CAMBRIDGE LIBRARY COLLECTION

Books of enduring scholarly value

Botany and Horticulture

Until the nineteenth century, the investigation of natural phenomena, plants and animals was considered either the preserve of elite scholars or a pastime for the leisured upper classes. As increasing academic rigour and systematisation was brought to the study of 'natural history', its subdisciplines were adopted into university curricula, and learned societies (such as the Royal Horticultural Society, founded in 1804) were established to support research in these areas. A related development was strong enthusiasm for exotic garden plants, which resulted in plant collecting expeditions to every corner of the globe, some-times with tragic consequences. This series includes accounts of some of those expeditions, detailed reference works on the flora of different regions, and practical advice for amateur and professional gardeners.

Flora Capensis

This seminal publication began life as a collaborative effort between the Irish botanist William Henry Harvey (1811–66) and his German counterpart Otto Wilhelm Sonder (1812–81). Relying on many contributors of specimens and descriptions from colonial South Africa – and building on the foundations laid by Carl Peter Thunberg, whose *Flora Capensis* (1823) is also reissued in this series – they published the first three volumes between 1860 and 1865. These were reprinted unchanged in 1894, and from 1896 the project was supervised by William Thiselton-Dyer (1843–1928), director of the Royal Botanic Gardens at Kew. A final supplement appeared in 1933. Reissued now in ten parts, this significant reference work catalogues more than 11,500 species of plant found in South Africa. Containing orders of the Calyciflorae with a monopetalous corolla and an inferior ovary, Volume 3 covers Rubiaceae to Campanulaceae.

Cambridge University Press has long been a pioneer in the reissuing of out-of-print titles from its own backlist, producing digital reprints of books that are still sought after by scholars and students but could not be reprinted economically using traditional technology. The Cambridge Library Collection extends this activity to a wider range of books which are still of importance to researchers and professionals, either for the source material they contain, or as landmarks in the history of their academic discipline.

Drawing from the world-renowned collections in the Cambridge University Library and other partner libraries, and guided by the advice of experts in each subject area, Cambridge University Press is using state-of-the-art scanning machines in its own Printing House to capture the content of each book selected for inclusion. The files are processed to give a consistently clear, crisp image, and the books finished to the high quality standard for which the Press is recognised around the world. The latest print-on-demand technology ensures that the books will remain available indefinitely, and that orders for single or multiple copies can quickly be supplied.

The Cambridge Library Collection brings back to life books of enduring scholarly value (including out-of-copyright works originally issued by other publishers) across a wide range of disciplines in the humanities and social sciences and in science and technology.

Flora Capensis

*Being a Systematic Description
of the Plants of the Cape Colony,
Caffraria & Port Natal*

VOLUME 3:
RUBIACEAE TO CAMPANULACEAE

WILLIAM H. HARVEY *ET AL.*

CAMBRIDGE
UNIVERSITY PRESS

CAMBRIDGE
UNIVERSITY PRESS

University Printing House, Cambridge, CB2 8BS, United Kingdom

Cambridge University Press is part of the University of Cambridge.
It furthers the University's mission by disseminating knowledge in the pursuit of
education, learning and research at the highest international levels of excellence.

www.cambridge.org
Information on this title: www.cambridge.org/9781108068086

© in this compilation Cambridge University Press 2014

This edition first published 1894
This digitally printed version 2014

ISBN 978-1-108-06808-6 Paperback

FLORA CAPENSIS.

FLORA CAPENSIS:

BEING A

Systematic Description of the Plants

OF THE

CAPE COLONY, CAFFRARIA, & PORT NATAL.

BY

WILLIAM H. HARVEY, M.D., F.R.S.
PROFESSOR OF BOTANY IN THE UNIVERSITY OF DUBLIN, ETC., ETC., ETC.

AND

OTTO WILHELM SONDER, Ph. D.
OF HAMBURGH.
MEMBER OF THE IMPERIAL LEOP.-CAROLINE ACADEMY NATURÆ CURIOSORUM, ETC , ETC., ETC.

VOLUME III.
RUBIACEÆ TO CAMPANULACEÆ.

L. REEVE & CO., Ltd.,
THE OAST HOUSE, BROOK
Nr. ASHFORD, KENT

1894

PREFACE.

This third volume contains the Orders of CALYCIFLORÆ with a monopetalous corolla and an inferior ovary. The fourth volume (shortly to be in preparation for press) will, it is hoped, include the Heaths *(Ericeæ)* and all the *Monopetalæ* with superior ovaries, *i. e.* the COROLLIFLORÆ proper. The fifth volume will probably take in all the MONOCHLAMYDEÆ, and the sixth the MONOCOTYLEDONES and *Ferns,* completing the work.

At this stage of their publication the authors feel themselves unwillingly compelled to fix the price of the *present* and *future* volumes at eighteen shillings, instead of selling them at twelve shillings, as charged for Vols. I. and II. Those volumes, having been published at the lower price, will continue to be sold at it. The advance is only to be charged on Vols. III., IV., V., VI. This measure has become necessary in self defence, owing to the price of the earlier volumes having been fixed much too low; for, after paying expenses and allowing for interest of money, although a considerable part of the edition has been sold, the profits are still in expectancy. Nor do they anticipate that the purchasers of this work, on reflection, will consider the price henceforward to be charged as unreasonably high. On the contrary, it will be found rather *under* than over that charged for similar works. The Australian Flora of Mr. Bentham, a work on a similar plan, and one that will include fully as many volumes, sells at twenty shillings per volume, and no one complains that this price is excessive. The volumes of the Cape Flora are of equal size, and equally closely printed. Both works stand on a somewhat similar footing, namely, that they enjoy government grants of £150 each volume, in aid of publication.

The authors have again to express to their many kind friends in South Africa their renewed thanks for collections of specimens sent during the publication of the present volume. Among their older correspondents, already mentioned in Vols. I. and II. those who have most constantly contributed specimens during the publication of Vol. III. have been MRS. F. W. BARBER and her brother HENRY BOWKER, Esq., who continue unremittingly and successfully their joint explorations; J. SANDERSON, Esq., of Natal; and Messrs. GERRARD and McKEN, of the same colony. These last-named gentlemen have contributed, within the last two years, over one thousand species of plants, including many novelties.

In the preface to Vol. II. mention was made of a collection of Damaraland plants received from "*Miss Elliott:*" the author's thanks should have been offered to that lady by her married name, Mrs. KOLBE.

The authors now present their thanks to the following new correspondents for their respective contributions:

To GEORGE FANNIN, Esq., of the Dargle, Maritzburg, Natal, for very interesting collections of plants, including many novelties, made in the neighbourhood of his residence. Among the more noticeable is a noble new species of Anemone (*A. Fanninii,* H.), with flowers fully larger than those of *A. Capensis,* and broad, digitately many-lobed, thick, and softly-velvetty leaves; a most desirable plant to introduce to gardens.

To PETER MAC OWAN, Esq., Principal of Shaw's College, Grahamstown, for several hundred species of the plants of his district, most carefully and beautifully dried. From none of their correspondents have the authors received more admirably prepared specimens, and though the immediate neighbourhood of Grahamstown is not particularly rich, and has already been well beaten over, Mr. Mac Owan has already detected more than one new species, and has added to the Flora the *Nuxia congesta,* of Abyssinia. A greater service to South African Botany has also been rendered by Mr. Mac Owan, in that he has succeeded in introducing among the pupils under his care a taste for Botany, which may lead to great results in the next generation. Among his most promising botanical pupils is Mr. R. W. READE, who has contributed many interesting species, especially of *Compositæ*, and whose well dried specimens do credit to his teacher.

To J. CHAPMAN, Esq., for a very considerable collection of the plants of the regions to the north of the Colony, in the direction and neighbourhood of Lake Ngami. This collection unfortunately suffered much in its long transit from the interior, and afterwards lay for a couple of years among lumber in a store in Capetcwn, from which it was rescued through the exertions of —— LAYARD, Esq., of the South African Museum, who most obligingly sought it out and forwarded it to Dublin. Though many of the specimens had perished, and others were much injured, a considerable number are in a state fit for examination.

To Mrs. OCTAVIUS BOWKER, through her sister-in-law Mrs. F. W. BARBER, for an interesting parcel of plants collected in the Orange Free State.

To Colonel BLAGRAVE for specimens of the *Hymenophylla* found about Table Mountain, &c.

To Captain BULGER for specimens collected at Windvogelsberg, and some Orchids from the neighbourhood of Capetown.

Lastly, to the Rev. Dr. J. CROUMBIE BROWN, Colonial Botanist, for his unremitting kind attention to the interests of this work, and for the zeal which he has shown, since his appointment, in endeavouring to promote the study of botany in all parts of the Colony, and among the neighbouring extra-colonial missionaries. The authors trust that his untiring efforts to diffuse botanical information throughout the Colony will in due time yield an abundant harvest.

To our friends in Europe, already mentioned in Vol. I, we have again to offer our best thanks for their continued support and assistance. And more especially our obligations are due to Sir W. J. HOOKER, for the continued free use of the noble Kew Herbarium ; and to Professors FRIES and ARESCHOUG for the equally free use of such portions of Thunberg's Herbarium as we desired to consult. The examination of the Thunbergian *Compositæ* has enabled Dr. Harvey to verify many of the obscure species of older authors, and in many cases to correct the synonymy. And Dr. Sonder has similarly profited, in his troublesome task on the synonymy of the *Campanulaceæ*.

Since the publication of our last volume South African Botany has lost in Dr. LUDOVIC PAPPE, late Colonial Botanist, a most diligent, devoted and successful explorer. His large herbarium, the result of many years labour, has been purchased by the Colonial Government for £200, and is now in course of arrangement. A portion containing the orders from *Ranunculaceæ* to *Umbelliferæ* inclusive, is already accessible to the student, and it is hoped that before the end of the present year the portion available for study will extend to the end of *Campanulaceæ*. When the whole shall have been completed the South African student will possess, at Capetown, an aid to study better than any book of reference.

Nor can the authors close this preface without a tribute of gratitude to RAWSON W. RAWSON, Esq., late Colonial Secretary, and now Governor of the Bahamas. To the active interest which from the first Mr. Rawson took in their undertaking, and to his powerful advocacy in the Colonial Parliament, are greatly due the very existence of the Flora. Nor was his assistance limited to the greater acts of patronage which became his station, but extended to the smallest details, such as the forwarding of parcels, &c.: in every detail and on every occasion he was invariably kind and considerate.

Trinity College, Dublin,
24th Feb., 1865.

SEQUENCE OF ORDERS CONTAINED IN VOLUME III.
WITH BRIEF CHARACTERS.

———◆———

FLORA CAPENSIS.

ORDER LXXIII. RUBIACEÆ, Juss.

(By W. Sonder).

Flowers regular, mostly bisexual, complete. *Calyx*-tube adhering to the ovary; the limb 4–6 lobed or toothed, sometimes obsolete. *Corolla* monopetalous, epigynous, 4–6 lobed, deciduous, either valvate or spirally-imbricate in æstivation. *Stamens* inserted in the tube of the corolla, as many as its lobes, and alternate with them. *Ovary* inferior, of two or several cells, crowned with an epigynous disc, and usually with the sub-persistent calyx tube; *ovules* one or many in each cell; *style* single, filiform, simple or, in the *Stellatæ*, bifid; *stigma* mostly thickened, bifid or bilamellate; capitate in *Stellatæ*; long and hairy in *Anthospermeæ*. *Fruit* either a capsule, berry, or drupe, or a pair of dry nuts cohering by their faces. *Seeds* with copious, fleshy, or horny albumen. *Radicle* next the hilum.

A very large Order, consisting of trees, shrubs, half-shrubs, and minute herbs dispersed over the globe, but chiefly abundant in tropical and sub-tropical regions. Leaves opposite, simple, quite entire, midribbed, and mostly penninerved. Stipules constantly present, interpetiolar, usually small, and connate into a toothed or awned lamina; in the *Stellatæ* leaflike and separate. Many important products are derived from this Order, as Peruvian Bark from various species of *Cinchona* and allied genera; Ipecacuanha from the roots of *Cephaelis* and others; Gambir, an intensely astringent substance from *Nauclea*; and above all Coffee, from the berries of *Coffea Arabica*. The wood of the arborescent kinds is hard, close, and heavy, and that of several of the S. African species in use for making yokes, axles, fellies, ploughs, &c. (See *Pappe, Sylva Capensis, p. 18–19*.)

TABLE OF THE SOUTH AFRICAN GENERA.

* Cells of the fruit many-seeded.

Tribe 1. GARDENIACEÆ. *Fruit* indehiscent, fleshy, 2- (rarely by abortion 1-) celled. *Seeds* not winged.—Shrubs.

I. **Burchellia.**—*Corolla* clavato-infundibuliform, limb short. *Stamens* about the middle of the tube.

II. **Oxyanthus.**—*Corolla* with a very long, slender tube. *Stamens* exserted. *Style* clavate. *Fruit* berry-like.

III. **Stylocoryne.**—*Corolla* salver-shaped or funnel-shaped, with a cylindrical tube. *Stigma* clavate. *Berry* dry.

IV. **Gardenia.**—*Corolla* salver-shaped, with a long tube; limb spreading, 5–9-cleft. *Stamens* in the throat of the tube. *Berry* fleshy.

V. **Randia.**—*Corolla* funnel-shaped, with a short tube. *Stigmas* 2. *Berry* nearly dry.

Tribe 2. HEDYOTIDEÆ. *Fruit* dehiscent, capsular, 2-celled. *Seeds* not winged.

VI. **Hedyotis.**—Small herbs.

** Cells of the fruit 1-seeded, or rarely 2-seeded.

Tribe 3. GUETTARDACEÆ. *Fruit* drupaceous, with 2–10 pyrenæ (or nuts). *Seeds* terete (without longitudinal furrow).—Shrubs or small trees.

VII. **Vangueria.**

Tribe 4. ALBERTIEÆ. *Fruit* dry, 10-ribbed and furrowed, 2-celled. *Seeds* convex at back, flat in front, without furrow. *Albumen* fleshy. Shrub or tree.

VIII. **Alberta.**—*Calyx* 5-lobed, 3 lobes small, 2 larger, ear-shaped, much enlarged in fruit.

Tribe 5. COFFEACEÆ. *Fruit* a 2-celled berry. *Seeds* convex at the back, flat, and with a longitudinal furrow in front. *Albumen* horny.—Trees or shrubs; very rarely herbs.

IX. **Canthium.**—*Corolla* with a short tube, 4–5-lobed, bearded or naked in the throat; æstivation valvate. *Style* filiform, exserted. *Stigma* undivided, thick, ovato-globose or mitræform.

X. **Plectronia.**—*Corolla* short, funnel-shaped, 5-lobed, bearded or naked in the throat. *Style* short. *Stigma* bilamellar.

XI. **Pavetta.** *Limb* of calyx 4-toothed. *Corolla* salver-shaped, with a long, slender tube, and 4-parted limb, imbricate in bud. *Style* much exserted.

XII. **Grumilea.**—*Limb* of calyx 5-toothed. *Corolla* with a short tube, villous at the throat; valvate in bud. *Stigma* bifid.

XIII. **Kraussia.**—*Limb* of calyx 5-cleft. *Corolla* funnel-shaped, with a short obconical tube, 5-cleft, imbricate in bud. *Style* short. *Stigma* clavate, striato-lamellate, bifid or bidentate.

XIV. **Bunburya.**—*Limb* of calyx bilabiate, 6-toothed. *Corolla* funnel-shaped, 6-lobed, imbricate in bud. *Style* filiform. *Stigma* bilamellate, lamells short, acute.

Tribe 6. SPERMACOCEÆ. *Fruit* nearly dry, with 2–6 pyrenæ or nuts. *Stigma* bilamellar, or globose, undivided.—Herbs, rarely shrublets. Stipules commonly split into several bristles.

XV. **Spermacoce.**—*Flowers* axillary, crowded. *Corolla* salver or funnel-shaped, with a short tube, naked inside. *Capsule* 2-celled, nuts divisible into 2 parts from the apex, one open, the other closed.

XVI. **Pentanisia.**—*Flowers* terminal, capitate-spiked. *Corolla* very long, naked inside. *Capsule* bipartible into 2 indehiscent nuts.

XVII. **Mitracarpum.**—*Flowers* in axillary and terminal heads. *Corolla* with a circular line of hairs inside. *Capsule* membranous, circumcised.

XVIII. **Hydrophylax.**—*Flowers* axillary, solitary or geminate. *Stigma* roundly 2-lobed. *Fruit* dry, corky, angular.

Tribe 7. ANTHOSPERMEÆ. *Fruit* nearly dry, bipartite (rarely fleshy and 2-celled). *Stigmas* 1 or 2, much elongated, hairy. *Albumen* fleshy.—Small shrubs or herbs. Stipules small, 1–3-toothed. Flowers often diœcious.

XIX. **Galopina.**—*Limb* of calyx very minute. *Fruit* of 2 mericarps; commissure flattish. *Stigmas* 2.—*Large terminal panicle.*

XX. **Anthospermum.**—*Limb* of calyx 4–5-toothed. *Fruit* of 2 mericarps; commissure concave. *Stigmas* 2.—*Flowers axillary or subpanicled.*

XXI. **Carpacoce.**—*Limb* of calyx 4–5-lobed. *Fruit* subdidymous, 2-seeded, or by abortion nearly terete, 1-seeded. *Stigma* 1.—*Flowers axillary, solitary.*

XXII. **Ambraria.**—*Fruit* spuriously 3–4-celled. *Stigmas* 2.

Tribe 8. STELLATÆ. *Fruit* dry, biparted (a cremocarp), rarely fleshv and 2-celled. *Style* bipartite. *Stigmas* capitate.—Herbaceous plants, with (spuriously) whorled leaves.

XXIII. **Rubia.**—*Corolla* 5-parted. *Fruit* fleshy.
XXIV. **Galium.**—*Corolla* 4-parted. *Fruit* dry.

TRIBE I.—GARDENIACEAE (Gen. I.–V.)

I. BURCHELLIA, R. Brown.

Calyx-tube obovate; limb equally 5-cleft beyond the middle, and drawn out above the ovarium. *Corolla* clavately funnel-shaped; the inner surface of the tube glabrous, except a bearded circle near the bottom; throat naked; lobes imbricated and twisted in æstivation. *Stamens* 5, inserted in the middle of the tube; filaments very short; anthers inclosed. *Stigma* oblong-clavate, bearing 5 convex crests, and a series of tufts of hairs. *Berry* sub-globose, crowned by the calyx, 2-celled. *Placentas* adnate to the dissepiment. *Seeds* angular, embryo slender. *DC. l. c.* 4, *p.* 368. *Endl. Gen. n.* 3315.

A shrub with petiolate, ovate, coriaceous leaves, intrapetiolar, cuspidate, deciduous stipulae, and terminal, capitate, sessile, scarlet flowers. Named in honour of *William Burchell,* the zealous and meritorious investigator of South Africa.

1. B. Capensis (R. Brown in Ker. Bot. Reg. t. 466); leaves on short petioles, ovate, acute, subcordate, entire, downy. *Pappe Sylv. cap. p.* 18. *E. Z. n.* 2285. *Lonicera bubalina, L. fil. suppl. p.* 146. *B. Kraussii, Hochst! Flora,* 1842, *p.* 237. *Zeyh.* 2702.

VAR. β. **parviflora**; leaves oval-lanceolate, smoothish, flowers a little smaller. *B. parviflora, Lindl. Bot. Reg. t.* 891. *E. Z. n.* 2286. *B. bubalina, Sims. Bot. Mag. t.* 2339. *Lonicera bubalina, Thunb. ! fl. cap. p.* 187.

HAB. Forests of Swellendam, George, Uitenhage, Albany, Caffraria, and Port Natal. Oct. Nov. (Herb. Thunb., D., Sd.)

A tree 12–14 feet high, called *Buffeldoorn.* Branches opposite, erect. Leaves about 3–5 inches long, 1½–3 inches broad, mostly subcordate at the base, pilose or pubescent beneath ; in var. γ, usually 2–3 inches long, 1 inch broad and rarely subcordate, nearly glabrous or with a few hairs on the middle nerve. Flowers 9–10 lines long, of a deep scarlet colour, adpressed hairy outside, in var. β. 8–9 lines long, orange-coloured. Limb of the corolla ovate or ovate-lanceolate, acute. Anthers pointed, nearly sessile in the upper part, or near the middle of the corolla. Style as long as the corolla. Fruit 4 lines long, shorter than the calyx lobes.

II. OXYANTHUS, DC.

Calyx-tube obovate, limb short, acutely 5-toothed. *Corolla* with a very long tube, a glabrous throat and a 5-parted, regular limb, with oblong and acuminated segments. *Stamens* 5, rising from the throat of the corolla, free, exserted ; anthers acute. *Style* filiform, clavate at the apex. *Fruit* baccate, 2-celled. *Seeds* numerous. *Albumen* cartilaginous. *Cotyledons* foliaceous. *DC. l. c. p.* 374. *Endl. gen. n.* 3307.

Shrubs with elliptic, acuminated, short-petioled leaves, oblong, triangular, deciduous stipules, and axillary, racemose-corymbose peduncles. Name from ὀξὺς, sharp, and ἀνθος, a flower, in reference to the acute teeth of the calyx and segments of the corolla.

Leaves elliptic-oblong or ovate-lanceolate: tube of the corolla
one half shorter than the leaves　...　...　...　...　...　... (1) **Natalensis.**
Leaves elliptic-lanceolate, acuminate : tube of the corolla 6–8
times shorter than the leaves　...　...　...　...　...　... (2) **Gerrardi.**
Leaves ovate-cordate, cuspidate ...　...　...　...　...　... (3) **latifolius.**

1. O. Natalensis (Sond. in Linn. vol. xxiii. p. 50); branches, leaves, and calyx quite glabrous ; leaves elliptic-oblong or ovate-lanceolate, *shortly acuminate,* quite entire; stipules ovate-acuminate, nearly thrice longer than the petiole ; racemes axillary, corymbose, *lax-flowered;* teeth of calyx subulate, shorter than the tube; tube of corolla *one-half* shorter than the leaves; style exserted. *Megacarpha pyriformis, Hochst! pl. Krauss (fructifera).*

HAB. Wet places in woods, Port Natal, *Krauss.* 110. *Gueinzius,* 432. *Cooper,* 1227. July. (Herb. Sd., D.)

Leaves penninerved, in our specimens 7–8 inches long, 3 inches broad, not cordate. Raceme about 16–20-flowered. Calyx 3–3½ lines, teeth 1½ line long. Tube of the corolla 2½ inches, lobes 7–8 lines long, subulate. Fil. short; anthers exserted. Berry pomiform, 1 inch or more long, glabrous, crowned by the calyx lobes, 2-celled, many-seeded. Seeds ovate, compressed. Very similar to *O. speciosus,* DC.

2. O. Gerrardi (Sond.); branches and calyx glabrous ; leaves elliptic-lanceolate, *acuminate*, quite entire, glabrous, *but bearded in the axils of the nerves beneath ;* stipules ovate, acuminate, twice longer than the petiole; racemes axillary, corymbose, *dense-flowered ;* teeth of calyx sub-ulate, shorter than the tube ; tube of corolla 6–8 *times shorter* than the leaves ; style exserted.

HAB. Natal, *Gerr. & M'K.* 1359; *Cooper,* 1102; *Sander.* 616. Feb.–Mar. (Hb. D.)
A large shrub. sometimes a low tree. Leaves 6–8 inches long, 2–2½ inches broad, shining, penninerved, nerves in the axils with a tuft of whitish hairs. Racemes 25–30-flowered, flowers white. Calyx 2 lines long. Tube of corolla 1 inch, the long pointed lobes 4–5 lines long. This comes very near *O. Natalensis*, but differs in the short tube of the corolla.

3. O. latifolius (Sond.); branches, leaves, and calyx quite glabrous; leaves *ovate-cordate, cuspidate,* quite entire; stipules 3-angular-ovate, acu-minate, twice longer than the petiole; racemes axillary, *lax-flowered*; teeth of calyx subulate, about equalling the tube ; tube of corolla one-half shorter than the leaves, style exserted.

HAB. Umgena, Port Natal, *Gerr. & M'K.* 718. Dec. (Herb. D., Sd.)
A very handsome shrub or low tree, with fragrant white flowers. Leaves 6–8 inches long, 4–5 inches broad, penninerved, paler on the underside. Flowers as in *O. Natalensis*, from which it is distinguished by the much broader, subcordate leaves.

III. STYLOCORYNE, Cavan. (Richd.)

Calyx-tube ovate-globose ; limb short, tubular, 5-toothed. *Corolla* salver-shaped, with a cylindrical tube, and a 5-parted limb. *Stamens* 5, inserted in the mouth of the tube of the corolla; anthers linear, very long. *Style* exserted ; stigma clavate, undivided, or the lobes are closely consolidated. *Berry* globose, crowned by the calyx, dry, 2-celled ; placentas spongy, adnate to the dissepiment. *Seeds* numerous, angular, rugulose. *Albumen* rather cartilaginous. *Embryo* long. *Stylocoryne et Cupia, DC. prod.* 4, *p.* 377, 393. *Ceriscus Nees ab Escnb.*

Trees or unarmed shrubs. Leaves opposite, oval or oblong, petiolate. Stipules broad at the base, acuminated at the apex, 1-nerved in the middle. Peduncles axillary, corymbose. Name from στυλος, a style, and κορυνη, a club, in reference to the cla-vate stigma.

1. S. cuspidata (E. Mey.) ; shrubby, unarmed, glabrous ; branches subangular; leaves petiolate, oval-oblong, narrowed to both ends, mem-branous; corymbs axillary on longish peduncles, trichotomous; pedicels and calyx adpressed-puberulous, at length glabrous, bracteated; berry globose, 6–8 seeded, shorter than the pedicels.

HAB. Port Natal, *Drege, Gerr. & M'K.* 136, April. (Herb. D., Sd.)
A large shrub. Stipules with a 3-lines long point. Leaves 6–8 inches long, 2–3 inches broad, penninerved and strongly veined, green on both sides, tapering at the base, cuspidate, acuminate at the apex. Petiole ½–1 inch. Peduncles several inches long. Corymb with spreading branches and pedicels. Flowers not developed in our specimens. Berry about the size of a pea, black, shining, the inner structure quite as in *S. Webera*, Rich.

IV. GARDENIA, Ellis.

Calyx tube ovate, often ribbed ; limb tubular, truncate, toothed, cleft or parted. *Corolla* tubulose or subcampanulate, hypocrateriform; limb twisted in æstivation, but afterwards spreading, 5–9 parted. *Anthers* 5–9, linear, nearly sessile in the naked throat of the corolla, or exserted. *Style* long; stigma clavate, bifid, or bidentate. *Ovary* 1-celled, half divided by 2–5 incomplete dissepiments. *Berry* fleshy, crowned by the calyx, imperfectly 2–5 celled. *Seeds* minute, immersed in fleshy parietal placentæ. *DC. l. c.* 4, *p.* 379. *Endl. Gen. n.* 3305.

Trees or shrubs unarmed or spinescent. Leaves opposite or rarely in whorls, ovate or oval. Flowers generally white, solitary, axillary, or terminal, sweet scented. Named after *Alexander Garden, M.D.* of Charlestown, South Carolina, one of the correspondents of Ellis and Linnæus.

1. **Eu-Gardenia.** *Calyx*-tube ribbed. *Corolla* tube cylindrical (1) **florida.**
2. **Kumbaga.** *Calyx*-tube not ribbed : limb equally toothed or lobed, rarely truncate. *Cor.* tube cylindr. or campanulate.
 Corolla-tube bell-shaped, villous ... ... (2) **globosa.**
 Corolla salver-shaped :
 Leaves obovate... ... ... (5) **Neuberia.**
 Leaves elliptical-lanceolate :
 Cor.-tube 2ce as long as calyx ... ... (3) **citriodora.**
 Cor.-tube 4-times as long as calyx ... ... (4) **Gerrardiana.**
3. **Piringa.** *Calyx*-tube not ribbed ; limb tubular, leafy, cleft down one side. *Cor.* salver-shaped ... ... (6) **Thunbergia.**
4. **Rothmannia.** *Calyx*-tube ribbed or angular. *Corolla* dilated upwards, or nearly funnel-shaped ... ... (7) **Rothmannia.**

1. G. florida (Linn. spec. 305); shrubby, unarmed, erect; leaves elliptical, acute at both ends ; flowers solitary, almost terminal, sessile ; calycine segments vertical, lanceolate-subulate, equalling the tube of the corolla in length; berry elongated, turbinate, ribbed. *Bot. Reg. t.* 449. *G. jasminoides, Sal. Phil. Trans.* 52, *t.* 20. *Jasm. Capense, Mill. t.* 180.

Hab. Cultivated, native of China, Nov. Dec. (Herb. Th., Sd.)
Shrub 2-6 feet. Flowers large, white, 5-9-parted, often double. Berry 5-6 angled, 5-6 celled at the base, 1-celled at the apex, orange coloured, size of a pigeon's egg; the pulp is used for dyeing yellow.

2. G. globosa (Hochst Flora, vol. 25, p. 237); leaves shortly petiolate, lanceolate, quite glabrous ; stipules acute, short ; flowers terminal or subaggregated, subsessile ; *limb of calyx shortly 5-toothed ;* tube of corolla campanulate, villous on both surfaces, lobes broadly ovate, acute, spreading; ovary oblong, fruit globose. *B. M. t.* 4791. *Harv. Thes. t.* 5.

Hab. Sea coast, Natal, to 1200 ft. *Krauss,* 467 ; *Gueinzius,* 130, 546. *Gerr. & M'K.* 714. *J. Sanderson,* 700. (Herb. D. Sd.). Aug. Oct.
A low tree or shrub, unarmed, glabrous, except in the leaf buds, young leaves, and inflorescence. Leaves 2-4 inches long, about 1 inch broad, acute at each end, or bluntish, glossy, penninerved, nerves reddish. Flowers on very short pedicels. Calyx tube 2 lines long, minutely pubescent, limb campanulate with rigid teeth. Corolla white, striated with pink lines (*Gerrard*), the tube 1 inch long, ½ inch broad, lobes 4 lines long. Fruit as large as a walnut.

3. G. citriodora (Hook. Bot. Mag. t. 4987); leaves petiolate, elliptical-lanceolate, subacuminate, glabrous ; *stipules from a broad base subulate-acuminate ;* corymbs axillary, shorter than the leaves; *lobes of calyx lanceolate, acuminate ;* corolla salver-shaped ; tube twice longer than the calyx, glabrous ; lobes obovate, obtuse ; fruit oval. *Mitriostigma axillare, Hochst ! l. c.*

HAB. Near D'Urban, Natal, *Krauss; Gueinzius; Gerrard,* 716. (Herb. D.)
A glabrous unarmed shrub several feet high, with green spreading branches.
Leaves 2–4 inches long, ¾–1¼ inch broad, acute or bluntish, penninerved. Stipules
broad at the base, 3 lines long. Flowers white. Calyx with subulate bracts at the
base, turbinate, limb campanulate, lobes ciliate. Tube of corolla ½ inch, lobes 4–5
lines long. Style exserted, stigma mitræform. Fruit as large as a small cherry.

4. G. Gerrardiana (Sond. & Harv.); leaves shortly petiolate, *elliptical-lanceolate*, acute or subobtuse, glabrous ; *stipules short, acute;* flowers
terminal, solitary ; tube of calyx campanulate, with short, acute, ciliate
teeth ; corolla salver-shaped, *tube cylindrical, nearly 4 times longer than
the calyx,* lobes obovate, obtuse.

HAB. Near D'Urban, Port Natal, *Gerr. & M'K.* 715. (Herb. D., Sd.)
Ultimate branches greyish, very short. Leaves coriaceous, glossy, scarcely paler
on the lower surface, and obsoletely penninerved, about 2 inches long, 9 lines broad.
Calyx turbinate, minutely bracteolate at the base, 5 lines long, the broad acute teeth
1 line long. Cor.-tube 1½ inch long, lobes 10 lines long, 4–5 lines broad, minutely
downy on both surfaces. Anthers and stigma subexserted.

5. G. Neuberia (E. Z. ! 2289); glabrous, *spinescent;* branches and
spines opposite, divergent ; leaves opposite or pseudo-fasciculate, subsessile, *obovate*, acute, or bluntish, shining above ; stipules very short,
acute; flowers terminal, solitary, sessile ; tube of calyx campanulate,
with short, acute, ciliate teeth; corolla salver-shaped, tube *clavate,
twice longer* than the calyx, lobes obovate, obtuse; fruit oval. *Hyper-
anthus floridus, E. Mey.*

HAB. Woods of the Zuureberge mts. near Enon, and on Grasrugg Uit. *E. & Z.
Drege.* Howisonspoort, *Zeyh.* 2703. Fl. Nov. Fr. Mart. (Herb. Sd.)
Branches greyish-white, terete, the ultimate very short. Spines from ½–1 inch.
Leaves numerous on the short branches, tapering into the very short petiole, 1–2
inches long, ½–1 inch broad, a little paler and scarcely penninerved beneath. Calyx
with 2 very small basal bracts, 5 lines long, turbinate at base, teeth 1 line long.
Flowers uncial, the lobes a little shorter than the tube, minutely downy on both
surfaces. Anthers and stigma subexserted. Fruit as large as a cherry, crowned by
the cylindrical, 5-toothed calyx. Near *G. amœna. Sims. Bot. Mag. t.* 1904.

6. G. Thunbergia (L. fil. suppl. 162); unarmed, much divided ;
leaves elliptic, tapering at both ends, petiolate, veined, glabrous ;
flowers terminal, solitary, sessile ; calyx tubular, foliaceous, cleft at
one side, with leaf-like appendages ; corolla tubular, salver shaped,
lobes 8, obtuse ; fruit ovate. *Thunb. diss. Gardenia, n. 3. Bot. Mag.
t.* 1004. *Thunbergia Capensis, Montin. act. Holm.* 1773, *t.* 11.

HAB. Forests of the Krakakamma and Oliphant's Hoek, Uitenhage. *E. Z.* 2287.
Zeyh. 2704. Natal, *Gerr. & M'K.* 717. Jan.–Feb. (Herb. D., Sond.)
Height of trunk, from 8–10 feet; diameter from 10–12 inches. Bark smooth,
greyish white. Wood hard. Leaves 4–6 inches long ; on the flowering branches
smaller, and often as broad as long. Flowers large, white, fragrant. Tube of calyx
1 inch, of the corolla 2½–3 inches long ; lobes of corolla 1 inch or more long. Anthers and stigma exserted. Fruit 2–2½ inches long, smooth, white, very hard,
almost woody, imperfectly 5 celled, many seeded.

7. G. Rothmannia (L. fil. suppl. p. 165); unarmed ; branches angular, rough ; leaves on very short petioles, oblong, acute, glabrous ;
flowers terminal, solitary, sessile ; calyx cylindrical, ribbed, hairy within, 5-cleft, its segments filiform, acute ; corolla with an obconical tube,

a campanulate throat, and spreading acute segments; fruit ovate, fleshy, costate, 2-valved, 2-celled, many seeded. *Thunb. diss. Gd. n. 6. Bot. M. t. 690. E. Z. 2288. Rothmannia Capensis, Th. act. Holm. 1776, p. 65, f. 2.*

HAB. Woods, Swellendam, George, Uitenhage; Natal. Jan.–Feb. (Hb. D., Sd.) Tree; stem 15–30 feet high. Wood very hard. Leaves 3–4 inches long, 1–1½ inch broad, coriaceous, veiny. Flowers white, spotted with red, sweet scented, calyx 1 inch and more, the corolla 3 inches long. Fruit nearly 1½ inch in diameter.

V. RANDIA, Linn.

Calyx-tube obovate, limb 5-lobed. *Corolla* funnel-shaped, with a short tube; limb 5-parted, twisted in æstivation. *Anthers* sessile within the tube of the corolla, included. *Stigmas* 2, thick. *Berry* nearly dry, crowned by the calyx, corticate, 2-celled. *Seeds* many in each cell, fixed to a central placenta, wingless, imbedded in the pulp, or imbricated downwards. *Albumen* cartilaginous. *Embryo* straight, with a terete radicle, and orbicular flat cotyledons. *DC. l. c. 4, p. 384. Endl. Gen. n. 3304.*

Much branched small trees or shrubs, with axillary opposite or subverticillate thorns. Leaves sessile or subsessile. Flowers almost sessile; usually solitary, rising from the axils of the leaves. Named after Isaac Rand, once a demonstrator of botany at the Chelsea botanic garden.

Peduncles 1-flowered, shorter than calyx; anth. linear, acute ... (1) **rudis.**
Pedunc. 2–3 fl, longer than calyx; anth. oblong ... ... (2) **Kraussii.**

1. R. rudis (E. Meyer); unarmed, young branches minutely downy; leaves fasciculate, obovate, suborbicular or elliptic, obtuse or subacute, cuneate at the base, glabrous or minutely downy beneath, on very short petioles; peduncles axillary, *one-flowered, shorter than the calyx*; tube of calyx obconical, lobes obovate, obtuse; corolla funnel-shaped, limb erect, spreading with oblong flat lobes; anthers *linear*, acute; fruit sub-globose, 10-ribbed. *Harv. Thes. t. 34. Heinsia Capensis, Buek in herb. Ecklon. Gardenia microcarpa, Hochst. in pl. Krauss.*

VAR. α. Calyx glabrous.
VAR. γ. Calyx setulose or scabrous. *R. parvifolia, Harv. Thes. t. 35.*
HAB. Mts. near Enon, near Grahamstown, Karregarivier, and between Hoffmann's Kl. and Driefontein, *Drege, Zeyh.* 2710, *H. Hutton, E. & Z.* Var. β. Natal, *Krauss,* 314, 347. *Gueinzius,* 117, 555. *J. Sanderson,* 266, 199. Sept. (Herb. D., Sd.) A rigid shrub, with erect, spreading, greyish branches. Stipules on the young twigs broadly ovate. Leaves ½–1 inch long, when young often minutely downy. Peduncles very short. Calyx 2 lines long, tube about as long as the spathulate, obtuse, or mucronulate lobes. Corolla ½ inch long. Fruit the size of a pea, black. I cannot regard Var. γ. as a distinct species, the leaves being often larger than in Var. α. In many specimens collected by *Ecklon* and *Zeyher,* the calyx lobes are evidently mucronulate and not quite glabrous.

2. R. Kraussii. (Harv. Thes. t. 33.); unarmed or rarely spinous; young branches minutely downy; leaves obovate or elliptic-obtuse, cuneate at the base, glabrous, netted beneath, on very short petioles; peduncles axillary, 2–3 *flowered, pedicels longer than the calyx;* tube of calyx ovate, glabrous, lobes obovate, obtuse; corolla minutely downy, tube very short, lobes broadly obovate, obtuse, undulate; anthers *oblong;* fruit globose. *Randia Lachnosiphonium, Hochst. Flora,* 1842, 237. *Lachnosiphonium obovatum, Hochst. ! l. c. p. 238.*

Hab. Natal, *Krauss*. 129. *Sanderson*, 200. *Gueinzius*, 120 ; *Gerr. & M·K.* 864. (Herb. Sd., D.)

A shrub 8–12 feet high. Axillary spines ½–1 inch long. Leaves 1½–2 inches long, 1 inch or more in width, tapering into a short petiole, glabrous or nearly so. Stipules short, acute, very broad at the base. Pedicels 6–8 lines long. Tube of calyx 3 lines, the veined limb 2 lines long. Corolla white, tube shorter than the calyx, with a hairy ring in the throat, segments patent, 6–8 lines in diameter. Stigma bifid. Fruit when dry very hard, the size of a small walnut, crowned by the limb of the calyx.

TRIBE II.—HEDYOTIDEÆ (Gen. VI.)

VI. HEDYOTIS, Lam.

Calyx-tube ovate or subglobose, limb 4-, rarely 5-toothed. *Corolla* tubular, limb 4-, rarely 5-lobed, throat villous or glabrous. *Stamens* exserted a little, or sessile within the tube. *Stigma* simple or bifid. *Capsule* subglobose, membranous, crowned with the distant calycine lobes, 2-celled, dehiscing at the summit. *Seeds* very numerous, minute, affixed to a subglobose placenta. *Wght. et Arn. Prodr. Fl. penins. Ind. or.* 1, 407.

Small herbaceous plants with opposite leaves, setulose stipules, and axillary or terminal, 1–2 or many-flowered, corymbose or somewhat spiked, often long peduncles. Name from ἥδυς, sweet, ous, an ear ; the leaves are oval, soft, and firm, and have been compared to ears.

Leaves suborbicular or spathulate ... ... (18) **tenella.**
Leaves small (ovate, lanceolate, linear.)
   a. Corolla with short tube and 4-cleft limb :
      Flowers densely corymbose ... ... (1) **chlorophylla.**
      Flowers capitate ... ... (2) **cephalotes.**
      Fl. cymose; perennial ; stem and lvs. pubescent ;
        in fl. glabrous ... ... (3) **Gerrardi.**
      Pedicels axillary, 1–2 flowered.
        Annual, pedicels 3 times longer than the calyx (5) **scabrida.**
        Annual, pedicels as long or shorter than the
          calyx, leaves six lines long ... ... (4) **Capensis.**
        Suffruticose, pedicels geminate; lvs. 1½ inch long (6) **geminiflora.**
   b. Corolla with a long terete tube and 4-lobed limb :
      1, Flowers terminal, umbellate ... ... (12) **Amatymbica**
      2, Flowers terminal, tufted or densely panicled :
        Fruit papillose-scabrous ... ... (11) **thymifolia.**
        Fruit quite smooth ... ... (16) **Natalensis.**
      3, Flowers axillary, or loosely panicled, fruit smooth.
        Annual, scabrous; leaves linear-lanceolate ... (9) **brachyloba.**
        Annual, glabrous, or nearly so :
          Lvs. lance-linear, flat ... ... (8) **Heynei.**
          Lvs. narrow-linear, with revolute margins (10) **stricta.**
          Lvs. ovate subcordate, flat ... ... (14) **rupicola.**
        Perennial : leaves glabrous, lanceolate (1 inch) (13) **decumbens.**
        Perennial; lvs. hairy-scabrous, ovate ... ... (15) **hirtula.**
        Suffruticose ; leaves with a recurved mucro ... (7) **setifera.**
   c. Corolla 5-lobed ... ... (17) **pentamera.**

a. **Oldenlandia**; corolla with a short tube and a 4-cleft limb. (Sp. 1–6.)

1. H. chlorophylla (Hochst. in Flora, 1844); suffruticose, erect, stem terete or subangular, smooth or a little scabrous; leaves lanceolate or ovate-lanceolate, 5-nerved, yellowish-green, subscabrous; stipules on both sides of 5 or 7 bristles ; *flowers aggregated in a terminal dense corymb;* calyx 4-fid, lobes subulate, recurved at the apex ; corolla equalling the calyx, throat hairy.

HAB. Between Natalbai and Umlaas River, and on the Table Mts. Nov.–Dec. *Krauss.* 39, 53. *J. Sanderson,* 399. *Gueinzius,* 125. *G. & M.'K.* 1579, *pte.* (Hb.Sd.,D.)
Stem 1 foot or higher, simple or a few branched. Leaves sessile, 1½–2 inches long, 6–9 lines broad, the upper more distant and smaller. Corymb many flowered, 1 or 2 inches in diameter. Pedicels with subulate bracts, as long as or shorter than the calyx. Tube of the calyx 1 line long, equalling the lobes. Fruit unknown.

2. H. cephalotes (Hochst. ! l. c.); herbaceous, ascending, quite glabrous and smooth ; stem nearly terete, branched ; leaves small, ovato-lanceolate, 3-nerved, with revolute margins; stipules bifid, ciliate; *flowers terminal, capitate ;* calyx 4-fid, lobes mucronate, as long as the hispid tubes; corolla equalling the calyx; capsule subglobose, pilose.

HAB. D'Urban, Natal, *Krauss.* 111 ; *J. Sanderson,* 321 ; *Gerr. & M'K.* 588, 589. Nov. Dec. (Herb. D., Sd.)
Habit of Epilobium. Root creeping. Stem 1–2 feet. Leaves acute at both ends, 8–10 lines long, 2–3 lines broad. Capitulum as large as a small hazel nut. Calyx 1 line long ; tube as long as the teeth.

3. H. Gerrardi (Sond.); perennial; stems erect, as well as the leaves scabrous, with short spreading hairs ; leaves linear, mucronate, with revolute margins; stipules of many bristles; flowers disposed in a cymose panicle ; peduncle, pedicels, and flowers glabrous ; calyx-lobes ovato-lanceolate, equalling the tube of the shortly tubulose corolla ; style as long as the corolla ; anthers linear-oblong.

HAB. Zulu, Natal, *W. T. Gerrard,* April; *Gerr. & M'K.* 1366. (Herb. D., Sd.)
Stem very leafy, from minute branchlets in the axils of the leaves, 3–4 inches high. Leaves 8–10 lines long, 1 line broad. The upper or flower-bearing part of about the same length as the leafy stem, but not hairy. Panicle erect. Pedicels unequal, 1–4 lines long. Calyx 1 line long ; tube subglobose, a little shorter than the acuminate lobes. Corolla 4-fid; lobes short, acute. Stamens inserted in the middle of the tube. Style glabrous. Fruit unknown.

4. H. Capensis (Lam. ill. n. 1425); annual; stem angular, much branched, decumbent, scabrous, hairy, or villous ; leaves linear, acute, with revolute margins ; stipules on both sides, with 2 or 3 setæ; pedicels axillary, numerous, 1-flowered, shorter than the leaves, and as long or shorter than the calyx; calyx 4-, rarely 5-fid, lobes subulate, ciliated, nearly as long as the tube; corolla a little longer than the calyx; throat villous ; fruit glabrous. *Oldenlandia Capensis, Thunb. ! fl. cap. p.* 147.

HAB. Sandy places near Koopmansrivier, *Thunb. ;* on the Witrivier, near Enon, *Drege,* 7659 ; Campground, *Sir O. Bunbury;* Paarl. Dec.–Feb. (Herb. Th., D., Sd.)
Branches 4–6 inches long, diffuse. Leaves about 6 lines long, hairy, sometimes subglabrous. Pedicels shorter or as long as the calyx. Corolla white. Stigma clavate. *(Thunb.)* Capsule 1 line long, subglobose.

5. H. scabrida (Sond.); *annual;* stem quadrangular, branched, suberect, scabrous on the angles ; leaves linear or linear-lanceolate, acute, subglabrous; stipules on both sides with 3 or 4 setæ; pedicels axillary, *solitary or subaggregated,* 1-flowered, shorter than the leaves, but 3 *times longer than the calyx ;* calyx 4-fid, lobes subulate, subciliated, nearly as long as the tube ; corolla a little longer than the calyx, throat hairy ; fruit glabrous. *Oldenlandia scabrida, DC. prodr.* 4, *p.* 425. *Kohautia Dregeana,* Presl. *in herb. Drege ? (imperfect specimen.)*

HAB. Port Natal, *Gueinzius*, 123. (Herb. Sond.)
Very nearly allied to the preceding, but easily distinguished by the lax habit, longer branches, glabrous, somewhat longer and broader leaves (8–10 lines long, 1½–2 lines broad), and the pedicels. The flower and fruit are not different. Stigma bifid.

6. H. geminiflora (Sond. in Linn. xxiii. p. 51); *suffruticose*, stem and branches angular, scabrous; leaves linear-lanceolate, acute, somewhat scabrous, with revolute margins; stipules membranous, with 4–6 setæ; *pedicels axillary, in pairs*, 1-flowered, in fruit deflexed, about as long as the calyx; lobes of the calyx subulate, a little longer than the tube, scabrous; corolla equalling the calyx; fruit glabrous.

HAB. Grassy places on the Magalisberg, Nov. Zeyh. 756. (Herb. Sd., D.)
A foot and more high, much branched. Leaves 1½ inch long, 1½–2 lines wide. Pedicles scabrous, 2 lines long. Calyx 4-fid. Lobes of the corolla nearly as long as the tube. Capsule subglobose.

b. **Kohautia.** Corolla with a long terete tube, and a 4-lobed limb. (Sp. 7–16.)

7. H. setifera (Sond.); *suffruticose*, erect, terete, branched; branches subcompressed or rib-furrowed, puberous, or a little scabrous; leaves linear or linear-lanceolate, with revolute margins, cuspidately mucronate, mucro recurved, glabrous; stipules cleft into 6 spreading bristles on both sides, which are *longer* than the membrane; corymbs few-flowered, laxly panicled; pedicels erect, in fruit 2–4 times longer than the calyx; lobes of the calyx *a little longer* than the tube, smooth; tube of the corolla elongate, terete, with short, oval lobes; fruit glabrous. *Oldenlandia Caffra, E. Z.! 2291. Kohautia setifera, DC. l. c. 4. p.* 430.

VAR. β. **pubescens**, leaves and the lower part of the stem pubescent.
HAB. Stony places near the Zwartkopsrivier, *E. & Z.; Zeyh.* 2705; Natal, *J. Sanderson*, 400, 138, 270, *T. Williamson*, var. β. near D'Urban. *Gerr. & M'K.* 582. Oct. (Herb. D., Sd.)
Stem ½–1 foot high, with erect-spreading branches. Leaves ½–1 inch long, 1–2 lines broad. Pedicels rarely solitary, mostly geminate, or 3–4 aggregated in the upper part of the branches, unequal, glabrous. Calyx 1 line long, with subulate teeth. Corolla nearly 3 lines long. Capsule globose, smooth.

8. H. Heynei (R. Br. in Wall. cat. 867); annual or biennial, quite glabrous; stem erect, branched; leaves lance-linear, acute; stipules of 2 or 3 very short bristles; pedicels axillary, 1-flowered, capillary, about as long as the leaves; corolla funnel-shaped; tube 3–4 times longer than the calycine lobes; styles subexserted, a little longer than the anthers; capsule subglobose. *Wght. & Arn. Prod. p.* 416.

HAB. Zulu, Natal, *Gerr. & M'K.* 1365. (Herb. D., Sd.)
About 1 foot high; branches erect, spreading. Leaves 6–8 lines long, ¼ line broad. Stipules cupular, with minute bristles. Pedicels spreading. Flowers white, 2 lines long. Limb of corolla 3 times shorter than the tube. Anthers oblong, sessile in the throat. Capsule 1 line long. The leaves are somewhat narrower than in East Indian specimens collected by Dr. Hooker and Thompson, but there are no other differences.

9. H. brachyloba (Sond. in Linn. vol. xxiii. p. 50); *annual, scabrous*, branches subcompressed; leaves *linear-lanceolate*, with subrevolute margins; stipules cleft *into 3–4 bristles* on both sides, which are *shorter* than the membrane; panicle lax, dichotomous; flowers sessile and pedicellate; teeth of the calyx *twice shorter* than the tube; corolla with very long tube and linear-oblong lobes; fruit smooth.

HAB. Komseep, near Springbokkeel. Feb. *Zeyh.* 761. (Herb. Sd.,D.)
A span high, beset with minute tubercles. Root longish, simple. Leaves 1 inch long, 2 lines broad, 1-nerved. Panicle subcorymbose, 3-4 inches long. Flowers geminate, one sessile or nearly so, the other on a 2-4 lines long pedicel. Bracteoles minute. Corolla 5-6 lines long; the lobes 1 line long, ¼ line broad. Capsule globose, crowned by the minute calyx-teeth, 1 line long.

10. H. stricta (Smith, Rees's Cyclop. vol. 17, no. 21); annual, *smooth* or a little tubercled; branches erect, subcompressed; leaves narrow-linear, with revolute margins; stipules joined to the petioles by a very narrow margin, furnished *each with 2 short bristles;* flowers trichotomously corymbose; pedicels unequal; teeth of the calyx shorter than the tube; corolla with very long tube and oblong lobes, fruit *smooth.* *Kohautia stricta, DC. l. c. p.* 430. *K. thymifolia, E. Mey. in Hb. Drege. b. non a.*

HAB. Winterveld and Nieuweveld, *Drege.;* Taba Uncha, Gamkariver, Springbokkeel, *Zeyh.* 759, 760. Namaqualand, *A. Wyley,* Dec.-Jan. (Herb. D., Sd.)
Very nearly allied to *H. brachyloba,* but differs in the more straight habit, narrower leaves (1 inch long, ½ line broad), which are quite smooth and glabrous, somewhat larger flowers and fruit. The cyme or corymbose panicle is about 1 inch long. Calyx 1 line, the tube of corolla 6 lines, the lobes 2 lines long. Capsule 1½ lines long.

11. H. thymifolia (Presl.! Bot. Bemerk. p. 85); suffruticose, *scabrous;* branches erect, quadrangular; leaves linear, acute, with revolute margins; bristles of stipules 2, setaceous; panicle 2-3-chotomous, many-flowered; flowers sessile and pedicellate; teeth of the calyx as long as the tube; corolla with a very long tube and oblong lobes; fruit subglobose, *papillate-scabrous. H. thymifolia, E. Mey. in Herb. Dr. a, c. non. b.*

HAB. On the Zwartkops. and the Garip, *Drege;* Vaalriver and Taba Uncha, *Zeyh.* 757; Namaqualand, *A. Wyley, Rev. H. Whitehead.* Sept.—Dec. (Herb. Sd., D.)
1-1½ foot high; stem, branches, leaves, pedicels, and calyx scabrous from minute, acute tubercles. Leaves 1-1½ inch long, 1 line broad. Bristles of the stipules as long or longer than the membrane. Pedicels 3-4 times longer than the fruit. Corolla 7-8 lines long. Capsule the size of a small pea, crowned by the subulate calyx-lobes.

12. H. Amatymbica (Hochst. in pl. Krauss.); perennial, glabrous, glaucescent; *stem erect,* much compressed, simple or branched; leaves *narrow linear,* cuspidate, caniculate above; bristles of stipules 3, setaceous; flowers *terminal, umbellate;* teeth of the calyx as long as the tube; corolla with a very long tube and oblong lobes; fruit subglobose, smooth. *Kohautia Amatymbica, E. & Z.!* 2292. *K.* 7657. *Hb. Drege.*

HAB. Katriviers and Winterberg, Keyrivier, Caffraria; Zwartehoogdens, Albany, *E. & Z. Zeyh.* 758, *Col. Bolton, Zuurebergen, Drege;* Port Natal, *Gueinzius,* 35. *J. Sanderson,* 245. Magalisberg, *Burke & Zeyh.* Apl. Jul. Oct. (Herb. Sd., D.)
Root perpendicular. Stem 1-1½ foot. Leaves 1½-3 inches long, 1 line broad. Umbel 6-10-flowered. Calyx about 2 lines long, the teeth subulate. Tube of the corolla 1 inch, lobes 3-4 lines long. Capsule as large as a pea, didymous, crowned with the 4, rarely 5, rather shorter calyx-teeth.

13. H. decumbens (Hochst.! l. c. p. 67); *perennial, glabrous; stem decumbent,* hexagonal; branches erect, dichotomous; leaves *lanceolate,* acute at the base, scabrous, hairy above; stipules joining the petioles, truncate, ciliate; peduncles *solitary, axillary* and terminal, 1-flowered, often trifid, pedicels 4–5 times, tube of the corolla 3 times longer than the calyx, expanded at the throat; lobes acute; fruit globose, smooth. *Kohautia longiflora, E. Mey. non DC.*

HAB. Natal, *Drege, Krauss,* 305. *Gueinzius,* 470. *R. W. Plant,* 64. *Gerr. & M·K.* 583. Jul. (Herb. D., Sd.)

Branches ½–1 foot or longer. Leaves about 1 inch long, 3–4 lines broad, adpressed-hairy, sometimes glabrous above, ciliated at the base; upper ones smaller, bract like. Peduncles longer than the leaves, ½–1 inch long, the pedicels spreading, about twice shorter. Corolla 2 lines long. glabrous in the throat, the 4 lobes acute, anthers subexserted. Capsule crowned with the short calyx-teeth.

14. H. rupicola (Sond.); annual, glabrous; stems ascending; leaves shortly petiolate, *ovate, subcordate,* acute, a little scabrous above; stipules of many bristles; cymes corymbose, terminal, few-flowered; pedicels short; calyx scabrous; lobes ovate, acute, 4 times shorter than the corolla; style subexserted.

HAB. Rocky places, Tagoma, Natal, *Gerr. & M·K.* 1364. Feb. (Herb. D.)

A small herb, about a span high, with the habit of *H. (Lucya) tuberosa,* Linn. Leaves 5–6 lines long, 4 lines broad, more or less cordate at base; the margins revolute. Cymes 9–12-fl. Corolla white, 4 lines long; the lobes oblong; the throat hairy. Anthers inclosed. Style bifid at the apex. Capsule nearly 1½ line long, crowned by the calyx-lobes. Seeds black, punctate.

15. H. hirtula (Harv.); perennial? decumbent or prostrate, closely much-branched; leaves *ovate,* acute, cuneate at base, petiolate, flat, *roughly hispid with short, swollen hairs;* stipules with 2–4 setæ; pedicels axillary, *very short;* calyx-lobes ovate, hispid, equalling the tube; corolla-tube 2–3 times as long as the calyx, lobes oblong, hispid on both sides; fruit hispid.

HAB. On trees, Krauss Kloof, *J. Sanderson,* 605. (Herb. D.)

A small plant, 3–4 inches long, with the aspect of *H. ternervia,* but rough with white, swollen, papillæform hairs. Leaves 3–4 lines long, 2–3 wide; petiole 1–1½ line long. Pedicels 1–1½ line long. Corolla-tube 3–4 lines long. Fl. white.

16. H. Natalensis (Hochst. l. c. p. 67); perennial? *stem erect,* tetragonal, few-branched, *pubescent;* leaves lanceolate, acuminate, scabrous above, pubescent, at length glabrous beneath; stipules with 2 short deciduous bristles; flowers terminal, *fasciculate;* pedicels much shorter, tube of the corolla 3 times longer than the calyx, lobes ovate; fruit ovato-globose, smooth. *Crusea? acuminata, E. Mey.*

Between Morley and Omtata, and near Port Natal, *Drege, Krauss., Cooper,* 1083; Kreilis Country, Caffraria, *H. Bowker.* Feb.–Apl. (Herb. D., Sd.)

More than a foot high. Leaves about 1½–2 inches long, 5–6 lines broad. Three flowers aggregated at the top of the terminal branchlets. Calyx with glabrous oval tube 1 line long, terminated by the 3 times longer, linear, rigid-ciliate teeth. Corolla a little hairy, swollen at the throat. Stamens subexserted. Capsule the size of a pea. Seeds angular, punctate.

c. **Pentodon**; flowers pentamerous. (Sp. 17.)

17. H. pentamera (Hochst.!); quite glabrous, stem decumbent or creeping, tetragonal, dichotomously branched; leaves ovato-lanceolate, acute; stipules with 1 or 2 bifid bristles; flowers axillary, racemose, pedicels geminate, in fruit deflexed, 4–5 times longer than the shortly toothed calyx; corolla funnel-shaped, hairy inside, lobes acute; fruit ovato-globose, smooth. *Pentodon decumbens, Hochst. l. c. p. 67.*

HAB. Natal, *Drege*, 4887; *Krauss*, 332, *J. Sanderson; Guienzius*, 471; *Ger. & M'K.* (Herb. D., Sd.)

Stem 1 or 2 feet long. Leaves 1½–2 inches long, ½ inch broad, acute at both ends. Racemes simple, 5–10-flowered, or compound 12–16-flowered, on a peduncle as long or longer than the leaf. Pedicels with a minute bract. Calyx 1 line long, the teeth equalling the tube. Corolla 4 lines long, white? limb ovate. Capsule truncate, didymous, crowned by very short calyx teeth.

Doubtful species.

H. tenella (Hochst. l. c. p. 68); glabrous, stem delicate, branched; leaves spathulate, suborbicular; stipules with many bristles; peduncles axillary and terminal, 2–3-flowered; tube of the corolla nearly twice longer than the 4-cleft calyx, lobes ovate-elliptic, obtuse; capsule ovate, crowned with the triangular-ovate, recurved, spreading calyx-lobes.

HAB. Table Mountains, near Port Natal. Dec. *Dr. Krauss.*

TRIBE III.—GUETTARDACEAE (Gen. VII.)

VII. **VANGUERIA**, Comm.

Calyx-tube short, obovate or hemisphærical; limb spreading, 5, rarely 4-toothed or parted. *Corolla* campanulate, 5, rarely 4-cleft, hairy inside, lobes lanceolate, acute, reflexed. *Stamens* 5, or 4, with very short filaments, and scarcely exserted, oblong acute anthers. *Stigma* capitate, cylindrical, thick, or truncate at both ends, obtuse or toothed. *Drupe* pomiform, when ripe not crowned with the calyx lobes, but marked by a sinuated areola at the summit, containing 5, or by abortion 4–2 bony, 1-seeded nuts, which are obtuse at the base and acute at the apex. *Seed* affixed to the middle of the inner angle of the cell, oblong. *Albumen* fleshy. *Embryo* large, inverted. *DC. l. c.* 4, *p.* 454. *Endl. gen. n.* 3191. *Pachystigma, Hochst. Regenst. bot. Zeit.* 1842, *p.* 234. *Lagynias, E. Mey. in herb. Drege.*

Shrubs or small trees with ovate or oblong leaves, acute or lanceolate stipules, and cymose or somewhat panicled axillary white or reddish flowers. Name from Voa Vanguer, the Madagascar name of V. edulis.

Calyx 5-toothed, teeth about as long as the tube.
  Leaves reticulated-veined; stigma obtuse   ... ... (1) **infausta.**
  Leaves not retic.-veined; stigma 4–5-toothed   ... ... (2) **parviflora.**
Calyx 4–5-lobed, lobes 2–4 times longer than the tube.
  Leaves tomentose   ... ... (3) **macrocalyx.**
  Leaves glabrous.
    Corolla glabrous.
      Leaves obovate or obovate-lanceolate; cymes few-flowered   (4) **venosa.**
      Leaves suborbicular or broad ovate; peduncle 3-flowered   (6) **latifolia.**
      Leaves oblong-lanceolate-acuminate; flowers cymose ...   (7) **Zeyheri.**
    Corolla closely hairy   ... ...   (5) **lasiantha.**

1. V. infausta (Burch. Trav. II. p. 258 et 259, c. icon.); leaves shortly petiolate, ovate or suborbicular, acute or subacuminate, tomentose, *netted-veined*; cymes axillary, panicled; calyx-teeth tomentose, bluntish, 4 times shorter than the corolla; *stigma obtuse. Sond. Linn. vol.* 23, *p.* 57. *V. tomentosa, Hochst! Regenst. bot. Zeitg.* 1842, *p.* 238.

VAR. β. **virescens,** leaves green, glabrous or nearly so.

HAB. Forests in Caffraria, *H. Bowker ;* Natal, *Krauss,* 219. *J. Sanderson,* 306. *T. Williamson, Drege, Gueinzius,* 56. Vaalriver and Magalisberg, *Zeyher,* 764. Nov. (Herb. D., Sd.)

A tree about 10 feet high, unarmed, with terete and subcompressed, opposite branches. Leaves quite entire, densely clothed with a yellowish tomentum, penninerved, 4–5 inches long, 3–4 inches broad. Stipules large, cuspidate. Cymes shorter than the leaves, tomentose. Calyx subsessile ; the teeth acute or bluntish, shorter than the tube. Corolla tomentose, 3 lines long, the lobes glabrous above. Ovary 5-celled, style filiform, equalling the stamens ; stigma obtuse. Fruit round, glabrous, an inch in diameter, edible. The variety is nearly allied to *V. edulis,* Vahl. This is the "*Wild Medlar*" of the colonists.

2. V. parvifolia (Sond. l. c. p. 58) ; young branches reddish, as well as the leaves and flowers velvetty-tomentose ; leaves subsessile, orbicular, subacute, penninerved, *not netted-veined* ; cymes axillary, sessile; calyx-teeth acute, 4-times shorter than the corolla; ovary 5-*celled, stigma* 4–5-toothed.

HAB. Magalisberg, Nov. Dec. *Burke & Zeyher,* 765. (Herb. D., Sd.)

Shrub with spreading terete branches. Leaves 12–14 lines long, 12 lines broad, coriaceous. Cymes shorter than the leaves. Tube of calyx as long as the limb. Corolla 2½ lines long ; the lobes glabrous above. Style equalling the stamens ; stigma thick, cylindrical, a little larger at the minutely toothed apex. Fruit (half ripe) as large as a small cherry, crowned by the rudiments of calyx teeth.

3. V. macrocalyx (Sond. l. c. p. 59) ; *young branches, leaves, and flowers tomentose ;* leaves shortly petiolate, ovate, acute, at length subglabrous ; cymes axillary, subsessile, few-flowered, shorter than the leaves; calyx teeth linear-oblong, obtuse, as long as the corolla; throat of the corolla naked or nearly so ; ovary 5-celled ; stigma obtuse.

HAB. On the Zwartkopsrivier and Winterhoeksberge, Uit., *E. & Z., Ebenac,* 17; Kreili's country, *H. Bowker ;* Port Natal, *Gerr. & M'K.* 1344. Nov. (Herb. D., Sd.)

A shrub, 6–8 feet high. Leaves 1½–2 inches long ; quite entire or subdenticulate ; the petiole 1–2 lines long. Peduncles bracteolate. Calyx tube 1 line long. the lobes 3 lines long, 1 line broad. Corolla 2½ lines long, the tube nearly glabrous outwards, but silky near the base within, the throat quite glabrous, or with a few hairs; lobes glabrous above, acute. Ovary 5-celled. Stigma as in *V. infausta.* The ripe fruit as large as a common plum, 3–4 seeded, delicious eating. *(W. T. Gerrard).*

4. V. venosa (Sond.); leaves shortly petiolate, obovate or obovatelanceolate, obtuse, or subacute, pale and much veined beneath, as well as the branches quite glabrous; cymes shortly pedunculate, few flowered ; calyx lobes linear, obtuse ; corolla *glabrous* outside. *Pachystigma venosum, Hochst. l. c. Lagynias discolor, E. Meyer.*

HAB. Port Natal, *Krauss,* 131. *Gueinzius,* 553. (Herb. D., Sd.)

Branches alternate and opposite, terete. Leaves 1½–2 inches long, 8–10 lines broad, tapering into the petiole, green above, whitish and netted-veined beneath. Cymes shorter than the leaves ; peduncles and pedicels bracteate, hairy. Calyx lobes 3 lines long. Corolla 3–4 lines long, lobes acuminate. Fruit nearly 1 inch in diameter, globose, glabrous, with 5, or by abortion with 4 one-seeded nuts. Seed oblong, curved, affixed to a central placenta.

5. V. lasiantha (Sond.); leaves petiolate, *oblong* or *ovate-oblong,* subacute, discoloured, green and pellucid-veined above, greyish-white beneath, glabrous, the young ones hairy on the nerves ; cymes shortly pedunculate, branches divaricate, *many-flowered ;* calyx-lobes linear-spathulate, obtuse ; corolla *appressed-hairy. Pachystigma lasianthum,* Sond. *Linnæa, Vol.* 23, *p.* 55.

HAB. Natal, *Gueinzius,* 119, 549. *J. Sanderson,* 520, 658. *Gerr. & M'K.* 851. Dec. (Herb. Sond., D.)

Very nearly allied to *V. venosa,* but differs by larger, ovate, or oblong leaves on longer petioles, many-flowered divaricate inflorescence, spathulate calyx-lobes and hairy corolla. Leaves 2 inches long, about 1 inch wide ; petiole 3–4 lines long. Cymes 12–14 flowered, 1 inch long. Calyx lobes 2½ lines long, ⅓ line broad. Corolla 4¼ lines long, glabrous in the throat, but with reversed hairs in the tube ; lobes lanceolate-acuminate. Ovary 5-celled. Stigma thick, truncate at both ends.

6. V. latifolia (Sond.); leaves shortly petiolate, *suborbicular* or *broad ovate,* discoloured, pale and netted-veined beneath, glabrous ; *peduncles axillary, 3-flowered,* bibracteate, glabrous ; calyx glabrous or subpilose, lobes lanceolate; corolla *glabrous* outside. *Pachyst. latifolium, Sd. t. c. p.* 56.

HAB. Natal, *Gueinzius,* 115. *T. Williamson.* (Herb. Sond., D.)

Branches glabrous, terete or subcompressed. Leaves quite entire, 2 inches long, 1½ inch wide, the petiole 1 line long. Stipules as in *V. venosa* and *V. lasiantha.* Peduncle 6 lines long, at the summit with 3 flowers on very short pedicels. Bracts ovate. Corolla nearly as in *V. venosa.*

7. V. Zeyheri (Sond.) ; leaves *oblong-lanceolate, acuminate,* attenuate at the base, quite glabrous, green, with prominent nerves and veins on both sides ; cymes axillary, shorter than the leaves, glabrous as well as the flowers; calyx lobes lanceolate; corolla glabrous outside. *Pachystigma Zeyheri, Sond. l. c.*

HAB. Stony places in woods, Magalisberg. Nov. *Zeyher,* 766. (Herb. Sd., D.)

Shrub with opposite, glabrous branches. Stipules broad, cuspidate, about 2 lines long. Leaves quite entire, coriaceous, 4–5 inches long, 12–14 lines broad, tapering into a petiole 3–6 lines long. Cymes opposite, 2 inches, the primary peduncle 1 inch long; the pedicels bracteolate. Calyx 3 lines long; tube hemisphærical, lobes twice longer. Corolla 4 lines long, with a series of silky hairs in the tube ; the throat and lobes glabrous. Ovary 5-celled, by abortion 2–3-celled. Style filiform ; stigma truncate at both ends. Drupe (unripe) ellipsoidal, glabrous, crowned with the vestiges of the calyx-lobes.

TRIBE IV.—**ALBERTIEÆ** (Gen. VIII.)

VIII. ALBERTA, E. Meyer.

Calyx-tube turbinate, 10-ribbed; limb 5-fid, 3 lobes short, acute, 2 oblong, obtuse, in fruit much larger. *Corolla* elongated, tubular, incurved; throat naked; limb 5-fid, with very short, erect, imbricate, acute lobes. *Stamens* 5, included ; anthers sessile. *Ovary* 2-celled ; ovules solitary. *Style* filiform, exserted; *stigma* attenuated, very shortly bifid. *Fruit* dry, ovate-oblong, 10-ribbed and furrowed, crowned with the 2 opposite, foliaceous, dilated, membranaceous, reticulate-veined, coloured calyx lobes, 2-celled ; cells 1-seeded. *Seeds* oblong, convex at back, flat in front. *Albumen* fleshy. *Endl. gen. n.* 3327.

A shrub or small tree with subsessile, oblong, obtuse, coriaceous, glabrous, glossy leaves, cuplike aristate stipules, terminal panicle, with cymose branches and purplish, silky pubescent corollas. It is named in honour of Albertus Magnus, of the house of Bolstadt, a famous philosopher and theologian of the thirteenth century, who wrote a treatise in seven books " De Vegetabilibus et Plantis."

1. A. magna (E. Mey. in Herb. Dreg.) *Harv. Thes. Cap. p. 29, t. 45.*

HAB. On hills between Omsamcabe and Omsamculo, 1000–2000 feet, *Drege ;* Port Natal, *Gerrard & M'Ken,* No. 1358, Jan.—Feb. (Herb. Sd., D.)

Young branches reddish brown, glabrous. Leaves shortly petiolate, oblong or elliptical-oblong, 4–5 inches long, 1½–2 inches broad, flat, with recurved margins, penninerved. Panicle large. Pedicels pubescent, bracteate at the base. Calyx shortly tomentose. Corolla 1 inch long, the segments obliquely ovate, acute. Stamens inserted in the middle of the tube. Fruit oval, nearly 3 lines long, its scarlet, winglike, larger, much veined lobes, 10–12 lines long, 2–3 lines broad.

TRIBE V.—COFFEACEÆ (Gen. IX.–XIV.)

IX. CANTHIUM, Lam.

Calyx-tube ovate, limb short, 4–5 toothed. *Corolla* with a short tube, a bearded throat, and 4–5 spreading lobes, with valvate æstivation. *Anthers* 4–5 in the throat, scarcely exserted. *Style* filiform, exserted. *Stigma* undivided, thick, ovato-globose, or mitre-formed. *Berry* globose, or didymous, crowned with the calyx teeth, fleshy, 2-celled. *Seeds* solitary in each cell. *Albumen* fleshy. *DC. l. c.* 4, *p.* 473.

Shrubs with spinous or unarmed branches. Leaves opposite, coriaceous. Stipules solitary. Peduncles axillary, short, many-flowered. *Canti* is the Malabar name of one of the species.

1. C. Gueinzii (Sond. Linnæa, vol. 23, 1, p. 54); unarmed, *branchlets pubescent ;* leaves shortly petiolate, *oblong, cuspidate-acuminate, subcordate,* netted-veined, veins pellucid, glabrous, shining above, a little hairy on the nerves beneath ; cymes pedunculate, bifid, many-flowered, 4-times shorter than the leaves; *peduncles, pedicels as well as the 5-toothed calyx pubescent ;* corolla glabrous outside, lobes revolute, as long as the tube ; style exserted, stigma mitre-formed.

HAB. Natal, *Gueinzius,* 71, 576. *Gerr. & M'K.* 709. (Herb. Sd., D.)

Stipules broad, cuspidate-acuminate, hairy. Leaves reddish, 3¼–4 inches long, ¾–1½ inch broad ; the petiole 2 lines long. Peduncle of the cyme half an inch ; the pedicels bracteolate. Calyx about 1 line long, with very short teeth. Corolla 2½ lines long; throat and tube bearded within; lobes acute, revolute, as long as the tube. Distinguished from the similar *Canth. Cornelia,* Cham. and Schl. (*Pavetta Cornelia, Reichb. in Sieb. pl. Senegal)* by the much larger leaves, petioles and calyx not being villous, shorter calyx-teeth, and smaller flowers.

2. C. obovatum (Klotzsch. in E. Z. Enum. n. 2293); unarmed, *glabrous ;* leaves shortly petiolate, *obovate or elliptical, subacute at the base,* shining above, paler below, few-nerved ; cymes axillary, trichotomous, many-flowered, shorter or equalling the leaves; *peduncles, pedicels, and calyx glabrous;* corolla glabrous, lobes revolute ; style exserted ; stigma mitre-formed. *Mitrastigma lucidum Harv.!* in *Lond. Jour. Bot.* vol. 1, p. 20. *Thes. Cap. t.* 22. *Phallaria lucida, Hochst.! Flora,* 1842, p. 238. *Psychotria obtusifolia, E. Mey. et Psychot,* 2361. *Herb. Drege.*

VAR. β. pyrifolium ; leaves elliptic or suborbicular, obtuse at the base. *C. pyrifolium Klotzsch, E. Z.* 2294, et *Ehretia ?* 62.

HAB. In distr. Uitenhage, near Adow, Winterhoeksberg, *E. & Z. ;* Glenfilling and near Strandfontein and Matjesfontein, *Drege ;* Van Stadesberg, *Zeyh.* 2706 ;

Natal, *Krauss,* 178.   *Gueinzius, T. Williamson, Gerrard & M'Ken.*   Var. β. in Krakakamma, *E. & Z.*   Jan.–Feb.   (Herb. D., Sd.)

Shrub or tree 12–15 feet high, with tetragonal subcompressed branches. Stipules cuspidate. Leaves 1–2 inches long, with revolute margins, and a gland at the base of the lateral nerves. Cymes pedunculate. Calyx 1 line long, with very short teeth. Corolla 2 lines long, 5, rarely 4-parted, the lobes acute ; the inner surface of the tube and throat hairy. Anthers subsessile, exserted. Style much exserted. Fruit the size of a large pea, roundish, subcompressed, 1–2-seeded.

X. PLECTRONIA, L.

Calyx-tube obovate or oblong ; limb with 5 short teeth. *Corolla* subinfundibuliform, with a 5-parted limb, acute, reflexed segments valvate in bud, and a hairy or naked throat. *Stamens* 5, in the throat, subexserted. *Style* short. *Stigma* subcapitate, of 2 approximate lamellæ. *Berry* fleshy, obovate-oblong, compressed, didymous, of 2 pyrenæ. *Pyrenæ* indehiscent, 1-seeded. *Albumen* fleshy. *Lam. ill. t.* 146. *Plectronia et Psilostoma, Klotzsch. in E. Z. enum.*

Shrubs or small trees with subspinous branches, ovate leaves, and axillary racemose or corymbose peduncles, with small flowers. Name from πληκτρον, a whip, in allusion to the square branches. [Too nearly allied to *Canthium,* W. H. H.]

Throat bearded.
Leaves pale beneath ; petioles glabrous ... ...   (1) **ventosa.**
Leaves one-coloured ; petioles pubescent ... ...   (2) **Mundtiana.**
Throat naked.
Pedicels axillary, 2-flowered; tube of corolla as long as the limb   (3) **parviflora.**
Pedicels axill., 1–2-flowered ; tube of corolla longer than the limb   (4) **ciliata.**
Pedicels axill., 5–12-fl.; tube of corolla shorter than the limb   (5) **spinosa.**

1. P. ventosa (Linn. Mant. p. 52); *glabrous,* branches brachiate, patent, spiny ; twigs quadrangular ; leaves opposite, petiolate, ovate or ovate-oblong, entire, subcoriaceous, *pale on the lower surface ;* corymbs 10–20-flowered; tube of the corolla conical, longer than the limb ; throat bearded. *Serissa Capensis, Thunb. l. c. p.* 193, *excl. syn. Cruse Rubiac. cap. p.* 20, *t.* 2, *E. Z.!* 2297. *Canthium Thunbergianum, Cham. & Schl. in Linnæa,* 4, *p.* 130. *Burm. Afr. n.* 257, *t.* 94.

HAB. In woods and ravines throughout the Colony. Oct.–Nov. (Herb. D., Sd.)
A tree, from 10–20 feet high, diameter from 6–10 inches. Bark white. Wood hard, close, and tough. Spines of the branches horizontal. Leaves quite glabrous, obtuse, or subacuminate, 1–2½ inches long. Petioles 4–6 lines long, glabrous. Corymbs shorter than the leaves. Corolla about 3 lines long. Style equalling the throat. Fruit about 5–6 lines long, 4 lines broad. Seeds oblong. *Plectronia ventosa,* Thunb. fl. cap. p. 195, is *Apodytes dimidiata, Benth.* (Conf. vol. 1, p. 235).

2. P. Mundtiana (Pappe Sylv. cap. p. 19); branches erect-spreading, a little spiny ; twigs subangular ; leaves opposite, petiolate, ovate, or elliptical, entire, membranaceous, *one-coloured, petioles and middle rib pubescent* on the upper surface ; corymbs 5–10-flowered ; tube ot the corolla cylindrical, as long as the limb, *throat bearded. Canthium Mundtianum Cham. et Schlecht. in Linnæa,* 4, *p.* 131. *E. & Z.!* 2295. *Herb. Un. Itin. No.* 637. *Drege,* 6747.

VAR. β. **pubescens** (Cham. and Schl. l. c.); branches and leaves softly pubescent. *P. parvifolia, Mundt. in Herb. E. & Z.*
HAB. On the Table and Devil's Mountain, Simon's Bay, Grootvadersbosch, Plet-

tenbergsbay, and near Gnadenthal.   Var. *β.* near Witte-boome, Table Mountain. Nov.–Dec. (Herb. Sd., D.)
    Arborescent shrub, 5–8 feet high.   Bark and wood white. Very similar to the preceding, but branches more straight, less spiny, spines much shorter, not dis-coloured leaves with hairy petioles, and smaller flowers.

3. P. pauciflora (Klotzsch. ! E. Z. n. 2300); *glabrous, branches and twigs terete;* leaves opposite, petiolate, ovate, acute, few-nerved, entire, membranaceous, *one-coloured;* peduncles opposite, axillary, 2-flowered; tube of the corolla *as long as the limb; throat naked;* tube hairy in the middle inside.   *Randia ? triplinervis, E. Meyer. in Herb. Drege.*

HAB. W., Plettenbergsbay, *Mundt.;* Natal, *Drege. Cooper,* 1165.   Dec.–Jan. (Herb. D., Sd.)
    A shrub, probably unarmed, with erect, spreading, greyish-green, virgate branches. Leaves 1 inch long, 8 lines broad, with 2 or 3 pairs of lateral nerves, not veined. Petiole 2 lines long. Stipules cuspidate, short.   Peduncle ¼ inch ; the 2 pedicels 4 lines long. Calyx 1 line. Corolla 2 lines long ; the lobes acute; the throat quite naked, but there is a hairy ring in the tube below the anthers. Fruit black, didy-mous, but by abortion of the one cell, obliquely oblong, nearly ½ inch long.

4. P. ciliata (Sond.) ; branches spiny, terete, glabrous ; twigs pu-bescent ; leaves *opposite,* shortly petiolate, ovate or subcordate-ovate, subacute, penninerved, *pale and veined beneath,* appressed-hairy, or glabrous above; peduncles opposite, axillary, 1-, rarely 2-fl.; tube of corolla longer *than the limb,* throat naked ; tube with scattered hairs inside.   *Psilostoma ciliata, Kl. ! E. Z.* 2296.   *Randia ? venosa, E. M.*

VAR. *β.* **mollis** ; leaves softly pubescent on both sides; corolla hairy outside, *Pl. mollis, Gerr. MS.*
VAR. *γ.* **glabrata** ; leaves quite glabrous on both sides, or with scattered hairs beneath ; corolla glabrous.
HAB. On the Katriver, near Philipstown and Nieuweport, *E. & Z.;* Howisons-poort, *Zeyh.* 2708 ; Buffel River, Gekau, Morley, and near Port Natal, *Drege,* var. *β.* Natal, *Gueinzius,* 554.   *J. Sanderson,* 444, 281.   *Gerr. & M'K.* 5, et 1345 ; var. *γ.* Port Natal, *Gerr. & M'K.* 1346. ‑Oct.–April. (Herb. Sond., D.)
    A large shrub with horizontal opposite spines.   Leaves opposite, rarely fasciculate, 8–12 lines long, subcoriaceous ; petioles 1–2 lines long.   Stipules subu-late, equalling the petiole.   Peduncles capillary, ½–1 inch long, bibracteate in the middle.   Corolla 3 lines long.   Anthers subsessile.   Style subexserted ; stigma capitate.·   Fruit didymous, one cell often wanting by abortion ; about 4 lines long.

5. P. spinosa (Klotzsch.! E.Z. n.2298); nearly glabrous, branches very spinous, terete ; leaves shortly petiolate, *fasciculate, oval or obovate, ob-tuse,* attenuate at the base, quite entire, penninerved, paler and veined beneath; peduncles axillary, solitary or fasciculate, racemose or corym-bose-panicled, 5–12-flowered; tube of the corolla *shorter than the limb ;* throat and tube naked.

HAB. Woods near Adow and Olifantshoek, *E. & Z.;* Stadensriver, *Zeyh.* 674 *Drege,* 6750; Natal, *Gueinzius,* 403; *Gerr. & M'K.* 534. Dec.–Jan. (Herb. Sd., D.)
    Habit of a Celastrus ; shrub 5–8 feet in height.   Branches spreading, greyish-white.   Spines ½–1 inch long.   Stipulæ subulate.   Leaves rising from an axillary bud ; petiole 1–2 lines long.   Young leaves as well as the twig sometimes appressed-hairy, old leaves quite glabrous, 1–1½ inch long.   Racemes or panicles shorter than the leaves ; pedicels about 2 lines long.   Calyx with very short teeth.   Corolla 1¼ lines long.   Fruit didymous, one cell often abortive.

XI. **PAVETTA**, Linn.

Calyx-tube ovate; limb 4-toothed or cleft. *Corolla* salver-shaped, with a long slender tube, and a 4-parted spreading limb. *Anthers* 4, sessile in the throat. *Style* much exserted beyond the tube of the corolla, stigma clavate, entire or nearly so. *Berry* drupaceous, 2-celled, crowned with the limb of the calyx. *DC. l. c.* 4, *p.* 485.

Shrubs or small trees, with opposite leaves and corymbs of white or yellow flowers. Pavetta is the vernacular name of P. indica in Malabar.

(1) Calyx-teeth setaceous, 2-4 times longer than the tube.
     Leaves obovate, almost sessile, glabrous ... ...   (1) **Caffra.**
     Leaves elliptic, petiolate, pubescent... ... ...   (2) **Cooperi.**
     Leaves ovato-lanceolate, glabrous, hairy on the midrib
       beneath; panicle pubescent ... ...   (5) **Gerrardi.**
     Leaves oblong-lanceolate, tomentose beneath; panicle
       glabrous ... ...   (4) **Bowkeri.**
     Leaves lanceolate-acuminate, petiolate, glabrous ... ...   (3) **Natalensis.**
(2) Calyx-teeth as long as the tube.
     Leaves obovate ... ...   (6) **obovata.**
     Leaves oblong-lanceolate, attenuate at the base ... ...   (9) **Zeyheri.**
(3) Calyx-teeth minute, shorter than the tube.
     Leaves obovate, limb of calyx with minute, acute teeth   (7) **assimilis.**
     Leaves lanceolate-acute; limb of calyx repando-dentate   (8) **edentula.**
     Leaves lanceolate-attenuate; teeth of calyx subobtuse,
       half as long as the tube ... ...   (10) **lanceolata.**

1. P. Caffra (Thunb.! fl. cap. p. 146); branches terete, as well as the leaves *glabrous;* leaves *obovate, almost sessile;* stipules broad, cuspidate; corymbs dense-flowered; teeth of calyx setaceous, twice longer than the tube. *E. Z.!* 2302. *P. corymbosa, Houtt. Nat. Hist. t.* 10; *Ixora caffra, Poir. suppl.* 3, *p.* 209.

VAR. β. **pubescens**; branches, leaves, and flowers shortly pubescent.

HAB. In woods, Dists. of Uitenhage, Albany, Caffraria and Port Natal, *Thunb. E. & Z. Drege, Zeyh.* 2711. Dec.—Jan. (Herb. Thunb., Sd., D.)

Shrub 3–6 feet high, with whitish branches. Leaves with slightly recurved margins, the lateral nerves often not conspicuous, quite glabrous, or in Var. β. pubescent, 1½–2 inches long, about 1 inch broad, attenuated at the base, sometimes in a short petiole. Teeth of calyx 2 lines, tube of corolla 6 lines long. Fruit black, shining, as large as a pea crowned with the calyx.

2. P. Cooperi (Harv. & Sond.); branches terete, *very shortly pubescent;* leaves *petiolate, elliptic,* acute at both ends, hairy-scabrous above, softly pubescent and strongly nerved beneath; stipules broad, cuspidate; corymbs capitate; calyx-teeth twice as long as the tube, setaceous.

HAB. Natal, *T. Cooper,* 1175; *Gerr. & M'K.* 1349. (Herb. D., Sd.)

A large shrub. Leaves 1½–2 inches long, 1–1½ inch broad, with very prominent middle and divergent lateral nerves. Petioles 3–6 lines long. Corymb globose, peduncles and pedicels very short, pubescent as well as the 2 lines long calyx-teeth. Corolla pilose, or nearly glabrous, white; length of the tube and size of the fruit as in *P. Caffra,* from which this species, at first sight, differs by the elliptical, nerved leaves.

3. P. Natalensis (Sond.); quite glabrous; young branches compressed; leaves on longish petioles, *lanceolate, acuminate,* attenuate at the base, shining; stipules cuspidate-acuminate; corymb loose, peduncle naked at the base, pedicels 1-flowered; teeth of calyx three times as long as the tube, setaceous.

HAB. Port Natal, *Gerrard & M'Ken.* (Herb. D.)
Very different from the preceding. The leaves are 3-4 inches long, 7-10 lines broad, with subrecurved margins, much acuminated at the apex. The corymb or racemose panicle, in the only fruit-bearing specimen I have seen, is 1 inch high and 2 inches wide. Calyx-tube 1 line long. Fruit somewhat larger than in *P. Caffra.*

4. P. Bowkeri (Harv. Thes. t. 131); twigs compressed, clothed with a very short bifarious pubescence; leaves *oblong-lanceolate,* tapering into a pubescent petiole, with recurved margins, quite glabrous above, *tomentose beneath;* corymbose panicle loosely trichotomous, *glabrous;* teeth of calyx 3-4 times longer than the tube, setaceous.

HAB. Mouth of Bashee R., Caffr., in swamps. Feb. *H. Bowker,* 459. (Hb. D., Sd.)
Leaves 3-4 inches long, 1-1½ inch broad, shortly acuminated into a bluntish point, penninerved, softly tomentose beneath; petiole ½ inch long. Stipules cuspidate or aristate. Panicle spreading; bracts very small. Calyx 3 lines long. Tube of the glabrous white corolla ½ inch, lobes 3 lines long, acute.

5. P. Gerrardi (Harv. MSS.); twigs compressed, minutely downy; leaves *long-petioled,* ovato-lanceolate, acute or subacuminate, tapering at the base, with subrecurved margins, glabrous, but hairy on the middle nerve beneath; corymbose panicle trichotomous, *shortly pubescent;* teeth of calyx twice longer than the tube, setaceous.

HAB. Tugela River, Natal, *Gerr. & M'K.* 1350. (Herb. D.)
A bushy shrub, 2-4 feet high, with light green foliage and large white flowers, nearly allied to *P. Bowkeri,* from which it chiefly differs in the longer, broader, glabrous, and green leaves. Leaves 5-6 inches long, and nearly 2 inches broad. Petioles pubescent, ½-1 inch long. Stipules cuspidate. Panicle many-flowered. Corolla glabrous, the size as in the preceding.

6. P. obovata (E. Mey. in Dreg. enum.); quite glabrous, twigs compressed; leaves obovate, tapering into the petiole, obtuse or acute, with revolute margins, subcoriaceous, penninerved; panicle corymbose, loosely trichotomous; teeth of calyx *acute, as long as the tube. P. revoluta, Hochst.! Flora,* 1842, *p.* 237.

HAB. Port Natal, *Drege, Krauss.* 98. Kowie sand hills, *P. Mac Owan,* 420. *J. Sanderson. Plant.* (Herb. Sd., D.)
Leaves 2-3 inches long, 12-15 lines broad, paler beneath. Petiole 1-4 lines long. Panicle spreading; bracts minute. Calyx 1 line long. Corolla white, the tube 6 lines, limb nearly 3 lines long. Style very long. Fruit as large as a large pea, shining, crowned by the rigid calyx.

7. P. assimilis (Sond.); quite glabrous; twigs subcompressed; leaves *obovate,* tapering in the petiole, obtuse or subacute, with recurved margins, subcoriaceous, penninerved; panicle corymbose, trichotomous; limb of calyx cupshaped, *with very minute, acute teeth.*

HAB. Port Natal, *Gerrard & M'Ken,* 1355. Dec. (Herb. D., Sd.)
Closely resembling *P. obovata* in habit and in the size and form of the leaves, but well characterized by the calyx. Panicle 1-1½ inch long. Calyx 1 line long, the teeth like a mucro. Tube of corolla 4 lines, the limb 2 lines long. Style one half longer than the corolla.

8. P. edentula (Sond.); quite glabrous; twigs terete; leaves *lanceolate* acute, tapering in a longish petiole, flat, coriaceous, shining, penni nerved, and reticulate; panicle corymbose, much shorter than the leaves; limb of calyx *cupshaped, repando-dentate.*

HAB. Port Natal, near Ingoma. March. *Gerr. & M'K.* 1353. (Herb. D.)

A shrub, 4–5 feet. Leaves 6–7 inches long, 1¼–1½ inch broad. Petiole 1 inch long. Stipules very broad, aristate. Panicle 1½–2 inches long, with minute bracts. Calyx 1½ line long, repand, without teeth. Corolla white, tube 6 lines, limb 2 lines long. Exserted part of the style as long as the corolla.

9. P. Zeyheri (Sond.); quite glabrous; twigs terete or subcompressed; leaves (small) *oblong-lanceolate* or oblong, *tapering at the base,* nearly sessile; corymb abbreviate, 8–12 flowered; teeth of calyx *as long as the tube,* acute. *P. lanceolata, Zeyh.* 768, *not of E. & Z.*

HAB. Magalisberg, Vaalriver, *Zeyher & Burke.* Dec. (Herb. Sd., D.)
Shrub with widely spreading, opposite branches. Leaves often tufted, 10–12 lines long, 2 lines broad. Corymb as long or shorter than the leaf. Calyx nearly 1 line long. Corolla wanting. Fruit as large as a pea, black.

10. P. lanceolata (Eckl.! in South Afr. Quart. Journ. 1830, p. 374); quite glabrous; branches subcompressed; leaves petiolate, *lanceolate,* attenuate at both ends, with recurved margins, coriaceous, shining above; panicle corymbose, trichotomous; teeth of the calyx bluntish, *twice shorter than the tube. E. Z.* 2303. *P. lanceolata, E. Mey. in* H*b. Drege.*

HAB. Forests in Olifantshoek, Uitenhage; near Grahamstown; Fort Beaufort, Katriver, and Natal, *E. & Z., Drege, Gerr. & M'Ken.*—Jan. (Herb. D., Sd.)
A shrub, 6–8 feet high, with yellowish or grey branches. Leaves 2½–3 inches long, 5–9 lines broad. Petioles short, in the specimens from Natal often 4–6 lines long. Stipules aristate. Corymbs terminal, lax, many-flowered. Calyx 1 line, tube of corolla 4 lines, lobes 2 lines long. Fruit black, globose, as large as a small pea.

XII. **GRUMILEA**, Gaertn.

Calyx-tube urceolate, limb with 5 very short teeth. *Corolla* with a short tube, villous in the throat; limb 5-parted, reflexed, valvate in bud. *Stamens* sub-exserted; anthers oblong. *Ovary* 2-celled, cells 1-ovulate. *Style* exserted; stigma bifid. *Berry* globose, 2-celled, 2-seeded. *Seed* planoconvex. *Albumen* cartilaginous, ruminate, or grumose from chinks and fissures. *Embryo* erect, small. *DC. prod.* 4, *p.* 495.
Shrubs. Leaves opposite, petiolate, elliptic or obovate, coriaceous. Stipulæ acute, broad at the base, deciduous. Flowers terminal, cymose. Name from *grumula,* a little heap; in reference to the albumen, which is grumose.

1. G. Capensis (Sond.); glabrous; branches subtetragonous or compressed; leaves shortly petiolate, obovate or oblong-ovate, subacute, attenuated at the base, with revolute margins; corymbs pedunculate, trichotomous; bracts and pedicels pilose. *Logania capensis, Eckl.! in South Afr. Quart. Journ.* 1. *p.* 371. *Plectronia citrifolia, E.Z.! n.* 2299. *P. coffeacea, E. Z.! pl. exsics. Grumilea cymosa, E. Mey.! G. globosa, Hochst.! Flora XXVII. II. p.* 554.

VAR. β. **angustifolia**; leaves lanceolate-oblong, obtuse, mucronulate or subacute.
VAR. γ. **pubescens**; leaves obovate, or obovate-oblong, pubescent beneath, at length glabrous.

HAB. Woods, Zuureberge, Drege; Olifantshoek, on Bosjesmannsrivier, *E. & Z.;* Howison's Port, *H. Hutton, Zeyh.* 2707; Port Natal, *Drege, J. Sanderson; Krauss.* 428. *Gerrard & M'Ken,* 830. Var. β. near Grahamstad, *Zeyh.* Var. γ. Port Natal. Sept.—Jan. (Herb. Sond., D.)
Evergreen shrub or tree. Leaves penninerved, 3–5 inches long, 1½–2 inches broad, shining above, a little paler beneath, quite glabrous, or with short adpressed hairs on the middle nerve; in var. γ. evidently pubescent, especially on the nerves. Pe-

tiole 2-4 lines long. Stipules coriaceous, very broad, mucronulate. Corymb trifid, on an uncial peduncle; pedicels appressed-hairy, at length glabrous, bracteated at the base; the ultimate pedicel terminated by a 6-12-flowered umbel. Calyx ½ line long, shorter than its pedicel, sinuated, but acutely toothed. Corolla yellow *(Gerrard)*, 3 lines long; lobes equalling the tube. Filaments glabrous. Lobes of stigma thickened. Fruit the size of a peppercorn, black, shining, 2-seeded. Seeds ruminate.

XIII. KRAUSSIA, Harvey.

Calyx-tube ovate; limb short, 5-cleft. *Corolla* funnel-shaped, tube short, obconical, limb 5-cleft, lobes oblong, with imbricate æstivation; the throat densely hairy. *Stamens* 5, in the throat, exserted; filaments short. *Anthers* erect, attenuate. *Ovary* 2-celled, cells 1-ovulate. *Style* very short. *Stigma* clavate, fusiform, striato-lamellate, bifid, lobes erect or revolute. *Berry* globose, crowned with the calyx limb, 2, or by abortion, 1-seeded. *Harv. in Lond. Journ. Bot. vol.* 1, *p.* 20.

Glabrous shrubs with lanceolate or ovate-lanceolate, acute, quite entire glossy, shortly petiolate leaves, broad, abruptly subulate-acuminate stipules and axillary cymes of white(?) flowers. Named in honour of *Dr. Ferdinand Krauss, professor at Stutgardt,* who travelled and collected in South Africa.

Stigma club-shaped, undivided or bidentate, erect :
    Lvs. sub-petiolate, oblong or lance-oblong, subacute ... (1) **floribunda.**
    Lvs. petiolate, broadly-oblong, much acuminate ... ... (2) **pavettoides.**
Stigma deeply 2-lobed, the lobes revolute :
    Lvs. lanceolate-acuminate ... ... (3) **lanceolata.**
    Lvs. ovate-oblong, subacute ... ... (4) **coriacea.**

1. **K. floribunda** (Harv. 1. c.); branches reddish brown, quadrisulcate; leaves *ovate-oblong or lanceolate-subacute;* cymes axillary, paniculate, nearly as long as the leaves, pedicels spreading; limb of the calyx *obtuse;* anthers subsessile ; stigma very large, *clavate, as long or a little longer than the style. Harv. Thes. t. 21. Coffea Kraussiana, Hochst. in Flora,* 1842, *p.* 237. *Psychotria oblongifolia, E. Mey. a. (non. b.)*

HAB. Port Natal, *Krauss,* 121; *Gueinzius,* 109, 111, 551. *Drege, Gerr. & M'K.* 706. (Herb. Sond., D.)
Leaves 2½-3 inches long, 1 inch broad, penninerved and with a gland at the base of each nerve below. Stipules mucronulate. Cymes trichotomous, the peduncle nearly an inch long; the pedicels with a minute lanceolate bracteole at the base, and about twice as long as the obtuse calyx. Corolla 3-4 lines long. Filaments very short. Fruit the size of a garden pea.

2. **K. pavettoides** (Harv.) ; twigs dark-colored, bluntly 4-angled ; leaves obovate-oblong, *acuminate, tapering at base into a petiole;* cymes terminal and axillary, peduncled, trichotomous ; pedicels spreading, puberulous as well as the calyx; calyx-lobes obtuse; anthers subsessile; style hoary in the middle, twice as long as the club-shaped, bidentate stigma.

HAB. Ravines on Field-Hill, Natal, *J. Sanderson,* 656. (Herb. D., Sd.)
A tree or large shrub, 10-20 feet high, with the habit of a *Pavetta.* Leaves 4-5 inches long, 1½-2½ inches broad, of thin substance, turning very dark in drying, narrow cuneate at base, and passing into a ½-¾ inch long petiole, distantly penninerved and netted beneath ; the midrib puberulous. Cymes mostly sub-terminal, from the axils of the smaller leaves. Peduncles uncial; pedicels much-branched, corymbose. Stipules cuspidate. Tube of corolla ¾ as long as the limb. Throat *minutely,* but closely *pubescent* within : by which character this differs from the other species.

3. K. lanceolata (Sond. in Linn. xxiii. p. 53); branches yellowish quadrisulcate; *leaves lanceolate, acuminate;* cymes axillary, many-flowered, much shorter than the leaves; limb of the calyx *acute;* filaments exserted, nearly as long as the anthers; stigma *bifid, 3 times shorter than the style,* with revolute lobes. *Carpothalis lanceolata, E. Mey. Psychotria saligna, E. Mey. (fruct.)*

HAB. Natal, *Drege, Gueinzius,* 68, 556. *Gerr. & M'K.* 713. (Herb. Sd., D.)

Leaves subcoriaceous, nerved and veined on both surfaces, 3–3½ inches long, 8–10 lines broad; petiole 4 lines long. Cymes 6–8 lines high; the bracteoles sheathing. Calyx acutely 5-toothed. Corolla 3 lines long, lobes oblong, reflexed. Fruit globose, the size of a small pea, crowned by the calyx, with 2 pyrenæ.

4. K. coriacea (Sond. l. c.); branches pale, subangulate; leaves coriaceous, very shortly petiolate, *oblong or ovate-oblong,* subacute at both ends, with recurved margins; cymes axillary, 4–8 flowered, three times shorter than the leaves; calyx acutely 5-toothed; filaments very short; stigma bifid, nearly 4 times shorter than the style, with revolute lobes.

HAB. Natal, *Gueinzius,* 100; *Gerr. & M'K.* 712. (Herb. Sd., D.)

Branches greyish white. Leaves 3 inches long, 1 inch broad, reticulated veined; the petiole about 2 lines long. Pedicels 4–5 lines long, with 2 alternate bracteoles. Corolla 4 lines long, the reflexed lobes nearly as long as the tube; the throat very villous with white hairs. Filaments affixed below the middle of the thrice longer linear anthers. Ovary 2-celled. Near the preceding, but easily distinguished by the thickish, coriaceous, not acuminate leaves, and the densely-bearded throat of the corolla.

XIV. **BUNBURYA**, Meisn.

Calyx-tube subglobose; limb shortly bilabiate, 6-toothed. *Corolla* funnel-shaped, limb 6-lobed, lobes lanceolate-oblong, twice shorter than the tube, with imbricate æstivation. *Stamens* 6 in the throat, exserted; filaments short, affixed in the middle of the linear-oblong anthers. *Ovary* bilocular, cells 1-ovulute. *Style* filiform, glabrous. *Stigma* bilamellate, lamells short, acute. Fruit ... *Meisn. in plant. Krauss, Reg. Bot. Zeitg.* 1844, *non. Harvey. Natalanthe, Sond. in Linnœa, vol. xxiii. p.* 52.

A shrub with the habit of *Coffea Arabica;* branches opposite, twigs mostly pubescent; stipules broad, with subulate acumen; leaves shortly petiolate, ovate-lanceolate, tapering to a bluntish point, obtuse or acute at the base; cymes axillary 3–4-flowered on very short peduncles; calyx subtended by a cup-like toothed bract. Named after *Sir Charles J. F. Bunbury, Bt. F.R.S.* who travelled and botanized in South Africa.

1. B. Capensis (Meisn. l. c.); *Harv. Thes. Cap. t.* 132. *Natalanthe floribunda, Sond. l. c.*

HAB. Near Port Natal, *Krauss, Drege, Gueinzius,* 69, 560. *Gerrard & M'K.* 1368; Kreili's Country, Caffraria, *H. Bowker.* July. (Herb. Sd., D.)

Large leaves 3–4 inches long, 1½ inch broad, penninerved and veined, glabrous, except on the nerves beneath. Petioles nerved, and calyx adpressed- or subsilky-pubescent. Corolla ½ inch long, with 6, rarely 7 lobes, white.

TRIBE VI.—**SPERMACOCEÆ** (Gen. XV.–XVIII.)

XV. **SPERMACOCE**, Meyer.

Calyx with an ovate or turbinate tube, and a 2–4-lobed limb, and sometimes accessory teeth in the recesses of the primary lobes. *Corolla*

salver-shaped or funnel-shaped, 4 lobed. *Stigma* bifid or undivided.
Capsule crowned, 2-celled ; nuts 1-seeded, dividing into 2 parts from
the apex ; the one part closed by the adnate dissepiment, the other
open. *Seed* oval-oblong, marked inside by a longitudinal furrow. *DC.
prod.* 4, *p.* 552.

Herbs or sub-shrubs, stems or branches tetragonal. Leaves opposite. Stipules
combined with the petioles, sheathing, fringed by numerous bristles. Flowers axillary, sessile, crowded. Name from σπερμα, a seed, and ακωκη, a point, in allusion
to the capsule being crowned by the calycine points.

1. S. Natalensis (Hochst. in pl. Krauss.) ; stem erect, glabrous ;
leaves linear-lanceolate or linear, with revolute margins, glabrous,
sometimes scabrous on the middle nerve beneath ; stipules rigid, with
5 bristles on each side ; flowers glomerate, semi-verticillate ; stamens
sub-inclosed ; capsule crowned by the calyx-teeth. *Diodia elongata,
E. Meyer. in Herb. Drege.*

HAB. Natal, *Krauss.* 328; *Gueinzius,* 516; *Gerr. & M'K., J. Sanderson,* near Glenfilling, on the Key and Omblas, *Drege.* Dec.–April. (Herb. Sd., D.)
Stem 1–2 feet high, simple or branched. Leaves 1–2 inches long, 2–3 lines broad,
paler beneath, the upper ones smaller. Heads of flowers the size of a large pea, or
a small hazel nut. Flowers 1 line long.

XVI. PENTANISIA, Harv.

Calyx tube obovate or turbinate, limb with 2 or 3 (rarely 4) elongate,
linear-subulate lobes, and some (3–4) small accessory ones. *Corolla*
salver-shaped, with a long tube, a bearded throat, and a 5-, rarely
4-lobed limb. *Stamens* 5 or 4, inserted in the throat, subexserted ; anthers oblong. *Ovary* 2-celled, cells 1-ovulate. *Style* filiform ; stigma
bifid, with linear-lobes. *Capsule* coriaceous, didymous, crowned by the
calyx-lobes, bipartite into 2, monospermous, indehiscent nuts. *Seed*
subtriquetrous, plano-convex. *Lond. Journ. Bot.* 1842, *p.* 21. *Diotocarpus, Hochst. in Reg. Bot. Zeitz.* 1843, *p.* 70.

Perennial herb, with a tuberous root, a tetragonal or compressed stem, opposite
nerved leaves, on each side with 3–4 fid setaceous stipules ; and pedunculated,
terminal, capitate, sometimes spiked flowers. Name from πεντε, *five,* and ανισος,
unequal; alluding to the unequal calycine lobes.

1. P. variabilis (Harv. l. c. p.)

VAR. α. latifolia; stem erect or ascending, hairy or hirsute; leaves elliptic, ovate
or oblong, acute, hairy or nearly glabrous; flowers disposed in umbel-like spikes.
P. variabilis, β. *latifolia, Hochst. l. c. P. variabilis, Harv.*

VAR. β. intermedia; stem erect, downy, or as well as the leaves glabrous or nearly
so ; leaves oblong-lanceolate or lanceolate, narrower at the base ; flowers disposed
in umbel-like, rarely interrupted spikes. *P. variabilis, Harv. Crusea variabilis et
lanceolata, E. Mey. in Herb. Dreg. Declieuxia prunelloides, Klotzch. E. Z. n.* 2301.
Diotocarp. prunelloides, Hochst. l. c.

VAR. γ. glaucescens; stem erect, as well as the leaves glabrous ; leaves lanceolate
or linear-lanceolate, glaucous-green, obtuse or narrower at the base ; flowers disposed
in oblong, interrupted spikes. *P. glaucescens, Harv. l. c. Crusea glaucescens, E. Meyer.
Pentanisia et Diotocarpus angustifolius, Hochst.*

HAB. Mountains, var. α., near Port Natal, *Drege, Gueinzius,* 513. *Plant,* 33. *Dr.
Sutherland, J. Sanderson, Gerr. & M'Ken, Krauss.* Var. β. Winter- and Katrivier-
berge, *Caffraria, E. & Z., H. Bowker,* near Port Natal, *Krauss. Drege.* Var. γ. near

D'Urban, *Gerrard & M'Ken*, 34, 580. *Krauss.* 93, *Drege;* Magalisberg, *Zeyh.* 762. Oct.—Nov. (Herb. D., Sd.)

Root 2-3 inches long, nearly 1 inch broad, brown. Stem ½-1 foot, quadrangular or compressed, reddish. Stipules variable, 2-3 lines or nearly 6 lines long, 3-fid or 5-fid, the middle lobe often larger. Leaves sessile, or on very short petioles; in var. *a.*, 1-1½ inch long, 6-9 lines broad; in var. *β.* et *γ.* narrower, and often 2 inches long or longer. Flowers beautiful lilac, on a longish naked peduncle, in var. *a.* and *β.* aggregated in a head-like, dense spike, involucrated by the outer calyx-lobes, pubescent; in var. *γ.* forming an interrupted spike ½-2 inches long, glabrous. Calyx-lobes about 3-4 lines long. Corolla nearly 1 inch long. Fruit appressed-hairy or glabrous, subglobose, shorter than the calyx-lobes.

XVII. MITRACARPUM, Zuccarini.

Calyx with an ovate tube, and a 4-toothed, permanent limb; 2 of the teeth usually larger than the others. *Corolla* salver-shaped, with a terete tube, which is furnished near the base inside with a circular line of hairs, a glabrous throat, and a 4-lobed limb. *Stigma* bifid. *Capsule* membranous, crowned, 2-celled, circumcised round the middle. *Seeds* solitary in the cells, basilar, partly fixed to the dissepiment. *Roem. et Schult. syst. 3, p. 210. A. Rich. mem. soc. hist. nat. par. 6, p. 151, t. 14, f. 4.*

Herbs or subshrubs, with the habit of *Spermacoce*. Stipules combined with the petioles a little way, and ending in many bristles. Flowers in dense, verticillate, axillary and terminal heads, the latter always involucrated by 4 leaves. Name from μιτρα, a girdle or ring, and καρπος, a fruit, in allusion to the fruit being cut round about in the middle.

1. M. Dregeanum (E. Mey. in Herb. Drege); stem erect, branched, as well as the leaves hairy; leaves sessile, lanceolate, narrower at the base, with revolute margins; stipules with 5-6 bristles on each side; flowers disposed in capitate whorls; calyx with 4 teeth, 2 a little smaller, ciliated, somewhat shorter than the corolla.

HAB. Near Port Natal, *Drege, Gerrard & M'Ken.* April. (Herb. Sd. D.)

Stem 1-2 feet high, subhirsute. Leaves 1½-2 inches long, 4-6 lines broad, acute, scabrous, hairy. Bristles of the stipules 2-3 lines long. Calyx-teeth subulate, about 1½ lines, corolla 3 lines long, the latter with 4 oblong, bearded lobes. Stamens subexserted. Capsule hairy, ovate. Seeds black, oblong, with a longitudinal furrow inside.

XVIII. HYDROPHYLAX, Lin. fil.

Calyx with an ovate, angular tube, and a 4-parted permanent limb; teeth acute, erect. *Corolla* campanulately funnel-shaped, 4-lobed. Anthers 4, sessile in the throat, exserted, linear-oblong. *Stigma* roundly 2-lobed. *Berry* dry, corky, angular, lanceolate, or ovate-oblong, crowned by the calyx, 2-celled. *Seeds* solitary in the cells, oblong, bisulcate inside, with hard albumen and a straight embryo. *Lam. ill. 76, f. 1. Serissus, Gœrtn. fruct. 1, p. 118, t. 25.*

Glabrous, creeping herbs. Stems terete. Leaves ovate-oblong, fleshy, joined with the stipules into a cupular, toothed sheath at the base. Flowers axillary, sessile, solitary or by twos. Name from ὑδωρ, water, and φυλαξ, a keeper or guardian. The plant always grows by the sea-side.

1. H. carnosa (Sond.); quite glabrous; stem decumbent; branches ascending; leaves ovato lanceolate, acute, fleshy; stipules 3-dentate; flowers axillary, solitary; limb of calyx shortly tubular, cleft on the outer side, with eroso-denticulate margin ; corolla 4-lobed ; capsule oblong-ovate, angulate-striate, compressed, crowned by the connivent calyx. *Diodia carnosa, Hochst. pl. Krauss.*

HAB. Sandy sea-coast of Zitzikamma, *Krauss.* 1667 ; Port Natal, *Gerrard and M'Ken,* 350. Mar.–Jul. (Herb. D. Sd.)

Habit of *Arenaria peploides,* but larger. Stipules 3–4 lines long. Leaves ½ inch long, 2 lines broad, flat above, with a minute point. Flowers white, ½ as long as the leaf. Fruit 3 lines long, 2 lines broad. Nearly allied to *H. maritima,* L.

TRIBE VII.—**ANTHOSPERMEÆ** (Gen. XIX.–XXII.)

XIX. **GALOPINA**, Thunb.

Flowers hermaphrodite, dioecious or polygamous. *Calyx-tube* obovate, limb 4–parted, very small, scarcely any. *Corolla* subrotate, 4–5-parted ; lobes oblong, lanceolate, patent-reflexed. *Stamens 3–5,* inserted in the base of the corolla. *Anthers* oblong, slightly exserted. *Styles* 2, with long hairy stigmas. *Fruit* obovato-didymous, consisting of 2 mericarps, at length separating, indehiscent, warted at back, 1-seeded, with a flat or flattish commissure. *DC. l. c.* 4, *p. 579. Anthospermi spec. Thunb. Prod. p. 23. Phyllis spec. Cruse, Linn. 6, p. 19. Galopina et Oxyspermum, E. Z. enum. p. 364.*

Perennial herbs, with opposite ovate or ovato-lanceolate leaves, and terminal lax panicles of small flowers. Name unexplained.

1. G. circæoides (Thunb. Diss. Nov. Gen. 1, p. 3) ; *glabrous;* leaves ovato-lanceolate, acuminate, tapering into the petiole, scabrous on the margins ; panicle expanded ; pedicels capillary, elongated ; flowers hermaphrodite. *E. & Z.* 2304. *Zeyh.* 2712. *Anthospermum galopina, Thunb. Fl. Cap. p. 156. Phyllis galopina, Cruse, l. c.*

HAB. Shady, moist places from Swellendam to Albany, and in Caffraria. Jan.–Feb. (Herb. Thunb., Sd., D.)

Stem rather angular, with 2 elevated lines which run from the stipules, 2 feet and more in height, branched, quite glabrous, very rarely in the lower part a little downy. Stipules 3-forked. Petiole 3–6 lines long. Leaves 1½–3 inches long, ½–¾ inch broad. The trichotomous panicle often 1 foot in diameter. Bracts subulate. Pedicels ½–1 inch long. Ripe fruit about ¾ line long and broad; mericarps convex at back.

2. G. aspera (Sond.) ; *tomentose ;* leaves ovate or ovate-lanceolate, acute, subcordate at the base, on very short petioles ; panicle oblong ; flowers dioecious or polygamous. *Oxyspermum asperum, E. & Z.* 2305. *G. hirsuta, E. Mey. G. tomentosa, Hochst.*

HAB. Katriviersberg, near Philipstown, *E. & Z., H. Hutton;* Natal on the Umlaasriver, and near D'Urban, *Drege, Krauss., J. Sanderson, Gerr. & M'Ken.* 828. Mar.–Jun. (Herb. D., Sd.)

Stem 2–3 feet, stiff, simple, 4-cornered, as well as the leaves and panicle densely covered with short greyish hairs. Leaves with some smaller in the axils, 1–1½ inch long, 6–10 lines broad, quite entire, when old often subscabrous. Stipules 3–6-parted. Panicle ½–1½ ft. long, trichotomously compound, leafy or nearly naked ; the branches and pedicels erect, spreading, rough-hairy in fruit, much shorter and more rigid than in *G. circæoides.* The specimens from Caffraria are polygamous (and hermaphrodite, *E. & Z.*), those from Natal mostly dioecious. Fruit obovate, densely papillate, ½–¾ line long.

XX. **ANTHOSPERMUM**, Linn.

Calyx-tube obovate ; limb 4–5-toothed. *Corolla* tubular, limb 4–5-

parted, lobes linear or lanceolate, spreading, revolute, valvate in æstivation. *Stamens* 4–5, inserted in the lower part of the tube. *Anthers*
versatile, exserted. *Styles* 2, mostly very short, rarely connate; stigmas
2, very long, hairy. *Fruit* composed of 2, easily separable, indehiscent,
1-seeded mericarps, which are a little compressed at the raphe, and joined
together by a concave commissure. *Seeds* erect, affixed at the base
Embryo dorsal, erect. *Albumen* subcartilaginous. *Cruse, Dissert. Rubiac.
Cap. p.* 7, *t.* 1, *f.* 1, 2. *Linn.* 6, *p.* 4. *DC. Prod.* 4, *p.* 579.

Small shrubs or herbs. Stems branched. Leaves opposite or ternately verticillate,
linear or lanceolate, rarely ovate or oblong, coriaceous. Stipules adhering to the
petioles at the base, drawn out into a single tooth, rarely into 2 or 3 teeth in the
middle. Flowers axillary, sessile, rarely panicled, small, with 2–3 bracts at the base
of the ovary, dioecious or polygamous or hermaphrodite. Name from ανθος, *a flower,*
and σπερμα, *a seed.*

  1. Flowers dioecious, axillary, subsessile (1–11).
       Stipules simple (1–9).
           Flowers pentandrous :
               Lvs. whorled, linear-lanceolate, ciliated   ...   (7) **Bergianum.**
               Lvs. opposite, lanceolate, ciliated when young   (9) **hirtum.**
           Flowers tetrandrous :
               Branches and leaves hirsute :
                   Stem procumbent ; lvs. longer than inter-
                       nodes, spreading   ...   .   (5) **hispidulum.**
                   Stem erect ; lvs. shorter than internodes,
                       erect-incurved ...   .   (6) **Burkei.**
               Branches minutely downy or glabrous :
                   Lvs. elliptical, ovate or oblong ...   .   (8) **Dregei.**
                   Lvs. linear or lanceolate.
                       Fruit sharply 3-ribbed   ...   .   (3) **tricostatum.**
                       Fruit smooth or obsoletely ribbed.
                           Stem rooting at the nodes ...   ...   (2) **prostratum.**
                           Stem not rooting.
                               Fruit crowned by calyx-teeth   (1) **Æthiopicum.**
                               Fruit destitute of calyx-teeth   (4) **ciliare.**
       Stipules 2–4-parted (10–11).
           Lvs. lanceolate, 1–1½ inch long...   ...   .   (10) **lanceolatum.**
           Lvs. ovate or ovato-lanceolate, 4–6 lines long ...   (11) **hedyotideum.**
  2. Flowers dioecious, panicled ...   .   (12) **paniculatum.**
  3. Flowers hermaphrodite, rarely polygamous (13–18).
       Fruit crowned by one leafy calycine lobe   ...   ...   (13) **calycophyllum.**
       Fruit with or without minute calyx-teeth :
           Fruit glabrous :
               Lvs. erect, linear-lanceolate, bluntish ; an-
                   thers oblong   ...   .   (14) **pumilum.**
               Lvs. spreading, linear-lanceolate, acute ;
                   anthers linear ; corolla glabrous   ...   ...   (15) **rigidum.**
               Lvs. reflexed, oblong-lanceolate, mucronate ;
                   anthers linear ; corolla hairy ...   .   (16) **Ecklonis.**
           Fruit hirsute :
               Flowers axillary-sessile   ...   .   (17) **Lichtensteinii.**
               Flowers cymoso-panicled   ...   .   (18) **Crocyllis.**

1. A. Æthiopicum (L. Spec. 1511) ; stem erect ; young branches
downy ; leaves opposite or ternate and whorled, linear-lanceolate or
subspathulate, glabrous ; flowers *aggregated,* verticillate-spiked, tetrandrous ; fruit *oblong, nearly cylindrical ;* mericarps elliptic, convex,
obsoletely-ribbed and minutely-dotted at back, crowned by the calyx-

teeth. *Hort. Cliff.* 455, *t.* 27. *Thunb.! Fl. Cap. p.* 157. *Cruse, Diss. p.* 10. *E. & Z.!* 2307.

VAR. *a.* **ternifolium** (Cruse! l. c. p. 11); leaves 3 in a whorl, linear-lanceolate, acute; branches straight. *A. æthiop., var.* γ. *verticillata, herb. Thunb.; var. a. et β. E. & Z. l. c. herb. Un. Itin.* 24.

VAR. *β.* **oppositifolium** (Cruse! l. c. p. 11); leaves opposite, linear-subspathulate, sometimes bluntish; branches virgate, erect-spreading. *A. æthiop. a. et β. Herb. Thunb. A. spathulatum, Spreng. Syst. Veg.* 1, *p.* 399. *Var.* γ. *et* δ. *E. & Z. l. c. Herb. Un. Itin.* 4, 5, 7, 9, 11, 25, 26. *Drege,* 7661, 7666, 9551. *Zeyh.* 2714.

VAR. γ. **Ecklonianum** (Cruse! Linn. 6, p. 10); leaves opposite, linear-lanceolate, acuminate. *Var.* ε. *E. & Z. l. c. A. spath. var. longifolium, E. Mey.*

VAR. δ. **montanum;** leaves opposite, lanceolate, bluntish, mucronulate; branches shorter, very leafy. *Var.* ζ. *alpinum, E. & Z. l. c.*

HAB. On plains and mountains, common. Aug.–Jan. (Herb. Thunb., D., Sd.) A very variable woody shrublet, ½–3 ft. high, much branched, greyish-brown. Branches opposite or ternate. Leaves aggregated, 3–8 lines long, ½–1¼ line broad, with revolute margins, shining and dotted above, pale with prominent middle nerve beneath. Stipules minute. Flowers 2–4 times shorter than the leaves, usually 2–6 or more in a tuft. Cor. with a 4-parted limb, equalling the tube. Fruit 1 line long.

2. A. prostratum (Sond.); stem elongate, *prostrate, rooting;* branches short, downy or glabrous; leaves opposite, lanceolate or subspathulate, mucronate, glabrous, with revolute margins; flowers *solitary,* tetrandrous; fruit *obcordate;* mericarps *roundish* on the back, minutely downy or glabrous; calyx-teeth nearly obsolete.

VAR. *a.* **velutinum;** stem, branches, and fruit minutely-downy.

VAR. *β.* **glabrum;** stem, branches, and fruit quite glabrous.

HAB. Cape flats, var. *a. C. Wright,* 491; var. *β. Ecklon.* (Herb. D., Sd.) Root woody, ½–1 foot, stem 1–3 feet high, rooting at the internodes. Branches 1 inch or a finger long. Leaves nearly as in *A. spathulatum,* Spr. or a little larger, not whitish, but mostly rufous beneath. Female flowers as in *A. æthiopicum,* stigmas very long, hairy. Fruit as long, but a little broader and more emarginate than in the preceding.

3. A. tricostatum (Sond.); stem erect, branched; branches slender, downy; leaves opposite, linear-lanceolate, with revolute margins, glabrous; flowers solitary, tetrandrous; fruit *oblong;* mericarps cuneate, prominently 3-ribbed at back, glabrous, minutely-dotted; calyx-teeth nearly obsolete.

HAB. Rietvallei, *E. & Z.;* between Driekoppen and Bloodriver, *Drege,* 9550. April–Nov. (Herb. Sd.) A woody shrub, with the habit of *A. æthiopicum, var. β.* Leaves aggregated, 3–4 lines long, acute. Fruit 1½–2 lines long; the mericarps with 2 prominent marginal and a dorsal rib; the bipartite spine, separating the mericarps, about half as long as the fruit. Stigmas very long and hairy.

4. A. ciliare (Linn. Spec. 1521); stem decumbent or suberect; branches downy; leaves opposite, lanceolate or linear-lanceolate, acute, ciliated, when old often without cilia, with revolute margins; flowers solitary, or 2–4 together, tetrandrous; fruit *obcordate;* mericarps round-ish at back, glabrous, shining, *destitute of calycine limb. Thunb.! l. c. p.* 157. *Herb. ex pte. Cruse, Diss. p.* 13. *Plukn. Mant.* 51, *t.* 344, *f.* 5. *A. galioides, Reichb. Spreng. Syst. Veg. c. p, IV.* 2, *p.* 338. *E. & Z.* 2308. *A.*

spathulatum? E. M. in Herb. Drege. Herb. Un. Itin. 12, 14, 27, 191. *Zey.*
2718, 2719. *Drege,* 9548. *Sieber. Fl. Cap.* 88. *A. ciliare, α. et β. fem. (non
mascul.) et γ. mas. Herb. Thunb.*

VAR. *β.* **papillatum**; fruit densely papillate, at length glabrous or nearly so.

VAR. *γ.* **glabrifolium**; leaves subglabrous, without cilia, or scabrous on margins.

HAB. Stony places, plains and mountains, common; var. *β.* Simon's Bay and
Rietvalley. May–Sept. (Herb. Thunb., D., Sd.)

Stem 4–10 inches, much branched, procumbent, rarely erect. Leaves 3–4 lines
long, aggregated, lanceolate or ovate-lanceolate (*var. latifolium,* E. Z.), in other
specimens linear-lanceolate (*var. angustifolium et scabrum,* E. Z.), glabrous and
shining above, pale or rusty beneath. Corolla a little hairy outside; the 4 lobes
longer than the tube. Stigmas very long, hairy. Fruit ¾ line long and broad, brown.

5. A. hispidulum (E. Mey.); stem procumbent, as well as the short
branches and leaves *clothed with spreading white hairs;* leaves opposite,
longer than the internodes, spreading, lanceolate, acute, with revolute
margins; flowers solitary, tetrandrous; fruit *oblong-cylindrical;* mericarps convex outside, hairy, crowned by the *subulate* calyx-teeth.

HAB. Rocky places, Omsamwabo to Omsamcaba. May. *Drege,* Natal; *Gerr.
& M'K.* 1361. (Herb. D., Sd.)

A low shrub, with greyish stem and branches, glabrous at the base. Leaves 4–6
lines long, 1 line broad, hispidulous. Stipule subulate. Fl. very small. Stigmas very
long, hairy. Fruit 1 line long, nearly cylindrical, as long as the ciliate bracts.

6. A. Burkei (Sond.); stem erect; branches long, hirsute; leaves
opposite, *shorter than the internodes, erect-incurved,* lanceolate, acute,
with revolute margins, densely clothed with short hairs; flowers aggregate, *tetrandrous;* fruit oblong; mericarps convex outside, pubescent,
crowned by the *minute, acute,* calyx-teeth.

HAB. Magalisberg, June, *Burke & Zeyh.* (Herb. Sd., D.)

Primary branches 1–2 feet long; ultimate branchlets very short. Internodes about
¼ inch. Leaves with many smaller axillary leaves or short branchlets, 2–3 lines long,
pointed. Flowers very small. Fruit scarcely 1 line long; the calycine lobes much
shorter than in *A. hispidulum,* to which it is closely allied by the leaves and flowers,
but not in habit.

7. A. Bergianum (Cruse! Diss. p. 8); stem erect; branches clothed
with whitish pubescence; leaves 3 in a whorl, connately perfoliate,
imbricated, *linear-lanceolate, ciliated;* flowers *pentandrous;* subverticillate, spiked; fruit obovate; mericarps convex outside, glabrous, minutely
punctate. *Linn.* 6, *p.* 7, *E. Z.!* 2306. *Drege,* 7668.

HAB. Cape flats, common. July–Aug. (Herb. Sd., D.)

A very leafy, more or less branched shrub, about 1 foot high. Leaves longer than
the internodes, densely imbricated, pale green, the margins and keel ciliated with
white hairs, 3–5 lines long, acute. Flowers in 3-flowered whorls. Male and female
flowers 5-parted, with short tube. Stigmas long. Fruit 1 line long.

8. A. Dregei (Sond.); stem erect; branches spreading, minutely
downy; leaves opposite, *elliptic, ovate or oblong,* mucronulate, glabrous,
when young minutely downy, rusty beneath, with slightly recurved
margins; flowers tetrandrous, solitary; fruit (when young) obovate;
mericarps glabrous, punctulate.

HAB. Betw. Koussie and Zilverfontein, 2000 ft. Aug. *Drege,* 3016. (Hb. Sd., D.)

A foot or more high, with a long, nearly simple, woody root. Twigs reddish or brown. Internodes longer than the leaves. Stipules very small. Leaves shortly petiolate, 3–4 lines long, 2 lines broad. Corolla with a short tube. Anthers linear, white. Stigmas long, hairy. Ripe fruit unknown.

9. A. hirtum (Cruse! Diss. p. 11); stem erect; branches erect-spreading, hairy as well as the leaves at the base ; leaves opposite, lanceolate, acute, ciliated when young; stipules simple, subulate; flowers solitary or 2–5 aggregated, axillary, pentandrous; fruit obovate; mericarps roundish outside, glabrous, papillate. *E. Z.* 2311. *A. ciliare, α et β, mas. herb. Thunb. non fem. A. rubiaceum, Reichb.! in Spreng. syst.* 4, *p.* 338. *Herb. Un. Itin. n.* 29. *A. lanceolatum, Sieb.! fl. cap. n.* 90. *A. hirsutum, DC. prod.* 4, *p.* 580.

HAB. Mountains near Capetown, *Bergius, Ecklon, Sieber, W. H. H.* Piquetberg, *Drege,* 7677. Sept.–Jan. (Herb. Th., Sd., D.)
Stem 1–2 feet, purplish brown, branched. Leaves 1 inch long, 1–2 lines broad, acute or acuminate, with slightly recurved margins, pale or whitish beneath. Stipules 1 line long. Flowers hairy outside. Tube of the corolla shorter than the lobes. Female flower solitary. Fruit 1 line long, destitute of the calycine limb.

10. A. lanceolatum (Thunb. prodr. p. 32); stem procumbent or erect, much branched; branches glabrous or downy; leaves opposite, lanceolate, acute, glabrous, or puberulous; stipules 2–4-parted, lobes subulate; flowers axillary, subverticillate, 4–5-androus; fruit obovate-oblong; mericarps convex at back, glabrous, or papillate. *Thunb. herb. α, et β, non γ. Cruse! Diss. p.* 12, *E. Z.* 2310. *A. herbaceum, L. fil. Suppl. p.* 440. *A. nodosum,* E. Mey.

VAR. β, latifolium; branches downy or densely hairy; leaves ovate-lanceolate, mostly rusty beneath, downy, at length glabrous; fruit papillate. *A. latifolium, E Mey. A. ferrugineum, E. Z.!* 2309.

HAB. Cape dist., *Thunb. Sieber,* 239; Kleinfontein, *Mundt. & Maire;* Klynriviersberg, and in distr. Uitenhage, *E. Z.;* Plettenberg Bay, *Dr. Pappe;* Howisonspoort, *H. Hutton;* Natal, *Drege.* VAR. β. in dist. Uit., Albany, and in Caffraria, *E. & Z., Zeyh.* 2716, ex parte ; near Natal, *Drege, Gueinzius,* 468. Oct. (Herb. Th., Sd., D.)
Stem 1–3 feet; as well as the branches, reddish-brown. Leaves sessile, 1–1½ inch long, 2–3 lines long, in var. β, ¾–1 inch long, about 4 lines broad, with slightly recurved margins, acute or acuminate, paler beneath, in var. β, on the under side with brown resinous dots. Corolla with conical tube, and lanceolate-linear lobes. Anthers oblong. Fruit 1 line long. Very similar to *A. hirtum,* of which it has the habit and foliage ; but the stipules are very different, and the leaves somewhat broader. In this species I often found female flowers on the lower twigs.

11. A. hedyotideum (Sond.); perennial; stems short, simple, or a little branched, as well as the leaves, minutely downy; *leaves opposite, ovate, acute, or ovate-lanceolate,* with revolute margins; stipules 3-fid, the lateral lobes smaller; flowers solitary, tetrandrous; fruit elliptic; mericarps convex at the back, minutely papillate; limb of calyx nearly obsolete.

HAB. Kreili's Country, Caffraria, *H. Bowker;* Keiskamma, *Drege.* (Herb. D., Sd.)
Stems many from the root, 3–4 inches high. Petiole nearly 1 line long. Leaves at length subglabrous, acute at both ends, pale and with some lateral veins beneath. 4–6 lines long, 2 lines broad. Stipules small. Calyx minute, 4-toothed. Lobes of the corolla shorter than the tube. Fruit 1 line long, the mericarps usually with a line on the back

12. A. paniculatum (Cruse! Diss. p. 15, t. 1, f. 2); stem erect, branched from the base, branches downy; leaves opposite, linear-sub-spathulate; stipules short, simple; flowers panicled, tetrandrous; fruit obovate; mericarps 3-ribbed, glabrous. *E. Z.* 2314. *A. Æthiopicum, fem. herb. Thunb.*

HAB. Houtniquas and Hanglip, *Mundt and Maire;* Winterhoeksberg, Kraka-kamma and Stadensriviersberg, Uit., Grahamstown, and in Caffraria, *E. & Z. Zeyh.* 2715; Howisonspoort, *H. Hutton.* Oct.–Jul. (Herb. Thunb., D., Sd.)
Root woody. Stem 1–1½ ft.; branches reddish, very leafy. Leaves aggregate, quite glabrous, with revolute margins, narrowed at base, 4–5 lines long, ½ line broad. Panicle terminal, narrow, 2–6 inches long. Flowers dioecious, but in a few speci-mens I observed polygamous flowers. Corolla with linear-lanceolate lobes longer than the short tube. Fruit about 2 lines long, crowned by the 4 erect, acute, caly-cine lobes.

13. A. calycophyllum (Sond.); stem *shrubby*, erect, branched; twigs pubescent; leaves opposite, petiolate, ovate or ovate-lanceolate, glabrous; flowers hermaphrodite, pentandrous, disposed in a short, terminal spike; calyx 5-toothed, teeth unequal, 4 very minute; fruit elliptic-oblong, downy; mericarps convex at back, one of them tipped by the *large leafy calycine lobe.*

HAB. Magalisberg, *Burke and Zeyher;* Natal, *Dr. Sutherland, J. Sanderson.* (Herb. Sd., D.)
One or several feet high, much branched. Stipules minute, simple, subulate. Leaves 4–5 lines long, 1½–2 lines broad, coriaceous, acute, with slightly revolute margins. Terminal spike about ½ inch long, the flowers mostly geminate on very short pedicels, in the axils of the soon deciduous, ciliate leaves. Fl. hermaphrodite, rarely polygamous. Four of the calyx-teeth nearly obsolete, the fifth expanded to a leaf. Tube of the corolla twice longer than the lobes. Anthers oblong. Stigmas elongated. Fruit 1 line long. The leafy calyx-lobes oblong, acute, equalling or a little longer than the easily separated mericarps.

14. A. pumilum (Sond.); *perennial;* stems numerous, erect, simple minutely downy; leaves opposite, erect, linear-lanceolate, *bluntish,* cili-olate-scabrous on the revolute margins; stipules simple; flowers herma-phrodite, tetrandrous, axillary, solitary or geminate; anthers *oblong;* fruit obcordate; mericarps convex at the back, glabrous, shining.

HAB. Caledon River, *Zeyher.* Jan. (Herb. Sd.)
3–4 inches high, with the habit of *A. Æthiopicum.* Leaves 2–3 lines long, aggre-gate, convex above by the revolute margins. Flowers small; calyx 5-toothed, equal. Fruit 1 line long; mericarps crowned by the minute or rudimentary calyx-teeth.

15. A. rigidum (E. Z.! n. 2315); suffruticose; stems erect, woody, much branched; branches minutely downy; leaves opposite, *spreading, linear-lanceolate, acute,* a little scabrous, with revolute margins; stipules simple; flowers polygamous, tetrandrous, axillary, aggregate; corolla *glabrous outside;* anthers *linear;* fruit obovate-oblong; mericarps con-vex at back, glabrous, minutely punctate, *crowned by the minute calyx-teeth.*

HAB. Karroo, near the Gauritzriver, Swell., *E. & Z.;* Nieuwejaars-pruit, between Garip and Caledonriver, at the foot of the Witbergen, *Zeyh.* Oct.–Dec. (Herb. Sd.)
Root thick, woody, about 1 foot long, simple. Stems numerous, 1 span high. Leaves 4–6 lines long, about ¾ of a line broad, pointed, shining above. Flowers 2–5-together. Calyx minute, equal. Corolla white, lobes about as long as the tube. Stigmas long, hairy. Fruit 1 line long.

16. A. Ecklonis (Sond.); shrubby; stem erect; branches and leaves *pubescent-scabrous;* leaves opposite, *reflexed, lanceolate or oblong-lanceolate,* mucronulate, with slightly recurved margins, whitish beneath; stipules simple; flowers hermaphrodite, tetrandrous, axillary, subaggregated; corolla *scabrous-hairy outside;* anthers linear; fruit obovate; mericarps convex on the back, glabrous, *destitute of calycine limb.*

HAB. On the Olifantriver and near Villa Brakfontein, Clanw., *Ecklon.* (Herb., Sd.) Several feet high. Leaves 6 lines long, 1–1½ line broad. Flowers white. Calyx minute, equal. Calyx-lobes longer than the tube. Fruit 1 line long.

17. A. Lichtensteinii (Cruse! Diss. p. 15); stem erect, branched; branches erect, downy at top; leaves opposite, *linear,* acute, *keeled, ciliated;* stipules simple, ciliated; flowers hermaphrodite, tetrandrous, *axillary,* whorled; style *very short;* stigmas long, hairy; fruit oblong; mericarps *hairy. Linn.* 6, *p.* 16. *E. Z.* 2316. *Spermacoce ericæfolia, Lichtst. in R. & Sch. syst. veg.* 3, *p.* 281.

HAB. Sandy places near Capetown and in Hottentottsholland, *Lichtenstein, E.&Z. Drege.* Nov.–Dec. (Herb. Sd.)

Shrub, 1 or 2 ft., with glabrous, slender, terete branches, naked at the base. Internodes about as long as the leaves. Stipules minute. Leaves 2–3 lines long in the axils, with some smaller or equal leaves. Flowers 2–5, verticillate. Calyx with 4 or 5 short, acute teeth. Limb of corolla longer than the tube. Stigmas long, hairy. Fruit 1½ line long; mericarps convex at back, densely clothed with white, short hairs; the commissure a little concave. One of the mericarps is often abortive; the fertile cell separated from the sterile by a deeply bi-parted spinule.

18. A. Crocyllis (Sond.); stem erect; branches spreading, glabrous, minutely downy at top; leaves opposite, linear, *obtuse, fleshy,* glabrous; stipules minute, bifid; flowers hermaphrodite, pentandrous, *paniculate;* style *bifid from the middle,* stigmas papillate; fruit (when young) obovate; mericarps hirsute. *Crocyllis anthospermoides, E. Meyer.*

HAB. Stony places near Verleptpram on the Garip, *Drege;* Namaqualand, *A. Wyley.* Sept. (Herb. Sd., D.)

At first sight known from *A. Lichtensteinii* by the white, spreading branches and panicled inflorescence. Seemingly a large shrub. Leaves 3–6 lines long, about 1 line broad, a little convex above, sulcate beneath, solitary or tufted, the ultimate short branches forming a cymose panicle. Flowers 2 lines long; calyx 5-fid, with acute lobes; corolla appressed-hairy outside; tube conical, about as long as the lanceolate lobes. Anthers linear. Ovary didymous, clothed with long, white hairs, 2-celled; cells 1-ovuled. Style glabrous; the spreading stigmas sub-exserted. Ripe fruit unknown.—Perhaps to be separated from *Anthospermum,* as a distinct genus?

XXI. CARPACOCE, Sond.

Flowers polygamous; hermaphrodite and male on the same plant. Herm. Fl. : *Calyx* 5-fid; lobes subulate, persistent, equal or 1 longer. *Corolla* funnel-shaped, tube short, limb 5-parted, lobes spreading, linear-lanceolate, with a reversed tooth above the thick-pointed apex. *Stamens* 5, inserted in the tube; filaments capillary; anthers linear-oblong. *Ovary* cuneate, 2-celled, 2-ovuled. *Style* very short, terminated by a very long, hairy stigma. *Fruit* crowned by the calyx, 2-seeded, didymous, easily separated when ripe; or 1-seeded by abortion. *Seeds* obovate or oblong, affixed at the base. *Commissure* subconcave. *Albumen* subcartilaginous. Male flowers : *Calyx* as in the hermaphrodite flower. *Corolla* tubular, 5-toothed. *Stamens* 5. *Style* none.—*Anthospermi spec. auct.*

Small shrubs. Stems branched. Leaves opposite, linear or lanceolate. Stipules adhering to the petioles at the base. Flowers axillary, solitary, bracteated.—Name from καρπος, fruit, and ακωκη, a point, in allusion to the fruit being terminated by the calycine point.

Calyx-lobes equal, subulate. Fr. 2-seeded. Lvs. linear ... ... (1) **scabra.**
Cal.-lobes uneq., 1 much longer. Fr. 1-seeded. Lvs. lanceolate (2) **spermacocea.**

1. C. scabra (Sond.); stems much branched, branches minutely downy at top; leaves opposite, *linear*, subtrigonal, acuminated, scabrous on the margins and keel; flowers axillary, sessile; fruit crowned by the subulate, calycine limb. *Anthospermum scabrum. Thunb.! Fl. Cap. 158. Cruse! diss. p. 14. E. Z.! n. 2313.*

HAB. Mountains near Capetown, *Thunb.*, *Bergius;* Middelenfontain, *Mundt and Muire;* Hott.-holld. and Van Stadensbg., *E. & Z.* July–Sept. (Hb. Th., Sd., D.)
Stem very short; branches slender, erect, 1 foot high. Leaves ½–1 inch long, narrow-linear, with a cartilaginous point, ciliolato-scabrous on the margins, aggregate at the top of the branches Calyx-lobes acuminate, nearly 2 lines long. Lobes of corolla 3–4 times longer than the short tube. Fruit quite glabrous, cuneate, about 1 line long and broad, sub-compressed; the calyx-lobes spreading, longer than the fruit. Seeds obovate, rugose.

2. C. spermacocea (Sond.); stem herbaceous, flexuous, branched, glabrous, but minutely downy at the apex; leaves opposite, lanceolate, with scabrous margins; stipules entire; flowers axillary, solitary, pedicellate, glabrous; fruit crowned by the calycine limb. *Anthospermum spermacoceum, Reichb. in Spreng. syst. 4, p. 338. Cruse! Linn. 6, p. 17. E. Z. n. 2312. A. fœtidum, Eckl. in Herb. Un. itin. n. 30. Lagotis spermacocea, E. Meyer.*

HAB. Cape flats; Table and Devil's Mts.; near Zwellendam, and in Caledon, *E. Z.;* Dutoitskloof *Drege.* Simon's Bay, *C. Wright,* 487. Sept.–Oct. (Herb. D., Sd.)
Habit of *Anthosp. lanceolatum.* Stem 1 or more feet high, erect or decumbent, rather angular. Leaves lanceolate or oblong-lanceolate, with incurved point, paler beneath. Stipules pilose, cleft into bristles. Peduncles 3–6 lines long. Calyx 1 line long, nearly cylindrical; one of the lobes nearly 3–4 times longer than the others, lanceolate or ovate-lanceolate. Tube of corolla short; lobes lanceolate. Fruit conical or subcylindrical, 2-celled, but 1 cell always abortive, minute.

XXII. AMBRARIA, Cruse.

Flowers diœcious, exactly as in *Anthospermum.* *Capsule* indehiscent, pseudo-3-locular, the intermediate cell vacant, the lateral ones 1-seeded, or 4-locular; 2 of the cells fertile, 1-seeded. *Cruse dissert. Rub. Cap. p. 16, t. 1, f. 3 & 4. Nenax, Gœrtn. de fruct. et sem. pl. 1, p. 165, t. xxxii. f. 7.*

Small shrubs, with linear leaves and axillary flowers.—Name derived from Cape Ambra? in Madagascar.

Fruit glabrous.
  Leaves linear, ciliate, ternately verticillate ... ... (1) **glabra.**
  Leaves linear, glabrous, opposite ... . (2) **acerosa.**
  Leaves (minute) ovate, acute, glabrous, verticillate ... ... (3) **microphylla.**
Fruit tomentose ... ... (4) **hirta.**

1. A. glabra (Cruse! l. c. p. 17, t. 1, f. 3); stem ascending or sub-erect; branches quadrangular, glabrous; leaves 6–12 in a whorl, connate at base, linear, acute, subtriquetrous, *margins and keel ciliated ;* fruit obovately subglobose, glabrous. *Linnæa* 6, *p.* 18. *E. Z.! n.* 2317.

VAR. β. **Tulbaghica**; branches virgate; fruit elliptic-oblong.

VAR. γ. **papillata** ; fruit small, subglobose, papillate.

HAB. Cape flats; var. β. near Waterfall, Tulbagh. Nov.–Dec. (Herb. Sd., D.)
Stem branched, 1–2 feet high ; young branches reddish, shining. Leaves flat above, keeled beneath, 3–6 lines long, ½–1 line broad. Stipules obtuse, not toothed. Fl. very small, axillary, subverticillate. Calyx 4 . rarely 5-toothed. Fruit 3- or 4-locular, variable in size, usually 2 lines long, subangular, in some specimens only 1 line long, nearly globose; in var. β. 2½ lines long, obsoletely angular, crowned by the short, acute or obtuse calycine limb.

2. A. acerosa (Sond.) ; stem erect, much branched; branches terete, glabrous ; leaves *opposite, linear,* subtriquetrous, *quite glabrous;* fruit elliptical, glabrous. *Nenax acerosa, E. Z.* 2319.

HAB. Stony places near Tulbaghskloof, Worcester. Sept. *E. Z.* (Herb. Sd.)
An erect greyish shrub, very different in habit from *A. glabra.* Ultimate branches very short. Leaves spreading, 2–3 lines long, ½ line broad, acute. Male flowers tetrandrous, exactly as in *Anthospermum.* Female flowers with long, hairy stigmas. Fruit 2 lines long, obsoletely angular, 4-locular, terminated by the nearly obsolete calyx-teeth.

3. A. microphylla (Sond.); stem woody, with short, glabrous, at length subspinous branches ; leaves sessile, whorled, *ovate-acute,* subconcave above, nerve-keeled beneath, glabrous; fruit globose, glabrous; 4-locular.

HAB. Rocky places ; Sandriver, *Zeyh.* 769. *Burke,* 506. Jan. (Herb. Sd., D.)
A dwarf, grey shrub, with numerous short branches. Leaves spreading, recurved, ½–1 line long, fleshy, carinate or concave above, a little scabrous on the margins. Flowers very small, axillary, sessile. Stigmas very long, hairy. Fruit red, about 1 line in diameter, 4-celled, 2 cells sterile.

4. A. hirta (Cruse! Diss. p. 17, t. 1, f. 2); stem woody, much-branched, densely hairy at top; leaves 3 in a whorl, linear, acute, subtriquetrous, margins and keel ciliated; fruit obovate or subglobose, tomentose, 3–4-locular. *Linn.* 6, *p.* 19. *E. & Z.* 2318.

VAR. β. **macrocarpa** (E. & Z. l. c.) ; leaves longer and fruit larger.

HAB. Devil's Mt. near Drieankerbay, *Bergius;* mts. near Capetown and near the cataract of Worcester, *E. & Z.;* var. β. on Breederiver, *Mundt;* Bosjesveld on the Doornriver, *Drege.* Sept.–Oct. (Herb. D. Sd.)
Shrub about 1 foot high. Branches numerous from the base, with short branchlets at top. Leaves much aggregated in the axils of the verticils, 2–3 lines, in var. β. 3–5 lines long, equalling the internodes. Flowers 3–6, verticillate-spiked. Calyx 4–5-toothed. Corolla funnel-shaped ; lobes ovate-lanceolate. Stamens exserted. Stigmas very long, hairy. Fruit 1–1½ lines, in var. β. 2–2½ lines long, obsoletely angular, crowned by the acute, nearly glabrous calycine limb.

Drege's No. 7660, 7661, 7664, 7665 cannot be satisfactorily determined. No. 1144, of which the ripe fruit is wanting, seems to be a very distinct species.

XXIII. **RUBIA**, Linn.

Calyx-tube ovate-globose ; limb 4-toothed. *Corolla* 4–5-parted, rotate. *Stamens* 4–5, short. *Styles* 2, short. *Fruit* didymous, nearly globose, baccate, juicy. *Lam. ill. t.* 60. *DC. prod.* 4, *p.* 588.

Herbs or sub-shrubs. Stems diffuse, much branched, tetragonal. Leaves opposite, usually furnished with 1-2, rarely 3-4, stipules on both sides, which are very
like the leaves, forming 4-10-leaved whorls. Flowers small, greenish-white or pale-
yellow. Berries black, rarely red or white.—Name from *ruber*, red ; in allusion to
the red colour of the roots, which yield the dye called *Madder.*

Leaves *petioled*, membranaceous, cordate-acuminate ... ... (1) **cordifolia.**
Leaves *petioled*, subcoriaceous, cordate-acute ... ... (2) **petiolaris.**
Leaves *sessile*, lanceolate or elliptic-lanceolate, coriaceous ... ... (3) **peregrina.**

1. R. cordifolia (Linn. mant. p. 197); leaves 4 in a whorl, petiolate,
cordate-acuminate, 5-nerved, *membranous*, beset with prickles on the
middle nerve, margins, petioles, and angles of stems; panicle longer than
the leaves; flowers tetramerous or pentamerous. *R. cordifolia, et mun-
jista, DC. l. c. p.* 588. *R. petiolaris, E. Z. n.* 2320, *ex pte. Drege,* 7669.

HAB. In Caffraria, *E. & Z.; Drege.* Natal, *Drege, Gueinzius,* 405, *Gerr. & M'K.*
561, 562, *Krauss.* Dec. (Herb. D., Sd.)
Stem several feet long. Larger leaves 1-1½ inch long, ¾-1 inch broad, not coriaceous, much veined. Panicle 3-4 inches long, trichotomous. Lobes of corolla
ovate-acuminate. Berries black.

2. R. petiolaris (DC. l. c. p. 588); leaves 6-8 in a whorl, petiolate,
cordate, acute, 3-nerved, or the upper ones oblong-lanceolate, 1-nerved,
rather coriaceous, beset with prickles on the middle nerve, margins, petioles and angles of stem; panicle abbreviate; flowers tetramerous or
pentamerous.

VAR. α. **isophylla**; all the leaves cordate-acute. *R. cordifolia, Thunb. ! fl. cap.
p.* 151, *non Lin. Zeyh. n.* 2721. *R. petiolaris, E. Z. ex pte.*
VAR. β. **heterophylla**; lower leaves cordate, upper ones oblong-lanceolate or lanceolate. *R. petiolaris, DC. l. c.*
HAB. Among shrubs on the sea shore near Algoa Bay, Port Elizabeth, and at
the mouth of the Vanstadensrivier, *Thunb. E. & Z. Drege.* Dec. Var. β. on the great
Vetriver, *Burke & Zeyh.* 774. March. (Herb. Thunb., D., Sd.)
Stem ascending, about 1 foot—in Var. β. apparently 2 feet and more high, much
branched. Leaves serrato-scabrous, 4-6 lines long, 3-5 lines broad, the upper ones
in var. β. ½-1 inch long, 1-1½ lines broad. Petioles the length of the leaves or
longer, tetragonal. Panicle mostly few-flowered, in var. β. more expanded ; the
pedicels glabrous, spreading. Flowers as in *R. cordifolia.* Berries black.—It is
distinguished from the preceding by much smaller, coriaceous, not conspicuously
veined leaves, and a different habit.

3. R. peregrina (Lin. Spec. p. 158); leaves 4-6 in a whorl, permanent, sessile, *lanceolate* or *elliptic-lanceolate*, 1-nerved, veinless, shining
above, smooth, but scabrous from hooked prickles on the margins and
along the angles of the stem; peduncles axillary, 3-chotomous. *Engl.
Bot. t.* 851. *R. anglica, Huds. R. lucida, Linn. R. lœvis, Thunb. fl. cap.
p.* 151. *R. lucida, var. β. herb. Thunb.*

HAB. Cape, *Thunberg.* (Introduced from Europe.)

XXIV. GALIUM, Scop.

Calyx with an ovate-globose or oblong tube, and hardly any limb.
Corolla 4-parted, rotate, rarely 3-parted. *Stamens* 4, short. *Styles* 2,
short. *Fruit* didymous, roundish, rarely oblong, dry, composed of 2
indehiscent 1-seeded mericarps. *DC. prodr.* 4, *p.* 593.

Branched herbs.   Leaves forming whorls along with the leaf-like stipules.   Inflorescence variable.   Name from γαλα, *milk;* some species are used for curdling milk.

1, Annual (a weed of cultivation)	(10) **Aparine.**
2, Perennials.	
Fruit glabrous, granulated	
Pedicels very long, villous	(13) **tomentosum.**
Pedicels short.	
quite glabrous	(11) **glabrum.**
branches and peduncles densely hairy	(12) **asperum.**
Fruit smooth, glabrous or hairy, *hairs not hooked.*	
Stem or branches smooth or hispid, not prickly.	
Flowers solitary, subsessile	(1) **Amatymbicum.**
Flowers peduncled.	
Stem and lanceolate lvs. villous ; *fruit* glabrous	(2) **monticolum.**
Stem, lin-lanceolate leaves and fr. glabrous (or downy)	(3) **Capense.**
Stem, oblong lvs., *pedicels* and fruit densely hairy	(9) **subvillosum.**
Stem, lanc. lvs. and fr. hairy ; *pedicels glabrous*	(8) **Dregeanum.**
Stem or branches scabrous or serrated by reflex prickles.	
Leaves linear, subtriquetrous ; flowers panicled	(6) **horridum.**
Leaves lanceolate ; peduncles solitary, 2-flowered	(7) **mucroniferum.**
Fruit hispid, with *hooked* hairs or bristles.	
Leaves linear :	
Stem erect, downy	(4) **Wittbergense.**
Stem climbing, weak, glabrous	(5) **Garipense.**
Leaves roundish-ovate	(14) **rotundifolium.**

1. G. Amatymbicum (E. & Z.! n. 2328); whole plant hispid; stems cæspitose, diffuse, filiform, 4-sided, much branched; leaves 6 in a whorl, lanceolate, acute; peduncles axillary, solitary, very short; fruit hispid.

HAB. Acacia fields near the Key R., Tambukiland, *E. & Z.* Dec. (Hb. D., Sd.)
A small prostrate herb, with the habit of *G. helveticum.* Leaves 2–3 lines long, acute at both ends, ciliate-hairy. Peduncle ½ line long, in fruit recurved. Flowers white. Fr. densely hispid, ⅓ line in diameter.

2. G. monticolum (Sond.); stem erect, branched, 4-sided, villous; leaves 6–8 in a whorl, linear-lanceolate, mucronate, nearly terete by the revolute margins, villous on both sides; peduncles axillary and terminal, 1- or few- flowered ; corollas and fruit glabrous.

HAB. Mountains near Capetown, *Ecklon.* (Herb. Sd.)
Habit of *G. maritimum,* L., and agreeing in the pubescence and leaves, but differing by glabrous pedicels, fl., and fr. ; branches 2 inches long, with very short lateral branchlets, bearing the flowers. Leaves 3–4 lines long, reflexed, with incurved tips. Fl. very small, pedicels 1 line long.

3. G. Capense (Thunb.! Prod. p. 30); stem ascending or erect, downy or subscabrous, branched; leaves 6–8 in a whorl, linear-lanceolate or linear, mucronate, quite glabrous, rarely ciliolate on the revolute margins; peduncles lateral, 1–3-flowered; floriferous branches panicled; fruit glabrous, smooth, or a little downy. *G. expansum, E. Z.,* 2326, *ex pte., E. Mey. in herb. Dreg.*

VAR. β. **minus**; stems ascending, short, weak. *G. mucronatum, Thunb.! l. c. Zeyh.* 773, *ex pte.*

VAR. γ. **expansum**; stem glabrous, downy or subscabrous ; panicle larger ; fruit glabrous or a little downy. *G. expansum, Thunb.! l. c. E. Z. ex pte. G. mucronatum, E. Z.* 2327. *Drege,* 7675, 7678, 7680, 7681, 7685. *Zeyh.* 1773, *ex pte.*

Var. δ. **scabrum**; stem, branches and leaves scabrous-hairy; fruit a little downy. *G. Namaquense, E. Z.!* 2322. *Drege,* 7682, 7683.

Hab. Sandy spots, and on mts. throughout the colony. Var. δ. high mts. near Heerelogement, Clanw., *E. & Z., Drege.* Sept.–Dec. (Herb. Thunb., Sd., D.)

Polymorphous; var. α. resembling *G. saxatile,* var. β. *G. uliginosum.* Stems from ½–2 feet high, often numerous from the perennial roots; sometimes terete at the very base, but usually 4-sided as well as the mostly simple branches. Leaves 4–6 lines long, about ½ line broad, spreading or reflexed. The axillary inflorescence longer than the leaves. Flowers white. Fruit very small.

4. G. Wittbergense (Sond.); stem erect; branches spreading, *downy;* leaves 6–8 in a whorl, erect-spreading, linear, mucronate, glabrous, ciliate-scabrous on the revolute margins; peduncles axillary, solitary, cymose-tripartite; cyme equalling the leaves; fruit hispid, with hooked hairs.

Hab. Rocky, wet places in the Wittbergen, 6–7000 ft. Jan. *Drege.* (Herb. Sd.)

Very like *G. Capense;* differing by cymose peduncle and hispid fruit Lower branches ½ foot long. Leaves 5–6 lines long, with a longish mucro, peduncle at the apex 3-parted, bracteated, the pedicels bifid. Fruit very small, didymous.

5. G. Garipense (Sond.); stem climbing, much-branched, 4 sided, as well as the branches *quite glabrous,* smooth or a little scabrous from scattered, reversed, minute prickles; leaves 4–6 in a whorl, much spreading or reflexed, linear, mucronate, glabrous; peduncles axillary, solitary, cymose-tripartite, 2–3 times longer than the leaves; fruit hispid with hooked hairs.

Hab. On the Garip near Buffelvallei, 4000 ft. Dec. *Drege;* Buffaloeriver, *Gerr. & M'K.* (Herb. D., Sd.)

Habit of *G. uliginosum;* prickles on the stem and branches very minute, sometimes wanting. Leaves shorter than the internodes, 3–4 lines long, ½ line broad. Peduncles and divaricate pedicels with 1 or 2 leafy bracts. Fruit as in the preceding, from which this species is distinguished by the more diffuse habit, glabrous, often prickly stem, smaller spreading or reflexed smooth leaves, and longer cymes.

6. G. horridum (Thunb.! Fl. Cap. p. 152); stem erect, 4–6-sided, prickly along the angles; leaves 8 or more in a whorl, elongated-linear, subtriquetous, reflexed, serrated from reversed prickles on the margins and keel; panicle axillary, elongate, trichotomous; fruit glabrous.

Hab. Cape, *Masson in herb. Thunb.;* Port Natal, *Gerr. & M'K.* 1339; Magalisberg, Dec. *Zeyh.* 771. (Herb. Thunb., Sd., D.)

Like *G. Aparine,* but is a stronger perennial herb, armed by larger prickles, and easily known by the long 3-angled leaves. Stem at the base nearly as thick as a goose's quill. Leaves 2–4 inches long, 1½ line broad, channelled above, sharply keeled beneath. Panicle in our specimens 2–4 inches long; the flower bearing pedicels 1–2 lines long. Flowers glabrous. Ripe fruit didymous, about 1½ line in diameter.

7. G. mucroniferum (Sond.); stem erect or ascending, 4-sided, glabrous, prickled on the angles; leaves 6 in a whorl, lanceolate, with a longish mucro, prickly along the margins, shining; peduncles axillary, solitary, 2-flowered; pedicels as long as the peduncle, and fruit glabrous. *G. mucronatum, E. Mey. non Thunb. Drege,* 7677, 7684, 7686, 7689.

Hab. Wet rocky places, near Gnadenthal, Dutoitskloof, *Drege;* in Caffraria, *E. Z.* Oct.–Jan. (Herb. D., Sd.)

Stems ½–1½ foot, minutely prickled. Internodes as long or longer than the leaves, shining. Leaves 4–5 lines long, 1 line broad, with recurved prickles on margins and middle nerve. Pedunc. and pedicels capillary, bracteated at the division. Flowers glabrous. Ripe fruit the size of a small peppercorn, shining; one of the mericarps often abortive. It varies with downy ovaries.

8. G. Dregeanum (Sond.); stem erect, 4-sided, densely beset with spreading hairs, at length subscabrous; leaves 8 in a whorl, linear-lanceolate, mucronate, ciliate-hairy on both sides, with revolute margins; peduncles axillary, solitary, 2-flowered; pedicels about as long as the peduncles, glabrous; fruit hispidulous.

HAB. Dutoitskloof, *Drege*, 7688. (Herb. D., Sd.)
Very similar to *G. mucroniferum*, but more robust, and well distinguished by the hirsute stem and hairy, more revolute, leaves. Peduncles as long as the leaves or longer. Fruit didymous.

9. G. subvillosum (Sond.); stem ascending, 4-sided, densely beset with short, spreading hairs, subscabrous when old; leaves 6 in a whorl, oblong, obtuse, mucronate, ciliate-hairy on both sides, with recurved margins; peduncles axillary, solitary, 3-flowered; pedicels as long as the peduncle or shorter, hairy, as well as the young fruit.

HAB. Dutoitskloof, 1–2000 ft. Oct.–Jan. *Drege*, 7687. (Herb. Sd.)
Of this I have only seen a single specimen. It resembles exactly *G. villosum*, Lam. Leaves 3 lines long, 1 line broad, with short recurved mucro, peduncles equalling the leaves, sometimes longer, with a leafy bract at top; pedicels 2–6 lines long, spreading, hairy. Ripe fruit unknown.

10. G. Aparine (L. Spec. p. 157); annual; stems weak, 4-sided, scabrous from reversed prickles; leaves 6–8 in a whorl, linear-lanceolate, apiculate, 1-nerved, scabrous from reversed prickles along the margins and keel; peduncles axillary, simple and bifid, sometimes panicled, scabrous; fruit didymously globose, very hispid from hooked bristles. *Engl. bot.* 816. *G. horridum, E. Z.!* 2329, *non Thunb. Drege*, 7670, 7673. *Zeyh.* 2722, 2723, 2724.

HAB. Shady places near Capetown, and throughout the Colony. Sept.–Oct. (Herb. Sd., D.)
" Common cleavers" or " goose-grass." Stem climbing, villous or scabrous at the nodes. Leaves sometimes larger, obovate-lanceolate. Fl. white or greenish. The small specimens from Hassaquaskloof, *Zeyh.* 2722, agree perfectly with *G. tenerum*, Schleich. from Switzerland. [Probably introduced from Europe.]

11. G. glabrum (Thunb.! Fl. Cap. p. 152); stem erect, flexuous, tetragonal, *glabrous*, prickly along the angles; leaves 6 in a whorl, obovate-oblong, shortly-pointed, *glabrous*, serrated by reversed prickles along the margin; peduncles capillary, *glabrous*, terminal and lateral, panicled. *E.Z.!* 2325. *G. uncinatum, Licht. Bart. et Wendl. Beyt.* 2, *p.* 12.

HAB. In woods, Duyvelsbosch and Voormannsbosch, near Puspas valley, Swell., *E. & Z.*; Buffeljagdrivier, *Zeyh.* 2725. Oct. (Herb. Thunb., Sd.)
Stem 2 ft. and more, alternately branched, resembling *G. sylvaticum, L.* Leaves 8–12 lines long, 4 lines broad, minutely punctate, obtuse, with short mucro-like point. Peduncles longer than the leaves. Fl. small, white. Fr. 1 line long, densely granulated.

12. G. asperum (Thunb.! Fl. Cap. p. 152); stem erect, flexuous, 4-sided, *as well as the branches and peduncles, densely clothed with white* *hairs*, the old stems sub-glabrous and scabrous along the angles; leaves 6 in a whorl, obovate-oblong, mucronate, glabrous, serrated by reversed prickles on the margins; peduncles capillary, terminal and lateral, disposed in a panicle; *pedicels 2-4 times longer than the flower. E. Z. !* 2323. *G. tomentosum, E. Meyer, var.*

HAB. Among shrubs in Stellenbosch, Worcester, Swellendam, and Uitenhage, *Thunb. E. & Z.; Drege,* 7671, 7672; Namaqualand, *A. Wyley.* Oct.-Jan. (Herb. Thunb., D., Sd.)

Nearly intermediate between *G. glabrum* and the following, differing from the first by the densely hairy branches and panicle, from *G. tomentosum* by the short pedicels. In some of the larger specimens the lower part is quite glabrous, but the branches and peduncles are clothed with short, usually reversed hairs. Flowers glabrous, white. Fruit as in *G. glabrum.*

13. G. tomentosum (Thunb.! Fl. Cap. p. 151); stem climbing, 4-sided, much branched, with the angles scabrous, hairy, or glabrous and prickled; upper branches densely clothed with white hairs; leaves 4-6 in a whorl, obovate, or obovate-oblong, glabrous, prickled on the margins; peduncles dichotomous, capillary, villous, forming a large panicle; pedicels *elongated,* 10-20 *times longer than the flower. E. Z.!* 2324. *G. maritimum, Thunb. Prodr., not of Linn. G. asperum, var. β. villosum, E. Z.* 2323. *Zeyh.* 2720.

HAB. Among shrubs, through the colony. Oct.-Jan. (Herb. Thunb., D., Sd.)
Well distinguished by the long (1 inch and more) villous pedicels. Leaves smaller than in *G. asperum,* 5-6 lines long, 2 lines broad. Fl. glabrous. Fr. 1 line in diam.

14. G. rotundifolium (Linn. Spec. p. 156); stem diffuse, glabrous or hispid; leaves 4 in a whorl, roundish-ovate, 3-nerved, ciliated; peduncles axillary and terminal, loose, elongated, naked, trichotomous at the apex; fruit nearly globose, beset with hooked bristles. *Bocc. Sic. t. 6, f. 1. Jacq. Astr. t. 94. G. Thunbergianum, E. Z.* 2321.

VAR. *β.* hirsutum; stems, leaves, and peduncles densely hairy. *G. rotundifolium, Thunb. fl. cap. p.* 151.

HAB. Mountains; Katriviersberg, *E. & Z.;* Dornkop near Vetrivier, *Zeyh.* 772. Natal, *Gerr. & M'K.* 1338. Var. *β. Masson in herb. Thunb.* Feb.-March. (Herb. Thunb., Sd., D.)

Stem procumbent or suberect, 1 foot high, glabrous or hispid. Leaves 4-6 lines long, 3-5 lines broad. Fl. white or greenish. I cannot find any difference between the South African and European plants.

ORDER LXXIV. **VALERIANEÆ,** DC.

(By W. SONDER).

Flowers mostly bisexual. *Calyx*-tube adnate; limb 3-4-toothed, often enlarged after flowering. *Corolla* epigynous, tubular, usually 5-, rarely 3-4-lobed, subunequal, the lobes obtuse, imbricate in bud. *Stamens* 1-5, inserted in the tube, alternate with the lobes of the corolla; anthers separate, 2-celled, introrse. *Ovary* inferior, 1-3-celled; ovules solitary, pendulous; style filiform. *Fruit* dry (like an achene), crowned with the often enlarged calyx-limb, 1-3-celled; 2 cells abortive. *Seed* pendulous. *Embryo* straight, without albumen.

Herbaceous, rarely half-shrubby plants, with opposite, often cut or pinnatisect, exstipulate leaves. Flowers in cymes or fascicles, or solitary in the forks of the branches, small. A small Order, chiefly from the temperate zones of both hemispheres ; abundant in the Andean region of South America.

TABLE OF THE SOUTH AFRICAN GENERA.

I. **Valerianella.**—Calyx small, unequally toothed, crowning the fruit.

II. **Valeriana.**—Calyx a thickened margin, crowning the ovary, at length unfolding into a feathery pappus.

I. VALERIANELLA, Poll.

Limb of calyx toothed. *Corolla* regular, 5-lobed. *Stamens* 3. *Stigma* almost undivided, or trifid. *Fruit* 3-celled, rather membranous, generally 2 of the cells fertile. *Pollich palat.* 1, 29. *Endl. gen.* 2181.

Annuals. Stems dichotomous at the top. Leaves oblong or linear, undivided or toothed, or the upper ones pinnatifid. Flowers solitary in the forks, or in fascicles or corymbs, bracteate, small, white, rarely rose-coloured.—Name, a diminution of *Valeriana*.

1. **V. eriocarpa** (Desv. Journ. Bot. 2, p. 314, t. 11, f. 2)'; fruit ovate, obsoletely ribbed, convex at back, and flattish in front ; limb of calyx as broad as the fruit, campanulate, obliquely truncate, reticulate-veined, minutely 6–8-toothed ; flowers densely corymbose. *Koch, Syn. ed.* 2, 372. *Reichb. icon. fl. germ.* 1406, t. 713. *Fedia campanulata, Presl. sic.* 11.

HAB. Near Grootvadersbosch. Oct. *Zeyh.* 2726. (Herb. Sd.)

4 inches to 1 foot high, downy ; branches spreading. Leaves spathulate, entire. Corymbs nearly capitate. Flowers very small. Limb of calyx erect. Fruit hispid, or glabrous at the base. The 2 abortive cells in the front of the fruit form 2 projecting lines or ribs ; the perfect cell is terminated by the broad, acute tooth.

II. VALERIANA, Linn.

Limb of calyx involute at the time of flowering, but at length unfolding into a deciduous pappus, composed of many plumose bristles. *Corolla* monopetalous, 5-cleft, gibbous at the base. *Stamens* 3. *Fruit* 1-celled and 1-seeded at maturity. *L. gen.* 44. *DC. prod.* 4, *p.* 632. *Endl. gen.* 2186.

Herbs or sub-shrubs. Leaves variable even in the same plant. Flowers corymbose, capitate or panicled, white, rarely blue, rose-coloured or yellow.—Name from *valere*, to be powerful, on account of the medical virtues of *V. officinalis.*

1. **V. Capensis** (Thunb. Fl. Cap. p. 33) ; stem erect, striated ; leaves imparipinnate ; leaflets opposite or alternate, ovate, acute, toothed, the terminal one largest ; corymb panicled ; flowers triandrous. *E. Z.* 2330.

HAB. In kloofs and moist valleys ; distr. of Cape, George, Albany, Uitenhage, and in Caffraria. Dec.–Feb. (Herb. D., Sd.)

Habit of *V. officinalis.* Stem 2-3 feet high, glabrous or hairy on the joints. Leaves with 4–8 pairs of glabrous or pilose leaflets, the odd one ovate or acuminate, 1 inch or more long, the lateral ones gradually smaller. Lower leaves on long footstalks, upper ones sessile and with narrower leaflets. Panicle, flowers, and fruit as in *V. officinalis.*

ORDER LXXV. **DIPSACEÆ,** Vaill.

(By W. SONDER).

Flowers perfect, crowded in heads, on a common receptacle, surrounded by a general involucre; each flower also seated in a calyx-like, dry, persistent involucel. *Calyx*-tube adnate; limb cup-like, subentire or split into several naked or feathery bristles, often enlarged after flowering. *Corolla* epigynous, tubular; limb oblique, somewhat 2-lipped, 4–5-lobed, imbricate in bud. *Stamens* 4, inserted in the tube, alternate with the lobes of the corolla, two mostly longer; filaments exserted; anthers 2-celled, introrse. *Ovary* inferior, 1-celled; ovule solitary, pendulous; style filiform; stigma simple. *Fruit* a dry utricle, crowned by the enlarged, persistent calyx-limb, and enclosed in the cup-like involucel. *Seed* pendulous. *Embryo* straight, in the axis of fleshy albumen.

Herbs or rarely suffrutices, with opposite or whorled, simple, often pinnatisect or lyrate, exstipulate leaves. Pubescence mostly copious, rough or silky. A small Order, chiefly from the warmer parts of the temperate zones; abundant in the Mediterranean region. The type of the Order is *Dipsacus*, the "*Teasle*," whose old and prickly flower-heads are used in carding wool.

TABLE OF THE SOUTH AFRICAN GENERA.

I. **Cephalaria.**—*Inv.* scales imbricate in several rows, shorter than the paleæ. *Calyx*-limb cup-like or discoid.

II. **Scabiosa.**—*Inv.* scales sub-biseriate. *Calyx*-limb crowned with 5 bristles.

I. **CEPHALARIA,** Schrad.

Involucre of many imbricated leaves, shorter than the paleæ. *Involucel* 4-angled, 8-furrowed, terminated by a 4–8-toothed crown. *Limb* of calyx rather cup-shaped or discoid. *Corolla* 4-cleft. *Stamens* 4. *Fruit* tetragonal, crowned by the limb of calyx and the involucel. *Schrad. Catal. Sem. Hort. Goett.* 1814. *Scabiosæ, spec. Linn. et others. Endl. Gen. n.* 2192.

Perennial herbs. Leaves toothed or pinnatifid, rarely entire. Heads of flowers terminal, globose; paleæ imbricated, outer ones sterile. Corollas white, cream-coloured or lilac. Name from κεφαλη, a *head*, the flowers are disposed in round heads.

Lvs. linear, quite entire	(1) **lavandulacea.**
Lvs. oval-oblong, serrated	(2) **rigida.**
Lvs. elongate-lanceolate, 3-fid or pinnate; lobes quite entire	(3) **attenuata.**
Lvs. lyrately-pinnatifid or inciso-serrate; lobes toothed; stem not scabrous	(4) **ustulata.**
Lvs. pinnatifid or bipinnatifid; lobes linear, incised; stem scabrous	(5) **scabra.**

1. C. lavandulacea (Sond.); glabrous; stem terete; leaves linear; 1-nerved, quite entire, with revolute margins; peduncle elongate; heads subglobose; scales of involucre ovate, obtuse, ciliate; paleæ acute; flowers pubescent, 4-fid.

HAB. Houhoeksbergen, Stellenbosch., 1000–3000 ft. July. *E. & Z.* (Herb. Sd.)
Plant 1 foot or more high, erect or ascending. Leaves sessile, approximate, three

times longer than the internodes, 1-1½ inch long, 1 line broad ; middle nerve impressed on the upper side. Peduncle 3-4 inches. Head of flowers as large as a hazel nut. Involucre of about 10 scales. Corollas equal, white? or pale rose-coloured, nearly ½ inch long, downy outwards, lobes ovate, a little shorter than the tube. Fruit unknown. Habit of *Lavandula (Lavender).*

2. C. rigida (Schrad. l. c.) ; stem terete ; branches elongate ; leaves rigid, ovate-oblong, serrated, often eared at base, upper ones lanceolate ; heads nearly globose ; scales ovate, obtuse, ciliated, appressed-pubescent, at length glabrous. *Scabiosa rigida, L. Mant. p.* 328. *Thunb.! Fl. Cap.* 143. *Sieb. Fl. Cap. exs.* 252. *Ceph. rigida et scabra, E. Z.!* 2331, 2332. *Herb. Un. Itin.* 727. *Commel. Hort. Amstel. t.* 93.

VAR. *a.* **glabra** ; quite glabrous.

VAR. *β.* **scabra** ; scabrous with short, rigid hairs.

HAB. Mts. near Capetown, and in Hottentottsholland ; Natal, *Krauss.* Dec.-Feb˙ (Herb. Th., Sd., D.)
Stem 2-3 ft. high ; branches elongate, opposite, leafless in the upper part ; lower lvs. close, 1½-3 in. long, ½-1 in. broad, upper ones smaller, lanceolate, with revolute margins. Heads the size of a walnut. Flowers white, downy outside, 4-fid ; lobes ovate, twice shorter than the tube. Fruit quadrangular, cuneate, silky-pubescent.

3. C. attenuata (R. & Sch. Syst. Veg. III. p. 44) ; stem striated or angular ; leaves oblong, lanceolate or sublinear, entire, trifid or pinnatifid ; lobes divergent, quite entire ; heads globose ; scales obtuse, villous ; paleæ acute.

VAR. *a.* leaves oblong, glabrous or hairy, with ciliate margins ; cauline ones entire or pinnatifid, lobes linear. *C. decurrens, E. Z.!* 2333. *C. ustulata, E. Mey.* and 9554, *Herb. Drege. S. rigida, Krauss, Herb.* 320.

VAR. *β.* glabrous ; leaves elongate-lanceolate, attenuated at both ends, entire or pinnatifid at the base, lobes spreading, acute, quite entire ; cauline leaves entire or more or less pinnatifid. *Scab. attenuata, Lin. fil. Suppl.* 118. *Willd. Spec.* 1, 546. *S. trifida, Thumb. Fl. Cap.* 144. *Zeyh.* 2333. *Ceph. attenuata, E. & Z.!* 3234. *C. longifolia, E. Mey.*

VAR. *γ.* glabrous or hairy ; lower leaves pinnatifid ; lobes decurrent, linear, acute, diverging, quite entire ; cauline leaves entire or pinnatifid. *Scab. decurrens, Thunb. l. c. Sc. humilis, Thunb. l. c. Ceph. decurrens, R. & Schult. C. attenuata, E. Z. ex pte.*

HAB. Wet places and grassy hills ; var. *a.* in Keiskamma, between Stadesmountains and Krakakamma, near Grahamstown, and Port Natal ; var. *β.* in distr. Uitenhage, George and Port Natal ; var. *γ.* Albany. Dec.-Feb. (Herb. Thunb., Sd., D.)
Stem 1-2 feet. Branches angle-furrowed, leafless or nearly so. Petioles of the radical leaves 1 inch-1 foot long. Leaves in var. *a.* about 2 inches long, ½-1 inch broad, in var. *β* sometimes 1 foot long, 4 lines-1½ inch broad, coriaceous ; rhachis and lobes of var. *γ* 2-3 lines broad. Lower stem leaves sessile, attenuate at the base. Heads as large as a small walnut, or smaller. Scales and paleæ purplish or blackish at the apex.

4. C. ustulata (R. & Sch. l. c. p. 43) ; glabrous or hairy ; stem striately angular ; lvs. lyrate-pinnatifid or serrate-incised, lobes toothed ; heads globose ; scales villous, obtuse ; inner paleæ acute.

VAR. *a, glabrous ;* stem branched ; lower leaves lyrate-pinnatifid ; lobes ovate, toothed ; scales and paleæ purplish at the apex. *Scab. ustulata, Th. Fl. Cap.* 144.

VAR. *β, pilosa ;* stem simple ; lower leaves serrate or lyrate-incised at the base ; lobes serrate ; upper leaves pinnatifid.

HAB. Var. *a*, Bocklandsberg, Nov.-Dec., *Thunb.*; var. *β*. Caledonriver and Magalisberg, and stony places near Commissiepoort, *Zey.* 779. (Herb. Th., D., Sd.) Stem 1-2 feet. Radical leaves petiolate, in var. *a*, nearly 6 inches long, 1½-2 inches broad; lobes decurrent, upper ones larger, the terminal ovate, incised-toothed; in var. *β*, 3-4 inches long, 6-9 lines broad, sharply-toothed or pinnately-incised; lobes ovate-oblong. Heads nearly the same as in the preceding, subglobose in var. *β*, with more downy but not sphacelate apex. Corolla pubescent.

5. C. scabra (R. & Sch. l. c. 44); stem erect, terete, rigid, branched, as well as the leaves scabrous-hairy; leaves stiff, pinnatifid or bipinnatifid; lobes linear, dentately-cut, acute, with revolute margins; scales villous, obtuse; paleæ acute. *Sc. scabra, Th.! Cap.* 144. *Linn. f. Sup.* 118.

HAB. Hills near Zoetemelksvalley, *Thunb.*; Winterhoeksberg, 1000-2000 feet, *Drege.* Jan.-Feb. (Herb. Th., D., Sd.) Plant 2-3 feet high; branches elongated, rigid. Leaves sessile, 2 inches long, lobes about 4-6 lines long, as well as the rhachis, 1, rarely 2, lines broad. Heads the size of small walnuts. Flowers pubescent, white.

II. SCABIOSA, Linn.

Involucre of many leaves, nearly biseriate. *Receptacle* chaffy. *Involucels* usually cylindrical, with 8 foveolæ, terminated by a campanulate or rotate, scarious limb. *Limb* of calyx tapering into a neck at the base, and ending in 5-awned bristles, or 1-4 by abortion. *Corolla* 4-5-cleft. *Stamens* 4. *Roem. et Schult. Syst. III.* 2. *Asterocephalus, Vaill. Endl. Gen.* 2195.

Perennial or suffruticose herbs. Leaves variable. Heads of flowers depressed. Outer flowers of heads usually radiant. Name from *scabies*, the itch; which disorder the common sort is said to cure.

Stem herbaceous; lvs. lyrately-pinnatifid or bipinnatifid ...	(1) **Columbaria.**
Stem herbaceous; lvs. obovate or obovate-oblong, irregularly toothed	(2) **Africana.**
Stem suffruticose; lvs. spathulate, serrato-dentate at the apex	(3) **Buekiana.**

1. S. Columbaria (Lin. Spec. p. 143); stem branched, smoothish; radical leaves petiolate, spathulate, *crenate or inciso-serrate*, hairy on both surfaces; cauline leaves pinnate-parted, with flat, linear lobes or undivided; peduncles elongated, 2-3-fid; corollas radiant, 5-fid, rarely 4-fid; heads of fruit ovate-globose; crown 20-nerved, *one-half shorter than the tube;* bristles of calyx 5, 2-4 times longer than the crown. *Engl. Bot. t.* 1311. *Reichb. Pl. Crit.* 4, *t.* 354. *E. Z.!* 2338. *S. acaulis, maritima et ochroleuca, Thunb.! Fl. Cap.* 145. *S. Africana, E. Z.*, *Thunb. Herb. S. pallida, E. M. c. Herb. Un. Itin.* 725. *Zey.* 778, 2727, 2728, 2729.

VAR. *β*. **dissecta;** glabrous or downy; lower leaves lyrate, unequally and bluntly serrated; upper ones 2-3-pinnatifid; lobes linear, incised. *S. laciniata, Licht. R. & Sch.* 3, 87. *E. Z.* 2339. *S. anthemifolia, E. Z.* 2337. *S. pallida, E. Mey. excl. c. S. ochroleuca, β. Thunb. Hb. Hb. Un. Itin.* 726. *Drege,* 9552. *Zey.* 2730 *et* 2731.

VAR. *γ*. **elata;** stem elongated, thick, fistulose; radical leaves on long petioles, pinnati-partite; lobes cuneiform. *S. crassicaulis, E. Mey. S. columbaria, Hb. E. Z. expt.*

VAR. *δ*. **simplicior;** glabrous; radical leaves on elongated petioles, lanceolate, entire or toothed or pinnatifid, with a few, distant, much-spreading, linear-cuneate, entire lobes, the terminal oblong, acute; cauline leaves linear; heads smaller. *Drege,* 9553.

HAB. Sandy places throughout whole colony, at Natal, and in Namaqualand,

Magalisberg, *Zey.* 776; var. *β.* dist. of Cape, Swellendam, Uitenhage, and Albany; Walwekop, *Zey.* 775; var. *γ.* Nieuweveldsbergen near Beaufort, 3–5000ft. *Drege*, v. δ.

A very polymorphous plant. Flowers blue, violet or purplish, and white according to Thunberg.

2. S. Africana (Lin. Spec. 145); whole plant *villous or pubescent;* stem *herbaceous,* erect, terete or subangular at the base, branched; leaves obovate or obovate-oblong, eroso-dentate or lyrato-pinnatifid at the base; upper leaves much smaller, undivided or pinnatifid; peduncles elongated; heads of fruit globose; corollas radiant, 5-fid; crown 20–24-nerved; bristles of calyx 5, twice longer than the crown. *Thunb.! Fl. Cap.* 145. *Herrm. Parad. t.* 219. *Herb. Un. Itin.* 724. `*Coult. Dips.* 37, t. 2, f. 12. *E. Z.* 2335. *S. indurata, Lin. Mant.* 196. *S. altissima, Jacq. Hort. Vind. t.* 185. *S. maritima, E. Mey.*

HAB. Hills near Capetown, Capeflats, &c. Jul.–Oct. (Herb. Th., Sd., D.)
Remarkable for its large leaves, which are 4–6 inches long, 1–2 inches broad, and clothed with soft and short tomentum; the lower ones aggregated, irregularly-toothed or incised at the narrowed base. Heads about the size of a walnut. Leaves of involucre lanceolate, silky, shorter than the pubescent corolla. Crown equalling the pubescent fruit, or a little shorter.

3. S. Buekiana (E. Z.! 2336); whole plant clothed *with appressed silky hairs;* stem *suffruticose, branched and leafy at the base;* leaves sessile, lyrately- or cuneate-spathulate, serrato-dentate from the middle, quite entire at the base; peduncles elongated; heads of fruit subglobose; corollas radiant, 5-fid; crown about 20-nerved; bristles of calyx 5, 2–3-times longer than the crown. *S. tomentosa, E. Mey.*

VAR. *β.* **virescens**; tomentum greenish; heads often smaller.
VAR. *γ.* **angustiloba**; leaves pinnatifid; lobes lanceolate.

HAB. Grassy hills near Olifantshoek on Bosjesmansriver, Uitenh., Sept., *E. Z.;* near Salomonstown, May, *Zey.* 2732; var. *β.* in Caffraria, *Eck. Herb., T. Cooper,* 303; var. *γ.* near Grahamstown and Glenfilling, Dec., *Drege.* (Herb. Sd., D.)
Stem short, but branches 1 foot and more high Leaves 1–1½ inch long, 6–8 lines broad, obovate-cuneated, bluntly-toothed or serrated, or in var. *γ.* deeply-pinnatifid; lobes about 4 lines long, 1 line broad. Heads nearly as in *S. columbaria.* Peduncles 1 foot long. Leaves of involucre lanceolate, twice shorter than the radiant outer flowers. Lobes of the pubescent corolla twice shorter than the tube. Fruit and involucel as in *S. columbaria;* but the bristles are generally somewhat shorter.

ORDER LXXVI. **COMPOSITÆ**, Juss.

(By W. H. HARVEY).

Flowers (of minute size) arranged in *heads (capitula)* on a general *receptacle,* surrounded by an *involucre,* composed of several, separate or cohering, dry and membranous or green and leaf-like, *scales.* The *heads* are either many- or few-flowered, or one-flowered; separate, or aggregated into compound clusters called *glomerules. Flowers* sessile on a disc-like, flat or convex, receptacle. *Calyx*-tube adhering to the ovary; limb (called *pappus*) either obsolete, annular or ring-like, coroniform, toothed, scaly, bristle-shaped or feathery, usually enlarged as the ovary swells, and more or less persistent. *Corolla* epigynous, tubular, either regular and 4–5-toothed, or uni-labiate and strap-shaped *(ligulate)*, or rarely bi-labiate. *Stamens* inserted in the tube of the corolla, alternate with its lobes; anthers united by their edges into a tube, surrounding the style. *Ovary* with a single, erect ovule; *style* filiform, forked in the fertile flowers; simple in the abortive. *Fruit* a dry, small nut or *achene;* seed without albumen.

This Order, synonymous with Linnæus's 19th class, *Syngenesia*, is by much the largest in the vegetable kingdom, comprising over 1,000 genera and perhaps 13,000 species, of which numerous examples may be found in most countries. In S. Africa their number is considerably over 1,500, and easily found examples are *Aster, Chrysocoma, Helichrysum* (or Everlasting), *Senecio* (grounsel), *Osteospermum, Gazania, Cryptostemma, Sonchus* (sow-thistle), &c. But conspicuous as are the flowers of these and kindred plants, a little examination shows that each apparent *"flower"* is in reality an *inflorescence* made up of a multitude of small flowers, each perfect in itself, arranged on a common platform or disc *(receptacle)* surrounded by numerous involucral scales. To understand the structure, which is general throughout this Order, the student should take some common example,—which may be picked up on any roadside or in any waste ground,—make a vertical section through the centre of one of the "flowers" (properly *flower-heads* or *capitula*), and compare what he will there find with the above diagnosis. In order further to assist him, I shall now briefly explain some of the terms used in the following descriptions.

Each **capitulum** or flower-head is composed of an **involucre** or outside covering, resembling a calyx; a **receptacle** on which the flowers stand; and the **flowers** themselves.

Involucre.—The small, leafy or scale-like parts of which the **involucre** is composed are called its **scales** ("scales of the involucre"). These may be in one, or in several rows, separate or cohering into a tube, and are of various textures and forms. In some genera, as in *Senecio*, there are frequently a number of smaller scales at the base of the proper involucre, which form a sort of supplementary or external involucre; this is called a **calyculus**, and the scales which compose it, for distinction sake, are called its **bracteoles**. Involucres which have a calyculus are said to be **calycled**.

Receptacle.—The **receptacle** is the summit of the flower-stalk, and is usually flattened out into a disc, which may be either nearly flat, hemispherical, or conical. In a flower-head which has gone to seed it is usually exposed, from the opening out or deflection of the involucral scales. The receptacle is said to be **nude**, when its surface is bare, with bald interspaces between the scars left by the fallen flowers; **honeycombed**, when it is minutely-pitted, like honeycomb; and **fimbriate** or **fimbrilliferous**, when the margins or walls of the honeycomb-cells are jagged or fringed. In some cases (as in *Geigeria*, and in the section "Lepicline" of *Helichrysum*) the fringes or fimbrils are separate, and resemble teeth or scales. Such must not be confounded with the true scales of the receptacle, which are called **paleæ**. These latter are modified bracts, and are generally similar in appearance to the innermost involucral scales, but are more rigid and horny; there is never more than one **palea** to each flower; it is placed on the outer side of the flower, and very generally its base wraps round the ovary or the young achene. **Paleæ** are always important generic characters; they occur in *Amellus, Athanasia, Œdera, Eriocephalus, Sphenogyne*, &c.

Flower-heads.—When all the flowers in a flower-head are tubular, and of nearly equal length, the head is said to be **discoid** (as in *Chrysocoma, Athanasia, Pteronia*, &c.). When the flowers in a head are of two kinds, the central ones tubular and 4–5-toothed, the marginal ones longer, strap-shaped, and one-sided, such a head is said to be **radiate**; the central portion is called *the disc*, and the spreading, marginal flowers *the rays* (as in *Aster, Senecio, Osteospermum, Gazania*, &c.). When all the flowers in a head are strap-shaped and one-sided, the head is **semiflosculose** (as in *Sowthistle, Dandelion*, &c.).

When all the flowers in a head are similar and *perfect*, each having stamens and an ovule-bearing pistil, the head is **homogamous** (as in *Chrysocoma*). When the flowers are of different sexes, or some perfect and some imperfect, the head is **heterogamous**. All *radiate* capitula are **heterogamous**; their disc-flowers being either perfect, as is commonly the case, or *male*, with perfect anthers and an abortive pistil;

their ray-flowers either female or neuter. Some heads are **monœcious**; when, on the same root, the heads are diverse, some containing only male flowers, others only female. And some are **diœcious** (as in *Brachylœna* and *Tarchonanthus*); when, on different roots, male and female flower-heads occur. And lastly, a head is **hetero-monœcious**, if the marginal flowers be female, the disc-flowers completely male, with abortive stigmas (as in *Osteospermum, Othonna,* &c.) Some heads are **glomerulate**, or aggregated in a *glomerule:* this is when a number of small heads (each with its proper involucre) are crowded together on a common receptacle, surrounded by a general involucre. A *compound*-head is thus formed (as in *Sphœranthus* and *Œdera*).

Flowers.—The *corolla* is either **tubular**, equally 4–5-toothed; **filiform**, or thread-shaped, and usually truncate ; **ligulate** or strap-shaped, one-sided ; or **bilabiate**, i.e., ligulate with a minute lobe or pair of lobes opposite the strap-shaped limb.

Pappus.—This term is given to the limb of the calyx, which takes a great variety of forms. It is usually an important generic character, and therefore to be closely attended to. Though the pappus exists when the flowers first open, it frequently enlarges after flowering, and is therefore best observed on the mature or nearly mature ovary. It is either **paleaceous**, formed of flat scales; **setaceous**, of slender bristles; **plumose** or feathered, when the bristles are plumed, like a feather, with slender hairs; **barbed**, when the bristles are set with short processes or imperfect plumes; it is **tooth-like** when formed of a definite number of small teeth ; **coroniform**, or crown-shaped, when the teeth are partly confluent; and **annular**, or like a ring. An ovary is said to be **calvous**, or bald, when there is no obvious pappus.

Anthers.—It is often of importance to examine the anthers, in order to observe whether they are **tailed**, i.e. produced at base, on each side of the filament, into a bristle-shaped point ; or **tailless** i.e., rounded or truncate at base. Tailed anthers alone separate the *Gnaphalieæ* (everlastings) from other Senecionideæ ; and *Inuleæ* from other Asteroideæ. When they occur, they are always of systematic value.

Style.—When the ovary contains an ovule, the *style* is constantly 2-lobed, or at least 2-toothed at the extremity ; its lobes are called " **style-branches.**" When the ovary is abortive, the style is, with a few exceptions, quite simple or unbranched. In the systematic arrangement of the Compositæ the *style-branches* afford characters of the highest value, and therefore must be closely and carefully observed by the student, either with a strong pocket lens or a simple microscope. The Tribes, except the *Cichoraceæ*, into which the Order is divided, are solely distinguished by certain characters of the style-branches, as indicated in the Table annexed. Unless these distinctions are mastered by the student, he will find it impossible to proceed in the study of Compositæ ; but, in most cases, the distinctions are easily seen, if carefully looked for, and then what follows is comparatively easy. He must bear in mind. however, that, for systematic purposes, it is the style of *bisexual* or perfect flowers which affords the indicated characters : the styles of *male* flowers being usually simple, and those of exclusively *female* flowers, though branched, having similar branches in all the Tribes. A difficulty, therefore, exists in the classification of such plants as *Osteospermum, Othonna,* &c. in which there are no bisexual flowers. Having no certain guide to follow in their flowers, we are forced to place them next those genera to which, by their general aspect, they appear nearest of kin.

Achenium or Achene—This is the name applied to the small, seedlike nut or fruit of the Compositæ. It often affords generic characters in its shape and markings. Thus it may be cylindrical, angular, flattened, winged, &c.; or it may be glabrous, hairy, woolly ; smooth, scabrous, granulated, echinated, &c. In some it is prolonged into a tapering or thickened point, when it is said to be beaked. It is usually sessile on the receptacle, but is sometimes (as in *Cotula*) raised on a short pedicel.

TABLE OF THE TRIBES.

**Tubuliflorœ.* Heads either discoid or radiate : the disc flowers tubular, regular, 4–5-toothed.

Tribe 1. VERNONIACEÆ. *Style-branches* long, much exserted, filiform, acute, *equally hispid or bristly* on the outer surface.—Heads always discoid. Leaves alternate. (Gen. 1–6.)

Tribe 2. EUPATORIACEÆ. *Style-branches* long, much exserted, terete, obtuse or thickened at the point, *minutely granulated* on the outer surface.—Heads always discoid. Leaves *mostly* opposite. (Gen. 7–10.)

Tribe 3. ASTEROIDEÆ. *Style-branches* linear or lance-linear, *flattish* or flattened, mostly acute, *minutely and equally downy* on the outer surface.—Heads radiate or discoid. Leaves various. (Gen. 11–38.)

Tribe 4. SENECIONIDEÆ. *Style-branches* long, linear, flattish, truncate, *bristly at the apex only*, or tipped with a short, bristly cone.—In the male flowers the styles are simple, more or less bristly or brush-like at the point.—Habit various. (Gen. 39–124.)

Tribe 5. CYNAREÆ. *Style* suddenly thickened towards the apex, and often hispid at the thickening; its *branches* convex, either partially cohering or separate, minutely downy on the outer surface.—Habit various. (Gen. 125–146.)

** *Ligulifloræ.* Heads semiflosculose, all the flowers strap-shaped and bisexual.

Tribe 6. CICHORACEÆ. *Style-branches* long, subobtuse, filiform, equally pubescent on the outer surface.—Juice milky, very bitter. (Gen. 147–154.)

Note.—Tables of the genera will be found under each of the Tribes respectively.

TRIBE I.—VERNONIACEÆ.

Flower-heads discoid, all the flowers tubular. *Style-branches* long, much exserted, filiform, sharp-pointed, *equally hispid or bristly* on the outer surface. Leaves alternate. (Gen. I.–VI.)

Heads several-flowered ; invol. of many imbricating scales :
  Pappus none, or a small fleshy ring :
      All the fl. perfect, bi-sexual, with bell-shaped corollas (1) **Ethulia.**
      Central fl. *male*, funnel-shaped ; marginal *female*, fili-
          form, in many rows ... ... (2) **Litogyne.**
  Pappus bristle-shaped, in 2 or more rows :
      Inv. scales and leaves spinous-pointed ... ... (4) **Hoplophyllum.**
      Inv. scales not spinous-pointed ... ... (3) **Vernonia.**
      Pappus of 7–9 spreading, flat, white scales ... ... (5) **Platycarpha.**
  Heads 1-flowered; invol. of 2 opposite scales, compressed ... (6) **Corymbium.**

I. ETHULIA, Cass.

Heads many-flowered, homogamous. *Involucre* imbricate, scales herbaceous, acute. *Receptacle* naked. *Corolla* with a slender tube, and bell-shaped, equally 5-fid limb; lobes lanceolate, margined. *Anthers* short, included. *Style* shortly exserted, its branches subulate, bristly. *Achenes* inversely pyramidal, 4-angled, 4-ribbed, glabrous, glandular between the ribs, truncate, and 4–5-angled at the summit. *Pappus* an entire, thickened, marginal rim. *DC. prodr. 5, p. 12.*

Erect, branching, leafy, herbaceous plants. Leaves alternate, entire or serrated, pellucid-dotted. Flower-heads small, in a much-branched corymb. Flowers red or purple. Name invented by Linnæus, who has not explained its meaning. *E. Gariepina*, DC., and *E. alata*, Sond., constitute our genus LITOGYNE.

1. E. conyzoides (Linn. Sp. 1171) ; closely puberulous ; branches corymbose, somewhat spreading ; leaves lanceolate, acuminate, distantly serrated towards the summit. *DC. l. c. 12, also, D. Kraussii, Sch. B.! in Walp. Rep. 2, p. 945, D. gracilis, DC., and D. angustifolia, Boj.!*

HAB. Common about Port Natal. (Herb. D., Sd., Hk., &c.)

2 or more feet high, thinly clothed with minute hairs. Stem and branches furrowed. Leaves pellucid-dotted, 2–3 inches long, in the Natal specimens seldom ½

inch wide, tapering to both extremities. Corymb densely much branched ; flower-
heads 2 lines across. Fl. purple, fading to white. A common weed in the tropics
of Asia and Africa. The leaves vary somewhat in breadth and serratures, whence
the 4 sp. of the authors above quoted.

II. LITOGYNE, Harv. (n. gen.)

Heads hetero-monœcious, many-fl., discoid ; the *male fl.* central, funnel-shaped, 5-toothed ; *female fl.* shorter, filiform, 3-toothed, in many
rows round the margin. *Invol.* imbricated, the scales acute. *Recept.*
naked, depressed. *Pappus* none. MALE : *anthers* linear, partly exserted, acute at base. *Style* simple, emarginate, much exserted, the
exserted portion hispid all round. *Ov.* abortive. FEM. : *Anth.* none.
Style exserted, deeply bifid, the arms glabrous, blunt, spreading. *Ovary*
glabrous, minute, ovule-bearing. *Achenes* unknown.

S. African, much branched suffrutices, with entire or denticulate, alternate leaves,
decurrent in narrow wings along the stem. Heads small, corymbulose or tufted at
the ends of the branches. *Fl.* pale ?—Name from λιτος, thin or mean, and γυνη, a
female; from the very slender, thread-like, female flowers.

1. L. glabra (Harv. Thes. t. 155); *glabrous,* smooth. *Ethulia ? Gariepina, DC.! prodr.* 5, *p.* 13.

HAB. On the Gariep, *Drege !* Namaqualand, *A. Wyley !* (Herb. D., Sd., Hk.)
Erect, perhaps 2-3 feet high, paniculately branched, sparsely leafy ; stems angular, minutely rough in the furrows. *Lvs.* 1-1½ inch long, 1-2½ lines wide, broadly
linear, sub-acute, smooth, entire or denticulate, decurrent in narrow wings along
the stem. Heads 3-6 together, on short pedicels, or subsessile. Invol. scales oblong-acuminate, smooth.

2. L. scabra (Harv.) ; *very scabrous. Ethulia alata, Sond. in Linn.*
xxiii. 60.

HAB. Bloomspruit, Fals river and Aapjes R., *Burke & Zeyh.!* (Hb. Sd.,Hk.)
Stem ascending, 1-2 ft. long, much branched, the branches erect and closely leafy,
either rough throughout or smoothish below. Leaves linear or linear-lanceolate, acute
or subacute, entire or remotely denticulate, decurrent in narrow wings along the
stem very rough with raised points on both sides. Heads tufted, 3-6 together,
subsessile. Invol. scales ovate-acuminate, rough at back.—Chiefly distinguished
from the preceding by its very scabrous surface.

III. VERNONIA, Schreb.

Heads few- or many-fl., homogamous. *Invol.* imbricated, shorter than
the flowers ; scales obtuse or acute, the inner longer. *Recept.* naked or
honeycombed. *Cor.* deeply 5-fid, the lobes lanceolate. *Filaments* smooth.
Anth. sagittate. *Achenes* linear or prismatic, glabrous, pubescent or silky-
villous, striate or rib-furrowed. *Pappus* biseriate, the *inner* of many
long, serrated bristles, the *outer* of very narrow, short scales ; rarely
both series nearly alike, subequal ; or the outer series of few bristles.
DC. Prodr. 5, *p.* 15. *Decaneuron, DC. l. c. p.* 66. *Webbia, DC. l. c. p.* 72.
Vernonella, Sond. in Linn. 23, *p.* 62.

A vast, tropical, and sub-tropical genus, common to both hemispheres, and consisting of herbs, climbers, shrubs, and even trees of various aspect. Lvs. alternate,
petiolate or sessile, often gland-dotted. *Infl.* various, either corymbose, spiked, or
with solitary heads. Fl. purple-rosy or white. *Decaneuron,* DC., to which is attributed a single pappus, has in several of its acknowledged species a distinctly double

one, and the 10 ribs of its achenes are slight exaggerations of the 10 striæ found on many acknowledged *Vernoniæ;* nor is there any distinctive habit to mark its species, at least those of the section *Gymnanthemum.* In *Webbia,* DC., which is founded on its unisexual flowers, I find (as C. H. Scultz, Bip. has already pointed out) *bisexual* flowers in all the Cape species. Nor can I keep *Vernonella,* Sond., separate without also founding genera upon our *V. monocephala* and *V. stahelinoides,* which differ as widely in habit both from each other and from the species of the first section.—Vernonia is named in honour of W. Vernon, a botanist of the 17th century, who travelled in North America in search of plants.

a, Erect or half-climbing much-branched shrubs or hf.-shrubs.
　　Heads corymbose or panicled.
　　　Lvs. minutely pubescent or glabrous :
　　　　Panicles divaricate-spreading ; invol. scales *acute*　　　(1) **anisochætoides.**
　　　　Corymbs erecto-patent ; invol. scales obtuse.
　　　　　Invol. scales powdery ; pappus white　...　...　(2) **angulifolia.**
　　　　　Invol. scales glabrous ; pappus reddish　...　...　(3) **mespilifolia.**
　　　Lvs. densely tomentose or hoary beneath :
　　　　Lvs. subsessile ; invol. sc. tomentose ; heads 4–5-fl.　(4) **corymbosa.**
　　　　Lvs. long-petioled ; inv. sc. glabrous ; hds. 15–20-fl.　(5) **Senegalensis.**
b. Herbs, with simple, erect stems. *Heads* corymbose (*Webbia,* DC.)
　　Stems uniformly leafy :
　　　Lvs. *petioled,* ovate, silvery beneath, penninerved　(6) **Kraussii.**
　　　Lvs. *sessile,* silky-silvery on one or both sides :
　　　　Lvs. lanceolate, penninerved ; inv. sc. aristate　(7) **Natalensis.**
　　　　Lvs. linear, 1-nerved ; inv. sc. mucronate　...　(8) **pinifolia.**
　　　Lvs. *sessile,* roughly pubescent (not silvery) ;
　　　　Lvs. cordate at base ; hairs simple, curled　...　(9) **hirsuta.**
　　　　Lvs. acute at base ; hairs forked　...　...　...　(10) **Sutherlandi.**
　　Stems nearly naked ; radical leaves obovate, large　(11) **Dregeana.**
c. Herbs or suffrutices.　Branches 1-headed :
　　Inv. scales oblong, broad, obtuse or mucronate :
　　　Suffruticose, much branched, closely pubescent　...　(12) **staehelinoides.**
　　　Herbaceous, glabrous ; lvs. subspathulate　...　...　(13) **Vernonella.**
　　Inv. scales lanceolate, much acuminate :
　　　Glabrous ; achenes densely silky　...　...　...　...　(14) **Gerrardi.**
　　　Hairy ; achenes 10-ribbed, quite glabrous　...　...　(15) **monocephala.**

1. V. anisochætoides (Sond. in Linn. 23, p. 61); shrubby, the branches, inflorescence, and young leaves minutely pubescent, becoming glabrous; branches flexuous, angularly striate ; leaves conspicuously petioled, cuneate at base, broadly obovate, sharply and angularly 3–5-toothed at the summit, or the upper ones subentire, acute, 3-nerved at base, pellucid-dotted ; panicle terminal, much branched, leafy, its branches *widely spreading or deflexed, loosely divaricating, multifid;* invol. scales ovatolanceolate, *acute,* nerve-keeled, woolly ; heads 12–13-flowered ; pappus pale.

HAB. Near D'Urban, Natal, *Gueinzius,* 320, 591. *Gerr. & M°K.* 335, 336. (Herb. Sd., D.)
A large, half-climbing shrub. Lower leaves 2½ inches long, and equally wide, with 5 sharp, toothlike lobes, with wide interspaces ; upper 1½–2 inches long, 1–1¼ inch wide, mostly entire. Invol. scales with a dark central line, woolly at the sides, becoming glabrate. Fl. purple. Outer pappus of few scales.—Readily known, among S. African species, by its divaricate inflorescence.

2. V. angulifolia (DC. Prodr. 5, p. 29); subherbaceous, the branches, inflorescence, and foliage minutely puberulous; branches angularly striate; leaves shortly petiolate, cuneate at base, *deltoid-ovate*, acute, with prominent lateral angles, or shortly 3-lobed or coarsely toothed, 3–5-nerved at base; panicle terminal, compound, leafy, its branches loosely corymbose; invol. scales ovate-oblong, obtuse, nerve-keeled, minutely scurfy, membrane-edged; heads 6–9-flowered; pappus white.

HAB. Near D'Urban, Natal, *Drège, Krauss* 223, *Gueinzius* 587, *Plant* 40, *Gerr. & M'K.* 334. (Herb. D., Sd., Hk.)

Much branched and tall, imperfectly woody (!), drying blackish. Leaves membranous, 1–1½ inches long, 1–1½ wide, the upper ones trowel-shaped, the lower more or less toothed or lobed. Scales of the invol. pale horn-coloured, with a dark, central line, under a lens minutely rough. Fl. purple. Achenes thinly silky.

3. **V. mespilifolia** (Less. in Linn. 1831, p. 641); shrubby, the young branches, inflorescence, and foliage minutely puberulous; branches round, striate; leaves shortly petioled, much attenuated at base, *broadly obovate*, coarsely 3–7 or several-toothed, or angularly lobulate at the summit; panicle terminal, compound, its branches densely corymbose; invol. scales ovate oblong, obtuse, round-backed, *glabrous, ciliolate;* heads 3–5-flowered; pappus fulvous. *DC. l. c.* 29.

HAB. Swellendam, *E. Z.* Forests of Uitenhage, *Burchell, Zeyher!* Common in Albany and Kaffirland, and at Natal, *E. Z., Drège, Bowker*, &c. (Hb. D., Sd., Hk.)

A half-climbing shrub, drying dark. Leaves membranous, 2–3 inches long, 2 in. wide, thinly puberulous, minutely gland-dotted beneath. Invol. scales pale-horny, neither keeled nor nerved, very obtuse or subacute, either quite glabrous or downy at top. Achenes with silky striæ. Outer pappus very narrow.

4. **V. corymbosa** (Less. l. c. p. 647); shrubby, the branches, inflorescence, and under sides of leaves hoary or tomentose; branches rodlike, rigid, angularly striate; leaves scarcely petioled, cuneate-oblong or obovate, coarsely few-toothed at the summit or beyond the middle, coriaceous, *tomentose beneath,* the older ones becoming glabrate above; panicle terminal, leafy, corymbose; invol. scales oblong, obtuse, round-backed, *tomentose;* heads 4–5-flowered; pappus fulvous. *DC. l. c.* 29. *Stœhelina corymbosa, Th. / Cap.* 628.

VAR. β, mespilioides (DC.); leaves larger, sharply 7–9-toothed, less hoary beneath. *E. Z.* 269; *Drège,* 5036; *Gerr. & M'K.* 344.

HAB. From Uitenhage to Natal, *Burchell, E. Z. Drège, Gerr. & M'K.* &c. (Herb. Th, D., Sd., Hk.)

A large, much branched shrub. Twigs straight. Leaves very variable in size and shape; sometimes narrow-cuneate, subentire, 1½ in. long, ½ in. wide; sometimes broadly cuneate, 1½ inch long, 1 inch. wide; and in var. β 2–2½ in. long, 1½ in. wide. Invol. scales rufous brown, not keeled. Achenes with pubescent striæ. Outer pappus narrow, brittle.

5. **V. Senegalensis** (Less. in Linn. 1829, p. 265); shrubby, the branches, infl., and under sides of leaves *tomentose;* branches flexuous, round, striate, sprinkled with glands; leaves *on longish petioles,* slightly cuneate at base, broadly ovate or oblong, obtuse, subentire, undulate, scabrous above; panicle terminal, leafy, compound, its branches widely spreading, loosely corymbose; inv. scales broadly ovate, mucronulate, round-backed, glabrous, ciliolate; heads 15–20-flowered; pappus fulvous. *Decaneurum Senegalense, DC. l. c.* 68. *Eupat. coloratum, Willd.*

HAB. Zululand, *Rev. Mr. Hewitson;* on the Shiré, *Dr. Kirk.* (Herb. D., Hk.)
A large, half-climbing shrub, 10–12 ft. high. Lower leaves not seen; upper on uncial petioles, 2–3 in. long, 1½–2 in. wide, not much attenuated at base, irregular in outline, but not toothed. Outer pappus of few and narrow scales. Achenes 10-rib-striate, the striæ glabrous, with gland-downy interspaces. Corolla pilose.

6. V. Kraussii (Sch. Bip. in Walp. Rep. 2, p. 947); herbaceous; the simple, striate stem, the inflorescence, and the *under surface* of the leaves silky-canescent; leaves shortly petioled, ovate, acute or acuminate, flat, entire, becoming glabrous above, penninerved; corymb of few or many heads, flat-topped; invol. scales lanceolate, acute, mucronate, canescent; achenes silky. *Webbia elæagnoides, DC. l. c. 73.*

VAR. β. **oligocephala;** dwarf (6–8 inches high), slender; heads 3–10. *Webbia oligocephala, DC. l. c.*

HAB. Between Omsamwubo and Omsamculo, *Drege!* Magallisberg, *Burke & Zey.!* Natal, *Krauss, Sanderson,* &c. Var. β. with var. α. *Drege, Gueinzius,* 339. (Herb. D., Sd., Hk.)
Stem 1½–3 feet high. Leaves 1–1½ inch long, ⅜–1 inch wide, at first thinly silky above, the adult quite glabrous, white and silky beneath. Heads many-flowered. Corymb 2–4 inches across. Corolla pilose. Decandolle's *W. oligocephala* seems to me to be merely a starved variety.

7. V. Natalensis (Sch. Bip. in Walp. Rep. 2, p. 947); herbaceous; the simple, striate stem, the inflorescence, and *both surface* of the leaves silky canescent; leaves sessile, crowded, *lanceolate,* acute or acuminate, *narrowed to the base,* penninerved, entire, flat; corymb of many heads, flat-topped; invol. scales lanceolate, much-attenuated, bristle-pointed; achenes silky. *Webbia aristata, DC. l. c. 73.*

HAB. Katberg, and at the Key and Basche rivers, *Drege;* Kaffirland, *E. Z.!* Natal, *Krauss,* 443; Winterberg, *Mrs. F. W. Barber,* 233; near Maramballa, *Dr. Kirk!* (Herb. D., Sd., Hk.)
This resembles *V. pinifolia,* but has much broader, strongly-nerved, and canous leaves, and long-pointed involucral scales, &c. Leaves 2–3 inches long, ¼–¾ inch wide, erect. Heads many-flowered. Corymb 2–5 inches across. Corolla pilose.

8. V. pinifolia (Less. in Linn. vol. 2, p. 257); herbaceous; the simple, rigid, striate stem, the inflorescence, and the *under* surface of leaves more or less canescent or pubescent; leaves sessile, crowded, linear, acute, with revolute margins, one-nerved, becoming glabrous above; corymb mostly compound, of many heads, flat-topped; invol. scales lanceolate-oblong, mucronate, canescent; achenes densely silky. *Webbia pinifolia, DC. Prod.* 5, *p.* 72. *Bot. Mag. t.* 5412. *Conyza pinifolia, Lam. C. canescens, Th.! Cap.* 665. *Erigeron Capense, Houtt.*

VAR. β. **glabrata;** stem weak, nearly glabrous; leaves few, glabrous and dotted; heads few.

HAB. Langekloof and Krum river, *Thunb. E. Z.!* Georgetown, *Pappe!* Uitenh. Albany, Caffraria, and Natal, *Drege, Hutton, Sander.,* &c.; var. β. Howisonspoort, *Hutton!* (Herb. Th., D., Sd., Hk.)
Stems 1–2 feet high, mostly closely-leafy throughout. Leaves 1–4 inches long, ½ line to 3–4 lines wide, the wider ones penninerved. Pubescence very variable, in var. β. scarcely any save on the involucre. Corymb 3–8 inches across. Flowers purple. Heads 20 or more flowered. Corolla pilose, bright purple.

9. V. hirsuta (Sch. Bip. l. c.); herbaceous; the simple, striate stem

the inflorescence, and under surface of leaves, hairy with rough, curled hairs; leaves sessile, *cordate at base,* oblong or obovate, mucronate, acuminate *(or obtuse),* subentire or with inflexed serratures, thinly-hairy above, netted-veined beneath; corymb densely many-headed, flat-topped; invol. scales lanceolate, acute or acuminate, woolly; achenes silky. *Webbia hirsuta, DC. 5, p. 73.*

VAR. β. obtusifolia ; leaves, save the uppermost, broadly-obovate, bullated, very obtuse or shortly-cuspidate.

HAB. Katberg and Tambukiland, *E. Z.! Hutton! Bowker;* Natal, *Drege, Krauss,* 293; var. β. Natal, *T. Williamson, Gerr. & M'K.* 343. (Herb. D., Sd., Hk.)

Stems 1–3 feet high, the whole plant coarsely-hairy, with greenish-yellow hairs. Leaves 1½–2½ inches long, ¾–1½ inch wide, very variable in shape, but always cordate at base. In var. β. the leaves are thicker, and much more strongly-netted. Corymb dense, 2–4 inches across. Corolla pilose.

10. V. Sutherlandi (Harv.); herbaceous; the simple, striate stem, inflorescence, and leaves thinly-hispid *with rigid, mostly forked hairs;* leaves sessile, *acute at base,* oblongo-lanceolate, acute, mucronulate, few-toothed beyond the middle, or subentire, thinly-hispid on both sides, dotted beneath, penninerved; corymb pedunculate, cymose, 4–6- or 12-headed, the heads on longish, simple pedicels; invol. scales lanceolate, acuminate, hispid; achenes densely-silky.

HAB. Klip river, Natal, 3500–4500 feet, *Dr. Sutherland!* (Herb. D , Hk., Sd.)

Stem 1–2 feet high, the lower half closely-leafy, the upper naked, with 2–3 distant, small, linear-subulate, depauperated leaves. Lower leaves 1½–2 inches long, ½–¾ inch wide, the uppermost entire, the lower coarsely few-toothed. Pubescence in most places *forked* or stellate, rather rigid. Pedicels of the central head ¾ inch, of the lateral 1½–2 inches long. Heads many-flowered. Corolla bright purple, pilose.

11. V. Dregeana (Sch. B. l. c.); herbaceous; the simple, angularly-striate stem, inflorescence, and foliage *nearly glabrous* or thinly-silky ; *radical* leaves obovate, narrowed at base, mucronulate, penninerved, sub-denticulate, with slightly-inflexed margins, nearly glabrous ; *cauline* few, remote, linear, erect, with involute margins ; corymb closely many-headed ; invol. scales oblongo-lanceolate, mucronate, woolly; achenes silky. *Webbia nudicaulis, DC. Prod. 5, p. 73.*

HAB. Betw. Key and Bashe, *Drege!* near Grahamstown, *Genl. Bolton, H. Hutton;* Natal, *T. Williamson, Sanderson, Gerr. & M'K.* 316. (Herb. Sd., Hk., D.)

Stem 1–2 feet high. Radical leaves several, rosulate, 2–2½ inches long, ½–1 inch wide; cauline 1 or 2 inches apart, 1–3 inches long, 1–2 lines wide. Heads on longish pedicels, many-flowered. Corolla pilose.

12. V. stæhelinoides (Harv. Thes. t. 156); suffruticose, paniculately much-branched ; the rigid, striate branches, peduncles, and foliage thinly appressed-pubescent, sub-canescent ; leaves sessile, narrow-linear, obtuse, with revolute margins, one-nerved beneath ; heads terminal and axillary, solitary, pedunculate, about 10-flowered ; invol. oblong, the outer scales roundish or ovate, obtuse, the inner oblong, subacute, mucronulate, all close-pressed, at first silky, becoming glabrate; achenes minutely-pubescent. *Zey.! 1027.*

HAB. Magalisberg and Crocodile river, *Burke & Zey.!* (Herb. Hk., D., Sd.)

Stem erect, 1–2 feet (or more?) high, much-branched, the branches erect, ramulose and somewhat corymbose. Leaves 1-1¼ inch long, not a line wide, very blunt, exactly linear. Peduncles 1-1½ inch long, mostly from the axils of upper leaves, naked. Heads, like those of a *Stæhelina* or *Serratula*, 5 lines long, 2 lines across, about 10-flowered. Pappus biseriate, the outer very short, the inner 5-6 times as long, with rough, dentato-plumose bristles. Recept. honeycombed. A very remarkable species.

13. **V. Vernonella** (Harv.); herbaceous, branched from the base; branches simple, furrow-striate, *glabrous*, leafy; leaves sessile, spathulate-linear, obtuse, narrowed to the base, with recurved edges, glabrous; heads *solitary*, terminal, pedunculate, many-flowered; invol. scales *broadly-oblong*, mucronate, with scarious, wavy, lacerate edges, glabrous; achenes *densely-silky*. *Vernonella Africana, Sond. in Linn.* 23, *p.* 62.

HAB. Port Natal, *Gueinzius,* 334! (Herb. Sond.)
A glabrous, many-stemmed perennial, 6-8 inches high. Branches or stems erect, simple, leafy beyond the middle. Leaves 1-2 inches long, 2 lines wide, slightly narrowed to the base, conspicuously 1-nerved. Heads ⅔ inch across. Invol. scales thin, membranous, purplish, ragged at the edges, loosely imbricated. Fl. purple. Pappus discoloured.

14. **V. Gerrardi** (Harv. Thes. t. 157); herbaceous, branched below; branches simple, furrow-striate, dotted, glaucous, leafy; leaves sessile, linear, acute, 1-nerved, dotted, flat, glabrous; heads solitary, terminal, pedunculate, many-fl.; inv. scales lanceolate, cuspidate-acuminate, with very long, squarrose points, cobwebbed, woolly-ciliate; achenes ribbed, densely silky.

HAB. Bigarsberg, Zululand, *W. T. Gerrard,* 1060. (Herb. D.)
A foot or more high, rigid, nearly glabrous and glaucous, conspicuously gland-dotted. Leaves 1-1½ inch long, 1 line wide, thickish. Heads ⅔ inch across. Inv. scales rigid, with long, squarrose points, like those of an *Athrixia.* A very distinct and remarkable species, allied to *V. Vernonella* and *V. monocephala.*

15. **V. monocephala** (Harv.); herbaceous, branched from the base; branches sub-simple, striate, *roughly hairy* or pubescent, leafy; leaves sessile, oblongo-lanceolate or lanceolate, mucronate, dotted, glabrous (or sparsely setose), ciliate on the margin and mid-rib beneath; heads *solitary*, terminal, pedunculate or sessile, many-flowered; invol. scales *linear-lanceolate*, much acuminate, subpungent, scabrous and setose; achenes *quite glabrous*, 10-ribbed, and furrowed.

HAB. Magalisberg, *Burke & Zeyher!* 801; Zululand, *Gerr. & M'K.* 1011. (Herb. Hk., Sd., D.)
Roughly hairy, perennial, 1-4 ft. high, either erect and much branched, or ascending from a decumbent base, branched from the crown or the lower part of stem. Branches pale, variably hairy, sometimes leafy to the summit, sometimes ending in a naked peduncle, 4-6 inches long, each branch usually bearing a fl.-head. Heads ¾-1 inch across, with very many flowers. Pappus dirty-white, the outer of a few short lin.-lanceolate bristles. Corolla externally scabrous, especially on the reflexed, linear-lanceolate lobes.—This has the achenium of a *Decaneurum*, DC., and the habit of *V. simplex*, Less.

IV. HOPLOPHYLLUM, DC.

Heads few- (5-6) flowered, homogamous. *Invol.* ovate-oblong, closely imbricate, its scales broad, obtuse, spinous-mucronate, scarious or horny. *Recept.* narrow, fimbrilliferous. *Cor.* glabrous, tubular, deeply and equally 5-cleft, the lobes linear. *Filaments* smooth. *Anthers* linear, sagittate. *Stigmata* 2, filiform, very long, bristly throughout. *Achenes* thick, subtrigonous, densely hairy, slightly contracted under the pappus. *Pappus* in many rows, persistent, scabrous, the outer paleæ bristle-shaped, the inner subulate-acuminate. *DC. Prodr.* 5, *p.* 73.

Rigid shrubs, with pungent, linear or subulate leaves. Heads sessile, axillary or terminal.—Name compounded of ὅπλον, a *dart*, and φυλλον, a *leaf;* from the sharp pointed leaves.

Leaves semiterete, entire; heads crowded; invol. scales glabrous...   (1) **spinosum**.
Leaves flat, with spinous, marginal teeth; heads solitary; invol. sc.
   powdery ... ...   (2) **ferox**.

1. H. spinosum (DC. l. c. p. 73); glabrous; leaves horizontally spreading, *subulate, semiterete,* furrowed, *quite entire;* invol. scales broadly oblong, very obtuse, shortly mucronate, quite glabrous; heads sessile in the upper axils and several-crowded at the ends of the branches.

HAB. Olifants R., and Bokkeland, *Thunb.;* Clanwilliam, *E. Z.!* Gariep, *Drege!* (Herb. D., Sd., Hk.)
A much-branched, rigid suffrutex : branches and leaves divaricate. Leaves 1-2 inches long, pungent, 1-2 lines wide. Invol. ½ inch long, of many close-pressed, horny, wide-margined scales. Pappus fulvous. Corolla pale.

2. H. ferox (Sond.); subglabrous; leaves spreading, *linear-mucronate, flat, armed with 1-2 pairs of spinous, marginal teeth,* furrowed; invol. scales ovate, strongly mucronate, dorsally glauco-pulverulent; heads terminal, solitary sessile.

HAB. Gamke R., *Burke & Zeyher;* Krails R. Beaufort, *Zeyh.!* (Herb. Hk., Sd.)
Very rigid, much branched, and spiny as the last, but with *flat* leaves, margined, with a few spinous teeth, and solitary, terminal heads. Leaves 1-1¼ inch long, 1 line wide. Pappus fulvous.

V. PLATYCARPHA, Less.

Heads densely crowded, sessile, many-flowered, homogamous. *Invol.* scales in many rows, lanceolate, entire, scarious, pungent-mucronate, the inner narrow, resembling paleæ. *Cor.* with a long, slender, hispid tube, and equally 5-parted limb, the lobes linear, with recurved, gland-bearing points. *Fil.* smooth. *Anthers* linear, obtuse, subequally sagittate at base. *Stigmata* 2, filiform, diverging, equally papillate throughout. *Achenes* glabrous, oblong, bluntly 5-angled. *Pappus* persistent, of 7-9, white, linear-acuminate, entire, spreading scales. *DC. Prodr.* 5, 71.

Only one species. The name is compounded of πλατυς, *flat,* and καρφος, a *scale;* referring to the flat scales of the pappus.

1. P. glomerata (Less. in Linn. 6, p. 688); *DC. l. c. Cynara glomerata, Th.! Cap.* 141. *Stobœa glomerata, Spr. Syst.* 3, *p.* 394.

HAB. In saline spots, Zwartkops R., *E. Z.!* Drege. (Herb. Th., D., Sd., Hk.)
A stemless perennial. Radical leaves numerous, spreading like a star on the ground, petiolate, pinnately divided, albo-tomentose beneath, cobwebby, but becoming glabrate above, the lobes coarsely toothed, the teeth pungent; the lower lobes short and distant, the upper close together and larger. Flower-heads numerous, densely crowded in a sessile tuft, covering the crown of the root. Invol. scales very sharp-pointed, rigid, spreading. Cor. purple. Much white wool about the insertion of the leaves.

VI. CORYMBIUM, Linn.

Heads 1-flowered. *Invol.* cylindrical, of two opposite channelled scales, one clasping round the other, with 2-3 very small, exterior bracts at base. *Cor.* salver-shaped, deeply and equally 5 cleft. *Anthers* included, simple at base. *Stigmata* 2, filiform, equally hispid. *Achenes* attenuated at base, densely clothed with long, straight, silky hairs. *Pappus* short, crown-like, irregularly cleft or fimbriate. *DC. Prodr.* 5, 88.

Herbaceous, nearly stemless plants, all natives of S. Africa, their thick rootstock clothed with long, soft, silky hairs. Leaves radical, linear, parallelly many-nerved, rigid. Flowering stems nearly naked, with a few clasping, small leaves often reduced to scales, corymbose at the summit. Corymbs loose or close, compound, each branch ending in an involucre. Flowers pink or white, rarely yellow.—Name from κορυμβος, *a corymb*, because the inflorescence is corymbose.

1. *Stems, leaves, and involucres glabrous :*
   Nerves of the leaves prominent; margin thickened, smooth :
   Nerves only 3, distant ... ... (1) latifolium.
   Nerves numerous, close-placed ... ... (2) nervosum.
   Nerves of the leaves immersed or scarcely prominent :
   Nerves 1-3; corymb loosely branched ... ... (3) glabrum.
   Nerves several; corymb very densely crowded ... ... (4) cymosum.
2. Stems and involucres scabrous; leaves glabrous ... ... (5) scabrum.
3. Stems and *leaves* hairy and glandularly muricate :
   Leaves linear, acute, elongate ... ... (6) villosum.
   Leaves broadly lanceolate, tapering to both ends ... ... (7) congestum.

1. C. latifolium (Harv.); glabrous; stem compressed, smooth, much longer than the leaves ; leaves *broadly* lanceolate, linear or oblong-linear, with 3 *distant, prominent* nerves and thickened margins; corymb loosely much branched; invol. scales smooth.

HAB. Vanstaadensberg, Uit., *Zeyher!* No. 2737 (and 303). (Herb. Sd., Hk., D.) This has broader and shorter leaves, and a laxer inflorescence, than *C. nervosum*, from which it more especially differs in never having more than 3 nerves, even in its broadest leaves. Leaves 6-8 inches long, ¼-1 inch wide, sometimes falcate.

2. C. nervosum (Th. Cap. 729); glabrous; stem angular, compressed, smooth ; leaves broadly-linear or lanceolate-linear, with *many strongly prominent, close-placed nerves,* and *thickened* margins; corymb *loosely* much-branched; invol. scales smooth. *DC. l. c.* 89. *C. glabrum, Linn. f. Lam. ill. t.* 723, *f.* 2. *Drege! Herb. and E. & Z.! ex pte.*

VAR. β. **subulifolium**; leaves very narrow-subulate, convolute-terete. *Zey.!* 2736.

HAB. Moist, sandy places, from Capetown to Uitenhage, common; var. β. Uit. *Zey.!* Grahamstown, &c. *Mrs. F. W. Barber.* (Herb. D., Sd., Hk., &c.) Root stock robust, densely woolly. Leaves 10-14 inches long, 2-5 lines wide, microscopically granulated, yellow-green. Stems scarcely taller than the radical leaves, with 2-3 clasping, cauline leaves. Flowers pink or purple. Var. β. has leaves almost as narrow as those of *C. scabrum, β. filiforme.*

3. C. glabrum (Thunb. 729); glabrous; stem angular, smooth ; leaves broadly-linear, subacute, flat, with 1-3 *impressed or immersed, stria-like nerves,* and *rough* margins; corymbs loosely much-branched; invol. scales smooth. *DC. l. c.*

HAB. Cape, *Th.; nr.* Capetown, *E. Z.!* Wynberg, *Dr. Wallich!* (Hb. Sd., D., Hk.) Often confounded with the preceding, from which its immersed (not prominent) nerves readily distinguish it. *Drege* and *E. & Z.'s* specimens are indiscriminately named. Lvs. 6–8 inches long, 3–4 lines wide, often falcate; the narrower *one*-nerved, the broader faintly 3-nerved. Invol. 4 lines long.

4. C. cymosum (E. Mey.); glabrous ; stem bluntly angular, smooth, mostly *shorter* than the leaves ; leaves *very long*, linear, acute, with many, scarcely prominent nerves, and a roughish margin ; corymb *very dense*, fasciculate ; the glabrous involucres subsessile. *DC. l. c. p.* 89. *(excl. sp. ex Drege).*

HAB. Hottentotsholland, *Mundt!* Eastern Districts, *Burchell,* fide DC. Stellenbosch and Swellendam, *E. & Z.!* Stellenbosch, *W. H. H.* (Herb. Hk., Sd., D.)
Readily known from *C. glabrum* by its densely crowded inflorescence. Leaves 8–16 inches long, 3–4 lines wide. Scales of calyculus pointed. *Drege's* specimens marked " *C. cymosum,*" so far as I have seen, belong to *C. scabrum,* β.

5. C. scabrum (Linn. f. Suppl. 392); stem terete, *densely muricated;* leaves linear or filiform, flattish or convolute, glabrous, minutely shagrined ; corymb dense ; invol. scales scabrous and glandular, viscidulous. *DC. l. c.* 89. *Thunb. Cap.* 729. *Lam. ill. t.* 723, *f.* 1. *Burm. Afr. t.* 70, *f.* 1.

VAR. β. filiforme (Th.); leaves very narrow, filiform, semi-terete. *C. filiforme, Linn. f. DC. l. c. C. cymosum, Drege! in Herb. Hk., D., Sd. Zey.!* 2739.

VAR. γ. luteum ; corolla (said to be) yellow. *C. luteum, E. M.! DC. l. c. p.* 89.

HAB. Round Capetown and Stellenbosch ; at the Paarl, *Drege;* Worcester, *E. Z.;* Caledon, *Pappe,* &c. ; Uit., *Zey.!* Var. γ. at the Paarl, *Drege!* (Hb. D., Sd., Hk.)
Leaves 4–12 inches long, ¼ line to 2–3 lines wide, variable in the same tuft. Stems mostly taller than the lvs. very rough. Flowers pink or pale ; in var. γ "yellow" (perhaps *primrose* colour?). I find *Drege's* specimens of *C. luteum* quite as scabrid as those of our var. β, and in the dry state undistinguishable.

6. C. villosum (Less. in Linn. 1831, p. 691); stem terete, villous and glandularly muricated ; leaves *linear*, acute, flat, sub-falcate, on both sides pilose and glandularly muricated ; invol. scales glandularly muricated, bearded and pilose. *DC. l. c.* 89. *C. villosum and C. hirtum, Th. Cap.* 729, 730. *C. hirsutum, Eck.*

HAB. Cape, *Thunb., E. & Z.!* Dutoitskloof and Piquetberg, *Drege!* Stellenbosch, *W. H. H.* Bamskloof, *Hutton!* Zwartberg, Caledon, *Pappe!* (Hb. D., Sd., Hk., &c.)
Leaves 4–6 inches long, 3–4 lines wide ; cauline long or short, but always linear. Corymb dense. Exterior scales of the invol. nearly half as long as the inner ones.

7. C. congestum (E. Mey.); stem somewhat angular, hispid and glandularly muricated ; leaves *broadly-lanceolate, tapering to both ends,* acute, on both sides glandularly pubescent ; corymb dense ; invol. scales densely glandular, sharply 3-toothed. *DC. l. c.* 89. *Harv. Thes. t.* 69.

HAB. Drakeensteinberg, 2–300 ft. *Drege!* Worcester, *E. & Z.!* (Hb. D., Sd., Hk.)
Leaves broader and more membranous than in other species, 5–6 inches long, ¾–1 inch wide. Cauline leaves clasping, oblongo-lanceolate. Division of the corymb few-flowered.

TRIBE II.—EUPATORIACEÆ.

Flower-heads discoid, all the flowers tubular. *Style-branches* long, exserted, terete, *obtuse,* or thickened at the point, *minutely granulated* on the outer surface.—Leaves mostly opposite, rarely alternate. (Gen. VII.—X.)

Leaves alternate.   Pappus of 3–5 unequal scales   ... ...   (7) **Anisochæta.**
Leaves opposite :
   Heads many-flowered :
      Pappus of 5–10 broad, toothed, pointed scales   ...   (8) **Ageratum.**
      Pappus of 3–5 gland-tipped bristles   ... ...   (9) **Adenostemma.**
   Heads 4-flowered.   Pappus of many, slender bristles ...   (10) **Mikania.**

VII. ANISOCHÆTA, DC.

Heads many-flowered, homogamous. *Invol.* ovate, imbricated, its scales lanceolate, appressed, shorter than the disc. *Recept.* naked. *Corolla* tubular, deeply and equally 5-cleft, the outer flowers curved downwards towards the periphery. *Anthers* sagittate. *Style* shortly exserted ; stigmata 2, cylindrical, obtuse, nearly smooth. *Achenia* (unripe) oblong, striated, scarcely downy. *Pappus* of 3–5, narrow-subulate, much acuminated, unequal, erect scales. *DC. Prod. 5, p. 109.*

A half climbing, or scrambling shrubby plant.   The name is compounded of *ανισος, unequal,* and *χαιτη,* a *bristle* ; alluding to the pappus.

1. A. Mikanioides (DC. l. c. p. 110.)

HAB. Between the Omsamwubo and Omsamcaba, *Drege!* Natal, *Gueinzius!* 542. (Herb. D., Sd., Hk.)
Stem terete, flexuous, striate, pubescent; branches widely spreading or deflexed. Leaves alternate, petioled, broadly-ovate, 3–5-nerved and subtruncate or deltoid at base, coarsely and bluntly incised, thinly downy, becoming glabrous, except on the nerves and petiole. Panicle ending the branches, its divisions spreading; ultimate pedicels longer than the heads, with 2–3 scattered bractlets. Invol. scales in several rows, minutely gland-dotted and rough-edged.

VIII. AGERATUM, Linn.

Heads many-flowered, homogamous, subglobose. *Invol.* loosely imbricated, its scales lanceolate, acuminate, equalling the disc. *Recept.* naked. *Cor.* tubular, equally 5-fid. *Style* exserted ; stigmas 2, cylindrical, obtuse. *Achenes* with a basal callus, 5-angled, narrowed to the base. *Pappus* of 5–10, lacerate or pectinate, bristle-pointed (or obtuse) scales. *DC. Prod. 5, p. 108.*

Herbaceous plants, chiefly American. *A. conyzoides* is commonly dispersed throughout the tropics of both hemispheres.   Name unexplained.

1. A. conyzoides (Linn. Sp. 1175); leaves ovate, rhomboid or cordate, on longish petioles; scales of the pappus 5, broad and serrated at base, their long bristle-points equalling the small corolla. *DC. l. c. p.* 108. *Schk. Hanb. t.* 238. *Hook. Exot. Fl. t.* 15.

HAB. About Port Natal, *Drege, Krauss,* 233; *Gueinzius, Gerr. & M'K.* 337, &c. (Herb. D., Sd., Hk., &c.)
Variable in pubescence and the shape of leaves, &c. Stem 2–3 feet high, terete. Leaves mostly pubescent, 1½–2 inches long, crenate; the petioles mostly very hairy. Corymbs dense, terminal. Flower-heads 2–3 lines across. Corollas lavender-blue or white, small.

IX. ADENOSTEMMA, Forst.

Heads many-flowered, homogamous. *Receptacle* flat, naked, honey-combed. *Invol.* campanulate, rather shorter than the flowers, reflexed after flowering, its scales in a single or double row, connate at base, oblong, herbaceous. *Cor.* shortly tubular, funnel-shaped, 5-toothed, villous externally. *Stigmata* 2, much exserted, linear, dilated, coloured. *Achenes* oblong, bluntly angular, crowned with 3–5 short, rigid, spreading, gland-tipped bristles. *DC. Prodr.* 5, *p.* 110.

Tropical and subtropical herbaceous plants, natives of both hemispheres. Leaves opposite, petiolate, toothed, 3-nerved at base. Heads panicled, pedicellate. Flowers white.—Name from αδην, a *gland*, and στεμμα, a *crown*.

Stem pubescent; leaves coarsely and sharply toothed ... ... (1) **Caffrum.**
Stem nearly glabrous; leaves with blunt, shallow teeth or crenate... (2) **Dregei.**

1. A. Caffrum (DC. l. c. 112); stem erect, 4-angled, thinly puberulous or scabrid below, more densely pubescent on the panicle; leaves ovato-rhomboid, acuminate, broadly cuneate or truncate at base, coarsely callous-toothed, nearly glabrous above, pubescent, especially on the nerves, beneath; panicle loosely dichotomous; invol. scales linear-oblong, obtuse, thinly downy at back; achenes granulato-muricated.

HAB. Caffraria and Natal, *Drege, Williamson, Gueinzius,* 347; *Bowker,* 34; *Gerr. & M'K.* 313. (Herb. D., Sd., Hk.)
A tall, coarse-growing weed. Leaves 3–5 inches long, 2–3 inches wide, herbaceous. Panicle terminal, irregularly forked, covered with rusty, subappressed hairs. Bracts small, few, linear. Petioles ¾–1 inch long. Invol. partially biseriate.

2. A. Dregei (DC. l. c. 111); stem sub-erect (or erect), sparingly muricated, especially about the nodes, otherwise glabrous; leaves conspicuously petioled, ovato-deltoid, cuneate at base, acute, bluntly callous-toothed, sparsely hispidulous on both sides, scabrid on the nerves beneath; panicle loosely forked, the pedicels long, slender, puberulous; invol. scales linear-oblong, obtuse, hispidulous; achenes granulato-muricated. Also *A. Natalense, DC. l. c.* 112.

HAB. Near Natal, Drege! *Gerr. & M'K.* 313. (Herb. D., Sd., Hk.)
A weaker, more slender, and more glabrous plant than *A. Caffrum,* with larger, thinner, more evidently petioled and less coarsely serrate leaves, and very lax inflorescence. Leaves 4–6 inches long, 3–4 inches wide. Stem rooting at the base, 2 ft. high. I cannot distinguish *A. Natalense* specifically.

X. MIKANIA, Willd.

Heads 4-flowered. *Recept.* naked, narrow. *Invol.* 4-leaved, bracteolate at or below the base. *Cor.* with a short tube, and bell-shaped, 5-cleft limb. *Anthers* sub-exserted. *Achenes* angular. *Pappus* of many, rough bristles, in one row. *DC. Prod.* 5, 187.

Tropical and sub-tropical, mostly climbing shrubs or herbs, chiefly American, with outliers in India, trop. Africa, and Madagascar. Leaves opposite, petioled, mostly cordate or sagittate. Heads in the S. African species corymbose. Corolla pale-coloured. Name in honour of Prof. Mikan, formerly of Prague.

All parts pubescent; leaves velvetty beneath ... ... (1) **Natalensis.**
Stem glabrous; leaves either glabrous, or hispidulous beneath (2) **Capensis.**

1. M. Natalensis (DC. l. c. 198); herbaceous, voluble, slender, *more or less pubescent* in all parts; leaves petioled, *cordato-sagittate*, acuminate, sinuato-dentate, 5-nerved, densely velvetty beneath, thinly pubes-

cent above, corymbs trichotomous, on peduncles longer than the leaves; invol. scales oblong-linear, pubescent, acute ; achenes glabrous, thinly glandular, pappus rufous.

HAB. Natal, *Drege,Sanderson*, 23 ; *Krauss,* 289 ; *Williamson,* &c. (Hb., D.Sd., Hk.) More slender than *M. Capensis,* with much more copious pubescence and narrower leaves, one-third shorter invol.-scales and smaller flowers.

2. M. Capensis (DC. l. c.); stem suffruticose, voluble, *glabrous ;* leaves petioled, cordate-acuminate or hastate-triangular, denticulate, glabrous above, hispidulous beneath or glabrous, 5-nerved ; corymbs trichotomous ; invol.-scales oblong-linear, acuminate ; achenes glabrous, thinly glandular; pappus rufous. *Eupatorium scandens, Thunb. Cap.* 627.

VAR. *a.* **cordata;** leaves *cordate*-acuminate, hispid beneath. *M. Capensis, DC. l. c.*

VAR. *β.* **oxyota;** lvs. *hastate* triangular, *mostly* glabrous. *M. oxyota, DC. l. c.*

HAB. Eastern districts of Caffraria, frequent. *β.* Vanstaadensberg, Drege ! Albany, *Mrs. F. W. Barber !* (Herb. D., Hk., Sd.)

A much branched climber, 20-30 f. long. Petioles 1-3 inches long. Leaves 2-3 inches long, 1½-2 in. wide, taper-pointed. Invol. 3-3½ lines long. Flowers pale or greenish. The leaves vary in shape and pubescence. This species scarcely differs from *M. chenopodifolia,* Willd.

TRIBE III.—ASTEROIDEÆ.

Flower-heads either discoid, all the fl. tubular ; or more frequently radiate, the central fl. tubular, 4–5 toothed, the marginal ligulate. *Style-branches* linear or lance-linear, flattened, mostly acute, their outer surface equally and minutely downy.—Leaves commonly alternate, rarely opposite. (Gen. XI.—XXXVIII.)

(1). FL-heads *conspicuously* radiate ; the rays blue, pink, or white (not yellow) :
　Recept. covered with rigid scales (*paleæ*) ... ... (12) **Amellus.**
　Recept. naked :
　　Ray and disc-fl. both with bristle shaped pappus :
　　　Pappus-bristles feathered (*plumose*) ... (13) **Mairea.**
　　　Pappus-bristles rough, but not feathered :
　　　　Pappus uniform, of many similar bristles ... ... (17) **Aster.**
　　　　Pappus double ; the *outer* bristles very short ... ... (18) **Diplopappus.**
　　Ray-fl. without pappus ; disc-fl. with bristle-pappus :
　　　Disc-fl. fertile, with *well-feathered* pappus ... ... (16) **Charieis.**
　　　Disc-fl. sterile, with *serrated* papp. bristles ... ... (14) **Gymnostephium.**
　　All the fl. without pappus. Leaves pinnatifid (21) **Garuleum.**
(2). Fl.-heads *conspicuously* radiate ; rays yellow :
　Recept. nude, smooth or minutely rough :
　　Pappus well-feathered (*plumose*) ... ... (13) **Mairea.**
　　Pappus bristle-shaped, roughish, *uniseriate :*
　　　Anthers not tailed at base ... ... (20) **Nidorella.**
　　　Anthers tailed at base ... ... (33) **Inula.**
　　Pappus bristle-shaped, *in many rows,* uniform-copious ... ... (11) **Alciope.**
　　Pappus double ; *outer* of short scales ; *inner* of bristles ... ... (34) **Pulicaria.**
　　Pappus none ... ... (15) **Anaglypha.**
　Recept. conspicuously honey-combed or fimbrilliferous :

Rays neuter, with glabrous, abortive achenes   (36) **Cypselodontia.**
Rays female; all the achenes hairy.   Pappus double:
     Pappus of several toothed bristles, the
         outer shorter ... ...   (37) **Minurothamnus.**
     Pappus of broad scales, all or half of
         them awned ... ...   (38) **Geigeria.**
(3). Fl. heads *heterogamous,* either discoid or very imperfectly radiate ; the marginal-fl. female, filiform, or with *very short,* unilabiate or bilabiate corollas (imperfect rays):
     Pappus double ; *outer* of short scales, *inner* of
         bristles ... ...   (34) **Pulicaria.**
     Pappus single, of few or many slender bristles :
         Marginal-fl. imperfectly ligulate (unilabiate):
             Rays white ; disc yellow   ... ...   (19) **Erigeron.**
             Rays yellow, as well as the disc   ... ...   (20) **Nidorella.**
         Marg.-fl. filiform, in a single row. Lvs. heath-
             like ... ...   (25) **Leptothamnus.**
         Marg.-fl. filiform, in many rows. Lvs. broad.
             Anthers without tails   ... ...   (26) **Conyza.**
             Anthers tailed at the base   ... ...   (32) **Blumea.**
     Pappus none, or of very minute, flat scales :
         Recept. flat.   Disc-fl. 5-toothed, sterile   ...   (31) **Denekia.**
         Recept. conical.   Disc-fl. 4-toothed   ... ...   (27) **Dichrocephala.**
(4). Fl.-heads *dioecious,* discoid ; male and female flowers in *separate* heads, on
distinct roots.   (Balsamic shrubs or trees) :
     Achenes woolly, without pappus ... ...   (30) **Tarchonanthus.**
     Achenes pubescent, with copious, bristle-shaped
         pappus ... ...   (29) **Brachylæna.**
(5). Fl.-heads *homogamous,* discoid ; all the flowers tubular and perfect:
     Recept. covered with rigid scales *(paleæ)*   ... ...   (12) **Amellus.**
     Recept. nude :
         Anthers not tailed at base :
             Pappus bristle-shaped, *uniform :*
                 Pappus uniseriate, very slender, de-
                     ciduous   ... ...   (23) **Chrysocoma.**
                 Pappus in many rows, rigid, per-
                     sistent ... ...   (24) **Pteronia.**
             Pappus double, the *outer* of short, narrow
                 scales, *inner* of long, rough bristles   ...   (22) **Fresenia.**
         Anthers tailed at base : pappus copious, of
             many bristles, or of scales and bristles   ...   (35) **Pegolettia.**
(6). Fl.-heads *compound ;* many small, discoid heads united on a common receptacle, with a general involucre.   Pappus none ... ...   (28) **Sphæranthus.**

Sub-Tribe 1. ASTEREÆ.   Heads sometimes homogamous and discoid, but mostly
heterogamous, radiate ; the ray-fl. female ; disc-fl. bisexual, rarely abortive. *Anthers
without tails.* (Gen. 1–24.)

XI. ALCIOPE, DC.

Heads many-flowered, radiate ; *ray fl.* ligulate, uniseriate, female,
with abortive stamens; *disc fl.* tubular, 5-cleft, regular, perfect. *Recept.*
broad, naked.   *Invol.* imbricate, its scales linear, appressed. *Style*
deeply bifid, its branches linear, divergent, semi-terete, obtuse, dorsally
puberulous.   *Achenes* linear-oblong, angular, sub-compressed, pubescent. *Pappus* bristle-shaped, in many rows; the bristles slender, rough,
deciduous. *DC. Prodr.* 5, *p.* 209.

Branching, erect undershrubs, natives of S. Africa.   Stems covered with dense,
white wool.   Leaves alternate, petioled, entire or denticulate, woolly beneath.

Heads large, terminating the branches, solitary or 2–3 in a corymb. Flowers yellow. The name is that of one of the nymphs, fancifully applied to this genus.

Invol. scales thinly cobwebby, becoming glabrous ... ...   (1) **Tabularis.**
Invol. scales densely and persistently woolly ... ...   (2) **lanata.**

1. A. Tabularis (DC.l.c.p.210); stem shrubby at base, erect, branched; branches tomentose, elongate, bearing 1–3 flower heads; leaves short-petioled, ovate or oval, entire or sinuato-denticulate, cobwebby, becoming glabrous above, woolly tomentose beneath; invol. scales oblong-linear, *soon glabrous*, sub-biseriate, the outer ones few and loose. *Arnica Tabularis, Th.! Cap.* 668. *Ligularia Tabularis, Less. Celmisia rotundifolia, Cass.*

HAB. Table Mt., *Thunberg, Drege;* Muysenberg, *W. H. H.* (Hb. Th., D., Sd., Hk.)
1–2 feet high, the branches with deciduous tomentum. Leaves 1½–2 inches long, 1½ inch wide, an inch or two apart. Branches naked at the summit, with a few linear, scattered bracts.

2. A. lanata (DC. l. c.); stem shrubby, erect, subsimple, *very densely woolly;* branches few, woolly, prolonged into long peduncles, bearing 1–2 fl.-heads; leaves petiolate, cordate-ovate, obtuse, with subrevolute margin, coriaceous, laxly woolly, becoming glabrous above, densely woolly beneath; inv. scales *densely and persistently woolly,* 2–3-seriate. *Arnica lanata & A. grandis, Th.! Cap.* 667, 668. *Ligularia lanata, Less.*

HAB. Roodesand, near Winterhoek, *Th.;* near Tulbagh, *Drege!* (Herb. Th., D., Sd., Hk.)
Much more densely and persistently woolly than the preceding; with larger lvs. and longer peduncles. Leaves 3–4 inches long, 2½–3 inches wide. Peduncles 12–15 inches long.

XII. AMELLUS, Cass.

Heads many-flowered, mostly heterogamous; *ray fl.* ligulate, in one series, female and fertile; *disc fl.* hermaphrodite, tubular, 5-toothed:—rarely *heads* homogamous and discoid. *Recept.* convex, covered with *paleæ* between the flowers. *Invol.* scales imbricated, rigid, acuminate, the inner ones assimilated to the paleæ. *Achenes* wedge-shaped, compressed, those of the *ray* somewhat 4-angled, scabrous, of the *disc* smooth, scabrous edged. *Pappus* of the ray of a few very short, unequal scales; of the *disc* double, the outer similar to that of the ray, the inner of 4–5 scabrous, deciduous bristles. *DC. Prodr.* 5, *p.* 213.

Small suffrutices or herbs, all S. African. Lower leaves opposite, upper alternate, oblong, entire or few-toothed, strigillose or canescent. Peduncles terminal, 1-headed. Ray fl. blue; disc yellow. Name adopted from the ancients.

A. **Perennes;** *Root* perennial. *Stems* rigid, half-woody. (Sp. 1–3.)
  Thinly silky, with close-pressed hairs; leaves broad or narrow, entire ... ...   (1) **Lychnitis.**
  Hispid with spreading bristles and hairs; lvs. very rough, entire or toothed ... ...   (2) **hispidus.**
  Sparingly setulose, with subappressed hairs; heads small   (3) **scabridus**
A. **Annui;** *Root* annual. *Stems* herbaceous. (Sp. 4–8.)
  Paleæ of recept. *acuminate,* as long as the disc fl., or longer:
    Rays conspicuous, *broad,* numerous:
      Stem erect, simple, with minute, appressed pubescence, and long hairs... ... ...   (4) **anisatus.**

Stem diffusely branched : *peduncles* hollow and
   swollen upwards ... ...   **(6) coilopodius.**
Stem diffusely branched ; pedunc. solid, filiform   **(5) strigosus.**
Rays very small and *narrow ;* pubescence copious
   and long ... ...   **(7) microglossus.**
Paleæ of recept. *club-shaped,* shorter than disc, few ; heads
   subsessile ... ...   **(8) nanus.**

A. **Perennes** *; Stem perennial, rigid, half-woody.* (Sp. 1-3.)

1. A. Lychnitis (Linn. Sp. 1276) ; stem suffruticose, erect or diffuse,
alternately branched, the branches, foliage, and involucres *minutely
appresso-pubescent ;* leaves alternate or opposite, either obovate, lanceo-
late-oblong-obtuse, or linear, very entire. *DC. l. c.* 214. *Lam. ill. t.* 682,
f. 1, *Jacq. coll.* 5, *t.* 10. *A. tenuifolius* (with narrow leaves), *Burm. DC.
l. c. Zey.!* 799.

VAR. β, flosculosus (Benth.) ; heads discoid, homogamous. *Hænelia Capensis,
Walp. Rep.* 2, *p.* 974. (Herb. D.)

HAB. Cape flats, &c., Riebeckskasteel, *Drege!* Riet Valley, *E. Z.!* Half way to
Stellenbosch, *W. H. H.; Namaqualand, Zeyher!* (Herb. Th., Hk., D., Sd.)
1-1½ foot high, pale, much branched. Stem sometimes sparsely pilose, as well as
silky. Leaves 1-1½ inch long, 1-3 lines wide, rigidly and closely puberulous. Heads
many-fl., 5-6 lines across. *A. tenuifolia* merely differs in its narrower, more exactly
linear leaves, varying on the same bush. Var. β. is in all respects identical with
the broad-leaved *A. Lychnitis,* minus the rays.

2. A. hispidus (DC. Prodr. 5, 214) ; stem suffruticose at base,
branched, erect, *hispid with spreading bristles and hairs ;* leaves alternate, linear, subacute, *very rough with rigid bristles and hairs,* some
quite entire, others with 1-2 small, lateral lobes or teeth ; invol. scales
densely setose at back, acuminate ; heads radiate. *Agathæa spathulata,
Kze.! Pugil.* 1, 9 *(fide Hort. Hamb.).*

VAR. β, angustissimus (DC.) ; slender, more densely covered with *long, spreading hairs* mixed with short ones ; leaves very narrow, and quite entire.

Var. γ, flosculosus ; leaves entire ; heads discoid. *A. flosculosus, DC. l. c.*

HAB. Kamiesberg, 300-400 ft. and near the Gariep, *Drege!* β, Oliphants River,
Drege! γ. Little Namaqualand, and Gariep, *Drege!* (Herb. D., Hk., Sd.)
Like *A. Lychnitis,* but readily known by its coarse, spreading pubescence. Var.
γ is a form precisely analogous to the flosculose var. of *A. Lychnitis.* A specimen
of *Agathæa spathulata,* Kze. in Herb. Sond., on the authority of the Bot. Gard. of
Hamburg, is an *Amellus!* and seems to be a garden state of this species.

3. A. scabridus (DC. l. c. p. 214); "stem suffruticose, branched,
sparingly setulose, with subappressed hairs ; leaves linear, acute, very
entire, sprinkled on both sides and at the margin with subappressed,
rigid bristles ; invol. scales acuminate, rough-hairy at back." *DC. l. c.*

HAB. Zwellendam, *Burchell;* Karroo, *E. Z.* (Unknown to us.)
"Stem rigid, slender. Branches not striated. Leaves 7 lines long, scarcely a
line wide. The heads when rubbed smell of aniseed, and are about half the size of
those of *A. Lychnitis.* Rays pale. Achenes mostly compressed, with callous margin,
at length nearly calvous." *DC.*

B. **Annui** : *Herbaceous, annual.* (Sp. 4-8)

4. A. anisatus (Cass. Dict. 37. p. 489) ; "stem herbaceous, erect,
simple, covered with minute, appressed pubescence, and sprinkled with

long, jointed bristles; leaves opposite or alternate, linear, very entire, on both sides roughly pubescent." *DC. l. c. p.* 214.

HAB. Cape. (Unknown to us.)

5. A. strigosus (Less! Syn. 162); annual, diffuse, ascending or depressed, many branched from the crown, rough with spreading hairs; leaves mostly alternate, roughly pubescent, linear, lanceolate or spathulate, narrowed to the base, either entire, or unidentate at each side beyond the middle, or shortly 3-lobed; paleæ acuminate.

VAR. *α.* **Thunbergii;** leaves mostly entire, sometimes denticulate. *Aster strigosus, Thunb.! Cap.* 687, *ex pte. Amellus strigosus, DC., l. c.,* 214.

VAR. *β.* **Wildenovii;** leaves mostly unidentate at each side. *A. annuus, Willd. Sp.* 3, 2215, *DC., l. c.* 215. *Kaulfussia amelloides, Zey.! Exsic.* 99. *Zey.!* 785.

VAR. *γ.* **tridactylus;** most of the leaves 3-lobed, the upper entire. *A. tridactylus, DC.! l. c.*

HAB. Groenekloof, *Thunb.!* Beyond the Gariep, *Burchell.* Gauritz R., *E. and Z.! Drege!* Bosjisveld, Swell., *Dr. Pappe!* Sneeweberg, *A. Wyley! β.* Zwartkop, R. *Zeyher!* Albany, *T. W.* Nieuweveld, and Beaufort, *Drege! γ.* Sneeweberg, *Drege!* Zwartkops R. *Zey!* (Herb. Thunb., D., Hk., Sd.)

A diffuse or often nearly prostrate annual, variable in size, and in the toothing of the leaves, but none of the above forms are constant to their assigned characters. Leaves ½-1-2 inches long, acute or obtuse, 1-3 lines wide. *Zey!* 2754 is intermediate between *β* and *γ*.

6. A. coilopodius (DC. l. c. 214); "root subsimple; stems numerous, herbaceous, terete, pubescent, the middle one erect, the lateral ascending; leaves linear, very entire, appressedly pubescent; *peduncles hollow and swollen under the heads;* invol. scales appressed, pubescent.'

HAB. Near the Gariep, *Drege.* (Unknown to us.)

"Stems 5-6 inches long. Leaves 1-1¼ inch long, a line wide. Rays many, blue. Chaff of recept. longer than the disc." *DC.* The hollow peduncles, if any thing, seem to mark this species.

7. A. microglossus (DC. l. c. 215); annual, diffusely much branched, pilose with long, spreading hairs; leaves linear-spathulate, subobtuse, narrowed at base, very entire, pubescent, invol.-scales infolded, acuminate, squarrose, hispid at back; *rays very small and narrow;* achenes of the ray pubescent, of the disc *thick-edged,* ciliate, central ones abortive.

HAB. Olifant's R., and Zilverfontein, Little Namaqualand, *Drege!* (Herb. D., Hk., Sd.)

Readily known by its very small ray-florets. Leaves 1-1½ in. long, 1½-3 lines wide, roughly pubescent. Heads on short branches, leafy nearly to the summit. Pappus very deciduous, shorter than the flowers; corona persistent.

8. A. nanus (DC. l. c. 215); annual, much branched from the base, depressed, in all parts rigidly and shortly appresso-pubescent, scabrid; leaves linear-lanceolate, obtuse, narrowed to the base, entire; heads subsessile, subtended by 1-2 leaves, becoming lateral; invol.-scales lanceolate, the outer keeled and infolded; rays ample, obtuse, twice as long as the involucre; *paleæ of the recept. shorter than the flowers, few, club-shaped, torulose;* achenes *silky-ciliate;* pappus similar in disc and ray.

HAB. Between Kaus, Natvoet, and Doornpoort, near the Gariep, *Drege !* (Herb. D., Hk., Sd.)

A depressed or prostrate, closely leafy annual, with large, blue-rayed, nearly sessile fl. heads, and crowded leaves. Lvs. 1-1½ in. long, 1-2 lines wide. The paleæ of recept. and achenes are different from those of other species.

XIII. MAIREA, DC.

Heads many-flowered, heterogamous; *ray-fl.* ligulate, female and fertile; *disc-fl.* tubular, 5-toothed, hermaphrodite, either fertile or sterile. *Recept.* naked. *Invol.* subimbricate. *Anthers* simple at base. *Achenes* plano-compressed, glabrous or sparsely pilose. *Pappus* in one series, of *feathered* bristles. *DC. Prod. 5, p.* 217.

Herbaceous perennials, or small, slender, shrubby plants, associated by an artificial character, *the feathery pappus.* Leaves alternate. Stems or branches naked at the summit, peduncular, one-headed, rarely corymbose. Disc yellow; ray purple-blue or white, in one species *yellow.* The name is in honour of *M. Maire,* a Prussian explorer, who accompanied. *Mundt* in his South African travels.

1. **Pteropappus** (Less.): Disc-flowers fertile. Herbaceous plants, with expanded leaves. (Sp. 1-3).

Stemless. Rad. leaves obovate-oblong, cobwebby, at length
glabrous, crenate ... ...   (1) **crenata.**
Caulescent. Lvs. obl.-obovate; *densely hairy beneath,* crenate   (2) **hirsuta.**
Nearly stemless. Lvs. linear-spathulate, hairy, subentire ...   (3) **perezioides.**

2. **Zyrphelis** (Cass.): Disc-flower sterile. Rays blue or white. Mostly suffrutices. Leaves linear or linear-lanceolate, quite entire, rigid, ciliate. (Sp. 4-9).

Nearly stemless. Rad. lvs. linear-spathulate, 4-5 inches long   (4) **Burchellii.**
Branching undershrubs, with linear or subulate, ciliate lvs. :
Heads terminating leafless branches (or *peduncles*) :
Achenes glabrous (or microscopically setulose) ...   (5) **taxifolia.**
Ach. densely silky; heads many-fl., invol. acuminate   (6) **lasiocarpa.**
Ach. pilose; heads small, few-fl.; inv. scal. subacute   (7) **microcephala.**
Heads terminating leafy branches, small and few-fl. ...   (8) **foliosa.**
Heads corymbose, on bracteate pedunc., ach. canescent   (9) **corymbosa.**
3. **Homochroma** (DC.): Rays yellow. Disc sterile. A scabrous suffrutex ... ...   (10) **Ecklonis.**

SECT. I. **PTEROPAPPUS**: *Disc-fl. fertile* (Sp. 1-3).

1. M. crenata (Nees. Ast. 348); stemless; leaves crowded round the crown of the root, obovate-oblong, tapering into a petiole, thickish, 1-nerved, bluntly toothed, cobwebby, becoming glabrous; peduncle scape-like, woolly, with 2- distant, linear bracts; invol. scales pilose, membrane-edged. *DC. l. c.* 217. *Arnica crenata, Th. ! Cap.* 668. *Gerbera crenata, Bot. Reg. t.* 855. *Aster crenatus, Less.*

HAB. Langekloof, *Thunb.;* Table Mnt. summit, common, *Mundt and Maire, Drege, W. H. H.,* &c.; Swellendam, *E. & Z.* (Herb. Th., D., Sd., Hk.)

Root-stock thickish. Radical leaves numerous, tufted, including the imperfect petiole, 2-3 inches long, ½-¾ inch wide, their teeth callous-tipped. Peduncles 2-10 inches high, at first densely covered with coarse, whitish, curly hairs, afterwards almost nude. Flowers large and handsome, the disk an inch across; rays nearly uncial, violet-purple.

2. M. hirsuta (DC. l. c. 217); stem ascending, subsimple, very hairy, leafy to near the middle; leaves oblong-obovate, tapering into a petiole, 1-nerved, bluntly toothed, pilose becoming glabrous above, *densely* *hairy beneath;* peduncle terminal, sometimes branched, very hairy, with 2–3 distant, linear bracts; invol. scales hairy.

HAB. Swellendam, in mts. near Puspas valley, *E. Z.! Zey.! 3071.* (Hb. D., Sd.) Differs from *M. crenata* in its longish, sometimes branching stem, and copious *rusty* pubescence, less thick leaves, and rather smaller flowers.

3. M. perezioides (Nees, Ast. 249); cæspitose, scarcely caulescent; leaves subradical, linear-spathulate, acute, with recurved margins, subentire or here and there toothed, pilose; pedunc. scapelike, with a few scattered, depauperated leaves on its lower half, densely glandular-pubescent; invol.-scales rather longer than the disc, pluriseriate, linear-acute, hairy and glandular. *DC. l. c. 217. Aster perezioides, Less! Syn* 167. *Leyssera pilosella, Thunb! Cap. and Herb. ex pte.*

HAB. Groenekloof and Zwartland, *Thunberg!* (Herb. Thunb.) Stems densely tufted, 1–2 inches long, naked below, densely leafy at the extremity. Leaves 2–2½ inches long, 2–3 lines wide, strongly midribbed beneath. Pedunc. 4–6 inches long; its leaves ¾ inch long, 1 line wide; its pubescence copious, of gland-tipped, short hairs. Rays purple. Pappus feathery. Achenes black, hairy on the face, margined and ciliate.—A distinctly marked species. which I have only seen in Herb. Thunb. A very imperfect specimen from *Mundt*, in Hb. Hooker, comes near this, but is more rigid and scabrous, with smaller flower-heads; it may however be a mere variety.

SECT. 2. **ZYRPHELIS**: *Disc-flowers sterile,* rays blue or white. (Sp. 4–9.)

4. M. Burchellii (DC. l. c. 218); "nearly stemless, radical leaves oblong-linear, glabrous, rigid, very entire, tapering at base and fringed with hairs; stems nearly leafless, sparingly branched, somewhat hairy; invol.-scales acuminate, ciliate." *DC.* (Disc-fl. sterile).

HAB. Eastern districts, *Burchell*, No. 8200. (Unknown to us.) "Leaves 4–5 inches long, below with a subprominent nerve, above half infolded. Stems rather shorter than the leaves. Invol. in two rows. Achenes subcompressed. very villous, narrowed at base." *DC.*

5. M. taxifolia (DC. l. c. 218); stem half-shrubby, erect, branched; leaves linear, acute, rigidly ciliate to the middle or beyond it; pedunc. softly hairy; invol.-scales acuminate; achenes *quite glabrous* or hispidulous. *Aster taxifolius, Linn.—Leyssera ciliata, Th! Cap. 690. Aster filiformis, Eckl.*

VAR. β. **pinifolia**; achenes sparingly and minutely hispidulous. *M. pinifolia,* Sch. B. *M. lasiocarpa, Drege! in Herb. Hook.*

HAB. On and about Table Mt., common: β, with the common variety. (Herb. Th., D., Sd., Hk., &c.) A slender suffrutex, 6–15 inches high, much-branched; branches erect, subsimple, leafy for ¾ of their length. Leaves ½–1 inch long, ½ line wide. Disc 3–4 lines across; rays 4 lines long, purple. The stem is sometimes decumbent, with ascending branches and falcato-secund leaves.

6. M. lasiocarpa. (DC. l. c.); stem half-shrubby, erect, branched; leaves linear, acute, rigidly ciliate; pedunc. glandularly bristly; invol. scales acuminate; achenes *densely silky.*

HAB. Near Caledon, *E. Z.!* Witsenberg, and Zwarteberg, *Zeyher! 2751.* (Herb., Sond., Cap.) Very like *M. taxifolia,* but easily known by its silky achenes. *Drege's* specimens distributed under this name belong (so far as I have seen) to *M. taxifolia β.* The pedunc. in this species is rough with short, spreading, glandular bristles ; in *M. taxifolia,* thinly clothed with soft, cobwebby hairs.

7. M. microcephala (DC. l. c.); stem half-shrubby, erect, branched ; leaves linear, ciliate or pubescent ; pedunc. glandularly bristly ; heads small ; invol.-scales few, subacute ; achenes pilose. *Elphegea micro-cephala Less. Syn.* 184.

HAB. Cape, *Mundt & Maire, Burchell; Zeederberg, Drege ! Zwellendam, Ecklon; Zwarteberg, Zey./ 2751.* (Herb. Hk., Sd.)
Very similar to the preceding, but smaller in all parts, with much smaller heads. Peduncles naked, 2–4 inches long.

8. M. foliosa (Harv.) ; stem half-shrubby, corymbosely much-branched, branches spreading, *leafy to the summit;* leaves linear, ob-tuse, ciliate and pilose ; heads small, *subsessile,* ending leafy branches ; invol.-scales few, subacute ; achenes pilose.

HAB. Seekuvallei, *E. Z.!* Hott. Hollandsberg, *Zey. ! 2752.* (Herb. Sd., Hk.)
Nearly related to *M. microcephala,* but with quite a different aspect, 3–5 inches high, very much branched and ramulous, all the flowering branches leafy to the summit. The dried specimens are glaucous.

9. M. corymbosa (Harv.); shrubby, robust, erect, branched ; leaves broadly linear-subulate or lanceolate-subulate, acute, flat, 3-nerved, ciliato-serrate ; pedunc. numerous, *corymbose* at the ends of the branches, scabrous, pluribracteate ; involucral scales 3-seriate, lanceolate, acute, glabrous, ciliate, 3-ribbed ; achenes of ray densely cano-pubescent.

HAB. Near the mouth of the Klyn R., *Zey.! 2742.* (Herb. Sd., Hk., D., Cap.)
Two feet high or more; the most robust and woody of the genus. Leaves 1–1½ inch long, 1–3 lines wide at base, the young ones erect, the old deflexed, all rather rigid, with immersed glands. The branches are leafy to the summit, and from the axils of the uppermost leaves spring many short (2–3 inch long) peduncles, each bearing several depauperated, scattered leaves or bracts. Invol. scales each with 3 glandular rib-striæ. Achenes of the ray with copious, plumose pappus. Almost identical in aspect and foliage with *Gymnostephium corymbosum,* but with a very different involucre, and copious feathery pappus !

SECT. 3. **HOMOCHROMA** : *Rays yellow.*

10. M. Eklonis (Sond.); half-shrubby, erect, branched, all parts *very rough with glandular bristles,* and sparingly pilose ; leaves linear-lanceolate, 1-nerved, acute, flat, on both sides scabrid, the younger also ciliate with long, soft hairs ; peduncles short; heads many-flowered ; invol. scales acuminate ; achenes densely silky. *Homochroma Ecklonis, DC. l. c.* 324.

HAB. Near Tulbagh, Worcester, *E. & Z.!* Brackfontein, *Zey.!* (Herb. Sond.)
More rigid and robust than any of the preceding section, with broader, flatter, and very scabrid leaves. The rays are *yellow,* on which character De Candolle founds his genus *Homochroma:* in all other respects this plant agrees with *Mairea.*

XIV. **GYMNOSTEPHIUM**, Less.

Heads many-fl., radiate ; *ray-fl.* female, ligulate, in one row; *disc-fl.* tubular-funnel-shaped, 5-toothed, sterile (with perfect, bifid styles, but abortive ovaries). *Invol.* scales imbricate. *Achenes* of the ray-fl. plano-compressed, rib-margined, beakless, glandularly-scabrous on the disc;

of the disc linear, empty. *Pappus* caducous, of few setæ, either shortly-feathery or barbato-serrate; ray-fl. either wholly without pappus, or with 1–2 bristles. *DC. l. c. p.* 300.

South African suffrutices or half-herbaceous plants, with slender, rod-like, leafy branches. Leaves alternate, linear or subulate, acute, entire, smooth or ciliate at the margins. Heads pedunculate, solitary, small; the ray blue, the disc yellow. Invol. scales mostly marked with long, glandular ribs. Nearly related to *Mairea*, from which it scarcely differs except by the scanty and less plumose pappus; I find the ray-flowers with or without pappus in species which are otherwise very closely allied. Name from γυμνος, *naked*, and στεφος, a *crown;* alluding to the want of pappus in the ray-flowers.

Invol. scales glabrous or nearly so, flat, marked with 3 *glandular* rib-striæ.
  Lvs. with minutely revolute margins, one-nerved :
    Very slender; pedunc. glabrous; invol. scales few (1) **gracile.**
    Robust; pedunc. villous; invol. scales many ... (2) **fruticosum.**
  Lvs. channelled and keeled; pedun. glandularly scabrous (3) **angustifolium.**
Invol. scales hispid or pubescent, keeled :
  Pedunc. elongate, *one-headed;* without bracts :
    Stems and leaves roughly hispid and scabrous ... (4) **hirsutum.**
    Stems glabrous; leaves glabrous, but ciliate ... (5) **ciliare.**
  Pedunc. short, *corymbose,* pluri-bracteate; leaves ciliato-serrate ... . (6) **corymbosum.**

1. G. gracile (Less ! Syn. 186); quite glabrous, very slender, diffuse; branches flexuous, laxly leafy ; leaves scattered, narrow-linear, acute, with revolute margins, one-nerved and veiny ; heads shortly peduncled, small; invol. triseriate, scales few, linear-oblong, flat, glabrous, ciliolate, with 3 glandular ribs. *DC. l. c. p.* 300.

HAB. Tradouw, *Mundt and Maire!* (Herb. Sond.)
A very slender, scarcely lignescent plant. Branches filiform. Leaves about 1 in. long, ½ line wide, thin. Pappus of disc of one or two caducous, plumose bristles ; of the ray none. Ripe achenes not seen.

2. G. fruticosum (DC.! l. c. 300); nearly glabrous, shrubby ; branches rather closely leafy, glabrous ; leaves alternate, linear, acute, *with revolute margins,* one-nerved, *glandularly-dotted,* quite glabrous ; pedunc. short, villous, with many barbato-ciliate bracts ; invol. pluriseriate, scales numerous, oblongo-lanceolate, acute, glabrous, ciliate, with 3 glandular ribs.

HAB. Near Swellendam, *Drege!* (Herb. Sond., Hook.)
Much more robust than *G. gracile,* with larger fl. heads and involucres. Leaves 1 inch long, 1-1½ lines wide. Our specimens are very imperfect, and I have seen neither pappus nor achenes.

3. G. angustifolium (Harv.) ; erect, nearly glabrous ; branches rod-like, glabrous, rather laxly leafy ; leaves subulato-filiform, channelled and keeled, acute, glabrous, very entire or the uppermost ciliate ; pedunc. glanduloso-muricate; invol. triseriate, scales several, oblongo-lanceolate, acute, minutely glandular and ciliate, with 3 glandular ribs.

HAB. Near the River Zonder Einde, *Zeyh.!* 2743. (Herb., Sd., Hk., Cap.)
1-1½ f. high, woody below, corymbosely branched upwards. Branches erecto-patent. Leaves an inch or rather more long, not ¼ line wide, the younger ones rigidly ciliolate, especially near their base, the older quite smooth-edged, all semi-amplexicaul. Pappus of disc of 2-3 caducous, barbellate bristles, of ray none. Ripe achenes glandularly-granulate on the disc. This has quite the aspect of *Aster simulans.*

4. G. hirsutum (Less.! Syn. 185); suffruticose; branches leafy, *roughly hispid* with short, spreading hairs ; leaves linear-subulate, acute, flat, one-nerved, *on both sides glandularly hispid*, ciliolate ; pedunc. elongate, glandularly scabrous; invol. 3-seriate, scales many, linear-lanceolate, acute, keeled, one-ribbed, the outer hispid. *DC. l. c.* 300.

HAB. Steenbock's Riv., at Zwarteberg, *Ecklon !* (Herb. Sond.)
1–2 f. high, lignescent below. Leaves 1–1½ inch long, 1–1½ line wide at base, gradually narrowing to the acute point, minutely gland-dotted, spreading. Achenes black, glandularly scabrid on the disc.

5. G. ciliare (Harv.) ; suffruticose, branches leafy, *glabrous ;* leaves linear-subulate, acute, flat, one-nerved, glabrous, *setoso-ciliate ;* pedunc. elongate, glandularly pubescent (or glabrous, *DC.*) ; invol. pluriseriate, scales numerous, one-nerved, pubescent. *Felicia ciliaris, DC.! l. c. p.* 221. Also *Heteractis falcata, DC. Prodr. VI.* 468.

HAB. Hills near Swellendam, *Drege !* Genadendahl, *Dr. Pappe !* (Herb. Sd., D., Hk.)
Very similar to the preceding, except in pubescence. and perhaps it is only a glabrous variety. Pappus plumoso-barbellate. Achenes black, 2-ribbed, the disc glandularly granulated.

6. G. corymbosum (Harv.) ; shrubby, robust ; branches glabrous, densely leafy ; leaves linear-subulate, acute, flat, 1–3-nerved, ciliato-serrate, *shortly decurrent ;* pedunc. short, *densely corymbose,* pluri-bracteate, hispido-scabrous and glandular ; invol. pluriseriate, scales many, linear-lanceolate, acuminate, roughly pubescent ; achenes of ray with 1–3 barbellate bristles. *Agathœa corymbosa, Turcz. Bull, Mosc. v. XXIV. II. p.* 60. *Walp. Ann.* 5, *p.* 172.

HAB. Near the River Zonder Einde, *Zeyher !* 2741. (Herb. Sd., Hk.)
This has the foliage of *G. ciliare,* but is much more robust and woody, with a very different inflorescence. Stems 1–2 f. high, branched above. Branches corymbose, leafy to the summit. Lvs. 1–1½ inches long, 2 lines wide at base, somewhat gland-dotted, the wider ones obviously 3-nerved. Pedunc. many from the axils of the uppermost leaves, simple or branched, each bearing several subulate, scattered bracts. Achenes of ray black, 2-ribbed, glandularly granulated on the disc. Very similar in aspect to *Mairea corymbosa,* which has however a very different involucre and pappus.

XV. ? **ANAGLYPHA**, DC.

Heads many-fl., radiate ; *ray-fl.* 1-seriate, strapshaped, female ; *disc-fl.* tubular, 5-toothed, hermaphrodite. *Invol.* 2-seriate, scales equal, rather longer than the disc, much acuminate. *Recept.* flat, honeycombed. *Tube* of corolla hairy. *Anth.* . . . . *Style* . . . . *Achenes* obovate, downy, without pappus. *DC. Prodr.* 3, *p.* 311.

A suffrutex with the aspect of *Mairea taxifolia,* or of *Gymnostephium ciliare;* the stem and ends of the branches velvetty with short, glandular down. Leaves alternate, linear, recurved at the acuminate point, on both sides striato-sulcate, very rough with rigid, spreading cilia along the margins and nerves beneath, toward the apex. Heads small, terminal, solitary. Fl. of disc and ray yellow.—Name, from ava, *with,* and γλυφη, *sculpture,* from the furrowed leaves.

1. A. aspera (DC. l. c.):—
HAB. Between Coega and Zwartkops Rivers, under 1000f., *Drege.*

Of this I have only seen (in Herb. Sond.) what purports to be a leafy twig, without flowers; in this state not distinguishable from *Osteospermum scabrum*, Th. De Candolle had seen only one imperfect specimen.

XVI. CHARIEIS, Cass.

Heads many-fl., radiate; *ray-fl.* female, in one row; *disc-fl.* hermaphrodite, 5-toothed, tubular-bellshaped. *Recept.* honey-combed. *Invol.*-scales bi-serial, the outer ones few, narrow, infolded and keeled, the inner membrane-edged, with a green keel. *Style* of the disc-fl. with flattened branches and deltoid stigmas. *Achenes* obovate, compressed, girt with a thickened rim; those of the *ray-fl.* frequently empty and always without pappus. *Pappus* of the *disc-fl.* of feathered-bristles in a single row. *DC. Prodr. 5, p. 300.*

A small, pubescent annual. Lower leaves opposite, upper alternate, membranous, attenuate at base. Peduncles terminal, leafless, 1-headed. Ray-fl. blue; disc either blue or yellow. The stigma has something the character of that of a *Senecionea*, but the habit and general character are those of the herbaceous *Agatheœ* or *Amelli*. Name, χαριεις, *elegant*.

1. C. heterophylla (Cass.); *DC. l. c.* 300. *Kaulfussia amelloides*, Nees, *Hort. Ber. t.* 11. *Leyssera pilosella, Th.!* Cap. 691 (*ex pte.*)

HAB. Wet spots on the Cape Flats, and other places in the Western Districts, common. (Herb. Th. D., Sd., Hk.)

Stem erect or diffuse, branching, 4–6 inches high, hispid and scabrous. Leaves 1–2½ inches long, 3–5 lines wide, oblongo-lanceolate, tapering at base, hispid on both sides. Peduncles elongate, glandular and hispid. Disc either blue or yellow; rays blue.—Cultivated in English gardens.

XVII. ASTER, Linn.

Heads many-fl. heterogamous; *ray-fl.* ligulate, female, in 1 row; *disc-fl.* 5-toothed, perfect, rarely sterile. *Recept.* naked, areolate or honeycombed. *Invol.* scales imbricated, in few or several rows, appressed or loose. *Achenes* compressed. *Pappus* of many serrulated, caducous or subpersistent, uniform bristles, uniseriate (or pluriseriate?) *Felicia, Munychia, Agathœa, Bellidiastrum, Aster., Tripolium, DC. Prodr. 5, pp.* 218, 222, 223, 226, 253.

A vast cosmopolitan genus of annual or perennial herbs, suffrutices, or small shrubs. Leaves alternate or opposite, entire or toothed, sessile or petiolate. Fl. heads (in the Cape species) solitary, terminal, mostly peduncled. Disc yellow, very rarely purple; rays blue, white, or pinkish, never yellow. Name from *aster*, a star; alluding to the radiate fl.-heads.

1. **Felicia.**—Inv. scales imbricated in 3 or more rows, the outer scales short, the next longer, the innermost longest. (Sp. 1–24.)

A. Achenes more or less pubescent or hispidulous. (Sp. 1–18).

  (1) Herbaceous: annual or perennial.

Annual or biennial: lvs. linear; branches 1-headed	(1) **tenellus.**
Annual; lvs. linear; branches corymbose at top	(2) **microspermus.**
Perennial; with a thick, woody root; lvs. spathulate	(3) **macrorhizus.**

  (2) Shrubby or half-shrubby, branching; erect or diffuse:

    (a) Pappus white; achenes thinly hispid or puberulous (*rarely silky*):

      Heads discoid; leaves linear-spathulate ... ... (15) **discoideus.**

      Heads radiate;—

        Leaves glabrous or ciliate:

Lvs. subulate or filiform, without obvious nerve :
   Lvs. subterete, glabrous or ciliate ... (4) **muricatus.**
   Lvs. subulate, channelled, glabrous ;
      Achenes puberulous ; pappus
        deciduous, flexuous ... ... (6) **simulans.**
      Achenes densely silky ; pappus
        persistent, straight ... ... (5) **Bowiei.**
Lvs. nerved, smooth-edged ... ... (7) **angustifolius.**
Lvs. nerved, ciliate or serrulate :
   Pedunc. terminal, 1-headed :
      Lvs. lance-linear, acute... ... (8) **serrulatus.**
      Lvs. oblong-obovate, obtuse... (9) **Zeyheri.**
   Peduncles lateral, in a panicle or
      thyrsus ... ... (10) **erigeroides.**
Lvs. pubescent, canescent, or scabrid glandular :
   Lvs. minutely-pulverulent or velvetty-pulvt.:
      Lvs. obl.-spathulate, obtuse, entire (11) **Dregei.**
      Lvs. 3–5-*toothed* at point, *velvetty* (11) **Dregei,** β.
      Lvs. 3–5-*lobed* at point, *scabrid*-pul-
        verulent ... ... (12) **scabridus.**
   Lvs. roughly-pubescent or canescent :
      Lvs. *spreading*, appressed-pubescent
        or canous ... ... (13) **hyssopifolius.**
      Leaves *erect*, roughly-hairy ; twigs
        rod-like ... ... (14) **confusus.**
 (b) Pappus reddish, rigid ; achenes densely silky or hairy :
   Glaucous ; lvs. filiform, rigidly ciliate, glabrous... (16) **rufibarbis.**
   Robust, divaricate, green ; lvs. linear, glabrous or
      pilose ... ... (18) **Burkei.**
   Glandular ; lvs linear, entire, the up. glandular (17) **lasiocarpus.**
B. Achenes perfectly glabrous. (Sp. 19–24).
   Leaves opposite, linear, fleshy, obtuse, entire ... ... (19) **ficoideus.**
   Leaves scattered, linear or lanceolate :
      Lvs. margined with stalked glands ... ... (20) **Caffrorum.**
      Lvs. glabrous, revolute-reflexed ; invol. scales
        squarrose ... ... (21) **retortus.**
      Lvs. ciliate, imbricate ; inv. scales erect ... ... (22) **imbricatus..**
   Leaves broad-based, oblong, ovate, or ovato-lanceolate :
      Branches rod-like ; lvs. imbricate ; heads sub-
        corymbose ... ... (23) **echinatus.**
      Branches curved ; lvs. strongly recurved ; heads
        terminal ... ... (24) **reflexus.**

2. Agathæa.—Invol. campanulate, the scales either in 1 row, or, if in 2 or 3 rows, nearly of equal size, the outer scales being as long, or nearly as long, as the inner, and mostly keeled. (Sp. 25–46).

A. Annuals :
   Disc-fl. abortive. Pappus of few, barbellate bristles.
      Invol. scales linear, hairy at back ... ... (25) **Bergerianus.**
      Invol scales broadly-oblong, quite glabrous ... (26) **annectens.**
   Disc. fl. generally perfect. Pappus of many bristles.
      Leaves mostly alternate (save the lower) :
        Inv. scales 2–3-seriate, acute or acuminate :
          Root slender. A *slender*, branching,
            hairy annual ... ... (27) **adfinis.**
          Root thick. A *robust*, very hairy, and
            glandular annual ... ... (28) **Namaquanus.**
        Inv. scales uniseriate, obtuse ; lvs. roundish-
          oblong... ... ... (44) **Capensis,** β.
      Leaves mostly opposite (some alternate) :
        Lvs. *close-set*, linear or spathulate, ¾–1 in. long (29 **Pappei.**

Lvs. *distant*, ovate or obovate, 2–3 lines long (30) **demissus.**
Lvs. *distant*, lance-oblong, 3-ner., 1½–2 in. long (31) **tener.**
B. Perennials: either shrubby, half-shrubby, or herbaceous :—
  Leaves distinctly petiolate, *opposite*, ovate, crenate :
    Ray-fl. 3-toothed. Disc fertile. Stem woody ... (32) **hirsutus.**
    Ray-fl. subentire. Disc sterile. Stem herbaceous (33) **Cymbalariæ.**
  Leaves distinctly petiolate, *alternate*, ovate, toothed (34) **petiolatus.**
  Leaves sessile or subsessile :
    Stems herbaceous, simple (or very short) ; lvs. large:
      Stems ascending, leafy; peduncles pubescent,
        elongate ... ... (35) **Candollei.**
      Stemless, with rosulate radical lvs.; pedunc.
        with 6–8 scattered, leafy bracts ... ... (36) **Natalensis.**
    Stems much branched, ligneous or shrubby :—
      Leaves mostly opposite :
        Achenes quite smooth and glossy ... (45) **Ethiopicus.**
        Achenes pubescent or puberulous :
          Lvs. roundish-oblong, obtuse, with
            reflexed edges ... ... (44) **Capensis.**
          Lvs. subacute, *erect*, very hairy, ob-
            long or lanceolate ... ... (42) **strigosus.**
          Lvs. blunt, linear-oblong, thick,
            beneath bristly ... ... (39) **barbatus.**
          Lvs. spreading, oblong or obovate,
            hispid with long, spreading hairs (40) **elongatus.**
      Leaves mostly alternate :
        Ripe achenes quite glabrous and glossy :
          Lvs. oblong or obovate, obtuse,
            entire ... ... (45) **Ethiopicus.**
          Lvs. cuneate, sharply 3–5-toothed,
            acute ... ... (46) **leiocarpus.**
        Ripe achenes pubescent or puberulous :
          Leaves glabrous, serrulate :—
            Lvs. linear-lanceolate, tapering
              at base ... ... (38) **linifolius.**
            Lvs. oblong-linear, broad-based (37) **serratus.**
          Leaves hairy, scabrid or glandular.
            Lvs. scattered, oblong, obovate,
              lanceol. or spathulate, hairy
              and glandular ... ... (39) **elongatus.**
            Lvs. oblong-linear, ciliate with
              swollen, jointed hairs ... (41) **hirtus.**
            Lvs. crowded, narrow with re-
              volute margins, above rough
              with prominent glands ... (43) **Ecklonis.**

SECT. 1. **FELICIA.** (Sp. 1–24).

1. A. tenellus (Linn. Mant. 471); annual or biennial, herbaceous or
at base lignescent, erect, diffuse or prostrate, simple or much branched,
sparsely pilose or nearly glabrous, or glandular-pubescent ; leaves *narrow-linear*, callous tipped, nerveless, *mostly* rigidly ciliate ; fl. branches
peduncular, one-headed; invol.-scales acute or subobtuse, margined,
hispidulous or glabrous ; achenes minutely pubescent. *Bot. Mag. t. 33.*
A. tenellus and *A. dentatus, Thunb.! Felicia tenella, DC. l. c.* 219. *F.
fragilis, Cass. Kaulfussia ciliata, Spr. Cineraria tenella, Link.*

VAR. *a.* **pusillus**; very small and slender, annual, erect, hispid, with long peduncles ; lvs. ciliate.

VAR. *β.* **cotuloides**; very slender, annual, erect, much-branched, *sparsely* hispid or glandular, with long peduncles, and small fl. heads ; lvs. very narrow, *now and then* ciliate. *F. cotuloides, DC.! l. c. Zey.!* 782.

VAR. γ. **glaber** ; annual, subsimple, erect, nearly glabrous ; lvs. short, sparingly ciliate; inv. scales broad,flat,obtuse,glabrous(or pilose). *F. tenella, litt. a, in Hb.Drege.*

VAR. δ. **robustus** ; annual, erect, *strong*, much branched, copiously hispid and glandular, or nearly glabrous. *Zey.* 783.

VAR. ε. **longifolius** (DC.) ; biennial or perennial (?), branches rooting at base, ascending or prostrate ; lvs. long or short, ciliate, otherwise glabrous. *DC.! C. Wright,* 386, 395. *Sieb. Fl.* 253.

HAB. Damp spots throughout the western districts, common. (Hb. D., Hk., Sd.,Th.) Very variable in size and pubescence. Stems 2–14 inches high. Roots either threadlike and simple, or of many robust fibres, the prostrate portions of the branches rooting at base. Lvs. ¼ inch to 2 inches long. Pubescence hairy or glandular and viscidulous. None of the above varieties are constant to the given characters.

2. A. microspermus (DC.) ; annual, herbaceous, many stemmed ; stems diffuse, simple below, *branched and corymbose, and glandularly pubescent near the summit ;* leaves linear, semiterete, obtuse, callous tipped, minutely glandular, (not ciliate); invol. scales linear, subacute, glandular ; achenes " glabrescent " DC. *Felicia microsperma, DC. l. c.*

HAB. Near the Gariep, *Drege !* (Herb. Sond., Hook.)
Perhaps a mere var. of *A. tenellus,* from which it differs chiefly in ramification. "Stems 8–12 inches long. Lvs. 9–10 lines long, ½ line wide. Heads of *A. tenellus,* but corymbulose. Achenes sprinkled with a few appressed hairs." *DC.*

3. A. macrorhizus (Thunb. ! Cap. 687) ; root *very thick and woody,* elongate ; stems many, *short, densely tufted,* simple ; leaves alternate, very closely set, from broad-clasping bases linear-spathulate, much attenuated towards the base, entire, pilose ; heads peduncled ; invol. scales multiseriate, imbricate, flat, glanduloso-scaberulous, lanceolate-linear, acuminate ; achenes reddish-brown, pilose. *Less. Syn.* 175. *Felicia macrorhiza, DC. l. c.* 219.

HAB. Cape, *Thunberg !* (Herb. Thunb.)
Root ½ inch in diameter, 8–12 inches or more long, probably growing in fissures of mountain rocks. Stems 2–3 inches long, densely imbricated with leaves. Bases of the leaves ovate, glabrous, remaining as persistent scales after the fall of the lamina. Leaves about an inch long, 1–1½ lines wide at top, thinly sprinkled with long, white hairs. Pappus sulphur-coloured. Achenes rather copiously furnished with long, straight hairs. A very distinct species.

4. A. muricatus (Less.! Syn. 176) ; shrubby, erect, much branched, thinly hispid or glabrescent ; leaves linear-terete, furrowed above, spreading, obtuse, callous tipped, entire, either glabrous or sprinkled with hairs or ciliate with rigid bristles ; axils often bearing leaf-tufts, heads peduncled ; invol. scales acute, linear, minutely glandular or glabrous ; achenes puberulous. *A. filifolius, Zey.!*

VAR. α. **fascicularis** (E. M.) ; leaves mostly tufted, glabrous or sparsely pilose. *Felicia fascicularis, DC. l. c.* 220. *Zey.!* 790.

VAR. β. **ciliatus** ; all the young leaves rigidly ciliate. *Zey.!* 793.

VAR. γ. **chrysocomoides** (Sond.) ; leaves mostly scattered, viscidulous-subglandular. *Zey.!* 797. *Drege !* 9096.

HAB. Carroo places in Uitenhage, Albany, and Br. Caffraria, *Krebs, E. Z.! Drege !* &c. β. Modder River Spruit, *Burke and Zeyher !* γ. Cape, *E.Z.!* Zuureberg, *Zey.!* (Herb. Th., D , Hk., Sd.)

A small, slender shrublet, 3–8 inches high. Leaves 3–8 lines long, very slender, conspicuously channelled. Pedunc. 2–3 inches long, glabrous or puberulous. Inv. scales flattish. Variable chiefly in its pubescence.

5. A. Bowiei (Harv.); suffruticose, glabrous, flexuous; branches closely leafy; leaves filiform-subulate, channelled, very long, acute (the younger bristle-pointed), quite entire; peduncles elongate, quite glabrous; invol. scales 3-seriate, imbricate, flat, lanceolate, acuminate, 1-nerved; achenes *densely silky;* pappus of many, *sub-persistent, straight,* scaberulous bristles.

HAB. Cape, *Bowie.* (Herb. Hk.)

Near *A. simulans,* but with much longer lvs., densely silky achenes, and persistent, copious, and straight pappus-bristles. Lvs. 1½–2 inches long, ⅓ line wide, involute, indistinctly nerved at base, tapering to a setaceous point. Peduncles 4 inches long. Head many-flowered. Of this I have only seen a solitary specimen in Hb. Hk.

6. A. simulans (Harv.); shrubby, erect, *glabrous* (save the peduncles); branches rod-like, closely leafy; lvs. *linear-subulate, channelled,* callous-tipped, acute, quite entire; peduncles *thinly pubescent;* involuc. scales pluriseriate, imbricate, flat, lanceolate-linear, acute, 1-nerved, ciliolate; achenes puberulous; pappus of many slender, subflexuous, deciduous, rough bristles.

HAB. Berg River, *Zey./* Sept. (Herb. Sond.)

One to two feet high, slender, but woody. Leaves an inch or rather more long, erect, closely overlapping, not half a line in diameter, broad-based. Peduncles 2–3 inches long. Heads rather small. Almost identical in aspect with *Gymnostephium angustifolium,* but differing in pappus and involucre.

7. A. angustifolius (Jacq. Schoenbr. t. 370); "shrubby, erect; branches slender, flexuous, glabrous, the younger leafy; leaves membranous, scattered, sessile, *obovate-lanceolate, linear,* acuminate, *smooth-edged,* tapering at base, 1-nerved, at length quite glabrous; heads pedunculate; invol. pluriseriate, scales flat, glabrous, fimbriate; achenes puberulous." *Less. Syn. p.* 178. *F. angustifolia, Nees. DC. l. c.* 220.

HAB. Cultivated from Cape seeds.

One to two feet high. Leaves 6–12 lines long, ½–1 line wide. Heads with blue rays. Said by *Lessing* to be scarcely distinguishable from *A. hyssopifolius,* save by its glabrous leaves, slender branches, and leaves without leaf-tufts. It is probably a mere garden variety.

8. A. serrulatus (Harv.); suffruticose or shrubby, much-branched; branches glabrous or pubescent; leaves *lanceolate-linear, flat* or concave, acute, 1-*nerved* or nearly nerveless, *calloso-serrulate or rough-edged,* the younger ciliate (sometimes setose), the adult glabrous; peduncles elongate, pubescent near the summit; invol. scales pluriseriate, flat, lanceolate, glabrous, ciliolate; achenes puberulous.

VAR. *a.* **polyphyllus**; a rigid, much-branched, fastigiate shrub; leaves crowded, 3–5 lines long, concave, with *indistinct* nerve. *Zey./* 2749, 2729.

VAR. *β.* **densus**; similar to *a,* but more glabrous, with *distinctly* nerved leaves, 3–4 lines long. *Zey./* 791.

VAR. *γ.* **glaber**; slender, suffruticose, nearly glabrous; leaves less closely set, 6–8 lines long. *Felicia angustifolia, β. glabra, DC. l. c.* 220.

VAR. *δ.* **setosus**; suffruticose, pilose; leaves *setoso-ciliate,* the younger ones also pilose. *Zey./* 2729, *b.*

HAB. Vars. β and δ, Rietkuil and Grootvadersbosch, *Zey.!* β. Wolvekop, *Burke and Zey.!* γ. Zeederberg and Stormberg, *Drege!* δ. Vanstaadensberg, *Zey.!* (Herb. D., Hk., Sd.)

When fully grown this is a woody and scrubby, much-branched bush, 10–15 inches (or more) high. Leaves 3–8 lines long, more or less attenuated to the base, the adult (save in var. δ.) nearly glabrous; but even then the stumps of the cilia remain as serratures: by which character this seems to differ from *A. angustifolius.*

9. A. Zeyheri (Less. Syn. 177); shrubby, flexuous; branches pilose, becoming glabrous; leaves *oblong-cuneate or obovate,* flat, one-nerved, *obtuse, glabrous, closely and rigidly ciliate;* heads on long, glabrous or pilose peduncles; invol. scales glabrous, flat, linear-lanceolate, subacute; achenes thinly pubescent. *F. Zeyheri, Nees. DC. l. c. p.* 220. *Zey.!* 2756.

HAB. Dry places near the Commandokraal, Zondag river, and on Vanstaadens-bergen, *Zey.!* (Herb. Sd., Hk., D.)

A small, scrubby plant, 6–8 inches high, rigid and irregularly branched, the adult parts becoming glabrous. Leaves crowded, ½–1 inch long, 2–4 lines wide, greatly narrowed to the base, and fringed with short, rigid bristles. Peduncles 3–4 inches long, slender. Invol. 3-seriate.

10. A. erigeroides (Harv.); shrubby; branches virgate, appressedly pubescent, leafy; leaves obovate, elliptic-oblong or broadly linear, obtuse, *mostly* 3-nerved, narrowed to the base, sparsely hispid becoming glabrous, serrulato-ciliolate; axils bearing leaf-tufts; lateral twigs naked toward the summit, ending in a fl.-head, forming a thyrsus or panicle of fl.-heads; invol.-scales flattish, acute, ciliate, glabrescent; achenes pubescent, narrow-cuneate.   *Felicia erigeroides, DC. l. c.* 219.

VAR. β. **Schultesii** (Harv.); more branched, less virgate; leaves narrower, tapering much to the base. *F. Natalensis, Sch. Bip. Walp. Rep.* 2. 956.

VAR. γ. **trinervius**; less virgate, more ramulous; the twigs longer, bearing 2–3 heads; leaves larger and more obovate. *Zey.* 2740. *F. trinervia, Turcz!* *Bull. Mosc.* 1851. *vol.* 24. 2. *p.* 60.

HAB. Near Natal, both α and β., *Drege! Krauss! Sanderson! Ger. & McK.* 274. γ. Uitenhage, *Zeyher!* (Herb. D., Sd., Hk.)

One to two feet high, robust, becoming quite woody. Branches 12–18 inches long, rodlike, closely leafy, bearing toward the summit many short, half-leafy flowering twigs. Leaves ¾–1¼ inch long, 2–4 lines wide: the lateral nerves *not always* apparent, but often visible even in the narrowest leaves of var. β. The ray-fl. are pink or light purple.

11. A. Dregei (Harv.); shrubby, erect, branched; *the branches, leaves and peduncles closely and minutely velvetty-pubescent;* the branches and peduncles angularly striate; leaves oblong-spathulate, obtuse, narrowed much to the base, entire or nearly so, one-nerved; invol.-scales flattish, linear, acute, fringed and bearded above; achenes pubescent and ciliate. *Zey.!* 795. *Fel. Dregei, DC. l. c.* 221.

VAR. β. **dentata** (DC.); lvs. obovate, obtuse, sharply 3–5 toothed at the summit.

HAB. Little Namaqualand, *Drege!* Heerelogement, *Zey.!* (Herb. D., Hk., Sd.)

One to two feet high. Leaves 6–8 lines long, 2 lines wide, often with axillary leaf-tufts. Peduncles 3–4 inches long. Disc fl. yellow "changing to purple"; ray blue. The pubescence is remarkably close and knappy, formed of minute, prostrate, crossed hairs. In general habit this species resembles *A. erigeroides*, but is much more closely allied to *A. scabridus.*

12. A. scabridus (E. Mey.!); suffruticose, the old stems glabrate,

twigs puberulent ; leaves save the lowest alternate, scabrido-pulveru-
lent, cuneate, narrowed at base, 3–5 *toothed or lobed at the apex,* the
lobes obtuse ; pedunc. terminal, glandularly puberulous ; invol.-scales
linear-lanceolate, flattish, imbricate, in 3–4 rows ; ripe achenes sparsely
pubescent. *Aster scabridus, E. Mey.! Agathæa scabrida, DC. l. c. p.* 225.

VAR. β. **brevifolius**; old twigs spinescent ; lvs. shorter. *Ag. brevifolia, DC. l. c.*

HAB. Between Hexriviersberge and the Bokkeveld, *Drege!* β. near Zilverfontein,
Drege! (Herb. Hk., D., Sd.)

A rather slender suffrutex, but rigid, 1–1½ ft. high, with spreading branches.
Leaves in var. *a*, 6–8 lines, in β, 2–3 lines long. Pappus rigid, as long as the fl.,
discoloured. Disc fl. reddish upwards. This comes very near *A. Dregei*, β. *dentatus,* from which, except by the pubescence it scarcely differs.

13. A. hyssopifolius (Berg.! Cap. 287) ; shrubby, much branched ;
branches pubescent or canous ; leaves linear or spathulate, flat, obtuse,
thick, nerveless, spreading, strigoso-pubescent or canescent ; heads
peduncled, the peduncle pubescent or canous; inv.-scales acute or subobtuse, pulverulent, or canous; achenes minutely pulverulent. *Less.!
Syn.* 175.

VAR. *a.* **hirtus** ; leaves thinly strigose, green. *Aster hirtus, Thunb.! Cap.* 688.
F. angustifolia, a., DC.!

VAR. β. **rigidulus**; leaves copiously strigose or rigidly pubescent, green. *Felicia
rigidula, DC.! l. c.* 220.

VAR. γ. **linifolius**; lvs. very long (½–1 in.) tapering at base, rigidly pubesc. *Zey.!* 2747.

VAR. δ. **canescens**; foliage silvery canescent, with rigid appressed hairs. *F.
rigidula, var. sub-canescens, DC.!*

VAR. ε. **canus**; all parts clothed with *soft*, dense, white hairs. *Felicia cana, DC. l. c.*

VAR. ζ. **Wallichii** ; silvery canescent ; inv.-scales *oblong-linear, obtuse! A. Wallichii, Harv. MS.*

HAB. Var. *a.* Zwartland and Groenekloof, *Th.!* Olifant's R., *Mundt and Maire.*
N. West regions, *Drege!* Springbokkeel, *Zey.!* 798, 3089. β. Uitenhage and Albany,
common. γ. Kommando Kraal, *Zey.!* δ. Grassrugg, Uit., *E. Z.! Drege!* Grahamstown, *Burke!* ε. Swellendam, *E. Z.* ζ. Cape, *Dr. Wallich!* (Hb. Th., D., Hk., Sd.)

A much branched, at length woody and rigid, small *scrub*, 8–15 inches high :
branches erect or widely spreading. Pubescence very variable, but never quite
absent ; sometimes very white and copious. Leaves commonly ½ inch long, varying
from 3–9 or in γ, 12 lines long, 1–1½ lines wide. Rays blue. Var. ζ. differs from
the rest in its broader and much more obtuse invol.-scales.

14? A. confusus (Harv.) ; "stem shrubby, branches rod-like, hairy-
canescent ; leaves linear, flat, sub-obtuse, on both sides strigose-hairy ;
heads solitary, on short peduncles ; invol. scales in 2–3 rows, the outer
ones hairy." *DC. F. hirsuta, DC. l. c.* 220.

HAB. Beyond the Gariep, *Burchell;* Ceded Territory and Graaf Reynet, *E. & Z.;*
Klaap river, *Drege!*

"Leaves 5–6 lines long, erect, distributed along the whole branch, except a short,
naked, peduncular extremity."—*DC.* I have not seen *Burchell's* or *E. & Z.'s* specimens. *Drege's* specimens (in Herb. Hk., D., Sd.) under this name may be thus
described :—"Shrubby, much-branched, flexuous or divaricate, *the twigs glandularly
pubescent and viscidulous;* leaves narrow-linear, channelled, obtuse, erect or spreading,
on both sides roughly setose or (in Herb. D.) setoso-ciliate; heads solitary, on short
or long, glandular peduncles; invol. scales in 2–3 rows, the outer minutely glandular."
These are very like some forms of *A. muricatus,* but differ in pubescence ; I fear a
very variable character. A specimen in Hb. Sd. from *Drege,* named "*F. imbricata*,"
agrees better with Decandolle's *F. hirsuta,* but has long peduncles !

15. A. discoideus (Sond.); shrubby, rigid, divaricately much-branched; twigs closely leafy, viscidulous ; leaves *linear-spathulate*, flat, obtuse, nerveless, glabrous ; heads shortly pedunculate, *discoid;* invol. scales loosely imbricate, linear-lanceolate, glabrous, acuminate, at length squarrose and reflexed ; achenes puberulous.

HAB. Cape, *Ecklon and Zeyher!* (Herb. Sond.)
A very rigid, depressed shrub, the old twigs hardening and almost spiny. Leaves 2–3 lines long, ⅓ line wide, sensibly narrowed to the base. Peduncles very short, or almost none. Heads globose. This is readily known by its discoid flowers.

16. A. rufibarbis (Harv.); shrubby, much-branched, slender, *glaucous*, sparsely setose ; leaves filiform, slender, scattered, *spinoso-ciliate*, pale ; heads on slender peduncles ; invol. scales linear, acute, triseriate, glabrous ; achenes *densely silky ;* pappus copious, *rusty-reddish ;* rays scarcely longer than the pappus. *Felicia,* 103, 4, *in Hb. Eckl.*

HAB. Cape, *Ecklon!* (Herb. Sond.)
A slender, pale shrub, 6–8 inches high. Branches striate, straw-colour, erect. Leaves 5–7 lines long, ¼ line in diameter, margined with stiff bristles, otherwise glabrous. Peduncles 3 inches long. Heads small. Pappus a bright brick-dust colour, serrulate. Achenes clothed with long, silky, white hairs. A very distinct species, but apparently allied to the following.

17. A. lasiocarpus (Harv.); "stem shrubby, dwarf, glabrous, *glanduloso-puberulent* in the upper parts ; leaves linear, *very entire*, acute, tapering at base and semi-amplexicaul, the lower ones glabrous, the upper with the stem and involucres glandular; achenes *densely silky*, with appressed hairs."—*DC. Felicia lasiocarpa, DC. l. c.* 221.

HAB. Between Hexrivier and the Bokkefeld, 3000–4000 feet, *Drege.*
"Stem 4 inches high. Leaves uncial, a line wide. Pappus rufous, densely serrulate. Heads 9–10 lines in diameter." This I have not seen. A specimen from *Drege* in Herb. Sond. marked "*F. lasiocarpa,* DC." is merely a fragment of *Mairea taxifolia!*

18. A. Burkei (Harv.); rigidly shrubby, divaricate; twigs glabrous, the older spinescent ; leaves narrow-linear, flattish or channelled, thick, obtuse, scattered, glabrous or sparsely setose ; heads on short, pubescent peduncles ; invol. scales oblong-linear, acute, keeled, 3-seriate, glabrous, subciliate ; achenes clothed with long hairs; pappus rufous, barbellate; rays not much longer than the pappus. *Zey.!* 784.

HAB. Gamke river, *Burke & Zey.!* (Herb. Hk., Sd., D.)
A rigid, scrubby, thick-stemmed bush, 6–8 inches high, divaricately much-branched. Leaves 4–5 lines long, ¼ line wide, slightly tapering to the base. Peduncles 2 inches long. Possibly only a scrubby, starved state of *A. rufibarbis*, but the habit is somewhat different.

19. A. ficoideus (Harv.) ; quite glabrous, stem *decumbent, rooting ;* leaves *opposite, fleshy,* linear-semiterete, obtuse, quite entire, slightly narrowed at base ; peduncles terminal, not much longer than the leaves; invol.-scales oblong, obtuse ; pappus fugacious, of 5–6 setæ. *Burch. Cat. n.* 5311. *Fel. ficoidea, DC. l. c.* 221.

HAB. Eastern Districts, *Burchell.* Wet spots near the mouth of the Zwartkops R., *E. Z.!* (Herb. D., Sd., Hk.)
A half prostrate, fleshy plant, like a small *Mesembryanthemum.* Stems striate, 1½ lines in diam. Leaves 1–1½ in. long, 1½ line wide. Fl. 4–5 lines across ; rays white!

20. A. Caffrorum (Less. Syn. 179); shrubby, much branched, nearly glabrous, twigs densely leafy ; leaves lanceolate-linear, flat, one-nerved, much attenuated at base, mucronate-acuminate, glabrous, spreading, *margined with prominent glands ;* heads on short, scabrous peduncles, corymbulose at the ends of leafy twigs ; invol.-scales pluriseriate, flat, *with a glandular-reddish midrib* and minutely lacerate, scarious margins; achenes quite glabrous. *Felicia Caffrorum, Nees. Ast.* 214. *DC. l. c.* 221.

HAB. Caffraria, *Krebs,* No. 178. (Herb. Sond.)
Of this I have seen only the flowering twigs. Leaves nearly uncial, 1½–2 lines wide, thinnish, pierced with pellucid dots, either quite entire or with 1–2 lateral teeth. Heads 16–20 fl., numerous, crowded round the ends of the twigs.

21. A. retortus (DC.); "quite glabrous ; stem shrubby, the branches leafy to the summit ; leaves linear, acute, *revolute-reflexed* with the nerve at base produced into a decurrent prominence ; outer invol. scales *revolute-reflexed,* inner longer, erect." *DC. Felicia retorta, DC. l. c.* 221.

HAB. Near Gnadendahl, *Burchell.* 7793. (Unknown to us.)

22. A. imbricatus (DC.) ; " shrubby; branches densely covered with imbricating leaves, only at the extremity scarcely nude, one-head-ed ; leaves linear, subacute, not tapering at base, thickish, with a prominent nerve, nearly trigonous, at the margins and sometimes on the midrib ciliate with bristles ; outer inv.-scales linear, hairy at back, the inner ones glabrous and scarious," *DC. F. imbricata, DC. l. c.* 221.

HAB. District of Worcester, *E. Z.* (Unknown to us).
" Lvs. 4 lines long, a line wide. Hds. nearly as in *A. echinatus.* Ach. glabrous." *DC.*

23. A. echinatus (Less.! Syn. 177); stem shrubby, branched, either hairy, thinly pilose, or quite glabrous ; branches virgate, closely set with leaves throughout ; leaves broad-based, sessile, ovate-oblong or lanceolate-oblong, pungent mucronate, spreading or deflexed, glabrous, often ciliato-serrate ; pedunc. one or several, subterminal, corymbose, short, hairy, with scattered bractlike-leaves ; inv.-scales dentato-ciliate, acuminate, either glabrous at back or rough with toothlike-bristles. *Pteronia echinata, Thunb.! Cap.* 629.

VAR. *a.* echinata ; stem hairy ; leaves ciliato-serrate ; inv.-scales very rough dorsally. *Felicia echinata, DC.! l. c.* 222.

VAR. *β.* Paralia ; stem glabrous or thinly pilose ; lvs. entire-edged or sparingly ciliate ; inv.-scales dorsally either glabrous or sparingly scabrous. *Fel. Paralia, DC. l. c.*

HAB. Plettenburg Bay, *Bowie! Mundt!* Outenaqualand, *Burchell, Drege. !* Knysna, *Pappe!* Port Elizabeth, *Mrs. Holland,* 135. β. Uitenhage, *Burchell,* Albany, *E. Z.! Mrs. F. W. Barber!* Pt. Elizabeth, *Dr. Pappe!* (Hb. Th., D., Hk., Sd.)
A stout, small shrub, 1–2 f. high, with crowded, simple, erect branches, densely leafy. Leaves 5–10 lines long, 2–3 lines wide, at length deflexed. Heads large ; rays either blue or white. Variable in pubescence. I find intermediate states connecting the two varieties.

24. A. reflexus (Linn. Sp. 1225); stem shrubby, fasciculately branched, glabrous or setose ; branches simple, flexuous, closely leafy throughout ; leaves broadbased, sessile, oblong or sublanceolate, *strongly recurved,* acute, glabrous, rigid, *serrato-ciliate or entire;* heads subsessile or shortly pedunculate, pedunc. sparsely leafy, hispid ; invol.-scales *oblong, subacute*, glabrous or with a few dorsal, toothlike bristles. *Andr. Rep. t.* 93. *Bot. Mag. t.* 884. *Thunb. Cap.* 689. *Chrysocoma scabra, Th.! Cap.* 627. *C. Wright*, 335. *Felicia reflexa*, DC. l. c. 222. *Sieb. Fl.*, 27.

Var. β. **brachyphyllus** (Sond.) ; slender, with very small and nearly entire lvs.

Hab. Mts. round Capetown, common. (Herb. Th., D., Sd., Hk.)

Very variable in pubescence and in the marginal teeth of leaves, &c. A *straggling under shrub*. Stems and branches curving. Leaves 2–4, rarely 6–9 lines long, 1–1½ lines wide. Rays white, purple on underside.

Sect. 2. AGATHÆA. (Sp. 25–45).

25. A. Bergerianus (Harv.);

annual ; stem and branches *densely hairy* with long, spreading hairs ; leaves opposite or alternate, obovate-oblong, denticulate, subacute, 1-nerved on both sides, hairy with long, sparse hairs ; peduncles long, pubescent ; invol. scales *uniseriate, linear-oblong*, obtuse, with a *hairy keel* and membranous margins ; *disc fl. sterile;* achenes of ray whiteish-hairy, 2-ribbed, obovate, flattish. *Elphegea Bergiana, Less.! Syn.* 182. *Cineraria Bergeriana, Spreng.! Agathœa Bergeriana, DC. l. c. p.* 223.

Var. β, **alternifolia**; leaves mostly alternate.

Hab. Lion's Mt., Capetown, *Eckl.! Pappe!* β, Cape, *Vilette!* (Herb. Sd., Hk., D.)

An upright annual, simple or branched, 3–8 inches high, very hairy, with long hairs. Leaves 1–1½ inch long, 3–5 lines wide, not much narrowed at base. Heads smaller than in *A. adfinis;* the disc sterile.—Var. β (in Herb. Hook.) differs merely by its alternate leaves.

26. A. annectens (Harv.);

annual, slender ; stem and branches pilose with long, spreading hairs ; leaves mostly alternate, oblong-obovate or oblongo-lanceolate, denticulate, subacute, tapering at base, 1-nerved, on both sides hairy ; pedunc. long, pubescent ; invol. scales uniseriate, *broadly-oblong, obtuse, quite glabrous*, with wide, membranous margins ; disc. fl. mostly sterile ; achenes of ray pilose, wide-margined ; pappus scanty, barbellate.

Hab. Groenekloof, *Thunberg!* at the Paarl, *W. H. H.*, 1838 ; Riv. Zonder Ende, Zey.! 2759, in part. (Herb. Th., D., Hk., Sd.)

A small annual, 3–6 inches high. Leaves tapering much at base, 1–1½ inch long, 2–3 lines wide. Pedunc. 2–4 inches long, naked or unibracteate. Inv. scales with a green dorsal line and pinkish or silvery, scarious, slightly jagged margins, about twice as long as broad. Rays strongly revolute. Readily known from *A. Bergerianus* by its involucre. It is also allied to *Charieis heterophylla*, with which it is confounded in Herb. Thunb., and its pappus is of intermediate character.

27. A. adfinis (Less.! Syn. p. 174);

annual, with a slender root ; stem and branches pubescent and pilose with long, spreading hairs ; leaves mostly alternate, obovate-oblong, tapering at base, subacute, 1-nerved, entire, on both sides hairy with long and short hairs, the lower leaves opposite ; peduncles long, pubescent ; invol. pluriseriate, the outer scales keeled, hairy, acute, the inner flattish, glabrous, membranous ; achenes pubescent, 2-ribbed, oval, brown. *Felicia adfinis, Nees. DC. l. c.* 219. *Charieis heterophylla, Drege! Herb. (ex pte.)*

Var. β, **strictus** ; stem panicled ; leaves obovato-lanceolate, oblongo-lanceolate, or linear-lanceolate, remotely subdenticulate. *Agathœa stricta, DC. l. c.* 224.

HAB. Lion's Mountain, Capetown, *Ecklon!* Namaqualand, *A. Wyley!* Modder-fontein, *Rev. H. Whitehead!* β, Riebechskasteel, *Drege!* Steendal, *Dr. Pappe!* Riv. Zonderende, *Zey.!* 2759 (2750). (Herb. Sd., D., Hk.)

A branching, hairy annual, 3–12 inches high. Leaves 9–15 lines long, 2–4 lines wide, the upper smaller. Pedunc. 2–4 inches long. Heads many-flowered, the disc fertile. Invol. in 2 or 3 rows, the outer scales nearly as long as the inner. Variable in size, branching, and pubescence as are most annuals. Var. β seems to me scarcely worth separating even as a variety, much less generically and specifically from *Lessing's* plant. *Drege's* distributed specimens of *Charieis heterophylla* partially belong to this.

28. A. Namaquanus (Harv.); annual, with a *thick* simple root; stem branched from the base, glandularly pubescent and pilose ; leaves mostly alternate, oblongo- or obovato-lanceolate, tapering at base, obtuse or subacute, on both sides roughly hairy and glandular, papulose above, 1-nerved ; peduncles terminal, glandular ; invol. bi-seriate, the outer scales keeled, glandular, green, acute, the inner lanceolate with wide membranous edges and a green back, all barbellate; achenes pubescent, 2-ribbed, obovate, black. *Zey.!* 796.

HAB. Bitterfontein, Namaqualand, *Zeyher!* (Herb. Sd., Hk., D., Cap.)

Seemingly an annual, but more robust and glandular than any of our other annual *Asters.* Rootstock 1–1½ lines in diameter, woody, deeply descending. Stems 4–10 inches high, erect, branched from the crown, closely leafy. Lvs. very rough, the older ones almost echinated with the stumps of fallen hairs. Heads many-flowered ; invol. of about 20 scales. Disc partly sterile. This looks different from *A. adfinis*, but it is not easy to express the difference in words, except by saying that it is more robust, rough, and glandular.

29. A. Pappei (Harv.); annual, root fibrous; stems simple or divided near the base into many ascending, leafy, appressedly-pubescent branches; leaves mostly opposite, close, much longer than the internodes, linear or spathulate, thickish, with recurved margins, gland-scabrid and appressedly-pubescent and ciliate; pedunc. long, shortly and softly pubescent; inv. uniseriate, of about 20 linear, obtuse, scabrous and barbellate scales ; rays 18–20, revolute ; disc fl. fertile ; achenes minutely hispidulous ; pappus of many rigid, serrulate, fulvous bristles.

HAB. Sandy spots on the Camp-ground, Rondebosch, *Dr. Pappe!* (Herb. Cap., D.)

Root much branched. Main stem erect, 1–3 inches high, either simple or throwing out many opposite, spreading, ascending-erect, simple branches. Leafpairs ¼–½ inch apart. Leaves ¾–1¼ inch long, 1–2 lines wide, mostly tapering to the base. Pubescence short and close-lying, rather stiff: surface minutely glandular and rough. Heads many-fl., showy, closely resembling those of *Charieis heterophylla*, with which this species was inadvertently mixed in Herb. Pappe, but from which it differs in gen. char. and in foliage. It appears to me to be a very distinct species, though hitherto overlooked, within an half-hour's drive of Capetown. Rays apparently violet-purple.

30. A. demissus (Harv.); annual, slender, diffusely branched, densely hairy with short, spreading hairs ; leaves opposite, distant, oval or obovate-oblong (small), entire, densely hairy; peduncles long, pubescent; invol. scales linear, acute, keeled, hairy at back ; "mature achenes nearly smooth on both sides."—*DC. Agathœa diffusa, DC. l. c.* 224.

HAB. Draakensteensbergen, *Drege!* (Herb. Sond.)

A slender, hairy, spreading annual, of which I have only seen a single, imperfect specimen. Leaves 2–3 lines long, 1–1½ line wide.

31. A. tener (Harv.); annual; stem erect, simple below, oppositely or alternately branched above, sparsely pilose; leaves, save the uppermost, *opposite*, remote, sessile, 3-nerved, oblong or lanceolate-oblong, distantly callous-serrate, membranaceous, thinly hispid; pedunc. elongate, shortly hairy and scabrous; invol. scales lanceolate, the outer hispid, inner membranous, glabrous. *Agathœa tenera, DC. l. c.* 224.

HAB. Cape, *Th / near the Gariep, Drege!* Groenekloof, *Pappe!* (Hb. D., Th., Hk., Sd.) 12–15 inches high, sometimes much-branched, leaves 1½–2 inches long, half-clasping, 3–5 lines wide. Pedunc. 3–6 inches long, pale. Heads 4–6 lines across. Rays blue. Confounded, in Herb. Thunb. with *Amellus strigosus* (*Aster. strigosus*, Th., ex pte.). Dr. Pappe's specimens are much stronger and more branched than *Drege's*; their branches and upper leaves frequently alternate. Root much branched.

32. A. hirsutus (Vent.); shrubby, with decumbent branches; leaves petiolate, ovate, sparingly sinuate-toothed, hispidulous; invol. 2–3 seriate; disc. fl. fertile, rays deeply 3-toothed at the point. *Munychia hirsuta, DC., l. c.* 222. *Cineraria hirsuta, Vent. Malm. t.* 95.

HAB. Cultivated from the Cape. (Unknown to us).

33. A. Cymbalariæ (Thunb.! Cap. 689); herbaceous or suffruticose, diffuse, pubescent or hairy; leaves opposite, long-petioled, ovate or oblong, 3–5-crenate or toothed; peduncles lateral and terminal, elongate, 1-headed, hairy; invol. scales 1–2-seriate, linear-acute; achenes sparsely hispid. *Munychia Cymb. Nees, Ast.* 218. *DC. Prod.* 5, *p.* 223.

VAR. β. **Ionops;** pubescence shorter and rougher; disc-fl. violet-purple.

HAB. Among stones about Table Mt., near the summit, *Thunb., Mundt, Ecklon, Drege, &c.*; β. Mitchell's Pass, *A. Wyley!* (Herb. Th., D., Sd., Hk.) Old stems woody at base, much-branched, 2–12 inches long or more. All parts clothed with spreading, long or short hairs. Peduncles ½–1½ inch long. Leaf like that of *Veronica hederœfolia*, 4–8 lines long and wide, sometimes cordate at base. Pedunc. 3–5 inches long.

34. A. petiolatus (Harv. Thes. t. 154); many-stemmed, procumbent or pendulous; stems very long, subsimple, pubescent; leaves alternate, sub-distant, *petiolate*, ovate, coarsely 3–5-toothed, membranous, veiny, dotted; pedunc. terminal, naked, 1-headed; invol. scales sub-biseriate, equal, pilose, lanceolate; ovary puberulous; pappus rough.

HAB. District of Albert, and in Basutuland, hanging over precipitous rocks, *T. Cooper!* 658 and 727. (Herb. D., Hk.) Stems 1–2 feet or more long. slender, terete, simple, with or without a few short, lateral ramuli, ending in a naked, 1-headed peduncle. Leaves ¾–1 inch long, 4–6 lines wide, on each side 2-toothed, with a deltoid, terminal tooth, the teeth mucronulate. An interesting species, with the general aspect of *A. Cymbalariæ*, but with alternate, distant leaves.

35. A. Candollei (Harv.); perennial, many-stemmed; stems ascending, subsimple, closely leafy at base, hairy with long, soft hairs; leaves lanceolate-oblong, acute, tapering much at base, denticulate beyond the middle, thinly-membranous, hairy and ciliate with long, soft hairs, 3-nerved at base; peduncles long (scapelike) pubescent; invol.-scales subuniseriate, keeled and hairy-backed, acute; achenes obovate, black, sparsely hispidulous, and with a hairy margin. *Munychia Cymbalarioides, DC. l. c. p.* 222.

HAB. About Table and Devil's Mountain, *Ecklon!* Waterfall, on E. side of Devil's Mt., *W. H. H.* (Herb. D., Sd.)

Stems 2–6 inches long, closely leafy, ending in a long, naked peduncle, 3–8 inches long, and sometimes having a single, slender bract in the middle. Leaves 1½–2 in. long, ½–¾ inch wide, drying very pale, and in that state translucent and netted with slender veins ; the base tapering into a cuneate, 3-nerved, imperfect petiole. *DC.* states the disc-fl. to be sterile ; but some of the outer ones at least are fertile. I venture to alter the specific name, because the term "*cymbalarioides*" serves to mislead. The leaves may be much more justly compared to those of *Silene Capensis* than to those of *A. Cymbalariæ*.

36. A. Natalensis (Harv.) ; herbaceous, nearly stemless, perennial ; radical leaves sub-rosulate, opposite, oblongo-lanceolate, denticulate, on both sides coarsely hairy ; flowering-stem scapelike, hirsute, one-headed, with several scattered linear-oblong or linear bracts (depauperated leaves) ; heads many-flowered ; invol.-scales biseriate, subequal, flat, one-nerved, lanceolate-linear, sub-acute, densely hairy, the inner membranous ; achenes (young) pubescent ; pappus copious, uniseriate, rough. *Agathœa Natalensis, Sch. B. in Walp. Rep. 2. p. 956.*

HAB. Summit of the Tafelberg, Natal. *Krauss!* No. 381. (Herb. Hook., D.)

Root woody, cylindrical, præmorse, with long, thickish fibres. Leaves crowded round the crown, 2–3½ inches long, ¾–1 inch wide, subacute, shaggy with long, whitish hairs. Scape 3–6 inches high with 6 or 8 leafy bracts, ½–1 inch long, and 1–2 lines wide. Heads resembling those of *A. alpinus*. Pappus of about 20 setæ.

37. A. serratus (Thunb! Cap. 689) ; stem ascending, branched, leafy, glabrate or scaberulous ; leaves alternate, subamplexicaul, broad-based, oblong-linear, acute, 3-nerved, erecto-patent, glabrous, calloso-serrulate ; flowering branches from the upper axils, scabrous, with 2–3 small, narrow leaves, one-headed ; invol.-scales linear, acuminate, hispidulous ; ovaria sparingly hispidulous. *Less! Syn. p. 173. Agathœa serrata, Nees.—DC. l. c. 223.*

HAB. Cape, *Thunberg!* (Herb. Thunb.)

Root, according to *Thunberg,* "annual ;" but I think this doubtful. Stem robust, a foot or more high, branched from the lower part, the branches long and simple, erect. Leaves ½–1 inch apart, 1–1½ inches long, 2–3 lines wide, sometimes sprinkled with rough, callous points, otherwise glabrous. I cannot understand why *Lessing* calls them "4–5 in a whorl :" they are not only all scattered, but rather distantly so. Heads small.

38. A. linifolius (Harv.) ; shrubby, scaberulous ; leaves mostly alternate, linear-lanceolate, acute, thin (sub-pellucid veiny), setoso-serrulate, with subreflexed margins, on both sides glabrous or nearly so ; pedunc. setose upwards ; invol.-scales sub-biseriate, pubescent and glandular, 3-ribbed ; achenes hispidulous. *Agathœa Zeyheri, Nees.—DC. l. c. 224. Cineraria linifolia, Zey. MSS.*

HAB. Uitenhage, *Zeyher!* (Herb. Sond.)

Of this I have only seen an imperfect specimen : it comes near *A. serratus.* Leaves 10–15 lines long, 1–2 lines wide, tapering at base, when held to the light sub-pellucid between the closely-netted veins, but not "pellucid-dotted" in the usual sense of that term.

39. A. barbatus (Harv.) ; shrubby, divaricately much-branched ; twigs set with long bristles ; leaves mostly opposite, crowded, linear-oblong, blunt, thick, on the under side and margin bearing long, stiff

bristles ; peduncles elongate, glandularly pubescent ; invol.-scales linear, acute, setose and barbato-ciliate, glandular ; achenes hairy. *Agathœa barbata, DC. l. c.* 224.

HAB. Graaf Reynet, *Drege!* Riv. Zonderende and Zwarteberg, *Zey.!* 2755. (Herb. Hk., Sd.)
A rigid, scrubby bush, about a foot high. Lvs. 5–8 lines long, 1–2 lines wide, the longer ones tapering to the base. Pubescence very rigid, except on the pedunc. and invol.

40. A. elongatus (Thunb.! Cap. 688); suffruticose at base, diffusely much branched ; stem, branches, leaves, and peduncles hispid with long, spreading or deflexed, rigid hairs ; leaves opposite or alternate, sessile, lanceolate-linear, linear or oblong or ovate-oblong, acute or obtuse, midribbed, membranous, quite entire, often glandular-pubescent ; pedunc. long, naked ; invol.-scales biseriate, equal, the outer hispid and glandular, the inner glabrous or nearly so; achenes hispidulous. *Less.! Syn.* 170. *Agathœa hispida, DC.! l. c.* 224.

VAR. *a.* **Thunbergii** ; leaves lanceolate-linear, subacute, mostly opposite. *A. elongatus, Hb. Thunb.! Agathœa hipida, DC. l. c.* 224.

VAR. *β.* **barbiger** ; leaves broadly linear-oblong, obtuse, opposite. *Ag. hispida, var. barbigera. DC.*

VAR. *γ.* **crassifolius** ; lvs. opposite or alternate, thickish, obovate or spathulate.

VAR. *δ.* **spathulæfolius** ; leaves mostly alternate, crowded, glandular and pilose, obovate or spathulate.

VAR. *ε.* **Pappei** ; leaves mostly alternate, lin.-lanceolate, acute, glandular and ciliato-pilose ; heads small.

VAR. *ζ.* **Candollei** ; leaves alternate, distant, ovate-oblong, acute. *Agathœa elongata, DC. in Herb. Drege! (an sp. propr.?)*

HAB. Zwartland and Saldanha Bay, *Th.!* Cederberg, *Drege!* Stellenbosch Mt., *W. H. H. β.* near the Gariep, *Drege! γ.* Albany and Somerset, *Mrs. F. W. Barber! δ.* Berg Riv., *Zey.!* 787. *ε.* Steendahl, Tulbagh, *Dr. Pappe! ζ.* near the Gariep, *Drege!* (Herb. Th., D., Hk., Sd.)
Many-stemmed, often forming large spreading tufts, slender, ascending. Leaves in vars. *a.* and *β.* mostly opposite ; in the other vars. alternate, variable in shape, texture, and indument ; in *δ.* and *ε.* mostly, but not constantly, well covered with minute, raised glands, mixed with the scanty or copious hairs, ¾–1–1½ inches long, 1–3–4–5 lines wide, either broadbased (in *ζ.*) or usually tapering. Pedunc. 5–10 inches long. Ripe achenes black, sparsely bristled. Of the above varieties the last (*ζ.*) is the most distinct looking, but being only known to us by the fragments distributed by *Drege,* we cannot venture to separate it specifically : its *distant* and broad-based leaves are remarkable.

41. A. hirtus (Harv.); suffruticose, ascending, branched ; branches elongate, leafy, terete, with minute powdery pubescence mixed with rigid jointed bristles ; leaves oblong-linear, sessile, entire, with subrevolute margin, rigidly ciliate on margin and nerve, the lowest opposite, the rest *alternate;* peduncles terminal, long, glandularly pubescent and setose; invol.-scales rough with swollen-jointed bristles, and glandular; achenes puberulous. *Agathœa hirta, DC. l. c.* 224.

HAB. Breedriver, *Drege!* (Herb. D., Hk., Sd.)
Branches from near the root, curving upwards, 12 inches long or more. Leaves 6–8 lines long, not tapering at base, rigid in substance, and rough with very stiff, subulate, jointed bristles. Pedunc. 6–10 inches long. Rays blue. The bristles are often coloured red or purple.

42. A. strigosus (Licht. Herb.) ; shrubby, much branched, erect ; branches hirsute, densely leafy ; leaves opposite, half-amplexicaul, *erect,* oblong or lanceolate-oblong, subacute, on both sides densely covered with long, rigid, white hairs ; pedunc. elongate, glandular and scabrid ; inv.-scales 2-seriate, narrow-lanceolate, the outer scabrid and pilose, 1-nerved ; pappus barbellate. *Aster Charieis, Less. Syn.* 171. *Leyssera ovata, Thunb.! Cap.* 692. *Kaulfussia strigosa, Spr. f. suppl.* 25. *Agathœa strigosa, Nees. DC. l. c.* 225.

HAB. Roggeveld, *Thunb.*, *Lichtenstein*, Olifant's R., *Mundt amd Maire.* Sandhills between Klein R. and Cape L'Agulhas, *E. Z.!* Uitenhage, *Zey.!* (Hb. Sd , Th.!)

A small bush, 1 or more feet high, the old branches bare and ash-coloured, the younger closely leafy. Foliage pale, almost canous with copious long hairs. Leaves 6–8 lines long, 1–2 lines wide, thick, 1-nerved, slightly recurved at the top. Pedunc. 3–6 inches long. Achenes hispidulous. Pappus *almost* plumose. This has nearly as much right to range with *Mairea* as with *Aster.*

43. A. Ecklonis (Less.! Syn. 170) ; shrubby ; branches glandular and pilose, closely leafy ; leaves mostly alternate (but crowded), narrow-oblong, broad-based, sessile, subacute, entire, with revolute margins, above rough with prominent glands, dotted beneath, the young ones somewhat pilose ; pedunc. elongate, glandular ; invol.-scales sub-biseriate, lanceolate, acute, 3-nerved, glandular ; achenes black, puberulous. *Agathœa Ecklonis, Nees. DC. l. c.* 225. *Cineraria trachyphylla, Spr. Syst.* 3, 547.

HAB. Between Kraderivier and Duyvenhoeks R., *Ecklon!* (Herb. Sond.)

44. A. Capensis (Less. Syn. 168) ; shrubby or suffruticose ; leaves mostly opposite, oblong-obovate or roundish-elliptical, obtuse, either sessile or tapering into a short, imperfect petiole, roughly hairy or pilose, with reflexed margins ; inv. scales 1-nerved, hispid-strigose, lanceolate ; achenes *minutely* scaberulous. *Agathœa amelloides, DC. l. c.* 225. *Cineraria amelloides, Linn. (ex pte.) Agathœa cœlestis, Cass.*

VAR. β, rotundifolia ; root annual ? (fide Thunb.) *Aster rotundifolius, Thunb.! Cap.* 689. *Less. Syn.* 172. *Agathœa rotundifolia, Nees. DC. l. c.* 223.

HAB. Table Mountain, *Thunberg;* Potberg, *Eckl.;* Vanstaadensberg, *Zey.!* 2758 ; Knysna, *Dr. Pappe!* Howison's Poort, Albany, *Mr. Hutton!* Var. β, Cape, *Thunberg!* Zwartkops R., *Zey.!* 2757. (Herb. Th., D., Hk., Sd.)

Often confounded, especially in gardens, with *A. Æthiopicus*, but usually much more roughly hairy, and always distinguishable by its hispidulous achenes. Except in its reputed annual root, I cannot distinguish *A. rotundifolius;* the specimens so-named in Herb. Thunb. appear to me to be certainly plants of the first year, but *doubtfully* annual.

45. A. Æthiopicus (Burm.) ; shrubby, erect, variably pubescent, or glandular or subglabrous ; branches rodlike, closely leafy ; leaves opposite or alternate, spreading or deflexed, subsessile, oblong or obovate, obtuse, mostly rigidly ciliate on margin and midrib (or glabrous), sometimes thinly hispid on one or both sides ; inv. scales 3-nerved, glandular ; achenes *quite smooth and glossy. Less. Syn. p.* 169. *Agathœa microphylla, Cass. DC. l. c.* 225. *Cineraria microphylla, Vahl.*

VAR. β, glandulosa ; branches and foliage glandular ; leaves linear-oblong or lanceolate, scabrous.

HAB. Platteklip, Table Mt., *Ecklon*, &c.; Simon's Bay, *C. Wright*, 324 ; Caledon,

E. Z.! Zuureberg, *Drege!* Uitenhage, *Zey.!* Albany, *Burchell.* β, Hemel and Arde, *Zey.!* 2760. (Herb. Sd., Hk., D.)

A slender shrub, 1–2 ft. high. Lvs. 6–8 lines long, 2–3 lines wide, in β, 1 line wide. Stem either glabrous, pubescent, hispid or glandular. Achenes always glabrous and very smooth.—Often cultivated in England for *A. Capensis.*

46. A. leiocarpus (Harv.); suffruticose at base; stems diffuse, much branched below (often trailing), roughly pubescent; leaves alternate, oblong-cuneate or spathulate, *sharply 3–5 toothed at the apex* or entire, sessile, on both sides hairy, ciliate below; pedunc. slender, puberulous, elongate; invol. scales 2–3-seriate, lin.-lanceolate, acute, 3-nerved, puberulent, ciliate, the outer shorter; *ripe achenes quite glabrous. Agathæa leiocarpa, DC. l. c.* 225.

HAB. Drakensteenberg, *Drege!* W. H. H. (Herb. D., Hk., Sd.)
Slender, the diffuse branches 12–15 inches long, sparsely leafy. Lvs. ¾–1½ inch long, 1–2½ lines wide, the lateral teeth-lobes spreading. Heads small, the disc-fl. sterile, downy. Ray achenes dark brown, soon calvous, glossy, and quite glabrous.

Doubtful Species.

A. tricolor (Agathæa tricolor, Nees. Ast. 267); "shrubby? leaves opposite, oblong-lanceolate, entire, on both sides patently hirsute; inv. scales narrow, 1-nerved; achenes setulose; rays 2-coloured." *(Perhaps a var. of A. elongatus?)*

A. amoenus (Agathæa, *Sch. B.*); "root annual; stem branched, a span long; leaves cuneate, linear, ciliate chiefly below, glabrous, with axillary tufts, of shorter, linear leaves; pedunc. long, naked, as well as the *uniseriate,* 21-leaved invol., densely pubescent; disc fl. mostly fertile; achenes pubescent; pappus of 20 bristles. *Walp. Rep.* 2, 956.

HAB. Sandy shore at Hout Bay, *Krauss.*

A. Kraussii (Agathæa, *Sch. B.*); "shrubby; leaves opposite, oblong, acute, sessile, very entire, as well as the invol., densely covered with short, glandular hairs; achenes quite glabrous, black. *Walp. Rep.* 2, 956.

HAB. Near Constantia, *Krauss.* (Probably a var. of *A. Æthiopicus.*)

XVIII. DIPLOPAPPUS, DC.

Heads many flowered, radiate; *ray fl.* in one row, female; *disc-fl.* hermaphrodite, 5 toothed. *Recept.* flat, somewhat honeycombed. *Invol.* imbricate. *Achenes* oblong, compressed. *Pappus* of disc and ray double, the *outer* of short, the *inner* of long, rough bristles. *DC. Prod.* 5 p. 175.

Shrubs or herbs, natives of both hemispheres, various in habit. Leaves alternate or fascicled. Ray fl. blue or white; disc yellow. Name, from διπλοος, *double* and παππυς, *pappus.*

Sect. 1. FRUTICOSÆ. Much branched, glabrous shrubs, with linear or filiform, mostly tufted leaves.
Leaves *flat* or flattish, basally attenuated ... ... (1) **fruticulosus.**
Leaves linear-filiform, somewhat channelled ... ... (2) **filifolius.**

Sect. 1. HERBACEÆ. Simple-stemmed, perennial herbs, with 3–5-nerved, scattered leaves.
Scabrous and hairy ... ... (3) **asper.**
Glabrous: edges of the oblong leaves perfectly smooth ... (4) **lævigatus.**
Glabrous: edges of the lanceolate leaves cartilagineo-serrulate (5) **serrulatus.**

Sect. 1. FRUTICOSÆ.   Glabrous shrubs, with narrow, tufted leaves.   (Sp. 1-2).

1. D. fruticulosus (Less.); a much branched shrub ; leaves *linear, basally attenuate, flat,* quite entire, 1-nerved, gland-dotted, glabrous ; lateral ramuli short, peduncular, 1-headed ; invol. scales glabrous, 1-nerved, the nerve ending in an oblong gland. *DC. l. c. p.* 275. *Aster fruticosus, Linn. Th. Cap.* 687. *A. fruticulosus, Willd. Bot. Mag. t.* 2286. Also *D. extenuatus, Nees. DC. l. c.,* and *D. obtusatus, Less. DC. l. c. Aster obtusatus, Willd. A. rosmarinifolius, Spr.*

HAB. Very common round Capetown, and throughout the W. districts. (Herb. D., Sd., Hk. &c.)

A bush, 2-3 feet high and across, densely branched and twiggy. Flowering twigs very numerous and closely set, lateral, the leafy part ½-1½ inch long, the peduncle 1-3 inches. Rays blue. Achenes villous.

2. D. filifolius (DC. l. c. 276); similar to *D. fruticulosus* except that the leaves are *linear-filiform, very narrow, channelled. Aster filifolius, Vent. Malm. t.* 82. *Diplostephium fil. Nees. Ast.* 197. *Diplopappus tereti-folius, Less. Syn.* 165.

Var. β. **elongatus** (DC.); leaves not tufted, uncial, without glands ; peduncles longer, *D. elongatus, DC. l. c.*

HAB. Throughout the Colony. Tulbagh, *Pappe.* Uitenhage, *Zey!* Albany, *Williamson.* Natal, *Miss Owen,* &c. (Herb. Th., D., Sd., Hk.)

Only known from the preceding by its narrow leaves. The peduncles are most variable in length. *Zey.* 794, from Groot-valleyberg (Hb. Sond.); seems to be a bad specimen of this species.

Sect. 2. HERBACEÆ.   Simple stemmed herbs, with 3-5-nerved leaves.   (Sp. 3-5).

3. D. asper (Less.); stem herbaceous, ascending-erect, simple, *scabrous or rigidly-setose,* ending in a long or short, 1-headed, leafless peduncle ; leaves ovate-oblong, oblongo-lanceolate or broadly linear, 3-5-nerved, sessile, entire or coarsely few-toothed, *on both sides scabrous, hispid and ciliate;* invol.-scales acuminate, flat, dorsally pubescent, margined. *DC. l. c.* 276. *Calendula hispida, Th.! Cap.* 704. *Dipl. Natalensis, Sch. Bip. in Hb. Krauss,* 399! *Zey. !* 2763.

Var. β, **pleiocephalus**; heads sometimes several ; rays white or pale ; invol.-scales nearly glabrous.

HAB. Throughout the Eastern Districts, Caffraria and at Natal. β, near Ladysmith and Greytown, Natal, *Gerr. and MᶜK.,* 340,1009. (Herb. Th., D., Sd., Hk., &c.)

Stems 6-12 inches high. Leaves very variable in length, breadth and shape, always *harsh* to the touch, 2-5 inches long, 3 lines to an inch or more wide. Peduncle 2-12 inches long. Heads an inch or more across ; the rays uncial, bright blue. *D. Nata-lensis, Sch. B.,* which I have from *Dr. Sutherland, and Mess. Gerr. and MᶜK.,* as well as from *Krauss,* varies in all its attributed characters.

4. D. lævigatus (Sond.); stem herbaceous, ascending-erect, *glabrous,* rib-striate, ending in a 1-headed, leafless, 1-2 bracteate peduncle; leaves *ovate-oblong* or obovate-oblong, obtuse or subacute, 3-nerved, half-amplexicaul, very entire, with a thickened, *perfectly smooth margin,* glabrous ; invol.-scales acuminate, flat, glabrous ; achenes *densely silky.*

HAB. Vanstaadensberg, Uit., *Zey. !* 2764. (Herb. D., Sd., Hk.)

This resembles *D. Asper*, but is perfectly glabrous and smooth in all parts, with smaller flower heads. Leaves 2–2½ inches long, ½–1 inch broad.

5. D. serrulatus (Harv.); stem herbaceous, ascending-erect, glabrous, rib-striate, ending in a 1-headed, 2–3 bracteate peduncle; leaves *lanceolate or linear-lanceolate, acuminate,* 3-nerved, half-amplexicaul, with a thickened, *cartilagineo-serrulate margin,* glabrous; invol. sc. acuminate, flat, glabrous; achenes *thinly pilose. Zey.! 800.*

HAB. Magalisberg, *Burke and Zey.!* (Herb. D., Hk., Sd.)
Similar to *D. lævigatus*, but with much more acuminate and narrower, scabroserrate, rigid leaves, and much less hairy achenes. Leaves 2–3½ inches long, 2–5 lines wide.

[Doubtful genus—allied to Diplopappus ?]

ASTEROSPERMA, Less.

Heads radiate. *Style*-branches in the disc-fl. tipped with a short cone, and pubescent with suberect hairs. *Achenes* wingless, beakless, flattened, 2-ribbed. *Pappus* similar in disc and ray, in two rows, the outer bristles short, inner long. *Less. Syn.* 389. *DC. Prodr. VII. p.* 299.
A much-branched shrublet, resembling *Aster hyssopifolius*, hairy canescent, with scattered, linear leaves. Ray-fl. blue; disc-fl. yellow. Heads solitary, on the ends of leafless branches. Name from Aster, and σπερμα, a seed; the achenes are like those of an Aster.

1. **A. chrysocomoides** (Less. l. c.); *DC. l. c. p.* 300.
Cape, *Lessing.* (Nothing further known of this plant.)

XIX. ERIGERON, L.

Heads many-fl., heterochromous, radiate; *ray-fl.* female, in many rows, linear, equalling the disc (or longer); *disc-fl.* tubular, 5-toothed, either all perfect, or the outer female, the inner hermaphrodite, or all abortive. *Recept.* naked, honey-combed, fimbrilliferous. *Achenes* compressed, beakless. *Pappus* bristleshaped, rough, in one row. *DC. Prodr.* 5. *p.* 283.
A large and widely spread genus, chiefly from the Northern Hemisphere. The only S. Afr. species is a weed, originally of N. American origin, but now found in most settled countries. Name from ηρι, *early*, and γερων, an *an old man:* soon grows seedy.

1. **E. Canadense** (Linn.); stem erect, hispid, paniculately much-branched; leaves linear-lanceolate, ciliate. Heads small, very numerous; rays scarcely longer than the cylindrical involucre, very narrow; achenes oblong, shortly pubescent. *DC. l. c.* 289.

HAB. Eastern Districts and Natal, *Drege, Hutton, Gerr. & McK.,* 323, &c. (Herb. D., Sd. &c.)
One to two feet high, annual. Lvs. crowded, 2–2½ in. long, 1–2 lines wide. Rays white.

XX. NIDORELLA, Cass.

Heads many-fl., heterogamous, homochromous, radiate; *ray-flowers* female, very shortly strapshaped or bilabiate, in one, two or several rows; *disc-fl.* tubular, 5-fid, hermaphrodite, the central ones sometimes sterile. *Invol.* campanulate, imbricate. *Recept.* honeycombed. *Anth.* without tails. *Achenes* oblong, terete or subcompressed, mostly downy. *Pappus* in one row, of roughish bristles, slightly connate at base. *DC. Prod.* 5, *p.* 321.

Herbs or half-shrubby plants, natives chiefly of S. Africa. Leaves alternate, toothed or entire. Heads corymbose, rarely solitary. Flowers all yellow, the rays sometimes paler than the disc. This genus differs from *Erigeron* (to which it is re-united by Schultz Bip.) in having ray-fl. "*homochromous*," or of the same colour as those of the disc. Name from *nidor*, a strong smell, which these plants have.

Sect. 1. GENUINÆ. Heads densely corymbose. (Sp. 1–14).
  A. Lvs. either pinnatifid, lobed, or toothed :
    Lvs. on *slender petioles*, ovate or cordate, coarsely toothed   (1) **mespilifolia.**
    Lvs. *tapering* at base, lobed or pinnatifid :
      Deeply pinnatifid ; lobes lanceolate, acute...   ...   (2) **pinnatilobata.**
      Lobed or pinnatifid ; lobes few, obtuse :
        Stem and lvs. sparsely bristly and rough   ...   (3) **resedæfolia.**
        Stem and lvs. roughly and copiously hairy   (4) **hirta.**
    Leaves *tapering* at base ; the upper sessile, auricled,
      obovate or oblong, toothed or entire   ...   ...   (5) **auriculata.**
    Leaves linear-lanceolate, acuminate, nearly glabrous,
      distantly toothed   ...   .   (6) **linifolia.**
  B. Lvs. quite entire or obscurely repand :
    Pappus copious. Ray-fl. flat, obtuse or 3-toothed :
      Pubescent ; hairs short, scabrous or glandular :
        Lvs. acute or acuminate :
          Subsimple. Ray-fl. sharply 3-toothed   (7) **solidaginea.**
          Much-branched. Ray-fl. bluntly toothed   (8) **polycephala.**
        Lvs. obtuse, linear-cuneate :
          Lvs. scattered, distant...   ...   .   (3) **resedæfolia,** *β.*
          Lvs. crowded, often tufted ...   .   (9) **foetida.**
      Woolly or villous ; hairs long and curled ...   ...   (10) **Hottentotica.**
      Glabrous or nearly so ; leaves long :
        Frutescent. Cauline lvs. lanceolate, petiolate :
          Lvs. quite entire, gland-dotted   ...   ...   (11) **punctulata.**
          Lvs. callous-toothed, gummy   ...   ...   (12) **conyzoides.**
        Herbaceous. Rad. lvs. very long ; cauline
          half-clasping, sub-decurrent, undulate   (13) **undulata.**
    Pappus of few, small bristles. Ray-fl. 2-lipped   ...   (14) **depauperata.**
  Sect. 2. SPURIÆ. Heads solitary, on terminal peduncles. (Sp. 15–16).
    Herbaceous, erect, subsimple. Lvs. glandular-linear   (15) **exilis.**
    Shrubby, divaricately much-branched, hispid and setose   (16) **Gariepina.**

1. N. mespilifolia (DC. l. c. 321) ; suffruticose, loosely-branched, glabrescent ; leaves on slender petioles, cordate or ovate, coarsely incise-toothed ; heads corymbose. *Aster mespilifolius, Less. Syn.* 180. *N. Natalense, Sch. B. Mss. Erigeron (Nid.) clematideus, Sch. B. Walp. Rep.* 2, *p.* 958.

HAB. Districts of Stellenbosch, *Eck.;* Uitenhage and Albany, Br. Caffraria and Natal, common. (Herb. D., Hk., Sd.)
  A half-climbing or scrambling plant, some feet in height, with spreading branches. Pubescence very scanty and minute. Petioles sub-uncial, very slender. Leaves 1–1½ inch long, ¾–1½ inch wide, acuminate, with 3–6 large teeth at each side. Peduncles ending the branches, naked, striate, either being a single, dense corymb, or panicled, each branch corymbose. Achenes very minutely-downy.

2. N. pinnatilobata (DC. l. c. 321) ; suffruticose, sub-simple or corymbose at top, the whole plant scabrido-pubescent ; leaves tapering at base into an imperfect petiole, inciso-pinnatifid or pinnato-partite, the lateral lobes 2–4, lanceolate, acute ; corymb compound, many-headed.

VAR. β. **glabriuscula** ; thinly-pubescent ; the leaves sub-glabrous, narrow, with involute margins and channelled lobes.

HAB. Between the Zwartkey and Basche, *Drege!* Caledon, *Dr. Pappe!* β. Cape, *Bowie!* in Hb. Hk. (Herb. Hk., Cap.)

Two feet high or more, rather robust, the tall, simple stem ending in many leafy flower-branches, nearly of equal height. Leaves 1–1¼ inch long. Invol. scales linear, subacute. Rays very short, obovate.

3. N. resedæfolia (DC. l. c. 322); herbaceous, erect, the stem and petioles sparsely bristly; leaves scabrido-pubescent, amplexicaul, taper-ing at base into a petiole, the uppermost entire, the lower pinnate-parted, lobes in 1–2 pairs, oblong, blunt, mucronulate, the terminal broader, obovate-oblong ; corymb compound ; invol.-scales oblong, glabrous, ciliolate. *Zey.!* 804.

VAR. β. **rapunculoides** ; all the leaves entire, or the lowest inciso-lobulate. *N. rapunculoides, DC. p.* 323. .

HAB. Little Namaqualand, near the Gariep, *Drege!* Caledon R., *Burke and Zeyher!* Zulu land, *Miss Owen!* Natal, *Sanderson!* β. Neuweveld, Beaufort, and Zondag and Koega Rs., *Drege!* (Herb. Hk., D., Sd.)

2 to 3 feet high, simple or branched ; the branches spreading, somewhat corym-bose. Leaves from a broad base, 1½–2½ inches long, the terminal lobe 3–5 lines wide. All the lobes suddenly mucronulate. Fl.-heads crowded. *Drege's* specimen of *N. rapunculoides,* DC., in Hb. Hook. has the lower leaves inciso-lobulate.

4. N. hirta (DC. l. c. 322); "herbaceous, subsimple, erect; stem roughly hairy; leaves hairy scabrid, the lower oblong, tapering at base into a long petiole, entire or here and there lobulate, the rest dilated at base and semiamplexicaul, pinnate-partite, the pinnæ in 1–2 pairs, oblong-linear ; corymb compound, crowded; invol.-scales linear, acute, omewhat hairy at back." *DC. l. c.*

HAB, On the Stormberg, 5–600 feet. *Drege.* (Unknown to me).

" Allied to *N. resedæfolia.* Cor. of ray deep yellow. Ligulæ very small. Lvs. often fascicled." *DC.*

5. N. auriculata (DC. l. c. 322); herbaceous, erect, subsimple or corymbose, the whole plant shortly and closely pubescent or scabrous-pubescent ; leaves more or less distinctly eared at base, oblong or ob-ovate, the upper subsessile and entire or nearly so, the rest tapering at base into a long or short petiole and coarsely toothed, veiny beneath ; corymb compound, dense ; invol.-scales obtuse, thin, glabrescent and ciliolate. *Erigeron Kraussi, Sch. B.*

VAR. β. **obovata**; ears small; lvs. shortly petiolate, obovate. *N. obovata, DC. l. c.*

VAR. γ. **senecionidea**; ears small; lower leaves on long petioles, oblong. *N. sene-cionidea, DC. l. c. Zey.!* 2765.

HAB. Districts of Uitenhage and Albany ; Kaffraria and Natal, common. (Hb. Th., Hk., D., Sd.)

A coarse plant, 2 f. high, variable in the breadth and toothing, shape and pubes-cence of the leaves, sometimes with a large, rounded, earlike lobe at the base of the leaf, sometimes with a small and angular, or with an obsolete ear. Pubescence rigid, though short. I find it impossible to separate, by persistent characters, the 3 species of *De Candolle* here united. One of the folia of *"Inula fœtida"* in Herb. Th. belongs to this.

6. N. linifolia (DC. l. c. 322); herbaceous, erect, simple, nearly glabrous, stem scaberulous; leaves linear-lanceolate, acuminate, amplexicaul, the upper sessile, entire, the lower tapering to the base, distantly callous-toothed above the middle; corymb compound, dense; inv.-scales linear-oblong, subacute, glabrous, ciliolate.

HAB. Near Port Natal, *Drege!* (Herb. Sond., Hk.)

A slender, virgate herb, with long, narrow leaves. Lower leaves 3–4 inches long, 3–4 lines wide; upper 1–1¼ inch long, 1–2 lines wide. Rays very minute.

7. N. solidaginea (DC. l. c. 322); root woody; stems herbaceous, erect, subsimple, below sparingly setose, above closely and minutely scabrid-pubescent; leaves *linear or lanceolate-linear, entire, acute,* scabrid-pubescent, the upper ones broad-based, the lower tapering at base, setose beneath and more or less ciliate; corymb compound; the pedicels and acute, 1-nerved invol.-scales scabrid-pubescent; rays narrow, sharply 3-toothed, longer than the involucre.

HAB. Witbergen and Stormberg, *Drege!* (Herb. Hook.)

Stems 2 feet high, rodlike, rib-striate. Lvs. 1½–2 inches long, 2 lines wide.

8. N. polycephala (DC. l. c. 323); "herbaceous, erect, paniculately much branched, the whole plant covered with short, greyish pubescence; cauline-leaves half-clasping, elongate, linear-lanceolate, acuminate, 3-nerved; rameal linear, 1-nerved, all quite entire, or 1–2 toothed; corymb of each branch compound, dense; invol.-scales acute, puberulous." *DC. l. c.*

HAB. Betw. Omtendo and Omsamculo, *Drege!* Orange State, *Cooper,* 1035. (Hb. D.)

"Two f. high. Cauline leaves 3 inches long, 3 lines wide; rameal 1 inch long, 1 line wide. Rays small, obtusely 3-toothed, paler than the disc." *DC. Mr. Cooper's* specimen agrees very well with the above description of *DC.* save that the leaves are now and then unidentate.

9. N. fœtida (DC. l. c. 323); stem erect, herbaceous, simple, hispid; leaves densely crowded, linear-cuneate, tapering to the base, obtuse or subacute, very entire, scabrid on both sides; axils mostly leaf-tufted; corymbs panicled, dense; invol. scales linear, puberulous or glabrate; rays very minute, obtuse. *Erigeron fœtidum, Linn. Inula fœtida, Th.! Cap.* 667. *Nid. foliosa, Cass.*

VAR. β, **hyssopifolia**; more slender, with less crowded leaves and smaller heads. *N. hyssopifolia, DC. l. c.*

HAB. Dry hills round Capetown and Kamp's Bay; Verloren Vlay, *Wallich.* β, round Capetown, and in Uitenhage, Albany, and Kaffraria, *E. Z.* (Herb. Thunb., D., Hk., Sd.)

1–2 ft. high, simple, or divided from the base, the lower part of the stems densely covered with patent, tufted leaves; the upper part rather bare. Leaves 1–1½ inch long, 2 lines wide, pale green, thick, with an immersed nerve. Pubescence rough and short. Rays paler than the disc, and scarcely so long. Var. β, which I have seen in Herb. Sond. and received from *Gerr. & M'K.* 261, does not materially differ from the normal form.

10. N. Hottentotica (DC. l. c. 323); stem lignescent at base, erect, villous or woolly; leaves entire, thinly or densely woolly, above sometimes becoming glabrate, the lowest obovate, tapering into a petiole; the medial oblong-cuneate, obtuse, narrow at base, the uppermost broad-

based, mucronate; corymb compound; inv. scales linear, glabrous or
thinly villous, pale.

VAR. β, lanata; all parts *very densely woolly*, with long, curled hairs.

HAB. Beyond the Gariep, between Klaarwater and Litaku, *Burchell;* Mooje R.
Zey.l 904; Zululand, *Miss Owen !* Var. β, Wonderfontein, Bechuanaland, *Zeyher!*
(Herb. D., Hk., Sd.)
Distinguished from other species by its woolly pubescence, which in var. β. is
very copious. The inflorescence resembles that of *N. foetida.*

11. N. punctulata (DC. l. c. 323); "quite glabrous; stem shrubby,
with viscidulous branches; leaves tapering into a petiole, elongate-
lanceolate, acuminate, quite entire, *on both sides marked with minute,
glandular dots;* corymb compound; pedunc. compressed; inv. shorter
than the pappus, and even than the disc, its scales obtuse." *DC. l. c.*

HAB. Beyond the Gariep, between the Asbestos Mts. and Wittewater, *Burchell.*
" Lvs. 3-4 in. long, 3-5 lines wide, thickish. Ray fl. female, minutely ligulate." *DC.*

12. N. conyzoides (Harv.); suffruticose, nearly glabrous, viscidulous;
leaves tapering at base into a short petiole, broadly lanceolate, callous-
toothed, penninerved, acute or acuminate; heads densely corymbose,
pedicels puberulous; inv. scales oblong, obtuse, quite glabrous; ray-fl.
very minute, ovate-spoonshaped, denticulate.

HAB. Delagoa Bay, *Capt. Owen !* (Herb. Sond.)
This has much the habit of *Conyza ivæfolia,* but with broader leaves and more
densely crowded heads: the marginal flowers are truly *ligulate,* though the expanded
portion is so short as to be overlooked without close examination. Leaves 3-4 in.
long, 1-1½ in. wide, rather closely penninerved, the nerves proceeding at an angle
of about 75° to the margin; veinlets netted.

13. N. undulata (Sond.); herbaceous, quite glabrous, ascending-
erect; stem angular, simple; leaves entire or remotely repand-toothed,
the *radical* very long, linear-oblong, strap shaped or lanceolate, much
attenuated at base, the *cauline* at base cordate, semi-amplexicaul, subdecurrent, lanceolate-linear, acute or subacute, undulate, the margin
slightly recurved; corymb panicled, dense; invol. scales linear, obtuse;
rays very minute and narrow. *Chrysocoma undulata, Th.! Cap. p.* 627.
Nidorella amplexicaulis, DC. & N. longifolia, DC. l. c. 324. *Zey.!* 2766.

HAB. Cape, *Thunb.!* Eastern districts, *Burchell;* Zuureberg, *Drege!* Vanstaadens Mts., Uit., *E. Z.!* Howison's-Poort, *Hutton!* (Herb. Th., D., Hk., Sd.)
Root thick. Stem rising from a curved or deflexed base, 2-3 feet high, robust.
Rad. leaves 8-14 inches long, ¾-1 inch wide, obtuse or acute; cauline 3-6 inches
long, 4-5 lines wide, the upper ones smaller. They vary in having a widely-cordate,
clasping base, or a slightly-cordate one, as *N. auriculata* does. Heads very small
and numerous.

14. N. depauperata (Harv.); herbaceous, erect, viscidulous; stem
very straight, simple or branched, scabrous; leaves erect, the lowest
linear-oblong, obtuse, much-attenuated at base, scabrous and pilose;
upper linear, rigid, one-nerved, scabrous and ciliate, flat or involute,
the very uppermost broadbased, short and ovato-lanceolate; heads
small, in compound corymbs; invol.-scales shorter than the disc,
broadly oblong, obtuse, glutinous; ray-fl. very minute, *bilabiate,
pappus of very few, small, unequal, rough bristles. Zey.!* 803.

HAB. Wolvekop, Caledon R., and Magalisberg, *Burke & Zey.*/ Mohlamba Range, Natal, *Dr. Sutherland*/ 5–6000f. (Herb. Hk., Sd., D.)

Stem 1–2f. high, either quite simple, or split into many very erect appressed branches. Lowest leaves (including the long, tapering petiole) 2–3 inches long, spathulate; upper ones ¾, 1, 1¼ inch long, 1–2 lines wide; those of the corymb ¼–¾ in. long, ovato-lanceolate. All parts rough to the touch, exuding gummy matter. Heads minute. The bilabiate corollas of the ray-fl., and few scales of pappus are irregular in the genus, but the habit and other characters are those of *Nidorella.* Disc-fl. sterile, their anthers scale-tipped, broad; stigmata perfect.— *Dr. Sutherland's* specimen has rather broader and more acute leaves, but does not otherwise differ.

15. N. exilis (DC. l. c. 324); "stem herbaceous, lignescent at base, erect, depauperated, naked above; branches few, long, one-headed; leaves linear, subacute, quite entire, all, as well as the branches, covered with *minute, glandular pubescence*, the lower longer, and pilose with long, scattered, white bristles; heads solitary, on long peduncles; inv.-scales subimbricate, flat, linear, acute, as long as the disc; rays 8–10, in the dry state revolute; achenes villous." *DC.*

HAB Clanwilliam, *E. & Z.*/ (Herb. Sond.)
Of this I have only seen a specimen from which the fl.-heads have disappeared. Stem 6–12 inches high, slender. Leaves 2–3 inches long, 1 line wide, thickish, rigid, one-nerved.

16. N.? Gariepina (DC. l. c. 324); suffruticose, divaricately much-branched, terete, hispid with spreading bristles; leaves linear, obtuse, one-nerved, on both sides *hispid and setose;* fl.-branches shortly naked at the summit; inv.-scales linear, acuminate, pluriseriate, hispid, scarious and ciliate at the margin; achenes slightly downy; rays minute.

HAB. Dry places near the Gariep, *Drege*/ (Hb. Hk., Sd., D.)
Very diverse in habit from others of this genus, resembling an *Aster* of the sub. gen. *Felicia*, except in its ray-flowers. Leaves spreading, 5–7 lines long, 1 line wide. Recept. quite naked. Invol. at length reflexed. Pappus copious.

Doubtful Species.

N. floribunda (Lehm. Hort. Hamb. 1851. Linn. XXV. p. 310); "stem shrubby, erect, much-branched, pubescent; leaves short-petiol-ed, oblong-lanceolate, serrulato-dentate, veiny, glabrescent, beneath and on the margins rough with short pubescence; corymb many-headed, fastigiate; rays equalling the disc, bluntly 3-toothed; young achenes silky." *Walp. Ann. 5. p. 192.*

Cult. at Hamburg, from Cape seeds. A garden var. of *N. auriculata?*

N. Sprengelii (Sch. B. in Walp. Rep. 2. 958); half-shrubby, roughly pubescent; stem corymbose, densely leafy; lvs. oblong-linear-cuneate, 3-nerved, the lower toothed, upper entire, with axillary leaf-tufts; heads numerous; inv.-scales oblong.

HAB. Near Melkout-Kraal, Hb. *Zeyher.*

N. Kraussii (Sch. B. l. c.); corymb many-headed, compound; stem suffruticose; lvs. ovate-elliptical, very entire or rarely toothed; the whole plant canescent with short hairs.

HAB. Cape, *Krauss.*

XXI. GARULEUM, Cass.

Heads many-flowered, monoecious, radiate ; *ray-fl.* strap-shaped, female ; *disc-fl.* tubular, 5-toothed, male, with abortive ovules. *Recept.* convex, naked. *Invol.* scales biseriate. *Style* of the disc-fl. with divergent branches, externally hairy, glandular at the margins inside. *Ovaries* of the disc oblong. *Achenes* of the ray obovate-oblong, 3-4-ribbed, beakless, roughish ; of the disc plano-compressed, smooth, empty. *Pappus* none. *DC. Prod. 5, p. 309.*

S. African half-shrubby plants. Leaves alternate, pinnatifid, with toothed lobes. Heads terminal, pedunculate ; the ray blue, disc yellow. Name unexplained.

Pubescent :
Lvs. incised, the lobes with *hook-pointed*, subulate teeth	(1) **latifolium.**
Lvs. pinnate-parted, the lobes bluntly-toothed	(2) **pinnatifidum.**
Glabrous, but scabrid ; lvs. sub-bipinnate, with subulate lobes	(3) **bipinnatum.**

1. G. latifolium (Harv.); densely scabrido-pubescent; leaves broadly-oblong, *inciso*-pinnatifid, the lobes coarsely toothed, *serratures callous, subulate, hook-pointed.*

HAB. Tongaat river, Natal, *W. T. Gerrard.* (Herb. D.)
" Two to three feet high, with blue flowers." Stem and foliage densely and roughly, but minutely pubescent. Upper leaves only seen by me ; these are 1½-2 inches long, 1 inch wide, cleft *half-way* to the midrib, the lobes 3-4 lines wide. The hooked serratures are remarkable. Peduncles woolly.

2. G. pinnatifidum (DC. l. c. 309) ; minutely velvetty ; leaves *pinnati-partite,* the lobes oblong, toothed, *serratures blunt or subacute, callous-tipped. Osteosp. pinnatifidum, Th. Cap. 717. L'Her. Nov. St. t. 11. O. cœruleum, Ait. Jacq. Ic. Rar. t. 179. Gar. viscosum, Cass.*

HAB. Sneeuwberg, *Drege! A. Wyley!* Seven Fountains, *Burke & Zey.!* (Herb. D., Hk., Sd.)
One to two feet high, much-branched, minutely glandular and viscid, all parts downy. Leaves 1-1½ inch long, ¾ inch wide, pinnately cleft to the midrib or nearly so ; lobes 1-1½ line wide, few-toothed. Pedunc. glandular and downy ; serratures sometimes broad and shallow, sometimes narrow and deep.

3. G. bipinnatum (Less.); *scabrous,* rigid ; leaves pinnate or *bipinnati-partite,* the lobes *subulate,* acute, few-lobed or simple. *DC. l. c. 309. Osteosp. bipinnatum, Th. Cap. 717. Zey. 1011.*

HAB. Hantom, *Thunb.; in the Karroo, Drege! E. & Z.!* Zondag river and Graaf Reinet, *Drege!* Jackal's Fontein, *Burke & Zey.!* (Herb. D., Hk., Sd.)
About a foot high, the whole plant *glabrous,* but sprinkled with rough points, rigid. Branches flexuous. Leaves uncial, once or twice pinnately parted ; the lobes awl-shaped, sub-pungent.

XXII. FRESENIA, DC.

Heads many-fl. (15-16 fl.), homogamous. *Invol.* scales 3-seriate, imbricate, linear, submembranous at margin, bearing a dorsal row of 1-3 glands. *Recept.* areolate, narrow. *Cor.* tubular, glabrous, 5-toothed, not wider at the throat. *Anth.* without tails. *Style* as in *Chrysocoma. Achenes* compressed, beakless, silky-villous. *Pappus* double, the outer of short, narrow scales, the inner of long, rough bristles. *DC. Prod. 5, 328.*

Glabrous suffrutices, with linear, opposite or alternate, entire leaves. Peduncles

1-headed. Fl. pale-yellow. Name in honour of Fresenius, a Frankfort botanist, who wrote on Arabian and African plants. I have not seen *Drege's Fresenia?* 5843, from the Kromriver.

1. F. leptophylla (DC. l. c.); leaves *opposite*, linear-subulate, *glabrous.*

HAB. Cederberg, *Drege!* (Herb. Hk., Sd.)
Much-branched, woody, 6–12 inches high. Leaf-pairs nearly an inch apart. LVS. 3–6 lines long. Nerve of the invol. scales ending in a long, prominent gland.

2. F.? scaposa (DC. l. c.); "leaves *alternate*, crowded at the base of the branches, linear, *villous* with appressed hairs."—*DC.*

HAB. Near Driekoppen, Karroo, *Drege.*
"Stem woody, dwarf : old branches clothed with the remains of leaves. Leaves 6–8 lines long, either quite linear or wider near the tip. Fl. branches 3 inches long, 1-headed, softly hairy. Invol. scales linear, villous. Cor. yellow, glabrous. Achenes oblong, silky-villous. Pappus double, rough, the outer short."—*DC. l. c.*

XXIII. CHRYSOCOMA, Cass.

Heads many fl. homogamous. *Recept.* naked, somewhat honey-combed. *Invol.* campanulate, shorter than the flowers, imbricate, of oblong-lanceolate scales. *Achenes* laterally flattened, beakless, hispidulous. *Pappus* in one row, of slender, rough bristles. *DC. Prod.* 5, *p.* 353.

S. African shrublets, with linear, scattered, mostly entire rarely lobed or pinnatifid leaves. Branches ending in long or short, 1-headed peduncles, rarely corymbose. Heads globose, yellow. Name, from χρυσος, *gold,* and κομη, a *head of hair;* alluding to the numerous bright yellow fl. heads.

1. Eu-Chrysocoma.—Shrubs, with narrow, sessile leaves. Heads solitary.
Leaves linear or filiform ;
　Leaves rigidly ciliate ; setose beneath... ... ... (2) **ciliata.**
　Leaves glabrous, entire :
　　Leaves flat, at least ½ line wide ; heads stalked ... (1) **Coma-aurea.**
　　Leaves linear-filiform, about ¼ line wide.
　　　Flowering branches leafy to the summit or near it (3) **tenuifolia.**
　　　Fl. branches ending in a long or short peduncle :
　　　　Lvs. 2–5 lines long ; pedunc. elongate ... (4) **peduncularis.**
　　　　Lvs. 1–2 inches long ; pedunc. short ... ... (5) **longifolia.**
　Lvs. elliptic-oblong, flat, rigidly ciliate ... ... (6) **oblongifolia.**
　Lvs. lobed or pinnatifid :
　　Lvs. 2–3-lobed, the lobes linear, blunt ... ... (7) **tridentata.**
　　Lvs. pinnatifid, lobes *on each side* 2–3, blunt ... ... (8) **pinnatifida.**

2. Heteropsis.—Herbs, with decurrent, pinnatifid leaves. Heads corymbose.
　Stem 2 ft. high, scabro-pubescent; lvs. lyrate or pinnatifid (9) **decurrens.**

Sub-genus I. **Eu-Chrysocoma.**—*Cor.* narrow, funnel-shaped, with recurved lobes. *Style* deeply bifid. *Setæ* of pappus straight, rather broader at the points. *Small shrubs, with narrow, entire or lobed, sessile leaves. Heads solitary* (Sp. 1–8).

1. C. Coma-aurea (Linn. Sp. 1178); shrubby, *glabrous or nearly so;* leaves scattered, linear, subacute, spreading or recurved, quite entire; heads shortly pedunculate, 120–150 flowered. *DC. l. c. p.* 353. *Pluk. Alm. t.* 327, *f.* 2. *C. aurea,* Th. *Cap. p.* 626. *C. patula, Linn.* Th. *Cap. l. c. C. cernua, Linn.* Thunb. *! Cap. l. c. (ex pte.). Sieb. fl. Cap. n.* 32.

HAB. Western districts, common. (Herb. Th., Sd., Hk., &c.)
A much branched shrub, 1–1½ feet high, the branches corymbose, fastigiate,

curved, erect, glabrous, or the younger parts sub-pubescent. Leaves 7–10 lines long, ½-line wide, obtuse or sub-acute, squarrose. Flowering branches 3–6 inches long, sparsely leafy, the apex bare of leaves. Heads 5–7 lines long. The leaves are occasionally sparsely ciliate.

2. C. ciliata (Linn. Sp. 1177); shrubby, the twigs *pubescent*; leaves scattered, linear, obtuse or subacute, erect or erecto-patent, *rigidly ciliate* and *setose beneath ;* heads shortly pedunculate. *Thunb. Cap.* 627. *DC. l. c. 353. C. ciliaris, Willd.*

HAB. Cape, *Thunberg, E. Z.!* Kaus, near the Gariep, *Drege!* (Herb. Sd., Hk.)
Similar to *C. Coma-aurea,* except for the pubescence, which though variable in amount is very copious, especially on the young parts. The branching is more diffuse, and flexuous ; the leaves more erect and the peduncles shorter. The peduncles in *Drege's* specimen are nearly glabrous ; in *E. & Z.'s,* densely pubescent and much shorter.

2. C. tenuifolia (Berg. Cap. 285); shrubby, glabrous or nearly so ; leaves scattered, linear-filiform, short, obtuse, spreading or squarrose, or erect with spreading points, quite entire ; fl. twigs leafy nearly or quite to the summit; heads 80–90-flowered. *DC. l. c. 353. C. cernua, Th. (ex pte.) Ch. microphylla, Thunb ! Cap. 626. Zey.! 2769.*

Var. β. microcephala, leaves shorter (1–3-lines long) and more erect ; heads 50–60-flowered. *Ch. microcephala, DC. l. c. Zey!* 2767, 2798.
HAB. Distr. of Worcester, Swell., Uit., Albany, and in Kaffraria and Natal. (Herb. Th. D., Sd., Hk.)
Smaller in all parts than *C. Coma-aurea,* with narrower, less flat leaves, and smaller and more sessile heads, and ⅓ shorter flowers. Leaves 3–7-lines long, ¼ line wide. I find at least 50 flowers in the heads of var. β., which grows in the same localities as the normal state, and which chiefly differs by its shorter leaves,—a very variable character.

4. C. peduncularis (DC. l. c. 353); shrubby, the young twigs sub-pubescent; leaves scattered or opposite, linear-filiform, short, obtuse, erect or squarrose, quite entire; heads *on longish, naked peduncles,* about 50-flowered.

HAB. Saldanha Bay, Namaqualand, *E. Z.!* Stormberg and Camiesberg, *Drege!* (Herb. D., Sd., Hk.)
About the size of *C. tenuifolia, var. β,* with similar but more erect leaves. Pedunc. 1–2 inches long, thinly puberulous.

5. C. longifolia (DC. l. c. 354); shrubby or suffruticose, glabrous ; leaves scattered, 1–2 inches long, narrow-linear, obtuse, erect or erecto-patent, quite entire ; heads shortly pedunculate, about 50-flowered.

HAB. Near Ebenezer, Oliphant R., and Nieuvalei, *Drege!* (Herb. D., Hk.)
Less branched than the preceding. Lvs. 1–2–2½ inches long, ¼ line wide. Pedunc. scarcely uncial. *Drege's* specimens of var. β, patula, *DC.,* marked "*a*" seem to me to belong to *C. tenuifolia ;* those marked "*b,*" to the present. Or perhaps this is an extravagant form of *C. tenuifolia.*

6. C. oblongifolia (DC. l. c. 354); shrubby, dwarf, divaricate; lvs. *elliptic-oblong,* flat, spreading, *rigidly ciliate,* glabrous ; heads pedunculate; inv. scales linear-lanceolate, acute, imbricate, subscarious at margin. *C. ciliata, E. Mey. fide DC.*

HAB. Between Hexrivier and the Bokkeveld, *Drege !* (Herb. Sond.)

A small scrubby shrublet. The leaves on the younger branches and twigs are 3–4 lines long, and 1–2 lines wide ; those on the older branches, which spring from the buds of old leaf-scars, are often fascicled, and linear or linear-spathulate, with or without cilia.

7. C. tridentata (DC. l. c. 354); shrubby, glabrous or nearly so ; leaves thickish, linear, 3-lobed or trifid, the upper ones sometimes linear, *entire* or with one lateral lobe ; heads shortly pedunculate ; inv. scales linear-lanceolate, with scarious edges.

HAB. Zwarteberg, and near Kendo, *Drege!* (Herb. Hk., Sd.)

A foot or so in height, erect, with rod-like branches. Leaves 3–6 lines long, the lobes ½ line wide, either opposite each other or alternating, or one or both abreast. Heads nearly as in *C. Coma aurea.*

8. C. pinnatifida (DC. l. c. 354); "shrubby, glabrous; leaves sessile, oblong-linear, thickish, *pinnatifid*, lobes on each side 2–3, entire, obtuse, the margins subrevolute ; heads peduncled." *DC.*

HAB. Zwarteberg, *Drege.*

" Stem 2 feet high, simple, shortly panicled at the apex. Inv. purplish. Fl. pale-yellow," *DC.*—Unknown to us.

Sub-genus 2. HETEROPSIS. *Cor.* campanulate, with a slender tube and erect lobes. *Style* shortly bifid. *Setæ* of pappus flexuous, taper-pointed. *A tall herb, with simple stem, ending in a corymbose inflorescence. Leaves decurrent in narrow wings along the stem.*

9. C. decurrens (DC. l. c. 354); stem angular, herbaceous, sub-simple, thinly pilose ; leaves alternate, sub-distant, elongate, roughly hispid, inciso-pinnatifid, *decurrent* at base, the lateral lobes alternate, short, toothlike ; corymb panicled, fastigiate, the subdivisions dense ; invol. scales scabrous.

VAR. β, **pterocaula** ; lower leaves lyrate, the lateral lobes short and blunt, the upper lanceolate. *C. pterocaula, DC. l. c.*

HAB. Witberg, at the foot of the mountains, in upland valleys and on flats, 4500–5000 ft., *Drege!* Var. β, Leeuwenspruit, between Kraai River and the Witberg, 4500 ft., *Drege!* (Herb. D., Hk., Sd.)

Stems 2 ft. or more high, 2–3 lines in diameter ; the leaves 1–2 inches apart, 2–4 inches long, 2–4 lines wide, the lobes nearly ½ in. apart and 1–2 lines long. Flowering branches corymbose, each division corymbulose, 4–6-headed. Heads 60–80-flowd. Achenes minutely pubescent. Var. β, from nearly the same locality as α, differs so slightly that I cannot keep it specifically apart. Of it, however, I have only seen an imperfect specimen. *Drege's* specimens of var. α. vary in the shape of the leaves.

XXIV. PTERONIA, L.

Heads discoid, homogamous, many-fl., rarely 3–1-flowered. *Recept.* honeycombed, the margins of the cells either short or fringed. *Invol.* imbricated in many rows. *Achenes* compressed or top-shaped, glabrous or villous, rarely shortly-tapering at the summit. *Pappus* in many rows, bristle-shaped, bristles thickish, closely barbed, often concrete at base. *DC. Prodr.* 5, *p. 356.*

A large and natural genus of S. African small, dry, or glutinous shrubs. Leaves opposite or rarely alternate, mostly entire, glabrous or hairy, often ciliate. Heads terminal, solitary or corymbose. Invol. scales scarious, often shining. Flowers yellow, rarely purple.—Name, from πτερον, *a wing.*

Excluded.

Pteronia, 2772 in Herb. Drege,=**Ambraria.**

(Artificial Analysis of the Species).

A. Heads 3 or more, in tufts or corymbs at the ends of the branches :
  (*a*) Leaves alternate, linear-subulate :
        Heads few-flowered, subsessile ; achenes *hairy*...   (3) **baccharoides**, β.
        Heads many-fl., ending leafy branches ; achenes
        glabrous ... ...   (51) **camphorata.**
  (*b*) Leaves opposite, *sheathing at base*, linear or subulate :
        Heads 1–3, many-flowered ; lvs. filiform ... ...   (20) **flexicaulis.**
        Heads 3, sessile, 3–5-fl.; lvs. short ... ...   (21) **paniculata**, β.
        Heads corymbose, pedicelled, 3–5-fl.; lvs. filiform   (21) **paniculata.**
        Heads densely fascicled, sessile, 1-fl.; lvs. subulate   (22) **fasciculata.**
  (*c*) Leaves opposite or ternate, *not sheathing* at base :
      Leaves linear-trigonous, glabrous :
        Leaves in threes, hook-pointed ; heads co-
        rymbose ... ...   (4) **uncinata.**
        Leaves opposite, obtuse ; heads 1–3, pedi-
        cellate, clammy   ... ...   (13) **pallens.**
        Leaves opposite, hook-pointed ; heads in
        threes, sessile ... ...   (14) **tricephala.**
      Leaves linear-oblong, *flat*, nerved, powdery ...   (39) **quinqueflora.**
      Leaves broadly ovate or oblong, sub-petiolate :
        Leaves *flat*, pale, powdery ... ...   (5) **divaricata.**
        Leaves *wavy and curled*, glandular ... ...   (24) **undulata.**
B. Heads solitary at the ends of the branches.

  1. Inv. sc. *close-pressed*, often viscid, mostly broad, blunt, bluntish or mucronate.
  (*a*) Leaves alternate (at least the upper ones).
      Glabrous ; leaves smooth-edged ; achenes hairy   (16) **gymnocline.**
      Glabrous ; lvs. rigidly ciliate ; achenes *glabrous*   (49) **onobromoides.**
      Hispid and scabrous ; lvs. linear ; achenes hairy   (32) **villosa.**
      Pulverulent-scabrous ; achenes hairy :
        Inv. scales membr.-edged, horny ; lvs. acute   (31) **centauroides.**
        Inv. scales not bordered, straw-colour ; lvs.
        blunt ... ...   (10) **lucilioides.**
  (*b*) Leaves opposite, *not glabrous.*
      Softly-tomentose or canescent :
        Lvs. lin.-oblong, 1–2 lines wide, 4–5 l. long   (9) **cinerea.**
        Lvs. elliptic-oblong, nerved, 3–5 lines wide,
        8–10 lines long... ... ...   (25) **ovalifolia.**
      Pulverulent-scabrous or glandular :
        Inv. scales round-topped, membr.-edged ;
        lvs. acute, faintly nerved... ... ...   (31) **centauroides.**
        Inv. scales ovate, subacute, not bordered ;
        lvs. acute, midribbed   ... ...   (7) **oppositifolia.**
        Inv. scales straw-colour, obtuse ; lvs. blunt,
        nerveless ... ...   (10) **lucilioides.**
      Rigidly pubescent or hispid :
        Leaves ciliate, linear-lanceolate ; fl. purple   (23) **hirsuta.**
        Leaves oval or oblong-lanceolate ; fl. yellow   (6) **inflexa.**
  (*c*) Leaves opposite, *glabrous* (sometimes ciliate) :
      Leaves oval or round ; inv. scales with a white
      border   ... ...   (36) **leucoloma.**
      Lvs. (small) oblong-linear or oblong, nerved :
        Heads 4-flowered ; inv. scales with a broad,
        membr. border ... ...   (29) **Candollei.**
        Heads several-fl. ; inv. scales rigid, opaque :
          Achenes densely hairy :
            Leaves sub-connate ; inv. scales
            gummy, ciliolate   ... ...   (12) **glomerata.**
            Leaves not connate ; inv. scales
            glossy, *softly fringed*   ... ...   (17) **ciliata.**
          Achenes sparsely pilose ; inv. scls. viscid   (45) **viscosa.**

Leaves linear or lin.-lanceolate, *flat*, not ciliate :
  Flowers yellow :
    Inv. scales softly fringed, callous tipped　(33) callosa.
    Inv. scales entire-edged, rigid　... ...　(19) glabrata.
  Flowers purple　... ... ...　(23) hirsuta, γ.
Lvs. linear-terete, trigonous or filiform, *not ciliate :*
  Leaves sheathing at base, linear-filiform ...　(20) flexicaulis.
  Leaves not sheathing at base :
    Lvs. from 5 lines to an inch or more in length :
      Inv. scales dry, quite entire, membr.
        edged... ... ...　(18) succulenta.
      Inv. sc. dry, rigid, opaque, serrato-
        lacerate　... ... ...　(38) tenuifolia.
      Inv. sc. clammy, ciliolate, mem-
        brane-edged, lacerate　... ...　(13) pallens.
      Inv. sc. sub-viscid, horny, silky-
        ciliate　... ... ...　(35) turbinata.
    Leaves 1–3- rarely 4- lines long :
      Twigs scabrous ; heads obovoid,
        viscid, many-fl. ... ...　(46) elongata.
      Glabrous; heads cylindrical; scales
        memb. edged　... ... ...　(37) cylindracea.
      Glabrous, *dwarf;* heads oblong-top-
        shaped ; scales horny　... ...　(43) empetrifolia.
Leaves rigidly *ciliate :*
  Glabrous :
    Inv. scales mucronate ; achenes villous　(34) mucronata.
    Inv. sc. subacute; achenes sparsely pilose　(45) viscosa.
    Inv. sc. obtuse ; achenes glabrous, glan-
      dular... ... ...　(48) adenocarpa.
  Twigs scabrous :
    Achenes sparsely pilose　... ... ...　(46) elongata.
    Achenes quite glabrous　... ... ...　(47) stæhelinoides.

2. Invol. scales *loosely imbricated*, dry, linear, lanceolate or acuminate, often with wide, membranous edges.

  Leaves alternate (at least the upper ones).
    Leaves, involucres, and twigs rough, with minute,
      harsh points　... ... ...　(50) scabra.
    Leaves glabrous or nearly so :
      Achenes *glabrous;* leaves filiform　... ...　(51) camphorata.
      Achenes pubescent, or densely hairy :
        Leaves linear-trigonous, crowded, fur-
          rowed beneath ... ...　(3) baccharoides.
        Lvs. lin.-spathulate ; inv. scales linear,
          obtuse... ... ...　(2) glaucescens.
        Lvs. linear, obtuse, 2–4 lines long ; inv.
          scales *acuminate*　... ... ...　(28) leucoclada.
        Leaves oval or oblong :
          Obtuse ; inv. scales broadly ovate,
            with a *wide*, membr. border　...　(26) scariosa.
          Acute ; inv. scales lin.-lanceolate,
            acuminate... ... ...　(1) aspalatha.
        Lvs. lanceolate, 10–12 lines long ; inv.
          scales lanceolate, acuminate ... ...　(30) acuminata.
  Leaves opposite, or mostly so :
    Achenes hairy or villous :
      Inv. sc. ovate, with a *wide*, membranous border :
        Pappus *bright red* (lvs. glaucous, small,
          ovate or oblong)　... ... ...　(11) erythrochæta.
        Pappus pale, whitish or fulvous :

Glabrous (leaves oval or oblong, obtuse)	(26)	**scariosa.**
Powdery or sub-canous on twigs and young leaves :		
Heads 5-fl. Leaf-pairs distant	(39)	**quinqueflora.**
Heads several-fl. Lvs. close.		
Leaves 4–5 ls. long, 1–1½ line wide	(27)	**membranacea.**
Lvs. 1–2 ls. long, ½ in. wide	(15)	**beckioides.**
Inv. scales obtuse, with a *very narrow*, membranous border:		
Inv. scales oblong or ovate-oblong ; leaves 3–6 lines long	(8)	**incana.**
Inv. scales linear ; leaves 1–3 lines long	(41)	**leptolepis.**
Inv. scales opaque, entire or ciliolate, *acute*:		
Glabrous. Invol. sc. lanceolate-oblong, entire	(40)	**leptospermoides.**
Minutely canous. Inv. scales broadly-ovate, downy, ciliolate	(42)	**glauca.**
Outer achenes hairy ; inner *glabrous ;* inv. scales oblong, obtuse	(44)	**heterocarpa.**
Achenes glabrous ; inv. sc. lanceolate, acuminate	(51)	**camphorata,** ε.

Sect. 1. SCEPINIA. *Heads* many or few-fl. *Recept.* honey-combed, shortly fimbrilliferous. *Achenes* terete or subcompressed, villous, not tapering into a neck. (Sp. 1–22)

1. P. aspalatha (DC. l. c. 356) ; " glabrescent ; leaves scattered, sessile, elliptical or oblong, nerveless, rather fleshy, acute, the lower with 1–2 marginal teeth ; heads shortly pedunculate, terminal, many fl. ; inv.-scales linear-lanceolate, *acuminate*, as well as the twigs and upper leaves minutely puberulous." *DC.*

HAB. Betw. Karroe R. and Dwaal, *Burchell,* 1427. Nieuweveld, *Drege?* (fide DC.)
A scrap without flowers from *Drege*, in Herb. Sond, said to belong to this species, seems referable to *Pegolettia polygalæfolia.* We therefore give the present species wholly on *De Candolle's* authority.

2. P. glaucescens (DC. l. c. 357) ; the young parts downy, otherwise glabrous; leaves alternate, close, linear-spathulate, glaucous, somewhat fleshy, quite entire, obtuse; heads ending leafy branches, subsessile, 10–12 fl. ; inv.-scales linear, *obtuse*, loosely imbricated; achenes silky ; pappus tawny.

HAB. Nieuweveld, *Drege !* (Herb. Hk., D., Sd.)
A scrubby, rigid, robust, small shrub ; the old twigs hardening and spinescent, divaricate. Leaves 2–3 lines long, 1 line wide, narrowed to the base. Heads small. Invol. loosely 3-seriate ; the scales dry but not membranous. Our specimens are far from perfect.

3. P. baccharoides (Less.! Syn. 196) ; glabrous ; leaves alternate, crowded, spreading, linear-trigonous, keeled, when dry furrowed beneath, acute or subacute ; heads sessile or subsessile, 3–12-flowered, either corymbose or solitary ; inv.-scales varying from narrow-oblong to ovate, 1-nerved, entire, subacute or acute, the inner longer; achenes topshaped, densely silky. *Osteospermum teretifolium, Thunb.! Cap.* 713. *O. trigonum, Spr. Syst.* 3, 628.

VAR. α. **Lessingii** ; heads 8–12-flowered, the inv.-scales ovate. *P. baccharoides, Less.! Pt. acerosa, var. multiflora, DC. l. c. and P. baccharoides, DC.*

VAR. *β.* **acerosa**; heads 3-5-flowered, mostly corymbose ; inv.-scales narrow-oblong. *Pt. acerosa, DC. l. c.* 357. *Vars. 3-flora and 5-flora.*

HAB. Districts of Uitenhage and Albany, common. (Herb. Th., Hk., D., Sd.)
A much-branched, upright bush, 1-2 f. high ; branches closely leafy to the summit. Leaves 4-7 lines long. Heads variable in size and number of flowers, and in the breadth and shape of the invol.-scales, even on the same branch ! Vars. *a.* and *β.* therefore, are only casually distinct.

4. P. uncinata (DC. l. c. 357); glabrous ; leaves 3 *in a whorl,* crowded, spreading, linear-trigonous *with a hooked point,* acute ; heads 4-5-fl., corymbose, shortly pedicellate ; invol.-scales linear-oblong, sub-acute, entire, the inner ones longer, acute ; achenes topshaped, densely silky. *P. verticillata, DC. l. c.?*

HAB. Lange Valei, *Drege !* Verloren Valei, among shrubs near the sea shore, *Dr. Wallich !* (Herb. D., Hk., Sd.)
Very similar in aspect and flowers to *P. baccharoides,* but differing in foliage. I do not see how DC.'s *P. verticillata* (by description) materially differs. Its fl. are said to be purplish, those of the present sp. yellow.

5. P. divaricata (Less. Syn. 196); branches divaricate ; twigs and leaves scabrous with a very minute, powdery and glandular pubescence; leaves opposite, elliptic or ovate, obtuse, shortly petioled ; heads 5-6-fl., corymbose, pedicellate ; invol.-scales glabrous, shorter than the disc, oval-oblong, obtuse, with membranous, entire margins ; ach. obovate, compressed, pilose. *DC. l. c.* 357. *Eupatorium divaricatum, Th.! Cap.* 628, *Chrysocoma oppositifolia, Linn.*

HAB. Cape, *Burman.* Groenekloof, *E. Z., Drege !* Modderfontyn, *Rev. H. White-head !* Namaqualand, *A. Wyley!* Blauweberg, *Zey.!* 818. (Hb. Th., D., Hk., Sd.)
A scrubby bush. Twigs opposite, horizontal or widely spreading. Lvs. ½-¾ in. long, 3-5 lines wide, pale, 1-nerved, sometimes mucronulate. Pedicels opposite, ½-1 in. long. Pappus twice as long as the invol., reddish.

6. P. inflexa (Thunb.! Cap. 633); twigs pubescent; leaves opposite, sessile, oval, oblong or oblongo-lanceolate, obtuse, thickish, faintly 1-nerved, on both sides rigidly pubescent, or glabrescent ; heads terminal, sessile, solitary, oval, nearly glabrous; inv. equalling the disc, scales closely appressed, very broad and obtuse, roundish-oblong, convex, with a membranous, entire margin ; the lowest almost obovate. *Burch. Cat. Geogr.* 1495. *P. lupulina, DC.! l. c.* 357.

HAB. Hantum, *Thunb.!* Sack Riv., *Burchell.* Nieuweveld, *Drege!* Bitterfontyn, *Zey.!* 3097. (Herb. Sd., Th.)
A very rigid, robust, scrubby bush, 6-12 inches high, divaricately much-branched ; old twigs spinescent. Leaves variable in shape and pubescence, the smallest 1-2 lines long, others 4-7 lines long and 1-2 l. broad. Heads 6-8 lines long, 4-5 l. in diam., on our specimens immature.

7. P. oppositifolia (Linn. Syst. Veg. 614); twigs and leaves covered with a minute, powdery-scaberulous indument, cinereous ; leaves oppo-site, lanceolate-oblong, rather longer than the internodes, sessile, acute, strongly midribbed ; heads terminal, solitary, sessile, oval, many-fl. ; inv.-scales ovate, obtuse or subacute, appressed, rigid, glabrous, the inner subviscid, all minutely ciliolate. *DC. l. c.* 358. *Th.! Cap.* 632.

HAB. Cape, *Thunberg! Burchell,* Cat. 6837. (Herb. Thunb.)

Much branched, the old branches glabrate. Leaves 4–5 lines long, 1½ wide, often with axillary tufts. Heads ½–¾ inch long, ½ inch diam., the inv.-scales horn-coloured at the edge, dark-centred. Pappus rufescent.

8. P. incana (Less.! in Hb. Thunb.); divaricate ; twigs and leaves thinly tomentose-canescent ; leaves opposite, crowded, linear-oblong, obtuse, sessile ; heads terminal, solitary, sessile, 6–9 fl. ; invol.-scales glabrous, oblong, subobtuse, entire, with a very narrow membranous margin ; achenes topshaped, densely and rigidly hairy. *Zey.! 2771. DC. l. c.* 358. *Pt. xantholepis, DC. l. c. Eupatorium cinereum, Thunb.! Cap.* 627. *Chrysocoma incana, Burm.*

HAB. Caledon, *Mundt!* Attaqua's Kloof, *Dr. Gill!* Swellendam and Uitenhage, *E. Z.!* Albany, *T. W.* Tulbagh, *Pappe!* Kaus, *Drege!* (Herb. Th., D., Hk., Sd.)
A scrubby, much-branched, canescent bush, the short ramuli tipped with sessile, shining fl.-heads. Invol.-scales at first greenish-yellow, then fulvous and finally brown in centre. Pappus foxy. Lvs. 3–6 lines long, 1–1½ line wide.

9. P. cinerea (Linn. f. Supp. 356) ; young twigs and leaves tomentose-canescent ; leaves opposite, oblong-linear, sessile, flat, thickish, obtuse or subacute, crowded, somewhat nerve-keeled ; heads oval, pluri-flowered, terminal, solitary, sessile ; invol.-scales *dorsally cobweb-tomentose*, ovate-oblong or ovate, obtuse, membrane-edged and ciliolate-lacerate ; achenes silky-villous. *Thunb.! Cap.* 632. *DC. l. c., also Pt. canescens, DC.! l. c.* 358.

HAB. Cape, *Thb.!* Onderbockeveld and Zeederberge, *Drege!* (Hb. Th., D., Hk., Sd.)
A small, scrubby, hairy bush ; the older branches becoming glabrous. Lvs. 4–5 lines long, 1–2 l. wide. Heads ¾ in. long, 3–4 l. in diam. DC.'s *P. canescens* quite agrees with the original specimens of *Pt. cinerea*, in Hb. Thunb.

10. P. lucilioides (DC. l. c. 358) ; slender, the twigs and leaves canescent with very short, close, powdery (swollen) scabrid hairs ; lvs. opposite, linear or spathulate, obtuse, narrowed to the base, entire, nerveless ; heads terminal, solitary, sessile, 7–9-fl.; inv. oblong ; its scales oblong, obtuse, dry, glabrous, entire, closepressed ; achenes densely silky-villous.

VAR. β. sparsifolia ; rameal leaves often alternate or imperfectly opposite.

HAB. Near the Gariep in Little Namaqualand, *Drege!* β. Namaqualand, *A. Wyley!* (Hb. D., Hk., Sd.)
A slender, much-branched, pale shrub ; the twigs at length glabrescent. Leaves 3–6 lines long, ½–1 line wide, subdistant or close, the rameal often bearing leaf-tufts. Pubescence scabrous, composed of minute, conical, vertical hairs. Invol. glossy, pale straw-colour. Specimens of this plant, from *Drege*, in Hb. Hk. and Sd. are marked "*P. gymnocline!*"

11. P. erythrochæta (DC. l. c. 358) ; divaricate, nearly glabrous (the young parts minutely pubescent) ; leaves opposite, crowded, sessile, ovate or oblong, obtuse, glaucous, thick and nerveless ; heads terminal, solitary, sessile, 4–5-flowered ; invol.-scales shorter than the disc, ovate-oblong, acute, glabrous, dorsally lined, membrane-edged ; pappus bright red ; ach. rigidly hairy.

HAB. Sack R., *Burch.* Zw. Ruggens, *Drege!* George, *Pappe!* (Herb. D., Hk., Sd.)
A very rigid, robust, divaricate or procumbent, scrubby shrub. Twigs densely leafy, 1–2 inches long. Lvs. 3–4 lines long, 2 l. wide, the upper imbricating.

12. P. glomerata (Linn. f. Suppl. *356*); glabrous, leaves very short, opposite, sessile, subconnate, thick, oblong-carinate, obtuse, usually with axillary leaf-tufts ; heads 10–12-fl., terminal, solitary, sessile, oboval, viscid ; invol.-scales ciliolate, appressed, the outer short, ovate, the inner oblong, nearly equalling the disc, obtuse, rigid ; achenes rigidly hairy. *DC. l. c. 359. Pt. microphylla, DC. l. c. 359.*

HAB. In the Karroo, *Th.! Drege! Wallich!* Zwarteberge, and Camisberge, &c., *Drege!* Stellenbosch, *E. Z.* (Herb. Th., D., Hk., Sd.)
A strong, rigid, scrubby bush, 6–12 in. high. Lvs. 1–2 lines long, 1 line wide, erecto-patent. Heads 8–9 lines long, very clammy. *DC.'s "P. microphylla,"* of which I have seen *Ecklon's* specimen in Hb. Sd., scarcely differs by any tangible character.

13. P. pallens (Linn. f. Suppl. *357*); glabrous, much-branched; lvs. opposite, linear-trigonous, elongate, obtuse ; heads 12–15-fl., solitary or in threes at the ends of the branches, shortly pedunculate, ovate-turbinate; inv. clammy, its scales close-pressed, oblong, obtuse, ciliolate, the inner with lacerate-membranous edges ; achenes rigidly hairy. *DC. l. c. 359. Thunb. Fl. Cap. 630.*

HAB. Cape, *Thunberg!* In the Karroo, *E. Z.* (Herb. Sd., Th.)
Robust, densely much-branched, corymbose, 6–12 inches high, with pale twigs, leaves and involucres. Lvs. 1 inch long, ½ line wide. Heads 4–5 lines long, acute or conical at base. Inv.-scales very blunt.

14. P. tricephala (DC. l. c. *359*); glabrous, somewhat viscid, much-branched; leaves opposite, linear-trigonous, hook-pointed, subconnate; heads 4–5-flowered, *sessile in threes* at the ends of the twigs, oblong ; invol. scales shorter than the disc, narrow, acute, ciliolate ; achenes densely hairy.

HAB. Nieuweveld, *Drege!* (Herb. D., Sd.)
A small scrub, 5–6 inches high. Leaves 5–6 lines long, ½ line wide, scarcely so hooked as those of *P. uncinata.* Invol. and pappus pale.

15. P. beckioides (DC. l. c. *359*); young twigs and leaves minutely pulverulent ; twigs slender ; leaves opposite (small), linear, subacute, thickish, somewhat keeled, entire, powdery ; heads terminal, solitary, sessile, oblong-turbinate, conical at base, glabrous; invol. scales ovato-lanceolate, acute or acuminate, with membranous and torn edges, the outer short ; achenes densely hairy.

HAB. Swellendam, *Drege!* (Herb. Sd.)
A slender, twiggy shrub; the older twigs quite glabrous. Leaves 1–2 lines long, ½ line wide. Heads 8–10 lines long; the invol. scales in many rows, glaucous-green when dry, with pale edges. Pappus foxy.

16. P. gymnocline (DC. l. c. *359*); glabrous; twigs slender, striate, pale ; lower leaves opposite, upper alternate, sessile, linear-spathulate, obtuse, thickish, narrowed at base, nerveless, minutely punctulate ; heads terminal, solitary, sessile, oblong, obtuse at base, 12–15-flowered ; invol. scales appressed, the outer oblong, obtuse, the inner oblongo-lanceolate, subacute, glabrous and glossy, quite entire ; recept. naked ; achenes densely hairy. *P. calvescens, Drege in Herb.!*

HAB. Near the Gariep, in Little Namaqualand, *Drege!* (Herb. Hk., D., Sd.)

A slender, small and pale, or glaucous shrub. Leaves 4–5 or 6 lines long, scarcely 1 line wide, somewhat channelled and bluntly keeled. Heads pale straw-colour, dry. The "*P. gymnocline*" of *Drege*, so far as I have seen, belongs to "*P. lucilioides.*"

17. P. ciliata (Thunb.! Cap. 632); glabrous, much-branched, rigid; leaves opposite (but not connate), crowded at the ends of the twigs, oblong, oblong-linear or linear, rigid, thick, with a broad and prominent keel-like rib, very entire, glaucous ; heads terminal, solitary, elliptic ; inv. scales closely imbricating, glossy, broadly oblong, obtuse, rigid, fringed with soft hairs ; achenes densely hairy. *DC. l. c. 359.*

VAR. *α*. **Thunbergii**; leaves linear-trigonous, 4–6 lines long.

VAR. *β*. **Ecklonis**; leaves oblong, 1½–2 lines long, 1 line wide.

HAB. Namaqualand, *E. & Z.!* Olifantsriver, *Drege!* (Herb. Sd., D., Th.) A small scrubby, intricately-branched bush. Leaves separate at base, spreading. Heads 7–8 lines long, 3–4 in diameter, the outer scales brownish or greenish. Invol. shorter than the disc. *E. & Z.'s* plant has much shorter and proportionably broader leaves than *Thunberg's*, but variably so ; in other respects the plants agree.

18. P. succulenta (Thunb.! Cap. 630); glabrous, much-branched; leaves opposite, linear-trigonous, fleshy, subacute, spreading or recurved; heads terminal, campanulate, many-flowered; invol. rarely as long as disc, scales broadly ovate, very obtuse, dry, entire, membrane-edged; achenes densely hairy. *DC. l. c. 360.*

HAB. Carroo and Hantum, *Th.!* Olifantriver, *Drege!* (Herb. Th., D., Hk., Sd.) Leaves granulated when dry, 4 lines to an inch long. Heads ¾ inch long, 4–5 lines diameter. Fl. yellow. Branches of style much protruded, the stigmas long and hispid. Invol. scales nerveless, with a linear gland beneath the apex.

19. P. glabrata (Linn. f. Suppl. 358); glabrous, much-branched; leaves opposite, broadly linear, *flat*, thickish, obtuse (or acute), 1-*nerved*, spreading; heads terminal, campanulate, many-flowered; invol. scarcely as long as disc, scales broadly ovate or oblong, very obtuse, dry, *many-striate*, membrane-edged, entire. *Thunb.! Cap.* 630. *P. sesuviifolia, DC.! l. c. 358.*

HAB. Carroo, below Bockland, *Thunb.!* Kaus and Olifants river, *Drege!* Heerelogement, Clanw., *E. & Z.!* (Herb. Th., D., Hk., Sd.) Very near *P. succulenta*, but with flat more or less nerved leaves, and the invol. scales marked with several longitudinal, coloured lines. Leaves 1–1¼ inch long, 1½–2 lines wide. Fl. yellow. Perhaps merely a more luxuriant state of *P. succulenta*.

20. P. flexicaulis (Linn. f. Suppl. 355); glabrous, very viscid above; young leaves sheathing at base, the sheath afterwards splitting, linear-filiform, channelled above, incurved, obtuse; heads 15–25-flowered, solitary or in threes at the ends of the branches, subsessile; inv. oblong, viscid, nearly equalling the disc, scales appressed, obtuse, the outer oval, the inner oblong. *Th.! Cap.* 629. *Pt. connata, DC. l. c.* 360.

HAB. Carroo, *Thunb.!* Swellendam, *Dr. Thom, E. & Z.!* in the Karroo, *Drege, E. & Z.!* (Herb. Th., D., Hk., Sd.) This has nearly the habit and foliage of the following, but a very different inflorescence. It is also more gummy.

21. P. paniculata (Th.! Cap. 629); glabrous, viscid above; young leaves sheathing at base, the sheath afterwards splitting, linear-semi-terete, incurved, obtuse; heads 3-flowered, densely corymbose, pedicellate, about 3 on every pedicel; invol. cylindrical, glabrous, shorter than disc, scales oblong, obtuse; achenes hairy. *P. flexicaulis, DC. l. c. 360. P. fastigiata, Zey.*

VAR. β. **fastigiata**; leaves closely set, 3–5 lines long; heads sessile, mostly in threes, 3–5-flowered; invol. scales ovate-oblong, subacute. *P. fastigiata, Thunb./ Cap. p. 629.*

HAB. Carroo, *Thunb.!* Underbokkeveld and Zeederbergen, *Drege!* Elandsberg, *Wallich!* Namaqualand, *Rev. H. Whitehead!* Swell., Uit., &c. *E. & Z.!* Slaaykraal, *Burke & Zey.!* (Herb. Th., D., Hk., Sd.)

Very much branched, scrubby, with incurved branches, twigs and leaves. Leaves ¾–1¼ inch long, their sheath, after splitting, forming a membranous base, broader than the lamina. Heads yellow, densely corymbose, fastigiate. Var. β. has much shorter and more closely set leaves, and somewhat larger and less corymbose heads: varying in this respect as *P. baccharoides* more notably does: I have only seen it in Herb. *Thunb.!*

22. P. fasciculata (Linn. f. Suppl. 357); glabrous, viscid above; leaves opposite, shortly connate at base (the sheath at length splitting), narrow-lanceolate, acute or acuminate, viscid, concave above, keeled beneath, densely set, scabrous-edged; heads numerous, *one-flowered, densely tufted at the ends of the branches, sessile;* invol. linear, compressed, glabrous, scales lanceolate, acute. *DC. l. c. 360. Th./ Cap. 630. Henanthus fasciculatus, Less. Syn. 195.*

HAB. Carroo near Hexriver, *Thunb.!* Gauritz R., *E. & Z.!* Zeederbergen, *Drege!* (Herb. Th., D., Hk., Sd.)

A strong growing bush, 1–2 feet high. Branches naked below, very closely covered with leaves above. Leaves 1–1½ inch long, 2 lines wide, their short sheath on bursting forming a broad, membranous base to the leaf. The dense tuft of yellow flower-heads is 1–1½ inch diameter.

Sect. 2. **Pachyderis.** *Heads* many-fl. *Recept.* honeycombed. *Achenes* compressed (the outer sometimes 3-cornered), tapering into a short, thick, glabrous neck (which is often hidden by the copious hairs of the lower portion), sometimes quite glabrous. (Sp. 23–50).

23. P. hirsuta (Linn. f. Suppl. 356); branches scabrous; hispid or pilose (rarely glabrous); leaves opposite, linear-lanceolate, somewhat keeled, acute, rigidly ciliate (or smooth), scabrous or pilose beneath; heads terminal, sessile, solitary, ovoid, many-flowered; invol. glabrous or pilose, viscid; scales broadly ovate or oblong, the inner longer, membranous at edge, ciliolate, and often curled. *P. retorta, Linn. f. Suppl. 356.*

VAR. a. **vera**; stems and undersides of leaves very rough with *long, white hairs. P. hirsuta,* L. *DC. l. c. p. 360. Th. Cap. 681. Zey./ 815.*

VAR. β. **Cephalotes**; stems and undersides of leaves very rough with rigid points, but *not pilose. Pt. Cephalotes, Linn. f. DC. l. c. 361. Thunb. p. 631. Zey./ 2775. Pt. pseudo-cephalotes, Less.!*

VAR. γ. **glabra** (DC.); branches *glabrous;* leaves without cilia, folded or flat, glabrous on both sides. *DC. l. c. 361.*

HAB. Districts of Cape, Stellenb., Tulbagh, Worcester, Clanw., and Swellendam, *Th., E. & Z., Drege,* &c.; var. β. the common form, round Capetown; var. γ. in the Karroo, *E. & Z.!* (Herb. Th., D., Sd., Hk., &c.)

A rigid, usually very rough, much-branched, small shrub, 12–18 inches high, variably pubescent. Leaves closely placed, 5–7 lines long, 1–2 lines wide, thick and stiff. Heads 1–1¼ inch long, ½ inch diameter. Flowers *purple*, the corollas curved outwards.. Except in pubescence, which varies much in different specimens, I cannot distinguish *P. Cephalotes* from *P. hirsuta. P. pseudo-Cephalotes*, Less.! is founded on a specimen in Herb. Thunb., having the characters of var. β., with the invol. scales *remarkably* fimbriate and curled ; I find this character occasionally in the other varieties.

24. P. undulata (DC. l. c. 361) ; twigs and foliage very minutely glanduloso-pulverulent, scaberulous ; leaves opposite, *tapering at base into a petiole*, oval or oblong, obtuse, *wavy and curled;* heads pedicellate, from the upper axils, corymbose at the ends of the branches, 4–5-flowered ; invol.-scales erect, oblong, glabrous, entire, obtuse, with a narrow membr. margin ; achenes thinly villous.

HAB. Zilverfontein, Little Namaqualand, *Drege !* (Herb. D., Hk., Sd.)
A slender shrub, with opposite, erect branches, and remarkably curled, glandular foliage. Leaves 7–8 lines long, 3–4 l. wide. Pedicels ½–1 inch long ; heads 3, 5, 7, &c. in a corymb, small.

25. P. ovalifolia (DC. l. c. 361) ; twigs cobwebby, becoming glabrous ; leaves opposite, ellipt.-oblong or oval, sessile, obtuse, very entire, tomentose-canescent with appressed, woolly hairs, the older ones becoming almost nude ; heads oval-oblong, sessile, solitary, many-flowered; invol.-scales cobwebbed, appressed, ovate or oblong, the inner with a narrowish, membranous margin, ciliolate ; achenes silky villous.

HAB. Olifant's R., *Drege !* Worcester, *E. Z.* (Herb. D., Hk., Sd.)
A slender shrub, with opposite brachiate twigs. Leaf-pairs subdistant. Leaves 8–10 lines long, 3–4 lines wide, leathery. Heads 1¼ in. long, ½ inch diam. Fl. yellow. Habit of *P. cinerea*, but with larger and more distant leaves, larger heads, &c.

26. P. scariosa (Linn. f. Suppl. 356) ; divaricately much-branched ; rigid, at length spinescent, glabrous ; leaves opposite, or the upper alternate, subsessile, oval or oval-oblong, obtuse, thick, nerveless, glabrous ; heads terminal, solitary, sessile, oblong, many-fl. ; invol.-scales loosely imbricated, broadly ovate, with a very wide, membranous, lacerate margin and an opaque ovato-lanceolate, pungent-mucronate middle. *Thunb.! Cap. 633. DC. l. c. 361.*

HAB. Near Hantum, *Thunberg !* Betw. Koussia and Pedroskloof, and at Zilverfontyn, *Drege !* (Herb. Th., Sd., Hk., D.)
A robust, scrubby bush, very rigid. Leaves 1½–2½ lines long, 1–1½ line wide, acute at base or sub-petiolate. Heads 9–10 lines long, 4–5 l. wide. Inv. very membranous, shining.

27. P. membranacea (Linn. f. Suppl. 357) ; twigs and foliage minutely pulverulent-canescent ; leaves opposite, sessile, crowded, linear-oblong or linear, acute ; heads terminal, solitary, sessile, campanulate, many-flowered; invol. glabrous, scales oblong, with a lanceolate, opaque middle, and a very wide, membranous, pellucid, lacerate margin. *Th. ! Cap. p. 633. DC. l. c. 361.*

HAB. The Karroo, *Thunb.! Mundt.* About the Zwartkops and Koega Rs., Uit., *E. & Z.!* Boschesmans Riv. and Zuureberge, *Drege !* Zwarteberg and R. Zonder Eiende, *Zey.!* 2773. (Herb. Th., D., Hk., Sd.)
A rather slender, much branched, thinly pulverulent shrub, 1–2 f. high. Leaves 4–5 lines long, 1–1½ wide, 1-nerved, imbricating. Invol. shining, the scales loose, not in the least viscid. Fl. yellow.

28. P. leucoclada (Turcz.!); much branched, the twigs rigid, divaricate, glabrous, white-barked ; leaves alternate (small), linear-oblong or linear, obtuse, thick, nerveless, sub-sessile, glabrous and glaucous; heads terminal, solitary, sessile, several-flowered; inv.-scales loosely imbricated in many rows, ovato-lanceolate, much acuminate, the innermost tapering into a long, awnlike point, all with a broad, membranous, entire margin ; achenes densely silky.

HAB. Bitterfontyn, *Zeyher!* 811. (Herb. Sond., D.)
A rigid shrub, 1–2 f. high, with brachiate branches and twigs, and remarkably white, smooth bark. Leaves scattered, subdistant, 2–4 lines long, scarcely 1 line wide, sometimes fascicled. Heads 4–5 lines long, 3 lines diam., pale. A remarkable species ; apparently very nearly related to *P. acuminata.*

29. P. Candollei (Harv.) ; " quite glabrous, leaves opposite, oblonglinear, thickish, with a prominent midrib beneath, glaucous, crowded about the apices and in the axils'; heads terminal, solitary, sessile, 4-flowered ; scales of the glabrous invol. oval-oblong, with a wide, membranous margin, acuminate-submucronate." *DC. P. glabrata, DC. l. c.* 362, *non Linn. f.*

HAB. Karroo River, *Burchell,* Cat. 1410. (fide DC.)
Unknown to us. *P. glabrata,* L. f. (by Herb. Thunb.!) is *P. sesuviifolia,* DC.!

30. P. acuminata (DC. l. c. 361); "glabrous; leaves alternate, lanceolate, tapering to each end, mucronate, nearly nerveless ; heads terminal, subsessile, oblong, 10–12-fl.; invol. scales lanceolate, acuminate, scarious at the margins." *DC.*

HAB. Beyond the Gariep, *Burchell,* 1587 (fide DC.)
" Leaves 10–12 lines long, 3 lines wide. Achenes compressed, very villous, attenuated under the reddish pappus." *DC.*

31. P. centauroides (DC. l. c. 362); branches and adult leaves glanduloso-pulverulent, scaberulous, twigs and young foliage glandulosovelutinous and canescent; leaves opposite and alternate, sessile, entire, acute, the cauline oval-oblong or lanceolate, midribbed, those of the twigs linear; heads terminal, solitary, oval, many-flowered; invol. scales appressed, entire, very obtuse, glabrous, with a very narrow membranous edge, the outermost short, canescent at base.

HAB. Dutoitskloof, *Drege!* (Herb. Hk., Sd.)
Apparently a shrub, 2–3 feet high, with erect, ramulous branches. Cauline leaves 7–12 lines long, 2–3 lines wide; rameal 5–6 lines long, scarcely a line wide. Heads an inch or more long, 6–8 lines in diameter.

32. P. villosa (Linn. f. Suppl. 356) ; twigs scabrous and pilose ; leaves mostly alternate, linear, obtuse, flat, thick, nerveless, on both sides rigidly pilose with swollen, jointed hairs; heads terminal, solitary, sessile, many-fl., ovato-globose ; invol. scales appressed, glabrous, broadly oblong, very obtuse, membrane-edged, the inner equalling the disc, scarious and undulate at apex ; achenes densely hairy, with a short, thick neck. *Thunb.! Cap. p.* 631.

HAB. Hantum and Carroo below Bockland, *Thunberg!* (Herb. Thunb.)
A scrubby, flexuous bush, a foot or so in height. Pubescence very coarse and rough. Leaves ½–1 inch long, 1 line wide, by *Thunberg* said to be " opposite," but

on his own specimen they are alternate. The young leaves are bristly all over, the older more or less glabrate and ciliate : but very few remain on the only specimen I have seen.

33. P. callosa (DC. l. c. 362); "glabrous, much branched; leaves linear, thickish, subobtuse, very entire, not ciliate, indistinctly-nerved; heads terminal, solitary, sessile, about 15-fl. ; inv. oval-oblong, scales appressed, delicately and softly ciliate, obtuse and callous at the apex." *DC.*

HAB. Zwart-Ruggens, on the Karroo, *Drege.*
"Suffrutex, 6 inches high. Leaves 5-7 lines long, scarcely more than ½ line wide. Heads 9 lines long. Inv. pale-greenish, more intensely coloured at the tips of the scales. Cor. yellow. Achenes very villous, compressed, shortly beaked. Recept. nearly concave, areolate, shortly honeycombed."—*DC.* Unknown to me. It seems to come very near the following.

34. P. mucronata (DC. l. c. 362); glabrous, much-branched; leaves opposite, *connate at base*, crowded, linear, bluntly-keeled, fleshy, obtuse, *rigidly-ciliolate;* heads terminal, solitary, sessile, 12–15-fl. ; invol. oval-oblong, somewhat viscid, scales appressed, horny-membranous, broadly oblong, with a sharp, often patent mucro, and lacerate-ciliate margin.

HAB. Zwart-Ruggens, on the Karroo, 2-3000 ft., *Drege!* Bitterfontyn, *Zey./* 817. (Herb. D., Hk., Sd.)
A scrubby, flexuous, small, intricately branched shrub. Leaves 3-8 lines long, ½ line wide, sometimes scabrous at back. Heads 7-8 lines long, 3-4 lines wide. Fl. yellow. Achenes densely villous.

35. P. turbinata (DC. l. c. 362); glabrous, much-branched; leaves sub-opposite, *distinct at base*, crowded, linear, thickish, round-backed, obtuse, quite entire, the upper smaller; heads terminal, solitary, sessile, about 15-fl.; invol. oval-oblong, slightly viscidulous, scales appressed, horny, oval-oblong, obtuse, silky-ciliate.

HAB. Little Namaqualand, *Drege!* (Herb. D., Hk., Sd.)
A small, rigid scrub, intricately branched and flexuous. Leaves 4-5 lines long, ½ line wide, not strictly opposite, though very nearly so, very close-placed. Invol. scales rigid and opaque, pale. Our specimens from *Drege* do not quite accord with *DC.'s* description.

36. P. leucoloma (DC. l. c. 362); "glabrous; leaves opposite, sessile, orbicular or oval, very entire, 1-nerved ; heads terminal, solitary, sessile, many-fl.; invol. cylindrical, scales appressed, obtuse, with a white, very entire, scarious margin." *DC.*

HAB. Kamiesberg, *Drege.*
"Leaves 2-3 lines long, 2 lines wide, longer than the upper internodes. Heads 10 lines long, 4-5 lines diameter. Achenes compressed, very hairy, with a short, glabrous beak."—*DC.* Unknown to me.

37. P. cylindracea (DC. l. c. 363); "glabrous ; leaves opposite, crowded on the ramuli, spreading, lin.-trigonous, slightly hook-pointed, subobtuse ; heads terminal, solitary, sessile, cylindrical; inv. scales appressed, broadly ovate, with membranous, ciliate-lacerate margins, and mostly tipped with a callous mucro." *DC.*

HAB. In the neighbourhood of the Gariep, *Burchell*, Cat. 1603. (fide DC.)

"Leaves subglaucous, 3 lines long. Heads 10 lines long, 3 l. wide, 9-fl. Ovary compressed, sparsely hairy, contracted under the pappus."—*DC.* Unknown to me.

38. P. tenuifolia (DC. l. c. *363*); glabrous; stems ascending-erect, subsimple or branched, suffruticose; leaves opposite, elongate, filiform-subulate, channelled, entire, acute; heads terminal, solitary, sub-pedunculate, 8–12-fl.; invol. oval-oblong, scales close-pressed, glabrous, broadly oblong, very obtuse, opaque and rigid, with a very narrow, serrulate-lacerate margin; achenes compressed, hairy, minutely rostrate.

HAB. Near Caledon, *E. Z.!* (Herb. Sond., D.)
6–12 inches high, either with several simple stems, or branched below, with several erect, simple branches, each bearing a fl. head. Leaves 1–1½ inch long, ½ line wide, very erect, longer than the internodes. Heads 8–10 lines long, 3–5 lines in diam. Pappus rufous.

39. P. quinqueflora (DC. l. c. *363*); shrubby; twigs slender, rod-like, minutely canescent, becoming glabrate; leaves opposite, oblong-linear, subacute, flat, thickish, nerved at back, grey, the younger canescent with very minute, powdery, appressed hairs; heads at the ends of the branches, solitary or in threes, very shortly pedunculate or sub-sessile, 5-flowered; invol. oval-oblong, scales ovate, acute, 1-nerved, submucronulate, powdery-canescent in the middle, with a broad, membranous, glabrous, subciliate margin; achenes compressed, hairy.

HAB. Graaf Reynet, *E. Z.!* (Herb. Sond.)
8–12 inches high, woody below. Internodes on the branches an inch or more apart. Leaves 4–5 lines long, 1–1½ lines wide, spreading. Heads about ½ inch long. Invol. glaucous.

40. P. leptospermoides (DC. l. c. *363*); glabrous, much branched, slender; leaves opposite, distinct at base, *minutely woolly in the axil*, crowded, linear, subacute, narrowed to the base, thickish, round-backed, quite entire, punctate; heads terminal, solitary, sessile, oblong, 6–7-flowered; invol. glabrous, dry, scales horny, lanceolate-oblong, acute, quite entire.

HAB. Zilverfontein, *Drege!* (Herb. D., Hk., Sd.)
A slender, twiggy bush, with opposite, short twigs. Leaves 6–7 lines long, not a line wide, the base on the inside covered with short, white wool. Invol. 7–8 lines long, 2–3 lines in diameter, pale testaceous. Achenes sparingly pilose.

41. P. leptolepis (DC. l. c. *363*); twigs canescent; leaves opposite, densely crowded at the ends of the twigs, linear-oblong, obtuse, thick-ish, flattish, canescent, nerveless but slightly keeled; heads solitary, terminal, sessile, 5–7-fl.; inv. scales loosely imbricated, linear-oblong, 1-nerved, obtuse, dry, rigid, with a very narrow, membranous, torn margin; achenes villous. *P. oppositifolia, E. M.! in Hb. Drege (non L.).*

HAB. Zwart Ruggens in the Karroo, *Drege!* (Herb. Sond.)
A very small, much-branched ramulous bush; twigs short and curved. Leaves 2–3 lines long, not 1 line wide. Heads 3–5 lines long, in 4–5 rows, obconic.

42. P. glauca (Thunb.! Cap. *631*); much branched; twigs and foliage minutely and thinly canescent; leaves opposite, crowded, linear-oblong, short, thick, somewhat keeled, obtuse, entire; heads terminal, solitary, cylindr., 4–7-fl.; invol. scales broadly ovate, acute, appressed, 3–5-nerved, dorsally downy, ciliolate. *P. latisquama, DC. l. c.* 363.

HAB. Hantum, *Thunberg!* Kamiesberg and Onder Bottkeveld, *Drege!* (Herb. Th., D., Hk.)
Scrubby, depressed, rigid, and small. Twigs very short and closely leafy; old branches glabrous. Invol. either pale-horny, or the scales glaucous-green at back, with horny edges. I find 7 fl. in some heads.

43. P. empetrifolia (DC. l. c. 363); very dwarf, ramulous, glabrous, *viscidulous;* leaves opposite, densely crowded on the short twigs, spreading, linear, keeled, flattish above or channelled, obtuse, thick, entire ; heads terminal, solitary, sessile, oblong-turbinate, many-fl. ; inv.-scales oblong, appressed, obtuse, opaque, horny, with a very narrow ciliate-lacerate margin ; achenes villous.

HAB. Cape, *Drege!* no station given. (Herb. Sond.)
A very dwarf plant, 3–4 inches high. Internodes scarcely a line long. Leaves 1–1½ lines long. Heads 7 lines long. Inv. yellowish. Fl. purplish. I have only seen a fragment.

44. P. heterocarpa (DC. l. c. 364); glabrous, decumbent; leaves opposite, connate, woolly internally at base, linear-elongate, obtuse, entire, fleshy ; heads term., solitary, sessile, campanulate, many-fl. ; invol. (shorter than the uppermost leaves) glabrous, dry, scales oval-oblong, obtuse, with a wide, membranous, lacerate margin ; outer achenes hirsute, inner quite glabrous.

HAB. Olifant's R., *Drege!* (Herb. D., Hk., Sd.)
This looks like a salt-marsh plant, and is much less woody than most other species. Lower leaf-pairs distant, upper approximate. Lvs. 1–1½ inch long, 1–1½ line wide. Heads 8–9 lines long and wide. Pappus foxy.

45. P. viscosa (Thunb.! Cap. 632); divaricately much-branched ; twigs glabrous; leaves opposite, but not connate, oblong or ovate-oblong, obtuse or acute, thick, flattish, with a medial furrow beneath, either rigidly ciliate, or sparsely bristly on one or both sides, or quite glabrous; heads terminal, solitary, sessile, oblong, several-flowered ; inv.-scales viscid, appressed, horny, the outer ovate-oblong, acute, the inner oblongo-lanceolate; achenes compressed, sparsely hispid. *Pt. viscosa, litt. b , Herb. Drege!*

HAB. Hantum, *Thunberg!* Zwart Ruggens, *Drege!* (Herb. Th., Hk.)
A very rigid, scrubby bush, 6–12 inches high; branches spreading widely. Leaves 3–4 lines long, 1–1½ line wide, on the same branch varying as above noted. Invol. uncial, the inner scales much longer and narrower than the outer, very acute, almost acuminate. By its achene it is related to *P. elongata,* but has a very different involucre.

46. P. elongata (Thunb. Cap. 631); twigs scabrous; leaves opposite, sessile, crowded, linear-trigonous, keeled, acute, the keel and margin rigidly ciliate (or glabrous); heads terminal, sessile, obovoid, many-flowered ; invol. glabrous, viscid, scales broadly oval or oblong, appressed, callous tipped, with a narrow, membranous, lacerate margin. *DC. l. c.* 362.

HAB. Cape, *Thunb.!* Uitenage, *E. & Z.!* Koega and Zondag R., *Drege!* Boschman's R., *Zey.!* 2774. (Herb. Th. D., Sd., Hk.)

This has quite the aspect of *P. hirsuta* β., but *yellow* flowers, &c. Leaves 3–4 lines long, 1 line wide, somewhat channelled above, with prominent keel. Heads 1–1¼ inch long, half an inch or more across. Achenes compressed, very sparingly pilose.

47. P. stæhelinoides (DC. l. c. 364); much-branched, ramulous; twigs scabrous-echinulate; leaves opposite, sessile, linear or linear-oblong, acute, rigid, keeled or infolded, rigidly ciliate on the margin (and sometimes on keel and upper surface), glaucous; heads terminal, solitary, sessile, ovate-oblong, many-fl.; inv.-scales at first powdery scaberulous, then glabrous, appressed, broadly oblong, obtuse or mucronulate, with a narrow, membranous and torn margin; achenes quite glabrous, not glandular.

VAR. β. **stenocephala**; fl.-heads narrower, with fewer flowers. (Hb. Sond.)

HAB. Graaf Reynet, *E. Z.!* *Drege!* Karroo, and Namaqualand, *E. Z.!* (Herb. Sond., D., Hk.

A rigid, scrubby bush, with large fl.-heads. Leaves 4–5 lines long, 1 line wide. Heads an inch long, ¼ inch in diameter. Scales dry.

48. P. adenocarpa (Harv.); robust, much-branched; twigs glabrous; leaves opposite, sessile, obovate-oblong, subacute, recurved at point, spreading, thick, nerveless, rigidly ciliato-serrulate at the sub-reflexed margin, glabrous; heads terminal, solitary, oblong, many-fl.; inv.-scales viscid, appressed, broadly oblong, very obtuse, opaque, entire, the edge scarcely sub-membranous; achenes compressed, covered with raised glands, glabrous. *P. viscosa, DC. l. c.* 364.

HAB. Winterhoek, *E. Z.!* Betw. Zwarteberg and Aasvogelberg, *Drege!* Somerset, *Mrs. F. W. Barber!* (Herb. Sd., Hk., D.)

A rigid shrub, 1–2 f. high, with flexuous branches and often opposite twigs. Leaves 4–6 lines long, 2–3 lines wide, scarcely connate at base, the younger balsamic. Heads an inch long, ¼ inch wide. Pappus fulvous. Achenes obovate, with a marginal rib, glabrous, except a tuft of hairs at their base. *Drege's* specimens marked *P. viscosa* "*a*," belong to this; those marked "*b*" to *Pt. viscosa*, Th.

49. P. onobromoides (DC. l. c. 364); glabrous; leaves *alternate*, closely approximate, sessile, sub-lanceolate-linear, elongate, concave above, round-backed, *rigidly ciliate* at the margin; heads term., solitary, sessile, ovate (*large*), many-flowered; invol. scales oval-oblong, very obtuse, viscidulous, horny, appressed and quite entire; achenes glabrous, glandular.

HAB. Olifant's R., *Drege!* Namaqualand, *E. & Z.!* Dr. *Atherstone!* Elandsberg, and towards Verloren Vlay and near S. Helena Bay, *Dr. Wallich!* Vogel-fontyn, *Zey.!* 816. (Herb. D., Hk., Sd.)

2–3 feet high, the lower branches bare, rugose, the upper densely leafy. Lvs. 1½ inch long, 2–3 lines wide, acute, rather rigid, somewhat involute when dry. Heads 12–15 lines long, 10–12 l. wide. Pappus pale. "The leaves are succulent and very aromatic, used by the native Namaquas and Bastards as a perfume, mixed with fat, under the name *Buchu*. It is called *Sâb* in the Namaqua language, and is dried and collected for sale." *Dr. Atherstone.*

50. P. scabra (Harv.); *all parts closely rough with minute, harsh points;* leaves mostly alternate (the lowest sometimes opposite), linear, acute, sessile, thickish, flat, strongly-nerved, pale or glaucous; heads terminal, solitary, sessile, ovate or oblong; invol. multiseriate, loosely

imbricate, scales ovato-lanceolate, acuminate, rigid, opaque, with spreading points, quite entire, scarcely or not at all margined; (flowers and
achenes not seen).

HAB. Howhoek Pass, *Zeyher!* (Herb. Sond.)

Suffruticose or shrubby, erect, not much branched, the whole plant pale or greyish,
and rough like a file to the touch. Leaves 7–8 lines long, 1 line wide. Heads
(immature) about an inch long, and 4–5 lines wide at base, the scales with free points,
though closely imbricated. This seems to be a distinctly marked plant, but unfortunately the only specimen I have seen is immature.

Sect. 3. **Pterophorus.** *Heads* many fl. *Recept.* fimbrilliferous. *Achenes* beakless,
plano-compressed, quite glabrous. *Pappus* concrete in a ring, sub-uniseriate.

51. P. camphorata (Linn. Sp. 1176); stem *mostly* scabrous; leaves
scattered or tufted (*rarely opposite*), linear-filiform, acute, rigidly ciliate
or glabrous; heads terminal, solitary or sub-corymbose, sessile or on
nearly naked (very sparsely leafy) branches; invol. scales lanceolate,
acuminate, keeled, serrulato-ciliate, shorter or longer than the disc;
recept. fimbriate and deeply pitted; achenes flattened, glabrous.

VAR. *a.* **armata**; stem very rough with simple or jointed hairs, or rough points;
leaves rigidly ciliate and often bristly, ½–1 inch long; invol. shorter than disc. P.
camphorata, Thunb. Cap. 629, *DC. l. c.* 364. *Lam. Ill. t.* 667. *F.* 1. *Zey.!* 809 (*ex pte.*)

VAR. *β.* **stricta**; stem villous-hirsute; leaves very densely crowded, uncial, ciliate
at base; invol. scales much acuminate, often longer than the disc. *P. stricta, Ait.
DC. l. c.* 364.

VAR. *γ.* **longifolia**; branches minutely-scabrous; leaves 1–2 inches long, remotely
ciliate or glabrous, very densely crowded; heads sessile, fl. branches corymbose, leafy.
P. aspera, DC.! l. c. 364. *Zey.!* 2776.

VAR. *δ.* **aspera**; branches minutely scabrous; leaves 4–5 lines long, laxly scattered,
glabrous; fl. branches long and nearly bare of leaves; invol. scales recurved-squarrose,
P. aspera, Thunb.! Cap. 631.

VAR. *ε.* **lævigata**; branches and leaves quite glabrous and glossy; leaves mostly
opposite! glabrous, 5–6 lines long. *Zey.!* 810.

HAB. About Capetown and in the Western districts generally. β. Caledon, *Mundt!*
Zwarteberge, *Drege!* γ. Stellenbosch and Swellendam, *E. & Z.!* Paarl, *Drege!*
Banks of Erste Riv., *W.H.H.* Riv. Zonderende, *Zey.!* δ, only seen in Hb. *Thunb.!*
ε, 24-rivers, *Zey.* (Herb. Th., D., Sd., Hk.)

A shrub, 1–3 feet high, much branched and twiggy, and usually closely covered
with filiform leaves; but very variable in pubescence, and in the length and frequency
of the leaves. Involucres constantly pale straw colour or horny, the scales rigid,
with a more or less developed reddish gland along the keel. I find no clear limits
between the four first varieties; var. ε, by its opposite leaves and glabrous stems
and branches, has more character; but even on it the upper leaves are occasionally
scattered, and here and there may be seen a minute rough point.

Doubtful Species.

Pt. elegans (Sch. B. mss.); "glabrous; stem (in the only specimen
seen) one-headed, a span long, densely leafy; leaves opposite connate,
thick, linear trigonous, acute, sub-ensiform, 1–1½ inch long, sometimes
with leaf-tufts; head 25-fl.; invol. 8 lines long, ovate-oblong, scales
5-seriate, appressed, ovate-oblong, rounded, viscid at back, obscurely
greenish-brown, towards the edge paler, and lacero-ciliate." *Walp. Rep.*
2. *p.* 970.

HAB. Near Hemelendarde, Zwell.

Judging by the description this is probably only *P. tenuifolia*, DC.

Sub-Tribe 2. BACCHARIDEÆ   *Heads* (in the S. African genera) heterogamous, the female marginal flowers pluriseriate, with filiform corollas. *Anthers* without tails. (Gen. 25-28).

XXV. LEPTOTHAMNUS, DC.

Heads many-fl., heterogamous; *ray-fl.* filiform, truncate or bi-dentate, female, in one row; *disc-fl.* tubular, hermaphrodite, 5-toothed. *Recept.* flat, naked, somewhat honeycombed. *Invol.* scales imbricate, acuminate, in few rows. *Style* of ray bifid, with subulate branches; of the disc bifid, with flattened branches, and compressed, conical, pubescent stigmas. *Pappus* bristle-shaped, deciduous, of the ray uniseriate, of the disc sub-biseriate (?) or at least more copious. *Achenes* obovate-oblong, compressed, silky (of ray and disc both ovuliferous). *DC. Prod.* 5, 367.

Slender suffrutices. Leaves scattered, linear-subulate, prominently nerved below, rigidly ciliate. Branches ending in simple, 1-hoaded peduncles. Flowers yellow, turning red above. Name from λεπτος, *slender*, and θαμνος, a *shrub*.

Leaves closely-set, imbricating ... ... (1) **ciliaris.**
Leaves few and distant   ... ... (2) **rarifolius.**

1. L. ciliaris (DC. l. c.); stems branched from the base; leaves closely placed, *imbricating;* ray-flowers very slender, *minutely bidentate. Burch. Cat.* 1839 *and* 2512.

HAB. Beyond the Gariep at Klaarwater and Kosifontein, *Burchell;* Klipplaat R., *Drege;* Wolvekop, *Burke & Zey.!* Zulu-land, *Miss Owen!* (Herb. D., Hk., Sd.)
Root very thick and woody; many subsimple stems rising from the crown, 6–10 inches high, erect. Leaves 3–6 lines long, ½ line wide, strongly ciliate, very erect and closely imbricating. Peduncles 1–2 inches long, minutely bristly. Inv. scales glabrous, linear-lanceolate, acute, bistriate. I find no difference in the achenes of disc and ray, both are silky.

2. L. rarifolius (Turcz. Bull. Mosc. xxiv. 2, p. 65); stems simple below, corymbose above, angular-striate; leaves *few, distantly scattered,* much shorter than the internodes; ray-fl. *sub-ligulate, shortly bifid. Walp. An.* 5, *p.* 195.

HAB. Magalisberg, *Burke & Zeyher!* 802 (also 103). (Herb. Hk., D., Sd.)
Root woody. Stems 15–18 inches high, simple from 4–6 inches above the base, then corymbose, with several long, erect branches. Leaves 1–1½ inch apart, each leaf 4–6 lines long, close-pressed, less strongly ciliate than in *L. ciliaris.*

XXVI. CONYZA, Less.

Heads many-fl., heterogamous, all the fl. tubular; marginal fl. female, in many rows, with very slender, filiform, truncate or 2–3-toothed corollas; central few, male, tubular, 5-toothed. *Recept.* flat or convex, naked or fimbrilliferous. *Invol.* scales in many rows. *Anth.* without tails. *Achenes* flattened, mostly glabrous, narrowed at base. *Pappus* uniseriate, of slender, scarcely rough bristles. *DC. Prod.* 5, *p.* 377.

Herbaceous or suffruticose plants, almost all from the Eastern hemisphere. Stems erect, terete, branched. Leaves various, often toothed or incised. Heads pedunculate, corymbose or panicled. Flowers yellow. Pappus rufescent. Differs from *Nidorella* by its filiform (not shortly ligulate) marginal flowers. Name from κωνωψ, a *gnat:* said to drive away gnats!

Stem herbaceous, simple or branched; heads laxly corymbose or subsolitary.
  Lvs. sub-glabrous, clasping, oblong, remotely denticulate (1) **Caffra.**
  Lvs. scabrid, obovate-cuneate, coarsely toothed; pedunc.
    long... ... ... (2) **podocephala.**
  Lvs. (cauline) sessile, pinnatifid: the lobes shortly oblong :
    Pubescence scanty; stems weak and laxly leafy ... (3) **pinnatilobata.**
    Pubescence copious; stems robust ... ... (4) **obscura.**
  Lvs. pinnati-partite, hairy, lobes lanceol.-linear ... ... (5) **pinnatifida.**
  Lvs. (cauline) petioled, eared at base, ovate, toothed,
    gland-pubescent ... ... (6) **incisa.**
  Lvs. sessile, oval, mucronate, entire, hispid-villous ... (7) **pusilla.**
Stem shrubby, with rodlike branches; heads densely corymbose.
  Lvs. petiolate, lanceolate or linear-lanceolate, serrate :
    Leaves glabrous, and exuding gum copiously ... (8) **ivæfolia.**
    Lvs. scabrido-puberulous, scarcely gummy ... ... (9) **ivæfolia β.**
  Lvs. (and stems) glabrous, subsessile, oblong-lanceolate,
    with the midrib beneath prominent and pale ... ... (9) **costata.**

1. C. Caffra (DC. l. c. 381); stem herbaceous, erect, glabrous or hispidulous; leaves *half-clasping, oblong tongue-shaped,* subobtuse, mucronulate, remotely calloso-denticulate, membranaceous, glabrous on both
sides or scaberulous; pedunc. loosely corymbose, softly hairy; invol.-
scales linear-subulate, softly pubescent, equalling the disc; male fl.
about 25, hairy at the summit, fem. innumerable, filiform; recept. flat,
naked. *C. (Blumea) Natalensis, Sch. Bip.! in Wal. Rep.* 2, *p.* 971.

HAB. Near the Key R., *Drege!* Natal, *T. Williamson! Krauss!* 227. *Gerr. and
M'K.* 313. Namaqualand, *A. Wyley!* (Herb. D., Hk., Sd.)
  One and a half to 2 f. high : stems simple below, corymbose above. Leaves
1½–2½ inches long, 4–6 lines wide, the teeth minute or obsolete. Heads 5–8 in each
corymb, 5–6 lines across.

2. C. podocephala (DC. l. c. 387); stem scarcely woody at base, erect
or ascending, herbaceous, sparsely setose; leaves *obovate, cuneate at base,*
the lower ones tapering into a petiole, coarsely toothed above the middle,
sparsely scabrido-hispid, the uppermost depauperated; branches *naked
at the summit,* 1-headed, thinly hispid; invol.-scales linear-subulate,
glabrescent, equalling the disc; male fl. numerous, glabrous, fem. innumerable; recept. honeycombed and toothed.

HAB. At Zw. Stroom and Witbergen, *Drege!* Caledon R., *Burke and Zey.!* 807.
(Herb. D., Hk., Sd.)
  Many-stemmed, diffuse, 1–2 f. high, harsh to the touch. Lower leaves 2¼, upper
1–1½ inches long, 3–6 lines wide, the teeth broad and shallow, blunt or sharp.
Pedunc. 3–8 inches long, with 2–3 distant bracts.

3. C. pinnatilobata (DC. l. c. 387); stem erect or diffuse, herbaceous,
weak, sparsely setulose or glabrous; radical and lower leaves petioled,
cauline sessile, half-clasping, all scaberulous, sparsely-hispidulous or
subglabrous, *pinnatifid,* the lobes shortly oblong or cultrate, toothed
or entire, subacute, scabrous on the edge and nerves; heads loosely
corymbulose, pedunc. pubescent; invol.-scales glabrescent; male fl.
30–40, glabrous, fem. innumerable; recept. honeycombed and toothed.
Erigeron pinnatum, L. f. Thunb.! Cap. 666. *Baccharis leucanthemifolia,
Burm. C. obscura, β. calvescens, DC. l. c.* 387 *(ex pte.)*

HAB. Cape, *Thunb., Burchell, &c.* Kochmans Kloof, *Mundt!* Zwartkops R.,

Zey.! 2780.   Natal, *Miss Owen!* *Gueinz.!* 321.   *Gerr. and M'K.*, 318.   Detroit's kloof and Paarl, *Drege!* (Herb. Th., Hk., D., Sd.)
    Stems 2 f. or more in height, weak, soon broken. Lower leaves (including petiole) 3–4 inches long, upper 1–2½ inches, the incisions not reaching the midrib. Heads numerous, 4–5 lines across. Achenes compressed.

4. C. obscura (DC. l. c. 387); herbaceous, all parts clothed with spreading, white hairs ; leaves half-clasping, narrowed at base, oblong, pinnatifid, the lobes shortly oblong, toothed or entire, subacute; heads corymbose ; invol.-scales hairy ; achenes hispidulous ; recept. naked. *C. obscura, DC. Conyza,* 2779, *Zey.!*

    VAR. β. **calvescens** (DC. ex pte); lower leaves hairy, the upper and the stems scabrous. *DC. l. c.*
    HAB. Districts of Uitenhage, Albany and Kaffraria, *Drege!* *E. Z.!* Somerset, *Mrs. F. W. Barber!* (Herb. D., Hk., Sd.)
    A stronger growing plant than *C. pinnatilobata*, of which it has nearly the foliage, and from which it chiefly differs in its copious pubescence ; in this latter character it agrees with *C. pinnatifida.* Some of *Drege's* distributed specimens of var. β. belong to *C. pinnatilobata.*

5. C. pinnatifida (Less. Syn. 204); herbaceous, all parts densely clothed with spreading, white hairs; cauline leaves pinnate-parted, the lobes lanceolate-linear, obtuse, spreading, entire or few toothed ; heads 2–3 at the ends of the branches; invol.-scales hairy ; achenes downy. *DC. l. c.* 387. *Erigeron pinnatifidum, Th.! Cap.* 666.

    HAB. Cape, *Thunb.!* Stellenbosch, *Ecklon!* Estuary of the Kowie, *H. Hutton!* Betw. Omtendo and Omsamculo, *Drege!* (Herb. D., Sd., Th.)
    Stems loosely branched, all parts very hairy. Leaves 1–2½ inches long. ; their lobes 5–6 lines long, 1–1½ wide, narrowed at base, nearly horizontal.

6. C. incisa (Ait. Kew, 3, p. 184); stem herbaceous, branched, pubescent ; cauline leaves petiolate, the petiole eared (as if stipuled) at base, ovate, oblong or subcordate, coarsely toothed, on both sides glanduloso-pubescent ; heads loosely corymbose ; invol.-scales puberulous; achenes glabrous ; recept. honeycombed and toothed. *DC. l. c.* 387. *Erig. incisum, Th. Cap.* 666. *Erig. hirtum, Th.! Cap.* 665. *Zey.!* 2778.

    HAB. Distr. of Stellenbosch, Tulbagh, Clan-William and Uitenhage, *Thunb.*, *E. Z.! Drege! &c.* Natal, *Gerr. and M'K.*, 338, 339. (Herb. Th., D., Hk., Sd.)
    Diffusely much branched, glandular and somewhat viscid, shortly, but fully pubescent. The ear-shaped bases of the petioles are sometimes very small or obsolete, and sometimes there is an additional pair, higher up on the petiole. Leaves 1–2 in. long, ½–1 in. broad. Axils obtuse ; branches spreading, curved upwards.

7. C. pusilla (Houtt. Hist. Nat. x. bl. 618, t. 69, f. 1); " hispid-villous, cinereous, herbaceous, branched; upper leaves sessile, oval, mucronate, either quite entire or with 1–2 apical, point-like teeth, 3-nerved at base; pedicels 1-flowered, as long as the leaves, laxly race-mose; inv. scales linear, acuminate, puberulous, at length reflexed." *DC. l. c.* 388.

    HAB. Cape. (Unknown to us).

8. C. ivæfolia (Less. Linn. 1831, p. 138); shrubby, much branched ; branches terete, puberulous ; lvs. lanceolate, acute or acuminate, *taper-ing at base into a short petiole,* sharply serrate, 3-nerved at base, glabrous,

viscidulous; corymb. compound, many-headed; inv. scales broadly
linear, obtuse, glabrous. *DC. l. c.* 388. *Baccharis ivæfolia, Linn. Thunb.!
Cap.* 666.

VAR. β, **scabrida**; *leaves* (as well as stems) densely *scabrido-puberulous*, 3-nerved
at base, and also somewhat penninerved as well as netted-veined. *C. scabrida, DC.
l. c.* 387.

HAB. Throughout the Colony, and on to Natal. Var. β, at Uitenhage, *E. Z.!*
(Herb. D., Hk., Sd., Th.)

A much-branched under-shrub, 2–3 feet high and more, densely leafy, and exuding resinous matter. Leaves distinctly petioled, 1½–3 inches long, ½–1 inch wide,
with callous-tipped serratures. Heads very numerous, smaller than in the other
species. Var. β is by no means constant to its characters, and grows in localities
where α is also common. I cannot keep it apart specifically.

9. C. (?) costata (Harv.); quite glabrous, shrubby; branches rod-
like, rib-striate; leaves oblongo-lanceolate, acute, sessile or nearly so,
obtuse or acute at base, sharply serrate, *the midrib beneath very prominent and thick* (pale); the lateral veins netted, immersed; inflorescence
unknown.

HAB. Magalisberg, *Zeyher!* (Herb. Sond.)

This has strongly the aspect of *C. ivæfolia*, but can hardly be a variety of it. It
may not even be a *Conyza*, but as it is a remarkable-looking plant, from a famous
locality, I am unwilling to omit it altogether. It may perhaps be a *Nidorella*, near
N. conyzoides.

[*Doubtful genus—probably a* Conyza.]

WEBBIA, C. H. Schultz, Bip.

Heads many-fl., heterogamous, all the flowers tubular; those of the circumference
filiform, in many rows, slender, truncate, female; of the centre numerous, hermaphrodite, 5-toothed. *Inv.* imbricate, the scales acuminate, hairy. *Recept.* honeycombed,
the cells toothed. *Anthers* without tails. *Achenes* of the female fl. glabrous, flattened, margined; of the disc terete-compressed, margined. *Pappus* in the marginal
fl. uniseriate, of scabrous bristles; of the disc biseriate, bristle-shaped, the outer very
short, inner of 15–16 barbellate bristles. *Sch. B. in Walp. Rep.* 2, *p.* 971.

Named in honour of Phillip Barker Webb, author of a Nat. Hist. of the Canary
Islands, and other works. It is to be regretted that the name of so distinguished a
naturalist should have been bestowed on this obscure and weed-like plant. Does it
really differ from *Conyza?*

1. W. Kraussii (Sch. B. l. c.). *Conyza Kraussii, Sch. B. MSS.*

HAB. Cape, *Krauss.* (Unknown to me.)

Said to be a hairy, subcanescent suffrutex, ascending-erect, 1 foot high, densely
leafy, corymbose. Leaves alternate, linear, inch long, 1 line wide, distantly toothed;
the teeth 1–2 lines long, 1 line wide, hairy-strigose, often with leaf-tufts. Heads
corymbose, hemispherical, erect; fl. yellow.—I have abridged the author's long
description above quoted. Of the plant intended to be indicated I know nothing.

XXVII. DICHROCEPHALA, DC.

Heads many-fl., heterogamous, all the fl. tubular; marginal-fl. female,
in many rows, 3–4-toothed, slender; central by abortion male, few,
campanulate, 4-toothed. *Recept.* naked, conical. *Invol.* expanded,
nearly uniserial, the scales ovate, subequal. *Style* included. *Achenes*
compressed, without beak, the marginal without pappus, the central
each with 1–2 bristles. *DC. Prodr.* 5, *p.* 371.

Annuals, natives of the warmer parts of Asia and Africa. Leaves alternate, toothed. Heads globose, small, in racemes or panicles. Name apparently from δις, *twice;* χροω, to *colour,* and κεφαλη, a head.

1. D. latifolia (DC. l. c. 372); stem diffuse or ascending, sparingly pilose; leaves ovate, on winged petioles or lyrato-pinnatifid, with an ovate terminal lobe and one or two pair of small lateral lobes, coarsely toothed, membranous, sparsely setulose or glabrous; heads on longish pedicels, in terminal racemes. Also *D. Capensis* and *D. sonchifolia, DC.! Ethulia auriculata, Thunb.! Cap.* 623. *Centipeda Capensis, Less. Syn.* 201.

HAB. Cape, *Thunberg.* Eastern Districts, *Burchell.* Natal, *Gueinzius!* 347. *T. Cooper!* 1206. (Herb. Sd., Th., D.)

A coarse, weedy annual, 6-12 inches or in rich soil more in length. Leaves tapering at base into winged petioles, the lamina 1-2 inches long, ¾-1½ inches wide. Heads as large as peas. Female corollas very minute.

XXVIII. SPHÆRANTHUS, Vaill.

Heads crowded into a globose glomerule (or compound-head) surrounded by a common involucre; each partial head sessile in the axil of a bract, on a convex common-receptacle. *Partial* heads few-flowered, monœcious, all the fl. tubular; *female* filiform, 3-toothed, subdilated and indurated at base; *males* swollen, 5-toothed, with abortive styles. *Partial* recept. nude. *Partial* invol. of several close-pressed, imbricated scales; *genl.* various. *Anth.* without tails. *Styles* of female fl. shortly bifid; of males undivided. *Achenes* without pappus, not beaked, pilose or glabrous. *DC. Prodr. 5, p. 369.*

Tropical and subtropical herbs, Asiatic and African. Leaves decurrent, serrate or subentire. Peduncles terminal or opposite the lobes. Name, from σφαιρα, a *sphere,* and ανθος, a *flower.*

1. S. peduncularis (DC. l. c. 370); scaberulous; leaves broadly linear, acute, serrate, decurrent in long, narrow, serrate wings along the stem; peduncles as long as the leaves or longer, terete; glomerule globose, subtended by broadly ovate, mucronate, invol.-bracts.

HAB. Near Natal, *Drege! T. Williamson! Krauss!* 224. *Gerr. & M'K.,* 355 Kreili's Country, near Butterworth, *H. Bowker!* (Herb. D., Hk., Sd.)

Root perennial. Stems 1-2 f. long, erect or diffusely branched, the branches spreading. Leaves 1½-2½ inches long, 2-3 lines wide. Flowers purplish.

Sub-tribe 3. TARCHONANTHEÆ. ¦*Heads* (in the S. African genera) diœcious; flowers tubular, 5-toothed. *Anthers* tailed at base. (Gen. XXIX.—XXX.)

XXIX. BRACHYLÆNA, R. Br.

Heads many fl., diœcious. *Recept.* naked. *Invol.* scales imbricated, dry, shorter than the flowers. *Cor.* tubular, unequally 5-toothed. *Male-fl.: anth.* tailed at base, connate, exserted. *Style* filiform, simple. *Ovary* hispid, abortive, sparingly pappous. *Fem. fl.: anth.* abortive, separate. *Style* bifid, the branches short and broad. *Achenes* glandular-pubescent. *Pappus* in two rows of rough bristles. *DC. Prodr. 5, p. 430.*

Strongly scented shrubs or small trees, natives of S. Africa. Leaves alternate, coriaceous, shortly petiolate, entire or toothed, glabrate above, often tomentose beneath. Heads in branching racemes or panicles. Fl. yellow. Name from βραχυς, *short,* and κλαινα, a *cloak,* alluding to the invol. shorter than the flowers.

Adult leaves glabrous on both sides or *nearly so :*

Leaves lanceolate or linear-lanceolate	(1) **nereifolia.**
Lvs. oblong or lanceol.-oblong, cuneate at base, subacute ...	(2) **dentata.**

Adult leaves above glabrous, beneath *densely tomentose :*

Lvs. oval or oblong, subentire, mucronate; infl. axillary ...	(3) **racemosa.**
Lvs. oblong or cuneate-oblong, toothed, simple or lobulate ; male infl. panicled, the inv.-scales glabrous, in few rows	(4) **elliptica.**
Lvs. oval-obovate, obtuse, repand or denticulate; male infl. panicled, the inv.-scales woolly-edged in many rows ...	(5) **discolor.**

1. B. nereifolia (R. Br.); leaves *lanceolate or linear-lanceolate,* acute or acuminate, very entire (or *sharply few-toothed* near the apex), the young ones rusty beneath, the adult quite glabrous. *DC. l. c.* 430. *Baccharis nereifolia, Linn. Tarchonanthus lanceolatus, Th.! Cap.* 638. *T. dentatus, Eckl. Zey.!* 2784.

HAB. Western Districts generally. Swellendam, *Mundt,* and Plettenberg's Bay, *Pappe!* (Herb. Th., Hk., D., Sd.)
A large, leafy shrub, with leaves like those of *Oleander.* Leaves 4-5 inches long, ½ in. to ¾ in. wide, mostly quite entire. Young leaves and inflorescence rusty with minute powder. Invol.-scales broadly ovate, longitudinally striate.

2. B. dentata (Less.! non DC.); leaves *oblong or lanceol.-oblong,* cuneate at base, subacute, very entire or *sinuate-toothed* near the apex, the young ones rusty beneath, the adult at length nearly glabrate. *Tarch. dentatus, Thunb.! Cap.* 638. *Br. grandifolia, DC.! l. c.* 430.

HAB. Eastern Districts. Zuureberg, Alexandria, *Drege!* Vanstaadensberg, Uit., *Zey.!* 2785. (Herb. Th., D., Hk., Sd.)
Very similar to *B. nereifolia* but with much broader, proportionably shorter and blunter leaves, and (apparently) quite an eastern plant. Leaves 3-4 inches long, 1-1½ in. wide. Fl. heads similar to those of *B. nereifolia.*

3. B. racemosa (Less.! Syn. 208) ; leaves elliptical or oblong, entire or denticulate, mucronate, glabrous and netted above, tomentose beneath; racemes axillary and terminal, shorter than the leaves, few-headed ; invol. nearly glabrous. *DC. l. c.* 430. *Tarchon. racemosus, Thunb.! Cap.* 638. *Zey.!* 2787.

HAB. Distr. of Uitenhage and Albany, *Thunb., Drege, E. Z., &c.* Weenen Country, Natal, *Dr. Sutherland!* (Herb. Th., D., Hk., Sd.)
A rigid, divaricately branched shrub. Leaves 1-1½ inches long, 4-7 lines wide. Invol.-scales varying from ovate to oblong, blunt. The tomentum on the under sides of the leaves varies from whitish, to fulvous or ferruginous.

4. B. elliptica (Less. Syn. 208); leaves oblong, linear-oblong or cuneate-oblong, *mostly* cuneate at base, often shortly 3-lobed at the summit, denticulate, sinuato-dentate or subentire at the margin, glabrous and netted above, whitish tomentose beneath, undulate; racemes of both sexes in a terminal thyrsus; male invol. nearly glabrous, the scales in few rows, broadly ovate. *Tarch. ellipticus, Thunb.! Cap.* 638. *Zey.!* 2786. *Br. dentata, DC. l. c.* 430 *(non Less.!)*

VAR. β. **salicina** (DC. l. c.) ; leaves elliptic-lanceolate, acute, sharply denticulate, *evidently penninerved* beneath.

HAB. Districts of Uitenhage, Albany and Br. Kaffraria, *Thunb., E. Z.! Drege! &c.* Olifant's Hoek, *Pappe!* Natal, *Gerr. and M'Ken,* 349. β. Br. Kaffraria, *Ecklon!* Omtata and Omsamwubo, *Drege!* (Herb. Th., D., Hk., Sd.)
A much-branched shrub. Leaves very variable in shape and comparative width ;

1⅓-3½ inches long, ¼-1 inch wide, except in var. β. (whose female flowers are still unknown) tapering much to the base. The nervature is sometimes obvious, sometimes nearly hidden.

5. B. discolor (DC. 1. c. 430); leaves elliptical or elliptic-obovate, obtuse or subacute, repand-toothed or denticulate, cuneate at base, glabrous and glossy above, whitish tomentose and strongly penninerved beneath ; racemes axillary and terminal, the males forming a thyrsus or panicle ; invol. turbinate, scales in many rows, woolly at the edges, the inner scales of the female invol. lanceolate, glabrous. *B. Natalensis, Sch. Bip., in Hb. Krauss,* 243.

HAB. Uitenhage, *Burchell, E. Z.!* Somerset, *Mrs. F. W. Barber.* Natal, *Drege, Krauss, Gerr. and M'K.* 348. Delagoa Bay, *Forbes!* Karrega R., *Zey.!* 2783. (Hb. D., Hk., Sd.)

A large shrub or small tree, 15-20 f. high. Leaves 3-4 inches long, 1-1½ in. wide. Fl. dirty yellowish white. "The Dutch prepare an alkali from the ashes, for the manufacture of soap." *Mr. Gerrard.*

Imperfectly known species.

B. ?? uniflora (Harv.); female fl. heads 1-*flowered,* spicato paniculate, subsessile ; invol.-scales at first cobwebby, then glabrous, lanceolate, sub-acute, keeled, dark brown ; pappus white ; *leaves unknown.*

HAB. Natal, on Mt. ranges 2000-3500 f., from 30-60 miles from the sea, *Dr. Sutherland!* (Herb. Hook.)

A scrap of female inflorescence, without leaves, is all that we yet know of this plant, which, whether a *Brachylæna* or not, seems to have marked characters.

Excluded from the genus.

B. trinervia (Sond.! in Linn. 23, p. 63).

XXX. TARCHONANTHUS, L.

Heads diœcious, few or several-fl. (rarely one-fl.). *Invol.* of the male fl. of 5 scales, connate to their middle; of the female, of many separate scales, in a double row. *Recept.* hairy. *Cor.* tubular-campanulate, 5-toothed, externally hairy and viscid, glabrous within. *Male fl.* : *anthers* exserted, connate, with long, setose tails, and glabrous filaments. *Ovary* abortive. *Nectary* (within the corolla) of large size, callous, hollow at top, simulating an ovary; *style* filiform, scarcely 2-lobed at the point. *Female: stam.* abortive. *Nect.* none. *Style* exserted, bifid ; the lobes revolute. *Achene* very woolly, without pappus. *DC. Prodr. 5, p.* 431.

Strongly resin-scented S. African shrubs. Leaves alternate, petioled, coriaceous, netted-veined and at length glabrous above, tomentose beneath. Heads small, in terminal and axillary panicles, rarely subsolitary and axillary. Name from ταρχος, pl. ταρχεα, *funeral-rites,* and ανθος, a *flower.*

Sub. gen. 1. EUTARCHONANTHUS. Fem. fl.-heads 3-5-flowered, with broadly-ovate inv.-scales. Achenes very woolly.—*Leaves undivided.*—(Sp. 1-2).

Leaves lanceolate-oblong, tapering at base, 3-5 inches long; fl. panicled ... ... (1) **camphoratus**.
Lvs. 1-1½ inch long, oblong or obovate; fl.-heads subsolitary or racemose ... . (2) **minor**.

Sub. gen. 2. HENOTOGYNA. Fem. fl.-heads one-flowered, with linear inv.-scales. Achenes sparingly woolly.—*Leaves rugose, 3-lobed at the summit.*

     LVS. 5–7 inches long, trifid ; panicles axillary, shorter
        than leaves   ...   .   (3) **trilobus.**

Sub. gen. 1. EUTARCHONANTHUS. (Sp. 1–2).

1. T. camphoratus (Linn. Sp. 1179); leaves lanceolate-oblong (or obovate), acute at base, subacute or obtuse at apex, quite entire or denticulate, the adult glabrous and tessellato-reticulate above, tomentose and penninerved beneath ; panicle terminal, many-headed. *DC. l. c.* 431. *Th.! Cap.* 637. *Lam. Ill. t.* 671.

VAR. β. **Litakunensis**; leaves *pitted*-reticulate above. *T. Litakunensis, DC. l. c. Zey.!* 829.

HAB. Throughout the Colony, common. β. Litaku, *Burch.* Rhinoster Kopf, *Burke!* Omsamculo, *Drege!* Hopetown, *A. Wyley!* (Herb. Th., D., Hk., Sd.)
A large shrub, with a strong, balsamic odour. Leaves 3–5 inches long, ½–1½ in. wide, the young ones densely velvetty above, the old glabrate and finely reticulate, the reticulations either flat or hollow in the middle, as if pitted. I cannot distinguish *T. Litakunensis,* so far as Drege's specimens are authority. The pitted and non-pitted reticulation varies in the same specimen, and seems to arise either from some condition of the leaf, or irregularity in drying. The leaves also vary much in size.

2. T. minor (Less. Syn. p. 208); leaves (small) obovate, oblong, or linear-oblong, obtuse, entire, the adult glabrous and tessellated above, thickly tomentose, and indistinctly nerved or nerveless beneath; heads axillary and terminal, either solitary or shortly racemose, the racemes few-headed.—*T. obovatus, DC.* and *angustissimus, DC. l. c.* 431.

HAB. Beyond the Gariep, *Burchell.* Zuureberg and Nieuweveld, *Drege!* Albany and Tambukiland, *Ecklon!* Zululand, *Miss Owen!* Cape, *Bowie!* Basutuland, *T. Cooper!* 708. (Herb. Hk., Sd., D.)
Smaller in all parts than *T. camphoratus,* with fewer fl.-heads and less branching inflorescence ; but possibly merely a dwarf variety. Leaves commonly 1–1½ inch long, 3–5 lines wide. The fl.-heads are similar to those of *T. camphoratus,* and are sometimes sessile, or subsessile, sometimes distinctly pedicellate, the pedicel 2–4 lines long. I find the leaves to vary in form on the same specimen.

Sub. gen. 2. HENOTOGYNA, DC. (Sp. 3).

3. T. trilobus (DC. l. c. 432); leaves cuneate, shortly 3-lobed at the abrupt summit, sinuato-dentate at the sides, becoming glabrous and corrugated above, penninerved, reticulated and hirsuto-tomentose beneath ; panicles axillary, shorter than the leaves ; scales of the fem. invol. linear, elongate.

HAB. Omtendo, *Drege!* Port Natal, *Gueinzius!* (Herb. D., Hk., Sd.,)
Branches bare of leaves below, closely leafy at the summit. Leaves petioled, sub-horizontal, 5–7 inches long, 1½–2¼ inches wide at the trifid point, green and wrinkled above, whitish hairy beneath. Racemes slightly branching, 2–4 inches long, woolly. Young achenes much less woolly than in other species. Very distinct in habit from the other species.

Subtribe 4. INULEÆ. *Heads* never diœcious ; if heterogamous, the marginal fl. female, with ligulate (sometimes very short) corollas. *Anthers* tailed. (Gen. 31–38).

XXXI. DENEKIA, Thunb.

Heads many-fl., heterogamous ; *marginal fl.* in several rows, female, equalling the disc, with bilabiate corollas, the lips oval, equal, entire ; *disc-fl.* funnel-shaped, 5-fid, hermaphrodite (but sterile). *Recept.* naked, flat. *Invol.* campanulate, biseriate. *Anthers* tailed. *Style* of disc-fl. bifid, with flattened branches. *Achene* oblong, sessile, beakless. *Pappus* of the ray-fl. none ; of the disc of a single, slender scale, palmato-fimbriate at the summit, and of a few small, simple scales. *DC. Prodr.* 5, *p.* 462.

S. African herbs. Leaves alternate, semi-amplexicaul, oblong or lanceolate, denticulate or undulate. Heads small, densely corymbulose or subfasciculate, the corymb in a panicle. Flowers white.—Name in honour of some botanist ?

1. D. Capensis (Thunb. Prodr. 177) ; leaves green above, *canescent-tomentose beneath,* subentire or here-and-there toothed. *Thunb. Fl. Cap.* p. 665. *DC. l. c.* 462. *Selloa Capensis, Spr. Syst.* 3, *p.* 496.

HAB. Wet spots in the Langekloof, *Thunb. !* Eastern Districts, Kaffraria, and Natal, *E. Z. ! Drege! Bowker, Sanderson, Gerr. & M'K.* 268. *Zey./* 901, 902. (Herb. Th., D., Hk., Sd.)
Stems from a perennial root, ascendent-erect, 6–18 inches high, leafy. Radical and lower leaves 5–6 inches long, lanceolate-oblong, mucronate, tapering to the base and subpetiolate ; cauline amplexicaul, oblong or ovate, sometimes subruncinate, the young ones cobwebby above, all white beneath. Corymbs pedunculate, very dense. Heads 1½ lines across. Invol. and pedicels glandular.

2. D. glabrata (DC. l. c.) ; leaves *green on both sides,* either minutely glandular-pubescent or glabrate, subentire or sharply toothed.
HAB. Klipplaat R., Zuureberge, &c , *Drege!* Uitenhage and Tambukiland, *E. Z.!* (Herb. D., Hk., Sd.)
Similar to the preceding, except in the pubescence. Perhaps a green variety ?

XXXII. **BLUMEA**, DC.

Heads many-fl., heterogamous ; *marginal fl.* in many rows, female, very slender-filiform, truncate or 2–3-toothed ; *disc-fl.* few (5–25), hermaphrodite, cylindrical, 5-toothed, scarcely dilated in the throat. *Recept.* flat, mostly naked. *Invol.* in few rows, subimbricate, the scales linear, acuminate. *Anthers* with slender tails. *Achenes* terete. *Pappus* in one row, of many, roughish bristles. *DC. Prodr.* 5, *p.* 432.

Herbs or half-shrubs, chiefly from tropical Asia, a few African. Lvs. alternate, mostly villous. Heads panicled or laxly corymbose. Fl. yellow or purplish.—These have the aspect of *Conyza,* from which they differ by the tailed anthers and terete achenes. The name is in honour of C. L. Blume, a celebrated Dutch botanist, and author of many works on the Indian Flora.

Leaves sessile (not decurrent), obovate, tapering much at base	(1) lacera.
Leaves decurrent, as narrow wings to the stem :	
Glandularly pubescent, scaberulous :	
Leaves elliptic-oblong ; pedicels equalling the invol., cernuous … … … … … … … … …	(2) alata.
Leaves acuminate ; pedicels much longer than the invol., slender … … … … … … … …	(3) pterodonta.
Softly and thickly silky-canescent ; leaves broadly-linear, entire … … … … … … … … …	(4) Gariepina.

1. B. lacera (DC. l. c. p. 436) ; stem herbaceous, erect, softly villous, lowest leaves petiolate ; cauline leaves obovate, sessile, but *much attenuated to the base,* in the upper-half denticulate, pubescent above, villous beneath, the younger ones on both sides silky; corymbs few-headed, subterminal, or from the axils of the upper leaves, forming an interrupted thyrsus; pedunc. and invol. densely silky; invol. scales narrow-linear, equalling or overtopping the disc; corolla of the disc-fl. externally glandularly tuberculated. *B. Dregeana, DC. l. c. Also B. Wightiana, DC. p.* 435. *B. Dregeanoides, Sch. Bip.*

HAB. Omtata and Omsamwubo, *Drege!* Magalisberg, *Burke & Zeyher!* 806; Umgeni, Natal, *Gerr. & M'K.* 301; Natal, *J. Sand.*, 369, *Krauss!* 216. (Hb. D., Hk., Sd.)
Stem 1–2 feet high, subsimple. Radical leaves and lower stem-leaves tapering into a slender petiole, 2–3 inches long, 1–1½ wide, exactly obovate; upper leaves more sessile, oblong or lanceolate, sometimes acute, 1–2 inches long. Invol. scales often tipped with crimson. Style of male fl. bifid.—I gladly follow *Bentham* (Hong Kong Fl. p. 178–179) in uniting the *Bl. Dregeana* of S. Africa with the common Indian weeds above quoted.

2. B. alata (DC. l. c. p. 448); stem herbaceous, erect, branched, clothed, as well as the leaves, with a short, glandular, rufous pubescence; leaves elliptic-oblong, denticulate, *decurrent along the stem* as narrow, entire or distantly-toothed wings; peduncles short, axillary, one or few-headed, disposed in a panicle or raceme; heads cernuous; inv. scales in several rows, the outer lanceolate, leafy, squarrose-reflexed, pubescent; inner linear-acuminate, equalling the flowers; cor. of disc-fl. glabrous. *Conyza alata, Roxb. Erigeron alatum, Don. Prodr.* 171.

HAB. Natal, *Drege! J. Sanderson,* 289, *Gerr. & M'K.* 300; Magalisberg, *Burke & Zeyher!* 912. (Herb. D., Hk., Sd.)
2–3 feet high, more or less densely glandular-pubescent, rusty-coloured. Leaves 2–3 inches long, ½–¾ in. wide, veiny, rather harsh to the touch. Heads in a simple or branched leafy raceme, ending the branches; invol. scales remarkably squarrose, with similar rough gland-pubescence as the leaves.—Sent also by *Mr. Mann* from the Cameroon Mts.—I find the wings frequently toothed on the Cape specimens.

3. B. pterodonta (DC. l. c. 448); stem herbaceous, erect, branched, clothed, as well as the leaves, with short, glandular pubescence; leaves ovate-oblong, *acuminate,* denticulate, decurrent along the stem as broadish, distantly-toothed wings; heads *on long, slender, naked pedicels,* disposed in a loose panicle; invol. scales pluriseriate, the outer squarrose and minutely glandular, the inner erect, linear-acuminate, scarious; cor. of disc-fl. glabrous.

HAB. Magalisberg, *Burke & Zey.!* 912. (Herb. Hk., D.)
2–3 feet high, loosely branched. Lower leaves not seen; medial oblong, upper ovate, all sharply taper-pointed. Pedicels 1–1½ inch long, thread-like, straight or curved. Invol. pale, all the scales very narrow.—Differs from *B. alata* in being green, not rusty; with shorter and less copious pubescence; broader wings to stem; acuminate leaves; and long, slender pedicels.

4. B. Gariepina (DC. l. c. 448); suffruticulose, much branched, *all parts densely silky-canescent;* leaves broadly-linear, entire, subacute, decurrent along the stem as narrow, entire wings; heads shortly pedicellate, corymbulose at the ends of short branchlets, forming a leafy panicle; invol. scales pluriseriate, linear-subulate, erect, imbricated, silky, equalling the disc; cor. of disc fl. setulose on the teeth.

HAB. Gariep, *Drege!* Namaqualand, *A. Wyley!* (Herb. D., Hk., Sd.)
Readily known by its linear leaves and soft and copious pubescence. Lvs. 1–1½ inches long, 2–4 lines wide. Heads like those of *Erigeron Canadensis.*

[*Species ? ? Unknown to us.*]

B. (Conyza) Natalensis (Sch. Bip. in Walp. Rep. 2, p. 971); an annual, with oblong-linear, toothed, sessile, half ear-clasping, glabrescent leaves.

Hab. Natal, *Krauss.*

B. (Conyza or Pluchea) Kraussii (Sch. Bip. l. c. p. 972); shrubby; cauline leaves 1–2 inches long, 2–5 lines wide, sessile, linear-lanceolate, acute, toothed or entire, glabrous.

Hab. Natal, *Krauss.*

XXIII. INULA, Gaertn.

Heads many-fl., heterogamous ; *ray-fl.* in one row, female (or by abortion sterile), mostly ligulate, rarely subtubular, 5-fid ; *disc-fl.* hermaphrodite, tubular, 5-toothed. *Invol.* imbricate, in several rows. *Recept.* flat or flattish, naked. *Anthers* tailed at base. *Achene* beakless, subterete (very rarely 4-angled). *Pappus* in one row, of roughish bristles. *DC. Prodr. 5, p. 463.*

A large genus, chiefly European and Asiatic, herbaceous, mostly perennial. Cauline leaves alternate, often amplexicaul, undivided, entire or serrated. Heads corymbose or solitary, and peduncled, yellow. Name of uncertain derivation.

1. I. Africana (Lam. dict. 3, p. 256); "stem herbaceous, slender, scarcely hairy, branched ; leaves amplexicaul, subscabrid, gland-dotted beneath, the lower oblong, acute, serrated, the uppermost lanceolate, shorter, nearly quite entire ; branches 1-headed, some leafless, others sparingly leafy ; inv.-scales linear-setaceous." *DC. l. c.* 469. *I. punctata, Less. Syn.* 193.

Hab. Cape, *Sonnerat in Hb. Lam.* (fide DC. l. c.)

XXXIV. PULICARIA, Cass.

Heads many-fl., heterogamous ; *ray-fl.* female, in one row, nearly always ligulate ; *disc-fl.* tubular, 5-toothed, hermaphrodite. *Recept.* naked, areolate, flattish. *Invol.* laxly imbricate in few rows, scales linear. *Achen.* beakless, downy, terete (not compressed). *Pappus* in two rows, the outer very short, coronæform, toothed; the inner of 10–20 rough bristles. *DC. Prodr. 5, p. 477.*

Herbaceous plants, chiefly European, a few African, growing in moist places, villous, erect, branched. Cauline leaves cordate-sagittate at base, entire, or toothed, lanceolate. Pedunc. 1-headed. Flowers yellow. Name from *pulex*, a flea : the popular name is "*Flea-bane.*"

1. P. Capensis (DC. l. c. 479); stem erect, simple below, corymbose above, silky-villous ; leaves (especially the younger) on both sides densely silky-villous, cordate-amplexicaul at base, oblongo-lanceolate, callous-mucronate and calloso-serrulate ; invol.-scales linear-acuminate, loosely imbricate, very villous ; ray-fl. narrow-ligulate, slightly longer than the involucre, 3-toothed; achenes thinly pubescent; outer pappus 10-toothed, inner of about 16 setæ. *Erigeron scabrum, Th.! Cap.* 665. *Inula Capensis, Spr.! Herb. Un. Itin.* 431.

VAR. β. **erigeroides**; stem and foliage thinly pubescent, or glabrescent; invol. scales scabrous-pubescent; achenes puberulous. *P. erigeroides, DC. l. c. p.* 480.

HAB. Both varieties throughout the Colony, and on to Port Natal, *Mundt! E. Z.! Drege! Gerr. and M'K.! 275, Pappe!* &c. (Herb. Th., D., Hk., Sd.)

One to 2 ft. high, robust, the villosity sometimes much less copious. Leaves 1–2½ inches long, ¼–½ inch wide. Corymbs few or many-headed. Except in being more slender and less hairy β. does not differ.

XXXV. PEGOLETTIA, Cass.

Heads many-fl., homogamous. *Recept.* naked, dotted, flat. *Invol.* imbricate in 2–3 rows, shorter than the flowers; scales acute. *Corolla* tubular 5-fid, regular (or sub-ringent). *Anthers* 2-tailed at base. *Achenes* cylindrical, rib-striate, beakless. *Pappus* in two or three rows, the inner of long, rigid, straight, serrato-ciliate (rarely plumose) bristles; the outer much shorter, either of flat, entire or lacerate, unequal scales, or of bristles nearly similar to those of the inner pappus. *DC. Prod. 5, p.* 481. *Pegolettia and Carphopappus, Sch. Bip.*

Small, rather rigid African half-shrubs or herbs, with something the aspect of *Pteroniæ.* Leaves alternate, dotted, elliptical or linear. Heads solitary at the end of the branches, yellow. In the few species of which this genus consists the pappus varies so much, from species to species, that if strictly insisted on as of generic value, the genus might be broken up into almost as many genera as there are species. In *P. Senegalensis* the inner bristles are quite plumose; in all the S. African species on the contrary, they are simply ciliate-barbellate: but the outer pappus presents even wider discrepancies. Nevertheless, I do not think it advisable to follow Schultz Bip. in breaking up the genus. The name is in honour of some botanist?

Leaves toothed:
  Herbaceous, viscid-pubescent: lvs. sessile, oblong-lan-
    ceolate, acuminate ... ... (1) **oxyodonta.**
  Shrubby, nearly glabrous; lvs. petioled, oval or oblong,
    subacute, thick ... ... (5) **baccharidifolia.**
Leaves quite entire:
  Shrubby, much-branched; lvs. petioled, oval or obovate,
    subacute ... ... (4) **polygalifolia.**
  Suffruticose; lvs. oblong, tapering to both ends, glan-
    dularly pubescent ... ... (2) **acuminata.**
  Subherbaceous, scabrous; lvs. linear-lanceolate, hispid;
    achenes pubescent ... ... (3) **lanceolata.**
  Herbaceous, slender, hispid; lvs. elliptic-oblong, taper-
    ing to both ends ... ... (6) **tenella.**

1. P. oxyodonta (DC. l. c. 481); herbaceous, stem glabrous below, glandular and viscidulous above; leaves sessile, oblongo-lanceolate, acuminate, sub-pungent, sharply few-toothed at the sides, glanduloso-pubescent and viscid; invol.-scales pluriseriate, acuminate, the inner broader and ciliate, the outer glandular-pubescent; heads many-fl.; pappus biseriate, the outer of many short, flat, obtuse, unequal scales.

HAB. Little Namaqualand, near the Gariep, *Drege! A. Wyley!* (Hb. D., Hk., Sd.)
Stems several from a woody base, 6–9 inches high, simple or slightly branched, erect. Leaves 1–2 inches long, 4–7 lines wide, the teeth subulate, projecting. Heads 30–40-fl. Achenes sprinkled with gland-tipped hairs.

2. P. acuminata (DC. l. c.); "suffruticose, glandular toward the apex; leaves oblong, entire, acuminate at both ends; inv.-scales much acuminate, loosely imbricate; pappus biserial, the outer shorter." *DC.*

HAB. Beyond the Gariep, *Burchell*. Cat. Geogr. 2396 (fide DC.)
Twigs and lvs. glandularly puberulous. Achenes glabrate, striate, subterete. *DC.*

3. P. lanceolata (Harv.); stems subherbaceous, branched near the
base ; branches elongate, simple, angular, rigidly pubescent ; leaves
sessile, linear-lanceolate, acute, quite entire, callous pointed, dotted,
sparsely hispid ; inv.-scales pluriseriate, acuminate, roughly pubescent;
heads many-fl. ; pappus pluriseriate, of many rough, unequal, long
bristles ; achenes rib-striate, pubescent.

HAB. Dry plains in Zululand, *W. T. Gerrard*, 1044. (Herb. D.)
Stems 1 foot or more high, subsimple or branched from below ; branches rather
closely leafy below, sparsely leafy or subnude above. Leaves 1–1½ inches long, 1–2
lines wide. Pappus nearly like that of a *Pteronia*. Anthers with very long tails,
bearded at the extremity.

4. P. polygalæfolia (Less. Syn. 200); shrubby, much-branched,
minutely viscoso-scaberulous, otherwise glabrous ; branches tortuous
or divaricate ; leaves tapering at base into a short petiole, oval or
obovate, subacute, fleshy, entire ; invol. scales biseriate, lanceolate,
acuminate ; pappus biseriate, the outer of many, narrow-subulate,
acuminate, serrated scales. *DC. l. c. Eupatorium retrofractum, Thunb.!*
Cap. 628. *Vernonia polygalæfolia, Licht.*

HAB. Cape, *Thunb.*; Kl. Fische R. and Zwartberg, *Drege! E. & Z.!* Cradock,
Burke! Springbokkeel, *Zey.!* 813. (Herb. Th., D., Hk., Sd.)
Woody, intricately ramulous, about 1 foot high. Leaves 5–8 lines long, 2–3 lines
wide, gland-dotted. Heads few-flowered, subsessile.

5. P. baccharidifolia (Less. Syn. 200); shrubby, minutely scaberulous or glabrous; branches curved ; leaves shortly petioled, oval or
oblong, subacute, fleshy, coarsely few-toothed, viscidulous; invol. scales
3-seriate, oblong-lanceolate, acute, pubescent ; pappus pluriseriate, all
of barbellate bristles, the inner long, the outer gradually shorter. *Car-*
phopappus baccharidifolius, Sch. Bip. in Walp. Rep. 2, *p.* 973. *Pteronia*
dentata, Spr.

HAB. Cape, *Krebs.*; Fisch R. and Zuureberg, *Drege!* Somerset, *Dr. Atherstone!*
Matavisberg, Albert, *T. Cooper!* 568. (Herb. D., Hk., Sd.)
Very like the preceding except in the rather larger and toothed leaves; the broader
and less acuminate invol. scales and pappus. It is perhaps less strongly woody and
more diffusely branched. Leaves 4–10 lines long, 2–4 lines wide.

6. P.? tenella (DC. l. c. 482); "herbaceous, erect, slender, sparingly
hispid; lvs. elliptic-oblong, tapering to both ends, very entire; branches
ending in long, leafless, pubescent, one-headed peduncles ; invol. scales
biseriate, acuminate ; pappus biseriate, the outer very short." *DC.*

HAB. Between Omtata and Omsamwubo, *Drege*, 5159 (fide DC.)
"Flowers unknown ; the genus therefore doubtful. Recept. naked, punctate.
Achenes terete, substriate. Pappus-bristles thick, white." DC.

XXXVI. CYPSELODONTIA, DC.

Heads many-fl.; the *ray-fl.* in one row, ligulate, neuter; *disc-fl.* deeply
5-fid, hermaphrodite. *Invol.* imbricate in several rows, the scales linear,
the outer squarrose. *Recept.* honeycombed, the cells toothed at edge.

Stam. filaments short, hairy; anthers prolonged into a coriaceous appendix, 2-tailed at base. *Style* of the disc-fl. with obtuse, nearly cohering lobes, downy at back near the summit. *Achenes* of ray abortive, glabrous, with a tuft of hairs round the base, of the disc obovate, very villous with rufous hairs. *Pappus* with thick, white, toothed bristles, few in the ray-fl., in many series in the disc-fl. *DC. Prod.* 7, *Suppl. p.* 286.

" A much-branched suffrutex; the branches striate, shortly nude and one-headed at the summit. Leaves alternate, obovate-oblong, subobtuse, cuneate at base, mucronate, very entire, canescent-villous beneath, glabrous and one-nerved above Heads solitary. Corolla yellow. Apparently related to *Geigeria* and *Pegolettia.* The name is compounded of κυψελη, a *bee-hive,* and οδον, a *tooth;* in allusion to the toothed margins of the cells of the honeycombed receptacle." *DC. l. c.*

C. Eckloniana (DC. l. c.)

HAB. District of Uitenhage, *Ecklon* (fide DC.)

XXXVII. MINUROTHAMNUS, DC.

Heads many-fl., heterogamous; *ray-fl.* ligulate, female, in one row; *disc-fl.* tubular, 5-toothed, hermaphrodite. *Inv.* scales biseriate, the outer shorter, acuminate, inner longer, obtuse, membrane-edged. *Recept.* . . . . *Stam.* filaments glabrous; anthers shortly tailed. *Style* bifid, with spreading branches. *Achenes* obovate ? clothed with very abundant, white woolly hairs. *Pappus* in two rows, of about 20 straight, yellow, toothed bristles, the outer 10 rather shorter, the inner more rigid. *DC. Prodr.* 7, *Suppl. p.* 286.

" A suffrutex, with the aspect of *Cypselodontia;* branches terete, naked and one-headed at the summit. Leaves alternate, linear, very entire, glabrous above, appressedly tomentose beneath. Ends of the peduncular branches and involucres clothed with very short, rufous toment. Corolla yellow. The name is compounded of μινυρος, *slender,* and θαμνος, a shrub." *DC. l. c.* Said to be related to *Pegolettia.*

M. phagnaloides (DC. l. c.)

HAB. District of Caledon, *Ecklon* (fide DC.)

XXXVIII. GEIGERIA, Griesselich.

Heads many-fl., radiate; *ray-fl.* uniseriate, female, ligulate; *disc-fl.* tubular, perfect, 5-lobed, the lobes lanceolate, erect, externally gland-scabrous. *Recept.* convex, piloso- or paleaceo-fimbrilliferous. *Inv.* closely imbricate, the outer scales mostly leaf-tipped, the inner horny, acute or acuminate. *Anthers* tailed. *Style*-branches of the disc-fl. linear-lanceolate, flattened. *Achenes* subtrigonous, hairy or pilose, beakless, narrowed to the base. *Pappus* biseriate, of 10–16 scales, either all aristate, or the outer ones blunt; or all of them toothed. *DC. Prodr.* 5, *p.* 482. *Geigera, Less. Syn.* 199.

Suffrutices, glabrous or scabrous, with rigid, leafy stems, or nearly stemless. Leaves alternate, linear or linear-subulate, 1-nerved, quite entire or denticulate, gland-dotted on each side of the rib. Heads sessile, either terminal or in the forks, or secundly pseudo-lateral on lengthening branches. Flowers yellow. Name in honor of Prof. Geiger, of Heidelberg. These plants have many characters of *Polychætia,* Less., with which genus several of them have been confounded; but they differ not merely by the double pappus, but by the asteroid style. The pappus in all is strongly hygrometric and expands almost immediately in warm water.

Stemless or with *very short, tufted*, closely branched stems :
- Stemless ; all the pappus-scales bristle-pointed ... ...   (1) **Africana.**
- Short-stemmed ; the *outer* pappus-scales pointless   ...   (2) **passerinoides.**

Stem elongate, simple or branched, sparsely leafy :
- Leaves 2–4 inches long, grass-like :
  - Stem depressed, trichotomous, laxly leafy ... ...   (3) **protensa.**
  - Stem erect, simple or branched at top only, closely leafy :
    - Glabrous ; leaves *flat*, entire, narrow   ... ...   (4) **Burkei.**
    - Glabrous ; leaves *linear-involute*, entire ... ...   (5) **Zeyheri.**
    - Hispid with swollen hairs ; lvs. flat, broadly lin.   (6) **aspera.**
  - Leaves less than 1 in. long, blunt; stem much branched   (7) **brevifolia.**
  - Leaves uncial, flat, midribbed, veiny, pungent-mucronate   (8) **pectidea.**

1. G. Africana (Gr. in Linn. V. p. 411); stemless or forked; leaves crowded at the crown, linear-elongate, acute, tapering much at base, entire or denticulate, the younger ones setoso-ciliate ; heads sessile, seemingly tufted (from the non-development of internodes); inner inv.-scales acuminate, ciliate ; *all the scales of the pappus bristle-pointed. DC. l. c.* 482. *Geigera Africana, Less. Syn.* 200. *Zeyheria acaulis, Spr. Dizonium longifolium, Willd. Herb.*

HAB. Karroo, near Beaufort, *Zeyher!* (Herb. Sond.)
A minute, scarcely branched plant, 1–3 inches diameter ; the leaves 1–2 inches long, 1 line wide. Several crowded, but not truly fascicled heads in the forks, or on the sides of scarcely developed branches. Recept. densely clothed with setaceous fimbrils. Disc-fl. with narrow-lanceolate, externally scabrous, erect lobes, and strong marginal nerves. Pappus of 8–16 scales, all of them bristle-pointed. Dwarfer than *G. passerinoides*, which it much resembles, but has a different pappus.

2. G. passerinoides (Harv.) ; stem shrubby, very dwarf, multifid at the crown, the branches very short, tufted ; leaves from a dilated, amplexicaul base linear, very long, one-nerved beneath, punctate, glabrous or sparsely hispid ; heads terminal or lateral, sessile, ovate, subtended by leaves; inner inv.-scales acuminate, ciliate; outer pappus of oblong, blunt, inner of bristle-pointed scales. *Relhania passerinoides, L'Her. Sert. Angl. p.* 22. *Polychætia passerinoides, DC. Prodr. 6, p.* 285.

HAB. Cape, *L'Heretier.* Nieuwevelde, *Drege!* Brack R. and Orange R., *Burke & Zeyher.* Zululand, *Gerrard,* 1024. (Herb. Hk., D., Sd.)
Root thick and woody. Stem shortly multifid, the crowded branches 1–2 inches long. Leaves 1–2–2½ inches long, 1 line wide, flat or revolute, thickish, subacute. Rays taper-pointed, long. Recept. bluntly conical, thickly clothed with bristle-shaped fimbrils. Outer pappus of 5 or 6 oblong, blunt, nerveless scales; inner of as many nerved, aristate scales. Style-branches in the disc-fl. deeply divided, the arms exserted, linear, flattened.

3. G. protensa (Harv.); suffruticose, glabrous ; stem depressed, subtrichotomous ; leaves linear, tapering to each end ; heads axillary, sessile, solitary ; recept. densely clothed with long, linear-subulate, flat fimbrils ; pappus of 10–12 scales, the 5–6 inner aristate, the outer obtuse or mucronate.

HAB. Zululand, *Miss Owen!* Interior of the Colony, Herb. *Hooker!* (Hb. D., Hk.)
Stem woody at base, rigid, flexuous, forked or zig-zag, almost prostrate, a foot or more in length. Leaves 3–4 inches long, 1–2 lines wide in the middle, thickish in substance, with subinvolute margins, one-nerved, dotted, glabrous, the lower scattered, the upper in tufts at the ends of the branches. Heads sessile in the forks of the stem, a few of the outer invol.-scales with leafy points ; the next broadly

ovate, acuminate ; the inner oblong, mucronate ; all rigid and glabrous. Recept. conical; its fimbriæ longer than the achenes, almost equalling the invol., resembling narrow paleæ, many surrounding each flower. Disc-fl. much longer than the invol., deeply 5-lobed. Achenes obconical, hairy. Branches of the style linear-lanceolate, flattened. Anth. with long tails.

4. G. Burkei (Harv.) ; stem erect or diffuse, shrubby, branched at the base ; branches erect, simple or corymbose at the summit, rufo-puberulous, subscabrid, closely leafy throughout ; leaves alternate, sessile, linear, elongate, subobtuse, *flat*, one-nerved, impress-dotted, scaberulous, glabrous ; heads terminal, solitary, sessile, subtended by many leaves ; fimbrils setaceous ; outer inv.-scales with short, leafy-tips, inner much acuminate, subulate, pungent ; outer pappus of oblong, blunt, inner of bristle-pointed scales.

VAR. β. diffusa; dwarf, stem and branches diffuse; younger heads pseudo-lateral. (Hb. Hk.)

HAB. Schon-Stroeme and Mooje Rivers, *Burke & Zeyher! Zey.! 915*, ex pte., 916. Zululand, *Gerr. & M'K.,* 1024, β. Doorn Kop, *Burke & Zey.!* (Hb. Hk., D., Sd.)

Stems 6–18 inches high, divided near the base ; branches at first quite simple, one-headed, afterwards by the sprouting of twigs from the axils of the uppermost leaves, close under the heads, becoming corymbose or fasciculately branched at the summit. Leaves close or subdistant, 1½–2½ inches long, a line wide. Heads ¾ inch long, many-fl., ½ inch across, ovoid. Inner inv.-scales much acuminate. Fimbrils setaceous, longer than the hairy achenes. Pappus of 5–6 oblong, short, blunt scales, alternating with as many bristle-pointed ones. Cor.-lobes subulate, acuminate, with strong marginal ribs. Style-branches long, linear, flattened.

5. G. Zeyheri (Harv.) ; stem shrubby, branched at base ; branches erect, simple or cymose at the summit, rufo-tomentulose, leafy throughout ; leaves alternate, sessile, *linear-involute*, nerve-keeled, impress-dotted, glabrous ; heads terminal, solitary, sessile, subtended by many leaves ; fimbrils setaceous ; outer inv.-scales short and broad, with short, leafy-tips, inner lanceolate, acute, spreading, subpungent ; outer pappus of oblong, blunt, inner of bristle-pointed scales.

HAB. Magalisberg, *Burke & Zeyher! Zey.! 915*, ex pte. (Herb. Hk., D., Sd.)

Very like *G. Burkei*, but the leaves are involute ; the fl.-heads smaller ; and the inner inv.-scales broader and less acuminate. Stems 12–18 inches high ; leaves 2–3–4 inches long, not ½ line wide. Young stems slightly tomentulose.

6. G. aspera (Harv.); stem erect or spreading, branched, leafy, rough with rigid, swollen hairs ; leaves alternate, sessile, broadly-linear, elongate, subacute, flat, calloso-denticulate or entire, sprinkled above with rigid, swollen hairs, conspicuously gland-dotted ; heads terminal or pseudo-lateral, sessile, subtended by leaves ; outer invol.-scales with short, leafy tops, inner lanceolate, acute ; outer pappus of oblong, blunt, inner of bristle-pointed scales.

HAB. Sand River, *Burke and Zeyher!* (Herb. Hk., Sd.)

Stem 6–8 inches (probably more) in height, closely leafy. Leaves 2–2½ inches long, 1½–2 lines wide. Stem and leaves rough to the touch, with short, thick hairs. Cor. lobes subulate, strongly ribbed. Style branches long, linear, flat. Pappus of 6 outer, blunt scales, and as many bristle-pointed inner ones. Known by its broader, subdenticulate leaves and rough pubescence.

7. G. brevifolia (Harv.) ; stem erect, much-branched, angularly ribbed and furrowed, gland-dotted; leaves shortly linear (4–5 times as long as wide), flat, thick, midribbed, blunt, submucronulate, gland-dotted; heads solitary in the forks and terminal, sessile; inv.-scales ovate-oblong, mucronate, the inner tomentose above; fimbrils longer than the achenes, multifid, concrete; achenes pubescent; pappus of many, short, toothed scales. *Polychætia brevifolia, DC. l. c.* 285.

HAB. Beyond the Gariep, *Burchell,* 2171. Cape, Hb. *Hooker!* (Herb. Hk.)
A foot or more in height, loosely much-branched, rigid, the branches straight, spreading, ramulous. Leaves subdistant, shorter than the internodes, 3–4 lines long, ¼–¾ line wide, conspicuously multipunctate (as well as the branches) on each side of the midrib; the very young leaves cobwebbed. Heads ovoid, 3 lines long, the younger woolly at top. Fimbrils broad, involving the achenes. Lobes of disc-fl. externally glandular, with thick nerves. Rays short, 3-toothed. Style branches of disc long, linear, flattened, puberulous. The *pappus* is somewhat different from that of other species, and apparently in a single row.

8. G. pectidea (Harv.); stem shrubby, terete, glabrous; leaves broadly linear, or linear oblong, narrowed to the base, cuspidate-mucronate, pungent, flat, midribbed and margined, subdenticulate, impress-dotted; heads 4 or more together, cymoso-glomerate, sessile; outer inv.-scales shortly leaf-topped, inner ovate, cuspidate, pungent, outer pappus of oblong, blunt, inner of bristle-pointed scales; achenes very villous. *Polychætia pectidea, DC. l. c.* 285.

HAB. Region beyond the Gariep, *Burchell.* Gariep, near Verleptpram, *Drege!* 2894. (Herb. Sond.)
I have only seen a flowering branch, 3–4 inches long. Leaves nearly uncial, about 2 lines broad, rigid, strongly midribbed, somewhat veiny and remarkably pungent-mucronate. It has all the gen. characters of a *Geigeria.*

Tribe 4. SENECIONIDEÆ. *Flower-heads* heterogamous, or homogamous, or monœcious (*disc* ♂, *ray* ♀). *Style-branches* (in the hermaphrodite flowers) long, linear, flattish, obtuse or truncate, pencilled at the apex or tipped with a bristly cone; in the male-fl. the styles are simple or emarginate, more or less pencillate or bristle-tipped. *Habit various.* (Gen. 39–124).

Sub-tribe 1. HELIANTHEÆ. *Leaves* opposite, expanded, mostly petiolate. *Recept.* bearing paleæ among the disc-flowers. *Anthers* without tails, short, dark-coloured. *Pappus* of a few stiff, persistent awns or none. (Gen. 39–44).

Achenes beakless, without pappus:
  Recept. flat. Inv. scales biseriate, ovate, acute ... (39) **Eclipta.**
  Recept. flattish. Inv. double; *outer* of 5 spathulate,
     spreading, leafy scales ... ... (40) **Siegesbeckia.**
  Recept. conical. Inv. scales short, appressed ... (44) **Spilanthes.**
Ach. beaked, bearing 2–5 rigid, retrorsely hispid bristles (42) **Bidens.**
Achenes beakless, with pappus:
  Pappus cup-shaped, toothed ... ... (41) **Wedelia.**
  Pappus of 8–10, rigid, unequal, rough bristles ... (43) **Lipotriche.**

Sub-tribe 2. HELENIEÆ. *Leaves* mostly alternate (in *Oedera* opposite). *Recept.* either nude or bearing paleæ among the disc-flowers. *Anthers* without tails. *Pappus* of several flat, dry, separate, conspicuous scales in a single row (occasionally with a few slender bristles as an inner pappus). (Gen. 45–49).

Recept. without paleæ.
  Inv. uniseriate. A glabrous, aquatic herb ... ... (45) **Cadiscus.**
Recept. bearing paleæ among the disc-flowers:
  Leaves opposite, linear. Heads conglomerate, in-
     volucred... ... ... (46) **Oedera.**

Leaves alternate. Heads separate.
> Rays female. Pappus of 2–3 acuminate, un-
> equal scales ... ..: ... ...   (47) **Callilepis.**
> Rays neuter. Pappus of 5 obovate, obtuse
> scales. *Achenes* with a tuft of soft hairs at
> base, sub-cylindrical ... ...   (48) **Sphenogyne.**
> Rays neuter. Outer pappus of 5 obovate scales;
> inner of 5 slender bristles. Achenes obovate,
> quite glabrous ... ...   (49) **Ursinia.**

Sub-tribe 3. ANTHEMIDEÆ. *Leaves* mostly alternate, rarely opposite. *Heads* discoid or radiate. *Recept.* either nude or bearing paleæ. *Anthers* wholly without tails, blunt at base. *Pappus* either none, or coroniform, or consisting of a few *minute* scales or jointed hairs. (Gen. 50–75).

(1). Receptacle bearing paleæ among the flowers.
> Heads radiate. Achenes glabrous. Leaves opposite   (50) **Eumorphia.**
> Heads heterogamous, more or less radiate. Achenes woolly.
>> Disc-fl. perfect. Leaves pinnatisect. Herbs or
>> suffrutices ... ...   (51) **Lasiospermum.**
>> Disc-fl. male. Leaves simple or 3-fid. Rigid
>> shrubs ... ...   (75) **Eriocephalus.**
> Heads homogamous, discoid. Pap. either minutely
> scaly, or of a few hairs, or none ... ...   (74) **Athanasia.**

(2). Receptacle nude (not bearing paleæ among the flowers).
> *Heads discoid :
>> Corolla of disc-fl. 5-toothed :
>>> Heads *homogamous*, (all the fl. similar and perfect) :
>>>> Pappus none : (shrubs or half shrubs) :
>>>>> Corolla glabrous :
>>>>>> Leaves linear, entire ... ...   (73) **Stilpnophytum.**
>>>>>> Lvs. cuneate, 3–5 lobed or toothed   (66) **Pentzia.**
>>>>> Cor. glandular. Achenes terete ...   (68) **Adenosolen.**
>>>>> Cor. pubescent. Achenes scabrid, sub-
>>>>> terete ... ...   (60) **Brachymeris.**
>>>> Pappus membranous, oblique, ear-shaped :
>>>>> Small shrubs or half-shrubs ... ...   (66) **Pentzia.**
>>>>> Annuals, with pinnatisect leaves ...   (61) **Matricaria.**
>>>> Pappus of several small, equal scales ...   (67) **Marasmodes.**
>>> Heads *heterogamous;* marg.-fl. filiform, female.
>>>> Disc-fl. male, with abortive ovaries...   (65) **Hippia.**
>>>> Disc-fl. fertile, ovuliferous ... ...   (64) **Artemisia.**
>> Corolla of disc-fl. 4-toothed :
>>> Tube of corolla roundish (not flattened) :
>>>> Shrublets or half-woody perennials :
>>>>> Achenes glabrous. Fl. all perfect ...   (62) **Tanacetum.**
>>>>> Marginal achenes villous ; of the disc
>>>>> glabr., abortive ... ...   (63) **Schistostephium.**
>>>> Annuals with pinnatisect leaves ... ...   (61) **Matricaria.**
>>> Tube of corolla flattened or 2-winged :
>>>> Annuals, or small, tufted perennials :
>>>>> Cor. not spurred at base ... ...   (71) **Cotula.**
>>>>> Cor. with a broad spur, infolding the
>>>>> ovary ... ...   (70) **Otochlamys.**
>>>> Virgate shrubs, with entire, sessile leaves   (69) **Peyrousea.**
> **Heads radiate ; the disc flowers fertile :
>> Corolla of disc-fl. 4-toothed :
>>> Herbs with multifid leaves :
>>>> Disc-fl. flat-tubed. Achenes flat ... ...   (72) **Cenia.**
>>>> Disc-fl. terete. Achenes angular ... ...   (61) **Matricaria.**

Small shrubs or half-shrubs :
    Rays female. Lvs. linear or 3-fid ...   ...   (58) **Adenachæna.**
    Rays neuter:
        Lvs. lobed or pinnatifid ...   .   (52) **Lidbeckia.**
        Lvs. linear, entire, silky ...   .   (53) **Thaminophyllum.**
Corolla of disc-fl. 5-toothed:
    Inv. scales in 1-row, connate or concrete.
        Disc fertile; all the achenes glabrous   ...   (54) **Gamolepis.**
        Disc sterile; ray-achenes silky, fertile   ...   (55) **Steirodiscus.**
    Inv. scales imbricate in few or many rows:
        Rays white:
            Small shrubs or suffruticose:
                Achenes papillose or glandular:
                    Recept. flat.   *Ray-fl.* some
                        filiform, some ligulate   ...   (57) **Phymaspermum.**
                    Recept. convex. Ray-fl. all
                        ligulate ...   .   (58) **Adenachæna.**
                Achenes angular, smooth...   ...   (59) **Chrysanthemum.**
            Herbs with multifid leaves   ...   (61) **Matricaria.**
        Rays yellow:
            Heads corymbose. Rays 1–2   ...   (56) **Iocaste.**
            Heads solitary, terminal: (small shrubs
            or herbs ; sometimes annual.)   ...   (59) **Chrysanthemum.**

Sub-tribe 4. GNAPHALIEÆ. Leaves alternate. *Heads* homogamous, heterogamous or hetero-monœcious, the marginal-fl. female, or rarely neuter, tubular or ligulate. *Anthers tailed at the base! Inv.-scales* mostly membranous, shining, not withering. *Foliage* very frequently woolly. *Pappus* various. (Gen. 76–106).

A. Heads one-flowered, either in spikes or glomerules :
    Pappus feathery   ...   .   (92) **Stoebe.**
    Pappus none   ...   .   (93) **Perotriche.**
B. Heads several-flowered, *discoid :*
    (1). *Homogamous ;* all the flowers perfect, 5-toothed :
        Pappus of several *feathery* bristles :
            Heads *large,* showy, solitary or corymbose   (79) **Helipterum.**
            Heads *small,* sessile, in tufts or spikes :
                Pappus-plumes *concrete* at base, fea-
                thered in the upper half only   ...   (87) **Elytropappus.**
                Pappus-plumes *separate,* feathered
                throughout ...   .   (88) **Pterothrix.**
        Pappus of several simple or serrate bristles :
            Achenes glabrous, smooth or granulated :
                Pappus biseriate   ...   .   (77) **Leontonyx.**
                Pappus uniseriate :
                    Achenes sessile, granulated. *(Leaves* neither *pungent nor*
                    spirally twisted):
                      Pappus-bristles rough or ser-
                      rate...   ...   .   (78) **Helichrysum.**
                      Pappus bristles beaded near
                      the tip ...   .   (82) **Eriosphæra.**
                  Ovaries stipitate (tapering at both
                  ends), smooth. (Leaves pun-
                  gent, often spirally twisted) ...   (84) **Metalasia.**
            Achenes hairy or woolly :
                Achenes beakless   ...   .   (85) **Lachnospermum.**
                Achenes with a glabrous, thick beak   (86) **Pachyrhynchus.**
        Pappus of 5 ovate, short, toothed scales   ...   (104) **Oligodora.**
    (2). *Heterogamous:* disc-fl. perfect; marginal-fl. filiform, female, few or many :
        Pappus of all the flowers amply feathered   ...   (83) **Lasiopogon.**
        Pappus bristle-shaped :
            Marginal female-fl. without pappus   ...   (81) **Amphidoxa.**

Marginal-fl. with pappus, as the disc-fl. :
 Recept. bearing paleæ among the fl. (76) **Rhynea.**
 Recept. nude or fimbrilliferous :
  Pappus biseriate ... ... (77) **Leontonyx.**
  Pappus uniseriate :
   Female-fl. few ... ... (78) **Helichrysum.**
   Female-fl. in 2 or more rows (80) **Gnaphalium.**
(3). *Hetero-monœcious :* disc-fl. abortive or male ; the marginal female, fili-
 form, few or many :
  Heads *large,* solitary, inv. radiating, lustrous (95) **Phænocoma.**
  Heads *small,* corymbose ; inv. radiating :
   Recept. very woolly, without paleæ (97) **Anaxeton.**
   Recept. nude in the centre, with mar-
    ginal paleæ ... ... (96) **Petalacte.**
  Heads *small,* in tufts or spikes ; inv. not
   radiating ... ... (94) **Trichogyne.**
C. Heads several-flowered, *radiate ;* rays ligulate :
 (1). Ray-flowers white or purple *(not yellow).*
  Pappus of several feathered bristles :
   Heads 2-fl. ; pappus-bristles feathered in
    the upper half only ... ... (91) **Disparago.**
   Heads 6–15-fl. ; pap. feathered throughout (89) **Amphiglossa.**
  Pappus of rough bristles, or of alternate bristles and short scales :
   Inv.-scales in few rows. (Moss-like, densely
    tufted) ... ... (90) **Bryomorphe.**
   Inv.-scales in many rows, bristle-pointed.
    (Suffrutices or shrubs, laxly leafy) ... (98) **Athrixia.**
  Pappus of several, very short scales ... ... (105) **Osmites.**
  Pappus none; ray neuter ... ... (106) **Osmitopsis.**
 (2). Ray-flowers yellow (often coppery beneath).
  Recept. nude. Pappus of many rigid bristles (99) **Antithrixia.**
  Recept. nude. Pappus in the ray of short
   bristles ; in the disc of feathered bristles ... (100) **Leyssera.**
  Recept. honey-combed or fimbrilliferous :
   Pappus of several short, free or connate scls. (102) **Nestlera.**
  Recept. bearing paleæ between the flowers :
   Pappus uniform, of minute separate or con-
    nate scales, or coroniform, subentire or
    crenate ... ... (103) **Relhania.**
   Pappus of the disc-fl. double ; the outer of
    short, broad scales ; the inner of 2 long
    bristles ... ... (101) **Rosenia.**

Sub-tribe 5. SENECIONEÆ. *Leaves* alternate. *Heads* radiate or discoid. *Recept.*
nude. *Anthers* without tails, blunt at base. *Pappus* of many or few slender, hairlike
bristles; the marginal fl. sometimes calvous. *Inv.* mostly uniseriate, either nude at
base or nearly so. (Gen. 107–119).
 Heads discoid :
  Homogamous : all the flowers perfect, 5-toothed :
   Inv. scales 5, winged or crested at back ... (111) **Lopholæna.**
   Inv. scales neither winged nor crested :
    Style-branches of disc-fl. truncate ... ... (117) **Senecio.**
    St. branches of disc-fl. minutely cone-tipped :
     Achenes flattened or winged ... ... (110) **Cineraria.**
     Achenes not flattened :
      Pappus in several rows ... ... (113) **Kleinia.**
      Pappus uniseriate ... ... (112) **Cacalia.**
  Heterogamous or hetero-monœcious : the marginal-fl.
   filiform, female :
    Shrubby or perennial ; fl. with copious pappus :
     Disc-fl. fertile. Leaves fleshy, thick ... (113) **Kleinia.**

Disc-fl. sterile. Lvs. expanded, glabrous (114) **Doria.**
Annual. Fl. with scanty pappus ... ... (107) **Stilpnogyne.**
Heads radiate:
Achenes flattened or winged ... ... ' ... (110) **Cineraria.**
Achenes terete or angular:
Pappus copious, in many rows:
Disc-fl. fertile:
Inv. scales separate, often calycled... (117) **Senecio.**
Inv. sc. connate, not calycled, valvate (118) **Euryops.**
Disc-fl. sterile:
Pappus of curved, barbed, deciduous
bristles ... ... (119) **Ruckeria.**
Pappus of straight, rigid bristles ... (115) **Othonna.**
Pappus of few bristles, uniseriate: (sometimes
wanting in disc-fl.):
Disc-fl. fertile:
Pappus in the disc-fl. of several, in
the ray of few bristles ... ... (109) **Mesogramma.**
Pappus of 4-5-curved, barbed
bristles ... ... (108) **Oligothrix.**
Disc-fl. sterile without pappus... ... (116) **Gymnodiscus.**

Sub-tribe 6. CALENDULEÆ. *Leaves* mostly alternate, rarely opposite. *Heads* radiate, hetero-monœcious, the ray-fl. female; the disc-fl. male, with abortive simple styles; or rarely (in *Dimorphotheca*) some or all of the disc-fl. fertile. *Anthers* acute or cuspidate at base or very minutely tailed. *Pappus* none. (Gen. 120–124).

Achenes quite beakless:
Inv. uniseriate:
Marginal achenes 3-cornered, either smooth,
tubercled, or sharply toothed at the angles (120) **Dimorphotheca.**
Marg. achenes covered with sharp thorny
points ... ... (124) **Xenismia.**
Inv. in few rows, 2-3 seriate:
Marg. achenes nutlike, thick-shelled, either
smooth, wrinkled, 3-cornered or 3-winged (123) **Osteospermum.**
Achenes produced into a beak:
Achenes 3-cornered, wingless or minutely winged (122) **Oligocarpus.**
Achenes broadly 3-winged ... ... (121) **Tripteris.**

Sub-tribe 1. HELIANTHEÆ. (Gen. 39–44).
XXXIX. **ECLIPTA**, Linn.

Heads many-fl., heterogamous; *ray-fl.* in few rows, female, short and narrow-ligulate; *disc-fl.* tubular, 4-toothed, hermaphrodite. *Recept.* plano-convex, with linear-filiform paleæ, ciliate at the apex and equalling the achenes. *Invol.* in two rows, of 10–12 ovate-lanceolate, acuminate scales. *Style-branches* in the disc-fl. linear, flattened, obtuse. *Achenes* of the ray 3-cornered, of the disc flattened, without pappus, either pointless or crowned with 1-3 minute teeth, the younger downy. *DC. Prodr.* 5, *p.* 489.

Tropical and subtropical herbs, erect or prostrate, scabrous or hairy. Leaves opposite, entire or serrate, penninerved. Pedunc. axillary, solitary or in pairs, one-headed. Flowers white. Name, from εκλειπω, *to be deficient*; it wants pappus?

1. **E. erecta** (Linn.); stem and leaves appressedly strigose; leaves oblongo-lanceolate, or lanceolate, tapering to each end, pedicels solitary or in pairs, 2-4 times as long as the lvs. *DC. l. c. p.* 490. *Dill. Elth. fig.* 137.

HAB. Port Natal, *Drege, Gerr. & M'K.* 299. (Herb. D., &c.)
A common tropical weed, variable in size and habit, erect or prostrate. Most if not all of the so-called species of *Eclipta* seem to be merely varieties of this universally diffused plant.

XL. SIEGESBECKIA, L.

Heads few-fl., heterogamous; *ray-fl.* uniseriate, female, ligulate or irregular; *disc-fl.* 3–5-toothed, tubular, hermaphrodite. *Invol.* biseriate, the scales covered with gland-headed bristles, the 5 exterior linear-spathulate, spreading; the inner ones half-clasping round the ray flowers. *Recept.* flat, bearing oval-oblong paleæ wrapping round the achenes. *Style-branches* in the disc-fl. short, somewhat flattened, very obtuse. *Achen.* obovate-oblong, somewhat 4-angled, arching inwards, without pappus. *DC. Prodr. 5, p.* 495.

Annuals with weak, forked stems, glandular and viscidulous above. Leaves opposite, mostly tapering at base into a petiole, ovato-triangular, coarsely toothed, pubescent. Peduncles from the forks of the branches, 1-headed, small, yellow. Name in honour of Dr. J. G. Siègesbeck, author of a "Flora of St. Petersburgh," 1736.

1. S. orientalis (Linn. Sp. 1269); leaves ovate, cuneate at base, acuminate at apex, coarsely toothed, the uppermost oblong-lanceolate; outer invol.-scales twice as long as the inner. *DC. l. c. Lin. Hort. Cliff. t.* 23.

HAB. Near Durban, Natal, *Gerr. & M'K.* 833. (Herb. D., Sd.)
A common tropical weed. 1–2f. high, the lower branches opposite, spreading, the upper forked. Leaves 5–6 inches long, 2–3 in. wide, thin, sparsely pubescent, with 3 principal nerves and laxly netted. Pedunc. hairy, 1–1½ in. long. Outer invol.-scales very narrow, 5–7 lines long.

XLI. WEDELIA, Jacq.

Heads many-fl., heterogamous; *ray-fl.* ligulate, female, in one row; *disc-fl.* hermaphrodite, tubular, 5-toothed. *Inv.* in 2–3 rows, the outer scales leaf-like, the inner membranous. *Recept.* somewhat convex, covered with paleæ. *Style* branches in the disc-fl. tipped with a short cone. *Achenes* obovate or compressed, beakless with a crown-like or cup-like, often substipitate pappus, consisting of concrete, toothed and ciliate scales. *DC. Prodr. 5, p.* 538.

Suffrutices or herbs, chiefly American; rare in Asia, Africa and Australia. Leaves opposite, shortly petiolate, serrate or trifid. Pedicels solitary, terminal, often in the forks of the branches. Corolla yellow. Name in honour of Professor Wedel, of Jena, author of many botanical dissertations.

1. W. Natalensis (Sond.! in Linn. 23, p. 63); herbaceous, very scabrous with short, rigid bristles; leaves subsessile, ovate or lanceolate, acuminate or acute, distinctly serrate, 3-nerved, with subrevolute margins; pedunc. about as long as the leaves; outer scales of the invol. ovate, or oblong, leafy, subacute, strigose; inner smaller, scarious, ciliate; achenes compressed, 4-angled, pilose, crowned with a hardened, toothed cup. *Sond. l. c.*

HAB. Port Natal, *Williamson! Gueinzius! Dr. Sutherland.* (Hb. D., Sd., Hk.)
2 to 3 feet high, coarse and weedlike. Leaves 2–5 inches long, 1–1½ wide, variable in form; sometimes quite lanceolate, when it seems to come very near *W. Africana*, Pers. It is also very near *W. calendulacea.*

XLII. BIDENS, L.

Heads many-fl., either homogamous and discoid, or more frequently radiate, the *ray-fl.* ligulate, neuter, the *disc-fl.* perfect, 5-toothed. *Inv.* scales in a double row, similar or dissimilar. *Recept.* flattish, bearing paleæ. *Style* branches tipped with a short cone. *Achenes* more or less compressed, aculeate, tapering into a beak, tipped with 2–5 rigid, retrorsely hispid, scabrous bristles. *DC. Prodr. 5, p. 594.*

A large and widely-dispersed genus, most numerous in America. Stems herbaceous, square or round, mostly with opposite branches. Leaves opposite, either undivided, cut or pinnate-partite; lobes cut or serrate. Flowers mostly yellow, rarely with white rays.—Name, *bidens*, 2-toothed, in allusion to the bristles of the pappus.

1. B. pilosa (Linn. Sp. 1166); stem erect, 4-angled, striate, glabrous or pilose; leaves pinnati-partite, the uppermost trifid or tripartite, their lobes ovate, acute or acuminate, sharply serrate, glabrous or pubescent, equal at base or the lateral oblique; heads subcorymbose, long-peduncled, either discoid or with white rays; achenes linear-angular, glabrous or pubescent, with 2–4 bristles, unequal. *DC. l. c. p. 597.*

VAR. β, leucantha; heads radiate, the ray white. *B. leucantha, Willd. DC. l. c.* 598. *B. Wallichii, DC. B. Sundaica, Bl. DC. l. c.*

HAB. Eastern Districts and Port Natal: also in waste ground. (Herb. D.,Sd.,Hk.) A common tropical and sub-tropical weed. Pubescence variable.

XLIII. LIPOTRICHE, R. Br.

Heads many-fl., radiate; *ray-fl.* in 1 row, female, broadly ligulate, 3-toothed, externally pubescent; *disc-fl.* hermaphrodite, tubular, 5-toothed. *Inv.* 2–3-seriate, loosely imbricate; the scales lanceolate, leafy. *Recept.* convex, covered with flattish, acuminate leafy paleæ. *Style* branches of the disc-fl. tipped with a hispid cone. *Anthers* nigrescent. *Achenes* turbinate-prismatic, 3–4-angled, those of the disc subcompressed. *Pappus* of 8–10, short, rigid, unequal, deciduous, rough bristles. *DC. Prodr. 5, p. 544.* Also *Psathurochæta, DC. l. c. p. 609.*

A coarse-growing, scabrous herb, with the aspect of *Wedelia*. Leaves opposite, long-petioled, the lower 3-lobed, the upper hastate-lanceolate, 3-nerved, serrate. Pedunc. long, 1-headed, solitary or subcorymbose. Fl. bright yellow. Name from λειπω, *to cast away*, and θριξ, *a bristle;* because the pappus is deciduous.

1. L. Brownii (DC. l. c. 544); *Sch. B. in Walp. Rep. 6, p. 157. Psathurochæta Dregei, DC. l. c. 609. Trigonotheca Natalensis, Sch. B. MSS. Zey.!* 1042.

HAB. On the Congo, *Chr. Smith;* Natal, *Drege! Krauss,* 31; *Gerr. & M'K.* 298. Crocodyle R., *Burke!* Kreili's Country, *H. Bowker!* 385. (Herb. D., Hk., Sd.) Stem 2–3 feet high, 4-angled, appressedly strigose. Leaves on long petioles, the lower 5–6 inches long, 3 inches wide at base, 3-lobed, the terminal lobe very long, trowel-shaped, the lateral short, all sharply serrate; upper leaves simple, all very scabrous. Pedunc. 4–6 inches long. Ray-fl. rich, crocus-yellow.

XLIV. SPILANTHES, Jacq.

Heads many-fl., either heterogamous, with *ray-fl.* female, ligulate, often short, in a single row; or homogamous, all the fl. perfect and tubular, 4–5-toothed. *Inv.* scales in 2 rows, appressed, shorter than the disc, the outer scales green, the inner submembranous, folded. *Recept.* convex, covered with membranous paleæ, enwrapping the flowers. *Style* branches of disc-fl. truncate and pencilled. *Anthers* nigrescent. *Achenes* of disc compressed, beakless, often ciliate at the sides or naked, of the ray 3-cornered or subcompressed. *Pap.* none. *DC.Prodr.* 5, *p.* 620.

Annual or perennial herbs, mostly tropical. Leaves opposite, petioled, subentire. Peduncles 1-headed, terminal or from the forks of the branches. Cor. mostly yellow ; heads ovate or conical.   Name, σπιλος, *a spot*, and ανθος, *a flower*.

1. **S. Africana** (DC. l. c. 623); stem rooting at base, ascending, sparingly pubescent ; leaves opposite, petiolate, ovate or ovato-lanceolate, coarsely toothed, minutely scabro-ciliolate at the margin, glabrous or sparsely setulose ; pedunc. 2–3 times as long as the leaves ; heads conical-ovate ; rays not much longer than the involucre ; invol. scales oblong, obtuse ; achenes glabrous, without cilia or awn.

HAB. Natal, *Drege! Gerr. & M'K.* 342, 816, *Krauss*, 303, &c. (Hb. Hk., D., Sd.)
Seemingly perennial. Stems 1–2 feet long, numerous, not much branched. Lvs. 1–1½ inches long, ¾–1 inch wide, unequally toothed. Pedunc. 2–4 inches long. Heads conical.—Is this different from *S. caulirhiza* ? The species seem to require revision and much condensation.

Sub-Tribe 2. HELENIEÆ.   (Gen. 45–49.)

XLV. CADISCUS, E. Mey.

Heads many-fl., radiate ; *ray-fl.* broadly ligulate, female, in one row ; *disc-fl.* campanulate, 5-toothed, fertile or the inner sterile. *Inv.* scales 8–10, in a single row, concrete into an 8–10-toothed, cup-like involucre. *Recept.* honeycombed, convex. *Anthers* not tailed, with a large, apical scale. *Style*-branches truncate, penicillate at the apex. Fertile *achenes* terete, ribbed and furrowed, slightly rostrate, villous at base, pubescent ; *sterile* linear, smooth. *Pappus* of the fertile fl. of 10–12 rigid, subulate-acuminate, persistent scales ; of the sterile, of more slender bristles. *DC. Prodr.* 7, *p.* 254.

A glabrous, aquatic herb. Stems long, weak, vaguely and distantly branched, floating, or rooting in the mud, at the nodes. Leaves alternate, stem-clasping, distant, linear tongue-shaped, entire, subacute, not obviously nerved. Peduncles opposite the leaves, short, 1-headed. Fl. white or yellowish. Name unexplained.

1. **C. aquaticus** (E. Mey.! in Hb. Drege); *DC. Prodr. v.* 7, *p.* 255.

HAB. Near Dassenberg, and between Groenekloof and Saldanha Bay, *Drege! Zey!* 1044. (Herb. D., Hk., Sd.)
Stems 1–2 feet long, 2 lines diameter, succulent. Leaves 2–3 inches long, 2–4 lines wide. Pedunc. ¾ inch long, cernuous, in fruit pendulous. Achenes nearly 4 lines long ; pappus 1 line.

XLVI. ŒDERA, Linn.

Heads crowded within a bracteated cluster, cylindrical, few-fl., radiate ; *ray-fl.* few, female, elongate toward the circumference of the cluster, short toward the centre ; *disc-fl.* tubular, 5-toothed, hermaphrodite. *Inv.* scales appressed, scarious, in few rows. *Recept.* paleaceous. *Anthers* tipped with a truncate appendage, without tails. Branches of the *style* truncate. *Achenes* wingless, angular-cylindrical, glabrous. *Pappus* of several minute or longish, equal or unequal, semi-lanceolate scales, or crown-like, toothed. *Œderia, DC. Prodr. 6, p.* 1.

Small, densely leafy, slightly branched, rigid, S. African shrubs. Leaves opposite or ternate, or spirally imbricate, rough-edged. Glomerules or head-clusters terminal, sessile. Flowers yellow. Named in honour of Professor Geo. Œder, of Copenhagen, author of *Flora Danica.*

1. EU-ŒDERA: Glomerule of heads sessile :
   Leaves on both sides clothed with rigid, gland-tipped, short
   hairs ... ... (1) **hirta.**
   Leaves glabrous, ciliate-serrate or entire :
   Lvs. ovate or ovato-lanceolate ; fl. leaves broad... ... (2) **latifolia.**
   Lvs. linear-lanceolate or subulate ; fl. leaves narrow... (3) **prolifera.**

2. ERIOPODA (DC.): Glomerule on a woolly peduncle ; ray-fl.
   very small ... ... (4) **lævis.**

1. Œ. hirta (Thunb.! Cap. 725); leaves broadly ovate or ovato-lanceolate, flat, on both sides densely clothed with rigid, gland-tipped short, spreading hairs. *Less.! Syn.* 248, *DC. l. c. p.* 2.

HAB. S. Africa, *Thunb! E. Z.!* (Herb. Th., Sd.)
All parts clothed with short, very rigid, glandular hairs. Stems a foot long, slightly branched, imbricated with spreading leaves. Leaves ¾ inch long, 4–5 lines wide, thick, obscurely midribbed. Pappus of several unequal scales.

2. Œ. latifolia (Less. Syn. 247); leaves either roundish-cordate, ovate, or ovato-lanceolate, or lanceolate, spreading or reflexed, flat, glabrous, the margin (and midrib) calloso-serrate ; outer floral leaves broadly-ovate, many-nerved, inner ovato-lanceolate or lanceolate, on both sides villous below; inv. scales linear, entire or lacerate. *DC.! l. c.* 2. *Œ. prolifera, γ. Hb. Th.! Œ. imbricata, Lam. Dict.* 2, 345. *Œ. obtusifolia, Cass. Œ. intermedia, DC. l. c.*

HAB. Cape, *Thunb.!* Muysenberg, *Eckl.! W.H.H.! C. Wright,* 378; Groenekloof and Frenchhoek, and Zwarteberg, *Drege!* Stellenbosch and Caledon, *E. Z.!* near Grahamstown, *T. Williamson, Genl. Bolton!* (Herb. Th., D., Hk., Sd.)
Stem robust, 1–2 feet high. Leaves very variable in shape, sometimes as broad as long, 5–6 lines each way ; again 5–6 lines long, 2–3 lines wide, in one var. 1½ in. long, 2–3 lines wide! mostly recurved. I find the invol. scales as often fimbriate-lacerate as not; they are even so in the original specimen marked by *Lessing* in Herb. Thunb.!

3. Œ. prolifera (Linn. f. Suppl. 391); leaves linear-lanceolate or subulate, spreading, subconcave or channelled, glabrous, the margin ciliato-serrate (or entire and softly ciliate!); floral leaves oblong-lanceolate or lanceolate, calloso-serrate or ciliate, more or less villous below; invol. scales lacerate. *Thunb.! Cap.* 724. *Less.! Syn.* 247. *DC. l. c.* 2.

VAR. β, **integrifolia**; leaves subulate, channelled, quite entire, *softly* ciliolate.

HAB. Cape, *Burm., Thunb.!* Groenekloof, *Drege!* Caledon and Stellenbosch, *E. Z.!* about Capetown, *W.H.H.* β, Steendal, Tulbagh, *Dr. Pappe!* (Hb. Th., D., Hk., Sd.)
About a foot high, ascending or decumbent, branched from below. Lvs. spreading or recurved or arcuate, 7–10 lines long, 1–2 lines wide. Floral leaves more glabrous than in *Œ. latifolia.* Var. β. has the leaves narrower, more channelled than usual, and *quite entire*, though fringed with soft, minute, white hairs; but there are intermediate stages connecting it with the normal form.

4. Œ. lævis (DC. l. c. 2); "leaves spirally disposed, lanceolate-linear, acuminate, concave, on both sides smooth, the margin serrulato-subcili-ate; *glomerule on a woolly peduncle;* floral leaves rather wider than the cauline, at base villoso-ciliate." *DC. l. c. Burch. Cat. Geogr.* 6870.

HAB. S. Africa, *Burchell!*
"Leaves nerveless, finely striate lengthwise, shining, 6 lines long, 2 lines wide. Branches woolly, especially near the apex. Heads within bracts on a woolly pe-duncle. Outer inv. scales very short. Ligules of each head 1-2, nearly bi-ligulate, very short, whence the head is pseudo-discoid. Pappus of the ray-fl. very short or none; of the disc-fl. membranous, toothed at the summit, longer than the tube of corolla." *DC. l. c.* Unknown to us.

XLVII. CALLILEPIS, DC.

Heads many-fl., heterogamous; *ray-fl.* numerous, in one row, ligulate, female; *disc-fl.* tubular, 5-toothed, perfect. *Invol.* campanulate, scales imbricated in many rows, lanceolate, subequal. *Recept.* flat, covered with hard and dry, infolded, acuminate paleæ, clasping the flowers. *Style*-branches tipped with a short cone. *Achenes* glabrous, those of the ray 3-angled-compressed; of the disc flattened. *Pappus* of 2-3 scarious acuminate, unequal, persistent, keeled scales; sometimes in the disc-fl. 1-2 minute additional ones.—*DC. Prodr.* 5. *p.* 671.

Glabrous or pubescent, rigid undershrubs, with subsimple stems or branched from near the base. Leaves rigid, undivided, the lowest opposite, the rest alternate, linear or lanceolate. Heads terminal, solitary, of large size. Pappus of the ray of 3, of the disc usually of 2 scales. Rays numerous, white(?). Name καλος, *beautiful*, and λεπις, *a scale;* in allusion to the paleæ of the receptacle, or the pappus.

Leaves broadly lanceolate, 3-nerved, entire or subdenticulate (1) **Laureola.**
Leaves linear-subulate, one-nerved ... ... (2) **leptophylla.**

1. C. Laureola (DC. l. c. 671); leaves opposite or alternate, *broadly lanceolate*, 3-nerved, margined, quite entire or minutely denticulate.

VAR. a. **glabra** (Sond.); glabrous in all parts. *C. glabra, DC. l. c.*

VAR. β. **intermedia** (Sond.); thinly pilose on the stem. *C. laureola, DC. l. c.*

VAR. γ. **hispida** (Sond.); densely villous on the stem and young parts. *C. hispida, DC. l. c.*

HAB. About Natal, all the varieties, *Drege! Krauss! Sanderson! Plant! Gerr. & M'K.* 325, &c. (Herb. D., Hk., Sd.)
Many-stemmed. Stems simple or branched near the base, with many erect, vir-gate branches, 1-2 f. high, either glabrous or villous, sometimes densely so. Lower leaves opposite, 1-1½ inch long, 3 lines to ½ inch wide, rigid, acute or acuminate, mostly denticulate. Heads terminal, the peduncle laxly leafy nearly to the summit. Rays very numerous, nearly uncial. Achenes quite smooth, with a marginal, entire wing. Var. β. scarcely differs tangibly from a.; I find its leaves (in *Drege's* specimens) as strongly toothed, and the pubescence very scanty: var. γ., by its copiously hairy stem, is more clearly marked.

2. C. leptophylla (Harv.); leaves scattered, *linear-subulate*, elongate, one-nerved, quite entire. *Zey.!* 1043.

HAB. Magalisberg, *Burke & Zey.!* (Herb. Hk., D., Sd.)
Stems 1-2 f. high, closely leafy and branched below, laxly leafy above, glabrous, or the younger parts softly villous. Leaves 1-1½ inches long, not a line in diameter, acute, flexuous, somewhat margined, quite entire. Flowering branches a foot long, sparsely leafy to near the summit, rib-striate. Heads very similar to those of *C. Laureola*, but the rays are rather smaller and less numerous. Achenes glabrous, with a marginal, ciliate wing.

XLVIII. **SPHENOGYNE**, R. Br.

Heads many-fl., radiate, *ray-fl.* in one row, neuter ; *disc-fl.* tubular, 5-toothed, perfect. *Recept.* paleaceous, the paleæ scarious, clasping the flowers, truncate or very rarely appendiculate at top. *Inv.*-scales imbricate in many rows, the inner ones larger, amply membranous at the apex. *Style*-branches truncate. *Anthers* tipped with a cordate appendage. *Achenes* at the base girt with a circle of long, soft hairs, cylindrical, scarcely narrowed below. *Pappus* of about 5, obovate or cuneate, obtuse, broad scales, spirally rolled round each other before the flowers open, much enlarged on the mature fruit, milk white. *DC. Prodr.* 5. *p.* 681.

A large African genus of suffrutices or herbs, strongly scented, with gland-dotted, mostly pinnati-partite, rarely serrate or quite entire, alternate leaves. Branches mostly ending in naked, one-headed peduncles, in *S. pinnata* panicled. Corollas mostly yellow ; in many species the rays are coppery or dark brown on the lower surface, then said to be "*discoloured* ;" when both sides are of the same colour, they are "*concolourous*." The name is compounded of σφην, a *wedge*, and γυνη, here meaning a *flower*.

Div. 1. THELYTHAMNOS. Outer involucral scales either without any membranous edge, or with a very narrow one. (Sp. 1–38).

* Herbaceous annuals, with pinnati-partite leaves :
　　Rays dark-coloured on the lower side ... ...　(1) **anthemoides**.
　　Rays concolourous (yellow on both sides) :
　　　Outer inv. scales oblong, obtuse, membr. tipped　(2) **calendulacea**.
　　　Outer inv. scales acute :
　　　　Inv. scales ovate. Pedunc. solitary　...　(3) **fœniculacea**.
　　　　Inv. sc. lanceolate. Pedunc. sub-corymbose　(4) **pusilla**.
** Herbaceous (annual?). Leaves tongue-shaped, inciso-
　　serrate. Pedunc. very slender, thread-like　... ...　(5) **filipes**.
** Suffrutices, or small shrubs. Leaves various.

(*a*) Leaves hairy, silky or tomentose, pinnate or inciso-serrate :
　　Lvs. oblong, sharply inciso-serrate, very woolly　(6) **coronopifolia**.
　　Lvs. pinnati-partite : lobes linear.
　　　Pedunc. very short, or scarcely any :
　　　　Rays concolourous. Lvs. crowded ...　(7) **hispida**.
　　　　Rays discoloured. Lvs. scattered ...　(8) **brachypoda**.
　　　Pedunc. long, scape-like. Stems short, tufted :
　　　　Loosely hairy. Inv. scales ovate-
　　　　　acuminate　... ... ...　(9) **Dregeana**.
　　　　Closely silky-canescent. Inv. scales
　　　　　lin.-lanceolate... ... ...　(10) **macropoda**.
　　　Pedunc. long. Stems shrubby, branched :
　　　　Closely and softly silky-tomentose in
　　　　　all parts ... ...　(11) **sericea**.
　　　　Hairy with short, curled, woolly hairs :
　　　　　Inv. sc. densely tomentose, acute　(12) **abrotanifolia**
　　　　　Inv. scales glabrescent, obtuse,
　　　　　　membr. tipped　... ... ...　(13) **subhirsuta**.
(*b*) Leaves glabrous, pinnate, each lobe tipped with a bristly hair :
　　Stems glabrate. Rays discoloured ... ...　(14) **dentata**.
　　Stems hirsute. Rays concolourous ... ...　(15) **concolor**.
(*c*) Leaves nearly glabrous, pinnate-parted, the lobes acute or mucronate.
　　Stems short, tufted. Pedunc. elongate, scape-like :
　　　Rays concolourous :
　　　　Leaves naked at base, pinnate beyond the
　　　　　middle ... ...　(16) **nudicaulis**.

Lvs. pinnate throughout their whole length   (17) **scapiformis.**
Rays discoloured :
  Lvs. pinnate throughout their whole length   (18) **chamomillæfolia.**
  Lvs. pinnate beyond their middle ... ...   (19) **Natalensis.**
Stems erect, branching, suffruticose or shrubby :
  Rays concolourous :
    Leaves 1–2 inches long; lobes filiform :
      Lvs. pinnate throughout, lobes 6–8
        pair ... ... (21) **anethoides.**
      Lvs. pinnate above, in few pair or trifid   (22) **crithmifolia.**
    Leaves ½ inch long or less; lobes very short :
      Leaf-lobes several, the upper 2–3-fid   (23) **punctata.**
      Leaf-lobes few, entire, subobtuse ... (24) **pauciloba.**
  Rays discoloured :
    Pedunc. one-headed, elongate, erect :
      Outer inv. scales broad-based, oblong or acuminate :
        Inv. scales acuminate: leaf-lobes
          long ... ... (20) **anethifolia.**
        Inv. scls. aristate; leaf-lobes lin.   (27) **chamæmeloides.**
        Inv. scales oval-oblong, obtuse :
          Glabrous. Leaf-lobes 6–8 pair   (25) **rigidula.**
          Cobwebby. Lf.-lobes 2–3 pair   (26) **discolor.**
      Outer inv. scales very narrow-linear,
        acuminate ... ... (28) **paleacea.**
    Pedunc. panicled, spreading or divaricate   (29) **pinnata.**
(*d*) Leaves nearly glabrous, 3–5-lobed or parted :
  Virgate shrubs, with straight, densely leafy branches : heads subsessile.
    Leaves 5-parted nearly to base ... ... (30) **quinquepartita.**
    Leaves 3-parted nearly to base ... ... (31) **tripartita.**
  Diffusely branched : heads pedunculate :
    Lvs. 3–5 lines long, shortly 3-lobed; lobes blunt   (32) **trifida.**
    Lvs. ½ inch long, unequally 3-lobed; lobes acute   (33) **heterodonta.**
    Lvs. 1–1½ inch long, *ciliate below;* lobes *long,*
      acute ... ... (34) **trifurca.**
(*e*) Leaves quite entire, glabrous or nearly so :
  Heads subsessile. Lvs. 5 lines long, *linear* ... ... (37) **leptoglossa.**
  Heads pedunculate. Lvs. 1–3 inches long, *linear :*
    Lvs. rigidly ciliate in their lower half ... ... (35) **ciliaris.**
    Lvs. quite smooth-edged, very long ... ... (36) **tenuifolia.**
  Heads pedunc. Lvs. *lanceolate,* 1–3-nerved ... ... (38) **Eckloniana.**

Div. 2. XEROLEPIS. All the involucral scales with broadly membranous edges and membranous expanded apices. (Sp. 39–44).

Leaves serrate or pinnatifid. Rays concolourous :
  Lvs. oblong, coarsely serrate; serratures deltoid   (39) **serrata.**
  Lvs. oblong, inciso-serrate; serratures acuminate   (40) **incisa.**
  Lvs. oblong, pinnatifid, lobes 2–3-toothed ... ... (41) **tridentata.**
Leaves inciso-pinnatifid. Rays discoloured   (42) **odorata.**
Leaves pinnati-partite, slender. Rays discoloured :
  Thinly pilose. Lf.-lobes 7–12 pairs, bristle-pointed   (43) **pilifera.**
  Glabrous. Leaf-lobes simple, mucronulate ... (44) **scariosa.**

Div. 1. **THELYTHAMNOS** (Sp. 1–38).

1. S. anthemoides (R. Br.); annual, erect, glabrous or puberulous; leaves either pinnati-partite or sub-bipinnati-partite, the lobes linear-filiform, acute or mucronate, the lower shorter or very small, the upper bi-trifid or pinnulate, spreading; pedunc. long, naked, cernuous; outer inv. scales ovate or oblong, subacute, glabrous or puberulous; paleæ uniform, sheathing, truncate; rays discoloured; achenes scaberulous. *Less. Syn. p.* 242. *DC. l. c.* 681. *Arctotis pinnata, and A. cernua, Th.! Cap.* 711–712. *Arct. anthemoides, L. Bot. Mag. t.* 544. *Zey.!* 2787, 2788, 2791.

VAR. β. **versicolor;** glabrescent; leaves mostly simply pinnate; rays purple at base, pale above, at length becoming purple. *S. versicolor, DC. l. c.* 681.

VAR. γ. **adonidifolia;** glabrous or nearly so; leaves multijugate, the lower lobes short, the upper long and compound; paleæ golden yellow. *S. adonidifolia, DC. l. c.* 682.

VAR. δ. **microcephala;** leaves sub-simply pinnate; fl. heads half the usual size; the whole plant *starved.* *S. microcephala, DC. l. c.* 682. *Drege,* 9101.

HAB. Western and N. Western districts, common. Zwartkops R., Uit. *Zey.!* 2788. β., Camiesberg, *Drege!* γ, Paarl, *Drege!* Lion's Mt., Capetown, *Ecklon.!* Tulbagh, *Pappe!* δ, Berg River, *Drege.* (Herb. Th., D., Sd. Hk.)

Stems 3–12 inches high, branched from near the base, the branches mostly straight and erect. Leaves variable in section, even in the same plant. Paleæ of the fruit twice as long as the involucre, either dirty straw colour or golden. Var. δ looks like a starved plant. I find no permanent marks between any of the above varieties; the inv. scales are identical in form in all, but variable in colour.

2. S. calendulæflora (DC. l. c. 682); herbaceous, annual, branching, glabrous; leaves bipinnati-partite, lobes linear, flattish, divaricate, subacute; pedunc. long, naked; outer inv. scales oblong, obtuse, membrane-edged, and tipped with small scales; rays concolourous, at least four times as long as the involucre; paleæ uniform, truncate.

HAB. Silverfontein and Knakarberg, N. W. coast, *Drege!* (Hb. D., Sd., Hk.)

A much branched annual, 4–10 inches high, the branches ending in peduncles 4–5 inches long. Rays bright orange yellow, about an inch long : by which character and that of the inv. scales this species is well marked.

3. S. fœniculacea (Less. Syn. 242); herbaceous, nearly glabrous, erect; leaves bipinnati-partite, the lobes linear, acute; pedunc. elongate, naked; outer inv. scales ovate, acute, glabrous; paleæ uniform, sheathing, truncate; rays concolourous. *DC. l. c. p.* 682. *Arctotis fœniculacea, Jacq. Schoenb. t.* 156.

VAR. β. **pallida;** slender; leaves either pinnati-partite or sub-bipin.; inv. scales pale. *Sph. pallida, DC.*

HAB. Cape, *Burch. Cat.* 6166 (fide DC.). *Eckl.!* β. near Clan-William, *Eckl.!* Camiesberg, *Drege.* (Herb., Sd., D., Hk.)

Except for its concolourous rays, I fear a character of little value, this scarcely differs from S. *anthemoides.* I can by no means distinguish S. *pallida* specifically.

4. S. pusilla (DC. l. c. 682); "herbaceous, branched, glabrous; leaves thickish, pinnate-parted, the lobes linear, sub-obtuse, either entire or 2–3 parted; pedicels numerous, one-headed, sub-corymbose, nearly as long as the stem; outer inv. scales lanceolate, subacute; rays concolourous; paleæ truncate; achenes scaberulous." *DC.*

HAB. Olifant River, *Drege!* (Hb. Sond.).
Only known to me by a very imperfect specimen.

5. S. filipes (E. Mey.!); herbaceous, (annual?) tall, branching, glabrous; leaves sessile, oblong, tongue-shaped, one nerved, acutely and deeply inciso-serrate or pinnatifid, the marginal lobes subulate, the three terminal bifid or trifid; pedunc. axillary and terminal, very long and slender ; heads small ; outer inv. scales lanceolate, acute, minutely membrane-edged ; rays concolourous. *DC. l. c. p.* 687.

HAB. Draakensteenberg, *Drege!* (Herb. D., Hk., Sd.)
A straggling herb, possibly annual, 1–2 ft. high or more, branching above, pale. Leaves 1–1½ inch long, the undivided portion 2–3 lines wide, the lateral teeth or lobes 2 lines long, spreading or recurved, acute. Pedunc. 4–8 inches long, curved, threadlike. Heads 3–4 lines wide.

6. S. coronopifolia (Less. Syn. 244); suffruticose, erect, every part densely clothed with curled, woolly hairs ; leaves crowded, oblong-ligulate, subtruncate, sharply inciso-serrate along the sides and at the broad point, the serratures broadly subulate, mucronate ; uppermost leaves linear, acute, entire ; peduncles elongate, terminal, one-headed ; outer invol.-scales densely woolly, acuminate ; rays concolourous ; paleæ erose ; pappus sc. subdenticulate. *DC. l. c. p.* 683.

HAB. Districts of Tulbagh and Worcester, *Ecklon!* (Herb. Sond.)
Stem a foot or more high, subsimple, with 2–3 short, erect, subterminal flowering branches. Leaves 1–1½ inch long, 4 lines wide ; the serratures spreading, with wide, rounded interspaces ; the summit of the lower leaves shortly 3-lobed, the middle lobe 3-toothed, the lateral 2-toothed. Pedunc. 5–6 inches long, densely woolly. Outer invol.-sc. hidden in their copiously woolly covering.

7. S. hispida (DC. l. c. 683) ; suffruticulose, erect, softly hairy ; leaves pinnatisect, the segments linear, mucronate-acute, entire ; heads *nearly sessile* above the uppermost leaves ; outer invol.-scales acute ; paleæ 3-toothed at apex ; rays concolourous, short. *DC. l. c.*

HAB. Cape, *Burchell,* 6998 and 7035 (fide DC.) Cape, *Bowie* (in Hb. Hook).
I have not seen *Burchell's* specimens on which this species was founded. One from *Bowie,* in Hb. Hook, is a branch about 5–6 inches long, closely leafy to the summit. Leaves imbricated, ¾ inch long, with about 3 pair of shortly mucronate pinnæ, densely villous. Heads small, terminal, sessile, the invol. densely villous ; inner scales with an oblong membranous apex of moderate size.

8. S. brachypoda (Harv.); suffruticose, erect, slender, much-branched, softly hairy ; leaves laxly set (short), pinnatisect, the segments 2–3 pair, linear, mucronate, entire, the lowest short ; heads few-flowered, on peduncles scarcely longer than the leaves ; invol. woolly, the outer scales blunt ; paleæ bluntly 3-toothed ; rays coppery at back, short, bidentate.

HAB. Voormansbosch, Swellendam, *Zey.!* 2798. (Herb. Sond., Hook.)
Twelve to 18 inches high, the young stems thinly villous, the older glabrate. Leaves ½ inch long, the upper lobes 3 lines, the lowest ½–1 line long. Heads 3–4 lines in diameter ; the rays not twice as long as the involucre. The habit is that of *S. dentata,* save the very short, scarcely uncial peduncles.

9. S. Dregeana (DC. l. c. 683); stems numerous from a woody crown, short, tufted, subsimple, tomentose ; leaves densely crowded, clothed with long, soft hairs, pinnate-partite, the lobes linear, mucronate, entire (or the uppermost 2–3-fid), on each side several, the lowest shortest ; peduncles very long (8–12 inches), glabrescent, one-headed ; outer inv.-scales ovate, acuminate, thinly pilose ; rays concolourous ; paleæ bluntly truncate.

HAB. On the Gift-berg, *Drege!* 1500–2500 f. (Herb. D., Hk., Sd.)

Stems 2–4 inches high, in pulvinate tufts. Leaves 1½–2 inches long, the lobes not ¼ line wide, sometimes all short and simple, sometimes ¾ inch long and bifid or trifid. Pedunc. reddish, at length glossy and glabrous.

10. S. macropoda (DC. l. c. 683); stems numerous, suffruticose, short, tufted, subsimple, appressedly silky-canescent; leaves crowded, the younger appressedly silky, the older glabrescent, petiolate, pinnate-parted above the middle, the lobes few (2–3 pair), linear, elongate, thickish, mucronulate, simple; pedunc. very long, glabrous; outer inv.-scales linear-lanceolate, thinly silky or glabrate, acuminate; rays concolourous; paleæ subdenticulate.

HAB. On the Giftberg, 1500–2500 f.; on the Kandeberg, 3–4000 f. and Ezelbank, *Drege!* (Herb. Hk., Sd., D.)

Stems 2–6 inches high, closely leafy, ending in a one-headed peduncle, 10–15 inches long. Leaves 2–3 inches long, more than half the rachis petiolar; pinnæ 2–3 pair, ¾–1 inch long, erecto-patent. Membranous tops of the inner inv.-scales fewer and less expanded than usual. Rays pale yellow. With the habit of *S. Dregeana* this differs in pubescence, foliage, and the much narrower and longer invol.-scales.

11. S. sericea (Less. Syn. 243); suffruticose, erect, branched, densely clothed with short, soft, close-lying, silky and silvery tomentum; branches densely leafy; leaves on longish petioles, pinnatisect or bi-pinnatisect above the middle, the lobes long or short, narrow, blunt or acute, simple or divided; peduncles elongate, tomentose, one-headed; invol. thinly tomentose, the outer scales ovate-oblong, obtuse or subacute; rays concolourous; paleæ denticulate. *DC. l. c.* 683. *Arctotis sericea, Thunb.! Cap. p.* 713.

HAB. Cape, *Thunberg!* Cederberg, *Drege!* Tulbagh, *Eckl.!* 24 Rivers, *Zey.,* 825. Alexander's Kloof, *Dr. Wallich!* Winterhoeksberg, *Pappe!* (Herb. D., Sd., Hk., Th.)

Stems diffuse at base, or erect, simple or much-branched, 6–12 inches long. Leaves either simply or doubly pinnati-parted, always with the lower half bare of lobes or *petiolar.* Pubescence very soft and close, pale or white. Pedunc. 6–15 inches long. The inv.-scales vary in being acute or obtuse.

12. S. abrotanifolia (R. Br.); shrubby, erect, branched, clothed with soft, short, curled, pale, woolly hairs; branches elongate, densely leafy; leaves bi-tri-pinnatisect, the segments narrow-linear, divergent, acute, the lowest short and subsimple; peduncles elongate, tomentose, one-headed; outer inv.-scales densely tomentose, acute; rays concolourous; paleæ toothed at the apex. *Less. Syn. p.* 243. *DC. l. c.* 683.

HAB. Paarl-berg, *Drege! Dr. Alexander Prior!* (Herb. D., Hk., Sd.)

A shrub, 1–2 f. high; the erect, curved, leafy branches 10–12 inches long, exclusive of peduncle. Leaves 1½–2 inches long, about an inch wide, the multifid lobes ½ line wide, flat and thinnish. Pubescence greyish, villous on the leaves, shorter on stem and pedunc. Pedunc. 6–10 inches long.

13. S. subhirsuta (DC. l. c. 683); shrubby, flexuous, branched, thinly clothed with soft, curled, woolly hairs; leaves sub-bi-pinnatisect, the lowest 4–5 pair of segments short, simple, the upper 3–4 pair longer, and pinnati-partite, lobes and lobules linear, acute; pedunc. villous, 2–3 times as long as the leaves, one-headed; invol. glabrescent, outer invol.-scales oblong, obtuse, membrane-tipped, brown; rays concolourous; paleæ truncate.

HAB. Near Swellendam, *Mundt! Ecklon!* Grootvadersbosch, *Zey./* 2800. Riv. Zondereinde, *Zey./* 2801. Baviaan's R., Gnadendahl, *Dr. Pappe!* (Hb. Hk., Sd., D.) A shrub, 1–2 f. high. Leaves 2 inches long, 1 inch wide; the lower lobes (pinnæ), 2–3 lines long, the upper 6–8 lines, ¼ line to 1 line diam. Pedunc. 3–4 inches long. Near *S. abrotanifolia,* but much less hairy, with differently cut leaves, and a very differeǹt involucre; the outer scales have a narrow membranous border as well as expanded apex, but the membrane varies much in different specimens, as does also the breadth of the leaf-lobes. *Zey./* 2800 has broader leaf-lobes and more membranous scales than the other specimens seen. *Zey./* 2802 partly belongs to this, partly to *S. grandiflora, DC.*

14. S. dentata (R. Br.); shrubby, much-branched, nearly glabrous; leaves (short) pinnati-partite, the lobes short, entire or trifid, the teeth tipped with a bristle; peduncles elongate, one-headed; heads rather small, outer inv.-scales ovate, subacute or obtuse; rays coppery underneath. *DC. l. c.* 684. *Arctotis dentata, Linn. (non Thunb.) Arct. pectinata, Thunb.! Cap.* 710. *Arct. piligera, Berg.*

VAR. β. **setigera;** stem slightly villous; pedunc. with a few, distant, simple, depauperated leaves; inv.-scales dark-edged, ovato-lanceolate, some of them membrane-tipped. *Sph. setigera, DC. l. c.* (Herb. Hk., Sd.)

HAB. On Table Mt., Capetown and near Simon's Bay, common. Var. β. at Cape L'Agulhas, *Drege!* (Herb. Th., D., Sd., Hk.)
A slender shrub, 1–2 f. high, with curved, closely leafy branches, ending in naked peduncles, 6–9 inches long. Leaves ½–¾ inch long, the lobes 2–3 lines long, rachis broadly linear, all conspicuously gland-dotted and mostly quite glabrous. Outer inv.-scales generally with a few dorsal hairs below the tip, sometimes quite obtuse with a dark border. Invol. ½ inch diameter; rays twice as long. β. scarcely differs from the state of *S. dentata* with dark-edged inv.-scales, which often has scattered leaves on the peduncle, puberulous stems and equally long bristles to the leaves.

15. S. concolor (Harv.); suffruticose, much-branched; branches roughly setose; leaves (short) pinnati-partite, the lobes dilated, broadly linear, short, the lower 2–3-toothed or lobed, the upper 5–7-lobed, all the teeth and lobes tipped with a bristle, otherwise quite glabrous; pedunc. elongate, one-headed; heads rather small, outer inv.-scales ovate, acute, pale; rays concolourous.

HAB. Sneewekop, *Dr. Wallich!* (Herb. D.)
Allied to *S. dentata* which it much resembles in aspect, but from which it differs in the concolourous (pale yellow) rays, more compound leaves, with thinner and broader lobes and hirsute stem. Leaves pale green, ¾–1 inch long, gland-dotted. Pedunc. 4–6 inches long. Heads 3–4 lines diameter.

16. S. nudicaulis (Less.! Syn. 243); glabrous; stems very short, tufted, simple, densely leafy; leaves sub-radical or cauline, petiolate, pinnate-partite beyond the middle, the segments 5–7 pair, alternate, erecto-patent, simple or the uppermost rarely 2–3-fid, apices obtuse or subacute; peduncles very long, one-headed, nude or with 1–2 minute bracts; outer inv.-scales lanceolate, acute, puberulous, medial ovate; rays concolourous. *DC. l. c. Arctotis nudicaulis, Thunb.! Cap.* 711.

VAR. β. **gracilior;** leaves twice as long, their lobes acute. (Hb. Thunb., Sd., D.

VAR. γ. **alpina;** dwarf; stems procumbent, throwing up many short, crowded, leafy branches, prolonged into *sparsely leafy* peduncles; leaves small (¾-inch long); inv.-scales ovate; rays concolourous?

HAB. Summit of Table Mt., common, *Th., Burch,, E. Z.! W. H. H., &c.* Var.

β, Lambert's Kloof, *Dr. Wallich* ; Cape, *Thunb.!* Var. γ, Summit of Sneewekop, *Dr. Wallich !* (Herb. D., Sd., Th.)

Stems decumbent at base, ascending, densely tufted, the leafy portion 1–2 inches long: an underground or trailing leafless trunk, sometimes 6–8 inches. Leaves 1½–2 inches, in β. 3–4 inches long, the lobes ¼–½ inch. Peduncles 8–10 inches long, with a few slender, filiform, scattered bracts. Inv.-scales rigid, dark coloured. Rays yellow, twice as long as the inv. Var. γ. may be a species. Its achenes are powdery and shortish.

17. S. scapiformis (DC. 1. c. 684, ex pte.) ; glabrous or nearly so ; stems short, tufted, simple, densely leafy ; leaves subradical and cauline, pinnati-partite throughout their whole length, segments 10–12 pair, the lowest short, subulate, acuminate, the upper longer, mostly 2–3-lobed, mucronate ; pedunc. very long, slender, one-headed, nude ; outer inv.-scales ovate or oblong, obtuse or subacute ; rays concolourous, not much longer than the inv.

HAB. Langekloof, *Drege!* Descent from Sneewekop, *Dr. Wallich!* (Hb. Sd., Hk., D.)
A smaller and slenderer plant than *S. nudicaulis* (to which it is closely allied)> with more numerous and taper-pointed leaflobes, and much smaller flower-heads· It differs from the following by the concolourous rays, &c.

18. S. chamomillæfolia (DC. 1. c. 684) ; glabrous or thinly downy on the young parts ; stems decumbent or ascending, shortish, simple (or branched) ; leaves scattered, pinnati-partite throughout the segments, in 8–10 pairs, the lowest short and subulate, acuminate, the upper longer, 2–3-lobed, mucronate ; pedunc. mostly very long, slender, one-headed ; outer inv.-scales ovato-lanceolate, acute, dark-edged ; rays dark coppery beneath, twice as long as the involucre. *Also S. scapiformis,* *DC. quoad Sp. Eckl.!*

VAR· β. **elongata** ; stems 1 foot long, branched, suberect ; pedunc. 4–6 uncial. *S. chamomillæfolia, Herb. Eckl.!*

HAB. Tulbagh and Worcester, *Eck.!* Waterfall, Tulbagh, *Dr. Pappe !* β. Worcester, *Eck. !* (Hb. Sd., Hk., D.)
Stems 4–6 inches long ; in β. much longer and more erect, sparsely leafy. Leaves 1½–2 inches long, the upper lobes 3–4, the lower 1–2 lines long. Pedunc. (save in β.) 6–12 inches long. This chiefly differs from *S. scapiformis* by its discoloured rays and longer and less glabrous stems.

19. S. Natalensis (Sch.! B. in Walp. Rep. 6, p. 182) ; suffruticose, glabrous or nearly so ; stems decumbent or ascending, short, subsimple ; leaves pinnati-partite beyond the middle, the segments in 3–4 pairs, elongate, linear, apiculate, the upper sometimes 2-lobed ; pedunc. elongate, nude, outer inv.-scales ovate, subacute, dark-edged ; rays discoloured, twice as long as the involucre.

HAB. Port Natal, *Krauss!* 446. *Gueinzius,335, Dr. Sutherland!* (Hb. D., Hk., Sd.)
Very near *S. chamomillæfolia* from which it chiefly differs in its less compound foliage. Stems 3–4 inches long, closely leafy below, ending in a 3–5 inch long peduncle. Lvs. 1½–2 inches long, the lobes 4–6 lines long, mostly simple, rarely bifid.

20. S. anethifolia (Less. Syn. 243) ; erect or ascending, shrubby, glabrous or nearly so ; leaves pinnati-partite, the lobes linear-filiform, acute, the lower ones short (or obsolete), entire, the upper often 2–3 fid ; pedunc. long, one-headed ; outer inv. scales from a broad base, linear-acuminate, herbaceous ; ligules discoloured, 1–2ce as long as involucre. *DC. l. c. p.* 684.

VAR. a; leaves from the base pinnati-lobed ; upper lobes mostly entire.

VAR. β; leaves pinnati-lobed beyond the middle only; lobes long and often 2–3 fid.

HAB. Var. a, Zwarteberge, *Zey.!* 2804. Caledon, *Dr. Pappe!* Simon's Bay, *C. Wright!* 399. Var. β. Cape, *Verreaux! Thom!* Sweet Valley, *Wallich!* (Herb. D., Hk., Sd.)

A more robust, more woody, taller and more erect plant than *S. chamomillæfolia*, to which it is closely allied; it may also be compared with *S. anethoides* and *S. crithmifolia*, from which its discoloured rays separate it.   Leaves 1½–2 inches long, the lobes ¾–1 inch; in var. β. many of the leaves are only 3-forked, the lower lobes disappearing; *Mr. Wright's* specimen connects a and β.

21. S. anethoides (DC. l. c. 685); shrubby, erect, branched, glabrous or thinly cob-webby; leaves pinnati-partite throughout, the lobes 6–8 pair linear-filiform, mucronate, the lowest short, entire, the upper frequently 2–3 fid; pedunc. long or shortish, one-headed; outer inv. scales lanceolate, acute, medial oblong, obtuse, membrane-tipped; rays concolourous, about twice as long as the invol.; paleæ truncate.

VAR. β, **brachyglossa** (DC.); rays not much longer than involucre.

VAR. γ. **ramosissima** (DC.); slender, all parts smaller; leaves short; stem much branched.

VAR. δ. **brevipes**; pedunc. short (1–3-uncial); leaves closely set, mostly simply pinnate.

HAB. Breede River and Drachenstein, *Drege!* Worcester, *Ecklon!* Tulbagh and Caledon, and Gnadendahl, *Dr. Pappe!* Witsenberg, *Zeyher!* 826. β. Piquetberg, *Drege!* γ, Giftberg, *Drege!* Tzitsikamma, *Dr. Pappe!* δ, Albany, *T. Williamson! T. Cooper!* 1558. Grahamstown, *Genl. Bolton!* New Year's River; *Mrs. Barber* 257. (Herb. D., Sd., Hk.)

An erect shrub, 1–2 f. high or more, rather robust, the young branches mostly cob-webby. Leaves rather crowded 1–1½ inches long, the upper lobes 4–6 lines long, (except in β. and γ.), semiterete, channelled above, the lower much shorter. Pedunc. commonly 6–8 inches long, sometimes 4–2 inches, erect, sometimes minutely bracteate. Inner invol. scales with a very ample, silvery-membr. apex. Ray bright golden yellow. Variable in size, the heads in γ. very small; in a, nearly an inch in diameter. Var. δ has uniformly much shorter peduncles and closer leaves ; the latter ¾ inch long.

22. S. crithmifolia (R. Br.); shrubby, glabrous or nearly so, erect, branched ; leaves pinnati-partite (or trifid), the lobes linear-filiform, semi-terete, acute, the lower lobes short or obsolete, the rest elongate ; pedunc. unusually elongate, bearing a few, simple, erect, scattered leaves below ; outer invol. linear, acute, or subacute, or obtuse, the medial lanceolate, the inner amply membrane-tipped; rays concolourous; paleæ truncate. *Less. Syn.* 244. *DC. l. c.* 686. *Arctotis crithmifolia, Berg. Cap.* 326.

VAR. β, **grandiflora**; more luxuriant; larger in all parts, especially the fl. heads, leaf-lobes more numerous, scarcely otherwise different. *S. grandiflora, DC.! l. c.* 685. Zey.! 2802 (ex pte.), 2803. *S. nudicaulis, Hb. Eckl.!* 1706.

VAR. γ., trifurcata (DC.); leaves (mostly) trifurcate, petiolate. *Arctotis trifurcata, Burm.*

HAB. Mountains near Capetown and Simon's Bay, *E. & Z. Drege! W.H.H. &c.* Gnadendahl, *Dr. Pappe!* β. Hott. Hott.-berg, Herb. *Deless.* Draakensteenberg, *Drege!* Stellenb., *W.H.H.* γ, Capetown, *W.H.H.*, M'*Gillivray.* (Hb. D., Hk., Sd.)

A stout shrub, 1–2 ft. high, densely leafy. Leaves 1½–2 inches long, the upper lobes ¾–1 inch long, ½ line diameter, the lower lobes in γ obsolete or wanting, in β 5–7 pair, in a 2–4 pair, always much shorter than the upper. Pedunc. 3–8–10 inches long. Heads 1–1½ diam., with long, bright-yellow, gland-dotted rays.—A variable plant, but I cannot draw a line between var. a and β.

23. S. punctata (Less.! Syn. 243); shrubby, slender, erect, virgate, nearly glabrous; leaves (small) pinnati-partite, the lower and medial throughout their whole length, the lower lobes simple, upper bifid or trifid, all mucronate and very short; upper leaves lobed beyond the middle only, pinnate or trifid; pedunc. elongate, slender; heads small; outermost inv.-scales very few, small, acute, the medial oblong, obtuse, broad; rays concolourous, short; paleæ truncate. *DC. l. c. p.* 685. *Also S. gracilis, DC. l. c.* 684. *Arctotis punctata, Thunb.! Cap.* 711.

HAB. Cape, *Thunberg!* Zeederberg, *Drege!* (Herb. Th., Hk., Sd.)
A slender shrub, 1-1½ ft. high; branches rodlike, 6-8 inches long, ending in pedunc. 4-6 inches long. Leaves about ½ inch long, the lobes 1-2 lines long; all the upper leaves are petioled, or nude in their lower half and mostly trifid at the apex. Heads few-fl., 3-4 lines across. Leaves gland-dotted as in other species. I cannot find any difference between *Thunberg's* original specimens and *Drege's* "*S. gracilis.*"

24. S. pauciloba (DC. l. c. 686); "shrubby, branched, erect, glabrous; leaves (short) coriaceous-fleshy, pinnatisect, the lobes mostly 5, subobtuse, entire, the 3-terminal near together, the uppermost linear, undivided; the moderately-elongate pedunc. and the young branches somewhat cobwebby; outer inv.-scales subobtuse, dark-edged; rays concolourous, not much longer than the invol.; paleæ truncate." *DC.*

HAB. Swellendam, on the Mountain, 4-500 f., *Drege*. (Hb. Sond., sine fl.)
Judging by a scrap, without flowers, in Hb. Sond., this seems near *S. punctata*. The leaves are ½-¾ inch long, the lobes 2-3 lines long, mucronate.

25. S. rigidula (DC. l. c. 685); suffruticose, erect or ascending, glabrous or nearly so; leaves pinnati-partite, rigid, the lobes short, 6-8 pair, rigidly mucronate, the lowest shortest, entire, the upper either entire or 2-3-fid; pedunc. elongate, naked; outer inv-scales oval-oblong, obtuse, with a narrow scarious edge and small membr.-tip, obtuse; rays discoloured.

HAB. Piquetberg, *Drege*. Hassaquaskloof and Buffeljads R., *Zey.!* 2799. (Hb. Sd.)
Of *Drege's* plant I have only seen a scrap without flowers. *Zeyher's* specimen, which seems the same, is diffusely branched, with subdistant leaves, the leafy portion of the branches 6-8 inches long, the pedunc. about 6 inches. Leaves 1-1½ inch long, the upper lobes 3-4 lines long, the lower 1-2 lines, each pair 2-3 lines apart. Inv.-scales glabrous, brown, with a narrow, white border. Pappus not seen, the disc-fl. having been insect-eaten; the genus therefore uncertain.

26. S. discolor (Less.! Syn. 243); suffruticose, diffusely branched, thinly cobwebby, becoming glabrescent; leaves (small) pinnati-partite, glaucous, the lobes in 2-3 pairs mucronate, lowest lobes very short (or none), upper longer; some leaves simply trifid; pedunc. elongate, often with a few linear, entire, appressed leaves at base; heads small; outer inv.-scales oval-oblong, somewhat membrane-edged and tipped, obtuse; rays coppery beneath; paleæ truncate. *DC. l. c.* 685. *Also S. brevifolia, DC.! l. c. Arctotis dentata, Thunb.! non L. Sph. pilifera, γ. sub-canescens, DC.! in Hb. Drege, Prodr.* 5, *p.* 688.

HAB. Cape, *Thunb.!* Uitenhage and Caledon, *Ecklon!* Swellendam, *Dr. Pappe! Drege! Zey.!* 2795. (Herb. Th., D., Hk., Sd.)
A diffuse or divaricate, much-branched half-shrub, 1-1½ ft. long, all the younger parts covered with prostrate, deciduous, cobwebby hairs. Leaves ½-¾ inch long,

the upper ones often only 3-fid, the lobes often hook-pointed. Heads 4–5 lines diam., the inv.-scales reddish-brown, with scabrous spots under the membranous tip, and a very narrow membr.-edge. Rays twice as long as the invol.

27. S. chamæmeloides (DC. l. c. 685); "suffruticose, glabrous; leaves pinnatisect, the lobes linear, acute, the (lower) entire, the (upper) mostly trifid; outer inv.-scales triangular, acuminate-aristate; rays discoloured." *DC. l. c.*

HAB. Algoa Bay, *Forbes* (fide DC.)
"Fl. branches sparingly leafy, 1–2 inches long. Most of the inv.-scales scarious-membranous. Ligules 5–7-nerved," *DC.* In translating the specific char. from DC. I have ventured to transpose the words "upper" and "lower," which seem to be accidentally misplaced in the Prodr. text. Possibly an *Ursinea;* and if so, it may be a mere variety of *U. chrysanthemoides?*

28. S. paleacea (Less. Syn. 244); shrubby, nearly glabrous; branches slender, virgate; leaves somewhat fleshy, filiform, the lowest pinnate-partite, 2-jugate, the medial trifid at the apex, the uppermost linear, quite entire; apices bluntish; pedunc. long, 1-headed, with a few scattered, linear leaves at base, outer and medial invol. scales very narrow, linear-acuminate or subulate, rigid, innermost with a very small, membranous tip; rays discoloured, scarcely as long as the involucre; paleæ truncate, subdenticulate. *DC. l. c. 685. S. subflosculosa, E. Mey.! DC. l. c. 686. Arctotis paleacea, Th.! Cap. 712.*

HAB. Cape, *Thunb.!* Cederberg, *Drege!* (Herb. Th., D., Sd. Hk.)
A slender shrub, 1–2 feet high. Leaves laxly set, ½–¾ inch long, the lobes 2–3 lines long, mostly blunt. Pedunc. 3–4 inches long, erect. Heads 4–6 lines in diameter, the rays scarcely as long as the small-topped inner scales. *Thunberg's* specimens precisely accord with *Drege's.*

29. S. pinnata (Less.! Syn. 243); shrubby, glabrous, much branched; branches virgate, spreading; leaves (small) sessile, half-clasping, pinnati-partite, the lobes linear-filiform, mucronate, in 2–4 pair; upper leaves tri-partite or simple; inflorescence *panicled*, the fl. branches rigid, spreading, sprinkled with a few linear, simple leaves; heads small; outer invol. scales linear-oblong, obtuse; rays very short, discoloured; paleæ obtuse. *DC. l. c. 686. Relhania pinnata, Thunb.! Cap. 641. Arctotis pinnata, Thunb.! ex pte. Thelythamnos filiformis, Spr. f. Suppl. 25.*

HAB. Cape, *Thunb.!* Drachensteinberg and Giftberg, *Drege!* Caledon, *E. & Z.!* Baviaan's Riv., *Pappe!* Cape, *Wallich!* (Herb. Th., D., Hk., Sd.)
2–3 feet high, the stout main trunk breaking into many spreading, rodlike, simple or again divided, rigid branches, which are closely covered with leaves in their lower half, becoming nearly naked or with a few distant, depauperated leaves upwards for 8–10 inches below the summit, then *panicled;* the rigid, divaricating arms of the panicle 1-headed. Leaves ½–¾ inch long, the lobes 4–6 lines long, slender.— This is readily known by its panicled inflorescence.

30. S. quinquepartita (DC. l. c. 686); shrubby, becoming nearly glabrous, the straight, rodlike branches closely covered with leaves to the very apex; leaves half-clasping, pinnately 5-partite nearly to the base, the lobes linear-filiform, erect, acute, the undivided part broadish, flattened, persistent on the older branches; heads terminal, solitary, subsessile; outer inv. scales ovato-lanceolate, subacute; rays discoloured. *Chronobasis quinquepartita, Hb. Eckl.*

HAB. Stellenbosch, *E. & Z.!* (Herb. Sond.)

2–3 feet high, rigid, with very straight stem and erect branches, imbricated with leaves. Leaves ¾–1 inch long, the lobes 8–9 lines long, the 3 terminal slightly removed from the basal pair. Heads 5–6 lines diameter, the rays about twice as long as the invol., dark-coloured beneath. Young branches and leaves pubescent.

31. S. tripartita (DC. l. c. 686); shrubby, at first pubescent, becoming nearly glabrous, the straight, rod-like branches leafy to the apex; leaves half-clasping, 3-parted nearly to the base, the lobes linear-filiform, erect, acute, the undivided part broadish, flattened, persistent as a scale on the old branches; heads terminal, solitary, subsessile; outer inv. scales oblongo-lanceolate, obtuse, puberulous; rays discoloured. *Chronobasis tripartita, Hb. Eckl.*

HAB. Cape, Hb. *Lambert* (fide DC.); Swell., *E. Z.!* Hott.holl., *Drege!* (Hb. Sd.) Precisely like *S. quinquepartita,* save that the leaves are *tripartite.* The pubescence is either a little more copious or more persistent.

32. S. trifida (Less.! Syn. 244); shrubby, slender, much-branched; leaves linear-cuneate, either all shortly three-lobed at apex or some quite entire, the lobes flat, bisulcate beneath, obtuse, much shorter than the undivided portion; peduncles short, slender; heads small; outer inv. scales linear-oblong or lanceolate, acute or obtuse; rays concolourous. *DC. l. c.* 687. *Also S. triloba, DC. l. c. Arctotis trifida, Th.! Cap.* 711.

VAR. β. **gracilis**; more slender, many of the leaves simple; inv.-scales narrower and more acute.

HAB. Cape, *Thunb.! Bowie!* Swellendam, *Burchell, E. Z.!* Voormansbosch, *Zey.!* 2797. β., Cape, *Bowie!* in Hb. D. (Herb. Th., D., Sd., Hk.)

1–2 feet high, distinctly woody, though slender, the old branches bare of leaves. Leaves 3–5 lines long, either all shortly trifid, or trifid and simple intermixed, conspicuously dotted. Pedunc. 1–1½ inch long. Heads 3–4 lines diam. The rays in *Thunberg's* specimen, though some are faded, do not appear to have been originally discoloured—on one or two flowers they still retain their yellow. There seems therefore no reason for keeping *S. triloba,* DC. apart, on account of a supposed difference in colour of ray.

33. S. heterodonta (DC. l. c. 687); suffruticose, glabrous (or cobwebby), laxly leafy; leaves mostly trifid at top, petiolate, the two lateral lobes usually longest, linear-filiform, acute, the medial lobes short, toothlike and somewhat hooked; rarely all three nearly equal, often the upper leaves simple, entire, or with toothlike lateral lobes; pedunc. elongate, one-headed; outer inv.-scales acuminate, medial membrane-tipped; rays discoloured.

HAB. Eastern Districts, *Burchell, Verreaux!* Uitenhage, *E. Z.!* *Zey.!* 2795 β. (Herb. D., Sd., Hk.)

About a foot in height, slender, laxly leafy. Leaves ½–¾ inch long, the lower and medial ones mostly trifid, the uncleft portion twice as long as the lobes, the medial lobe *mostly* very short. Heads many-fl., ½–¾ inch across.

34. S. trifurca (Harv.); suffruticose, glabrous; stems decumbent, subsimple; leaves half-clasping at base, thickish, mostly 3-forked, the uppermost simple, rigidly ciliate in the lower half, the lobes subequal, linear terete, acute; pedunc. terminal, elongate, with a few scattered, small leaves; outer inv.-scales linear-oblong, brownish, medial slightly membr. tipped; rays discoloured.

10*

HAB. Near the mouth of the Potrivier, *Zey.!* 2805. (Hb. Hook., Sd.)

Stems 3-8 inches long, prostrate, except at the upturned extremity, densely leafy. Leaves 1-1½ inches long, curved upwards. Rays yolk-yellow, brownish beneath. With the habit of *S. nudicaulis,* this chiefly differs from *S. ciliaris* by the 3-forked leaves ; it may perhaps be a variety ?

35. S. ciliaris (DC. l. c. 687) ; suffruticose, glabrous ; stems erect or ascending, subsimple (tufted ?) ; leaves half-clasping at base, linear-subulate, elongate, thickish, glaucous, acute, *rigidly ciliate* in the lower half, entire above ; pedunc. terminal, elongate, nude or with 2–3 scattered, small leaves, one-headed; outer inv.-scales linear-oblong, brown-edged, medial membr. tipped ; rays elongate, discoloured.

HAB. At Cape L'Agulhas, near the sea, *Drege!* (Herb. Hook., Sd.)

Stems 6-8 inches long, closely leafy. Leaves very erect, rigid when dry, 1½ inch long. Pedunc. 8-10 inches long, scapelike. Heads ¾ inch diam., the rays at least twice as long, golden-yellow above, dark coppery beneath.

36. S. tenuifolia (DC. l. c. 687); suffruticose, many-stemmed, tufted ; stems short, ascending, subsimple ; leaves clasping at base, *very long* (2–3 inches), narrow-linear, flattish, acute, glabrous ; pedunc. elongate, one-headed ; outer inv.-scales narrow, medial ovate-oblong, membrane-tipped, inner amply membranous ; rays concolourous. *Arctotis tenuifolia,* Linn. *Mant.* 288. *Ursinia tenuifolia, Poir.*

VAR. β. **heterochroma**; rays discoloured !

HAB. Cape, *Eckl.!* Cape Flats, *W. H. H.* Witkamp, *Zey.!* 3112. β. near Simonstown, *C. Wright,* 372. (Herb. D., Sd.)

Stems numerous, 4-8 inches long, decumbent at base, closely leafy. Leaves 2-3 inches long, ½-¾ line wide, almost always quite simple, very rarely a pinnati-parted leaf occurs intermixed. Pedunc. 6-10 inches long, naked. Rays in α, pale golden yellow on both sides; in β, which in other respects is identical, orange-yellow above, coppery beneath. The habit is that of *S. nudicaulis.*

37. S. leptoglossa (DC. l. c. 687); "suffruticose, slender, erect, branched; branches sub-dichotomous, glabrescent; leaves linear, acute, punctate, glabrous, quite entire (5 lines long, ½ line wide); heads terminal, *subsessile,* small ; outer inv. scales oblongo-linear ; rays very narrow, subentire, sparingly glandular." *DC. l. c.*

HAB. District of Swellendam, *Burchell,* No. 6897 (fide DC.)

38. S. Eckloniana (Sond.! in Linn. 23, p. 64); shrubby, robust, erect, densely leafy; leaves sub-imbricating, linear-lanceolate, acute, flat, quite entire, 3–nerved, sessile, or both sides impressed-dotted, minutely downy beneath; pedunc. 2–4 times longer than the leaves, one-headed, pilose; outer inv. scales lanceolate, acute, hispidulous; rays concolourous.

HAB. Kleinriviersberge, 4th height, *Ecklon!* (Herb. Sond.)

Stems 2-4 ft. high, branches as thick as a goose quill, the young parts pubescent. Leaves 1½-2 inches long, 3-5 lines wide, broad and half clasping at base. Pedunc. 2-4 inches long; outer inv. scales herbaceous, brownish-edged, achene cylindrical, glabrous, furrowed, girt with soft, white hairs at base. Pappus of 5-6 obovate, obtuse, white scales.—A remarkable species, with the habit of *S. serrata,* but very different involucre and foliage.

Div. 2. **XEROLEPIS** (Sp. 39–44).

39. S. serrata (DC.! l. c. 688); suffruticose, robust, puberulous, becoming glabrate; leaves crowded, half-clasping, oblong or lanceol.-oblong, nerved, coarsely and sharply serrate; pedunc. terminal and axillary, numerous, long, sub-corymbose, with a few small, subulate leaf-scales; the invol. scales *amply* membrane tipped; rays concolourous; achenes quite glabrous. *Arctotis serrata, Thunb.! Cap. 705. Ursinia serrata, Poir. Sph. Sonneratii, Cass.*

HAB. Cape, *Thunb.! Burchell,* 5965. *Bowie!* Swellendam, *Mundt.!* Grootvadersbosch, *Zey.!* 2806. (Hb. Th., D., Hk., Sd.)
Stems 2–3 ft. high, 3–4 lines in diameter, but full of pith, imbricated with leaves throughout. Leaves 1–1½ inches long, 3–5 lines wide, obtuse, equally serrate throughout, the serratures obliquely deltoid. Invol. ½–¾ inch diam., all the scales amply membranous and convex. Rays golden yellow. *Thunberg's* specimens are more slender than the others cited above, with narrower and longer and more pubescent leaves, but otherwise the same.

40. S. incisa (DC. l. c. 688); "suffruticose, puberulous; leaves elongate-oblong, dilated upwards, inciso-serrate, the serratures acuminate, the lower ones small and distant, the upper close and longer; pedunc. four times longer than the leaves; rays concolourous." *DC. l. c.*

HAB. Swellendam, *Burchell,* No. 7008 (fide DC.)
"Leaves 1½ inch long, 2–3 lines wide at base, 4–5 lines at apex." *DC.*

41. S. tridentata (Harv.); suffruticose, villoso-pubescent; leaves sessile, oblong-tongue-shaped, one-nerved, inciso-pinnatifid, *the marginal and terminal lobes sharply 2–3 toothed at the truncate apex,* those of the upper leaves simple or bidentate; pedunc. elongate, numerous, with a few small, subulate leaf-scales; all the inv. scales *amply* membrane-tipped; rays concolourous; achenes with a few basal hairs.—*Sph. serrata, Herb. Eckl.!*

HAB. Cape *Ecklon!* (Hb. Sond.)
Much more villous than *S. serrata,* less robust, with differently cut leaves. Leaves 1 inch long, the undivided portion 1–2 lines wide, the lobes 1–2 lines long. Heads smaller than in *S. serrata.*

42. S. odorata (R. Br.); "shrubby, glabrous; leaves flat, inciso-pinnatifid at the apex; pappus obsolete; rays discoloured." *Less. Syn.* 244. *DC. l. c.* 688. *Anthemis odorata, Ait. Kew. 3. p. 238.*

HAB. Formerly cultivated in England from Cape seeds: now lost in gardens.

43. S. pilifera (Less.! Syn. 244, non DC.?); suffruticose, thinly pilose; leaves pinnati-partite, the lobes 7–12 pairs, narrow-linear, the lower simple, the upper 2–3 fid, all tipped with a long, white bristle, dotted, nearly glabrous; pedunc. elongate, pubescent near the top; all the invol. scales *amply* membrane edged and tipped; ligules discoloured; paleæ lacerate-toothed. *Arctotis pilifera, Thunb.! Cap.* 712.

HAB. Cape, *Thunberg!* Riv. Zonder Einde, *Zey.!* (Herb. Th., Sd.)
Slightly woody at base, simple or branched from the base, the leafy portion 6–8 inches long, ending in a striate peduncle 6–9 inches long. Leaves 1–1½ inch long, the lower lobes 3–4 lines long, the upper rather longer, and except on the upper leaves, 2–3 fid, all dotted, tapering to the point. Heads an inch in diameter, the rays white above, coppery beneath.

44. S. scariosa (R. Br.); "shrubby, glabrous; leaves pinnatisect, the segments linear-filiform, very entire, mucronulate; pedunc. 2–3 times longer than leaves, outer inv. scales altogether scarious, subrotund; rays sub-discoloured; paleæ truncate; achenes softly and thinly villous." *DC.l.c.p.*688. *Less.Syn.*244. *Arctotis scariosa, Willd. excl. syn. Th. Ursinia scariosa, Poir.*

VAR. *a,* brevipes (DC.); peduncles very short.
HAB. Cape, *Burchell,* No. 5451. β, Cape, *Drege* (fide DC.)

Doubtful Species.

S. pilifera (DC. l. c. p. 688, non Less.); "shrubby, diffuse; leaves pinnati-sect, fleshy, spreading and shortly hispid, lobes linear-tipped with a bristle; pedicels long, sub-hispid; outer inv. scales obtuse, with a narrow membranous margin; rays discoloured; paleæ truncate; pappus scales obovate, rather shorter than the achene," *DC.—Bot. Reg. t.* 604, *and Jacq. Schœnbr. t.* 155, *fide DC.*

XLIX. URSINIA, Gærtn.

Character the same as that of *Sphenogyne,* except: 1, *Achenes* obovate or pear-shaped, oblique, distinctly tapering to the base, quite glabrous, with 5 obscure ribs; 2, *Pappus* biseriate, the outer (as in *Sphenogyne*) of 5 broad, obovate, white, spirally rolled scales, the inner of 5 slender, white bristles. *DC. l. c.* 688.

Herbs or suffrutices with completely the habit of *Sphenogyne,* and undistinguishable except by the above characters. Leaves pinnate-partite. Name unexplained.

Root perennial. Stem suffruticose or herbaceous. Rays mostly discoloured.
  Stems short, densely tufted. Pedunc. scape-like.
    Leaf-lobes thickish, bluntly mucronulate ... (1) **montana.**
    Leaf-lobes tipped with a white, subulate point (2) **apiculata.**
  Stems elongate, diffuse, branched.
    All the invol. scales membr. tipped, obtuse ... (3) **paradoxa.**
    Outer inv. scales ovate, subacute: paleæ toothed:
      Quite glabrous; stem shorter than pedunc. (4) **tenuiloba.**
      Glabrescent; stem 6–12 inches long ... (5) **affinis.**
    Outer inv. scales ovate-acuminate, very acute (6) **chrysanthemoides.**
Root annual. Rays concolourous.
    All the inv. scales membrane-tipped, obtuse... ... (7) **speciosa.**
    Outer inv. scales ovato-lanceolate, very acute ... (8) **cakilefolia.**
    Outer inv. scales ovate, obtuse or subacute ... ... (9) **annua.**
    Outer inv. scales ovate, setaceo-acuminate. Paleæ
    tipped with a reniform scale ... ... (10) **pygmæa.**

1. U. montana (DC. l. c. 688); shrubby, ascending, depressed, scarcely branched, the short stem and branches pubescent; lvs. densely crowded, pinnati-partite, pubescent, becoming glabrate, the lobes thickish, terete, "the lowest very minute, and acute" (*DC.*), the upper linear or bifid, bluntly mucronulate; pedunc. naked, elongate; outer invol. scales glabrous, ovate-oblong, obtuse, dark-edged and membrane-tipped; rays small, discoloured; paleæ truncate.

HAB. On the Witbergen, 5–8000 ft., *Drege!* (Herb. Hook.)
I have only seen a small, imperfect specimen. The leaves are scarcely uncial, pinnated *above the middle,* the lowest lobes (above described) not visible; several of the upper lobes bifid.

2. U. apiculata (DC. l. c. 689); shrubby, tufted, depressed, glabrous; leaves pinnati-partite, all the lobes tipped with a white, conico-subulate, cap-like mucro, the lower lobes very short, subulate, the upper longer, linear or 2–3-fid; pedunc. naked, elongate, outer inv. scales ovate, obtuse, dark-edged and slightly membrane-tipped; rays discoloured; paleæ obtuse.

HAB. On the Sneeuwebergen, 5–6000 feet, *Drege!* also collected by *Ecklon!* Elandsberg, *T. Cooper!* 218. Kreilis country, *H. Bowker!* 281. Natal, *J. Sanderson!* (Herb. Hk., Sd., D)

Root thick and woody. Stems numerous, 3–4 inches long, subsimple, forming a dense tuft. Leaves about uncial, the lobes below the middle reduced to teeth, the upper 3–4 lines long. In habit it is like the foregoing, but differs from that and the following by the remarkable, ivory-white tips to the leaf-lobes. These are, however, variable, and much less evident in the specimens from Kreilis country and Natal, which in other respects are very similar.

3. U. paradoxa (Gærtn. Less.! Syn. 245); suffruticose at base; stems diffuse, ascending, nearly glabrous; leaves pinnati-partite, the lower lobes short, subulate (or tooth-like), entire, the upper linear, simple or 2–3-fid, all sharply mucronate; peduncles longish; all the invol. scales membrane-tipped, obtuse; rays strongly discoloured; paleæ subtruncate (not appendiculate). *DC. l. c.* 689. *Arctotis paradoxa, Linn. Th. Cap.* 712.

HAB. Cape, *Thunberg !* (Herb. Thunb.)

Stems 6–12 inches long, simple or branched, laxly leafy. Leaves thickish, 1½ in. long, the upper lobes 3–5 lines long, all tapering to a very acute point, but scarcely setaceous. Heads at least an inch across, with long rays; outer inv. scales brown, rigid, with an expanded, membrane tip. Rays coppery beneath.—I have only seen this in Hb. Thunb.; *Drege's* specimens, quoted by DC., so far as I have seen, belong to *U. chrysanthemoides* as does also one of the specimens, marked "*paradoxa*," in Herb. Thunb.—These species are easily distinguished by their inv. scales and paleæ.

4. U. tenuiloba (DC. l. c. 689); "quite glabrous; stem suffruticulose, subdecumbent, short, sparingly branched; leaves pinnate-parted, the lobes linear, aristate-acuminate, the lower shorter, entire, the upper 2–3-lobed; pedunc. longer than the stem; paleæ obtusely toothed; inner pappus a little shorter than the outer." *DC. l. c.*

HAB. Katriversberg, *Drege* (fide DC.).

Ecklon's "*Sph. fœniculacea,*" from Uitenhage, referred to by *DC.*, seems to me to belong to *U. annua.*

5. U. affinis (Harv.); perennial (?), herbaceous, many-stemmed, diffuse or procumbent, glabrescent or hispidulous; leaves pinnati-partite or sub-bipinnati-partite, the lower lobes short or tooth-like, subulate-acuminate, the upper 2–5-fid; ped. elongate; rays short, discoloured; outer inv. scales ovate, subacute, dark-edged; paleæ obtusely 3-toothed; inner pappus rather longer than the outer.

HAB. Cape, *E. Z.!* 1494; between Hassaquaskloof and Riv. Zondereinde, *Zey.!* 2793. (Herb. Sond.)

Possibly the same as *DC.'s U. tenuiloba.*" Stems 6–12 inches long, laxly leafy. Leaves 1–1½ inch long, the lower lobes sometimes wanting, often reduced to teeth. Heads 3–4 lines diameter, smaller than in *U. annua*, all the scales black-edged, pale-green on the disc; the innermost with a dark-coloured scale. The stems are more rigid than in *U. annua*, the rays discoloured; the root very fibrous, and *possibly* perennial.

6. U. chrysanthemoides (Harv.); half-herbaceous or suffruticose, glabrescent; stems diffuse or ascending, often rooting from the lower nodes; leaves sub-bipinnati-partite, the lobes linear, acute or mucronate, spreading, either all compound or the lower ones simple, the upper 2–3-fid or pinnate; pedunc. elongate; outer inv. scales ovate-acuminate or sublanceolate, very acute, glabrous; paleæ tipped with a roundish-reniform scale, the outer ones sheathing, the inner narrow, linear-spathulate; rays usually *more or less* discoloured; inner disc-fl. often (if not always) abortive. *Sphenogyne chrysanthemoides, Less.! in Hb. Thb.! DC. l. c.* 682. *Arctotis anthemoides (ex pte.) and Arct. paradoxa (ex pte.). Thunb.! in Herb. Ursinia albicaulis, U. pedunculosa, U. radicans, and U. filicaulis, DC. l. c.* 689. *Zey.!* 2796, 2794.

HAB. Wet places, throughout the Colony. Cape Flats, *E. Z.! Drege!* Simonstown, *C. Wright!* 398 ; Steendal, Tulb., *Dr. Pappe!* Swellendam, and on the Gariep, Namaqualand, *Drege!* Algoa Bay and Uitenhage, *Drege! Zey.!* &c. (Hb. Th., D., Hk., Sd.)

6–12 inches high, diffusely branched. Stems sometimes cobwebby *(U. albicaulis).* Leaves 1–2 inches long, mostly bipinnate, sometimes very slender *(U. filicaulis).* Pedunc. 3–6 inches long or longer *(U. pedunc.).* Rays ¾ inch long, sometimes *concolourous,* but more commonly faintly, or fully discoloured. It is known from *U. paradoxa* by the inv. scales and paleæ. I can find no limits between the four species of DC. here united. *Lessing's* plant (in Hb. Thunb.) is evidently the same, though not being in fruit, he has mistaken its genus.

7. U. speciosa (DC. l. c. 690); herbaceous, annual, erect, branching, glabrous ; leaves pinnati-partite or sub-bi-pinnati-partite, the lobes linear, flat, midribbed, the lower short and simple, the upper pinnatifid and toothed, all bristle-pointed; pedunc. elongate; all the inv.-scales tipped with a broad, obtuse membranous-scale; rays concolourous; paleæ truncate.

HAB. Between Kaus, Natvoet and Doornpoort, and Silverfontein, Namaqualand, *Drege!* (Hb. D., Sd., Hk.)

About a foot high, with pale, straw-like stems. Leaves 1½–2 inches long, the lobes nearly 1 line wide, mostly compound. Pedunc. 4–6 inches long. Invol. ¾–1 inch across, the rays fully an inch long.

8. U. cakilefolia (DC. l. c. 690); herbaceous, annual, erect, branching, glabrous ; leaves somewhat fleshy, pinnati-partite, the lobes linear, acute, the lowest very short or reduced to subulate teeth, the upper long, simple or 2–3-fid; pedunc. elongate; rays concolourous; outer inv.-scales ovato-lanceolate, much acuminate, very acute, subscarious at tip; outer paleæ tridentate.

HAB. Giftberg, *Drege!* (Herb. D., Sd., Hk.)

Six to twelve inches high. Leaves more simple than in *U. speciosa,* from which this specially differs in its involucre. The scales are rather rigid, purple-tipped, and remarkably acuminate. Rays about an inch long, deep yellow or orange.

9. U. annua (Less. Syn. 245) ; annual, diffuse, much-branched, glabrous or nearly so ; leaves somewhat fleshy, pinnati-partite, the lobes linear, obtuse or mucronulate, the lowest short (or obsolete), entire, the upper bi-trifid; pedunc. shortish; rays concolourous; outer inv.-scales ovate, subacute or obtuse; paleæ tridentate.

VAR. *α.* **indecora**; invol.-scales all concolourous, pale. *U. indecora, DC. l. c.* 690.

VAR. *β.* **nana**; outer inv.-scales dark-edged (variably so). *U. nana, DC. l. c.*

HAB. Var. *a.,* Nieuweveld, betw. Brakrivier and Uitvlugt, 3–4000 f. *Drege'* Bitterfontein, Namaqualand, *Zey.! 822.* Var. *β.* Silverfontein, *Drege!* Zwartkops River, Uitenhage, *Zey.! 2792.* Kreili's Country, *H. Bowker!* (Herb. D., Hk., Sd.) Two to twelve inches high, branched from the base. Leaves 1–1½ inch long, the lobes ½–¾ line diameter. Pedunc. 1–2 inches long. Heads 5–6 lines across. I adopt *Lessing's* name for the two sp. of DC. above indicated, and which I cannot satisfactorily separate. To these might perhaps be added *U. abyssinica,* Sch. Bip.! (Schimp! 1238). Nor can I distinguish "*Sphenogyne fœniculacea,*" Eck.! No. 312 (non Less.) which is by DC. referred to his "*U. tenuiloba*" from the present: it has every appearance of having an annual root.

10. U. pygmæa (DC. l. c. p. 690); annual, dwarf, corymbosely branched, glabrous; leaves pinnati-partite beyond the middle, the lobes in few pair, linear-filiform, acute; pedunc. shorter than the leaves; outer inv.-scales ovate, setaceo-acuminate, glabrous; paleæ dissimilar, the outer sheathing, the inner very narrow, all tipped with a roundish-reniform, membranous scale; rays (concolourous?).

HAB. Valleyfontein, *Zeyher!* (Herb. Sond.)
The specimens are 1–2 inches high, including the short peduncle, branched from the base; leafy nearly to the summit. The leaves are about ¾ inch long, some 2–3-jugate, others three-parted. Heads 2–3 lines across. The rays are withered.

L. **EUMORPHIA**, DC.

Heads many-fl., radiate, solitary; *ray-fl.* uniseriate, female; *disc-fl.* hermaphrodite. *Invol.* imbricate. *Recept.* convex, covered with paleæ, inrolling the flowers. *Cor.* of disc tubular, 5-toothed, cylindrical, not glandular, dilated at base above the ovary. *Anthers* without tails. Branches of the disc-*styles* divergent, exserted, obtuse; of the ray longer, subacute. *Achenes* glabrous, prismatic, 3–4-angled, without pappus. *DC. Prodr. 6, p. 2.*

A small, glabrous shrub. Leaves opposite, imbricated, linear, very short. Heads terminal, sub-pedicellate, with white rays and a yellow disc. Name from εν, *well,* and μορφη, *a form;* alluding to the neatness of foliage.

1. E. Dregeana (DC. l. c. p. 3); *Harv. Thes. Cap. t. 70. Zey.! 960.*

HAB. Sneeuweberg, 4–5000 ft., *Drege! Dr. Wallich!* Africa's Hoogde, *Burke and Zeyher!* (Herb. Hk., Sd., D.)
A rigid, woody, erect shrub, 1–2 feet high, much-branched and ramulous. Leaves 2 lines long, linear-terete, very blunt, decussate, closely imbricated, or, on the flower-branches more distant, when dry with a dorsal furrow. Pedunc. half an inch long, mostly with 1 or 2 pair of leaves. Heads small. Invol. scales linear, blunt, shorter than the disc.

LI. **LASIOSPERMUM**, Lag.

Heads many-fl., heterogamous; *ray-fl.* female, either ligulate or minute and tubular; *disc-fl.* tubular, 5-toothed, hermaphrodite. *Recept.* broad, paleaceous. *Invol.* broadly campanulate, the scales imbricated, shorter than the disc. Branches of disc-style truncate. *Achenes* wingless, without pappus, the younger hairy, the old envolved in very thick wool. *DC. Prodr. 6, p. 37.*

Glabrous, S. African herbs or half-shrubs. Leaves alternate, pinnatisect, with linear lobes. Peduncles long, terminal, one-headed. Ray-fl. when present white; disc yellow. Name from λασιος, *hairy,* and σπερμα, *a seed.*

Heads discoid; the marginal female fl. minute, tubular ...   (1) **pedunculare.**
Heads radiate; the marginal-fl. ligulate :
   Perennial. Rays linear, thrice as long as involucre...   (2) **radiatum.**
   Annual. Rays oblong, very short ... ...   (3) **brachyglossum.**

1. **L. pedunculare**(Lag. Nov. Gen. p. 31); herbaceous, perennial; leaves " membranaceous," pinnati-partite, the lobes linear, here and there lobulate; invol. expanded, the scales oblong, obtuse; heads discoid, the marginal female flowers minute, tubular. *DC. l. c. p.* 38. *Santolina erecta, Lam. Ill. t.* 671. *t.* 4 (*excl. syn.*). *Sant. eriosperma, Pers. Schr. Hort. Monach. t.* 67.

HAB. Sneeweberg, 5-6000 ft.; and Winterveld, near Groot-Tafelberg, 3-4000 f., *Drege!* (Herb. Hk., Sd)
Known from the following chiefly by its discoid capitula.

2. **L. radiatum** (Trev. Nov. Act. 13. 1. p. 205); herbaceous, perennial; leaves somewhat fleshy, pinnati-partite, the laciniæ linear, sub-falcate, simple or with 2-3 linear lobes, all the apices sharply albo-mucronate; inv. expanded, the scales broadly oblong, obtuse; heads radiate, the rays linear, thrice as long as the involucre. *DC. l. c. p.* 38. *Lidbeckia bipinnata, Thunb.! Cap.* 694. *Lancisia bipinnata, Pers. Matricaria bipinnata, Spr.* Also *Mataxa Capensis, Spr.*

HAB. Langekloof, *Thunberg!* Zachriver, *Burchell,* No. 1404. Roggeveld, *Eckl.! Smaldeel* and Uitvlugt, *Zey.!* 821; Albany, *Mrs. F. W. Barber!* (Herb. Th., Hk., Sd., D.)
2 or more feet high, diffusely branching, pale. Leaves 1-2 inches long, the upper laciniæ ½-1 inch long, ½ line wide, the lower shorter and simpler, all erecto-patent. Upper leaves often quite entire, pedunc. 6-10 inches long, straw-colour. Rays white, ¼ inch long or more.

3. **L. brachyglossum** (DC. l. c. 38); annual; somewhat fleshy bipinnati-partite, the lobes or lobules spreading, linear, acute or mucronate ; inv. shorter than the disc, the scales oblong, obtuse; heads radiate, the rays oval-oblong, scarcely longer than the involucre.

HAB. Silverfontein, on the flats, 2000 f. Sep.-Oct. *Drege!* (Herb. D., Hk, Sd.)
About a foot high, branched chiefly from the base. Leaves 1½-2 inches long, the lobes divaricating. Pedunc. 2-3 inches long. Heads smaller than in the preceding, and the rays very much shorter.

LII. **LIDBECKIA,** Berg.

Heads many-fl., radiate ; *ray-fl.* uniseriate, ligulate; neuter ; *disc-fl.* tubular, hermaphrodite, 4-toothed. *Recept.* flattish, bearing a few filiform shreds. *Inv.* 2-3-seriate, the scales as long as the disc. Branches of the disc-*style* truncate. *Ovaries* of the ray-fl. abortive. *Achenes* of of disc fertile, wingless, glabrous, without pappus, longitudinally ribbed, crowned by a cylindrical nectary. *DC. Prodr.* 6, *p.* 39.

Suffruticose or half herbaceous. Leaves alternate, lobed or pinnatifid, the lobes oblong or linear, entire, mucronate. Branches ending in one-headed peduncles. Rays white ; disc yellow. Name in honour of Eric Gustavius Lidbeck, formerly Professor of Botany at Lund, Sweden.

Glabrescent, glaucous ; leaves subsessile, oblong, pinnatifid ...   (1) **pectinata.**
Copiously hairy ; leaves petioled, palmately 3-5-lobed ... ...   (2) **lobata.**

1. **L. pectinata** (Berg. Cap. 307, t. 5, f. 9); glabrate; leaves subsessile, oblong, pinnatifid, glaucous beneath, punctate, sparsely pilose, the lobes

on each side 3–5 broadly linear. *Thunb.! Cap. p.* 694. *DC. l. c.* 39. *L. Capensis. Gmel. Lancisia, Lam. Ill. t.* 701, *f.* 2. *Lancisia pectinata, Pers.*

HAB. Rodesand R., below Winterhoek, Tulbagh, *Thunb.! Pappe!* Cape, *Drege! E. Z.!* 24-Rivers, *Zey.!* 827. (Herb. Th., D., Hk., Sd.)

2–3 f. high, sparingly pubescent or glabrous; branches long, straight, rodlike, leafy. Leaves 1–1½ inch long, the lobes ½–¾ inch long, 2 lines wide, the lower smaller, all mucronate. Pedunc. 1–3 inches long, sometimes bearing 1–2 small leaves. Heads like those of *Chrysanth. leucanthemum.* (*Ox-eye-daisy*).

2. L. lobata (Thunb.! Cap. 694); silky-villous; leaves petiolate, sub-palmately 3–5-lobed, punctate, villous, the lobes broadly oblong, short, mucronate. *DC. l. c.* 39. *Lidbeckia quinqueloba, Cass. Cotula quinqueloba, Linn. f. Suppl.* 377. *Lancisia, Lam. Ill. t.* 701, *f. 3. L. lobata, Pers.*

HAB. Olifant's Riv., *Thunb.!* Clan William, *E. Z.!* Zeederberg, *Drege!* (Herb. Th., Sd., D., Hk.)

1–2 f. high, copiously clothed with long, silky hairs, the young parts quite shaggy. Petioles ½–1 inch long, about equalling the usually 5-lobed lamina. Heads smaller than in the preceding; invol. very hairy.

LIII. THAMINOPHYLLUM, Harv.

Heads few or many-fl., heterogamous; *ray-fl.* few or several, ligulate, neuter; *disc-fl.* hermaphrodite, acutely 4-toothed. *Recept.* naked, narrow or conical. *Invol.*-scales linear, sub-biseriate, loosely imbricating, herbaceous. *Anthers* without tails. *Style*-branches truncate. *Achenes* oblong, subcompressed, glabrous, crowned with a hardened, conical style-base or nectary. *Pappus* none.

Much-branched, ramulous, closely leafy, silky-pubescent small shrubs, with the general aspect of *Phylica.* Leaves alternate, crowded, spreading or squarrose, linear, acute, quite entire, one-nerved, with reflexed margins. Heads subsessile or shortly pedicellate, terminal or in leafy racemes. Rays white; disc-fl. tinged with purple. Name from θαμινος, *close* or *thick-set* and φυλλον, a *leaf.* Very unlike *Lidbeckia* in habit, but closely approaching it in generic character.

Heads racemose, few-flowered, with two rays ... ... (1) **Mundtii.**
Hds. solitary, short pedicelled, many-flowered, with several rays (2) **multiflorum.**

1. Th. Mundtii (Harv.); flower-heads in leafy, terminal racemes, pedicellate; ray-fl. two; disc-fl. 7–8.

HAB. Cape, *Mundt!* (Herb. Hook.)
A robust, much-branched, woody shrub, with the aspect of a *Phylica.* Leaves 5–6 lines long, ½ line wide, horizontally spreading or deflexed, linear, acute, silky. Ray-fl. narrow obovate or spathulate, entire, without style, their ovary filiform. Lobes of disc-fl. remarkably acuminate.

2. Th. multiflorum (Harv.); fl.-heads terminal, solitary, subsessile or shortly pedicelled, many-flowered; rays 10 or 12; receptacle conical.

HAB. Cape, *Hb. Hook.!* Distr. of George, *Dr. Alexander Prior!* (Hb. Hk. Cap.)
Foliage and habit exactly as in *Th. Mundtii,* but the fl.-heads are solitary, many-flowered, resembling those of an *Osmites.* I have seen perfect specimens in Hb. Cap., collected by *Dr. Alexander Prior.* The species was first named however from a small fragment, marked "834" in Hb. Hook., without locality or ray flowers.

LIV. GAMOLEPIS, Less.

Heads many-fl., radiate; *ray-fl.* ligulate, female; *disc-fl.* tubular,

5-toothed, hermaphrodite. *Recept.* convex, naked or slightly honey-combed. *Inv.*-scales numerous, uniseriate, connate into a cup for one-third or half their length, or more. Branches of the disc-*style* tipped with a cone. *Achenes* glabrous, wingless, without pappus, with a ter-minal areole. *DC. Prodr. 6, p. 41. Psilothamnus, DC. l. c. 41.*

Small shrubs or herbs, mostly glabrous. Leaves alternate, pinnatisect, rarely entire. Pedunc. one-headed. Flowers yellow. Naturally allied to *Euryops*, from which it is known by the want of pappus, and the appendaged styles. Name from γαμεω, *to marry or unite*, and λεπις, *a scale;* alluding to the connate invol-scales.

Annual. Lvs. pinnati-partite. Inv. ovate-urceolate (1) **Tagetes.**
Shrubs or suffrutices. Invol. expanded:
   Pedunc. long and slender (*at least* 4 times as long as leaves, usually many-
    times):
     Leaves pinnatifid or pinnati-partite:
      Lvs. crowded, 1–2 inches long; lobes
       broad, flat, oblong, nerved .. ... (2) **chrysanthemoides.**
      Lvs. imbricate, ½ inch long; lobes shortly
       subulate, pungent ... ... (10) **munita.**
      Lvs. sparse, 1 inch long; lobes narrow-
       linear, semiterete ... ... (5) **debilis.**
     Leaves 3-lobed or 3-forked:
      Pedunc. many times longer than the leaves:
       Achenes quite glabrous. Leaves glabrous or powdery.
        Stem much-branched, flexuous,
         closely leafy; lvs. 4–9 lines
         long ... ... (7) **trifurcata.**
        Stem sub-simple, straight, laxly
         leafy; lvs. 1½ inches long ... (6) **laxa.**
       Achenes velvetty.   Stem short,
        simple. Leaves pilose ... ... (9) **hebecarpa.**
      Pedunc. about 4 times as long as the
       leaves. St. and lvs. quite glabrous ... (8) **euryopoides.**
     Leaves simple, short, subulate, imbricate or
      appressed ... ... (11) **ericoides.**
   Pedunc. short (twice as long as leaves *at most*, usually much shorter):
    Leaves pinnati-partite:
     Achenes tuberculated ... ... (3) **pectinata.**
     Achenes smooth ... ... (4) **brachypoda.**
    Leaves 3-forked (or mostly so):
     Axils woolly. Pedunc. scarcely
      longer than leaves ... ... (4) **brachypoda** γ.
     Axils glabrous. Pedunc. twice as
      long as leaves ... ... (7) **euryopoides.**
    Leaves entire, linear, mucronate ... ... (4) **brachypoda** β.

1. G. Tagetes (DC. l. c. 40); annual, slender, corymbosely branched; leaves pinnati-partite, the lobes linear, on each side 5–7, either quite entire or with 1–2 lobules, obtuse; invol. ovate-urceolate, the scales 10–15, concrete much beyond their middle. *Othonna Tagetes, Linn. Sp.* 1309. *Thunb.! Cap.* 723. *Gamolepis annua, Less.! Syn.* 251. *Psilo-thonna Tagetes, E. Mey.!*

HAB. On the Cape Flats, *Thunb.! E. Z.! W.H.H.!* Groenekloof, *Drege!* (Herb. Th., D., Hk., Sd.)

A slender, wiry, glabrous, mostly branching annual, 3–10 inches high, the branches flexuous, ending in fl.-heads on nearly the same level. Leaves 1–1½ inch long, the lobes ⅓–¾ inch. Rays broad, revolute, deep and bright yellow or full-orange.

2. G. chrysanthemoides (DC. l. c. 40); glabrous, dichotomous;

leaves crowded towards the ends of the branches, spreading, subpetiolate, pinnati-partite, or pinnatifid, the lobes on each side 3–5, broadly
linear or oblong-linear, flat, acute, nerved, the lower shorter or tooth-
like ; pedunc. slender, elongate ; invol. campanulate, its scales ovate
or ovato-lanceolate, acuminate, several-nerved; achenes obovate, bluntly
rib-striate.

HAB. Eastern Districts, *Burchell.* 3641. Buffalo River, *Drege, Pappe!* Bosjesman's R., Uit., *E.&Z.!* Somerset, *Mrs. F. W. Barber!* (Herb. D., Hk., Sd.)
　Apparently a shrub 2–3 ft. or more in height, the old branches denuded, the upper
closely leafy. Leaves 1½–2 inches long, the lobes 1–2 lines wide, the upper often
confluent, erecto-patent. Pedunc. 4–5 inches long. Achenes dark-purplish.—The
leaves are much more membranous and expanded than in other species ; not unlike
those of *Chrysanth. pinnatifidum.*

3. G. pectinata (Less. Syn. 251); glabrous, erect, closely leafy ; leaves
coriaceous, nerveless, pinnati-partite, the segments on each side 2–3–4,
linear, semiterete, mucronate; pedunc. about twice as long as the leaves ;
invol. campanulate, its scales lanceolate-acuminate, faintly 3–4-nerved,
concrete to the middle ; achenes *densely tuberculated*, subglobose-obo-
vate. *DC. l. c.* 40.

HAB. Cape, *Thunberg!* (Herb. Th., marked " *Othonna abrotanifolia,* 1 ").
　This has the aspect of *G. brachypoda,* but the leaves are uniformly pinnatisect,
with longer lobes, and much more closely set ; not to speak of the tuberculated
achenes. I have only seen the solitary specimen in Hb. Th.

4. G. brachypoda (DC. l. c. 40); erect, subdichotomous, flexuous,
closely leafy, with *woolly tufts* in the axils of the young leaves, otherwise glabrous ; leaves coriaceous, thick, nerveless, mucronate, *polymorphous*, either linear-entire, or with 1–2 lateral teeth; or 3-forked, with
subulate, lateral lobes ; or pinnati-partite, with 2–3 pair of pinnules ;
pedunc. scarcely longer than the leaves ; invol. campanulate, its scales
lanceolate-acuminate, 3–4-nerved, concrete to the middle ; achenes
striate. *Osteospermum heterophyllum, DC. l. c.* 467.

VAR. β, **integrifolia**; most of the leaves quite entire, a few unidentate.
VAR. γ, **tridens**; most of the leaves 3-forked or 3-fid. *Zey.!* 861.

HAB. Districts of Albany, *Drege, Burchell;* Mts. near Grahamstown, in several
places, *Zey.!* 2801. Var. β, from *Dr. Pappe!* without Hab. (Herb. D., Hk., Sd.)
　2–3 feet high, irregularly much branched, the old branches bare of leaves, the
upper densely covered with them. Leaves extremely variable, even on the same
branch, from ⅓–¾ inch long, about ½ line wide, erecto-patent or squarrose. Pedunc.
rarely an inch long. Rays broad, twice as long as invol. or more. Achenes with
many slender, longitudinal, raised striæ. β. is more robust and more densely leafy,
with almost all the leaves linear. γ. may be mistaken for *G. euryopoides,* but is known
by the axillary wool-tufts, and thicker, narrower, sharply-mucronate leaf-lobes.

5. G. debilis (Harv.) ; stems many from a thick woody root, short,
subsimple, slender; leaves rather sparse, pinnati-partite, spreading, the
rachis and lobes narrow-linear, thickish, nerved beneath, subacute, the
lobes 2–3-pair; pedunc. elongate; inv. scales lanceolate, acute, connate
at base only; ovary glabrous, smooth. *G. pectinata, var. Natalensis,
Sch. B. (not of Less.)*

HAB. Port Natal, *Krauss!* 461. (Herb. D., Hk.)
Stems, in the only specimens seen (which appear to be a growth after burning)

4–6 inches high, sparsely leafy below, closely leafy above. Leaves uncial, the lobes 4–5 lines long, ½ line wide. Pedunc. 3–4 inches long. Heads small. Fl. yellow.

6. G. laxa (Harv.); glabrous; stems slender, subsimple, erect, laxly leafy; leaves thickish, linear, *flat*, trifid near the apex, the lateral lobes alternate, linear, nerved, acute, callous-tipped, straight-pointed; pedunc. very long, naked; inv. campanulate, its scales broad, oblong, acuminate, pluri-nerved; achenes obovate, smooth.

HAB. Zululand, *W. T. Gerrard*, 1015. (Herb. D.)

Stems from a woody crown, several, 1–2 feet high, slender, simple or with few, very erect branches. Leaves 1½ inch long, ½ line wide, the undivided portion an inch or more in length, the lobes 5–6 lines long. Pedunc. 8–12 inches long, terminal or pseudo-lateral. Inv. scales semi-connate, broader than in *G. trifurcata*, from which this is known by its erect habit, long, straight, laxly leafy branches, and much longer leaves.

7. G. trifurcata (Less. Syn. 251); at first cobwebby, afterwards glabrous; branches dichotomous, ramulous; leaves thickish, linear-terete, mostly trifid beyond the middle, sometimes 5-lobed, sub-pinnatisect, the lobes linear, mucronate, recurved at point; ped. very long, naked; invol. campanulate, its scales lanceolate, acuminate, 3-nerved; achenes obovate, bluntly 10-ribbed, smooth. *DC. l. c.* 40.

HAB. Between Graaf Reynet and Kamdekaka, *Burch.* 2961; Zondags Riv. and Zuureberg, *Drege!* Uit. and Clanw., *E. Z.!* Bushman's R., Albany, *Zey.!* Koega Kamma Kloof, *Zey.!* 2808. (Herb. Hk., Sd., D.)

A scrubby, much-branched bush, about a foot high, the lower branches nude. Leaves 4–9 lines long, the lobes 1–3 lines, generally much shorter than the undivided portion of the leaf. Pedunc. 8–12 inches long, pale, 1-headed. Rays yellow, 2–3 times as long as invol. Achenes pear-shaped, ribbed and furrowed.

8. G. euryopoides (DC. l. c. 41); glabrous, erect; leaves crowded, mostly trifid to the middle, some entire, the rachis and lobes linear, concave, thickish, subacute; pedunc. 2–4 times longer than the leaves; inv. scales lanceolate-oblong, subacute; achenes obovate, 10-ribbed and furrowed, the ribs finely cross-striate.

HAB. Cape, *Bowie!* Vanstaadenberg, Uit., *E. Z.!* *Zey.!* 2810; Grahamstown, *Sir. C. F. Bunbury! Genl. Bolton!* Sandfontein, *Burke!* Port Elizabeth, *Dr. Pappe!* (Herb. D., Hk., Sd.)

An erect, leafy bush, 2 feet high, with the aspect of *Euryops abrotanifolia*, in all parts, even the young shoots, quite glabrous. Leaves 1–1½ inch long, the lobes about ½ inch, ½–1 line wide. Pedunc. 2–4 inches long. Ripe achenes strongly ribbed and furrowed.

9. G. hebecarpa (DC. l. c. 41); stem short, sub-simple, robust, ligneous; leaves crowded, 3-forked or simple, the rachis and lobes linear, fleshy, sub-terete, rigid, mucronate, the younger ones pilose with long, white hairs, the older spreading or reflexed, glabrate; pedunc. elongate; invol. flattish, the scales 20–24, lanceolate, ciliolate; achenes turbinate, velvetty-canescent.

HAB. At Cape L'Agulhas, *Drege!* (Hb. D., Hk., Sd.)

Stems 3–6 inches long, densely leafy, at length denuded below. Leaves 1–1½ inches long, a line in diameter, the young softly pilose; perfectly simple and entire leaves occur here and there among the 3-forked. Pedunc. 12–14 inches long. Invol. 1 inch diameter; rays 1 inch long, bright yellow. The largest flowered species, and well marked by its velvetty fruit, &c.

10. G. munita (Less.! Syn. 251); quite glabrous, erect, sub-dichotomous, branches closely imbricated with short leaves; leaves erect-incurved, sessile, rigid, strongly midribbed, pinnatifid, the lobes on each side three, shortly subulate, acuminate, nerved; pedunc. elongate, wiry, from the forks, or the ends of the branches; invol. campanulate, its scales about 12, concrete to about their middle; achenes smooth. *DC.l.c.p.* 40. *Othonna munita, Linn. f. suppl.* 388. *Th.! Cap.* 722. *Osteospermum fallax, Spr.!—DC.! l. c. p.* 467.

HAB. Cape, *Thunberg!* Worcester, *Eckl.* Gauritz R., *Burchell.* Kromriver, Langekloof, *Drege!* Vanstaadenberg, *Zey.!* 2809. (Herb. Th., D., Hk., Sd.)

A very rigid small shrub, 1–2 f. high; the forked branches clothed throughout with close lying pale-green neat foliage. Leaves about ½ inch long, the lobes 1–2 lines long, broad or narrow, very acute, keeled. Pedunc. 6–8 inches long, quite naked. Invol. scales acuminate. Rays yellow, obtuse, 2–3 times as long as involucre.

11. G. ericoides (Less.! Syn. 251); stems shrubby, di-tri-chotomons, much branched, closely imbricated with short leaves, sparingly cobwebby in the axils, otherwise quite glabrous; leaves trigono-subulate or subulate, rigid, very acute, erect or erecto-patent, more or less incurved or closely appressed; pedunc. elongate; invol. scales connate for about a third their length, lanceolate, acute; achenes (not quite ripe) ridged and furrowed, the ribs warted. *DC.l.c.c.*41. Also *Psilothamnus ericifolius, DC. l. c.* 41. *Othonna ericoides, Th.! Cap.* 717.

VAR. β, **adpressifolius**; rather more slender; leaves very *closely appressed*, the branches rope-like. *Psilothamnus adpressifolius, DC.! l. c.*

VAR. γ. **incurvifolius**; more robust; leaves hooked inwards, trigono-subulate, keeled, on each side bisulcate. *Othonna ericoides, Thunb.! in Herb.*

HAB. Cannaland, *Thunberg!* Swellendam and Albany, *Drege!* Uitenhage, *E. & Z.* Amsterdamvlacte, mouth of Zwartkops Rr. *Zey.!* β and γ, in the same localities. (Herb. Th., D., Hk., Sd.)

A small, much branched, erect, rigid shrub; branches many times irregularly forked and closely imbricated with leaves throughout. Leaves 1½–2½ lines long, from broadish, clasping bases, sometimes trigono-subulate, with a very thick midrib, margined on each side by an evident lamina (as in *Thunberg's* and some of *E. & Z.'s* specimens); but oftener reduced to the terete, subulate midrib. The inflexion of the points varies greatly, as does also the degree of appressedness or the contrary. Pedunc. wiry, 6–8 inches long. Heads small. Invol. scales certainly connate, though to a rather less extent than in other species.—Nearly allied to *G. munita,* chiefly differing in foliage.

LV. **STEIRODISCUS**, Less.

Heads several-fl., heterogamous; *ray-fl.* female, ligulate, *disc-fl.* hermaphrodite but sterile, terete, 5-toothed. *Recept.* naked. *Invol.* campanulate, of many scales in a single row. *Ovaries* of disc linear, glabrous, without pappus, always sterile. Branches of disc-*style* tipped with a cone. *Achenes* of ray-fl. obovoid, silky. *DC. Prodr.* 6. *p.* 74.

Small annuals with the aspect of *Gamolepis Tagetes* in miniature. Leaves pinnatipartite; the lobes linear-filiform, either quite entire or with 1–2 lobules, obtuse. Heads terminal, solitary, with 5–10 rays. Name from σтεıρος, *sterile* and δισκος, the *disc;* because the disc-flowers are sterile.

1. S. capillaceus (Less. Syn. 251); inv. scales green (*not-dotted*), with *narrow* margins, subacute; rays 5–6, *pale yellow* as the disc. *DC. l. c.* 74. *Cineraria capillacea, Th.! Cap.* 672.

HAB. In sands. Zwartland, *Thunberg.* Paarl, *Drege!* (Herb. D., Hk., Sd.)·
2–3 inches high, very slender, glabrous. Leaves uncial, their lobes 2–4 lines long,
½-line wide. Heads few-fl., 2-lines diameter. Rays revolute.

2. S. linearilobus (DC. l. c. 74); inv. scales *black-dotted*, with *wide*
margins, acuminate; rays 8–10, *deep orange*, rather darker than the disc.

HAB. Olifant's River, *Drege!* (Herb. D., Hk.)
4–6 inches high, very similar to the preceding, but the heads are larger (3 lines
diam.), the rays more numerous and of deeper colour, and the inv. scales dotted.

LVI. IOCASTE, E. Mey.

Heads several-fl., heterogamous; *ray-fl.* 1–2, ligulate, female; *disc-fl.*
12–15, tubular, 5-toothed. *Inv.* 2–3-seriate, imbricate, the inner scales
scarious-tipped. *Recept.* naked. Branches of *style* truncate. *Anth.* not
tailed, tipped with a scale. *Achenes* terete, with 8–10 obtuse, raised
striæ, glandularly scabrid, without pappus, truncate. *E. Mey./ MSS.
in Hb.Drege. Oligoglossa,DC.! Prodr. 6, p. 76. Jacosta. Endl.Gen.* 2680.

A glabrous, many-stemmed suffrutex, the branches closely leafy at base, laxly
leafy, or half naked upwards, and corymbosely panicled at top. Leaves linear-subulate, erect, quite entire, keeled, pungent. Heads pedicellate, fl. yellow. *E. Meyer's*
gen. name, here preserved, is mythological.

1. I. acicularis (E. Mey.!); *Oligoglossa acicularis, DC.! l. c.*

HAB. Oudeberg, 3–4000 f.; and Nieuwveld, betw. Rhinosterkop and Ganzefontein, 3500–4500 f., *Drege!* Uitenhage *E. Z.!* Cradock, *Burke!* (Herb. Hk., D., Sd.)
Woody at base and dividing into several stem-like branches, 1–1½ foot high, the
lower part, for 2–4 inches long, closely leafy. Leaves ¾–1 in. long, not 1 line diam.,
the upper small and distant, pale green. Heads laxly corymbose, on slender pedicels, 2–3 lines diam. Inv.-scales ciliate at top. Tube of corolla glandularly scabrous.

LVII. ? PHYMASPERMUM, Less.

Heads many-fl., heterogamous; *ray-fl.* of two forms, some ligulate,
some filiform; *disc-fl.* hermaphrodite, terete, 5-toothed. *Recept.* naked,
flat. *Inv.* campanulate, closely imbricated. *Ovary* wingless, without
pappus, papillose-canescent, elliptical. *Style* of disc-fl. with truncate
branches. *DC. Prodr. 6, p.* 44.

A much-branched shrub. Leaves scattered, sessile, leathery, linear, keeled, with
very entire, revolute margin, at length shining. Heads solitary, small, terminal ;
rays white. Inv.-scales tomentose, obtuse, mucronate. Achenes obovate. Name
from φυμα, a *swelling*, and σπερμα, a *seed ;* the achenes are papillated.

1. Ph. junceum (Less.! Syn. 253) ; *DC. l. c.* 44. *Osteospermum junceum, Thunb.! Cap.* 714.

HAB. Cape, *Thunberg!* (Herb. Th.)
Stem diffusely branched, angularly ribbed, minutely puberulous, reddish. Leaves
5–7 lines long, quite glabrous, rigid, erect or erecto-patent, the revolute margin concealing the lower surface. Of this a very poor specimen, from which all the flowers
have dropped, alone remains in Herb. Thunb. It is glued on a sheet with a specimen of *Osteosp. spinosum.* I have been unable to verify the generic character. The
aspect is quite that of a genuine *Osteospermum.*

LVIII. ADENACHÆNA, DC.

Heads many-fl., heterogamous ; *ray-fl.* ligulate, 1-seriate, female;
disc-fl. hermaphrodite, 4–5-fid, with a terete tube. *Inv.*-scales shorter

than the disc, in 2–3 rows, subequal, linear-lanceolate, the inner obtuse. *Recept.* plano-convex, at length subglobose, minutely papillose. *Achenes* all subterete, without pappus, truncate and denticulate at the expanded summit, rib striate, the striæ covered with papillæ or glandular granules. *DC. Prodr. 6, p.* 49.

Erect, much-branched, glabrous or pubescent, S. African small shrubs. Pedunc. terminal, one-headed, short. Leaves alternate, linear, trifid, or on each side one-toothed. Rays white, reflexed. Name from αδην, a *gland*, and *achenium;* from the glandular fruit.

Glabrous ; leaves sharply 3-lobed and simple ; rays numerous   (1) **leptophylla.**
Thinly silky or glabrous; lvs. blunt, simple or 3-lobed; rays few   (2) **parvifolia.**

1. A. leptophylla (DC. l. c. 49); quite glabrous ; leaves linear, with strongly revolute margins (5–12 lines long) acute, sometimes trifid ; heads very many-flowered ; rays about 15 ; recept. convex.

HAB. Swellendam, *Drege, Dr. Pappe!* Karroo, *Ecklon!* (Herb. D., Sd.)
A rigid, erect or spreading, small bush, with pale stems and foliage. Leaves from ½–1 inch long, not a line wide, erect or spreading. Heads like those of *Chrysanth. leucanthemum*, but smaller. Recept. at length hemispherical. Achenes densely glandular.

2. A. parvifolia (DC. l. c. 50); appressedly pubescent or glabrous ; leaves linear, with subrevolute margins, obtuse or mucronulate, often tufted, either quite simple, or obtusely 3-lobed ; rays 6–9; recept. flat ; achenes striate, papulose ; corolla viscidulous.

VAR. α. **pubescens**; twigs and leaves closely pubescent. *A. pubescens, DC. l. c.* 49.
VAR. β. **glabra**; quite glabrous. *A. parvifolia, DC. l. c.*

HAB. Nieuwe Hantam, *Drege!* Wolfkop and Cradock, *Burke!* Albany, *Mrs. F. W. Barber!* 410. β. Uitenhage, *Ecklon.* Albert, *T. Cooper!* (Herb. Hk., D., Sd.)
A rigid, robust, much-branched and ramulous scrubby bush, 12 inches or more high. Leaves mostly tufted, very variable in length, from 2–3 lines to an inch long, less than a line wide, mostly thinly silky. Heads on short, silky peduncles, about 15-flowered ; the disc-fl. reddish or yellow. " This is known in the Colony as 'the good Karroo' ; it is one of the most valuable plants, as pasture, for Merino sheep, and where it grows plentifully the farms are most favourable for woolled sheep."—*Mrs. F. W. Barber.*

LIX. CHRYSANTHEMUM, L.

Heads many-fl., heterogamous ; *ray-fl.* ligulate, uniseriate, female or rarely neuter, disc-fl. hermaphrodite, 4–5-toothed, with a compressed, fleshy, 2-winged tube. *Inv.* imbricate, campanulate, the scales membrane-edged. *Recept.* naked, flat or convex. *Style*-branches truncate. *Achenes* dissimilar : those of the *ray* 3-angled or 3-winged, 2 angles or wings lateral, the third on the inner face ; of the *disc* compressed or subterete, with a short wing on the inner face. *Pappus* none, or coroniform. *DC. Prodr. 6. p.* 63.

Herbs or small shrubs, European and African : habit various. Leaves alternate. Rays yellow or white, or white with a yellow base.

Annual. Rays yellow. Leaves oblong, sharply toothed or cut   (1) **segetum.**
Shrubs, with linear-filiform, simple or trifid, or pinnately 5-lobed leaves :
  Peduncles very long, filiform :
    Leaf-lobes long, slender, erect, acute (or none) ...   (2) **nodosum.**
    Leaf-lobes short, fleshy, spreading (rarely none) ...   (3) **Thunbergii.**

Peduncles very short:
  Leaves mostly shortly trifid at the apex, semiuncial   (4) **carnosulum.**
  Leaves quite simple, uncial (very rarely 3-lobed) ...   (4) **carnosulum** β.
Perennial, sub-simple. Rays white. Lvs. lanceolate, sharply
  serrate ... .   (5) **osmitoides.**

1. C. segetum (Linn. Sp. 1254); annual, glabrous, branching; leaves amplexicaul, oblong, some sharply and coarsely toothed, others laciniate; pedunc. terminal, one-headed; inv.-scales with very wide, membranous edges; achenes without pappus. *DC. l. c. 64. E. Bot. t. 540.*

HAB. A corn-weed, introduced from Europe. It has already penetrated 'the bush,' 50 miles beyond the Kei, *H. Bowker!* 335. (Herb. D.)
Flower-heads large, handsome, bright golden yellow. One of the plants called "*Gowan*" in Scotland.

2. C. nodosum (DC. l. c. 65); shrubby, quite glabrous; leaves from callous, swollen bases, linear-filiform, slender, either quite entire or beyond the middle trifid or pinnately 5-lobed, the lobes elongate, acute; pedunc. terminal or axillary, very long, naked, one-headed; outer inv.-scales broadly ovate, acute, striate, inner widely membrane-edged; rays concolourous; disc-fl. 5-toothed; achenes amply winged; pappus coroniform, membranous. *Arctotis nodosa, Th.! Cap.* 711. *Pinardia nodosa, Less.! in Linn.* 1831, *p.* 169. *Ismelia nodosa, Less.! Syn.* 255. Also *Chrys. leptophyllum, DC.! l. c.* 65. *Argyranthemum leptophyllum, Fenzl.*

HAB. Patrysberg, *Thunberg!* Olifant's R. and Kl. Namaqualand, *Drege!* (Herb. Th., Hk., D.)
A subdichotomous shrub, 1–2 f. high, the old branches bare and rough with the wartlike swollen bases of the fallen leaves. Leaves 1–1½ inch long, not ½ line diam., simple and partite on the same branch. Pedunc. 8–12 inches long. Heads ⅔ inch diam. or more. The habit is that of a *Sphenogyne.* I have compared *Drege's* "*Ch. leptophyllum*" with *Thunberg's* original specimen, and find them identical in all respects.

3. C. Thunbergii (Harv.); shrubby, quite glabrous; leaves linear-terete, fleshy, either entire or near the summit shortly trifid or pinnately 5-fid, the lobes short or tooth-like, spreading, subacute; pedunc. terminal, very long, naked, 1-headed; outer invol. scales short, ovate, inner widely membrane-edged and tipped; rays linear, concolourous (?); disc fl. 5-toothed; achenes amply winged; pappus coroniform, membranous. *C. frutescens, Thunb.! Cap.* 693 *(non Linn.). Pyrethrum frutescens, Hb. Th.! n.* 2. *Ismelia frutescens, Less.! Syn.* 255. *Pinardia frutescens, Less. in Linn. l. c.*

HAB. Cape, *Thunberg!* (Herb. Thunb.)
With the foliage of *C. carnosulum* this has the peduncle and involucre of *C. nodosum.* Two specimens marked "*Pyrethrum frutescens*" exist in Hb. Thunb.; one of them, marked "No. 2," belongs to this species; the other, "No. 1,"to *C. carnosulum.*

4. C. carnosulum (DC. l. c. 65); shrubby, quite glabrous; leaves linear-terete, fleshy, dotted, some quite entire, most of them near the summit shortly trifid, the lobes patent or recurved, mucronate; pedunc. very short (2–3 times longer than the leaves), 1-headed; outer invol. scales oblong, obtuse, the inner membrane-tipped, with a narrow, membrane-border; rays short, concolourous; disc-flower 5-toothed; pappus coroniform, membranous, ample. *Pyrethrum frutescens, No.* 1, *Th.! Herb. Pentzia frutescens, Fenzl.*

VAR. β, **filifolium**; leaves uncial, almost *all* of them quite entire, 1 or 2 occasionally 3-lobed.

HAB. Cape, *Thunberg!* East of Capetown, *Burchell!* 6578 ; Karmelksriver, on hills, Swell., *Drege! Zey.!* 2831 ; at Gauritz R., Swell, *Dr. Pappe!* β, Swellendam, *Dr. Pappe!* (Herb. Th., D., Hk., Sd.)

2 feet high, much-branched and twiggy, the branches woody, rigid, slender, angle-striate. Leaves about ¼ inch long, the lobes 1 line long. Pedunc. 1–1¼ inch long or less. Heads ¼ inch diameter. Var. β. has terete leaves, an inch long, mostly quite simple ; but in one sp. (in Herb. D.) a few leaves are trifid, as in the ordinary state.

5. C.? osmitoides (Harv. Thes. t. 152); stems erect, suffruticose at base, subsimple, rib-furrowed, puberulous, becoming glabrate, leafy; leaves alternate, sessile, lanceolate, acuminate, sharply serrate, glabrous, the uppermost reduced, subentire; heads shortly pedicellate, terminal, or 3-subcorymbose, many-fl.; invol. pauciseriate, the scales with membranous, torn edges; rays numerous, creamy white; disc-fl. 5-toothed, subterete; pappus none; achenes (immature) of disc and ray obovate, compressed, 10-ribbed.

HAB. Omgati, Natal, *W. T. Gerrard*, 1026. (Herb. D.)

Stems 2–3 feet high, several from the crown ? simple, 1- or few-headed, leafy nearly to the summit. Leaves 1–1¼ inch long, 3–4 lines wide, coarsely and sharply serrate. Heads like those of *Chrysanth. leucanthemum.* Invol. scales 2–3-seriate, nearly equal, almost entirely scarious. Recept. flat and broad, quite naked. Corollas of ray and disc continuous with the ovaries, strongly adnate. Ripe achenes not seen. Until the ripe fruit shall have been seen the genus must be doubtful.

LX. **BRACHYMERIS**, DC.

Heads 8–16-fl., homogamous. *Invol.* campanulate-ovate, imbricate, the scales appressed, oval-oblong. *Recept.* naked. *Corolla* pubescent externally, with short tube and 5-fid limb. *Anth.* not tailed. *Style-* branches bearded at summit, scarcely exserted. *Achenes* terete, sub-compressed, striate, beakless, scabrid, without pappus. *DC. l. c.* 76.

A very rigid, much-branched, scrubby bush, the younger parts thinly silky ; twigs ribbed and furrowed. Leaves very small, thick, oblong or linear, obtuse, entire, appressedly silky. Heads subsessile, from the upper axils, somewhat racemose, crowded. Name from βραχυς, *short*, and μερις, *a part;* parts of the flower and the leaves very short.

1. B. scoparia (E. Mey.); *DC. l. c.* 76. *Brachystylis scoparia, E. Mey.!* in *Herb. Drege.*

HAB. Winterfeld, near Limoenfontein and Groot-Tafelberg ; Zeekoe-river and Sneeuwberg, *Drege!* Klein Tafelberg, *Burke & Zeyher!* (Herb. Hk., D., Sd.)

1–2 feet high, densely much branched and twiggy. Leaves 2–3 lines long, 1 line wide. Heads about 2 lines diam.

LXI. **MATRICARIA**, Linn.

Heads many-fl., homogamous or heterogamous ; *ray-fl.* ligulate, uniseriate, female ; *disc-fl.* tubular, hermaphrodite, 4–5-toothed, with a terete tube. *Invol.* imbricate in few rows, the scales membrane-edged, subequal. *Recept.* ample, naked, globose or ovate-conical ! Branches of the disc-*style* truncate. *Achenes* wingless, angular, glabrous, similar in disc and ray, with a large epigynous areole, either without pappus, or with a short, toothed, or entire crown-like or earshaped pappus. *DC. Prodr. 6, 50.*

Herbaceous, mostly annual, glabrous, branching, chiefly from the Northern Hemisphere. Leaves multipartite with narrow linear lobes. Heads solitary or corymbose. Rays when present white ; disc yellow.

I. EU-MATRICARIA. Hds. radiate; rays ligulate, white, long or short. (Sp. 1–5.)
 Ach. without pappus. Lvs. glaucous. Rays short, oblong (1) **nigellæfolia.**
 Achenes with an *entire-edged*, truncate, frill-shaped pap-
  pus. Rays long, slender ... ... (3) **tenella.**
 Ach. of disc with a *sharply-toothed*, unequal sided pappus.
  Diffuse. Lower leaves bi-pinnatifid, upper pinnatifid.
   Sparsely pilose or glabrous. Recept. globose ... (2) **Capensis.**
   Leaves copiously hispid-pilose. Recept. conical (5) **hirta.**
  Erect. Leaves glabrous, simply pinnatisect ... ... (4) **glabrata.**
II. ANACTIDEA. Heads discoid. (Sp. 6–11.)
 * Pappus rim-like, subentire or denticulate, equal-sided : Cor. 4-toothed.
  Pedunc. terminal, one-headed, scattered or subcorymbose :
   Inv.-scales obtuse, glabrous. Heads globose ... (6) **globifera.**
   Inv.-scales obtuse, tomentose. Heads depressed (7) **grandiflora.**
   Inv.-scales acuminate, glabrous. Cor. lobes taper-
    pointed ... ... (9) **acutiloba.**
  Infl. densely corymbose ; corymbs branching, many-
   headed ... ... (8) **multiflora.**
 * Pappus unilateral, ear-shaped, nearly as long as the achene. Cor. 5-toothed.
  Leaves bi-pinnatisect ; lobes and lobules narrow ... (10) **albida.**
  Leaves simply pinnatisect, lobes flat, linear ... ... (11) **dichotoma.**

Sect. 1. EUMATRICARIA. (Sp. 1–5).

1. M. nigellæfolia (DC. l. c. 50); glabrous or hispidulous, glaucous, diffuse or procumbent ; leaves sessile, bi-pinnipartite above, with multifid, linear-lanceolate, acute, nerved lobes, the lowest lobes simple, short ; pedunc. not much longer than the leaves ; inv.-scales obtuse ; rays oblong, entire; recept. globose ; pappus none ; disc-fl. 4-toothed, their achenes 4-angled.

VAR. β. **tenuior** (DC.) ; inv.-scales subacute ; rays bilabiate, each lip entire. *DC. l. c.*

HAB. Zondag R., *Drege!* Swellendam, Worcester and Kaffraria, *E. Z.! Zey.!* 2835. Wittedrift, Plettenbergs-bay, *Pappe!* Near Maritzburg, Natal, *Gerr. & M'Ken!* 341. *T. Cooper!* 1146. β. Dumpfel, Albany, *Drege!* (Herb. Hk., D., Sd.)
Stems many from the base, 1–2 ft. long, ascending or decumbent, often rooting where they touch the soil, and possibly therefore perennial, leafy throughout. Lvs. 1–1½ inch long, their multifid lobes from ½– 1 line wide. Pedunc. 1–2 inches long. Heads globose, with a very prominent disc and short rays. Achenes crowned with a callous, thickened margin ; those of the ray-fl. flattened. Var. β. is more slender, with heads half the ordinary size.

2. M. Capensis (Linn. Mant. 115); erect or diffuse, much-branched, thinly pilose or glabrous ; lower leaves bipinnatisect, the upper simply pinnatisect, the lobes linear, flattish, fleshy, acute ; pedunc. elongate ; invol. scales obtuse, with a wide, scarious-membranous margin and tip ; disc-fl. 4-toothed ; pappus toothed, unilateral or unequal-sided. *Th.! Cap.* 693. *DC. l. c.* 50. *M. Africana, Berg. Cap.* 296. *Cotula Capensis, L. Sp.* 287.

HAB. Cape, *Thunberg! Verreaux!* (Herb. Th., D., Sd.)
1–2 feet high, much-branched, annual. Leaves 1–2 inches long ; the lobes ½–1 line wide. Pedunc. 3–6 inches long. Disc-fl. reddish or yellow.

3. M. tenella (DC.! l. c. 51); glabrescent or thinly puberulous, simple or branched; leaves petiolate, bipinni-partite, the pinnæ horizontally patent, lobules short, subacute or mucronulate; pedunc. elongate; inv. scales obtuse, with a wide, membranous margin and tip; disc-fl. 4-toothed; pappus entire, truncate or wavy, unilateral or split at one side.

HAB. Langevaley, *Drege!* Cape, *Zey.!* 829.   (Herb. Hk., D., Sd.)
6–8 inches high, slender, either many-stemmed or simple, laxly leafy. Leaves 1–2 inches long, the laciniæ uncial, the lobules 1–3 lines long, ½ line wide. Pedunc. 3–5 inches long. Heads smaller than in *M. Capensis,* and whole plant more slender.

4. M. glabrata (DC. l. c. 51); stem erect, glabrous, simple or branched, branches angular, one-headed; leaves glabrous or nearly so, pinnati-partite, the lobes linear, subacute, entire or rarely one-toothed; invol. scales obtuse, with wide membranous margin and tip; rays broadly oblong or obovate, reflexed; achenes of the disc angular, with a sharply toothed, unilateral pappus, of the ray sub-abortive, with a shorter, more equally-toothed pappus. *Chrysanthemum glabratum, Thunb.! Cap.* 693. *Pyrethrum glabratum, Less.! Syn.* 254.

HAB. Cape, *Thunberg! E. & Z.!* Piquetberg, *Drege!* (Herb. Th., Hk.)
A foot or more high, rigid, sparsely leafy. Upper leaves often quite simple; lower 1–1½ inch long, their lobes 3–5 lines long, ½ line wide. Peduncle 2–3 inches beyond the uppermost leaf. Heads smaller than in *M. Capensis,* with much broader and shorter rays. Recept. very convex, globose.

5. M. hirta (DC. l. c. 51); stem diffusely much branched or prostrate, nearly glabrous, angular; pedunc. terminal and opposite the leaves, striate, and thickened under the fl.-heads; leaves roughly hispid, the lower bi-pinnati-partite, the upper pinnatifid, lobes linear, subacute, thickish; inv.-sc. glabrous, with membranous margins; recept. ovate-conical; rays linear-oblong, reflexed; achenes of disc bluntly angular, with a sharply 5-toothed, unequal or unilateral, ear-like pappus; of the ray flattish, without pappus. *Chrysanthemum hirtum, Th.! Cap.* 693. *Pyrethrum hirtum, Less.! Syn.* 254.

HAB. Wet places near Capetown, *Th.!* Marshes near the Salt R., Cape Flats, *W. H. H.* (Herb. Th., D.)
Stems 2 ft. long or more, robust, many spreading from the crown, irregularly forked. Young stems sparsely hispid. Leaves copiously hispid with brittle hairs.

Sect. 2. ANACTIDEA. (Sp. 6–11.)

6. M.globifera (Fenzl!); annual, much-branched, glabrescent or sparsely pubescent; leaves petioled, bi-tri-pinnatisect, the lobules linear, short, subobtuse; fl. branches pedunculoid, naked, 1-headed; heads globose, discoid; inv. scales oblong, obtuse, glabrous, membrane-edged; corollas 4-lobed; pappus denticulate, short. *Cotula globifera, Th.! Cap.* 696. *Tanacetum globiferum, DC. in Deless. Ic.* 4, *t.* 48. *Prodr.* 6, *p.* 132. *T. obtusum, Th. Cap.!* 641. *Cenocline globifera, Koch. Bot. Zeit.* 1, *p.* 41.

HAB. Cape, *Thunberg!* &c. A common weed throughout the colony. (Herb. Th., D., Hk., Sd.)
This is the "Stink-net" of the colonists. Stem 1–1½ ft. high, often excessively branched, but not ligneous, robust or slender, purplish or pale, quite glabrous or hairy. Leaves 1–1½ inch long, including the petiole, finely divided. Pedunc. 1–3 inches long. Recept. very convex, at length quite globose, inflated. Cor. lobes short, ovate. The fl. heads vary much in size.

7. M. grandiflora (Fenzl) ; annual, erect, pubescent, corymbose at
the summit ; leaves petioled, half-clasping and often toothed at base of
petiole, bi-tri-pinnatisect, the pinnæ several, lobules linear, callous mucronate ; fl. branches pedunculoid, corymbose, more or less fistular upwards ;
heads flattish-convex, discoid ; inv. scales oblong, obtuse, ribkeeled tomentose ; the innermost membrane-tipped ; corolla 4-lobed ; pappus obsolete,
sub-entire. *Tanacetum grandiflorum, Th. ! Cap.* 642. *DC. l. c.* 132. Also,
Cotula tripinnata, Th. ! Cap. 696. *Cenocline grandiflora, Koch.*

VAR. β. lobes of corolla taper-pointed, very acute.   (Hb. Hk.)

HAB. Cape, *Thunberg!* Olifant's R., *Drege!* Brackfontein and Kapallsloot, *E. & Z.*
(Herb. D., Hk., Sd.)

Stem robust, 1–1½ ft. high, branched chiefly towards the summit, the branches
level-topped. Leaves 1½–2 inches long, with basal stipulæform lobules, more divided
than in *M. globifera.* Pedunc. 2–3 inches long, swelling upwards. Heads 6–8
lines across, much flatter than in *M. globifera.* β. has flowers nearly as in *M. acutiloba*, but the invol. of *M. grandiflora.*

8. M. multiflora (Fenzl !); annual, corymbosely much branched, erect
or diffuse, thinly pubescent ; leaves subpetiolate, bipinnatisect, the
lobules linear, short or toothlike, subacute ; corymbs branching, many-
headed, leafy at base ; heads (small) subglobose, discoid ; inv. scales
broadly oblong, very obtuse, membrane-edged, nearly glabrous ; pappus
obsolete, rim-like, subentire. *Tanacetum multiflorum, Th. ! Cap.* 642.
DC. l. c. 132. *T. fruticosum, Linn. Sp.* 1183. *Cotula tanacetifolia, Willd.*

VAR. β. leptoloba; more lax and slender, with longer leaves. *Tan. leptolobum,
DC. ! l. c.* 133.

HAB. Cape, *Thunb. ! E. & Z. ! Drege!* &c. From Capetown to Namaqualand, and
throughout the Western Districts, common. Very common near Capetown. (Herb.
Th., D., Hk., Sd.)

Stems strong, the larger ones almost ligneous at base; those from the sea-side
(near Greenpoint) especially stout, depressed or widely spreading, rigid and closely
much-branched. Leaves commonly uncial, in β. 1½–2 in. long. Heads 2–3 lines
across, bright yellow.

9. M. acutiloba (Harv.); annual, rigid at base, thinly pubescent, erect,
much-branched; leaves sessile, bipinnatisect, the lobes linear, obtuse,
2–4 of the basal ones stipular; branches ending in nude, one-headed
peduncles; heads flattish-convex; inv. scales pluri-seriate, subulate-
acuminate; corolla urceolate, 4–lobed, the lobes taper pointed; pappus
rim-like, subentire. *Tanacetum acutilobum. DC. ! l. c.* 132. *Cenocline
acutiloba, Koch, l. c. Chamœmelum acutilobum, Fenzl.*

HAB. Verleptpram, on the Gariep, *Drege!* (Herb. D., Hk., Sd.)

More rigid than *M. globifera*, almost ligneous at base, with smaller leaves, flatter
fl. heads, and specially acuminate inv. scales. The corolla has a cylindr. tube, and
a much inflated, narrow mouthed, sharply 4-toothed limb. Achenes white, obconic,
sub-compressed and sharp-edged, with a prominent ventral angle. Pappus obvious
as a marginal rim, but very short. Pedunc. 1–2 inches long. Heads 6–7 lines wide.

10. M. albida (Fenzl); annual, dichotomously much-branched, diffuse;
stems and branches glabrescent; leaves petiolate, bi-pinnati-partite,
pubescent or villous, the lobes and lobules linear, obtuse or acute;
peduncles long, glabrate, one-headed, filiform; invol. scales oval-oblong,
obtuse, pubescent, membrane edged, the outer subacute; pappus ear-

shaped, unilateral, nearly as long as the achenes. *Tanacetum albidum,*
DC. l. c. 132. Also *Pentzia annua, DC. l. c.* 138.

HAB. Near the Gariep, *Drege! Zey.!* 837. Namaqualand, *A. Wyley!* (Herb. D.,
Sd., Hk.)
Leaves nearly 1-inch long, including the ½ inch of petiole, copiously pubescent,
the lobes 3–4 lines long. Pedunc. 5–8 inches long, terminal, becoming lateral as
the stem lengthens out. Corolla 5-toothed. Pappus as in *Pentzia,* to which genus
this and the following species are rather artificially referred by De Candolle.

11. M. dichotoma (Fenzl); annual, dichotomously-branched, diffuse;
stems and branches glabrescent; leaves cuneate at base or sub-petiolate,
pubescent, some trifid, others pinnati-partite, the lobes simple, linear,
flat, in few pair, mucronate; pedunc. long, glabrate, one-headed, filiform;
inv. scales oblong, obtuse, pubescent, widely membrane-edged, sub-
biseriate, equal; cor.-lobes acuminate; pappus earshaped, unilateral,
nearly equalling the achene. *Pentzia dichotoma, DC. l. c.* 138. *Cotula*
Teesdaliæ, Drege! Hb. ex pte, nec DC.

HAB. Olifants R., *Drege!* (Herb. Sd., Hk.)
Near *M. albida,* but the leaves are less compound.

LXII. **TANACETUM**, Linn. (ex pte).

Heads discoid, either homogamous or heterogamous, having a few
marginal, female, 2–3–4-toothed flowers. *Recept.* convex, naked. *Inv.*
imbricate. *Corolla* of disc terete, 4-toothed (in the S. African species).
Achene sessile, angular, glabrous, with a large epigynous disc. *Pappus*
either none, or minute and coroniform, entire or somewhat toothed,
often oblique. *DC. Prodr.* 6, *p.* 128.

A large (and rather ill-constructed) genus, chiefly from the Old World, consisting
of shrubs, half-shrubs, herbaceous perennials and annuals. Leaves alternate, variously divided. Heads solitary or corymbose. The garden herb *Tansy* is the type of
the genus, but belongs to a section with 5-toothed corollas. Name said to be a
corruption of *Athanasia.*

Inflorescence corymbose; corymb few or many-headed:
    Lf.-lobes in several pair, sharply 2–3-toothed or some entire   (1) **hippiæfolium.**
    Leaf-lobes in 2–4 pair, entire, or the uppermost 1-lobed   (2) **heptalobum.**
Peduncles simple, one-headed; leaf-lobes in 2 pair, linear, with
    revolute margins ... ... (3) **griseum.**

1. T. hippiæfolium (DC. l. c. 133); stem herbaceous, erect, angular,
glabrous, resin-dotted; leaves subpetiolate, pinnatisect, gland-dotted,
glabrous, the lobes in several pair, sharply 2–3-toothed or some entire,
acutely mucronate; corymb compound, many-headed; pedicels short;
heads subglobose; inv.-scales glabrous, in several rows, loosely imbricating; fl. glandular; achenes without pappus.

HAB. Cape (Eastern Districts ?) *Krebs, E. Z.!* (Herb. Sond.)
Stems 1–2 ft. high, subsimple or much branched, branches erect, virgate. Leaves
1–2¼ inches long, tapering at base into a narrow-cuneate, imperfect petiole, the lobes
in 5–8 pairs, close-set, cuneate or linear, mostly toothed. Heads 2 lines diameter.
Drege's specimens (Hb. Hk., Sd.) under this name (from Natal) rather belong to
T. heptalobum.

2. T. heptalobum (DC.! l. c. 133); stem herbaceous, erect, nearly
terete, silky; leaves sessile or subsessile, pinnatisect, gland-dotted, more

or less silky, the lobes in 2–4 pair, oblong, acutely mucronate, entire or
the uppermost with a lateral, spreading lobe, the terminal sometimes
trifid ; corymb simple or compound, few or several-headed ; pedicels
short ; heads subglobose ; inv.-scales glabrate, in several rows, loosely
imbricating ; fl. glabrous ; achenes without pappus.

HAB. Betw. Omcomas and Omblas, *Drege!* Natal, *Krauss,* 149 ; *Gueinzius,* 333,
590. (Herb. D., Sd., Hk.)
Very similar to *T. hippiæfolium,* but with more terete stems, a copious or scanty
silky pubescence, and fewer leaf-lobes, which are commonly quite entire, with thickened edges, sometimes lobed (but not sharply toothed). Hds. nearly as in *T. hippiæfol.*

3. T. griseum (Harv.); suffruticose, erect, all parts densely silky-
canescent; branches virgate, closely leafy below, pedunculoid upwards,
one-headed ; leaves sessile, pinnati-partite, the lobes in 2 pair, linear,
acute, one-nerved, with revolute margins ; peduncles elongate, nearly
nude; inv.-scales linear-lanceolate, acute, silky ; flowers 4-toothed, all
perfect; pappus none; achenes glabrous.

HAB. Umboti District, Natal, *W. T. Gerrard,* 1050. (Herb. D.)
1–2 ft. high, branched in the lower half; the branches straight, erect, 10–18 inches
long, closely leafy in the lower half, pedunculoid upwards. Pedunc. 3–6 inches
long, one-headed. Heads nearly ½ inch across, homogamous. Corolla-tube terete,
not winged ; lobes short, bluntish. The foliage is something like that of *Schistost.
cratægifolium,* but the inflorescence is quite different, and the heads are homogamous.

LXIII. SCHISTOSTEPHIUM, Less.

Heads many-fl., heterogamous, discoid; marginal-fl. female, with very
short, bilabiate corollas ; disc-fl. bisexual (but commonly abortive),
4-toothed. *Invol.* imbricate. *Recept.* convex, naked. *Achenes* sessile ;
of the disc glabrous, ovuliferous, with bifid styles; of the margin silky-
villous, obovate. *Pappus* none. *Less. Syn. p.* 251. *DC. Prodr. 6, p.*
74, also *Tanacetum, Sec. V. Hippioides, DC. l. c.* 134.

Shrubby, suffruticose or half-herbaceous perennials, silky or tomentose in all parts.
Leaves alternate, flabelliform and many-nerved, or pinnatisect. Heads corymbose.
Flowers yellow: the marginal female-fl. not longer than the disc, bilabiate, the inner
lip toothlike, the outer deeply 3-toothed. Name from σχιζω, *to cut* and στεφος, *a
crown ;* alluding to the deeply-toothed marginal or ray-flowers.

Leaves petiolate, roundish or flabelliform :
  Shrubby, silvery ; lvs. lobulate ; inv.-scales obtuse   ...   (1) **flabelliforme.**
  Herbaceous, green ; lvs. toothed ; inv.-scales lanceolate (2) **rotundifolium.**
Leaves sessile, pinnatifid, the upper segments toothed or cut (3) **cratægifolium.**

1. S. flabelliforme (Less. Syn. 251) ; shrubby, erect, branches and
leaves silvery-canescent, with short, close, silky pubescence ; leaves
petiolate, roundish-flabelliform, 3–5-nerved, cuneate at base, crenate
or shortly and bluntly 3–5–9-lobed ; corymb terminal, branched, densely
many-headed ; inv. silky, scales ovate-oblong, shorter than the flowers;
marginal-flowers female, with very short corollas, and flatted, pilose
achenes. *Schistostephium argyreum, Fenzl. Tanacetum argyreum, DC.!
l. c.* 134.

HAB. Cape, *Krebs.* Near Grahamstown, and betw. Morley and Omtata, *Drege!*
Grahamstown, *Genl. Bolton!* Howison's Poort, *H. Hutton!* *Zey.!* 2833. (Herb.
D., Hk.. Sd.)
1–2 feet high, virgate, with the habit of *Pentzia flabelliformis.* Leaves ½ inch

long, rather wider than their length, more or less deeply toothed or lobulate. Heads 1–1½ line diam., several-flowered, 20–30 in a dense corymb. Disc-fl. glandular, 12–15; ray-fl. 5–6, with very short, imperfect corollas. I have not seen any specimen of *Lessing's Schistostephium flabelliforme,* but, by description, it can hardly be different from *Tanacetum argyreum, DC.*

2. S. rotundifolium (Fenzl. Ms.); herbaceous, erect, tall, branched; stem and branches terete, villoso-tomentose; leaves on slender petioles, those of the sterile twigs sub-opposite, the rest alternate, roundish, 3–7-nerved, coarsely toothed, closely silky on both sides; upper branches ending in many-headed, branching corymbs; heads globose; inv.-scales lanceolate, acute, silky, shorter than the flowers; achenes without pappus; marginal female flowers with short, bilabiate corollas, and flattened, pilose achenes. *Tanacetum rotundifolium, DC.! l. c.* 133.

HAB. Betw. Gekau and Basche, and betw. Omsamwubo and Omsamcaba, *Drege!* Natal, *Gueinzius,* 332,592. (Herb. D., Hk., Sd.)

2–3 ft. high, robust, densely leafy. Petioles ½ inch long, very slender. Leaves ¼–1 inch long, and equally wide, truncate or broadly cuneate at base, toothed above, the teeth shallow, very broad, mucronate. Fl. branches in a terminal panicle, each corymb 10–20-headed. Recept. conical. Some of the central flowers abortive.

3. S. cratægifolium (Fenzl, Ms.); stem half-shrubby, branched, erect or ascending; branches appressedly silky-villous; leaves sessile with a stipular lobe at base, pinnatifid, the lower lobes short and simple, lanceolate, acute, the upper oblong, often toothed or trifid, all copiously silky, especially beneath; fl. branches shortly naked at top, bearing a few-headed corymb; pedicels long or short; heads globose, many-flowered; inv.-scales pluriseriate, silky, linear, acute; marginal-flowers female, with short corollas, and flattened, pilose achenes. *Tanacetum cratægifolium, DC. l. c.* 134. Also *T. consanguineum, DC. l. c. fide sp. ex Drege!*

HAB. Cape, *Burchell, Krebs.* (fide DC.) *Mundt! Verreaux!* Witbergen, *Drege!* Adow, *Zey.!* 2835. Albany, *T. Williamson!* Grahamstown, *Sir C. J. F. Bunbury!* Doorn Kopf, *Burke & Zey.!* 1041. Near Ladysmith, Natal, *Gerr. & M'K.!* 264. Albert Distr., *T. Cooper!* 619. (Herb. D., Hk., Sd.)

A robust, branching, half shrub, 1–2 ft. high, with copious, straight, silky, whitish pubescence. Leaves 1–1½ inches long, deeply pinnatifid or pinnati-partite, when young on both sides silky, the older ones becoming naked above, lobes 1–1½ line wide, 2–3 lines long. Inv. somewhat silvery. Heads 4–12 in a corymb, 2–4 lines across. I find marginal female flowers in *Drege's* specimen of " *T. consanguineum,*" and therefore venture to combine that with the present; the size of the fl.-heads and length of pedicels are characters too inconstant to be trusted. The marginal fl. are less clearly bilabiate, and their teeth much smaller than in the other species.

LXIV. **ARTEMISIA,** Linn.

Heads discoid, homo- or heterogamous, the *marginal* flowers in one row often female, 3-toothed, with a long, exserted, bifid style; *disc fl.* 5-toothed, hermaphrodite, or abortive, or male. *Invol.* imbricate, the scales with membranous borders. *Recept.* without paleæ, flattish or convex, naked or hairy-fimbrilliferous. *Achenes* obovate, with a small epigynous disc. *Pappus* none. *DC. Prodr. 6, p.* 93.

A very large genus of bitter or aromatic herbs or suffrutices, almost exclusively natives of the northern hemisphere. Leaves alternate, variously pinnatisect. Heads in spikes or racemes, or paniculate, of small size, several-flowered. Fl. yellow or purple. The common garden plants, *Wormwood, Southernwood,* and *Tarragon* are familiar examples of this genus.—Named from *Artemis,* the Grecian Diana.

A. Afra (Jacq. Hort. Schoenbr. t. 467); suffruticose, leaves becoming glabrous above, canescent beneath, interruptedly bi-pinnati-partite, the pinnules linear-acute, entire or toothed, with recurved margins, rachis pectinate with similar lobules, the petiole leaf-stipuled at base; stipules simple or divided; heads racemoso-paniculate, secund, hemispherical, cernuous; inv. scales oblong, obtuse, scarious, canescent at back, with a green medial line; corollas naked; recept. conical, pilose! *DC. Prodr.* 6, *p.* 106. *A. pontica, Thunb.! Cap.* 643.

HAB. Cape, *Thunberg! Bowie! Thom! E. Z.! Drege!* Gnadendal, *Dr. Roser!* Grahamstown, *Genl. Bolton!* Bushman's R., Natal, *Gerr. & M'K.* 345 ; Keiskamma Hoek, and Baillie's Grave, *W. S. M. D'Urban!* (Herb. Th., D., Hk.)

Stems robust, 2–3 feet high, much branched and leafy, tomentose, rib-striate. Leaves 2–3 inches long, petioled, bipinnate, the pinnæ ½–1 inch long, pinnules 3–5 lines long, ½ line wide. Racemes in a much-branched, terminal panicle. Heads 1½–2 lines diameter. *A. vestita,* Wall! from India, is very near indeed to this, if it be permanently distinct.

LXV. **HIPPIA**, Linn.

Heads many-fl., discoid, heterogamous; *ray-fl.* in 1 row, female, with a filiform corolla continuous with the ovary; *disc-fl.* male, 5-toothed. *Recept.* small, convex, naked. *Invol.*-scales sub-biseriate, elliptical or ovate, with a scarious margin. *Achenes* of ray, roundish, flattened, with marginal ribs or wings; of disc abortive. *Pap.* none. *DC. Prodr.* 6, *p.* 144.

Herbs or suffrutices, all natives of S. Africa, with the odour of chamomile. Lvs. alternate, mostly pinnately lobed, rarely entire. Heads small, yellow, corymbose. Name in honour of some botanist ?

Leaves pinnati-partite or deeply pinnatifid :
 Shrubby, robust, with rodlike branches; lvs. 2 inches long (1) **frutescens.**
 Slender, filiform, diffuse, flexuous ; lvs. ½–¾ inch long :
  Branches pubescent, villous near the top ... ... (2) **gracilis.**
  Branches very hairy, with long, spreading hairs ... (3) **hirsuta.**
Leaves oblong, entire; stem slender, decumbent, hairy ... ... (4) **integrifolia.**

1. H. frutescens (Linn. Mant. 261); suffruticose or shrubby; branches rodlike, densely pubescent; leaves pectinato-pinnatipartite, the lobes in 5–10 pairs, opposite, linear-oblong, or linear, mucronulate, veiny, pubescent, the lowest small, at the base of the petiole like a pair of stipules; heads corymbose. *DC. l. c.* 144. *Th.! Cap.* 723. *Tanacetum frutescens, Linn. Sp.* 1183.

VAR. β, **Thunbergii** ; leaf-lobes narrow-linear, with revolute margins, 6–7-pair, acute or acuminate. *Tanacetum suffruticosum, Th.! Cap.* 541, *ex pte.*

HAB. Cape, *Thunberg,* &c.; Paarl, *Drege!* Cape Flats, *W. H. H.;* Gnadendal, *Dr. Roser!* Plettenberg's Bay, *Mundt.!* (Herb. D., Hk., Sd.)

Tall, straggling, not much branched; the branches 12–18 inches long, leafy. Lvs. 2–2½ inches long, their lobes 5–8 lines long, 1–2 lines wide. Heads sometimes in a lax panicle of small corymbs, sometimes in a dense, branched corymb. Var. β, which I have only seen in Hb. Thunberg, has leaf-lobes 6 lines long, not ½ line wide, much more rigid and pointed than in the normal state.

2. H. gracilis (Less.! Syn. 268); suffruticose, slender, diffuse; branches filiform, curved, pubescent, with scattered, longer hairs towards the apex ; leaves pinnati-partite, the lobes in 2–5 pairs, linear, nerved, acute, closely pubescent, no stipular lobes; heads corymbose. *DC. l. c.* 144. *Tanacetum pilosum, Berg. Cap.* 244. *T. cotuloides, Linn. Mant.* 282. *Cotula fimbriata, Spr.*

VAR. **repens**; root creeping ; less subsessile ; heads few. *H. repens,* DC. l. c. 144.
HAB. Cape, *Burman.* &c. Stellenbosch and Draakenstensberg, *Drege!* W. H. H. Zwarteberg, Pappe! Bamskloof, Worcester, *H. Hutton!* Table Mt. summit, *Ecklon!* W. H. H. Simon's bay, *C. Wright!* 346, 347. *Zey.!* 2837. β. at Gnadendal, *Drege!* (Herb. D., Sd., Hk.)
Much more slender, branching, and diffuse than *H. frutescens,* with much smaller and less divided leaves. Leaves ¼–½ inch long, the lobes 1–2 lines long, ¼–1 line wide. Var. β. is scarcely different from the common form.

3. H. hirsuta (DC. l. c. 144) ; "suffruticulose ; *branches very hairy;* leaves puberulous, cuneate, pinnatifid ; lobes acuminate, in 3–4 pairs, the lower shorter ; heads corymbose."

HAB. Bot River, *Ecklon.* Swellendam, *Burch.,* 6981. (Unknown to me).
"The branches, especially the younger ones, hispid with long, soft, spreading hairs. Leaf-lobes less deep and much more acuminate than in *H. gracilis.*" *DC.*

4. H. integrifolia (Less. Syn. 268) ; suffruticose, slender, diffuse or procumbent ; branches filiform, patently hairy ; leaves sessile, oblong or obovate-oblong, mucronate, quite entire or unidentate, hairy, one-nerved, spreading or deflexed, gland-dotted ; heads small, terminal, solitary or 2–3 together, sub-corymbose. *DC. l. c.* 144.

HAB. Cape. Herb. *Vahl., E. Z.!* (Herb. Sd., Cap., D.)
Stems 1 foot or more long, flexuous, as are the long filiform branches. Whole plant hairy with fulvous, soft, spreading hairs. Leaves squarrose, ½–¾ inch long, 2–3 lines wide, slightly narrowed to the base.

LXVI. **PENTZIA**, Thunb.

Heads many-fl., homogamous. *Recept.* narrow, flat, then convex, sparingly fimbrilliferous. *Invol.* obovate, equalling the disc ; its scales imbricated, scarious. *Corolla* 5, rarely 4-toothed. *Achenes* angular, wingless, sessile. *Pappus* membranaceous, shortly tubular, irregularly torn, oblique or ear-shaped. *DC. Prodr.* 6, 136.

Small, rigid, much branched, mostly cano-pubescent shrubs or half shrubs. Lvs. alternate, variously toothed, cut or pinnatisect. Heads terminal, yellow, corymbose or solitary. Name in honour of some botanist ?

Heads corymbose ; corymbs simple or compound :
    Leaves pinnatisect, or twice trifid :
        Outer inv.-sc. herbaceous, keeled, pubescent; inner
            scarious ... ... (1) **Burchellii.**
        All the inv.-scales shining, scarious-membranous (2) **elegans.**
    Leaves cuneate or fan-shaped, toothed at the extremity :
        Lvs. canescent, petioled, fan-shaped, 3–5-nerved
            beneath ... ... (3) **flabelliformis.**
        Lvs. glabrous, dotted, narrow-cuneate, tapering at
            base, sessile ... ... (4) **Cooperi.**
Heads solitary, at the ends of nude, pedunculoid branches :
    Old twigs and pedunc. changing to rigid spines ... (9) **spinescens.**
    Unarmed :
        Leaves cuneate, truncate, 3–5-toothed or lobed at the apex.
            Erect, with virgate branches. Pappus mem-
               branous, ear-like, toothed ... ... (5) **quinquefida.**

Dwarf, flexuous, much-branched. Pappus none   (10) **tortuosa.**
Leaves pinnati-partite or trifid,
 Erect, with virgate branches: hds. very many-fl.   (6) **sphærocephala.**
 Divaricately much-branched; heads rather small.
  Lvs. petioled ; inv.-scales obtuse, the
   outer keeled ... ...   (7) **virgata.**
  Lvs. sessile; inv.-scales taper-pointed   (8) **globosa.**

1. P. Burchellii (Fenzl); shrubby, much-branched, spreading, thinly tomentulose ; leaves petiolate, glabrous or puberulous, punctate, in the upper half pinnatisect or bipinnatisect, the pinnæ in 2–3 pairs, either 3–5-lobed or entire, linear, thickish, mucronulate; corymbs subsimple, few or several-headed, the pedicels twice as long as the leaves, tomentulose ; heads flattish, many-fl. ; inv.-scales oblong, obtuse, the outer herbaceous, keeled, puberulous, inner membranous ; pappus earshaped, irregularly torn or cut. *Tanacetum Burchellii, DC. l. c.* 132 ? *Pentzia punctata, Harv. (in Hb. Hk., D.)*

HAB. Betw. Capetown and Zacksriver, *Burchell; Graafreynet, E. Z.!* Zuureberg, *Burke & Zey.!* Albany, *Mrs. F. W. Barber.* Sneeberg, *Dr. Wallich.* (Hb. Sd., D., Hk.)
I adopt this sp. on the authority of *E. & Z.'s* specimens in Hb. Sond., which were seen and marked by De Candolle. Yet I can hardly persuade myself that DC.'s character, taken probably from *Burchell's* plant, can refer to the same species. DC. expressly says that his "*Burchellii*" has the aspect of a *Matricaria*, and is wholly without pappus *(achæniis omnino calvis)* ; Ecklon's plant here described is a small shrub, with every character, including the conspicuous pappus and flat receptacle of *Pentzia.* In aspect and foliage it resembles *P. elegans,* but differs in the involucre and pappus. Leaves ½–¾ inch long ; the uppermost sometimes subsimple ; some occasionally with basal, stipuloid lobes. Corymbs 8–12 or more headed ; heads 3–4 lines wide.

2. P. elegans (DC. l. c. 136); shrubby, much branched, unarmed, minutely canescent ; leaves at the apex shortly pinnatisect, or biternate, the lower pair of lobes either bifid or trifid or entire, the terminal always trifid, lobules linear, thick, obtuse ; corymbs many-headed, heads on short pedicels, oblong; inv.-scales all shining, scarious, obtuse; pappus white, tubular, unequal-sided, toothed.

HAB. Kendo, in stony places, 3000–4000 f., *Drege !* (Herb. D., Hk., Sd.)
A foot or more high, the older branches leafless, nodulose, twigs closely leafy. Leaves 3–5 lines long, the lobes 2 lines long, ½ line wide. All young parts, save the involucres, minutely and closely pubescent, with whitish or foxy hairs. Heads 2 lines long, 1 line wide, 10–20 in a corymb, scarious.

3. P. flabelliformis (Willd. Sp. 3, 1808); shrubby, much branched, greyish-canescent ; leaves petiolate, cuneate, tapering at base, truncate and toothed at the extremity, 3-nerved beneath ; corymbs many-headed, heads on short pedicels, obovate-oblong; inv.-scales oblong, obtuse, the outer dorsally pubescent, inner scarious, glabrous ; pappus white, tubular, unequal-sided, toothed. *Gnaphalium dentatum, Linn. Tanacetum flabelliforme, L'Her. Sert. Angl.* 21, *t.* 27. *Pentzia crenata, Th.! Cap.* 637. *Balsamita flabelliforme, Pers.*

HAB. Cape, *Thunberg ! Thom !* Nieuweveld, betw. Brakriver and Uitvlugt, and betw. Zwarteberg and Kendo, *Drege !* Fish River, *Burke! Zey.!* 856, 855. (Herb. Th., D., Hk., Sd.)
1–2 f. high, erect. Leaves 3–6 lines long, 2–4 lines wide. Corymbs 10–20-headed. Heads 2½ lines long, 1½ wide, pale straw or buff colour.

4. P. Cooperi (Harv.); shrubby, erect, unarmed, the young parts cobwebby-flocculent, the adult becoming glabrous; leaves narrow-cuneate, tapering much at base, truncate and sharply 3-toothed or trifid at the summit, punctate, glabrous; corymbs compound, many-headed; heads on long or short pedicels, globose; inv. oblong or ovate-oblong, obtuse, the outer herbaceous, inner scarious, glabrous; pappus white, rather short, unequal-sided, toothed.

HAB. District of Albert, near the Witberg, and in Basutuland, *T. Cooper!* No. 628, 711. (Herb. D.)

1–2 feet high, erect, with erect rodlike branches. Young twigs and leaves hoary with cobwebby hairs; older twigs minutely puberulous. Adult leaves glabrous, 5–8 lines long, 2 lines wide at the 3-toothed summit. Corymbs 6–20-headed, flat-topped. Uppermost leaves often entire, round-topped.

5. P. quinquefida (Less.! Syn. 266); shrubby, erect, virgate, silky-canescent or cinereo-pubescent, unarmed; leaves cuneate, tapering at base into a petiole, truncate at the apex, and 3–5-toothed or lobulate, the teeth or lobes obtuse, closely silky-canescent, the uppermost leaves linear, entire; pedunc. one-headed, long or shortish; inv.-scales obtuse, the outer strongly keeled, silky. *Cotula quinquefida, Th.! Cap. 695. Pentzia microphylla, DC. l. c. 137.* Also *P. cinerascens, DC.! l. c. 138.*

VAR. β. **nana**; dwarf; leaves coarsely 5-toothed, with close, silky pubescence; peduncles long. *Pentzia nana, Burch. Cat.* 1731. *Trav. t. p.* 400.

HAB. Cape, *Thunberg!* Plettenberg's Bay, *Burchell.* Kendo and Adow, *Drege!* Koegakamma Kloof, *Zey.!* 852. β. Ky-Gariep, *Burchell.* (Herb. Thunb.)

A foot high, all the young parts silky-canescent. Leaves 5–6 lines long, 2–3 lines wide at top, sometimes merely crenate, sometimes deeply toothed or lobed, varying in the same plant. Pedunc. 2–6 inches long. I have not seen Burchell's *P. nana*, but by the characters assigned to it, it can scarcely differ from the plant collected by Thunberg and here described. *Drege's* specimen of *P. cinerascens* (Hb. Sd.) is also identical. Except in the *cuneate* and shortly-lobed leaves this scarcely differs from *P. sphærocephala.*

6. P. sphærocephala (DC.! l. c. 138); shrubby, rigid, branching; branches erect, virgate, thinly cinereo-canescent; leaves petiolate, in the upper half pinnati-partite, the lobes linear, obtuse, in 2–3 pairs, the upper leaves trifid or entire, closely and thinly canescent; branches ending in a long one-headed peduncle; heads flattish, very many flowered; inv. scales thinly pubescent or nearly glabrous; oblong, obtuse, scarious at margin, the outer keeled; pappus unequal-sided, toothed.

HAB. Near the Zwartkey and Fish R., *Drege!* Uitenhage, *Ecklon.* Koegakamma Kloof, *Zey.!* Rhinoster River, *Burke!* District of Albert, *T. Cooper!* 578. (Herb. D. Hk.)

A more robust, less divaricate bush than *P. virgata*, with much longer flowering stems and larger heads. Fl. branches, often a foot or more long, naked for 3–6 inches below the head. Leaves 1 inch or more long, the lobes 4–5 lines long; or, in starved specimens, of half that size. Heads half an inch or more across: the largest in the genus.

7. P. virgata (Less.! Syn. 266); unarmed, shrubby, rigid, divaricately much branched; twigs thinly canescent; leaves petiolate, at the apex trifid, or pinnately 5–lobed, the lobes linear, obtuse, with revolute margins, more or less canescent; heads terminal, solitary; inv. scales nearly glabrous, oblong, obtuse, scarious at the margin, the outer keeled; pappus

white, very unequal-sided, toothed. *DC. l. c.* 137. *Chrysanthemum inca-*
num, Thunb.! Cap. 693. *P. cotuloides, DC. l. c.* 138, *fide Sp. ex Drege!*

VAR. β. **gracilior;** slender, leaves an inch long or more, ¾ of this length simple,
3–5 lobed; outer inv. scales keeled, pubescent.

VAR. γ. **microcephala** (Fzl.); diffuse or procumbent; heads on short lateral twigs,
small, few-flowered. (Hb. Sd., D., Hk.)

HAB. Cape *Thunberg!* Kochman's Kloof, *Mundt.!* Knysna, *Dr. Pappe!* Uiten-
hage, *Zeyher*, 1849. Betw. Uitkomst and Geelbekskraal; on the Zwarteberge; betw.
Hex Riversberg and Bokkeveld, and on the Zondag and Koega R., *Drege!* β., Natal,
Miss Owen! γ. Rhosterkop, *Zey.!* 854. (Herb. Th., D., Hk.)

A much branched, rigid little bush, more or less clothed with grey pubescence,
1–2 f. high. Leaves commonly 4–5 lines long, in β. twice that length at least, the
lobes 1–2 lines long, blunt. Heads variable in size, 2–4 lines diam. I can find
nothing whereby to distinguish *P. cotuloides* from *Thunberg's* plant, which agrees
well with recently collected specimens from all the above localities. DC. states that
Drege's "*P. virgata,*" came from Kl. Namaqualand, but no habitat is given in the
"*Documente.*"

8. P. globosa (Less. Syn. 266); unarmed, shrubby, rigid, divaricately
much branched; twigs thinly canescent, glabrate upwards; leaves sessile,
pinnati-partite from the base or near it, the lobes in 3–4 pairs, short,
linear, obtuse (or mucronate), sub-glabrous; heads terminal, solitary;
inv. scales glabrous, lanceolate, taper-pointed, the inner minutely membrane tipped; pappus unequal-sided, toothed. *DC. l. c.* 137.

HAB. Cape, *Lichtenstein.* Winterveld, near Groot Tafelberg, and betw. Zilverfontein and Kaus. *Drege.* Gamke River, Zuureberg and Vat Rivier, *Burke! Zey.!*
850,851,853. Roggevelt, *A. Wyley!* (Herb. D., Sd., Hk.)

Similar to *P. virgata,* but readily known by the leaves being pinnatisect from the base,
or very generally so, and especially by the narrow, long-pointed invol. scales. The
leaf-lobes are sometimes very short, under 1 line long. Heads 2–4 lines wide, convex.

9. P. spinescens (Less. Syn. 266); shrubby, rigid, divaricately much
branched, the old twigs and peduncles hardening into spines; leaves
petiolate, at the apex 3-lobed or pinnately 5-lobed, the segments shortly
linear, obtuse, furrowed beneath, glabrescent; pedunc. widely spreading,
short incurved, somewhat racemose. *DC. l. c.* 137. *Osteospermum spines-*
cens, Thunb. in Herb. ex pte.

HAB. Cape, *Thunberg!* (Herb. Thunb.)

The specimen in Herb. Thunb. has suffered much from insects, and I have not
examined a flower. It is a more robust plant than *P. virgata.* The twigs are closely
silky. The inv. scales obtuse, with a brown, membranous border. Leaves 2–4 lines
long.

10. P. tortuosa (Fenzl); shrubby, dwarf, much branched, flexuous,
glabrous (microscopically scaberulous); leaves narrow-cuneate, thickish,
punctate, shortly and bluntly 3–5 lobed; pedunc. naked, slender, one-
headed, the head at first cernuous, then erect; inv. scales broad, obtuse,
the outer green at back, the inner membranous, pappus none. *Tana-*
cetum tortuosum, DC.! l. c. 133.

HAB. Witbergen, 7–8000 f., *Drege!* (Herb. D., Hk.)

3–4 inches high, woody, many times forked. Leaves 3–4 lines long, 1½ line wide
at top, pale. Pedunc. threadlike, 2 inches long. Heads 2 lines across. Corolla
5-toothed. Pappus none.—I follow *Fenzl* in referring this to *Pentzia,* of which genus
it has the habit, 5-toothed corolla, and flat receptacle, but no pappus! at least, none
developed on the fl. when it first opens. I have not seen mature flowers.

LXVII. MARASMODES, DC.

Heads several or many-fl., homogamous. *Inv.* ovate, imbricate, the outer scales scarious at apex. *Recept.* narrow, naked. *Cor.*-tube short, glandular, limb 5-toothed, not much dilated in the throat. *Anthers* not tailed. *Style*-branches capitellate. *Achenes* terete, beakless. *Pappus* of several separate, membranous, obtuse scales. *DC. Prodr.* 6, *p.* 136. *Oligodorella, Turcz.?*

Rigid, branching, divaricate, glabrous half-shrubs. Leaves alternate, sessile, thickish, dry, small, quite entire, heathlike, mucronulate. Heads at the ends of the branches in clusters, sessile or pedicelled, solitary. Cor. yellow. Distinguished from *Adenosolen* and *Pentzia* by the pappus. Name from μαρασμος, a *wasting away;* because these plants look lean or ill-fed.

Heads few-fl., several together at the ends of the branches ... (1) **polycephala.**
Heads many-fl., solitary, peduncled; lvs. hook-pointed ... ... (2) **Adenosolen.**

1. M. polycephalus (DC.! l. c.); scrubby, much branched, leaves linear-terete, short, obtuse or mucronulate; heads 5–7, crowded at the ends of the branches, few-flowered; corolla with a glandular tube; pappus of about 5, short, roundish, blunt, membranous scales. *Oligodorella teretifolia, Turcz. Mull. Mosc. XXIV. p.* 187 ?

VAR. β. **oligocephalus;** heads 1–3 at the ends of the branches. *M. oligocephalus, DC. l. c.* 136.

HAB. Stellenbosch, *Ecklon!* β., Groenekloof, *Eklon.* (Hb. Sd.)

Root thick and woody. Stems much branched from the base, flexuous, 6–12 in. high, bushy. Leaves 2 lines long. Heads 5–8-fl. I have not seen var. β., but, by description, it can scarcely be different specifically.

2. M. Adenosolen (Harv.); shrubby, much branched; leaves linear-terete, dotted, hook-pointed, calloso-mucronulate, glabrous; heads terminating the upper ramuli, pedicellate, many-flowered; recept. convex; corolla cylindrical, very short, with a glandular tube; pappus of about 5 short, roundish, blunt, membranous scales.

HAB. Cape, *Ecklon & Zeyher!* (Herb. Sond.)

A small shrub, with the aspect of *Aster muricatus.* Leaves 5–6 lines long, not ½ line diameter, recurved at the point. Heads 2 lines across, on peduncles ½–¾ inch long, solitary or few approximate, ending the uppermost twigs, and so corymbulose. This so nearly agrees in habit (judging by DC.'s description) with "*Adenosolen tenuifolius*" DC., that I am tempted to believe it to be identical. Yet I find a most evident pappus, which could hardly have been overlooked by De Candolle. No habitat is given in Herb. E. Z.

LXVIII.? ADENOSOLEN, DC.

Heads many-fl., homogamous. *Inv.* imbricate, in about 3 rows. *Recept.* convex, naked. *Cor.* tube glanduliferous, dilated at base and closely adhering to the achene, the throat dilated, campanulate, limb 5-fid. *Anth.* tailless, exserted in the sterile, sub-included in the fertile flowers. *Style* branches exserted, capitellate. *Achene* terete, without pappus. *DC. Prodr.* 6, *p.* 136.

A glabrous, branching half-shrub. Leaves alternate, sessile, linear-subulate, acuminate, quite entire. Heads subcorymbose at the ends of the branches, few, pedicellate. Name from αδην, *a gland,* and σωλεν, *a tube.*

1. A. tenuifolius (DC. l. c.)

HAB. Near Uitenhage, *Ecklon.*
"Leaves 7–8 lines long, nearly 1 line wide." *DC.* Unknown to me; unless
our *Marasmodes Adenosolen* be the same.

LXIX. PEYROUSEA, DC.

Heads many-fl., discoid, homogamous. *Recept.* flat, naked. *Inv.* campanulate, the scales in 2–3 rows, of sub-equal length, longer than the
disc. Tube of the *corolla* flattened, 2-winged, the limb short, 4-lobed.
Style-branches truncate. *Achene* flattened, with a thick marginal rim,
glabrous, without pappus, similar. *DC. Prodr. 6, 76. Lapeyrousia, Th.!
Cap.* 700. *Less. Syn.* 260 *(non Pourr.)*

Virgate, sparingly branched shrubs, all parts clothed with appressed, silky, and
silvery hairs. Leaves sessile, alternate, imbricate, oblongo-lanceolate, quite entire,
callous-mucronate, midribbed, silky-tomentose. Heads in a terminal, leafy corymb,
on short pedicels. Fl. golden yellow. Name in honour of La Peyrouse, the celebrated and unfortunate circumnavigator.

1. P. calycina (DC. l. c. 77); inv. scales obtuse, with an evident
scarious margin, even to the summit. *Osmites calycina, Linn. Lapeyrousia calycina, Thunb.! Cap. p.* 700. *Relhania calycina, Poir. Lapeyrousia Thunbergii, Cass.*

HAB. Houteniquas, *Th.! Bowie! Drege!* Swellendam, *E. & Z.*, Georgetown, *Dr.
A. Prior!* Tzitsikamma, *Dr. Pappe!* (Herb. Th., D., Hk., Sd.)
A tall shrub, 2–3 f. high, closely leafy to the summit; branches straight, a foot
or more long. Leaves 1–1½ inches long, silky, 4–5 lines wide. Heads ½ inch or
more across, the inv. over-topping the disc.

2. P. oxylepis (DC. l. c.); "inv. scales acuminato-subulate, scarcely
membrane-edged." *Cotula umbellata, Linn. f. suppl.* 378.

HAB. Eastern districts, *Burch. Cat.* 5185 (fide DC.)
"Heads few or sub-solitary." Possibly a starved condition of *P. calycina.*

LXX. OTOCHLAMYS, DC.

Heads many-fl., discoid, heterogamous, all the flowers on toothlike,
compressed pedicels; marginal fl. uniseriate, few, female, without corolla;
most of the flowers hermaphrodite. *Corolla* of the hermaphr. flowers
tubular, 4-toothed, its tube widely winged, and produced at base into
a broad, concave, ear-like spur, which completely enwraps the ovary
and finally the achene! *Inv.* double, the outer of few, broad, herbaceous, loose scales; the inner of few or many scarious, appressed scales.
Recept. flat, without paleæ, but covered with the persistent, toothlike
pedicels of the flowers. *Achenes* of the female, marginal flowers flattened,
and bordered with a wide, membranous wing; of the disc-fl. oblong,
wingless, but completely enwrapped in the earlike appendage of the
corolla. *Pappus* none.—*DC. Prodr. 6, p.* 77.

A small annual with opposite, entire, or pinnati-partite, slender lvs., and peduncled,
terminal fl. heads. Fl. yellow. The habit that of *Cotula,* from which, however, the
very remarkable form of the disc-flowers sufficiently separates it. Name from *ous,
otos, an ear,* and χλαμυς a *cloak;* from the earlike appendix to the corolla, which
wraps round the achene like a cloak.

1. O. Eckloniana (DC.! l. c. 77).

VAR. *a*, **filifolia**; leaves filiform-subulate, quite entire, slender.

VAR. *β*. **majuscula**; leaves pinnati-partite, few lobed, or trifid. *O. majùscula, Harv. in Hb. Hk. and D.*

HAB. Sandy fields near the sea, in wet spots. Riet Valley and Greenpoint, *Ecklon!* β. Greenpoint and Camp's Bay, *Pappe!* Cape Flats, *W.H.H.* (Herb. D., Sd., Hk.)

Annual, very variable in size, 1–6 inches high, slender or robust, simple or branched from the base, the young parts appressedly hispid, older glabrous. Leaves opposite, in *a* quite simple; in β more or less pinnate-parted. Sheaths of the leaves ciliate. Inv. glabrous, the outer scales few, inner either few or many, according to the size of the head, which varies from 1½ to 4–5 lines diameter. Disc-fl. mostly fertile; ray-fl. often abortive.

LXXI. COTULA, Gaertn.

Heads many-fl., discoid, heterogamous or rarely homogamous ; marginal fl. in one or several rows, female, either without corolla or with a 2-toothed or filiform one; disc-fl. with a flattened or winged tube, sometimes shortly 2-eared at base, and a 4-lobed limb. *Recept.* flat, often papillate, without paleæ. *Inv.* 2–3-seriate, of few or several, subequal, blunt scales. *Achenes* plano-compressed, often wing-margined, glabrous or hispid, without pappus; those of the ray mostly stipitate, of the disc narrower, less winged, and often subsessile. *DC.Prodr.6,p.77. Strongylosperma, Less. Pleiogyne, Koch. in Bot. Zeit. 1, p. 40.*

Small branching annuals, or rarely perennials, chiefly African, but found also in the warmer parts of the temperate zone. Leaves rarely opposite, or spuriously whorled, often sheathing at the base, toothed, incised or pinnatisect, with narrow lobes. Branches ending in naked, 1-headed peduncles. Heads small, hemispherical, yellow. Name from κοτυλη, *a cup;* the involucres are somewhat cuplike. Of several species I have seen either no specimen or very imperfect ones ; some are ill-defined, and probably to be rejected on further examination.

I. EUCOTULA. Marginal female flowers few, either in 1, or very rarely in 2 rows. (Sp. 1–16).

Leaves opposite, sheathing, multipartite; lobes filiform　...　(1) **myriophylloides.**
Leaves alternate, sheathing at base (glabrous) :
　Lvs. flat, toothed or cut ; inv. scales *linear*　...　...　(2) **coronopifolia.**
　Lvs. pinnati-partite, narrow ; inv. sc. ovate-oblong　...　(3) **pusilla.**
　Lvs. linear, quite entire ; inv. sc. ovate-oblong :
　　Peduno. filiform, nude　...　...　...　...　...　(4) **filifolia.**
　　Peduno. leafy nearly to the summit　...　...　...　(5) **bracteolata.**
Leaves alternate, clasping or scarcely clasping, pinnatisect :
　Glabrescent or sparsely hispid :
　　Lf.-base wide, toothed, *clasping;* lobes rigid, *pungent*　(6) **bipinnata.**
　　Lf.-base scarcely half-clasping ; lobes flaccid, not pungent:
　　　Leaves simply or sub-simply pinnatisect :
　　　　Peduncles nude, long, slender :
　　　　　Diffusely branched ; leaf-lobes few,
　　　　　　broad, veiny...　...　...　...　...　(7) **laxa.**
　　　　　Dwarf, with radical leaves ; lobes
　　　　　　narrow, thickish　...　...　...　...　(13) **Teesdaliæ.**
　　　　Ped. bearing a few linear, undivided leaves:
　　　　　Erect, glabrous ; lobes in 7–8 pairs　(14) **pterocarpa.**
　　　　　Many-stemmed, branching, sparsely
　　　　　　pilose　...　...　...　...　...　...　(11) **tenella.**
　　　Leaves bipinnatisect ; upper lobes pinnulate　(10) **multifida.**
　Pilose or villous (copiously or more thinly) :
　　Leaves bipinnatisect, or partially so :
　　　Thinly pilose ; inv. scales broad.
　　　　Diffusely branched, 3–4 inches long　...　(8) **sororia.**

Minute, erect, 1–2 inches high   ... ...   (9) **leptalea.**
Villous or pilose, with long, soft hairs :
   Lvs. scarcely clasping ; their lobes short-
    ish, broad ... ...   (15) **heterocarpa.**
   Lvs. clasping ; their lobes very narrow,
    straight... ... ...   (16) **ceniæfolia.**
Leaves simply pinnatisect ; lobes linear   ... ...   (12) **Zeyheri.**

II. PLEIOGYNE. Marg. fem. fl. in many rows ; disc-fl. few, and often sterile. (Sp. 17).
   Diffuse, pilose ; lvs. bipinnatisect ... ...   (17) **anthemoides.**

III. DISCO-COTULA. All the fl. perfect, with 4-toothed corollas. (Sp. 18–22).
   Root annual. Peduncles scapelike.
    Leaves glabrescent. Achenes winged at sides
     and front ... ...   (18) **nudicaulis.**
    Leaves silky-villous. Achenes not winged :
     Inv. scales *silky,* narrow, brown-edged...   (20) **Thunbergii.**
     Inv. sc. *glabrous,* widely membrane-edged   (19) **barbata.**
   Perennial, lignescent at base ; invol. scales glabrous, membr.-edged.
    Pedunc. scape-like, leafless ... ...   (21) **sericea.**
    Pedunc. ending leafy stems or branches   ...   (22) **hispida.**

I. EUCOTULA (Sp. 1–16).

1. C. myriophylloides (Harv. in Hk. Ic. t. 335) ; glabrous ; stem creeping (under water) ; leaves opposite, connate and sheathing at base, digitately partite into many slender, filiform, elongate, entire lobes ; pedunc. slender, axillary and terminal ; invol. uniseriate, of 4–5 broad, elliptic-oblong, obtuse scales ; ray-fl. stipitate, with winged achenes ; disc-fl. subsessile, with compressed corollas and wingless fruits.

HAB. In shallow pools at Greenpoint, near Capetown, *W. H. H.* (Herb. D., Hk.)
A very remarkable species with the aspect of a *Myriophyllum.* At first sight the leaves appear to be whorled, many thread-like leaves in each whorl ; but the whorl is really composed as above described, of two opposite, connate leaves, cleft to the base into many shreds.   Leaf-lobes 1–1½ inch long, not ¼ line diam.   Heads 3–4 lines diam.

2. C. coronopifolia (Linn. Sp. 1257) ; ascending or decumbent, glabrous ; leaves loosely sheathing at base, broadly linear or sublanceolate, irregularly toothed or pinnatifid, rarely subentire, obtuse ; sheath not bearded, with a reflexed, often toothed margin ; pedunc. naked or with 1–2 small leaves ; inv.-scales numerous, linear, obtuse, 3-nerved, membr. tipped ; ray-fl. uniseriate, stipitate. *Th! Cap.* 695. *DC. l. c.* 78. *Lam. Ill. t.* 700, *fig.* 1. *Dill. Elth. fig.* 26.

HAB. Wet spots throughout the Colony. (Herb. Th., D., Hk.)
Stems numerous, 10–12 or more inches in length, the longer ones trailing and rooting ; root sometimes perennial. Leaves 1½–2 inches long, 2–4 lines wide, very variable in shape, sometimes with a few long lobes. Heads bright yellow, 3–4 lines diameter.

3. C. pusilla (Th. ! Cap. 695) ; glabrous, erect, slender ; leaves alternate, loosely sheathing at base, pinnati-partite, the lobes linear, acute ; pedunc. naked or with 1–2 small, simple leaves ; inv.-scales few, broadly elliptical or oblong, obtuse, membrane-edged ; ray-fl. uniseriate, stipitate ; achenes hispid on the inner face. *DC. l. c.* 78.

VAR. β. **subindivisa** (DC.) ; an inch high ; some leaves quite entire, others sparingly pinnate-lobed.

HAB. Sandy places, Zwartland, *Thunberg!* a. and β., Greenpoint, *Ecklon!* (Hb. Th., Sd., D.)

Thunberg's specimens in his Herbarium are about 2 inches high; DC. has seen one 6 inches. The fl.-heads are 2–3 lines diam. The inv.-scales resemble those of *C. filifolia;* the leaves those of *C. coronopifolia;* from both which the hispid achenes separate this species.

4. **C. filifolia** (Thunb.! Cap. 696); glabrous, erect or decumbent; leaves linear or filiform, quite entire, shortly sheathing at base, the sheath ciliate or bearded; pedunc. slender, naked; inv.-scales broadly ovate, very obtuse; ray-fl. uniseriate, stipitate. *DC. l. c. 77. Zey.* 838.

VAR. β. **decumbens;** decumbent, stems rooting from the lower nodes; leaves broader and flatter, their sheaths more bearded. *C. barbellata, Fenzl!*

HAB. Wet spots, in sandy places, about Capetown, &c., *Th.! Drege, E. & Z.! W. H. H., &c.* (Hb. Th., Sd., D., Hk.)

A small, slender annual, 2–6 inches high, including the peduncle.. Leaves 1–1½ inch long, not half a line wide. Also a native of Pt. Phillip, Australia, *Dr. F. Mueller!* Var. β., from very wet spots *(Dr. Wallich! E. & Z.!)* is of much stronger growth, resembling some states of *C. coronopifolia,* but easily known by its broad invol.-scales.

5. **C. bracteolata** (E. Mey.); " glabrous, sub-decumbent; leaves sheathing at base, linear, thickish, quite entire, obtuse; *fl.-branches bearing smaller leaves nearly to the fl.-heads;* inv.-scales ovate, obtuse, hyaline at point." *DC. l. c. p.* 78. *C. integrifolia, Burch. Trav.* 1. *p.* 61?

HAB. Paarden Eyland, *Drege.* (Unknown to me.)

6. **C. bipinnata** (Th.! Cap. 696); sparingly pilose, at length glabrous, branching, erect; leaves with a wide, half-clasping, inciso-dentate base, pinnati-partite, the lobes linear-subulate, rigid, pungent-mucronate, either pungently few-toothed, or entire; pedunc. short, one-headed; ray-fl. uniseriate, shortly stipitate; invol.-scales numerous, oblong, membrane-tipped, obtuse. *Less.! Syn. p.* 261. *C. oxyodonta, DC.! l. c.* 78. *C. tenella, E. M.! ex pte. DC. l. c.* 80.

HAB. Swartland, Piquet Berg and Verloren Valley, *Thunberg!* At the Paarl, *Drege!* (Herb. Th., Hk., D., Sd.)

From 2 to 12 inches high, more rigid than most, the larger specimens very much branched. Leaves about uncial, strongly nerved, with reflexed edges, all their lobes remarkably acute, and often hook-pointed. Pedunc. scarcely 1 inch long, bearing a small leaf or two. Heads 2–3 lines diameter. *Thunberg's* specimens are much larger and more branching than *Drege's,* but in all other respects identical. *Drege's* specimens of " *C. tenella*" marked " b" belong to this (Hb. D., Hk.); those marked " a" are retained as a species; see next page.

7. **C. laxa** (DC.! l. c. 78); sparsely pilose or sub-glabrous, diffuse, slender; leaves scarcely half-clasping at base, the lowest petiolate, the upper sub-sessile, all pinnati-partite, the lobes broadly linear or oblong, veiny, mucronate, the lateral in few and distant pairs or alternate, terminal 3–5 together; pedunc. filiform, elongate; inv.-scales about 10, ovate-oblong, shorter than disc, membrane-edged; ray-fl. uniseriate, stipitate.

HAB. Zilverfontein, 2–3000 f. *Drege!* (Herb. D., Hk., Sd.)

Very slender and weak-growing. Leaves pale green, 1–1½ inch long, their lobes 2–3 lines long, 1 line wide. Heads not 2 lines diameter. Disc-fl. plano-compressed, widely winged; ray-fl. without corolla.

8. C. sororia (DC. l. c. 79); thinly pilose, slender, diffuse or erect; leaves scarcely half-clasping at base, petiolate, the lower bi-pinnati-partite, the upper pinnatifid, lobes linear, mid-nerved, acute; pedunc. filiform, short or longish, appressedly bristly under the fl.-head; inv. scales about 10, ovate-oblong, shorter than the disc, membr. edged; ray-fl. uniseriate, stipitate, without corolla.

HAB. Graafreynet, 3–4000 ft., *Drege!* Uitenhage, *E. & Z.!* (Herb. D., Sd.)
A slender, weak-growing annual. Leaves conspicuously petiolate, pinnatisect above the middle, 1½ inch long, the lobes scarcely 1 line wide. Heads 2–3 lines diam. or pedunc. 1–2 inches. The aspect is nearly that of *C. australis* (Hook. f. Fl. Tasm. t. 50, A.), but the female flowers are fewer. More hairy than *C. laxa,* with narrower, more divided, and longer leaf-lobes: *Fenzl* suggests that it may be a mere variety of that species.

9. C. leptalea (DC. l. c. 80); minute, thinly pilose, slender, suberect; leaves scarcely half-clasping at base, petiolate, bi-pinnati-partite, the lobes linear, mucronate, the lower undivided, the upper pinnate or ternately parted; pedunc. slender, filiform, appressedly bristly under the head; inv. scales about 10, oval-oblong, membr. edged; "ray-fruits flattened, winged, disc-fruits oblong, wingless." *DC.*

HAB. Between Zilverfontein, Kooperberg and Kaus, *Drege!* (Herb. D., Hk.)
To me this appears like a starved form of *C. sororia.* By *DC.* it is said to be very near "*C. tenella.*"

10. C. multifida (DC. l. c. 80); "glabrescent, ascending, branched; leaves semi-amplexicaul at base, pinnati-partite, lobes on each side 2–3, linear [the medial* naked, undivided, the uppermost pinnate-parted], the lobes trifid, the lobules linear, acute; branches at the apex naked, one-headed; inv. scales oblong, obtuse, membrane-edged." *DC.*

HAB. Algoa Bay, *Forbes.* (Unknown to me).

11. C. tenella (E. Mey.); annual, slender, erect, branching, sparsely pilose or glabrescent; leaves pinnate-parted from the half-clasping base, the lobes linear, acute, one-nerved, entire or occasionally 1–2-lobed; peduncles terminal, filiform, with a few distant, simple leaves; inv. scales oblong, glabrous, brown-edged; ray-achenes flattened, winged, of the disc oblong, wingless. *DC. l. c. p.* 80.

HAB. Roodesand, between Nieuwekloof and Schlangenheuvel, *Drege!* Brackfontein, *E. & Z.!* (Herb. Sond.)
Drege's specimens marked "b," from Ebenezer, Olifant R., belong to the very different *C. bipinnata,* Th. Root very slender. Stems pale straw-colour, weak, filiform. Leaves mostly quite simply pinnatisect, ½–¾ inch long, the lobes thin, ¼ line diameter. Heads 3 lines across, globose. I have not seen ripe achenes.

12. C. Zeyheri (Fenzl! in Hb. Sond.); annual, slender, erect, branching, variably pilose; leaves half-clasping at base, petiolate, pinnate-partite, the lobes narrow-linear, entire, callous-mucronate; pedunc. terminal, filiform, with a few distant, simple leaves; inv. scales oblong, glabrous, broadly membrane-edged; ray-achenes with a very wide,

* *DC.'s* words are "*lobulis utrinque 2–3 linearibus, medio subnudis indivisis apice pinnati-partitis:*" I have ventured, in rendering them, to substitute *mediis* for "*medio,*" and *summis* for "*apice;*" whether advisedly or not, I cannot say.

membranous wing; disc-fl. hispidulous, with oblong achenes. *C. villosa,* *Zey.! (ex pte.) No.* 2820. *C. tenella, Zey.! (ex pte.) in Herb.*

HAB. Hassaquaskloof and near the Zwartkops River, *Zeyher!* (Herb. Sond.)

Very near *C. tenella,* if it be more than a variety, with narrower leaf-lobes and more copious pubescence. It differs from " *C. villosa*" (if I rightly understand that plant) in the few female marginal flowers, &c.

13. C. Teesdaliæ (DC.! l. c. 80); dwarf, erect, sparsely pilose or glabrescent; leaves subradical, petiolate, somewhat fleshy, pinnati-partite, the lobes linear, callous-mucronate, the upper ones longer; pedunc. scapelike, 2–3 times as long as the leaves, appressedly puberulous or glabrous; inv. scales broadly oval, obtuse, rufous, membrane-edged, glabrous; ray achenes (fide DC.) " compressed, somewhat winged; of the disc compressed, the younger ones with a narrow wing."

HAB. Near Holriver, on Karroo-like hills, 1000 f., *Drege!* (Herb. Sd.)

Drege's specimens are very dwarf, the stem ½ inch long; pedunc. 1½–2 inches long. Leaves about uncial, pinnate above the middle. Heads 3–4 lines in diameter. Marginal female flowers few (I have failed to find them on the specimen examined).

14. C. pterocarpa (DC. l. c. 80); " dwarf, glabrous, erect; leaves sessile, pinnati-partite, with linear rachis, lobes on each side 7–8, the lower toothlike, medial linear, entire, uppermost toothed, all acute; branches bearing linear, entire leaves, one-headed; inv. scales oval-oblong, obtuse, with hyaline margins; ray-fruits pedicellate, elliptical, much flattened, winged." *DC.*

HAB. Between Paardeneiland, Blauwberg and Tygerberg, on the flats. *Drege.* (Unknown to me.)

15. C. heterocarpa (DC. l. c. 80); annual, diffuse or ascending, many stemmed, villous with long, soft hairs; branches or stems long, laxly leafy; leaves scarcely half-clasping at base, pinnati-partite from the base, the lowest pair of lobes simple, the rest pinnulated; segments linear, acute or mucronate; pedunc. terminal, elongate, filiform, naked or bearing 1–3 distant, abortive leaves; inv. scales numerous, 2–3 seriate, glabrous or pubescent, oblong or ovate-oblong, the inner with membranous tips; achenes of the ray-fl. broadly winged, scaberulous on both faces; of the disc oblong, sub-compressed. *C. ceniæfolia, Drege! ex pte.* *(in herbariis pluribus).*

HAB. Cape, *Drege!* *E. & Z.!* Albany, *T. Williamson! T. Cooper,* 8! Elandsberg, *T. Cooper,* 246! (Herb. D., Sd., Hk.)

Stems 6–12 inches long, sometimes densely, sometimes more sparingly villous and silky, the whole plant with the habit of *Cenia turbinata,* and by *Drege* distributed under the name " *C. ceniæfolia,*" which, however, (fide *E & Z.!* specimens) is different.

16. C. ceniæfolia (DC.! l. c. 79); annual, diffuse or erect, (or stemless and stoloniferous), villous with long, soft hairs; leaves with a broad, membranous, clasping base, sub-bipinnati-partite from the base, the lowest 2–3 pair lobes simple, the rest pinnulated, segments narrow-linear, rigid, callous-mucronate; pedunc. terminal, elongate, filiform, nude; inv. scales ovate-oblong, brown-edged, obtuse, glabrate; marginal-fl. uniseriate, female, with compressed, 2-toothed corollas; their achenes broadly membrane-winged, glabrous; of the disc oblong, with a very narrow, membranous border. *C. poecilophylla, C. Koch? Bet. Zeit.* 1. 39.

Var. β, **scaposa** (DC!); almost stemless, but throwing out stolons; leaves sub-radical; pedunc. scape-like.—*C. stolonifera, Harv. MSS. olim, in Hb. Hook.*

Hab. Stellenbosch and Swellendam, *Ecklon!* Zondag R., *Burchell*, fide DC. β., Caledon, *Zey.!* 2821. (Herb. D. Hk.)

Stems long or short. Leaves more rigid, with narrower leaf-lobes than in *C. hetero-carpa*, their bases much spread. Unripe achenes with a very wide, delicately mem-branous wing. I think *C. poecilophylla*, C. Koch, must be intended for a var. of this plant.

II. PLEIOGYNE. (Sp. 17).

17. C. anthemoides (Linn. Sp. 1256); variably pilose or hirsute, diffusely much branched; leaves half-clasping and often toothed at base, sub-bipinnatifid, the lower lobes short, simple or toothed, the upper longer, inciso-dentate or pinnulate, the lobes and lobules broadly linear, nerved, acute; pedunc. lateral or terminal, shorter or longer, filiform; inv.-scales numerous, narrow-oblong, obtuse, glabrous, with hyaline or white edges; ray-fl. in many rows, their achenes winged, with minutely granulated or smooth faces. *DC. l. c. 79. Lam. Ill. t. 701, f. 3. Dill. Elth. t. 23, f. 25. C. abyssinica, Sch. B. Pl. Schimp.* 137 *(non* 1381).

Var. a. **pilosa**; copiously pilose or hirsute.   *C. anthemoides*, and *C. villosa, DC.?*

Var. β. **glabriuscula**; sparingly pilose or glabrescent.   *C. microcephala, DC.! l. c. p. 79. C. sororia, Drege (ex pte.) non DC.*

Hab. Betw. Beaufort and Rhinosterkop, *Drege!* Natal, *Miss Owen! J. Sanderson*, 78. β. Gariep, near Verleptpram, *Drege!* Litaku, *Burchell*, 2323 (fide DC.) (Hb. D., Sd., &c.)

Stems evidently spreading, 4–6 inches long or more, repeatedly subdivided. Leaves 1–1½ inch long, generally conspicuously clasping at base, their lobes ⅛–¾ inch long, ½–1 line wide. Pubescence very variable in amount. Pedunc. 1–3 inches long. A native of Egypt and Nubia, &c. *C. villosa, E. Z.!* in Hb. Sond. appears to me to be identical with this. I have not seen Burchell's plant.

III. DISCOCOTULA. (Sp. 18–22).

18. C. nudicaulis (Thunb.! Cap. 695); annual, stem multifid at the crown; branches short, tufted, closely leafy, sparsely pilose or glabrate; leaves *opposite*, clasping at base, pinnati-partite above the middle, sparsely pilose or glabrous, lobes linear, thickish, entire, callous-tipped; peduncles elongate, scapelike, glabrous; inv.-scales about 10, roundish-oblong, very obtuse, biseriate; heads homogamous, but the outer fl. generally abortive, all stipitate with similar corollas; corolla tube very short, widely 2-winged, limb 4-toothed; achenes flattened, with a mar-ginal, membranous wing, and *two connivent, inrolled wings on its inner face. Cenia discoidea, Less.! Syn.* 261. *DC. l. c.* 83.

Hab. Cape, *Thunberg!* (Herb. Thunb.)

A very curious and distinctly marked species, which might almost be made the type of a genus, in some measure connecting *Cotula* with *Otochlamys*. The structure of the achene is very curious, reminding one in some degree of that of an *Arctotis*. Viewed on its dorsal side, it shows merely the flat, wing-edged, oblong, disc-like form of an ordinary achene in *Cotula*, but viewed in front, it presents, within the marginal wing, the aspect as of a grain of wheat, with a slit more or less open down the middle: this is formed by two intra-marginal, longitudinal, inrolled wings. The marginal flowers, though formed like the rest, are generally abortive, on longer pedicels than the central ones. The old receptacles are almost echinate with the persistent fl.-pedicels. The achenes are pale. In general aspect it resembles *C. barbata*, but is quite different in structure.

19. C. barbata (DC. l. c. 79); annual, cæspitose, villous, with long silky hairs, the crown dividing into several very short, rudimentary stems; leaves subradical and tufted, softly villous, petiolate, pinnatipartite above the middle, the lobes few, narrow-linear, simple or forked, callous-tipped; pedunc. very long, scapelike, glabrous, heads many-flowered, homogamous; inv.-scales 10–12, the outer broadly ovate with membranous edges, the inner scarious, very obtuse; corolla tube longer than the bell-shaped limb, widely 2-winged, limb 4-toothed, the teeth granulated; ovaries granulated (achenes not seen).

HAB. Betw. Zilverfontein and Kaus, 2–3000 f., and near Lelifontein, *Drege!* (Hb. D., Hk., Sd.)

United by *Fenzl* (Hb. Sd.) with *C. nudicaulis*, of which it has the habit; but besides the copious pubescence, this differs in the form of the corolla and the surface of the ovary. I have not seen the achenes, but, judging by the ovary, they must differ greatly from those of *C. nudicaulis.* Tufts subglobose, 2 inches in diameter. Pedunc. scapelike, 5–6 inches long. Heads 4–5 lines in diameter. I cannot find any female marginal flowers.

20. C. Thunbergii (Harv.); annual, stem multifid at the crown; branches short, tufted, closely leafy, pubescent; leaves alternate, clasping at base, long petioled, at the apex trifid or shortly pinnatisect, densely silky with soft, spreading hairs, the lobes linear, simple or forked, callous-tipped; pedunc, elongate, scapelike, glabrous; inv.-scales numerous, narrow-oblong, silky, one-nerved, brown-edged, with a narrow membranous border; heads homogamous (the outer fl. commonly abortive), all fl. subsessile, with similar 4-toothed corollas; cor. tube narrowly winged, *expanding at base into a short, spreading frill, which overhangs the ovary;* achenes compressed, wingless. *C. sericea, Thunb.! in Herb. No.* 2 *(excl. Sp. No.* 1 *).*

HAB. Cape, *Thunberg!* (Herb. Th., No. 2).

With the aspect and foliage of *C. barbata*, this has a very different involucre and structure of flowers; the frill at base of the corolla is evidently an appendage similar to that which envelopes the ovary in *Otochlamys.* Stems 1–2 inches long, densely tufted. Leaves about uncial, for ¾ their length simple and petiolar, lobes 3–4 lines long. Pedunc. 4–6 inches long. Heads 4 lines diameter.

21. C. sericea (Thunb.! Cap. 696); perennial? lignescent at base, many-stemmed, tufted; leaves on long clasping petioles, bi-pinnatipartite, densely silky with appressed hairs, lobes and lobules linear-terete, entire, obtuse; peduncles long, scape-like, nude, 1-headed; heads flattish-convex; inv. scales longer than the disc, glabrous, oblong, obtuse, nerved, with a wide, hyaline, membranous border; heads homogamous, all fl. subsessile, with similar, 4-toothed corollas; cor.-tube compressed, slightly winged, shortly ear-frilled at base; achenes compressed, wingless. *C. sericea, Thunb.! in Herb. No.* 1 *(excl. sp. No.* 2 *). Tanacetum linearilobum, DC. l. c.* 133.

HAB. Cape, *Thunberg!* Witbergen, 7–8000 ft., *Drege!* (Herb. Th., Hk.)

Stems 2–3 inches high, ascending, tufted. Leaves 1½–2 inches long, including the petiole. Pedunc. 3–12 inches long. In Thunberg's herbarium are two sheets marked "*Cotula sericea:*" that marked No. 1 belongs to this species; that marked No. 2 to our *C. Thunbergii.*

22. C.? hispida (Harv.); perennial, lignescent at base, many-stemmed, tufted; stems ascending-erect, simple, pilose, ending in a nude, one-headed peduncle; radical and lower leaves petiolate, bipinnati-partite, the lobules linear, callous-tipped; cauline sessile, simply pinnate; all villous, with long, soft hairs; pedunc. elongate, glabrous, somewhat fistular under the head; invol. scales numerous, bi-triseriate, oblong, 1-nerved, glabrous, obtuse, with a hyaline, membranous border; heads homogamous, all fl. subsessile, with similar 4-toothed corollas; cor.-tube narrowly winged, shortly ear-frilled at base; achenes compressed, wing-less. *Tanacetum hispidum, DC. l l. c. 133. Matricaria hispida, Fenzl! MSS*.

HAB. Witberg, *Drege! T. Cooper,* 660; Draakensberg, *Cooper,* 1022. (Hb. D., Hk., Sd.)
Stems 6–12 inches long or more, closely leafy in the lower half, pedunculoid up-wards; the upper leaves shorter, simpler, and laxer, gradually passing into scales. Lower and radical leaves, including the petiole, 2–2½ inches long; upper ½–¾ inch. Heads depressed or flat, 5–8 lines diameter. This has the aspect of a *Cenia.* Its evidently winged corolla excludes it from *Tanacetum* and *Matricaria,* and brings it near *C. sericea* and *C. Thunbergii,* from both which it differs in having its flowering stems leafy to or beyond the middle.

Doubtful Species.

C. villosa (DC. l. c. 79); "whole plant villous-hispid, erect, branched; leaves sessile, pinnati-partite, the lobes of the lower leaves trifid, of the upper entire, oblong-linear, mucronate; branches leafless at top, 1-headed, pubescent; inv. scales membr.-edged, elliptic-oblong, obtuse; achenes very thick, muriculate on the disc." *DC. l. c.*

HAB. Cape, *Burchell,* 6462.
I have not seen *Burchell's* plant, but specimens in Hb. Ecklon (Hb. Sond.), named "villosa," seem to me not to differ materially from *C. anthemoides.*

C. stenophylla (C. Koch. Bot. Zeit. 1, 40); "villous; stem branched at base; branches simple, elongate, below set with pinnate leaves, above leafless; inv. scales oblong; ray-fl. stipitate, in 2–3 rows." *Walp. Rep.* 2, *p.* 991.
HAB. Cape, *Ecklon.*
Is this the same as *C. Zeyheri, Fenzl?*

LXXII. CENIA, Commers.

Heads many-fl., heterogamous; *ray-fl.* female, 1–2 rows, some shortly ligulate or bilabiate, (the outer lip ligulate, the inner much shorter 3-toothed), some (in *C. turbinata*) without corolla; rarely all the flowers tubular. *Disc-fl.* compressed, 4-toothed. *Recept.* convex, naked, crown-ing *the top-shaped hollow apex of the peduncle. Inv.* scales biseriate. *Achenes* compressed, wingless, but margined, oblong or obovate, without pappus. *DC. Prodr.* 6, 82.

Small, hairy, annual or perennial plants, with pinnatisect, or bipinnatisect, alternate leaves, and terminal, 1-headed, naked peduncles. Distinguished from *Cotula* by the marginal rays (not always present), and specially by the hollow, top shaped apex of the peduncle, which is very conspicuous in the fruiting heads: hence the name, from κενος, *empty* or hollow

Inv. scales marked with 3–5 nerves:
      Rootannual. Stems slender, diffusely branched or simple (1) **turbinata.**
      Root perennial. Stems suffruticose, ascending or procumbent;
        Leaves with slender lobes. Pubescence close-lying,
          silky, pale ... ... (2) **sericea.**

Lvs. with short, broader lobes. Pubescence loose,
　　very copious, foxy … … … … … … (3) **pectinata.**
Inv. scales one-nerved. Annual, softly villous with spread-
　ing hairs … … … … … … … … … … (4) **microglossa.**

1. **C. turbinata** (Per. Ench. 2. 465); diffusely much branched, copiously pubescent or pilose; leaves from a broad-toothed base, pinnatisect
or bipinnatisect, the lobes linear, spreading, callous-mucronate, very acute;
pedunc. elongate, naked, thinly and appressedly pilose; inv. sc. 8–12,
ovate or oblong, 3–5-nerved, glabrous; rays slightly longer than the
involucre; achenes obovate, margined, minutely granulated or nearly
smooth, some of the marginal (which are generally abortive and have
no corolla) muriculated. *Less. Syn.* 261. *Cotula turbinata, Linn. Sp.* 1258.
Lidbeckia turbinata, Th. Cap. 694. *Lam. Ill. t.* 701, *f.* 1.

VAR. *a,* **concolor;** rays pale yellow on both sides. *C. turbinata, DC. l. c.* 82.
C. anthemoidea, DC. l. c.

VAR. β. **discolor;** rays yellow above, purplish beneath. *C. pruinosa, DC. l. c.*
Zey.l 2819.

VAR. γ. **debilis;** diffuse, weak; leaf-lobes shorter and broader than in *a* and β.
C. debilis, DC. l. c.　C. sub-heterocarpa, Less?　E. & Z.! in Hb. Sd.

HAB. A common weed throughout the colony. (Herb. D., Hk., &c.)
Very variable in size, (2–15 inches long, erect or procumbent) ramification and
pubescence, also in the leaves, which are 1–3 inches long, simply or doubly pinnatisect, with narrow or broadish lobes on the same plant. Nor do I think the discoloura-
tion of the rays a more certain character. No one seems to have noticed that *some*
of the marginal flowers are constantly destitute of corolla, exactly as in *Cotula,* while
others have a bilabiate-ligulate corolla; those destitute of corolla, though with well
formed styles, are often abortive; when fertile their achenes are muriculated, and
such plants become *C. anthemoidea,* DC. I can only distinguish *C. debilis,* DC. by
its somewhat broader leaf-lobes, a char. which he admits to be variable. The number
of inv. scales, by DC. said to be 8, are surely often 12 or more; and possibly *C.
sub-heterocarpa,* Less. should be added to the above varieties: *E. & Z.'s* specimens
at least are by me undistinguishable.

2. **C. sericea** (DC. l. c. 82, excl. syn.); suffruticose at base, ascending;
multipartite; branches tufted, short, robust, closely leafy; leaves from
a broad, toothed or pectinated base, pinnatisect or bipinnatisect, *appressedly* silky-tomentose, the lobes linear, callous-mucronate, acute; pedunc.
elongate, naked, appressedly silky or glabrate; inv. scales 8–12, ovate,
3-nerved, glabrate; rays equalling or scarcely equalling the involucre;
achenes obovate, margined, smooth, uniform.

VAR. *a.* **concolor;** rays yellowish; pubescence copious; leaf-lobes broader. *C. sericea,*
DC. l. c.

VAR. β. **discolor;** rays coppery beneath; pubescence more scanty; leaf-lobes
slender. *C. discolor, DC. l. c.*

HAB. Var. *a,* Zuureberg and Zondag R., *Drege.* Zwartkops River, *E. & Z.!* near
Pt. Elizabeth, *Zeyher!* β. Albany, *Verreaux!* near Grahamstown, *Genl. Bolton!*
(Herb. D., Hk., Sd.)
Seemingly perennial, rigid and ligneous at base, many stemmed; stems 2–8 inches
long, sub-simple, ascending, the older ones closely imbricated with the broad, pectinated bases of broken leaves, the young parts closely leafy. Leaves 1–1½ inch long,
copiously or more sparingly silky, the lobes simple or pinnulate. Pedunc. 3–6 inches
long. Marginal fl. all ligulate-bilabiate.—*Thunberg's* " *C. sericea,*" composed partly
of our "*Cotula sericea*" and partly of "*Tanacetum linearilobum,* DC.," is quite different
from the plant here described.

3. C. pectinata (DC.! l. c. 83); suffruticose at base, robust, diffuse, decumbent or ascending, *densely fulvous-silky* and shaggy ; branches robust, closely leafy ; leaves from a broad pectinated base, bipinnatisect, *roughly* fulvo-sericeous, the lobes short, incurved, very hairy, obtuse ; pedunc. elongate, naked, appressedly silky or glabrate ; inv.-scales ovate, bluntly acuminate, biseriate, numerous, faintly plurinerved, glabrous ; rays . . . ; achenes obovate, smooth, uniform. *Lidbeckia pectinata, Eckl.! in Hb.*

HAB. Cape Recief, near Algoa Bay, *Ecklon! Zey.!* 2816. (Herb. Sd.)

Very similar in character to *C. sericea*, but much more robust, with more copious, looser, longer, and tawny pubescence. Stems a foot in length, 2 lines diameter at base, very closely leafy, the leaves spreading.

4. C. microglossa (DC.! l. c. 83); diffusely much-branched from the crown; branches elongate, leafy at intervals, softly and patently villous; leaves scattered or tufted, from broad, pectinate bases, bipinnatisect, softly villous with long, spreading, very slender hairs, the lobes linear, thickish, callous-mucronate ; pedunc. very long (the medial one scape-like, rising from the crown), villous with spreading hairs; inv.-scales 20–30, biseriate, *one-nerved*, the outer ovate, the inner oblong, all glabrous and membrane-edged ; rays shorter than the invol. (or none) ; achenes oblong-obovate, smooth, uniform.

VAR. **flosculosa**; ray-fl. tubular, female or sterile. *C. flosculosa, DC.! l. c.*

HAB. Betw. Pedroskloof and Liliefontein, and on the Roodeberg, *Drege!* Specktaakel, Namaqualand, *Rev. H. Whitehead!* β. Betw. Zuurebergen and Kl. Bruintjeshoogte, *Drege!* (Herb. Hk., D., Sd.

Stems a foot long, purplish, interruptedly leafy. Leaves 1½–2 inches long, pale, very villous, all the hairs widely spreading. Pedunc. 8–12 inches long, in fruit widely turbinate under the head. Inv.-scales with a single, dark-coloured midnerve. Except in the absence of the imperfectly ligulate rays, I find nothing to distinguish *C. flosculosa,* DC.

LXXIII. **STILPNOPHYTUM**, Less.

Heads many-fl., discoid, homogamous. *Recept.* flat, naked. *Inv.*-scales imbricated, pluriseriate, dry. *Achenes* oblong, cylindrical, angularly ribbed or striate and furrowed or cuneate, subcompressed. Pappus none. *DC. Prodr. 6. p.* 92.

Glabrous shrubs with the habit of *Athanasia*, from which genus this is distinguished solely by the want of paleæ between the flowers. Leaves alternate or rarely opposite, linear, entire. Heads corymbose, rarely solitary. Flowers yellow. Name from στιλπνος, *shining*, and φυτον, *a plant ;* the involucres are polished and glossy.

Sub-genus 1. EU-STILPNOPHYTUM. *Achenes* cylindrical, rib-striate. *Leaves* alternate. *Heads* corymbose.

     Heads many-flowered. Inv.-scales acute, serrulato-ciliate   (1) **linifolium**.

     Heads 15–20-flowered. Inv.-scales oblong, membr. edged   (2) **oocephalum**

Sub-genus 2. ASÆMIA (Harv.) *Achenes* cuneate, sub-compressed. *Leaves* opposite *Heads* solitary, terminal, on minute, axillary ramuli   ...   .   (3) **axillare**.

1. S. linifolium (Less.! Syn. 264); leaves linear-filiform or subulate, acute ; corymb simple or slightly compound, 5–10–20-headed ; inv. ovato-globose, shining, the scales ovate, acute, rigid, serrulate ; heads many-flowered.

VAR. *a.* **longifolium**; leaves few or many times longer than the internodes. *S. longifolium, Less.! DC.! l. c.* 92. *Tanacetum longifolium, Th.! Cap.* 642.

VAR. *β.* **brevifolium**; leaves scarcely longer than the internodes. *S. linifolium, Less! Syn. DC.! l. c.* 92. *Tanacetum linifolium, Th.! Cap.* 642.

HAB. Cape, *Thunberg! Burchell,* 5721, *Mundt!* Outenequaland, *Drege!* George, *Dr. Prior! β.* Langekloof, *Th.!* Clanwilliam, *Ecklon.* (Herb. Th., Hk., D., Sd.)

A tall, virgate, strong-growing, slightly branched suffrutex; branches 1–2 ft. long, closely or sparsely leafy. Leaves in *a,* 2–3 inches, in *β,* 8–12 lines long, scarcely 1 line wide. Inv.-scales thick and glossy, finely serrato-ciliate, broadly ovate. Heads 50–100-fl. Var. *β* merely differs in its shorter leaves, a most variable character, when many specimens are seen. *Drege's " S. linifolium"* in some Herb. is *" Athanasia flexuosa."*

2. S. oocephalum (DC. 1. c. 93); slender; leaves scattered, linear, flattish, obtuse, entire, obliquely erect, above slightly concave; corymb compound, loosely many-headed; heads oblong, 15–20-flowered; inv. scales oblong, obtuse, scarious, membrane-edged; outer achenes sharply 5-angled, inner abortive.

HAB. Near Swellendam, *Burchell,* 7462. Hassaquaskloof, *Zey.!* 2822. (Herb. Hook., Sd., D.)

A slender suffrutex with the habit of *Athanasia linifolia.* Branches virgate. Leaves half an inch apart, 6–8 lines long, 1 line wide. Corymb 30–40-headed. Heads 2–3 lines long, pale straw-colour, the inv. scales thinnish and very blunt. *Zeyher's* plant (here described) seems to agree with the description of *Burchell's* original, which I have not seen.

3. S. (Asæmia) axillare (Less.! Syn. 264); diffusely much-branched, flexuous; leaves opposite, connate and slightly sheathing at base, linear-filiform, acute; heads solitary, terminating minute, axillary, leafy ramuli; invol. ovoid, its scales appressed, broad, obtuse, with a narrow, membranous border; achenes cuneate, sub-compressed, smooth. *DC. l. c. p.* 93. *Tanacetum axillare, Th.! Cap.* 642. *Asæmia axillaris, Harv. MSS. in Hb. Thunb.!*

HAB. Cape, *Thunberg!* (Herb. Thunb.)

Root thick and woody. Whole plant glabrous, pale. Stems 1 foot or more in length, according to *Thunberg* "erect;" but, judging by the aspect of his dried specimens, seemingly very diffuse or prostrate. Leaves 1–1¼ inch long, not a line wide. Invol. 3–4 lines long, shorter than the flowers. Flowers yellow. Anthers tailless. Style-branches truncate. Very unlike other *Stilpnophyta* in habit, and perhaps better considered as the type of a separate genus, distinguished by the connate leaves, solitary heads, and sub-compressed (young) achenes. The name *Asæmia,* proposed above, is derived from ασημος, *obscure* or *ignoble:* in contrast to the shining *Stilpnophyta.*

LXXIV. **ATHANASIA**, Linn.

Heads few- or many- flowered, discoid, homogamous. *Receptacle* bearing paleæ between the flowers. *Invol.* imbricate, scarious (except in *A. capitata,* where the outer scales are large and leafy) in several rows, the outer short. *Cor.* 5-toothed. *Anth.* without tails. *Achenium* oblong, sharply 5-angled or winged. *Pappus* either of several short, flat, unequal scales; or of swollen, jointed, short, deciduous hairs; or none. *Hymenolepis, Holophyllum, Athanasia, and Morysia, DC. Prodr. pp.* 85, 86, 90.

Small shrubs or half-shrubby plants, strongly scented and glandular. Leaves scattered, either entire, toothed, lobed or pinnati-partite, varying in shape on the same branch or root. Pubescence very variable. Heads corymbose, rarely solitary. Flowers yellow. There is no difference in habit among the species of the above combined genera, except in *Holophyllum*, which consists of a single species having a leafy involucre. The species are very variable, and some of them difficult to fix natural limits to; perhaps I have retained too many. The name is compounded of *a, negative,* and θανατος, *death;* because the flower-heads are dry and persistent.

I. HYMENOLEPIS (Cass.).   *Inv. scales* imbricated, scarious, the outer ones small. *Pappus* of several short, flat, acute scales. (Sp. 1–8.)

Flower-heads small, each containing 3–5–7 flowers :
 Leaves pinnatisect, *lobes long,* linear ; inv. glabrous (1) **parviflora.**
 Leaves some entire, some 1–2-toothed or pinnatisect, *lobes short :*
  Inv. pubescent ; pappus very short, ciliate ... (2) **incisa.**
  Inv. glabrous ; pappus of palmatifid, flat scales (3) **schizolepis.**
 Leaves all undivided, linear ... ... (4) **indivisa.**
Flower-heads many-flowered (20–50-fl.) :
 Recept. flattish.—Lvs. oblong, conspicuously gland-
  dotted ... ... (5) **punctata.**
 Recept. conical :
  Leaves oblong-lanceolate, softly hairy, entire or
   toothed ... ... (6) **Dregeana.**
  Leaves ovate, sharply serrate, glabrous ; twigs
   cobwebby ... ... (7) **leucoclada.**
  Lvs. pinnati-partite, lobes linear; twigs pubescent (8) **coronopifolia.**

II. HOLOPHYLLUM (Less.).   Outer *inv. scales* leaf-like, as long as the inner scales and the disc-flowers. *Pappus* of many short, swollen, brittle, jointed hairs.

More or less villous ; leaves oblong, obovate or roundish ;
 heads few, many-flowered ... ... (9) **capitata.**

III. EU-ATHANASIA.   *Inv. scales* imbricated, scarious, the outer ones small. *Pappus* of many short, swollen, brittle, jointed hairs. (Sp. 10–30.)

Fl.-heads oblong or oval, 10–25- (rarely 35-) flowered :
 Corymb much branched, densely many-headed :
  Tomentose ; leaves flat, 3–5-forked or pinnatifid,
   the upper entire ... ... (10) **hirsuta.**
  Tomentose or glabrous ; lvs. linear, semi-terete,
   mostly entire ... ... (11) **filiformis.**
 Corymb slightly compound or simple, loosely few-headed :
  Lvs. linear, glabrous ; corymb slightly compound (12) **flexuosa.**
  Lvs. linear, velvetty; corymb simple; inv. tomen. (13) **sertulifera.**
  Lvs. oblong or lin.-oblong, glabrous and rugulose
   beneath ; twigs and invol. hairy ... ... (14) **rugulosa.**
  Lvs. short, ovate or oblong, or 3-lobed, glabrous ;
   invol. glabrous ... ... (15) **microphylla.**
  Lvs. orbicular, glabrous ; inv. scales downy ... (16) **rotundifolia.**
  Lvs. obovate or cuneate, entire or 3-toothed,
   glabrous; inv. glabrous ... ... (17) **dimorpha.**
Fl.-heads globose or obovate, many-flowered (40–50–100-fl.).
 Leaves narrow-linear, entire (or 3-toothed or lobed) :
  Corymb few-headed ; inner inv. scales scarious
   and torn at edge ... ... (18) **oligocephala.**
  Corymb 10–20-headed ; inv. scales oblong, obtuse,
   quite entire ... ... (19) **glabrescens.**
  Slender ; corymb few-headed ; outer inv. scales
   ovate, acute, inner scarious ... ... (20) **eriopoda.**
 Lvs. oblong or ovate-obl., entire (or 3–5-toothed or lobed).
  Branches, leaves, pedicels, and inv. *softly hairy;*
   corymb many-headed ... ... (21) **pubescens.**

Branches hairy; leaves glabrescent; corymb panicled, branched ... ... (22) **crenata.**
Branches, leaves, ped., and invol. *glabrous;* corymb simple, many-headed ... ... (25) **trifurcata, ζ.**
Branches pilose; leaves fleshy, wrinkled when
dry; heads solitary, sessile ... ... (23) **scabra.**
Twigs, leaves, pedunc., and invol. hairy; heads
subsolitary, pedicillate ... ... (24) **pachycephala.**
Leaves obovate-cuneate, entire, glabrous, closely im-
bricating ... ... (27) **imbricata.**
Leaves oblongo-cuneate, *tapering at base*, 3-5-lobed
or toothed; invol. globose ... ... (25) **trifurcata.**
Leaves *very short*, recurved, 3-lobed or toothed; inv.
tapering at base ... ... (26) **Mundtii.**
Leaves *short*, recurved, obovate-orbicular, entire or
3-5-toothed; invol. globose ... ... (28) **quinquedentata.**
Lvs. 3-5-lobed to the middle; the lobes long and narrow:
Heads globose; inv. scales broad, obtuse, inner
lacerate ... ... (25) **trifurcata, β., γ.**
Heads ovoid; inv. sc. narrow, acute, rigid, entire (29) **crithmifolia.**
Lvs. palmatifid, the medial lobe entire, lateral deeply
bifid; all parts of plant silky ... ... (30) **palmatifida.**

IV. MORYSIA (Cass.). *Invol. scales* imbricated, scarious, the outer ones small.
Pappus none. (Sp. 31–40.)

Leaves linear, quite entire:
All parts tomentose ... ... (31) **tomentosa.**
Glabr. Corymb densely much branched, many-headed:
Heads 7-9-fl. Leaves crowded, mucronulate ... (32) **fasciculata.**
Heads 15-20-fl. Lvs. scattered or close, acute,
narrow ... ... (33) **linifolia.**
Heads 20-25-fl. Leaves *distant*, broadly linear (34) **juncea.**
Glabr. Corymb simple, few-headed. Heads 25-40-fl. (35) **longifolia.**
Leaves ovate or obovate, or oblong, toothed, or lobulate:
Lvs. half-clasping, *broad-based,* squarrose or reflexed (36) **dentata.**
Lvs. *tapering at base;* tube of corolla hairy! ... ... (37) **spathulata.**
Leaves pinnati-partite (or some simple); lobes linear:
Glabrous; leaf-lobes channelled; invol. scales obtuse (38) **pectinata.**
Twigs and pedunc. hairy; lvs. glabrous, lobes with
recurved margin; inv. scales very narrow, glandular (39) **acerosa.**
All parts velvetty-tomentose, canescent ... ... (40) **pinnata.**

Sect. I. HYMENOLEPIS. (Sp. 1–8).

1. A. parviflora (Linn. Sp. 1182); leaves from the middle pinnati-
partite, lobes few, distant, linear, elongate, sub-acute; corymb much-
branched, spreading, very many-headed; heads 3–4-fl.; inv.-scales
glabrous, oblong, concave, sub-acute; pappus of several, unequal, narrow, minute, flat scales. *Thunb.! Cap.* 636. *Jacq. Schoenbr. t.* 149.
Burm. Afr. t. 68, *f.* 4. *Hymenolepis parviflora, DC. l. c.* 85. *H. lepto-
cephala, Cass.*

HAB. Western Districts. Mts. round Capetown, common. Paarl, *Drege!* Tulbagh, *Dr. Pappe!* Elandsberg, *Dr. Wallich.* (Herb. Th., D., Hk., Sd)
A robust, densely leafy shrub, 2–4 ft. high, corymbosely branched, the young
branches and leaves flocculent, the old glabrous. Leaves 2–2½ inches long, the
lobes 1–1½ inch long, 1 line wide. Inv. sulphur-yellow, glossy; heads 1 line diam.,
2 lines long. Corymbs densely much-branched.

2. A. incisa (Harv.); "glabrous, sub-glaucous; some leaves linear, quite entire, some trifid, others pinnately 5-lobed, all the lobes with sub-revolute margins, mucronate; heads oblong, 4–5-flowered, crowded in a very dense corymb; inv.-scales *pubescent,* appressed; pappus very short, scarcely membranous, ciliate." *DC. Ath. pinnata, E. Mey. in Coll. Drege, fide DC. Hymenolepis incisa, DC. l. c.* 85.

HAB. Betw. Hexriver and Bokkeveld, on table land and stony hills, 3-4000 f. *Drege.*

Said to resemble *A. pinnata,* but to differ in the glabrous leaves, and minute pappus.

3. A. schizolepis (Harv.); erect, glabrous; branches striate; leaves erect, linear, callous-mucronate, some quite entire, some on each side unidentate, some sparingly pinnatisect, the lobes short, mostly alternate; heads oblong, 4–5-fl., densely corymbose; inv.-scales glabrous, membrane-edged; paleæ lacerate; pappus of several, *palmately-multifid,* flat scales. *Morysia, Sp.* 84, *Hb. Eckl.!*

HAB. Cape, *Ecklon!* (Herb. Sond.)
A single specimen, without locality. Leaves scarcely inch long, ½ line diam., thickish, the lower mostly quite entire, the upper more or less inciso-dentate or pinnatisect, the lobes 1–2 lines long, erect, acute. Invol. 3–4 lines long, sub-compressed, turning blackish in drying. By its foliage this seems near *A. incisa,* but if *DC.'s* account of the pappus be correct, it cannot be the same. The pappus in our plant is more developed than in most of this section, and each scale is deeply cut into many narrow lobes.

4. A. indivisa (Harv.); glabrous, leaves linear-filiform, nearly terete, undivided, quite entire, sub-acute; corymb compound, many-headed; heads 3–4-fl.; inv.-scales glabrous, oblong, concave, obtuse; pappus of several, unequal, narrow, minute, flat scales. *Hymenolepis indivisa, Drege!*

HAB. On the Zuureberg, betw. Enon and Driefontein, *Drege!* (Hb. D., Hk., Sd.)
This has the habit, and nearly the foliage and inflorescence of *A. fasiculata,* for which it may readily be mistaken; but has a pappus precisely like that of *A. parviflora.* It is either omitted by *De Candolle* or combined with his *Hymenolepis incisa,* which (by the char. given) seems to be different, and comes at least from a very different locality. Leaves 1–1½ inch long, not 1 line diam., crowded. Heads as in *A. parviflora.*

5. A. punctata (Harv.); virgate, erect, simple or branched; branches terete, pubescent, robust; leaves densely crowded or tufted, sessile, oblong or cuneate-oblong, sharply serrate or entire, midribbed, on both sides conspicuously gland-dotted, glabrous; corymb compound, fastigiate, the pedicels pubescent; heads 20–25-flowered; inv.-scales oblong or ovato-lanceolate, loosely imbricated, glandular; recept. flattish, with many narrow-linear paleæ; pappus of several, unequal, broadish, flat, entire or toothed scales. *Hymenolepis punctata, DC. l. c.* 85.

HAB. Gauritz R., *Burchell,* 5055. Kromriver, *Drege!* Vanstaadens Mts., Uit., *E. & Z.! Zey.!* 289. Near Grahamstown, *Genl. Bolton!* Natal, *T. Cooper!* 1155. Maritzberg to Ladysmith, *Gerr. & McKen!* 280. (Herb. D., Hk., Sd.)
2–3 f. high, strong-growing, the young plants sub-simple, branched at the summit only. Leaves ½–¾ inch long, 2–3 lines wide, mostly toothed, the upper ones often entire; dots oblong, resinous. Corymbs 2–3 inches across. Heads 2–3 lines diam. Pappus of broadish scales. *A. punctata,* Th.! is a var. of *A. capitata.* The anthers are minutely produced at base.

6. A. Dregeana (Harv.); stem erect; branches terete, densely and softly hairy; leaves crowded, sessile, oblongo-lanceolate, acute, on both sides softly hairy, midnerved, entire or denticulate; corymb compound, fastigiate, the pedicels hairy; heads 20–25-fl. ; inv.-scales lanceolate, serrulate, loosely imbricated, glandular; recept. conical ; pappus of about 8 tooth-like scales. *Hymenolepis Dregeana, DC. l. c.* 85.

HAB. Betw. the Omsamcaba and Omsamwubo, *Drege !* (Herb. Sond.)
I have only seen a small branch. Leaves ½–¾ inch long, 3–4 lines wide, mostly entire. Whole plant save the involucres densely and softly hairy.

7. A. leucoclada (Harv.); "branches terete, virgate, with a whitish, depressed, cobwebby covering; axils without ramuli; leaves sessile, erect, coriaceous, subpunctate, glabrous, ovate, acute, sharply serrate ; corymb simple, dense, few-headed, the pedicels bracteolate, shorter than the subglobose, many-flowered fl.-head ; inv.-scales oblong, truncate (recept. conical) ; achenes rib-striate, crowned with a minute, scaly, many-toothed pappus." *DC.—Hymenolepis leucoclada, DC. l. c.* 86.

HAB. Betw. the Omsamwubo and Omsamcaba, 1–2000 f., *Drege.*
Unknown to me.

8. A. coronopifolia (Harv.) ; erect ; branches virgate, terete, striate upwards, pubescent with whitish, short hairs ; leaves tufted, crowded, sessile, pinnati-partite, the younger hoary, the older glabrate, lobes in few pairs, short, linear, subacute ; corymb subsimple or not much branched, few-headed ; pedicels longer than the subglobose, many-fl. fl.-heads; inv.-scales lanceolate, serrulate, loosely imbricated; recept. conical ; achenes rib-striate, pappus of 8–10 toothlike scales.

HAB. Betw. Maritzburg and Ladysmith, Natal, *Gerr. & M'K.!* 279. (Hb. D.)
Over 2 feet high, our specimen quite simple, leafy throughout and ending in a corymb of 20–25 heads. It seems to be a year's shoot, from a thick woody crown, whose stems had been burned the previous year. Leaves about ¾ inch long, the lobes 2–4 lines long, and not a line in diameter. The scales of the pappus are broadish.

Sect. 2. HOLOPHYLLUM. (Sp. 9).

9. A. capitata (Linn. Sp. 1181) ; diffuse or ascending, more or less villous; leaves loosely imbricating, oval, oblong, obovate or subrotund, acute or obtuse, entire or sharply few-toothed, villous or glabrous, nerveless, thickish ; heads 2–7 at the ends of the leafy branches, subsessile or shortly pedicelled, many-flowered; invol. mostly very villous, campanulate, the outer scales *leaflike*, as long as the inner and as the disc, oblong or lance-oblong ; achenes sharply 5-angled and furrowed ; pappus of a few, fragile, jointed, short hairs. *Thunb.! Cap.* 635. *Holophyllum capitatum, Less.! Syn.* 262. *DC. l. c. Also Hol. lanuginosum and H. scabrum, DC. l. c.*

VAR. β. **glabrata**; leaves subrotund, very obtuse, glabrous or nearly so. *A. punctata, Berg. Cap.* 238. *Thunb.! l. c.*
HAB. Sea-shores near Capetown and Greenpoint, common. Eastern Districts, *H. Hutton!* (Herb. Th., D., Sd., Hk.)
A small, many-stemmed bush, 1–2 ft. high, not much branched. Leaves very variable in shape and in pubescence, sometimes very densely clothed on both sides with long, soft hairs ; occasionally sharply and deeply few-toothed, the apex often

recurved. Heads sometimes solitary, mostly several together, 5 lines long, 4 lines diam. Fl. bright yellow ; invol. leaflike, green.

Sect. 3. EU-ATHANASIA.   (Sp. 10–30).

10. A. hirsuta (Th.! Cap. 655); branches virgate, densely tomentose, leaves crowded, erect, sessile, flat, dimorphous, the lower 3–5-forked or inciso-pinnatifid beyond the middle, the upper entire, linear-acute, the adult at length becoming glabrate on the under, but remaining densely tomentose on the upper surface, callous-mucronate; corymbs very much branched, dense, fastigiate; heads oblong, 12–15 fl.; invol. tomentose, the scales obtuse ; pappus of squarrose, swollen hairs. *DC. l. c.* 90.

HAB. Cape, *Thunberg! Dr. Thom!* (Herb. Thunb., Hook).

Branches 1–1½ ft. long, robust. Leaves 1–1½ inch long, 1–1½ line wide, the young ones densely woolly on both sides, the lowest simple for ¾ or ⅔ of their length, thence shortly 3–5-lobed or pinnatifid, upper quite simple, tapering to a sharp point. Infl. very much branched, on woolly, multifid peduncles, round-topped. Heads 2 lines long, 1 line wide, 100 or 200 in each corymb.

11. A. filiformis (Linn. f. Suppl. 361); much-branched, glabrous or tomentose ; leaves linear-semiterete, thickish, acute, crowded, glabrous or tomentose, some occasionally 3-forked; corymb compound, fastigiate, densely many-headed ; heads oblong, 10–15-fl. ; inv. scales oblong, obtuse, appressed, shorter than the disc ; achenes sharply angled ; pappus of many squarrose, swollen hairs. *Thunb.! Cap.* 634.

VAR. *a.* **glabra**; whole plant glabrous or nearly so. *A. filiformis, DC.! l. c.* 87. *Burch. Cat.* 4962.

VAR. *β.* **cinerea**; more or less tomentose, old parts becoming glabrate. *A. tomentosa, DC.! l. c. excl. syn. Thunb.! A. cinerea, Linn. f.? Suppl.* 361. *Burch. Cat.* 6829.

HAB. Cape, *Thunberg!* Var. *a.* and *β.* Kochman's Kloof, *Mundt! E. & Z.!* Betwn. Breede and Gauritz R., *Drege! Pappe!* Cypher Font., *P. MacOwan.* (Herb. Th., D., Sd., Hk.)

1–2 feet high or more. Leaves 1–1½ inch long, not a line wide, spreading. Corymbs 1½–2 inches across, of 20–50 or more heads. Heads 2–3 lines long, 1 line diam. Both varieties occur together, and differ merely by the varying amount of tomentum, which is often very scanty on var. *β. Thunberg's ' A. tomentosa' is Morysia velutina, DC.! Paleæ,* occasionally *very few* and slender.

12. A. flexuosa (Thunb.! Cap. 634, non DC.) ; much-branched, glabrous, flexuous ; leaves linear-flattish, mucronulate or obtuse, scattered or somewhat crowded; corymb slightly compound, fastigiate, loosely many-headed ; heads ovoid, 25-35-fl. ; inv. scales appressed, oblong, obtuse, shorter than disc ; achenes angled and furrowed; pappus of many jointed, swollen hairs. *A. affinis, Sond. in Hb.*

HAB. Cape, *Thunberg!* Olifant's R. and Nieuweveld, Driefontein, *Zeyher!* 843. (Herb. Sd., Hk.)

With much of the habit of *A. filiformis* this has shorter, broader, and flatter leaves, and larger flower-heads. Leaves scarcely uncial, 1 line wide.

13. A. sertulifera (DC.! l. c. 87); all parts tomentose ; leaves linear, flat or concave beneath, entire, mucronulate, velvetty-tomentose ; corymb simple, few-headed ; heads half as long as the pedicels, oval-oblong, 25-fl ; inv. densely tomentose, the scales linear, obtuse, the outermost subacute.

HAB. Kochman's Kloof, *Mundt!* Between Bergevalei and Langevalei, near Zwartbastkraal, *Drege!* (Herb. Hk., Sd.)

Foliage nearly as in *A. filiformis,* β; the older leaves partly glabrate ; fl. heads larger and fewer.

14. A. rugulosa (E. Mey.!) ; much-branched ; branches and twigs densely hairy; leaves scattered, sessile, oblong or oblong-linear, acute, erect, thick, glabrous on the under, hairy on the upper surface, when dry rugulose, quite entire ; corymb simple, few-headed ; heads oval-oblong on short, hairy pedicels, 20–25-fl. ; inv. scales oblong, hairy, the outer subacute ; achenes angled and furrowed, pappus of many-jointed, swollen hairs. *DC.! l. c.* 87. *A. pubescens, Th. ! Cap. 635, non Linn.*

HAB. Cape, *Thunb.!* Groenekloof, *Drege!* Between Klipfontein and Predikstoel, *Zey.!* 844. (Herb. Th., D., Hk., Sd.)

Much-branched and ramulous, 1–2 feet high. Leaves 3–6 lines long, 1–3 lines wide, varying from linear to oblong, slightly concave, the upper surface almost always hairy. Corymbs 4–12-headed, in the latter case slightly compound.

15. A. microphylla (DC. l. c. 87); twigs minutely tomentulose, otherwise glabrous; leaves crowded, sub-imbricate, short, sessile, ovate or oblong, entire or *three-lobed*, thick, subacute, glabrous; corymbs simple or slightly compound, few-headed, fastigiate ; heads oblong, 12–15 fl., inv. scales oblong, obtuse, glabrous ; achenes angled, pappus of many swollen hairs. *A. eriopoda, Herb. Drege (in pluribus herbariis).*

HAB. Honigvalei and Kaudeberg, and Ezelbank, *Drege!* (Herb. D., Sd., Hk.)

A rigid, erect, virgate shrub, 1–2 f. high. Leaves 2–3 lines long, 1–1½ wide, on the same branch trifid and entire, sometimes all entire. Heads 6–15 in a corymb, 2–3 lines long. All the specimens we have seen from *Drege* are by him marked " *A. eriopoda,*" but they agree (except as to the trifid leaves) with *DC.*'s description of *A. microphylla.*

16. A. rotundifolia (DC. l. c. 87. excl. sp. Eckl.); "leaves orbicular, entire, glabrous as the branches; corymb 7–8 headed, on downy pedicels ; heads about 15-fl.; inv. obovate; scales obtuse, sub-scarious, downy at back, the innermost a little shorter than the disc." *DC. l. c.*

HAB. Eastern Districts, *Burchell, No.* 6509.

Unknown to us. *Ecklon's* specimens referred to by *DC.* do not agree with the character given, and seem to belong to *A. dimorpha.*

17. A. dimorpha (DC.! l. c. 87) ; twigs and pedicels tomentulose ; leaves obovate or oblong, linear-oblong or cuneate, entire or shortly 3-toothed at the apex ; glabrous, the younger (fide *DC.*) often tomentulose ; corymbs simple or slightly branched, 10–20 or many headed ; heads obovate, 20–30-fl.; inv. scales glabrous, obtuse, the outer ones small, acute ; achenes wing-angled, pappus of swollen hairs.

VAR. β. **obovata**; leaves obovate, mostly entire. *A. rotundifolia, Hb. Eckl.!*

VAR. γ. **minor**; leaves shortly obovate or cuneate, 3-toothed or entire. *A. rotundif. var. minor, Pappe!*

HAB. Vanstaadens Riv., Uit., *Drege!* β. Krakakamma Forest, Uit., *E. & Z.!* Knysna, *Dr. Pappe!* γ. Swellendam, *Dr. Pappe.* (Herb. D., Hk., Sd.)

A small shrub, glabrous, except on the young parts. Leaves 4–6 lines long, 2–4 lines wide, variable in form and margin, crowded, spreading, more or less concave above. Heads 3–4 lines long, the outer scales slightly scattered on the pedicel. Invol. pale yellow.

18. A. oligocephala (DC. in Hb. Eckl.); variably tomentulose, or hairy, or glabrous; leaves linear, thickish, subacute or obtuse, quite entire or some 3-toothed or trifid; corymb simple, few-headed (or heads solitary); heads on longish pedicels, with or without bracts, ovoid-globose, many-flowered; inv. scales appressed, oblong, obtuse glabrous or villous, the inner ones scarious and lacerate at the apex; pappus of many swollen hairs. *A. flexuosa, DC. l. c.* 87, *excl. syn. Thunb. A. ebracteata, E. Mey! DC. p.* 86.

Var. *a*, **glabra**; glabrous, or the young parts pilose. *A. flexuosa, DC.*

Var. *β*. **adenatha**; twigs and invol. sub-persistently villous; cor. lobes with 1-2 reddish glands. (*Hb. Eckl.*)

Var. *γ*. **brachypoda** (DC.); lvs. and twigs thinly tomentulose; pedicels short, ebracteate.

Var. *δ*. **araneosa** (DC.); "lvs. glabrous; twigs cobwebby, at length glabrous; pedicels twice as long as the heads, densely and shortly hairy." *DC. l. c.*

Var. *ε*. **hirsuta** (*E. M.*); twigs densely clothed with long hairs; young leaves somewhat hairy, adult glabrous; pedicels hairy, twice as long as the heads.

Hab. Cape, Hb. *Lambert* (DC.) Kl. Draakenstein, and the Giftberg, &c., *Drege!* Cape, *Eckl.!* Swellendam, *Pappe!* Zwartland, *Wallich!* Predikstoel, *Zey.!* 845. (Herb. D., Hk., Sd.)

A virgate shrub, 1-3 f. high, very variable in pubescence. Leaves ½-1 inch long, scarcely 1 line wide, rather closely set. Heads sometimes solitary, or 2-3-8-10 in a lax corymb. This does not occur in Hb. Thunb., whose '*A. flexuosa*' is very different. I cannot keep *A. ebracteata* apart by any persistent character.

19. A. glabrescens (DC.! l. c. 88); twigs and young leaves stellulato-tomentulose, older glabrate; leaves linear, mucronulate, quite entire; corymb simple, fastigiate, 10-20-headed, pedicels with 1-3 bracts, elongate; heads obovate, many-flowered; inv.-scales oblong, obtuse, powdery-stellulate, quite entire, outer subacute; stamens gland-tipped; pappus of swollen hairs.

Hab. Olifant's River, *Drege!* (Herb. D., Hk.)

Similar in foliage to *A. filiformis*, but with a different inflorescence, and much nearer *A. oligocephala*. Heads 40-50-fl. Leaves 1½ inch long, scarcely a line wide, curved or squarrose.

20. A. eriopoda (DC. l. c. 89); slender, virgate, young twigs and pedicels stellato-pubescent, leaves short, linear, subacute, entire or on each side 1-2-lobed, convex beneath and (when dry) rugulose, the younger ones stellate-pubescent above; corymb simple, few-headed; pedicels naked or few-bracteolate; heads many-fl., globose; inv. pubescent, the scales broadly ovate, acute, the inner scarious; pappus of many swollen hairs. *A. microphylla, Hb. Drege! (in pluribus herbariis).*

Hab. Ezelsbank, Zeederbergen, 3-4000 f., *Drege!* (Herb. Sd., D., Hk.)

1-1½ ft. high, erect, with a few erect branches and twigs. Pubescence mostly deciduous. Leaves 2-4 lines long, not a line wide, simple and 2-3-lobed on the same twig. Heads solitary or 3-5 in a corymb, the outer scales broader and more acute than in most. A much more slender plant than *A. oligocephala*, with different invol.-scales.

21. A. pubescens (Linn. Sp. 1182); branches densely and softly hairy; leaves oblong or lance-oblong, sessile, mucronulate, entire (or tridentate), on both sides softly hairy, at length becoming sub-glabrous;

corymbs simple, many-headed; heads on hairy, bracteate pedicels, sub-globose, many-fl.; inv. hairy, the outer scales lanceolate, acute, the inner oblong-lanceolate, bearded at the membranous apex. *DC.! l. c.* 88. *A. canescens, Thunb.! Cap.* 634.

HAB. Picketberg, *Thunberg! Drege!* Betw. Groenekloof and Saldanha Bay, *Drege!* Bergvalley, *E. Z.! Klipfontein, Zey.!* 846. (Herb. Th., D., Hk., Sd.)
Densely and softly hairy, with longish hairs. Leaves 1–1½ inch long, 2–3 lines wide, the margins slightly revolute. Heads 3 lines diameter, hairy.

22. A. crenata (Linn. Sp. 1180); "branches virgate, hairy, sub-striate; leaves oblong, acute, glabrescent, entire; corymb panicled, branched; heads globose, many-fl.; inv. glabrous, the outer scales lanceolate, acute, middle oblong, obtuse, innermost produced into a scarious, roundish, serrato-fimbriate appendage." *DC. l. c.* 88.

HAB. Cape.
Not in Herb. Thunb. Unknown to us. Possibly *A. trifurcata*, var. *glabra!*

23. A. scabra (Thunb.! Cap. 634); rigid, branched and ramulous; twigs short, pilose, closely leafy; leaves loosely imbricating, short, ovate-oblong, acute, thick and fleshy (shrinking and wrinkled when dry), glabrous, nerveless; heads solitary, terminal, sessile, many-flower-ed; invol. glabrous, ovate-oblong, the scales appressed, the outer short, inner successively longer, oblong, obtuse, horny, the innermost with membranous, lacerate tips; paleæ lacerate; achenes . . . ?; pappus of a few, fragile, jointed, short hairs.

HAB. Cape, *Thunberg!* (Herb. Thunb.)
A much branched, small bush, with leaves not unlike those of *A. rugulosa*, E. Mey., and sessile involucres. Leaves 2–3 lines long, 1–1½ line wide, rugulose when dry ("*rugoso-scabrida*," Th.), but probably quite smooth when living. Heads 4 lines long, 2½–3 lines wide, invol. glossy.

24. A. pachycephala (DC. l. c. 88); "branches glabrous; twigs, leaves, peduncles and involucres hairy or somewhat woolly; leaves crowded, erect, oblong-linear, entire, or oblong, or cuneate-incised, the lobes 3–5, obtuse; uppermost leaves linear, acute, distant; heads soli-tary on the ends of the short, sub-corymbose twigs, globose, nearly 100-flowered." *DC. l. c.*

HAB. Zeederberg, *Drege.*
Unknown to us. Strongly marked by its large fl. heads.

25. A. trifurcata (Linn. Sp. 1181); stellulato-tomentulose or gla-brescent or quite glabrous; leaves oblongo- or obovato-cuneate, much narrowed at base, rarely entire, commonly 3–5-toothed or lobed *(very variable in degree of incision)*, the teeth or lobes mucronulate, mostly stellulate on both sides; corymb simple or sub-simple, racemulose, fastigiate; heads on long, naked or 1–2 bracteolate pedicels, many-fl.; invol. globose, glabrous or nearly so, the outer scales acute, medial obtuse, innermost mostly scarious and lacerate at margin; pappus of many swollen hairs.

VAR. *a.* **Linnæi**; densely stellulate; lvs. broadish, sharply 3–5-toothed or incised. *A. trifurcata, DC.! l. c.* 89. *Thunb.! Cap.* 635 (ex pte.) *A. scariosa, DC.! l. c.* 88. *Zey.!* 2828.

VAR. β. **tricuspis** (DC.); stellulate; leaf-lobes long, narrow, linear-acute. *DC. l. c.*

VAR. γ. **virgata**; thinly stellulate or glabrescent; leaves 3–5-fid, lobes obl.-linear, acute. *A. virgata, Jacq. Sch. t.* 148. *DC. l. c.* 89. *Th.! Cap. (ex pte.)*

VAR. δ. **Thunbergii**; glabrescent or minutely stellulate; lvs. 3–4 lines long, ob-ovate or broadly cuneate, entire or bluntly 3-crenate. *A. trifurcata, Th.! Hb.(ex pte.)*

VAR. ε. **cuneiformis**; glabrescent; leaves elongate-cuneate, entire or sharply 3-5-toothed. *A. cuneiformis, DC.! l. c.*

VAR. ζ. **glabra**; quite glabrous; lvs. elongate-cuneate, acute, quite entire. *A. glabra, Th.! Cap.* 634.

HAB. On dry hills throughout the colony, common. (Herb. Th., D., Hk., Sd.)

2–3 f. high, robust. Exceedingly variable in pubescence and foliage, but more constant in the scarious and lacerate involucres. Leaves commonly ¾–1 inch long, in β. 1–1½ inch, in δ. very small.

26. A. Mundtii (Harv.); young twigs and leaves minutely stellulate, older glabrous; leaves *(very short)*, broadly cuneate, *recurved or reflexed*, 3-lobed or 3-toothed, the lobes oblong or deltoid; corymb simple, few-headed; heads on bracteolate pedicels, many-fl.; invol. *obovate-turbinate, tapering at base,* scales obtuse, the innermost scarious and lacerate at margin; pappus of many swollen hairs.

HAB. Grassy hills near Swellendam, *Mundt!* (Herb. Hook.)

A diffusely branched, small shrub, with flexuous branches. Leaves 2 lines long, 1½ line wide, strongly recurved or squarrose, crowded. Heads 4–6 in a corymb, 4–5 lines long, 2–3 lines wide. Distinguished from all forms of *A. trifurcata* by the shape of the involucres and the strongly recurved leaves, scarcely at all narrowed at base.

27. A. imbricata (Harv.); glabrous or nearly so; branches virgate; leaves closely imbricating, erect, obovato-cuneate, entire, mucronulate, the apices sub-recurved; corymb simple, racemulose, many-headed; heads on long, bracteolate, stellulate pedicels, very many-flowered; invol. globose, glabrous, the outer scales sub-acute, medial oblong, ob-tuse, serrulate, inner somewhat scarious-lacerate at the margin; pappus of many, long swollen hairs. *Herb. Eckl.* 89, 9.

HAB. Cape, *Ecklon!* (Herb. Sond.)

Leaves 6–8 lines long, 3–4 lines wide. Pedicels 1–2 inches long, with several narrow linear bracts. Heads 15–20 or more, 4–5 lines diameter, among the largest of the genus, except (perhaps) *A. pachycephala.*

28. A. quinquedentata (Thunb.! Cap. 635); twigs and peduncles stellulate; leaves crowded, shortly obovate-orbicular, squarrose or re-flexed, glabrous, entire or 3-5-toothed, the teeth short and broad; co-rymb simple or sub-simple, many-headed; heads on long, bracteolate or naked pedicels, many-fl.; invol. globose, glabrous, scales oblong, obtuse, the innermost scarious and lacerate at the margin; pappus of many swollen hairs. *DC. l. c.* 90.

HAB. Cape, *Thunberg! Dr. Thom! Mundt!* E. & Z.! Rietkuil and Grootvaders-bosch, *Zey.!* 2827. (Herb. Th., Hk., Sd.)

Nearly related to some forms of *A. tridentata*, especially to our var. δ., but with much shorter and broader and more patent eaves. *Zeyher's* specimens (2827) pre-cisely agree with *Thunberg's;* the others are of stronger growth, with larger heads and less scarious inv.-scales. Leaves 3–4 lines long, 3–4 lines wide, nerveless, when entire mucronulate. Pedicels 1–1¼ inch long; heads 3 lines diam.

29. A. crithmifolia (Linn. Sp. 1181); adult leaves glabrous, 3–5-fid to the middle (rarely shortly trifid or entire!), the lobes linear, elongate, calloso-mucronate ; corymb compound (sometimes panicled), many-headed, fastigiate; the pedicels bracteolate; heads ovoid, many-fl.; inv. scales glabrous, in many rows, narrow, acute or acuminate, rigid, entire; pappus of many swollen hairs.   *Thunb.! Cap. 636. Burm. Afr. t. 69, f. 1. DC. l. c.* 89. *Sieb. Cap.* 418.

HAB. Common round Capetown and in the Western Districts.   Berg R., *Drege!* Winterhoek, Tulbagh, *Pappe!*   (Herb. Th., Hk., D., Sd.)
A densely leafy shrub, 2–3 feet high, mostly glabrous, the young parts sometimes clothed with woolly hairs, sometimes stellulate.   Leaves 1–2½ inches long, the lobes ½–1 inch or more long, not a line wide; rarely (in the Winterhoek specimens) varying from linear-lanceolate, entire, to narrow-cuneate, tridentate or shortly 3-lobed, 1–1½ line wide.   Corymb usually compound, of very many heads.   Invol. scales variable, sometimes very narrow and sharp, sometimes broader and blunter.

30. A. palmatifida (DC. l. c. 89); all parts densely clothed with soft, short, silky hairs; leaves densely imbricating, broadly linear to the middle, then palmately cleft, the medial lobe linear, entire, the lateral lobes deeply bifid, all callous-mucronate; corymb (undeveloped) many-headed ; heads many-fl., globose, hirsute, (fl. not seen).

HAB. Gnadenthal, in clefts of mountain rocks, 2–3000 feet, *Drege!* (Herb. Sd.)
A remarkable species, densely tomentose in all parts.   Leaves about uncial, the undivided portion ⅓ inch long, 1–1½ line wide; the lobes about as long, and ⅓ line wide; margins slightly recurved.   The inflorescence in the only specimen I have seen is only about half-opened, the corymb being half hidden among the uppermost leaves.

§ 4. MORYSIA.   (Sp. 31–40.)

31. A. tomentosa (Thunb.! Cap. 634.); virgate, all parts tomentose; leaves crowded, linear, flat, midribbed, mucronulate; corymb compound, densely many-headed, fastigiate ; heads oblong, 10–12 fl.; invol. scales obtuse, with deciduous toment.; pappus none; achenes (ripe) crowned with a few minute teeth.   *Morysia velutina, DC.? l. c.* 91.

HAB. Cape, *Thunberg! Mundt!*   (Herb. Th., Hk.)
Branches long, straight, simple, densely tomentulose and velvetty.   Leaves 1–1½ inch long, 1 line wide, with an evident midrib, sub-obtuse.   Heads 50–10 in a much-branched corymb.   There is no pappus, but each rib of the achene is produced upwards into a minute tooth.—I suppose this to be DC.'s "*Morysia velutina*," found at Swellendam by *Ecklon*, but having seen no specimens so marked, I forbear to quote *Ecklon's* habitat.   Possibly *Ecklon* received his specimen from *Mundt*, who collected largely at Swellendam.

32. A. fasciculata (Harv.); glabrous, flexuous; leaves crowded, linear, semi-terete (when dry channelled above), obtuse or mucronulate; corymb much branched, dense, very many-headed; heads 7–9-fl.; invol. scales obtuse, oblong, in few rows ; pappus none.   *Morysia fasciculata, Less.! Syn.* 262. *DC. l. c.* 91.   *Pteronia fasciculata, Willd. fide Less. Tanacetum vestitum, Thunb.! Cap.* 642.

HAB. Cape, *Thunberg!* Langekloof and Uitenhage, *Ecklon!* Swellendam? *Mundt! Thom!*   (Herb. Th., Sd., Hk.)
Very similar in aspect to *A. indivisa*, and also, though less so, to *A. filiformis,* but differing from both in wanting pappus.   Leaves about 1 inch long, ½ line wide,

fleshy. Heads very small, like those of *A. parviflora*. Inv. scales very obtuse, not "*acute*," as in DC.

33. A. linifolia (Harv. non Linn.); glabrous, virgate; leaves scattered or approximate, linear, thickish (when dry often channelled above), acute, callous-mucronate; corymb. much branched, dense many headed; heads 15-20-flowered; inv. scales obtuse, oblong, in few rows; pappus none. *Morysia pauciflora and M. microcephala, DC. l. c. 91.*

HAB. Ezelbank; and Pedroskloof to Leliefontein and Koussie, Kl. Nam.; also on Camdeboberg and Zwartberg, 4–6000 ft., *Drege!* Kamiesberg and near Worcester, *E. & Z.!* Camp-ground, near Rondebosch, *W.H.H.* (Herb. D., Sd., Hk.)

1–2 f. high, with few rod-like branches. Leaves ¾-1 inch long, ½ line wide, erect or spreading or squarrose, callous-pointed; less terete, more acute-pointed and less crowded than in *A. fasciculata*, but otherwise very similar. Fl. heads according to *DC.* "5–7-*flowered*" but I find them in both *Drege's* and *Eckl.* specimens at least 15-fl. and often 20-fl. The heads are *larger*, and the fl. *more numerous* than in *A. fasciculata*; hence I am compelled to reject both of *DC.'s* specific names. The only sp. I have seen of *M. pauciflora* has undeveloped fl.-heads.

34. A. juncea (DC.); "glabrous; branches terete, virgate; leaves distant, sessile, broadly-linear, subacute, quite entire, not tapering at base; corymb compound, fastigiate, dense; pedicels bracteolate; inv. scales oblong, appressed, obtuse; heads oval-oblong, 20–25-fl." *Morysia juncea, DC. l. c. 91.*

HAB. Betw. Breederiver and Gauritz R., Swell., *Drege.*—(Unknown to me).

35. A. longifolia (Lam. dict. t. 326); "glabrous; leaves linear-lanceolate, erect, acuminate, broad-based, with a decurrent midrib; corymb simple, few-headed; heads globose, glabrous, 25–40-fl.; pappus none." *DC. Morysia longifolia, Less. Syn.* 262. *DC. l.c.92. Ath. linifolia, Linn. f. Suppl. 361, fide Less.*

HAB. Cape. (Unknown to us.)

36. A. dentata (Linn. Sp. 1181); glabrous; leaves ovate or oblong, acute, half-clasping, squarrose or reflexed, more or less serrate, *rarely subentire*, teeth sharp; corymb compound, densely many headed; heads oblong, 15–20 fl.; inv. scales oblong, obtuse; pappus none. *Thunb.! Fl. 636. Morysia dentata, DC. l. c. 91. M. diversifolia, Cass. Less. Syn.* 262. *A. aspera, Th.! Cap.* 636.

VAR. β. **subintegrifolia**; leaves sub-entire, acute or acuminate, or lanceolate.

VAR. γ. **pachyphylla**; leaves reniform-orbicular, entire obtuse. *Morysia pachyphylla, Sch. B. Walp. Rep.* 6. p. 211.

HAB. Dry rocky places throughout the colony, common. (Herb. Th., D., Hk. Sd.)

2 feet or more high, laxly branched, the branches long, simple, closely set with short and broad reflexed leaves. Leaves 3–7 lines long, 3–4 lines wide, varying in shape and degree of serration; sometimes quite entire. Heads several flowered. *Thunberg's* specimen marked "*A. aspera*" is not distinguishable even as a variety. *Schultz's* (Bip.) *M. pachyphylla* seems to be a var. with entire leaves.

37. A. spathulata (Harv.); quite glabrous; leaves cuneato-spathulate, tapering to the base, more or less sharply serrate or lobulate beyond the middle; corymb compound, many headed; bracts small, subulate;

heads 5–6 flowered ; tube of corolla very hairy ; inv. scales obtuse ; pappus none. *Morysia spathulata, DC. l. c.* 91.

HAB. Near Groenriver and Waterval, 2500–3000 f., *Drege!* (Herb. D., Hk., Sd.)
About 1–1½ f. high, with slender, virgate, leafy branches. Leaves erecto-patent, nearly ¾ inch long, 4–5 lines wide at top, closely placed below, sub-distant toward the ends of the branches. Invol. pale straw-colour. A well marked species.

38. A. pectinata (Linn. f. Suppl. 361); glabrous or nearly so ; leaves pinnati-partite, or deeply 3–5 lobed, lobes narrow-linear, channelled, acute ; corymb compound, very dense, many headed ; bracts subulate; heads 9–12-fl.; cor. glabrous; inv. scales obtuse; pappus none. *Thunb.! Cap.* 636. *Morysia lineariloba, DC.! l. c.* 92.

HAB. Cape, *Thunberg!, Burch.* 6507. Gauritz R., *Drege!* Swellendam, *E. & Z.! Pappe!* (Herb. Th. D., Hk., Sd.)
Leaves on the same twig some pinnatisect, with 3–4 pinnæ on each side, some trifurcate or 5-lobed, all from ¾ to 1 inch long, the lobes ½ line wide, corymb very close, round-topped. *Thunberg's* specimens quite agree with *Drege's*.

39. A. acerosa (Harv.); twigs and peduncles pubescent or villous ; leaves glabrate, crowded or tufted, some pinnati-partite, some 3–5 lobed, some entire and acerose, lobes narrow-linear, with recurved margins, acute ; corymbs compound, densely many headed ; bracts subulate ; heads 7–9-fl.; cor. glabrous; inv. scales very narrow, gland-bearing ; paleæ few; pappus none. *Morysia acerosa, DC. l. c.* 92.

HAB. Betw. Omcomas and Omsamculo, *Drege!* Natal, *Plant, No.* 100. Between Maritzburg and Ladysmith, *Gerr. & M'K.* 266. *T. Cooper,* 1139. (Herb. D. Hk., Sd.)
Branches virgate, densely leafy to the summit, the leaves mostly with axillary tufts. Leaves ¾–1 inch long, like those of *A. pectinata.* Recept. with marginal paleæ only. Inv. scales narrower than in any other species.

40. A. pinnata (Linn. f. Suppl. 361) ; all parts velvetty-tomentose, canescent ; leaves crowded, pinnati-partite, the lobes linear, with sub-revolute margins, callous-mucronate ; corymb much-branched, densely many headed ; pedunc. hairy ; heads 15–20-fl. ; cor. glabrous ; inv. scales oblong, pubescent, sub-obtuse, outer acute. *Thunb.! Cap.* 637. *Morysia pinnata, DC. l. c.* 92.

HAB. Cape, *Thunberg!* Langekloof, *Drege!* Zwarthoogte, Uit. *Zey.!* 839 and near Grahamstown, *E. & Z.! Genl. Bolton!* (Herb. Th., D., Hk., Sd.)
2–3 feet high, robust, densely clothed with grey-downy leaves. Leaves 1–1½ inch long, the lobes ½–¾ line wide. Corymbs 2–3 inches across, very dense and compound.

Doubtful Species.

A. aspera (Linn.); "villous-scabrid; leaves ovate, 5-toothed; corymb panicled, sub-globose." *DC. Morysia aspera, Less. Syn.* 263. *DC. l. c. Ath. lævigata, L.*

HAB. Cape.
Probably a scabrid var. of *A. dentata*: it is said to vary with *glabrous* leaves.

LXXV. ERIOCEPHALUS, Linn.

Heads 10–15-fl., heterogamous ; the *ray-fl.* female, with a bifid style and mostly ligulate, the ligule broadly-obovate, toothed, sometimes very short and small ; *disc-fl.* tubular, 5-toothed, male, with perfect stamens and a simple, club-shaped, truncate style. *Recept.* bearing paleæ. *Inv.* double, the outer campanulate, of 4–5 ovate, free scales ; the inner of cohering scales, very woolly externally. *Achenes* of the ray flattened, wingless, without pappus. *DC. Prodr. 6, p.* 145.

Much-branched, rigid, mostly silky or silvery shrubs. Leaves alternate or opposite, or fascicled, small, mostly linear, simple or trifid, rarely glabrous. Heads pedicellate or sessile, racemose, umbellate or solitary, subglobose, after flowering becoming very woolly. Name from εριον, *wool,* and κεφαλος, *a head : woolly-head ;* the colonial name is *Capok-bosch.* The species have a great resemblance to each other, and perhaps I have retained too many.

Sect. 1. PHÆNOGYNE (DC.). *Corolla* of the ray conspicuously ligulate, longer than the style and involucre, obovate or cuneate, bluntly 3-toothed. (Sp. 1–12.)

```
 * Heads pedicellate, sub-corymbose :
     Leaves (or many of them) 3–5-lobed :
         Silky or silvery-pubescent :
             Inner invol. septiferous within :
                 Leaves 6–12 lines long, silky      ...  ...   (1) Africanus.
                 Leaves 3–5 lines long, silvery     ...  ...   (2) septulifer.
             Inner invol. not septiferous :
                 Silky-villous ; heads loosely corymbose      (3) sericeus.
                 Silvery ; hds. sub-sessile, in close corymbs  (4) capitellatus.
         Glabr.  Lvs. 3-fid, the lateral lobes bifid, mucronate (5) petrophiloides.
     Leaves all quite entire ;  (rarely minutely unidentate) :
         Leaves glabrous, impress-dotted or smooth :
             Outer inv.-scales with a membr. border   ...   (6) punctulatus.
             Outer inv.-scales not bordered    ...  ...  ...  (7) pteronioides.
         Leaves glabrous, rough with raised tubercles   ...  (8) tuberculosus.
         Leaves pubescent or silky-silvery :
             Hds. mostly racemose; ray-fl. cuneate, elongate  (9) scariosus.
             Hds. umbellate; ray-fl. broadly obovate    ...  (10) umbellulatus.
** Heads solitary, sessile at the ends of the twigs :
         Heads bearing pencils of long, straight, silky hairs ...  (11) eximius.
         Heads bearing curled, woolly hairs (as in most species)  (12) spinescens.
```

Sect. 2. CRYPTOGYNE (DC.). *Corolla* of the ray inconspicuous, nearly hidden within the involucre, shorter than the style, tubular, subentire or obliquely truncate, or imperfectly ligulate. (Sp. 13–17.)

```
Leaves silky-villous or conspicuously pubescent :
     Not spinescent. Heads in leafy, unilateral racemes...  (13) racemosus.
     Subspinescent.  Leaves short, 2–5 lines long.
         Heads solitary, shortly pedicellate, terminal   ...  (14) aspalathoides.
         Heads on slender pedicels, axillary or subumbell.   (15) pubescens.
Leaves glabrous or minutely puberulous :
     Lvs. 1–2 lines long.  Heads several-flowered, racemose
         or spicate  ...  ...  ...  ...  ..  ...  ...  ...  ...  (16) glaber.
     Lvs. 2–3 lines long.  Hds. few-fl., racemose, very small  (17) microcephalus.
```

§ 1. PHÆNOGYNE, DC. (Sp. 1–12.)

1. E. Africanus (Linn.! Sp. 1310) ; leaves opposite or tufted, silky-pubescent, thickish, obtuse, linear or trifid, channelled ; heads umbellate at the ends of the branches, the inner involucre having three projecting ridges (septa) within. *Less.! Syn.* 269. *E. septifer, Cass. DC. l. c.* 145. *Dill. Elth. t.* 110.

HAB. Cape, *Thunberg!* (Herb. Thunb.)
Of this I have only seen a specimen without flowers, in Hb. Th. Leaves 6–12 lines long, thinly silky, ¼ line thick.

2. E. septulifer (DC.! l. c. 145); leaves alternate, tufted, silky-silvery, thickish, obtuse, shortly linear, or very rarely trifid, channelled ; heads umbellate at the ends of short, lateral twigs, the pedicels longer than the leaves ; "the inner invol. having very short projecting ridges (septa) within." *DC.*

HAB. Cape Flats, *Drege!* *W. H. H.* (Herb. D., Sd., Hk.)
Leaves 3–5 lines long, ½ line thick, silvery. Wool of the invol. coarse and fulvous.

3. E. sericeus (Gaudich.) ; leaves often opposite, linear, entire or trifid, silky-villous ; heads sub-corymbose at the ends of the branches, inner invol. not septiferous within. *DC. l. c.* 145. *E. Africanus, Burm.*

HAB. Cape : not uncommon round Capetown, according to De Candolle.
Of this I know nothing, unless a specimen in Hb. Hook. without flowers and probably from a garden, be referable to it.

4. E. capitellatus (DC.! l. c. 146) ; leaves alternate, tufted, linear, entire and trifid, silky-silvery, obtuse, sub-terete; heads at the ends of short, lateral twigs, sub-sessile, small ; rays 1–3, short.

HAB. Zwarteberg and Zwart Ruggens, *Drege!* Hexrivier, *Burke! Zey.!* 860. (Herb. Hk., D., Sd.)
A slender, much-branched bush ; twigs filiform, closely covered with leaf-tufts. Leaves 2–4 lines long, ¼ line thick, several in almost every tuft trifid. Heads 1 line long, on pedicels 1–2 lines long or less. Wool pale fulvous.

5. E.? petrophiloides (DC.! l. c. 146); stem and leaves quite glabrous; leaves alternate, trifid above the middle, the medial lobe linear, entire, the lateral lobes mostly bifid, all callous-mucronate ; pedicels axillary, solitary, filiform, longer than the leaves ; outer inv. of 6–8 oblong, glabrous, basally connate scales. *Eriocephalus, 446, Hb. Eckl.*

HAB. In S. Africa (no station given), *Drege! Eckl.!* (Herb. Sond.)
A most remarkable species, unlike any other and possibly not of this genus ; but the fl.-heads, in the only specimens I have seen, have had their contents eaten by insects, leaving merely the outer invol. and a dense tuft of discoloured wool. What may be the origin of this wool, whether from an inner invol. or from the achenes, remains undetermined. *DC.'s* specimens seem to have been equally imperfect. Leaves 1 inch or more long, the lobes 4–5 lines long, ½ line wide, all tipped with a white, hard point. Pedicels 1½ inch long.

6. E. punctulatus (DC. l. c. 146) ; leaves opposite and alternate, scattered or crowded, linear-semiterete, acute or obtuse (occasionally unidentate at each side), thickish, glabrous, more or less impress-dotted; pedicels in terminal umbels and axillary, short or long ; outer invol. scales glabrous, widely membrane-edged, inner not septiferous ; rays obovate, 3-toothed.

VAR. α, **pedicellaris**; pedicels uncial ; leaves scattered. *E. pedicellaris, DC. l. c.*
VAR. β, **punctulatus**; pedicels semi-uncial; leaves crowded or scattered. *E. punctulatus, DC. l. c.*
VAR. γ, **tenuifolius**; pedicels mostly shorter than the leaves ; leaves crowded. *E. tenuifolius, DC. l. c.*
VAR. δ, **filifolius** (DC.) ; leaves linear-filiform, elongate, remote ; pedicels 2–3 times longer than the heads. *E. tenuifolius* ; β, *filifolius, DC. l. c.*

HAB. α, Little Namaqualand, *Drege!* β, Slopes of the Kamiesberg, near Kaspars-

kloof and Geelbekskraal, *Drege!* Modderfontein, *Rev. H. Whitehead!* Basutu-land,
T. Cooper! 692 ; γ, Witbergen and Sneeuwbergen, *Drege!* Stormberg, *Wyley!* δ,
Onder Bokkeveld, *Drege.* (Hb. D., Hk., Sd.)
  Our var. β seems exactly intermediate between a and γ. The species is best
marked by its glabrous, dotted leaves and membrane-edged invol. Leaves 3–4–6
lines long, ⅓ line wide. I have not seen var. δ. On *Mr. Cooper's* specimens the
leaves are occasionally unidentate about the middle.

  7. E. pteronioides (DC. l. c. 146); "leaves opposite and alternate,
linear-oblong, obtuse, thick, glabrous; pedicels thrice as long as the heads,
sub-umbellate at the ends of the branches; outer inv. scales dorsally
glabrous, not membrane-edged; inner not septiferous; ligules small."
DC. l. c.

  HAB. Olifant's R., *Drege.* (Unknown to me.)

  8. E. tuberculosus (DC. l. c. 146); twigs glandular; leaves opposite
and alternate, short, linear, subtrigonous, thickly sprinkled with prominent, glandular tubercles, otherwise glabrous; heads at the ends of
the branches spicate; outer inv. scales 5, albo-membranous, with reddish margin; rays 2–3-lobed.

  HAB. Modderfonteinsberg, Namaqualand, *Drege!* (Herb. Sond.)
  At once known by its curiously gland-tubercled leaves. I have not seen the fl.-
heads, and copy *De Candolle's* description of them. Leaves 2–3 lines long, very
slender, closely placed.

  9. E. scariosus (DC. l. c. 147); leaves alternate, linear, incurved, subacute, entire, thinly and closely pubescent; pedicels 2–3 times as long
as the fl.-heads, the lower axillary, sub-racemose, the uppermost subumbellate; outer inv. scales glabrous or silky with a green middle and
broadly membranous, white margin; inner not septiferous; rays cuneate,
elongate, 3-toothed.

  HAB. Betw. Natvoet and the Gariep, *Drege!* Namaqualand, A. Wyley! (Herb.
D., Hk., Sd.)
  Leaves about an inch long, scattered. Rays longer and proportionably narrower
than in other species. The pubescence is sometimes copious, sometimes scanty.

  10. E. umbellulatus (DC. l. c. 147); leaves opposite and alternate,
mostly tufted, linear, entire, silky-villous or silvery; heads umbellate at
the ends of the branches and twigs, pedicels 1½–3 times as long as the
fl.-heads; outer inv. scales densely silky, membrane-edged; inner not
septiferous; rays broadly obovate, 3-toothed. *E. Africanus, Sieb.! Cap.
No. 35. Krebs. No.* 758. *E. racemosus, Gaertn.* 2, *t.* 168. *Lam. Dict.* 717,
f. 2, *Th.! Cap. p.* 724 *(ex pte.) E. paniculatus seu umbellulatus, Cass.*

  VAR. β. **glabriusculus** (DC.); leaves pubescent, the young ones silky; pedicels
twice as long as head; inv.-scales glabrescent.
  VAR. γ. **argenteus** (DC.); leaves silvery; pedicels longish; inv. scales villous.
  HAB. Rocky, dry places and roadsides, throughout the colony, common; γ. in
Namaqualand, *Drege! Wyley!* (Herb. Th., D., Hk., Sd.)
  Often confounded in Herbaria with *E. racemosus,* which is much less common.
Many of *Drege's* distributed specimens marked "*racemosus*" belong to this; also one
of the specimens in Hb. *Thunb.!* It is readily known by its inflorescence and ray-
flowers from all states of *racemosus.*

11. E. eximius (DC.! l. c. 147); twigs spinescent; leaves opposite and alternate, closely imbricated on the twigs, entire, short, silky-silvery; heads terminal, sessile, solitary; outer inv.-scales broadly ovate, flat, silky, with very long, straight, silky hairs from their axils, the inner invol. glabrescent, 3-lobed, septiferous within, the septa linear, alternating with the scales, free above; rays broadly oblong, 3-toothed.

HAB. Sneeuwbergen, 4–5000 f., *Drege!* Stormberg, *A. Wyley!* Witteberg and Basutuland, *T. Cooper*, 627,741. (Herb. D., Hk., Sd.)
A rigid, robust, scrubby bush, with very silvery leaves, and long, silky and silvery involucral wool. Leaves 2–3 lines long, mostly opposite, as I find them.

12. E. spinescens (Burch. Trav. 1.272); divaricately much-branched, spinescent; leaves opposite and alternate, tufted and densely imbricated on the twigs, entire, linear-trigonous, short, closely silky or silvery; heads at the ends of minute, lateral twigs solitary, sessile; outer inv. scales oblong, silky (inner not septiferous?); rays small, 3-toothed.

HAB. Near Karroo River, *Burchell,* No. 1419. Sneeuwberg, *Drege!* Sand River, *Burke!* 279. *Zey.!* 858. (Herb. D., Hk., Sd.)
A rigid, robust, dwarf scrubby bush. Leaves less silvery than in *E. eximius,* 2–3 lines long. Heads smaller, wanting the long, silky wool, with shorter and smaller rays, often disposed in a spurious spike towards the ends of the spinescent branches. Flowering twigs 2–3 lines long.

§ 2. CRYPTOGYNE, DC. (Sp. 13–17.)

13. E. racemosus (Linn. Sp. 1311); leaves opposite and alternate, often tufted, linear, entire, silky-villous; pedicels axillary; heads forming leafy racemes toward the ends of the branches, often unilateral and cernuous; ray-fl. few, short, tubular. *Th.! Cap. p.* 724 *(ex pte.) DC. l. c.* 147. *E. spicatus, Burm.*

VAR. β. **affinis**; pedicels twice as long as the head; heads a little larger. *E. affinis, DC. l. c.*

HAB. Cape, *Thunberg,* &c., *Drege!* (ex pte.) Langevalley, *Zey.!* 857. β. on the Cape Flats, *Drege!* (Herb. Th., D., Hk., Sd.)
Less common than *E. umbellulatus,* which is often found mixed with it in Herbaria, and was by Thunberg and older authors confounded with it. Twigs somewhat virgate, often arching. Leaves ½–¾ inch long. Pedicels long or short, very variable. I cannot separate *E. affinis,* DC. by any valid character.

14. E. aspalathoides (DC.! l. c. 148); divaricately much-branched, sub-spinescent; leaves opposite and alternate, crowded or tufted, linear-terete, short, entire, closely silky; heads *solitary at the ends of short, lateral twigs, shortly pedicellate;* outer inv.-scales silky, with a narrow membranous border; ray-fl. minute, obliquely truncate, shorter than the style.

HAB. Betw. Zwarteberg and Aasvogelberge, and near Mierenskasteel, Namaqualand, *Drege!* (Herb. D., Hk.)
Very similar in aspect to *E. spinescens,* but the flower-heads are pedicellate and the rays minute, almost hidden. Leaves 2 lines long. Pedicels 2–3 lines. *Drege's* specimens so marked in some Herbaria (as in Hb. Sond.!) belong to *E. pubescens,* which is a slenderer plant, with different inflorescence; again, in Herb. D., one of *Drege's* specimens marked "*pubescens*" belongs to *aspalathoides.*

15. E. pubescens (DC. l. c. 148); divaricately much-branched, sub-

spinescent; leaves alternate and opposite, crowded, linear-filiform, obtuse, entire, closely pubescent or silky ; heads on slender pedicels, 2–3
times as long as the leaves, either axillary or sub-umbellate or solitary
at the ends of the twigs ; outer inv.-scales silky with a narrow membranous margin ; ray-fl. very minute.   *E. aspalathoides, Drege! ex pte.*

HAB. Mierenskasteel, Kl. Namaqualand, *Drege* / Gamke R., *Zey.!* 859. (Herb.
D., Hk., Sd.)

Very like the last, but with longer leaves and different inflorescence. Leaves 3–5
lines long, slender. Pedicels nearly ¾–1 inch long. In Hb. Sond. a specimen of
this species from *Drege* is marked *aspalathoides*. *Zeyher's* plant (859) is more robust,
with solitary pedicels, nearly 1 inch long, densely silky.

16. E. glaber (Th.! Cap. 724) ; leaves opposite and alternate, very
short, linear, sub-trigonous, patent, obtuse, glabrous or puberulous, punctate; heads at the ends of minute twigs solitary or few together, but forming leafy racemes or spikes toward the ends of the branches, sub-sessile or
pedicellate; outer inv.-scales (mostly purple) glabrescent, with a narrow
membranous border ; ray-fl. very minute.   *Less. Syn.* 269.   *Tarchonanthus ericoides, Linn. f.*

VAR. *a.* **glaber**; adult leaves quite glabrous, or nearly so ; pedicels short. *E.
glaber, DC.! l. c.* 148.

VAR. *β.* **pubescens**; leaves puberulous ; pedicels 1–2 together, scarcely ½-uncial.
E. microphyllus, DC.! l. c

VAR. *γ.* **laricinus**; leaves gland-dotted, glabrescent; pedicels scarcely longer than
leaf. *E. laricinus, DC.!*

VAR. *δ.* **sessiliflorus** (Sond.) ; heads sub-sessile, in terminal spikes. *E. sessiliflorus,
Sond. MS.*

HAB. Cape, *Thunberg* / Little Namaqualand, and near the Gariep, *Burchell,
Drege! Wyley!* Var. *γ.* near the Fish River, *Drege!* Albany, *Bowker!* Var. *δ.*,
Cape, *E. & Z.!* Z. n. N. n. E. 23.   (Herb. Th., D., Sd., Hk.)

Leaves very minute, commonly 1–2 lines long, with occasional subtending leaves
3–4 lines long, the young ones and twigs minutely puberulous ; the older either glabrate or retaining their downy surface, all more or less dotted. Pedicels 1–6 lines
long. I cannot keep the above varieties specifically apart. The *aspect* of all is the
same, and the pubescence most inconstant ; the length of pedicel partly depends on
the advanced or nascent state of flowering

17. E. microcephalus (DC.! l. c. 148) ; leaves opposite, tufted on
the sterile twigs, linear-filiform, slender, glabrous or puberulous; pedicels solitary in the axils of the upper leaves, longer than the leaf, forming a leafy raceme ; heads very small, 5–8-fl. ; outer inv.-scales oblong,
glabrescent (purple) with a narrow border ; ray-fl. minute.

HAB. Little Namaqualand, *Drege!* (Herb. D., Hk., Sd.)

More slender than *E. glaber*, with much smaller fl.-heads, but otherwise very
similar, and perhaps a mere starved variety. Leaves 2–3 lines long, on the fertile
twigs in sub-distant pairs. Heads 1 line long.

Sub-Tribe 4. GNAPHALIEÆ. (Gen. 76–106.)

LXXVI. RHYNEA, DC.

Heads many-fl., heterogamous, discoid ; the disc-fl. (about 15) perfect,
5-toothed ; the marginal (about 5) filiform, female. *Recept.* bearing
linear, deciduous paleæ between the flowers. *Inv.* imbricate, the outer
scales downy, obtuse, the inner produced into an oblong, spreading,

obtuse, white appendix. *Anthers* shortly tailed. *Style-*branches trun-cate. *Achene* obovate-oblong, beakless, minutely downy. *Pappus* in one row, bristle-shaped. *DC. Prodr. 6, p.* 154.

A tall, branching undershrub, with rod-like, canous branches. Leaves alternate, decurrent, lanceolate, mucronate, green above, white beneath, one-nerved, with sub-recurved margins. Heads in branching, subsessile corymbs, the branches bracteate. Inv. oblong, 2½ lines long, the inner scales opaque, snow-white, as are also the paleæ. The aspect is that of a *Helichrysum*, from which genus this is readily known by the paleæ of the receptacle. The name is in honour of *W. v. Rhyne*, a traveller in S. Africa of the 17th century.

1. R. phylicæfolia (DC. in Del. Ic. 4, t. 52); *DC. Prodr. l. c.*

HAB. Various localities near Natal, *Drege! Plant, Gueinzius, Gerr. & M'Ken,* &c. (Herb. D., Hk., Sd.)

Stem 2–3 ft. high, much branched. Lvs. ½–1 inch long, 2–4 lines wide, coriaceous.

LXXVII. **LEONTONYX**, Cass.

Heads 20–30 fl., homo- or heterogamous, a few of the marginal fl. female, the rest perfect. *Recept.* flat and naked. *Inv.* imbricated, the outer scales very woolly, the inner elongate, rigidly membranous, hooked or straight. *Achenes* sessile, oblong, granulated. *Pappus* in a double row, copious, of very slender, caducous bristles, equalling the flowers. *DC. Prodr. 6, p.* 167.

Herbaceous or suffruticose, densely woolly, S. African plants. Leaves alternate, sessile, obovate or spathulate. Heads mostly crowded at the ends of the branches, rarely solitary, closely subtended by a false involucre of leaves. This genus scarcely differs from *Helichrysum;* the character of the double pappus is not very definite or easily seen ; but the habit is peculiar. The generic name is compounded of λεων, a lion, and ονυχ, a claw; alluding to the hooked invol.-scales. Lessing's *L. pusillus* will be found under *Helichrysum (H. pachyrhizum).*

Inner inv.-scales acute or acuminate, hooked or straight :
  Inv.-scales *remarkably* hook-pointed :
    Stem robust, ascending ; lower lvs. larger and closer (1) **squarrosus.**
    Stem slender, diffuse or prostrate, equally leafy throughout :
      Inv.-scales rufous, rosy, or pale straw-colour ... (2) **glomeratus.**
      Inv.-scales yellow, with a bright red medial spot,
        elongate ... ... (3) **bicolor.**
  Inv.-scales straight or slightly recurved, not hooked.
    Leaves obovate, flat ... ... (2) **glomeratus β.**
    Leaves linear-spathulate, snowy white ... ... (4) **angustifolius.**
Inner inv.-scales obtuse or bidentate, lvs. obovate ... ... (5) **spathulatus.**

1. L. squarrosus (DC. l. c. 167); densely white-woolly, many stem-med, with rosulate radical leaves ; stems ascending, simple or branched, densely leafy below, more laxly upwards ; leaves obovate-oblong, acute, or obtuse, immersedly 3-nerved ; heads crowded, many together in densely-woolly tufts, subtended by several floral-leaves, heterogamous ; inner inv. scales taper-pointed, strongly reflexed, white, rosy, or deep purple. *Gnaph. squarrosum, Linn. Sp.* 1197. *Th.! Cap.* 656. *Jacq. Fragm. t.* 3. *f.* 4. *Leontonyx tomentosus, Cass.*—*Less.! Syn. p.* 325. Also *L. tinctus, DC.! l. c. L. coloratus β. contractus, Less.! Syn.* 327. *Gnaph. tinctum, Thunb.! Cap.* 656.

VAR. β. **discretus**; heads separate, shortly pedicelled, in dense cymules (not tufts).

HAB. Cape-Flats and hills in dry ground; and similar situations, throughout the colony. Natal, *Gueinzius! Gerr. and M'K.!* 271. β. Natal, *Gerr. and M'K.!* 272. (Herb. Th., Hk., D., Sd.)

Perennial. Radical leaves numerous, 1–2½ inches long, ⅓–⅔ inch wide. Cauline leaves smaller and narrower, sometimes densely, sometimes laxly set, membranous (when stripped of wool), in the weaker varieties spathulate. Stems 6–12 inches high, simple or corymbose. Heads glomerated in wool, a few fl. female. Inv. scales much acuminate and recurved, very variable in colour. *Thunberg's G. tinctum* is very dwarf; *DC.'s*, at least *Ecklon's* quoted specimens, immature, but, as I think, otherwise undistinguishable. Our var. β. looks much more like a species, all the heads being quite separate. But it grows with the ordinary form, and I have a nearly intermediate specimen from *Mr. Hutton.*

2. L. glomeratus (DC.! l. c. 168); closely or loosely woolly; stems prostrate or diffuse, slender, much branched from the base, the branches decompound, laxly leafy throughout, white-woolly; leaves obovate or spathulate, narrowed to the base, obtuse; heads 3–6 together, sessile on short lateral twigs, subtended by floral leaves, heterogamous; inner inv. scales lanceolate, taper-pointed, reflexed, rufous or rosy. *Gn. glomeratum, Linn.—Th.! Cap.* 656. *Leont. coloratus, Cass.—L. col., var. gracilis, Less.! Syn.* 326. *Leont. angustifol. var. diffusus, ex pte. DC.*

VAR. *a.* **verus**; inv. scales narrow, much acuminate and strongly hooked, whitish or rosy.

VAR. β. **intermedius**; inv. scales broad, acute or sub-acute, spreading, scarcely recurved.

VAR. γ! **stramineus**; leaves very ⁻⁻oolly; inv. scales acuminate, hooked, straw-coloured. *L. stramineus, DC.! l. c.* 168.

HAB. Mts. round Capetown, and on the flats. Uitenhage, *Zeyher!* Olifants R., and near the Gariep, *Drege!* Swellendam, &c., *E. & Z.! Zey.!* 2882. var. γ. Piquetberg, *Drege!* (Herb. Th., Hk., D., Sd.)

Stems 6–12 inches long, the larger specimens prostrate, variably woolly. Leaves ¼–½ inch long, 2–4 lines wide, sometimes closely, sometimes loosely woolly or shaggy, white or whitish. Infl. mostly on short, lateral branchlets. Tips of the invol. varying much in degree of acumination, in breadth and in colour; sometimes barely acute, not acuminate. *DC.'s L. stramineus*, of which I have seen but imperfect specimens, seems chiefly to differ by its yellowish inv. scales.

3. L. bicolor (DC.! l. c. 168); white-woolly, dwarf; stems diffuse, branching, tufted, leafy throughout; leaves obovato-spathulate, concave, obtuse; heads sub-solitary, sessile, subtended by floral leaves; inner inv. scales *very long*, linear-lanceolate, acuminate, hooked, yellow, with a bright rosy spot in the middle.

HAB. Camiesberg, *Drege!* (Herb. Sond.)

Known by its long, distinctly 2-coloured inv. scales, and mostly solitary fl.-heads

4. L. angustifolius (DC.! l. c. 168); wholly covered with a snow-white, close toment.; stems diffuse, slender, much branched, leafy throughout; leaves linear-spathulate, obtuse, infolded, spreading, straight or recurved; heads sub-solitary or 2–3 together, sessile on short lateral twigs, subtended by floral leaves; inner inv. scales lanceolate, taper-pointed, straight, rufous.

HAB. About Capetown and at George, *E. & Z.!* Lange Valley, *Drege!* (Herb. Hk., Sd., D.)

Except by its whiter and closer indument, narrower leaves and straight inv. scales this scarcely differs from *L. glomeratus. DC.'s* var. β. *diffusus*, by the distributed specimens, belongs rather to *L. glomeratus.*

5. L. spathulatus (Less.! Syn. p. 327); closely or loosely white-woolly (sometimes snow-white); stems prostrate, slender, much-branched and decompound, leafy throughout; leaves obovate or spathulate, narrowed to the base, obtuse; heads 3–6 together, sessile on short, lateral twigs, subtended by floral leaves; inner inv.-scales linear or oblong, straight, obtuse or minutely bidentate, rosy-tipped. *Gnaph. spathulatum, Th.! Cap.* 656.

VAR. *a.* **hirsutus**; toment loosely woolly. *L. spathulatus, DC.! l. c.* 168. *L. pusillus, Hb. Drege!*

VAR. β. **candidissimus**; toment very white and close-lying. *L. candidissimus, DC.! l. c.* 169.

HAB. Table Mountain, &c. and on the Flats, common; also in Worcester, Stellenbosch and Uitenhage, *E. & Z.!* Near the Gariep, *Drege!* Zey.! 2880, *ex pte.*, 2879, &c. (Herb. Th., Hk., Sd.)

Only to be known from *L. glomeratus*, with which it grows intermixed, and from whose var. β. it scarcely differs, by the *obtuse* (not acuminate) and very generally bidentate inv.-scales. Var. β. is snow-white in all parts, with *closely* interwoven wool, but both these characters vary greatly. Some of *E. & Z.*'s distributed specimens marked "*L. angustifolius*" belong to this.

Doubtful species.

Eriosphæra dubia (DC. l. c. 167); "stems scarcely suffruticose at base, numerous, erect, cobwebby-woolly; leaves obovato-spathulate, on both sides woolly-tomentose, nearly nerveless, obtuse, exapiculate; heads crowded, tufted-corymbose, subtended by a few floral leaves, about 20-fl.; inv.-scales acuminate, submembranaceous." *DC.*

HAB. Wittberg, *Drege!* (Herb. Hk., D.)

Of this I have only seen bad specimens, which I am unable to separate from *Leontonyx squarrosus.* Possibly *Drege* may have in this instance (as in several others he certainly has) distributed a wrong plant.

LXXVIII. HELICHRYSUM, Vaill.

Heads many or few-fl., discoid, either homogamous, all the fl. tubular, perfect, 5-toothed; or heterogamous, the marginal-flowers filiform, female, very few or in a single row. *Inv.* imbricating, dry and membranous. *Recept.* without paleæ, either quite naked, honey-combed, toothed or fimbrilliferous. *Achenes* beakless, sessile, very generally minutely granulated, (in *H. ericoides*, silky). *Pappus* in one row, of many or few, slender, scabrous or serrulated bristles. *DC. l. c. p.* 169.

A vast genus of herbs or suffrutices, very various in aspect, abundant in S. Africa and Australia, rare in S. Europe, N. Africa and Central Asia; not found in America. Stems and leaves mostly woolly. Inv. white, rosy, horn-coloured or yellow. Cor. yellow, rarely purple. This genus only differs from *Gnaphalium* by the fewer filiform, marginal-flowers; from *Helipterum* by the bristle-shaped pappus. The name is derived from ἥλιος, *the sun*, and χρυσος, *gold;* applicable to such species as *H. fulgidum.*

Sub-genus I. EU-HELICHRYSUM. Receptacle naked (neither fimbrilliferous nor toothed) or scarcely honey-combed. (Sp. 1–80.)

A. ARGYREIA. Inv. scales white or reddish, rarely tawny, horn-colour or brownish *(not yellow).** (§ 1–9.)

§ 1. LEPTORHIZA. Slender annuals. Heads small, either tufted at the ends of the branches or rarely corymbose. Inv. scales either white, brown, horn-colour, or pale straw-colour. (Sp. 1–8.)

Heads several together, sessile or subsessile in terminal tufts :
    Inv. scales brown or horn-colour. Lvs. mucronate   (1) **leptorhizum.**
    Inv. scales straw-colour or purplish. Lvs. obtuse :
        Inv. scales acuminate, hook-pointed ... ... (2) **Leontonyx.**
        Inv. scales obtuse or subacute, *straight :*
            Leaves oblong or obovate :
                Heads 25–30-flowered ... ... (3) **alsinoides.**
                Heads about 10-flowered ... ... (4) **micropoides.**
            Leaves linear, woolly. Heads 5–8-fl. ... (5) **filagineum.**
    Inv. scales *snow-white*, obtuse ... ... (7) **herniarioides.**
Heads pedicellate, racemose or axillary ; lvs. petioled ; inv. scales whitish ... ... (6) **capillaceum.**
Heads corymbose, panicled ; lvs. spathulate ; inv. snow white ... ... (8) **expansum.**

§ 2. ERICÆFOLIA. Small shrubs or suffrutices. Leaves *linear* or *minute*, and scale-like. Heads small, terminal, *sessile*, solitary or in tufts or spikes. Inv. scales horn-colour or white, rarely rufous. (Sp. 9–16.)

Leaves linear, with revolute margins :
    Heads numerous, in a terminal spike. Lvs. woolly (9) **spiciforme.**
    Heads solitary, terminal :
        Branches and leaves woolly ... ... (10) **cerastoides.**
        Branches and leaves glabrous ... ... (11) **paronychioides.**
        Br. and lvs. gland-scabrid, the young cobwebby (12) **oxybelium.**
    Heads 3–6, in terminal tufts :
        Loosely branched, erect or spreading *(very variable)*... ... ... (13) **ericæfolium.**
        Prostrate, densely ramulous, level-topped ... (14) **cæspititium.**
    Leaves linear, quite flat; heads solitary, terminal ... (15) **simulans.**
    Leaves minute, scale-like, ovate, appressed ; ach. silky (16) **ericoides.**

§ 3. DECLINATA. Herbaceous or suffrutices, sometimes nearly stemless. Leaves *(not heath-like)* either spathulate, oval or obovate, or linear-oblong. Heads small, corymbose, rarely solitary. Inv. scales white. (Sp. 17–21.)

Stems loosely branched, diffuse or ascending :
    Herbaceous. Leaves linear-spathulate, acute ... (17) **declinatum.**
    Suffruticose. Leaves oval or obovate; inv. scales linear-oblong ... ... (18) **serpyllifolium.**
Dwarf, tufted ; branches short, closely leafy :
    Heads corymbose, on laxly leafy flower-branches :
        Leaves obovate or oval, short ... ... (19) **Sutherlandi.**
        Leaves lance-linear, elongate ... ... (20) **Ernestianum.**
    Heads solitary, at the end of leafy short branches ... (21) **sessile.**

§ 4. IMBRICATA. Much branched (mostly *small*), half-herbaceous or suffruticose perennials. Leaves obovate or spathulate. Heads 10–60-fl., either solitary or 2–3-together, ending short branches, pedicellate. *Inv.* imbricated in several rows, glabrous, its scales rarely acute, concave or squarrose, reddish-brown, tawny, or horn-colour, the innermost sometimes white or creamy. (Sp. 22–28.)

---

* In two or three species of § 7 the invol. varies from pure white to *pale lemon-colour :* in a few other cases to *straw-colour.*

All the inv. scales of one colour (rufous or horn-colour) :
  Heads pedicellate, corymbose or panicled :
    Depressed, fastigiate. Heads 10–12-fl. ... ...   (22) **fastigiatum.**
    Diffuse. Heads 40–50-fl. ... ... (23) **cochleariforme.**
  Heads 2–3-together, subtended by ovate, leafy bracts (24) **pulchellum.**
Inner inv. scales white or creamy ; outer tawny, rufous, or rosy :
  Inv. scales squarrose or spreading ; heads oblong :
    Inv. scales *acute;* medial purple ; inner bright
      white ... ... (25) **rubellum.**
    Inv. scales *obtuse;* medial tawny ; inner spoon-
      like, creamy ... ... (26) **cylindricum.**
  Inv. scales erect or inflexed ; heads globose :
    Inv. scales concave, *very obtuse;* inner spoon-
      like, snow-white ... ... (27) **imbricatum.**
    Inv. scales ovate, *acute,* flattish ; inner white-
      limbed ... ... (28) **stellatum.**

§ 5. OBVALLATA. Root woody, long unbranched. Stems many from the crown, half-herbaceous or suffruticose, *depressed* or *decumbent.* Leaves oblong or obovate, narrowed to the base. Heads few or many-flowered, globose or oval, one or *several in a tuft* at the ends of the branches, *sessile (except in No. 29),* subtended by leaves. Inv. snow-white or rosy. (Sp. 29–35.)

Heads *globose,* 1–4-together, *pedicellate,* corymbulose ... (29) **chionosphærum.**
Heads globose, solitary, sessile ; inv. *quite glabrous,* silver-
  white ... ... (30) **argyrosphærum.**
Heads oblong, in terminal tufts, sessile ; inv. scales acute or acuminate.
  Leaves linear, spathulate, or oblong, sessile :
    Heads many-flowered, 1–3-together ; lvs. very
      woolly ... ... (31) **obvallatum.**
    Heads 20–25-flowered, *several* in a tuft :
      Root thick and woody ; stems much and
        closely branched ... ... (32) **pachyrhizum.**
      Root slender ; stems slender, trailing or
        prostrate... ... ... (33) **leptolepis.**
    Lvs. oval or oblong, petiolate, nerved, *thinly* woolly (34) **Gariepinum.**
Heads in terminal tufts, sessile ; inv. scales very obtuse (35) **catipes.**

§ 6. SPHÆROCEPHALA. Erect or diffuse suffrutices. Leaves sessile, ovate, oblong, or (in No. 41), lance-linear. Heads few or many-fl., in *dense,* mostly pedunculate, closely-branched cymes or tufts. Invol. scales broadly ovate or obovate, obtuse (rarely *acute*), erect, not radiating, opaque-white or rosy. (Sp. 36–41.)

Stems diffuse or procumbent, mostly branched :
  Inv. scales *acute;* leaves oval or oblong, mucronate (36) **diffusum.**
  Inv. scales obovate, *very obtuse;* lvs. obovate or spathulate.
    Stem alternately branched ; lvs. spathulate-obl. (37) **crassifolium.**
    Branched at the crown ; lvs. roundish-obovate (38) **rotundifolium.**
Stems erect or suberect (1–4 feet high) :
  Leaves broadly ovate or obovate, 3–5-nerved :
    Inv. scales cream or pale straw-colour ... ... (39) **grandiflorum.**
    Inv. scales snow-white .. .. (40) **fruticans.**
  Leaves ovato-lanceolate or lanceolate, *acuminate* ... (41) **felinum.**

§ 7. PANICULATA. Erect or ascending suffrutices. Leaves sessile, acute or acuminate, rigid. Heads many-flowered, in simple or few-headed corymbs, mostly pedicellate. Inv. scales spreading, often radiating, mostly acute (in No. 44 obtuse), white or rosy, or *pale lemon-colour.* (Sp. 42–47.)

Leaves lanceolate-acuminate, or subulate :
  Branches and leaves *green,* glabrous or villous (not
    silvery) ... ... (42) **striatum.**
  Branches and leaves *silvery-white,* silky :
    Leaves linear-lanceolate ; inv. scales acute ... (43) **paniculatum.**

| Leaves subulate ; inv. sc. obtuse (heads small) | (44) **mucronatum.** |
| Leaves imbricated, ovate or ovato-lanceolate, *broad-based* | (45) **chlorochrysum.** |

Leaves *recurved,* obovate or spathulate, *narrow*-based :

| Villous, dirty-whitish ; inv. scales lanceolate, dull ... | (46) **sordescens.** |
| Silvery-white ; invol. scales lustrous white or rosy, lanceolate, the innermost short and obtuse   ... | (47) **argenteum.** |

§ 8. **XERANTHEMOIDEA.** Diffuse or procumbent shrubs or suffrutices. Leaves sessile, mostly recurved, rigid, acute or acuminate, crowded. Heads many-flowered, *solitary* at the ends of the branches. Invol. scales *radiating,* acute or acuminate, white or purplish or rosy. (Sp. 48–52.)

Flowering branches ending in a nearly nude *peduncle:*

| Leaves silvery ; inv. scales very long, narrow, acuminate ... ... | (48) **lancifolium.** |
| Leaves glabrous above, silky beneath ; invol. scales ovato-lanceolate ... ... | (49) **recurvatum.** |

Flowering branches closely leafy to the very summit :

| Stem dwarf, umbellately much branched   ... ... | (50) **ramulosum.** |

Stem diffuse or prostrate, or trailing :

| Leaves linear, acute, 1–1½ line wide, squarrose | (51) **stoloniferum.** |
| Lvs. oblong or ovate, 2–4 lines wide, recurved | (52) **retortum.** |

§ 9. **ELEGANTISSIMA.** Erect *herbs,* with crowded radical leaves, and simple or corymbose, leafy stems. Leaves sessile, oblong, of thin substance. Heads very many-fl., radiating, corymbose or solitary. Inv. scales acute or acuminate, lustrous white or rosy. (Sp. 53–55.)

| Heads solitary ; lvs. nude, with woolly margins ... ... | (53) **marginatum.** |

Heads corymbose :

| Lvs. copiously woolly (especially the radical)   ... | (54) **adenocarpum.** |
| Lvs. thinly cobwebby, becoming nude above .. ... | (55) **elegantissimum.** |

B. **DASYLEPIDEA.** Inv. scales dorsally *densely pubescent* or *shaggy,* not radiating, and neither conspicuously white or yellow. (Shrubs and suffrutices. Heads of small size, corymbose or glomerate.) (§ 10.—Sp. 56–61.)

| Shrub. Leaves on long petioles, cordate-ovate or subrotund ... ... | (56) **populifolium.** |

Suffrutices. Leaves *sessile,* oblong, obovate or roundish :

Heads sessile, several crowded together in tufts :

Diffuse or prostrate, slender	(57) **marifolium.**
Erect or suberect. *Pedunc. long.* Heads 4–5-together	(58) **rotundatum.**
Erect or suberect. Heads 20–30, in subsessile tufts	(59) **coriaceum.**
Heads pedicellate, in closely-branched cymes. Stem virgate	(60) **umbellatum.**
Suffrutex. Leaves sessile, *linear.* Heads corymbose ...	(61) **Lambertianum.**

C. **CHRYSOLEPIDEA.** Inv. scales yellow (usually *golden-yellow,* but in § 11 varying to *straw-colour;* and in a few other cases *horn-colour* or colourless). (§ 11–12.)

§ 11. **XEROCHLÆNA.** Herbs or suffrutices. Heads corymbose or solitary, usually many-flowered. Invol. glabrous and glossy, *radiating,* scales lanceolate, acute or acuminate, straight, golden-yellow or *straw*-colour. (Sp. 62–69.)

| Stem herbaceous. Cauline lvs. *decurrent,* scabrous. Hds. very many-fl. ... ... | (62) **Cooperi.** |

Stem herbacous, erect ; cauline leaves broad, clasping ; heads *very many*-flowered :

| Leaves glandular and setose on both sides, setoso-ciliate ... ... | (63) **setosum.** |
| Lvs. scabrous above ; more or less woolly and whitish beneath ... ... | (64) **fœtidum.** |

Lvs. scabrous and woolly above; white-woolly beneath (65) **decorum.**
Lvs. scab., erect, with woolly margins; heads solitary (66) **fulgidum.**
Stem suffruticose, erect; lvs. sessile, broad, *very woolly;*
heads corymbose, *few-flowered* ... ... (67) **lanatum.**
Stem suffruticose, diffuse; lvs. obovate-spathulate, silvery;
heads corymbose ... ... (68) **argyrophyllum.**
Stem suffruticose; lvs. linear, acute, with revolute mar-
gins; heads solitary ... ... (69) **squamosum.**

§ 12. STŒCHADINA. Herbs, suffrutices, or small, rigid shrubs. Leaves various. Heads of small size, mostly few-flowered, cymoso-corymbose or densely aggregated. Inv. not radiating, its scales imbricating, either all glabrous or the short, outer ones thinly cobwebbed, golden-yellow (in No. 75, 78, 79, *horn-colour* or pellucid). (Sp. 70-80.)

Sub-radical leaves *petiolate*, oblong-oval, penninerved;
stem simple ... ... (70) **acutatum.**
Leaves neither petioled nor penninerved:
Leaves linear, with *revolute* margins ... ... (71) **splendidum.**
Leaves *flat* (margins not revolute):
Annual, diffuse. Inv. bright yellow ... ... (72) **rutilans.**
Perennial, herbaceous or suffrut., with virgate branches:
Heads 20-50-fl., cymose or corymbose:
Lvs. one-nerved:
Lvs. spathulate, obtuse; inv. sc.
lemon-yellow ... ... (73) **adscendens.**
Lvs. lance-lin., acuminate; inv.
sc. membr., yellow-tipped ... (74) **psilolepis.**
Lvs. 3-5-nerved, lanceol.; inv. scales
horn-colour ... ... (75) **hebelepis.**
Heads 5-10-fl., congested with wool into flat-topped disc-like cymes:
Lvs. sessile; cymes subglobose ... (76) **subglomeratum.**
Lvs. tapering much at base; cymes
disc-like, rayed beneath ... ... (77) **umbraculigerum.**
Small, rigid, much-branched *shrubs*, with obovate leaves:
Inv. scales erect, obtuse, horn-colour:
Heads pedicellate; inv. 1½ line long (78) **pentzioides.**
Heads sessile; inv. 3 lines long ... (79) **lucilioides.**
Inv. scales squarrose, acute, golden-yellow (80) **excisum.**

Sub-genus II. LEPICLINE. *Recept.* fimbrilliferous, or at least covered with tooth-like prominences between the flowers. (Sp. 81-137.)

§ 13. PLANTAGINEA. *Stems* herbaceous or scarcely suffruticose, mostly simple, erect. *Leaves* 3-5- or many- nerved, either all radical, or the lowest much larger than the rest. *Heads* small or of middle size, densely corymbose or glomerate. *Inv. scales* not radiating, mostly yellow or yellowish, *rarely* brown, horn-colour, or red and white. (Sp. 81-96.)

Stem leafy *at base* or to *the middle;* pedunculoid at top:
Inv. scales *membranous*, brown, purple or horn-colour:
Leaves pilose-scabrid above:
Rad. lvs. *sessile;* infl. globose; inv. scales
obtuse ... ... (81) **latifolium.**
Rad. lvs. *petioled;* infl. diffuse; inv. sc.
acuminate ... ... (82) **griseum.**
Leaves quite smooth above ... ... (83) **pedunculare.**
Inv. scales *opaque*, white, purple, or creamy; leaves
woolly beneath ... ... (84) **undatum.**
Inv. scales yellow or yellowish, or straw-colour:
Leaves quite smooth (not *scabrous*) on the upper surface:
Leaves on both sides *silvery;* inv. scales
acute, wavy ... ... (85) **coriaceum.**

Lvs. quite *glabrous* above, woolly or nude below; inv. sc. obtuse :
Cymes very dense; inv. scales flat,
    pale yellow or reddish ... ...   (86) **allioides.**
Cymes loose; inv. sc. golden-yellow (87) **leiopodium.**
Leaves scabrous on the upper surface; nearly nude :
  Cauline leaves lanceolate or lance-linear
    (rarely obovate) ... ... (88) **nudifolium.**
  Cauline leaves broadly ovate, 5-nerved,
    strongly decurrent   ... ... (89) **quinquenerve.**
Stem bearing leaves to the summit, or very near it : ·
  Cauline leaves with *revolute* margins, rather rigid :
    Cauline leaves quite *smooth* above; inv. scales
      acute, golden... ... ... (90) **miconisefolium.**
    Cauline leaves scabrous above :
      Cauline leaves lance.-linear, acute; rad.
        petioled; inv. scales straw-colour ... (91) **Krebsianum.**
      Cauline leaves subulate, hook-pointed; inv.
        sc. tawny-horn-colour ... ... (92) **subulifolium.**
      Cauline leaves lance.-linear, 3-nerved; outer
        inv.-sc. purple, inner white-tipped ... (93) **oxyphyllum.**
  Cauline leaves flat or undulate (not *revolute*-edged) :
    Heads golden : upper cauline leaves *not* membrane-tipped :
      Root lvs. woolly on both sides; cymes
        globular   ... ... ··· ... ... (94) **cephaloideum.**
      Root lvs. glabrous above, very long; cymes
        becoming loose ... ... (95) **longifolium.**
    Heads pale-yellow or purplish; upper caul. lvs.
      tipped with a membranous scale ... ... (96) **appendiculatum.**

§ 14. DECURRENTIA. Stems herbaceous or fruticose, simple or branched, leafy throughout, or pedunculoid upwards. Leaves *strongly decurrent*, at least the upper ones. Heads in No. 97, rather large and corymbose; in the other species small and cymose. Inv. scales yellow, rufous, or whitish. Receptacle honey-combed or fimbrilliferous. (Sp. 97–104.)

Heads very many-fl., corymbose; inv. sc. golden, rigid (97) **Xerochrysum.**
Heads few-fl. (5–12, rarely 20–25-fl.), cymose, subsessile :
  Invol. scales creamy white :
    Leaves glabrous above; inflorescence panicled (98) **Mundtii.**
    Leaves white-woolly on *both* sides; infl. cymose (99) **Natalitium.**
  Inv. scales rufous; lvs. scabrid, decurrent in a
    broad wing ... ... (100) **platypterum.**
  Outer inv. scales fulvous, inner yellow; stem branched :
    Leaves nude above, with revolute margins :
      Lvs. glabrous, quite smooth above ... (101) **Gerrardi.**
      Lvs. tomentose beneath, scabrous above (102) **stenopterum.**
    Lvs. flat, on both sides woolly :
      Heads 5–6-fl.; peduncles long ... ... (103) **gymnocomum.**
      Heads 10–12-fl.; pedunc. short or long (104) **odoratissimum.**

§ 15. APTERA. Suffrutices or small shrubs. Leaves mostly *sessile*, or *half-clasping*, rarely *petiolate*, but not decurrent, or very minutely so. Heads small or mediocre, cymose. Inv. scales imbricated, not radiating, either yellow, rufous, or pure white. (Sp. 105–127.)

* Inv. scales *yellow*, at least the inner ones :
  Leaves flat or flattish (not strongly revolute-edged) :
    Shrub, much branched and ramulous; leaves
      3–5-lined above ... ... (105) **trilineatum.**
    Suffrutices, with virgate branches :
      Leaves obviously 3-nerved, half-clasping :
        Inv. scales erect; lvs. becoming nude above :
          Robust; leaves rigid, oblong or
            linear, mucronate ... ... (107) **cymosum.**

Slender; lvs. membr., lanceolate-
acuminate ... ...　(106) **tenuiculum.**
Inv. scales squarrose ; lvs. woolly on
both sides　...　... ... ...　(108) **simillimum.**
Leaves one-nerved :
Inv. scales glabrous and glossy :
Lvs. on both sides nearly equally woolly :
Heads 10–12-fl. in globose,
few-headed tufts ...　...　(109) **melanacme.**
Heads 5-flowered, in much
branched cymes :
All the flowers perfect;
pubescence thin　...　(110) **parviflorum.**
2 or 3 fl. perfect ; 2 fe-
male; pub. woolly...　(111) **callicomum.**
Heads 40–45-fl., in globose,
few-headed tufts ...　...　(112) **aureonitens.**
Lvs. thinly cobwebbed, becoming
nude above :
Lvs. linear, *smooth* above ..　(113) **subdecurrens.**
Lvs. broadly oblong, undu-
late, *scabrous* above　...　(114) **intricatum.**
Inv. scales dorsally silky...　...　...　(115) **maritimum.**
Stem simple ; lvs. cordate-clasping, very woolly,
large　...　... ... ...　(116) **pannosum.**
Lvs. linear or oblong, with *strongly revolute margins :*
Inv. scales dorsally silky...　...　...　...　(117) **plebeium.**
Inv. scales glabrous ; heads many-flowered ...　(118) **anomalum.**
Inv. scales glabrous ; heads 5-flowered :
Leaves hooked ; flowers all perfect　...　(119) **hamulosum.**
Leaves straight ; 4 fl. perfect, 1 female ...　(120) **Kraussii.**
** Inv. scales rufous or horn-colour, or brown :
Inv. sc. loosely imbricated, squarrose, inner spreading :
Shrub ; leaves linear-filiform, glabrate above...　(121) **teretifolium.**
Half-shrubs ; lvs. 3-nerved, cobwebby or woolly above :
Heads many-fl.; fimbrils short　...　...　(122) **rugulosum.**
Heads 12–15-fl.; fimbrils long　...　...　(123) **capitellatum.**
Inv. scales *erect* (not reflexed or spreading) :
Leaves *smooth* (not *scabrous*) above, glabrate or woolly :
Leaves sessile :
Inv. scales glossy, rufous, sub-opaque,
rigid. (*Shrub*). ...　...　...　...　(124) **erosum.**
Invol. scales membranous-pellucid.
(*Slender half-shrub.*)　...　...　...　(125) **Dregeanum.**
Leaves minutely petiolate, lance-oblong,
3-nerved　...　...　...　...　...　(126) **tricostatum.**
Leaves scabrous or gland-hispid above :
Lvs. copiously woolly ; inv. scales rufous,
glossy　...　...　...　...　...　...　(127) **revolutum.**
Lvs. undulate, glandular, nearly nude ;
inv. scales pellucid, pale　...　...　...　(128) **scabrum.**
*** Inv. scales opaque-white or creamy (rarely *primrose*-col.) :
Heads subglobose, 12- or many-fl. ; inv. loosely imbricate :
Lvs. petiolate, ovate, woolly ...　...　...　...　(129) **petiolatum.**
Lvs. on eared-petioles, cordate, green above,
white beneath　...　...　...　...　...　(130) **hypoleucum.**
Lvs. eared at base, panduriform :
Heads in densely much-branched cymes ;
inv. scales obtuse ...　...　...　...　...　(131) **crispum.**
Heads in loose cymes, many-fl. ; inv. scales
acute　...　...　...　...　...　...　...　(132) **auriculatum.**

Heads cylindrical, 5-fl.; inv. closely imbricate ... (133) **Zeyheri.**

§ 17. CHIONOSTEMMA. Very woolly suffrutices. Leaves sessile, oblong or linear. Heads large, many-fl., solitary or aggregated. Inv. scales lustrous, white or whitish, lanceolate, acuminate, radiating. Recept. covered with lanceolate-acuminate, scale-like fimbrils, longer than the ovary. (Sp. 134-135.)

Tall; lvs. oblong or linear; inv. lustrous-*white;* corollas
purple above ... ... (134) **vestitum.**
Dwarf; lower lvs. tapering at base; inv. whitish; corollas
yellow ... ... (135) **Ecklonis.**

§ ·18. EDMONDIA. Slender, glabrate suffrutices. Leaves subulate or filiform, glossy. Heads solitary, terminal. Inv. scales lustrous, either white, rosy, purple-brown or yellow, lanceolate, acuminate, radiating. Recept. covered with long or short fimbrils. (Sp. 136-137.)

Lvs. keeled. Inv. campanulate. Fimbrils longer than
the ovaries ... ... (136) **sesamoides.**
Lvs. round-backed. Inv. turbinate. Fimbrils very short (137) **humile.**

Sub-genus 1. **EU-HELICHRYSUM.** (Sp. 1-80.)
A. *ARGYREIA.* (§ 1-9.)
§ 1. LEPTORHIZA. (Sp. 1-8.)

1. H. leptorhizum (DC. l. c. 169); root annual, slender, simple; stems many from the crown, diffuse, simple, slender, woolly like the leaves; leaves obovate-oblong, *mucronate*, tapering at base; heads 3–5, in the upper axils and terminal, sub-sessile, about 16-fl., ovate; outer inv.-scales loosely woolly, the rest glabrous, scarious, *brown and horn-colour*, oblong, *obtuse*, straight.

HAB. Kaus Mt., Little Namaqualand, 3-4000 f., *Drege!* (Herb. Hk., Sd.)
A minute annual. Stems 1-2 inches long, laxly leafy. Leaves 4-5 lines long, 2 lines wide. Heads 2 lines long. Female-fl. about 8.

2. H. Leontonyx (DC. l. c. 169); root annual, slender, simple; stems many from the crown, diffuse, simple, slender, woolly like the leaves; leaves obovate, *very obtuse*, tapering at base, nerveless, the cauline remote, the radical and those under the fl.-heads crowded; heads 3–6, at the ends of the branches, subtended by shorter floral leaves, 14–16-fl., heterogamous; outer inv.-scales short and woolly, the rest elongate, glabrous, scarious, rigidly *acuminate*, purplish and straw-colour, hook-pointed. *Hel.* 641, *Drege! in Hb.*

HAB. Kaus Mt., Little Namaqualand, 3-4000 f., *Drege!* Also between Dweka and Zwartbulletje, *Drege!* (Herb. Hk., Sd.)
A minute annual, resembling *H. leptorhizum,* but differing in the very obtuse leaves and acuminated, hook-pointed inv.-scales. The invol. is quite that of a *Leontonyx,* but the pappus is uniseriate, with slightly clavate bristles. Fem.-fl. 5-6.

3. H. alsinoides (DC. l. c. 169); root annual, slender, simple; stems many from the crown, diffuse, forked, cobwebby on the younger parts; leaves sub-distant, oblong or spathulate, obtuse, tapering at base, faintly nerved, slightly decurrent; heads crowded-subracemose at the ends of the branches, sessile, about 25-30-fl.; inv.-scales biseriate, the outer 7-8, short, broadly oblong, somewhat cobwebby at back, the inner 10-15, linear, glabrous, straw-colour, hyaline and obtuse at the tip.

HAB. S. Africa, *Drege!* (Herb. Sond.)

Stems (fide DC.) 3-4 inches long: I have seen but a branch. Leaves 4-6 lines long, 1-2 wide. Old involucres stellately patent; the inner scales twice as long, and half as broad as the outer.

4. H. micropoides (DC. l. c. 170); root annual, slender, simple; stems many from the crown, sparingly branched, glabrate below, cobwebby above; leaves scattered, obovate-oblong or spathulate, obtuse, tapering at base, thinly woolly, the uppermost close under the crowded, sessile, 10-fl. heads; inv. scales pale straw-colour, biseriate, linear, the outer woolly, the inner glabrate, subacute.

HAB. Nieuweveld, betw. Zakrivier'spoort and Sionfnt., 3-4000f., *Drege!* (Hb. Sd.) Only known to me by an imperfect specimen in Hb. Sd.

5. H. filagineum (DC.! l. c. 170); root annual, simple; stem branched from the base; branches erecto-patent, angularly bent, cobwebby; leaves linear, obtuse, scarcely narrowed at base, on both sides woolly, whitish; heads 2-4, in sessile tufts, opposite the leaves or ending the branches, subtended by floral leaves, about 5-8 flowered; inv. scales biseriate, pale straw-colour, the outer short, acuminate, inner linear, obtuse, thinly cobwebbed at back.

HAB. Cape, *Drege!* Springbokkeel, *Zeyher!* (Herb. Sond., D.)
A slender, divaricately branched, rigid annual, 2-4 inches high. Leaves nearly ¼ inch long, about 1 line wide. Lateral tufts of heads quite sessile. Inner inv. scales twice as long as the outer.—A distinctly marked little species.

6. H. capillaceum (Less.! Syn. 275); root annual, slender; stem slender, diffusely much branched or sub-erect, thinly cobwebby or glabrous; leaves scattered, obovate or ovate, petioled, mucronate, thinly membranous, cobwebby or woolly, at length more or less glabrous above; heads pedicellate, racemose or corymbulose, or axillary, about 20-25 fl., heterogamous, inv. biseriate, cylindrical, the outer scales short, inner linear, whitish, obtuse, glabrous, at length widely spreading. *DC.! l. c.* 170. *Gnaph. capillaceum, Th.! Cap.* 660.

HAB. Mountains, &c., throughout the colony, common. Mohlamba range, Natal, *Dr. Sutherland!* (Herb. Th., Hk., D., Sd.)
Very variable in size; stems 2-12 inches long, the larger forms alternately much branched, weak and often interwoven. Leaves distinctly petiolate, at first white on both sides, then green and naked above, 3-10 lines long, 2-8 lines wide. Heads small, almost always pedicellate; the inv. scales a dull, whitish-horn colour.

7. H. herniarioides (DC. l. c. 170); root annual, slender, simple; stem very short, branched from the base; branches many, diffuse, simple, cobwebby; leaves oblong or obovate, obtuse, tapering much at base, cobwebby, afterwards somewhat naked; heads crowded in terminal glomerules, subtended by a few leaves, 25-40 fl.; inv.-scales loosely imbricate, the outermost woolly, with narrow, short points, the rest ovate-oblong, broad, radiating, snow-white, obtuse.

HAB. Betw. Roggeveld and the Gariep, *Burchell*, Cat. 1582. Zilverfontein, *Drege!* Springbokkeel, *Zeyher!* (Herb. Hk., D., Sd.)
Central stem erect, about an inch high, ending in a cluster of fl.-heads; branches from near its base, 2-4 inches long, nearly prostrate, each ending in a similar cluster. Inv.-scales conspicuously white, except the outer ones which are thinner and somewhat straw-coloured or occasionally tawny.

8. H. expansum (Less.! Syn. 276); root annual, simple ; stem herbaceous, diffusely branched, woolly; leaves oblong, spathulate or nearly linear, sessile, narrowed at base, nigro-mucronate, on both sides woolly; heads homogamous, 12–20-fl., corymboso-paniculate or crowded at the ends of short branches, pedicellate or subsessile ; inv. imbricate, the outer scales short, blunt, thinly cobwebby, pale horn-colour, the inner snow-white, opaque, broadly-oblong, very obtuse, glabrous. *DC. l. c.* 170. *Gnaph. paniculatum, Berg. Cap. p.* 256, *non Th. Gnaph. notatum, Th.! Cap.* 653. *Gn. expansum, Th.! Cap.* 651. *Gn. achillæoides, Sieb. Fl. Cap.* 201. *Heb. Zey.!* 2893.

HAB. Mountain sides, throughout the Colony. (Herb. Th., Hk., D., Sd.)
The largest and strongest of the annual species. Stems 8–18 inches long, sometimes erect and subsimple, sometimes breaking at base into very many diffuse branches. Leaves very variable in comparative length and breadth, the lowest largest. Lower leaves often 1½–2 inches long, 4–6 lines wide ; upper ½–1 inch long, broad or narrow. Inner inv.-scales remarkably white.

§ 2. ERICÆFOLIA. (Sp. 9–16.)

9. H. spiciforme (DC. l. c. 171); " stem suffruticose, erect, branched at top ; branches bearing lateral, subsessile fl.-heads arranged as in a spike ; leaves sessile, linear, obtuse, with revolute margins, densely woolly ; outer inv.-scales leaf-like, inner lengthened into a scarious, lanceolate, acuminate limb." *DC. l. c.*

HAB. Beyond the Gariep, *Burchell*, Cat. 2247.
" A suffrutex, 1–1½ f. high. Leaves 4 lines long, 1 line wide. Heads about 30 in the spike."

10. H. cerastioides (DC. l. c. 171); " stem suffruticose, branches tomentose, bearing at the summit a solitary, sessile capitulum surrounded by leaves ; leaves sessile, lanceolate-linear, with sub-revolute margins, densely woolly; outer inv.-scales leaf-like, inner lengthened into a scarious, lanceolate, acuminate limb." *DC. l. c.*

HAB. Beyond the Gariep, near Litaku, *Burchell*, Cat. No. 2225.
" Leaves nearly as in *Cerast. tomentosum*, but shorter and alternate. Scales of the invol. sometimes sub-rufescent." *DC.*

11. H. paronychioides (DC. l. c. 171); " stem shrubby, procumbent, much-branched, twigs leafy to the very summit; leaves sessile, linear, with sub-revolute margins, sub-acute, glabrous (as are the branches) ; heads solitary, terminal on the branches and twigs, 12-fl.; inv.-scales nearly all scarious, lengthened into a lanceolate, subacuminate limb." *DC. l. c.*

HAB. Regions beyond the Gariep, *Burchell*, Cat. No. 2234.
" Remarkable for its glabrous foliage, but with fl.-heads very similar to the preceding. Leaves 2 lines long." *DC.*

12. H. oxybelium (DC. l. c. 171); shrubby, rigid, divaricately much-branched ; branches *glandularly scabrid*, becoming glabrous, twigs sub-spinescent; leaves linear or lance-linear, sessile, acute, one-nerved, with sub-revolute margins, glandularly-scaberulous, the very young ones thinly cobwebby, the uppermost subtending the fl.-heads; fl.-heads terminal, solitary, 15–20fl.; outer inv.-scales short, leaf-like, inner oblong, acuminate, whitish or purplish.

HAB. Kamiesberg, *E. & Z.!* Konstapel, in the Carroo, *Drege!* (Hb. Sd., Hk., D.) An intricately-branched, slender shrublet, a foot high, the twigs horizontally spreading. Branches and twigs dark-brown, covered with short-stalked glands, which rub off, leaving them bare. Lvs. 4–6 lines long, not 1 line wide, mostly naked.

13. H. ericæfolium (Less. Syn. 314); stem suffruticose or shrubby, loosely or divaricately much branched, naked or woolly; leaves linear, sessile, spreading, crowded or scattered, obtuse, with revolute margins, woolly or glabrous; heads sessile, aggregated, 3–6 together at the ends of the twigs (rarely solitary), surrounded by leaves, 8–12 fl.; inv. scales in few rows, oblong, subacute, obtuse or denticulate, pale-horn colour, pure white or reddish. *DC. l. c.* 172. *Gnaph. ericoides, Linn.—Th.! Cap.* 647, *and Stoebe aspera, Th. Cap.* 728.

VAR. a. **vulgare**; shrubby, much branched; adult leaves glabrous; inv. scales horn-colour, subacute or toothed. *H. ericæfol., DC. l. c.—Zey.!* 2903, *a. and* δ.

VAR. β. **albidulum** (DC.); leaves and branches white-woolly; inv. scales often purple. *DC. l. c. Zey.!* 2903 β. *and* 2904.

VAR. γ. **metalasioides**; shrubby, adult leaves glabrous, crowded; inv. scales snow-white, subacute. *H. metalasioides, DC. l. c.* 171. *Zey.!* 2905. *H. callunoides? Sch. B.*

VAR δ. **laxum**; half-shrubby, straggling, loosely and divaricately branched; leaves sub-distant, woolly or glabrate; heads 2–3 together or solitary; inv. scales horn-colour, subacute. *H. laxum, E. Mey. DC. l. c. p.* 171.

VAR. ε. **lineare**; habit and foliage of var. a, but the inv. scales more obtuse. *H. lineare, DC. l. c. p.* 172, *excl. var.* β. *H. comosum, Sch. B. in Hb. Krauss! No.* 421.

HAB. Mountain places throughout the colony and in Caffraria, common. Port Natal, *Krauss!* 421. (Herb. Th., Hk., D., Sd.)

Certainly a very variable plant in ramification and indument, yet I find it difficult to draw clear lines between the above forms. Of *H. metalasioides,* DC. I have only seen Ecklon's specimens, which seem to differ chiefly in the brighter white of their inv. leaves. *H. laxum,* E. M.! seems to be merely a weak growing plant, and *H. lineare* (excl. var. β.), only differs from *H. ericæfolium* (*verum*) by its blunter invol. scales, surely a very inconstant character. *H. callunoides,* Sch. B. (Comp. Kraus. p. 15) by description, seems referable to one or other of our vars.; perhaps to var. γ.

14. H. cæspititium (Sond.); dwarf, shrubby, prostrate, excessively branched and ramulous; twigs very short, erect, densely leafy; leaves linear, spreading, obtuse, with revolute margins, glabrous (the younger cobwebby); heads 3–6 together (or solitary) at the ends of the twigs, sessile, surrounded by leaves, homogamous, 12–15-fl.; inv. scales loosely imbricated, squarrose, snow-white or rosy, broadly oblong, obtuse or subacute. *H. lineare β. cæspititium, DC. l. c.* 172.

HAB. Tambukiland, *Drege! Ecklon!* Aapjes R., *Burke!* Queenstown District, on iron-stone hillocks, in the flats, *Mrs. F. W. Barber, No.* 314. Natal, *W. T. Gerrard!* (Herb. D., Hk., Sd.)

Except by its very peculiar, prostrate habit and crowded, minute leafy twigs, this scarcely differs from *H. ericæfolium,* of which it may possibly be merely an extreme variety. Specimens from different habitats have all the same characters.

15. H. simulans (Harv. and Sond.); shrubby or suffruticose, divaricately much branched; twigs canous; leaves sessile, linear, *flat,* obtuse, on both sides silvery with close-lying, short indument; heads sessile, solitary at the end of short twigs, subtended by leaves, 12–15-fl.; inv. scales imbricated in few rows, the outer ones short, obtuse, inner acute or acuminate, all rufous-horn colour, scarious, semi-pellucid.

HAB. Cape, *Ecklon!* (Herb. Sond.)
Very similar in aspect to *H. ericæfolium* vars. *laxum* and *albidulum*, but differing in its *flat* (not revolute-edged), slightly adnate leaves and solitary fl.-heads. The single specimen which I have seen was separated from *H. ericæfol.* by *Dr. Sonder.*

16. H. ericoides (Pers. ench. 2, p. 415); shrubby, depressed, much-branched and ramulous; leaves very minute, imbricating, appressed, coriaceous, ovate, obtuse, glabrous without, concave and tomentose within; heads solitary at the ends of the branches, 10–12-fl.; inv.-scales at the base fulvous, otherwise white, glabrous, lanceolate, acuminate, spreading; achenes densely silky. *DC. l. c. 172. Less. Syn. 324. Argyrocoma ericoides, Lam. Ill. t. 693, f. 2. Stoebe nivea, Th. Cap. 729.*

HAB. Near Verkeerde Valley, *Thunberg!* Among rocks on the summit of the Sneeuwekop, *Dr. Wallich!* (Herb. Th., D., Hk., Sd.)
Root thick and woody. Stems many from the crown, ligneous, spreading or prostrate, very much and closely branched, and twiggy, every twig closely imbricated to the summit. Leaves precisely as in *Elytropappus Rhinocerotis*, ⅓ line long. Heads 3 lines long. A very distinctly marked, neat little plant : its silky achenes very unusual in *Helichrysum.*

§ 3. DECLINATA. (Sp. 17–21.)

17. H. declinatum (Less.! Syn. 278); perennial, herbaceous; stems many, diffuse or ascending, branched toward the extremities, woolly; leaves sessile, linear or linear-spathulate, acute, crowded, sub-erect, nerved, on both sides albo-tomentose; heads corymbulose at the ends of the branches, shortly pedicelled or sub-sessile; invol. glabrous, radiating, the outer scales and bases of the inner testaceous, tips of the inner scales oblong, sub-acute or denticulate, snow-white. *DC. l. c. p. 172. Gnaph. declinatum, Linn. f. Gn. pygmæum, Th! Cap. p. 652.*

HAB. East and West Districts. Drakeensteenberg and Albany, *Drege!* Caledon, Swellendam, and Uitenhage, *E. & Z.! Pappe! Mundt!* (Herb. Th., Hk., D., Sd.)
Resembling *H. expansum;* but the root is perennial, the crown somewhat woody, and the invol.-scales narrower, longer, and inclining to acute. Stems 3–10 inches long, mostly prostrate. Leaves ½ inch long or more, 1–2 lines wide. In *Drege's* specimens the heads are less obviously corymbose than usual, and the outer inv.-scales paler and less glabrous.

18. H. serpyllifolium (Less.! Syn. 277); suffruticose, loosely much-branched, the branches flexuous, spreading, tomentose; leaves oval or obovate, sessile, narrowed at base, nerved, flat or wavy, albo-tomentose beneath, green and either glabrous or cobwebby above; heads densely corymbose or fascicled at the ends of the branches, cylindrical, homogamous, about 12-fl.; outer inv.-scales woolly, inner narrow oblong, glabrous, spreading, milk-white, obtuse, *DC. l. c. 172. Gnaph. orbiculare, Th. Cap. 659, and G. polifolium, Th.! Cap. 656.*

HAB. Cape Flats, and throughout the Colony, to Natal. (Herb. Th., Hk., D., Sd.)
Stems 2–3 f. long, branching chiefly near the extremity: branches simple or again divided. Leaves 2–5 lines long, 2–3 lines wide, varying from nearly circular to obovate-oblong, snow-white beneath, more or less green above. Corymbs 1 inch across, many-headed.

19. H. Sutherlandi (Harv.); dwarf, tufted, the barren twigs short, densely leafy, rosulate at the extremity, fl.-branches sparingly leafy, erect, simple, short ; branches and leaves white-woolly ; leaves obovate or oval, narrowed to the base, flat, obtuse, (not obviously) 3-nerved ; heads in a loosely branched, terminal, corymbose-cyme, subpedicellate, 20–25-fl.; inv. campanulate, radiating, outer scales woolly, inner elongate, linear, spreading, milk-white, obtuse or subacute; recept. nude, areolate.

HAB. Rovelo Hills, Natal, 7000 f., *Dr. Sutherland!* Basutu Land, *T. Cooper!* 709. (Herb. Hook., D.)

Stems woody, robust, but very short, multifid, throwing up many 1–2 inches long erect branchlets, which form a leafy *tuft* or cushion. Flowering branches 2–3 inches long, rising from the tuft. Leaves 4–6 lines long, 2–3 lines wide. The habit is quite peculiar, or most like that of *H. Ernestianum* (which greatly differs in foliage) ; the fl. heads are not unlike those of *H. serpyllifolium*, but rather larger, *Mr. Cooper's* specimens, probably from a lower elevation, are much more luxuriant than *Dr. Sutherland's*, with fl. branches 8–10 inches long, and leaves ½–¾ inch long, 3–4 lines wide, with the 3 nerves sometimes a little obvious through the dense wool. Except in luxuriance they do not differ.

20. H. Ernestianum (DC. l. c. 173) ; dwarf, tufted ; stems sparingly leafy, simple, scarcely twice as long as the radical leaves ; leaves lanceolate-linear, thick, subobtuse, clothed with very close-lying tomentum, nerveless, ˙erect, the radical longer and broader, cauline few, acute ; heads in a terminal, crowded corymb, pedicellate; "outer inv.-scales woolly, inner linear-oblong, white, subobtuse."

HAB. Camdebo-berg, *Drege!* (Herb. Sond.)

Rad. leaves 1–1½ in. long, 2–3 lines wide, narrowed to the base, obtuse or subacute. Stem 2–3 inches high, scape-like, with a few small, closely pressed leaves, and a terminal corymb. Flowers not seen by me. Is not this related to *H. chionosphærum?*

21. H. sessile (DC. l. c. 173); shrubby, dwarf, robust, branching, densely tufted ; branches crowded, clothed with old, persistent, withered leaves, the apices stellately leafy ; leaves broadly linear, thick, subacute, clothed with close-lying tomentum ; heads sessile at the ends of the leafy branches, solitary ; inv.-scales scarious, white, ovato-lanceolate.

HAB. Sneeuweberg, *Drege! Zeyher!* (Herb. Sond.)

A very dwarf, but stout, intricately-branched bush ; old branches denuded; the younger flexuous, imbricated with reflexed, withered leaves. Leaves at and near the apices crowded, or rosulate, ¾–1 inch long, 1–1½ line wide. Fl. not seen by me.

§ 4. IMBRICATA. (Sp. 22–28.)

22. H. fastigiatum (Harv.); stem suffruticose at base, depressed, paniculately much branched, fastigiate, and with the leaves white-woolly; leaves half-clasping, shortly obovate-oblong or narrow-oblong, subacute, nigro-mucronate ; heads 2–4 at the ends of the branchlets, corymbulose, pedicellate, 10–12-fl.; invol. cylindrical, not radiating, imbricate, inv.-scales oblong, very obtuse, horn-colour or rufescent, squarrose, scarious, subpellucid, with bullated tips, the innermost similar.

HAB. Near Riv. Zonder Ende and Buffeljagd River, *Zey.!* No. 2859. Zwarteberg, *Dr. Pappe!* (Herb. Hk., Sd.)

Excessively branched, subdichotomous, forming depressed, roundish tufts, 3–4 inches high and 6–8 in diameter, with circumscribed, floriferous apices. Leaves 3–5 lines long, 1–2 lines wide. Heads 1½ line long, the scales either brownish horn-colour or some of them rosy. Allied to *H. cochleariforme*, but with very much smaller fl. heads and a different habit.

23. H. cochleariforme (DC.! l. c. 185); herbaceous, branched from the base, diffuse, the branches paniculate near the summit, and with the leaves albo-tomentose; leaves half-clasping, linear-oblong or sub-spathulate, nigro-mucronate; heads loosely panicled, pedicellate, 40–50-fl.; inv. campanulate, quite glabrous, loosely imbricate, its scales oblong, obtuse, concave-bullated (spoon-like) at the apex, concolourous, all rufous horn-colour, subpellucid.

HAB. Cape, *Burchell,* Cat. 6491. Kamiesberg, *Drege!* (Herb. Hk., D., Sd.)
Root branching, somewhat woody, perennial? (suffruticose, *DC.*) Stems 6–8 inches high. Leaves 4–6 lines long, 1–2 l. wide. Heads terminating the spreading ramuli of the panicled branches, 2½ lines long. Recept. areolate.

24. H. pulchellum (E. Mey.!); "stem at the base only fruticose and much branched, branches subsimple, cobwebbed, and white; leaves half-clasping, oblong, obtuse, loosely woolly, whitish; heads 2–3 together at the ends of the branchlets, subtended by a few ovate, leafy bracts; inv.-scales scarious, glabrous, reddish brown, ovate, very obtuse, erect, concave but not squarrose." *DC. l. c. p.* 190.

HAB. Betw. Hexrivier and Bokkeveld, on table-land and stony hills, 3–4000 f., *Drege!* (Herb. Sond.)
Only known to me by a small specimen, with immature fl. heads, in Hb. Sond.

25. H. rubellum (Less. Syn. 280); stem suffruticose at base, paniculately much-branched, and as well as the leaves white-woolly; leaves half-clasping, obovate-oblong or spathulate, sub-acute; heads 2–3-co-rymbulose at the ends of the branchlets, shortly pedicellate or sub-sessile, 25–30-fl., heterogamous; invol. shortly radiating, campanulate, loosely imbricate, the scales squarrose, acute, the outer shorter, rufous or horn-colour, glabrous or cobwebbed at margin, the medial flesh-coloured or purplish, acute or acuminate, the innermost paler, with an ovate or lanceolate, spreading limb.

VAR. *a.* **Thunbergii**; inner inv.-scales with an ovate-acute spreading limb. *H. rubellum, Less!* *DC. l. c. p.* 191.

VAR. *β.* **incarnatum**; inner inv.-scales ovato-lanceolate, or lanceolate, acute or acuminate. *H. incarnatum, DC.! l. c.* 191.

HAB. Hexrivier Mountains, *Thunberg.* β. Breederiver, *Drege!* Blauwberg, *Zey.!* 892. Cape District, *Miller! W. H. H.!* (Herb. D., Hk., Sd.)
With the aspect of *H. cylindricum,* but differing in involucre, &c. The inner inv. scales vary much, in different specimens, in their comparative length and breadth, and except in extreme forms the two varieties above named are scarcely different.

26. H. cylindricum (Less. Syn. 281); stem suffruticose at base, paniculately much-branched, woolly; leaves half-clasping, oblong, sub-spathulate or linear, obtuse or sub-acute, white-woolly; heads 2–3 together at the ends of the branchlets, shortly pedicellate or sub-sessile, 15–20-flowered; invol. cylindrical-turbinate, imbricate, the outer scales short, obtuse, rufous or horn-colour, cobwebbed at margin, squarrose, the inner longer with an ovate, sub-acute, opaque mostly cream-colour limb; recept. punctate. *DC. l. c.* 190. *Also H. imbricatum, DC. l. c.* 191, *excl. Syn. Less. Gn. cylindricum, Linn. Sp.* 1194. *Gn. paniculatum, Th.! Fl. Cap.* 647, *non Berg. Burm. Afr. t.* 80, *f. 3 ?*

HAB. Western Districts. Worcester, Cl. William, and Swellendam, *Eck.! Zey.!* 891. Waterfall, Tulbagh, *Dr. Pappe!* (Herb. Th., Sd., Hk., D.)

Very variable in foliage and in the amount of pubescence. Smaller specimens are herbaceous, 6–8 inches high, larger distinctly ligneous 1–1½ foot high. Leaves 1–1½ inch long, 2–5 lines wide. Inner inv. scales at first erect, then spreading, always paler in colour than the outer, mostly creamy-white, more rarely pale horn-colour. All the specimens I have seen of *H. imbricatum* DC. (vars. β, γ, and ε) belong to this species; his sp. char. partly applies to this, and partly to the true *H. imbricatum*, Less.

27. H. imbricatum (Less. Syn. 279, non DC.); stem suffruticose at base, paniculately much-branched, as well as the leaves white-woolly; leaves half-clasping, oblongo-obovate or spathulate, sub-acute; heads solitary or 2–3 together at the ends of the branchlets, pedicellate, 50–60-fl., heterogamous; invol. shortly radiating, globose, imbricate, the outer scales rufous or horn-colour, cobwebbed at margin, glossy, concave-bullated, rounded and very obtuse, the innermost with a spoon-shaped, spreading, snow-white, obtuse limb. *Gn. imbricatum, Linn. Sp. 1195, fide Less.* Also *G. discolorum? Linn. Burm. Afr. t. 80, f. 2.*

HAB. Rietvalley, *Mundt & Maire.* Cape Flats, *Dr. Pappe!* Kuilsriver, *Zey.!* 870. (Herb. D., Hk., Sd.)

Root and base of stems distinctly ligneous. Stems about a foot high, paniculately branched from the base, the branches spreading. Leaves 1–1½ inch long, 2–3 lines wide above, tapering towards the broad and clasping base. Heads on longish, laxly leafy pedicels, 3 lines long and nearly as wide. Female-fl. 15–20; herm. fl. 35–40.

28. H. stellatum (Less.! Syn. 279); suffruticose, erect, branched from the base, woolly; leaves half-clasping, narrowed above the base, oblong-linear, acute, on both sides loosely woolly; heads roundish-ovate homogamous, many-fl., corymbose; invol. scales imbricated in many rows, glabrous, glossy, appressed, ovate, acute, rosy or fulvous at base, white-tipped, the innermost clawed, with an oval limb. *DC. l. c. 174, Gnaph. stellatum, Linn. Sp. 1198. Gn. discolorum and G. helianthemifolium, Th.! Cap. 650. Gn. pyramidale, Berg., Gn. fragrans, Sieb. No. 250. Zey.! 2858 and 869.*

VAR. β. **globiferum**; involucres globose, the scales loosely imbricated, concave, obtuse or the innermost only subacute; leaves densely woolly, obovato-spathulate.

HAB. Western Districts, frequent. β. Brederivier, Port Beaufort, *Mundt.!* (Herb. Th., D., Hk., Sd.)

A robust, woolly, corymbosely branched plant, woody at base, about 1 foot high. Leaves 1 inch long, 2–3 lines wide, equally distributed, callous-pointed. Heads 3 lines diameter, the inner scales white, the outer either rosy or fulvous, pedicellate.

§ 5. OBVALLATA. (Sp. 29–35.)

29. H. chionosphærum (DC.! l. c. 174); suffruticose, depressed, tufted, much branched; branches short, ascending, woolly-canescent; leaves sessile, linear-oblong or spathulate, obtuse or subacute, on both sides woolly-canescent, 3–5-nerved, the nerves parallel, raised on the under-side; heads terminal, 1–4, pedicellate, corymbulose, globose, very many flowered; inv. scales loosely imbricated, glabrous and glossy, snow-white, broadly oblong, obtuse or subacute.

HAB. Stormberg, and Witberg, *Drege!* Caffraria, *Ecklon!* Wolf Kop, *Burke!* Zwartkey and Klipplaat Riv., *Zey.!* 873. (Herb. Hk., D., Sd.)

Root woody. Primary stems robust, woody, prostrate, throwing up herbaceous, short, laxly leafy fl. branches. Leaves ¾-1 inch long, 2-3 lines wide, the nerves at first hidden by the thick wool. Heads occasionally solitary, mostly 3-4 in a corymb, 4-5 lines diam., the inv. satiny-white.

30. H. argyrosphærum (DC.l.c. 174); suffruticose, slender, branched from the base, diffuse or decumbent, branches cobwebby; leaves sessile, linear-oblong or spathulate, narrowed at base, submucronate, one-nerved, cobwebby-villous; heads globose, many-fl., solitary at the ends of the branches and branchlets, sessile, surrounded by leaves; inv. quite glabrous, imbricated in many rows, the scales silvery-white, ovate-oblong, subacute or obtuse, erect or spreading, scarcely longer than the flowers.

HAB. Beyond the Gariep and near Litaku, *Burchell*, 1645 and 2285. Near the Gariep, *Drege!* Natal country, *Miss Owen!* Aapjes River, *Burke and Zey.!—Zey.!* 872. (Herb. D., Hk., Sd.)

Root woody, many stemmed; stems or primary branches 6-10 inches long, simple or branched. Leaves ¾-1 in. long, 2-3 lines wide. Heads 4 lines diameter, very many flowered, much larger than in *H. gariepinum.* A specimen from *Drege*, in Hb. T.C.D., marked "*H. gariepinum, a.*" belongs to this species; but specimens similarly marked, in Hb. Hook and Sd., are true *H. gariepinum.*

31. H. obvallatum (DC. l. c. 174); suffruticose, depressed, much branched and ramulous, tufted; branches and leaves densely and softly woolly-canescent; leaves linear-oblong or oblong, tapering at base, obtuse or subacute, crowded, scarcely nerved, the uppermost surrounding (like an involucre) the 1-3 sessile, terminal, many-flowered heads; inv.-scales glabrous and glossy, white or rosy-tipped, oblongo-lanceolate, acute, erect.

HAB. Sneeuweberg and Klipplaat river, *Drege!* Cradock, *Burke & Zey.!* 871. Albany District, *Mrs. F. W. Barber!* (Herb. D., Hk., Sd.)

Root woody. Stems rather strong, 3-5 inches long, spreading or prostrate, with erect, very short, leafy fl. branches. Leaves 5-7 lines long, 1-2 lines wide. Toment. copious and white. Heads 3-4 lines diameter.

32. H. pachyrhizum (Harv.); root-stock thick and woody, deeply descending; stems many from the crown, spreading or prostrate, shrubby, branching and ramulous, tomentose; twigs erect, leafy to the summit; leaves sessile, spathulate or lance.-linear, narrowed to the base, acute or obtuse, white-woolly; heads one, or several-capitate at the ends of the branches, surrounded by several woolly leaves, 20-25-fl.; inv.-scales pluri-seriate, oblongo-lanceolate, acuminate, white or whitish, glabrous, glossy, scarious, straight or recurved.

VAR. β. **Thunbergii**; heads mostly solitary; branches erect; lvs. acute or acuminate. *Gn. pusillum, Th.!* *Cap.* 651. *Leontonyx pusillus, Less. Syn.* 327, *DC. l. c.* 168. HAB. Aapjes R., *Zey.!* 895 *and Burke.* β. Cape, *Thunberg!* Springbokkeel, *Zey.!* 897. (Herb. Th., D., Hk., Sd.)

Stock ½ inch diam., several inches long. Stems subhorizontally spreading, throwing up many tufted branches 1-2 inches long. Leaves ½-¾ inch long, very woolly. Heads solitary or several in a tuft. I venture to remove this plant from *Leontonyx*, and also to reject the very inapplicable name "*pusillus.*" It is nearly related to *H. leptolepis*, but differs in habit.

33. H. leptolepis (DC.? l. c. 170); stem suffruticose (at first herbaceous), branched from the base, the branches decumbent, simple or much divided, loosely woolly, becoming glabrate ; leaves sessile, oblong or lance.-oblong, obtuse or acute, loosely woolly ; heads several, densely aggregated (or *capitate*) at the ends of the branches, sessile, surrounded by several woolly leaves, 20–25-fl., homogamous ; inv. scales pluriseriate, lanceolate-acuminate, squarrose and spreading, glabrous and glossy, scarious, silvery-white.

HAB. Delagoa Bay, *Forbes! Capt. Speke!* Near Port Natal, *Mr. Hewitson!* (Hb. Hook., D.)

When it first begins to flower (the state described by *De Candolle?*) this has a slender, simple root, herbaceous, simple branches, 3–4 inches long, each ending in a globose cluster of fl. heads ; afterwards the branches become distinctly woody, 12–15 inches long, simple to beyond their middle, then alternately much branched, divergent, each branchlet ending in a cluster of heads. Leaves ⅔–1 inch long (the lower ones 1½ inch), 2–4 lines wide, variably woolly. Heads 12 or more in a tuft, distinctly radiating. If this be not the same as *De Candolle's* plant (a native of Madagascar) it is at least very closely related to it. *Capt. Speke* found it to be "common" at Delagoa Bay. The *habit* is that of *H. Gariepinum*, but the heads are fewer-flowered, and the leaves quite sessile. *Mr. Hewitson's* specimens are very woolly, with broad, subacuminate leaves.

34. H. gariepinum (DC. l. c. 174); half-herbaceous, branched from the base, diffuse or decumbent, flexuous, slender, cobwebby ; leaves oval or oblong, tapering much at base, subpetiolate, obtuse or acute, thinnish, nerved, loosely woolly, the older becoming nude ; heads ovato-globose, many-fl., two or more at the ends of the branches, subsessile, surrounded by leaves ; outer inv. scales few, woolly, narrow, inner imbricated in many rows, glabrous and glossy, oblong or lanceolate, acute, white or rosy.

HAB. Little Namaqualand, near the Gariep, *Drege!, A. Wyley!* (Herb. D., Hk. Sd.

Root ligneous, perennial. Stems many from the crown, 6–12 inches long, cobwebby at length glabrous, alternately branched. Leaves nearly uncial, 3–4 lines wide, the narrower ones spathulate. Heads 2 lines diameter ; 40–50 fl., inv. scales sometimes rosy-purple.

35. H. Catipes (Harv.); stem suffruticose, many branched from the base, decumbent, tufted; branches ascending, simple, slender, woolly, the flowering ones naked toward the summit and laxly leafy; lower leaves obovate-oblong, upper oblong, nigro-mucronate, all loosely woolly on both sides, more copiously beneath; heads 6–9, sessile, in a subglobose tuft, subtended by 4–5, nigro-cuspidate floral-leaves, 12–15 fl., outer inv. scales woolly, inner with an obovate, glabrous, very obtuse, radiating apex; pappus barbellate. *Eriosphæra Catipes, DC.! l. c.* 167.

HAB. On the Zeederberg, *Drege!* (Herb. Sond.)

Barren stems about 2, fertile, 4–5 inches long. Leaves 4–5 lines long, 2 lines wide. The habit, as *De Candolle* observes, is not unlike that of *Antennaria dioica.* Heads 2–2½ lines long; the inner inv. scales distinctly radiating and broad tipped.

§ 6. SPHÆROCEPHALA. (Sp. 36–40.)

36. H. diffusum (DC. l. c. 175) ; stems scarcely suffruticose at base, slender, elongate, sparingly branched, diffuse or ascending, white-woolly; leaves sessile, broadly oval or oblong, broad-based, mucronate, one-nerved, on both sides woolly and white; heads several, in a dense, sub-pedunculate, terminal corymb or fascicle, shortly pedicellate, many fl.; inv. scales glabrous, oblong-ovate, *acute*, imbricate, cream-white; recept. naked. *Gnaph. diffusum, EM. in Hb. Drege.*

HAB. Dutoitskloof, *Drege!* (Herb. D., Hk., Sd.)
Stems several from the crown. 10–18 inches long, simple or alternately branched, weak and straggling, leafy throughout, the uppermost leaves smaller and more distant. Leaves 1–1½ inch long, ½–¾ inch wide. Corymb globose, almost capitate, of 6–12 heads. This resembles *H. auriculatum*, but the leaves are not auricled, and the recept. is quite nude.

37. H. crassifolium (Less. Syn. 282); stem suffruticose, diffuse or procumbent, alternately branched, branches spreading, white-woolly; leaves sessile, obovate-oblong or spathulate, obtuse, not obviously nerved, thickly clothed with close, white wool; heads many, in dense, branching, corymbose-cymes, subsessile, 20–25-fl.; inv.-scales obovate, very obtuse, imbricated, undulate, cream-white, not radiating. *DC. l. c. 175. Gn. crassifolium, Linn. Mant. 112, Th. Cap. 658. Gn. multiflorum, Th.! Hb. (non Fi.) Hel. leucophyllum, DC.! l. c. 175.*

HAB. Cape, *Thunb.!* Cape Flats and near Worcester, *Ecklon!* Sands near Muysenberg and Simon's Bay, *W.H.H., Dr. Wallich! C. Wright!* 327, 333. (Hb. Th., D., Hk., Sd.)
Stems 1–2 feet long, spreading on the ground, the ends of the branches ascending; branches issuing at right angles, flexuous. Whole plant very white and softly woolly. Leaves ¾–1 inch long, 3–4 lines wide, thick. Inflorescence sometimes very compound, each short branch bearing a dense cyme. I cannot find any specimen in Hb. Th. marked "*crassifolium*," but one marked "*multiflorum*" quite agrees with our plant, and is identical with *H. leucophyllum, DC.*

38. H. rotundifolium (Less.! Syn. 277); suffruticose, branched from the crown; branches decumbent-ascending, simple, leafy, densely woolly; leaves half-clasping, roundish-obovate, very obtuse, indistinctly 3-nerved, densely clothed with whitish wool; heads corymboso-capitate at the ends of the branches, surrounded by leaves, subsessile, 20–25-fl., homogamous; inv.-scales glabrous, white, obovate, obtuse, loosely imbricated. *DC. l. c. 176. Gnaph. rotundifolium, Th.! Cap. p. 660.*

HAB. Cape Flats, *Bergius.* Draakenstenberg, *Drege!* (Hb. Th., D., Hk., Sd.)
Stems 2–4 inches long, leafy throughout. Leaves ¾–1 inch long, ½ inch wide or more, the nerves only visible on the older and partially denuded. Glomerules an inch across. Nearly related to the preceding, but differing in habit, with much broader leaves.

39. H. grandiflorum (Less.! Syn. 289); stems suffruticose and decumbent at base, ascending, elongate, subsimple, ending in a pedunculoid fl. branch, woolly; leaves crowded near the base of the stem, sessile, the lower obovate, the upper oval, ovate or oblong, 3–5-nerved, obtuse, on both sides woolly, those on the peduncle lanceolate or linear, acute, the uppermost bract-like; heads hemispherical, many-fl., homogamous, cymoso-corymbose; inv. scales glabrous and glossy, ovate, acute or subacute, imbricated, not radiating, cream-colour. *DC.! l. c. 175. Gnaph. grandiflorum, Linn. Sp. 1191, Th.! Cap. p. 659. Burm. Afr. t. 76, f. 1. Gn. fruticans, Willd. Sieb. No. 202, non Linn.*

HAB. Table and Devil's Mt., *Ecklon! Drege! Sieber! W.H.H.*, &c. (Herb. Th.! D., Hk., Sd.)
Stems many from the base, decumbent and nude at base, then ascending and

densely leafy, and ending in a more or less pedunculoid fl. branch, bearing depauper-ated leaves or bracts.   Lower leaves 3-4 inches long, 2-2½ inches wide.   Cymes branching, about 2 inches across.   The specific name "*grandiflorum*" is peculiarly ill-chosen : it is the leaves, not the fl. heads, which are large.   Fl. heads about 3 lines diam.

40. H. fruticans (Less.! Syn. 288); stem shrubby, robust, erect, sub-simple or branching, woolly, closely leafy below, ending in a laxly leafy, pedunculoid fl.-branch ; leaves half-clasping, elliptical or oblong, nigro-mucronate, 3-5-nerved, above cobwebby becoming nude, beneath densely woolly and white; heads in a much-branched, corymbose cyme, homogamous, many-fl., pedicellate ; inv. scales glabrous and glossy, loosely imbricating, snow-white, broadly ovate or oblong, sub-acute or obtuse. *DC. l. c.* 175. *Gnaph. fruticans, Linn. Mant.* 282. *Bot. Mag. t.* 1802. *Astelma fruticans, Bot. Reg. t.* 726. *Sieb. No.* 13. *Gn. grandi-florum, Willd. Andr. Bot. Rep. t.* 489. *Gn. fruticans and Gn. petiolatum, Th.! Cap.* 659. *and* 660.

HAB. Summit of Table Mountain, *Thunb.! E. & Z.! W. H. H., &c.* (Herb. Th.! D., Hk., Sd.)

Stem robust, 3-4 feet high, denuded of leaves below, densely leafy above to within 6-12 inches of the summit, where the laxly leafy or naked flowering portion starts. Leaves 2-2½ inches long, 1-1½ inch wide, green above, spreading. Cyme 4-5 inches across ; each head 5-6 lines diam. Scales radiating.—A much handsomer plant than *H. grandiflorum*, with which it has been confounded.

41. H. felinum (Less.! Syn. 287); stem erect, shrubby ; branches elongate, leafy, loosely tomentose; leaves broad-based, sessile, spreading or deflexed, varying from ovato-lanceolate to lanceolate-linear, acuminate, rigid, rugose and either naked or woolly above, albo-tomentose beneath; heads globose, homogamous, many-fl., loosely or densely corymbose or glomerate ; inv. globose, its scales glabrous, white or pink, loosely im-bricate, obtuse, broad. *DC. l. c.* 176. *Gn. felinum and G. serratum, Th.! Cap.* 648 *and* 659. *Gn. congestum and G. discolorum, Willd.*

HAB. Swellendam and the Eastern Districts, common. (Hb. Th., D., Hk., Sd.)

A very variable, but easily recognised species. Stem 2-3 ft. high, not much branched, branches 1-2 f. long. Leaves ½-1-1½-2 inches long, ¼-½ inch wide, the upper surface when young woolly, at length naked, and then either glabrous or densely papillate-scabrous, mostly wrinkled.   Branches either leafy to the summit, or more commonly ending in laxly leafy peduncles. Heads commonly crowded in a globose tuft, but sometimes cymose or loosely panicled. *DC.* enumerates 6 varieties.

§ 7. PANICULATA. (Sp. 42-47.)

42. H. striatum (Th.! Cap. p. 179); stem shrubby, erect, branched, glabrous or villous, leafy to the summit; leaves sessile, lanceolate-subu-late, channelled, very acute, rigid, the younger villous (but not canescent), the older mostly glabrous ; heads corymbose, on villous pedicels ; inv. ovate, radiating, the scales ovato-lanceolate, acute or acuminate, the innermost narrow, short, obtuse.   *Less. Syn.* 296. *DC. l. c.* 179. *Xeranth. rigidum, Andr. Rep. t.* 387.

VAR. β. villosum (DC.); branches and leaves persistently villous.

HAB. Districts of Uitenhage, Albany, and Somerset, *Ecklon! Drege! Zey.!* &c. (Herb. Th., D., Hk., Sd.)

1-2 feet high. Leaves 1-1½ inch long, 1-2 lines wide, green, either glabrous or
thinly clothed with long, silky, loose, yellowish hairs. Inv. scales glossy, either
pure white or primrose colour. Pedicels short, densely hairy. Readily known from
all forms of *H. paniculatum* by the want of silvery hairs.

43. H. paniculatum (Th. Cap. 661); stem shrubby, erect, corymbosely
branched, the virgate branches and leaves silky-canescent; leaves sessile, linear-lanceolate, or subulate, rigid, much acuminate, mucronate,
3-nerved, erecto-patent, flattish or concave; heads corymbose or corymbo-cymose, on long pedicels; inv. hemispherical, shortly radiating,
the scales glabrous, loosely imbricated, ovato-lanceolate, acute or acuminate, either snow-white, pale-citron, or purple, innermost linear.
Less. Syn. p. 294. *DC. l. c.* 179. *Xeranth. paniculatum, Linn. Burm. t.*
67, *f.* 1.

VAR. β, **intermedium**; leaves shorter and broader; heads globose, scarcely radiating; outer scales acute, inner obtuse. *H. intermedium, Less. Syn.* 295, *DC. l. c.* 179.

VAR. γ, **angustifolium**; leaves subulate; inv. scales very acute.

HAB. Districts of Swellendam and George, common. β, Kardow and Pikeneers-
kloof, *Zey.!* 868. (Herb. Th., D., Hk., Sd.)

1-2 feet high, robust or slender. Leaves very variable in breadth, sometimes as
narrow as in *H. mucronatum;* sometimes 1-2 lines wide. Infl. either simply corymbose, or the pedicels branching, each bearing 2-3 heads. Inv. scales varying much
in shape, sometimes taper-pointed, mostly acute; the innermost linear, either acute
or obtuse. Though generally known from *H. mucronatum* by its broader leaves,
acute inv. scales, and larger fl. heads, specimens occur which seem to connect the
extreme forms, as our vars. β. and γ.

44. H. mucronatum (Less.! Syn. 295); stem shrubby, erect, corymbosely branched; branches and leaves silky-canescent; leaves sessile,
linear-subulate, rigid, mucronate, erecto-patent; pedicels several, from
the upper axils, corymbose, 1-headed; heads hemispherical, not radiating; scales glabrous, loosely imbricated, broadly ovate, short, obtuse,
or subacute. *DC. l. c.* 178. *Gn. mucronatum, Berg. Cap.* 269. *Burm.*
Aft. t. 66, *f.* 3.

VAR. α. **citreum**; invol. pale-lemon or sulph. coloured.

VAR. β. **niveum** (DC.); inv. snow-white.

HAB. Tradow, *Mundt & Maire.* Caledon, *Eckl.!* β. Groote Howhoek, *Zey.!* 2856.
Zwarteberg, *Pappe!* (Herb. Sd., D., Hk.)

About a foot high, closely leafy, much branched. Leaves ¾-1 inch long, ⅓ line
diam., tipped with a long, slender mucro. Pedicels 1-2 inches long, slender, naked.
Heads 4 lines across; fl. yellow. Very like *H. paniculatum* in miniature, but the
leaves are narrower, heads smaller, and inv. scales shorter and blunter, &c. Perhaps
a mere variety.

45. H. chlorochrysum (DC. l. c. 179); stem shrubby, erect, branched;
branches densely leafy to the summit; leaves sessile, broadly ovate or
ovato-lanceolate, imbricated, subsquarrose, acuminate, several nerved,
closely silky-canescent; heads corymbose, ovate or subglobose; invol.
scarcely radiating, the scales ovate-acute, the innermost shorter, obtuse,
pale-lemon coloured (greenish when dry).

HAB. Bontebock'sjagt, at Karsrivier, *L. H. Beil,* in Hb. Eckl. (Herb. Sond.)

1-2-f. high, robust. Leaves ½-¾ inch long, 3-5 lines wide, closely many nerved
or striate, when the copious indument has been removed. Heads on short woolly
pedicels, 4-5 lines diameter. The leaves resemble those of a *Priestleya* in shape.

46. H. sordescens (DC.! l. c. 178); stem suffruticose, flexuous, branching, the branches and both surfaces of the leaves silky-villous; leaves linear-spathulate, acute, mucronate, 3–5 nerved at base, squarrose or recurved, flattish or convolute; pedicels several from the upper axils, corymbose, one-headed; heads hemispherical, many-fl., shortly radiating; scales glabrous, lanceolate, acute, appressed, whitish or pale horn-colour; pappus scabrous.

Hab. Cape, *Bowie!* Sandhills near Port Elizabeth, *Drege!* (Herb. D., Hk. Sd.)
Stem 1 foot or more long, diffuse or ascending, corymbosely branched. Pubescence long, soft, and loose, dull or dirty whitish (not silvery). Leaves 1–1½ inch long, their lower half appressed, adhering to the stem by the long, marginal hairs, the upper half strongly recurved. Pedicels 1–2 inches long, silky, slender, corymbose. Heads 4–5 lines diam, inv. greenish-white or dull whitish.

47. H. argenteum (Th.! Cap. 662); stem shrubby, erect, branched, the fl.-branches subpeduncular, tomentoso-canescent; leaves obovate-oblong, narrowed at base, plurinerved, rigid, spreading or squarrose, recurvo-mucronate, on both sides closely-silky-silvery; heads corymbose, on longish, naked or bracteate pedicels; invol. ovate, radiating, the outer scales ovate, acute, the medial lanceolate, the innermost much shorter, narrower, obtuse. *DC. l. c.* 179. *H. affine, Less.!* 292 *(fide Sond.) DC. l. c.* 178.

Hab. Districts of Uitenhage and Albany, *Th., Drege! E. & Z.! &c.* (Herb. Th., D., Hk., Sd.)
1–2 feet high, erect or ascending, irregularly branched. Leaves ¾–1¼ inch long, 3–5 lines wide; those on the fl.-branches much depauperated and distant. Heads rarely solitary; commonly on pedicels 1–2 inches long. Inv. glossy-white or rose-coloured. *Dr. Sonder* informs me that he has ascertained, by comparison with the original specimens in Herb. Berol., that *H. affine*, Less.! is identical with this species; and that *H. argenteum*, Less.! is a syn. of *Helipterum argyropsis*, DC.

§ 8. Xeranthemoidea. (Sp. 48–52).

48. H. lancifolium (Th.! Cap. 662); stem suffruticose, diffuse or decumbent, branching, branches and leaves closely silky-silvery; sterile branches closely leafy to the summit, fl.-branches leafy at base, prolonged into a sparsely bracteated, elongated, one-headed peduncle; leaves linear lanceolate, acuminate, squarrose or recurved, channelled or complicate, cobwebby and appressed at base; heads oblong-turbinate, acute at base, radiating; scales glabrous, acuminate, the outermost ovate, brownish, the medial ovato-lanceolate, more acuminate, purple, the innermost very long, narrow-lanceolate, white. *Hel. xeranthemoides. DC.! l. c.* 178.

Hab. Cape, *Thunberg! Ecklon!* Langekloofberg in several places, *Drege!* (Herb. Th., D., Hk., Sd.)
Stems 1–2-f. long, straggling. Leaves 5–6 lines long, 2–3-l. wide, closely set. Pedunc. 4–8 inches long, bearing several, narrow linear, acuminate, closely appressed bracts or depauperated leaves. Heads like those of an *Edmondia*, the outer inv. scales effused along the pedunc. Pappus of many setæ, subconcrete at base and barbellate near the summit. Inner inv. scales ¾ inch long, 1–2 lines wide, satiny within.

49. H. recurvatum (Th.! Cap. 662); shrubby, diffuse, much branched, the branches and under surfaces of leaves whitish-tomentose with long, appressed hairs; sterile branches closely leafy to the summit, fl. branches leafy below, prolonged into a one-headed peduncle; leaves half-clasping, linear-lanceolate, acuminate, recurved, complicate, glabrous above, densely hairy beneath, ciliate and appressed at base ; heads hemispherical, obtuse at base, radiating ; scales glabrous, silvery-white or purplish, the outer and medial ovato-lanceolate, acuminate, the innermost short, obtuse. *Less. Syn.* 294, *DC. l. c.* 178. *Xeranth. recurvatum, Linn. f.*

HAB. Cape, *Thunberg!* Zondag R., *Burchell,* No. 4216. Uitenhage and George, *Ecklon! Drege! Zey.! No.* 2854. (Herb. Th., D., Hk., Sd.)

About 1 foot high, robust, corymbosely branched. Pubescence of long and soft, but rather coarse hairs. Leaves ½-¾ inch long, 2-3 lines wide, infolded, the upper surface green and glabrous. Pedunc. 2-3 inches long, hairy, with 1-2 erect bracts, the uppermost scarious and silvery. Heads broader than long, about ½ inch across. Inv. scales often purple-tinted. Pappus barbellate throughout.

50. H. ramulosum (DC. l. c. 176); "stem shrubby, branches crowded, subumbellate, leafy nearly to the summit, one-headed ; leaves half-clasping, lanceolate-linear, acute, spreading, on both sides albo-tomentose; inv. campanulate, glabrous, the scales acute, the outer shorter, red-brown, loosely imbricated, inner sublinear, pale, radiating." *DC. l. c.*

HAB. Eastern Districts, *Burchell,* No. 7530.

"A much-branched, dwarf, erect shrublet. Leaves 3 lines long, 1 line wide. Heads small." *DC.*

51. H. stoloniferum (Thunb.! Cap. 662) ; stem shrubby, depressed, tufted, and trailing, branches flexuous, closely leafy throughout, canescent ; leaves crowded, sessile, linear or linear-oblong, acute, recurved-pointed, silky-silvery, at length becoming nude above ; heads terminal, sessile, solitary, many-fl. ; inv. campanulate, radiating, scales glabrous and glossy, imbricate, ovato-lanceolate, acute, the outer shorter and reddish-brown, the inner white, often rosy-tipped. *Less. Syn.* 291. *DC. l. c.* 175. *Xeranth. stoloniferum, Linn. f.*

HAB. Cape, *Thunberg!* Witbergen, *Drege!* (Herb. Th., D., Hk., Sd.)

Root deeply descending, woolly. Stems very many from the crown, some short and tufted, others long, trailing, flexuous, throwing up erect branchlets. Leaves 4-5 lines long, 1-1½ line wide, linear, squarrose. Heads 5-6 lines diameter. Habit of *H. retortum,* but with narrower leaves and much smaller heads. Recept. narrow, naked. *Thunberg's* original specimens are more silvery than *Drege's.*

52. H. retortum (Thunb. Cap.! 662) ; stems numerous, suffruticose, prostrate, rooting at intervals, branching, the apices erect, closely leafy throughout ; leaves oblong or ovate, broad-based, sessile, squarrose or recurved, subacute, closely silky-silvery on both sides ; heads terminal, solitary, sessile, oblong, obtuse, very many-fl.; inv. glabrous, radiating, longer than the disc, scales closely imbricating, acuminate, the outer ovate, fulvous, the inner lanceolate, white or purple tinted; bristles of the pappus numerous, subconcrete at base, barbellate at apex. *Willd. Sp.* 3, 1907. *DC. l. c.* 177. *H. radicans, Thunb.! Cap.* 663. *Less. Syn.* 292.

VAR. β, **minus** (DC.) ; smaller in all parts.

HAB. Sandy sea-shores, Table and Camp's Bay, and Simon's Bay. Klynriver mouth, *Zey.* / 2853. β, Roodeberg and Egelskop, 4-5000 ft.; and mountains near Welgelegen, *Drege!* (Herb. D., Hk., Sd.)

Stems 1 or several feet long, flexuous, forked or irregularly branched ; the short fl. branches erect, leafy to the very summit. Whole plant silvery. Leaves 4-7 lines long, 2-4 lines wide. Heads ¾ inch long, 5-6 lines wide.

§ 9. ELEGANTISSIMA. (Sp. *53–55.*)

53. H. marginatum (DC. l. c. 180); rhizome ligneous; stems several, ascending-erect, simple, 1-headed, leafy, woolly; radical leaves oblong-tongue-shaped, obtuse, 3–5-nerved, the cauline linear-oblong, obtuse, the uppermost acuminate, all (save the youngest, which are cobwebby) on both sides glabrous, with a woolly margin; heads sessile or sub-pedunculate, many-fl.; inv. scales scarious, snow-white, glossy, lanceolate, acuminate, twice as long as the disc.

HAB. Witbergen, 7–8000 ft., *Drege !* (Herb. Hk., Sd., D.)
Root thick and woody. Stems 3–6 inches high, closely leafy, the leaves smaller and narrower upwards. Root leaves 1–1½ inch long, 3–5 lines wide. Pappus rough, Recept. naked, flat.

54. H. adenocarpum (DC. l. c. 180); stem herbaceous, erect, simple or corymbosely branched; branches virgate, loosely woolly and glandular; radical leaves oval or oblong, obtuse, sessile, loosely woolly, cauline sessile, oval-oblong or linear-oblong, densely glandular and often very woolly, the uppermost narrower, acuminate; corymb loosely panicled; heads campanulate, radiate, very many fl.; invol. scales shining, silvery-white or rosy, lanceolate, acuminate, much longer than the disc; achenes glandular.

HAB. Common near Natal, from the sea level *(Drege !)* to 2–3000 feet, *Plant, Gueinzius, Grant, Sutherland, Gerr. & M'K. No.* 267, &c. Orange Free State, *T. Cooper,* 1121. (Herb. D., Sd., Hk.)
Stem 1–2 feet high, more woolly and branching than *H. elegantissimum,* with shorter and broader radical leaves and more lanceolate invol. scales. Invol. mostly rosy-purple externally.

55. H. elegantissimum (DC. l. c. 179); stem herbaceous, erect, simple, terete, softly hairy below, glandular-pubescent upward; radical leaves oval-oblong, thin, narrowed to the base, cobwebby-villous, becoming nude on the upper surface; cauline sessile, eared at base and stem-clasping, lanceolate-oblong, acute or acuminate, glandularly pubescent; corymb compound, loosely branched; heads campanulate, radiate, very many-fl.; inv. scales shining, silvery-white on the outer ray, oblongo-lanceolate, acute, longer than the disc.

HAB. Witbergen, 5–6000 ft., *Drege!* Basutuland, *T. Cooper /* 721. (Herb. Hk., D., Sd.)
Stem 1–2 feet high. Radical leaves numerous, 2½–3 inches long, 1 inch wide; cauline 1½–2 inches long, ½ inch wide, green. Recept. naked, 4–5 lines diam. Fl. 2–300 in a head. Inv. ¾ inch across.

B. *DASYLEPIDEA.* (§ 10.)

§ 10. ERIOCEPHALA. (Sp. *56–61.*)

56. H. populifolium (DC.! l. c. 180); a shrub; branches tomentose-canescent; leaves on long petioles, cordate at base, ovate or ovato-subrotund, acute or obtuse, 5–7- or many-nerved at base, cobwebby-canescent above, becoming bare, densely tomentose and white beneath; panicles terminal and axillary, decompound, divaricately much-branched; heads at the ends of its divisions in clusters of 3–5, subsessile, ovate, 12–16-fl.; inv. scales in few rows, the outer dorsally tomentose, the inner whitish, glabrous, obtuse, squarrose; recept. honey-combed.

HAB. Between Omsamculo and Omsamwubo, *Drege!* Natal, *Gueinzius!* (Herb. Hk., D., Sd.)

Seemingly a large shrub. Petioles 1–2 inches long; lamina 2–3½ inches long, 1½–3 inches wide, varying in amount and persistence of indument, and in shape. Panicles 5–6 inches long, very much-branched. Heads 2 lines long, 1 line diam. Fl. yellow. Not like any other S. African species. Lvs. like those of some *Abutilon.*

57. H. marifolium (DC.! l. c. 186); stems diffuse or prostrate, slender, flexuous, irregularly branched, woolly; branches sparsely leafy, naked towards the summit; leaves alternate, obovate-oblong, cuneate at base, nigro-mucronulate, thinly woolly (becoming nude) on the upper, densely woolly on the lower surface; heads 5–8, sessile, in peduncled tufts, about 20-flowered, homogamous; recept. naked; inv. scales imbricated, the outer densely shaggy externally, the inner hairy at the apex, linear, subacute; pappus plumoso-barbellate. Also *Eriosphæra Oculus Cati, DC.! Prod. 6, p.* 166 *(non Lessing);* and *Eriosph. apiculata, DC.?*

HAB. Table Mt., *Drege!* Swellendam, *E. & Z.!* (Herb. D., Hk., Sd.)

Stems 12–14 inches long, leafy nearly to the summit. Leaves about ⅓ inch apart, nearly ½ inch long, 2–3 lines wide, green (but woolly) above, foxy or dirty white, and thickly woolly beneath, much attenuated at base. Heads 2½ lines long; in *E. & Z.'s* specimens 20-flowered! Pappus armed with long, acute, slender, lateral cellules. *DC.'s* "*Eriosphæra apiculata,*" judging by his description, scarcely, if at all, differs. Of it he says :—"Stem diffuse? tomentose at the summit; leaves obovate, nigro-mucronate, the younger on both sides tomentose, the adult cobwebby above; the flowering branches subelongate, sparingly leafy, the floral leaves (?) oblong, woolly beneath and at the margin, glabrous above; heads 3–4-sessile, densely woolly, about 20-fl."—*DC. Eriosphæra apiculata, DC. l. c.* 166. (Eastern Districts, *Burchell.*)

58. H. rotundatum (Harv.); "stem erect? tomentose; leaves obovate-subrotund, very obtuse, obsoletely tipped with a black point, tomentose beneath, velvetty-cobwebbed and at length glabrate above; peduncles elongate; floral leaves oblong, acute, on the outside and at the margin densely woolly, within glabrescent; heads 4–5, ovate, sparingly woolly." *DC. Eriosphæra rotundifolia, DC. l. c.* 166.

HAB. Eastern Districts, *Burchell.*

Unknown to me. Said to be allied to "*Eriosphæra apiculata.*" Leaves 8–9 lines long, 6–7 wide, the older ones glabrate above, and then 3-nerved at base. Flowers not known.

59. H. coriaceum (Harv.); stem robust, suffruticose, erect or suberect, densely woolly, the woolly branches leafy nearly to the summit; leaves crowded, spreading, obovate-oblong, nigro-mucronulate, coriaceous, closely tomentose above, densely woolly beneath; heads numerous (20–30) densely crowded in sessile or shortly peduncled, branching, subcapitate cymes, 20–25-fl., homogamous; inv. scales imbricated, linear, all densely shaggy externally; pappus barbellate. *Eriosphæra coriacea, DC.! l. c.* 167. *Gnaph. coriaceum, E. Mey.! in Hb. Drege.*

HAB. Mountains near Swellendam, *Drege!* (Herb. D., Hk., Sd.)

Much more robust than *H. marifolium,* with more compound inflorescence. Leaves ¾–1 inch long, 4–5 lines wide. Heads campanulate, 3–4 lines long, clothed with fulvous wool; tips of the inv. scales scarcely protruding.

60. H. umbellatum (Harv.); stem erect, suffruticose; branches virgate, closely leafy nearly to the summit; leaves spreading or reflexed, shortly oblong, nigro-mucronulate, thickly woolly on both sides ; heads numerous, (12–20–30), pedicellate, in hemispherical, closely branched, woolly cymes, about 15–18-fl., homogamous ; inv. scales imbricated, linear, shaggy externally, the innermost subglabrous at tip; pappus barbellate. *Eriosphœra umbellata, Turcz.! in Bull. Mosc.* 1851. *Hol.* 24, *p.*

HAB. Riv. Zonder Einde, *Zeyher!* 2891. (Herb. Sond.)
Stem 1–2 feet high, distantly branched ; the branches subumbellate, simple, erect. Leaves squarrose, 4 lines long, 2 lines wide. Flowering branches laxly leafy toward the summit, for 1–1½ inch below the cyme. Cymes about 1 inch across.

61. H. Lambertianum (DC.! l. c. 190) ; stems suffruticose, erect, branched from the base, branches virgate, and, as well as the leaves and involucres white-woolly ; leaves sessile, linear, subacute, one-nerved, with slightly-reflexed margins, nigro-mucronate ; heads in a terminal, simple or branched, sparsely-leafy, crowded corymb, turbinate, homogamous, 20–25-fl.; inv. imbricate, its scales linear, pale, obtuse, externally densely woolly.

HAB. Cape, Herb. *Lambert., Drege.* Clanwilliam and Tulbagh, *Ecklon!* Above the Tulbagh Waterfall, *Dr. Pappe!* (Herb. Sd., D., Hk.)
Root and bases of the stems woody. Stems tufted, 10–12 inches high. Whole plant closely covered with white wool. Leaves ¾–1 inch long. 1 line wide. Corymbs 1–2 inches across. Heads 2½–3 lines long ; the narrow-linear, creamy scales, with a green, central line on the inside, all of them densely woolly externally.

C. *CHRYSOLEPIDEA.* (§ 11–12.)
§ 11. XEROCHLÆNA. (Sp. 62–69.)

62. H. Cooperi (Harv.) ; stem herbaceous, erect, robust, panicled, the branches leafy, densely glandularly setose and hairy, one or more headed ; cauline leaves clasping and *strongly decurrent*, oblong or lanceolate, acute, rameal sessile or but slightly decurrent, clasping, oblong, acuminate, all glandularly scabrous on both sides, woolly-edged, plurinerved, green ; heads very many-fl., shortly pedicellate; inv. hemispherical, stellate, radiating, its scales imbricating, glossy, scarious, oblong, acute, golden yellow.

HAB. Near Draakensberg, Orange Free-state, *T. Cooper!* 1117. (Herb. D.)
Stem 4–5 feet high, closely leafy. Leaves 3–4 inches long or more, 1–1½ wide, the medial and lower ones decurrent as a wing to the stem for 1–2 inches below their base. Heads as in *H. fœtidum,* from which this is readily known by its strongly decurrent leaves, green on both sides. A very fine species, worthy of cultivation.

63. H. setosum (Harv.); stem herbaceous, erect, paniculately much branched, the branches long, leafy, glandularly setulose, one-headed ; leaves cordate-clasping, oblongo-lanceolate, acute or acuminate, the uppermost narrow, sessile, concolourous, glandular and setulose on both sides, setoso-ciliate ; heads terminating long, leafy branches, sessile, subtended by 2–4 lanceolate, setose leaves, very many-fl.; inv. stellate, radiating, its scales imbricating, glossy, scarious, lance-oblong, acute, golden yellow.

HAB. On the Vaal River, *Burke and Zey.! Zey.!* 875. (Herb. Hk., D., Sd.)
Probably annual or biennial. Stem 1–2 f. high, pale, much branched. Leaves green, thin, glandular and viscidulous, strongly scented, not in the least woolly.

Heads nearly as in *H. fœtidum*, to which this is allied, but from which it differs in habit, the solitary fl.-heads, and pubescence.

64. H. fœtidum (Cass.); stem tall, herbaceous, erect, tomentose; leaves cordate-clasping, oblong or ovate, acute or acuminate, spreading, scabrous above, more or less woolly beneath; heads panicled or glomerate, on leafy branchlets, very many-fl., pedicellate; inv. hemispherical-stellate, radiating, its scales imbricated, glossy, scarious, lanceolate, or ovato-lanceolate, acute, golden-yellow or cream-colour. *Less. Syn.* 284. *DC. l. c.* 187. *Gn. fœtidum, Th.! Cap.* 653. *Bot. Mag. t.* 1987. *Anaxeton fœtidum, Lam. Ill. t.* 692, *f.* 1.

HAB. Throughout the Colony, and at Natal. (Herb, Th., D., Hk., Sd., &c.)
A strong growing biennial, 2–4 f. high, simple or corymboso-paniculate. Leaves variable in shape, the upper cordate-ovate, acuminate, the lower longer, and oblong or oblongo-lanceolate, 2–3 inches long, ½–1 inch wide, strongly and disagreeably scented. Heads very showy, ½–¾ inch across, sometimes in a loose, sometimes a much contracted panicle. Scales varying from cream-white to brilliant yellow.

65. H. decorum (DC.! l. c. 188); stem herbaceous, erect, simple, white-woolly, branched at the summit, subcorymbose, the branches leafy, one-headed; leaves sessile, the radical ovate-oblong, narrowed at base, 3–5-nerved, the cauline half-clasping, oblongo-lanceolate, acute, the uppermost acuminate, all scabrous and thinly woolly on the upper side, white-woolly beneath; inv. hemispherical-radiate, the scales imbricating, glossy, scarious, lanceolate-acute, bright-yellow.

HAB. Port Natal, *Drege! T. Williamson! Dr. Grant! Dr. Sutherland! Gerr. & M'K.,* 853. (Herb. Hk., Sd., D.)
Intermediate between *H. fœtidum* and *H. fulgidum,* with more the habit of the former, but more woolly. The root is said to be perennial. Stems 2–3 feet high. Leaves 1½–2 inches long, ¾–1 inch wide. Heads in a leafy corymb.

66. H. fulgidum (Willd.); stem herbaceous, erect, subsimple, loosely woolly; leaves half-clasping, oblong, erect, acute, concolourous, on each side scabrous and glandular, with woolly margins; heads very many-fl., mostly solitary at the end of the stem, or of the leafy branches, sometimes aggregated, sessile; inv. stellate, radiating, its scales imbricating, glossy, scarious, lanceolate, acuminate, bright yellow. *Th.! Cap. p.* 664. *Less. Syn.* 285. *DC. l. c.* 187. *Xeranth. fulgidum, Linn. f. Jacq. Ic. Rar. t.* 173. *Bot. Mag. t.* 414.

VAR. β. **angustifolium** (DC.); root-leaves long and narrow, on each side loosely woolly; upper-leaves wool-margined. *Zey.!* 2840. (Herb. Hk., D., Sd.)

VAR. γ. **heterotrichum** (DC.); root-leaves elliptic-oblong, 3-nerved, narrowed at base; medial wool-margined; uppermost glandular-pubescent. (Herb. D., Hk., Sd.)

VAR. δ. **subnudatum** (DC.); lower lvs. scarcely wool-margined, sparingly gland-pubescent above, hispid beneath along the nerves; root-lvs. ovato-lanceolate. (Hb. Hk.)

VAR. ε. **monocephalum** (DC.); stems sparsely woolly, simple, one-headed; root-leaves oblong, obtuse, narrowed at base, the younger on each side tomentose, adult scabrid; lower cauline wool-edged, upper glandular-scabrid. *H. decorum, Pl. Krauss! No.* 315. *Gueinz.* 595, 317. *Gerr. & M'K.! No.* 270. (Hb. Th., D., Hk., Sd.)

VAR. ζ. **nanum** (DC.); stem very short, woolly tomentose, one-headed; lower leaves oblong, obtuse, woolly on both sides; cauline linear-acuminate, wool-margined; adult glandularly scabrid. (Herb. Sd.)

HAB. Eastern Districts and Port Natal. (Herb. Th., D., Hk., Sd.)
Very variable in size and minor characters; stem 4 inches to 2–3 feet high. Root

woody, perennial ? Root-leaves 4–8 inches long ; cauline 1½–3 inches, mostly wool-margined. Heads very showy, generally larger than in *H. fœtidum*, often 1–1½ inch across, but in vars. ε. and ζ. much smaller, ¼–⅞ in. Scales very brilliant, deep yellow.

67. H. lanatum (Harv.); root woody ; stems erect, subsimple, leafy throughout, and, as well as the leaves, thickly coated with long, white, interwoven woolly hairs ; leaves sessile, obovate, callous-mucronate, beneath the wool penninerved, the upper ones narrower and oblong or elliptical; corymb subpedunculate, much branched, many-headed, with woolly branches; heads subsessile, 12–15-fl.; invol. campanulate, 3–4-seriate, the scales loosely imbricated, outer and inner of nearly equal size, all glabrous and glossy, lanceolate-acuminate, bright yellow ; recept. naked, narrow.

HAB. Sand River, *Burke & Zeyher ! Zey.!* 878. (Herb. Hk., D., Sd.)
Root thick and woody. Stems 1 or more, 12–15 inches high, densely leafy below, more laxly upwards, ending in a subpedunculate, spreading, corymbose cyme. All parts save the involucres very thickly white-woolly. Lower leaves 1½–2 inches long, 1 inch wide, upper 1–1½ inch long, ¼–⅓ inch wide. Inflorescence 2–4 inches across. Inv. scales 3–4 lines long, ½ line wide. Recept. quite naked. Pappus rough. Achenes granulated. A very handsome and distinct species.

68. H. argyrophyllum (DC. l. c. 186); stem suffruticose at base, diffuse, branches ascending, and as well as the leaves coated with a silvery-white, densely interwoven, somewhat glossy, felted indument ; leaves obovate or obovate-spathulate, hook-pointed, much attenuated to the base (3-nerved, *DC.*), the uppermost distant and narrow; corymb pedunculate, loosely few-headed ; heads on long, 1–2 bracteated pedicels, many-fl., stellato-campanulate; invol. imbricate, radiating, the outer scales short, very thinly cobwebby, fulvous, the rest oblong or oblongo-lanceolate, glabrous, glossy, opaque, yellow, the innermost longer than the disc, acute.

HAB. Kaffirland, *Ecklon !* (Herb. Sond.)
Stems decumbent at base, the barren branches short, densely leafy throughout; the fertile 6–12 inches long, nearly naked toward the summit, where they divide into 4–5 branchlets, each of which bears a head. Leaves 1–1½ inch long, 3–4 lines wide, the narrowed portion twice as long as the broad extremity. Heads 6–7 lines diameter, lemon-yellow.

69. H. squamosum (Th. ! Cap. 661); suffruticose at base, erect or ascending, branches virgate, closely leafy to the summit, woolly ; lower leaves oblong, medial and upper linear-subulate, with strongly revolute margins, sordidly woolly, acute, mucronate, at length often nude and then scabrous above, the uppermost often tipped with a membranous scale ; heads sessile, very many-flowered; inv. ovate-oblong, radiating, its scales lustrous, scarious, imbricating, oblong-lanceolate, acute, fulvous-yellow. *Less. Syn. p.* 286. *DC. l. c.* 185. *H. splendens, Bot. Mag. t.* 1773. *H. herbaceum, Andr. Bot. Rep. t.* 487. *Helipterum ferrugineum, DC. l. c.* 412 *(excl. syn.).*

HAB. Cape, *Thunb.!* Uitenhage and Albany, frequent, *E. Z.! Drege! Sir C. Bunbury !* &c.; Slaagekraal, *Burke !* Natal, *Gueinzius! Plant !* (Hb. Th., D., Hk., Sd.)
12–18 inches high, tufted ; stems simple, or corymbosely branched, the branches virgate, 1-headed. Leaves about uncial, the lower ones sometimes 4–5 lines wide, the rest 1–2 lines wide, erect or squarrose, several of those near the heads tipped

with a polished scale. Wool dirty whitish or fulvous. Pappus bristle-shaped, rough or serrulate. This has quite the habit of a *Helipterum*, but wants the plumose pappus. All *Ecklon's* and *Drege's* specimens of "*Hel. ferrugineum*" that I have seen, belong to this species.

§ 12. STOECHADINA. (Sp. 70–80.)

70. H. acutatum (DC.! l. c. 186); "stem erect, simple, cobwebby, the younger leaves cobwebby, at length on both sides rough with scattered, thick, glandular bristles," the lower leaves oval-oblong, penninerved, acute, tapering at base into a longish petiole, "the upper linear-elongate, much-acuminate; corymb compound, fastigiate; heads 8–12-fl.; inv. oblong, its scales lanceolate, acuminate, the outer dorsally woolly, the rest bright-yellow, shining." *DC. l. c.*

HAB. Between Omtendo and Omsamculo, on Grass-hills, under 500 feet, *Drege!* (Herb. Sond.)
One of the lower leaves and a few unopened fl. heads only seen by me. The leaf has a petiole about 3 inches long, and an ovato-lanceolate lamina 4½ inches long, 2½ inches wide, thinly cobwebby and rough with raised, glandular points. Fl. heads about 2 lines long, 1 line wide. The *habit* seems to be that of § 13, *Plantaginea;* but DC. says the recept. is "scarcely honey-combed."

71. H. splendidum (Less.! Syn. 286); stem shrubby or suffruticose, branches virgate, woolly; leaves sessile, linear or linear-oblong, with revolute margins, nigro-mucronate, on both sides white-woolly; heads subglobose, many-flowered, heterogamous, in densely-crowded, subglobose cymes, short pedicelled; inv. woolly at base, all the scales bright yellow, shining, opaque, oblong, obtuse, subconcave, scarcely radiant. *DC.! l. c.* 185. *Gn. splendidum, Th.! Cap.* 648. *Gn. strictum, Link.*

VAR. β. **xanthinum**; more slender, with laxer, less woolly leaves. *H. xanthinum, DC.! l. c.* 185.

VAR. γ. **montanum**; dwarf, depressed and tufted; branches erect, virgate; leaves crowded, broadish, the lower oblong-subspathulate, all very woolly. *H. montanum, DC.! l. c.* 186.

HAB. Cape, *Thunb.!* Eastern Districts, *Burchell*, 5047. Winterberg, *Mrs. F. W. Barber*, 239. Assagaysbosch, on mountains near Goerreshoogte, *Mundt!* Var. β. Zwarte, Storm, and Witberg, and on Sneeuweberg, *Drege!* Var. γ. Witberg and Camdebooberg, 6–7000 ft., *Drege!* Spitzkop, Somerset, *Dr. Atherstone!* (Herb. Th., D., Hk., Sd.)
Stems 2–5 feet high; in γ. 6–8 inches, closely or laxly leafy. Leaves ½–1½ inch long, 2–3 lines wide, the lowest flattish, all the rest with strongly reflexed edges. Cymes 1–1½ inch diam. Invol. very bright. I find filiform female fl. in all the vars.

72. H. rutilans (Less.! Syn. 275); stem herbaceous (annual), erect, simple or divided near the base into many simple branches, albo-tomentose; leaves half-clasping, spathulate, nigro-mucronate, spreading, on both sides densely albo-tomentose; heads in dense, branching, corymbose cymes, homogamous, subsessile; invol. cylindrical, loosely woolly at base, its scales opaque, imbricate, the outer shorter, oblong, the inner clawed, broadly obovate, very obtuse, undulate, bright-yellow. *DC. l. c.* 185. *Gnaph. rutilans, Linn. Sp.* 1291. *Th. Cap. p.* 630. *Gn. odoratum, Th.! Cap.* 654, *excl. syn.*

HAB. Sandy ground, chiefly in the Western Districts. (Herb. Th., Hk., D., Sd.)
A white-woolly annual, 10–15 inches high, with slender, subsimple roots. Leaves

subdistant, 1-1½ inch long, 2-4 lines wide near the summit, narrowed toward the base. Infl. generally much-branched. Involucral-scales rigid, bright-golden yellow.

73. H. adscendens (Less.! Syn. 274); stem herbaceous (perennial), ascending, simple, leafy at base, laxly leafy upwards; lower leaves spathulate, obtuse, very much attenuated towards the half-clasping base, upper linear, acute, all thinly and loosely woolly on both sides, one-nerved; heads subglobose, many-fl., in terminal, branching, corymbose cymes, on woolly pedicels; inv. campanulate, loosely imbricated, glabrous, all the scales pale-yellow, opaque, the medial and inner oblong, obtuse, spreading. *Gn. adscendens, Th.! Cap. 654.*

HAB. Cape, *Thunberg! Krebs*, No. 152, fide *Less. l. c.* (Herb. Thunb.)
Very similar to *H. rutilans*, but less woolly, and said to be perennial. Lower leaves 2-2½ inches long, 2-3 lines wide toward the apex, about 1 line wide toward the base; upper 1-1½ inch long, not a line wide. Corymb 1½ inch across, rather loose. Scales lemon-yellow. I have only seen *Thunberg's* specimen; *H. adscendens, DC.!* is very different (our *H. psilolepis*).

74. H. psilolepis (Harv.); stem herbaceous, sub-erect, simple or branched from near the base, together with the leaves thickly white-woolly; leaves half-clasping, broad-based, lanceolate or lance-linear, acute or acuminate, nigro-apiculate; heads subglobose, many-flowered, in densely congested, spheroidal cymes, shortly pedicellate or subsessile; invol. campanulate, woolly at base, the outer scales membranous, horn-colour, thin and pellucid, oblong or ovate,-the inner linear, with an oblong, obtuse, bright-yellow, spreading limb, subpellucid. *H. adscendens, DC. l. c.* 185.

HAB. Cape, *Ecklon!* between Zwartkey and Stormberg, also at Kl. Bruintjeshoogte, Zuureberg, *Drege!* (Hb. D., Hk., Sd.)
Perennial, many stemmed, very woolly. Leaves ¾-1½ inch long, gradually attenuated upwards, 1-2 lines wide, ending in a black, sharp point (often hidden under the wool). Infl. not much branched. Invol. scales very thin in substance, the apices only yellow.

75. H. hebelepis (DC. l. c. 187); stem suffruticose at base, ascending-erect, simple, or branched from the base, the branches virgate, woolly; leaves sessile, lanceolate, apiculate, loosely cobwebbed above, white-woolly beneath, 3-5-nerved, the lower acute, the upper acuminate; cymes subpedunculate, compound, flat-topped, densely many-headed, its branches woolly, leafless; heads subsessile, campanulate, about 20-25-flowered, homogamous; inv. scales dorsally thinly woolly, glabrous near the apex, linear, scarious, thin, the outer acute, pale horn-colour, the inner obtuse, fulvous-flavescent.

VAR. β, **angustius** (DC.); leaves narrower and more acuminate, the lateral nerves scarcely obvious, the corymb more crowded, and the heads shorter.

HAB. Kaus Mt. and Modderfontein, Namaqualand, and Hexrivierskloof, *Drege!* Modderfontein, *Rev. H. Whitehead!* β, Brackfontein, *Ecklon!* (Herb. Hk., D., Sd.)
Stems or branches 1-1½ ft. high. Leaves in α, 1½-2 inches long, 4-7 lines wide; in β, 3-6 lines wide, thinly woolly and pale-greenish above. Cymes corymbose, much branched. Heads 2-2½ lines long. Recept. quite naked! Pappus scabrous. The broad-leaved form is very similar in aspect to *H. tricostatum*, but differs in the acute inv. scales, and absence of fimbrils on the receptacle.

76. H. subglomeratum (Less. Syn. 283); stem shrubby at base, simple or divided into several, virgate branches; branches and leaves coated with a silvery-white, interwoven, felted indument; leaves sessile, either elliptical, oblong, or lingulate, subacute; heads very densely cymoso-corymbose, sessile, the tufts congested with wool underneath; inv. 8–10-fl., cylindrical, the outer scales testaceous, inner bright yellow, scarious, obtuse, subpellucid; pappus barbellate. *DC. l. c.* 186.

VAR. *a.* **lingulatum**; leaves linear-oblong or tongue-shaped, 1-2 inches long.

VAR. *β.* **imbricatum** (DC.); leaves elliptical or oblong, imbricated, ½-¾ inch long.

HAB. Uitenhage and Albany, *Krebs, Burchell, E. & Z.! Drege! H. Hutton! Genl. Bolton!* &c. (Herb. D., Hk., Sd.)

1-2 f. high, silvery; the indument usually membranous and smooth, but occasionally (in *Drege's* sp.) loose and woolly. Typical specimens of the two varieties look very distinct, but intermediate states are readily found. The woolly agglutination of the heads, though considerable, is less than in the following species.

77. H. umbraculigerum (Less. Syn. 284); stem suffruticose at base, erect or ascending, simple, tomentose-canescent; leaves membranous, oblongo-obovate or ovato-lanceolate, narrowed much at base (almost petiolate), calloso-mucronate, thinly woolly above, albo-tomentose beneath (or uniformly canous), 3-nerved; heads 5-fl., very numerous, densely congested in a flat topped cyme, the branches of the cyme so united by interwoven white wool as to form a disc-like, peltate pseudomembrane, marked with raised, radiating, nerves beneath, and completely clothed with fl. heads above; inv. scales bright yellow, obtuse. *DC. l. c.* 186.

HAB. Cape, *Krebs,* No. 147. Betw. Graaf Reynet and Komma-Dabka, *Burchell, No.* 3078. Betw. Omsamcuba and Omsamwubo, *Drege.* Near Maritzburg and Ladysmith, *Gerr. & M'K.,* 282. Lower Umcomas and Zululand, *Gerr. & M'K.!* 1033, 1034. (Herb. D., Hk., Sd.)

Stems 12-15 inches high. Leaves 1-1½ inch long, 4-6 lines wide. Inflorescence very peculiar; the flower-heads seemingly closely sessile on a broad, woolly, nerved disc. *Burchell's* MS. name *Gn. dorstenicæflorum* is very appropriate. The woolliness varies much in density and whiteness.

78. H. pentzioides (Less.! Syn. 282); stem rigid, shrubby, divaricately-branched, twigs canescent or nude; leaves obovate, tapering much at base, recurvo-mucronulate or very obtuse, thickish, on each side dotted, thinly canescent or glabrate; heads corymbulose, subpedicellate, 10–12-fl.; invol. cylindrical, not radiating, imbricate, its scales pale-horn-colour, dotted, obtuse, scarious, thin, appressed, subpellucid, the tips slightly bullated. *DC. l. c.* 192.

HAB. Kanna River, *Mundt & Maire.* Nieuweveld, betw. Brakriver and Uitflugt; and betw. Zondag riv. and Ado, *Drege!* Fishriver, *Ecklon!* Zuureberge, *Burke & Zeyher!* (*Zey.!* 887). (Herb. Hk., Sd., D.)

A stout, woody and scrubby bush, a foot or more high, much branched, sometimes nearly glabrous, sometimes canous in all parts, Leaves 4-6 lines long, 2-3 lines wide, rigid. Heads 1½ lines long.

79. H. lucilioides (Less. Syn. p. 290); stem rigid, shrubby, divaricately branched, twigs tomentose; leaves obovate or oblong, tapering much at base, recurvo-mucronate, thickish, impunctate, 1-nerved, thinly canescent; heads two or more together, sessile at the ends of the branches, subtended by leaves, 10-fl., homogamous; inv. cylindrical, not radiating,

imbricate, its scales pale-horn-colour, impunctate, obtuse, scarious, glossy, thin, appressed, semipellucid. *DC. l. c.* 191. *Gn. stœhelinoides, Th. !* *Cap.* 652.

HAB. Cape, *Thunb.! Ecklon! Burke & Zeyher!* Nieuwefeld, Beaufort, *Drege!* (Herb. Th., Hook., D., Sd.)

Very similar to *H. pentzioides* in habit and foliage, but with different inflorescence, longer and more shining involucres, and corollas and pappus fully twice as long. Invol. 3 lines long. Leaves 3-4 lines long, 1½-2 lines wide, whitish or yellowish.

80. H. excisum (Less. Syn. 282); stem shrubby, rigid, much branched, the twigs canescent ; leaves spathulate-cuneate, tapering much at base, recurvo-mucronate (and almost obcordate), thickish, on each side cano-tomentose; heads corymbulose, subsessile, 9-10-fl. ; inv. cylindrical, not radiating, imbricate, its scales golden-yellow, lanceolate, acute, squarrose, scarious, the outer ones short and membranous. *DC. l. c.* 192. *Gnaph. excisum, Th. ! Cap.* 655.

HAB. Cape, *Thunberg! Burchell,* 6508. *Dr. Thom!* Langekloof, betw. Grootfontein and Gangekraal, *Drege!* On the Karro, *Mundt!* Hassagay'skloof and Breederivers-poort, *Zey!* 2865. (Herb. Th., Hk., D., Sd.)

A densely-branched, rigid, small shrub, less divaricate than *H. pentzioides*, with nearly similar foliage, but very different involucres. Leaves 3-6 lines long, 2-3 lines wide, owing to the strongly recurved point nearly obcordate. Heads 1½ line long, the tips of all the scales very acute and strongly recurved.

Sub-genus II. **LEPICLINE**. (Sp. 81-137.)

§ 13. PLANTAGINEA. (Sp. 81-96.)

81. H. latifolium (Less.! Syn. 297); stem erect, simple, leafy at the base only, pedunculoid upwards and densely woolly; subradical leaves sessile, either ovate-oblong or ovato-lanceolate, generally acute-pointed and narrowed to the base, rarely obtuse at both ends, the cauline (1-2) stem-clasping, lanceolate-oblong, narrow, all 5-7-9-nerved, woolly beneath, piloso-scabrid and thinly cobwebbed or glabrate above; heads many in a densely-branched, flat-topped or globose cyme, many-fl., pedicellate ; inv. campanulate, imbricate, woolly at base, apices of the scales glabrous, membranous, *broadly-oblong, obtuse,* pale horn-colour or purplish. *DC. l. c. p.* 198. *Gnaph. latifolium, Th. ! Cap.* 660.

VAR. β. **reticulatum ;** cauline leaves ovate, clasping ; all the leaves *thinly* tomen-tose beneath, with dark-coloured, netted veins between the nerves.

HAB. Galgebosch, *Thunberg.* Gauritz R., *Burchell,* 4734. Zuureberg, *Drege!* Grahamstown, *Genl. Bolton!* Kaffirland, *Ecklon!* Natal, *Krauss,* 384, *Gueinzius, Sutherland!* β. Between Maritzberg and Ladysmith, *Gerr. & M'Ken.!* 281. (Herb. Th., D., Hk., Sd.)

Stem 12-15 inches high. Leaves 2-4 inches long, 1-2½ inches wide, becoming glabrate above, but persistently rough with swollen, jointed bristles ; the under surface variably tomentose. Condensed cyme 1-1½ inch diameter, at length flat-topped.

82. H. griseum (Sond.! in Linn. 23, p. 65); stem erect, simple, leafy at the base only, pedunculoid upwards, densely woolly ; subradical leaves shortly petiolate, ovate or oblong, acute at each end, the cauline one or two, stem-clasping, ovate or oblong, acute or acuminate, all faintly 5-7-nerved, densely woolly beneath, piloso-scabrid and (at first) thinly cobwebbed, becoming glabrate above ; heads very many in a

diffusely much-branched, woolly corymbose-cyme, many-fl., pedicellate;
invol. campanulate, imbricate, all the scales dorsally woolly below, their
apices glabrous, membranous, *lanceolate-acuminate*, squarrose, pale horn-
colour or purplish.

HAB. Natal, *Gueinzius!* 322, 590.   Near D'Urban, *Gerr. & M'Ken.!* 307.   (Herb.
Sond., D.)
    Allied to *H. latifolium*, from which it differs in the evidently petiolate root-leaves,
the much more diffuse inflorescence, and especially in the invol. scales.   Stem 1–1½
feet high.   Leaves 3–4 inches long, 1½–2 inches. wide.   There is usually a single,
narrow-lanceolate, acuminate floral-leaf about the middle of the peduncle, and one
at the base of the cyme.

83. H. pedunculare (DC.! l. c. 198); stem erect, simple, leafy at base
or to beyond the middle, usually prolonged into a nearly naked, woolly
peduncle; subradical leaves coriaceous, lance-oblong or obovate-oblong,
subacute, more or less basally attenuated or petiolate, above glabrous
and smooth, 5–7–9-nerved, beneath white-tomentose, with inconspicuous or faintly marked nerves; heads very many (20–60) in a dense,
subglobose, branching cyme, many-fl., subsessile; invol. campanulate,
imbricate, woolly at base, not radiating, the scales flat, membranous,
linear, obtuse, brownish horn-colour. *Gnaph. pedunculare, Linn. Mant.*
284, fide DC.

VAR. β. **pilosellum**; stem dwarf (2½ inches high); heads about 5 in a cluster.
H. pilosellum, Less. Syn. 297. *DC. l. c.* 198.

HAB. Swellendam, *Burchell!* 720. Uitenhage, *Ecklon!* Kango, *Dr. Thom.!* Beth-
elsdorf, *Zey.!* 879. Caledon R., *Burke!* Beaufort, *T. Cooper!* 415. Var. β. Kango,
Mundt and Maire. (Herb. Sd., Hk., D.)
    Stem 6 inches to 3 feet high, leafy chiefly near the base; the cauline leaves
clasping, oblong or oblongo-lanceolate; the peduncular extremity woolly, with a
few scattered, subulate, depauperated leaves.   Lower leaves 3–7 inches long, 1–2½
inches wide, the tomentum on the under surface closely interwoven, either copious
(concealing the nerves) or scanty.   Inv. scales testaceous.   The young inflorescence is nearly spheroidal, and very dense; the older flattens into a trichotomous
cyme. *H. pilosellum*, Less., by description, can scarcely be anything but a dwarf
specimen of this plant.   According to *Mr. Cooper*, this plant, called *Ery'kue* by the
Fingoes, is used to cure wounds, particularly the wounds of circumcision.

84. H. undatum (Less.! Syn. 298); stem erect, simple, leafy at base
or below the middle, pedunculoid upwards and densely woolly; subradical leaves tapering at base into a short or longish petiole, ovate-
oblong, acute or acuminate, 3–5-nerved, cauline (when any) sessile,
clasping, ovate or lanceolate, acuminate, all densely woolly beneath,
piloso-scabrid and thinly cobwebbed above, undulate; heads many in a
densely much-branched, woolly, flat-topped cyme, many-fl., pedicellate;
invol. campanulate, woolly at base; apices of the scales glabrous, glossy,
oval-oblong, concave and connivent, very obtuse, sub-opaque, the outer
purplish, inner white. *DC. l. c.* 198. *Gn. undatum, Th.! Cap. p.* 655.

VAR. β, **pallidum**; inv. sc. cream-colour or pale buff. *H. pallidum, DC. l. c.* 199.

HAB. Paardekop, *Mundt & Maire.* Cape, *Bowie!* Fish R. and Zuureberg, *Drege!*
Albany, *E. Z., T. W., Genl. Bolton.* Var. β, Albany, *T. Williamson!* Caledon R.,
Burke! Katriversberg, Zwartkey and Buffel River, and between Omsamculo and
Omtendo, *Drege!* (Herb. Th., D., Hk., Sd.)
    Stem 1–2 feet high, sometimes leafy in its lower half, sometimes at base only.
Rad. leaves more or less distinctly pedunculate, 4–6 or 8–10 inches long, 1–2–3 in.

wide, mostly acuminate; in the more luxuriant the forking of the nerves is much above the base of the leaf.   De Candolle's *H. pallidum* merely differs in the colour of the inv. sc.   I have both forms from the same locality, collected by *T. Williamson.*

85. H. coriaceum (Sond.! Linn. 23, p. 65); stem erect, simple, leafy at base or below the middle, prolonged into a nearly naked, woolly peduncle; subradical leaves coriaceous, lanceolate, acuminate, tapering at base, on both sides coated with a membranous, separable, silvery and rather glossy coat, 3–5-nerved, the upper surface at length denuded, lower persistently silvery; heads many in a dense, subglobose or branching and flat-topped cyme, many-fl.; invol. campanulate, imbricate, woolly at base, not radiating, the scales wavy, membranous, ovato-lanceolate, *acuminate* or acute, pale straw-colour, subpellucid.

HAB. Magalisberg and Aapjes Riv., *Burke & Zey.! Zey.!* 880.   (Hb. Hk., Sd., D.)
Stem 1–2 feet high, densely leafy at base.   Root leaves numerous, 5–6 inches long, ½–¾ inch wide, tapering to both ends.   Cauline leaves sessile, 2–3 inches long, 2–4 lines wide.   The coating of the leaves may be peeled off as a thin, membranous, white skin; it finally splits longitudinally and peels off the upper surface.   Mature inflorescence 2–3 inches across, umbellate, of 40–60 heads.   Inv. scales whitish.

86. H. allioides (Less. Syn. 299); stem erect, simple, leafy at the base or below the middle, pedunculoid upwards, tomentose; subradical leaves crowded, petiolate, varying from ovate to ovato-lanceolate, or linear-lanceolate, acute or acuminate at both ends, 3-nerved, cauline leaves sessile, 1 nerved, all white-woolly beneath, glabrous and smooth above, the younger cobwebby on the nerves; heads very numerous (50–100), crowded in a densely much-branched, flattened cyme, many-fl., pedicellate; invol. campanulate, imbricate, woolly at base, the scales flat, ovate, the outer subacute, inner *obtuse*, variable in colour (pale-yellow, reddish-brown or straw). *DC. l. c.* 198.

HAB. Cape, *Krebs,* 145; Stormberg and Zwartekey, *Drege!* Uitenhage, *Ecklon!* Natal, *Krauss,* No. 369, *Gueinzius! Sutherland!* &c.   (Herb., Sd., Hk., D.)
Stem 1–2 feet high. ̄Leaves extremely variable in shape and comparative length and breadth: in some ovate, 2–2½ inches long, 1½ inch wide; in some 4–6 inches long, 1–1½ inch wide; and in others 4–5 inches long, 3–4 lines wide.   Petioles vary from 1–5–6 ̇inches in length.   The nerves are generally obvious on both surfaces.   Inv. scales varying much in colour, not very glossy.

87. H. leiopodium (DC.! l. c. 200); stem erect, simple, leafy to or beyond the middle, tomentose; radical leaves elliptic-oblong or lanceolate, acute at both ends, tapering at base into a glabrate petiole, 3–5-nerved, smooth above, cobwebby-tomentose and whitish beneath; cauline, sessile, lance-linear, acuminate; corymb loosely much-branched; heads subsessile; inv. campanulate, imbricate, nearly glabrous, its scales shining, imbricate, roundish oval, very obtuse, erect, golden-yellow.

VAR. *β. denudatum*; radical leaves glabrate beneath.

HAB. Near Onzer, and betw. Gekau and Basche, *Drege!* Caledon, Uitenhage, Albany and Tambukiland, *Ecklon!* Var. *β.* Bosjesveld, *Mundt!* Thaba Unka, *Burke & Zeyher!* 381. Vanstaadensberg and near Grahamstown, *Zey.!* 2862.   (Hb. D., Hk., Sd.)
Very like *H. nudifolium* (especially var. *β.*) but more tomentose, with the upper surfaces of the leaves quite smooth, not scaberulous, their bases not decurrent, &c. I fear however that the characters are not constant.

88. H. nudifolium (Less. Syn. 299) ; stem erect, simple or branched, leafy to or beyond the middle, scabrous or thinly woolly below, peduncular and thinly woolly above ; root-leaves numerous, ovato-lanceolate or linear-lanceolate, elongate, tapering to both ends, prolonged at base into a petiole, 3–5-nerved, concolourous, above and on the prominent nerves and veins beneath scaberulous, not woolly or cobwebbed; cauline lanceolate or lance-linear, more or less decurrent, acuminate; corymb much branched, spreading, thinly tomentose ; heads many-fl., subsessile ; invol. campanulate, nearly glabrous, its scales broadly oblong, very obtuse, connivent, membranous, subpellucid, the outer rufo-fulvous, inner yellow, or all yellow. *DC. l. c.* 200. *Gnaph. nudifolium, Linn. Sp.* 1196. *Th.! Cap.* 658.

VAR. β. **obovatum**; lower and medial cauline leaves 3–4 inches long, 1½ inch wide, obovate, suddenly acute, tapering at base, decurrent, 3-nerved above the base, piloso-scabrid on both sides and on the margin ; upper leaves passing from obovato-lanceolate to lanceolate, linear-acuminate and subulate along the peduncle, which is thinly tomentose upwards ; *immature* involucres fulvous. (Herb. Sd.)

HAB. Throughout the Colony and at Natal, common. (Herb. Th., D., Hk., Sd.) Commonly known as "*Hottentot Tea,*" a name given to the allied species also. Stem 1–2 feet high or more, brown, very thinly tomentose or nude. Root and lower leaves 5–6 inches long, ½–1½ inch wide, very harsh to the touch, quite green, with prominent ribs and veins. Cauline leaves mostly very narrow, but varying from lanceolate-acuminate to linear. Var. β. is founded on a single specimen collected by *Ecklon,* and now in Dr. Sonder's possession, by whom it is regarded as a form of *H. quinquenerve.* To me it seems rather referable to *H. nudifolium,* if it do not show the necessity of uniting those species in one.

89. H. quinquenerve (Less. Syn. 300) ; stem erect, simple, tomentose, leafy to or beyond the middle, pedunculoid-upwards ; root-leaves subpetiolate, oblong, attenuate at base, acute or acuminate, cauline broadly ovate, strongly decurrent, above and along the margin scabrid, beneath thinly cobwebbed, becoming glabrate-scaberulous, 5-nerved, the nerves prominent, scabrid, connected by prominent, netted veins; corymb very much branched, spreading, woolly ; heads many-fl., subsessile ; inv. campanulate, nearly glabrous, its scales broadly oblong, very obtuse, connivent, membranous, subopaque, primrose or lemon-coloured. *H. nudifolium.* γ. *plantagineum, DC. l. c.,* also *H. multinerve, DC.! l. c.* 199. *Gnaph. quinquenerve, Thunb. Cap.* 658.

HAB. Near George, *Mundt & Maire! Dr. Thom! Dr. Pappe!* Vanstaadensberg, *Zeyher! Drege!* Stormberg and Witberg, and in the Langekloof, *Drege.* Betw. Maritzberg and Ladysmith, *Gerr. & M'K.* 265. (Herb. Th., D., Hk., Sd.) Nearly allied to *H. nudifolium,* but with much broader, more ovate and clasping stem leaves, 5 or several nerved, more strongly decurrent ; a more woolly stem, and more rigid and opaque, less yellow inv. scales. De Candolle seems to have described it under two names. Both his plants grow in the same localities.

90. H. miconiæfolium (DC.! l. c. 200) ; stem erect, simple, leafy and tomentose throughout ; radical leaves oval or ovato-lanceolate, 3-nerved, acute at both ends and tapering at base into a short or long, hispid petiole, the younger ones cobwebby above, then glabrate-scaberulous, beneath tomentose (or glabrate); cauline sessile, lance-linear, acuminate, erect, with revolute margins, quite smooth above, cobwebbed or tomentose beneath ; corymb loosely much-branched ; heads subsessile ; inv. woolly at base, its scales glabrous, shining, loosely imbricated, undulate, ovate, acute, subacuminate, golden or lemon-yellow.

HAB. Witberg, *Drege!* Albany Distr., *Ecklon! Mrs. F. W. Barber*, 506. (Herb. Sond., D., Hk.)

Stem white-woolly throughout, 12–15 inches high, sometimes laxly leafy upwards or imperfectly pedunculoid ; sometimes densely leafy to the corymb. Root leaves sometimes on petioles 4–5 inches long; the lamina 3 inches long, 1–1½ inch wide : Sometimes on petioles scarcely an inch long. Usually the lower leaves are persistently tomentose on the under side ; but in some specimens they become glabrous. —Nearly allied to *H. Krebsianum*, but differing in the involucre, &c. This is also called "Kaffir Tea" (fide *Mrs. F. W. B.*)

91. H. Krebsianum (Less.! Syn. 308); stem erect, simple, leafy throughout, setoso-scabrid and angular below, tomentose above ; subradical leaves oblongo-lanceolate, tapering at base into a ribbed petiole, on both sides piloso-setose, prominently 3-nerved, the cauline sessile, broad-based, lance-linear, acuminate, with revolute margins, piloso-scabrid above, beneath cobwebby ; corymb loosely much-branched ; heads pedicellate, globose ; inv.-scales nearly glabrous, loosely imbricated, oval, concave, obtuse, straw-colour, connivent. *DC. l. c.* 204, and *Hel. crassinerve, DC.! l. c.* 199.

HAB. Cape, *Krebs*, No. 154. Caffraria, *Ecklon!* (Herb. Sond.)

Stem 1–1½ f. high. Root-leaves few, 2–3 inches long, ½–⅔ inch wide : cauline numerous, 1–1½ inch long, 1–3 lines wide. Heads 3–4 lines diam.—Compared by *Dr. Sonder* with the original specimens in Hb. Berol., described by *Lessing*, who had not seen the radical or lower cauline leaves.

92. H. subulifolium (Harv.) ; stem erect, simple, tomentose, leafy nearly to the summit ; radical leaves . . . . ; cauline crowded, from a broadish base linear-attenuate, hook-pointed, with strongly revolute margins, the lower ones 3-nerved, all piloso-scabrid above and on the nerves beneath, at first thinly cobwebby-tomentose, becoming glabrate ; heads very many, in a much-branched, spreading corymb, many-fl., subsessile ; inv. glabrous (or thinly cobwebbed at base), imbricate, its scales glossy, membranous, oblong, obtuse, concave, tawny or horn-colour.

HAB. Mooye River, *Burke & Zeyher! Zey.!* 882. (Herb. Hk., D., Sd.)

Stem 1–2 feet high; the base and root-leaves not seen. Cauline leaves 4–5 inches long, 1–2 lines wide, strongly revolute at margin, tapering to a sharp, recurved point, the young ones hoary, the old glabrate, but very scabrid. Upper leaves gradually smaller. Corymb 3–5 inches in diameter, its branches tomentose. Invol. almost glabrous. Pappus pale. Fimbrils fulvous, subulate, longer than the achenes.

93. H. oxyphyllum (DC. l. c. 199) ; "leaves along the lower part of the stem sessile, half-clasping, from a broad base lanceolate-linear, the lowest obtuse, the rest much acuminate, with subrevolute margins, 3-nerved, somewhat scabrid above, beneath (as well as the simple stem) villoso-tomentose, whitish ; heads 50–60 in a compound corymb; inv. woolly at base, the outer scales purplish, the innermost scales oblong, white-tipped, subacute." *DC. l. c.*

HAB. Near the Gauritz River, *Burchell*, No. 4733.

"Leaves 4 inches long, 6–7 lines wide. Stem 1½ f. high. Inv. woolly at base, the outer scales purplish, the inner white and subacute at apex. Cor. yellow. Pappus white. Fimbrils golden." *DC. l. c.*

94. H. cephaloideum (DC.! l. c. 197); stem erect, simple or branched, white-woolly, leafy throughout; leaves on both sides white-woolly, the radical oblong or lance-oblong, elongate, acute or obtuse, not much narrowed at base, 3-nerved, the cauline sessile, lance-linear, acuminate, erect, the uppermost subulate; heads in a dense, subglobose fascicle or densely corymbose cyme, campanulate, many-fl.; inv. scales glabrous and glossy, flat, imbricated, scarious, lanceolate, acute, yellow.

VAR. β. **polycephalum** (DC.); stem branched, branches erect, straight; heads and clusters smaller; inv. scales rather blunter and less bright.

HAB. Kaffirland, *Ecklon!* Betw. Omsamwubo and Omsamcaba, and at Natal, *Drege!* Natal, *Dr. Grant! Gueinzius!* 355, 593. *Dr. Sutherland!* Betw. Maritzburg and Ladysmith, *Gerr. & M'K.* 290. Var. β. Albany, *Ecklon!* Karrega river, *Zey.!* 2877. Betw. Maritzburg and Ladysmith, *Gerr. & M'K.*, 258, 263. (Hb., D., Hk., Sd.)

Stems 1-2 feet high, quite simple or branched below the middle. Rad. leaves 3-6 inches long. Lower cauline 3-4 inches long, 3-5 lines wide; upper gradually smaller, all very woolly and sharp pointed. Heads in var. *a.* very brilliant golden or orange-yellow, with very sharp-pointed scales; in β. smaller, much less bright, in smaller clusters and with blunter scales. This last variety is most common in the Colony, and the former seems prevalent in Kaffraria and Natal, but I have both forms collected together by *Gerr. & M'Ken.*

95. H. longifolium (DC.! l. c. 198); stem erect, simple, white-woolly, leafy throughout; radical leaves linear-lanceolate, acute or acuminate, elongate, glabrous and 5-nerved on the upper, white-woolly on the under surface; cauline erect, linear-acuminate or subulate, woolly on one or both sides; heads campanulate, very many-fl., in a close or open corymbose-cyme, pedicellate; inv. scales glabrous and glossy, flat, imbricated, scarious, oblongo-lanceolate, acute or acuminate, bright yellow.

HAB. Between Omsamculo and Omtendo, *Drege!* Natal, *T. Williamson! Gerr. & M'K.!* 855, *Dr. Sutherland!* Kaboosie Mts., Br. Kaffraria, *W. S. M. D'Urban!* 52.

Very nearly related to *H. cephaloideum*, but differing in the discoloured root-leaves and larger fl. heads. In an early stage the inflorescence is nearly as much condensed as in *H. cephaloideum*, but afterwards becomes open, the heads on long pedicels. Heads not unlike those of *H. fœtidum.*

96. H. appendiculatum (Less.! Syn. 308); stem erect or ascending-erect, simple, loosely woolly or cobwebbed, leafy throughout; leaves loosely woolly and pilose on both sides, the radical elongate, 3-5-nerved, the cauline lanceolate-oblong, half-clasping, acute or mucronate, the uppermost erect, lance-acuminate, tipped with a scarious, glabrous and withered point; heads densely corymboso-cymose, crowded, pedicellate, campanulate, many-fl.; inv. scales glabrous, pale-yellow or purplish, flat, loosely imbricated, ovate, acute or acuminate, scarious, but not glossy. *DC. l. c* 208. Also *H. folliculatum, DC.! l. c.* 197. *Gn. appendiculatum, Th.! Cap.* 649 *and Gn. humile, Th.! p.* 655.

VAR. β. **discolor** (DC.); leaves green and very thinly woolly above; invol. scales *very much acuminate, squarrose. H. discolor, DC.! l. c.* 197.

HAB. Cape, *Thunberg,* &c. Common in Uitenhage and Albany, *E. & Z.! Drege! &c.* Var. β. Near Port Natal, Drege! *Krauss,* 398*! Plant! Gerr. & M'Th.!* 291, 854, 857, &c. (Herb. Th., D., Sd., Hk.)

Stems 1-2 f. high, simple or branched below the middle, loosely woolly. Leaves on both sides sprinkled with jointed, swollen, persistent bristles, and clothed with partially deciduous, loose, silky-woolly hairs in variable quantity, faintly 3-5-nerved. Var. β. chiefly differs in its much more taper-pointed, curved or recurved inv. scales ; but this character is far from being constant, even on the same specimen. The leaves, though *glabrous and smooth above* on some of *Drege's* specimens, are in others variably *hispid and cobwebby*, exactly as in var. α.

§ 14. DECURRENTIA. (Sp. 97–104.)

97. H. xerochrysum (DC.! l. c. 201); stem erect, simple, leafy in the lower half, pedunculoid upwards, and as well as the leaves clothed with rufous or fulvous wool; lower leaves broadly oblongo-lanceolate, acute, narrowed to the base, upper lanceolate or lance-linear, acuminate, strongly decurrent; corymb few-headed, subsimple; heads (rather large), long pedicelled, woolly at base, campanulate, very many-fl.; inv. scales loosely imbricate in many rows, glabrous and glossy, rigidly scarious, golden-yellow, lance-oblong, acute or acuminate.

HAB. Between Zandplaat and Komga, *Drege!* (Herb. Hook., Sd.)

Stem 12–15 inches high; the upper half peduncular, with 1–2 narrow, very erect, depauperated leaves. Lower leaves 3-4 inches long, ¾–1 inch wide, gradually shorter, narrower, and more decurrent upwards. Heads 8–10 lines diameter; outer scales fulvous. Fimbrils whitish.

98. H. Mundtii (Harv.); stem erect, simple, leafy, tomentose, panicled at the summit; lower leaves oblongo-lanceolate, subacute, very much attenuate to the base, 3-nerved above the middle, upper linear-lanceolate or linear, acute or acuminate, all glabrous on the upper, white-woolly on the lower surface, and decurrent at base into a long, narrow stem-wing; inflorescence corymboso-paniculate, much and loosely branched; heads aggregate, subsessile, 20–25-fl.; inv. glabrous, imbricated, its scales erect, oblong, obtuse, cream-coloured, subopaque; recept. honeycombed and toothed.

HAB. Wagenmaker's Bosch, Swell., *Mundt!* (Herb. D., Hk.)

Stem 2–3 feet high, herbaceous, simple except for the panicled inflorescence. Lower leaves 6-8 inches long or more, about ⅔ of the length forming the tapering base, below the forking of the nerves: the expanded portion ¾–1½ inch wide. Upper leaves shorter and narrower, less tapering at base. Panicle 6–12 in. long and wide.

99. H. Natalitium (DC. l. c. 201); stem erect, simple, elongate, leafy, shortly pedunculoid at the summit, and, as well as the leaves, loosely (or thinly) whitish-woolly; leaves lanceolate or linear-lanceolate, 3-nerved, tapering to each end, long-decurrent in a narrow wing; inflorescence corymbo-cymose, branching; heads very densely crowded, sessile, 10–12-fl.; inv. woolly at base, its scales loosely imbricate, oblong, subacute, wavy, cream-coloured; recept. shortly honey-combed; pappus of very few, slender bristles.

HAB. Port Natal, *Drege! T. Williamson! Gueinzius!* No. 350, *Dr. Sutherland!* (Herb. Hk., D., Sd.)

Stem 1–2 feet high. Root leaves 5–6 inches long, ¾ inch wide, much acuminate and tapering at base; cauline leaves 2-4-6-8 inches long. Peduncle often branched, each branch bearing 2-3 very dense cymes. Heads 2 lines long. Pappus very scanty, sometimes none.

100. H. platypterum (DC.! l. c. 201); "stem erect, simple, below

rough with thick, glandular bristles, above loosely tomentose ; leaves
on the upper surface sparingly, on the lower along the nerves, and
specially at the margins, rough with thick, glandular bristles, oval-
lanceolate, on each side decurrent in a long, broad wing ; corymb compound ; heads 12–15-fl. ; inv. ovate, the scales membranous, oval, very
obtuse, glabrous, rufescent ; recept. scarcely honey-combed. *DC. l. c.*

HAB. Between Omsamwubo and Omsamcaba, 1–2000 f., *Drege!*
Of this I have only seen a frustule in Hb. Sond.

101. H. Gerrardi (Harv.) ; stem erect, branching, laxly leafy, glabrous and striate, smooth ; branches long ; leaves linear-lanceolate,
acuminate, narrowed to the base, with revolute margins, quite smooth
and even on the upper, glabrous on both surfaces, all 3-nerved, decurrent at base into a long, narrow stem wing ; infl. corymboso-cymose ;
heads (not mature) crowded, sessile, few-fl. ; inv. glabrous, glossy, its
scales erect, oblong, obtuse, lemon-yellow.

HAB. Nototi, Natal, *W. T. Gerrard!* 1001. (Herb. D.)
Very like *H. stenopterum,* but quite glabrous, with perfectly smooth (not scabrous)
upper surfaces of leaf. Flowers not seen.

102. H. stenopterum (DC.! l. c. 201) ; stem erect, branching, laxly
leafy, cobwebby ; branches long ; leaves linear-lanceolate, acuminate,
narrowed to the base, with revolute margins, scabrous on the upper,
white-woolly on the lower surface, the lowest 3- the upper 1-nerved,
all decurrent at base into a long, narrow stem-wing ; inflorescence
corymbo-cymose, branching ; heads very densely crowded, sessile, 5–6-fl. ;
invol. glabrous and glossy, its scales membranous, erect, oblong, acute,
the outer tawny, inner lemon-yellow, all semi-pellucid.

HAB. Betw. Omsamculo and Omcomas, and at Natal, *Drege!* Natal, *Dr. Grant!*
Gueinzius! Betw. Maritzburg and Ladysmith, *Gerr. & M'K.!* 286. (Hb. D., Hk., Sd.)
Stems straggling or subscandent, several (?) feet long, slender ; branches 12–18
inches long. Rameal leaves an inch or more apart, 2–4 inches long, 2–4 lines wide,
the young ones thinly cobwebby, older glabrate, but scaberulous, deep green.
Heads innumerable, very small, rufous, tipped with gold. Recept. honey-combed,
the cells toothed. Pappus rough.

103. H. gymnocomum (DC.! l. c. 202) ; stem rigid, erect or diffuse,
branching, branches ending in a long, leafless peduncle, tomentose ;
leaves long-decurrent, the lowest obovate, 3-nerved, subobtuse, mucronate, upper oblong, the younger on both sides loosely woolly, the old
becoming nude above, setoso-scabrid ; cymes long-peduncled, very
densely much-branched, leafless ; heads oblong, 5–6-fl. ; inv. scales
erect, glabrous and glossy, oblong, concave, acute or subacute, fulvous-
yellow.

VAR. β. **acuminatum** (DC.) ; leaves longer, acuminate, more scabrid above ; inv.
scales less acute.

HAB. Betw. Bushman's and Gauritz R., *Burchell,* 4334, 4385. Cape Recief,
Algoa Bay, *Ecklon!* Var. β. Vanstaadensberg, *Drege!* Albert, *T. Cooper!* 620.
(Herb. Sd., Hk., D.)
Very nearly allied to *H. odoratissimum,* from which it chiefly differs in its rather
longer involucre and flowers, and the *fewer* number of flowers in each head, and
subacute inv. scales. β. is slender, and less woolly.

104. H. odoratissimum (Less.! Syn. 301); stem suffruticose at base, erect, or diffuse, branching, branches tomentose, ending in a leafless peduncle; leaves long-decurrent, spathulate, oblong or lanceolate, 1–3-nerved, membranous, white-woolly, the older becoming nude above, setoso-scabrid; cymes more or less peduncled, densely much-branched, the branches unibracteate; heads subsessile, oblong, 10–12-fl.; inv. scales erect, glabrous, oblong, obtuse, the outer shorter and fulvous, inner yellow. *DC. l. c.* 202. *Gnaph. odoratissimum, Linn. Sp.* 1196. *Gn. strigosum, Th.! Cap.* 654. *Gn. aureofulvum, Berg. p.* 257.

VAR. *a.* **acuminatum** (DC.); leaves linear or lanceolate, acuminate.

VAR. *β.* **undulæfolium** (DC); leaves much more woolly, oblong or obovate-oblong, subacute, undulate shortly decurrent; cymes dense, depressed-globose.

VAR. *γ.* **lanatum** (Sond.); stems trailing; leaves very woolly, white, oblong and undulate.

HAB. Common throughout the Colony. *β.* Caledon and Swellendam. *γ.* Uitenhage. (Herb. Th., D., Hk., Sd.)

Stem ½–1½ f. high, erect or spreading, much branched. Leaves ½–1½ inch long, 2–4 lines wide. Cymes 1–3 inches across, variably compound, always very dense.

§ 16. APTERA. (Sp. 105–133.)

105. H. trilineatum (DC.! l. c. 192); stem shrubby, much-branched, branches tomentose; leaves sessile, crowded, linear or oblong, very obtuse, recurvo-submucronulate, on the upper surface longitudinally 3–5 lined, woolly beneath; heads in dense tufts or corymbs at the ends of the branches, subsessile or shortly pedicellate, ovate, 30–40-fl.; inv. scales oblong or oblongo-lanceolate, obtuse or subacute, bright-yellow, the outer cobwebby or woolly at base; recept. fimbrillate.

VAR. *a.* **glabriusculum**; leaves glabrous above, thinly tomentose beneath; heads corymbose; inv. scales obtuse. *H. trilineatum, DC.! l. c.*

VAR. *β.* **tomentosum**; younger leaves white-woolly on both sides, older glabrous above; heads densely tufted, subsessile; inv. scales sub-acute. *H. alveolatum, DC. l. c.*

VAR. *γ.* **brevifolium**; leaves shortly oblong (4 lines long, 1½–2 l. wide), thinly tomentose above, more densely beneath, the hairs yellowish; heads densely tufted, subsessile; outer inv. scales rufous, acute, inner bright-yellow, very obtuse. (Hb. Hk.!)

HAB. Witberg, 7–8000 f., *Drege!* Lostafelberg, 6–7000 f.; and Compasberg, *Drege!* Somerset, *Dr. Atherstone! β.* Witbergen, 6–8000 f., *Drege!* Basutuland, *T. Cooper,* 714. Var. *γ.* Mts of Natal, 100 miles inland, at 6500 f., *Dr. Sutherland!* (Herb. Hk., Sd.)

A stout, much-branched, small shrub, 6–12 inches high, variable in pubescence. Var. *β.* is dwarfer and much more copiously woolly, but too similar in general characters and habitat to be regarded other than a more alpine variety. Leaves 6–8 lines long, 1–2 lines wide. Heads 2–2½ lines long. The recept. is completely that of a *Lepicline.*

106. H. cymosum (Less.! Syn. 302); stem suffruticose at base, branching, branches virgate, thinly canescent; leaves half-clasping, *slightly* decurrent at base, linear-oblong or linear, mucronate, 3-nerved, spreading, woolly beneath, cobwebby or naked above, not scabrid; cymes densely much-branched; heads subpedicellate, separate, 10–12-fl.; inv. cylindrical, glabrous and glossy, the scales closely imbricating, golden-yellow, obtuse, the outer shorter. *DC. l. c.* 202. *Gnaph. cymosum, Linn. Sp.* 1195. *Th. Cap.* 649. *Berg. Cap.* 258. *Gn. tricostatum, Sieb. No.* 206. *Dill. Elth. fig.* 128. *Burm. Afr. t.* 76, *f. 3.*

VAR. β, minus (Sond.); leaves more closely set, much narrower and shorter, linear; cymes more contracted; heads shorter and narrower. *Zey.!* 883.

VAR. γ, pauciflorum; leaves close-set, shortly oblong, obtuse or sub-acute, 3-nerved, spreading or recurved, glabrous above; heads 5-flowered. *H. parviflorum,* var. *aureum, DC.!* (quoad exempl. *Eckl.!*)

HAB. Throughout the Colony, and in Caffirland; vars. α. and β., Natal, *Gerr. & M'K.* 293. Var. γ. Cape, *Burchell,* 63,315, and 4216, *Ecklon!* (Herb. Th. D., Hk., Sd.) Stems 1-2 feet hight, woody below, much-branched; branches long and leafy nearly to the cymes. Leaves in α. subdistant, 1½ inch long, 2-3 lines wide; in β. ½-¾ inch long, 1 line wide, but obviously 3-nerved. Invols. very bright, the heads not massed together and often shortly pedicelled. VAR. γ., referred to by DC. to *H. parviflorum,* has all the essential characters of *H. cymosum,* except the smaller fl. heads.

107. H. tenuiculum (DC. l. c. 203); stems suffruticose at base, very slender, flexuous, diffuse or ascending, branches long, suberect, tomentose-canescent; leaves half-clasping, subdecurrent, lanceolate-acuminate, 3-nerved, flat, thinly membranous, soon becoming glabrous above, whitish woolly beneath; heads 12-20 in dense, slightly-branched cymes, short pedicelled, oblong, 10-12-fl.; invol. scales obtuse, glossy, close-pressed, the outer fulvous, inner golden-yellow.

HAB. Dutoitskloof, 3-4000 f., *Drege!* (Herb. D., Sd., Hk.) Stems weak and straggling, 1-2 feet long, in tufts; branches 8-10 inches long, erect, laxly leafy upwards. Leaves horizontally spreading or deflexed, green above, of thin substance, ¾-1 inch long, 1-2 lines wide. Nearly allied to *H. cymosum,* but very much more slender, with thinner leaves.

108. H. simillimum (DC.! l. c. 203); stem suffruticose, slender, branched, the branches virgate, erect, white-woolly; leaves half-clasping, imbricate, erect, on both sides whitish-woolly, ovato-lanceolate, nigro-mucronate, 3-nerved, with subrevolute margins; heads many, in crowded, branching, subglobose cymes, short-pedicelled, 12-15-fl., heterogamous, oblong; inv. scales glabrous, oblong, yellow-fulvous, with squarrose-reflexed, obtuse tips.

HAB. Cape, *Krebs,* 211, fide *Sond.!* Kaffirland, *Ecklon!* between the Key and Gekau, *Drege!* near Butterworth, Kreili's Country, *H. Bowker!* (Herb. Sond. D.) Stems branched from the base; branches 8-10 inches long, closely leafy throughout. Leaves ½ inch long, 2½ lines wide at base, of thin substance, *evidently 3-nerved* (when viewed by transmitted light) and also *mucronate* in *Ecklon's* specimens. Heads 1½ lines long; inv. squarrose. According to *Sonder,* this agrees exactly with *Krebs'* 211, alluded to by *Lessing* in an obs. under *H. capitellatum* (Syn. p. 305). I have not seen *Drege's* specimens, which may come nearer to *H. melanacme* than those here described.

109. H. melanacme (DC.! l. c. 203); stem suffruticose at base, very slender, diffusely-branched, branches long, suberect, tomentose; leaves half-clasping, subdecurrent, lanceolate, one-nerved, nigro-mucronate, with subrevolute margins, both sides albo-tomentose, the younger erect, the old spreading or reflexed, at length denuded above; heads 12-20, in dense subglobose, nearly sessile cymes, short-pedicelled, oblong, about 12-fl.; inv. scales obtuse, glossy, close-pressed, the outer fulvous, inner golden-yellow.

HAB. Witberg, 6-7000 ft., and Camdebooberg, 4-5000 ft., *Drege!* (Herb. D., Sd., Hk.) A slender, apparently trailing, closely-leafy, thinly-woolly plant; young parts whitish. Leaves ½ inch long, 1 line wide. Heads 1½ line long, like those of *H. cymosum. DC.* says the heads are " 25-fl. ;" not in the Camdeboo specimens at least.

110. H. parviflorum (DC.! l. c. 203); stem shrubby, robust, much-branched, branches virgate, thinly canous ; leaves sessile, broad-based, linear or oblong, or lanceolate, one-nerved, on both sides canous-tomentose, often recurvo-mucronate, and concave ; heads very nume-rous, in densely much-branched, flat-topped cymes, subsessile, 5-fl., homogamous, cylindrical ; inv. scales glossy, semi-pellucid, imbricate in many rows, appressed, oblong, obtuse, pale-lemon coloured. *H. niveum, Less.! Syn.* 302. *Gn. niveum, Linn. Sp.* 1192. *Gn. cernuum, Th.! Cap.* 654. *Gn. parviflorum, Lam.*

Var. β. **longifolium** (DC.) ; leaves very long and narrow, acute, linear.

Var. γ. **latifolium** (DC. ex pte.) ; leaves oval-oblong or elliptical, obtuse or mucronulate.

Hab. Western Districts, common. Vanstaaden's River, *Drege.* Hassagayskloof, *Zey.!* 2873, 2870. Var. γ. Uitenhage, *Eckl.!* Between the Koega and Zwartkop's Riv., *Zey.!* 2871. (Herb. Th., D., Hk., Sd.)

A small shrub, 1-2 feet high ; branches 10-12 inches long. Leaves very variable in length and breadth ; commonly ¾-1 inch long, 1-2 lines wide ; in β. 1-1½ inch long, ½ line wide ; in γ. sometimes 1 inch long, 4-6 lines wide ! Heads constantly 5-fl. and *pale*-lemon colour. *Drege's* plant from Sneeuweberg, referred by DC. to var. γ., is (in some Herbaria at least) *H. subglomeratum, Less.*

111. H. callicomum (Harv.) ; stem shrubby at base, branches virgate, closely set with leaves throughout, and, as well as the leaves, densely woolly ; leaves sessile one-nerved, the lower obovato-spathulate, obtuse, mucronulate, the upper passing from linear-oblong to lance-linear, acute or acuminate, with subrecurved margins ; cymes loosely panicled, flat-topped, much branched ; heads 2-4 at the ends of the pedicels, subsessile, cylindrical, 4-5-fl., (2-3 herm., 2 fem.) ; invol. glabrous and glossy, its scales appressed, imbricate in many rows, acute, pellucid, straw-colour, the outer short, inner oblongo-lanceolate ; recept. narrow, short-toothed. *Zey.!* 886.

Hab. Mooye River, *Burke and Zeyher.* Basutuland, *T. Cooper!* 730. Zululand *Gerr & M'K.* 1035. (Hb. Hk. D., Sd.)

Stems 2-3-ft. high. Branches 1-1½-f. long. All parts thickly woolly, greyish. Lower leaves ¾-1 inch long, 2-3 lines wide at lip ; upper ¼-¾ inch long, the younger erect, old spreading or reflexed. Panicle corymbose, 2-4 inches across. Inv. with a glassy lustre.

112. H. aureo-nitens (Sch. Bip.) ; stems slender, subsimple, erect, canous, leafy throughout ; leaves half-clasping, on both sides tomentose, with subrecurved margins, one-nerved, the lowest obovate-spathulate, the upper oblong, mucronulate ; heads 12-20, in dense, subglobose cymes, 40-45-fl., subsessile ; inv. campanulate, imbricate, the scales glabrous, semi-pellucid, very broad and blunt, yellow or tawny ; recept. honey-combed and toothed. *Sch. Bip. Comp. Krauss, p.* 16.

Hab. Port Natal, *Dr. Krauss!* 280. Ranges 3-4000 f. *Dr. Sutherland.* (Hb. D. Hk., Sd.)

Probably growing in tufts, many stems from a woody crown. Stems 6-12 inches high, the taller disposed to branch below the short-peduncled cyme. Leaves 1 inch long, 2-3 lines wide, white, closely woolly. Inv. 1¼ line long.

113. H. subdecurrens (DC. l. c. 202); "stem suffruticose at base; leaves half-clasping, subdecurrent, linear, one-nerved, villous-tomentose beneath, the lowest subacute, uppermost mucronate-acuminate; branches leafless near the extremity, tomentose; heads ovate, 8–9-fl., crowded in a subglobose, compound cyme; invol. glabrous, not radiating, the scales closely imbricating, obtuse, fulvous-yellow." *DC. l. c.*

HAB. Cape, *Burchell*, 5970.
Said to differ from *H. cymosum* in the narrower, more acuminate, one-nerved leaves; ovate fl.-heads; and very short fimbrils.

114. H. intricatum (DC.! l. c. 204); stem suffruticose at base, slender, nude; branches long, laxly leafy, tomentose and pedunculoid upwards; leaves half-clasping, sub-decurrent, broadly oblong, obtuse, membranous, undulate, subrepand, setoso-scabrid, becoming nude above, loosely woolly beneath; cymes pedunculate, leafless, crowded, branched, hemispherical; heads oblong, 7–9-fl., subsessile, woolly at base; inv. scales shining, oblong, obtuse, the outer fulvous, inner golden-yellow.

HAB. Swellendam and George, *Ecklon!* (Herb. Sond.)
Stems weak and straggling, 1–2 f. long; branches 10–12 inches long. Leaves an inch apart, spreading, thin in substance, at first cobwebbed above, then green, but scabrous, ¾–1 inch long, 2–4 lines wide. Cymes 1 inch across. Head 1½ line long.

115. H. pannosum (DC.! l. c. 204); the subsimple stem, leaves, and peduncles densely clothed with interwoven, white wool; leaves distant, the lowest cordate-clasping at base, oval-oblong, acute, the upper half-clasping, linear-oblong, or lance-linear, acuminate; corymb loosely compound or panicled; heads clustered at the end of the pedicels, sessile, woolly-based, about 10-fl., cylindrical; inv. scales glabrous, ovate-acute, closely imbricate, fulvous, with yellow tips or lemon-yellow.

HAB. Between Omsamwubo and Omsamcaba, and Omcomas and Omblas, *Drege.* Umginto, Natal, *Gerr. & M'K.* 1036. (Herb. D., Hk., Sd.)
Stem 2 feet high, flexuous, ending in a many-forked corymbose-cyme. Leaves 1½–2 inches apart; the lower 2½ inches long, 1 inch wide; upper as long, only half as wide, complicate. All parts clothed with white, closely interwoven wool. The lower leaves are *several*-nerved; but the nerves completely hidden under the wool.

116. H. maritimum (Less.! Syn. 304); stem shrubby, branched; branches long, woolly; leaves half-clasping, oblong or lance-oblong, acute, nigro-mucronate, with subrecurved margins, softly hairy above, woolly and whitish beneath; heads many, in dense, flat-topped branching, bracteated cymes, 20–25-fl., campanulate; outer invol. scales herbaceous, dorsally silky, inner oblong, obtuse, scarious, opaque, with spreading, straw-coloured tips. *DC. l. c.* 204. *Gn. maritimum, Linn.* Th. Cap. 649, and *G. molle, Th.!* 653. *Gn. dasyanthum, Willd. Gn. serratum, Linn. Sp.* 1194. *Burm. Afr.* 76, *f.* 3.

VAR. β. **microphyllum** (DC.!); leaves much smaller, blunt, undulate, concolourous, thickish, densely and equally tomentose on both sides.

HAB. Sandy ground, near the coast in the Western Distr. Stellenbosch, *Ecklon!* Swellendam and Outeniquasland, *Drege!* Var. β. Kamiesberg, *Drege, Ecklon!* (Hb. Th., D., Hk., Sd.)
Stem woody, erect or straggling, distantly branched. Leaves spreading, ½–1 inch long, 3–4 lines wide, pale-green above, grey-white beneath. Inner inv. scales shortly radiating, pale straw-colour, not "golden."

117. H. plebeium (DC.! l. c. 206); stem shrubby, branches virgate, closely leafy below, laxly leafy and tomentose upwards; leaves sessile, linear or oblong-linear, narrowed above the base, mucronulate, with revolute margins, the younger on both sides woolly, the older cobwebby above or glabrate; heads many, in densely-branched, hemispherical cymes, 12–15-fl.; inv. campanulate, the outer scales dorsally silky, the inner longer, spreading, obtuse, tawny; fimbrils long, white.

HAB. Cape, *Burchell!* 6568; Swellendam, *Mundt ! Ecklon !* (Herb. Sd., Hk.)
A small shrub with something the aspect of *H. anomalum,* but having longer leaves and hairy involucres, something like those of *H. maritimum.* Leaves ¾–1 inch long, 1–2 lines wide. Cymes an inch diameter, subcapitate. Inv. 2 lines long.

118. H. anomalum (Less.! Syn. 303); stem shrubby, much-branched, branches canescent; leaves sessile, linear (rarely oblong), with revolute margins, erect or erecto-patent, obtuse, on both sides albo-tomentose; heads many-fl., pedicellate, in simple or branching corymbs; inv. ovate or obovate-turbinate, outer scales closely imbricate, rufous-brown, membranous, glabrous, very obtuse, inner with spreading, golden-yellow, opaque, obtuse tips; pappus of one or two short bristles. *DC. l. c.* 204. *Gn. niveum, Th.! Cap. 654, non Linn. Manopappus anomalus, Sch. B.*

VAR. *a.* **turbinatum;** inv. obovate-turbinate, tapering at base, descending along the pedicel; leaves linear, erect.

VAR. *β.* **ovatum;** inv. ovate, obtuse at base; leaves linear, erect or erecto-patent.

VAR. *γ.* **brevifolium** (DC.); inv. ovate, obtuse at base; leaves elliptic-oblong, 3-4 lines wide, with subrevolute margins, scabrous and densely woolly, with interwoven white hairs on both sides. (Herb. Sond.)

HAB. Eastern Districts, common. Var. *γ.* near Grahamstown, *Ecklon!* (Herb. Th., D., Hk., Sd.)
A rigid, much-branched bush, 1–1½ f. high. Leaves 6–8 lines long, not 1 line wide. Heads 2 lines long in *β.*; in *a.* tapering into the pedicel indefinitely.

119. H. hamulosum (E. Mey.!); stem shrubby, branching, branches tomentose; leaves crowded, sessile, linear, with a recurved point and revolute margins, glabrous above, white-woolly beneath; heads in dense, branching or simple, terminal corymbose-cymes, subsessile, about 5-fl., homogamous; invol. oblong, not radiating, imbricate, its scales pale-yellow, glabrous and glossy, erect, lanceolate, acute or acuminate, all similar. *DC. l. c.* 192.

HAB. Nieuweveld, on hills, &c., near Bokpoort, 3500-4500 f., *Drege !* Cape, *Ecklon !* (Herb. Sond.)
A small shrub, about a foot high, the lower part of stem nude; the branches and twigs leafy and woolly. Leaves ½–1 inch long, ½ line wide, one-nerved. Recept. fimbril-toothed, not "*naked,*" as stated by *DC.* Very near *H. Kraussii,* but the involucres are longer (3 lines long), and scales more acuminate; the flowers all hermaphrodite, and the leaves very constantly hook-pointed.

120. H. Kraussii (Sch. Bip.! Bot. Zeit. 1844, p. 679); stem shrubby, robust, branches virgate, thinly canous; leaves sessile, linear, with strongly revolute margins, mucronate, at first cobwebby, afterwards glabrate above, tomentose beneath; heads very many, in densely much-branched, flat-topped cymes, subsessile, 5-fl. (4 herm., 1 fem.), cylindrical; inv. scales glossy, semi-pellucid, appressed, imbricated, oblong, subacute, pale lemon-yellow or straw-colour. *Sch. Bip. Comp. Krauss, p.* 15.

HAB. Natal, *Krauss!* 459, *Mr. Hewitson!* Delagoa Bay, very common, *Capt. Speke!* (Herb. D., Hk.)

1–2 feet high, with many long, virgate, closely leafy branches. Leaves spreading or reflexed, 7–9 lines long, ½–¾ line wide, the whole under surface often completely hidden by the strongly revolute margins. Heads 1½–2 lines long, not unlike those of *H. parviflorum*, but heterogamous, with rather more acute scales. Still more nearly allied to *H. hamulosum.*

121. H. teretifolium (Less.! Syn. 312) ; stem shrubby, much branch-ed, branches closely leafy ; leaves sessile, linear, rigid, with revolute margins, recurvo-mucronate, glabrate above, tomentose beneath ; heads many-fl., subsessile or pedicellate, in simple or branching corymbs ; invol. subglobose, glabrous, loosely imbricate, the outer scales tawny or rufous, short, membranous, inner paler (creamy), undulate, spreading, very obtuse ; fimbrils twice as long as the ovary. *DC.! l. c.* 205. *Gn. teretifolium, Linn. Th.! Cap.* 644. *Gn. ericoides, Lam. Burm. Afr. t.* 77, *f.* 3.

VAR. β. **Natalense**; inv. scales ovate, acute or subacute, all rufous, or the inner paler. *Krauss!* 291.

HAB. Throughout the Colony and in Caffraria. β. Natal, *Krauss! Sutherland!* (Herb. D., Hk., Sd.)

A much-branched, ramulous bush, a foot or more high. Leaves 4–8 lines long, ½ line wide, spreading, mostly hook-pointed. Heads sometimes on long pedicels, commonly subsessile. Inv. sometimes very pale, creamy ; sometimes cinnamon-brown ; commonly 2-coloured, the outer scales deeper in colour. Var. β. only differs in its acute inv. scales. *Gn. heterophyllum,* Th.! referred here by *Lessing,* belongs, according to Herb. Thunb. to *H. erosum.*

122. H. rugulosum (Less.! Syn. 307); stem suffruticose at base, branches long, simple, densely-leafy, whitish ; leaves sessile, wrinkled and cob-webby above, white-woolly beneath, three-nerved, the lower oblong, acute, the upper lanceolate, acuminate, with subrevolute margins ; heads many-fl., in simple or branched corymbose-cymes, subsessile or pedi-celled ; inv. slightly cobwebbed at base, campanulate, loosely imbricate, the scales broadly oblong, membranous, undulate, obtuse, rufous, squar-rose, the innermost paler, spreading ; fimbrils short. *DC. l. c.* 205.

HAB. Plettenbergs Bay, *Mundt & Maire.* Assagaysbosch, Albany, *Zeyher.* Uit-enhage, *Ecklon!* Bushman's River, Ado, and the Zuureberg, &c., *Drege!* (Herb. D., Hk., Sd.)

Stems branched near the base; branches 10–15 inches long, leafy throughout. Leaves 1–1¼ inch long, 2 lines wide ; sometimes the upper ones very long and nar-row, ⅓ line wide. Inv. either pale, tawny, reddish, or brownish-purple. Less woody and branching than *H. teretifolium*, with broader leaves. Sometimes confounded with *H. erosum*, but always to be known by its involucre. Heads 35–40-fl.

123. H. capitellatum (Less.! Syn. 305); stem suffruticose at base, branches long, erect, tomentose, laxly leafy ; leaves broad-based, half-clasping, oblongo-lanceolate, acuminate, 3-nerved, loosely woolly on both sides (the old becoming naked above) ; heads in densely much branched, subglobose cymes, 12–15-fl., heterogamous ; inv. glabrous, campanulate, scales loosely imbricated, squarrose, undulate, submem-branous, obtuse, rufous or cream-coloured ; fimbrils long. *DC. l. c.* 206. *Gnaph. capitellatum, Th.! Cap.* 653.

HAB. Table Mt. sides, *Bergius, Eckl.!* Cape Flats, *Wallich!* French Hoek, *Thunb.* Common in the Western Districts, Swellendam, *Mundt,* &c. (Hb. Th., D., Hk., Sd.)

Allied to *H. rugulosum* in foliage and involucres ; but with much smaller fl. heads, longer fimbrils, &c. Leaves spreading or deflexed, ¾-1 inch long, 2-4 lines wide at base. Heads 1½ line long. *Drege's* specimen of "*H. capitell. β. mollius,* DC." in Hb. Hook. is a more woolly state of this species ; but specimens also from *Drege,* and similarly marked in Hb. Sond., belong to *H. erosum,* γ. *concolorum.*

124. H. erosum (Harv.) ; stem shrubby, much branched, branches whitish, leafy ; leaves sessile, linear or oblong, the broader 3-nerved, the narrow 1-nerved, all subacute or obtuse, whitish-tomentose on the under, glabrous or cobwebby on the upper surface (in var. γ. woolly on both sides); heads numerous, in simple or branching corymbs, subsessile, 20-fl., heterogamous; inv. oblong, glabrous or glossy, not radiating, the scales closely imbricated, concave, erect, obtuse, cream-colour or rufous ; fimbrils short. *Hel. rosum, Less. Syn.* 306. *DC. l. c. p.* 205. *Gn. rosum, Berg. Cap.* 260. *Gn. stoechas, Th.! Cap.* 647, *non L.,* also *Gn. hetero-phyllum, Th. ! Cap.* 644.

Var. *a,* **angustifolium**; lowest leaves narrow oblong, upper linear.

Var. *β.* **latifolium**; lower and medial leaves oblong, distinctly 3-nerved; upper lanceolate.

Var. γ. **concolorum** (Sond. !); leaves *on both sides* densely white-woolly. *H. con-colorum, DC.! l. c. p.* 206.

Hab. Cape, *Thunberg, Krebs.* 150,209. Paarlberg and Bosjesveld, *Drege!* Stellen-bosch, Swellendam, and Uit., *Ecklon!* Uitenhage and Albany, common. Var. γ. George, Swell., and Uit., *Eckl.!* Vanstaadensberg, *Zey.!* (Herb. Th., D., Hk., Sd.)
Very variable in foliage and pubescence: constant in its involucre. Except in being copiously woolly, *H. concolorum* does not differ, and this character varies very much in different specimens. Dr. Sonder informs me that a specimen in Hb. Berol. marked by Lessing as "*H. rosum*" entirely agrees with DC.'s *concolorum.* I venture to substitute "*erosum*" for the unmeaning "*rosum*" of authors. Bergius, with whom this latter word originated, says of the leaves " *margine reflexo-undulato dentatis ;*" a character nearly synonymous with *erosis.* The " e" apparently dropped out by a printer's error.

125. H. Dregeanum (Sond. and Harv.); stem half-shrubby, slender, diffuse, branching, branches closely leafy; leaves sessile, lance-linear or linear, one-nerved, with revolute margins, on both sides loosely woolly or nude above ; heads 3-4 or several, corymbulose, subsessile at the ends of the branches 20-25-fl., heterogamous; inv. glabrous, campanu-late, its scales imbricate, membranous, pellucid, very obtuse, rufous; recept. with short, toothlike prominences. *H. scabrum, DC.! non Less. (quoad specim. Drege!) p.* 205.

Hab. Cape, Hb. *Ecklon!* Stormberg, 5-6000-f., *Drege!* Witteberg, Albert, *T. Cooper,* 617. (Herb. Sd., D., Hk.)
A small, slender shrublet, with diffuse or trailing stems, and erect branches. Leaves imbricating, 4-6 lines long, ½-1½ line wide. Heads 2 lines long. *De Candolle's* descript-ion does not agree with *Drege's* specimens, in which the leaves are neither "densely punctato-scabrid " nor " hispidulous " nor "acuminated." *Mr. Cooper's* specimens are rather less woolly, more luxuriant, and with as many as 12-20 heads in a corymb.

126. H. tricostatum (Less. Syn. 310); stem shrubby, branches spread-ing, virgate, tomentose; leafy nearly to the summit; leaves tapering to the base, minutely petiolate, oblong or lanceolate-oblong, acute, with subrecurved margins, 3-nerved, thinly tomentose, the uppermost narrow and one-nerved; heads in dense, branching, flat-topped cymes, 20-fl., cylindrical; inv. scales imbricate, erect, bullated, obtuse, pale horn-colour or creamy, semipellucid. *DC. l l.c.* 209. *Gn. tricostatum, Th. l Cap. p.*657.

HAB. Picketberg, *Thunberg!* Olifantriver, *Drege!* (Herb. Th., Hk., D., Sd.)
Branches 1–2 feet long, straight, laxly leafy, the upper leaves narrower, continued to the base of the inflorescence. Stems white. Leaves ¾–1 inch long, 2–4 lines wide; the petiole 1 line long. Heads 2½ lines long, narrow, woolly at base. Recept. fimbrilliferous.

127. H. revolutum (Less.! Syn. 305); stem shrubby at base, branches woolly, leafy and ramulous, leaves half-clasping, oblong or linear-oblong, above cobwebby and scaberulous, beneath densely woolly, with revolute margins, mucronate, heads numerous in much-branched, loose, corymbose-cymes, about 25-fl., heterogamous; inv. campanulate, glabrous, scales closely imbricated, glossy-subpellucid, erect, broadly oblong, very obtuse, rufous or horn-colour; fimbrils short. *DC. l. c.* 206. *Gn. revolutum, Th.l Cap.* 652. Also *H. leiolepis, DC.l l. c.* 187.

HAB. About Table Mt., *Bergius.* Rietvalley, *Mdt. & Mre.* Groenekloof and Paarl, *Drege!* Zwartland, *Ecklon!* Brackfontein, *Ecklon!* Tulbagh, *Dr. Pappe!* (Hb. Th., D., Hk., Sd.)
Stems 1–1½ f. high. Leaves spreading or reflexed, 1–1½ inch long, 2–4 lines wide, at length generally denuded above but persistently rough, with small, raised points; sometimes densely woolly on *both* sides. Inv. 2½ lines long. I have compared *Ecklon's* original specimens of "*H. leiolepis*" DC. and find them absolutely identical with ordinary *H. revolutum*, which varies in the paler or deeper colours of the inv. scls.

128. H. scabrum (Less.! Syn. 312 non DC.); stem suffruticose, much-branched, the branches spreading, thinly tomentose and glandular; leaves half-clasping, slightly decurrent, oblong or oblongo-lanceolate, the lowermost tapering at base, obtusely repand, undulate, with revolute margins, above glandularly-hispid or very scabrid, beneath glandular, cobwebbed or naked; heads in simple or branching, loose corymbs, pedicellate, campanulate, 20–25-fl.; inv. scales glabrous, oblong, very obtuse, pellucid, pale horn-colour, fulvous at the extremity. *Gn. scabrum, Th.! Cap.* 655. *Hel. repandum, DC.! l. c.* 203.

VAR. β. **scaberrimum**; stems and leaves echinate-scabrous, very rough, glandular.
VAR. γ. **microphyllum** (DC.); more woolly and less glandular; leaves shorter, narrower, more strongly repand, with fewer gland-bristles.

HAB. Caledon and Worcester, *Ecklon!* Namaqualand, *V. Schlicht!* β. Caledon, *Ecklon!* γ. at Kamiesberg, *Drege!* (Herb. Sd., D., Hk.)
A bush, about 1 foot high, ramulous, and resinous gland-dotted. Leaves ½–1 inch long, 1–3 lines wide, spreading, the upper ones with strongly revolute margins. Heads 2½ lines long.

129. H. petiolatum (DC. l. c. 208); stem shrubby or suffruticose, branches woolly, naked and pedunculoid at the extremity; leaves petiolate, ovate, obtuse or subacute, woolly-tomentose; heads in densely-branched, woolly corymbose-cymes, subglobose, glabrous; inv. scales oblong, obtuse, cream-white. *Gn. petiolatum, Linn.,* non *Th.*

HAB. Essenbosch, *Burchell,* 4825. Swellendam and George, *Ecklon!* Sneeweberg, *Drege!* (Herb. D., Hk., Sd.)
Stems 2 or more feet long, flexuous, loosely branched. Petioles ¼–⅓ inch long, slender, clasping at base; lamina ½–1 inch long, 4–8 lines wide, greenish above, the nerves not obvious through the wool. Corymbs 1–2 inches across. Allied to *H. crispum* but with distinctly petioled, flat leaves.

130. H. hypoleucum (Harv.); stem shrubby or suffruticose, branches widely spreading, quite woolly, more or less pedunculate at the extremity; leaves on narrowly winged petioles which expand at base into clasping auricles, cordate-ovate, acute or acuminate, 3–5-nerved, the young ones cobwebby, the adult nude and glandularly scabrous above, all very white-woolly beneath, membranous; heads in dense, subglobose, glomerate, woolly cymes, sessile, about 12-fl., heterogamous; inv. woolly at base, the inner scales oblong, obtuse, cream-white.

HAB. Drakensberg, Orange Free State, *T. Cooper*, 1025. Umgate and Ingoma, Natal, *Gerr. &. M'K*. 1005. (Herb. D.)

Stem slender, straggling, 2–5 feet high. Branches terete, flexuous, snow-white. Leaves on winged, clasping petioles, ½–1 inch long; the lamina 1–1½ inch long, ¾–1 inch wide, full green above, snow-white beneath. Heads few-fl. in globose tufts, nearly ¼ inch diam. Female fl. 3–4. Teeth of recept. subulate, yellow.

131. H. crispum (Less. Syn. 310); stem shrubby or suffruticose, branches widely spreading, woolly, more or less pedunculate at the extremity; leaves half-clasping and eared at base, fiddle-shaped, obtuse, undulate, thinly woolly above, white-woolly beneath; heads in densely-branched, woolly, corymbose cymes, subglobose, glabrous; inv. scales oblong, obtuse, cream-white or primrose colour. *DC. l. c.* 208. *Gn. crispum, Linn. Gn. divaricatum, Berg. Cap.* 250, *Th.! Cap.* 657. *G. polyanthos, Th. l. c. (but not p.* 643*). Gn. auriculatum, Lam. non Th.*

VAR. β. **citrinum**; inv. scales pale citron or primrose-colour. *Zey.!* 2875.

HAB. Round Capetown and on the Flats. Paarl, *Drege! Zey.!* 2861. Var. β. on the Cape Flats, *Wallich! Zeyher!* (Herb. Th., D., Hk., Sd.)

Stems 2–4 feet long, flexuous, with widely spreading branches, all parts woolly. Leaves ½–1½ inch long. very wide at the extremity, contracted in the middle, expanded and stem-clasping at base. Heads as in *H. petiolatum.*

132. H. auriculatum (Less.! Syn. 311); stem shrubby or suffruticose, branches widely spreading, woolly, pedunculoid at the extremity; leaves half-clasping and eared at base, fiddle-shaped, obtuse, subundulate, thinly woolly above, white-woolly beneath; heads in rather loosely branched, woolly (commonly few-headed) corymbose-cymes, many-fl., subglobose, glabrous; invol. scales ovate, acute or subacute, somewhat radiating, milk-white or rufous. *DC. l. c.* 209. *Gn. auriculatum, Th.! Cap.* 657. *Gn. patulum, Berg.*

VAR. β. **panduratum**; leaves very large, much less constricted below the middle, the wide extremity 3-nerved, mucronate, thinly tomentose; cymes very much branched, many-headed; heads smaller, inv. scales less acute, *G. & M'K.!* 256, 273, 856.

HAB. Table Mt. and neighbouring hills, *Thunb., Burchell, Ecklon! W. H. H.!* Simon's Bay, *C. Wright!* 320, 361; Winterhoek, Tulbagh, *Pappe!* R. Zonderende, *Zey.!* 2860; Omsamwubo and Omsamcaba, *Drege!* Var. β. near D'Urban, and also betw. Maritzburg and Ladysmith, Natal, *Gerr. & M'K.!* (Herb. Th., D., Hk., Sd.)

Very like *H. crispum*, but with somewhat larger fl. heads, laxer inflorescence, and more spreading and acute inv. scales. Var. β. has the dense, many-headed inflorescence of *H. crispum*, but larger heads, and very large and broad leaves, rather slightly contracted below the middle. Stems 2–4 feet long. Leaves ½–1½ inch long; in β. 1½–2½ inches long, 1–1½ inch wide above, ½–¾ inch wide in narrow part.

133. H. Zeyheri (Less.! Syn. 309); stem suffruticose, branching; branches virgate, white-woolly; lowest leaves obovate, narrowed at base sub-petiolate; cauline and rameal leaves sessile, oblong, ovate-oblong or obovate, acute or acuminate, callous-mucronate, one-nerved, on both sides closely and densely white-woolly; heads cylindrical, 5-fl., in branching, terminal corymbose cymes, subsessile; invol. glabrous, its scales erect, 3–5-stichous, close-pressed, not radiating, oblong, obtuse or truncate, snow-white; recept. narrow, fimbrilliferous.

VAR. a. **Zeyheri** (DC.); root leaves obovate; cauline subundulate, broad-based, oblong or obovate-oblong, obtuse or acute; wool copious. *H. Zeyheri, DC.! l. c.* 196.

VAR. β. **intermedium** (DC.); lower cauline leaves obovate, tapering much at base, upper subundulate, obovate-oblong or oblong, obtuse, or acute; wool rather copious. *H. Burchellii,* β. *intermedium, DC. l. c.* 196.

VAR. γ. **Burchellii** (DC); lower leaves cuneate, upper oblong or lance-linear, mostly acute, partly infolded, scarcely undulate; wool less copious and closer. *H. Burchellii,* a. *complicatum, DC.! l. c.* 196.

HAB. Var. a. Winterfeld, Beaufort, *Ecklon!* Heerelogement, *Zey.!* 888. Dutoits Kloof and Paarl, *Drege!* Paarl, *Rev. W. Elliott!* Winter Hoek's Berg, *Dr Pappe!* Lamberts Kloof, *Wallich!* β. Karroo and Zwarteberg, *Drege!* George, *Ecklon!* Algoa Bay, *Zeyher!* Var. γ. Beyond the Gariep, *Burchell,* 2506. Uitenhage, *E. & Z.!* Brackriver, *Zey.!* 884 and Rhinosterkop, *Zey.!* 885. (Herb. D., Hk., Sd.)

Stems woody at base, 8–12–18 inches high, branched in the lower part; the branches erect, elongate, mostly virgate, leafy throughout. Leaves rather close, thickish, ¾–1 inch long, 2–6 lines wide, very variable in shape in different specimens. Lower leaves almost always obovate; medial varying much; uppermost narrow, oblong or linear. All parts are canescent, but the wool is sometimes close, sometimes more loose. Heads 2–3 in a cluster, collected in branching cymes, 1–1½ inch across; each head 2 lines long, not a line diam.—I find it impossible to keep *H. Burchellii, DC.* separate by any tangible character.

§ 17. CHIONOSTEMMA. (Sp. 134–135.)

134. H. vestitum (Less.! Syn. 314); shrubby below, suffruticose above, stem and leaves very thickly clothed with whitish, interwoven wool; leaves oblong or linear-oblong, or tongue-shaped, nigro-mucronate, the uppermost on the fl. branches tipped with a scabrous-membranous scale; heads *(large)* terminal, solitary or aggregated, very many-fl.; invol. globose, radiating, glabrous and glossy, scales snow-white, loosely imbricating in very many rows, all lanceolate, acuminate; recept. clothed with acuminate, scale-like fimbrils, longer than the ovaries; pappus bristles concrete at base, minutely rough. *DC.! l. c.* 209. *Hel. speciosum, Thunb.! Cap.* 664, *excl. syn. pl. Xeranth. vestitum, Linn. Sp.* 1201, *non Th. Hel. lanatum, Schrank. Xeranth. variegatum, Andr. Bot. Rep. t.* 384, *non Berg. Burm. Afr. t.* 66, *f.* 1. *Leucostemma lingulatum, Don.* (leaves wider and less densely woolly).

HAB. Rietvalley, *Bergius.* Drakensteinberg, *Drege!* Near Capetown, also in Worcester and Caledon, *Eck.!* Hott.-holland, *Mundt!* Simon's Bay, *C. Wright,* 374, 376, *W. H. H., Dr. Hooker,* &c. Zwarteberg, *Zey.!* 865. (Herb. Th., D., Hk., Sd.) Stem 1–2 feet high, robust, densely leafy; branches short or long, erect, virgate, laxly leafy, the upper leaves smaller, and degenerating to bracts, tipped with a white scale. Leaves 2–2½ inches long, ¼–1 inch wide. Heads an inch diameter.

135. H. Ecklonis (Sond.!); stems short, herbaceous, ascending-erect, with the leaves thickly clothed with *loosely* interwoven wool; subradical leaves lance-oblong, tapering much at base, nigro-mucronate, cauline linear or lance-linear, acute, the uppermost sometimes tipped with a scale; heads (large) terminal, solitary, very many-fl.; inv. globose, radiating, glabrous and glossy; scales whitish ("pale straw," *Sond.*), loosely imbricating in many rows, all lanceolate, acuminate; pappus bristles scarcely concrete at base, serrulate along the whole bristle, and barbellate near the extremity; corolla yellow.

HAB. Cape, *Ecklon !* (Herb. Sond.)

Stems 4–5 inches long, simple, incurved. Radical and subradical leaves 2½–3 in. long, 4–5 lines wide; cauline shorter, narrower, and more acuminate. *Dr. Sonder* has the following note on this plant :—" De Candolle considers this as falling under *H. vestitum*, β. *ligulatum*, but from other specimens of that variety, as well as from the primary form, this differs by its humbler habit, much looser wool, and invol. scales pale-straw (not shining-white). But the specific differences become much more evident, when the flowers are accurately examined. For in *H. vestitum* the fl. are yellow in their lower, purple in their upper half; in *H. Ecklonis* altogether yellow. In *H. vestitum* the pappus bristles are united at base into a broad, glabrous ring, the serratures remote and very short, and toward the apex the bristle is evidently clavate; in *H. Ecklonis* the bristles are scarcely united at base, along their whole length armed with close-set serratures, more than thrice as long as the diameter of the bristle, and towards the apex scarcely thickened. Achenes in both plants similar. This therefore, certainly, is a distinct species." [I admit the differences, as pointed out by my colleague, but do not regard them of *specific* value.—*W.H.H.*]

§ 18. EDMONDIA. (Sp. 136–137.)

136. H. humile (Andr. Bot. Rep.t. 652); ascending, branches curved upwards, leafy throughout; leaves from a clasping base, subulate, rigid, pungent or obtuse, glossy, round-backed, the uppermost (on the flower branches) scale-bearing or reduced to membranous scales; inv. turbinate, the scales rosy, lanceolate-acuminate, radiating, the medial stipitate; fimbrils linear, very short; achenes smooth; pappus clavulate, finely serrated. *Less.! Syn. p.* 322. *Helipterum humile, DC. l. c.* 322. *Aphelexis humilis, Don. Helichrysum sesamoides, Curt. Bot. Mag. t.* 425. *H. spectabilis, Lodd. Cat. t.* 59.

HAB. Crevices of rocks on mts., Table Mt., and Stellenbosch Mts.; Simon's Bay, *Milne !* Baviansberg, Genadendal, *Dr. Pappe !* (Herb. D., Hk., Sd.)

Stems diffuse, 6 inches to 2 feet high. Leaves very variable in length, sometimes uniformly 2–3 lines, sometimes ½–1 inch long, shining, glabrous, with wool in the axil. Invol. scales varying from pale flesh-colour to deep rose-red. The pappus is scarcely serrulate, but gradually thicker upwards.

137. H. sesamoides (Thunb. ! Cap. 661); erect, suffruticose, laxly branched; branches virgate, incurved, leafy throughout; leaves half-clasping, subulate, rigid, glossy, acute or obtuse, keeled, the lowest long and filiform, spreading, the rest short, close-pressed, those under the heads scaly membranous; invol. campanulate or subturbinate, all the scales lance-acuminate, except the outermost shortly stipitate or subsessile, lustrous; fimbrils longer than the ovaries; achenes mostly granulate; pappus barbellate. *Xeranth. sesamoides, Linn.*

VAR. α. **Wildenowii**; inv. scales white or rosy; fimbrils bristle-shaped, very slender. *Hel. sesamoides, Less. ! Syn.* 322. *Helipterum sesamoides, DC. l. c.* 214. *Burm. Afr. t.* 67, *f.* 2.

VAR. β. **heterophyllum**; invol. scales white, or the outermost brown-tipped; fimbrils broadish, flat or conduplicate and keeled (variable in breadth). *Helipt. heterophyllum, DC. l. c.* 214. *Xeranth. fasciculatum, var. Andr. Bot. Rep. t.* 279.

VAR. γ. **fasciculatum**; inv. scales pale lemon-yellow or primrose, the outer fulvous; fimbrils as in β. *Helipterum fasciculatum, DC.! l. c.* 214. *Xer. fasciculatum, Andr. Rep. t.* 242.

Var. δ. filiforme ; inv. scales pale lemon-yellow, the outer fulvous, "inner scales more or less obtuse ' *(Less.);* fimbrils as in β. *Helipt. filiforme, DC. l. c.* 215. *Helichr. filiforme, Less.!*

Hab. Rocky, mountain situations, in the Western Districts, common. (Herb. Th., D., Hk., Sd.)

Stems 1–2 feet high, slender. Lower leaves loosely set, 1–2 inches long, flexuous; upper 2–3 lines long, closely applied to the stem. Invol. lustrous, satiny, variable in colour, white, rosy, pale-lemon, or mottled. Heads 1–1½ inch across, very handsome. The above four vars. are precisely similar in all respects, except in the trivial differences assigned to each : and these I find variable. The fimbrils are not to be depended on ; all breadths may be found, if a sufficient number of heads be examined. I have no hesitation therefore in restoring the old Linnæan species in place of the four into which it has been split.

LXXIX. HELIPTERUM, DC.

Character the same as in *Helichrysum,* except : *Pappus* plumose ! *Heads* in the Cape species homogamous. *Recept.* honeycombed. *DC. Prodr. 6, p.* 211.

Suffrutices or herbaceous plants, natives of Africa and Australia. Stems and leaves woolly. Leaves alternate, simple, entire. Heads showy, 10 or many-fl., with white, yellow, red or purple, glossy scales. The name is a contracted expression for a *Helichrysum* with a plumose pappus. As a generic group it must be regarded as *purely* artificial; there is no difference in habit from *Helichrysum,* and, among our Cape species, *Helichrysum argenteum* is closely similar to *Helipterum argyropsis;* and *Helichrysum vestitum* to *Helipt. speciosissimum.* DC.'s section " *Edmondia*" having *serrate,* not *plumose* pappus, is removed to *Helichrysum ;* many species of *Helichrysum* have similar pappus.

Sect. 1. ASTELMA. Inv. scales glabrous and glossy, ovate, oblong or lanceolate, spreading or close-lying. (Sp. 1–10.)

Heads 1–3 at the ends of simple branches :
  Inv. scales not yellow :
    Heads nodding. Inv. scales obtuse ... ... (2) **variegatum.**
    Heads erect. Inv. scales acuminate :
      Heads on leafless peduncles ... ... (1) **speciosissimum.**
      Heads on closely leafy branches :
        Lvs. small (2–10 lines long) ... ... (4) **canescens.**
        Lvs. 1½–2 inches long, 3–4 lines wide (3) **Zeyheri.**
  Inv. scales lemon-yellow :
    Fl. branches leafy throughout. Inv. sc. acute... (5) **ferrugineum.**
    Fl. branches pedunculoid at top :
      Heads 12–15-fl. Inv. scales acute ... ... (6) **virgatum.**
      Heads many-fl. Inv. scales obtuse ... ... (7) **citrinum.**
Heads corymbose, on branching peduncles :
  Inv. scales spreading, white, acute. Lvs. spathulate (8) **argyropsis.**
  Inv. scales close-lying, obtuse. Lvs. ovate or lanceolate.
    Heads oblong, 10–12-fl., innumerable ... ... (9) **phlomoides.**
    Heads globose, 50–60-fl., ruby-red ... ... (10) **eximium.**
Sect. 2. SYNCARPHA. Inv. scales woolly at base, close-lying and erect, ending in a reflexed or spreading, membranous point. (Sp. 11–12).
  Lvs. obovate or oblong. Tip of the inv. scales *broad,* obtuse, ruby-red ... ... (11) **Dregeanum.**
  Lvs. linear. Tip of the inv. scales very *narrow,* acuminate, strongly reflexed ... ... (12) **gnaphaloides.**

1. **H. speciosissimum** (DC.! l. c. 211) ; stem robust, suffruticose, simple or sparingly branched ; branches closely leafy for about half their length, pedunculoid upwards, densely and softly white-woolly as well as the leaves; leaves elliptic-oblong or oblong, half-clasping, acute,

one-nerved, thick; heads solitary, peduncled, very many-fl.; inv. campanulate, loosely radiating, the outer scales sessile, ovate or ovato-lanceolate, inner stipitate, elongate, lanceolate-acuminate. *Helichr. speciosissimum, Willd. Less./ Syn.* 315. *Hel. Stœhelina, Thunb./ Cap.* 664. *Xeranth. speciosissimum, Linn. Andr. Bot. Rep. t.* 51.

VAR. β, **angustifolium** (DC.); more slender, with longer peduncles; leaves linear, 2–3 lines wide. *DC. l. c.* 212. *H. seminudum, Sch. Bip. Bot. Zeit.* 27, 692.

VAR. γ, **glabratum**; toment yellowish; leaves glabrous beneath or nearly so, laxly woolly above; inner inv. scales less acuminate. *H. glabratum, Sond/ in Linn.* 23, *p.* 66.

VAR. δ, **polycephalum**; stem divided at the summit into many short, scaly, subpedunculoid, one-headed branches; leaves lance-linear, woolly on both sides; inv. scales narrow-lanceolate, acuminate (whiter than in α), the inner shorter and bluntish.

HAB. Table Mt., and the Cape Flats, common. Simon's Bay, *C. Wright,* 373. Kirstenbosch, *Zey./* β. Klynrivierkloof, *Zey./* Genadendal, *Pappe/* &c. γ. Klynrivierkloof, *Zey./* 2839. δ, near Swellendam, *Ecklon!* (Herb. Th., D., Sd., Hk.)

Stems 6–12 inches to 2 ft. high, branched chiefly near the base; branches erect, simple, in δ. multifid at the summit. Leaves 1½–2 inches long, ½–¾ inch wide, mostly acute. Heads 1½ inch across. Inv. scales a dull, ochraceous white, brighter in δ. Recept. deeply honey-combed and toothed.

2. H. variegatum (DC.! l. c. 212); stem robust, suffruticose, simple or sparingly branched; branches imbricated with leaves throughout, the upper leaves gradually smaller, and membrane-tipped, the lower narrow-oblong, sessile, subacute, on both sides densely and softly woolly; heads solitary, nodding, very many-fl.; inv. campanulate, loosely radiating, the scales obtuse, oblongo-lanceolate, the very-innermost much shorter than the medial. *Helichr. variegatum, Thunb./ Cap.* 663. *Less. Syn.* 316. *H. speciosum, Th.! ex pte. Hel. spirale, Andr. Rep. t.* 262.

HAB. Cape, *Thunberg!* Tulbagh, *Drege!* Worcester and Stellenbosch, *Ecklon!* Howhoek, *Munat!* Simon's Bay, *C. Wright,* 375. (Herb. Th., D., Hk., Sd.)

1–2 f. high; branches, when any, virgate. Whole plant closely shaggy, with whitish or fulvous wool. Leaves 1–1½ inch long, 3–4 lines wide, very closely set, gradually smaller upwards, all the upper ones tipped with a lanceolate, membranous, glabrous scale. Heads 1½–2 inches across. Inv. either white, or the scales tipped with brown, or rusty, glossy.

3. H. Zeyheri (Sond. in Linn. 23, p. 66); stem robust, shrubby; branches closely leafy throughout; leaves oblongo-lanceolate, narrowed to the base, the uppermost lance-linear, imbricated, erect, obtuse or subacute, very densely and softly woolly, one-nerved; heads terminal, 1–3 together, subsessile or short peduncled; inv. scales radiating, rosy-purple, lanceolate, acute or acuminate, the outer and inner short. *Dietr. Fl. Univ. Comp. t.* 5.

HAB. Grassy places, on hills at mouth of Klynrivier, *Zey./* 2845. (Hb. Sd., D.)

Stems short, 6–8 inches high, divided into several erect branches. Leaves 2 inches long, 3–4 lines wide, the upper shorter and narrower. Heads in the specimens seen, which are immature, subsessile, ending leafy branches; but probably, in mature specimens, shortly peduncled, the pedunc. sparsely scaly. To the foliage of *H. variegatum* this joins the invol. nearly of *H. canescens;* and in its 1- or 3-headed branches it unites the corymbose with the one-headed species.

4. H. canescens (DC.! l. c. 212); stem suffruticose, sparingly branched; branches virgate, closely leafy throughout; leaves (small), oblong

or obovate, acute or obtuse, callous-tipped, woolly canescent on both
sides, imbricating and very erect, or squarrose, the uppermost often
bearing membranous scales or reduced altogether to scales; heads solitary, erect; inv. turbinate or campanulate, radiating, scales rosy or
red-brown and white, the outer ovate, sessile; medial stipitate, lanceolate-acuminate, very innermost short, obtuse, truncate or bidentate.
*Helichrysum canescens, Willd. Thunb.! Cap. 663. Less.! Syn. 318.
Xeranth. canescens, Linn. Bot. Mag. t. 420.*

VAR, β. imbricatum; dwarf, ramulose; leaves very small: outer and medial inv.
scales red-brown ; inner white. *H. imbricatum, Th.! Cap. 663. H. canescens, var.
microphyllum, Sond.! in Linn. 23, p. 66.*

HAB. Western and N. W. districts, frequent. β. Cape, *Thunberg!* Klynriver's
Berg, *Zey.! 2852.* (Herb. Th., D., Hk., Sd.)
Stems 3 inches to 1-2 ft. high, straggling, variably branched. Leaves rarely 1
inch, commonly 3-4 lines long, very variable in size and shape, sometimes spreading,
but most commonly erect and close-pressed, the uppermost depauperated. Heads
when fully open an inch across. Inv. scales either pure white, rosy, deep red or
mottled with red, brown and white. De Candolle enumerates 6 varieties, differing
slightly in foliage and colour of involucre. The most distinct is Thunberg's *H. imbricatum,* with brown and white involucres and very small leaves.

5. H. ferrugineum (Sond. and Harv., nec Less.! nec DC.!); stem
robust, suffruticose, sparingly branched; branches virgate, closely leafy
throughout; leaves (rather small), oblong, subacute, callous-tipped,
woolly canescent on both sides, imbricating, erecto-patent, the uppermost reduced to membranous scales; heads solitary, erect, many-fl.;
inv. campanulate, radiating, the scales lemon-yellow or tawny, lanceolate-acuminate, the innermost short, acute or acuminate. *Xeranthemum
ferrugineum, Lam. Enc. III. p. 237, fide Willd.! Herb. No. 15530. Helichrysum ferrugineum, Pers. Syn. 2, 414? (not of Less. Syn.) " H. canescens, an var. imbricat.? b." Herb. Drège. Helipterum spinulosum, Turcz.!
Bull. Mosc. 1851, p. 197.*

HAB. Cape, *Mundt! E. & Z.! Drege!* (Herb. Sond., Hook.)
Nearly allied to *H. canescens,* from which it differs by its rather larger fl. heads,
yellow involucre, and all the scales, even the innermost acuminate. I adopt this as
the true *ferrugineum* on the authority of sheet No. 15,530 of Herb. Willd., compared
by *Dr. Sonder* with *E. & Z.'s* specimens. *H. ferrugineum,* Less.! Syn. 317, described
evidently from a specimen in Herb. Thunb., and at first named " *Helichrysum citrinum*" by Lessing, is our *H. citrinum. H. ferrugineum, DC.!* (according to the distributed specimens of *E. & Z.* and *Drege*) is *Helichrysum squamosum,* Thunb.

6. H. virgatum (DC.! l. c. 213); stem suffruticose, much branched;
branches virgate, closely leafy beyond the middle, pedunculoid upwards,
densely woolly as well as the leaves; leaves narrow-oblong, half-clasping,
subacute, callous-tipped, concave or infolded, one-nerved, thick, the
uppermost minutely scale-tipped; heads solitary, peduncled, 12–15-fl.,
erect; inv. campanulate, loosely imbricate, radiating, the scales shining-
lemon-coloured, the outer broadly ovate, short, obtuse, the inner oblongo-
lanceolate, acute or subacure. *Helichrysum virgatum, Willd. Less.! Syn.
318. Th.! Cap. 663, ex pte. Pteronia pauciflora, Sims, Bot. Mag. t. 1697.*

HAB. Cape, *Thunberg!* Bosjesvelde, *Mundt!* Swellendam, *Thom!* Ezelbank and
Piquetberg, &c., *Drege!* (Herb. Th., D., Hk.)
1–1½ feet high, sometimes very bushy. Leaves 1–1¼ inch long, 1-2 lines wide,

commonly infolded, whitish or rusty at tip. Heads scarcely half inch in diameter, few-flowered. Scales sometimes rusty.

7. H. citrinum (Harv. and Sond.); stem suffruticose, sparingly branched; branches virgate, closely leafy to or beyond the middle, pedunculoid upwards; densely woolly as well as the leaves; leaves oblong, half-clasping, obtuse, scarcely narrowed to the base, one-nerved, thick, the uppermost tipped with a minute scale; heads solitary, peduncled, many (25–40) fl., erect; invol. campanulate, imbricated, all the scales very obtuse, the outer oval, sessile, inner stipitate, elongate, oblong. *Helichrysum ferrugineum, Less.! Syn. p. 317. Helichrysum citrinum, Less.! MSS. in Herb. Thb.!*

Hab. Cape, *Thunberg!* Langekloof, *Mundt. and Maire., Bowie!* Pikenierskloof, *E. & Z.!* Betw. Omtata and Omsamwubo, *Drege!* (*Hel. virgatum,* c.), Kardow, *Zey!* 863, ex pte. (Herb. Th., Hk., D., Sond.)

Very similar to *H. virgatum,* but more robust, with flatter and broader leaves, larger fl. heads, and much blunter inv. scales. It was confounded with *H. virgatum* by Thunberg and De Candolle.—*Zeyher's* specimen (863) exactly agrees with the specimen in Hb. Thunb. quoted by Lessing, and marked by him "*citrinum, n. sp.? a virgato diversum.*" Some of *Zey.!* 863 specimens belong to *H. virgatum.*

8. H. argyropsis (DC.! l. c. 211); stem suffruticose, corymbosely branched; branches leafy throughout, clothed, as well as the leaves with close-lying, white, interwoven wool; leaves oblongo-spathulate, obtuse or subacute, greatly narrowed to the base, erecto-patent, one-nerved, thick; heads corymbose, pedicellate, many-fl.; invol. campanulate, loosely imbricate, scales shining-white, the outer ovato-lanceolate, sessile, the inner on a gradually longer, coloured (dark-red) stipe, bearing an ovato-lanceolate, acute, snow-white limb.

Hab. Swellendam, *Dr. Thom!* Caledon and Swellendam, *Ecklon! Dr. Pappe!* Cape L'Aguillas, *Drege!* (Herb. Hk., D., Sd.)

Robust 2–3 ft. high, much branched. Toment. thick, interwoven, snowy. Corymbs subsimple, 3–5–8 headed. Heads about ½ inch across. Inv. externally quite white; the coloured claws of the inner scales hidden by the outer ones. Pappus plumose. Recept. fimbrilliferous.—Very similar in aspect to *Helichrysum argenteum,* but differing in generic character.

9. H. phlomoides (DC.! l. c. 213); stem robust, shrubby; branches closely leafy throughout; leaves ovate or oblongo-lanceolate, acute or acuminate, densely and softly woolly, half-clasping, one-nerved; heads extremely numerous, oblong, 10–12-fl., in a much branched, corymbose panicle, bearing many membranous bracts (depauperated leaves); inv. not radiating, its scales oblong, erect, flat, obtuse, white or pale-rosy; recept. covered with rigid, subconcrete fimbrils, longer than the achenes. *Helichrysum phlomoides, Spreng. Less.! Syn. 320. Gnaph. milleflorum, Linn. Thunb.! Cap. 658. Gn. phlomoides, Lam.*

Hab. Uitenhage and Albany, *Thunberg!, E & Z.! Genl. Bolton, &c.* Zuureberg, *Drege!* (Herb. Th., D., Hk., Sd.)

Stem and leaves as in *H. eximium.* Inflorescence 6–8 inches in diameter, excessively branched. Heads ⅛ inch long, ⅓ inch diameter, quite white or pale-rosy at tip. Pappus united in a ring at base, feathery.

10. H. eximium (DC.! l. c. 213); stem robust, shrubby; branches closely leafy throughout; leaves (large) broadly ovate or elliptical, acute or acuminate, very densely and softly woolly, half-clasping, 1-nerved;

heads globose, corymbose; invol. not radiating, its scales concave, appressed, bright crimson, very obtuse. *Helichrysum eximium, Less.! Syn.* 319. *Gnaph. eximium, Linn. Bot. Mag. t.* 300. *Andr. Bot. Rep. t.* 654. *Bot. Reg. t.* 532.

HAB. Districts of Caledon, Swellendam and George, *Thunberg! Drege! Pappe!* &c. (Herb. Th., D., Hk., Sd.)

A very woolly small shrub. Leaves 2–3 inches long, 1½–2 inches wide, very thick and soft. Corymb sessile or subsessile, many-headed, dense, not much branched. Heads ½–¾ inch diameter, ruby-red. Flowers yellow, about 60. Much used at the Cape as an "*everlasting*" for winter nosegays, &c.

11. H. Dregeanum (DC.! l. c. 214); suffruticose, irregularly branched, tomentose; branches leafy to the summit; leaves half-clasping, obovate or oblong, obtuse, callous-tipped, recurved at the point, often infolded; heads terminal, solitary, sessile, oblong; invol. scales woolly at base, imbricated, erect, produced into a broad, lance-oblong, obtuse or subacute, ruby-red, torn-edged, spreading or reflexed appendage.

HAB. Cape, *Dr. Thom!* Zwellendam, Onder-Roggeveldt and Zeederberg, *Drege!* (Herb. Hk., D.)

1–2 feet high, rather robust, not much branched. Leaves 1–1½ inch long, 3–5 lines wide, more or less obovate, obscurely 1-nerved. Heads ¾ inch long, nearly ½ inch across. Tips of the inv. scales as bright as in *H. eximium.*

12. H. gnaphaloides (DC.! l. c. 213); suffruticose, much branched, snowy-white tomentose; branches ascending, leafy, the flowery ones naked at top; leaves narrow-linear, long-channelled, obtuse; heads solitary, peduncled, oblong, many-fl.; inv. scales woolly at base, imbricated, erect, produced into a very taper-pointed, narrow, red-brown, reflexed appendage. *Stæhelina gnaphaloides, Linn. Leyssera squarrosa, Th.! Cap.* 691. *Helichrysum stæhelinoides, Less.! Syn.* 321. *Gnaphalium modestum, Hook. Bog. Mag. t.* 2710.

HAB. Sandy sea-shore and hill-sides near Capetown and Simonstown. Also in Worcester and Stellenbosch, common. (Herb., Th., D., Hk., Sd.)

Stems 1–1½ foot high, woody at base; the branches long, simple, curving upwards, somewhat corymbose. Leaves 2–2½ inches long, scarcely 1 line wide, thickish. Heads about ½ inch long, ⅓ inch across. Readily known by its strongly reflexed, taper-pointed inv. scales.

LXXX. GNAPHALIUM, Linn.

Character the same as in Helichrysum, except, *marginal* female flowers filiform, in several rows, or much more numerous than the central, perfect flowers. *Recept.* nude. *DC. Prodr.* 6. *p.* 222.

Herbaceous mostly annual plants, with the aspect of the small-flowered species of *Helichrysum.* Foliage woolly. Inv. scales, in the Cape species, either white, or pale-horn or straw-colour. Name from γναφαλον, *soft down* or wool.

Leaves *decurrent*, lance-linear, green above; stem tall   ...   (1) **undulatum.**
Leaves *sessile* or amplexicaul, not decurrent:
   Inv. scales with *snow-white*, spreading tips:
      Lvs. oblong-lanceolate or obovate; inv. scales short   (2) **candidissimum.**
      Lvs. linear or narrow-spathulate; inv. scales longer   (3) **micranthum.**
   Inv. scales pale straw-colour or horn-colour:
      Stems prostrate or trailing, rigid:
         White-woolly; leaves linear spathulate   ...   (4) **repens.**

Cobwebby, or nude; lvs. obovate-oblong ... (5) **prostratum.**
Stems erect, or ascending: (annual).
  Tall; cauline leaves half-clasping ... ... (6) **luteo-album.**
  Minute annuals; cauline lvs. not clasping:
    Leaves narrow-linear, or subspathulate:
      Lvs. thickish; inv. scales subacute (7) **pauciflorum.**
      Lvs. thin; inv. scales emarginate ... (8) **glomerulatum.**
    Leaves obovate or obovate-spathulate:
      Inv. scales oblong, very obtuse ... (9) **parvulum.**
      Inv. scales lanceolate, acute ... ... (10) **confine.**

1. G. undulatum (Linn. Sp. 1197); stem herbaceous, tall, erect, branched above, cobwebbed; leaves *decurrent*, lance-linear, green and scabrous above, white-woolly beneath; heads tufted, sessile, the tufts numerous, in a much branched, corymbose panicle; heads 30–40-fl.; inv. scales glabrous, pale yellowish, oblong, obtuse or subacute. *Less. Syn. 331. DC. l. c. 226. Dill. Elth. t. 108. f. 130.*

HAB. Caledon, Swell., Uitenhage and Albany, *E. & Z.!* Paarl, *Drege!* Natal, *Drege! Gerr. and M'K. No.* 260. Zwartkops R., *Zey.!* 2883. (Herb. D., Hk., Sd., Th.)
Stem 1–2 ft. high, almost woody at base. Leaves 1½ inch long, 2–3 lines wide, acute, their base decurrent as a wing to the stem. Panicle much branched. Heads 2 lines long. Achenes minutely granulated. Female fl. numerous.

2. G. candidissimum (Lamk.); very white, softly and closely silky-tomentose; stems erect or diffuse, branched from the base, leafy throughout; leaves sessile, oblongo-lanceolate or spathulate, tapering much at base, obtuse, concolourous; heads densely glomerate at the ends of the branches, sunk in wool, about 20-fl.; inv. scales linear, with spreading, short, obtuse, snow-white tips. *Less. Syn. p.* 328. *DC.! l. c.* 229. *G. dealbatum and G. maculatum, Thunb.! Cap.* 652, 651.

HAB. Western and midland districts, *Thunberg, Mundt, E. Z.!* *Drege!* &c. Sandy Flats near Capetown and Simonstown; Zwartkops R., *Zey.!* 2886. (Herb. Th., Hk., D., Sd.)
Root fibrous, perennial (?). Stems 3–12 inches long or more, mostly decurrent at base, then erect, simple or branched. Leaves 1–1½ inch long, 2–3 lines wide. Heads small, like those of *Helichr. declinatum.* Female fl. numerous. Pappus serrulated.

3. G. micranthum (Th.! Cap. 651); albo-tomentose; stems sub-erect or diffuse, slender, branching, leafy throughout; leaves sessile, linear-lingulate or spathulate, mucronate, acute or subacute, concolourous; heads glomerate at the ends of the branches, sessile, woolly at base; inv. scales linear, with radiating, oblong, obtuse, snow-white tips. *Less.! Syn. p.* 329. *DC.! l. c.* 229.

HAB. Cape, *Thunberg!* Plettenberg's Bay, *Mundt & Maire;* Worcester, Stellenbosch, and Uitenhage, *E. Z.!* ·(Herb., Th., D., Hk., Sd.)
Very near *G. candidissimum,* but more slender, with narrower leaves, and longer inv. scales, the radiating apex 2–3 times as long as broad.

4. G. repens (Linn. Mant. 283); white-woolly; stems many from the crown, slender, suffruticose, decumbent or prostrate, branching, leafy throughout; leaves sessile, crowded, short, linear-spathulate, mucronate; heads 2–5-together, in tufts at the ends of the twigs, sessile; invol. scales spreading, linear, elongate, horn-colour, opaque, obtuse or subacute. *Thunb.! Cap. p.* 647. *Less.! Syn.* 330. *DC. l. c.* 230. *G. acilepis, DC. l. c.* 230.

HAB. Sea-shore near Capetown, *Thunberg !* about Port Elizabeth, *Zeyher !* 2890, ex pte. (Herb. Th., D., Hk., Sd.)

Stems 6-12 inches long or more, trailing. Leaves close, 2-3 lines long, 1 line wide, slightly narrowed at base. Heads 2½ lines long. *G. acilepis,* DC., quite agrees with Thunberg's *G. repens.*

5. **G. prostratum** (Thunb.! Cap. 652); stems prostrate, slender, very much and closely branched, cobwebby, becoming nude, leafy throughout; leaves sessile, obovate-oblong, obtuse, thinly tomentose, becoming nude; heads 3-6, glomerate at the ends of the short branchlets, woolly at base, few-fl.; inv. scales linear, spreading, elongate, horn-colour, obtuse. *Less.! Syn.* 330, *DC. l. c.* 230. *G. prostratum and G. debile,* Th.!

HAB. In the Carroo, Onder Bokkeveldt, *Thunberg!* Cango, *Mundt & Maire.* (Hb. Th., Sd.)

Root annual ? Stems many from the crown, 3-6 inches long. Leaves 2-3 lines long, 1-1½ line wide. Much less woolly than *G. repens,* much more branching, with broader and more obovate leaves. The root, too, is said to be annual.

6. **G. luteo-album** (Linn. Sp. 1196); annual, herbaceous, variably woolly or cottony ; stem *tall,* erect, subsimple or diffusely branched ; lower leaves obovate or oblongo-spathulate, tapering at base, obtuse, cauline half-clasping, linear or lance-linear, acute, with subrecurved margin ; heads tufted, the tufts either clustered at the ends of the branches or disposed in cymes or panicles; inv. scales scarious-hyaline, straw-colour, obtuse or subacute. *DC. l. c.* 230. *E. Bot. t.* 1002. *Fl. Dan. t.* 1763.

HAB. Eastern Districts and Natal, frequent. *Zey!* 2885. (Hb. Th., D., Hk., Sd.)

Very variable in size and woolliness. Stems 6 inches to 2 ft. high, the taller ones weak and straggling. Leaves 1-1½ inch long, 2-4 lines wide. Heads several in a tuft, each head 2½ lines long. Inv. scales glossy and very thin. A widely dispersed plant, found in Europe, Asia, Africa, Australia, St. Helena and Brazil.

7. **G. pauciflorum** (DC.! l. c. 229); annual, stem very dwarf, branching, woolly ; leaves narrow-linear, elongate, coriaceous, acute, white-woolly; heads sessile in the upper axils and terminal, somewhat tufted; inv. scales horn-colour, hyaline, subacute.

HAB. Round Capetown, *Ecklon!* (Herb. Sond.)

1-2 inches high, white-woolly, simple or branched, erect. Root small, fibrous. Leaves ¼-¾ inch long, ⅓ line wide, of thickish substance, opaque. Heads with very few perfect and many female flowers. Pappus white.

8. **G. glomerulatum** (Sond.! MSS.); annual, many-stemmed ; stems erect, filiform, loosely woolly; leaves linear or spathulate, membranous, 4-nerved, subacute or obtuse, loosely woolly ; heads sessile in terminal clusters surrounded by many linear floral leaves (like those of the stem); inv. scales horn-colour, hyaline, lance-linear, emarginate, subuniseriate; perfect fl. 6-8 ; females very numerous.

HAB. Cape, *Ecklon & Zeyher!* (Herb., Sd.)

A small annual, with the habit of *Lasiopogon micropioides.* Stems 2-3 inches long. Leaves 4-6 lines long, ⅓ line wide, of thin subpellucid substance. Inv. scales flat, ribbed below, varying from linear to lance-linear, always minutely and bluntly bidentulate at the apex. What seems a *starved* condition of this, scarcely an inch high and very slender in all parts, occurs also in Hb. Sond.

9. **G. parvulum** (Harv.); annual, slender, erect or diffuse, **branching,**

white-woolly; branches threadlike, flexuous; leaves scattered, obovate, tapering much at base, acute or mucronate; heads tufted at the ends of the branches, or subsessile or pedicelled, in the upper axils; inv. cylindrical, few-fl., woolly at base, the inner scales oblong, obtuse, hyaline, glabrous; female fl. 15–16, perfect 2–3.

HAB. Cape, *Drege!* Steendahl, Tulbagh, *Dr. Pappe!* Hassaquas Kloof, *Zey.!* 2887. (Herb. D., Hk.)
Root fibrous. Stems 3–6 inches high, irregularly branched. Leaves nearly ½ inch long, 2–3 lines wide. Heads about 20 fl., very few hermaphrodite flowers. A specimen in Hb. Hook. from Drege, marked " *Gn. micranthum, Th.! a.*" belongs to this species.

10. G. confine (Harv.); annual, slender, erect, many stemmed, branching, loosely white-woolly; branches corymbose upwards; leaves scattered, spathulate-obovate, tapering much at base, acute, mucronulate; heads tufted at the ends of the branches, or axillary, 40–50-fl.; inv. spreading, the scales pluriseriate, loosely imbricate, horn-coloured, hyaline, lanceolate, acute; fem. fl. numerous.

HAB. Linde, *Zeyher!* (Herb. Sond.)
Very like *G. parvulum*, but with different inv. scales and more flowers in the head. Stems 2–3 inches high, divided near the base, ramulous near the summit. Heads copious. Leaves 4–6 lines long, 1–2 lines wide.

Doubtful Species.

G. lasiocaulon (Link, Enum. h. her. 2, 317); "tomentose; leaves linear, acute, curved; heads fasiculate-corymbose, peduncled; inv. scales white, acute, shrubby. *DC. l. c. p.* 237, *No.* 102.

G.? cicatrisatum (Vahl.); "leaves linear-lanceolate, keeled; corymb compound, dense." *DC. l. c. p.* 237, *No.* 101. Genus and habitat alike uncertain.)

G.? acuminatum (Link); "shrubby; branches pubescent; leaves linear lanceolate, with very long sharp points, glabrous above, thinly tomentose beneath; heads corymbose, cylindrical; inv. scales very obtuse." *DC. l. c.* 236, *No.* 90. (Doubtful whether a Cape or an Australian species; its genus also uncertain!)

LXXXI. AMPHIDOXA, DC.

Heads many-fl., heterogamous, discoid, the marginal fl. very slender, female, numerous (in several rows); disc fl. perfect. *Recept.* flat, naked. *Inv.* imbricate, campanulate, the outer scales close-pressed, inner longer, radiating. *Achenes* oblong, granulated, those of the marginal fl. without pappus; of the disc with few (5–6) caducous bristles, barbellate at the apex only. *DC. Prodr.* 6, *p.* 246.

A single species, with the aspect of a small *Helichrysum* (such as *H. declinatum*) or of a *Gnaphalium*, but known from both genera by the want of pappus to the marginal flowers. The name is compounded of αμφι, *round about*, and δοξα, *opinion;* because the affinities are various with several genera.

1. A. gnaphaloides (DC. l. c.).

HAB. Near Uitenhage, *Ecklon!* Port Elizabeth, *Zey.!* 2890, ex pte. (Herb. D., Hk., Sd.)
Stems decumbent or trailing, turning up at the extremities, 8–12 inches long,

subsimple, suffruticose, flexuous, thinly woolly, closely leafy throughout. Leaves sessile, oblongo-spathulate, mucronate, undulate, thinly silky-canescent or cobwebby, 1-nerved. Heads shortly pedicelled, in terminal, close corymbs. Invol. campanulate, glabrous, 2 lines long, the outer scales brownish, the inner milk-white, obtuse, radiating.

LXXXII. ERIOSPHÆRA, Less.

Heads collected in globose, densely woolly glomerules, about 10-fl., homogamous, the fl. 5-toothed, terete, glabrous. *Recept.* flat, naked. *Inv.* scales in few rows, linear, enveloped in wool. *Achenes* narrow-obovate, angular, minutely granulated. *Pappus* caducous, of few very slender, hair-like bristles, smooth below, toward the apex beaded with swollen. roundish cells. *Less.! Syn. p. 270 (non DC. Prodr. 6, p. 166).*

Only one species. A small, many-stemmed herb. Name from εριον, *wool,* and σφαιρα, a *globe;* woolly head.

1. E. Oculus Cati (Less.! Syn. p. 270). *Harv. Thes. Cap. t.* 149. *Gn. Oculus Cati, Linn. Suppl. 364. Gn. Oculus, Thunb.! Cap. 657.*

HAB. Cape, *Thunberg!* (Herb. Thunb.)

Root annual? Stems slender, filiform, prostrate, dichotomous, cobwebby, at length glabrate. Leaves scattered, obovate, narrowed at base, membranous, obtuse, loosely woolly, the older becoming nearly nude. Glomerules sessile, surrounded by several floral leaves, similar in shape and size to the cauline, but more woolly. Heads from 12-20 in each cluster, their discs separated by broad rims of white wool. Inv. scales in 1-2 rows, narrow-linear, acute or obtuse, sometimes toothed at the truncate apex. Pappus of a very few, flexuous, slender hairs, quite naked for at least ¾ their length, near the summit on each side crenulated (or beaded) with opposite, swollen, pearly cells.—Of this curious little plant there are good specimens in Herb. Thunberg, from which *Lessing's* description is obviously taken. *De Candolle's* " *E. Oculus-Cati*" is quite different in habit, inflorescence, and pappus, and is, as well as the other species of *Eriosphæra,* DC., not only a true *Helichrysum,* but identical with *H. marifolium,* DC.!

LXXXIII. LASIOPOGON, Cass.

Heads glomerated, many-fl., heterogamous, the marginal-fl. filiform, females in one or more rows, disc-fl. 5-toothed, perfect. *Recept.* naked, flat. *Inv.* in few rows, linear, scarious, immersed in wool. *Achenes* obovate, subcompressed, glabrous. *Pappus* in a single row, feathery, with very long pinnules. *DC. Prodr.* 6. 246.

Minute, woolly annuals. Heads in globose clusters at the ends of the branches. Leaves spathulate or obovate, alternate, variably woolly. Pappus with very long plumes to its bristles. Name from λασιος, *woolly* and πωγων, a *beard.* These little plants can only be known from *Eriosphæra Oculus Cati* by the pappus and heterogamous heads; in all external characters they are almost identical.

1. L. micropoides (DC. l. c. 246); heads with about 20-25 perfect, and 15-18 female flowers. *Harv. Thes. Cap. t.* 150.

VAR. α; copiously woolly; leaves obovate or spathulate. *L. micropoides, DC.! l. c.*

VAR. β; cobwebby, the stems glabrate; leaves narrower. *L. molluginoides, DC. l.c.*

HAB. Betw. Zilverfontein, Kooper-bergen and Kaus, *Drege!* β. near Verleptpram, on the Gariep, *Drege.* Modderfontein, *Rev. H. Whitehead!* (Herb. Hk., D., Sd.)

Branched from the base, 1-3 inches high, suberect, the branches ending in globose, woolly glomerules of 8-12 heads. Leaves 3-5 lines long, 1-2 lines wide, thinly membranous. Closely resembling *L. muscoides* with larger fl.-heads, and fewer female flowers in proportion to the perfect ones, but more erect.

2. L. muscoides (DC.! l. c. 246); heads with few (5–6) perfect, and many (30–50) female, marginal flowers in several rows. *Gnaph muscoides, Desf. Atlant.* 2, *p.* 267. 231. *Lasiopogon lanatum, Cass. Bull. Philom.* 1818, *p.* 75.

HAB. Tarka, *E. & Z.!* S. Africa, *Drege.* (Herb. Sd., D.)

Stems 1–3 inches long, many from the crown, diffuse or prostrate, closely much branched, very woolly. Leaves spathulate, 3–5 lines long, 1–2 lines wide, very woolly. Tufts of heads ending all the branches. Inv. scales obtuse, testaceous. This little plant, besides being found in North Africa, occurs in several parts of Central Asia and in S. Europe. It is much more closely branched, more procumbent and woolly than *L. micropoides*, with which, in *Drege's* distribution, it is confounded. It is easily known by the *few* perfect and *many* female fl. in each head.

LXXXIV. METALASIA, R. Br.

Heads few or many-fl., homogamous; all the fl. tubular, 5-toothed, perfect. *Recept.* naked. *Inv.* scales closely imbricated, the outer often acute or acuminate, inner mostly dilated and coloured at the apex. Young *achene* attenuated towards each end, somewhat stipitate and beaked; ripe fr. sessile. *Pappus* uniseriate, of several rather broad, serrulate or clavate bristles. *DC. Prodr.* 6, *p.* 249. Also *Erythropogon, DC. l. c. p.* 254.

Erect or spreading, small shrubs. Branches mostly tomentose, closely leafy throughout. Leaves alternate, sessile, coriaceous, linear or oblong, with involute margins, concave and tomentose above, convex beneath and either shining-glabrous or clothed with deciduous wool, very generally spirally twisted. Axils very frequently leaf-bearing. Heads cylindrical or turbinate, rarely solitary, mostly corymbose or fascicled. Inner inv. scales petaloid, white or purple, rarely yellow. Corollas purple or whitish. Name from μετα, on the *other side*, and λασιος, *hairy;* alluding to the foliage. The species are difficult to characterize.

Subgenus 1. ERYTHROPOGON. Hds. 100–200-flowered. Pappus rufous. (Sp. 1–2.
Leaves lanceolate, straight, erect, not gemmuliferous.
 Inner inv. scales acute, radiating, white .. .. (1) **imbricata.**
Lvs. linear, spirally twisted, gemmuliferous. Inner inv.
 scales obtuse, rosy ... ... (2) **umbellata.**

Sub-genus 2. EU-METALASIA. Heads 3–10-flowered. Pappus white. (Sp. 3–20.)

Glomeratæ; Heads sessile in tufts, matted together by woolly hairs :
 Petaloid inv. scales imbricated in 3–4 rows, obtuse :
  Lvs. lance-linear and linear, spiral, squarrose (3) **Cephalotes.**
  Lvs. obovate-cymbiform, very convex at back,
   erecto-patent, tapering at base ... ... (4) **cymbæfolia.**
 Petaloid inv. scales uniseriate, acute :
  Dwarf, prostrate, divaricately much branched;
   lvs. ovato-cymbiform, convex backed ... (5) **depressa.**
  Diffuse, slender ; lvs. linear or lance-linear,
   twisted ... ... (6) **divergens.**
Capitatæ; Heads sessile in tufts, but not matted together :
 Lvs. tufted, spirally twisted ... ... (7) **capitata.**
 Lvs. solitary, hook-pointed, not twisted ... (8) **adunca.**
Condensatæ ; Heads subsessile or shortly pedicellate, in dense, roundish or
 imperfectly capitate tufts :
 Heads 8–10-fl., corymbulose ; petaloid inv. scales
  pluriseriate ... ... (9) **Lichtensteinii.**
 Heads 3–5-flowered :
  Petaloid inv. scales imbricated in 3–4 rows :
   Stems robust, erect, much branched; inv.
    scales obtuse ... ... (10) **nitidula.**

Stems slender, diffuse, straggling; inv.
     scales oblong, subacute   ...   .   (11) **erubescens.**
Stems slender. erect; *lvs. bristle-shaped ;*
     inv. scales lanceolate, acute   ...   ...   (12) **seriphiifolia.**
   Petaloid inv. scales sub-uniseriate :
     Hds. 3-fl.; petaloid inv. sc. concave, erect   (13) **tenuifolia.**
     Hds. 5-fl.; pet. inv. scales acute, recurved   (14) **quinqueflora.**
Fastigiatæ ; Heads individually subsessile or pedicelled, arranged in a branch-
     ing, hemispherical, rather loose corymb :—
   Petaloid inv. scales uniseriate or sub-biseriate :
     Pet. inv. scales *yellow*, scarcely exserted   ...   (15) **aurea.**
     Pet. inv. scales white or tawny, much exserted :
       Lvs. lanceolate, erect, straight, convex-
         keeled, solitary ... ...   16) **pulcherrima.**
       Lvs. mostly tufted, squarrose or spiral, various in shape :
         Inv. scales appressed, the outer blunt
           ormucronate, or awned, inner white   (17) **muricata.**
         Inv. scales *squarrose,* the outer taper-
           pointed, recurved, inner fulvous   (18) **stricta.**
   Petaloid inv. scales imbricated in 3-4 rows ;
     Leaves 6-8 lines long, subulate, spreading ...   (19) **concinna.**
     Lvs. 1-3 lines long, linear, much tufted   ...   (20) **fasciculata.**

Sub-genus 1. ERYTHROPOGON. (Sp. 1-2.)

1. M. imbricata (Harv.); leaves lanceolate, pungent-mucronate,
straight, erect, close-set, convex-keeled and shining beneath, concave
and woolly above, not gemmuliferous ; heads very many-fl., solitary or
3-5 in a corymb, on longish, leafy, tomentose pedicels; inv. campanulate,
the outer scales cobwebbed, with glabrous, acuminate-pungent points;
the inner lanceolate, acute, radiating, mostly white. *M. uniflora, Don.
Less.! Syn. 334. Erythropogon imbricatum, DC. l. c. 255. Stœhelina imbricata, Berg. Cap. 233. Th.! Cap. 628. Xeranth. imbricatum, Burm.*

HAB. Cape, *Bergius, Thunb.!* Groenekloof, *Drege!* Stellenbosch and Swellendam,
Ecklon, Zey.! 900. (Herb. Th., Hk., D., Sd.)
1-2 f. high, slightly branched ; branches long and simple, erect, curved, woolly.
Leaves 2-3 lines long, precisely similar to those of *M. pulcherrima.* Heads 5 lines
long, 4-5 in diam., 200-fl. Pappus purple-red.

2. M. umbellata (Don. l. c. 556) ; leaves linear, mucronate, spirally
twisted, spreading, with leaf-tufts ; heads very many-fl., umbellate, on
longish, leafy pedicels; invol. campanulate, the outer scales cobwebbed,
with glabrous, ovate, free points, the inner ovate-oblong, radiating, obtuse, rosy. *Less.! Syn. p. 335. Erythropogon umbellatum, DC.! l. c. 255.
Gnaph. umbellatum, L. Th.! Cap. 643. Xeranth. squarrosum, Lam. Ill.
t. 693, f. 3.*

HAB. Cape, *Thunberg!* Caledon, *Mundt! Ecklon.* Baviansberg, Genadendahl,
Pappe! Zey.! 2913. (Herb. Th., D., Hk., Sd.)
Erect, 12-18 inches high, simple or slightly branched, the branches long, curved
and closely leafy throughout. Leaves 3-4 lines long, convolute, with crowded tufts
in the axils. Umbel terminal, simple, the rays ½-1 inch long, with a few, leafy,
scattered bracts. Heads 4 lines long, 5 lines diam., with 100-200 fl. Pappus rufous.
The foliage is very similar to that of *M. fasciculata.*

Sub-genus 2. EU-METALASIA. (Sp. 3-20).

3. M. Cephalotes (Less.! Syn. 338); suberect or diffuse, robust ;

leaves lance-linear and linear, involute, spirally twisted, spreading or
squarrose, the axils gemmuliferous or nude; heads 5-fl., glomerated,
many in the glomerule; outer inv. scales very woolly, with lanceolate,
white or red glabrous points; inner (petaloid) imbricated in 3-4 rows,
the outer acute or subacute, the inner very obtuse or truncate; pappus
club-shaped, obtuse. *M. glomerata, M. rosea and M. muraltiæfolia,*
DC.! l. c. 252-253. *Gnaph. capitatum* and *Gn. Cephalotes, Thunb.! Cap.*
646, 643.

HAB. About Capetown and in the Western Districts, frequent. (Herb. Th., D.,
Cap., Hk., Sd.)
More robust than *M. divergens,* with much larger and more compound head-clusters
and a different involucre. Leaves 4-7 lines long, squarrose, crowded; or erecto-
patent, without gemmules. Thunb.'s "*G. capitatum*" is precisely DC.'s "*M. rosea.*"
"*M. muraltiæfolia*" seems to be merely a luxuriant form. "*M. Cephalotes,*" DC.
is quite a different species, our *M. concinna.*

4. M. cymbæfolia (Harv.); stem robust, erect, branched and ramulous, the twigs, woolly; leaves erecto-patent, straight or scarcely twisted,
obovate-cymbiform, tapering to the base, very convex and glossy at back,
concave and woolly above, pungent-mucronate; heads 3-4-fl. glomerated;
outer inv. scales very woolly, inner (petaloid) imbricated in 3-4 rows,
concave or spoonshaped, obtuse, dorsally apiculate, rosy; pappus lanceolate, flat, acute, serrulate. *M. divergens, var. γ. nudiuscula, DC. ! l. c. p.* 252.

HAB. In the district of Caledon, *Ecklon!* (Herb. Cap., Sd., D.)
A strong growing, twiggy bush, 1-2-ft. high, the younger branches and twigs
closely leafy. Leaves of the branches sparingly gemmiferous; of the twigs with
nude axils, 3-4 lines long, 1-1½ line wide, boatshaped, remarkably convex at back
and almost petiolate at base. Glomerules globose. Mature fl. heads not seen. The
pappus, even in the half-formed flowers is remarkably broad and flat, tapering at base,
lanceolate above. If this be a mere form of *M. divergens,* it is indeed a memorable one.

5. M. depressa (Harv.); dwarf, procumbent or prostrate, divaricately
much branched, the twigs woolly; leaves squarrose, ovate-cymbiform,
sessile, convex and cobwebby or glabrate at back, concave and woolly
above, mucronate, mostly gemmuliferous; heads 5-fl., glomerated, few
together; outer inv. scales very woolly, with lanceolate, brown, glabrous,
acuminate points; inner (petaloid) uniseriate, spreading, flat, subacute,
white; pappus clubshaped, obtuse. *M. divergens, Eckl. Herb., ex pte.*

HAB. Summit of Table Mountain, *Ecklon!, Pappe!, W. H. H.* (Herb. Cap., D.,
Sd., Hk.)
A prostrate shrublet, many-stemmed, spreading over the rocky soil, closely twiggy,
the twigs short, leafy. Leaves 2-3 lines long, almost always with small ones in the
axils. Heads 3-6 in each cluster. Allied to *M. cymbæfolia* in foliage, but much
smaller, and with very dissimilar involucres and pappus; it differs from *M. divergens*
chiefly in its foliage and prostrate, ramulous habit.

6. M. divergens (Don, l. c.); divaricately branched, diffuse; leaves
linear, or lanceolate, mucronate, spirally twisted, squarrose or reflexed,
the axils almost always gemmuliferous; heads 5-flowered, glomerated;
outer inv. scales very woolly, with subulate, brown, glabrous, acuminate,
recurved points; inner (petaloid) uniseriate, spreading, flat, lanceolate,
acute; pappus clubshaped, obtuse. *Less.! Syn.* 329. *DC. l. c.* 252, *ex pte.*
Gnaph. divergens, Th.! Cap. 645.

VAR. *a.* **linearifolia**; lvs. linear-involute; heads several in the glomerule.

VAR. *β.* **gracilis**; lvs. linear; heads about 3 in a tuft; stems very slender. *C. Wright,* 354, 368.

VAR. *γ* ? **purpurascens** (DC.); "lvs. lanceolate or oblongo-lanceolate, open, twisted; stems robust; inv. scales purple." *M. divergens, var. purpurascens, DC. ex pte.?*

HAB. About Capetown, and generally in the Western districts, frequent. (Herb. Th., Cap.; D., Hk., Sd.)

A very straggling bush, 12–18 inches long, distantly branched. Leaves 3–5 lines long, ½–1 line wide, widely spreading. Young plant often all over woolly. Many, at least, of Ecklon's distributed specimens of var. *purpurascens,* belong to *M. Cephalotes;* e. g. those in Hb. Cap. and D.; that in Hb. Sond. which is in a very bad state, is doubtful.

7. **M. capitata** (Less.! Syn. 337); divaricately branched, branches tomentose; leaves linear, involute, spirally twisted, mucronate, gemmuliferous, glabrous externally; heads numerous (20–30), cylindrical, 3-flowered, sessile in a terminal, simple, exinvolucrate tuft; invol. glabrescent, the outer scales appressed, ovate, acute, the inner uniseriate, linear-oblong, often brown-tipped, obtuse. *DC.! l. c.* 253. *M. ericoides, Sieb.! 15. Gnaph. capitatum, Lam. fide Less.! non Thunb.*

HAB. On the Cape Flats, *Sieber! Ecklon!* (Herb. Hk., Sd., Cap.)

Intermediate in habit between *M. cephalotes* and *M. divergens,* but with much less glomerated inflorescence; the heads, though sessile and closely tufted, stand apart from each other, and are not woolly, and the leaves immediately under the tuft are either obsolete or very small.

8. **M. adunca** (Less.! Syn. 341); leaves linear, involute, hook pointed, erecto-patent, not spirally twisted, without axillary leaf-tufts; heads cylindr., 5-flowered, sessile or subsessile in bracteated tufts or capitate corymbs; inv. glabrescent or slightly cobwebbed below, the outer scales close-pressed, oblong, mucronate, inner narrow-oblong, obtuse, uniseriate, white, spreading. *DC.! l. c.* 253. *Gnaph. muricatum, Th.! Cap.* 644, *ex pte.*

HAB. Cape, *Thunberg!* near Constantia, *Mundt!* Cape Flats, *Ecklon, Drege!* (Herb. Th., Hk., Sd.)

Readily known by its hook-pointed, untwisted, scattered leaves. Leaves 4–5 lines long, ⅓ line diam. Inv. white, the petaloid scales much longer than the calycine.

9. **M. Lichtensteinii** (Less. Syn. 335); robust, thinly woolly or glabrate; leaves linear, rigid, pungent-mucronate, spirally twisted, crowded, spreading, gemmuliferous; heads oblong, 8–10-fl., in a simple or compound corymb, the pedicels mostly shorter than the heads, woolly; outer inv. scales cobwebbed, acuminate, rufous-tipped, inner pluri-seriate, rosulate, subequal, oblong, spreading, milk-white, acute or subacute, or keeled and mucronulate; pappus obtuse, club-shaped. *DC. l. c.* 253.

HAB. In the interior, *Lichtenstein; Burchell,* No. 8182, *Bowie!* Hott.-Holl.-berge, *Eckl.! Zey.!* 2910. (Herb. Hk., Sd., Cap., D.)

A robust, erect, branching bush. Leaves shining, very pungent and closely set, 4–5 lines long, strongly spiral. Heads 10–20 or more in a corymb; in the luxuriant specimens the pedunc. are 2–3-headed.

10. **M. nitidula** (Harv.); stems erect, rigid, much branched; branches glabrous or nearly so; leaves linear-involute, narrowed at base, mucronate, twisted, glabrous and glossy, rather closely set, mostly without

axillary tufts; heads few or many, subsessile in a compound, umbellate tuft, subtended by a few leaves; inv. glabrate, the outer scales subcanescent, mucronate, the inner oblong, obtuse, rosy or white, imbricated in 3–4 rows. *M. erubescens, var. β, rigidula, DC. l. c.* 254.

HAB. Cape, *Mundt.!* Caledon, *Ecklon!* Babylonschetoornberg, *Zey.!* 2901. (Herb. Cap., Hk., Sd., D.)

1–1½ ft. high, robust, much branched, bushy, erect. Leaves 5–6 lines long, erecto-patent, curving outwards, occasionally gemmuliferous. The habit is very different from that of *M. erubescens,* which however it very closely approaches in character.

11. M. erubescens (DC.! l. c. 252); stems diffuse or procumbent, tomentulose or glabrate; leaves linear-involute, mucronate, twisted, glabrous, squarrose or recurved, mostly (except in var. β.) without axillary tufts; heads several together, crowded in a condensed, subcompound corymb, cylindrical, 3-fl.; inv. glabrate, the outer scales acute or mucronate; inner (petaloid) bi-triseriate, imbricate, oblong, acute, erect, concave, rosy or white.

VAR. β.?, **gemmulifera**; leaves shorter, more closely set, more or less gemmuliferous. *M. intermedia, DC. l. c.* 253.

HAB. Stellenbosch, *Ecklon!* Swellendam, *Mundt.!* Zwarteberg, *Dr. Pappe!* β. in Caledon, *Ecklon! Pappe!* (Herb. Cap., D., Sd., Hk.)

The normal form has the aspect of *M. tenuifolia,* from which it chiefly differs in the involucres. Its leaves are 4–6 lines long. β. has leaves from 2–4 lines long, almost always with leaf-tufts in the axils. Its involucre offers no character. I have some hesitation in uniting it to this species.

12. M. seriphiifolia (DC.! l. c. 255); slender, flexuous-erect, the branches tomentose; leaves linear-setaceous, involute, spirally twisted, mucronate, squarrose, very gemmuliferous in the axils; heads 3-fl., in a densely fascicled, globose, branching corymb, the pedicels very short; inv. cobwebbed or woolly at base, the outer scales subulate-acuminate, the inner pluriseriate, imbricate, lanceolate, acute, rosy-purple.

HAB. District of Caledon, *Ecklon!* (Herb. Cap., Sond.)

A slender, distantly branched bush, the branches flexuous, erect. Leaves very slender, as thick as bristle, 3–5 lines long, with dense leaf-tufts in the axils. Capitate-corymb ¾ inch diameter. Apparently a distinct species.

13. M. tenuifolia (DC.! l. c. 253); stems diffuse, flexuous, glabrate; leaves linear-involute, mucronate, twisted, glabrous, without axillary tufts; heads few together, subsessile in a subsimple, condensed corymb, cylindrical, 3-flowered, subtended by 2–3 leaves; inv. glabrate, the outer scales subulate, acuminate; inner (petaloid) subuniseriate, narrow-oblong, concave, subacute, erect.

HAB. Distr. of Caledon, *Ecklon! Drege!* (Herb. Hk., D., Sd.)

Stems 1–2 ft. long, diffuse or procumbent. Leaves 5–7 lines long. Heads 3–4 in a tuft, 4 or more such tufts in a nearly capitate fascicle. Nearly allied to *M. erubescens,* from which it differs chiefly in the few, subuniseriate petaloid scales of the involucre.

14. M. quinqueflora (DC.! l. c. 254); stems diffuse or tufted, flexuous, at first tomentose, then glabrate; leaves linear-involute, mucronate, twisted, glabrous, with or without axillary tufts, heads few or many, subsessile in a slightly compound, subumbellate tuft, cylindrical, 5-fl., subtended by a few leaves; inv. glabrate, the outer scales subcanescent, acute or mucronate, close-pressed, the inner sub-biseriate, oblong, acute, curving back at the extremity, whitish ; pappus club-shaped, obtuse.

HAB. Hott -Holl.-berge, *Mundt!* Stellenbosch, *Ecklon.* (Herb. Hk., D.)

Very like *M. tenuifolia* in habit, but the heads are constantly 5-flowered, and the inner inv. scales longer, spreading or recurved at the point. The leaves are either approximate and slightly gemmuliferous ; or widely apart, quite nude. The inner inv. scales are like those of *M. Lichtensteinii.*

15. M. aurea (Don. Wern. Trans. 5, p. 558) ; leaves linear-involute, rigid, squarrose, pungent-mucronate, spirally twisted, with axillary leaf-tufts; heads cylindrical, 4–5-fl., crowded in a branching, hemispherical corymb; inv. glabrescent, the outer scales appressed, sharply mucronate, inner uniseriate scarcely exserted, round-topped, mucronulate, dull-yellow; pappus bristles slender, serrulate, slightly thickened at the point. *DC. l. c.* 250.

HAB. Uitenhage and Albany, *Mundt! E. & Z.! Drege!* Elands river, *Zey.!* 2896. (Herb. Cap., D., Hk., Sd.)

Erect, 1–2 feet high, corymbosely branched; branches virgate. Leaves commonly ½ inch long, rarely 2–3 lines, horizontally spreading or deflexed, shining, very sharp. Heads rather few or very numerous, shortly pedicelled, 3 lines long, not a line diam.; the outer scales silvery, with a brownish or yellowish rib, ending in a mucro. Pappus finely serrulate, with a lance-linear tip. A small specimen from *Drege*, in Herb. Sond., marked "*M. octoflora*" DC., has but 5 fl. in the head, and in no respect differs from *M. aurea.* Is this really *M. octoflora*, DC. ?

16. M. pulcherrima (Less.! Syn. 340) ; leaves lanceolate, pungent-mucronate, straight, erect, close-set, convex-keeled and shining beneath, concave and woolly above, not gemmuliferous; heads cylindr., 3–4–5-fl., crowded in hemispherical branching corymbs ; outer inv. scales cobwebby, tawny, erect or spreading, aristate-acuminate, inner white or yellowish, acute. *DC. l. c.* 253, also *M. lanceolata, DC. p.* 251. *Stoebe gnaphalodes, Thunb.!* *Cap.* 726.

VAR. α. pallescens; invol. scales creamy white or whitish. *M. lanceolata, DC.!*

VAR. β. lutescens; inv. scales straw colour or pale yellowish. *M. pulcherrima, DC.!*

HAB. Cape, *Thunberg!* Langekloof, *Drege!* Vanstaadensberg, *E. & Z.!* β. Eastern Distr. *Burchell.* Drakensteenberg, *Drege!* Swellendam, *Ecklon.* (Herb. Th., D., Hk., Sd.)

This species is readily known by its erect, boat-shaped leaves, without axillary tufts. Leaves 2–4 lines long, 1 line wide. Heads 3 lines long. *Thunberg's* specimen has the heads 5-fl. ; *Ecklon's* mostly 3-fl., and in *Drege's* I have observed 3, 4, and 5 fl. in a head. There is no external char. of foliage, and only the colour of the invol. to separate "*M. lanceolata*" from this. *Ecklon's* specimens, marked "*M. pulcherrima*," in Hb. Sond. and Pappe, belong to *Stoebe gomphrenoides.*

17. M. muricata (Less.! Syn. 356) ; leaves linear or lance-linear (in β. and ε. elliptic-oblong), mucronate, with leaf-tufts, more or less spirally twisted, spreading ; heads cylindrical, 3–4–5-fl., crowded in a hemispherical, branching corymb ; inv. scales glabrescent, appressed, the outer either obtuse, acute, or cuspidate-mucronate ; inner milk-white or rosy, sub-biseriate, exserted, obtuse or acute. *M. muricata, fastigiata, polyanthos, and pungens, Don. Gnaph. muricatum, Linn. Gn. polyanthos and G. fastigiatum, Th.!* *Cap.* 643. *Sieb. no.* 17.

VAR. α. obtusiuscula; lvs. linear-involute spiral; outer inv. scales subacute or mucronate; inner obtuse or subacute. *M. muricata, DC.! l. c.* 249. *Zey.!* 2898, 2899.

VAR. β. phylicoides; lvs. elliptic-oblong, flattish, twisted at base only, white beneath, the margin only involute; heads as in α. *M. phylicoides, Don, l. c.* 559. *DC. l. c.* 250.

VAR. γ. aristata; lvs. as in α.; outer inv. scales cuspidate-acuminate; inner acute or subacute. *M. aristata, DC.! l. c.* 250.

VAR. δ. pungens; lvs. as in α, but more pungent; outer inv. scales taper-pointed, aristate, pungent; inner acute or acuminate. *M. pungens, Don. DC. l. c.* 251.

VAR. ε. tomentosa; lvs. as in β, but very woolly; inv. scales nearly as in γ. *M. lanceolata, var. tomentosa, DC.! l. c.*

HAB. Common throughout the Colony, extending eastward to Natal. (Herb. Th., D., Hk., Cap.)

Very variable in size, branching, and woolliness, and in the shape of the involucral scales, especially the outer ones. It is commonly erect, with level-topped branches, sometimes diffuse or straggling. Leaves 3–6–8 lines long, ½ line wide, or in β. and ε. 1–2 lines wide, mostly glabrous externally, but varying to cobwebby or very woolly. Heads 3 lines long, ½ line diameter. I can find no absolute and *permanent* characters between the above varieties; all grades of bluntness or sharpness may be found in the inv. scales, and in the expanded or involute, blunt or pungent leaves. The more numerous the specimens examined, the less possible is it to distinguish the varieties from one another.

18. M. stricta (Less. Syn. 336); branches tomentose; leaves linear, pungent-mucronate, involute, spirally twisted, with leaf-tufts; heads very many in a branching, woolly, hemispherical corymb, 3–5-fl.; inv. thinly cobwebbed, the outer scales rufous, taper-pointed, aristate, *squarrose-reflexed;* inner sub-biseriate, tawny-yellowish or dirty-white, acute or acuminate. *DC. l. c.* 251.

VAR. β. Dregeana; more branching, more glabrous, sometimes with whiter inv. scales. *M. Dregeana, DC. l. c.*

HAB. Hangklipp, *Mundt & Maire; Paarlberg, Drege! β.* also at Paarlberg, *Drege! Doornkopf, Burke!* Cape, *Dr. Wallich!* (Herb. Hook., D., Sd.)

Robust, either virgate or ramulous, with thinly or thickly woolly branches. Lvs. as in *M. muricata.* Outer inv. scales with strongly recurved points, by which character, and the tawny inner, this is alone distinguishable from var. γ. of *M. muricata.* I retain it, with some doubt, as a species.

19. M. concinna (Harv.); diffusely branched; leaves subulate-terete, mucronate, spreading, subspiral, gemmuliferous; heads cylindrical, 5-fl., sessile in few-headed, pedunculate tufts, disposed in a branching, close corymb; inv. cylindrical, cobwebbed at base, all the scales obtuse, the outer short, the inner with oblong, exserted, pluriseriate, convex, milk-white or rosy tips; pappus clavate. *M. cephalotes, DC.! l. c. p.* 252, *not of Lessing.*

HAB. Districts of Caledon and Worcester, *Ecklon!* Swellendam, *Dr. Thom!* Onderbokkeveldt, *Drege!* (Herb. Hk., Sd., D., Cap.)

This has the foliage and inflorescence of *M. muricata;* the habit of *M. cephalotes,* but not the inflorescence; and the involucre, but not the foliage of *M. fasciculata.* Leaves 6–8 lines long, ⅓ line wide, the axillary leaves much smaller. Stems slightly branched, the branches curved.

20. M. fasciculata (Don. l. c. 657); corymbosely branched; leaves short, linear-terete, mucronate, spirally twisted, squarrose, gemmuliferous; heads cylindrical, 3-fl., sessile, in few-headed, pedunculate tufts disposed in a branching, close corymb; invol. cylindrical, thinly cob-webbed externally, all the scales obtuse, spathulate, the outer with short, the inner with oblong, pluriseriate, milk-white tips; pappus acute. *Less.! Syn.* 330. *DC. l. c.* 253. *Gnaph. fasciculatum, Thunb.! Cap.* 645. Also *M. Eckloniana, DC.! l. c.*

HAB. About Capetown and in the Western Districts. (Herb. Th., D., Hk., Sd.) Much branched and twiggy. Leaves 1–3 lines long, crowded, fascicled. Heads 3 lines long, distinct, but close-lying in 3–5-headed tufts, each tuft on a short, common peduncle. Inv. scales milk-white. I cannot see by what character *M. Ecklo-niana,* DC.! which I have seen in Herb. Sond., differs from a common form of *M. fasciculata.*

Doubtful Species.

M. octoflora (DC. l. c. 250); "branches appressedly pubescent; lvs. linear, mucronate, glabrous at back, often gemmuliferous; heads ovate-cylindrical, pedicellate, 8-flowered, in a simple corymb; outer invol. scales appressed, acute, reddish; inner few, obtuse, scarcely exserted." *DC. l. c.*

HAB. Berg River, *Drege.* (But a specimen in Hb. Sond., sent by *Drege* as " *M. octoflora,*" differs in no respect from *M. aurea.*)

M. distans (DC. l. c. 254); "leaves linear, acute, twisted, glabrous outside, distant, not gemmuliferous; heads few at the ends of the branches, subumbellate; inv. terete, imbricate, outer scales sub-tomen-tose, mucronate, inner white, acuminate." *DC. l. c. Gn. distans, Schrank. Act. Acad. Morac.* 8, *p.* 160.

LXXXV. LACHNOSPERMUM, Willd.

Heads many-fl., homogamous, discoid. *Inv.* turbinate; the scales closely imbricate in many rows, with subpungent tips. *Recept.* with a few marginal paleæ. *Cor.* tubular, 5-toothed. *Style*-branches pencilled at the summit. *Achenes* oblong, angular, the younger ones villous at the angles, the full grown densely hairy. *Pappus* biseriate, of slender, serrulated bristles, rather longer than the corolla. *DC. Prodr.* 6, 255.

A slender, straggling shrub, with spreading, canous, subsimple branches, leafy to the summit. Leaves minute, linear-terete, obtuse or mucronate, slightly twisted, with axillary leaf-tufts; involute, as in *Metalasiæ.* Heads solitary, or 3–5 in a corymb on laxly leafy peduncles. Invol. 4–5 lines long, 3 lines wide, woolly, bristling with the slender, bristle-tips of the outer scales: inner scales linear; acute. Name from λαχνηεις, *woolly,* and σπερμα, *a seed.*

1. L. ericoides, (Willd.); *Less.! Syn.* 342. *DC.! l. c.* 255. *Stœhelina fasciculata, Th.! Cap.* 628. *Serratula fasciculata, Poir. Carpholoma rigidum, Don.*

HAB. Cape, *Thunberg! Mundt.! Burchell.* Namaqualand, *Ecklon.* Kl. Draken-stein, *Drege!* (Herb. Th., Hk., D., Sd.) Stems diffuse, 1–2 ft. long or more. Leaves 1–1½ line long. Whole plant thinly canous.

LXXXVI. PACHYRHYNCHUS, DC.

Heads about 10-fl., homogamous, all the fl. 5-toothed, perfect. *Recept.*

nude, flat. *Inv.* at first terete-oblong, acuminate, with short, leafy, villous, accessory scales or bracts, the rest (or true scales) elongate, scarious, shining, glabrous, linear at length spreading. *Achenes* ovate, very villous, sessile, crowned with a glabrous, thick beak. *Pappus* pilose, the bristles scarcely rough, in several rows, longer than the corolla. *DC. Prodr.* 6, 255.

" A branching, erect suffrutex, all parts, except the heads, villous-tomentose and canescent. Leaves sessile, lanceolate or oblong, callous-tipped, entire, crowded, sub-imbricate. Invol. straw colour." *DC. l. c.* Name from παχυς, *thick*, and ρυγχος, *a beak*.

1. P. xeranthemoides (DC. l. c.)

HAB. Cape, *Bruguiere*, in Herb. L'Her. (Unknown to us.)

LXXXVII. **ELYTROPAPPUS**, Cass.

Heads few (2–8) flowered, homogamous. *Recept.* naked. *Inv.* scales oblong, in few rows, imbricate, horny. *Style*-branches pencilled at the summit. *Achenes* beakless, sessile. *Pappus* of several, broad-based bristles, united in a ring at base, plumose at the summit, with a very minute, or rarely, a cupshaped, external rim. *DC. Prodr.* 6, 256.

Much branched shrubs, growing in dry ground. Leaves minute or heathlike, mostly spirally twisted, more or less gladular and strongly scented. Heads subsessile, in the axils of the upper leaves, solitary or few together, as if spicate-racemose. Name from ελυτρον, an *involucre*, and παππος, *pappus*; alluding to the minute rim surrounding the pappus, which is probably a mere expansion of the epignyous disc.

I. CYATHOPAPPUS (Sch. B.) *Heads* in terminal, globose tufts. *Outer-pappus* cupshaped, half as long as the ovary. *Achenes* wrinkled across :—
    Leaves 4–5 lines long, glabrous and glossy, gland-bristled ... (1) **spinellosus.**
    Lvs. 1–2 lines long, woolly-villous, sparsely gland-bristled ... (2) **cyathiformis.**
II. ELYTROPAPPUS. *Heads* scattered along the branches, or aggregated in spikes. *Outer-pappus* ring-like, very short. *Achenes* longitudinally striate :
    Leaves linear or subulate. Inv. scales acute or acuminate :
      Leaves *gland-bristled*, glabrous or cobwebbed ... ... (3) **glandulosus.**
      Lvs. *glandless*, densely white-woolly ... ... (4) **canescens.**
    Leaves minute, appressed.
      Inv. scales very obtuse ... ... (5) **Rhinocerotis.**
      Inv. scales acuminate ... ... (6) **adpressus.**

1. E. spinellosus (Cass.); leaves linear, involute, mucronate, twisted, erecto-patent, rigid, on the outside glossy and sparsely setose ; heads 6–8-fl., several together in terminal tufts, surrounded by several floral leaves ; inv. scales acuminate ; ovaries scaberulous ; outer pappus cupshaped, entire. *DC. l. c.* 256. *E. spinulosus, Less.! Syn.* 343. *Gnaphalium hispidum, Linn. Th.! Cap.* 645. *Metalasia hispida, Don.*

HAB. Bockland, *Thunberg!* (Herb. Thunb.)
About a foot high, erect, not much branched, glabrous or nearly so. Leaves closely set, erecto-patent, 4–5 lines long, pungent, most of them, but especially the upper ones, sprinkled with spreading, gland-tipped bristles. Tufts of heads sessile, equalling the surrounding leaves. The cup-like rim of the pappus is about half as long as the ovary ; the pappus is beautifully feathered. This has quite a distinct habit; no recent collector seems to have met with it.

2. E. cyathiformis (DC.! l. c. 257); leaves, especially the younger

ones villoso-canescent, woolly, linear, twisted, mucronate, with a few
scattered, pedicellate glands ; heads 3–4-fl., several together in terminal
tufts, surrounded by a few floral leaves; inv. scales puberulous, brown-
tipped, oblong, acute or subacute; achenes wrinkled across; outer
pappus cup-shaped, entire. *Cyathopappus metalasioides, Sch. B. Cass.
Unifl. p.* 27.

HAB. Cederberge, *Drege!* (Herb. Sond.)
Allied to *E. spinellosus* of which it has the remarkable cup-like outer pappus, but
differing in foliage. Leaves 1½–2 lines long, the young ones white-woolly, the older
gray and cobwebbed at least ; the upper ones chiefly gland-bearing. Heads several
in a tuft. I have only seen a small frustule, but sufficient for identification.

3. E. glandulosus (Less.! Syn. 343); leaves linear, terete-involute,
mucronate, straight or twisted, erect or spreading, on the outside bearing
few or many stalked glands; heads 2–4-fl. in small, sessile or pedunculate tufts; inv. scales acute; achenes smooth or striate lengthwise. *DC.!
l. c.* 256, also *E. ambiguus*, DC.! *Stoebe scabra, Linn. Th.! Cap.* 728.

VAR. *a*, **longifolius** (DC) ; lvs. 6–8 lines long, crowded, spreading, very glandular ;
tufts of heads sessile or peduncled (on the same root !).

VAR. *β*. **ambiguus** ; lvs. 2–4 lines long, straight or twisted, more or less glandular ;
tufts of heads as long as, or longer than the floral leaf. *E. ambiguus, DC.! l. c. Stoebe
cinerea, Sieb. no.* 20 *(not Th.) Achyrocome ambigua, Schrank. Stoebe mucida, E.M.*

VAR. *γ*. **microphyllus** (DC.) ; lvs. 2–3 lines long, suberect or spreading ; tufts of
heads shorter than the floral leaf.

HAB. Western Districts, *Thunberg! E.&Z.! Drege.* Elandsberg, *Wallich!* (Herb.
Th., D., Hk., Sd.)
A much branched bush, variable in branching, length of leaves, glands, and
inflorescence. Ramuli sometimes densely scabrous-glandular, sometimes smooth
and cobwebby. Glands sometimes closely fringing the leaves, sometimes on the
back also; sometimes very few and scattered. I find it impossible to separate *E.
ambiguus*, DC., from the var. "*microphyllus*" by any of the characters attributed to
it, all of which vary in different specimens.

4. E. canescens (DC.! l. c. 256); leaves linear, terete-involute, mucronate, straight or twisted, spreading, *not glandular*, densely white-woolly;
heads 2–3 fl., in a long, terminal spike; outer inv. scales woolly, thinly
villous, acute.

HAB. Clanwilliam, *Ecklon!* (Herb. Sond.)
Perhaps a mere glandless and woolly var. of *E. glandulosus*, which varies greatly
in woolliness and in copiousness of glands. I have only seen a specimen with young
inflorescence.

5. E. Rhinocerotis (Less.! Syn. 341) ; leaves minute, appressed, obtuse, smooth externally ; heads 3-fl., much longer than the floral-leaf ;
inv. scales obtuse ; achenes smooth, furrowed. *DC.! l. c.* 256. *Stoebe
Rhinocerotis, Linn. f. Th.! Cap.* 728. *Stoebe cernua, Th.! l. c.*

HAB. Dry ground, throughout the colony, common. (Herb. Th., D., Hk., Sd.)
Well known as the *Rhinoster-bosch.* Stems 1–2 ft. high, excessively branched and
ramulous, the twigs closely covered with minute, scale-like leaves. Heads solitary,
sessile, or on very minute ramuli. Inv. scales horn colour.

6. E.? adpressus (Harv.); slender, much branched, virgate; branches
(with the leaves) filiform ; leaves *most closely appressed*, imbricated,
linear, minute, subacute, nude, woolly edged; heads solitary in the

upper axils, as if spicate, 3-flowered; inv. scales acuminate; pappus of many plumes united in a ring at base; no external annulus? An *Seriphium adpressum, DC. l. c. p.* 263 *(??).*

HAB. Cape. (Herb. Sond.)
A slender shrublet, with stems and foliage closely resembling those of *Stoebe micro-phylla*, but with different inflorescence, and at least 3-flowers in the capitulum. It may possibly be *Seriphium? adpressum*, DC., of which the flowers are unknown. Stems 6–8 inches high, branches 3–6 inches long, quite simple, or ramulous at the tips only. The aspect is that of a very slender *Passerina.*

LXXXVIII. **PTEROTHRIX**, DC. (reform.)

Heads 3–10-fl., homogamous. Other characters as in *Amphiglossa, DC. Prodr.* 6. *p.* 280 *(excl. Sect.* 2. *Lucilioides).* Also *Amphiglossa, Sect.* 1. *Aglossa, DC. l. c. p.* 258.

Small shrublets, spinous or unarmed. Leaves and infl. as in *Amphiglossa*, from which genus this differs solely by the homogamous heads. From *Elytropappus* it differs somewhat in habit, and also in the well-feathered pappus, and the want of an external annulus or outer pappus. Name from πτερον, *a wing*, and θριξ, *a hair;* referring to the plumose pappus.

Spiniferous; lvs. linear, mucronulate; inv scales subacute ...　(1) **spinescens.**
Unarmed; lvs. linear, obtuse; inv. scales taper-pointed, acute　(2) **perotrichoides.**
Unarmed; lvs. lance-oblong, concave, 3-nerved beneath　...　(3) **cymbæfolia.**

1. P. spinescens (DC. l. c. 280); many-stemmed; stems much-branch-ed and twiggy, suberect or divaricate, the upper twigs often spinous; leaves linear, convolute, mucronulate, scattered; heads 3–5-fl.; inner inv. scales acute or obtuse, rufous. Also *Amphidoxa triflora, DC.! l. c.* 258.

HAB. Near the Gariep and at Litaku, *Burchell,* Cat. 1845. Zwartekey, on the flats, *Drege!* Brack River, *Burke & Zey.! Zey.!* 907. (Herb. Hk., D., Sd.)
A dwarf shrublet, 6–8 inches high, tufted; branches and especially the spinous upper twigs widely spreading. Leaves 3–5 lines long. Heads, even in *Drege's* specimens of "*Amph. triflora*," as often 4-fl. as not, and sometimes 5-fl. Inv. scales variable, sometimes very obtuse.

2. P. perotrichoides (Harv.); cæspitose, diffuse or decumbent, much-branched and twiggy, not spiny; leaves linear, convolute, very obtuse, somewhat spiral and now and then tufted; heads 4–5-fl.; inner inv. scales much acuminated, rufous. *Amph. perotrichoides. DC.! l. c.* 258.

HAB. Cape. *Burchell, Mundt!* Swellendam, *Ecklon!* (Hb. Hk., Sd.)
Nearly related to *P. spinescens*, but not spinous, with very blunt leaves and much more taper-pointed inv. scales.

3. P. cymbæfolia (Harv.); stems shrubby, distantly and vaguely branched, unarmed; leaves alternate, sparse, lance-oblong or oblong, cuneate at base, concave and tomentose above, convex, quite glabrous and 3-nerved beneath; heads about 8–10-flowered; inv. scales pluri-seriate, lanceolate, acuminate.

HAB. Betw. Bitterfontein and Mierekskasteel, *Zey.!* (Herb. Sond.)
Stems woody, 1 foot or 2 in length, distantly branched. Leaves nearly an inch apart, ¾–1 inch long, 2–4 lines wide, almost boat-shaped, obtuse or acute, tomentose-canescent above, distinctly 3-nerved and glabrous beneath. Heads 3 lines long, inv. scales imbricate in several rows. Achenes glabrous. Pappus of many amply-feathered, slender, deciduous bristles. Tails of the anthers short, barbellate or

granulated with cellules. Style with a large, bulbous base ; its branches truncate, recurved.

LXXXIX. AMPHIGLOSSA, DC. (reform.)

Heads 6–20-fl., heterogamous, radiate ; ray-fl. ligulate, sometimes very minute and shorter than the involucre; disc-fl. perfect, 5-toothed. *Recept.* naked. *Invol.* cylindrical, imbricate, scarious. *Achene* glabrous. *Pappus* of many very slender, deciduous, separate bristles, closely and amply feathered through their whole length, with or without a minute, external annulus. *DC. prodr. 6, p.* 258 *(excl. sect.* 1, *Aglossa).* Also *Pterothrix, sect.* 2, *Lucilioides, DC. l. c. p.* 280.

Small, much branched, erect or decumbent shrublets. Leaves small, linear or subulate, glabrous and convex externally, woolly within, with inflexed or involute edges. Heads solitary, terminal, sessile. Name from αμφι, *both,* and γλωσσα, *a tongue;* alluding to the ray-fl. being different in the two sections of this genus. It differs from *Pterothrix* merely in having heterogamous fl. heads.

Sect. 1. LEPTOGLOSSA. (DC.) Heads 6–15-fl., heterogamous. Ray-fl. very slender, included or scarcely exserted. (Sp. 3–4.)
Ray-fl. included, obsolete ; all the leaves spreading, acute ...   (1) corrudæfolia.
Ray-fl. subexserted, linear, 3-toothed. Twig-leaves imbricate,
close-lying   ...   .   (2) callunoides.
Sect. 2. PHÆNOGLOSSA. (DC.) Heads 9–10-fl., heterogamous. Ray-fl. exserted, strap-shaped, revolute. (Sp. 5.)
Leaves linear-trigonous, mucronate, spreading   ...   .   (3) tomentosa.

1. A. corrudæfolia (DC.! l. c. 258) ; diffuse or trailing, branched and ramulous; rameal leaves broadly subulate, mucronate, keeled, with inflexed margins, spreading, the axils with leaf-tufts ; twig-leaves short, spreading, mucronate; heads 6–8-fl., heterogamous, on very short, axillary branchlets ; ray-fl. 2–3, very slender, with very minute, included ligules ; inner inv. scales linear, obtuse.

HAB. Boschjemanskarroo, 3000–4000 feet, *Drege !* (Herb. Hk., D., Sd.)
Stems several inches long, woody, glabrate, apparently trailing ; but DC. says " *erect.*" Rameal leaves 4–6 lines long, 1 line wide; those on the twigs much shorter. Heads 3–4 lines long.

2. A. callunoides (DC.! l. c. 259) ; diffuse or trailing, the long, simple branches closely ramulous; rameal leaves broadly-subulate, mucronate, keeled, with inflexed margins, the axils ramuliferous; twig-leaves very small, close-pressed, oval-oblong or linear, obtuse or mucronulate; heads 10–12-fl., heterogamous, on short, axillary twigs; ray-fl. 5–6, with narrow, linear, 3-toothed, purple, sub-exserted ligules ; inner inv. scales linear, obtuse.

HAB. Among stones, in the channel of the Zwartkops River, near Uitenhage, *E. Z.! Verreaux! Dr. Alexander Prior!* &c. (Herb. Hk., D., Sd.)
Very near the former in character, but with a different aspect, and much more evidently ligulate ray-flowers. Twig leaves shorter and more imbricating.

3. A. tomentosa (Harv.); divaricately much branched, spreading, slender; leaves linear-trigonous, mucronate, spreading, often with leaf-tufts ; heads terminal, 9–10-fl., heterogamous ; ray-fl. about 3, with strapshaped, revolute, exserted ligules, longer than the disc; outer inv. scales ovate, inner linear-oblong, acute, rufous, glossy. *A. nitidula, DC.!*

l. c. 259. *Relhania tomentosa, Th. Cap.* 640. *Athrixia tomentosa, Less. Pterothrix tomentosa, DC. / l. c.* 280.

HAB. Cape, *Thunberg!* Uienvalei ; also between Koussie and Pedroskloof, and near Kaus, in Little Namaqualand, *Drege!* Kamiesberg, Namaqualand, *Von Schlicht! Zey.!* 2907. (Herb. Th. Hk., D., Sd.)

Root thick and woody. Stems many from the crown, erect, tufted, slender, much branched and ramulous, the younger tomentose, older glabrous. Leaves crowded, linear-involute, obtuse or mucronulate, 2–3 lines long, tomentose on the hollow, upper surface, cobwebby, becoming glabrous without. Heads sessile at the ends of the twigs, solitary, or 2–3–5 in a tuft. Inv. cylindrical, its scales close-pressed, the outer ovate, acute, inner oblong, obtuse, all glabrous, purple-tipped, slightly viscidulous. Rays very short, purple, revolute. Pappus-bristles very slender, copiously plumose throughout. *Thunberg's* and *Drege's* specimens agree in every particular.

XC. **BRYOMORPHE,*** Harv.

Heads several-fl., heterogamous, the ray-fl. few, ligulate, female; disc-fl. fertile, 5-toothed. *Inv.* imbricate in few rows, the scales linear, separate, caducous. *Recept.* naked. *Anthers* tailed. *Style branches* truncate. *Achene* glabrous, beakless, sessile. *Pappus* of several, slender, scabrous bristles in a single row.

A very densely tufted, many stemmed perennial, forming close, moss-like, level-topped cushions; branches very erect, closely imbricated with subulate leaves to the very summit. Leaves linear-subulate, mucronate, on both sides closely silvery. Heads solitary, terminal, sessile, half sunk among the upper leaves. Disc and ray flowers purple. Name from βρυον, *a moss,* and μορφη, *a form;* because this plant grows in moss-like cushions. It is allied to *Amphiglossa,* from which it differs in habit, and by the bristle-shaped, not plumose pappus. From *Helichrysum* and all its immediate allies it is broadly separated by the ligulate, not filiform, marginal flowers.

1. B. Zeyheri (Harv. Thes. Cap. t. 151); *Helichr. arctioides, Turcz.! Bull. d. l. soc. imp. Mosc. XXIV. II. p.* 79. *Klenzea lycopodioides, Sch. B.*

HAB. Summits of Table, and Hott. Holl. Mountains, common, but rarely flowering, *Zeyher!* 2908. Summit of Genadendahl Mt., 5000 f., *Dr. Roser!* 42. (Herb. Hk., D., Sd.)

Stems 2–3 inches long, branching, erect, with lateral roots, densely cushioned. Leaves 4–5 lines long, ½ line wide, erect. Heads few-flowered. Inv. scales obtuse, whitish or purple tipped.

XCI. **DISPARAGO,** Gaertn.

Heads 2-flowered, one fl. ligulate, either female or neuter ; the other 5-toothed, perfect. *Recept.* narrow. *Inv.* oblong, scales horny, in few rows linear. *Achene* oblong (glabrous or woolly). *Pappus* of 5 or many bristles, naked below, plumose above, often wanting in the ray-fl. *DC. Prodr.* 6, 257.

Dwarf, much branched, heath-like shrubs. Leaves spirally inserted and twisted, crowded, sessile, linear or subulate, involute, within tomentose, without glabrate or cobwebbed. Heads in terminal, round or oblong, very dense glomerules. Corollas purple or white. Name from dispar, *unequal ;* alluding to the two dissimilar fl. in the head.

Achenes glabrous :
 Ray-fl. female, with 5 pappus plumes. Leaves crowded   (1) **ericoides**.

* *Bryomorpha,* Kar. and Kir. is the same as *Thylacospermum,* Fenzl.

Ray-fl. neuter, without pappus :
 Leaves scattered ... ... (3) **laxifolia.**
 Leaves crowded ... ... (2) **Kraussii.**
Achenes woolly or tomentose ; ray-fl. neuter :
 Ray-fl. with 4 or 5 plumes ; tubular-fl. with many plumes (4) **seriphioides.**
 Both ray and tubular-flowers with many plumes ... ... (5) **lasiocarpa.**

1. D. ericoides (Gaertn.) ; leaves crowded, spreading, aristate ; ray-flower female; achene glabrous; pappus of both flowers 5-plumed; paleæ between the flowers. *Less./ Syn. p. 363. DC./ l. c. 257.*

HAB. Eastern Districts, common. *Zey./ 2919, 2921, 2922.* (Herb. D., Hk., Sd.)
A small, much branched, densely leafy shrub. Leaves 2–3 lines long, spirally twisted, closely set, involute, tipped with a longish bristle. Heads crowded, very many together in globose or oblong tufts or spikes. Inv. scales horny, taper-pointed. Ray-fl. shortly and broadly ligulate, purple.

2. D. Kraussii (C. H. Sch. B. Flora, xxvii., 693); leaves crowded, spreading, obtuse-mucronulate; ray-fl. neuter, without pappus, 3-lobed (white); tubular-fl. fertile with 5 plumes and a glabrous ovary. *D. lasiocarpa, b. Herb. Drege!*

HAB. Cape, *Krauss, Drege!* (Herb. Sond.)
Confounded by *Drege* with *D. lasiocarpa,* of which it has the habit, but much more nearly related to *D. laxifolia,* though different in aspect. Leaves ½ a line apart, 2–3 lines long, broadly linear, spirally twisted, thinly canescent. Inv. scales yellowish horn-colour, much acuminate.

3. D. laxifolia (DC. l. c. 257); leaves distant, spreading, obtuse-mucronulate ; ray-fl. neuter, without pappus; tubular-fl. fertile, with five plumes, and a glabrous, angular achene.

HAB. Zwarteberg, and Klyn-rivier'sberg, Caledon, *Ecklon.* Near Simon's Bay, *C. Wright, No. 353.* (Herb. D.)
A slender, straggling fruticulus, 6–8 inches long, with filiform glabrate branches. Leaves 2–3 lines apart, 2–3 lines long, either spirally involute or flattish, the younger shortly mucronate. Glomerules 3–4 lines diam. Rays pale, much smaller than in *D. ericoides.*

4. D. seriphioides (DC. l. c. 257); "leaves scattered, mucronate; ray-fl. neuter; achene tomentose; pappus of the tubular, fertile-fl. of 15–20 plumes, of the ray-fl. with 4–5 plumes." *DC. l. c.*

HAB. Cape, *Burchell, No. 705.*
"Very similar in aspect to *D. ericoides,* but the leaves are scarcely mucronate, not aristate ; the achene villous, not glabrous, and the ray-fl. neuter, not female." *DC.*

5. D. lasiocarpa (Cass.); leaves crowded, spreading, obtuse or mucronulate; ray-fl. neuter; achene densely woolly; pappus of the fertile fl. of many plumes, of the ray-fl. none. *DC./ l.c.258. Wigandia disparaginoides, Less./ Syn. 362.*

HAB. Cape Flats ; also in Worcester and Stellenbosch, *Ecklon,* Drege.! Hout Bay, *W. H. H.* Tulbagh, *Dr. Pappe,!* (Herb. D., Hk., Sd.)
Much-branched, very ramulous, level-topped, about 1 foot high. Leaves 2 lines long, spirally twisted, slender. Heads in globose, terminal tufts. Inv. scales taper-pointed. Ray-fl. narrow-oblong, purple.

XCII. **STOEBE**, Linn.

Heads one-flowered. *Inv.* scales oblong, imbricate, dry-membranous, the outer ones short and often woolly. *Cor.* tubular, 5-toothed. *Style-*branches pencilled at the abrupt point. *Pappus* of 5 or many bristles, naked below, plumose above, slightly connected at base into a ring and falling off together; a small rim or annulus, in many species, exterior to the true pappus, and persistent on the oblong achene. *Achene* either glabrous or woolly, sessile, beakless. *Stoebe and Seriphium, DC. Prodr. p. 259, 261.*

Small, rigid shrubs or half shrubs, almost all S. African. Leaves crowded, narrow, often pungent, very entire, linear, oblong or subulate, often spirally twisted, woolly and concave within, glossy, becoming glabrate externally. Heads crowded together in dense tufts (glomerules) or dispersed in long spikes. Name from στοιβη a *crowding* or *surrounding :* alluding to the crowded leaves of many. I unite the genera *Stoebe* and *Seriphium* of most authors, because the species grouped under them agree in habit, foliage and floral characters; and merely differ by an obscure and variable character of the fruit; the *achene* in *Stoebe* possessing an annulus exterior to the pappus; that of *Seriphium* being perfectly round-topped. I find an annulus in some species said not to possess one; and the breadth of margin varies very much, from species to species.

(1.) Heads collected in terminal, *globose,* perfectly defined glomerules:
  Inv. scales white or cream-coloured:
    Leaves erect, imbricating or appressed:—
      Leaves oval-oblong, concave, woolly-edged   (1) **gomphrenoides.**
      Leaves *very minute,* scale-like, close-pressed (4) **microphylla.**
    Leaves recurved or squarrose:
      Stem and branches erect; lvs. broadly linear (2) **leucocephala.**
      Stem diffuse; lvs. narrow-linear, elongate (3) **squarrosa.**
  Inv. scales rufous or brown-tipped:
    Erect or spreading much branched shrublets:
      Pappus of many plumes:
        Lvs. subulate, spirally twisted and falcate (5) **æthiopica.**
        Lvs. lance-oblong, erecto-patent ... (6) **phylicoides.**
      Pappus of 5-plumes; lvs. spirally twisted... (7) **tortilis.**
    Prostrate, slender, distantly branched; lvs. flat (8) **prostrata.**

(2.) Heads either in small, terminal tufts (*not globose*) or in oval or oblong, more or less decurrent glomerules or imperfect spikes:
  Leaves on the back *pitted* on each side of the keel:
    Lvs. subulate, *ribbed;* inv. scales much acuminate (9) **nervigera.**
    Lvs. linear, keeled; inv. scales obtuse ... ... (10) **rugulosa.**
  Leaves not pitted or furrowed on the back:
    Lvs. straight or straightish, erect or imbricate, expanded:
      Lvs. lance-linear, cobwebbed or glabrous;
        inv. sc. acuminate ... ... (11) **phlæoides.**
      Lvs. lance-oblong, on both sides woolly; inv.
        sc. mucronate ... ... (12) **copholepis.**
    Lvs. spirally twisted, squarrose or spreading:
      Inv. scales oblong, obtuse or scarcely acute (13) **fusca.**
      Inv. scales acuminate, very acute:
        Outer inv. scales shorter and woolly;
        inner glabrous, longer than the flowers:
          Inv. scales brown, squarrose, *very*
            much acuminate ... ... (14) **incana.**
          Inv. scales yellow-horn colour,
            rigid, straight ... ... (15) **spiralis.**
        All the inv. scales glabrous, rufous,
          equalling the flowers ... ... (16) **capitata.**

(3.) Heads in cylindrical or elongated and interrupted, compound-spikes. Leaves fascicled.
   Leaves slender; heads in a long, dense, cylindr. spike.
      Achenes with many toothed ribs, and an annulus (17) **alopecuroides.**
   Leaves various; heads in an interrupted, elongate spike.
      Achenes angled and furrowed, wrinkled, across
      without annulus   .. .. (18) **cinerea.**

1. S. gomphrenoides (Berg.! Cap. 336); erect, subsimple; leaves elliptic-oblong, erect, imbricating, concave, villoso-ciliate; heads in a globose cluster; inv. compressed, cream-white, scales taper-pointed; achenium 4–angled, silvery-tomentose; pappus of 5–plumes. *Thunb.! Cap.* 726. *Less. Syn.* 346. *DC. l. c.* 259. *S. gnaphalioides, Houtt. t.* 34. *f.* 1. *non Th. Metalasia pulcherrima, Hb. Eckl.! (non Less.!)*

HAB. Cape, *Thunberg! Mundt.! Ecklon!* (Herb. Th., Hk. Sd.)
   A foot or more high, many-stemmed, not much branched, branches erect. Leaves 3–5 lines long, 1½–2 lines wide, glabrous and round-backed externally, woolly within, either very obtuse or mucronulate. Glomerules ¼ inch diameter, the inv. scales straw colour.

2. S. leucocephala (DC.! l. c. 259); branches erect, subcanescent; leaves broadly linear, recurvo-patent, spirally twisted, with inflexed edges, tomentose within, villoso-ciliate; heads in a globose cluster; inv. scales white, taper-pointed; achene woolly tomentose; pappus of 20–24 plumes.

HAB. Uienvalei, and between Pickenierskloof and Markuskraal, 1–2,000 f. *Drege.!* (Herb. Hk., D., Sd.)
   Suffruticose, not much branched, erect and virgate. Leaves 3–4 lines long, ¾ line wide, obtuse or mucronulate. Glomerules as in *S. gomphrenoides*, from which this differs in foliage and pappus.

3. S. squarrosa (Harv.); diffuse, with flexuous, spreading branches; leaves linear-terete, slender, involute, recurved-squarrose, acute, somewhat twisted, glabrate; heads in a globose cluster; inv.-scales whitish, taper-pointed; achenes woolly-tomentose; pappus of 18–20 plumes.

HAB. Lambert's or Alexander's Kloof, *Dr. Wallich!* (Herb. Hk., D.)
   Very near *S. leucocephala*, but with a different habit and much narrower, longer, and acute leaves. Stems diffusely much-branched, 12–15 inches long. Leaves 5–6 lines long, ¼ line wide, completely involute.

4. S. microphylla (DC.! l. c. 259); erect, much branched, slender; leaves minute, scale-like, most closely appressed, oblong, acute, keeled, glabrous; heads in a globose cluster; inv. scales white, taper-pointed; *(flowers not seen).*

HAB. Langekloof, in rocky mountainous places, between Grootfontein and Ganzekraal, *Drege! Verreaux!* (Herb. Hk., D., Sd,)
   Stems 6–12 inches high, shrubby, but very slender. Leaves 1 line long, ½ line wide, completely imbricating the twigs. Glomerules not fully grown, brobably ½ inch diameter.

5. S. æthiopica (Linn. Sp. 1315); shrubby, robust, diffusely much-branched, flexuous; twigs villous; leaves lance-subulate or subulate, pungent-mucronate, concave, sub-involute, woolly within, squarrose-recurved, somewhat spiral or falcate; heads in a globose cluster; inv.

scales rufescent, acute; achenes terete, narrowed at base, when mature minutely and appressedly canescent (ovaries almost glabrous); pappus of many plumes. *Thunb.! Cap:* 725. *Less.! Syn.* 317. *DC. l. c.* 260. *Lam. Ill. t.* 722.

HAB. About Capetown and in the Western Districts, common. *Zey.!* 2923. (Herb. Th., D., Sd., Hk.)

Stems 1–2 feet high, diffusely branched or ascending, ramulous. Leaves half-spirally twisted, very rigid and pungent, 3–6 lines long, 1–1½ line wide, the inflexed edges not covering the upper surface. Glomerules ¾–1 inch diam. Scales not very prominent. Flowers bright purple. The ovaries appear, even under a strong lens, glabrous; the very minute and appressed pubescence is only obvious in advanced growth, after flowering.

6. **S. phylicoides** (Thunb.! Cap. 726); shrubby, erect, with virgate branches; twigs tomentulose, becoming glabrate; leaves oblong or lance-oblong, pungent-mucronate, concave, woolly within, with inflexed edges, erecto-patent, straight; heads in a globose cluster; inv. scales rufous, acuminate; achenes terete, narrowed at base, when mature minutely and appressedly canescent (ovaries nearly glabrous). *Less.! Syn.* 346. *DC.! l. c.* 260.

HAB. Cape, *Thunberg!* Betw. Grasbergriver and Waterval, and on the Giftberg, *Drege!* (Herb. Th., D., Sd.)

Nearly related to *S. æthiopica*, but much more erect and virgate, with broader, and straighter leaves, and tomentulose, not villous twigs. Leaves 4–5 lines long, 1–2 lines wide. Heads ½–¾ inch diameter. I find the achenes precisely similar to those of *S. æthiopica*. Those authors who describe them as "glabrous" must have examined only half-grown fruits.

7. **S. tortilis** (DC. l. c. 260); diffuse, straggling, irregularly branched, slender; leaves linear-involute, mucronate, spirally twisted, close-set, glabrous externally; heads in globose, terminal tufts, subtended by many leaves; inv. scales glabrous, mucronate; achenes angular, glabrous; pappus of 5-plumes; annulus entire.

HAB. Uitenhage, *Ecklon*. Fissures of rocks on the Vanstaaden Mts., *Zeyher!* 2920. (Herb. D., Hk., Sd.)

Stems a foot long. Leaves 2½–3 lines long. Glomerules small. This is readily known from its allies by the 5-pappus plumes. In aspect it resembles some states of *Disparago ericoides*.

8. **S. prostrata** (Linn. Mant. 291); stems slender, flexuous, prostrate, cobwebbed; leaves linear-oblong, mucronate, flat, with subinflexed margins, white woolly within, glabrous, glossy and one nerved beneath, spreading; axils nude; heads in a hemispherical, simple glomerule, subtended by several leaves; inv. scales scarious, taper-pointed; fl. purple; achene without annulus. *Thunb.! Cap.* 726. *Seriphium prostratum, Lam. Less.! Syn. p.* 351. *DC. l. c.* 263.

Cape, *Thunberg, Drege!* Summit of Table Mt. *W. H. H.* (Herb. D., Hk. Sd.)

Stems many from the crown, thread-like, distantly branched, trailing. Leaves 3–4 lines long, nearly 1 line wide, white above, glossy-green beneath. Glomerules 3–4 lines in diameter, their subtending leaves longer than the flowers.

9. **S. nervigera** (Sch. Bip.); dwarf, much-branched and ramulous; leaves imbricated, erect or recurved at the points, rigid, linear-subulate, pungent mucronate, glabrous externally, with a prominent midrib, and

thickened margin, the space between midrib and margin pitted; axils nude; heads in terminal, simple or subsimple glomerules; inv. scales chesnut colour, glossy, much acuminate; achene glabrous, somewhat furrowed; pappus of many long plumes. *Seriphium nervigerum, DC. l. c.* 263.

HAB. Cape, *Burchell*, No. 7570. Zeekuvalley, *Mundt.!* Clanwilliam, *Ecklon.* Zwarteberg and Olifant R., *Drege!* (Herb. D., Hk., Sd.)
A small, twiggy flexuous bush, 3–8 inches high; the young twigs villous. Leaves very rigid, not at all spiral, woolly within and at the axils, pale green and glossy without. Inv. scales nearly 4 lines long, very taper-pointed. Flowers purple. There is a trace of an annulus to the achene, and consequently, as DC. observes, this species is doubtfully referable either to *Stoebe* (Less.) or to *Seriphium*, an excellent reason for combining these genera!

10. S. rugulosa (Harv.); dwarf, much-branched and ramulous; leaves imbricated, erect, rigid, linear, mucronulate, glabrate, keeled, and pitted-inv. rugulose beneath; axils nude; heads in terminal, oblong, short spikes; scales horn-coloured, obtuse; achenes villous; pappus of many plumes.

HAB. Brede River, *Mundt.!* (Herb. Hk., D.)
This agrees with *S. nervigera* in habit and the pitted backs to the leaves; but the leaves though keeled are not prominently midribbed, the inv. scales are *blunt*, of paler colour and *shorter*, about 2 lines long. The achenes too are somewhat villous, and the heads more spicate. Flowers purple. The inflor. is nearly that of *S. fusca*.

11. S. phlæoides (Sch. Bip.); suffruticose, erect, with virgate branches, villous; leaves lanceolate-linear, acuminate, sub-aristate, straight and erecto-patent, or the lower squarrose and twisted, those on the twigs more linear and involute; axils not leaf-tufted; glomerules in an oblong, cylindrical spike, the floral-leaves often longer than the flowers; inv. scales scarious, acuminate, horn-colour; ovaries glabrous, furrowed; pappus of many plumes. *Seriphium phlæoides, DC. l. c.* 262. *Seriph. candicans, Mundt.! MSS.*

VAR. β., **phyllostachya**; spike more lax and interrupted, floral leaves longer than the glomerules. *Seriph. phyllostachyum, DC. l. c.*

HAB. Cape, *Burchell*, 5052, 5131. Mountains near Swellendan, *Mundt.! Ecklon!* Knysna *Pappe.!* Herb. D., Hk., Sd.)
A foot or more high, not much branched. Leaves 3–7 lines long, the larger lanceolate expanded, and midribbed, the smaller more or less involute, and somewhat twisted, the younger cobwebbed or thinly tomentose, becoming glabrate. Spikes 1–2 inches long, ½ inch diameter. Flowers white. In uniting the two forms indicated by DC. I adopt his name "*phlæoides*" as most expressive, although I do not find the inv. scales to be "*velvetty*" as described; but then DC.'s specimen was either immature, or, as he suspects, a monstrosity. In our plant the spike is very dense, and the floral leaves sometimes longer, sometimes shorter than the glomerules.

12. S. copholepis (Sch. Bip. Cass. Unifl. p. 23); shrubby, branching-virgate; leaves lance-oblong, imbricating, erect, convex-backed, setaceo-mucronate, on both sides woolly, not involute; axils nude; heads in dense, oblong or ovate, terminal spikes; floral leaves short; inv. scales oblong, obtuse, mucronulate, horn-colour; achene angular, silky; pappus of many plumes.

HAB. Palmiet River, *Ecklon!* (Herb. Sond.)
8–12 inches high, erect, bushy. Branches imbricated with leaves throughout. Leaves 3–4 lines long, 1–1½ line wide, hoary, with black bristle-points. Inflorescence about an inch long. Inv. scales nearly as in *S. fusca*, but mucronulate.

13. S. fusca (Thunb.! Cap. 728) ; shrubby, much branched and ramulous ; leaves linear, involute, mucronate or obtuse, spirally twisted, cobwebby-canescent ; axils nude ; heads in small terminal tufts, or short, oblong spikes ; floral leaves shorter than the involucres ; inv. scales oblong, sub-obtuse, pointless ; achene angular, villous, pappus of many plumes. *Seriphium fuscum, Linn. Sp.* 1317. *Less. Syn.* 351. *DC.! l. c.* 262.

HAB. About Capetown, and in the Western Districts. (Hb. Th., D., Hk., Sd.)
A much branched, flexuous, small bush, 3–12 inches high or more, closely leafy, with greyish, cobwebby indument. Leaves variable in length, 1–5 lines long, the shorter ones commonly pointless. Inv. scales varying from horn-colour to rufous-brown, subobtuse, 2¼ lines long. I find the achenes, both in *Thunberg's* and subsequently collected specimens, *villous*. Flowers white.

14. S. incana (Thunb.! Cap. 725) ; much branched, albo-tomentose, the branches slender ; leaves rather laxly set, linear-involute, slender, mucronate, spirally twisted and squarrose, white-woolly, the very old partly nude ; heads in oblong or subglobose, decurrent tufts (sub-spicate) ; inv. scales chesnut-colour, lanceolate-*cuspidate*, much attenuate, squarrose ; achenes silvery-canescent ; pappus of many plumes. *Less.! Syn.* 348. *DC. l.c.* 260. *S. æthiopica, Sieb.!* 204, *non Linn.*

HAB. Summit of Table Mt., abundant, July and Aug., *Thunberg! Ecklon, Drege! W.H.H., &c.* (Herb. Th., D., Hk., Sd.)
12–18 inches high, much branched, erect, very woolly. Leaves 3 lines long, the longer ones evidently spiral, all strongly recurved. The very much attenuated, glossy-chesnut inv. scales strongly contrast with the whiteness of the leaves and branches, and distinguish this species from *S. spiralis.*

15. S. spiralis (Less.! Syn. 347) ; branched, tomentose, the branches slender ; leaves linear-involute, slender, mucronate, spirally twisted and squarrose-recurved, loosely woolly, in age denuded ; axils mostly nude ; heads in globose or oblong, subdecurrent glomerules or spikes ; outer inv. scales very short, woolly ; inner pale horn-colour or yellowish, rigid, straight, lance-acuminate, longer than the flower ; corolla-lobes short, erect ; pappus of many plumes, without annulus. *DC. l. c.* 260.

VAR. **flavescens**; glomerules commonly oblong ; inv. scales tawny or buff-coloured. *Seriphium flavescens, DC.! l. c.* 263.

HAB. Stellenbosch, *Mundt & Maire, Ecklon!* β. Cape, *Burchell,* 7617, 8166. Dutoit's kloof and Draakenstein, *Drege!* Palmiet R., and elsewhere, *E. & Z.!* (Herb. D., Sd., Hk.)
12–18 inches high, much or little branched, variably, but generally, copiously woolly-canescent. Leaves 3–4 lines long, very squarrose. Inflorescence varying from a nearly globose glomerule to an oblong, or cylindrical, or interrupted spike. Next *S. incana,* but with much less taper-pointed and paler inv. scales, and (so far as I can see) no annulus to the pappus.

16. S. capitata (Berg. Cap. 338) ; ascending or erect, flexuous, twiggy, with virgate flowering branches ; leaves linear-involute, aristato-mucronate, spirally twisted, erecto-patent, glabrate or woolly ; axils nude ; heads in globose or oblong, simple or sub-decurrent glomerules or spikes ; inv. scales glabrous, rufous, the inner acuminate or cuspidate ; achenes silvery ; pappus of many plumes, with an annulus. *Seriphium capitatum, Less.! Syn.* 352. *DC.! l. c.* 263. Also, *S. perotrichoides, Less. l. c. DC. l. c.* 263.

HAB. Cape, *Bergius.* Tradouw, *Mundt.!* Paarlberg, *Drege!* Seekuvalley, *Ecklon.*
Cape Flats, *W.H.H.*, *Burchell*, 707. (Herb. D., Hk., Sd.)

1–1½ ft. high, straggling, either robust and woody or slender, ramulous below,
with long, erect fl.-branches. Leaves 2-6 lines long, commonly glabrous, but sometimes loosely woolly on the outside. Infl. varying from globose to oblong. I find an
obvious annulus to the pappus, and the habit is more that of a true '*Stoebe*' than of
a '*Seriphium*.'

17. S. alopecuroides (Less.! Syn. 349); shrubby, robust, erect, with
virgate branches; leaves densely crowded or fasciculate, linear-involute,
slender, mucronate, appressedly canescent, spirally twisted and squarrose;
tufts of heads in a long, cylindrical spike, mixed with leaves; achenes
glabrous, rib-furrowed, the ribs toothed or tubercled; pappus of many
plumes. *DC.! l. c.* 260. *Seriphium alopecuroides, Lam. dict.* 1. 271.
Stoebe cinerea, fol. 3, Thunb.! in Herb.

HAB. Cape, *Thunberg!*, *Thom! Bowie!* Zwartekops R., *Dr. Pappe!* (Herb. Th.,
Hk., D., Cap.)

A robust, small shrub, 1–2 ft. high, branched in the lower part; the branches
6-18 inches long, erect, simple, very densely leafy. Leaves 4-5 lines long, very
generally with axillary leaf-tufts. Spikes of glomerules 2-3 inches long, ¾ inch
diameter. Inv. scales horn-colour, acuminate. Young achenes somewhat cobwebbed,
old glabrous, deeply furrowed, with toothed ribs; a small annulus outside the pappus.
Corolla white.

18. S. cinerea (Thunb.! Cap. 727); shrubby, much branched and
ramulous; leaves *polymorphous* (varying from minutely-oval or grain-
shaped and obtuse to linear, elongate, mucronate, in the latter case
squarrose and spirally twisted, glabrate or woolly); heads in roundish
tufts, collected in a long, lax or dense, sub-interrupted spike; inner
inv. scales chesnut colour, much acuminate; achenes 4–5 angled and
furrowed, transversely lamellated above; pappus of many plumes. *Less.!
Syn.* 350.

VAR. *a*, **plumosa** (Less.!); leaves woolly canescent, mostly granular or shortly
linear, very obtuse, but varying on the same branch to linear and mucronate. *Stoebe
plumosa, Thunb.! Seriphium plumosum, Linn.! DC.! l. c.* 262. *Ser. vermiculatum,
DC.* (fide sp. *Eclk.!, Drege!*)

VAR. *β.* **virgata** (Less.!); leaves glabrate, frequently linear and mucronate, but
varying on the same branch to obtuse or granular; spike lax. *Stoebe virgata, Th.!
Seriph. plumosum, β. DC.! l. c.*

VAR. *γ.* **cinerea** (Less!); leaves linear, mucronate, squarrose, spirally twisted,
the axils often nude; spike denser, more cylindrical. *Stoebe cinerea, Th.! Seriphium
cinereum, Linn.! DC.! l. c.* 262.

HAB. In dry ground and by road-sides, throughout the Colony; in the Orange
State; and at Natal. (Herb. Th., D., Hk., Sd.)

Most variable in foliage, but I fully agree with Lessing on the propriety of uniting
the above varieties under one specific name. Leaves varying from ½ line to 3-4
lines long; and any thing but uniform on the same specimen. Spikes 3-5 inches
long. If *Drege's* specimens are to be depended on, "*S. vermiculata*," DC. does not
differ from our var. *a.*

Doubtful species.

S. ? vermiculata (DC. l. c. 263); "stems erect, branching, glabrous,
twigs woolly; leaves sessile, short, spreading, concave and tomentose
above, glabrate beneath, often with leaf-tufts; heads glomerate, the
glomerules spicate." *Seriphium vermiculatum, DC. l. c.*

HAB. Rondebosch and Wynberg; and Vanstaadens R., Uitenhage, *Drege! Ecklon!* (Herb. D.)
How does this differ from a common form of *S. cinerea?*

S.? adpressa (DC. l. c.); "stem erect, much-branched; leaves sessile, very minute, strictly adpressed, and with the twigs cobwebbed; heads in the upper axils, solitary, and thus spicate, erect." *Seriphium? adpressum, DC. l. c.*

HAB. Cape, *Burchell,* 7573. (Perfect inflorescence unknown; therefore doubtful.)

S. filaginea (Sch. Bip.); "leaves linear-involute, mucronulate, sub-erect, the younger cobwebby beneath, at length glabrous; axils nude; heads crowded in a subglobose glomerule; outer inv. scales very short, woolly; inner scarious, elongate, very much acuminate, brownish-purple. Flowers not seen." *Seriphium filagineum, DC. l. c. 263.*

HAB. Cape, *Burchell,* 7764. (Unknown to us.)
Specimens distributed by *Drege* under this name appear to me to be monstrosities of some species, with diseased inflorescence.

XCIII. PEROTRICHE, Cass.

Heads one-fl., the fl. tubular, 5-toothed, perfect. *Recept.* naked. *Inv.* scales oblong, imbricated, chaffy, appressed. *Achenes* without pappus, beakless, glabrous, sessile, cylindrical. *Style*-branches pencilled at the apex only. *DC. Prodr.* 6. 264.

A small shrub with the aspect of *Stoebe perotrichoides;* differing generically from *Stoebe,* by the want of pappus. Name, πηρος, *wanting,* and θριξ, *a hair;* without pappus.

1. P. tortilis (Cass.); *Less. Syn.* 353. *DC.! l. c.* 264. *Gymnachæna bruniades, Reich. in Sieb. Fl. Cap.* 23.

HAB. Cape Flats, in heathy places; *Bergius, Ecklon, Pappe, W. H. H.* (Herb. Hk., D., Sd.)
About 8–12 inches high, much-branched, densely leafy; the fl. branches virgate, ending in a globose glomerule of many heads. Leaves linear involute, aristate-mucronate, spirally twisted, cobwebbed, becoming glabrous. Flowers purple. This plant is so very similar in aspect to *Stoebe perotrichoides,* that it may easily be mistaken for it, unless the achenes be examined!

XCIV. TRICHOGYNE, Less.

Heads several-fl., monœcious, all the fl. tubular; the *female* marginal, 1–6, among the innermost inv. scales or paleæ; *male* numerous, central, 5-toothed, tapering at base. *Recept.* bearing marginal paleæ, naked in the centre. *Inv.* scales loosely imbricate in several rows. *Style* in the male-fl. simple, dilated and downy at the apex; in the female bifid. *Pappus* none in the female fl.; in the males of slender bristles plumose at top, uniseriate. *DC. Prodr.* 6, p. 264.

Depressed, branching, densely leafy suffrutices. Leaves minute, linear, often involute, woolly on their upper, glabrate on the lower surface, often fascicled. Name from θριξ, *a hair,* and γυνη, *a female;* the female fl. are slender, like hairs.

Stems shrubby, much branched:—
  Inv. scales all dry and membranous:
    Heads in terminal tufts. Lvs. erect. Inner inv.
      scales white ... ...   (1) **decumbens.**

Heads in short or long spikes. Lvs. spreading.
Inv. scales foxy ... ... (2) **reflexa.**
Outer inv. scales herbaceous ; inner membranous :
Unarmed. Outer inv. scales close-lying. Hds. 7–8-fl. (3) **seriphioides.**
Branches spine-tipped. Outer inv. scales spreading.
Heads 5-fl. ... ... (4) **laricifolia.**
Stems simple, herbaceous ; root annual :
Heads 3–5, axillary ; much shorter than the fl. leaf ... (5) **verticillata.**
Heads in terminal tufts, subtended by many long leaves :
Leaves very slender, terete-involute ... ... (6) **paronychioides.**
Leaves linear, flat, or with sub-inflexed edges ... (7) **glomerata.**

1. T. decumbens (Less.! Syn. *359*); stem shrubby, depressed or prostrate; leaves linear-terete, obtuse, imbricate, erect, thinly silvery; axils
nude ; heads in terminal, globose tufts ; inv. cylindrical, the scales
membranous, the innermost white, obtuse, spreading; pappus feathered
for more than half its length, the plumes spreading or recurved. *DC.!*
l. c. 265. Gnaph. decumbens, Th.! Cap. 646.

HAB. Cape, *Thunberg!* Sneeuweberg, *Drege!* (Herb. Th., D., Hk., Sd.)
Root woody. Stems 3–6 lines long, prone, densely ramulous ; the twigs short,
erect, closely imbricate with leaves. Leaves 2–2½ lines long, slightly whitish and
glossy. Heads 3–8 in a tuft. Outer inv. scales brown.

2. T. reflexa (Less.! Syn. *265*); stem shrubby, procumbent or creeping ; leaves linear-terete, mucronulate or obtuse, slender, spreading, the
adult externally glabrous ; axils often gemmiferous ; heads either in
oblong or cylindrical, dense or lax terminal spikes, or in small subsessile tufts along the branches ; inv. cylindrical, all the scales membranous, rufous, obtuse, incurved ; pappus serrate below, barbato-plumose
toward the apex, the plumes *very short,* erecto-patent. *DC.! l. c. 265,*
also *T. radicans, DC.! l. c.. Stoebe reflexa, Linn. Th.! Cap. 727. Seri-*
phium reflexum, Pers.

HAB. Cape Flats and Simon's Bay, and in the W. districts, common. (Herb.
Th., D., Hk. Sd.)
Stems extensively creeping, 1 or more feet long. Leaves 2–4 lines long. Inv.
scales foxy or pale-chesnut. "Trichogyne seriphioides, litt. e," *Drege!* belongs to
this ; as does also the same collector's specimens of *T. radicans* DC., which certainly is not "erect." The inflorescence varies greatly, not only in different specimens, but on different parts of the same specimen.

3. T. seriphioides (Less.! Syn. *360*); stem shrubby, erect or diffuse,
the fl. branches long, curved; leaves minute or linear, squarrose or
reflexed, woolly within, thinly cobwebbed without; axils gemmuliferous;
heads (not glomerate) in long, distichous spikes, 7–8-fl.; inv. cylindrical, the outer scales leaf-like, short, obtuse, appressed ; inner membranous, horn-colour, obtuse, incurved ; pappus shortly plumose toward
the apex. *DC.! l. c. 265. Gnaph. seriphioides, Berg. Cap. 267. Th. Cap.*
646. Metalasia seriphioides, Don. Seriphium distichum Lam. Stoebe dis-
ticha, Linn. Th. Cap., 727. Also Stoebe fasciculata, Th. Cap. l. c.

HAB. Round Capetown and Simonstown, and in the W. Districts. (Herb. Th.,
D., Hk., Sd.)
A straggling shrub, 2–3 ft. high. The leaves vary in size and shape, much as do
those of *Stoebe cinerea;* from ½ line to 2–3 lines long, the larger more open. Spikes
of heads 3–6 inches long or more, each head separate, laxly or densely set. Invols.
pale horn-colour. Male-fl. 6–7, female 1.

4. T. laricifolia (Less.! Syn. 361) ; stem shrubby, erect or diffuse, the fl. branches long, curved, often spine-tipped, glabrate ; leaves linear-involute, keeled, slender, spreading, glabrate externally ; axils gemmiferous ; heads in long, subdistichous spikes, 5-fl. ; inv. cylindr., the outer scales leaf-like, linear, spreading, inner membranous, horn-colour, obtuse ; pappus flexuous, shortly plumose toward the apex. *DC. l. c.* 265. *Seriph. laricifolium, Lam. Artemisia ambigua, Sieb.! Fl. No.* 21. *Stoebe disticha, fol.* 2., *Herb Thunb.!*

HAB. Dry ground round Capetown, *Thunberg! Mundt.! Ecklon, Drege! Sieber!* (Herb. Th., Hk., D., Sd.)

Very like *T. seriphioides*, but more scrubby, often spinous ; with narrower and longer leaves and smaller fl. heads ; the outer inv. scales less close-pressed and longer.

5. T. verticillata (Less.! Syn. 361) ; annual ; stems many from the crown, or solitary, erect, simple, densely leafy ; leaves linear, elongate, with inflexed edges, much longer than the fl. heads ; heads 3–5-together, sessile in the axils, 6–8-fl.; inv. scales membranous, glossy, acuminate. *DC.! l. c.* 265. *Gnaph. verticillatum, Thunb.! Cap.* 648. *Ifloga polycnemoides, Fenzl.! (fide Drege!).*

HAB. Throughout the colony, in moist, sandy places. (Herb. Th., D., Hk., Sd.)

Root, slender, simple. Stems 1–6 inches high, 1 or many, erect, quite simple, closely leafy throughout. Leaves ½–1 inch long, ⅓ line wide. Fl. heads in the axils of most of the leaves.

6. T. paronychioides (DC.! l. c. 266); annual; stems many from the crown, very short, diffuse, leafy at the summit only ; leaves linear, involute, elongate, glabrate without; heads in tufts at the ends of the stems, surrounded by many leaves ; invol. scales membranous, glossy, acuminate.

HAB Silverfontein, Kl. Namaqualand, *Drege!* (Herb. D., Hk., Sd.)

A minute annual; stems ½–1 inch long. Leaves ½ inch long, very slender. Invol. pale.

7. T. glomerata (Harv.) ; annual; stems several from the crown, very short, diffuse, sparsely leafy ; leaves linear, flattish, or with sub-inflexed margins, woolly within, glabrate without, acute ; heads in large, globose tufts at the ends of the stems, surrounded by many leaves; inv. scales membranous, rufous, glossy, acuminate.

HAB. Cape, *Echl. & Zey.!* (Herb. Sond.)

Very like *T. paronychioides* (perhaps a mere variety), but with broader and much flatter leaves, and rufous involucres. Leaves 3–5 lines long, nearly a line to a line wide. Glomerules ½–1 inch diameter.

XCV. PHŒNOCOMA, Don.

Heads very many-fl., monœcious, all the fl. tubular, 5-toothed; *marginal* female, in a single row ; *central* male, with abortive stigma. *Recept.* nude. *Pappus* in 1 row, of many rough bristles, equalling the corolla ; in the female fl. variously cohering, in the male club-shaped. *DC. Prodr.* 6, *p.* 266.

A much-branched, robust, small shrub. Branches tomentose. Twigs very short, almost papillæform, closely imbricated with minute, scale-like, bluntly-ovate, glabrous leaves. Cauline leaves (subtending the twigs), acuminate, rigid, deciduous. Heads terminal, solitary. Inv. imbricate in many rows, the scales woolly at base; outer short, appressed, acuminate; inner very long, radiating, lanceolate, acuminate, rosy-purple, very showy. Name from φαεινος, *shining*, and κομη, *a head of hair.*

1. P. prolifera (Don.). *Less./ Syn.* 358. *DC./l.c. Xeranth. prolife-rum, Linn. Helichrysum proliferum, Willd. Bot. Mag. t.* 2365. *Bot. Reg. t.* 21. *Andr. Rep. t.* 374.

HAB. Mountains in Stellenbosch, Worcester, and Caledon. (Hb. Th., D., Hk., Sd.) Stem 1–2 feet high. Branches mamillated with minute twigs. Heads 1–1½ inch across, very splendid.

XCVI. PETALACTE, Don.

Heads 10–20-fl., monœcious, all the fl. tubular, 5-toothed; the marginal few (1–3) female, hidden among the innermost inv. scales; the rest male. *Inv.* imbricated, the outer scales scarious, very hairy; inner stipitate, tipped with petaloid, (white), radiating, obtuse laminæ. *Recept.* with marginal paleæ, naked in the middle. *Achenes* beakless, glabrous, the central ones abortive. *Pappus* slender, in one row, capillary, that of the male-fl. subplumose or clavate at the apex. *DC. Prodr.* 6. *p.* 267.

Small shrubs or half shrubs with white-woolly, entire, alternate, spathulate or obovate leaves. Heads in corymbose, shortly-peduncled glomerules. Outer inv. scales very shaggy, rufous or light brown; inner with woolly claws, at the apex expanded into a broad, petal-like spreading lamina. Name from πεταλον, a *petal,* and ακτιν, a *ray.* In habit they resemble *Helichrysa* of the section *Dasylepidea.*

Outer inv. scales very woolly, stipitate; inner with subrotund
petaloid tips ... ... (1) **coronata.**
Outer inv. scales laxly villous, broad-based; inner with oblong,
petaloid tips ... ... (2) **canescens.**

1. P. coronata (Don.); leaves spathulate-oblong, coriaceous, on both sides albo-tomentose, tipped with a black mucro; outer inv. scales stipitate, very woolly, with small tips; inner with broadly obovate, subrotund, white tips; pappus shortly plumose. *DC. l. c.* 267. *Peta-lolepis coronata, Less./ Syn. p.* 357. *Gnaph. coronatum, Linn. Th./ Cap.* 650. *Gn. Achilleæ, Sieb. Fl. No.* 14.

HAB. Western Districts, common. *Zey./* 893. (Herb. Th., D., Hk., Sd.) A woolly suffrutex, about a foot or 18 inches high, erect, not much-branched; the branches virgate. Leaves ¾–1½ inch long, 2–3 lines wide, tapering to the base. Heads several together in a tuft; the tufts in a branching corymb. Inv. scales with opaque white, rarely rosy or purple laminas. *P. discolor,* Don seems to be a trifling variety, with discoloured rays.

2. P. canescens (DC. l. c. 267); leaves obovate-oblong, thinnish, on both sides albo-tomentose, tipped with a brown mucro; outer inv. scales broad-based, loosely woolly, with brown, ovate, glabrous tips; inner with linear-oblong, white tips; pappus-bristles thickened towards the summit, serrulate.

HAB. Cape, *Burchell,* 7310. Mts. round Swellendam, *E. & Z./ Drege! Zey./* 2906. (Herb. D., Hk., Sd., Cap.) A straggling, or decumbent, slender suffrutex, much less robust and woolly than *P. coronata.* Leaves ½ inch long, 2–3 lines wide, flexible. Heads in a simple, few-headed corymb, subsessile. Outer inv. scales deep brown, inner with snow-white tips.

XCVII. ANAXETON, Cass.

Heads few-flowered, monœcious, all the fl. tubular, 5-toothed; one or two female, the rest male. *Recept.* flat, without paleæ, woolly or glabrous. *Inv.* scales in many rows, dry, loosely imbricate, the innermost clawed, spathulate, with a roundish (white) lamina. *Style* in the male-fl. quite simple. *Achenes* sessile, cylindrical, beakless, the fertile granulated or pubescent. *Pappus* of a few scabrous or shortly plumose bristles, shorter than the flower. *DC. Prodr. 6. p.* 267.

Small shrublets, erect or ascending, sparingly branched ; the fl. branches closely leafy below, somewhat pedunculoid upwards. Leaves alternate, coriaceous, quite entire, sessile, mucronate, one-nerved, nerve channelled above, with revolute margins. Heads small in branching corymbs ; inv. scales white or rosy-purple. The habit is something like that of *Metalasia*, but the leaves are contrariwise ; glabrate or glossy on their *upper*, tomentose on their *lower* surface. Name unexplained.

Upper surface of the leaves quite smooth and even :
　　Outer inv. scales quite glabrous, rosy or purple; heads
　　　5–6 fl. ... ... (1) **arborescens.**
　　Outermost scales woolly, medial glabrous, brown; heads
　　　2–3-fl. ... ... (2) **virgatum.**
　　Outer and medial scales, silky-villous; heads 5–6 fl.:
　　　Leaves narrow-linear, erecto-patent　... ... ... (4) **asperum,** β.
　　　Leaves oblong, patent reflexed; outer inv. scales
　　　　brown ... ... (3) **nycthemerum.**
Upper surface of the leaves rough with raised points :
　　Younger leaves cobwebby or tomentose; older nude :
　　　Erect; lvs. erecto-patent; cymes peduncled　... (4) **asperum,** α.
　　　Diffuse; lvs. falcate-recurved; glomerules on leafy
　　　　branches　... ... ... (5) **recurvum.**
　　Younger leaves pilose, with long silky hairs; outer inv.
　　　scales much acuminate; inner with ovate tips... ... (6) **hirsutum.**

1. A. arborescens (Cass.); flowering branches woolly, leafless at the summit; leaves linear, mucronate, patent, tomentose beneath, glabrous and glossy above; heads about 5-fl.; outer inv. scales acute, glabrous, glossy, reddish, or purple, much shorter than the oblong, obtuse, or emarginate white inner scales. *Less.! Syn.* 354. *DC.! l.c.* 268. *Gnaph. arborescens, Linn. Gn. arboreum, Linn. Th.! Cap.* 645. *Lam. Ill. t.* 692, *f.* 2.

HAB. Common on the hills round Capetown and in the W. districts. (Herb. Th., D., Hk., Sd.)

A foot or 18 inches high; branches erect, or curved, virgate. Leaves ½–1 inch long, 1–2 lines wide, horizontally spreading or recurved, pungent, rigid. Pedunc. 1–3 inches long. Corymb much branched; outer inv. scales often purple. Achenes enveloped in long wool.

3. A. nycthemerum (Less.! Syn. 355); flowering branches woolly, distantly leafy ; leaves oblong, mucronate, patent-reflexed, tomentose beneath with revolute edges, glabrous and smooth above ; heads in a very dense, glomerulated, hemispherical, branched cyme, 5-flowered ; outer inv. scales *brown*, villous, shorter than the oblong, obtuse, white inner scales ; recept. quite nude. *DC. l. c.* 268.

HAB. Southern slope of Devil's Mt., *Bergius.* Collected by *Mundt!* but the locality uncertain, *E. & Z.!* (Herb. Sond., D., Cap.)

Ascending-erect, about a foot high, branched chiefly near the base, the branches curved upwards, closely leafy for more than half their length. Leaves strongly reflexed, $\frac{1}{4}$-$\frac{3}{5}$ inch long, 2 lines wide, all mucronate (even in the original specimen in Hb. Berol.) Inflorescence at first hemispherical, almost capitate, afterwards expanding into a much and closely-branched corymb. Outer inv. scales deep brown, strongly contrasting with the white inner scales.

2. A. virgatum (DC. l. c. 268); stem subsimple, erect, slender, leafless at the summit; leaves linear, with revolute margins, erect or erecto-patent, quite smooth above, the younger ones woolly beneath; heads minute, 2–3-fl., densely crowded; base of the invol. woolly, medial scales ovate, acute, brown, glabrous, inner white, short, undulate; recept. nude.

HAB. Cape L'Aguillas, *Drege*. Cape, *E. & Z.!* (Herb. Sond.)
I have seen no specimen of *Drege's* plant, but describe from a sp. of *E. & Z.* in Hb. Sond., which answers pretty nearly to *DC.'s* description. It looks like a very slender depauperated form of *A. arborescens*, with much smaller involucres and fewer flowers in each.

4. A. asperum (DC.! l. c. 268); flowering branches woolly, leafless at the summit; leaves linear, mucronate, erecto-patent, tomentose beneath, glabrous or cobwebbed, and either scabrous or smooth above; outer inv. scales acute, reddish, villous or softly and densely hairy, shorter than the obtuse inner scales; recept. nude.

VAR. α. **asperum**; upper surface of the leaves rough with raised points; inv. scales very hairy. *A. asperum, DC. l. c. Gn. asperum, Th.! Cap*, 641.

VAR. β. **læve**; upper surface of leaves quite smooth; invol. scales villous.

HAB. Hills about Capetown and Simon's bay, *Thunb.!* Stellenbosch and Caledon, *Ecklon!* Drakenstein and Hott. Holland, *Drege! Zey.!* 2912. β. Table Mt., and Simonstown, *W.H.H., C. Wright, No.* 328, 359, 360. (Herb. Th., D., Hk., Sd.)
Very similar in aspect to *A. arborescens*, but readily known by its silky-villous involucre. Var. β. is intermediate between true "*arborescens*," of which it has the leaves, and true "*asperum*," of which it has the involucres. The leaves in both vars. are more erect than in *A. arborescens;* the inv. scales less rufous. *Zeyher's* 2912 has the leaves copiously cobwebbed with loose, deciduous wool. The receptacle is nude and the achenes pubescent, but not wrapped in wool as they are in *A. arborescens*.

5. A. recurvum (DC.! l. c. 268); robust, diffuse, much-branched, flexuous; branches very woolly, closely leafy to the summit or nearly so; leaves linear, with strongly-revolute margins, mucronate, scabrous, but nude above, woolly beneath, spreading or falcate, recurved; heads *(immature)* densely-crowded in a hemispherical glomerule *(afterwards probably expanding ?)* ; outer inv. scales ovate, acute, brown, villous; inner obtuse *(not fully grown)*. *Gnaph. recurvum, Lamk. Dict.* 2, p. 746, *fide DC.*

HAB. Stellenbosch District, *Ecklon!* Hott. Holl. Berg., *Zey!* 2911. (Herb. Sd., D., Cap.)
Stems robust, diffuse or procumbent, shaggy and closely leafy. Leaves like those of *A. asperum*, from which this seems chiefly to differ in habit. DC. describes the inv. scales as "*quite glabrous*," but I find them copiously silky on *Ecklon's* original specimen. Until the fully-developed inflorescence shall be found this sp. must remain doubtful.

6. A. hirsutum (Less.! Syn. 356); fl. branches hirsute, sparsely leafy at the summit; leaves linear-oblong, mucronate, erect, with strongly revolute margins, hairy beneath, *roughly-pilose* (becoming nude in age)

and scabrous above; heads subcapitate ; outer inv. scales much acuminate, brownish, very densely hairy except at the points, shorter than
the *ovate* inner scales. *DC. l. c.* 268. *Gnaph. hirsutum, Thunb.!* *Cap.*
646 *(but not G. asperum, pte. as stated by Lessing).*

HAB. Cape (no locality stated), *Thunberg!* (Herb. Th.)
About a foot high, not tomentose or woolly, but clothed with long, soft, silky
brown hairs. Leaves closely set, imbricating, 3–5 lines long, rigid. Heads 6–12
together in terminal tufts. Outer inv. scales dark-coloured. The left hand specimen of "*Gn. asperum*" *Thunb.*, in Herb Th., referred to by Lessing, exactly agrees
with *A. asperum var. a,* DC., and differs from the present in its *woolly toment* not
silky pubescence.

XCVIII. ATHRIXIA, Ker.

Heads many-fl., heterogamous ; *ray-fl.* in a single row, ligulate or bi-
ligulate, female ; *disc.-fl.* perfect, 5–toothed, tubular. *Recept.* naked.
Inv. turbinate, the scales closely imbricate in many rows, aristate-re-
curved at the points. *Achenes* oblong, beakless, sometimes with a tuft
of hairs at the base, sometimes naked, glabrous or pilose. *Pappus* consisting either wholly of filiform, rough bristles in a single row ; or more
frequently of bristles and short serrulate scales alternating. *DC. Prodr.*
6, *p.* 276.

African suffrutices with the aspect of some *Asters*. Leaves alternate, decurrent
or sessile, linear or narrow, mucronate, rigid, with revolute margins, tomentose
beneath, smooth or scabrous above. Heads terminal, solitary ; rays spreading,
white or purple ; disc yellow. Natives of Abyssinia and Madagascar, as well as of
the Cape. Name, possibly from αθηρ, an *awn ;* from the awned inv. scales ?

Leaves sessile, ovato-lanceolate, smooth above, woolly beneath　(1) **phylicoides.**
Leaves sessile, linear, *smooth* above.
　Stem much branched, tall ; branches spreading ; achenes
　　pilose　...　...　...　...　...　...　...　...　...　(2) **elata.**
　Stem subsimple ; branches erect ; achenes quite glabrous　(3) **angustissima.**
Leaves sessile, linear, *gland-hispid* above　...　...　...　...　(4) **Gerrardi.**
Leaves more or less decurrent, *scabrous* above ; the upper-
　most at least linear-subulate, the lower either similar or
　ovate-oblong or lanceolate :
　　Pappus consisting of bristles, with short, interposed scales　(5) **heterophylla.**
　　Pappus consisting of bristles only *(no short scales)*　...　(6) **Capensis.**

1. A. phylicoides (DC.! l. c. 277); stem erect, slender, fruticose, diffusely much branched, the twigs glabrate ; leaves sessile, ovato-lanceolate or lanceolate, acuminate, with slightly reflexed margins, white-
tomentose beneath, above dark-green and glossy, 3-nerved at base, and
veiny ; inv. cobwebbed, the scales lanceolate-acuminate ; pappus with
interposed scales; achenes pilose.

HAB. Tambukiland, *Ecklon!* betw. Omsamcubo and Omtata, *Drege!* Trans Kei
H. Bowker! No. 70! Natal, *Dr. Sutherland!* Bushman's River, Natal, *Gerr. & M'K.!*
547. (Herb. D., Hk., Sd.)
A slender, much-branched, twiggy shrub, "used in making brooms and its leaves
infused as tea"—*(Dr. Sutherland)*. Leaves ½–1 inch long, not unlike those of *Phylica paniculata,* but larger, the midrib alone visible beneath through the toment.,
lateral veins obvious on the glossy, upper surface. Heads terminating small twigs ;
the invol. scales variable in shape, more or less taper-pointed. I find *scales* among
the pappus bristles, and the achenes constantly clothed with erect, white hairs.

2. A. elata (Sond. in Linn. vol. 23, p. 67) ; stem tall, tapering upwards, very much branched, the branches lateral, spreading, filiform,
simple or ramulous ; leaves linear, sessile, acute, glabrous and smooth
above, tomentose with revolute margins beneath ; heads terminal and
lateral-secund in the axils of the upper leaves ; invol. scales bristle-
pointed, squarrose; achenes pilose ; pappus with interposed scales.

HAB. Magalisberg, *Burke & Zeyher! Zey.!* 911. Basutu Land, 735, and Albert,
No. 1754, *T. Cooper!* (Herb. Hk., D., Sd.)
Stems 2–3 feet high, rigid, ligneous, terete, straight and tapering to a fine point';
branches lateral, spreading or recurved, filiform, in large specimens emitting lateral,
spreading twigs; young parts cobwebbed or woolly, old glabrate and glossy. Leaves
about 1 inch long, ½ line wide. Heads sometimes ending the branches, but more
commonly subsessile, or on short, axillary twigs in the axils of the upper leaves,
and then secund. Achenes exactly like those of *A. phylicæfolia.*

3. A. angustissima (DC.! l. c. 277); stem slender, erect, slightly
branched, the branches erect; leaves narrow-linear, acute, sessile, quite
glabrous and smooth above, tomentose below with revolute margins ;
inv. scales setaceous, erect ; ovaries slender, elongate, glabrous at base;
pappus with interposed scales, the bristles very caducous.

HAB. On the Witberg, 6–7000 f., *Drege!* (Herb. D., Hk.)
4–6 inches high, probably tufted. Leaves 1–1½ inch long, ½ line diameter, rigid,
erecto-patent. Heads ending the branches, sub-peduncled, with a few scales on the
peduncle, smaller than in other species, the inv. scales very narrow and bristle-
pointed.

4. A. Gerrardi (Harv.); stem ascending-erect, sparingly branched,
robust; branches erect, closely leafy to the summit, pubescent ; leaves
sessile, not decurrent, linear, with recurved points and revolute margins,
rough with glandular bristles on the outer surface, tomentose beneath ;
pappus with interposed scales.

HAB. Dry plains in Zululand, *W. T. Gerrard!* 1029. (Herb. D.)
Stem 1–2 ft. high, much stronger and more densely leafy than in *A. Capensis*, very
rough with glandular hairs. The leaves are uniformly linear, 1–1½ inch long, ½
line wide, not in the least decurrent at base. Heads ¾ inch long, and wide ; inv.
scales bristle-pointed. Flowers blue. Achenes hispidulous at base. Quite unlike
A. sessilifolia, DC.

5. A. heterophylla (Less.! Syn. 366); stem ascending-erect, sparingly
branched, branches cobwebbed, simple or ramulous ; lower leaves oblong, upper linear-subulate, pungent, with revolute margins, strongly
decurrent at base, rough with glandular bristles outside; pappus with
interposed scales. *DC. l. c.* 277. *Aster heterophyllus, Th.! Cap.* 688.
Also *A. sessilifolia, DC.! l. c.*

HAB. Langekloof, Krumriver and trans Kamtousriver, *Thunberg!* Uitenhage and
Swellendam, *Ecklon!* Dutoitskloof and Drakenstein, *Drege!* Albany, *T. W., Mrs.
F. W. Barber!* &c. Brit. Kaffraria, *T. Cooper!* 287. (Herb. Th., D., Hk., Sd.)
Undistinguishable from *A. Capensis* by any character except that of the *pappus!*
the lower leaves are sometimes as narrow as the upper. *Drege's* specimens (in Herb.
D., Hk.) of *A. sessilifolia,* DC. have the leaves certainly decurrent, and a tuft
of hairs at base of achene, and do not at all differ from weakly specimens of *A.
heterophylla.*

6. A. Capensis (Ker. Bot. Reg. t. 681); stem erect, sparingly branched;

branches erect, virgate, leaves (except in var. β.) linear-subulate, pungent, with revolute margins, strongly decurrent at base, rough with glandular bristles on the outer surface; pappus wholly of rough bristles. *Less.! Syn. p. 365. DC.! l. c. 276. Aster crinitus, Thunb.! Cap. p.* 688.

VAR. β. **latifolia** (DC.); lower and medial leaves broadly oblong or lanceolate, expanded, nearly smooth above or scabrous along the edge merely, white-woolly beneath, with inflexed margin.

HAB. Cape, *Thunberg! Ecklon.* Paarlberg and Swellendam, *Drege!* β. In Worcester and Clanwilliam, *E. & Z.! Drege!* (Herb. Th., D., Hk., Sd.)

About a foot high, branched chiefly near the base, very rigid, canescent. Leaves an inch long, mostly linear, the lower ones occasionally broader, decurrent for a long way down the stem. Inv. scales with very long, narrow, bristle-shaped, recurved points. Rays purple. Achenes with a tuft of bristles at base. Var. β. has the foliage nearly of *A. heterophylla,* but not the pappus!

XCIX. **ANTITHRIXIA**, DC.

Heads many-fl., radiate, the ray-fl. ligulate, flat, female; disc-fl. tubular, 5-toothed. *Inv.* imbricate, the scales linear-oblong, with dry, obtuse points. *Recept.* nude. *Anthers* tailed. *Style*-branches truncate. *Achenes* terete, glabrous, with a small, basal, pubescent style, sub-rostrate. *Pappus* in one row, of very many, rigid, roughish bristles. slightly united at base. *DC. Prodr. 6, p.* 277.

A small, dwarf shrub, much branched and twiggy; the young twigs cobwebbed, becoming glabrous. Leaves opposite, slightly connate at base, linear, short, obtuse, woolly above, midribbed and glabrate beneath, the midrib scabrous. Axils often leaf-bearing. Heads terminal, solitary, yellow. The invol. is that of a *Leyssera;* the pappus of *Athrixia;* and the foliage that of some *Pteronia.* Name from αντι, *contrary to,* and Athrixia; *different from Athrixia.*

1. A. flavicoma (DC. l. c. 278).

HAB. Camiesberg, *Drege!* (Herb. D., Hk., Sd.)

A much branched scrub, about a foot high. Leaves 2–3 lines long. Heads about ½ inch long. Inv. glossy, the scales rigid below, membranous at the point, fulvous.

C. **LEYSSERA**, Linn.

Heads many-fl., radiate; ray-fl. ligulate, female, with shorter ovaries; disc-fl. tubular, 5-toothed, perfect. *Recept.* without paleæ, subfimbrilliferous. *Inv.* imbricate, of dry scales. *Achenes* terete, shortly beaked, with a terminal areole. *Pappus* in one row; in the ray-fl. of short scales; in the disc-fl. of long, plumose bristles, alternating with small scales. *DC. Prodr. 6, p.* 278.

Suffrutices or herbaceous plants, all African. Branches slender, leafy, ending in one-headed peduncles. Leaves scattered, sessile, linear, often with axillary leaf-tufts. Flowers yellow. Name, in honour of Leysser, author of the Flora Halensis.

Pappus-bristles in the disc-fl. feathered *completely* from top to base :

 Inv. scales very obtuse, in few or many rows   ..   ...   (1) **gnaphalioides.**
 Inv. scales lanceolate, acute or acuminate   ...   .   (2) **incana.**

Pappus-bristles in the disc-fl. feathered at the top only, naked

 at base   ...   .   (3) **tenella.**

1. L. gnaphalioides (Linn. Sp. 1249, non Th.); suffruticose, inv. scales all very obtuse. *Less.! Syn. p. 368. DC.! l. c. 278. Jacq. Ic.*

Rar. t. 588. *L. callicornia, Th.! Cap.* 690. *Callicornia gnaphalioides,*
Burm. *Asteropteris Callicornia, Gaertn.* 2, *p.* 460, *t.* 173.

VAR. β. **gracilis** ; inv. scales in few rows, rigid, less membranous. *Herb. Th.!*

HAB. Common near Capetown and in the Western Districts. Var. β. Steendahl,
near Tulbagh, *Dr. Pappe!* (mixed with a). (Herb. Th., D., Hk., Sd.)

Stem woody at base, about a foot high, corymbosely branched, woolly or glabrous,
more or less scabrous. Leaves crowded, 10–12 lines long, not ½ line broad, linear,
acute, with reflexed margins, tomentose or nude, more or less sprinkled or margined
with stalked glands. Pedunc. terminal, nude, glabrous, spreading, 2–3 inches long.
Heads many-fl. ; the rays yellow above, brownish-red on the under side. Inv. scales
in a. in very many rows, with widely membranous borders ; in β. (otherwise quite
similar) in much fewer rows, more rigid, with narrow, scarcely membranous edges.
All the specimens in Herb. Thunb. belong to this variety, which is by no means so
common as a. *Drege's* specimens of "*L. tenella, var. subcanescens*" (in Hb. Sd.!)
belong to it also.

2. L. incana (Thunb.! Cap. 691); suffruticose ; inv. scales (or at least
the outer ones) lanceolate, acuminate. *Less.! Syn.* 369. *DC. l. c.* 278.

HAB. Cape, *Thunberg!* Round Capetown, *Ecklon !* W.H.H. (Hb. Th., Hk., Sd.)
Only known from *L. gnaphalioides* by its taper-pointed inv. scales, a character
sometimes strongly marked, but very variable, when a large number of specimens
are examined, and I fear not to be trusted to as a specific difference.

3. L. tenella (DC.! l. c. 279); very slender, wiry ; leaves very slender ; inner inv. scales folding at base round the ovary and closely clasping
the marginal achenes, linear, subacute ; pappus of the ray of 5 oblong,
toothed scales, of the disc of 5 obtuse scales and 5 bristles feathery at
top only ; tubular fl. with a very scabrous tube.

HAB. Zach River, *Burchell.* Silverfontein, *Drege!* Namaqualand, *Zey!* (Herb. Sd.)
Similar in aspect to *L. gnaphalioides var.* β., but still smaller and more slender,
and always to be known by the pappus bristles (in the disc fl.) being feathered at
their apex only. Some (at least) of *Drege's* distributed specimens of "*L. tenella var.
subcanescens*" have the fully plumed pappus of *L. gnaph.* β. *gracilis,* to which therefore
they must be transferred.

CI. **ROSENIA,** Thunb.

Heads many-fl., radiate ; ray-fl. ligulate, female ; disc-fl. tubular, perfect, 5-toothed, the teeth erect. *Recept.* covered with conduplicate,
scarious paleæ. *Inv.* imbricate, the scales dry, membrane-edged. *Anthers* tailed. *Achenes* beakless, glabrous, of the ray 3-cornered, 3-ribbed ;
of the disc terete, furrowed. *Pappus* of the disc-flowers in two rows,
the outer of many short, broad scales, the inner of 2 long bristles ; of
the ray of many short scales in a single row. *DC. Prodr.* 6. *p.* 280.

A rigid shrub, with the aspect of a *Pteronia.* Leaves minute, opposite, decussate.
Heads terminal, solitary. Name in honour of the brothers *Rosen,* eminent physicians
and botanists. DC's "*R. spinescens*" will be found under *Nestlera.* (*N. Dregeana,* H.)

1. R. glandulosa (Thunb.! Cap. p. 692); *Less.! Syn. p.* 370. *DC.! l. c.*
280.

HAB. Cape (probably some part of the Karoo), *Thunberg,!* (Herb. Thunb.)
A rigid, much-branched shrub, 1–2 f. high, subdichotomous, the old twigs indu-
rating. Twigs glandular. Leaves opposite, the pairs 2–3 lines apart, oblong or oblong-obovate, 2–3 lines long, 1 line wide, coriaceous, very obtuse, gland-viscid
externally, tomentose above, with subinflexed edges ; the younger tomentose be-

neath. Inv. loosely imbricate, oblong, the scales broadly oblong, obtuse, mucronulate, with a dark brown keel and very wide membranous edges. Paleæ of recept. keeled, equalling the disc., 1-nerved, acuminate. Pappus and fl. not seen by me. The specimen in Hb. Thunb. from which I describe is in a very imperfect state. A full account of it will be found in Lessing's work, above quoted.

CII. NESTLERA, Spreng.

Heads many-fl., radiate ; *ray-fl.* female, ligulate ; *disc-fl.* 5-toothed, tubular, perfect. *Recept.* without paleæ, either honey-combed or fimbriate, *Inv.* scales imbricate in several rows, the inner scales longer, membranous, arid. *Achene* beakless, sessile, glabrous or pubescent. *Pappus* short, either a toothed crown, or of several short, separate or connate scales. *DC. Prod. 6. p.* 283. Also, *Polychætia, Less. ! Syn.* 371. *DC. l. c., pro parte.*

Small, arid, rigid shrublets, rarely herbaceous biennials or annuals. Leaves alternate or opposite, sessile, linear or oblong, entire. Flowers yellow. Heads solitary, terminal, or in the forks of the branches. Name in honour of some botanist. Several of DC's. sp. of *Polychætia* have asteroid stigmas, and all other characters of *Geigeria,* under which genus they will be found.

Invol. scales acute or acuminate:
 Herbaceous ; leaves decurrent, obtuse ; inv. scales spreading (1) **biennis.**
 Shrubby, glabrous ; leaves sessile, subulate, pungent ... (2) **acerosa.**
Inv. scales obtuse:
 Pappus of several distinct scales :
  Erect, much branched shrubs :
   Leaves ovate-oblong, tomentose, fringed with
    glands, heads solitary ... ... (3) **Dregeana.**
   Lvs. linear-trigonous, canous or nude ; heads
    corymbose... ... ... (4) **Garnotii.**
   Prostrate ; lvs. linear convolute ; heads solitary ... (5) **prostrata.**
 Pappus coroniform, entire, subentire, or toothed :
  Heads 3-together at the ends of the twigs... ... ... (6) **tricephala.**
  Heads solitary in the forks or at the ends of the twigs :
   Leaves linear or spathulate, channelled above :
    Lvs. loosely woolly ; inv. rufous... ... ... (7) **humilis.**
    Lvs. glabrous ; inv. pale ; (stem dwarf) ... (8) **conferta.**
    Lvs. cobwebbed or glabrous, sparsely *spini-*
     ferous; branches spiny ... ... (9) **muriculata.**
   Leaves linear, flat, appressedly canescent... ... (10) **oppositifolia.**
  Heads subumbellate ; leaves linear-terete, closely
   involute ... ... (11) **tenuifolia.**

Sᴇᴄᴛ. 1. Pᴏʟʏᴄʜᴀᴛɪᴀ (Less.) : *Pappus* of several distinct scales (Sp. 1–5).

1. N. biennis (Spreng.); stem herbaceous, viscidulous and cobwebbed; becoming glabrate ; leaves decurrent, linear, obtuse, glandular and cobwebbed; inv. scales much acuminate, greatly longer than the disc-fl. spreading at the summit, the outer ones bristle-pointed, squarrose ; achenes elongate, 4-sided, longitudinally ribbed and furrowed. *Less. ! Syn. p.* 373. *DC. ! l. c.* 283, also *N. reflexa, DC. ! l.c. Columellea biennis, Jacq. Schoenb. t.* 301. *Stephanopappus reflexus, Less. ! in Linn.* 234. *Relhania reflexa, Thunb. ! Cap.* 640.

Hᴀʙ. Cape, *Thunberg!* Olifant's River, *Drege!* Bellfontein, Langekloof, *Dr. Wallich!* (Herb. Th., D., Hk., Sd.)
Root fibrous, annual or bennial. Stem erect, corymbosely branched, 6–12 inches high. Leaves decurrent, ¾–1 inch long, ½ line wide, the younger cobwebbed. All

parts glandular and viscidulous. Heads at first terminal, becoming lateral by the lengthening of the branch, just beneath the head. Invol. pale horn coloured. Pappus very white, of many short, unequal scales. *Drege's* specimen, referred by DC. to *N. biennis* precisely accords with *Thunberg's* "*R. reflexa*."

2. **N. acerosa** (Harv.); shrubby, erect, glabrous, viscidulous; leaves linear-subulate, pungent, round-backed, furrowed above, gland-dotted, crowded, imbricating; heads terminal, solitary, sessile; inv. scales lanceolate, acute, viscidulous; recept. honey-combed and toothed (scarcely fimbrilliferous); achenes long, slender, angular, glabrous; pappus of many short, slender, subulate scales. *Polychætia acerosa, DC.! l. c.* 285.

HAB. Witteberg, *Drege*. Stormberg, *A. Wyley!* Mts. near Mequeathing, Basutuland, *T. Cooper!* 733.

A strong growing small bush, 12–18 inches high; branches curved, with rounded axils, the older ones quite bare; upper branches and twigs closely covered with erecto-patent leaves. Leaves ⅓–1 inch long, not a line wide, rigid, emitting a balsam. Heads longer than the leaves, the inner inv. scales radiating. Ray-fl. not much longer than the scales, reddish brown beneath. Style-branches truncate, pencilled at the extremity. Anthers with long tails.—A very distinct species, especially by the characters of the receptacle.

3. **N. Dregeana** (Harv.); stems forked, flexuous, twigs spiniferous; leaves opposite, oblong-obovate, flattish, tomentose, the upper ones (at least) densely setoso-glandular along the margin; heads terminal, solitary, sessile; inv. scales oblong, broadly membrane-edged, very obtuse, the innermost elongate, at length reflexed; recept. fimbrilliferous; achenes linear-prismatic, glabrous; pappus of many short scales. *Rosenia spinescens, DC.! l. c.* 280.

HAB. Nieuweveld, betw. Rhinosterkop and Gangefontein, 3500–4500 f., *Drege!* (Herb. Sond.)

A scrubby, tomentose, glandular bush. Leaves ⅓ inch long, 2 lines wide, hoary above, deciduously woolly and gland-bearing beneath. This plant, erroneously referred to *Rosenia* by De Candolle, has all the gen. char. of a "*Polychætia*." The fimbrils of the recept. are much shorter than the achenes, slender and subulate.

4. **N. Garnotii** (Harv.); twigs tomentose, leaves linear, coriaceous, sub-trigonous, spreading or deflexed, crowded, obtuse, appressedly canescent or nude; heads several together in tufts at the ends of the branches; inv. oblong, few-fl., scales appressed, obtuse, shining; recept. fimbrilliferous; outer ovaries villoso-canescent; pappus of many minute, subulate scales. *Polychætia Garnotii, Less.! Syn.* 372. *DC. l. c.* 285.

VAR. β. denudata; leaves glabrous or nearly so; young twigs canescent.

HAB. Cape, *Garnot!* Zwartland, *Drege!* β. Cape, *Bowie!* (Herb. Sond., Hk.)

A small slender shrub, of which I have seen but small specimens. Leaves 2 lines long, ⅓ line wide, squarrose. Heads 3 lines long. Pappus very minute. Fimbrils of the recept. obtuse, connate into honeycomb-cells, obtuse. β. in Hb. Hook. agrees in all characters except pubescence,

5. **N. prostrata** (Harv.); stem shrubby, depressed or prostrate; branches elongate, glabrescent, viscidulous; leaves opposite, linear-convolute, obtuse, tomentose above, puberulous beneath; heads sessile at the ends of minute, lateral, densely leafy twigs, solitary; inv. scales oblong, obtuse; pappus of many subulate scales. *Polychætia oppositifolia, DC.! l. c.* 285.

HAB. Sneeuweberg, 3–4000 f., *Drege!* (Herb. Hk., D., Sd.)
Branches trailing, 3–6 inches long, leafy throughout; emitting short, erect, densely leafy ramuli. Leaves 2–3 lines long, not ½ a line diam. Heads 4 lines long. Inv. glabrous, pale horn-colour. Fimbrils subulate, rather longer than the achenes. Achenes of the ray pilose ; of the disc glabrous; pappus in both of many, narrow, toothed, acute, white, short scales. Style-branches truncate, bearded at the apex.

Sect. 2. NESTLERA (Less.) *Pappus* coroniform, toothed. (Sp. 6–11.)

6. N. tricephala (Harv.); stem slender, shrubby, dichotomous, nude, the twigs canescent ; leaves alternate, linear, obtuse, spreading, above concave and tomentose-canescent, beneath green, viscidulous, subpubescent ; heads 3 together at the ends of the branches, sessile, oblong ; inv. scales appressed, obtuse, brown-margined ; pappus *(fide DC.)* crown-like, toothed. *Relhania triflora, E. Mey. Polychætia tricephala, DC. l. c. 285.*

HAB. Roodezand, betw. Nieuwekloof and Slargenteuvel, *Drege!* (Hb. Sond.)
A small, slender shrub. Leaves 4–5 lines long, nearly 1 line wide. Heads 3 lines long. I have seen but a small specimen, and have not had an opportunity of examining the generic character. The foliage is not unlike that of *N. prostrata.*

7. N. humilis (Less. Syn. 372); shrubby, much-branched, subdichotomous; branches and twigs cobwebbed, becoming nude; leaves opposite or scattered, sessile, linear or spathulate, channelled, obtuse or subacute, more or less villoso-tomentose or woolly, at length nearly nude; heads ovate, terminal, and in the forks sessile, subtended by several closely-applied leaves; inv. scales close-lying, very obtuse, the inner spreading; achenes short, furrowed; pappus crown-like, toothed. *N. humilis, N. minuta,* and *N. rigida, DC. l. c.* 283, 284. *Relhania dichotoma, Willd. Pteronia minuta, Linn.? Sieb.! Fl. Cap.* 117. *Polychætia relhanioides, Less.! Syn.* 371.

HAB. Uitenhage and Graafreynet, *E. & Z.! Lichtenstein, Mdt. & Maire, Zey.!* 2929. (Herb. D., Hk., Sd.)
A rigid, much-branched, scrubby bush, a foot or more high ; twigs flexuous. Leaves ¼–¾ inch long, ½–1 line wide. Heads 4 lines long ; inv. sc. very rigid. Pappus minutely denticulate. A specimen (in Hb. Sond.) of " *Polychætia relhanioides,*" Less. marked by Lessing himself, has the *coroniform pappus* and all other characters of *N. humilis!* Lessing has therefore either confounded two plants together, or, more probably, erroneously given two names and characters to the same plant.

8. N. conferta (DC.! l. c. 284); stem shrubby, depressed, or prostrate, dichotomous, the branches very short, crowded ; leaves linear, glabrous, subobtuse, channelled, crowded ; heads sessile in the forks and terminal, subtended by leaves ; heads oval, the scales close-lying, very obtuse, pale, the inner membranous at the point and spreading ; achenes short, furrowed ; pappus crown-like, toothed.

HAB. In very dry, stony places. Betw. the Sack and Gariep R., *Burchell,* No. 1588. On the Karroo, and the Sneeuweberg, *Drege!* (Herb. D., Hk.)
Root deeply descending, woody. Stems very dwarf, 2–3 inches high, densely branched and leafy. Leaves 4–6 lines long, ⅓ line wide. Invol. pale straw-colour ; by which character and its dwarf habit and quite glabrous leaves it is known from *N. humilis.*

9. N. muriculata (DC. l. c. 284); stem shrubby, much-branched, subdicotomous, the lower parts of the branches and twigs armed with rigid, pungent spines (formed from the persistent bases of old leaves); leaves opposite, sessile, linear, channelled, subobtuse, soon glabrous, along the margin and midrib sparingly spinuliferous; heads oblong, sessile in the forks and terminal, oblong, subtended by leaves; inv. scales close-lying, very obtuse, the inner spreading; achenes short, furrowed; pappus crown-like, toothed.

HAB. Zwart Ruggens, *Drege.* Fish River, *Burke and Zey.!* Caledon R., *Zey.!* (Herb. D., Hk., Sd.)

Similar in habit to *N. humilis,* but known by the persistently spinous bases of the old decayed leaves and the spinous points on the young leaves. Inv. and its contents similar.

10. N. oppositifolia (DC. l. c. 283); "stem shrubby, dwarf, dichotomous, the younger pubescent; leaves linear, *flat, appressedly canescent,* the cauline tufted, the rameal opposite; heads ovate-oblong, solitary in the forks and at the apices of the twigs, sessile, subtended by leaves; inv. scales appressed, shining, very obtuse, even the innermost erect. *DC.*

HAB. Zwarteberg, *Drege.* (Unknown to us.)

11. N. tenuifolia (DC. l. c. 284); stem vaguely branched; leaves *linear-terete,* closely involute, furrowed above, tomentose in the furrow, glabrous without; heads subumbellate; invol. oblong, the scales appressed, the innermost obtuse." *DC.*

HAB. Cape, *Burchell,* 6760. (Unknown to us.)

CIII. **RELHANIA,** L'Her.

Heads many-fl., radiate; *ray-fl.* uniseriate, female, ligulate; *disc-fl.* tubular, 5-toothed, perfect or sterile. *Recept.* flat, bearing paleæ between the flowers. *Inv.* ovate or cylindrical, the scales hard and dry, closely imbricating. *Achenes* linear-prismatic, glabrous or pubescent, slender, sometimes shortly beaked. *Pappus* either crown-like and shortly toothed or subentire; or consisting of many small, sharp scales. *Relhania, Eclopes and Rhynchopsidium, DC.! Prodr.* 6, *pp.* 286, 287, 290.

Small, rigid, branching shrublets, or rigid annuals, often gummy, either glabrous or pubescent. Leaves alternate or opposite, quite entire, small. Heads terminal, either solitary or corymbose. Flowers yellow. Named in honour of Rev. R. Relhan, author of *Flora Cantab. Eclopes* differs merely by a minute character of pappus, not always definite; its species are undistinguishable to the eye from true *Relhania. Rhynchopsidium* differs from *Eclopes* by its *slightly* rostrate achenes, and annual roots. I do not think it answers any good purpose to retain such artificial genera.

Sect. 1. RELHANIA. Small shrubs. *Achenes* beakless. *Pappus* either coroniform or tubular, sub-dentate. (Sp. 1–3.)

Heads solitary, terminal, subsessile :
　　Leaves linear-involute, recurved-mucronate ... ... (1) **ericoides.**
　　Leaves oblong or obovate, very obtuse ... ... (2) **rotundifolia.**
Heads corymbose; leaves rigid, squarrose, oblong, acute ... (3) **squarrosa.**

Sect. 2. ECLOPES. Small shrubs. *Achenes* beakless. *Pappus* of many short scales, either separate or subconnate at base. (Sp. 4–13.)

Heads corymbose (or 2–3-together), pedicellate :
　　Inv. scales very obtuse; heads oval-oblong :
　　　　Leaves recurvo-mucronate, nerved ... ... (4) **genistæfolia.**

Leaves very obtuse, nerveless or nearly so ... ... (5) **affinis.**
Inner inv. scales long, narrow, acute ... ... (6) **multipunctata.**
Heads solitary, sessile, terminal : .
Leaves thickish, 1-nerved or nerveless :
    Leaves opposite, linear-trigonous, hook-pointed ... (7) **cuneata.**
    Leaves alternate, linear, thick, obtuse, dotted ... (8) **sedifolia.**
    Lvs. alternate, fleshy, acute; invol. scales lacerate (9) **centauroides.**
Leaves 3–5- or many-nerved, rigid, pungent :
    Ray-fl. not much longer than the involucre :
      Leaves *green* on both sides, dotted above ... (10) **quinquenervis.**
      Leaves *tomentose-canescent* on the upper surface:
        Lvs. with 3–5 subdistant ribs beneath ... (11) **trinervis.**
        Lvs. *closely* multi-striate beneath ... ... (12) **pungens.**
    Ray-fl. fully twice as long as the involucre; leaves
      glabrous, veiny beneath ... ... (13) **speciosa.**

Sect. 3. RHYNCHOPSIDIUM. Small annuals. *Achenes* shortly beaked. *Pappus* of many short scales. (Sp. 14–16.)

Heads ovate, few-flowered ; rays shorter than the involucre :
    Heads sessile or subsessile ... ... (14) **sessiliflora.**
    Heads on long peduncles ... ... (15) **pedunculata.**
Heads campanulate, many-fl.; rays much longer than the
    involucre ... ... (16) **pumila.**

Sect. 1. RELHANIA (Less.) (Sp. 1–3.)

1. R. ericoides (Cass. Dict. XLV. 30); leaves linear-involute, somewhat keeled, recurvo-mucronate; heads solitary, subsessile; inner inv. scales acuminate; pappus tubular, membranous, toothed. *Leyssera ericoides, Berg. Cap. p. 294, excl. syn.*

VAR. *a.* **santolinoides** ; leaves about ⅓ inch long, the younger canescent. *R. santolinoides, L'Her. Sert. Angl.* 24. *Thunb.! Cap.* 640. *Less.! Syn.* 374. *DC.! l. c.* 286.

VAR. *β.* **paleacea** ; leaves 3–4 lines long, mostly green. *R. paleacea, L'Her. l. c. Thunb.! Cap.* 640. *Less.! Syn.* 375. *DC. l. c. Leyssera paleacea, Linn. Syst.* 641.

HAB. Hills round Capetown, and in Worcester and Stellenbosch, *Thunberg! Drege! E. & Z.! Pappe! W.H.H.*, &c. (Herb. Th., D., Hk., Sd.)
About a foot high, corymbosely much branched, closely leafy. Leaves heath-like, 3–6–7 lines long, ⅓ line wide, green or canescent, spreading. Heads ending the branches. Inv. glossy, pale horn-colour. Pappus crenulate. Paleæ narrow-linear, conduplicate, one-nerved, entire.

2. R. rotundifolia (Less. Syn. *375*); twigs tomentose, leaves oblong or obovate, very obtuse, callous-tipped, flattish, with slightly inflexed margins, on both sides albo-tomentose, but more thickly on the upper; heads terminal, solitary, sessile, campanulate, the inv. scales broadly oblong, very obtuse, membrane edged; pappus tubular, membranous, crenate. *DC. l. c.* 286. *Gorteria ovata, Vahl., fide Less.*

HAB. Cape, Hb. *Vahl.* Groenekloof, *Eckl.* Klipfontein, *Zey.!* 917. (Hb. Hk., Sd·)
Shrubby, distantly branched ; the old branches naked, the younger closely leafy. Leaves ¼–⅓ inch long, 2–3 lines wide, canous on both sides. Inv. scales somewhat bullated. Rays short. Pappus tubular, half as long as the ovary, toothed. A very distinct species.

3. R. squarrosa (L'Her. Sert. Angl. 24); leaves oblong or oval, squarrose-recurved, acute-mucronate, subpungent, one-nerved, glabrous, punctate; heads corymbose, cylindrical, shortly pedicelled; inv. scales closely-imbricate, flat, oblong, the innermost spreading, obtuse; paleæ conduplicate, nerved, mucronate; pappus very short, crown-like, sub-dentate. *Thunb.! Cap.* 639. *Less.! Syn.* 376. *DC. l. c.* 286. *Athanasia squarrosa, Linn.?*

VAR. β. **brevifolia** ; leaves shorter ; heads on shorter pedicels. *R. recurva, DC. l. c. (excl. syn. Less.?)*

HAB. Karroo, *Thunberg!* Swellendam, *Mundt, Thom, E. & Z.!* Zederberg, *Drege! Zey.* 2826. Elandsberg, *Dr. Wallich!* (Herb. Hk., D., Sd.)

A foot or more high, robust, woody, corymbosely branched. Branches closely leafy, all the leaves recurved or squarrose, very rigid, glabrous. Heads several in corymbs, on pedicels as long as themselves or shorter. Whatever *R. recurva*, Less. may be, the plant distributed by *Drege* under that name is certainly not different from this species.

Sect. 2. ECLOPES (Gaernt.) (Sp. 4–13.)

4. R. genistæfolia (L'Her. Sect. 22); leaves linear or oblong-linear, punctate, viscidulous, recurvo-mucronate, more or less distinctly nerved; heads corymbose, oblong, on longer and shorter pedicels; inv. scales obtuse ; ray achenes 3-cornered, villous, of the disc slender, glabrous. *Thunb.! Cap.* 639. *Eclopes genistæfolia, DC. l. c.* 288. *Athanasia genistæfolia, Linn. Eclopes punctata, Cass. Less.! Syn.* 378.

VAR. a; **angustifolia** ; leaves squarrose or spreading, narrow-linear or oblong 2–4 lines long. *E. genistæfolia, DC.*

VAR. β; **glutinosa**; leaves erecto-patent, oblong-linear, 4–6 lines long, more gluti-nous, and rather more distinctly dotted. *E. glutinosa, DC. l. c.*

VAR. γ. **discoidea**; heads discoid; leaves obovato-linear, viscidulous. *E. viscida, Less. Syn. p.* 379.

HAB. About Capetown and in the Western and S. Eastern District, Common. Grahamstown, *Genl. Bolton! Zey.!* 2823, 2624. β, Piquetberg, *Drege! Zey.!* 842. (Herb. Th., D., Hk., Sd.)

A much branched, strongly scented, viscidulous and glabrous bush. Leaves vari-able in size and somewhat in form, always acute or mucronate, in β. erecto-patent and straighter. Heads several in a corymb. Var. γ. (from the eastern base of the Devil's Mt.) seems to be a mere accidental state, without ray-flowers.

5. R. affinis (Sond.! in Herb.) ; leaves linear-oblong or spathulate, punctate, viscidulous, *very obtuse*, nearly nerveless, spreading; heads corymbose, oblong, on very short pedicels; inv. scales and achenes as in *R. genistæfolia.*

HAB. Near the Berg River, *Zey.!* 842. (Hb. Sond., Cap., D.)

Very like *R. genistæfolia*, but more scrubby, with broader and very blunt, almost nerveless leaves. Perhaps a mere variety?

6. R. multipunctata (DC.! l. c. 286); stem shrubby; branches thinly cobwebby-canescent ; leaves lance-oblong, narrowed at base, flat, sub-acute, one-nerved, spreading; on both sides copiously gland-dotted; heads 2–3 together at the ends of the branches, pedicelled, cylindrical; inv. scales appressed, the outer short, obtuse, inner long, narrow, acute or acuminate ; paleæ setaceo-subulate ; pappus white, of many scarcely connate, subulate scales.

HAB. Zeederberg, *Drege!* (Herb. Sond.)

A slender, branching, small shrub. Leaves ½–¾ inch long, rather sparse, 1½–2 lines wide. Inv. 6–9 lines long, pale. Paleæ of the recept. very slender, much acuminate.

7. R. cuneata (Th. Cap. 639); quite glabrous, viscid; leaves opposite, the pairs decussating, linear-trigonous, hook-pointed, erect; heads ovate, terminal, solitary, sessile ; rays spreading ; inv. scales oblong, convex, all obtuse ; ray achenes compressed, ciliate; of the disc slender, terete, glabrous. *L'Her. Sert.* 25. *Rel-decussata, L'Her. p.* 24 ? *Eclopes cuneata, Less! Syn.* 378. *DC. l. c.* 288, also *E. decussata, DC. l. c.*

HAB. Cape, *Thunberg!* Eastern Districts, *Burchell,* 6763. Swellendam, *Dr. Thom! Ecklon.* Hassaquaskloof, *Zey.!* 2825. (Herb. Th., Hk., Sd.)
A foot or 18 inches high, corymbosely much branched ; branches erect. Leaves 3–4 lines long imbricating in alternate pairs, erect with recurved, callous points. Heads not many flowered, the invol. pale. Paleæ subulate, shorter than the flowers. *Lessing* omits to mention that the leaves are opposite. *Thunberg's* specimens entirely agree with those from *Dr. Thom.*

8. R. sedifolia (Harv.); glabrous ; leaves alternate, crowded, linear, thick, obtuse, impress-dotted, erect, the youngest subpubescent ; heads solitary (or few) at the ends of the branches, oblong, sessile; inv. scales narrow-linear, the outer obtuse, the inner acuminate ; paleæ slender, rigid, as long as the flowers. *Eclopes sedifolia, DC. l. c.* 288.

HAB. Zeederberg, *Drege!* (Herb. Hk., Sd.)
A small, rigid, nearly glabrous shrublet. Leaves 3–4 lines long, straight and erect, not curved-tipped. Achenes slender, glabrous. Pappus 2–5 parted. Heads few flowered.

9. R. centauroides (DC. l. c. 288); "branches irregularly forked, the younger hairy, the adult glabrous ; leaves alternate, scattered, close-together, subimbricate, carnoso-subtrigonous, acute ; heads solitary, sessile, discoid? ; inner inv. scales scarious and lacerate at the apices." *DC. l. c. Eclopes centauroides, DC. l. c. Stœhelina centauroides, Burm.*

HAB. Cape, *Burmann.*
18 inches high, erect. Leaves 3 lines long. Head ovate, 6 lines long. Recept. paleaceous. Pappus of many short scales. Achenes glabrous.

10. R. quinquenervis (Th.! Cap. 641) ; twigs hairy ; leaves lanceo-late, the younger on both sides green and softly hairy, older glabrate above, all concave above, grain-dotted, villous, and strongly or faintly 5–3 nerved beneath, pungent, spreading or squarrose ; heads solitary terminal, sessile; inv. scales oblong, obtuse. *Eclopes trinervis, Less.! Syn. (excl. syn. Th.!) DC. l. c.* 289. *R. lateriflora, Link.*

HAB. Cape, *Thunberg!* Uitenhage and Albany, *E. & Z.!* Slaay Kraal, *Burke!* (Herb. Th., D., Hk., Sd.)
A larger and stronger bush than *R. trinervis,* from which it is known by its villous and hirsute *green* pubescence ; never *albo-tomentose.* The young plants are densely villous all over, the older more or less glabrate. In Herbaria it is often marked "*subpungens.*" It varies with broader or narrower leaves, the narrow-leaved variety being most frequently mistaken for *R. pungens.*

11. R. trinervis (Th.! Cap. 641) ; twigs albo-tomentose ; leaves lance-subulate or lanceolate, concave and tomentose within, convex, cobwebby (becoming glabrate) and 3–5-nerved beneath, pungent, straight, spread-ing ; heads solitary, terminal, sessile; inv. scales oblong, obtuse or sub-obtuse. *Eclopes parallelinervis, Less. Syn.* 380. *DC. l. c. p.* 289.

HAB. Cape, *Thunberg!* Uitenhage and Albany, *Krebs, Ecklon and Zeyher. Zey.!* 2931. *Mrs. F. W. Barber, &c.* (Herb. Th., D., Hk., Sd.)

Similar in pubescence to *R. pungens*, but known at once by the *distantly* 3 (or rarely 5) nerved (not *closely multistriate*), leaves. The younger leaves are quite smooth (læves) externally; the older often very scabrous. Not an uncommon plant: most of *E. and Z.'s* and *Drege's* distributed specimens of "*E. subpungens*" belong to this. *Thunberg's* Herb. specimen of *R. trinervis* is unquestionably this species, and not, as stated by Lessing, "*quinquenervis.*"

12. R. pungens (L'Her. Sert. p. 24); twigs albo-tomentose; leaves lance-subulate, concave and tomentose above, convex, glabrous and closely multistriate beneath, pungent, straight, erect; heads solitary, terminal, sessile; inv. scales oblong, obtuse. *Eclopes subpungens. Thunb./ Cap.* 640. *Bot. Reg. t.* 2817. *Less./ Syn. p.* 380. *DC. l. c. (ex pte.)*

HAB. Cape, *Thunb./* Kochmans Kloof and Swellendam, *Mundt!* Near Grahamstown, *Genl. Bolton!* Buffelsjagdt, *Zey./* 2932. (Herb. Th., Hk., D.)
A rigid, erect bush, 12–15 inches high : older branches bare of leaves ; younger closely leafy. Leaves ¾–1 inch long, 1–2 lines wide, the under surface with many parallel, close-set rib-striæ, with furrows between. This seems a much less common species than *R. trinervis.*

13. R. speciosa (Harv.) ; twigs tomentose ; leaves broadly lanceolate or ovate-orbicular, acuminate, pungent, rigid, spreading, on both sides glabrous, scabrous-edged, flattish, 5-nerved and subreticulate beneath ; heads solitary, terminal, sessile, very many-fl., rays uncial ; inv. scales rather loosely imbricated, oblong, obtuse, with a broad, membranous margin. *Eclopes speciosa, DC. l. c.* 289.

VAR. β. **schizolepis** ; leaves shorter and broader. *Eclopes schizolepis, DC. l. c.* 289.

HAB. Near Genadendahl, 2–3000 f., and at Kendo, 3–4000 f., *Drege!* β. at Gnadendahl, *Burchell,* 7825. (Herb. Hk., Sd.)
A very handsome, large-flowered species, with heads resembling those of a *Sphenogyne.* Stem a foot or more high, erect, not much branched, closely leafy. Leaves ¼–¾ inch long, 2–3 lines wide ; like those of an *Oedera.* Ligules dark-red beneath, fully an inch longer than the involucre. The leaves vary much in shape, and β., which grows with the common form, if a specimen in Hb. Sond. be DC.'s plant, is scarcely distinguishable. The laceration of the membr. edges of the inv. scales depends on the age of the specimen.

Sect. 3. RHYNCHOPSIDIUM. (Sp. 14–16.)

14. R. sessiliflora (Th. Cap. 639) ; heads sessile or subsessile in the forks and at the ends of the branches, ovate, few-flowered ; rays shorter than the involucre. *Rhynchocarpus lateriflorus, Less ! Syn.* 383. *Relhania lateriflora, L'Her. Rhynchopsidium sessiliflorum, DC. l. c.* 290. *Athanasia sessiliflora, Linn. f.*

HAB. In moist, sandy ground, round Capetown and through the Western and S. Eastern Districts, common. (Herb. Th., D., Hk., Sd)
A small annual, 2–6–8 inches high, simple or much branched. Stems and leaves pilose with gland-tipped hairs. Leaves 4–7 lines long, 1 line broad, linear or spathulate, acute, midribbed, spreading. Heads quite sessile or *shortly* pedunculate, the peduncle rarely equalling the involucre. Outer inv. scales ovate, obtuse ; inner lanceolate, subacute. Achenes shortly rostrate, clothed with capitate hairs.

15. R. pedunculata (Harv.); heads on peduncles thrice their length, ovate, few-flowered ; rays shorter than the involucre. *Rhynchopsid. pedunculatum, DC. l. c.* 290.

HAB. Cape, *Burmann.* Olifant's R. and Kl. Namaqualand, *Drege!* Worcester and George, *Eckl.* Tulbagh, *Dr. Pappe!* (Herb. D., Hk., Sd.)

A small annual, more slender and divaricate than *R. sessiliflorum*, from which it is chiefly distinguished by its peduncled heads. Pedunc. ¾-1 inch long. Leaves sometimes tomentose, as well as gland-hispid.

16. R. pumila (Th.! Cap. 639); heads on peduncles thrice their length, campanulate, many-flowered; rays longer than the involucre. *Rhynchops. pumilum, DC.! l. c. 290. Athanasia pumila, Linn. f. Eclopes pumila, Less.! Syn. 382.*

HAB. Brandvalley, *Thunberg!* Clanwilliam, *Ecklon!* Kl. Namaqualand, *Drege! A. Wyley!* (Herb. Th., D., Hk., Sd.)

Annual, branched from the base, the lateral branches divaricate or decumbent, simple or branched near the apex. Leaves ¼-¾ inch long, flat or involute, linear, gland-hispid. Mature invol. broader than its length; the outer scales rounded, inner oblong, all very blunt and membrane-edged. Rays broad, bright yellow, twice as long as the involucre.

(Doubtful species of the section Eclopes.)

R. apiculata (DC.); "leaves elliptic-lanceolate, on both sides glabrescent, coriaceous, many-nerved, tipped with a rigid, subsphacelate mucro; achenes of the disc with the inner margin subciliate." *DC. Eclopes apiculata, DC. l. c. 289.*

HAB. Stony hills between Driekoppen, Bokkeveld and Hexrivier; and the Zwarteberg, near Klaarstroom, *Drege!* (Herb. Sond., *sine flore.*)

This looks like a more glabrous state of *R. quinquenervis*, so far as I can judge from a branch, without flowers, in Hb. Sd.

R. styphelioides (DC.); "quite glabrous; leaves lanceolate, acuminate, rigid, many-nerved beneath, the upper leaves equalling the head." *DC. Eclopes styphelioides, DC. l. c. 289.*

HAB. Near Gauritz R., *Burchell,* 4697. (Unknown to us.)

CIV. ? **OLIGODORA**, DC.

Heads 5-fl., homogamous, fl. 5-toothed, tubular, perfect. *Inv.* imbricate, subtrigonous, the scales appressed, the innermost with the margin clasping round the outer achenes. *Recept.* narrow, paleaceo-fimbrilliferous, the scales folding round the achenes. *Style*-branches included, obtuse and pilose at the apex. *Anthers* tailed? *Achenes* cylindrical, smooth. *Pappus* of 5, ovate, short, toothed scales. *DC. Prodr. 6. p. 282.*

An erect, branching, glabrous suffrutex. Leaves alternate, thickish, sessile, toothed, the teeth on each side 2-3 and a terminal one, mucronate. Heads corymbulose at the ends of the branches, on very short pedicels. Scales or bracteæ on the pedicels short, acute, spreading. Corolla (probably?) white. *DC.* The name is from ολιγος, *few,* and δωρον, *a gift;* alluding to the few flowers of the head.

1. O. dentata (DC. l. c.)

HAB. Zeederberg, *Drege.* (v. frust. in Hb. Sond.)

Of this I have seen only a minute fragment, and can add nothing to the above description, copied from *De Candolle.* Possibly it may be an *Athanasia;* it seems to differ merely in the anthers, if these be correctly described.

CV. **OSMITES**, Linn.

Heads many-fl., radiate; *ray-fl.* ligulate, female; *disc-fl.* tubular, 5-toothed, perfect. *Recept.* flat, covered with scarious paleæ, equalling the disc-fl. *Inv.* campanulate, the scales pluriseriate, herbaceous, sub-equal. *Anthers* tailed. *Achenes* sessile, beakless, glabrous or downy, somewhat 4-sided, compressed. *Pappus* of many short scales. *DC. Prodr.* 6, *p.* 290.

Cape suffrutices. Leaves alternate, crowded, sessile, oblong, obovate or linear; entire or toothed or pinnatifid, gland-dotted and strongly-scented. Heads solitary, terminal; disc yellow; rays white. Name from οσμη, *smell;* alluding to the balsamic odour of these plants.

Leaves sharply toothed, or the uppermost entire; pappus short:
    Stem erect (1–2 ft.); lvs. linear-lanceolate, acuminate,
      not veiny ... ...   (1) **Bellidiastrum.**
    Stems ascending, subsimple, pedunculoid upwards:
      Leaves obovate or oblong, glabrous, veiny, sharply
      toothed ... ...   (2) **dentata.**
      Leaves lance-linear, hirsute-subcanescent, toothed
      above ... ...   (3) **hirsuta.**
    Stems dwarf, branching, leafy to summit; leaves oblong,
      glabrous, deflexed ... ...   (4) **parvifolia.**
Leaves deeply pinnatifid, and lobed; lobes subulate; pappus bristle-pointed:
    Leaf-lobes 3–5 on each side ... ...   (5) **pinnatifida.**
    Leaf-lobes 1–2 on each side, near the summit ... ...   (6) **angustifolia.**

1. O. Bellidiastrum (Th. Cap. 701); stem shrubby, erect, branched; branches virgate, leafy below, pedunculoid at summit, striate, cobwebby-pubescent; leaves linear-lanceolate, acuminate, cuneate at base, sharply few-toothed near the summit, pale beneath, nearly glabrous, punctate, midribbed, not prominently veiny; invol. scales oblong, obtuse, the inner membranous and torn at edge; rays female. *O. anthemoides, DC. l c.* 291. *Bellidiastrum Osmitoides, Less. Syn.* 384. *Osmites lancea, Th.! in Hb.*

Hab. Cape, *Thunberg!* Kanna River, *Mundt.* Eastern Districts, *Burchell,* 6965. Swellendam, *Drege, Dr. Alexander Prior! Pappe! Zey.!* 2935. (Hb. Th., D., Hk., Sd.)
1–2 feet high, woody below; branches 10–12 inches long, erect, virgate. Leaves 1–1¼ inch long, 2–3 lines wide, with slightly revolute edges, quite entire except near the summit, where there are 2–5 sharp, taper-pointed teeth on each side. Heads about as large as in *O. hirsuta.*

2. O. dentata (Thunb! Cap. 701); stems subsimple, ascending-erect, below leafy, pedunculoid at summit, striate, cobwebby-puberulous; lower leaves obovate, medial oblong or lanceolate, upper depauperated, all but the uppermost sharply and deeply serrate throughout, glabrous, mid-ribbed and veiny; invol. pubescent, the scales acute or acuminate; rays female. *Less.! Syn.* 283 *DC. l. c.* 291. *Osmites camphorina, Gaertn. fr. t.* 74, *f.* 3, *non Linn.* Anthemis afra, Burm.

Hab. Table Mt., *Thunberg! E. & Z.! W.H.H.!* &c. Swellendam, Simon's Bay, *C. Wright* 392. (Herb. Th., D., Sd.)
Suffruticose at base, many stemmed, more or less decumbent, the fl. branches ascending erect,, 12–15 inches high. Lower leaves 1–1¼ inch long, ¼–1 inch wide, inciso-serrate; medial similar, but narrower; upper scattered on the peduncle, bract-like. Heads larger than in *O. hirsuta.*

3. O. hirsuta (Less. Syn. 385); hirsute, subcanescent; stems subsimple, ascending-erect, closely leafy below, shortly naked at the summit; leaves lance-linear, sharply toothed above the middle; inner inv. scales obtuse; rays female. *DC. l. c.* 291. also *O. Bellidiastrum, DC. l. c. excl. syn. omn.*

HAB. Fransche Hoek, *Ecklon!* W.H.H. Drakensteinberg and Dutoitskloof, *Drege!* Bavian'sberg, Genadendahl, *Dr. Pappe!* Table Mt., *Mundt.! Zey.!* 2936. (Herb. Th., D., Hk., Sd.)

Stems several from the crown, at first decumbent, then erect, 6–12 inches high, simple or branched near the summit. Leaves closely placed below, 6–8 lines long, 2 lines wide, becoming gradually smaller and more lax upwards, diminishing to scales below the sub-peduncled head. Heads like those of the Ox-eye Daisy (*Chrysanth. leucanthemum*). Invol. woolly. Scales broad, with jagged margins. Rays thick, opaque-white. *Drege's* specimens distributed under the name "*O. Bellidiastrum*" in no respect differ from the common *hirsuta.* I find the rays of this, in Fransche Hoek specimens (!), to be *female,* not "neuter" as stated by Lessing.

4. O. parvifolia (DC. l. c. 291); shrubby, dwarf, branching; branches densely and equally leafy to the summit; leaves oblong or obovate oblong, spreading or deflexed, glabrous, punctate, sharply toothed near the summit or throughout, teeth subulate-acuminate; inv. scales oblong, glabrous, serrato-ciliate; rays female.

HAB. Cape, *Mundt! Dr. Stanger!* Hott. Holl. Mts., *Drege!* (Herb. Hk., Sd.)

Stems woody, closely branched, 3–6 inches high, densely leafy. Leaves squarrose, 3–4 lines long, 2 lines wide. Inner inv. scales scarious at tip. Rays broad, white.

5. O. pinnatifida (DC.! l. c. 291); stems subsimple, ascending, leafy below, shortly nude at summit, glabrous; leaves deeply pinnatifid, glabrous, the lobes on each side 3–5, subulate, quite entire, the lower shorter; inv. scales oblong, the inner membranous and torn at edge; pappus of the ray of very minute scales, sub-coroniform; of the disc with 2 long bristle-pointed scales, the rest very small or obsolete.

HAB. Drakensteinberg, *Drege!* (Herb. Hk., Sd.)

With the habit of *O. dentata,* but very different foliage and pappus.

6. O. angustifolia (DC. l. c. 291); "leaves linear, quite entire or sparingly inciso-lobate near the apex; pappus of the ray and disc all with scales elongating into nearly equal bristles." *DC.*

HAB. Simonsberg, Stellenb., near the waterfall, *Drege.*

Of this I have only seen a few leaves (in Hb. Sd.): the leaf-lobes are subulate, 1–2 at each side near the summit.

CVI. **OSMITOPSIS**, Cass.

Characters as in *Osmites;* except; *Ray-fl.* neuter. *Pappus* none. *DC. Prodr. 6. p.* 292.

A closely leafy, strongly balsam-scented, erect shrub. Leaves lanceolate, scattered, dotted, one-nerved, sessile, entire or nearly so. Heads terminal, rays white. Name from *Osmites* and οψις, *resemblance;* like an *Osmites.*

1. O. asteriscoides (Cass. Dict. 37, p. 5). *Less. Syn.* 386. *DC. l. c.* 292.

VAR. *a,* **pubescens** (DC.); leaves on both sides pubescent, quite entire. *O. asterisc-oides, Less.! Osmites asteriscoides, Linn. Sp.* 1285. *Burm. Afr. t.* 58, *f.* 1.

VAR. *β,* **subdentata** (DC.); leaves pubescent, callous-denticulate near the summit. *Osmites camphorina, Linn., non Th.*

VAR. *γ,* **glabra** (DC.); leaves glabrous, quite entire. *O. camphorina, Less.!* 387. *Osmites camphorina, Thunb.! Cap.* 700.

Hab. Table Mt. summit, and similar situations in the Western Districts. (Herb. Th., D., Sd., &c.)

A much-branched, camphor-scented shrub, 4–5 feet high. Leaves 1¼ inch long, 2–4 lines wide, varying as above stated.

Sub-tribe 5. Senecioneæ. (Gen. 107–119.)

CVII. STILPNOGYNE, DC.

Heads 7–8-fl., heterogamous, discoid, all the fl. tubular, 3 marginal 3-toothed female, 4–5 central perfect, 5-toothed. *Inv.* scales 5–7 in a single row, equalling the disc, valvate, connate at base. *Recept.* naked, narrow. *Style* branches in the central flowers short, bearded at the summit only; of the fem. fl. longer, terete, downy. *Achenes* oblong, tapering to both ends, granulated; those of the fem. fl. without pappus. *Pappus* in the disc-fl. uniseriate, of many scabrid bristles. *DC. Prodr.* 6, *p.* 293.

A small, slender, glabrous annual. Leaves on long petioles, roundish, cuneate at base, bluntly 5–7-lobed or repand, sometimes with 1–2 runcinate lobes below a terminal, 5–7-toothed lobe. Stem branched, filiform, naked, divided into a few slender, 1–2 inches long, one-headed pedicels. Fl. yellow. Name from στίλπνος, *shining,* and γυνη, *a female.* It differs from *Kleinia* in habit, and the want of pappus to the fem. fl.

1. **S. bellidioides** (DC.! in Deless. Ic. Sel. 4, t. 54); *DC. l. c.* 294.

Hab. Little Namaqualand, *Drege!* (Herb. D., Hk., Sd.)

2–3 inches long. Heads oblong, 3 lines long. Leaves 4–5 lines diameter; their petioles 1–1½ inch long.

CVIII. OLIGOTHRIX, DC.

Heads many-fl., radiate; *ray-fl.* about 5, female; *disc-fl.* 5-toothed, perfect. *Inv.* uniseriate, campanulate, naked at base, the scales 12–15, connate below, striate. *Recept.* naked, flat. *Anthers* exserted, enclosing the style. *Achenes* obtusely 5-angled, beakless, granulated, especially on the angles. *Pappus* of about 5 very caducous, wavy, barbellate bristles. *DC. Prodr.* 6, *p.* 304.

A wiry, branching, glabrous annual. Stems erect, terete, 6–12 inches high, branched from the base, subcorymbose. Leaves small, eared and stem-clasping at base, oblongo-lanceolate, more or less toothed. Flowers yellow. Name, ολιγος, *few,* and θριξ, *a hair;* the obsolete pappus. The habit is that of *Gymnodiscus;* the heads like those of *Steirodiscus.*

1. **O. gracilis** (DC. in Deless. Ic. 4, t. 57); *DC. l. c.* 304.

Hab. Zeederberg, *Drege!* (Herb. D., Hk., Sd.)

CIX. MESOGRAMMA, DC.

Heads many-fl., radiate; *ray-fl.* uniseriate, ligulate, female. *Inv.* scales uniseriate, subcaliculate at base, the scales about 20, acuminate, each marked by two linear, intra-marginal glands (or *vittæ*). *Recept.* flat, naked. *Disc-fl.* tubular, 5-toothed, 5-lineate, the medial nerves strongly marked, intervals pellucid, or faintly nerved. *Anthers* without tails. *Style*-branches truncate, bearded at the apex. Young *achenes* compressed; adult 5-angled, tapering to each end, ciliate at the angles, with a very short beak covered with thick bristles, resembling an outer pappus. *Pappus* uniseriate, bristle-shaped, very slender, deciduous, in the central-fl. of many, in the marginal of few bristles. *DC. Prodr. 6, p.* 304.

A rigid, herbaceous, glabrous perennial. Stems several from the crown, rib-striate, branched above. Leaves on winged petioles, pinnatisect, the lobes lanceolate, toothed. Branches ending in naked, elongate, one-headed pedicels, subcorymbose. Corollas pale yellow, with red medial lines. Name from μεσος, the *middle*, and γραμμη, a *line;* alluding to the medial nerve of the corolla, being much more strongly marked than the marginal : a character also found in *Cineraria.*

1. M. apiifolium (DC.! in Deless. Ic. vol. 4, t. 58); *DC. l. c.* 304.

HAB. Near the Gariep, by Verleptpram, *Drege!* (Herb. D., Hk., Sd.)

CX. **CINERARIA**, Linn. (ex pte.)

Heads many-fl., rarely homogamous and discoid, mostly radiate ; *ray-fl.* ligulate, female. *Disc-fl.* tubular, 5-toothed, the medial nerve strongly marked. *Inv.* uniseriate, more or less caliculate at base, the scales membrane-edged. *Recept.* naked, flat. *Style*-branches in the disc-fl. tipped with a short cone, clothed with ascending pubescence. *Achenes* beakless, mostly flattened, often (at least the outer ones) winged at margin. *Pappus* in one or more rows, capillary, caducous. *DC. Prodr.* 6, *p.* 305.

Herbs or suffrutices, natives of Africa. Leaves alternate, various. Heads yellow. Name from cinereus, *grey* or ash-coloured ; the prevalent colour of the foliage in the original species. The habit is that of *Senecio,* from which *Cineraria* differs by the cone-tipped style and the *usually* flattened or many-angled achenes. Many species formerly comprised in it, including the parent of the garden "*Cineraria,*" have been removed to *Senecio;* others to other genera.

I. SENECIOIDES. *Achenes* quite glabrous, broadly 3–4-winged. *Leaves sessile,* clasping, woolly, entire or serrulate (not lobed). ... ... (1) **tomentosa.**

II. EU-CINERARIA. *Achenes* compressed or flattened, hispid or glabrous, with marginal border or wing (but no disc-wings). *Leaves* petioled, the petiole often eared at base, the lamina lobed, toothed or lyrate-pinnatisect, the terminal lobe often reniform. (Sp. 2–21.)

Stem and leaves more or less pubescent or woolly :
 Pedunc. elongate, *simple,* one-headed :
  Erect, branched ; branches ending in long peduncles (2) **pedunculosa.**
  Densely tufted, nearly stemless ; with sub-radical,
   lyrate or reniform leaves, and *scapelike* peduncles (3) **mollis.**
 Pedunc. corymbose or panicled, many-headed :
  Leaves sub-bipinnatisect, *multifid,* the terminal lobe
   deeply incised and toothed :
    Inv. campanulate, glabrous, 25–30-fl. ... ... (4) **aspera.**
    Inv. cylindr., cobwebbed, 12–15-fl. ... ... (5) **erosa.**
  Leaves either reniform or more or less lyrate, with
   a reniform, terminal lobe :
    Pubescence woolly or canescent :
     Stem rigid or suffruticose ; corymb compact, many-headed :
      Tomentose and canescent ; leaf-lobes
       sharply toothed and wavy, rigid (6) **canescens.**
      Cobwebbed ; leaves reniform, lobato-
       dentate... ... ... (8) **polycephala.**

Stem herbaceous ; corymb lax or panicled :
    Leaves amply auricled at base   ...   (7) **erodioides.**
    Leaves minutely auricled at base ... (9) **alchemilloides.**
    Pubescence hairy, not woolly or canous ... (10) **geifolia.**
Stem and leaves quite glabrous :
  Achenes hispid on surface ; ciliate at margin :
    Heads radiate :
      Suffrutices ; leaves reniform or fanshaped.
        Lvs. cordate-based ; pedunc. corymbose (11) **lobata.**
        Lvs. cuneate-based ; pedunc. subsimple (12) **Saxifraga.**
      Herbaceous; lvs. broadly hastate, sharply cut (16) **decipiens.**
      Herb. (annual); lvs. lyrato-pinnatisect ... (17) **platycarpa.**
      Herb.; lvs. oblong-lanceolate, toothed; rays
        minute ... ... (19) **microglossa.**
    Heads discoid, homogamous :
      Heads corymbose ; inv. scales appendiculate (20) **Dregeana.**
      Heads on simple pedunc. ; lower leaves ovate,
        tapering at base; upper oblong-linear ... (21) **exilis.**
  Achenes quite glabrous, not ciliate, wing-margined :
    Leaves reniform or lyrato-pinnatisect :
      Perennial ; lvs. reniform, on long petioles,
        the upper sometimes sub-lyrate ... ... (13) **geraniifolia.**
      Annual ; leaves lyrato-pinnati-partite, short-
        petioled, much cut and narrow ... ... (18) **lyrata.**
    Leaves hastate or deltoid, long-petioled :
      Lvs. deltoid, repand, amply auricled at base (14) **deltoidea.**
      Lvs. hastate, sharply cut, with narrow auricles (15) **atriplicifolia.**

III. OTHONNOIDES. *Achenes* obovate, subcompressed, densely clothed with short, rufous hairs ; the disc-achenes often abortive, but with perfect, bifid, cone-tipped styles. *Leaves* pinnati-partite. Fl. branches pedunculoid, one-headed. (Sp. 22.)

Annual ; cobwebby, subglabrous ; leaf-lobes linear ... ... (22) **othonnoides.**

§ 1. SENECIOIDES. (Sp. 1.)

1. C. tomentosa (Less. Syn. 391); stem robust, herbaceous, erect, simple, below densely white-woolly, glabrate above ; leaves sessile, half-clasping, cobwebby, becoming glabrous above, densely white-woolly beneath, the lower obovate, acute, calloso-serrate, with reflexed margins, the upper narrower and smaller, more oblong or linear ; corymb loosely branched, few-headed ; inv. glabrous, scales numerous, linear-acuminate ; rays several ; achenes glabrous, broadly 3- (rarely 4)- winged. *DC. l. c.* 308. *Senecio lanatus, Thunb. Cap.* 681.

HAB. French Hoek, *Thunberg.* Worcester, *Ecklon.* Drackensteinberg, *Drege!* Tulbagh, *Zey.!* (Herb. D., Hk., Sd., C.)

1–2 feet high, with the habit of *Senecio verbascifolius ;* quite unlike any other *Cineraria.* Lower leaves 1½–2½ inches long, ¾–1 inch wide, very white beneath ; upper leaves gradually smaller, diminishing to mere bracts.

§ 2. EU-CINERARIA. (Sp. 2–21.)

2. C. pedunculosa (DC.! l. c. 305); stem suffruticose at base, erect, short, closely branched, the branches ending in long, scaly, loosely woolly, one-headed peduncles ; leaves petioled, lyrato-pinnatifid, the woolly petiole not auricled at base, the lateral lobes in 1–3 pair, short, cuneate, bluntly toothed, the terminal 3–5-lobed, roundish or oblong *(not cordate at base)*, more or less deeply 3–5-lobed, the lobes toothed, glabrate above, woolly beneath ; inv. glabrescent, of 12–14 scales ; rays 12–14 ; achenes *(young)* compressed, hispid and ciliate.

HAB. Caledon, *Ecklon!* (Herb. Sond.)

Root fibrous, perennial. Stem curved at base, then erect, 6–8 inches high, the branches 2–3 inches long, closely leafy. Leaves 2 inches long; the terminal lobe not reniform. Pedunc. 6–8 inches long. I have only examined a single, *young* achene, but, judging from its characters, I doubt whether the mature fruit differs much from that of *C. mollis*, to which this seems to be allied. *Ecklon's* specimens, above quoted, are in poor condition.

3. C. mollis (E. Mey.!); stems very short, frutescent, branched at

the crown; leaves petiolate, lyrato-pinnatisect, the petiole woolly, half-clasping but not auricled at base, lateral lobes 2–4 pair, roundish, denticulate, the lower smaller, terminal reniform, callous toothed, all cobwebbed above, densely albo-tomentose beneath; peduncles elongate, loosely woolly, becoming nude, one-headed; inv. cobwebby, of 14–16 scales, equalling the disc; ligules 10–12; achenes oblong, minutely ciliate, sparingly hispid, the outer ones nearly glabrous on their inner face. *DC. l. c. p. 306.*

VAR. β. **polyglossa**; leaves either simply reniform, or lyrate-reniform, with one or two pair of small lateral lobes; inv. scales 20–22, rays 15–20. *C. polyglossa, DC.! l. c. C. arctotidea, Hb. Drege!*

HAB. Stormberg, *Drege!* Graaf Reynet, *Eckl.!* Elandslaagte, *Zey.!* (both varieties). β. Sneeuweberg, *Drege!* (Herb. D., Hk., Sd.)

Densely tufted; stems 1–3 inches long, closely leafy. Leaves 2–3 inches long, the terminal lobe ½–¾ inch wide, 4–6 lines long; lateral lobes very variable in number and size. Pedunc. 4–6 inches long. Intermediate forms connect α. and β.

4. C. aspera (Thunb.! Cap. 672); herbaceous, tall, much-branched,

scaberulous, the young parts cobwebby or woolly; leaves eared at base, petiolate, irregularly pinnate- or bipinnati-lobed, scabrous above, woolly beneath, the lobes horizontally spreading, few, distant, unequal, sharply toothed, the larger lobulate; corymbs compound, terminal; pedicels scaly; inv. campanulate, 8–10-leaved, glabrous; ray-fl. 3–5; disc-fl. 20–25; achenes flattened, pubescent, ciliolate. *DC. l. c. 306.*

HAB. Cape, *Thunberg!* Sak River, *Burchell,* 1477. Nieuweveld, *Drege.* Graaf Reynet, *Ecklon!* Queenstown and Cradock districts, flowering all the year round, *Mrs. F. W. Barber! No.* 320. Brit. Caffraria, *T. Cooper!* 195. (Herb. D., Sd.)

2–3 feet high, much branched, canous, becoming glabrate, but always scabrohispid. Leaves 1–2 inches long, the lobes and lobules 1 line wide. Corymbs many-headed, rather dense.—Differs from *C. oxyodonta* by its larger fl.-heads, &c.; it seems also to be exclusively an Eastern plant.

5. C. erosa (Harv.); herbaceous, tall, much-branched, scaberulous,

the young parts cobwebby; leaves eared at base, irregularly lyrato-pinnati-lobed, the lateral lobes unequal, oblong or obovate, coarsely and sharply-toothed, the upper ones confluent into a 3–5-parted, sharply-toothed limb; corymb compound, terminal, lax; pedicels bearing scales; inv. cylindrical, 5–7-leaved, shorter than the disc, cobwebby; rays 3–5; disc-fl. 9–12; achenes hispidulous and ciliate. *C. oxyodonta, DC. l. c. 306. Doria erosa, Th.! Cap. 674.*

HAB. Riebeckskasteel and Paardeberg, *Thunberg!* Paarlberg, *Drege!* Worcester, *Ecklon!* Cape, *Dr. Stuart!* Schoen Stromm, *Burke and Zeyher!* Stellenbosch Mt., *W. H. H.!* (Herb. Th., D., Hk., Sd.)

2 feet high, loosely branched. Leaves shortly petioled, multi-lobulate. Very like *C. aspera,* but the leaves are larger, with broader lobes the inv. is narrow, and

the heads much fewer-flowered : the leaves also show a disposition to form a terminal, reniform lobe, approaching that of *C. canescens.* The specimen in Hb. Thunb.
exactly agrees in foliage and involucres with *Drege's* (in Hb. D.) but the fl.-heads
are immature, and therefore *appear* rayless. Without doubt these plants are identical.

6. C. canescens (Wendl.); stem suffruticose, canescent, branched ;
leaves lyrate, tomentose and canescent beneath, cobwebbed, becoming
glabrate above, the lateral lobes in a single pair, oblong, toothed, the
terminal reniform, 5–7-lobed, the lobes sharply toothed and wavy ; petiole auricled at base ; corymb many-headed, compact, pedicels sparsely
leafy ; invol. calyculate, the scales 8–12, oblong, glabrate, 5-nerved ;
ligules 5–8 ; achenes hispidulous, ciliate. *DC. l. c. 307. C. parviflora,*
Ait. Kew. Ed. 2, v. 5, p. 72. C. Aitoniana, Spr.

VAR. β, **flabellifolia** ; lateral leaf-lobes wanting, terminal broadly reniform, lobed,
the lobes sharply and irregularly cut; corymb closely many-headed.

HAB. Cape, *Aiton.* Kamiesberg and Little Namaqualand, *Drege!* β at Modderfontein, *Rev. H. Whitehead !* (Herb. Hk., D., Sd.)
A foot or more high, whitish or greyish, with rather close toment. Leaves 1–1½
inch long, mostly lyrate, the lateral lobes sometimes abortive, terminal, ½–¾ inch
long, ¾–1 inch wide, rather rigid ; the teeth sharp. Pedicels cobwebbed or tomentose. β differs from the usual form in having no lateral leaf-lobes, a broader lamina,
and a denser and more glabrous inflorescence.

7. C. erodioides (DC. l. c. 307); stem herbaceous, erect, branching,
woolly below, pilose above ; leaves on long petioles, cordate-reniform,
shortly 5–7-lobed, the lobes sharply many-toothed, hispidulous above,
cobwebby-woolly and canescent beneath ; petioles amply auricled at
base ; corymb compound, many-headed, the pedicels about as long as
the involucre ; inv. scales 7–9, glabrous; ligules 5; achenes *almost*
glabrous, with a broad, subciliate marginal wing. *Cineraria tussilaginea,*
Thunb.! Cap. p. 671 (excl. syn. L'Her.).

HAB. Cape, *Thunberg!* Uitenhage, *E. & Z.!* (Herb. Th., Cap., Sd.)
This has the foliage nearly of *C. geifolia,* but a cobweb-woolly pubescence of
long, curled, white hairs, and achenes nearly like those of *C. geraniifolia.* The heads
are smaller and the pedicels much shorter than in *C. geifolia.* A specimen in Hb.
Thunb. marked "*Cineraria tussilaginis,*". exactly agrees with *E. & Z.'s* specimens,
and with *Thunberg's* description above quoted. I do not restore the name, lest it
may perpetuate a confusion with the *C. tussilaginis, L'Her.*

8. C. polycephala (DC. l. c. 307); "stem half-shrubby, erect ; branches
somewhat hairy or cobwebby ; leaves petioled, glabrate above, cobwebbed beneath ; the petioles of the lower leaves nude, of the upper
auricled at base ; limb orbicular, subcordate, lobato-dentate ; corymb
compound, many-headed, crowded ; pedicels tomentose, scaly ; inv.
cobwebbed or glabrous, subcalyculate, 12–13-leaved ; rays 8–10; achenes
on both sides hispid, ciliate. *DC. l. c.*

HAB. Mouth of Gauritz R., near Mossel Bay, *Drege!* (Herb. Sond.)
I have seen but a single poor specimen, from which it is difficult to form an opinion.
It is allied to *C. erodioides,* but has larger fl. heads, a closer panicle and a stronger
stem ; which may all be due to its littoral habitat.

9. C. alchemilloides (DC.! l. c. 307); stem herbaceous, erect, branching, loosely cobwebbed, becoming glabrous ; leaves on long petioles,
thinly cobwebbed or glabrous above, more persistently woolly and canous

beneath and on the petiole, which is *minutely* auricled at base, the limb cordate-reniform, sharply 5–7–9-lobed, the lobes toothed ; corymb lax, much divided or panicled, many-headed ; inv. scales 7–8 ; ligules few; achenes flat, puberulous and ciliate.

HAB. Tulbagh, *Ecklon!* (Herb. Sd.)
A loose growing, tall, straggling plant, resembling *C. lobata*, but tomentose. It is much more slender than *C. erodioides*, and wants the ample, leafy ear-lobes to the petioles, so conspicuous in that species. *Ecklon's* specimens are in poor condition.

10. C. geifolia (Linn. Sp. 1242) ; stem half-herbaceous, weak, flexuous, branched, pubescent; leaves on long petioles, cordate-reniform, bluntly or deeply lobed or coarsely toothed, the lobes 3–5-toothed, nearly glabrous above, beneath pubescent or villous ; petioles of the cauline leaves auricled at base; fl.-branches naked, laxly corymbose, few headed ; pedicels elongate, sparingly scaly ; inv. scales 9–12, oblong, acute, glabrous or puberulous, 3–5-nerved ; ligules 5–8; achenes on each side hispidulous, ciliate. *Th.! Cap.* 671, *DC. l. c.* 307. *C. distans, Kze. in Linn.* 17, *p.* 571?

HAB. Cape, *Thunberg!* Common in woody places, on the Mts. near Capetown and Simonstown. Kuilsriver, *Zey.! 920.* (Herb. Th., D., Hb., Sd.)
Stems 1–2 feet high, flexuous, vaguely and distantly branched. Petioles 1–2 inches long. Lamina ¾–1 inch long, 1–1½ inch wide, *more or less deeply* 7–9 lobed. Heads sometimes solitary, commonly 3–12 in a loose, irregular corymb. Inv. varying from glabrous to downy or hispid. Pubescence of branches copious or scanty. The leaves vary greatly in the depth or shallowness of the lobes. Judging by the author's description, *C. distans,* Kze. can hardly be any thing but this common plant; it was raised in Leipsic Bot. Garden, from seeds sent by *Gueinzius.*

11. C. lobata (L'Her. Sert. Angl. 26) ; stems suffruticulose, flexuous, much branched, glabrous; leaves on long petioles, roundish-reniform (rarely lyrate), shortly many lobed, the lobes 3–5-toothed, the teeth callous; petioles of the upper leaves eared at base; corymb loosely branched, many-headed ; pedicels scaly; inv. scales 6–18 glabrous ; achenes hispidulous and ciliate. *Th.! Fl. Cap.* 671. *DC. l. c.* 307. *Sieb.! Fl. Cap.* 368.

VAR. β. **Pappei**; heads very numerous, small, panicled; auricles of the petioles obsolete.

HAB. Uitenhage and Albany, *Bowie! E. & Z.! Drege! Sieber!* β, Winterhoek mountain, Tulbage, *Dr. Pappe!* (Herb. D., Hk., Sd.)
Stems 2–3 ft. high. Petioles slender, 1–2 inches long. Leaves 1 inch long, 1½-inch wide, occasionally with a pair of lateral lobes. Corymb. variable, very loose, few or many headed. Very like *C. geifolia,* except that it is in all parts quite glabrous, with a more branching inflorescence. The fl. heads vary much in size and number of inv. scales; our var. β. is an extreme state, in all respects, save pubescence, agreeing with *C. alchemilloides!*

12. C. Saxifraga (DC. l. c. 306) ; suffruticose, erect or diffuse, glabrous ; leaves on long petioles, roundish-reniform, truncate or cuneate at base, coarsely 5–9-toothed at the apex, many-nerved and netted-veined ; petiole not auricled at base, rarely with 1–2 lobules below the lamina ; fl.-branches nude, simple or branched, pedicels elongate, sparingly scaly ; inv. 8–12-scaled, sub-calyculate, scales glabrous, oblong 4–5-nerved ; ligules 3–5; achenes on each side hispidulous, ciliate.

HAB. Uitenhage and Albany, *E. & Z.! Verreaux! Genl. Bolton!* &c. Slaaye Kraal, *Burke and Zeyher!* (Herb. D., Hk., Sd.)

Stems much branched, slender, rigid, either erect or spreading, or trailing. Petioles 1–1½ inch long, slender; *lamina* 6–9 lines wide, 3–5 lines long. Peduncles 2–3 inches long, filiform. Inv. scales linear-oblong, with a triangular top.—Allied to *C. lobata*, but with much smaller and less reniform leaves, less cut.

13. C. geraniifolia (DC.? l. c. 308) ; quite glabrous ; stem herbaceous, erect, slender, laxly branched ; leaves on long petioles minutely auricled at base or nude, the lower leaves mostly reniform, 5–7-lobed and toothed, the upper lyrato-pinnatisect, with 1–2 pair of obovate, toothed, petiolulate, lateral lobes and a deeply 3–5-lobed terminal one ; flowering branches elongate, nude, laxly few-headed, the heads on very long nude pedicels ; inv. subcalyculate, glabrous, of about 12 scales ; rays 10–12 ; achenes quite glabrous, margined, or winged, not ciliate.

HAB. King William's Town, *Dr. Pappe!* (Herb. Cap., D.)
Stem straggling, 2 ft. high, weak. Leaves distant on 2–3 inch long petioles; most of the upper ones with lateral leaf-lobes ; varying considerably in depth of lobing and degree of dentition. Pedicels 2–5 inches long. Rays numerous, bright yellow. Not having seen authentic specimens of DC.'s plant I refrain from quoting his habitats. Specimens from *Ecklon* and *Drege* (in Hb. Sond.), purporting to belong to " *C. geraniifolia β. oligocephala*" DC. have hispidulous and ciliate achenes, and belong to *C. Saxifraga :* a specimen in Hb. Cap. from *Ecklon* has no fl. heads, and may be anything ! *Dr. Pappe's* plant, here described, is at least an entity.

14. C. deltoidea (Sond.! in Linn. 23. p. 68) ; herbaceous, erect, quite glabrous ; leaves petiolate, the petiole with reniform, toothed, ample auricles at base, the limb deltoid, truncate at base, many nerved, acute, repand-toothed ; heads loosely panicled or corymbose ; pedicels with few scales ; inv. subcalyculate, glabrous ; rays 4–5 ; achenes quite glabrous, compressed, narrow-margined.

HAB. Natal, *Gueinzius!* 343. (Herb. Sond.)
Very near *C. atriplicifolia*, from which it chiefly differs in the less deeply cut and jagged leaves, and the ampler and reniform auricles to the petioles.

15. C. atriplicifolia (DC.! l. c. 308) ; herbaceous, erect, quite glabrous ; leaves petiolate, the petiole bearing at base 1–2 narrow, small lobes, the limb truncate at base, hastate, sharply and coarsely incised or lobed, acuminate ; heads loosely panicled or corymbose, pedicels scaly ; inv. calyculate, glabrous ; rays 4–5 ; achenes flattened, smooth, wingless, quite glabrous, not ciliate.

HAB. Port Natal, *Drege! Dr. W. B. Grant.* Umzinto, *M. J. McKen!* (Herb. D., Hk., Sd)
Stems weak, 1–2 ft. high, green. Petioles occasionally bearing one or two lobes in the middle ; terminal lobe 1–1½ inch long, ¾–1¼ inch wide at base, sharply and unequally cut.

16. C. decipiens (Harv.) ; herbaceous, erect, quite glabrous ; leaves petiolate, the petiole eared at base, limb truncate and toothed at base, broadly hastate, sharply and coarsely incised and lobed, acuminate ; heads loosely panicled or corymbose, few-fl., pedicels filiform, nearly nude ; inv. of 8 scales, subcalyculate, glabrous ; rays about 4 ; achenes flattened, hispidulous, with a narrow, ciliate border.

HAB. Umvoti District, Natal, *W. T. Gerrard*, 1040. (Herb. D.)
This is so like *C. atriplicifolia* in foliage, that it can scarcely be known without an examination of the fruit : then there can be no mistake. It is a rather more

slender and branching plant, with smaller fl. heads, and larger and more leafy ears to the petiole.

17. C. platycarpa (DC. l. c. 305); annual, many-stemmed, diffuse, glabrous, pale; lower and medial leaves eared at base, petioled, lyrate, the lateral lobes 1–2 pair, oblong, dentate, the terminal large, roundish or 3-lobed, toothed; upper leaves sessile, eared at base, pinnatisect; fl. branches laxly panicled, the pedicels long, scarcely scaly; inv. 10–13-scaled, with few bracts at base; rays 6–10, twice as long as the disc; achenes broadly obovate, hispidulous on the disc, with a very wide, fringed margin.

HAB. Cape, *Drege.* Albany and Somerset, *Mrs. F. W. Barber.* (Herb. D., Hk.) 6–12 inches high or more, the stems straw-colour when dry. Leaves of very thin texture, veiny. Heads 3 lines long, and about as wide; rays oblong, spreading, yellow.

18. C. lyrata (DC.? l. c. 308); annual; stem branching, erect, glabrous, angular; leaves eared at base, lyrato-pinnati-partite, the lateral lobes in 2–3 pairs, obovate-oblong, toothed, the terminal larger, cuneate-obovate, 3–5-lobed, the lobes toothed; corymb compound, loose; inv. calycled, its scales 8–10, glabrous, oblong, acute; rays few; achenes flattened, quite glabrous, with a broad, perfectly entire, not ciliate marginal wing.

HAB. Caledon River, *Burke & Zeyher!* Nieuweveldt, *Drege?* (fide DC.) (Herb. D., Hk., Sd.) Stem 1–2 feet high, simple or branched. Leaves 1⅓–3 inches long, the lateral lobes 3–8 lines long, the terminal larger. Heads small. I describe from *Burke and Zey.* specimens, not having seen *Drege's* plant, which, if different, must be a nearly allied form.

19. C. microglossa (DC. l. c. 305); "herbaceous, glabrous, erect; branched from the suffruticose base; leaves petioled, oblong-lanceolate, tapering to both ends, coarsely toothed or sub-pinnatisect, the upper linear, sessile, subentire; branches bifid; pedicels elongate, nude; inv. 14–15-leaved; rays 10–12, scarcely longer than the involucre; achenes hairy, ciliate." *DC. l. c.*

HAB. Near the Gariep. *Drege.* (Unknown to me.) "Six inches high. Leaves 1–1½ inch long. Ray-achenes compressed." *DC.*

20. C. Dregeana (DC. l. c. 305); "glabrous; stem erect, herbaceous, terete, branched; lower leaves . . . . . ; upper with broad, roundish, clasping, sinuate toothed auricles at base of petiole; heads pedicellate, corymbose, discoid; inv. scales dilated into a minute, subscarious appendix; achenes subangular, on both sides whitish with short, close hairs." *DC. l. c.*

HAB. At the Gariep, *Drege.* (Unknown to me.) Said to resemble *C. geifolia,* but differing in fruit and the discoid heads.

21. C. exilis (DC. l. c. 305); "herbaceous, glabrous, erect; lower leaves ovate, toothed, tapering at base into a petiole, the rest oblong-linear, narrowed at base, somewhat toothed at apex; pedunc. few, elongate, scarcely scaly, one-headed; inv. nearly nude at base, equalling

the disc; heads discoid, homogamous; achenes compressed, somewhat
4-angled, the angles densely ciliate." *DC. l. c.*

HAB. Near Litaku, *Burchell*, 2274. (Unknown to me.)
"About 6 inches high, with few leaves. Inv. scales linear, about 18-20." *DC.*

§ 3. OTHONNOIDES. (Sp. 22.)

22. C.? othonnoides (Harv.); root fibrous, annual; stem erect, branching, woolly in the axils, glabrous or sparsely pilose; leaves sessile, half-
clasping, pinnati-partite, scaberulous, segments linear, subacute, entire
or 1-2-toothed, with recurved edges, decurrent; branches prolonged
into nude, one headed peduncles; inv. sc. 10–12, not calycled at base,
oblong, 3–5-lined; rays revolute; achenes rufo-pubescent. *Othonna
pinnatifida, Th.! Cap.* 721. *DC. l. c.* 482.

HAB. Cape, *Thunberg.* (Herb. Th., Sd.)
Root fibrous. Stem divided near the base into several erect, simple branches,
which are leafy in their lower part, nude and pedunculoid upwards. Leaves slightly
ear-clasping at base, 1-1½ in. long; their lobes 1 line wide. Pedunc. 4-6 in. long,
swollen under the head. Inv. without calycle, but the expanded apex of the peduncle
is minutely dentate, the teeth alternating with the inv. scales. Achenes narrow
obovate, copiously pubescent, their pubescence mucilaginous when moistened, emitting
spiral hairs as in *Ruckeria,* &c. The style of the disc fl. is branched; its branches
tipped with a short, hispid cone.

Doubtful species.

C. arctotidea (DC. l. c. 306); "herbaceous, erect, the whole plant
cano-tomentose when young, the adult cobwebby or glabrescent; lower
leaves equalling the stem, long-petioled, lyrate, the terminal lobe reniform, obtusely lobed, the lobes calloso-mucronate; the cauline shorter,
with petioles dilated at base; branches few, nearly leafless, one-headed;
inv. subcalyculate, a little shorter than the disc, subcanescent; rays
about 10; achenes compressed, downy, ciliate." *DC. l. c.*

HAB. Roggeveldt, *Burchell*, 1368.
Drege's specimens, from the Sneeuweberg, quoted by DC. are undistinguishable
from *C. mollis β.*; I have not seen Burchell's.

C. humifusa (L'Her. Sert. Angl. 25); "stem suffruticose at base,
ascending; radical leaves petioled, the petiole tomentose, not eared at
base, naked or with a lobe under the limb, limb ovate or subcordate,
or cuneate at base, lobulate-dentate, glabrous above, tomentose beneath;
pedunc. scapelike, sparingly scaly, one-headed; invol. calyculate, cobwebbed, of 10–12 scales; rays 8–10; achenes minutely downy, ciliate."
DC. l. c. 306.

HAB. Graaf Reynet, *Drege.* (v. frustulum in Hb. Sd.)
Is not this some form of *C. mollis?*

CXI. **LOPHOLÆNA**, DC.

Heads many-fl., all the flowers tubular, 5-fid, with a terete tube; the
marginal fl. mostly cleft on the inner side and with subabortive anthers,
thus polygamo-female; of the disc regularly tubular, the style sometimes
abortive. *Inv.* 5-leaved, the scales free, leaf-like, broadly keel-crested on
the back, winged at the sides. *Recept.* honey-combed. *Anthers* tail-less.

Style-branches in the perfect fl. elongate, downy at back, produced at the apex into a long, every where hispid, scarcely acute appendix. *Achenes* angular, beakless, glabrescent, ciliate at the angles, the central ones often abortive. *Pappus* multiseriate, bristle-shaped, scarcely rough. *DC. Prodr. 6, p. 336.*

A single species. The name is compounded of λοφος, *a crest*, and χλαινα, *a cloak;* alluding to the involucre.

1. L. Dregeana (DC. in Deless. Ic. 4. t. 59); *DC. l. c.*

Hab. Omsamculo and Omtenda, *Drege!* (Herb. D., Hk., Sd.)

A robust, glabrous, erect undershrub. Leaves 1-3 inches long, ½-¾ inch wide, alternate, sessile, half-clasping, oblong, obtuse, rigid, quite entire, 3-nerved, the upper narrower and more lanceolate. Branches leafy to the summit. Heads subcorymbose. Fl. seemingly yellow. A very remarkable plant.

CXII. CACALIA, Linn.

Heads several-fl., homogamous; all the fl. tubular, 5-fid, perfect. *Inv.* uniseriate, of 5-30-scales, with a very few bracteoles at base. *Recept.* without paleæ. *Style*-branches tipped with a short cone, hispid at base. *Achenes* oblong, beakless, glabrous. *Pappus* uniseriate, of many rigid, scabrous bristles. *DC. Prodr. 6, p. 327.*

Perennial herbs. Leaves alternate, mostly petiolate, toothed or lobed. Heads panicled or corymbose. Cor. white, rosy or yellow-white. The species are chiefly American and Asiatic; very few African, and only one from the Cape. Name used by Dioscorides for some similar plant.

1. C.? cissampelina (DC. l. c. 331); suffruticose, scandent, cobwebby, becoming glabrous; branches striate; leaves petiolate, bluntly 3-5-angled, 5-7 nerved and netted-veined, the angles mucronate; pedunc. axillary, shorter than the leaf, corymbose or subumbellate, several-headed; heads shortly pedicellate, 9-10-fl.; inv. scarcely calycled, of 6-8 scales much shorter than the flowers; corollas campanulate, with long, revolute lobes; anthers much exserted.

Hab. Ceded Territory, *Ecklon!* Katberg, *H. Hutton!* (Herb. Sd., D., Cap.)

With the habit of a scandent *Senecio*, but cobwebby. Petiole 1½-2 in. long. Leaf 2-2½ in. long and broad, pentagonal, rhomboid, or deltoid. Flowers pale. Corymb few or many headed, simple or branched. Scarcely generically separable from *Senecio*.

CXIII. KLEINIA, Linn.

Heads many-fl., discoid, almost always homogamous, (in 1-2 sp. heterogamous), all the fl. tubular, 5-toothed. *Recept.* flat, naked. *Inv.* uniseriate, many-leaved, with a few small bracteoles at base, rarely nude. *Style*-branches tipped with a short cone, ciliate at base. *Achenes* beakless. *Pappus* bristle-shaped, roughish, in many rows. *DC. Prodr. 6, p. 336.*

Fleshy, African shrubs or herbs, sometimes nearly stemless, often glaucous. Branches terete or angular. Leaves alternate, mostly quite entire. Flowers white or pale-yellow. Name, in honour of *J. Th. Klein*, a German zoologist. Better known by a peculiar habit than by differential characters. It differs from *Cacalia* by the more copious pappus; from *Senecio* by the conical tips of the style-branches, the constantly discoid heads, and white or very pale flowers. Several species unknown to me are given on the authority of De Candolle. They are difficult to discriminate in a dried state, and I fear have been needlessly multiplied.

Sect. 1. CACALIANTHEMUM. Heads homogamous, all the flowers perfect, 5-toothed, hermaphrodite. (Sp. 1-16.)

Leaves *glabrous*, fleshy, linear, subulate or fusiform :
  Lvs. minute, distant; stem much-branched, angular   (1) **longiflora.**
  Lvs. elongate, crowded, linear or fusiform :
    Lvs. fusiform, tapering *to base* and apex :
      Stem erect, fruticose :
        Lvs. laterally compressed :
          Powdery-glaucous; stem branched (2) **ficoides.**
          Scarcely glaucous ; stem very short (3) **aizoides.**
        Lvs. vertically depressed, glaucous ... (4) **repens.**
      Stem trailing, slender, rooting at intervals (5) **radicans.**
    Lvs. linear-semiterete, not tapering at base :
      Pedunc. much longer than the leaves :
        Pedunc. remotely forked, few-headed :
          Stem glabrous ; lvs. 2-2½ in. long (6) **crassulæfolia.**
          Stem villous; lvs. 1½ inch long ... (7) **pinguifolia.**
        Pedunc. closely-corymbose, very many
          headed ... ... (10) **talinoides.**
      Pedunc. shorter than the leaves :
        Pedunc. short, 3-4-headed ... ... (8) **breviscapa.**
        Pedunc. simple, one-headed ... ... (9) **Ecklonis.**
Leaves *tomentose*, terete, tapering to each end :
  Densely tomentose; leaves elongate ... ... (11) **Haworthii.**
  Closely tomentose; lvs. shorter and broader ... ... (12) **cana.**
Leaves glabrous, *expanded, flat*, quite entire :
  Lvs. wedge-shaped, obtuse or truncate ... ... (13) **cuneifolia.**
  Lvs. ovate-oblong; stem smoothish ... ... (14) **Anteuphorbium.**
  Lvs. lanceolate; stem rough with knobs ... ... (15) **papillaris.**
Leaves glabrous, *petioled, lobed or laciniate* .. .. (16) **articulata.**

Sec. 2. ERECHTHITOIDES. Heads heterogamous, the marginal flowers female, either tubular-truncate or minutely and very imperfectly radiate. (Sp. 17-18.)

  Stemless; lvs. radical, elongate; scape one-headed ... (17) **acaulis.**
  Branching, leafy; pedunc. short, corymbose ... ... (18) **subradiata.**

1. K. longiflora (DC. ! l. c. 337); glabrous; stem shrubby, fleshy, much-branched ; branches mult-angular (furrowed when dry) ; leaves distant, minute, linear-subulate, fleshy ; peduncles 3-5 at the ends of the branches, much shorter than the involucres, bearing 2-4 scattered, subulate bracts; inv. cylindrical, of 5, linear, acute, margined, rigid scales, half as long as the corollas ; flowers 5 in the head, perfect ; achenes elongate, cylindrical, multistriate, rigidly hispidulous between the ribs ; pappus longer than the corolla.

HAB. Beyond the Gariep, betw. Klaarwater and Nu-Gariep, *Burchell*, 1718. Nieuweveld, *Drege!* Cradock, *Burke & Zey.! Zey.!* 950. (Herb. Hk., D., Sd.)
Two feet or more high, much-branched ; the branches 3-4 lines diameter, prismatically angled. Leaves 3-4 lines long (perhaps longer on young shoots), scale-like. Peduncles ¼ and ⅓ inch long. Inv. scales 8-9 lines long. Pappus twice as long, pale-creamy, silky, copious.

2. K. ficoides (Haw. Succ. Pl. 313); "quite glabrous; stem fleshy-fruticose, erect, branched ; leaves fleshy, nerveless, compressed, acuminate, *powdery-glaucous;* pedunc. elongate, corymbose at the summit, many-headed; inv. cylindr., the scales 7-8; fl. 9-15, perfect; achenes downy. *DC. l. c. p. 337. Cacalia ficoides, Linn. Th. Cap. 623. DC. Pl. Grass. t. 90. Comm. Rar. t. 40.*

HAB. Cape, *Thunberg.* (v. v. cult.)

3. K. aizoides (DC. l. c. 337); "quite glabrous; stem very short, shrubby; leaves crowded on the crown, compressed, quite entire, callous-mucronate, acute; pedunc. twice as long as the leaves, nude, striate, bifid, 2–4-headed; pedicels elongate, scarcely scaly under the head; inv. sc. about 12, scarious edged, equalling the flowers; fl. 25–30; achenes terete, densely hairy." *DC. l. c.*

HAB. Zwarteberge, near Kendo, *Drege!* (Herb. Sond.)

Said to be allied to *K. ficoides*, but "scarcely glaucous, much smaller, with fewer fl. heads. Leaves 3–3½ inches long, 3 lines wide. Heads 6 lines long, and nearly as wide." *DC.* I give this on DC.'s authority; a specimen in Hb. Sond., with the fl. in bad condition, looks to me more like a *Doria.* Specimens collected by Dr. Pappe, at the Waterfall, Tulbagh (in Herb. Cap.) agree better with DC.'s character. Their leaves vary from 1½–4½ inches long, 2–3 lines wide. The scape, sometimes one-headed, is 6–12 inches long, in the larger plant forked and 2–3-headed; pedicels 2 inches long. Achenes densely and softly villous.

4. K. repens (Haw. Succ. Pl. 313); "quite glabrous; root creeping; stem fleshy-fruticose, erect; leaves fleshy, glaucous, oblong-acuminate, depressed, subconcave above; pedunc. naked, somewhat branched, corymbulose, few-headed; inv. campanulate, subcalyculate, of 5–6 (or 8–10) scales; flowers 15–16 (or 20–30); achenes glabrescent." *DC. l. c. p.* 337. *Cacalia repens, L. Th. Cap.* 623. *DC. Pl. Grass. t.* 42.

HAB. Cape, *Thunberg!* (Herb. Th., and v. v. cult.)

"Smaller and more glaucous than *K. ficoides*, which it closely resembles; it differs chiefly in the *depressed*, not *compressed* leaves.

5. K. radicans (DC. l. c. 337); stem herbaceous, decumbent, rooting at intervals, sharply wing-angled; branches short, erect; leaves fleshy, lanceolate or linear-lanceolate, tapering to each end, acute, scattered; pedunc. slender, naked, simple or bifid, pedicels elongate, unibracteate, one-headed; heads homogamous, 20–25-fl.; inv. 10–12-leaved, equalling the disc, acute at base, subcalyculate; scales linear-acuminate; achenes hispid, ribstriate. *K. gonoclada, DC. l. c.* 336. *Cacalia radicans, Th.! Cap.* 625.

HAB. At Saldanha Bay; on the Karroo, and elsewhere, *Thunberg!* Under bushes, in thickets by the Zwartkops R., *E. & Z.!* Fish R., *Burke!* (Hb. Th., D., Hk., Sd.)

Stems 1–2 feet long, prostrate, subsimple, rooting every few inches, secundly-leafy; branches 1–2 inches long, leafy, erect. Both stem and branches are wing-angled by ridges proceeding from the base of the leaves. Leaves ¾–1 inch long, 2–3 lines wide. Pedunc. 4–5 inches long. The leaves, which have a taste of turpentine, are eaten by the Hottentots. *Thunberg's* specimens quite agree with those of *E. & Z.* of "*K. gonoclada.*"

6. K. crassulæfolia (DC. l. c. 336); glabrous; stem fleshy-fruticose, short, simple, erect; leaves crowded on the upper half of the stem, linear-semiterete, acute or subulate, quite entire; peduncle elongate, naked, bifid or sparingly branched near the summit, the pedicels elongate, distantly scaly, one-headed; heads campanulate, homogamous, about 30-fl.; inv. bracteolate at base, scales about 12, membrane-edged, acuminate, with inflexed points not shorter than the disc; achenes pubescent; pappus white, equalling the flowers.

HAB. Uitenhage, *Eckl.* Stony places near the Zwartkops R., *Zey.!* (Hb. D., Hk.)

Stem 3–4 inches high, stipe-like, 4–5 lines diameter, rough with the bases of old leaves. Leaves 2–2½ inches long, 2–3 lines thick, quite glabrous. Pedunc. 6–8 inches long, slender, annual, terminal. Invol. 4–5 lines long. This comes near *K. repens*, but the leaves (in a dry state) seem different.

7. K. pinguifolia (DC. l. c. 336); "stem very short, fleshy-fruticose, *villous;* leaves subterete, acute, glabrescent, quite entire; pedunc. 3–4 times longer than the leaves, glabrous, remotely forked, leafless ; pedicels nude, much longer than the heads ; invol. scarcely calyculate, of about 12 acute scales; fl. about 20; achenes terete, downy." *DC. l. c.*

HAB. Cape, *Drege.* (Unknown to us.)
"Leaves 1½ inch long, 2 lines wide. Inv. 4 lines long, equalling the pappus," *DC.* It seems to come very near *K. crassulæfolia.*

8. K. breviscapa (DC. l. c. 337); "stem very short, fleshy-fruticose, irregularly terete, somewhat villous; leaves subterete, with a subspinous mucro, entire, glabrous ; pedunc. subdichotomous, 4-headed, scarcely equalling the leaves, scaly ; inv. calyculate, of about 12 leaves, as long as the disc ; fl. 25–30; achenes glabrous, subvillous on the angles." *DC. l. c.*

HAB. Ebenezar, Oliphant River, *Drege.* (Unknown to us.)
"Leaves 1½–2 inches long, often incurved. Heads 5 lines long and broad." *DC.*

9. K. Ecklonis (Harv.) ; rhizome creeping subhorizontally under ground ; stem short, ascending, fleshy, rough with broken leaf-bases; leaves crowded on the upper half of the stem, linear-terete, elongate, channelled above, attenuated, acute, glabrous ; scapes shorter than the leaves, slender, distantly scaly, one-headed ; inv. scales 10–12, acuminate, with broad margins equalling the disc; fl. all perfect (no female marginal fl.), 40–50; achenes (young), subglabrous, with a wide, epigynous disc. *K. acaulis, β. Ecklonis, DC. l. c.*

HAB. On dry hills at Adow, Uit., *E. Z., Zey. !* 2981. (Herb. D., Hk., Sd.)
Rhizome several inches long ; stem rising 2–3 inches above the soil, 4–5 lines diameter. Lvs. 2½–3¼ inches long, 1½–2 lines diameter. Scapes about 3 inches long.

10. K. talinoides (DC. l. c. 337); quite glabrous, fleshy-fruticose, erect; stem terete; leaves fleshy, elongate, 'terete-compressed' (?), quite entire, acute or acuminate; pedunc. much longer than the leaves, terete, nude, ending in a much-branched, many-headed, close, corymbose panicle; pedicels shorter than the invol. or about equalling it; invol. 1–3 bracteolate at base, of 5–7 scales, 5–7-flowered, shorter than the flowers; achenes glabrous.

HAB. Haazenkraal's River, near Uitkomst, *Drege !* (Herb. Sd., D.)
I have not seen the stem. Leaves 2–3 inches long, 2–3 lines wide, much compressed in drying. Pedunc. 6–9 inches long; corymb 2 inches across, 40–50-headed or more. Inv. 3–4 lines long, 2 lines wide.

11. K. Haworthii (DC. l. c. 338); "in all parts clothed with dense, white woolliness; stem fleshy-fruticose ; leaves terete, tapering to each end." *Cacalia tomentosa, Haw. Misc. p. 189, non Linn. C. Haworthii, Sw.*

HAB. Cape.
Cultivated in Europe (at least formerly); fl. undescribed.

12. K. cana (DC. l. c. 338); "in all parts clothed with close, whitish toment.; stem fleshy-fruticose, 5-angled; leaves thick, oval or obovate, tapering to each end; peduncle nude, one-headed; inv. about 7-leaved, tomentose, obtuse, subequalling the flowers." *DC. l. c.*

HAB. Camdebosberg, *Drege.*
Said to differ from *K. Haworthii* by its thicker leaves, 8–9 lines long, 3–4 lines thick.

13. K. cuneifolia (DC. l. c. 337); "glabrous; stem fleshy-fruticose, erect; leaves fleshy, cuneiform, veinless." *DC. l. c. Cacalia cuneifolia, Linn.—Th.! Cap.* 624.

Cape, *Thunberg!* (Herb. Th.)
A fragment without flowers alone exists in Hb. Thunb.

14. K. Anteuphorbium (DC. l. c. 338); "quite glabrous; stem fleshy-fruticose, erect; leaves ovate-oblong, flat; petioles decurrent in a triple line at base; heads shortly peduncled, solitary; inv. scales acuminate, equalling the flowers. *Cacalia Anteuphorbium, Linn. Dill. Elth. t.* 55, *f.* 2–3.

HAB. Cape. (Cultivated in Europe.)

15. K. papillaris (Haw. Succ. Pl. p. 313); "quite glabrous; stem fleshy fruticose, beset with cylindrical, truncate, petiolar prominences; leaves lanceolate, flat, subglaucous, mid-nerved." *DC. l. c.* 338. *Cacalia papillaris, Linn. Dill. Elth. t.* 55, *f.* 63.

HAB. Cape. (Cultivated in Europe.)

16. K. articulata (Haw. Succ. Pl. p. 315); quite glabrous; stem fleshy-fruticose, suberect; branches jointed, swollen; leaves petioled, flat, fleshy, glaucous, runcinate-laciniate, the lobes acuminate, the terminal larger; pedunc. elongate, nude, corymbose at summit, few-headed; pedicels longer than the heads; inv. scales 10–12, acuminate, scarcely calyculate, longer than the disc; fl. 15–20; achenes downy. *DC. l. c.* 339. *Cacalia articulata, Linn. Th.! Fl. Cap.* 624. *L'Her. Stirp. t.* 83. *C. laciniata, Jacq. Ic. Rar. t.* 68. *DC. Pl. grass. t.* 18. *C. runcinata, Lam.*

HAB. On the hill of Zwartkops Zoutpan, Uit., *Thunberg!, E. & K.! Zey.!* 2985. (Herb. Th., D., Hk., Sd.)
Long cultivated in Europe. Stems 1–2 f. high, branching, the branches constricted at intervals. Petioles 1 inch or more; lamina 1–2 inches long, deeply lobed. Pedunc. 6–8 inches long.

17. K. acaulis (DC.! l. c. 339); nearly stemless (with a thick rhizome); leaves from the crown numerous, aggregated, linear-semi-terete, elongate, channelled above, acute, mucronate; scape erect, one-headed, distantly scaly, rather longer than the leaves, striated; inv. scales 12–16, acuminate, with broad white margins, nearly equalling the disc; fl. 40–60, the outermost narrow-tubular, female, with elongated stigmas; achenes villous; pappus copious. *Cacalia acaulis, Linn. f. Th.! Fl. Cap.* 625.

HAB. Under Roggeveldt, *Thunberg!* (Herb. Th.)
Rhizome 4–5 lines diameter, scarcely protruding above the soil. Leaves 4–5–6 inches long, 1½–2 lines diameter, clasping at base, linear, acute, glabrous. Scapes

6–8 inches high, with a few slender, subulate, scattered leaves. Inv. ¼ inch long, among the largest in the genus, when mature nearly an inch across. Female marginal flowers, several, uniseriate, denticulate at summit, quite cylindrical. Achenes shortly hairy. *Thunberg's* original specimens are in excellent order; I have seen no modern ones agreeing with them. DC. quotes a specimen No. 6804 of Burchell's list from Swellendam. *Ecklon's* Uitenhage plant seems different; it is probably our *K. Ecklonis.*

18. K.? **subradiata** (DC. l. c. *339*); "shrubby, erect, branching, glabrous; branches terete; leaves sessile, fleshy, linear, thick, acute, quite entire; heads at the ends of the branches subcorymbose; pedunc. shorter than the leaves; pedicels scaly, shorter than the head; inv. scales 7–8; rays few, shorter than their own style; achenes softly pubescent." *DC. l. c.*

HAB. Graaf Reynet, *Ecklon.* (Unknown to me.)

Allied, according to DC., to *Senecio acutifolius,* but the leaves are 1½ inch long, and the expanded portion of the ligule so minute as to be scarcely obvious to the eye. Is the plant more properly a *Senecio?*

CXIV. DORIA, Less.

Character as in *Othonna,* except: *Heads* discoid. *Corolla* of the marginal, female flowers tubular, truncate, shorter than the involucre, *never ligulate. Less. in Linn.* 1831, *p.* 94. *DC. Prodr.* 6, *p.* 469.

Herbs or shrubs, with completely the aspect of *Othonna.* Name unexplained.

A. Rigid, ramulous shrubs or half-shrubs, with ligneous stems :
  Leaves lanceolate or linear-lanceolate :
    Lvs. lanc., entire or toothed; inv. sc. 8–9 ... (1) **acutiloba.**
    Lvs. linear-lanceolate, thick, entire; inv. sc. 5–6 :
      Pedunc. shorter than lvs.; inv. sc. acumin. (3) **cneorifolia.**
      Pedunc. 2–3-ce as long as lvs.; inv. scales
        subacute ... ... (4) **longipes.**
  Leaves oblong-linear, obtuse, *petiolate;* pedicels axill.,
    short ... ... (2) **lasiocarpa.**
  Leaves obovate, oblong or cuneate-oblong, entire or
    toothed : pedicels terminal.
    Heads solitary, on short or long pedicels :
      Branches divaricate, ending in hard spines (6) **rigida.**
      Branches not spinous :
        Pedicels very short or shortish :
          Lvs. obovate, obtuse, entire, drying
            pale ... ... (5) **cluytiæfolia.**
          Lvs. cuneate-oblong, 3–5-toothed,
            ciliolate, drying dark ... ... (7) **ciliata.**
        Pedicels 2–8 times longer than the
          decurrent leaves ... ... (8) **Kraussii.**
    Heads in subsessile, branching corymbs; lvs.
      decurrent ... ... (9) **alata.**

B. Fleshy or succulent, small shrubs ; branches mostly forked :
  Leaves pinnato-multifid, with linear-filiform lobes ... (14) **abrotanifolia.**
  Lvs. obovate or lanceolate, entire or toothed :
    Pedicels thrice shorter than the obovate, *serrulate*
      leaves ... ... (10) **nana.**
    Pedicels as long as leaves or longer :
      Leaves oblong-spathulate, obtuse, petioled ;
        achenes silky ... ... (11) **Ceradia.**

Lvs. lanceolate, the upper obovate; achenes
　　glabrous ... ...　(12) **Arbuscula.**
Lvs. oblong-lanceolate, 1-toothed or lobed
　　on each side; achenes glabrous ... ...　(13) **retrofracta.**

C. Herbaceous plants, mostly (or always?) with tuberous roots.
　Stem erect, *fleshy;* lvs. fleshy, pinnatifid :
　　Achenes glabrescent ; pappus scarcely longer
　　　than invol. ... ...　(15) **carnosa.**
　　Achenes hirsute; pappus twice as long as invol.　(16) **eriocarpa.**
　Stemless : rad. lvs. many, petioled, obovate, bluntly
　　pinnatifid or lobulate ... ...　(17) **taraxacoides.**
　Stem slender, simple or branched, weak or wiry :
　　* Achenes of the *disc* with pappus :
　　　Cauline lvs. conspicuously stem-clasping, cordate or cuneate :
　　　　Lower leaves subpetioled, cuneate or
　　　　　oblong, 3–5-lobed or toothed ... ...　(18) **digitata.**
　　　　Leaves cordate at base, obtuse or acuminate, entire :
　　　　　Ach. silky; pappus whitish or pale　(19) **perfoliata.**
　　　　　Ach. hispidulous ; pappus rufous　(20) **Lingua.**
　　　　　Ach. glabrous; pappus rufous ...　(21) **chromochæta.**
　　　Cauline lvs. linear or linear-lanceolate, sessile or somewhat clasping.
　　　　Marginal achenes glabrous :
　　　　　Lvs. linear, elongate ... ...　(22) **linearifolia.**
　　　　　Lvs. lanceolate, 1½ inch long, 4–5
　　　　　　lines wide... ... ...　(23) **Laureola.**
　　　　Marginal achenes pubescent: (stems branching) :
　　　　　Pappus whitish ; lvs. lanceolate,
　　　　　　undulate ... ...　(24) **ramosa.**
　　　　　Pap. rufous; lvs. linear-lanceolate;
　　　　　　stem twisted ... ...　(25) **tortuosa.**
　　** Achenes of the *disc* quite nude, *without pappus :*
　　　Lvs. sessile, linear-lanceolate ... ...　(26) **semicalva.**
　　　Lvs. cordate-oblong, clasping, subacute ...　(27) **Gymnodiscus.**

1. D. acutiloba (DC. l. c. 469) ; shrubby, glabrous, erect ; branches
terete; leaves lanceolate, acute, tapering much at base, coriaceous, with
immersed veins, sessile, some quite entire, some sharply few toothed,
others (occasionally) " toothed or cut along their whole margin ;"
pedunc. terminal, nude, long, 1-2-headed, rigid; inv. of 8–9 lanceolate,
membrane-edged, basally concrete scales; "marg. achenes terete, thick,
densely villous." *(DC.)*

HAB. District of Worcester, *Ecklon!* (Herb. Sond.)
　Leaves closely set, 1½–2 inches long, 2–6 lines wide, cuneate-attenuate at base.
Pedunc. 3–5 inches long, if branched, the pedicels as long. Fl. and fruit wanting
in the only specimens I have seen ; and which, to my eye, are undistinguishable
from *Othonna coronopifolia.*

2. D. lasiocarpa (DC. l. c. 469) ; " stem shrubby, terete, glabrous ;
leaves oblong-linear, quite entire, obtuse, glabrous, tapering at base
into a petiole; pedicels axillary, equalling the leaves; inv. scales 5,
oblong, subconcrete at the base only ; marg. achenes thick, obovate,
very hairy." *DC. l. c.*

HAB. Near Verleptpram, Gariep, *Drege.* (Unknown to me.)
　" Stem purplish. Leaves 2 in. long, 3 l. wide. Pappus dirty-reddish." *DC.*

3. D. cneorifolia (DC.! l. c. 470); shrubby, erect, glabrous ; branches terete; leaves linear-lanceolate, coriaceous, thick, quite entire, tapering at base, callous-tipped, subacute, faintly mid-nerved; pedicels axillary toward the ends of the branches, subracemose, a little *shorter* than the leaves; inv. sc. 5–6, concrete at base, acuminate, with broad, membranous edges; marg. achenes hairy-villous.

HAB. Kamiesberg, *Drege!* (Herb. Sond.)
Leaves 2 inches long, 2–3 lines wide.

4. D. longipes (Harv.);

shrubby, glabrous, erect; branches terete, ramulous; leaves linear-lanceolate, fleshy, quite entire, tapering at base, acute, almost nerveless; pedicels terminal and subterminal, slender, 2–3 times as long as the leaves; inv. scales 5–6, linear-oblong, subacute, with narrow, membr. edges; marg. achenes (immature) thinly silky?

HAB. Modderfontein, *Rev. H. Whitehead!* (Herb. D.)
Apparently a much-branched shrub. Leaves 1–1½ inch long, 2–3 lines wide, shrinking when dry. Pedicels 3–4 inches long, pale. Inv. sc. 7 lines long, 1½ line wide. The young achenes are compressed, and scarcely silky. Nearly allied to *D. cneorifolia,* but with different inflorescence, &c.

5. D. cluytiæfolia (DC.! l. c. 470);

shrubby, glabrous, erect, much branched; branches terete, flexuous, pale; leaves sessile, half-clasping and subdecurrent, obovate or oblong, obtuse, quite entire, rigid, nerveless, glaucous; pedicels terminal, as long or twice as long as the leaves; inv. sc. 5–7, broadly oblong or ovate, acute, concrete at base; marg.-achenes thick, very hairy.

HAB. Sneeuweberg and Nieuweveld, near Uitvlugt, *Drege!* (Herb. Hk., Sd.)
A very rigid, pale, glaucous bush. Leaves ¾ inches long, 4–6 l. wide, immersedly 3-nerved. Pedicels ½–1 inch long. Inv. scales (fide DC.) sometimes 9; they are commonly but 5, rarely 6 or 7.

6. D. rigida (Harv.);

shrubby, glabrous, divaricately much branched, the branches terete, often spine-pointed; leaves coriaceous, oblong-obovate, obtuse or mucronulate, glaucous, nerveless, some entire, others with two or few small, acute, lateral teeth; pedicels terminal, a little longer than the leaves; inv. scales 5, acute, concrete at base, with membranous edges; marg. achenes very hairy. *Cacalia rigida, Thunb.! Cap. p.* 624. *Doria spinescens, DC. l. c.* 470. *Kleinia rigida, DC. l. c.* 338.

HAB. Cape, *Thunberg!* Sneeuweberg, *Drege.* (Herb. Thunb.)
A very rigid, scrubby bush. the old twigs ending in spines. Leaves ¾ in. long, 3–4 lines wide, frequently bidentate near the apex. I have not seen *Drege's* plant, but DC.'s description agrees well with *Thunberg's* specimen.

7. D. ciliata (Harv.);

rigidly suffruticose, spreading, much branched; branches pale, angular; leaves sessile, half-clasping, cuneate-oblong, obsoletely 3–5-toothed at the extremity, cartilagineo-ciliolate, coriaceous, rigid, faintly three-nerved, glaucous (turning black when dry); pedicels short, one-headed, terminal or opposite the leaves; inv. scales 4–5, broadly oblong, connate at base, with membr. edges, acute; marg. achenes *glabrous* or nearly so (*ripe* not seen).

HAB. Vet Rivier, *Burke & Zeyher! Zey.!* 995. (Herb., Hk., Sd.)
Closely much branched, leafy; the pale twigs very rigid, but imperfectly ligneous.

Leaves 1–1½ inch long, 4–5 lines wide, with a deltoid extremity ; in the specimens seen turning quite black in drying. Pedicel shorter than the leaves. Inv. sc. 4–5 lines long, 2 l. wide.

8. D. Kraussii (Sch. Bip. in Bot. Zeit. xxvii. p. 769); half-shrubby, erect or diffuse, much-branched, glabrous; branches terete, flexuous; leaves obovate, obtuse, decurrent at base into a *narrow,* vanishing stem-wing, coriaceous, faintly 3-nerved, some quite entire, some few-toothed, callous-margined; pedicels sub-terminal, solitary or 2–3-together, one-headed, 2–8 times as long as the leaves; inv. scales 6–8, oblong, con-crete at base, membrane-edged, subacute; achenes silky-villous. *D. alata,* *DC. l. c. ex pte. Drege! litt. a. & b. Othonna frutescens, Eckl. (non L.)*

HAB. Aasvogelberg and Kopie, near mouth of Zwartkops R., *Drege! Zey.!* 3032. Uitenhage and Kaffirland, *Ecklon!* (Herb. D., Hk., Sd.)

Very similar in habit and foliage to *D. alata,* with which it was associated by *De Candolle;* but from which it differs in inflorescence, larger fl. heads, longer pedicels, more numerous and less connate inv. scales, and narrower stem-wings. Invol. scales 4 lines long. *DC.'s* description refers to *both* plants.

9. D. alata (Thunb. Cap. 673); half-shrubby, erect or diffuse, much branched, glabrous; branches terete, flexuous; leaves obovate, obtuse, strongly decurrent at base into a vanishing stem-wing, coriaceous, vein-less, entire, callous-margined; heads in subsessile, branching corymbs, the pedicels swollen upwards; inv. scales about 5, oblong, concrete to their middle, subacute; achenes densely hairy. *Less. Syn. p. 69. DC. l. c. (ex pte.), p.* 470.

HAB. Cape, *Thunberg!* Hills about Eland's R., Uitenhage, *Zey.!* 3033. (Herb. Th., D., Sd., Hk.)

Scrubby, divaricately branched, ligneous below, half-ligneous above; the bark dark-coloured. Leaves ¾ inch long, 3–6 lines wide, undulate, decurrent for 4–5 lines below the base as a broadish wing. Heads on short clavate pedicels, in a branching 10–20-headed corymb. Inv. sc. 3 lines long. *Zeyher's* specimens quite agree with *Thunberg's.*

10. D. nana (DC. l. c. 470); "glabrous; stem shrubby-carnose, dwarf, branched; branches terete, thick, leafy at the summit only; leaves somewhat tufted, sessile, obovate, tapering at base, obtuse, thickish, nerveless, serrulate or sublobulate at the margin; pedicels terminal, solitary, ⅓ as long as leaves; inv. sc. 5, ovate, concrete at base." *DC. l. c.* (Achenes not seen.)

HAB. Karroo, *Drege.* (Unknown to me.)

11. D. Ceradia (Harv.); shrubby-carnose, glabrous; branches repeat-edly forked, leafy at the tips; leaves tufted, oblongo-spathulate, obtuse or retuse, coriaceous, rigid, tapering at base into a 3-nerved petiole, obsoletely veiny; ped. 1 or several, terminal, 1-headed, equalling the leaves; invol. scales 5–6, oblong, acute, concrete at base only; achenes silky. *Ceradia furcata, Lindl. Bot. Reg.* 1845, *Pl. misc. p.* 12. *Walp. Rep.* 6, *p.* 727.

HAB. On the island of Ichaboe. (Herb. Hk., sp. cult. ?)

A fleshy shrub, with the general aspect and foliage of *Othonna arborescens.* In gen. char. it precisely agrees with other species of *Doria.*

12. D. Arbuscula (DC. l. c. 470); " shrubby-carnose, erect, di-trichotomous; branches terete; leaves lanceolate, flat, the uppermost obovate; pedicels 1–2-headed, terminal; inv. cylindrical, of 5 obtuse scales, concrete at base; achenes glabrous." *DC. l. c. Cacalia Arbuscula, Th.? Cap.* 624.

HAB. Roggeveld, *Thunberg?* Clanwilliam, *Ecklon.* (Herb. Th. ?)
I have not seen *Ecklon's* plant. In *Thunberg's* Herbarium are two specimens marked " *Cacalia arbuscula;*" that marked No. 2, which has no leaves, may belong to this; but No. 1, is a *Senecio*, and probably *S. aloides, DC.*

13. D. retrofracta (Less. Syn. 89); " shrubby-carnose, erect, glabrous; branches divaricate or bent back; leaves oblong-lanceolate, subobtuse, cuneate at base, in the middle on each side with a single large tooth or lobe; pedicels 1-headed, axillary, scarcely twice longer than leaves; inv. 5-parted; marg. achenes glabrous." *Othonna retrofracta, Jacq. Schoenbr. t.* 376.

Probably a garden variety of *D. Arbuscula.*

14. D. abrotanifolia (Harv.); glabrous, shrubby-carnose; twigs 4-angled, leafy at the extremity only; leaves tufted, petioled, *pinnato-multifid*, the pinnæ in few pairs, filiform, once or twice forked or multifid, the lobules slender, acute, the uppermost sometimes cuneate; pedicels terminal, 1 or more, equalling or scarcely exceeding the leaves; inv. scales 5, oblong, acute, membr.-edged, scarcely connate at base; achenes longitudinally striate, hispidulous. *Zey.?* 949.

HAB. Springbokfontein, Namaqualand, *Rev. H. Whitehead!* (Herb. D.)
A small, fleshy, glaucous bush, bearing tufts of leaves at the ends of the twigs. Leaves 1–1½ inch long, as much in expansion, divided, in a way between pinnate and dichotomous, into many filiform or very narrow, fastigiate lobes and lobules. Petiole springing from a hard white disc, channelled. Pedicels 1–1½ inch long. *Zey.!* 949 (Hb. Hk.) is either the same, or an allied species with nearly similar leaves; it is in too imperfect a condition for determination.

15. D. carnosa (DC. l. c. 471); herbaceous, glabrous, with tuberous root; stem *fleshy*, simple or branched, with woolly axils; leaves fleshy, sessile, stem clasping, narrowed toward the base, pinnatifid, the lobes on each side 2–3, oblong, entire, acute; pedunc. terminal, 1–3–5-headed, the heads pedicellate; inv. sc. 4–6, broadly oval, obtuse, scarcely connate at base; marg. achenes glabrescent, their pappus not longer than the invol., fulvous.

HAB. Grassy hills in Uitenhage, *E. & Z.!* Adow, *Zey.!* 3034. Hex R., *Drege.* (Herb. D., Hk., Sd.)
Tuber long, oval, or pear-shaped. Stem 3–6 inches high, simple or 2–3 times forked, thick, shrinking in drying. Leaves closely set, 2–3 inches long, 3–4 lines wide, the lobes 2–6 lines long. Pedunc. slender, longer than the leaves, *mostly* bearing 2–5 heads, rarely 1-headed.

16. D. eriocarpa (DC. l. c. 471); herbaceous, glabrous, with tuberous root; stem fleshy, simple or branched, with woolly axils; leaves fleshy, sessile, stem-clasping, tapering at base into a petiole, pinnatifid, the lobes on each side 2–3, oblong, acute; pedunc. terminal, corymbose, 3–5 headed, heads long-pedicelled; inv. sc. 5, oval oblong, connate at base, subacute; achenes hirsute, the pappus twice as long as the inv., fulvous.

HAB. Kendo; also at Ado, *Drege.* Zwartkops Zoutpan, *E. & Z.!, Zey.!* 3035. (Herb. D., Sd., Hk.)

Very similar in habit and foliage to *D. carnosa,* but differing in the very hairy achenes and the long, silky pappus.

17. D. taraxacoides (DC. l. c. 471); stemless, very woolly at the crown and in the axils; radical leaves glabrous, tapering into a broad petiole, obovate or spathulate, on each side bluntly 3-5 lobed or crenate; pedunc. scapelike, twice as long as the leaves, one-headed; inv. scales 13-14, concrete below, lanceolate, acute, membr.-edged; achenes densely hairy; pappus short, straw-colour.

HAB. Betw. Zilverfontein and Kaus, Namaqualand, *Drege!* (Herb. Hk.)

Crown of root thick. Leaves many, rosulate, 2 inches long, including the petiole. Pedunc. 3-4 inches long. Inv. campanulate, many-fl.; its scales connate nearly to their middle.

18. D. digitata (Less. Syn. 89); herbaceous, glabrous, or below pubescent, with tuberous root; stem short, simple or branched, with wool-tufts in axils of branches and leaves; branches sparsely leafy, ending in long, nude, one-headed peduncles; leaves polymorphous, all the cauline from broad, stem-clasping bases, the lowest tapering at base into a petiole, cuneate or oblong, or lanceolate, coarsely 3-5-lobed, or toothed, or entire; upper sessile, coarsely toothed or pinnatifid; heads many-fl.; inv. sc. 8-10, oblong, acute; pappus rufous; achenes closely silky. *DC. l. c.* 471. *O. digitata, Linn., Th.! Cap.* 720.

VAR. β. **lanceolata**; root leaves lanceolate, quite entire or here and there toothed or lobed. *D. campanulata, DC.! l. c.* 473. (Hb. Sd.)

HAB. Cape Flats, Blauweberg, Piquetberg, Hott. Holl., &c. *E. & Z.! Drege! W.H.H.* Var. β. Piquetberg, *Zeyher!* Table Mt., *Ecklon!* (Hb. Th., D., Hk., Sd.)

Bulb ovate, woolly at the crown. Root leaves 4-6 inches long, most variable, generally more or less toothed or lobed. The lower parts of stem and petioles are sometimes closely pubescent. Upper leaves oblong or spathulate, with very broad, clasping bases, incised or entire. Pedunc. 6-8 inches long. *D. campanulata,* DC., according to *Ecklon's* specimens in Herb. Sd., does not differ from our var. β., which I had brought hither before I knew it was DC.'s plant; it exists in Hb. Thunb also.

19. D. perfoliata (Th. Cap. 663); herbaceous, glabrous; root tuberous; stem elongate, branched, often climbing; radical leaves petioled, cordate or ovate; cauline cordate-amplexicaul, with ample basal lobes, oblong or lanceolate, or taperpointed, acute or acuminated, entire; pedunc. terminal or axillary, one-headed; inv. sc. 10-12, lanceolate, acute, connate at base; achenes closely silky; pappus pale, not much longer than the involucre. *DC. l. c.* 472. *Less. Syn.* 89. *Cineraria perfoliata, Lam.*

VAR. β. **angustifolia**; cauline leaves very narrow and much acuminate; whole plant smaller. *D. diversifolia, DC. l. c.* 472. (Herb. Hk.)

HAB. Sandy places on the Cape Flats, Langekloof and Olifant's River, *Drege!* Saldanha Bay, Worcester and Stellenbosch, *E. & Z.!* Kamp's Bay, *W. H. H.* β. Tigerberg, Blauweberg and near Konstapel, *Drege!* (Herb. Th., D., Sd., Hk.)

Stem 2-3 ft. high, supported among shrubs. Cauline leaves 2-2½ in. long, ½-1 in. wide; in β. narrow and smaller. Pedunc. 2-8 inches long. This has the habit of *Othonna amplexifolia,* but wants the rays.

20. D. Lingua (Less. Syn. 89); herbaceous, glabrous, often branched, erect or climbing ; root tuberous ; radical leaves petioled, ovate or lanceolate, entire; cauline cordate-amplexicaul, oblong or lanceolate, acute or acuminate ; pedunc. terminal and axillary, one-headed ; achenes hispidulous; pappus rufous or purplish. *DC. l. c.* 471. *Othonna Lingua, Jacq. Hort. Schœnbr. t.* 238.

VAR. *a.* **Jacquini**; cauline leaves lanceolate-taperpointed. (Hb. Sd.)

VAR. *β.* **latifolia**; cauline leaves oblong or ovate, or ovato-lanceolate. (Herb. D., Hk., Sd.)

HAB. Cape Flats and Western Districts, *a. Zeyher! β. E. & Z.! W.H.H., &c.*
Like *O. perfoliata*, but with reddish pappus and less pubescent achenes. The cauline leaves are, I fear, too variable to build a character upon ; but our vars. *a.* and *β.* look different in the specimens examined.

21. D. chromochæta (DC. l. c. 472); "herbaceous, glabrous, erect, much-branched; stem striate; cauline lvs. cordate-amplexicaul, broadly ovate, acutely mucronate, 3-nerved, entire; fl. branches forked, pedicels nude, one-headed; inv. sc. 10–12, concrete at base; ray-achenes glandular, minutely papulose, glabrous. *DC. l. c.*

HAB. Kl. Namaqualand, *Drege!* (Herb. Sd.)
"Pappus copious, dull yellowish-violet colour." I have only seen a fragment of this ; it differs from *D. perfoliata* chiefly in its coloured pappus and glabrous achenes.

22. D. linearifolia (DC. l. c. 471); "herbaceous, glabrous, woolly on the crown and in the axils ; root tuberous ; stem erect, simple, naked and one-headed at the apex ; leaves linear, quite entire, somewhat shorter than the stem ; marg. corollas tubular, truncate, very short ; achenes glabrous." *DC. l. c.*

HAB. Cape, in Hb. L'Her., fide DC. (Said to resemble *Othonna linifolia.*)

23. D. Laureola (DC. l. c. 472); herbaceous, glabrous, erect ; stem striate ; cauline leaves sessile, lanceolate, acute, somewhat shining, quite entire ; pedicels from the axils of the upper leaves, longer than the leaf, one-headed, in a simple corymb ; inv. sc. 10–12, connate at base, oblong ; marg. achenes striate, glabrous.

HAB. Mierenkasteel, *Drege!* (Herb. Sond.)
Of this I have seen only a leaf, and 2 fl. heads ; leaves 1½ in. long, 4–5 lines wide. Pedic. 2–2½ in. long.

24. D. ramosa (DC. l. c. 472); "herbaceous, erect, branched, glabrous, glaucescent ; crown woolly ; branches striate ; leaves sessile, obtusely eared and half-clasping at base, lanceolate, acute, subundulate, entire ; pedicels 1–3, at the ends of the branches, longer than the leaf ; inv. about 12-cleft ; marg. achenes pubescent." *DC. l. c.*

HAB. Hexriviersberg, *Drege.* (Unknown to me.)
"A foot high. Lvs. 2 in. long, 6 l. wide. Pappus whitish." A specimen from *Zeyher* (Hb. Sd.) from the Langevalley, seems very near this, but is unbranched and has fewer inv. scales. I forbear to describe it as different, and am not sure whether it be the same.

25. D. tortuosa (DC. l. c. 472) ; "herb., erect, branching, twisted ; collum and roots tuberous; stem compressed, subangular; leaves sessile, linear-lanceolate, the uppermost obtusely somewhat eared ; pedicels

2–3, at the ends of the branches; inv. ovate at base, 12-fid, the lobes acute; marg. achenes pubescent, their pappus shorter than the involucre; pappus of the disc-fl. of few bristles." *DC. l. c.*

Hab. Mierenkasteel, *Drege.* (Unknown to me.)
"6 inches high. Leaves 12–15 lines long, 2–3 l. wide. Fl. heads 4–5 l. long and wide. Pappus rufous." *DC. l. c.*

26. D. semicalva (DC. l. c. 473); "herbaceous, erect, weak, glabrous; branches striate; leaves sessile, linear-lanceolate, at each end tapering, quite entire; heads axillary, 2–3 on long pedicels; inv. 10–12-fid; [ray] achenes terete, closely silky-villous; those of the disc without pappus." *DC. l. c.*

Hab. Langevalley and Olifant's R., *Drege.* (Unknown to me.)

27. D. Gymnodiscus (DC. l. c. 473); herb., erect, glabrous; woolly at the crown; branches terete; cauline leaves sessile, round-eared and clasping at base, oblong or obovate, subacute, quite entire; branches 1–2-headed, heads ½ as long as the pedicels; inv. 10–12-fid; marg. achenes thick, terete, striate, closely silky, their pappus copious, reddish, twice as long as the involucre; disc-achenes without pappus, or with 1–2 bristles only.

Hab. Near Port Elizabeth, *E. & Z.!* (Herb. Sond.)
Very similar to small specimens of *D. Lingua,* but easily known by the calvous disc-achenes; I sometimes find one or two bristles on the outer ones.

(Imperfectly known.)

D. miser (Harv.); glabrous, stem shrubby-carnose, dichotomously much-branched, twiggy; leaves tufted at the ends of the twigs *(mature-leaves not seen),* obovate? tapering at base, lobulate? fleshy; pedicels terminal, very short, solitary; inv. sc. about 5, narrow-oblong, acute, connate to the middle; achenes closely-silky.

Hab. Rhinosterkop, Beaufort, *Burke & Zey.! Zey.!* 988. (Herb. Hk., Sd.)
A small, intricately-branched, fleshy bush; twigs forking, each prong ending in a tuft of leaves. Form of leaves uncertain. Pedicels ½ inch. Inv. 4–5 lines long, cylindrical, few-flowered. Seems to be near *D. Arbuscula,* but the achenes are silky.

D. undulosa (DC. l. c. 472); "herbaceous, glabrous; root?; stem terete, flexuous; leaves cordate, amplexicaul, acuminate, sinuate, undulate, the uppermost suborbicular, dentato-sinuate; pedicels nude, one-headed; marg. cor. truncate, very short; achenes multistriate, subpuberulous." *DC. l. c.*

Hab. Hexriviersberg, *Drege.* (Unknown to me.)

CXV. OTHONNA, Linn.

Heads radiate; *ray-flowers* ligulate, female; *disc-fl.* male, tubular, 5-toothed. *Recept.* convex or subconical, honey-combed, sometimes pilose. *Inv.* sc. uniseriate, never calycled, more or less concrete before opening, and strictly valvate. *Style* of male-flowers simple, tipped with a hispid cone. *Anthers* rounded at base. *Achenes* of the ray-fl. oval, hairy or glabrous, with very copious, bristle-shaped pappus in many

rows; of the disc-fl. abortive, slender, glabrescent, with uniseriate
pappus. *DC. Prodr. 6, p.* 473.

Small shrubs or herbaceous plants, chiefly S. African, mostly glabrous, often
glaucous. Roots in the herbaceous species often tuberous. Leaves entire or variously
cut, lobed or toothed; membranous, coriaceous, or fleshy. Heads either solitary or
corymbose. Name used by Pliny for some similar plant.

§ 1. FRUTICOSÆ. Rigid, ligneous or subligneous, branched and twiggy shrubs
or suffrutices : fl. branches mostly terminal and one-headed ; heads rarely subcorymbose. Rays yellow. (Sp. 1–11.)

Leaves pinnate-lobed ; the lobes narrow linear-acute :
    Heads loosely corymbose, on long pedicels; inv. sc. 5   (1) **Pavonia.**
    Heads on simple, long peduncles ; inv. sc. 8 ... ... (2) **leptodactyla.**
Leaves lanceolate, or linear-oblong or linear, acute or obtuse,
  entire or irregularly toothed, sessile :
    Achenes of ray-fl. *silky* or *pubescent:*
      Leaves lanceolate, often toothed, *tapering at base,*
        thick, nearly nerveless ; inv. sc. 7–8 ... ... (3) **coronopifolia.**
      Lvs. oblong or lance-oblong, rigid, one-nerved ;
        sc. 7–8 ... ... (4) **ramulosa.**
      Leaves linear, coriaceous, sub-obtuse, nerveless ;
        inv. sc. 5 ... ... (9) **pallens.**
    Achenes of ray-fl. *glabrous:*
      Spiniferous; lvs. linear, narrowed at base, nerve-
        less, glaucous ... ... (7) **spinescens.**
      Unarmed ; lvs. lanceolate, 3-nerved, entire or
        unidentate... ... ... (6) **trinervia.**
  Lvs. *minutely petiolate,* elliptic-oblong, obtuse ; twigs *spiny* (8) **pteronioides.**
  Lvs. obovate, or obovate-oblong, sessile or subpetioled :
    Lvs. obovate-oblong, *cusp.-mucronate,* 1-nerved, often
      toothed at base... ... ... (5) **mucronata.**
    Lvs. obovate, tapering into a petiole, glaucous, entire
      or 3-toothed ... ... (10) **osteospermoides.**
    Lvs. obovate, sessile, entire, 3–5-nerved, rigidly cori-
      aceous ... ... (11) **coriifolia.**

§ 2. SUFFRUTICOSÆ. Trailing or decumbent or diffuse suffrutices, lignescent at
base, the subsimple branches ascending, ending in long, one or few-headed peduncles. (Sp. 12–14.)

Leaves obovate, bluntly pinnatifid or subentire, copiously
  ciliate ... ... (12) **ciliata.**
Lvs. obovate, quite entire, or sparsely ciliolate ... ... (13) **frutescens.**
Lvs. linear or lance-linear, thick, entire or toothed ... (14) **multicaulis.**

§ 3. PANICULATÆ. Erect suffrutices or herbs. Stems virgate, simple and leafy
below, above more or less nude, ending in a panicled or much-branched, corymbose
inflorescence. (Sp. 15–21.)

Achenes of ray-fl. silky or villous :
  Lvs. *coriaceous, thick, veinless,* either cuneate-obovate,
    lanceolate, linear-lanceolate, or linear-elongate, en-
    tire or sharply few-toothed ... ... (15) **parviflora.**
  Lvs. *obviously midribbed and penni-nerved,* oblong or obovate, *broad :*
    Infl. loosely panicled or subumbellate ; bracts small and narrow :
      Leaves cuneate-oblong, half-clasping, obso-
        letely 5-toothed near the apex ... ... (16) **quinquedentata.**
      Lvs. oblongo-lanceolate, clasping, several-
        toothed beyond the middle - ... ... (17) **pluridentata.**
      Lvs. obovate, *obtuse,* sharply many-toothed,
        subpetiolate ... ... (20) **umbelliformis.**

Infl. densely corymbose ; bracts broad ; leaves
broadly obovate, cordate and ear-clasping at
base, entire or denticulate　...　...　...　...　(19) **amplexicaulis.**
Achenes of ray-fl. glabrous :
　Lvs. elliptical or obovate, subpetiolate; bracts broad　(18) **membranifolia.**
　Lvs. linear-terete, acute, quite entire　...　...　...　(21) **viminea.**

§ 4. CARNOSÆ. More or less fleshy or succulent shrubs, half-shrubs, or subherbaceous plants. Stems irregularly branched, the older parts commonly nude and
cicatricised. Leaves often crowded at the ends of the branches. Pedunc. subterminal, one or several-headed. (Sp. 22–35.)

Leaves linear or oblong, *fleshy*, semiterete or terete, nerveless and veinless :
　Stems shrubby and fleshy ; lvs. crowded at end of branches ; ped. corymbose :
　　Lvs. linear-elongate, 2–3 inches long, 1 line wide　(22) **cylindrica.**
　　Lvs. cylindrical, 1–2 inches long, 1½–2 l. diam.　(23) **carnosa.**
　Stems slender ; lvs. whorled at intervals, 1 in. long,
　　3 lines diameter　...　...　...　...　...　(24) **crassifolia.**
　Stems slender ; leaves opposite or alternate, ovoid or
　　shortly cylindr.; pedunc. mostly 1-headed　...　...　(25) **sedifolia.**
Leaves laterally lobed or pinnatifid :
　Leaves fleshy, obtuse, on each side 1–2-lobed ...　...　(26) **obtusiloba.**
　Lvs. subpetioled, pinnatifid, lobes on each side 3–4,
　　mucronate　...　...　...　...　...　...　...　(27) **quercifolia.**
Leaves obovate or oblong, entire or toothed :
　Pedunc. corymbose, several-headed : rays yellow :
　　Leaves tapering into a slender petiole, entire, 3-
　　　nerved from the middle ...　...　...　...　...　(28) **triplinervia.**
　　Lvs. sessile, toothed or subentire, faintly nerved,
　　　thickish　...　...　...　...　...　...　...　(29) **dentata.**
　　Leaves oblong, narrowed at base, fleshy, 1 inch
　　　long, 2–3 lines wide ; pappus rufous　...　...　(30) **macrosperma.**
　Pedunc. mostly forked, 2-headed ; rays purplish ;
　　leaves glaucous...　...　...　...　...　...　...　(31) **purpurascens.**
　Pedunc. simple, 1-headed, or rarely forked ; rays yellow :
　　Shrubby-carnose, much branched, 2–3 feet high ;
　　　pedunc. long　...　...　...　...　...　...　(32) **arborescens.**
　　Shrubby-carnose, branching ; leaves coriaceous,
　　　with *recurved margins*　...　...　...　...　(34) **cacalioides.**
　　Dwarf ; leaves somewhat 3-nerved, denticulate ;
　　　pedunc. very short ...　...　...　...　...　(33) **primulina.**
　　Very dwarf, unbranched ? woolly at crown ; lvs.
　　　petioled, cuneate at base, obovate, sharply
　　　toothed along the rounded apex　...　...　...　(35) **minima.**

§ 5. SCAPIGERÆ. Rootstock or caudex short, simple or multifid, woolly at the
crown. Leaves from the crown, numerous, rosulate. Peduncles scape-like, simple
and one-headed, or sub-corymbose, several-headed. (Sp. 36–41.)

Pedunc. several-headed ; lvs. oblongo-lanceolate, tapering
　at base, coriaceous, entire　...　...　...　...　(36) **Natalensis.**
Pedunc. one-headed ; leaves linear-lanceolate, coriaceous,
　quite entire　...　...　...　...　..　...　...　(39) **scapigera.**
Pedunc. several-headed ; lvs. spathulate, netted-veined, *ciliate:*
　Inv. sc. and rays about 8　...　...　...　...　...　(37) **retrorsa.**
　Inv. sc. and rays about 5　...　...　...　...　...　(38) **Zeyheri.**
Pedunc. 1-headed ; lvs. variably pinnatifid or subentire,
　or the lower petioled, cuneate-obovate ; fl. red or purple　(40) **auriculæfolia.**
Pedunc. rigid, tall, one-headed ; leaves lyrato-pinnatifid,
　the lobes in several pairs, lobulate and toothed ; fl. yellow ?　(41) **sonchifolia.**

§ 6. CAULESCENTES. Root mostly (or always ?) tuberous. Crown of the root
mostly woolly. Radical leaves (when present) few, petioled. Stem simple or

branched, alternately leafy; cauline leaves sessile or petioled.   Pedunc. axillary and
terminal, simple or branched.   (Sp. 42–58.)

* Ray and disc. fl. red or purple :
 Leaves pinnati-partite, the lobes few, linear, obtuse,
  entire ... ... (42) **cakilefolia.**
 Lvs. oblongo-lanceolate, coarsely toothed, teeth deltoid (43) **incisa.**
 Lvs. very glaucous, elliptic-oblong, entire ... ... (44) **rosea.**
** Ray and disc. fl. yellow :
 Rad. lvs. linear-attenuate or linear-lanceolate, entire (48) **linifolia.**
 Rad. lvs. (if any) roundish, ovate or ovato-lanceolate,
  entire or crenato-lobulate :
  Stem elongate, laxly leafy, branching, erect or climbing :
   Stem rigid ; invol. sc. 12–13 ; achenes rufo-
    pubescent ; pappus rufous ... ... (45) **rufibarbis.**
   Stem weak ; inv. sc. 9–10 ; achenes minutely
    hispid ; pappus whitish :
    Cauline leaves broadly cordate-ovate,
     acute or obtuse ... ... (46) **amplexifolia.**
    Cauline lvs. cordate at base, lanceolate,
     acuminate... ... ... (47) **filicaulis.**
  Stem short, flexuous, closely leafy, subsimple :
   Rad. lvs. ovate or ovato-lanceolate, acute,
    entire, glabrous ; cauline lvs. spathulate
    or obovate ; pedunc. 1-headed ... ... (49) **tuberosa.**
   Rad. lvs. broadly ovate or obovate, obtuse,
    repand or crenate, veiny beneath, often
    pilose ; cauline obovate, crenate ; pedunc.
    one-headed ... ... (50) **heterophylla.**
   Rad. lvs.? cauline *very large*, stem-clasping,
    ovate-oblong, acute or acuminate, subden-
    tate ; pedunc. branched, 8–10-headed ... (51) **macrophylla.**
   Rad. lvs. oblong, sinuate or crenate or erose,
    callous-toothed ; cauline tapering at base,
    subpetioled, obovate ; pedunc. branched,
    few-headed ; inv. sc. 8 ... ... (52) **petiolaris.**
  Rad. and lower cauline lvs. broadly cuneate, coarsely
   toothed at the apex; pedunc. branched, few-headed ;
   inv. sc. 7–8 ... ... (53) **cuneata.**
  Rad. lvs. pinnatifid, pinnati-partite or lyrate :
   Rad. and lower lvs. sinuato-pinnatifid, *netted-
    veined* ... ... (54) **reticulata.**
   Rad. lvs. pinnatifid, lateral lobes 4–5, term. lobe
    roundish, sinuate ... ... (55) **rotundifolia.**
   Rad. and lower lvs. *lyrate*, stem clasping ; lat.
    lobes 2–3, roundish, terminal very large,
    roundish-cordate, toothed or lobed ... ... (56) **lyrata.**
   Rad. lvs. *pinnate-parted ;* segments in several pair, rigid, roundish, oblong
    or lance-oblong, entire :
    Branches pedunculoid, one-headed ; inv. sc.
     12–13 ... ... (57) **pinnata.**
    Branches 2–4-headed ; inv. sc. about 8 ... (58) **tephrosioides.**

§ 1. FRUTICOSÆ. (Sp. 1–11.)

1. O. Pavonia (E. Mey.!); shrubby, erect, branching, glabrous; stem
terete; leaves sessile, pinnatipartite, the lobes linear, thick, acute, entire
or forked; corymbs peduncled, terminal and lateral, loosely few-headed,
pedicels long; inv. scales 5, oblong, concrete at base; rays 5; achenes
densely hairy. *DC. l. c.* 473. *Zey.! Fl. Cap. No.* 501, 145.

HAB. Klipplaat River, *Drege!* Graaf Reynet, *Zey.!* Brack R., Cradock, *Burke and Zey.! 990.* (Herb. D., Hk., Sd.)

1–2 feet high, with purplish stems and pale pedunc. and foliage. Leaves 1–1½ inch long, their lobes 6–8 lines long, 1–¾ line diameter. Pedunc. straw-colour, 2–3 times forked; pedicels 1½–2½ inches long.

2. O. leptodactyla (Harv.); shrubby, erect, glabrous; branches angular, striate, virgate; lateral twigs laxly leafy, ending in short or longish one-headed peduncles; leaves dimorphous, some linear-elongate, acuminate, quite entire, others at base on each side closely pinnated with 3–4 linear-subulate lobes, with a very long, linear-subulate, medial lobe; inv. lobes and rays about 8; achenes not seen. *Oth. frutescens?* *Drege! in Herb.*

HAB. Betw. Kromrivier, by Piquetberg and Bergvalei, *Drege!* (Herb. Hk.)

This has something the habit of *O. ramulosa*, but longer and more distantly leafy lateral twigs, and very dissimilar foliage. Most of the leaves are pinnate-lobed at base; a few here and there quite entire, 1–1½ inch long, ½ line wide. The pinnate portion is 3–4 lines long; each lobe 3–9 lines, and very patent; the terminal lobe 1–1¼ inch long. Heads as in *O. ramulosa*.

3. O. coronopifolia (Linn. Sp. 1310); stem shrubby, erect, glabrous, flexuous or forked; leaves lanceolate or linear-lanceolate, entire or irregularly few-toothed, tapering at base, acuminate, coriaceous, thick, nearly nerveless; pedunc. ending the branches, nude, one-headed, or branched and few-headed, the heads on very long pedicels; inv. sc. 7–8, connate at base, oblong; achenes silky-villous. *Th.! Cap. 719.* *DC.! l. c. 475. O. Bergeri, Spr.*

HAB. Cape, *Thunb.! Boivin,* 615. Near Capetown, *E. & Z.!* Salt River, W.H.H. Piquetberg, Olifants R., and Kl. Namaqualand, *Drege!* (Herb. Th., D., Sd., Hk.)

1–2 ft. high, bushy. Leaves 1½–2½ in. long, 3–5 l. wide, some entire, some with small, some with large lateral teeth. Pedunc. 3–6 inches long, mostly simple. Rays many-striate. There are 4 folia in Herb. Thunb.; of these, No. 1 belongs to *O. arborescens;* No. 2, 3, 4, to *O. coronopifolia.*

4. O. ramulosa (DC.! l. c. 477); shrubby, erect, glabrous, the axils woolly; branches subangular, virgate; lateral twigs leafy at base, ending in short or longish, nude, one-headed peduncles; leaves sessile, oblong or lance-oblong, one-nerved, subpungent mucronate, thick, rigid, quite entire or 1–2-toothed; inv. lobes and rays about 8; achenes puberulous.

HAB. Kendo and Konstapel, *Drege!* Driefontein, *E. & Z.!* Pinaars Kloof, *Burke & Zeyher! Zey.! 991.* (Herb. D., Hk., Sd.)

2–3 feet high, with very rigid stems and foliage; branches 1–2 ft. long, alternately leafy, with axillary, short, flowering twigs. Leaves ½–¾–1 in. long, 2–3 lines wide, either broad at base, or the lower ones tapering, and almost lanceolate. Pedunc. ½–1½ in. long. I have not seen ripe achenes.

5. O. mucronata (Harv.); shrubby, erect, glabrous, the axils woolly; branches subangular, virgate; lateral twigs leafy at base, ending in nude, simple or forked, 1–2-headed peduncles; leaves sessile, obovate-oblong, cuspidato-mucronate, narrowed and often sharply unidentate at each side at base, one-nerved, rigid, entire; inv. lobes and rays about 8; achenes pubescent.

HAB. Langvalley, *Zey.!* 998. (Herb. Hk., Sd.)
Very near *O. ramulosa*, but with much broader, and differently shaped leaves. Leaves 1 in. long, 5–6 l. wide, probably glaucous. Pedunc. 3–5 inches long.

6. O.? **trinervia** (DC. l. c. 477); " shrubby, glabrous; branches striate with lines decurrent from the leaves; leaves lanceolate, cuneate at base, 3-nerved, mucronate, entire or sharply unidentate at each side; pedicels subracemose at the apices of the branches, thrice as long as the leaves; inv. 7–8-lobed; [*ray-fl. unknown*]; achenes glabrous." *DC. l. c.*

HAB. Carroo, *Drege.* (Unknown to me).
DC. is uncertain whether this may not prove to be a *Doria*.

7. O. spinescens (DC. l. c. 476); shrubby, glabrous; branches terete, whitish, divergent, mostly spine-tipped; leaves sessile, linear, or linear-cuneate, tapering at base, mucronate, glaucous, nerveless, thick, entire, [or here and there subdentate]; fl. twigs attenuated upwards, subnude, one-headed; inv. lobes and rays about 8; ray achenes glabrous. *DC. l. c., excl., var. β.*

HAB. Betw. Droogeheuvel and Jackal'sfont., *Drege.* (Herb. Hk.)
Specimens from *Drege* (Herb: Sond.) marked " *O. spinescens a.* and *β*." from Sneeuwberg, Compasberg and Rhinosterberg, belong to *Doria rigida ;* except one, which I venture to describe as *O. pteronioides.* In Herb. Hk. is a specimen of the true *O. spinescens*, but the fl. heads have been eaten by insects; and also one of "var β.," which is merely *Doria rigida.*

8. O. pteronioides (Harv.); shrubby, glabrous, divaricately branched, the twigs ending in rigid spines; branches terete; leaves minutely petiolate, elliptic-oblong, very obtuse, entire or obscurely unidentate, thick, nerveless; fl. twigs one-headed; inv. lobes 6–8? ray achenes?

HAB. Cape, *Drege!* (Herb. Sond.)
Mixed with specimens of *Doria rigida.* One ligulate fl. remains to show that this is an *Othonna.* In foliage it differs so remarkably from *O. spinescens, DC.*, that I venture to keep it distinct. The achenes in the only specimen seen have been destroyed by insects.

9. O. pallens (DC.! l. c. 476); shrubby, erect, glabrous, much-branched and twiggy; branches terete, pale; leaves sessile, with a decurrent line from each side of the base, linear, coriaceous, glaucous, smooth, subobtuse, nerveless, quite entire; axils woolly; heads ending the twigs, solitary, very shortly pedicelled; inv. oblong, 5-lobed, lobes acute, membrane-edged; rays 5, very short, scarcely exserted; achenes silky-villous.

HAB. Sneeuwberg, *Drege! Dr. Wallich!* Orange R., *Burke & Zey.!* Queenstown Flats, *Mrs. F. W. Barber*, 229. (Herb. D., Hk., Sd.)
1–2 feet high, rigid, all parts very pale. Leaves ¾–1 in. long, 2 l. wide, slightly concave. Pedicels ¼–½ in. long; inv. 4–5 l. long. Rays inconspicuous. Intermediate in character between *Doria* and *Othonna.*

10. O. osteospermoides (DC.! l. c. 474); shrubby, erect, glabrous; branches terete, flexuous, cicatrised, the twigs leafy; leaves obovate or oblong-obovate, tapering at base into a petiole, either entire or coarsely 3-toothed at the apex, glaucous, coriaceous, glossy; the teeth callous-mucronate; pedicels from the uppermost axils, scarcely longer than the leaves, simple and one-headed, or forked at base, together imperfectly corymbose; inv. sc. 8, concrete below, acuminate, 3-nerved, scarious-edged; rays thrice as long as the inv., 5–8-nerved; achenes thinly pubescent.

HAB. Zwaanepoelspoortberg, 3–4000 ft., *Drege!* (Herb. Hk., Sd.)
A strong, flexuous, branching shrub, with brown bark. Leaves 2–2¼ in. long, 1 in. wide, often quite entire.

11. O.? coriifolia (Sond.); shrubby, erect, glabrous; branches flexuous, angular, closely leafy; leaves sessile, subdecurrent, obovate, entire, rigidly-coriaceous, 3–5-nerved, pale; pedicels from the uppermost axils, bibracteate *(undeveloped);* inv. sc. 5–6, oblong, valvate, barbellate at tips; *flowers not seen.*

HAB. Magalisberg, *Burke & Zeyher. Zey.* 944. (Herb. Hk., Sd.)
Very rigid in all parts, with pale stems and foliage. Leaves 1–1¼ in. long, ½–¾ in. wide, wavy. Infl. undeveloped, and genus therefore uncertain: it may be a *Doria?*

§ 2. SUFFRUTICOSÆ. (Sp. 12–14.)

12. O. ciliata (Linn. f. Suppl. 388); suffruticose, glabrous, decumbent, spreading widely, the fl. branches ascending, ending in long, nude, one-headed peduncles; leaves tapering much at base, oblong, obovate or spathulate, either undivided or (more frequently) inciso-pinnatifid, with short, broad lobes, all ciliate-toothed, the cilia slender, elongate; inv. sc. about 8, oblong, acute, barbellate at top, concrete at base; rays 8, long, broad, 4–5-nerved; ray achenes closely silky. *Th.! Cap.* 721. *DC. l. c.* 474.

HAB. Hills and Mt. sides near Capetown, frequent. Stellenbosch, *E. & Z.!* (Hb. Th!, D., Sd., Hk.)
Stems many from the crown, trailing. several feet long. Leaves 1½–2 in. long, 4–8 lines wide, mostly pinnatifid; the marginal cilia copious, 1–2 lines long. Pedunc. 6–8 in. long.

13. O. frutescens (Linn. Mant. 288); suffruticose, decumbent; branches subsimple, leafy; lower leaves obovate, mucronate, tapering much at base into a cuneate petiole, margined, quite entire or here and there ciliolate, midribbed; upper leaves sessile, oblong or linear, entire or toothed, small, passing upwards into scales; pedunc. terminal, 2–3-headed, pedicels very long; inv. sc. about 8, concrete at base, broadly oval, acute, barbellate; achenes glabrous. *Thunb.! Herb. DC.? l. c.* 478.

HAB. Cape, *Thunberg!* (Herb. Thunb)
Neither *Ecklon's* nor *Drege's* specimens (Hb. Sond.) quoted for this species by *DC.* bear any resemblance to *Thunberg's* plant, here described, which seems to me to have more of the *habit* of *O. ciliata* than of the fleshy species. *E. & Z.'s O. frutescens* is *Doria Kraussii; Drege's* (in Hb. Hk.) is our *O. leptodactyla.*

14. O. multicaulis (Harv.); rootstock woody; stems many from the crown, suffruticose, diffuse-ascending, laxly leafy, prolonged into simple or subsimple peduncles; leaves sessile, linear or linear-lanceolate, acute, entire or sharply few-toothed, coriaceous, thick, veinless and nerveless; pedunc. terminal, one-headed (sometimes emitting a lateral long pedicel from the axil of the uppermost leaf); inv. sc. about 8, oblong, acute; achenes silky.

Hab. Zwarteberg, *Zeyher!* 3048. (Herb. Sond.)
Notwithstanding its apparently distinct characters, I propose this species with much hesitation; the foliage closely resembling that of the narrow-leaved forms of *O. parviflora*, from which the habit and inflorescence separate it. Stems, including the peduncle, about 6 inches long. Leaves 1–1½ in. long, 1–3 lines wide. Heads as in *O. parviflora*.

§ 3. PANICULATÆ. (Sp. 15–21.)

15. O. parviflora (Linn. Mant. 89); suffruticose, erect or ascending, glabrous (or the axils woolly), subsimple ; stem leafy below, pedunculoid and loosely panicled above ; leaves *very variable in shape*, tapering much to the base, subdecurrent, sessile or subpetioled, *coriaceous, thick*, veinless or nearly so, cuneate-obovate, cuneate-lanceolate or linear-lanceolate, either entire or coarsely few-toothed toward the apex ; pedicels long ; inv. sc. 7–8, oblong, acute ; achenes silky-villous. *Th.! Cap. 720. DC. l. c. 476. Senecio rigens, Linn.*

Var. *a.* dentata ; leaves cuneate-obovate, coarsely 3–5-toothed. *O. quinquedentata, DC.! l. c.* 474, excl. syn. *O. Eckloniana, DC. l. c.* 476, ex pte. *Zey.!* 3049.

Var. *β.* bupleuroides ; lvs. oblong-cuneate or ellipt. lanceolate, entire. *O. bupleuroides, DC. l. c.* 475.

Var. *γ.* longifolia ; lvs. linear-lanceolate or linear-elongate, entire or 2–3-toothed. *O. longifolia, DC. l. c.* 476; and *O. leptophylla, DC. l. c.* 477.

Var. *δ.* depauperata ; dwarf, 6–8 in. high, leafy at base ; leaves linear-lanceolate, tapering at base, acute, entire or unidentate. *O. Eckloniana, ex pte. in Hb. Ecklon. O. nudata ??* DC. l. c. 475. (Herb. D., Sd.)

Hab. Common about Capetown and on the Cape Flats. Stellenbosch, Drakenstein, &c. *E. & Z.! Drege!* Uitenhage, *Zey.!* 3047. (Herb. Th., D., Hk., Sd.)
2–2½ f. high. Leaves 1½–2½ in. long, ½–⅔ in. wide ; in var. *γ.* 3–4 in. long, 4–5 lines wide, always opaque, thick, and drying dark. All the above forms grow together, and pass gradually one into the other.

16. O. quinquedentata (Th.! Cap. 720); suffruticose, erect, glabrous, virgate, subsimple ; stem closely leafy below, pedunculoid and loosely panicled at the summit; leaves cuneate-oblong, sessile, half-clasping, and subdecurrent, coarsely or obsoletely 5-toothed near the apex, subcoriaceous, obviously midribbed and penninerved, callous-margined ; pedicels elongate, subumbellate ; inv. sc. 7–8, oblong, acute ; achenes silky-villous. *O. subumbellata, DC.! l. c.* 474.

Hab. Cape, *Thunberg! Sieber!* Berg R., *Drege!* (Herb. Th., Hk., Sd.)
Stem 3–4 ft. high, robust, but not ligneous. Leaves 5–6 in. long, 1½–2 in. wide, of thinnish but tough substance, bluntish, mucronulate ; the shallow teeth callous-tipped. The lesser divisions of the panicles are imperfectly umbellate. A leaf and partial infl. of *O. subumbellata, DC.* (in Hb. Sd.) are undistinguishable from Thunberg's plant : " *O. quinquedentata,*" DC. is our *O. parviflora, var. dentata.*

17. O. pluridentata (DC.! l. c. 474) ; suffruticose, erect, glabrous, virgate, subsimple ; stem closely leafy below, pedunculoid and loosely panicled above ; leaves oblongo-lanceolate, acute, the upper ones clasping and adnate, the lower more and more strongly decurrent, coarsely few or several toothed beyond the middle, membranous, midribbed and penninerved, veiny, callous-edged ; pedicels long ; inv. sc. 7–8, oblong, acute ; achenes silky-villous. Also *O. grandidentata, DC. l. c.*

HAB. Dutoitskloof, *Drege!* (Herb. D., Hk., Sd.)

Very near *O. quinquedentata*, but with thinner, less cuneate, more taper-pointed leaves, strongly decurrent at base, with usually more teeth. *Drege's* specimens of *O. grandidentata* and *pluridentata* are from the same locality, and by me undistinguishable ; the number of teeth is a most uncertain character.

18. O. membranifolia (DC. ! l. c. 476); suffruticose, erect, glabrous; stem simple, terete, leafy, ending in a corymbose, *leafy* panicle; leaves subpetiolate, broadly elliptical or obovate, subacute or obtuse, membranous, midribbed and penninerved, entire or remotely toothed above the middle; floral leaves (on the panicle) ovate or roundish, cucullate, concave ; corymb laxly branched, its subdivisions few-headed, pedicels long; inv. sc. 8, oval, acute, connate below ; ray achenes glabrous.

HAB. Zwartkops R., Uit., *E. & Z.! Zey.* 3043. (Herb. D., Sd., Hk.)

In habit and foliage very like *O. amplexicaulis ;* but the leaves are not amplexicaul, their toothing (when present) is different, their substance thinner; the corymb is much less compound, more leafy, with fewer and larger heads, and the achenes are glabrous.

19. O. amplexicaulis (Th. ! Cap. 718); suffruticose, erect, glabrous, subsimple ; stem leafy, terete, ending in a much-branched, corymbose panicle ; leaves eared and clasping at base, sessile or adnate-subdecurrent, broadly obovate or obovate-oblong, or oblong, mucronate, coriaceous, penninerved, entire or denticulate ; corymb level-topped, very many-headed, pedicels long, slender; inv. sc. 8, oblong; achenes silky-villous.

VAR. *a.* **integrifolia**; leaves entire or subentire. *O. amplexicaulis, and O. senecionea, DC. l. c.* 475.

VAR. *β.* **denticulata**; leaves denticulate. *O. denticulata, Ait. Kew.* 3, *p.* 276. *Bot. Mag. t.* 1979. *DC.! l. c.* 475. Also *O. rigida, DC.! l. c.* 476.

VAR. *γ.*? **Dregeana**; leaves subdecurrent, entire. *O. Dregeana, DC. l. c.* 475.

HAB. Fields near the Berg R., at the Paarl; also in Worcester, Clanwilliam, &c., *E. & Z.! Drege! W. H. H. β.* on the Paarl Mt., *W. H. H.* Onder Bokkeveld, *Drege!* Simonstown, *C. Wright,* 296. *a.* near Tulbagh, *Dr. Pappe! β.* Zwartkops R., *Zey.!* 3044. (Herb. Th., D., Hk., Sd.)

Stem robust, 3–4 ft. high or more, leafy to the corymb. Lower leaves 6–8 inches long, 2–4 in. wide; upper gradually smaller. Except in the entire or denticulate leaves this species varies little. I have not seen *DC.'s O. Dregeana,* which I refer here, by description, and because it is stated to grow along with *O. amplexicaulis, DC.*, by the river side, at the Paarl.

20. O. umbelliformis (DC. ! l. c. 474); herbaceous (?), erect, glabrous ; stem simple or divided at base, leafy below, pedunculoid and umbellato-corymbose above; leaves obovate, obtuse, cuneate and tapering much at base, either sharply many-toothed or subentire, thinly coriaceous, midribbed and penninerved, the lowermost imperfectly petioled, the uppermost ear-clasping ; pedicels long, subequal; inv. sc. 7–8; achenes villous.

HAB. Berg River, *Drege!* Cape, *E. & Z.!* (Herb. Sd.)

Leafy portion of stem 3–4 inches long; pedunc. 6–8 inches, 5–6-headed. Leaves 1½–2 in. long, ¾–1 in. wide. *Drege's* specimens are immature ; but *Ecklon's* have ripe achenes. The root seems to be either thick or tuberous, judging from the stem-base of *Ecklon's* specimens.

21. O. viminea (E. Mey.); shrubby, erect, glabrous; branches virgate; leaves sessile, remote, linear-terete, acute, quite entire, erect; "panicle laxly branched, few-headed; inv. sc. 5–6, concrete at base; rays few, shorter than the involucre, and scarcely longer than their own styles; achenes glabrous." *DC. l. c.*

HAB. Uitvlugt, *Drege!* (Herb. Sond.)
An imperfect specimen, without flowers, only seen by me.   Leaves 1–1½ in. long, not ⅓ line diam.   Fl. heads very small.

§ 4. CARNOSÆ. (Sp. 22–35.)

22. O. cylindrica (DC.! l. c. 477, excl. syn. Th.); shrubby-carnose, glabrous; branches terete; leaves scattered, linear-elongate, semiterete, fleshy, glaucous, acute, tapering at base, spreading; pedunc. terminal, bearing a few- or several-headed, branching corymb, pedicels slender, long; inv. scales 8–9, narrow-oblong, membr. edged, separate nearly to the base; rays as long as invol., their achenes glabrous.   *Cacalia cylindrica, Lam. DC. Pl. grass. t. 48. O. tenuissima, Haw. Succ. p. 315.*

HAB. Olifant R., *Drege!* Clanwilliam, *Eck.* Heerelogement, *Zey.!* 948. (Herb. D., Hk., Sd.)
A loosely branched, forking shrub, with the habit of *Senecio longifolius*; older branches bare and cicatricised.   Leaves 1½–3 in. long or more, 1 line wide.   Pedunc. 4–6 in. long.   A specimen in Hb. Sd. from *Drege*, marked "O. cylindrica, b." belongs to *Senecio longifolius. Zeyher's* 948 is almost intermediate with *O. carnosa. Zeyher's* 947 (Hb. Sd.), in bad condition, seems to belong to this species.

23. O. carnosa (Less. Syn. 88); shrubby-carnose, glabrous; branches terete; leaves linear-terete, fleshy, acute or subacute, tapering at base; pedunc. terminal, elongate, ending in a few-headed, subcorymbose panicle, pedicels very long; inv. sc. 6–8, oblong, membr-edged; rays revolute; achenes glabrous.   *DC. l. c.* 478.   *Cineraria cacalioides, Linn. f. Th.! Cap.* 670.

HAB. Cape, *Thunberg!* Eastern Distr., *Burchell.* Uitenhage and Albany, *E. & Z.! Zey.* 3036, 3039. (Herb. Th., Hk., Sd.)
Similar to *O. cylindrica*, but with thicker, more fusiform, and usually shorter leaves, longer pedunc., and fewer fl. heads.   The ray-fl. said by *DC.* to be very short and narrow, must on his specimens have been immature: I find them nearly as in *O. cylindrica*, but commonly revolute.   Leaves 1½–2 in. long, 1½–2 l. diam.

24. O. crassifolia (Harv., non L., nec. E.M.); glabrous; stem slender, weak, suffruticose at base, irregularly branched; leaves fleshy, subterete? linear-oblong, mostly acute or mucronate, sometimes obtuse, either scattered or *crowded at intervals in imperfect whorls*, with bare spaces between; pedunc. terminal, subcorymbose, few-headed, pedicels very long and slender; inv. sc. about 8, oblong, margined; achenes glabrous.   *O. filicaulis, Eckl. Fl. Cap.* 295, *non Jacq.*

HAB. Uitenhage, *E. & Z.!* Bethelsdorp, *Zey.!* 3040. (Herb. Sond.)
Remarkable for its leaves being collected at intervals into whorls; they are probably *terete*, 1–1¼ in. long, 3 lines diam., with immersed nerves and veins.   All the younger seem to be acute or mucronate.   Pedunc. 3–6 in. long, very slender, 4–5-headed, pedicels 2–3 in. long.

25. O. sedifolia (DC.! l. c. 479); suffruticose, erect, forking, glabrous;

leaves either opposite or alternate, ovoid, ellipsoid or shortly cylindrical, fleshy, quite entire, obtuse or subacute, nerveless, minutely petioled; pedicels terminal, one-headed (or forked and 2-headed), much longer than the leaves; inv. sc. 8–9, oblong, membr. edged, connate at base; rays 10–12, longer than the inv.; achenes glabrous.

HAB. Namaqual., *Drege! A. Wyley!* Gamka R., *Burke & Zey.!* (Hb. D., Hk., Sd.) A minute shrublet. 3–6 inches high, repeatedly forked. Leaves 3–7 lines long, 2–3 in. diam. Pedunc. 3–4 in. long. In *Burke and Zeyher's* specimens the leaves are rather longer and more acute than usual.

26. O. obtusiloba (Harv.); half-shrubby, carnose; the branches nude below, with woolly cicatrices, twigs leafy; leaves linear, fleshy, obtuse, *on each side 1–2-lobed*, the lobes oblong or linear, very blunt, margins strongly reflexed, veins not visible; pedunc. terminal, forked or sub-corymbose, 3–5-headed; inv. scales 5, oblong, concrete at base; rays 5, multistriate; achenes?

HAB. Knakerberg, *Zeyher!* 997. (Herb. Sond., D.) The specimens are incomplete, but seem to indicate a small, succulent half-shrub or perhaps shrub. Leaves 1½–2 inches long, 1½–3 lines wide, mostly with a pair of opposite lobes in the middle, occasionally with a smaller pair of lobes or teeth. The lobes ¼–¾ in. long.

27. O. quercifolia (DC.! l. c. 478); glabrous, erect; stem suffruticose-carnose, leaves tapering at base into a cuneate petiole, obtuse, pinnatifid, the lobes on each side 3–4, oblong, mucronate; pedunc. terminal, bearing a compound, many-headed corymb; pedicels long, slender; inv. lobes 5–6, oblong; achenes glabrous.

HAB. Olifant R. and Brackfontein, *E. & Z.!* (Herb. Sond.) Leaves 2–4 inches long, ½–1 in. wide; the lobes 4–8 lines long. Pedunc. 3–6 in. long; pedicels about uncial. Rays many-lined, yellow.

28. O. triplinervia (DC.! l. c. 478); glabrous; stem shrubby-carnose, short, subsimple or forked, the lower part nude, cicatricised; cicatrices glabrous; leaves crowded toward the apices, obovate, obtuse, tapering at base into a slender petiole, drying thin, from the middle 3-nerved, netted-veined, quite entire or repand; pedunc. terminal, loosely corymbose, few or several headed, pedicels very long; inv. scales and rays about 5; achenes villous.

HAB. Konab and Kat R., *E. & Z.!* Howison's Poort, *Zeyher!* 3037, 3038, *H. Hutton!* (Herb. D., Sd.) Stem thick and fleshy. Leaves, with their petiole, 2½–3 inches long, 1 inch wide, slightly fleshy when recent, almost membranous, and pellucid if carefully dried. Pedunc. somewhat longer than the leaves; pedicels 3–4 in. long. Rays bright-yellow, conspicuous.

29. O. dentata (Linn. sp. pl. vol. 1, p. 926); glabrous; stem shrubby-carnose; leaves crowded at the ends of the branches, subrosulate, obovate, cuneate at base, sessile (not in the least clasping), either sharply and coarsely toothed, or denticulate or subentire, midribbed, faintly nerved; pedunc. terminal, elongate, nude, loosely corymbose or pani-cled at the apex, several headed; pedicels long, rising from minute, subulate bracts; inv. scales about 8, acute, membr. edged, connate; achenes villous. *DC. l. c.* 478. *Burm. Afr. t.* 59.

HAB. West side of Table Mt., *Drege!* Near Simonstown, *Mr. C. Wright,* 295, 296, 297. (Herb. D., Sd.)

Stem or branches fleshy, 3–4 lines diam., leafy at the extremities only; cicatrices nude. Leaves 1½–2 in. long, ¾–1¼ in. wide; very variable in dentition. Pedunc. 4–10 in. long, 3–12-headed; pedicels 1½–3 in. long. Pappus white. I omit *Ecklon's* habitat, quoted by *DC.,* because a specimen from *E. & Z.* in Herb. Sond., named "*O. dentata,*" is merely a dwarf, ill-grown plant of *O. amplexicaulis,* var. *denticulata;* having the amply ear-clasping leaves, infl. and fl. heads of that species.

30. O. macrosperma (DC. l. c. 478); "shrubby-carnose, branching, glabrous, ascending; cicatrices bearded; leaves fleshy, flat, oblong, obtuse, tapering at base, some entire, some about the middle bluntly one-lobed; fl. branches striate, subracemose, few-headed, nude; pedicels elongate; inv. cylindr., 5-parted; rays 5, ten-nerved; ray achenes silky with rufous, close-lying hairs; pappus rufous." *DC. l. c.*

HAB. Olifant's R., *Drege.* (Unknown to me.)
"Leaves 1 in. long, 2–3 l. wide. Ach. 3 l. long." *DC.*

31. O. purpurascens (Harv.); stem shrubby-carnose, the branches nude below with woolly cicatrices; leaves somewhat fleshy, flat, glabrous, glaucous, oblong or obovate-oblong, obtuse, narrowed to the base, entire or bluntly 1–2-toothed; pedunc. terminal, mostly forked, 2-headed, pedicels long; inv. sc. 5; rays *purplish;* achenes silky-villous.

HAB. Mastenberg, *Zeyher!* (Herb. Sond.)
The specimens are in very poor condition, but seem to belong to a well marked species. The foliage is very glaucous, and the rays seem to have been dull, purple-red. Leaves 1–1½ in. long, 4–5 lines wide.

32. O. arborescens (Linn. Sp. 1310); stem shrubby-carnose, the branches flexuous, nude and cicatricised below; cicatrices woolly; leaves obovate-oblong, tapering at base, obtuse or subacute, entire or sinuate-toothed, thickish, but netted-veined when dry; pedunc. terminal, elongate, simple or bifid, mostly one-headed; inv. sc. 5, broadly oval, concrete at base; rays multistriate; achenes villous. *DC. l. c.* 478. *Dill. Elth. t.* 103. *Burm. Afr. t.* 60. *f.* 1. *O. coronopifolia, fol.* 1. *Hb. Th.!*

HAB. Cape, *Thunberg!* Cape Flats, *Drege!* Table Mt., *W. H. H.!* Onrustrivier, *Zey.!* 3051. (Herb. Th., D., Sd., Hk.)
A fleshy shrub, 2–3 feet high. Axils of leaves and younger cicatrices with a tuft of short, white hairs. Leaves 1½–2 in. long, ¼–¾ in. wide. Pedunc. 3–4 in. long. *Zeyher's* 3051 is very robust, but does not materially differ from my Table Mt. specimens.

33. O. primulina (DC. l. c. 479); "shrubby, dwarf, quite glabrous; leaves crowded at the ends of the branches, obovate, obtuse, sessile, tapering at base [7–9 l. long, 3–4 l. wide], somewhat 3-nerved, denticulate; pedunc. scarcely longer than the leaves, one-headed; inv. lobes and rays 5–6; rays 5-nerved." *DC. l. c.*

HAB. Cape, *Krebs.* (Unknown to me.)

34. O. cacalioides (Linn. f. Suppl. 388); shrubby-carnose, glabrous, branching; branches very thick and fleshy, with white-woolly cicatrices; leaves tufted, obovate, tapering at base, very obtuse, thickish coriaceous,

faintly 3-nerved, entire or minutely denticulate, the margin thickened
and subreflexed; peduncle twice as long as the leaves, one-headed; inv.
lobes 5, very broad; rays?; achenes? *Th.! Cap.* 719. *DC. l. c.* 479.

HAB. Bockland'sberg, *Thunberg!* (Herb. Th.)
Stem and branches ¾ in. diameter, with a pale-brown or fulvous, glossy bark.
Leaves ¾ in. long, 4–5 lines wide. Pedunc. 1½ in. long. Heads small, few-flowered.

35. O. minima (DC.! l. c. 479); caudex very short, fleshy, woolly at
the crown; leaves from the crown, several, rosulate, obovate-flabelliform,
at base cuneate, entire, and tapering into a petiole, at the apex rounded
and sharply toothed; peduncle. scape-like, filiform, one-headed (or forked
and 2–3-headed), 2–4 times longer than the leaves; inv. sc. 5; achenes
hairy-villous.

HAB. Giftberg, *Drege!* (Herb. Sond.)
Allied to *O. cacalioides*, but differing in its very short, tuberous stem, and sharply
toothed leaves. Perhaps in more advanced specimens the stem might be longer.
Leaves on nearly uncial petioles, ¾ in. long, ½ in. wide. Peduncle. 2–4 in. long.
Raised in Hamburg Bot. garden from *Drege's* seeds.

§ 5. SCAPIGERÆ. (Sp. 36–41.)

36. O. Natalensis (Sch. Bip., in Bot. Zeit. XXVII. p. 771); herbaceous, glabrous; root tuberous, prolonged upwards into a thick rootstock, woolly at the crown; leaves from the crown, many, rosulate,
oblong or oblongo-lanceolate, obtuse or acute, much attenuated to the
base, with woolly axils, quite entire, coriaceous, midribbed, nerveless,
margined; stem scapelike, nude or with 1–2 bracts, once or twice forked
or corymbose at summit; pedicels long; inv. sc. about 8, lanceolate,
concrete at base; rays 8, yellow; achenes shortly pubescent; pappus
rufo-fulvous. *Walp. Rep.* 6, *p.* 275.

HAB. Natal, *Krauss!* 442. British Caffraria, *T. Cooper*, 299. (Herb. D., Hk.)
This has something the look of *Senecio coronatus.* Leaves 4–5 inches long, 1¼–2
in. wide, thickish, opaque. Peduncle. 6–12 inches high. Rays twice as long as the
involucre.

37. O. retrorsa (DC. l. c. 479); caudex suberect, simple or multifid
woolly and clothed with the reflexed bases of old leaves; axils very
woolly; leaves crowded round the apices, rosulate, oblongo-spathulate
or linear-spathulate, narrowed to the base, subacute, rigid, thin, prominently netted-veined on both sides, margined with spreading or recurved
cilia; scapes erect, striate, forked or loosely sub-corymbose, pedicels
long, nude; inv. scales and rays about 8; achenes silky villous; pappus
short.

VAR. β. **linearifolia**; leaves smaller and narrower, almost linear.

HAB. Little Namaqualand, *Drege!* (Herb. D., Hk., Sd.)
Caudices 2–4 inches high, nearly 1 in. thick. Leaves 2–3 inches long, 5–6 l. wide;
in β. about 2 lines wide. Scape 6–10 inches long, corymbose, few-headed.

38. O. Zeyheri (Sond.); caudex and leaves as in *O. retrorsa*; scapes
erect, scarcely striate, corymbulose, few-headed, pedicels *short*; inv. sc.
and rays 5; achenes?

HAB. S. Africa, *Zeyher*, 996. (Herb. Sond.)
Of this I have seen only very poor specimens. So far as I can judge, it merely

differs from *O. retrorsa* in the smaller fl. heads and fewer scales of the involucre, if these characters be constant.

39. 0. scapigera (Harv. Thes. t. 15); caudex short, simple or divided, woolly; axis woolly; leaves many from the crown, rosulate, linear or linear-lanceolate, acute, coriaceous, thick, one-nerved, veinless, quite entire and smooth-edged; scapes erect, simple, one-headed, with 1–2 leaf-scales; inv. scales and rays 8–9; achenes (young) glabrous.

HAB. Natal, *Dr. P. Sutherland!* (Herb. D., Hk.)
With the habit of *O. retrorsa*, this has very different leaves and 1–headed peduncles. Leaves 1½–2½ inches long, 2 lines wide. Scapes 6–8 inches high.

40. 0. auriculæfolia (Lichtenst.! in Less. Linn.!1831, p. 93); stemless (in β. sub-caulescent) woolly at the crown; radical leaves petioled or subpetioled, oblong, glabrous, glaucous, sinuato-pinnatifid or eroso-dentate, the lobes entire or denticulate; scapes one-headed, as long or twice as long as the leaves; inv. campanulate, the scales 12–14, concrete to the middle, 3-nerved, membr. edged; rays 12–14, blue; achenes silky, they and the pappus fulvous. *O. cyanoglossa, O. lactucæfolia and O. picridioides, DC. l. c.* 481–482.

VAR. β. **arctotoides** (Harv.); rad. leaves long petioled, cuneate-obovate, entire or repand; cauline variably incised or pinnatifid; heads as in α. (Herb. Hk., Sd.)

HAB. Cape, *Lichtenstein.* Sneeuwberg; Kendo; Pietermeintjesfont; near Graaf Reynet; and Hex Rivierberg, *Drege!* β. Wolve River, *Burke and Zeyher. Zey.* 992. (Herb. Sond., D.)
Rootstock very thick, tuberous at base. Leaves many, rosulate from the crown, 2–4 inches long, ¾–1 in. wide, variably incised or sinuate, of thickish, opaque substance. Pedunc. 2–5 inches long. Inv. ½–¾ in. diam. Rays deep blue when dry. *Dr. Sonder* has compared *Drege's* plant with the original sp. in Herb. Berol. described by *Lessing.* Var. β. has petioles 2–3 inches long, and much shorter and more entire lower and radical leaves; a 1–1½ in. long stem, and polymorphous cauline leaves; in its fl. heads it agrees with the common form. The rays dry blue; the disc.-fl. look as if they may have been red: quere, are both *red* when fresh?

41. 0. sonchifolia (DC.! l. c. 482); stemless; root tuberous, crown very thick, woolly; radical leaves shortly petioled, glabrous, lyrato-pinnatifid, the lobes in several pair, with rounded interspaces, shortly oblong, lobulate and mucronato-denticulate, the uppermost confluent; scapes rigid, striate, much longer than the leaves, one-headed; inv. lobes and rays about 8; achenes silky; pappus pale tawny.

HAB. Stellenbosch, *E. & Z.!* (Herb. Sond.)
Tuber globose or oblong; crown ¾ in. diam. Leaves 4–6 inches long, 1 in. wide; lobes ½–¾ in. long, 4–5 lines wide. Scapes 6–10 in. long. The rays, in *Ecklon's* specimens here described, seem to my eye to have been *yellow;* but *DC.* calls them *blueish (cærulescentes).*

§ 6. CAULESCENTES. (Sp. 42–58.)

42. 0. cakilefolia (DC. l. c. 482); herbaceous, erect, glabrous; root tuberous, somewhat woolly on the crown, stem nude at base, leafy and flexuous in the middle, branched above; leaves sessile, half-clasping, fleshy, pinnatipartite, the lobes in few pair, with rounded interspaces, broadly linear, obtuse, entire; upper leaves on each side one-lobed; branches pedunculoid, one-headed, subcorymbose; inv. campanulate, of 10–12 concrete, acute scales; ray achenes rufo-villous. *DC. l. c.,* *excl. var. β.*

HAB. Near Holrivier, in Karroid places, 1000 f., *Drege!* (Herb. Hook.)
Root an oblong-cylindrical tuber, 1 in. long, 2–3 l. diam. Bare portion of stem 2–3 inches long; leafy portion 4–5 inches. Pedunculoid branches 3–4 inches long. Rays and disc fl. crimson or purplish-red.

43. O. incisa (Harv.); herbaceous, erect, glabrous; root?; crown woolly; stem leafy in the lower half, pedunculoid, nude, and sub-corymbose above; leaves sessile, half-clasping, somewhat fleshy, ob-longo-lanceolate or lanceolate, coarsely several-toothed, the teeth deltoid, subacute, distant; upper leaves smaller, subentire or entire, acute; pedicels elongate, subcorymbose; inv. campanulate, of 8–10, connate, lanceolate scales; ray achenes tawny-villous; pappus whitish. *O. cakilefolia, var. β latifolia, DC.! l. c.* 482.

HAB. Betw. Uitkomst and Geelbekskraal, 2–3000 f., *Drege!* (Herb. D., Sd.)
This differs decidedly in foliage from *O. cakilefolia;* so much so that I must separate it from that species. The root-leaves, if any, and tuber are unknown; the cauline leaves are 6–8 inches long, 1–1½ wide. Pedunculoid stem 6–10 inches high. Both ray and disc fl. seem to have been purplish, much less bright than in *O. cakilefolia.*

44. O. rosea (Harv.); herbaceous, erect, branching, glaucous, quite glabrous; leaves sessile, stem-clasping, elliptic-oblong, mucronate, mar-gined, quite entire or subrepand, midnerved, with immersed veins; branches subpedunculoid above, forked or subcorymbose, few-headed; pedicels very long; invol. campanulate, of 10–12 oblong, semiconcrete scales; rays 4-lined; achenes (young) glabrous.

HAB. Modderfontein, Namaqualand, *Rev. H. Whitehead!* (Herb. D., Sd.)
Root and base of stem unknown. Leaves 3 inches long, 1–1½ inch wide, with the stem and inv. very glaucous. Pedicels subcorymbose, 2–3 inches long. Ray and disc. fl. bright, rosy-red. Pappus pale fulvous.

45. O. rufibarbis (Harv.); herbaceous, glabrous; root tuberous? woolly at the crown; stem erect, rigid, branching, leafy; radical leaves on long petioles, ovate, acute, 3–5-nerved at base, rigid, veiny; cauline sessile, ear-clasping, oblong or ovate-oblong, obtuse or acute, margined, entire, undulate; pedunc. terminal and axillary, long, one-headed; inv. scales and rays 12–13; achenes rufo-pubescent; pappus rufous.

HAB. Adow, Uit., *Zey.!* 3041. (Herb. D., Hk., Sd.)
Root not seen. Stem 1 foot long. Radical leaves on 4–5 inch long petioles, 1½–2 inches long, 1–1½ wide; cauline 1½ inch long, ½–¾ inch wide. Pedunc. 4–8 inches long. Pappus copious, dull purple.

46. O. amplexifolia (DC. l. c. 480); herbaceous, branching, half-climbing, glaucous, glabrous; root tuberous, woolly at the crown and in the lower axils; radical leaves?; cauline leaves broadly cordate, stem-clasping, acute or obtuse, undulate or repand, mucronate, soft, drying thin and veiny; pedicels terminal and axillary, slender, long, 1-headed; inv. sc. 9–10; achenes dark-brown, tapering, minutely hispid. *O. per-foliata, Jacq. Schoenbr. t.* 240 *(excl. syn.).*

HAB. Paarl, *Drege!* Tulbagh, *Pappe!* Piquetberg, *Zey.!* 1003. (Hb. Hk., D. Sd.)
This has quite the habit and foliage of *Doria perfoliata,* but differs in generic

character, and in the fewer inv. scales. Leaves 2½ inches long, almost equally wide, slightly fleshy, but pellucid and membranous when dry. Pedicels 3-5 inches long. Pappus whitish, short.

47. O. filicaulis (Jacq. Schœnb. t. 241); "herbaceous, glabrous, weak, branching; root tuberous; radical leaves petioled, ovate; cauline cordate-clasping at base, oblongo-lanceolate, acute; pedunc. nude, one-headed; rays narrow, long, bidentate; achenes glabrous." *DC. l. c. p.* 481.

HAB. Cape : formerly cult. in Europe.

Like *O. amplexifolia*, but with much narrower, taper-pointed leaves. The achenes, described as "*glabrous*," are perhaps, more probably, *hispidulous*.

48. O. linifolia (Linn. f. Suppl. 388); herbaceous, glabrous; root tuberous ; crown and axils woolly ; stems scape-like, bifid or 2-3 forked, nearly nude; radical leaves very long, linear-attenuate or linear-lanceolate, tapering to each end, rigid, ribbed and veiny or striate, quite entire; pedicels very long; inv. sc. and rays 9-10; ray achenes silky-villous. *Th.! Cap.* 718. *DC. l.c.* 479. *O. bulbosa, var. γ. Linn. Sp.* 1309.

HAB. Cape, *Thunberg!* Cape Flats, *E. & Z.! Dr. Pappe! W. H. H.* Berg. R., *Zey.!* 3042. (Herb. Th., Hk., D., Sd.)
Radical leaves 6-10 inches long, ½-3 lines wide. Stem 8-14 in. high, commonly forked in the middle, with 1-2 sessile, linear leaves.

49. O. tuberosa (Thunb.! Cap. 720); herbaceous, glabrous or sparsely pilose; root tuberous, woolly at the crown and axils; stems short, flexuous, simple or divided, leafy; radical leaves on long petioles, ovate or oblong or oblongo-lanceolate, entire, acute, membranous, veiny (when dry); cauline leaves sessile, half-clasping, oblong, spathulate or obovate, entire or repand; pedunc. very long, nude or nearly so, one-headed, curved, striate; inv. sc. 12-14, lanceolate, acute, concrete at base; achenes silky. *O. elliptica; O. adscendens; and O. oblongifolia, DC.! l. c.* 480. Also *O. Lingua, Herb. Th.!*

HAB. Common round Capetown and on the Cape Flats. *Zey.!* 3045, *pte.* (Herb. Th., D., Hk., Sd.)
Leafy portion of stem 2-3 inches long; pedunc. 6-8 in. long. Leaves mostly quite entire and glabrous ; rad. leaves 3-5 in. long, 1-2 in. wide; petiole 2-3 in. long. Some specimens of *Zey.* 3045 belong to *O. heterophylla.*

50. O. heterophylla (Linn. f. Suppl. 387); herbaceous, more or less copiously pilose, rarely subglabrous; root tuberous, woolly at the crown and in the axils; stems short, flexuous, simple or forked, laxly leafy; radical leaves petiolate, broadly ovate or obovate, or subrotund, obtuse, often cordate at base, repand or crenate, prominently nerved and veiny beneath, more or less pilose; cauline sessile, half-clasping, obovate or oblong, with recurved edges, crenate or sub-entire; pedunc. terminal, nude, one-headed; inv. sc. 13-14; achenes villous. *Th. Cap.* 721. *O. tuberosa, DC. l. c.* 479 *(excl. syn.)* Also *O. brachypoda, DC.! l. c.* 480.

HAB. Cape Flats, &c., *Thunberg! Drege, Eckl., W.H.H.* Voormansbosch, *Zey..* 3046, 3045 *(pte.)* (Herb. Th., D., Hk., Sd.)
Very near *O. tuberosa*, but more hairy, with thicker, more prominently nerved leaves, with recurved edges, usually crenate or repand; *Zeyher's* 3045 is partly made up of one, partly of the other form or species.

51. O. macrophylla (DC.! l. c. 480); herbaceous, glabrous; root tuberous, woolly on the crown; stem short, flexuous, simple or forked, leafy; radical leaves?; cauline leaves *(very large)* sessile, stem-clasping, soft (drying thin and veiny), ovate-oblong or oblong, the uppermost acuminate, subsinuate, entire or remotely callous-denticulate; pedunc. long, nude or nearly so, 3-4 times forked or loosely panicled, pedicels very long; inv. sc. 12-13, lanceolate, acute, connate; achenes velvetty.

HAB. Little Namaqualand, *Drege!* (Herb. D., Hk., Sd.)
Root leaves unknown; cauline 4-9 or 10 inches long, 2-5 in. wide, conspicuously midribbed and netted with slender veins when dry; probably somewhat fleshy. *Stem* and pedunc. together 2 ft. high, fistular, striate, pale; pedicels 3-6 inches long. Heads 8-12 in the inflorescence.

52. O. petiolaris (DC.! l. c. 480); herbaceous, glabrous; root tuberous, woolly at the crown; stem short, subsimple, few-leaved; radical leaves petioled, broadly oblong or ovate-oblong, sinuate or crenato-lobulate, or erose, or sublacerate, callous-denticulate, drying thin and veiny; cauline 2-3, tapering much at base or subpetiolate, obovate, crenate or lobulate; pedunc. long, nude, loosely panicled at summit, few-headed, the pedicels long; inv. sc. about 8, ovate-oblong, subacute; rays 8; achenes?

HAB. Little Namaqualand, *Drege!* Betw. Olifant's R. and Knakisberg, *Zey.!* 1001. (Herb. Hk., Sd.)
Radical leaves 4-6 inches long, on uncial or 2-uncial petioles, 2-3 in. wide; cauline 2-3 in. long, 1½-2 in. wide. Allied to *O. macrophylla,* but differing in the fewer inv. scales and the non-clasping cauline leaves.

53. O. cuneata (DC. l. c. 481); herbaceous, glabrous; root tuberous, woolly at the crown; stem erect, short, simple or branched, leafy; radical and lower cauline leaves petioled, broadly cuneate or subrotund, entire at base, dilated and coarsely several-toothed or lobulate at apex, thickish, faintly nerved and veined; upper leaves sessile or subsessile, toothed or entire; pedunc. long, nearly nude, once or twice forked, few-headed; pedicels very long; inv. sc. 7-8; achenes velvetty, with minute, reddish hairs.

HAB. Langvalei and Knakerberg; and near Holrivier, *Drege!* Betw. Olifant R. and Knakerberg, *Zey.!* 1000. (Herb. Hk., D., Sd.)
Petioles 2-4 inches long; lamina of lower leaves 1½-2 in. long, 2-2½ wide. Ped. 6-10 in. long, bearing 2-3 small leaves.

54. O. reticulata (DC. l. c. 481); herbaceous, quite glabrous, woolly at the crown; root tuberous? radical and lowest cauline leaves petioled, more or less deeply sinuato-pinnatifid, prominently netted-veined, the lobes ovate or oblong, entire, mucronulate; upper leaves sessile, stem-clasping, oblong or lanceolate, some pinnatifid, others toothed or entire; stem very laxly leafy, branched, its branches ending in very long, nude, one-headed pedicels; inv. lobes and rays about 8; achenes *(fide DC.)* "glabrescent."

HAB. Near Olifant R., and Brackfontein, *E. & Z.!* (Herb. Sd., D.)
1-2 feet high, the stem less pedunculoid than in others of this group; its pedunculoid branches 8-12 inches long. Leaves rather rigid, drying thin, 4-6 inches long, 1-2 inches wide.

55. O. rotundifolia (DC. l. c. 481); "herbaceous, quite glabrous, with bearded axils; radical leaves petioled, pinnatifid, the lobes on each side 4–5, with the terminal lobe roundish, obtusely sinuate, here and there submucronate; scape rather longer than the leaves, twice forked at the apex, pedicels longer than the heads; invol. 7–8-lobed; rays 7–8, mostly 5-nerved; ray achenes glabrous." *DC. l. c.*

HAB. Little Namaqualand, *Drege.* (Unknown to me.)
"Leaves 10 inches long, 2–2½ inches wide. Pappus reddish." *DC.*

56. O. lyrata (DC. ! l. c. 481); herbaceous, glabrous; root tuberous? the crown and lower axils woolly; stem short, flexuous, leafy; radical and cauline leaves lyrate, near or at the base of petiole dilated and stem-clasping, the lateral lobes in 2–3 pair, roundish subentire, broad-based or cuneate-attenuate; terminal lobe very large, roundish, often cordate at base, subentire, toothed, or lobulate; the blunt teeth or lobes very broad and shallow, mucronulate; pedunc. long, nude, repeatedly forked, pedicels long; inv. sc. and rays 7–8; achenes *(fide DC.)* "glabrous."

HAB. Clanwilliam and Worcester, *E. & Z.!* (Herb. Sond.)
Leaves 5–6 inches long, shortly petioled, the dilatation of the radical leaves shortly above the base of the petiole; that of the cauline, which are similar, at the base: their veins and nerves not prominent. Term. leaf-lobe 2–2½ in. long, 1½–2 in. wide. Pedunc. 10–12 in. high.

57. O. pinnata (Linn. f. suppl. 387); herbaceous, glabrous; root tuber-ous, the crown and lower axils woolly; stem short, flexuous, laxly leafy; radical and lower cauline leaves petioled, pinnatisect, the segments in several, close pairs, rigid, oblong, oval or subrotund, quite entire, margined, decurrent; branches pedunculoid, long, curved, nude, one-headed; inv. scales 12–13, lanceolate, concrete; achenes villous. *Bot. Mag. t.* 168. *Th. ! Cap.* 722. *DC. l. c.* 481.

HAB. Common on the Cape flats, and hills round Capetown, Kamp's Bay, &c. (Herb. Th., D., Hk., Sd.)
Leaves 4–6 inches long, including the petiole; lobes ¾–1 in. long, 3–4 lines wide. The rays are often red or purple beneath, but yellow above.

58. O. tephrosioides (Sond.); herbaceous, glabrous; root tuberous, the crown and lowest axil woolly; stem suberect, branching, laxly leafy; radical leaves petioled, pinnatisect, the segments in several, subdistant pairs, rigid, oblongo-lanceolate, subacute, quite entire, margined, decur-rent; branches 2–4 headed, pedicels 1–2-uncial; inv. sc. about 8, ovate, subacute; achenes?

HAB. Matjisfontein, *Zeyher.!* 999. (Herb. Hk., Sd.)
Root ovate-oblong, acute. Rad. lvs., with their petiole, 5–6 in. long; lobes 1–1½ in. long, 3–4 l. wide. Cauline lvs. sessile, the lowest pinnatisect, upper entire. Stems 1–1½ ft. high, forking.—Nearly allied to *O. pinnata*, but with taller, more erect and branching stem, more compound inflorescence and fewer inv. scales.

(Doubtful Species.)

O. nudata (DC. l. c. 475); "suffr., erect, glabrous; stem terete, leafy at base, otherwise nude, elongate, paniculato-corymbose; leaves lanceo-late, tapering at base, acute, with 1–2 large teeth on each side; pedicels rather shorter than the fl. head; inv. lobes and rays 7–8; achenes villous." *DC. l. c.*

HAB. Near Capetown, *Ecklon*. (Perhaps some form of *O. parviflora*.)

O. quinqueradiata (DC. l. c. 478); "suffruticose, glabrous, erect; stem short, terete, fleshy, sparingly branched; leaves crowded at the end of the branches, woolly at the axils, oblong, tapering at base, obtuse, submucronate, flat, quite entire; pedunc. terminal, twice as long as the leaves, one-headed, nude; inv. scales and rays 5, rays 7–9-nerved." *DC. l. c.* 479.

HAB. Cape, *Forbes*. (How does this differ from *O. arborescens?*)

Hertia crassifolia (Less. Syn. 88); *DC. Prodr.* 6, *p.* 483. *Othonna crassifolia, Linn. Mant.* 118 *(excl. syn. Comm.) Mill. Ic. t.* 245, *f.* 2, is omitted in this work, because, as I am advised by *Dr. Sonder*, the specimen, in Herb. Berol., on which *Lessing* founded his genus, is merely a cultivated state of the European *O. cheirifolia*. There is nothing like it in any collection of Cape plants known to me.

CXVI. **GYMNODISCUS**, Less.

Heads several-flowered, radiate; *ray-fl.* shortly ligulate, female; *disc-fl.* 5-fid, male. *Inv.* scales uniseriate, concrete at base, not calycled. *Recept.* nude. *Achenes* glabrous, smooth, ovate or obovate. *Pappus* uniseriate in the ray-fl.; none in the disc-fl. *DC. Prodr.* 6, *p.* 469.

Small annuals, natives of S. Africa. Root fibrous. Leaves chiefly radical, rosulate, lyrate or linear. Heads small, corymboso-paniculate. Name from γυμνος, *nude*, and δισκος, a *disc;* because the disc-fl. have no pappus.

Radical-leaves lyrate or sinuato-pinnatifid ... ... (1) **capillaris.**
Radical-leaves narrow-linear, quite entire ... ... (2) **linearifolia.**

1. G. capillaris (Less. Syn. 89); radical leaves obovate, lyrate or lyrato-pinnatifid; inv. 5–6-lobed; rays 5–6. *DC. l. c.* 469. *Othonna capillaris, Linn. f. Suppl.* 388. *Thunb. ! Cap.* 721.

VAR. β. latifolia (Sond.); rad. lvs. sinuato-pinnatifid; cauline very broad, oblong or obovate, obtuse, repand or subsinuate. (Hb. Sd.)

HAB. Cape Flats and throughout the Western Districts, in sandy ground. β. Saldanha Bay, *E. & Z.!* (Herb. Th., D., Hk., Sd.)
Habit of *Capsella Bursa Pastoris*. Radical-leaves many, rosulate, 3–5 in. long, petioled, the lat. lobes 2–3, rounded, terminal larger, bluntly ovate, entire or repand. Stem 6–12 in. high, forked or alternately divided, with small, sessile, toothed or incised leaves at the forks, diffusely panicled upwards; dwarf specimens simple and few-headed. Anthers rounded at base. Disc-style cone-tipped. β. is a very luxuriant state, with much larger and less cut or subentire stem-leaves.

2. G. linearifolia (DC. l. c. 469); leaves linear-elongate, fleshy, obtusely callous-tipped, veinless, quite entire; inv. 7–9-lobed; rays 6–7.

HAB. Boschjemans-karroo and Groenrivier, *Drege!* (Herb. D., Hk., Sd.)
Habit of *G. capillaris*, but with very different foliage.

(Doubtful Species.)

G. orbicularifolius (Sch. B., Bot. Zeit. 27, 768); "leaves orbicular, entire, 5–9 lines diameter, petioled; stem nude, corymboso-paniculate; the heads larger and rays paler than in *G. capillaris*." *Sch. B. l. c. Walp. Rep.* 6, *p.* 274.

HAB. Cape.

CXVII. SENECIO, L.

Heads either discoid-homogamous or radiate-heterogamous ; ray-fl.
ligulate, female.   *Invol.* uniseriate, rarely quite nude at base, usually
more or less calycled ; invol. sc. frequently with withered or dark tips,
membrane-edged, often 2-nerved.   *Recept.* nude or honey-combed.   *Style*
of the disc fl. with truncate branches, hispid at the tip.   *Achenes* terete,
truncate or slightly tapering at the summit.   *Pappus* pluriseriate, pilose,
caducous, the bristles straight, very slender, roughish.   *DC. Prodr.* 6,
341.   Also *Brachyrhynchos, Less. DC. l. c.* 437.

A vast, cosmopolitan genus, extremely varied in habit, consisting of annuals,
perennials, shrubs, or small trees.   Leaves alternate.   Flowers yellow or purple.
Inflorescence various.   Name from *senex,* an old man; alluding to the usually white,
hair-like pappus.   I have attempted a redistribution of the Cape species, in 12 sections, which are meant to be natural groups : they are not always easily definable.
The same indefiniteness attaches to the specific characters of many, and with fuller
materials I should probably have reduced to varieties several that are here retained.

§ 1. ANNUI.   *Root* annual.   *Stem* herbaceous, mostly branched, erect or diffuse.
Inflorescence diffusely panicled or subcorymbose.   *Heads* rarely discoid ; mostly
radiate, the rays yellow or purple.   (Sp. 1-27.)

(*a.*) Heads discoid.—(Leaves bluntly pinnatifid, glabrate)   (1) **vulgaris.**
(*b.*) Heads radiate.   Ray-flowers *yellow :*
    * *Glabrati :* glabrous or sparsely hispidulous :
        Leaves *sessile,* expanded, entire or pinnatifid (with *broad* lobes) :
           Inv. conspicuously *calycled* at base :
              Leaves fleshy, oblong-obovate, obtuse,
                denticulate   ...   .   (2) **maritimus.**
              Lvs. not fleshy, oblong, toothed or pinnatifid :
                Inv. 3-4 l. long ; achenes striate,
                  setulose   ...   .   (3) **littoreus.**
                Inv. 2 l. long ; ach. *canous,* notstriate (4) **laxus.**
              Lvs. lanceolate, tapering at base   ...   (5) **ruderalis.**
            Inv. nude at base, or nearly so :
              Leaves incised or pinnatifid, the lobes
                subentire ...   .   (6) **abruptus.**
              Lvs. deeply pinnatisect, lobes sharply
                toothed   ...   .   (7) **consanguineus.**
        Leaves (lower) *distinctly petioled,* simple or lyrate, with *broad* lobes :
            Inv. nude at base, or nearly so :
               Lower and medial lvs. petioled, orbi-
                cular, lobed or lyrate: upper sessile,
                pinnatifid, or entire   ...   ...   (8) **repandus.**
              Lower leaves petioled, cordate; upper
                cordate, clasping   ...   .   (10) **lobelioides.**
            Invol. calycled ; leaves obovate or oblong,
              crenate, long-stalked   ...   .   (9) **Lessingii.**
            Inv. calycled ; upper leaves pinnatifid :
               Leaf-lobes 5-9 pair, toothed ; rays *long* (11) **erysimoides.**
              Leaf-lobes 3-4 pair, entire ; rays short (12) **cardaminifolius.**
        Leaves either all *narrow-linear,* or pinnatisect or
        multifid, with *narrow-linear* segments and lobes :
            Inv. nude at base, or nearly so :
               Leaves bipinnati-partite ...   .   (13) **sophioides.**
              Leaves linear, quite entire   ...   ...   (14) **tenellus.**
             Inv. calycled :
               Erect : lvs. either linear, or pinnatisect (15) **lævigatus.**
              Diffuse : lvs. bipinnatifid or multifid   (16) **matricariæfolius.**

** *Semiglabri:* glabrous, except the densely *gland-*
pubescent peduncle, bracteæ, pedicels, and invo-
lucres ; lvs. pinnatifid ... ... (17) **trachylænus.**
*** *Pubescentes.* Branches and leaves, or whole plant *scabrid-pubescent :*
 Lvs. bipinnatisect or pinnatisect, with *linear* lobes :
  Diffuse or decumbent ; heads few, calycled (18) **diffusus.**
  Erect : upper lvs. pinnatisect ; heads pani-
   cled, nearly nude at base ... ... (19) **Paarlensis.**
 Lvs. lance-linear, entire or toothed ... ... (20) **puberulus.**
 Lvs. petioled, some ovate-oblong, some lyrate (21) **sisymbrifolius.**
 Lvs. oblong, obtusely pinnatifid or sinuate-toothed :
  Inv. quite glabrous, of 10–12 oblong scales (22) **glutinarius.**
  Inv. gland-scabrid, of many linear-subulate
   scales ... ... (23) **glutinosus.**
(*c.*) Heads radiate. Ray flowers *purple :*
 Inv. conspicuously calycled at base :
  Calycle of broad, *short*, ciliate bracteoles ... ... (24) **elegans.**
  Calycle of lanceolate bracteoles as long as inv. sc. (25) **multibracteatus.**
 Inv. nude at base or nearly so :
  Viscoso-pubescent : leaves polymorphous ... (26) **arenarius.**
  Glabrous, or sparsely hispidulous :
   Leaves pinnatipartite, with blunt lobes ... (27) **cakilefolius.**
   Leaves bipinnatipartite, with narrow-linear
   lobes ... ... (13) **sophioides.**

§ 2. SINUOSI. *Rootstock* perennial. *Radical* leaves several, larger than the cauline,
mostly sinuous-pinnatifid, pinnatifid, or incised, rarely entire, *very generally* pubescent
or viscidulous. *Stem* herbaceous, subsimple, nude or leafy, one or loosely several
headed. *Heads* many-fl., radiate or discoid ; flowers purple or yellow. (Sp. 28–49.)

(*a.*) Heads radiate ; disc and ray flowers *purple :*
 Radical and lower leaves *sessile ;* stem leafy, *racemose*
  above ... ... (28) **macrocephalus.**
 Rad. and lower leaves *petioled*; stem nude and
  forked above, *subcorymbose* ... ... (29) **concolor.**
(*b.*) Heads discoid ; the disc-fl. *purple :*
 Stems branching ; fl. branches nude, forked or corymbose ; leaves *rarely entire*,
  commonly lyrate or pinnatifid :
  Glandularly pubescent, viscidulous :
   Corymbs *laxly* few-headed ; achenes pubes. (30) **erubescens.**
   Corymbs *densely* many-headed, compound ;
    achenes glabrous;
    Medial and upper leaves *sessile* ... ... (31) **purpureus.**
    Medial and upper lvs. *decurrent* as a
     stem-wing ... ... (32) **odontopterus.**
  Glabrous : leaves pinnatisect, with cut lobes ... (33) **glabrifolius.**
 Stems simple, closely leafy nearly to the summit :
  Pubescence of long, jointed, white hairs :
   Heads corymbose ; upper lvs. crowded, 4–5
    lines wide, flattish ... ... (34) **barbatus.**
   Heads panicled ; upper lvs. lax, 1 line wide,
    revolute-edged ... ... (35) **thyrsoideus.**
  Pubescence of short, scabrid, close, knappy hairs (36) **Sandersoni.**
(*c.*) Heads discoid ; the disc-fl. *yellow :*
 Glabrous. Stem much branched, cauline lvs. coarsely
  toothed or cut ... ... (37) **incomptus.**
 Scabro-pubescent. Stem sub-simple, few-headed, laxly leafy :
  Lower-lvs. linear-lanceolate, acuminate, with
   recurved edges ... ... (38) **asperulus.**
  Lower-lvs. petioled, oblong or obovate, flat, toothed :
   Heads loosely corymbose, 40–50 fl. ... ... (39) **serratus.**

Heads solitary, or 2 (on long pedicels), about
100 fl. ... ... (40) **hypochœrideus.**
(*d.*) Heads radiate; disc and ray fl. *yellow :*
  Crown of rootstock clothed with long, silky-woolly hairs :
    Stem, bluntly-pinnatifid lvs., and invol. *glabrous* (41) **eriobasis.**
    Stem, bluntly-pinnatifid lvs., and invol. *scabro-
      pubescent* ... ... (42) **erosus.**
  Crown of rootstock *nude*, not silky or woolly :
    Scabro-pubescent, viscidulous ; or sparsely pubescent :
      Rad. leaves *inciso-pinnatifid*, or sinuate-pinnatifid :
        Inv. scales scabro-pubescent :
          Rad. lvs. inciso-pinnatifid, lobes
            blunt ... . (43) **hastulatus.**
          Rad. lvs. elongate-lanceolate, sin-
            uate-pinnatifid ... ... (44) **robertiæfolius.**
        Inv. scales *glabrous ;* lvs. pinnatifid,
          with many narrow lobes ... ... (45) **spiræifolius.**
        Rad. lvs. oblong, toothed ; heads corymbose (46) **hieracioides.**
    White-woolly, on leaves and involucre ; rad. lvs.
      pinnati-partite ... ... (47) **albifolius.**
    Glabrous *except* pedunc. and invol. ; rad. lvs.
      narrow, toothed, nerveless ... ... (48) **reptans.**
    Glabrous, subovoid ; rad. lvs. obovate, crenulate;
      scape nude, one-headed ... ... (49) **Bellis.**

§ 3. PLANTAGINEI.' *Rootstock* perennial. *Radical* leaves several, petioled, coriaceous, linear, lanceolate or ovato-lanceolate, glabrous or cottony (not viscoso-pubescent), entire (rarely *repand*), mostly callous-denticulate. *Stem* herbaceous, subsimple,
nude or leafy, its leaves clasping or sessile. *Heads* many-fl., conspicuously calycled,
radiate or discoid) ; fl. yellow, rarely purple. (Sp. 50–65.)

(*a.*) Heads radiate ; disc and ray-fl. *purple* ... ... (50) **Dregeanus.**
(*b.*) Heads radiate ; disc and ray-fl. *yellow :*
  Crown of rootstock very woolly ; lvs. ovate-oblong (51) **coronatus.**
  Crown of rootstock *nude*, not silky or woolly :
    Cauline lvs. sessile (not *decurrent* at base) :
      Achenes quite glabrous :
        Calycle of many bracteoles, much shorter than the invol. :
          Disc-fl. very many ; rays 10–15 (52) **Albanensis.**
          Disc-fl. 15 ; rays 5 ... ... (53) **inornatus.**
        Calycle of few, *very long* bracteoles ;
          rays very long, many ... . (54) **caudatus.**
      Achenes striate, scaberulous or pubescent :
        Cobweb-cottony; lvs. lanceolate, 10–18
          inches long ; achenes 4 l. long ... (55) **macrospermus.**
        Glabrous ; lvs. oblong, 3 inches long,
          8–9 l. wide ; ach. pubescent ... ... (56) **monticolus.**
        Glabr. or cobwebby; lvs. linear, 1 line
          wide ; ach. villous ... ... (57) **gramineus.**
      Achenes flattish, ciliate on the edges ... (58) **crispus.**
    Cauline lvs. *decurrent* at base, as a stem wing :
      Decurrent portion of leaf coarsely toothed (59) **decurrens.**
      Decur. port. of leaf crenato-dentate ... ... (60) **digitalifolius.**
(*c.*) Heads discoid ; flowers *yellow :*
  Inv. calycled, of about 20 narrow-linear, acuminate scales :
    Pedicels and involucre *scabro-pubescent ;* cauline
      leaves toothed ... ... (61) **polyodon.**
    Pedicels and invol. quite *glabrous :*
      Cauline lvs. clasping, toothed or denticulate (62) **crenulatus.**
      Cauline leaves sessile, elongate-linear, with
        revolute edges ... ... (63) **striatifolius.**

Inv. nude at base, of 10–12 broad, flat, glabrous scales :
>> Stem leafy below, nude and panicled upwards ;
>>> heads subcorymbose ... ... (64) **othonnæflorus.**
>> Stemless ; rad. lvs. rosulate, stem scapelike,
>>> scaly, one-headed ... ... (65) **petiolaris.**

§ 4. PAUCIFOLII. *Rootstock* perennial; roots *often,* (perhaps always) thickened or tuberous. *Radical* leaves none, or petioled. *Stem* rigid, glabrous or cobwebby, leafy at base, ending in a simple or corymbose peduncle. *Leaves* rarely all sessile, the lowermost commonly on long petioles, either quite glabrous or cottony (not scabro-pubescent), various in shape. *Heads* nearly nude at base, radiate or discoid; rays yellow or purple. (Sp. 66–80.)

(*a.*) Heads radiate ; disc and ray-fl. *purple* ... ... (67) **cymbalariæfolius.**
(*b.*) Heads radiate ; disc and ray-fl. *yellow* :
> Lower leaves *petioled,* either cordate, lyrate, reniform or multifid :
>> Lvs. lyrate, reniform or multifid ; achenes striate,
>>> setulose, tapering ... ... (67) **cymbalariæfolius.**
>> Lvs. cordate or ovate, repand-toothed ; achenes
>>> quite glabrous... ... ... (66) **cordifolius.**
> Lower leaves *sessile,* oblong or lanceolate :
>> Corymb few-headed ; heads many-fl. ; rays 8–10 (74) **paucifolius.**
>> Corymb many-headed ; heads few-fl. ; rays 3–5 :
>>> Rootstock not woolly (?) ; whole plant glabrous :
>>>> Cauline lvs. lance-oblong, shortly decur-
>>>>> rent ... ... (78) **adnatus.**
>>>> Cauline lvs. sessile, ovate, acute, stem-
>>>>> clasping ... ... (77) **glaberrimus.**
>>> Rootstock woolly at the crown :
>>>> Disc-fl. 20–30 ; lvs. acuminate, often
>>>>> falcate ... ... (76) **bupleuroides.**
>>>> Disc-fl. 10–12 ; lvs. broad and narrow (75) **latifolius.**
(*c.*) Heads discoid ; disc-fl. yellow :
> Heads subsolitary, many-fl. ; inv. scales 12–20 :
>> Glabrous :
>>> Lower leaves cordate-reniform or lyrate-
>>>> pinnatifid... ... ... (68) **tuberosus.**
>>> Lower lvs. bipinnati-partite, with linear lobes (70) **anthemifolius.**
>> Cobwebbed & *woolly*; lvs. polymorphous, mostly
>>> woolly beneath ... ... (69) **diversifolius.**
> Heads corymbose, few-fl.; inv. scales 5–9 :
>> Leaves more or less distinctly *peltate*:
>>> Lvs. cordate or hastate, toothed or angled (72) **oxyriæfolius.**
>>> Lvs. orbicular or oval, callous-toothod or
>>>> ciliate ... ... (71) **orbicularis.**
>> Leaves not peltate.
>>> Stem short; lvs. fleshy, rhomb-ovate; inv.
>>>> sc. 8–9 ... ... (73) **rhomboideus.**
>>> Stems tall, leafy ; lvs. coriaceous or mem-
>>>> branous; inv. sc. 5.
>>>>> Stem nude; lvs. membr., with slender
>>>>>> veins... ... ... (79) **isatideus.**
>>>>> Stem leafy throughout; lvs. rigid, with
>>>>>> strong, *prominent,* netted-veins ... (80) **venosus.**

§ 5. RIGIDI. *Shrubs* or *suffrutices,* rarely *herbaceous* perennials, commonly much branched, leafy. *Leaves* expanded, toothed or incised, or lyrato-pinnatifid or lyrate, variably pubescent or glabrous, rarely entire, midribbed and mostly closely netted-veined, *Heads* calycled, corymbose (rarely solitary), radiate or discoid. *Rays* yellow or purple. (Sp. 81–116.)

(*a.*) Heads discoid ; fl. yellow or sometimes "*blue*!" or white :
> Heads 80–100-fl., subsolitary ; leaves scabrous, obl.-
>> lanceolate, toothed... ... ... (81) **picridifolius.**

Heads 60-80-fl., corymbose; lvs. woolly, pinnate or
 pinnatifid... ... ...   (87) **cinerascens.**
Heads 8-20-fl., corymbose or panicled:
 Lvs. sessile, oblong-lanceolate, glabrous above
  cottony beneath ... ...   (82) **Thunbergii.**
 Lvs. sessile, lyrato-pinnatifid, scabr. above,
  canescent beneath ... ...   (83) **amabilis.**
 Lvs. on ear-clasping petioles, with an ovate limb:
  Lvs. *scabrous* on both sides, coarsely repand-
   toothed ... ...   (84) **Gerrardi.**
  Lvs. cobwebby or glabrate above, cottony
   beneath ... ...   (85) **pandurifolius.**
(*b*). Heads radiate; ray-fl. yellow.
 * Leaves distinctly *petioled*:
  Stem, petioles and one or both sides of leaves *white-woolly*:
   Lvs. cordate, denticulate, or the uppermost
    panduriform ... ...   (86) **verbascifolius.**
   Lvs. pinnatifid or pinnati-partite, lobes
    lance.-linear ... ...   (87) **cinerascens.**
  Twigs and *leaves* pubescent, or hispid-villous:
   Flexuous; lvs. cuneate-obovate, toothed;
    disc-fl. 30-40 ... ...   (88) **tortuosus.**
   Erect; lvs. ovate, deeply cut and toothed;
    disc-fl. 15 ... ...   (89) **blattarioides.**
  Stem and leaves quite, or nearly glabrous:
   Leaves *lyrate*, the petiole having 2-4 pair of small lobes:
    Heads 40-50-fl.; rays conspicuous;
     lvs. *finely* serrate ... ...   (90) **serratuloides.**
    Heads 12-15-fl.; rays *minute;* leaves
     coarsely serrate ... ...   (91) **microglossus.**
   Leaves ovate, obovate or lanceolate:—
    Shrubby:
     Leaves minutely and bluntly ser-
      rulated; corymb many-headed   (92) **crenatus.**
     Leaves sharply toothed or lobed;
      corymb lax, few-headed ...   (93) **oxyodontus.**
    Herbaceous:
     Lvs. cuneate-ovate or rhomboid,
      3-nerved at base ... ...   (94) **Zeyheri.**
     Lvs. lanceolate-acuminate, mem-
      branous, 1-nerved ... ...   (95) **pellucidus.**
 ** Leaves sessile, or decurrent:
  Lvs. stem-clasping, variably *lyrate* or *pinnatifid:*
   Heads 2-3-together, subsessile along the
    rod-like fl. branches; *calycle* woolly, sc.
    glabrous ... ...   (96) **pubigerus.**
   Heads corymbose or panicled; *calycle* and sc. glabrous:
    Leaves lanceolate, acuminate, sharply
     cut or pinnatifid; fl. branches rod-
     like, corymbose at top ... ...   (97) **incisus.**
    Lvs. lyrate, the upper oblong-toothed;
     panicle very diffuse, heads on long
     pedicels ... ...   (98) **expansus.**
    Lvs. lyrate, bluntly lobed and sharply
     toothed; corymbs densely many-
     headed, pedicels short ... ...   (99) **lyratus.**
  Lvs. sessile or clasping, or obscurely decurrent, *not pinnatifid*, netted-
   veined, cobwebby cottony or white-woolly beneath, mostly scabrous
   above:—
    Corymbs much branched, many-headed:
     Leaves broad-based, stem clasping:

Shrub, roughly hispid ; lvs. oblong
   or fiddle-shaped, obtuse, his-
   pid, thinly cobwebbed beneath **(**100**) rigidus.**
Suffrutices; lvs. white-woolly beneath :
   Lvs. oblong or lance-oblong,
     sharply toothed ... ... **(**103**) ilicifolius.**
   Lvs. linear or lance-linear,
     entire or sparingly toothed
     (or lobed) ... ... **(**104**) juniperinus.**
   Leaves tapering at base, subpetiolate,
     sub-decurrent, lanceolate, sharply
     serrate, white beneath ... ... **(**102**) pterophorus,**
   Corymbs few-headed, or heads solitary :
     Leaves clasping, ovate-oblong, rigid,
     sharply toothed, at first white-wool-
     led beneath ... ... **(**105**) oederiæfolius.**
   Lvs. obovate-oblong, subentire, cob-
     webbed beneath ... ... **(**106**) arnicæflorus.**
   Lvs. obovate-lanceolate, acuminate,
     quite entire ... ... **(**107**) microspermus.**
Lvs. (cauline) *decurrent* in a long stem wing ;
   netted-veined, denticulate, hispid beneath,
   cobwebbed above ; upper lvs. ovate, sessile **(**101**) caulopterus.**
Lvs. sessile, closely and *rigidly pubescent*, obo-
   vate, tapering at base, 5–7-toothed ... ... **(**108**) hirtifolius.**
Lvs. sessile or subsessile, quite *glabrous*, or the young parts *thinly* cob-
   webbed with *deciduous* web :
   Leaves broad-based more or less stem-clasping :
     Lvs. oblong or obovate, obtuse, sharply
     and finely serrate ... ... **(**109**) vestitus.**
     Leaves lanceolate, acuminate, rigid,
     coarsely toothed ... ... **(**110**) aquifoliaceus.**
     Leaves oblong-lanceolate, coriaceous,
     serrulate or entire ... ... **(**111**) lanceus.**
   Leaves narrowed to the base, not clasping :
     Lvs. subcuneate at base, ovato-lanceo-
     late, acute, serrulate ... ... **(**112**) scoparius.**
     Lvs. tapering at base, oblongo-lanceo-
     late, acute, serrulate ... ... **(**113**) Serra.**
     Lvs. cuneate-obovate, tapering at
     base, glaucous, coarsely few-toothed
     or subentire ... ... **(**114**) halimifolius.**
(*c.*) Heads radiate ; ray-fl. *purple :*
   Leaves oblong-ovate, with revolute margins, toothed,
     woolly beneath ... ... **(**115**) coleophyllus.**
   Leaves lance-oblong, flat, toothed, glabrous ... ... **(**116**) glastifolius.**

§ 6. MICROLOBI.   Erect *suffrutices*, more or less tomentose with short, soft hairs.
Branches closely leafy. *Leaves* sessile, ovate or lanceolate, few-toothed or entire,
with or without 1–3 pair of minute lobules at base, on one or both sides *clothed
with soft, greyish-white toment* (not cobwebby or loosely woolly). *Heads* sparingly
calycled, corymbose, radiate or discoid. Rays yellow. (Sp. 117–121.)

Heads discoid, 7–12-fl. ; lvs. ovate or oblong   ... **(**117**) oliganthus.**
Heads radiate ; rays 5–6 :
   Inv. scales tomentose ; lvs. toothed or entire :
     Lvs. ovate or oblong, 3–5 nerved, entire   **(**118**) quinquenervius.**
     Lvs. lanceolate, 3 nerved   ... ... **(**119**) lineatus.**
   Inv. scales glabrous :
     Lvs. 2–3 lobed at base, term. lobe few-
     toothed ... ... **(**120**) triplinervius.**
     Lvs. 3–4-lobed at base. term. lobe serrate
     throughout ... ... **(**121**) penninervius.**

§ 7. LEPTOLOBI. *Suffrutices,* or herbaceous *perennials,* commonly branched. *Leaves* pinnatipartite or bipinnatipartite, glabrous or tomentose, the lobes narrow-linear (in some species the leaves *vary* from *linear quite entire* to *pinnatipartite* with linear lobes). Heads corymbose, mostly calycled, radiate or discoid. Rays yellow, purple, or white. (Sp. 122–138.)

(*a.*) Heads radiate ; rays purple or white :
Suffr., glabrous ; lvs. *entire* or pinnate, with strongly
revolute edges ... ... (122) **umbellatus.**
Herb., pubescent ; lvs. pinnatipartite, lobes acuminate, entire or toothed, with subrecurved edges ;
rays *purple* ... ... (123) **grandiflorus.**
Herb., scabrous ; lvs. pinnatipartite, *very scabrous,*
lobes acute, with reflexed edges ; rays *white* ... (124) **leucoglossus.**

(*b.*) Heads radiate ; rays yellow :
Leaf-lobes very short and tooth-like ; all parts gla-
brous ... ... (126) **parvifolius.**
Leaf-lobes linear or subulate, *distant,* in few pairs :
Stem and leaves *very scabrous* with short, swollen
hairs ... ... (125) **muricatus.**
St. and lvs. glabrous or pilose, not scabrid :
Diffuse, herbaceous ; lvs. unequally pinnate (127) **Carroensis.**
Erect, suffrut. ; leaf-lobes linear-elongate (128) **pinnulatus.**
Leaf-lobes *closely set,* simple or lobulate :
Ascending-erect ; lvs. woolly, the lobes short,
obtuse, often 2–3-fid ... ... (129) **tanacetoides.**
Erect, virgate, glabrous or nearly so :
Upper leaf-lobes 2–3-fid or tooth-lobed :
Pedicels and invol. glabrous ; heads
subcorymbose ... ... (130) **achilleæfolius.**
Pedunc. and invol. scabrous ; heads
subsolitary ... ... (131) **serrurioides.**
All the leaf-lobes linear-elongate, quite
entire ... ... (132) **euryopoides.**

(*c.*) Heads discoid :
Leaves bipinnati-partite :
Heads 30–40-fl , in few-headed corymbs ... (133) **pinnatifidus.**
Heads 12–15-fl. :
Suffr. quite glabrous ; inv. sc. 7–8, broad,
subacute ... ... (134) **fœniculoides.**
Herb., viscidulous ; inv. sc. 10–12, taper-
pointed, narrow ... ... (136) **rhyncholænus.**
Heads 4–6-fl. in much branched, fastigiate
corymbs ... ... (135) **bipinnatus.**
Leaves pinnati-partite *(some simple, linear-elongate);*
Suffr. branching ; branches virgate, leafy ; leaf-
lobes few, long, slender ... ... (137) **paniculatus.**
Many-stemmed ; stems simple, ascending, leafy
at base, leaf-lobes short ; upper leaves linear,
entire ... ... (138) **multicaulis.**

§ 8. LEPTOPHYLLI. *Shrubs, suffrutices* or *herbaceous* perennials, mostly branching. *Leaves* linear, linear-filiform, or oblong-linear, mostly with revolute edges, entire or sparsely toothed, glabrous or variably pubescent or woolly. *Heads* mostly corym-bose, rarely solitary, radiate or discoid. *Rays* yellow or white. (Sp. 139–155.)

(*a.*) Heads discoid :
Leaves sessile, narrow-linear, not eared at base :
Glabrous in all parts ... ... (139) **angustifolius.**
Woolly on branches, leaves, and involucre ... (140) **niveus.**
Leaves eared or strongly clasping at base :
Leaves woolly or tomentose beneath :

Herbaceous; lvs. obtuse; corymbs dense;
  inv. woolly at base ... ...   (141) **persicifolius.**
Suffr.; lvs. pungent mucronate; corymbs
  loose; inv. glabrous ... ...   (142) **mucronatus.**
Leaves glabrous or cobwebbed:
  Suffr.; lvs. entire, the small ear toothed
    or entire ... ...   (143) **leptophyllus.**
  Herb.; lvs. denticulate, the ear deeply
    2-5-toothed or lobed ... ...   (144) **Diodon.**
(*b.*) Heads radiate; rays yellow, rarely white:
Herb.; lvs. serrate or unequally lobulate:
  Lvs. lanceolate, 10-12-toothed on each side ... (145) **Skirrhodon.**
  Lvs. elongate-linear, serrulate; heads racemose (146) **serrulatus.**
  Lvs. linear, with revolute edges, unequally
    toothed, shortly lobed or entire ... ... (147) **inæquidens.**
Suffr.; lvs. linear-filiform (some with filiform lateral lobes); heads on very
  long pedicels, or solitary:
  Rays white; heads corymbose) inv. of 12-13
    scales ... ... (148) **filifolius.**
  Rays yellow; hds. solitary; inv. of 18-20 scales (149) **debilis.**
Shrub.; glabrous; lvs. linear-elongate, quite entire,
  tapering to each end; corymbs compound; rays 5 (150) **longifolius.**
Shrub. or *suffr.;* lvs. linear or lance-linear, ear-clasping at base, 1-3 lines
  wide, with revolute edges:
  Corymb *densely* many-headed, compound; heads
    small ... ... '... ... ... (151) **rosmarinifolius.**
  Corymb loosely few-headed:
    Stem branching, often rough; lvs. close-set;
      achenes puberulous... ... ... (152) **Burchellii.**
    Stem virgate; lvs. glabrous, distant; ach.
      silky... ... ... (153) **vimineus.**
    Stem branching; lvs. sparse; ach. glabrous (154) **dracunculoides.**
  Heads solitary, on long peduncles; lvs. hairy... (155) **hirtellus.**

§ 9. PINIFOLII. Small, rigid, perfectly glabrous *shrubs.* *Leaves* needle-shaped,
crowded, rigid, pungent. *Heads* solitary or corymbose, radiate or discoid; rays
yellow. (Sp. 156-157.)

  Heads radiate ... ... (156) **pinifolius.**
  Heads discoid ... ... (157) **triqueter.**

§ 10. SCANDENTES. Climbing, half-climbing, or trailing *suffrutices* or *herbaceous*
perennials, mostly branching, and *quite glabrous.* *Leaves* petiolate, hastate, deltoid
or angle-lobed. *Heads* corymbose or solitary, radiate or discoid; fl. yellow. (Sp.
158-166.)

(*a.*) Heads discoid:
  Heads in branching corymbs, 5-15-flowered:
    Lvs. 5-7-angled or lobed; inv. of 8-9 scales ... (158) **mikanioides.**
    Lvs. deltoid-hastate, many-toothed; inv. of 5 sc. (159) **deltoideus.**
    Lvs. ovate, toothed; petiole channelled and
      winged; inv. sc. 12-15 ... ... (160) **canalipes.**
  Heads 1-5, amply calycled, 40-50-fl. ... ... (161) **quinquelobus.**
(*b.*) Heads radiate:
  Heads 1-5-7-together, *amply calycled,* 50-60-fl.:
    Leaves deltoid or hastate, with 3-5 acuminate
      angles ... ... (162) **macroglossus.**
    Lvs. bluntly 3-5-lobed, multidentate ... ... (163) **bryoniæfolius.**
  Heads corymbose, sparingly calycled, 10-30-fl.:
    Scandent: leaves hastate, unequally toothed or
      many-lobed, lobules acute ... ... (164) **tamoides.**

Half-erect, or scrambling ; lvs. ovate, tapering
at base, angle-lobed or subentire :
    Petioles 1-2½ inches long ; inv. sc. 10–12 ;
      disc-fl. 20, rays 4–6 ; achenes hispidulous (165) **angulatus.**
    Petioles short; inv. sc. 8 ; disc-fl. 6–8 ; rays
      3–4 ; achenes glabrous ... ... (166) **brachypodus.**

§ 11. KLEINOIDEI. *Shrubby* or *suffruticose*, with thick or succulent stems. *Leaves*
succulent, cylindrical, semi-cylindrical or flattish, mostly quite entire, rarely somewhat toothed or lobed. *Heads* radiate, yellow. (Sp. 167–176.)

Leaves flat, subsinuate or imperfectly pinnate-lobed :
    Lvs. lance-oblong or linear, entire or *bluntly* toothed;
      rays 5 ... . (167) **subsinuatus.**
    Lvs. *various*, sparingly and *sharply* toothed or lobed,
      rays 8–10 ... ... (168) **crassiusculus.**
Leaves quite entire, broadly linear, or semi-terete :
    Heads many, in a long-peduncled, thyrsoid raceme (169) **pyramidatus.**
    Heads one or few, on simple or branched peduncles :
      Stemless or short-stemmed ; ped. many times
        longer than the leaves ... ... (170) **scaposus.**
      Branching shrubs : pedunc. 2–3 times longer than the leaves :
        Inv. calycled ; ach. hispidulous on the striæ (171) **aloides.**
        Invol. scarcely calycled ; achenes densely
          pubescent ... ... (172) **bulbinefolius.**
    Heads in simple or branched corymbs :
      Leaves linear-elongate, tapering at base, *acute* (173) **corymbiferus.**
      Leaves linear-oblong, *obtuse* ... ... (175) **succulentus.**
    Heads on pedicels shorter than the mucronate leaves (175) **cotyledonis.**
    Heads *sessile*, ending leafy branches ; rays *very short* (176) **acutifolius.**

§ 12. APHYLLI. *Stem* branching, glabrous, succulent, *leafless*, herbaceous. *Heads*
corymbose, radiate ; rays yellow :
    Branches long and virgate, bearing terminal corymbs (177) **junceus.**

§ 1. ANNUI. (Sp. 1–27.)

1. S. vulgaris (Linn. Sp. 1216) ; annual, somewhat cobwebby or
glabrous ; stem erect, mostly branching ; leaves stem-clasping, bluntly
pinnatifid and toothed; heads corymbose, crowded or scattered, cernuous,
discoid; inv. ovoid, calycled, of many slender, tapering scales; achenes
striate, hispidulous on the striæ. *Eng. Bot. t.* 747. *Fl. Dan. t.* 513.
DC. l. c. 341.

HAB. A weed in cultivated ground, *E. & Z.!* (Herb. Sond.)
The common " Grounsel" of Europe, now dispersed over the globe.

2. S. maritimus (Linn.); annual, procumbent or prostrate, much
branched, glabrous ; leaves clasping at base, oblong-obovate, obtuse,
fleshy, veinless, calloso-denticulate ; corymbs few-headed, at the ends
of the branches ; heads radiate ; inv. calyculate, the scales glabrous,
oblong, acuminate, brown-tipped, membrane-edged ; achenes setulose.
Th.! Cap. p. 680. *DC. p.* 434. *S. telephiifolius, Jacq.? fragm. t.* 1, *f.* 3.
DC. l. c. 380. *S. ustulatus, Eckl. (ex pte.)*

VAR. β. minor; half the size of var. α, less succulent; stem ascending, paniculate;
leaves oblong, repand ; heads smaller than in α, with shorter rays.

HAB. Sea shores round Capetown, Greenpoint and Simon's Bay, *Thunberg, Pappe,
E. & Z., W.H.H., Wallich.* β. Kalk Bay, *Dr. Pappe!* (Herb. Th., D., Hk., Sd.)
Stems 1-1½ foot long ; in β. 5–6 inches, usually prostrate, alternately much branch-

ed, the branches long, subsimple, ending in a 3–5-headed corymb. Leaves rarely over an inch or 1¼ inch long, very blunt, obscurely denticulate. β. is much smaller in all parts and less fleshy ; its inv. scales scarcely ustulate.

3. S. littoreus (Thunb.! Cap. 681); annual, nearly glabrous ; stem erect, fistular, with spreading branches ; leaves eared and stem-clasping at base, oblong or oblong-hastate, or the lower ones pinnatifid, acute, unequally toothed or incised, the serratures deltoid, callous-tipped ; heads in a loose, corymbose panicle, radiate ; inv. calyculate, the scales glabrous, oblong-acuminate, brown-tipped, membrane-edged ; achenes striate, setulose. *DC. l. c. 379. S. arenarius, Th.! (ex pte.) p. 680. S. ustulatus, DC. l. c. 380. S. heterophyllus, Drege! (in Hb. Sd.)*

VAR. β. **hispidulus**; stem and lower surface of leaves more or less rough with swollen, white hairs; leaves often pinnatifid, sparingly eared and sometimes tapering to the base. *S. arenarius, Th.! (ex pte.) p.* 680. (Herb. Sd., D., Cap.)

HAB. Round Capetown, in sandy ground and near cultivation; Little Namaqualand and Saldanha Bay, *Drege!* (Herb. Th., D., Hk., Sd., Cap.)

1–2 ft. high, loosely much branched, erect or spreading. Leaves 1½–2 inches long, very variable, in β. the upper leaves are often obovate, in α. usually broad based, but all forms occur indiscriminately together. Inv. scales brown or black-tipped, those of the calycle either ciliate, pilose, or quite glabrous. *Drege's* " *S. ustulatus*" *DC.* is undistinguishable from this ; *Ecklon's* (in Hb. Cap.) is *S. maritimus*, Linn.

4. S. laxus (DC.? l. c. 381); annual, quite or nearly glabrous ; stem erect, fistular, angular, paniculately branched ; cauline leaves amply eared and amplexicaul at base, narrow, acute, toothed, incised or pinnatifid, the margin reflexed, lobes callous-denticulate ; panicle diffusely much branched, many-headed; pedicels slender, nude; heads *(small)* ; inv. 12–15-leaved, glabrous, calyculate ; disc-fl. 20–30 ; rays 5–6 ; achenes oblong, minutely downy-canescent, *not striate.*

HAB. Olifant's River, *Drege!* (Herb. Sd., D., Hk.)

2 ft. or more high, distantly branched, weak. Lower leaves unknown ; medial 1–2 inches long, distant, the lobes 1–2 lines wide. Heads 2 lines long, the disc 1½ wide. With the foliage of *S. littoreus*, this has a much more diffusely branched panicle, smaller fl.-heads; very narrow scales to the calyculus, and different achenes.

5. S. ruderalis (Harv.); annual, glabrous; stem erect, branched, rigid, angle-striate; leaves sessile, linear-lanceolate or lanceolate, acute, tapering to the base, distantly denticulate or subentire with subrevolute margins, sometimes slightly eared at base ; heads loosely corymbose, on long, sparsely scaly pedicels, radiate ; inv. calycled at base, glabrous, of 18–20 acute scales; rays 12–15, revolute; disc fl. 50–60; achenes closely puberulous.

HAB. Waste ground near Natal, *T. Williamson, J. Sanderson,* 326, *Gerr. & M'K.,* 324. *Dr. W. B. Grant.* Near Delagoa Bay, *Capt. Speke.* (Herb. D., Hk.)

1–1½ ft. high, diffusely branched, the stems green. Leaves 2 inches long, 2–4 lines wide, rather rigid when dry, on both sides green. Inv. campanulate, 2–3 lines long. Rays yellow, 4 lined.

6. S. abruptus (Thunb.! Cap. 685); annual, subglabrous or sparsely hispidulous; stem erect, weak, angular, paniculately branched; leaves half-clasping at base, oblong, deeply or sharply incised or pinnatifid, the lobes or teeth oblong, spreading, subacute; the larger ones dentate ; margins reflexed; infl. laxly panicled, the pedicels elongate, scarcely scaly; inv. naked at base or nearly so, glabrous, scales linear, acuminate, 2–3 ribbed; rays several, revolute; achenes striate, puberulous. *S. hetero-clinius, DC. l. c.* 380.

Var. β. **heterophyllus**; lower leaves toothed; upper narrower, linear, subentire. *S. heterophyllus, Th. ! Cap.* 679. *DC. ! l. c.* 379.

Hab. Sandy, moist ground in the Western Districts. *Thunberg!* Cape Flats, and Stellenbosch, *E. & Z.!* Paarl, *Drege!* Berg R., *Zeyher!* (Herb. Th., D., Sd., Hk.)

Habit of *S. littoreus,* but the leaves are commonly much more deeply pinnatifid, (though varying much in this respect) and always known from that species by the absence of calyculus, or the mere 1–2 minute bracteoles in its place. Leaves 1½–2 inches long, the lobes 2–5 lines long, 2–4 lines wide. Panicle diffuse.

7. S. consanguineus (DC. l. c. 381); "herbaceous, erect, branched; stem terete, minutely puberulous; leaves glabrous, membranaceous, sessile, ear-clasping, pinnatipartite, the lobes on each side 5–6, oblong, obtuse, sinuate-toothed, some of the lower ones petioled, exauriculate; corymb loosely few-headed; pedicels long, nude; inv. nude at base, 10-leaved, scarcely puberulous, not torulose, equalling the disc; disc-fl. about 20, rays 5, minute, revolute; achenes terete, pubescent." *DC. l. c.*

Var. β. **major**; "larger and more hispid; disc-fl. 30; rays 6–8." *DC. l. c.*

Hab. Near Verleptpram, on the Gariep, both vars. (Var. β. in Herb., Hk., Sd.) Stems in var. β. 1–1½ feet, fistular, striate, minutely scaberulous. Leaves 1½ inch long, subpetiolate, deeply pinnatifid, the lobes in several pairs, horizontal, oblong, sharply and unequally toothed or biserrate. Corymb rather close. Inv. scales sparsely hispid, flat. Achenes pubescent. I have not seen var. α.

8. S. repandus (Thunb. Cap. 683); annual, weak, erect; stem erect, scaberulous, with spreading branches; lower and medial leaves on long petioles, auricled at base, thinly membranous, veiny, ovate or orbicular, coarsely toothed, or lyrato-pinnatifid with a large terminal lobe and few lateral lobes, the upper sessile, pinnatifid or entire; inflorescence lax, the heads on long, slender pedicels; inv. nearly nude at base, the scales 12–13, linear-lanceolate, shorter than the disc, acuminate, scabrous, brown-tipped; rays 7–10, spreading; achenes minutely downy. *DC. l. c.* 381. Also *S. membranifolius, DC.! l. c.* and *S. inconstans, DC.! l. c.* 382.

Var. β. **erodiifolius**; dwarf, diffuse; leaves commonly lyrate; pedicels few. *S. erodiifolius, DC.! l. c.* 381. *Cineraria pumila, Thunb. Cap. p.* 672.

Hab. At the Paarl, *Thunberg, Drege!* Clanwilliam, *Ecklon!* Simonstown, *C. Wright,* 300 (inv. sc. *glabrate!*) (Herb. Th., D., Sd., Hk.)

3–12 inches high, very pale and tender. Leaves, with their petioles, 2–2½ inches long in the larger specimens, ¼ that size in the dwarf, varying from simple to lyrate-pinnatifid. Pedicels 2–3 inches long. Heads 4 lines long, 2½ broad : inv. usually quite nude, sometimes with one or two small bracteoles at base. No specimen of "*S. repandus*" exists in Thunberg's Herbarium, but his description agrees with our plant. I describe chiefly from *E. & Z.'s* specimens of "*S. membranifolius,*" as being the best in condition of those enumerated, which however all appear to belong to the same species. I have only seen a *scrap* of "*S. repandus*," DC.

9. S. Lessingii (Harv.); annual, erect, branching; stem and branches terete, striate, puberulous, especially the young parts; leaves on long petioles, oblong or obovate, repand or obsoletely crenate, membranous, veiny, glabrous or nearly so; petioles not eared at base, scarcely clasping;

heads loosely corymbose, on pubescent pedicels ; inv. of 12–14 broad, black-tipped, striate scales, calycled with several flat, ciliate, black-tipped bracteoles ; rays several ; achenes puberulous. *S. repandus, ex pte. Herb. Th.! S. petiolaris, Less.! in Hb. Th. MSS., non DC.*

HAB. Cape, *Thunberg.* (Herb. Th.)
Described from a specimen in Hb. Th. marked " *S. repandus,* 2," but well distinguished from that species by the amply calycled involucres, &c.   Petioles 1–1½ inch long ; lamina 1–1½ in. long, ¾–1 in. wide, more or less cuneate-attenuate at base. Heads rather large, like those of *S. littoreus.*

10. S. lobelioides (DC. l. c. 382); "herbaceous, erect, branched, quite glabrous ; stem terete ; lower leaves petioled, cordate, toothed; upper sessile, cordate-amplexicaul, toothed; corymb few-headed, the pedicels nude; inv. cylindrical, smooth, nearly nude at base, equalling the disc ; scales 9–10, barbellate at tip ; disc-fl. 25, rays 3, very small ; achenes pubescent." *DC. l. c.*

HAB. Zilverfontein, *Drege!* (Unknown to me.)
" 6–9 inches high. Heads 3–4 lines long. Fl. pale-yellow." *DC.*

11. S. erysimoides (DC. l. c. 382); annual, glabrous or nearly so ; stem erect, branched, angle-furrowed above; lower leaves petioled, ovate, sinuate, the rest sessile, ear-clasping at base, deeply pinnatifid or pinnati-partite, thin, the lobes in 5–9 pairs, spreading, some entire, some sharply toothed, acute, mucronate ; corymb lax, pedicels slender ; inv. glabrous, calycled with a few black-tipped, shortly subulate bracteoles, 13–15-scaled ; rays 8–10, spreading, 4-ribbed, longer than the involucre; achenes striate, minutely downy.

HAB. Zwarteberg, *Drege!* (Herb. Sd.)
Of this I have only seen two upper leaves and a few fl. heads. The latter differ from those of *S. cardaminifolius* by the longer rays, and black-tipped, broader, and shorter bracteoles. *De Candolle* also compares it with his *S. inconstans* and *S. consanguineus. Zeyher's* 2971, var. β. (Herb. Sd.), from Hassagay's Kloof, comes very near this.

12. S. cardaminifolius (DC.! l. c. 382); annual, glabrous or nearly so, erect, branched ; stem weak, rib-striate; leaves eared at base, of thin substance, the lower ones petioled, ovate or oblong, coarsely dentate, the upper subsessile, tapering to the base, deeply pinnatifid or pinnati-sect, the lobes in 3–4 pair, horizontally patent, subulate, acute, mucronate, quite entire; corymb rather close, pedicels short ; inv. glabrous, of 12–15 scales, calycled with a few very slender, setaceo-subulate bracteoles ; rays 6–8, shorter than the invol; achenes pubescent.

HAB. Zilverfontein, *Drege!* (Herb. D , Sd., Hk., Cap.)
A weak annual, 6–12 inches high or more, with thinly membranous, pale leaves. Leaves 1–1½ inch long, the lower ones ¾–1 inch wide. Inv. 2 lines long ; rays 1½, at length revolute.

13. S. sophioides (DC.! l. c. 382); annual, sparsely hispid or glabrescent ; stem erect, branching, angular ; leaves half-clasping at base, bipinnatipartite, shortly petioled, the lobes narrow-linear, with revolute margins, elongate, some simple, some laxly pinnulated, all callous-tipped; infl. subsimple, the pedicels terminal, elongated, slender, pubescent;

inv. of 12–13 acute, scabro-pubescent scales, nude or scarcely calycled; fl. 40–50; rays several, *purple!* (or "pale" *DC.;*) achenes striate, downy.

HAB. Districts of Caledon and Swellendam, *Ecklon!* (Herb. Sd., Cap.)
6–12 inches high, much branched. Leaves 1½–2 inches long, the lowermost bi-pinnate, the upper varying to pinnate; lobes ½ line wide, tipped with a remarkable callus. Rays in an original specimen, in Hb. Sd., marked by *De Candolle*, distinctly *purple;* in another in Hb. Cap., as clearly *pale*, and probably faded; or does it vary with purple and yellow rays? In other respects the specimens are identical.

14. S. tenellus (DC. l. c. 379); "annual, erect, branched, glabrescent, slender; leaves linear, acute, quite entire, subscabrid; fl. branches long, nude, one-headed; inv. uniseriate, of 12–13, linear, acute, membrane-edged scales." *DC. l. c.*

HAB. Cape, *Boissier.* (Unknown to me.)
" A minute annual (3–6 inches high?). Leaves 4–6 lines long, ½ line wide. Heads scarcely 3 lines diam., pale-yellow " Is not this a dwarf state of *S. lævigatus!*

15. S. lævigatus (Thunb.! Cap. 686); annual, nearly glabrous; stem erect, rigid, angle-striate, paniculately branched; leaves clasping at base, narrow-linear, with more or less revolute margins, the upper ones often entire, the medial and lower pinnati-partite, with few linear lobes, or inciso-lobulate; lobes quite entire; branches elongate, nearly nude, one-headed; inv. calyculate, the scales linear, acuminate, glabrous; rays several; achenes pubescent. *DC. l. c.* 382. *Also S. plebeius, DC.! l. c.* 379 *(fide Sp. Eckl.!)*

VAR. β, **integrifolius**; leaves undivided, linear, with revolute margins, glabrous. (Herb. D.)
HAB. Cape, *Thunberg!* Langekloof, *Burchell, Krebs, E. & Z.!* Simonsbay, *C. Wright!* 299. VAR. β. Albany, *T. Cooper*, 9, 12. (Herb. Th., Sd., D.)
About a foot high, rigid, and almost ligneous at base, paniculately much branched, slender. Leaves 1–1½ inch long, the lobes not a line diameter, rigid, glabrous or scaberulous. Fl. branches 2½–4 inches long, filiform. Heads 3 lines diam. I have not seen *Burchell's* plant, but *Ecklon's* is the same as *Thunberg's.*

16. S. matricariæfolius (DC. ? l. c. 383); annual, diffuse, glabrous or nearly so; branches angle-furrowed; leaves bipinnatisect, the lower petioled, the rest ear-clasping at base, lobes narrow, sharply pinnatisect or inciso-dentate, spreading (sparsely pilose beneath); pedicels few, long, slender, scaly; inv. glabrous, calycled with brown-tipped, ciliolate bracteoles, 10–12-scaled; rays several, spreading, longer than the invol., 4-nerved; achenes striate, cano-puberulous. *S. filipes, Turcz. Bull. Mosc. XXIV. II. p.* 87. *Walp. Ann. V. p.* 336.

HAB. Zwarteberg, near Klaarstroom, *Drege.* Hassaquaskloof, *Zeyher!* (Hb. Sd.)
I have not seen *Drege's* plant, described by *DC.;* but that of *Zeyher (S. filipes,* Tcz.!) so nearly agrees with *De Candolle's* description, differing seemingly in not being quite glabrous, that I venture to consider it the same. Stems many from the base, 6–8 inches long, widely spreading. Leaves 1–1½ inches long. their lobes not a line in breadth, sharply multifid. Pedicels 3 inches long.

17. S. trachylænus (Harv.); annual, nearly glabrous except on the inflorescence; stem erect, leafy; leaves sessile, half-clasping at base, glabrous, deeply pinnatifid, the lobes in 3–5 pair, broadly linear or oblong, subacute, entire or toothed with revolute margins; *apex of the* *stem, flowering branches, bracteæ, pedicels and involucres all equally and densely clothed with gland-tipped, rigid hairs;* inv. of 16–18 narrow, acute scales, nearly or quite nude at base, pale; rays 8–10, revolute, yellow; achenes striate, puberulous.

HAB. Bitterfontyn, *Zeyher!* (Herb. Sond.)
Probably branching, panicled or corymbose; perhaps 1–1½ ft. high. I have only seen a small and evidently dwarfed specimen, unbranched, about 5 inches high, bearing 2 fl. heads. The glandular pubescence on the inflorescence, combined with the glabrous stem and foliage, seems to me, however, to mark this plant so definitely that I venture to found a species on very imperfect data.

18. **S. diffusus** (Thunb.! Cap. 686); annual, diffuse, much-branched, aspero-pubescent, glandular; lower leaves bipinnatisect, the lobes multifid, divaricate, narrow linear; upper ear-clasping at base, inciso-pinnatifid, with entire or toothed lobes; heads few, on long, more or less scaly pedicels; inv. of 12–15, scabrous, linear-acuminate scales, calycled with few or several subulate, often ciliate bracteoles; rays several, spreading, scarcely equalling the inv., 4-nerved; achenes striate, the striæ minutely puberulous. *S. chamœmelifolius, DC.! l. c.* 383.

HAB. Zwartland, *Thunberg!* Konstapel, *Drege!* (Herb. Th., Sd.)
Stems many from the base, decumbent or ascending, 6–8 inches long, branched. Leaves 1–1½ inch long; the lobes ⅙ line wide. Rays pale yellow. *Thunberg's* original specimen so nearly accords with *Drege's* that I venture to refer *S. chamœmelifolius, DC.* to the older species; the imperfect specimen of it (in Hb. Sd.) which alone I have seen, chiefly differs in its less compound leaves and more copious calycle-bracts.

19. **S. Paarlensis** (DC! l. c. 383); annual, erect, much branched, aspero-pubescent; stem weak, angle-furrowed; leaves clasping at base, the lower petioled, sub-bipinnatisect, the upper sessile, pinnatisect, the lobes 4–5-pair, spreading, narrow-linear, acute, entire or toothed; infl. loosely panicled, the pedicels elongate, nude or 1–2-squamellate; inv. of 12–13 scabrous, linear, acute scales, nearly nude at base or 1–2-bracteolate; rays 8–10, spreading, 4-nerved, as long as the involucre; achenes striate, elongate, tapering upwards, cano-puberulous on the striæ.

HAB. At the Paarl, *Drege!* R. Zondereinde, *Zey.! 2971* (ex pte.) (Hb. Hk., Sd.)
6 inches to 1½–2 feet high, panicled. Leaves 1–1½ inch long, the lobes ¼–1 line wide. Inv. 3–4 lines long; rays deep or pale yellow, sometimes fulvous when dry. A specimen (Hb. Sd.) from *Drege* seems different, having a glabrous and calycled involucre; but it is in an immature state, and therefore doubtful: *Drege's* sp. in Hb. Hk. agrees with our plant.

20. **S. puberulus** (DC.! l. c. 380); "herbaceous, erect, sparingly branched, nearly all parts roughly pubescent; branches striate, the upper long, 1-headed, nude; leaves lance-linear, toothed or entire, acute, the lower tapering at base, the others sessile; heads on long pedicels; invol. sparingly calyculate, 16–20-leaved, glabrescent, scarcely shorter than the disc; disc fl. 30–40, rays about 10; achenes terete, striate, minutely pubescent." *DC. l. c.*

HAB. Albany, *Drege.* (Unknown to me.)

21. **S. sisymbrifolius** (DC. l. c. 382); annual, viscoso-pubescent; stem

erect, branching; leaves half-clasping at base, long-petioled, some ovate-oblong, sinuate toothed or incised, some pinnatipartite with a large terminal lobe; infl. laxly panicled, subcorymbose, the pedicels long and slender; inv. scarcely calyculate, 10–12-scaled, glandularly-pubescent; disc-fl. 20–30, rays 5, long, 4-nerved, spreading; achenes pubescent on the striæ.

HAB. Between Kaus and the Gariep, *Drege!* (Herb. Hk.)
6–10 inches high, slender. Petioles 1–1½ inch long, 1 line wide; lamina uncial. Inv. 3 lines long; the exserted portion of the yellow ray 4 lines.

22. S. glutinarius (DC.! l. c. 381); annual, viscoso-pubescent, pilose with swollen hairs; stem erect, fistular, panicled upwards; leaves eared and clasping at base, oblong, sinuate-toothed or bluntly pinnatifid, thickish, on both sides hairy; infl. subcorymbose, rather dense, the heads on short pedicels; invol. of 10–12 scales, subcalyculate, *glabrous*, the scales oblong, acute; disc fl. 20–30, rays few, short and revolute; achenes striate, puberulous. *S. glutinosus, E. Mey!* (non Th.)

HAB. Saldanha Bay, *Drege!* (Herb. D., Sd., Hk.)
A foot or more high, much branched, viscidulous. Leaves (not in good state in our specimens) 1–1½ inch long, ½–1 inch wide. Heads small; inv. 2¼ lines long. Rays very short. Known among allied species by its perfectly glabrous involucre and viscid pubescence of the parts.

23. S. glutinosus (Th.! Cap. 684); annual, in all parts viscoso-pubescent; stem erect, robust, fistular above, panicled upwards; leaves obtusely pinnatifid, elongate, the lowermost petioled, the upper amplexicaul and eared at base, sinuses rounded, lobes blunt, and bluntly toothed or repand, short; panicle lax, subcorymbose, the pedicels long, glandular; invol. either nude at base or 1–2-bracteolate, the scales numerous, linear-acuminate, glandularly pubescent; rays yellow, spreading or revolute; achenes striate, minutely pilose. *DC. l. c. p. 381.*

VAR. β. **inamœnus**; lvs. (in outline) oblong-obovate, their lobes deeper and longer than in α.; infl. corymbose. *S. inamœnus, DC.! l. c.* 382. *Zey.!* 2964 *(starved).*

VAR. γ. **brachyglossus**; stem more rigid, subsimple; lvs. narrower, with short lobes; inv. scales sparsely setulose; rays short, revolute. *S. brachyglossus, Turcz.! Bull. Mosc. XXIV. II. p.* 87. *Walp. Ann.* 5, *p.* 336. (Herb. Sd., D.)

HAB. Table Mountain and round Capetown, *Thunberg! Ecklon* (ex pte.), *W.H.H.* Var. β. Zwartkops R., Uitenhage, *E. & Z.!* γ. Caledon River, *Ecklon!* (Herb. Th., D., Hk., Sd.)
1–2 or 3 ft. high, or 2–3 inches when starved, covered in all parts with short, glandular pubescence. Lower leaves (in large specimens) 6–8 inches long, upper 2–3 inches; very variable in size, tolerably constant to an oblong form, deeply but bluntly cut, with *very round* spaces between the lobes. In foliage and other characters this greatly resembles *S. hastulatus*, with which it is often confounded in Herbaria; but that has a perennial, thick rhizome, and numerous radical leaves.

24. S. elegans (Linn! Sp. 1218); annual, viscoso-pubescent, erect or diffuse, paniculately branched; leaves ear-clasping at base, petioled, *polymorphous* (either oblong subentire or toothed, lyrate, pinnatifid, pinnati-partite or bipinnati-partite, with cuneate, toothed or incise lobes or lobules); corymb laxly several headed, the pedicels long and scaly; heads radiate, many-fl., the disc yellow, rays purple; inv. calycled with many broad ciliate black-tipped bracteoles, of 12–15 broadly linear, flat, glabrous, or pilose, black-tipped scales; achenes striate, puberulous. *Willd. Sp. 3.* 1990. *Bot. Mag. t.* 238. *S. pseudo-elegans, Less.! DC. l.c.* 407.

VAR. *a*, **erectus**; stem erect, slender, leaves more frequently pinnate or bipinnati-partite; inv. scales *glabrous! S. carnosus, ex pte., Th.! Herb. S. pseudo-elegans, Less.! Syn.* 391. *DC.! l.* c. 407. *S. elegans, Willd. Bot. Mag. t.* 238. *Seba. Thes.* 1. *t.* 22. *f.* 1. *S. elegans β. Linn.*

VAR. *β*, **diffusus**; stem diffuse, thick and rigid, roughly pubescent; leaves more frequently lyrate, with a large terminal lobe; inv. scales *piloso-scabrous! S. carnosus, ex pte. Th.! Herb. DC. l. c. p.* 409.

HAB. Sandy ground round Capetown and in the Western Districts, *Drege! Eckl.! Zey.! W. H. H., C. Wright* 312, 317. Cultivated in Europe. Var. *β.* near the sea shore, *Thunb.! Drege!* (Herb. Th., D., Hk., Sd.)

Stems 1–2 ft. high, the larger specimens much branched. Leaves 1½–3 inches long, extremely varied in shape and degree of incision. Inv. truncate at base; the numerous bracteoles of the calycle 1–2 lines long; the scales 5–6 lines long. Rays showy. A var. with double flowers is the well-known "American Grounsel," of Engl. gardens; why called "American" I know not. The wild plant seems constantly annual; the cultivated variety may be kept for 2 or more years.

Mr. Kippist has kindly furnished me with the following note, as the result of his examination of the Linnæan specimens of "*Senecio elegans* :" it seems to establish the incorrectness of Lessing's change of name to "*pseudo-elegans.*" "I find [in the Linnæan Cabinets] two plants pinned together, both marked *S. elegans* in Linnæus's own hand. There is no habitat to either, nor any information as to the history of the specimens. The first sheet, however, having the No. '13' (corresponding with the 1st Ed. of the Sp. Pl.) prefixed to the name *elegans* must be taken as the type specimen, and this is assuredly the *pseudo-elegans* of Less., the calyx being most distinctly bicalyculate, glabrous, with the outer scales serrulato-ciliate, and each of the pedicels bearing 4–5 squamæ. Why Lessing should have changed the name, I know not, especially as, in all three of the figures quoted by Linnæus, and even in Commelin's wretched plate, the invol. is clearly bicalyculate. * * * Linnæus subsequently introduced into his herbarium, with the name *S. elegans*,' a totally different plant, with the aphyllous pedicels and a uniseriate, hairy involucre; apparently agreeing with *S. elegans*, *γ. viscosissimus, DC.*"—The error perhaps originated with *Thunberg.* who gave the name *carnosus* to the plant with bicalyculate involucres; and the names *S. elegans, S. arenarius* and *S. myrrhifolius* to that with uniseriate involucres.

25. S. multibracteatus (Harv.); annual, sparsely pubescent, visciculous above, erect; stem terete; cauline leaves sessile, not auricled, lanceolate, coarsely few-toothed, tapering at base, glabrous or nearly so, with scarcely reflexed margins; rameal leaves stem-clasping, coarsely toothed at base, acuminate, more or less scabrid; heads subcorymbose, on very long, copiously scaly pedicels; inv. of many dark-tipped, glabrous scales, amply calycled with many imbricating, lanceolate bracteoles nearly equalling the inv. scales; disc yellow, rays purple.

HAB. Bethelsdorp, *Zeyher!* 2962. (Herb. Sond.)

Marked *perennial* by Zeyher; but the root looks like that of an annual. Stem 1–1½ f. high, branched above. Leaves 2–2½ inches long, 4–5 lines wide, the teeth deltoid, distant. The pedicels toward the fl. head are very scaly and the calycle nearly as long as the involucre, by which characters and the form of leaves this is known from *S. elegans.*

26. S. arenarius (Th.! Cap. 680, ex pte); annual, viscoso-pubescent, erect, paniculately branched; leaves petioled, more or less clasping at base, *polymorphous* (either oblong or obovate, subentire or toothed, lyrate, pinnatifid or bipinnatifid, with cuneate, toothed or incised lobes;)

corymb laxly several-headed, the pedicels long and scaly; heads radiate, 30–50-fl., the disc yellow, rays purple; inv. nearly nude at base, of 12–13 narrow, keeled, scabrous scales; achenes tapering, striate, puberulous. *S. elegans, Th.! Cap. 785, Less. Syn. 391. (excl. syn. Linn.) C.! l. c. 408. S. myrrhifolius, Th.! Cap. 685. S. Volkameri, Sch. Bip.*

HAB. Sandy ground near Capetown and in the Western Districts, common. (Herb. Th., D., Hk., Sd.)

A smaller and more slender plant than *S. elegans,* from which it is at once distinguished by its very dissimilar involucres. Stems 3–18 inches high, weak. *Thunberg,* both in his Herb. and Flora, confounds with it *S. elegans* and a hispidulous form of *S. littoreus:* and gives the name *myrrhifolius* to specimens with more divided leaves. Of the synonyms at my disposal, I adopt *arenarius,* as being most appropriate.

27. S. cakilefolius (DC. l. c. 408); "annual, herbaceous, glabrous, erect; leaves sessile, car-clasping, pinnatipartite, the lobes obtuse, the margin here and there revolute, fleshy; corymb subfastigiate, branched, pedicels nearly nude; inv. 1-seriate, not calycled, of 20–24 linear scales." *DC. l. c.*

HAB. Silverfontein, *Drege.* (Unknown to me.)

Allied to *S. arenarius,* from which it differs by its glabrous surface and larger fl. heads.

§ 2. SINUOSI. (Sp. 28–49.)

28. S. macrocephalus (DC.! l. c. 407); perennial, glandular and more or less pilose with long, soft, jointed hairs; radical and lower leaves obovate-oblong, obtuse, narrowed to the base, but not petioled, irregularly toothed or repand; stem herbaceous, ascending, leafy, subsimple, *racemose* at the summit; cauline leaves clasping, oblong, subdentate, obtuse or acute; heads in a simple or slightly branched raceme or thyrsus, pedicels not twice as long as the head; inv. sparingly calycled with a very few long, narrow bracteoles, of 18–20 narrow pilose or hirsute scales; heads radiate, 60–80-fl., disc and ray purple; achenes striate, pubescent.

β. hirsutissimus; upper part of stem, pedicels and involucres, shaggy with very copious, long, soft, jointed hairs. (Herb. D.)

HAB. Katberg and betw. Key and Buffel River, *Drege!* Natal, *J. Sanderson,* 56. *Dr. Sutherland!* β. Betw. Basche and Key, *Drege!* Kreili's Country, *H. Bowker!* (Herb. D., Hk., Sd.)

Very near *S. concolor,* from which it is chiefly known by its simpler and more leafy stems; racemose inflorescence, sessile (or very obscurely petioled) lower leaves, and somewhat larger fl. heads. Var. β. is remarkable for its excessive indument, resembling that of some of the Andean *Culcitia.* It only differs from *S. thyrsoideus* by its rayed fl. heads.

29. S. concolor (DC.! l. c. 407); perennial, glandular, and more or less copiously clothed with soft, jointed hairs; radical and lower leaves petioled, oblong, obtuse, either toothed, incised or lyrato-pinnatifid, with short, blunt lobes; stem herbaceous, ascending, weak, leafy below, nude and subdichotomous above; cauline leaves ear-clasping, oblong or linear, toothed or subentire; corymbs laxly few-headed, the pedicels long; inv. nearly nude at base, of 15–20 narrow, pilose scales; heads radiate, 50-fl., disc and ray purple; achenes striate, puberulous.

VAR. β. **hispidus**; copiously clothed with long, soft, jointed hairs; lower leaves oblong, toothed.

VAR. γ. **lyratus**; pubescence shorter and more scanty; lower lvs. lyrato-pinnatifid.

HAB. Near Tulbagh and in the Zuureberg, *Drege.* β. Abundant throughout Uitenhage and Albany; in Br. Caffraria, *Eckl.! D'Urban! Hutton! Genl. Bolton!* &c. Natal, *Gueinzius! Sanderson,* &c. γ. Simonstown, *C. Wright!* 309. Modderlagd, *Zey.!* 941. (Herb. D., Sd., Hk.)

Similar in habit to *S. erubescens,* from which it is known by its radiate heads, and *usually* by its more glandular and pilose pubescence. Var. β. is very abundant and ought perhaps to be regarded as the type of the species. Of *DC.'s typical* form I have only seen a single specimen (Hb. Hook.) of Drege's " *C.*" (from Katberg); it is incomplete, but looks as if it belonged to a *different* species.

30. S. erubescens (Ait. Kew 1, v. 3, p. 190); perennial, glandularly pubescent, viscidulous; radical and subradical leaves petioled, *polymorphous* (oblong, obovate or lyrato-pinnatisect, with short, toothed, lateral lobes and wide interspaces); stem herbaceous, ascending, weak, laxly leafy below, nude, pedunculoid and dichotomous above; cauline leaves ear-clasping, oblong or linear, toothed or subentire, or inciso-pinnatifid; corymbs laxly few-headed, the pedicels elongate; inv. very sparingly calycled, of 12–15 narrow, scabrid scales; heads discoid, 30–50-fl.; achenes striate, more or less pubescent or pilose. *DC. l. c.* 406. *S. vernonioides, Sch. B. Bot. Zeit.* 27, *p.* 701.

VAR. α. **lyratus**; lower leaves lyrato-pinnatifid; upper incised. (Hb. D., Hk.)

VAR. β. **dichotomus** (DC.); lower leaves shortly petioled, oblong or obovate, toothed, repand or subentire. *Jacobœa scabra, Thunb.! Cap.* 677. *Sen. adfinis, Less.! in Hb. Th. MSS.* (Hb., D., Th., Sd.)

VAR. γ. **crepidifolius** (DC.); leaves on longish petioles, oblong or obovate, more or less toothed. *Senecio variabilis, Sch.! Bip. Bot. Zeit.* 27, *p.* 700. (Hb. D., Sd., Hk.)

VAR. δ! **incisus** (DC.); stem ascending, leafy below, nude and subdichotomous above; leaves scabrous and pilose with jointed hairs, sharply toothed and incised. (Herb. D., Sd.)

HAB. Throughout the Colony, and in Caffraria and Natal. Var. *a,* near Capetown, *W. H. H.* β. Hott. Holland, *Eckl.!* Karrega R., Albany, *Zey.!* 942. Buffeljadgts R., *Zey.!* 2972, *ex pte.* γ. common at Natal, *Krauss! Gueinzius! Gerr. & M^cK.* 311, 842; *Sanderson!* 4. Ft. Bowker, *H. Bowker!* 566. Humansdorp, *Dr. Pappe.* (Herb. Th., D., Hk., Sd.)

Radical and lower leaves 2–3, rarely 4 inches long. Stem 1–1½ f. high, weak, nude for half its length or more; except in δ, which also differs from other varieties by its *pilose* indument, like that of *S. thyrsoides* in some respects. Corymbs 3–9-headed, very lax.—Like *S. purpureus* in miniature, especially our var. *a*; but much smaller, weaker, with more simple infl., rather longer involucres, and pubescent (not glabrous) achenes. The lower leaves vary much in form.

31. S. purpureus (Linn. Sp. 1215); perennial, glandularly pubescent, viscidulous; stem herbaceous, erect, tall, leafy, corymboso-paniculate at the summit; radical and lowest cauline leaves petiolate, elongate, lyrato-pinnatifid, tapering at base, the terminal lobe large and inciso-lobulate, the lateral short and broad, irregularly toothed or repand, with very wide rounded interspaces; medial and upper cauline leaves sessile, ear-clasping oblong, or lyrate or pinnatifid, toothed or repand; callous-denticulate; infl. compound, the partial corymbs densely many-headed; inv. nearly nude at base, of 10–13 scabrous scales, shorter than the fl.; heads discoid, 40–50 fl.; achenes striate, glabrous. *DC.! l. c.* 406 *(excl. syn. Thunb.*

et Berg?) *Jacq. Ic. Rar.* 3. *t.* 580.　*S. mucronulatus, Sch. Bip. Bot. Zeit.*
27. *p.* 701.

HAB. Moist places　Various localities on and round Table Mt., *Mundt.! Bowie!*
Eckl.! Drege! Cape Flats.　Dutoits Kloof, and Kl. Draakenstein, *Drege*.　Plettenberg Bay, *Dr. Pappe!* Vanstaaden Mts , Uit., *Zey.! 2972, ex pte.* (Herb. D., Sd., Hk.)
　Stem 3–4 ft. high, strictly herbaceous, fistular upwards.　Lowest leaves, with
their petiole, 12–18 inches long, the terminal lobe 3–4 inches wide.　Upper cauline
leaves 2–4 inches long, ½–1½ inch wide.　Heads very numerous, 3–4 lines long.
Thunberg's "*Jacobœa purpurea*" (by description) must be different; but I cannot
find it in his Herb.

32. S. odontopterus (DC.! l. c. 406); in all respects like *S. purpureus*, except that the cauline leaves are *more or less* prolonged at base, into a decurrent, toothed or incised, vanishing stem-wing.

HAB. On the Paarl Mt., *Drege!* (Herb. D., Hk., Sd.)
I have not seen the radical leaves, which seem, by description, to be identical
with those of *S. purpureus*: the *decurrent* character of the cauline leaves varies considerably, and I find no difference in the floral characters or in the achenes; this
probably, therefore, ought to rank as a mere *var.* of *S. purpureus*.

33. S. glabrifolius (DC. l. c. 406); "herbaceous, erect, branched, glabrous; fl. branches terete, nude, 1-headed, sparsely scaly under the head; leaves crowded at the base of the stem, petiolate, pinnatipartite, the lobes incised, often trifid, heads about 25-fl., discoid; inv. nude at base, of 12–15 scales shorter than the disc; achenes black, oblong, sub-compressed, with sparse, whitish pubescence; corollas tipped with red." *DC. l. c.*

HAB. Kamiesberg, *Drege.*
Unknown to me, except by a frustule, without fl. heads, in Hb. Sd.

34. S. barbatus (DC. l. c. 406); perennial, the stem, leaves, pedicels, and inv. scales pilose, *with long, jointed, soft hairs;* stem ascending, sim-ple, closely leafy throughout; lower leaves obovate-oblong, tapering at base, medial linear-oblong, sessile, both toothed; upper linear or lance-linear, subentire, with slightly revolute margins; pedicels several, sub-corymbose, terminal, rising among the upper leaves, and scarcely longer; heads discoid, many-fl.; inv. calycled with a few, very slender, filiform bracteoles, of 12–18 linear, scabrous, and pilose scales; achenes cano-pubescent.

HAB. Katberg, 4–5000 ft., *Drege:* same range, at 2000 ft., *H. Hutton!* (Hb. D.)
Stems several from the crown, at first decumbent, then suberect, 6–12 inches long.
Leaves closely imbricating, 2–3 inches long, 4–5 lines wide, gradually smaller and
narrower upwards; all but the uppermost ones toothed.　The hairs which are copiously scattered over all the young parts are very distinctly jointed, and nearly 1½
line long.　Invol. 5 lines long.　Flowers bluish-purple; according to Mr. Hutton,
"*blue.*"　A remarkable and seemingly rare species ; I have only seen the specimen
collected by Mr. Hutton, who found but one root.　It is nearly allied to *S. thyrsoi-*
deus, but looks different.

35. S. thyrsoideus (DC.! l. c. 406); perennial, the stem, leaves, pedi-cels, and inv. scales pilose with long, jointed, soft hairs ; radical leaves rosulate, numerous, lanceolate or oblongo-lanceolate, narrowed toward the base, repand-toothed or subentire; stem ascending-erect, ending in

a thyrsoid, many-headed panicle; cauline leaves scattered, linear, erect, with recurved margins; heads discoid, many-fl.; invol. calycled with a few very slender, filiform bracteoles, of 12–15 linear, scabrous, and pilose scales; achenes minutely puberulous.

VAR. β, **monticolus** (DC.); leaves more sparingly pilose; raceme simple; achenes more thickly downy.

HAB. Kaffirland, *Ecklon!* Winterberg, *Mrs. F. W. Barber*, 416. β, on the Witberg. 7–8000 ft., *Drege.* (Herb. Sond. D.)
Rootstock not woolly. Rad. leaves many, 2–3 inches long, 3–5 lines wide, mostly acute. Stem 10–12 inches high. Cauline leaves laxly set, 1–1½ inch long, 1 line wide. Invol. 5 lines long. Pubescence of the achenes close and short. I have not seen var β, which DC. suggests may prove a distinct species. *S. thyrsoideus* differs more by habit, and the very narrow and lax cauline leaves, than by any more definite character from *S. barbatus.* It is also like *S. macrocephalus,* but has discoid fl.-heads.

36. S. Sandersoni (Harv.);

perennial, glandularly pubescent, viscidulous; radical leaves none; stem herbaceous, thick and strong but fistular, closely leafy nearly to the summit, simple; leaves sessile, shortly adnate-decurrent at base, oblongo-lanceolate, subacute, spreading, coarsely and unequally toothed, *very scabrid;* heads several, in a subsimple, scarcely peduncled raceme, discoid, many-fl.; inv. sparingly calycled at base, of about 20 linear, scabrous scales equalling the disc; achenes copiously cano-villous.

HAB. Natal, *J. Sanderson!* 462. (Herb. D., Hk.)
Stem 2 ft. high or more, 4–5 lines in diam. below, 2–3 lines thick above, equally and closely leafy throughout. Leaves 2–2½ inches long, 4–5 lines wide, with broad, deflexed ears at base, which are adnate to the stem. Inv. ½ inch long, 4 lines diam. Heads with upwards of 100 fl. This has something of the habit of *S. barbatus,* with the inflorescence of *S. thyrsoideus;* but differs from both in foliage and pubescence. Its stems are very thick and strong for a herbaceous plant.

37. S. incomptus (DC. l. c. 386);

glabrous; rhizome ligneous, nude; stem herbaceous, rigid, terete, much branched from the base, subdichotomous; "lowest leaves petioled" (DC.); upper sessile, oblong, inciso-pinnatifid or coarsely serrate, the lobes or teeth on each side 3–5–9, short, acute, toothed or entire; infl. subdichotomous; heads on short, rigid pedicels, discoid, 12–20-fl.; inv. narrow, cylindrical, sparingly calycled, of 8–10, smooth, flattish, narrow scales; achenes striate, cano-puberulous.

VAR. β. **serratus**; upper leaves serrate, the teeth 7–9 on each side. *S. serratus, E. Mey., fide DC.*

HAB. Betw. the Zack and Gariep, *Burchell,* 1544. Near Cradock, on the Gt. Fish River, and near Uitvlugt, *Burke & Zey.! Zey.!* 1034. β. betw. Rhinosterkopf and Ganzefontein, *Drege!* (Herb. Hk., Sd.)
Stems 5–8 inches high, much-branched and rigid. Leaves small, 1–1½ inch long. Inv. ½ inch, 2–2½ lines across. Not like any of this section in habit.

38. S. asperulus (DC.! l. c. 386);

whole plant very scabrous, pubescent; rootstock not woolly; radical and lowest cauline leaves linear-lanceolate, elongate, acute or acuminate, tapering much at base and more or less petioled, rigid, calloso-denticulate, or serrate, with reflexed margins, one-nerved; stem erect, herbaceous, laxly leafy, few-headed; cauline leaves half-clasping, erect, linear-acuminate or subulate, with strongly revolute margins; heads discoid, 100-fl. or more; inv. scabrous, calycled, of 12–20 linear, acuminate, keeled scales; achenes terete, striate, puberulous.

VAR. β. **glabrior**; stem and leaves varying from scabrid-puberulous to nearly glabrous! (*Drege! in Hb. Hk.*)

HAB. Cape, *Bowie!* Albany, *Ecklon!* Cypher Font., Grahamstown, *P. McOwan*, 23. Zwartkey, *Drege!* Modder-rivier's Spruit, *Zey.!* (Hb. D., Sd., Hk., Cap.)

Rootstock thick and woody; stem rigid, 8–18 inches high, the smaller simple, the larger sparingly branched. Root leaves 3–10 inches long, 2–4 lines wide, varying from minutely calloso-denticulate to somewhat coarsely callous-serrate; the narrow ones almost entire. Heads much larger and leaves narrower and more rigid than in *S. serratus*, to which this is allied. *Drege's* specimens (var. β.) are very much less rough than *E. & Z.'s*.

39. S. serratus (Sond.!); scabrous-pubescent, viscidulous; rootstock not woolly; radical leaves long-petioled, oblongo or obovato-lanceolate, or lance-linear, tapering to the base, subobtuse or acute, one-nerved, unequally and rather coarsely callous-toothed; stem erect, herbaceous, laxly leafy, corymbose or panicled upwards; cauline leaves sessile, ear-clasping, oblong, toothed, the uppermost small and taper-pointed; infl. lax, few or many-headed, pedicels long or short; inv. campanulate, scabrous, with a few small basal bracteoles, of many linear, acuminate, keeled scales; heads discoid, 30–50-fl.; achenes terete, striate, puberulous. *Doria serrata, Th.!* Cap. 674. *Cineraria serrata, Willd.!* Senecio *leontodontis, DC.! l. c.* 386. *S. rudis, Turcz.!* Bull. Mosc. 1851.

HAB. Cape, *Thunberg! E. & Z.!* Camiesberg, *Drege!* Zululand, *Gerr. & M'Ken.!* 1013. (Herb. Sond., D.)

Root leaves many, 3–6 inches long, including the petiole; stem-leaves 2–3 inches long, 3–4 lines wide. Stem 6–18 inches high, the smaller specimens subsimple, 2–3-headed; the larger branching, 10–20-headed. All parts very rough. I do not find this in Hb. Thunb., but a specimen of *Thunberg's* collecting (Hb. Sond.) quite agrees with those of *Ecklon* and *Drege*. *Gerrard's* specimen has rather narrower leaves, but does not otherwise differ.

40. S. hypochoerideus (DC. l. c. 386); "herbaceous, erect, scabrid-pubescent; radical leaves oblong, obtuse, tapering into a petiole, coarsely callous-crenate; stems laxly leafy, simple, or divided into 1–2, one-headed, nude branches; heads nearly 100-fl., discoid; inv. sparingly calycled, of 20 scales; achenes terete, puberulous." *DC. l. c.*

HAB. Betw. Windvogelberg and Zwartkey, 2000–4000 ft., *Drege*.

Except for the larger fl. heads this seems to come very near *S. serratus*. De *Candolle* says it has the habit and roughness of *S. puberulus*; and the crenate leaves of *S. crenulatus*.

41. S. eriobasis (DC.! l. c. 388); woolly on the rootstock, nearly or quite glabrous; radical leaves petioled, oblong or elongate, obtuse, more or less deeply and bluntly inciso-pinnatifid, with short, toothed lobes and rounded interstices; stem erect, herbaceous, fistular, sparingly leafy, corymbose at the summit; corymb lax, the pedicels elongate; cauline leaves small, sessile, deeply incised; rays numerous, yellow; inv. calycled, of many linear, acute, glabrous scales; achenes terete, pubescent.

HAB. Worcester, *Eckl.!* Brackfontein, *Zey.!* (Herb. Sond., Cap.)

Root-leaves 6–10 inches long, of thin substance, 1–2 inches wide. Stem 1½–2 ft. high, branched beyond the middle. Heads several. A larger plant than *S. erosus,* from which it is known by its glabrous leaves, stems and involucres.

42. S. erosus (Linn. f. Suppl. 370); woolly on the crown of the root, viscoso-pubescent; radical leaves numerous, petioled, oblong or obovate-oblong, obtuse, bluntly inciso-pinnatifid, with short, toothed lobes and rounded interstices, or repando-dentate; stem scape-like, nude above, with a few distant, small, sessile, toothed or pinnatifid leaf-scales, simple and one-headed or branched, 2–3-headed; rays numerous, yellow; inv. sparingly calycled, of many linear, acuminate, scabrous-pubescent scales; achenes terete, minutely pubescent. *Thunb. Cap.* 684. *DC. l. c.* 388. *Cineraria pandurata, Th.! Cap.* 672, *and Doria incisa, Th.! Cap.* 675. *Senecio panduratus, Less. Syn.* 392 *(non DC.)*

VAR. β, **repandus**; leaves repand-toothed, not incised. *Cineraria crenata, Spreng. Sieb. Fl.* 284.

HAB. Cape, *Thunberg!* Kamp's Bay hills, *Eckl.! W.H.H.* Caledon, *Eck.!* Stellenbosch, *Dr. Prior!* Steendal, Tulbagh, *Dr. Pappe!* Babyl. Toorn, Kl. Rivier's Berg, *Zey.!* 2793. (Herb. Th., D., Hk., Sd.)

With the foliage of *S. hastulatus,* this has usually a simple, one-headed, scape-like stem, and may always be known from the dwarf states of that species by the woolly crown of the root.

43. S. hastulatus (Linn. Sp. 1218); rootstock thick and woody, sub-horizontal, not woolly at the crown; radical leaves numerous, petiolate, inciso-pinnatifid, viscoso-pubescent, the lobes in several pairs, short, blunt, eroso-dentate or coarsely toothed or lobulate, the sinuses rounded; stems herbaceous, leafy, corymboso-paniculate upwards, striate, pubescent; cauline leaves ear-clasping, inciso-pinnatifid, with short, toothed lobes; corymb laxly few or several-headed, the pedicels long; heads multi-radiate; inv. nearly nude at base, of many scabrid, linear-attenuate, round-backed scales; achenes finely striate, minutely downy. *DC. l. c.* 383. *Dill.! Elth. t.* 152, *f.* 184. Also *S. panduratus, DC.? l. c.* (*exlc. syn. Less.! and Th.!*). *E. & Z.! Drege! Herb. S. glutinosus, Eckl.!* (*ex parte*). *S.? brachyrhynchoides, DC.! l. c. ex Hb. Drege!*

VAR. β, **nudiusculus**; radical leaves obovate-oblong, tapering at base into a petiole, crenate or more or less deeply and bluntly inciso-pinnatifid; stems nearly leafless, simple and one-headed, or 2–3-headed. *S. nudiusculus, DC. l. c.* 384.

HAB. Round Capetown and Simonstown, *Bowie! W.H.H., C. Wright,* 306. Tulbagh, *Dr. Pappe!* Zwarteberg and Groenekloof, *Drege!* Caledon, *Eckl.!* Distr. of Queenstown, *T. Cooper!* 232. Buffelsjagt, *Gill!* Basutu Land, *T. Cooper,* 715 (depauperated). β, near Beaufort, *Drege.* Zwartkops R., *Pappe!* (Hb. D., Hk., Sd.,Cap.)

Rhizome ½ inch thick. Root leaves (from the crown) 4–7 inches long, 1–2 inches wide; their lobes ½–¾ inch wide. Stems 1–2 feet high, fistular, pale, corymboso-paniculate; pedicels 3–4 inches long. Heads 5 lines long, as many wide. Rays yellow, spreading. Known from *S. glutinosus,* which it resembles in foliage, by the rootstock; from *S. erosus* by the non-woolly crown, &c. *S.? brachyrhynchoides,* DC., according to a specimen from *Drege* (Hb. Hk.), is surely only a dwarf state of this species. Our var β. in its typical state, with obovate, crenate leaves, and nude, subsimple stems, looks different, but Dr. Pappe's specimens, all collected in the same spot, vary with deeply cut leaves, closely approaching those of the common variety. *Dillenius'* figure, above quoted, is an *excellent* representation of the normal form.

44. S. robertiæfolius (DC. l. c. 384); "stem herbaceous, sparsely·
pubescent, simple, leafy at base, nude above; radical leaves petioled,
elongate-lanceolate, sinuato-pinnatifid, subdentate, on both sides shortly
hairy; cauline sessile, acuminate, toothed; heads subsolitary, long-
pedicelled; inv. sparingly calycled; inv. of 20 torulose, hispid scales;
rays 10–12, narrow, long, flat, open; achenes quite glabrous." *DC. l. c.*

HAB. Camiesberg, *Drege*. (Unknown to me.)
Perhaps a mere variety of *S. hastulatus*.

45. S. spiræifolius (Thunb.! Cap. 684); rootstock oblique, not woolly
at the crown; radical leaves numerous, petioled, inciso-pinnatifid, scabrid-pubescent, the lobes in 10–15 or more pair, short, horizontal,
coarsely 3–5-toothed, with reflexed edges, the sinuses truncate; stems
herbaceous, nearly nude, or with 1–2 distant, pinnatisect leaves, or a
few small, clasping, taper-pointed leaf-scales, pubescent at base, glabrous
above; corymb 2–7-headed, pedicels long; heads multi-radiate; inv.
calycled at base, of many glabrous, linear-acute, flattish scales; achenes
striate, cano-pubescent. *S. pedicularifolius, DC.! in Hb. Eckl.! Drege!
(non H.B.K.) S. pseudo-Sceptrum, Steud. S. Ecklonianus, DC. l. c. 384.*

HAB. Cape, *Thunberg!* Worcester, *Ecklon!* Near Beaufort and betw. Paarl and
Frenchhoek, *Drege!* Steendahl, Tulbagh, *Dr. Pappe!* (Herb. Th., D., Hk., Sd.)
Rhizome ¼ inch thick. Rad. leaves 5–7 inches long. Cauline leaves often none.
Stem 10–18 inches high, sometimes 1-headed. A handsome species, with more
sharply cut and rigid leaves than *S. hastulatus;* nearly naked stems and *glabrous*
involucres. *Thunberg's* specimens, in excellent preservation, exactly agree with our
plant; his description only errs in calling the root *annual* and the inv. *hairy.*

46. S. hieracioides (DC. l. c. 384); "stem herbaceous, erect, branching, terete, striate, puberulous, viscidulous above; root-leaves petioled,
oblong, toothed, minutely puberulous; cauline sessile, ear-clasping,
oblong, incised, downy, the uppermost linear, viscidulous; corymb
panicled, the pedicels sparsely scaly; inv. of about 12 puberulous scales;
rays 8–10; achenes linear-compressed, striate, pubescent between the
striæ." *DC. l. c.*

HAB. Los Tafelberg, Kaffirland, in stony places 5–6000 f., *Drege!*
"Lower leaves 3 inches long, 4 lines wide. Stem 1–1¼ foot high." Of this I
have seen no specimen, unless *Zey.!* 938, from Vet River (Hb. Sd.), belong to it:
in it the radical and lower leaves are obovate-oblong, tapering into an imperfect
petiole, coarsely toothed, midribbed, with a thickened, subciliate margin, glabrous,
except for a few scattered bristles on the lower side; stem leaves few, clasping,
toothed, puberulous; infl. laxly few-headed; inv. nearly nude at base, of 12–15
narrow, scabrous scales; rays few, pale-yellow, shorter than the inv.; achenes
tapering upwards, striate, pubescent.

47. S. albifolius (DC.! l. c. 389); herbaceous, perennial, not woolly
on the crown of rootstock; radical leaves numerous, rosulate, petioled,
pinnati-partite, on both sides albo-tomentose, becoming glabrate above,
lobes on each side 4–6, linear, obtuse, with revolute margins; stem
scape-like, much longer than the leaves, nearly leafless, 2–3 headed,
the pedicels long; inv. white-woolly, scarcely calycled, of many-scales;
heads many-fl., rays 10–12, achenes striate, puberulous on the striæ.

HAB. Winterhoeksberg, Tulbagh, *Eckl.!* (Herb. Sond.)

A dwarf plant. Rootstock thick. Leaves 1½–2 inches long, the lobes 1 line wide, ¼–½ inch long, scape 4–6 inches long, sometimes with 1–2 leaves near the base ; pedicels 1–2 inches long. Inv. 4 lines long, 5 lines across. Rays 4-nerved, flat.

48. S. reptans (Turcz! Bull. Mosc. 1851, p. 207) ; nearly glabrous ; rhizome woody, nude, stoloniferous; stolons prostrate, rooting, with a tuft of leaves at the extremity; radical leaves numerous, petioled, linear-oblong, obtuse, inciso-dentate, or subpinnatifid, thickish, nerveless; stems ascending, simple, leafy below, ending in an erect, nude, gland-scabrid, one-headed peduncle; cauline leaves sessile, linear, entire or repando-crenate ; inv. *scabrous*, of 18–20 scales, many-fl. ; rays several, short ; achenes terete, with many slender, cano-puberulous striæ. *Walp. Ann.* 5, *p.* 335.

HAB. Cape, *Ecklon!* Caledon, R., *Burke and Zey.! Zey.!* 935. (Herb. Hk., Sd., D.)
Root woody; stolons numerous 2–3 inches long. Radical leaves, or those in the tuft at end of stolon, 1–1½ inch long, 1–2 lines wide. Cauline leaves ½ inch long, 1 line wide. Stems 3–4 inches long, the pedunc. about 2 inches. Compared by Turczanninow to *S. brachyrhynchoides, DC.* (which we consider a var. of *S. hastulatus*), but the resemblance is trifling ; and the present seems a well characterised plant.

49. S. Bellis (Harv.); glabrous or nearly so ; rootstock not woolly ; radical leaves numerous, rosulate, obovate-oblong, obtuse, tapering at base into a short petiole, one-nerved, thickish, subviscidulous, minutely callous-denticulate; stems simple, one-headed, angular below, striate above, *nude* or with a very few, distant, minute, subulate leaf-scales ; heads many fl., radiate; inv. glabrous, sparingly calycled at base, of about 20 linear, acuminate scales; rays numerous, revolute; achenes (*young*) short, compressed, glabrous.

HAB. On hill sides, Queenstown district, *Mrs. F. W. Barber,* 317. (Herb. D.)
Leaves 2–3 inches long, ½–¾ inch wide, shaped almost exactly like those of the *Daisy (Bellis perennis).* Stems 10–12 inches high. Inv. ½ inch long, 4–5 lines across. Apparently a very distinct and quite an elegant species.

§ 3. PLANTAGINEI. (Sp. 50–65.)

50. S. Dregeanus (DC.! l. c. 407); perennial, herbaceous; stem erect, ribstriate, rigid, glabrous and laxly leafy below, cobweb-woolly and pedunculoid above, corymbose at the summit; subradical and lower cauline leaves long-petioled, lanceolate, tapering at both ends, acute, glabrous, rigid, obliquely nerved, calloso-crenulate ; upper leaf-scales subulate ; corymb few-headed, the pedicels long, cobwebby, sparsely scaly ; heads radiate, many-fl. ; inv. campanulate, *imbricated in several rows*, the outer scales scarcely shorter, blending with the inner, all with long, glabrous, dark points, bearded at the summit; achenes glabrous.

VAR. β. **discoideus** ; rays none ; stems frequently simple, 1-headed. (Hb. Sd., D.)
HAB. Betw. Omsamculo and Omcomua, *Drege!* Near Umgena Waterfall, Natal, *Gerr. & M'Ken.!* 354. Var. β. Betw. Omsamwubo and Omsamcaba, *Drege!* Kreili's Country, *H. Bowker!* (Herb. D., Hk., Sd.)
Stem 2 feet high or more. Leaves 12–15 inches long, including the petiole, ½–¾ inch wide. Inv. very much cobwebbed, more like that of an *Athrixia* than of a *Senecio.* Disc and ray-flowers deep purple. β. is a smaller and weaker plant, with narrower leaves and often one-headed, simple stems.

51. S. coronatus (Harv.); rootstock very woolly; stem herbaceous,

erect, ribstriate, terete, loosely woolly below, cobwebbed, becoming
glabrous above, ending in a few-headed corymb; lower leaves broadly
ovate or oblong-obovate, tapering at base into a petiole, subacute,
penninerved, minutely calloso-crenulate, the younger loosely cobwebbed,
older glabrous, rigid; upper sessile, stem-clasping, ovate, oblong or
lanceolate, smaller upwards; corymb 3-20-headed, simple or branched,
the pedicels long, naked; heads many-fl. radiate; inv. calycled with
long, subulate bracteoles, glabrous or cobwebbed, of 20 or more, keeled,
subulate scales; achenes short, subcompressed, variably hairy. *Cineraria coronata*, Thunb.! Cap. 670. *Senecio lasiorhizus*, DC.! l. c. 387. *S.
lasiorhizoides*, Sch. B.

VAR. β. **minor**; radical leaves 3 inches long, ½-1 inch wide; cauline small and
narrow; heads few and of smaller size. (Herb. D.)

HAB. Camtous R., *Thunberg!* Uitenhage, Albany and Kaffraria, *Drege, Ecklon,
Zeyher*, 929, *Cooper*, 285. Kreili's Country, *H. Bowker!* Natal, *Miss Owen! Krauss*,
438, *Gueinzius!* 319, 594. β. Cafferland, *Cooper*, 150. Camperdowm and Bulu, Natal,
Gerr. & M'K. 1053. (Herb. Th., D., Hk., Sd.)
Rootstock thick, very woolly. Stem 6 inches to 2 feet high, sparsely leafy, nude
above. Root-leaves several, on longer or shorter petioles, 3-6 inches long, 1½-3
inches broad, varying considerably in comparative length and breadth. Cauline
leaves few and distant, broad or narrow. Young parts loosely cobwebbed. Heads
like those of *S. Albanensis*. Rays numerous and long, yellow. Achenes sometimes
densely, sometimes sparsely hairy. Var. β. is more slender, with smaller leaves
and heads; but otherwise the same.

52. S. Albanensis (DC.! l. c. 385); glabrous or cobwebby; rootstock not woolly; radical leaves several, long-petioled, elongato-lanceolate or lance-linear, tapering to each end, rigid, midribbed (the broader
penninerved beneath), callous-denticulate, with subrecurved margin;
stem erect, ribbed and furrowed, laxly leafy, corymbose at the summit;
cauline leaves sessile, narrow, taper-pointed, with revolute margins;
corymb lax, few or several-headed, the pedicels elongate, nude or scaly;
inv. of 15-20, glabrous or cobweb-woolly, narrow, acuminate scales,
calycled with several longish subulate bracts; rays numerous, spreading;
achenes glabrous, finely striate, narrowed to the base, slightly tapering
upwards. *Zey.!* 2948.

VAR. β. **doroniciflorus**; root-leaves 12-14 inches long, ¾-1 inch wide; heads few;
inv. quite glabrous; rays deep orange yellow. *S. doroniciflorus*, DC.! l. c.

VAR. γ. **angustifolius**; leaves very long and narrow, 6-15 inches long, 1½-5 lines
wide; heads several; inv. cobwebby; rays chrome yellow. *Zey.!* 2950.

VAR. δ. **brevifolius**; radical leaves 3-4 inches long or shorter, often very narrow;
infl. cobwebby; heads rather smaller than in α. and β. *Zey.!* 2949. *S. affinis,
DC.? l. c. (ex descript.) S. leiocarpus*, DC.! l. c. 390.

VAR. ε.? **leiophyllus**; perfectly glabrous in all parts; radical leaves lanceolate,
subobtuse; stem rigid, closely leafy, few-headed; cauline leaves linear-lanceolate,
narrowed at base, subobtuse, with revolute margins; heads rather smaller and rays
shorter than usual. (Herb. Hk., Sd.)
HAB. Common in Albany, Uitenhage, Queenstown, the Winterberg, and in
Kreili's Country, *E. & Z.! Drege, Mrs. F. W. Barber, &c.* Betw. Knysna and
Plettenberg Bay, *Dr. Pappe!* β. Buffelriver; betw. Gekau and Basche R.; and
betw. Omtendo and Omsamculo, *Drege!* Winterhoekberg, Krakakamma and Koega
R., *Zey.!* δ. Elandsriver, *Zey.!* Var. ε. Magallisberg, *Burke & Zey.! Zey.!* 928.
(Herb. D., Hk., Sd., Cap.)
Stem 8 inches to 2 ft. high. Leaves 3-15 inches long, narrow-lanceolate or

linear. Heads sometimes 2–3 on long, simple pedicels; sometimes 10–12, in a branching, very lax corymb. Pedicels 3–4 inches long. Heads ½–¾ inch across. Very variable in size, but generally recognizable by its leaves, involucre and achenes. I suspect that *S. affinis* and *S. inornatus*, both unknown to me, are varieties of this common plant. Our var. *ε. looks* different, but has scarcely characters sufficient to separate it specifically. *S. leiocarpus, DC.!* of which I have seen Ecklon's specimens in the Cape Govt. Herb. seems to me to be founded on a half-starved, narrow-leaved form of this species; the specimens, however, are in miserable condition, half decayed.

53. S. inornatus (DC. l. c. 385); "herbaceous, quite glabrous; stem terete, striate, erect, laxly panicled, sparingly leafy; leaves lance.-linear, subcoriaceous, the lower petiolate, dentato-serrate, upper sessile, quite entire; pedicels nude; inv. about 10–12 scaled, shorter than the disc, calycled with small bracteoles; disc 15-fl.; rays 5; achenes glabrous."

HAB. Betw. Grahamstown and Bothasberg, and betw. Zandplaat and Komga, in pastures, *Drege.* (Unknown to me.)

Except for the small size of the fl. heads this, by description, does not differ from *S. Albanensis,* var. *δ.* "Leaves 3–4 inches long, 3–4 lines wide. Stem 18 inches high."

54. S. caudatus (DC. l. c. 384); glabrous or nearly so; rootstock nude; stem herbaceous, erect, angular and furrowed, leafy to the middle, either one-headed, or dividing into 3–4 one-headed, nearly nude branches; radical and lowest cauline leaves oblongo-lanceolate or lanceolate, penninerved, membranous, acute, entire or calloso-denticulate; medial broad-based, clasping, many-striate at base externally, acuminate; upper leaves small, tapering into a long, narrow, acute, ciliolate, tail-like point; rameal leaves reduced to subulate, attenuate scales; inv. of about 20, glabrous, linear-subulate scales, calycled with a few subulate bracteoles of equal length; rays 12–13, twice as long as the involucre; achenes tapering upwards, glabrous.

HAB. Katriver'sberg, *Drege! H. Hutton!* (Herb. D., Hk.)

Stem (in *Mr. Hutton's* plant) 2–3 feet high, sharply angular. Lower leaves 6–7 inches long, 2–2½ inches wide; medial 3–4 inches long, 2½–3 inches wide; in *Drege's* specimens, which are very imperfect (Herb. D. Hk.) *much narrower,* scarcely ½ inch wide. Pedicels 6–16 inches long. Rays 1½ inch long. I describe from Mr. Hutton's complete and well-preserved specimens, gathered on the Katberg; and I omit all notice of *Ecklon's* (Herb. Sd.), because they are quite decayed; so much so that I cannot tell whether they belong to this species or not!

55. S. macrospermus (DC. l. c. 385); "herbaceous, erect, cobweb-cotonny, subcanescent; leaves callous-toothed, elongato-lanceolate, the lowest petioled, much attenuate at base" [and externally many-nerved]; "cauline sessile; corymb 6–7-headed, pedicels long, thick, nude" [rib-furrowed]; "inv. sparingly and shortly calycled, at length quite reflexed; recept. smooth; achenes terete-compressed" [rib-striate, *scaberulous* in the furrows] "tapering a little at each end." *DC. l. c.*

HAB. Witbergen, 6000 ft., *Drege!* (Herb. Sond.)

Of this I have only seen a leaf, an empty receptacle with its withered invol., and some ripe achenes. The leaves are 10–18 inches long (fide *DC.*); the heads an inch in diameter. Achenes 4 lines long, pale.

56. S. monticolus (DC. l. c. 384); "herbaceous, erect, glabrous; stem

simple, terete, nude above; leaves oblong, toothed, the lowermost tapering into a petiole, upper sessile, narrower; pedicels few, one-headed,
subracemose, 1–2-scaled at the summit, twice as long as the head; inv.
calycled, of 12 scales, equalling the disc; disc-fl. about 30, rays 8, long,
4-nerved; achenes tapering upwards, striate, pubescent." *DC. l. c.*

HAB. On the Klipplaat River and the Witberg, 3500–6000 feet, *Drege.* Albany
and Uitenhage. *Ecklon.* (Unknown to me.)
"A foot high. Lower leaves 3 inches long, 8–9 lines wide. Inv. 5 lines long." *DC.*

57. S. gramineus (Harv.)

; rootstock not woolly; radical leaves
many, narrow-linear, elongate, blunt, with strongly revolute margins,
glabrous or cobwebby below, cauline leaves similar, few and distant;
stem scapelike, filiform, rigid, cobwebby or glabrate, simple, 1-headed;
inv. calycled at base, of many narrow, keeled, taperpointed, cobwebby
scales; rays 6–8; disc-fl. 30–40; achenes (young) villous.

HAB. Vanstaadensberg, Uit., *Zey.!* (Herb. Sond.)
Rootstock slender. Radical leaves 3–4 inches long, ⅓ line wide, quite linear
throughout. Stem 6–8 inches high. Inv. 4 lines long. Of this I have seen but a
single specimen, and propose it as a species with some hesitation. But for its villous
ovaries it might be regarded as an extremely emaciated form of *S. Albanensis.*

58. S. crispus (Thunb.! Cap. 680)

; nearly or quite glabrous; root-
stock oblique, not woolly at the crown; radical leaves numerous, tapering at base into a longish petiole, oblong-lanceolate, obtuse, glabrous,
bluntly toothed or repand, with subreflexed edges; stems herbaceous,
rigid, ribbed and furrowed, erect, simple, corymbose at the summit,
leafy; cauline leaves sessile, erect, oblong, toothed, blunt, with revolute
edges; corymb closely several-headed; heads multiradiate, hemispherical; inv. glabrous, calycled, of many oblong-acuminate scales; achenes
(young) short, compressed, glabrous, ciliolate.   *S. sulcatus, DC. l. c.* 384.

HAB. Summit of Table Mt., *Thunberg! Drege, Ecklon! W.H.H.* (Herb. Th., D.)
Rhizome 4–5 lines diameter. Rad. leaves 3–7 inches long, the petiole 1–4 inches;
lamina ½–1 inch broad. Cauline leaves numerous, very erect. Stem 12–18 inches
high. Corymb closely branched, flat topped, 10–20-headed. Inv. 3 lines long, 4–5
lines wide. The ovaries look as if the achenes might be compressed or flattened, in
which case this species should be removed to *Cineraria.*

59. S. decurrens (DC. l. c. 388)

; stem herbaceous, erect, ribbed and
furrowed, cobwebby, virgate, corymboso-paniculate at the apex; leaves
oblongo-lanceolate, calloso-crenulate, above glabrate, beneath cobwebby-
canescent, "the radical on long petioles" *(DC.),* the cauline decurrent
in a coarsely toothed wing; corymb compound, fastigiate, the pedicels
long, nude; inv. cobwebby, of many scales, amply calycled with long,
2–3 seriate bracteoles; heads many-fl.; rays 10–12, yellow; achenes
quite glabrous, elongate.

HAB. Omsamcaba, *Drege!* Fort Bowker, *J. H. Bowker,* 603. (Hb. D., Hk., Sd.)
I have not seen the radical leaves, which are said to be nearly 2 ft. long. Stem
3–4 ft. high, leafy up to the base of the corymb. Lower leaves 6–12 inches, upper
3–5 inches long, 1–1½ inch wide. Heads 5–6 lines across, like those of *S. Albanensis,*
but with an ampler calycle.

60. S. digitalifolius (DC. l. c. 397)

; "herbaceous, erect, cobwebby
canescent; leaves crenato-dentate, acuminate, the" [radical and] "lower

tapering into a petiole, oblongo-lanceolate; cauline decurrent, somewhat eared, crenate even on the decurrent portion; corymb compound, crowded; pedicels scaly; inv. glabrous, calycled with several rows of long, imbricated bracteoles; rays narrow" [yellow]; "achenes glabrous." *DC. l. c.*

HAB. Betw. Omtata and Omsamwubo, *Drege.*

Of this I have seen a single radical leaf, in Hb. Sd., about 12 inches long and 1-1¼ inch wide. The cauline leaves are said to be about 3 inches long. By description, it must approach very closely to *S. decurrens;* though placed in a different § by DC.

61. S. polyodon (DC. l. c. 386); "stem herbaceous, erect, striate; below simple, rough with scattered pubescence, above panicled, *pubescent;* leaves oblong-linear, subscabrid, with 8-10 teeth on each side, the lower ones tapering at base, the upper sessile, shortly eared; heads hemispherical, about 80-fl., discoid; inv. of about 20 scales, sparingly calycled, equalling the disc; achenes glabrous." *DC. l. c.*

HAB. Near the Little and Great Fish River, 2-3000f., *Drege* (v. frust. in Hb. Sd.).

Said to be allied to *S. diodon,* DC. I can form no judgment of its characters from the frustule which alone I have seen. Cauline leaves obtuse, 1-1½ inch long, 4 lines wide. Inv. scales 3 lines long, *scabrous;* heads 3-4 lines diam.

62. S. crenulatus (DC.! l. c. 385); glabrous; rootstock not woolly; radical leaves long-petioled, varying from oblongo-lanceolate to linear-lanceolate, tapering to both ends, midribbed and penninerved, callous-denticulate, rigid; stem erect, rib-striate, leafy, corymbose at summit; cauline leaves sessile, amplexicaul, oblong, ovate-oblong, or linear, acuminate-denticulate; corymb rather close, few or many-headed; inv. hemispherical, glabrous, of about 20 narrow scales, calycled with a few subulate bracteoles; heads discoid; achenes silky-villous, canescent. *Zey./* 2951.

HAB. Districts of Uitenhage and Albany, frequent, *Ecklon! Drege! Zeyher!* &c. Zululand, *Gerr. & McK.* (Herb. Hk., Sd., D., Cap.)

In foliage this resembles *S. Albanensis,* but the stem leaves are strongly clasping, the fl. heads much smaller, and *always* (so far as I have seen, in the 28 specimens from various quarters that I have examined) *discoid;* and the achenes are very silky. *De Candolle,* however, describes the heads as having "6-8 rays," except in his var. β. "*discoideus.*" It is curious that in the four large Herbaria above quoted, no specimen with rays, from either Eckl., Drege, or Zeyher, should be forthcoming!

63. S. striatifolius (DC. l. c. 387); "herbaceous, erect, glabrous, stem terete, sub-striate; leaves sessile, elongate-linear, with revolute margins, nearly quite entire, 3-5 striate dorsally at base, one-nerved at apex; corymb few-headed; pedicels elongate; heads 50-60 fl., discoid; inv. of about 20 linear-acuminate, not wither-tipped scales, calycled; achenes pubescent." *DC. l. c.*

HAB. Witbergen, 6-7000 f. *Drege!* (Herb. Sd.)

Of this I have only seen a pair of leaves and two fl. heads. Stem said to be 1-2 ft. high. Leaves 5-6 inches long, 2-3 lines wide, tapering to a fine point. Heads like those of some vars. of *S. Albanensis,* but discoid. According to DC. it is allied to *S. reclinatus* and *S. diodon.*

64. S. othonnæflorus (DC.! l. c. 386); perennial, herbaceous, erect, quite glabrous; stem closely leafy at base, nude and panicled above, rigid, terete, striate; lowest leaves oblong or lance-oblong, narrowed at base into a short petiole, the rest sessile, amplexicaul, lanceolate, acuminate, entire or remotely and minutely callous-denticulate, one-nerved; panicle lax, subcorymbose, the pedicels long, swelling under the head; heads discoid, 50–60 fl.; inv. quite glabrous, and nearly or quite nude at base, of 10–13 broad, flat, oblong, acute, glaucous scales; achenes quite glabrous. *Doria denticulata, Th.! Cap. p. 674. Brachyrhynchos eupatorioides. Less. Syn.* 393.

HAB. Districts of Uitenhage and Albany, *Thunberg, Drege! E. & Z.! Zey.!* 2960. (Herb. Th. D., Hk., Sd.)

Rootstock somewhat woody. Stem 1–1½ f. high, divided beyond the middle into many corymbose, nude branches. Upper leaves small and much acuminate or subulate. Lower leaves 2–3 inches long, ½–¾ inch wide. Heads very like those of an *Othonna.* Some of *E. & Z.'s* specimens distributed as of this species belong to *S. Albanensis.*

65. S. petiolaris (DC.! l. c. 434); glabrous, or cobwebby; rootstock not woolly; radical or subradical leaves numerous, rosulate, on long, ribstriate and channelled petioles, oblong, obtuse, thick, nerved, undulate or repando-dentate, with subreflexed, thickened margins; stem scape-like, elongate, simple, one-headed, ribstriate, scaly; inv. of about 12–13 glabrous, oblong acuminate, flat scales, not calycled; heads many fl., discoid; achenes tapering upwards, hispidulous. *Doria undulata, Th.! Cap.* 673.

HAB. Bockland, *Thunberg!* (Herb. Th.)

Root seemingly perennial. Petioles 2–3 inches long; lamina 1½ inch long, 4–5 lines wide, somewhat fleshy! Stem 10–15 inches high, sparsely set with scale-like, subulate, depauperated leaves. Heads 5–6 lines long, and about as broad. This has the habit of *S. Bellis,* but different foliage and involucres, &c.; the heads resemble those of *S. othonnæflorus.*

§ 4. PAUCIFOLII. (Sp. 66–80.)

66. S. cordifolius (Linn. f. Suppl. 372); quite glabrous; root tuberous? stem herbaceous, slender, simple or sparingly forked, leafy below; leaves all petiolate or the uppermost subsessile, broadly cordate or ovate, obtuse, rigid, coarsely and bluntly toothed, or repand-denticulate, paler beneath, the margins more or less reflexed; pedunc. long, one or fewheaded, the pedicels long, sparsely and minutely scaly; heads many fl., radiate; inv. nearly nude at base, of about 12 flat, acute scales; rays 8–10, longer than the inv., 4-nerved; achenes quite glabrous. *Th.! Cap.* 683. *DC. l. c.* 394.

HAB. Top and sides of Table Mt., *Thunberg!* Drakensteinberg, *Drege!* (Herb. Th., Sd., Hk.)

This has quite the habit of *S. cymbalarifolius,* but different foliage and achenes. *Thunberg* says the root is annual, but probably by mistake; I conjecture from the aspect of the plant that it may be tuberous, and probably deep in the ground. Stems 3–6 inches high; the lower leaves with uncial petioles, and 1–1½ in. long, ½–1 inch wide laminæ of very rigid substance, not prominently veiny. Pedunc. 6–8 inches long, sometimes 2–3 times forked, with 3–5 inch long pedicels. Rays said to be yellow, but possibly sometimes purple.

67. S. cymbalarifolius (Less.! Syn. 391); root tuberous; stem erect, herbaceous, simple or branched, glabrous as well as the leaves and involucres; upper leaves amplexicaul, oblong or lance-oblong, passing upwards into subulate leaf-scales; lower leaves petiolate, *polymorphous* (either lyrate, cordate, reniform, subrotund or multifid), more or less toothed or lobed, sometimes sparsely setulose, more or less amplexicaul at base; pedunc. scaly, simple and one-headed, or laxly corymbose, several headed; inv. scales about 12, naked-based, glabrous; rays 8–10 *(yellow or purple)*, flat; achenes long and slender, ribstriate, setulose, slightly beaked. *Brachyrhynchos cymbalarif., DC. l. c.* 438. *Cineraria cymbalarifolia, Th. Cap.* 671.

VAR. *a.* **flavus** (DC.); leaves lyrate, with many lateral lobes; disc and ray yellow.

VAR. *β.* **purpureus** (DC.); leaves as in *a*; disc and ray purple.

VAR. *γ.* **rotundifolius** (DC.); leaves roundish-reniform, toothed or lobed, without lateral lobes; disc and ray purple.

VAR. *δ.* **hastifolius**; leaves somewhat reniform, deeply 3–5 lobed or pedatifid, without lateral lobes; disc and ray yellow. *S. hastifolius, Less.! Cineraria hastifolia, Th.! Cap.* 672.

HAB. Dry hills in the Western and S. Western districts, the several varieties often intermixed. *a.* and *β.*, Drakenstein, *Drege!* Round Capetown and Simonstown, *Eckl.! β. Wright*, 311, 316. Swellendam, *Zey.!* 2966. *γ.* Capetown to Swellendam, *E. & Z.!, Drege! Zey.!* 924. *δ.*, Hanglip, *Zey.!* 925, Simonstown, *Wright!* 310. (Herb. Th., D., Hk., Sd.)

Root a tuber about ½–1 inch diam., buried 2–3 inches. Stems 2–6 inches high, closely leafy at base, the upper half gradually passing into a scaly peduncle. Lowest leaves long-petioled, scarcely broad-based; the medial with broad, amplexicaul, leaf-like, toothed or cut bases; the upper sessile and amplexicaul. Lamina of the lower leaves very varied in its cutting, mostly quite glabrous, now and then setulose. Pedunc. 3–12 inches long, simple or branched; the branches corymbose, elongate, scaly, one-headed. Colour of flowers varying from bright purple to bright yellow, a very uncommon circumstance. Achenes with a cup-like disc.

68. S. tuberosus (Harv.); tuberous rooted, glabrous; stem erect, herbaceous, simple or branched, leafy at base, pedunculoid and one or several headed upwards; upper leaves sessile, amplexicaul, oblong or lanceolate, toothed; medial amplexicaul, petioled, lyrate-pinnatifid; lower on long petioles, either cordate-reniform, toothed, or lyrate with 1 or several lateral, toothed lobes, and a terminal reniform one; pedunc. laxly scaly; inv. scales 10–12, naked-based, glabrous; heads discoid, homogamous; achenes long, ribstriate, setulose, beaked. *Doria pinnatifida, Th.! Cap.* 675. *Brachyr. tuberosus, DC. l. c.* 438. *Cineraria tuberosa, E. Mey. Brachyrh. trachycarpus, DC.! l. c.* (excl. syn. Th.) *Senecio incertus, DC. l. c.* 433.

HAB. Paarl, *Drege!* Caledon, *Eckl.!* Tulbagh Waterfall, *Dr. Pappe!* (Herb. Th. D., Hk., Sd.)

Quite like *C. cymbalarifolius*, var. *a.*, except that the heads are discoid. Possibly a mere variety. I venture to unite DC.'s "*Br. trachycarpus,*" which, by description, can scarcely be different from the present.

69. S. diversifolius (Harv.); root . . . .; stem erect, herbaceous, simple or branched, leafy below, pedunculoid upwards, *more or less cobwebbed or woolly* as well as the leaves; upper-leaves sessile, lanceolate or subulate, passing upwards into scales; lower petioled, *polymorphous* (either pinnatisect, pinnatifid, or undivided) oblong-hastate, or oblong-lanceolate, acute, subglabrous or densely woolly beneath, the young ones often cobwebbed above; pedunc. elongate, scaly, simple or branched, the branches one-headed; inv. scales 14–20, glabrous or cobwebbed; heads many-fl., discoid; achenes long, angular, scabrous-muricated, somewhat beaked. *Brachyr. elongatus, Less.! DC. l. c.*

Var. *a.*, **pinnatifidus**; leaves in outline lance-oblong, acute or acuminate, deeply pinnatifid, the lobes in many pairs, toothed or incised, the uppermost 2–3 pairs, confluent into a pinnatifid terminal lobe. *Doria elongata, var. β., Hb. Th.! Brachyr. albicaulis, DC.! l. c.* 438, also *B. diversifolius, vars. a. and β, DC. l. c.* 439.

Var. *β.*, **integrifolius**; leaves oblong-hastate, acute, cuneate or truncate at base, toothed. *Doria elongata, vars. a. and γ., Herb. Th.! Brach. diversifolius, var. γ, cuneatus, DC. l. c.*

Hab. Western and S. Eastern Districts. *Thunberg!* a, Olifants R., and Saldanha Bay and Camiesberg. *Drege!* Stellenbosch and Swellendam. Worcester and Clanwilliam, *E. & Z.!* Groenekloof and Tulbagh, *Dr. Pappe!* β. *Hb. Thunb.!* Swellendam, *Mundt.!* Riebeckskasteel, *Drege.* Herb. Th., D., Hk. Sd.)
This varies quite as much in the cutting of its leaves as *S. cymbalarifolius* and *S. pinnatifidus*, but is to be known from the latter not only by its tendency to woolliness, but by the want of a terminal cordate-reniform leaf-lobe in the pinnatifid varieties, and the hastate-acute form of the simple lamina in var. *β.* I do not find it possible to distinguish as species the forms separated by De Candolle from *Thunberg's "Doria elongata"*; but I gladly adopt DC's trivial name *"diversifolius,"* as more expressive than that of Thunberg.

70. S. anthemifolius (Harv.); quite glabrous, crown not woolly; radical leaves long-petioled, bipinnati-partite, the lobes and lobules shortly linear, divaricate, flat, blunt; cauline leaves few, sessile, half-clasping, pinnati- or bipinnati-partite; stem simple, shortly peduncular at summit, one-headed; heads discoid; inv. scales 18–20, glabrous, slightly calycled; heads many-fl.; achenes long, tapering upwards, with many canescent striæ.

Hab. Voormansbosch, Swell., *Zeyher!* 746. (Herb. Sond.)
Root probably tuberous. Radical leaves numerous, 3 inches long, the pinnæ in several pairs, not a line wide, all remarkably patent, more or less compound. Stem 10–12 inches high, bearing 2–5 leaves, similar to the radical, but sessile. Inv. scales 6–7 lines long. Achenes tapering into a slender neck.

71. S. orbicularis (Sond.! in Herb.); root tuberous? all parts quite glabrous; stem shortly leafy, simple, ending in a long, nude, laxly corymbose peduncle; leaves on long petioles, fleshy, glaucous, peltate, orbicular, or oval, unequally callous-denticulate, or ciliato-dentate; inv. nearly nude, of 8–9 acute scales equalling the disc; heads discoid, 12–20-fl.; achenes hispidulous.

Hab. Magalisberg, *Burke & Zey!* 931. (Herb. Hk., Sd. D.)
Very near *S. oxyricæfolius*, but the leaves are almost exactly orbicular; the petiole nearly equidistant from the margin. The serratures vary much in degree. Leaves 1½–2 inches diameter; petioles 2–2½ inches long. Infl. as in *S. oxyricæfolius.*

72. S. oxyriæfolius (DC.! l. c. 405); root tuberous, all parts quite glabrous; stem erect, shortly leafy, simple, ending in a long, nude, corymboso-paniculate peduncle; leaves on long petioles, fleshy, glaucescent, *more or less peltate*, cordate or hastate, either repand, denticulate or angle-lobulate, the salient points acute; corymb or panicle laxly few-headed, pedicels elongate; inv. scarcely calycled, of 8–9 taper-pointed scales equalling the disc; heads discoid, 12–20-fl.; achenes hispidulous

on the striæ. *S. oxyriæfolius; S. peltatus; S. subnudus; S. peltiformis,
DC.! l. c.* 465.

HAB. Uitenhage, Albany and Caffraria, *E. & Z.! Drege!* Zwartkop Zoutpan,
Zey.! 2968. Albany, *H. Hutton!* Umcomas, Natal, *Gerr. & McK.* 1030. (Herb. D.,
Sd., Hk.)

The leaves vary a little in shape, in the incision of margin, and in amount of peltate character, the petiole being sometimes ½ an inch, sometimes 1 line within the
margin; but in all the specimens I have seen, including one of the *typical "oxyriæfolius"* DC., the peltate character exists. In *Mr. Hutton's* specimen this peltate char.
is at its maximum, and the margin nearly entire, or barely repand and minutely
denticulate. Petiole 2–3 inches long. Lamina 1–1½ inch long and broad. Peduncle
12–15 inches high. Fl. yellow.

73. S. rhomboideus (Harv.); root tuberous? all parts glabrous; stem
leafy at base, ending in a long, nude, corymbose or panicled peduncle;
leaves petiolate, fleshy, rhomboid or ovate, repand and denticulate, cuneate and tapering much at base, drying thin, closely netted-veined;
inv. scarcely calycled, of 8–9 taper-pointed scales, fl. . . ?

HAB. Eastern frontier? *Ecklon!* (Herb. Sd.)

Founded on a single imperfect specimen, mixed with "*S. subnudus,*" from which
the foliage so remarkably differs that I think it must be kept distinct. The fl. heads
would seem to be similar to those of *S. oxyriæfolius.*

74. S. paucifolius (DC.! l. c. 403); glaucous, quite glabrous; nude
at the crown; stem erect, short, simple, leafy below; leaves sessile, oval
or oblong, penninerved, quite entire, obtuse or acute, or mucronate,
either cuneate at base or rounded and half-clasping; pedunc. elongate,
nude, sparingly corymbose at summit, or forked, or simple; pedicels
2–8, long, sparsely scaly, one-headed; invol. not calycled, of 10–12
scales, equalling the disc; disc fl. 50–60, rays 8–10; achenes glabrous,
striate.

HAB. Swellendam, *Mundt.!* Uitenhage and Caffraria, *E. & Z.! Zey.!* 2967.
Zuureberg, *Drege.* Betw. R. Zondereinde and Zwartberg, *Zey.!* Vetrivier, *Burke &
Zey.!* 930. (Herb. D., Sd., Hk.)

75. S. latifolius (DC. l. c. 387); woolly on the crown of the root,
otherwise quite glabrous; stem erect, striate, tall, leafy, ending in a
much-branched, corymbose panicle; lower leaves oblong or obovate,
acute or acuminate, tapering to the base, entire or remotely denticulate;
upper numerous, cordate-eared and clasping at base, oblong or lanceolate, or linear-acuminate; partial corymbs many-headed, fastigiate,
pedicels short, nearly nude; inv. of 5–8 glabrous, nerved scales, nearly
nude at base; disc-fl. 10–12, rays 3–5; achenes quite glabrous.

VAR. α., *verus*; upper leaves broadly oblong or ovato-lanceolate, amply eared at
base, denticulate, or subentire. *S. latifolius, DC.! l. c.*

VAR. β. *retrorsus*; upper leaves varying from oblongo-lanceolate to linear-lanceolate, acuminate or attenuate, mostly denticulate. *S. retrorsus, DC.! l. c.*

VAR. γ. *barbellatus*; upper leaves still narrower and more acuminate, mostly
quite entire at edge. *S. barbellatus, DC. l. c.* 388.

HAB. Districts of Uitenhage and Albany; Caffraria and Natal. Var. α. Om-
samcaba, *Drege!* β. Zuureberg, *Drege!* Albany, *Ecklon!* Caledon R. and Magalisberg,
Burke, γ. Stormberg and Witberg, *Drege.* Albany, and in Caffraria, *Ecklon!* (Hb.
D., Hk., Sd.)

Stem 2-ft. high or more, leafy nearly to the base of the panicle, the upper leaves diminishing in size. Leaves very variable in size, and in the dentition of the margin, the wider ones penni-nerved, but even in the narrowest forms the upper leaves are amply cordate at base. Leaves in α, 4–6 inches long, 1–2 inches wide; in β, 3–4 inches long, ¾–1 inch wide; in γ, much narrower. Panicle 6–10 inches across. Inv. 2–3 lines long, much shorter than the disc.

76. S. bupleuroides (DC. l. c. 387); woolly on the crown of root, otherwise glabrous; stem erect, flexuous, striate, sparsely leafy, nude above, and ending in a lax, branching corymbose panicle; cauline leaves varying from broadly ovato-lanceolate to elongato-lanceolate-acuminate, straight or falcate, rigid, quite entire or subdenticulate, midribbed and obliquely penni-nerved; pedicels long, sparingly scaly; inv. of about 8 broad, glabrous, striate scales, nearly nude at base; disc-fl. 20–30; rays about 5; achenes glabrous, striate.

VAR. α, **latifolius**; leaves broadly ovato or oblongo-lanceolate, acute. *S. bupleuriformis, Sch. Bip.! Walp. Rep. 6. p. 263.*

VAR. β. **angustifolius**; leaves lanceolate-attenuate, 5–10 inches long, not ½ inch wide, often falcate. *S. bupleuroides, DC. l. c. S. bupl. var. falcatus, Sch. Bip.! l. c.*

HAB. Omsamcaba, *Drege!* Natal, *Plant! Gueinzius! &c.* β, Natal, *Krauss! Sutherland!* Nonote, *W. T. Gerrard!* (Herb. D., Sd., Hk.)

Very nearly related to *S. latifolius*, but with fewer leaves (usually 3–4), a flexuous stem, a laxer inflorescence, longer pedicels and more fl. in the heads. The leaves vary in shape quite as much as in *S. latifolius*: in α, 2–4 inches long, 1–1½ broad; in β. 5–10 inches long, 3–5 lines wide.

77? S. glaberrimus (DC. l. c. 403); "herbaceous? quite glabrous, rigid; stem simple, slender, angular, corymbose at summit; cauline leaves sessile, distant, ovate, acute, slightly cordate and clasping at base, quite entire, penni-nerved; corymb slightly compound, fastigiate, pedicels nearly nude, elongate; inv. half as long as disc, of 10 acute scales; disc-fl. 20–25, rays 4–5; achenes glabrous, striate.

HAB. Omsamcaba, Omtendo and Omsamculo, *Drege.*

"18 in. high. Radical lvs. unknown; cauline 3 in. long, 15 l. wide." Unknown to us. The above diagnosis answers equally well for *S. bupleuroides, DC.*; does it really differ?

78. S. adnatus (DC. l. c. 388); quite glabrous, rigid, erect, herbaceous; stem ribbed and furrowed, leafy, virgate, corymboso-paniculate at the apex; leaves elongato-lanceolate, quite entire, margined, midribbed and obliquely veined, acuminate, the lower (not seen), the medial stem-clasping, shortly decurrent or adnate at base, the uppermost sessile, gradually shorter and narrower; infl. level-topped, much divided, the partial corymbs 10–12-headed, with short, nude pedicels; inv. nearly nude at base, cylindrical, of 5–6 broad, flat scales; disc-fl. 8–12, rays 3, long, spreading; achenes quite glabrous, narrowed upwards.

HAB. Katberg, and betw. Windvogelberg and Zwartkey, *Drege!* Katberg, *H. Hutton!* (Herb. Hk., D., Sd.)

Stem probably 3–4 ft. high, the lower part not seen. Medial leaves 5–8 inches long, 1–1¼ inch wide. Inv. 4–5 lines long. Fl. yellow. Heads in the general infl. over 100.

79. S. isatideus (DC.! l. c. 387); quite glabrous, except for the woolly crown of the rhizome; stem erect, striate, tall, leafy below, nude above, ending in a much-branched, corymbose panicle; lower leaves oblong-obovate, tapering at base into a short, winged petiole, obtuse or mucronate, callous-denticulate, midribbed and penninerved; upper stem-clasping, oblong or lanceolate, much smaller; partial corymbs densely many-headed, pedicels short; heads 5-fl., discoid; inv. nearly nude at base, funnel-shaped, of 5 oblong, obtuse or subacute, glabrous scales; achenes glabrous. *Zey.! 933. S. marginatus, fol. 5 (excl. fol. 1, 2, 3, 4),·Herb. Th.! S. nitidus, Less.! MSS. in Hb. Th.*

HAB. Cape, *Thunberg! Bowie!* Near the Kraai R., 4500 f., *Drege!* Caledon R., *Burke & Zey.!* (Herb. Th., D., Hk., Sd.)

Stem woolly at the very base only, 2–3 ft. high, leafy for 6–12 inches below, ending in a much-branched inflorescence. Leaves 4–6 inches long, 2–3 inches wide. Upper leaves very erect, narrower, much smaller and ear-clasping. Inv.·3 lines long, 1–1¼ wide. Habit of a few-flowered *Othonna.*

80. S. venosus (Harv.); glabrous; stem erect, ribstriate, leafy to the base of the diffusely branched corymb (lower portion unknown); lower leaves . . . .; cauline leaves cordate at base and stem-clasping, lanceolate, acuminate, rigidly coriaceous, thick, ciliato-denticulate, glaucescent, conspicuously veiny on both surfaces, on the lower with *very prominent,* oblique, anastomosing veins and veinlets, many-veined at base; partial corymbs few-headed, pedicels short; heads 10-fl., discoid; inv. nude at base, funnel-shaped, of 5–6 oblong, obtuse, glabrous scales; achenes glabrous.

HAB. Magalisberg, Oct., *Zey.!* 936. (Herb. Sond.)

Apparently a tall plant, with the habit of *S. isatideus,* but leafy throughout, with close-placed, strongly and coarsely veiny, very rigid leaves. I have only seen a single specimen, about 14 inches long, being merely the flowering top of the stem, or of a branch: its leaves are 2–3 inches long, ½–¾ inch wide at base, very acute. Until the lower part of the stem shall be seen the characters cannot be fully given.

§ 5. RIGIDI. (Sp. 81–116.)

81. S. picridifolius (DC.! l. c. 386); perennial; stem herbaceous, weak, erect, sparingly branched, leafy, scabro-pubescent; cauline leaves membranous, ear-clasping and tapering much at base, oblongo-lanceolate, acuminate, coarsely and unequally toothed, scabrous above, paler and more or less glabrous beneath; fl. branches nude and pedunculoid at the apex, mostly one-headed, very scabrid; heads discoid, 80–100 fl.; inv. calycled, of many linear-acuminate, scabrous scales, achenes compressed, rib-striate, minutely scabrid.

HAB. Natal and Omsamculo, *Drege! T. Williamson! Cooper!* 1136. Magalisberg, *Burke & Zeyher! Zey.!* 940. (Herb. Sd., D., Hk.)

Stem 2–3 ft. or more high, fistular upwards, supporting itself among bushes (?) but scarcely climbing. Leaves scattered, tapering at base into an imperfect petiole, the lower ones 2–3 inches long, 1 inch wide. Upper leaves sessile, broad-based, often deeply incised, with larger ears at base. Peduncles mostly 1-headed, rarely branched. *Burke & Zeyher's* specimens from Magalisberg and Mooje R. seem to have grown in drier and more exposed places than those of Natal, being dwarfer, more erect, with more rigid foliage, &c. The natural affinities of this sp. are to me very doubtful.

82. S. Thunbergii (Harv.); stem robust, rigid-herbaceous, erect,

angle-striate, cobwebby, simple, imbricated with leaves throughout;
leaves half-clasping and adnate at base, oblongo-lanceolate, acuminate,
glabrous above, cobweb-tomentose beneath, distantly callous-denticulate
or entire, with recurved margins; heads in a much branched, dense,
thyrsoid-panicle, discoid, about 12-flowered; pedicels short, scaly, glabrate; inv. glabrous, calycled, of about 8 broadish, flat scales; achenes?
Cacalia tomentosa, Th.! Cap. 625 (non Jacq.)

HAB. Cape, *Thunberg!* (Herb. Thunb.)
Stem 1–2 ft. high, thick and strong, but full of pith. Leaves 1½–2 inch. long, 4–7
lines wide, erect, one-nerved, the young ones thinly cobwebbed above, adult quite
glabrous. Infl. not fully opened, very compound, forming (in the specimen seen) an
oblong, close panicle; at a later stage probably more corymbose; both pedicels,
bracts and invols. glabrous or nearly so. This is very different from *S. oliganthus*,
to which it is referred by DC.; I describe from Thunberg's specimens.

83. S. amabilis (DC.! l. c. 406); stem herbaceous, diffuse? branches
terete, scabro-puberulous, long, panicled or corymbose at summit; leaves
subdistant, ear-clasping at base, lyrato-pinnatifid, scabrous above, tomentose-canescent, and netted beneath, the lobes short, obtuse, in few-pair,
sharply denticulate, the terminal larger; corymbs densely several headed,
ending the longish branches of the panicle; pedicels very short; *heads
discoid*, 8–10 fl.; inv. glabrous, calycled, of about 8 acuminate scales,
achenes short, hispidulous; fl. white or whitish?

HAB. Langekloof, *Drege!* (Herb. D., Sd., Hk.)
Branches 2–3 ft. long or more, pale. Leaves 2–2½ inches long, the upper lobe
¾–1 inch wide; lateral lobes ½ inch long, 4 lines wide. Inv. 2–2½ lines long, the
scales pinkish at tip, as are the anthers and style. The habit is that of *S. incisus*,
from which the discoid fl. heads, &c. separate it.

84. S. Gerrardi (Harv.); herbaceous, tall, viscoso-puberulous; stem
. . ?; cauline leaves eared at base, petioled, ovate or ovate-oblong,
midribbed and penninerved, unequally and coarsely repando-dentate, the
imperfect lobules callous-toothed, on both sides closely scabro-puberu-
lent; rameal leaves sessile, obovate or oblong, toothed; infl. loosely
corymbose, many-headed; heads discoid, homogamous, 15–20-fl.; inv.
scabrous, calycled, of 10–12 linear acuminate scales; lobes of the cor.
barbellate; achenes terete, striate, glabrous or nearly so.

HAB. Umvote District, Natal, *W. T. Gerrard!* 1010.   (Herb. D.)
"Stems 4–5 ft. high." Petiole (of the only cauline leaf seen) about 1 inch long,
1 line wide, with a roundish, dentate lobe at base; lamina 2½–5 inches long (probably much more), 1½ inch wide. Of this plant I have received but a single leaf
and a branch of the inflorescence; these are quite unlike any species known to me.
Curiously enough, though the foliage is so very unlike that of *S. rhyncholænus*, the
flowers and fruit in both are almost identical.

85. S. pandurifolius (Harv.); stem? twigs and peduncles thinly cobwebbed; cauline leaves . . ; rameal leaves *panduriform*, or with a
broadly winged petiole amply eared at base, and an ovate, broadbased,
acute or acuminate limb, calloso-denticulate, penninerved and netted-
veined, cobwebby becoming glabrate above, cottony-canescent beneath,
the upper ones smaller, with shorter petioles; corymbs few-headed, panicled; inv. calycled, subglabrate, of 8–9 scales; heads discoid, about
12-fl.; achenes glabrous.

HAB. Highlands of Natal, *Gerr. & M'K.*, 1043. (Herb. D.)
"Herbaceous, 4–5 ft. high; fl. yellow or blue," *W. T. G.* Only known to me by a lateral twig, 14 inches long, with its subtending rameal-leaf. This latter is about 4½ inches long, of which the lamina is 2½ inches, the narrow or petiolar portion 1½– and the auricle about ¼ inch: lamina 2½ inches wide, winged petiole ¼ inch. Heads, small; the dried flowers are dark-tipped, and may have been blue. With the foliage of *S. verbascifolius*, this has widely different inflorescence. Fl. and achenes immature.

86. S. verbascifolius (Burm. Pr. Cap. 26); stem robust, ligneous at base, herbaceous, ascending, clothed, as well as the petioles and under surface of the leaves with a thick, snow-white, felted, woolly coating; leaves on long, linear, ear-clasping petioles, cordate, remotely callous-denticulate, cobwebby (becoming glabrate) above; petioles of the upper leaves broad and amply dilated at base; corymbs laxly few or several headed, the pedicels glabrate; inv. glabrous, calycled, of many scales; heads many-fl., radiate, the rays yellow, about 12; achenes glabrous. *DC.! l. c.* 389. *Cineraria tussilaginea, Thunb. Cap.* 671. *Cin. hypoleuca, Sieb.! Cap.* 33.

HAB. On Table Mountain, *Thunberg! Sieb.! Eckl.! Drege! Pappe! W. H. H.*, &c. (Herb. Th., D., Hk., Sd.)
Stems several, thick at base; 1–2 ft. high, leafy throughout, the lower leaves large and on long petioles; the upper gradually smaller, and the uppermost fiddle-shaped. Petioles 2–3 inches long; lamina 1½–2½ inches long, 1½–2 in. wide. Toment very thick and white. Inv. 4–5 lines long, as much in diam.

87. S. cinerascens (Ait.); stem suffruticose, erect, virgate, thinly white-woolly, leafy, corymbiferous; leaves more or less petioled, pinnati-partite or pinnatifid, the lobes on each side 2–4, linear or lance-oblong, obtuse, with revolute margins, albo-tomentose beneath, either tomentose or glabrate above; upper leaves sessile; corymb subsimple, few-headed, the pedicels long; inv. cobwebbed or glabrate, calycled, of many scales; heads many-fl., rays "3–5, flat, not revolute," or none. *Jacq. Schoenb. t.* 150. *DC.! l. c.* 389. *Jacobœa tomentosa, Th.! Cap.* 678. *Sen. tomentosus, Salisb. Prod.* 196.

HAB. Cape, *Thunberg!* Kamiesberg, *Drege!* (Herb. Th., D., Sd., Hk.)
Stem 1–2 ft. high, pale. Leaves 3–5 inches long, the rachis 1–4 lines wide; lobes ½–1 inch long, 1–2 lines wide. Corymb 3–8-headed. Heads 5–6 lines long, and nearly equally broad. I have not seen *rays* on the few specimens examined.

88. S. tortuosus (DC. l. c. 392); suffruticose or shrubby, erect, densely much-branched, in all parts (save the denuded stems) shortly pubescent; twigs flexuous, sparsely leafy; leaves cuneate-obovate, coarsely few-toothed, or somewhat lyrate, tapering at base into a petiole as long as the lamina or longer, not prominently veiny; pedunc. ending leafy twigs, short, few-scaled, one-headed; inv. nearly nude at base, gland-scabrid, 10–12-scaled; rays about 5, yellow, short, revolute, disc-fl. 30–40; achenes striate, minutely pubescent.

HAB. Olifant's R., *Drege!* Worcester, *E. & Z.!* Knakisberg, *Zey.!* 926. (Herb. D., Sd., Hk.)
About a foot high, woody in age, but slender, branches very flexuous or zigzag. All the younger parts minutely rough with subglandular, spreading hairs. Petioles ¾–1 inch long; lamina ½–¾ inch long, 4–6 lines wide, variable in shape. Pedunc. 1–1¼ inch long, slender. Inv. 4 lines long, 3 l. broad.

89. S. blattarioides (DC. l. c. *393*); "stem herbaceous? erect, simple, glabrescent; leaves all *petiolate*, the lower ovate, obtuse at base, acute at apex, deeply cut and toothed, both surfaces and the petiole hispid-subvillous, the uppermost oblong, tapering to both ends, toothed; corymb compound, loose, pedicels scaly, elongate; inv. scarcely calycled, glabrous, of about 12 scales; disc-fl. about 15; rays 5; achenes angular, glabrous." *DC. l. c.*

HAB. Betw. Omtata and Omsamwubo, *Drege.* (Unknown to me.)
" Stem 1½ ft. high. Lower leaves, with a 2-inch long petiole, 2 in. long limb, 15 lines wide." *DC.*

90. S. serratuloides (DC.! l. c. *395*); stem erect, herbaceous, puberulous or glabrous; branches rigid, striate, leafy, panicled upwards; leaves lyrate, the slender petiole with 2–4 pair of small, narrow, lance-linear, acuminate lobes, the terminal lobe very large, ovato-lanceolate or lanceolate, rigid, midribbed and penninerved, finely and sharply serrate, tapering to each end, glabrous, or the young ones flocculent; upper leaves small, sessile, inciso-pinnatifid below the middle; corymb branching, loosely many-headed, pedicels long, scaly; inv. calycled, of 10–12 oblong, acute, scales; disc-fl. 30–40; rays 8–10, spreading; achenes glabrous.

VAR. β, **gracilis**; branches laxly leafy, slender; terminal leaf-lobes lanceolate, serrulate.
HAB. Natal, *Drege! Gueinzius!* β, Natal, *Dr. W. B. Grant.* (Herb. Sd., Hk.)
Perennial, perhaps suffruticose at base. Branches 1–2 feet long. Leaves technically lyrate, but the terminal lobe is 2–3 inches long, ½–1 inch wide, and the lateral lobes, which are confined to the petiole, vary from 2–3 and from 5–12 lines long, and rarely exceed a line in width. Inv. scales dark-tipped. Rays pale.

91. S. microglossus (DC.! l. c. *395*); " quite glabrous; stem herbaceous, erect, angle-striate, corymbose at top; leaves petiolate, lyrate, the lateral lobes on each side 2–3, small, entire or few-toothed, the terminal lanceolate or oblongo-lanceolate, tapering to each end, *coarsely* serrate, obliquely penninerved, rigid; corymb compound, much-branched; pedicels slender, scaly; heads about 12-fl., the rays few and *very minute;* inv. calycled, of 8 scales; achenes glabrous.

HAB. Grass-fields, betw. Gekau and Basche, *Drege!* (Herb. Sd.)
Allied to *S. serratuloides*, but with much more coarsely serrate leaves, smaller fl. heads, and very minute, inconspicuous rays, shorter than their style. I have only seen a small fragment in Hb. Sd.

92. S. crenatus (Th.! Cap. 681); stem shrubby, erect, virgate, panicled at summit or simple, cobwebby, rib-striate, closely leafy; leaves shortly petioled, ovate or obovate, obtuse, rigid, conspicuously netted-veined, finely callous-serrulate, the younger cobwebbed, the older glabrous; corymb compound, many-headed, the pedicels scaly; heads radiate, 12–15-fl.; inv. calycled, of 7–9 broadish scales; rays 4–5, yellow, oblong; achenes striate, pubescent. *DC. l. c.* 392. *Sieb. Fl. Cap.* 243. *S. Kraussii, Sch. Bip. ?*

HAB. Eastern Districts, *Thunberg! Bowie!* Uitenhage, *E. Z.! Zey.!* 2940, *Drege!* (Herb. Th., D., Sd., Hk.)
Stems 2–3 feet high, strong, rod-like, subsimple, except at the flowering summit.

Leaves 1½-2 inches long, 1-1½ in. wide, tapering at base into a ½ inch petiole. Serratures minute and blunt. Invol. 3 lines long, much shorter than the flowers. Rays pale yellow.

93. S. oxyodontus (DC. l. c. 392); "shrubby, quite glabr.; branches terete; leaves petiolate, ovate, sharply toothed or lobed; corymb compound, few-headed; pedicels long, nearly nude; inv. of 12–13 scales, sparingly calycled, subturbinate, a little shorter than the disc; rays 5, 4-lined; disc-fl. 10–12; achenes angle-striate, hispidulous on the striæ.

HAB. Betw. Omsamcubo and Omtata, *Drege.*
"Petiole uncial. Leaf 6–12 lines long, 4–8 lines wide." Unknown to me.

94. S. Zeyheri (Turcz.! Bull. Moscw. xxiv. II. p. 89); glabrous, herbaceous; branches long, subsimple, striate, pale; leaves scattered, petiolate, glaucous, cuneate, and 3-nerved at base, ovate or rhomboid, unequally toothed, margined, netted-veined, the uppermost smaller, crenate; corymb very loose, branched, the divisions few-headed, pedicels long, minutely scaly; inv. slightly calycled, of 10–12 scales; disc-fl. about 20; rays 8, spreading; achenes striate, minutely hispidulous on the striæ. *Walp. Ann. Bot. 5, p. 337.*

HAB. Salem, *Zey.!* 2959. Near Grahamstown, *Genl. Bolton!* Olifant's Hoek, *Dr. Pappe!* (Herb. Sd., D., Hk.)
Stem probably suffruticose at base. Branches 1–2 feet long, laxly leafy, ending in a sub-peduncled, very loose corymb. Petioles slender, ½–1 inch long; lamina 1–1½ inch long, ½–¾ inch wide, acute or obtuse, sometimes obovate, coarsely toothed, of thinnish substance. Inv. shorter than the disc, 3 lines long; rays broad, 4-lined.

95. S. pellucidus (DC.! l. c. 380); stem herbaceous, diffusely much-branched; terete, striate, glabrous; leaves scattered, membranaceous, netted-veined, glabrous, lanceolate, tapering at base into a slightly decurrent petiole, acuminate, the lower more or less deeply toothed, the medial and upper denticulate or subentire; panicle loosely forked, widely spreading, the pedicels long, nearly nude; inv. calycled, its scales 12–20, glabrous, acuminate; disc-fl. 15–20; rays 5–6, yellow; achenes striate, minutely puberulous. *S. lancifolius, Turcz. Walp. Ann. Bot. 5, p. 337.*

HAB. Zuureberg, *Drege!* Adow, Uit., *Zey.!* 2958. Howison's Poort, *H. Hutton.* (Herb. D., Sd., Hk.)
Root perennial? stem at length suffruticose at base? Apparently a much and diffusely-branched plant, 3 feet or more in height, with rigid, slender branches, glabrous or nearly so in all parts. Leaves 2½–3 inches long, rarely ½ inch wide, pale green, tapering much to the base, imperfectly petiolate, the petiole shortly decurrent in a ciliate lamina. Panicle 1 foot wide; pedicels 1–1½ inch long. Inv. 4 lines long. The leaves on young plants are more often *cut* than on older, in which they are mostly minutely and distantly denticulate.

96. S. pubigerus (Linn. Amoen. 6, Afr. 66); stem ligneous at base, herbaceous upwards, diffusely much-branched; branches terete, striate, scabro-pubescent or subglabrous, widely spreading, alternately decompound; leaves stem-clasping, rigid, scabrous above, cobwebby beneath, the lower incised or pinnatifid, with few sharply-toothed lobes, the upper small, oblong or obovate, sharply few toothed; fl.-branches rod-like, scaly, bearing clustered heads in the axils of the scales, *in an im-*

perfect raceme; inv. with a many-leaved, woolly calycle, its scales 8–10, glabrous, flat, lanceolate, acuminate, equalling the disc-fl.; rays 5–6, yellow; achenes short, terete, closely hairy. *DC. l. c. 393. S. squamosus, Th.! Cap. 683. Prenanthes glomerata, Rchb. Sieb. Cap. 26. Breyn. Cent. t. 65.*

HAB. Roadsides and waste ground about Capetown and at the Paarl, very common. (Herb. Th., D., Sd., Hk.)

Root perennial. Stem 3–4 ft. high, bushy, quite woody at base; the branches widely spreading, slender, rod-like. Lower leaves 1–2½ inches long, ½–¾ inch wide, either entire or few-toothed, or deeply pinnatifid; upper leaves ½ inch—rarely 1 in. long; those on the fl. branches reduced to scales. Heads small; inv. scales 3–4 lines long.

97. S. incisus (Th.! Cap. 683); stem ligneous at base, diffusely much-branched, herbaceous upwards; branches rib-furrowed, scabro-pubescent and cobwebby, glabrous upwards; cauline leaves stem-clasping, rigid, scabrous above, cobwebby or cottony beneath, elongate-lanceolate, acuminate, sharply and unequally cut or pinnatifid, the lobes toothed, the upper leaves smaller, narrower and less deeply cut; fl. branches rod-like and racemose below, panicled and corymbulose at the apex, the heads partly clustered, partly pedicellate; pedicels short, scaly; inv. calycled, both calycle and scales glabrous, scales 8–12 flat, lanceolate-acuminate, equalling the disc-fl.; rays 5–6, yellow; disc-fl. 15–20; achenes pubescent. *S. phalachrolœnus. DC.! l. c. 393.*

HAB. Cape, *Thunberg!* East of Capetown, *Burchell*, 8026. Wagenmakers Bosch, Swell., *Mundt.!* Berg R., *Drege!* (Herb. Th., D., Hk., Sd.)

Very near *S. pubigerus* and somewhat intermediate between that and *S. expansus* From *S. pubigerus* it differs in the glabrous calycle and the tendency to a corymbose or panicled inflorescence; from *S. expansus* in general aspect, involucre, and a tendency to racemose infl. in the lower part.

98. S. expansus (Harv.); stem herbaceous, diffuse, terete, scabro-pubescent, the branches widely spreading, loosely panicled at summit; leaves alternate, sub-distant, stem-clasping, cordate at base, scabrous on both sides, paler and netted-veined beneath, the lower more or less deeply lyrate-pinnatifid, the medial oblong, distantly toothed and repand, the uppermost ovate-acuminate, small; panicle very loose, alternately compound, the pedicels filiform, scaly; inv. calycled, glabrous, of 10–12 narrow, acute scales equalling the disc; rays about 5, spreading; achenes minutely and closely puberulous. *S. lyratus, DC.! l. c. 382, excl. syn.*

HAB. Dutoitskloof; Simonsberg; and Kl. Draakenstein, near Berg R., *Drege!* (Herb. D., Sd., Hk.)

Root perennial? Branches 3–4 ft. long, very diffuse, the terminal, lax panicle 1 foot or more in length. Leaves 2–3 inches long, ½–1 inch wide, the uppermost diminished to scales. Pubescence very short, but rigid. Teeth of the leaves sharp, with wide, rounded interspaces; margin slightly reflexed. Inv. turbinate, 3 lines long. Allied to *S. pubigerus* and *S. incisus*, but differing in infl., &c. from both. DC. places it among the annual species, to none of which is it naturally allied.

99. S. lyratus (Linn. f. suppl.; non. DC.); herbaceous, erect or straggling, more or less scabrous, with minute, rigid pubescence; stem terete, striate, pithy; cauline leaves stem-clasping, *variable in shape*, lyrate or lacero-pinnatifid, acute, or the upper ones narrow-oblong, all sharply and unequally toothed, lobes short, unequal, the terminal larger; corymbs much-branched, many-headed, pedicels scaly; heads radiate, about 30-fl.; inv. calycled, glabrous, of many narrow scales; rays 6–8, yellow; achenes small, minutely hispidulous. *S. lyratus, Th.! Cap.* 684. *S. variifolius, DC. l. c.* 393.

VAR. β. **subcanescens** (DC.); leaves cobweb-canescent beneath, surface rough. *(E. & Z.!)*

VAR. γ. **pentaglottis** (DC.); "branches glabrous, striate and furrowed; lvs. cobweb-pubescent beneath; disc fl. 15–20; rays 5." *(Drege.)*

VAR. δ. **lacerus** (DC.); sparingly hispid; leaves deeply pinnatifid, the lobes incisoserrate; disc-fl. 15–20; rays about 5. *(Ecklon! Bowie!)*

HAB. Table Mountain summit, *Thunberg!* Swellendam and Uitenhage, *E. & Z.* Cape Flats, *W. H. H.* (Herb. Th. Sd., Hk., D.)

More herbaceous than *S. rigidus,* with (usually) more deeply-cut or pinnatifid leaves, but very variable. Var. β. (Hb. Sd.) looks like a weak state of *S. rigidus;* γ. is unknown to me; δ. (Herb. Sd., Hk.) ought perhaps to be made the type of the species. Leaves 2–6 inches long, ½–1½ wide. Heads not 2 lines long, campanulate. This is unquestionably the *S. lyratus,* Th., and therefore probably of Linn. f., but not of De Candolle. The specimens examined by me are not sufficiently copious to enable me to judge of its specific limits, or whether *all* the varieties might not be referred to *S. rigidus.*

100. S. rigidus (Linn. Sp. 1224); a large shrub; branches striate, terete, roughly hispid, the upper ones herbaceous, panicled at summit; leaves stem-clasping and often adnate-subdecurrent at base, oblong, or fiddle-shaped, obtuse or acute, the smaller often obovate, sharply and unequally toothed, scabrous above, more or less cobwebbed, and netted-veined beneath, harsh to the touch; corymb much branched, its divisions closely many-headed, pedicels scaly; heads radiate, 20–30-fl.; rays 5–6, yellow; inv. calycled, glabrous, of many narrow scales; achenes striate, hispidulous. *Th.! Cap.* 682. *DC. l. c.* 392.

VAR. β. **scaber**; rameal leaves conspicuously adnate-decurrent at base, obovate-oblong, obtuse, green on both sides, not cobwebbed. *S. scaber, Willd., DC. l. c.* 334. *Jacobæa scabra, Th.! Cap.* 677. (Herb. Thunb.)

VAR. γ. **serratus**; pubescence scanty, upper side of leaves minutely rough. *S. variifolius, var. serratus, DC. l. c.* (Herb. Sd.)

HAB. Cape Flats, common. Uitenhage, *Zey!* 2942. (Herb. Th., Hk., D., Sd.)

8–12 ft. high, much branched, scabrous and rough to the touch in most parts. Leaves 2–3 inches long, 1–2 in. wide, variable in shape and pubescence, sometimes almost woolly beneath. Infl. very compound, the heads small; inv. 1½–2 lines long.

101. S. caulopterus (DC.! l. c. 388); stem herbaceous, erect, rib-striate, minutely pubescent; branches corymbiferous; cauline leaves obovate or oblong, sharply calloso-denticulate, glabrous or cobwebbed above, minutely hispidulous, penninerved and closely reticulate beneath, decurrent at base into a long, toothed or subentire stem-wing; upper lvs. ovate, sub-sessile; corymb compound, laxly many-headed, the pedicels long, somewhat scaly; inv. calycled, of 9–12 glabrate scales; disc-fl. 10–12, rays 4–5; achenes (young) hispidulous.

HAB. Betw. Omsamwubo and Omsamculo, *Drege!* Inyatikab River, Natal, *W. T. Gerrard!* (Herb. Sd., D.)

Of *Drege's!* plant I have seen but the apex of a branch and one of the uppermost, non-decurrent leaves; my description therefore is chiefly adapted from DC. l. c.

The stem is said to be 2-3 ft. high. The leaves have the texture and close reticulation of those of *S. rigidus*, and are anything but glabrous beneath. *Mr. Gerrard's* specimen to which I refer here consists of a lateral, flowering branch, about a foot long, bearing many leaves similar to that on *Drege's*, but not quite sessile nor stem clasping.

102. S. pterophorus (DC.! l. c. 389) ; stem herbaceous, erect, ribstriate, cobwebbed or glabrate, virgate, corymboso-paniculate at the apex; leaves lanceolate, sharply and coarsely inciso-serrate, acuminate, tapering at base almost into a petiole, glabrate and green above, cano-tomentose beneath, the petiole produced at base (in var. *α.*) into narrow or obsolete, somewhat denticled stem-wings ; corymb very compound, pedicels scaly; inv. of many scales, amply calycled, glabrous ; disc-fl. 40-50, rays 8-10; achenes glabrous.

VAR. *α.* **verus**; wings of the stem more or less obvious. *S. pterophorus, DC. l. c.* *(Drege ex pte.)*

VAR. *β.* **apterus**; wings of the stem obsolete or none. *S. pterophorus, ex pte, Drege!* *S. polyanthemoides, Sch.! Bip. Bot. Zeit. 27, p. 697. S. polyanthemus, var. γ. subserratus, DC. l. c.*

HAB. Natal District. *α.* Betw. Omtendo and Omsamculo, *Drege!* *β.* Betw. Gekau and Basche, and betw. Omsamwubo and Omsamcaba, *Drege!* Natal, *Krauss!* 424. *Gueinzius!* (Herb. D., Hk., Sd.)

Stem 2-3 ft. high. Leaves 3-4 inches long, ¼-½ inch wide. The stem wing is very narrow at best, and often obsolete, if *Drege's* specimens all belong to the same species. This species is very closely related to *S. juniperinus*, but the leaves are not ear-clasping, and the fl. heads are smaller.

103. S. ilicifolius (Th.! Cap. 682); stem robust, suffruticose at base, erect, rigid-herbaceous, leafy, branched at the summit, the branches cobwebbed, corymbiferous ; leaves clasping and adnate-subdecurrent at base, oblong or oblongo-lanceolate, acute, sharply and coarsely toothed, with reflexed or revolute margins, glabrous and scabrous-dotted above, albo-tomentose beneath ; corymb much branched, spreading, many-headed, the pedicels scaly ; heads 50-60-fl., radiate ; inv. calycled, of 12-15 glabrous scales; rays about 12, yellow; achenes minutely granulated. *DC.! l. c. 391. S. quercifolius, Th.! Cap. 682.*

HAB. Cape, *Thunberg! Krebs, Mundt!* Uitenhage and Albany, *E. & Z.! Zey.!* 2945. Camtour's R., *Gill!* Knysna, *Dr. Pappe!* (Herb. Th., D., Hk., Sd.)

Stem 2-3 f. high, branched at the summit, closely leafy. Leaves 1-2½ inches long, 4 lines to 1½ in. wide, rigid, sharply indented like those of a holly, with round spaces between the teeth ; upper surface glossy green, lower white. Infl. 5- to upwards of 100-headed. A larger and coarser plant than *S. juniperinus,* with broader and more deeply toothed and rigid leaves. *S. quercifolius,* fide Herb. Th.! is merely the base of the stem, where the leaves are larger and more incised !

104. S. juniperinus (Linn. f. Suppl. 371); stem suffruticose, cobwebby-tomentose, branching, branches cobwebbed or nude, corymbiferous ; leaves half-clasping and often unidentate at base, linear-subulate or linear-lanceolate, subpungent-mucronate, with revolute or recurved margins, either entire or sparingly toothed, or some furnished with a pair of opposite, horizontal lobes, scabro-punctulate and glabrate (cobwebbed) above, albo-tomentose beneath ; corymb branching, loosely many-headed, the pedicels longish, scaly ; heads 40-60 fl., radiate ; inv. calycled, of 12-15 glabrous, barbellate scales; rays about 12, yellow ; achenes nearly glabrous, or minutely puberulous. *Thunb.! Cap. 679. DC.! l. c. 391*; also *S. polyanthemus, DC.! l. c.*

VAR. *a.* **genuinus**; leaves linear-subulate, with strongly-revolute margins, entire.

VAR. *β.* **salicinus** (DC.); leaves flattish, lanceolate, with recurved margins, entire or subentire; minutely glabrous-dotted above.

VAR. *γ.* **epitrachys** (DC.); leaves less expanded, with more strongly-revolute margins and more scabrid; entire, or here and there toothed.

VAR. *δ.* **cruciatus**; leaves (some at least) above the middle furnished with a pair of opposite, spreading lobes, like the arms of a cross. *S. cruciatus, Linn. f. Thunb.! Cap. 671.*

HAB. Cape, *Thunberg!* &c. Uitenhage, *E. & Z.! Zey.! 2945.* Zuureberg, Stormberg and Witberg, *Drege!* Hassaquaskloof, *Zey.! 2952.* (Herb. Th., D., Sd., Hk.)

Stems rigid and strong, but scarcely ligneous, except at base; the branches numerous, more herbaceous, but wiry. Branches and leaves more or less cobwebby, becoming nude. The leaves vary in length and breadth and degree of expansion, and in being sparingly toothed or entire; I find them constantly scabrous-dotted above, even in var. *salicinus* (*Drege's* specimens). *Drege's* var. "*subserratus*" appears to me to belong rather to *S. pterophorus.* Inflorescence generally widely spreading varying very much, however, in the number of heads. Var. *δ. cruciatus,* is rather a monstrosity than a genuine form.

105. S. œderiæfolius (DC.! l. c. 391); stem erect, rigid-herbaceous striate, scaberulous, the young parts cobwebbed, branches erect virgate; leaves sessile, half-clasping, and slightly adnate at base, ovate-oblong or ovato-lanceolate, acuminate, rigid, sharply and distantly toothed, with reflexed margins, scaberulous, glabrous and glossy above, white-woolly, (at length nude) beneath; heads either solitary, terminal or 2–5 in a corymb, radiate, over 100-fl.; rays 12–15; inv. amply calycled, flat-bottomed, of many glabrous or cobwebbed scales; achenes minutely granulated.

HAB. Vanstaadensberg, Uit., *E. & Z.! Zey.! 2944.* (Herb. D., Sd., Hk.)

A branching, bushy, but not ligneous, plant, 1-2-ft. high; with straight, erect, leafy one or few-headed branches. Leaves ¾-1¼ inch long, 5-6 lines wide, erect, very rigid. Heads 5-lines across. Similar in foliage to *S. ilicifolius,* from which it is known by its fewer and larger fl. heads.

106. S. arnicæflorus (DC.! l. c. 390); stem suffruticose, ascending, cobwebby, closely leafy below, subsimple, (either 1–3-headed or) corymbiferous and subpedunculoid at top; leaves oblong, sessile and sub-decurrent, rather narrowed to the base, entire or sparsely few-toothed, with revolute margins, mucronate, cobweb-woolly beneath, scaberulous and thinly cobwebbed, becoming glabrate above; upper leaves lax, small, linear, erect, acuminate; corymb simple or branched, few or several headed, the pedicels long, bearing 2–3 subulate leaf-scales; heads many-fl.; rays 10–12, spreading; inv. calycled, cobwebbed, of many narrow scales; achenes (young) pubescent.

HAB. Auteniqualand, *Burchell.* Cape, *Bowie!* Groenekloof, and near Cape l'Aquillas, *Drege!* Near Simonstown, *C. Wright,* 302, 303. (Herb. Hk., D.)

Stems 8-12 inches high. Leaves 1-2 inches long, 3-6 lines wide, with strongly revolute margins. Whole plant cottony-cobwebbed. Heads 4-5 lines across, yellow. I have not seen Burchell's plant, which may differ, for DC. speaks of a "valid mucro" to the leaves and "1-2 *sessile* fl. heads," characters which do not exist in *Drege's* specimen in Hb. Hook. The pedicels are corymbose, 1-1½ inch long; the mucro is small and not always obvious.

107. S. microspermus (DC. l. c. 391); "stem suffruticose, branching, *diffuse or depressed*, cobwebby; leaves crowded, sessile, half-clasping, ovato-lanceolate, acuminate, *quite entire*, scabrid above, cobweb-woolly beneath; branches scarcely nude at the summit, 1–2-headed; inv. of about 20 subulate, glabrous scales, shorter than the disc, amply calycled with cobwebby bracteoles; disc-fl. about 50, rays few? achenes *minute*, microscopically puberulous. *DC. l. c.*

HAB. Algoa Bay, *Drege.* (Unknown to me.)
" Leaves 8–9 lines long, 3 lines wide."

108. S. hirtifolius (DC.! l. c. 394); subherbaceous or suffruticose, all parts densely pubescent with short, spreading, rigid hairs; branches erect, angular; leaves crowded, sessile, oblong-obovate, cuneate and tapering at base, coarsely 5–7-toothed above, on both sides equally hairy, flat, rigid, thickish; pedunc. terminal, nude, simple or branched; corymbs few-headed, rather loose; inv. oblong, calycled, of about 12 scabrous scales, shorter than the disc; disc-fl. 15–20; rays about 5, reflexed; achenes striate, hispidulous.

HAB. Near the Zwartkops R., Uitenhage, *Ecklon!* (Herb. Sd., Cap.)
Stem 1–2 ft. high, the branches very erect, closely leafy. Leaves imbricate, 1½ inch long, ½–¾ inch wide, erect, of thickish substance, coarsely velvetty with short hairs. Pedunc. 4–6 inches long. Inv. scales 4–5 lines long. A well marked species.

109. S. vestitus (Berg. Cap. 282); quite glabrous and smooth; stem suffruticose, terete, finely striate, leafy; leaves stem-clasping, broad-based, oblong or obovate, obtuse, membranous, sharply and unequally serrate, netted-veined; corymb patently much-branched, many-headed, pedicels scaly, divaricate; heads radiate, 20–30-fl.; inv. calycled, glabrous, of many narrow scales; rays 6–8, yellow; achenes striate, hispidulous. *Jacobœa vestita, Th.! Cap.* 677. *S. rigidus, var. glabrescens, E. Mey.! in Hb. Drege. DC. l. c.* 393.

HAB. Cape, *Bergius, Thunberg!* Camiesberg and Modderfontein; also in Dutoit's Kloof, *Drege!* Knysna, *Dr. Pappe!* Mitchell's Pass, *A. Wyley!* (Herb. Th., D., Sd., Hk.)
Very similar to *S. rigidus* in the form of the leaves and habit; but perfectly glabrous and smooth in all parts, with thinner and softer leaves, and a more spreading panicle. If not a "species," it seems to be a well-marked and widely spread "form." Our specimens exactly agree with one in Hb. Thunb.!

110. S. aquifoliaceus (DC.! l. c. 394); stem herbaceous? terete, here and there setose; leaves very closely crowded, shortly adnate, lanceolate, sharply and coarsely toothed, penninerved, on each side glabrous, or the young ones setulose on the nerves, rigid; corymb compound, terminal, just overtopping the leaves; pedicels sparingly scaly; inv. calycled, glabrous, of 10–12 lanceolate scales, about equalling the disc; disc-fl. 30–40, rays 6–8; achenes nearly glabrous.

HAB. Betw. Sparrbosch and Tradow, Swell., *Drege!* (Herb. Sd.)
Of this I have only seen leaves and a division of the corymb. The leaves are 2–2½ inches long, 4–5 lines wide, like those of *S. glastifolius.* Fl. yellow.

111. S. lanceus (Ait. Kew. 3, p. 194); stem suffruticose, erect, sub-simple or panicled at top, glabrous or loosely cobwebbed, striate; leaves

stem-clasping and shortly adnate-decurrent at base, oblongo-lanceolate
or lanceolate, acute, coriaceous, glabrous, with subrecurved margins,
either sharply serrate, serrulate or subentire, the upper smaller and more
acute ; corymb compound, crowded, many-headed, the pedicels short,
scaly ; inv. calycled, glabrous, of 20 or more narrow-subulate scales ;
rays 10–12, shorter than the inv.; disc-fl. 40–50 ; achenes striate, glabrous. *Jacq. Schoenbr. t.* 304. *DC.! l. c.* 394. *S. marginatus, Th.! Cap.
p.* 680, *ex pte. majori.*

VAR. *a.* **serratus** ; leaves more or less sharply serrate. Vars. β, γ, є, *DC. l. c.*

VAR. β. **subinteger** (DC.) ; leaves nearly entire. *S. undulatus, Th.! Cap. p.* 680.

HAB. Cape Flats, and eastward through Swellendam to Uitenhage and Albany.
(Herb. Th., D., Hk., Sd.)
Stem strong, 3–4 ft. high, chiefly branched at the summit. Young parts often
cobwebbed with deciduous wool. Leaves 3–5 inches long, ¾–1½ in. broad, the upper
gradually smaller. Dentition of the margin very inconstant; the var. with nearly
entire leaves is frequent.

112. S. scoparius (Harv.) ; stem straight, erect, glabrous, striate,
simple or corymbose at summit ; leaves scattered, subsessile, subcune-
ate at base, ovate or ovato-lanceolate, acute, calloso-serrulate, netted-
veined, quite glabrous or the young ones cobwebbed, the upper smaller
and narrower; corymb very compound, flat-topped (broom-like), many-
headed, the pedicels long, filiform, sparsely scaly; inv. minutely calycled, glabrous, of 10–13 bluntish, keeled scales, shorter than the disc ;
rays 5–6, yellow, spreading ; disc-fl. about 20 ; achenes hispidulous.

HAB. Kreili's Country, about old Kraals, *H. Bowker,* 388, *Mrs. F. W. Barber,*
25. (Herb. D.)
Stem 2–3 ft. high, very straight, ending in a much-branched, broom-like corymb,
which is 4–6 inches diameter, compact and flat-topped. Leaves 1½–2 inches long,
¼–1 inch wide, the lowest quite sessile and bluntly wedge-shaped or rounded at
base, the uppermost often on 1 line long petioles, cuneate ; under surface rather
paler and closely netted. Pedicels 1–2 inches long. Heads 3 lines long, and equally
wide. Rays 3 lines long.

113. S. Serra (Sond. in Linn. 23, p. 68) ; stem herbaceous, erect,
ribbed and furrowed, glabrous, virgate, corymboso-paniculate at the
apex; cauline leaves oblongo-lanceolate, acute, cartilagineo-serrulate,
tapering and entire at base, decurrent in a very narrow (or obsolete)
entire stem-wing, the uppermost often auricled ; corymb compound,
flat-topped, the pedicels scaly; inv. of 10–12 scales, glabrous, calycled,
the scales dark-tipped ; disc-fl. about 20, rays 6–7; achenes glabrous.

HAB. Natal, *Gueinzius!* 358. (Herb. Sd., Hk.)
2–3 feet (or perhaps more) in height. Lower leaves not seen. Cauline leaves
3–5 inches long, ½–¾ inch wide, coriaceous, rigid, closely and finely serrulate. The
wings of the stem are very narrow and sometimes obsolete. Inv. 2 lines long.

114. S. halimifolius (Linn. Sp. 1223); shrubby, erect, glaucous, more
or less cobwebbed with woolly hairs; branches angular, ribbed, virgate,
closely leafy; leaves sessile, slightly ear-clasping at base, cuneate-obovate
or oblong, subobtuse, coarsely toothed, not conspicuously veiny, thickish, the adult glabrous, the younger cobwebbed ; corymb compound,
densely many-headed; pedicels short, scaly ; heads radiate, 30–40-fl.;

inv. calycled; rays 10–12, yellow, spreading; achenes striate, minutely pubescent. *DC.! l. c.* 392. *Dill. Hort. Elth. t.* 104, *f.* 124. *S. solidagi-noides, Berg. Cap.* 284. *Th. Cap.* 681. *S. sessilis, Th. Cap.* 682. *S. soli-dagineus, Spr. DC. l. c.* 435. *S. populifolius, Linn. Sp.* 1224.

VAR. *β.* **repandus** ; leaves *subentire,* slightly repand or angular. (Cape Flats, Herb. Hook.)

HAB. Cape Flats, *Thunberg! Eckl.! Wallich! W.H.H.* &c. (Herb. Th., Sd., Hk.)
A corymbose, pale green bush, 1-2 feet high, much branched, the branches end-ing in compound, often much divided, densely many-headed corymbs of yellow flowers. Young parts cobwebbed or woolly, older glabrate. Leaves 1½-2½ inches long, ½-1¼ wide, cuneate at base, with slightly recurved edges, coarsely callous-toothed, the teeth broad and shallow. Inv. 2-3 lines long, shorter than the flowers. Var. *β.* (Hb. Hk.) has the leaves nearly entire, or obsoletely angled and callous-denticulate.

115. S. coleophyllus (Turcz.! Bull. Mosc. xxiv. ii. (1851) p. 90); stem erect, robust, rigid-herbaceous, glabrous or cobwebbed, branches erect, ending in slender few-headed peduncles; leaves clasping and adnate-subdecurrent at base, ovate or oblong-ovate, acute, rigid, sharply and distantly toothed, with reflexed margins, spreading or deflexed, glabrous and glossy above, white-woolly beneath; heads either solitary or 2–5 subcorymbose, long-pedicelled, 80–100-fl. or more; rays 10–12 *purple;* inv. amply calycled, of many glabrous or cobwebbed, narrow scales; achenes minutely granulated. *Walp. Ann. V. p.* 337.

HAB. Mts. near Riv. Zondereinde, *Zey.!* 2953. (Herb. Sd., Hk.)
In foliage and general habit this nearly resembles *S. oederiæfolius,* but the pedun-culoid branches are slender and more nude, and the rays bright purple. Teeth of the disc-fl. papillose at the apex.

116. S. glastifolius (Linn. f. Suppl. 372); glabrous; stem suffruticose below, erect, branching; leaves oblong or lance-oblong, acute, coarsely and unequally toothed, half-clasping, the lower more or less decurrent, narrowed towards the base; panicle laxly corymbose, pedicels elongate; heads many-fl., radiate, the rays purple; inv. amply calycled with sub-ulate bracteoles, glabrous; achenes striate, puberulous. *DC.! l. c.* 408.

HAB. Table Mountain, *Thunberg.* East of Capetown, *Burchell.* Betw. Kabeljous Riv. and Gamtoos R., and Groenvalei, near Bosch Riv., *Drege!* (Herb. Th., D., Sd. Hk.)
Tall, strong-growing. Leaves 1-3 inches long, ½-1 inch wide, rigid, quite gla-brous, when dry punctulate. Panicle many headed, the partial corymbs 3-5 headed. Inv. campanulate, 4-5 lines diam. and about as long. Rays spreading, numerous, purple, fading to dirty whitish in the herbarium.

§ 6. MICROLOBI. (Sp. 117–121.)

117. S. oliganthus (DC.! l. c. 390. excl. syn. Th.!); suffruticose, villoso-tomentose, subsimple or branched, the branches virgate, closely leafy, sub-pedunculoid and corymbiferous at the summit; leaves sessile, ovate or shortly oblong, entire or 3–5 toothed, rigid, with subrevolute margins, tomentose on one or both sides, faintly 3-nerved; corymbs dense, the pedicels tomentose; inv. sparingly calycled, of 5–7 tomentose, membrane-edged scales; heads discoid, 7–12-fl.; achenes glabrous.

VAR. *a.* **discolor**; upper surface of leaves glabrous, green; lower tomentose.
VAR. *β.* **concolor** (DC.); both surfaces of leaves tomentose.

HAB. Cape, *Thunberg.* Gauritz R., *Burchell.* Vanstaadensberg and Langekloof, *Drege! E. & Z.! Zey.! 2939.* (Herb. D., Hk., Sd.)

1–2 ft. high or more, the branches 10–18 inches long, straight, imbricated with leaves for ¾ of their length or more, thence passing into a sparsely leafy peduncle. Leaves ½–¾ inch long, 4–5 lines wide, commonly 3-toothed, but often toothed and entire on the same branch. Toment. rusty, persistent. Inv. 2½ lines long, shorter than the flowers. *Cacalia tomentosa*, Th., referred to this by De Candolle, is a very different species, our *S. Thunbergii.*

118. S. quinquenervius (Sond.); stem erect, simple, suffruticose, villoso-tomentose, closely imbricated with leaves; leaves sessile, half-clasping at base, ovate or oblong-ovate, with reflexed edges, 3–5-ribbed and densely tomentose beneath, cinereo-tomentulose and impress-nerved above, either quite entire or 3–5-denticulate near the apex, *now and then with a minute, stipuloid lobule at base;* corymb compact, branching, several-headed, the pedicels short, tomentose; heads 25–30-fl.; inv. subcalycled, of 8–10, oblong, tomentose scales, shorter than the disc; rays 5; achenes glabrous.

HAB. Krakakamma, Uit., *E. & Z.!* (Herb. Sd., Hk.)

Very similar in general habit to *S. lineatus* but more woolly, with broader, commonly 5-nerved leaves, not tapering to the base, and (in our specimens) *very rarely* furnished with a single, minute, tooth-like basal-lobe, the margins more reflexed and the apical teeth smaller. A small specimen, without habitat, is in Herb. Hk., glued on the same sheet with *S. lineatus;* of which this is perhaps a mere variety.

119. S. lineatus (DC.! l. c. *395*); stem erect, simple, suffruticose, cinereo-tomentose; leaves crowded, sessile, at the very base furnished on each side with 1–3 narrow lobules (sometimes wanting), lanceolate, with recurved margins, either entire or coarsely few-toothed near the apex, tomentose and 3-nerved beneath, tomentulose, becoming glabrous above; corymb somewhat peduncled, compact, branching, several-headed, the pedicels tomentose, scaly; heads 20–25-fl.; inv. calycled, the scales 8–9, oblong, acute, tomentulose, shorter than the disc; rays 5–6, eq. the inv. scales; achenes glabrous. *Cineraria lineata, Linn. f. Th.! Cap. 670. S. persicifolius, Burm., fide DC.*

HAB. Cape, *Burman, Thunberg!* Round Capetown, *Drege! Ecklon! Pappe! W.H.H.* Koega R., *Zey.! 2938.* (Herb. Th., D., Hk., Sd.)

Stem 1–2 f. high, virgate, imbricated with leaves, the uppermost smaller, narrower, and passing into leaf-scales. All the young parts tomentose, the older leaves becoming nude above. Corymbs 12–30-headed. Inv. 4 lines long. Rays yellow.

120. S. triplinervius (DC. l. c. *395*); stem suffruticose, erect, cinereo-canescent, glabrescent at apex; leaves sessile, lyrate, having at base on each side 2–3 minute, approximate, lanceolate lobules, the terminal lobe (or principal lamina) oblong or lanceolate, with recurved margins, coarsely few-toothed near the apex, cinereo-tomentose and three-nerved beneath, glabrous, with impressed veins above; corymb lax, slightly compound; heads about 25-fl.; inv. minutely calycled, of about 8, oblong, acute, glabrous scales, shorter than the disc; rays 5–6, nearly as long as inv. scales; achenes glabrous.

HAB. Eastern Districts, *Burchell,* 5373. Kromrivier, *Drege!* (Herb. Sd.)

Only known to me by a small specimen, from *Drege,* in Hb. Sd. Except by the

glabrous inv. scales, if this be constant, I cannot distinguish it from *S. lineatus,*
which often has basal leaf-lobes.

121. **S. penninervius** (DC. l. c. *395*); "stem suffruticose, erect, cinereo-
puberulous, somewhat naked upwards; leaves sessile, lyrate, having at
base on each side 3–4 lanceolate lobules, the terminal lobe (or principal
lamina) lanceolate, serrate, penninerved and cinereo-pubescent beneath,
glabrous, with impressed veins above; corymb compound, many-headed;
heads 25–30-fl.; inv. calycled, glabrous, half as long as the disc-fl.;
rays twice as long as the involucre." *DC. l. c.*

HAB. Eastern Districts, *Burchell,* 5363.
Unknown to me. It is said greatly to resemble *S. triplinervius,* but to differ in
the more distant leaf-lobes, here and there toothed, in the terminal lobe which is
serrate nearly throughout, and in the more numerous heads. Both are probably
mere vars. of *S. lineatus.*

§ 7. LEPTOLOBI. (Sp. 122–138.)

122. **S. umbellatus** (Linn.); glabrous or nearly so; stem suffruticose
at base, erect, branching, striate; leaves linear, elongate, with strongly
revolute margins, acute, quite entire, or here and there sparsely and
irregularly toothed or pinnati-lobed, the teeth or lobes subulate, widely
spreading; corymb laxly few-headed, the pedicels long and nearly nude;
heads many-fl., radiate, the ray purple; inv. amply calycled, with many
subulate bracteoles, of about 20 glabrous, narrow scales; achenes striate, puberulous. *DC. l. c.* 408. *S. linifolius, Drege! ex pte. S. tenuilobus,
DC. l. c.* 398.

VAR. β. **latiusculus**; leaves shorter, broader (1–1½ inch long, 1½–2 lines wide),
with less reflexed margins, denticulate.

HAB. Cape Flats, *Wallich! W. H. H.!* Kamiesberg, *Drege!* β. Betw. Vanstaadensberg and Bethelsdorp, *Drege!* (Herb. D., Hk., Sd.)
Stem 1–2 feet high, forming a much-branched suffrutex, the lower parts ligneous,
the upper herbaceous. Leaves 2–3 inches long, ½–1 line wide, more usually quite
entire and simple; but varying on the same branch to toothed or lobed. Infl. like
that of *S. grandiflorus;* pedicels 2–3 inches long or more. Rays purple, but in Herbaria often faded to an uncertain, dingy whitish or buff-colour: such specimens have
been distributed as "*S. linifolius*" (*linoides? DC.*) by Drege (Hb. Sd.). *Ecklon's*
specimens (Hb. Sd.) are in a wretched condition, *quite scabrous,* and do not seem to
belong to our plant. *S. tenuilobus,* DC., unless it have *yellow* rays, which I cannot
determine from the imperfect specimens seen (Herb. D., Sd., Hk.), differs in no respect from ordinary *S. umbellatus.*

123. **S. grandiflorus** (Berg. Cap. 280); herbaceous, perennial; stem
tall, strong, virgate, striate, nearly glabrous, leafy for the greater part,
nude and panicled at the summit; cauline leaves sessile, half-clasping,
pinnati-partite, the lobes on each side several, linear, acuminate, entire
or toothed, with subrecurved margins, the lowest short or toothlike, all
sparsely pilose or sub-glabrous; panicle laxly corymbose, the partial
corymbs 3–5-headed, the pedicels long and nude; heads many-fl., radiate,
the rays purple, disc yellow; inv. amply calycled with many subulate,
spreading bracteoles, of 18–20 narrow, glabrous scales; achenes black,
striate, minutely puberulous in the furrows. *Less. Syn.* 301. *DC.! l. c.*
408. *S. venustus, Ait. Lindl. Bot. Reg. t.* 901. *S. pauciflorus, Th.! Herb.*

Var. β. **albidus**; disc and ray-fl. whitish or very pale. *S. albidus, E.M.! DC.
l. c.* 398.

Hab. Summit and East side of Table Mt., *Thunberg! Bowie! Eckl.! W. H. H.,
Wallich!* (Herb. Th., D., Hk., Sd.)

Stems 4–5 feet high, simple and closely leafy for ¾ of their length; then gradually
nude, and ending in a spreading panicle. Leaves 3–6 inches long; the lobes 1–2
inches long, 1–2 lines wide. Inv. campanulate, 4–5 lines diameter. The name
"*grandiflorus*" is inappropriate, as many species have larger fl. heads.

124. S. leucoglossus (Sond.); stem herbaceous, erect, straight, rib-
striate, scabrous below, glabrous above, simple, ending in a loose, corymbose panicle; leaves sessile, half-clasping, pinnati-partite, above very
scabrous with raised points, the rachis and lobes linear, with reflexed
edges, here and there sharply toothed, lobes few, distant, unequal, acute;
corymbs few-headed, pedicels spreading, smooth, nude; heads many-fl.,
radiate, *the rays* 10–12, *white;* inv. amply calycled with many subulate,
imbricating bracteoles, of 18–20 narrow, subulate, keeled, glabrous
scales; achenes hispidulous.

Hab. Witsenberg, *Zeyher!* 943. (Herb. Hk., Sd.)

Root ligneous. Stems 1–2 ft. high, pale, branched only at the summit. Leaves
scattered, 1–1¼ inch long, the lobes 1–1½ line wide, acute, very rough with sharp
points. The rays seem to be white; by which character and that of the involucre
this differs from the nearly allied *S. muricatus,* Th.

125. S. muricatus (Th.! Cap. 686); stem herbaceous, erect, ribstriate,
scabrous-pilose throughout, simple, ending in a loose, corymbose panicle;
leaves sessile, half-clasping, pinnati-partite, above very rough with
swollen, jointed hairs, the rachis and lobes linear, entire, with revolute
margins, acute; lobes few and distant; pedicels spreading, pilose, nude,
elongate; heads many-fl., radiate, rays 8–10, yellow; inv. calycled with
several lanceolate, flat, nerved bracteoles, of 12–13 broad, flat, nerved,
membrane-edged scales; achenes puberulous. *DC. l. c.* 435.

Hab. Cape, *Thunberg!* (Herb. Th.)

Very similar to *S. leucoglossus,* but with involucre and calycle of much fewer, flatter
and broader scales; *yellow* rays, fide Th.; and a more pilose pubescence.

126. S. parvifolius (DC.! l. c. 396); suffruticose ? erect, diffusely
much branched, quite glabrous; branches striate, pale; leaves scattered,
fleshy, small, irregularly inciso-serrate or pinnatifid with very short
lobes; fl. branches laxly subcorymbose, the pedicels very long, minutely
scaly; inv. calycled, of 12–13 black-tipped, acute scales; disc-fl. 40–50,
rays 8–10, revolute; achenes terete, pubescent.

Hab. Modderfonteinsberg, Namaqualand, *Drege!* (Herb. D., Hk., Sd.)

A very slender plant, 1–2 ft. high, much branched, with pale, straw-coloured stems
and branches, which seem ligneous at base. Leaves ½–¾ inch long, the lobes ½–1
line long, toothlike. Pedicels 3–4 inches long, incurved.

127. S. Carroensis (DC.! l. c. 396); herbaceous, diffusely branching,
glabrous; branches terete; leaves scattered, sessile, thickish, laxly
pinnati-partite, not eared at base, the lobes unequal, some short and
tooth-like, some subulate, acute, entire or denticulate; uppermost leaves
sometimes linear, entire; fl. branches forked or few-branched, pedicels
long, slender, with 1–2 minute scales; inv. calycled, of 10–12 glabrous,

nerved scales rather shorter than the disc; disc-fl. 25–30; rays 7–9, small, not much exserted; achenes minutely puberulous.

HAB. Kendo, in stony and rocky places, *Drege!* (Herb. D., Sd., Hk.)
Root possibly annual? Stem slender, subpaniculate. Leaves 1–1¼ inch long, the longer lobes 2–4 lines long, the shorter scarcely a line, ¼ line wide. Pedicels 2–3 inches long. Our specimens are very imperfect.

128. S. pinnulatus (Thunb. Cap. 679); stem suffruticose at base, erect, striate, simple or branched, pilose or glabrous; leaves sessile, often ear-lobed at base, approximate, distantly pinnati-partite, the lobes on each side 3–5, linear-elongate, acute, with reflexed or revolute margin, quite entire or here and there minutely toothed, mostly glabrous; fl. branches subnude upwards and loosely paniculate-corymbose, pedicels slender, sparingly scaly; inv. calycled, of 12–15 glabrous, black-tipped scales; disc-fl. 40–50, rays 8–10; achenes angular, pubescent. *DC. l. c.* 396. Also *S. scabriusculus, DC. l. c.*

HAB. Cape, *Thunberg!* Stellenbosch and Caledon, *E. & Z.!* Paarlberg, Drakensteinberg, Dutoitskloof, and about Table Mt., *Drege!* Steendaal, Tulbagh, *Dr. Pappe!* (Herb. Th., Sd., D.)
2–3 ft. high, with a subsimple, closely leafy lower stem, panicled and laxly leafy or nude above. Leaves 1–2–3 inches long, the rachis and ½–1¼ inch long lobes from ½–1 line wide. Pubescence scanty, white, sometimes absent. Pedicels 1–3 inches long, spreading. Inv. 4 lines long. Rays bright yellow, reflexed. I cannot distinguish *S. scabriusculus*, DC., which I have seen in Hb. Sd.

129. S. tanacetoides (Sond.!); stem suffruticose, diffuse or ascending; branches curved, woolly-canescent below, scabrous above, leafy; leaves sessile or petioled, on both sides woolly-tomentose, canous, closely pinnatisect, the lobes on each side 8–12, shortly linear, obtuse, simple or 2–3-fid; infl. loosely panicled, subcorymbose, the long, nearly nude pedicels very scabrous; inv. scarcely calycled, of 12–13 very scabrous, dark-tipped scales; disc-fl. 40–50, rays 12–13, 4-lined; achenes 5-angled, glabrous.

HAB. Langspruit and Dornkop, *Zey.! 934, and Burke!* Witteberg, *T. Cooper,* 626. (Herb. D., Sd., Hk.)
Stems numerous, woody at base, at first decumbent, then ascending-erect, 1–2 ft. long; the lower part nude and glabrate; the medial leafy and white-woolled; the upper green, scabrous, not woolly and sparsely leafy or nude. Leaves about uncial, sessile and petioled on the same specimen, the lobes 2–3 lines long, very close. Pedicels 2–3 inches long. Inv. scales 4 lines long. Rays pale yellow.

130. S. achilleæfolius (DC.! l. c. 396); quite glabrous or cobwebby; stem suffruticose, erect, virgate, branching; branches terete, leafy; leaves sessile, somewhat fleshy, closely pinnatisect, the lobes on each side 8–10, shortly linear, with recurved edges, subacute, the lower quite entire, the upper often bifid or 3–5-toothed; pedicels subcorymbose, elongate, scaly, or branches subpedunculoid and one-headed; inv. nude or nearly so, of 12–13 broad, flat, *glabrous,* membrane-edged scales; disc-fl. 50 or more, rays 10–12; achenes 5-angled, glabrous.

VAR. β. glaucescens; more slender, with less corymbose infl., the branches thinly cobwebbed. *S. glaucescens, DC.! l. c.* (Herb. Sond.)

HAB. Sneeuweberg, and Ondeberg, and in the Winterveld, 3–4000 f. *Drege!* Damp valleys of Graaf-Reinet, *Mrs. F. W. Barber.* β. Zwartbulletje, *Drege!* (Herb. D., Sd., Hk.)

Stems rigid and ligneous, but *Mrs. Barber* calls it "a branching *annual:*" possibly by inadvertence. Branches closely or sparsely leafy, 6–12 inches long. Leaves 1–1½ inch long, their lobes 2–3 or 4–5 lines long, simple or lobulate. Heads sometimes solitary, more generally in few-headed, long pedicelled corymbs. Inv. with 1–2 scales at base or nude; its scales 3–4 lines long, 1 line wide, green, white-edged, flat. Var. β. seems merely a weak-growing specimen.

131. S. serrurioides (Turcz.! l. c. p. 89); stem suffruticose, erect, branching, densely leafy, glabrous or scaberulous; branches angle-striate, leafy-below, subpedunculoid upwards; leaves sessile, glabrous or scaberulous, rigid, closely pinnatisect, the lobes on each side 8–12, linear-terete, acute, the lower entire, the upper bifid or lobulate; heads solitary or in pairs on long, sparsely leafy or nude, scabrous, pedunculoid branches; inv. nearly nude at base, of 12–15 flat, *acuminate, scabrous,* membrane-edged scales; disc-fl. 80–100; rays 10–12, longer than the inv., 4-lined; achenes 5-angled, glabrous. *Walp. Ann. Bot. 5, p. 337.*

HAB. Moist places on the Vanstaadensberg, Uit., *Zey.!* 2974. (Hb. D., Hk., Sd.)

Very near *S. achilleæfolius,* but much stronger, more densely leafy, with fewer, larger, and less corymbose fl. heads, and scabrous peduncles and involucre. Leaves imbricating, 1–1½ inch long, their lobes ½ line wide, 5–8 lines long; inv. scales 5–6 lines long, taper-pointed.

132. S. euryopoides (DC.! l. c. 396); glabrous or nearly so; stem suffruticose, erect, virgate, densely leafy, terete, corymbose at the summit; leaves imbricating, sessile, ear-lobed at base, pinnati-partite, the lobes on each side 6–8, *linear-elongate,* with recurved edges, acute, quite entire; corymb compound, the pedunc. long, nude, the pedicels scaly; inv. calycled, of 12–15 acuminate, glabrous scales; disc-fl. 35–40, rays 8–10, oblong, multistriate; achenes glabrous.

HAB. Between Welgelegen and Onzer, Langekloof, *Drege!* (Herb. D., Sd., Hk.)

Stem 2–3 feet high, closely invested with leaves throughout. Leaves 1½–2 inches long, the lobes 1–1½ inch long, not ½ line wide, flat. Peduncles 4–6 inches long; pedicels ½–1 inch. Inv. 3 lines long. Rays about as long, broad for their length.

133. S. pinnatifidus (Less.! Syn. 391, non DC.); quite glabrous; stem suffruticose, erect, striate, closely leafy, subsimple; leaves bipinnati-partite, the lobes narrow-linear, with recurved edges, acute, spreading, quite entire; fl. branches partly nude upwards, ending in a subsimple, few-headed lax corymb, or in a panicle of many such corymbs; pedicels elongate, scaly; inv. calycled, turbinate, of 12–13 scales, 30–40-fl., discoid, shorter than the fl.; achenes glabrous. *S. abrotanoides, E. Mey.! DC.! l. c. 397. Doria bipinnata, Herb. Thunb.! Th. Cap. p. 675. Cacalia pinnatifida, Berg. Cap.* 230, *fide Less. l. c.*

HAB. Hantam, *Thunberg!* Piquetberg, *Drege!* Winterhoek and at the Waterfall, Tulbagh, *Pappe!* (Herb. Th., D., Hk., Sd.)

Very similar in habit and foliage to *S. bipinnatus,* but readily known by its many-flowered heads, either loosely panicled or in a simple corymb. *Thunberg's* specimens quite agree with those of recent collectors.

134. S. foeniculoides (Harv.); quite glabrous; stem suffruticose, erect, striate, closely leafy; leaves bipinnati-partite, the lobes narrow-linear

with recurved edges, acute, entire; fl. branches partly nude upwards,
ending in a closely much-branched, many-headed, fastigiate corymb;
pedicels short, erect, scaly; inv. calycled, of 7–8 scales, 12–15-fl., shorter
than the fl.; achenes glabrous.   *S. pinnatifidus, DC.! l. c.* 397, *excl. syn.*,
also *S. bipinnatus, DC.! l. c., excl. syn.*

HAB. Cape Flats, *Ecklon! Zey.!* 1045, *W. H. H.* At the 24 Rivers, *Drege!* (Herb.
D., Sd., Hk.)

Very similar to *S. bipinnatus* in habit and foliage, but with larger fl. heads, and
more scaly pedicels, &c. *Ecklon's* specimens, quoted by DC. under his " *S. bipinnatus,*"
seem to me identical with *Drege's*, quoted under " *S. pinnatifidus*" by DC.

135. S. bipinnatus (Less.! Syn. 391, non DC.); quite glabrous; stem
suffruticose, erect, striate, densely leafy; leaves bipinnati-partite, the
lobes narrow-linear, with recurved edges, acute, spreading, quite entire;
fl. branches partially nude upwards, ending in a very compound, dense
corymb, pedicels fastigiate, erect, scaly; inv. sparingly calycled, cylindrical, of 5–6 oblong, scales, discoid, 4–6-fl., shorter than the flowers;
achenes glabrous.   *S. quinqueflorus, DC.! l. c.* 396.   *Cacalia bipinnata,*
Th.! Cap. 626, *excl. syn. Berg.*

VAR. β. *triflorus;* inv. scales 3! *lanceolate-acuminate;* fl. three.   (Hb. Hk., D.)

HAB. Summit of Table Mt., *Thunberg! W. H. H.!* Drakenstein, *Pappe!* β. Swellendam, *Mundt.!* (Herb. Th., D., Sd., Hk.)

Stems 1–2 ft. high, with long, virgate, densely leafy branches. Leaves 1–2 inches
long, decompound-pinnatisect, the lobes and lobules not ½ line wide. Whole plant
strongly scented. Corymb much divided, level-topped. Inv. scales and flowers
commonly 5–6; in β. but three. This is the true " *Cacalia bipinnata* " of Thunb.

136. S. rhyncholænus (DC.! l. c. 397); glabrous or nearly so, some-
what viscidulous; stem herbaceous, tall, erect, angle-striate, leafy; lower
leaves *on long petioles*, bipinnati-partite, the upper subsessile, ear-lobed
at base, pinnatisect, lobes of all linear, 1-nerved, with reflexed edges,
spreading, subacute, entire or toothed ; fl. branches loosely corymbose,
few-headed; pedicels sparingly scaly; heads oblong, discoid, 12–15-fl.;
inv. subcalycled, of 10–12 narrow, scaberulous, taper-pointed scales, as
long as the disc or longer ; achenes striate, sparsely hispidulous or
glabrous.

HAB. Omsamwubo to Omsamcaba, *Drege!* Zululand, *Gerr. & M'K.*, 1057. (Hb.
D., Sd., Hk.)

Root, according to *Mr. Gerrard*, annual. Stems 2 ft. high. Leaves 2–3 inches
long, multi-partite, 2–3-pinnatisect, viscidulous: lobes scarcely 1 line wide. Inv.
4–5 lines long. Lobes of the corolla barbellate. Ridges of the achenes minutely
rough, otherwise glabrous. *Mr. Gerrard* says that the Zulu Caffirs dry this plant
as a perfume.

137. S. multicaulis (DC.! l. c. 398) ; stems many from the crown,
ascending-erect, leafy at base, laxly leafy, becoming nude upwards,
loosely panicled at the summit, glabrous, striate ; leaves sessile, the
lower sparsely pilose, incise-toothed or shortly few-lobed, or pinnatifid
toward the apex, the upper linear, erect, glabrous, quite entire, all with
revolute margins, acute ; pedicels long, nearly nude ; heads discoid,
40–50-fl. ; inv. calycled with many subulate bracteoles, of 12–15 dark-
tipped, glabrous scales ; achenes puberulous.   *S. multicaulis, sp.* " *a,*"
Hb. Drege, excl. " *b.*"

Hab. Ezelsbank, 3–4000 f., *Drege!* (Herb. Sond.)

Root ligneous. Stems bent at base, 1–1½ ft. high, slender, pale, striate. Lower leaves close, 1–1¼ in. long, their lobes in 2–3 pair, 1–3 lines long; upper leaves few, very erect, 1–1½ inch long, ½ line wide. Panicle subcorymbose. *Drege's* specimens marked " b " (in Hb. Hk.) have *radiate* heads, and seem to belong to *S. umbellatus.*

138. S. paniculatus (Berg. Cap. 277); suffruticose, glabrous or nearly so; branches virgate, leafy below, pedunculoid above and loosely corymbose or panicled; leaves *polymorphous*, slender, elongate, with strongly revolute margins, either all quite entire or some entire and some sparingly pinnati-partite, or all pinnati-partite, with few and distant, entire or toothed lobes ; corymb diffuse, few-headed, simple or branched, the pedicels long and scaly ; heads discoid, many-fl.; inv. amply calycled with many subulate bracteoles, of 16–24 narrow, tapering, dark-tipped scales ; achenes hispidulous.

Var. *a.* **peucedanifolius**; leaves all pinnati-partite, often toothed at base, the lobes in few pair, entire or few-toothed. *S. peucedanifolius, Linn. f. DC. l. c.* 397. *Jacobæa peucedanifolia, Th.! Cap.* 677. *Cacalia peuced., Jacq. Ic. var.* 581. *Zey.!* 944.

Var. *β.* **intermedius** ; leaves some quite entire, linear-elongate ; some with a few short lateral lobes ; some pinnati-partite, with long lobes. *Zey.!* 2969.

Var. *γ.* **reclinatus**; usually more robust, with larger fl. heads, but variable; leaves all linear-elongate, entire or the revolute margins toothed. *S. reclinatus, Linn. f. DC. l. c.* 398. *L'Her. Stirp. Nov. t.* 5. *S. graminifolius, Jacq. Ic. Rar. t.* 180. *S. Chrysocoma, Meerb. Ic. t.* 156. *Jacob. reclinata, Th.! Cap.* 675. *J. bidentata, Th.! Cap.* 676.

Hab. Cape, *Thunberg!* Var. *a.* Clanw., Worcester and Swellendam, *E. & Z.!* Drakenstein and Paarl, *Drege!* Banghoek, *W.H.H.* Winterhoek, Tulbagh, *Pappe!* *β.* Riv. Zondereinde, *Zey.!* *γ.* Districts of Uitenhage and Albany, *Drege! E. & Z.! P. McOwan!* &c. (Herb. Th., D., Sd., Hk.)

Very variable in foliage and in the size of the fl. heads, but var. *β.* seems completely intermediate in character between *a.* and *γ.* Stem 2–3 ft. high, erect or diffuse, very pale, terete, with raised striæ, closely leafy below. Leaves 2–3 inches long or more, 1 line wide, the lobes, when present, equally narrow, all with strongly revolute margins and glabrous. Infl. widely spreading. Heads from 3 to 7 lines long, 3–6 lines diameter. Pedicels 1–3 inches long. Calycle 2–3-seriate, of subulate or lanceolate bracteoles. Except for the discoid heads it is very similar to *S. umbellatus.*

§ 8. Leptophylli. (Sp. 139–155.)

139. S. angustifolius (Willd. Sp. 3. p. 1973); suffruticose, glabrous; branches erect, slender, angle-striate, sparsely leafy ; leaves narrow-linear, with recurved edges, quite entire, mucronate ; infl. loosely panicled or subcorymbose, pedicels long, slightly scaly ; heads discoid, 30–40-fl. ; inv. sparingly calycled, of 12–14 narrow, glabrous scales ; achenes puberulous. *DC. l. c.* 399. *Jacobæa angustifolia, Th.! Cap.* 675.

Hab. Cape, *Thunberg!* Worcester, *E. & Z.!* Riv. Zondereinde, *Zey.!* 2970. Zeederberg, *Drege* (fide DC.) Tulbagh, *Dr. Pappe!* (Herb. Th., D., Sd., Hk.)

A much-branched half-shrub, 1–2 ft. high, woody at base, with many slender, pale, rigid, erect branches. Leaves 1–1¼ in. long, not ½ line diam. Corymbs 3–5 headed, the pedicels 1–1¼ in. long. Inv. pale, shorter than the fl., strongly reflexed in age. Calycle of very small, dark-tipped bracteoles, few or several.

140. S. niveus (Less.! Syn. 392); suffruticose, slender, cano-tomentose, branching, branches white-woolly, corymbiferous ; leaves sessile, linear, blunt, with revolute margins, fleshy, at first albo-tomentose, becoming glabrate ; corymbs few-headed, the pedicels mostly without scales ; inv. sparingly calycled, of 14–16 canous scales ; heads discoid, 30–40-fl. ; achenes striate, canescent. *DC.! l. c. 390. Jacobœa nivea, Th.! Cap.* 676, and *Doria nivea, Th.! Cap.* 673 *Cineraria nivea, Willd.*

HAB. Cape, *Thunberg!* Silverfontein, *Drege!* Namaqualand, *Eckl.* Betw. Zack and Gariep R., *Burchell.* Springbokkeel, *Zeyher!* 945. (Herb. Th., Hk., Sd.)

2–3 ft. high, straggling, with many long, lax, sparsely leafy branches, ending in a few corymbose heads. Tomentum sometimes very copious, sometimes scanty.

141. S. persicifolius (Linn. Amoen. 6, Afr. 82) ; stem herbaceous, rigid, angular and striate, glabrous or cobwebbed, branching ; branches corymbiferous ; leaves scattered, ear-clasping, linear, entire, with strongly revolute margins, obtuse, glabrate above, woolly beneath ; corymbs closely few-headed, the pedicels short ; inv. of 10–12 glabrous scales, calycled and woolly at base ; heads 20–30-fl., discoid ; achenes hispidulous. *DC. l. c. 389. Jacobœa persicifolia, Th.! Cap.* 676.

HAB. Cape, *Thunberg!* (Herb. Thunb.)

Stem 12–18 inches high, slender, laxly branched ; the branches long, simple or divided, floriferous. Leaves scattered, subdistant, 1–1½ inch long, not a line wide, but with well-developed basal ears. Pubescence cobwebby and fugacious. Heads 5–10, in subglobose, dense corymbs. Inv. 3 lines long, shorter than the disc ; scales of the calycle brown-tipped.

142. S. mucronatus (Willd. Sp. 1974) ; suffruticose, cobwebby-tomentose, much branched, the branches canescent, corymbiferous ; leaves half-clasping and often unidentate at base, linear-subulate, subpungent mucronate, with revolute margins, punctulate and glabrate (the younger cobwebby) above, tomentose beneath ; corymb lax, compound, the pedicels copiously scaly below the head ; inv. calycled, of 14–16 glabrous, barbellate scales ; heads discoid, 20–30-fl. ; achenes cano-pubescent. *DC. l. c. 390. Jacobœa mucronata, Th.! Cap.* 676.

HAB. Cape, *Thunberg!* Worcester, *Eckl.* Paarl and Kl. Drakenstein, *Drege!* Riebeck's Kasteel, *Zey.!* 946. Winterhoek, Tulbagh, *Dr. Pappe!* (Hb. Th., D., Hk., Sd.)

A much-branched bush, 1–2 feet high, all the young parts copiously flocculent. Leaves 1–2 inches long, 1 line wide, from a broadish base, which is very frequently produced at each side into a tooth-like ear. Corymbs spreading, 10–20-headed. Heads 3 lines long. It much resembles *S. rosmarinifolius*, but has discoid heads, and broad-based, often auricled leaves, &c. It is still nearer to *S. juniperinus*, from the narrow-leaved forms of which it only differs by the discoid heads.

143. S. leptophyllus (DC.! l. c. 400) ; suffruticose, erect, branching ; stem terete, sparingly cobwebbed ; leaves linear, with revolute margins, entire, *ear-clasping*, the auricle small, toothed or subentire ; corymbs laxly few-headed, pedicels nude ; heads discoid, 50–60-fl. ; inv. calycled, of about 15-margined scales ; achenes puberulous.

HAB. Uitvlugt, and near Kendo, *Drege!* Common on the Eastern Frontier, *Mrs. F. W. Barber!* 398. (Herb. Sd., D.)

A rigid suffrutex, 1–2 feet high, with the leaves and inflorescence of *S. Burchellii*, but discoid fl. heads. Leaves 1–1½ inch long, 1 line wide, the auricle small and sometimes obsolete. *Mrs. Barber's* specimens agree well with a poor one from *Drege*, in Herb. Sond.

144. S. Diodon (DC.! l. c. 399) ; stem herbaceous, erect, branched, sparingly hispid below, glabrous and fistular above; leaves lance-linear, elongate, glabrous, with subrecurved edges, denticulate, *ear-clasping*, the auricle deeply 2–5-toothed or lobed; corymbs few-headed, panicled; pedicels long, nearly nude; heads discoid, 80–100-fl.; inv. calycled with several lanceolate, black-tipped bracteoles, of 16–20 flat, glabrous, taper-pointed scales; achenes puberulous.

HAB. Clanwilliam, *Ecklon !* Brackfontein, *Zey./* (Herb. Sd., Cap.)
Stem tall, 2–3 feet high (or more ?), but weak and herbaceous. Lvs. 2–3 inches long, 2–3 lines wide, conspicuously eared at base, the ear deeply several-toothed; those of the upper leaves occasionally much more developed and pectinate-partite. Inv. shorter than the fl., the bracts of calycle broader than common.

145. S. Skirrhodon (DC. l. c. 401); " herbaceous? erect, quite glabrous; stem striate, naked at the apex and bifid, 2-headed; leaves sessile, lanceolate, tapering to both ends, on each side with 10–12 callous teeth; pedicels sparingly scaly; inv. calycled, of 15–18 scales, nearly equalling the disc; disc-fl. 100, rays 11–13; achenes glabrescent." *DC. l. c.*

HAB. Omsamwubo and Omsamculo, *Drege.* (Unknown to us.)

146. S. serrulatus (DC. l. c. 401); " herbaceous, sparingly branched, virgate, glabrous; branches long, striate, nude above; leaves sessile, elongate-linear, serrulate, acute; panicle narrow, racemose; bracteoles of the calycle linear, acuminate, subciliate; inv. of 12 scales as long as the disc; rays 10–12, oblong; disc-fl. 30; achenes glabrous." *DC. l. c.*

HAB. Witbergen, 6–7000 f., *Drege.* (Unknown to us.)
" 2 ft. high, suffr. at base. Lvs. 2½ in. long, 1 line wide. Thyrsus oblong." *DC.*

147. S. inæquidens (DC. l. c. 401); rigidly herbaceous, erect, paniculately branched, nearly glabrous; stem and branches angle-striate; cauline leaves clasping and often eared at base, linear, acute, with revolute margins, some pinnatifid with 2–4 pair of short, spreading lobes, some unequally and coarsely toothed, and others denticulate, the teeth inflexed; inf. very lax, subcorymbose, the pedicels very long and nearly nude, spreading; inv. calycled, glabrous, of about 20 scales; disc-fl. 60–70, rays 10; achenes closely puberulous.

HAB. Cape, *Drege.* Near the Zwartkops R., Uit., *Zeyher!* 2963. (Herb. Sd., D.)
Root fibrous, much-branched; possibly annual. Stems 1–2 ft. high, much-branched, the branches spreading. Leaves 1½ in. long, 2–3 l. wide, very variable in their toothing or lobes, but always with revolute edges. Of *Drege's* plant I have only seen a scrap (in Hb. Sd.) which seems the same as *Zeyher's*, here described.

148. S. filifolius (Harv., non Berg.); suffruticose, much-branched, quite glabrous, the branches slender, terete incurved; leaves half-clasping, the uppermost often eared and toothed at base, filiform, with strongly revolute margins, acute, either quite entire or *furnished with a few unequal, alternate, spreading, filiform, lateral lobes;* inf. loosely panicled, pedicels very long, subcorymbose, angular, sparsely scaly, curved; inv. calycled, of 12–13 narrow scales; disc-fl. 40–50, rays 6–8, revolute; achenes hispidulous on the striæ. *S. linifolius, Th./ Cap.* 678, excl. syn.

HAB. Cape, *Thunberg!* (fol. 2). Zwarteberg, Caledon, *Dr. Pappe!* (Herb. Th., D., Hk., Cap.)

Stems very rigid and wiry, ligneous at base, slender; the bark often purple or dark. Leaves 1½–2 inches long, not ½ line wide, appearing terete from the strongly revolute margins, perfectly simple and distantly and unequally pinnati-lobed intermixed on the same branch. Panicle widely spreading; pedicels 3–4 inches long. Rays creamy white. The leaves in *Thunberg's* specimen are quite entire; in other respects it agrees with *Dr. Pappe's.*

149. S. debilis (Harv.); suffruticose at base, nearly or quite glabrous; stem slender, angle-striate, diffusely branched, ascending-erect; leaves distant, narrow-linear, acute, with revolute margins, entire; pedunc. terminal, long, filiform, sparsely scaly, *one-headed*; inv. shortly calycled, of about 20 acute scales; disc-fl. 40–50; rays 10–12, revolute; achenes puberulous. *S. linariæfolius, Drege, ex pte.* and *S. paniculatus, Drege, litt. c. (excl. litt. a. and b.)*

HAB. Hexrivierskloof and Zuureberg, betw. Enon and Driefontein, *Drege!* Uitenhage, *Zey.!* (Herb. Hk., Sd.)

Perhaps 2 ft. high, laxly branched, very slender, with sparse foliage. Leaves 1½ inch long, ¼–1 line wide, either completely revolute at edges, and thus filiform, or flattish, with the margin reflected, and sometimes minutely callous denticled. Pedunc. 5–8 inches long, pale. Inv. 3–4 lines long, shorter than the disc; rays yellow, strongly revolute. Calycle not a line long. *Drege* seems to have confounded more than one species under his "*S. linariæfolius*;" the plant so named in Hb. Sond. is different from that here described, and may belong to *S. Burchellii.* Whether our plant be identical or not with *S. linoides*, DC., I cannot say.

150. S. longifolius (Linn.? Sp. 1222); quite glabrous; stem shrubby, much branched, flexuous, the older branches nude and cicatricised, the younger and twigs closely leafy; leaves linear-elongate, very narrow, quite entire, tapering to both ends, acute, *thickish* or fleshy (commonly terete-involute), with a slender midrib; corymbs peduncled, compound, spreading, the divisions few-headed, pedicels nearly nude; heads radiate, 25–30-fl.; inv. slightly calycled, of 10–12 narrow scales; rays oblong, about 5; achenes puberulous. *Cineraria filifolius, Th.! Cap. 669. S. longifolius. DC. l. c. 400, ex pte.*

HAB. Cape, *Thunberg!* Hill sides in Albany, Uitenhage, and Caffraria, *Mundt.! Eckl.! Zey.!* 2978; *T. Williamson, W. S. M. D'Urban,* 54; *MacOwan,* 145. Tzitzikamma, *Pappe!* (Herb. Th., D., Hk.)

A large shrub, with curved ash-coloured branches, nude and rough below. Leaves 3–6 inches long, 1–1½ line wide, apparently terete from the strongly involute edges, occasionally flat, becoming dark in drying and probably shrinking. Pedunc. terminal, nude, pale, 3–9 inches long, bearing a loosely much branched corymb. *Drege's* specimens in Hb. D., Hk., under this name belong to *S. Burchellii;* possibly also that in Hb. Sd.

151. S. rosmarinifolius (Linn. f. suppl. 369); stem shrubby, erect, branches virgate, leafy, cobwebbed or glabrate; leaves lance-linear or linear, entire or denticulate, with revolute edges, midribbed beneath, glabrous or cobwebbed, or scaberulous; corymb peduncled, branched, its divisions compactly many-headed; pedicels short, scaly; heads radiate, 50–60-fl.; rays 8–10, broad; invol. calycled, of 12–15 glabrous or cobwebbed, narrow scales; achenes hispidulous on the striæ. *Thunb.! Cap.* 678. *DC.! l. c.* 400. *Jacq. Ic. Rar. t.* 587. *S. asper, Ait. Kew.* 3, *p.* 196. *Jacob. mucronata, Th. ! Cap.* 676 (more cobwebbed, with looser inflorescence). *S. rigescens, Jacq. Coll.* 5, *t.* 6, *f.* 1. *S. longifolius, Berg.?* *Cap.* 279.

VAR. *β*, **cristatus**; leaves half-clasping and 2–3 toothed or ear-crested at the very base. *S. cristatus, Th. Cap.* 679.

HAB. Common round Capetown, and Eastwards to Uitenhage and Albany, *Zey.!* 2961. (Herb. Th., D., Hk., Sd.)

A bush, 1–3 ft. high, either whitish from cobwebby, loose hairs, or nearly glabrous. Leaves 1–3 inches long, 1–3 lines wide, either simple at base, or in *β*. somewhat eared and toothed. Heads like those of *S. rigidus*, about 2 lines long.

152. S. Burchellii (DC. l. c. 401); stem suffruticose, erect, scabro-pubescent or glabrous, striate; leaves *half-clasping and minutely eared and toothed at base*, linear, entire or denticulate, with revolute margins; fl. branches nude or sparsely leafy, loosely corymbose or panicled; pedicels elongate, scaly; inv. of about 12 scales, glabrous, calycled; disc-fl. about 40, rays 5–7, flat; achenes minutely puberulous. *Sieb.!* *Cap.* 34.

HAB. Cape Flats, *Burchell, Mundt! Sieber! Wallich! C. Wright*, 294. (Herb. Hk., Sd., D.)

A scrubby, half woody bush, with the habit of *S. rosmarinifolius*, but much looser inflorescence, the heads of twice larger size, on long pedicels, very few in the corymbs, or subsolitary. Stem more or less rough with minute, rigid hairs. Leaves 1–2 in. long, 1 line wide. Pedicels 1–3 in. long. *S. linariæfolius*, Drege! (Hb. Sd.) partly belongs to this; partly to *S. debilis* (Hb. Hk.).

153. S. vimineus (DC.? l. c. 400); suffruticose, quite or nearly gla-brous; main branches virgate, terete, striate, sparsely leafy; leaves (often with axillary leaf-tufts), narrow-linear, quite entire, with slightly revolute margins, stem-clasping at base, callous-tipped; fl. branches spreading, few-headed; heads subcorymbose, on long, scaly pedicels; inv. of 20–22, narrow, dark-tipped, acuminate scales, calycled with several flat, dark-tipped bracteoles; disc-fl. 80–100; rays 10–12, re-flexed; achenes silky.

HAB. Eastern Districts, *Burchell*. Basutuland, *T. Cooper!* 716. (Herb. D.)

Probably 2 ft. high, loosely branched; branches 1–1½ ft. long. very pale, simple or with a few subterminal, widely spreading branchlets, ending in a corymb of 3–8 heads. Leaves 1½–2 inches long, ⅓ line wide, pale-green, quite glabrous. Pedicels 2–4 inches long, bearing several scales. Inv. 4 lines long, 5 lines diameter, pale, of many narrow, white-edged, glabrous scales. Calyce ample, its bracteoles conspicu-ously dark-tipped. Whether this be *DC.'s* "*S. vimineus*" or not, I cannot tell: if not, it must be very near it. I describe from *Mr. Cooper's* specimens.

154. S. dracunculoides (DC.! l. c. 400); scarcely suffruticose, erect, much-branched, glabrous; branches angularly compressed, flexuous; leaves sparse, half-clasping and often eared at base, linear, subacute, quite entire, with reflexed margins; corymbs loosely few-headed, pedi-cels sparsely scaly; inv. calycled, of about 18 narrow, keeled, dark-tipped scales; disc-fl. 50–60, rays 10–12; achenes striate, *quite glabrous.*

HAB. Near the Zwartkop's R., Uit., *Ecklon!* (Herb. Cap.)

I describe from one of *Ecklon's* sp. in the Cape Govt. Herb., which agrees well with *DC.'s* character; other specimens distributed by *Ecklon* are diverse. One in Hb. Sd. seems to belong to *S. inæquidens*, a nearly allied sp., differing in foliage and in its pubescent achenes. The leaves are thickish, 1½ in. long, 2–3 lines wide.

155. S. hirtellus (DC.! l. c. 401); suffruticose, suberect or diffusely branched; stem copiously and rigidly pubescent, young parts also cobwebbed; leaves linear, entire, scabro-pubescent or rigidly hairy, with strongly revolute margins, stem-clasping at base; fl. branches pedunculoid upwards, the pedunc. long, nude, glabrate or hispid, one or two-headed; inv. calycled, glabrate, of 18–20 scales; disc-fl. 60–80; rays 10–12, revolute; achenes hispidulous.

VAR. β. **subcorymbosus** (DC.); leaves more hairy; branches 3–5-headed.

HAB. Uitenhage, *Ecklon!* β. in Albany, *Eckl.;* Howison's Poort, *H. Hutton!* (Herb. D., Cap.)

A small, weak, much branched, half-ligneous plant, 6–18 inches high. Lower half of branches closely leafy; pedunc. 4–6 inches long. Leaves 1–1¼ inch long, 1 line wide, like those of *S. rosmarinifolius,* either copiously or scantily clothed with short, curled, whitish hairs.

§ 9. PINIFOLII. (Sp. 156–157.)

156. S. pinifolius (Lam., dict. 3, p. 263); shrubby, glabrous; branches erect, imbricate with leaves to the summit, or scaly above, one-headed; leaves crowded, rigid, acerose-triquetrous, mucronate, erect or spreading; heads *radiate,* many-fl., rays 3–8, spreading; inv. calycled with several subulate bracteoles, of 12–18 flat, lanceolate, glabrous scales; achenes glabrous. *DC. l. c. 399. Inula pinifolia, Linn. Th.! Cap. p. 667. Cineraria laricifolia, Lam. dict. 2, p. 8. Hubertia pinifolia, Sieb. Cap. 31.*

HAB. About Capetown, common; extending eastwards to Uitenhage and Albany. (Herb. Th., D., Sd., Hk.)

Stem 6–18 in. high, variable in habit; branches curved or straight. Leaves ½–1 inch long, imbricating. Heads either solitary, corymbose or somewhat racemose, large or small. Distinguished from *S. triqueter* by its *radiate* fl. heads. It is a much commoner plant.

157. S. triqueter (Less.! Linn. p. 251); shrubby, glabrous; branches erect, virgate, imbricated with leaves, subpedunculoid, scaly and one-headed at apex; leaves crowded, rigid, acerose-triquetrous, mucronate; heads *discoid,* many-fl.; inv. slightly calycled, of 14–18 broad, flat, lance-oblong, glabrous scales; achenes glabrous, compressed. *DC. l. c. 399.*

HAB. Cape Flats, *Thunberg! Burchell, E. & Z.! W.H.H.,* &c. Hott. Holl. Bg. *Mundt!* (Herb. Th., D., Hk., Sd.)

Stem 1–1½ ft. high, rigid, nude and cicatricised below, closely leafy above. Leaves ¾–1 in. long, erecto-patent. Pedunc. short or long, more or less covered with reduced, scale-like leaves. Heads 5–6 lines long, equally broad. Very similar, except in its *discoid* heads, to *S. pinifolius,* with which it was associated by *Thunberg.* Both vary in the size of the head.

§ 10. SCANDENTES. (Sp. 158–166.)

158. S. mikanioides (Otto MSS.); scandent, quite glabrous; leaves petiolate, sharply 5–7-angled or lobed, hastate or cordate at base, the lobes deltoid or obsolete, with broad, shallow interspaces; fl. branches terminal and axillary, with a few depauperated leaves, bearing a compound, many-headed corymb; inv. very sparingly calycled, of 8–9 narrow scales, shorter than the flowers; heads discoid, 8–10-fl.; achenes hispidulous on the striæ. *S. scandens, DC.! l. c. 404, excl. syn. Th.*

Cacalia scandens, Zey.! Coll. 157. *Mikania senecioides, Hort. Delairea odorata, Lemaire. Breonia palmata, Hort.*

HAB. Forests of Uitenhage, Albany and Caffraria, *Burchell, E. & Z.! &c. Zey.!* 2977. Katberg, *H. Hutton!* Natal, *T. Cooper,* 1130. Cult. in Cape and European gardens. (Herb. D., Sd., Hk.)

A much branched, climbing suffrutex, many feet high. Petioles 1–1½ inch long, slender, sometimes auricled at base. Leaves somewhat fleshy, drying thin, *ivy-shaped,* variably lobed or angled. Corymb generally much branched, on weak shoots subsimple. Inv. scarcely 2 lines long, much shorter than the flowers. As this is not *Thunberg's* "*Cacalia scandens;*" and as the name "*scandens*" is already occupied for an Indian species, I am compelled to reject it.

159. S. deltoideus (Less.! Syn. 392); scandent, much branched, glabrous or nearly so; leaves petiolate, often ear-stipuled at base of petiole, deltoid-hastate, acuminate, unequally multi-dentate; corymbs from the upper axils or in a terminal panicle, divaricate, forked, several-headed; inv. scarcely calycled, of about 5 oblong, blunt scales; fl. 5–6; achenes hispid on the striæ. *DC.! l. c.* 404; also *S. Mikaniæ, DC.! and S. Mikaniæformis, DC.! l. c.* 405. *Cacalia scandens, Th.! Cap.* 625. *Mikania auriculata, Willd.*

HAB. Uit., Albany, Caffraria and Natal.; frequent. (Herb. Th., D., Hk., Sd.)

A much branched, angularly bent, slender climber, several feet long. Leaves on uncial petioles, 1–2 inches long, ½–1½ in. wide, trowel-shaped, more or less cordate at base, tapering to an acute point. Corymbs on very short peduncles, from the axils of the upper leaves or of leaf-scales, 6–12-headed; sometimes in a divaricate panicle. Inv. 2 lines long. *Cooper's* 1124, 1313 and 1513 belong to this.

160. S. canalipes (DC. l. c. 405); "suffrut., quite glabrous; branches terete; leaves ovate, acute, here and there toothed, acuminate, produced at base into a very narrowly winged, channelled petiole, which is angular at back; corymb dichotomous, panicled; heads discoid, 12–15-fl.; inv. of 12–15 scales, calycled; achenes terete, glabrous." *DC. l. c.*

HAB. Galgebosch, *Drege.* (Unknown to us.)

161. S. quinquelobus (DC. l. c. 404); scandent, quite glabrous; lvs. petiolate, sharply 5–7-angled or lobed, cordate at base, the lobes deltoid or obsolete, with broad, shallow interspaces; fl. branches with a few depauperated leaves, 3–5-headed; inv. amply calycled with broad, spreading bracteoles, of 12–15 scales; heads discoid, fl. 40–50; achenes glabrous. *Cacalia quinqueloba, Th.! Cap.* 626.

VAB. β. **helminthoideus** (Sch. B.); interspaces of the leaf-lobes denticulate; achenes hispidulous.

HAB. Forests of Uitenhage and Albany, *Thunberg! E. & Z.! Drege! Cooper,* 1514, MacOwan, 274. Katberg, *H. Hutton.* β. Natal, *Krauss,* 285, *Gerr. & McK.,* 1007. (Herb. Th., D., Hk., Sd.)

A climbing or trailing suffrutex, many feet long. Petioles slender, 1–1½ in. long. Lvs. 1–2 inches long, equally wide, shaped like the lower leaves of Ivy. Heads 6–7 lines long, many-fl., the inv. much shorter than the flowers. Leaves sometimes hispido-scabrous on the under surface!

162. S. macroglossus (DC. l. c. 404); scandent, quite glabrous; lvs. petiolate, hastate, with salient acuminate basal-lobes; or cordate, acute or acuminate, entire, or with 1–2 broad teeth or lobules on each side;

fl. branches bearing a few oblong or lanceolate leaf-scales, 1–3-headed ; disc-fl. 40–50, rays 12, multistriate. *Zey.! coll.* 70.

HAB. Bushman's R., Uitenhage, *Zey.!* 2986. Keiskamma Hoek, *W. S. M. D'Urban*, 43. Natal, *Drege! Gerr. & McK.*, 330. (Herb. D., Hk., Sd.)

Climbing up trees in woods. Leaves on uncial petioles, commonly deltoid-hastate, varying to cordate with imperfect lobes, 1¼ in. long and equally wide. Fl. branches 3–6 inches long, mostly 1-headed, each bearing 4–5 reduced leaves or scales. Bracts of the calyculus nearly as long as inv. scales. Except for the *rays* it comes near *S. quinquelobus*.

163. S. bryoniæfolius (Harv.); scandent, quite glabrous, leaves petiolate, 3–5-nerved and netted-veined, obsoletely 3–5-lobed, the lobes rounded or acute, denticulate; fl. branches lateral and terminal, corymbose or subumbellately 5–7-headed, pedicels elongate, filiform, laxly scaly; inv. calycled, of about 12 acuminate scales, equalling the disc; disc-fl. 30–40, rays 9–10, 8–10 striate; achenes glabrous.

HAB. Natal, *Gueinzius*. (Herb. D., Hk.)

Leaves 1½ in. long, about equally wide, drying thin, not unlike those of a Bryony; the lobes short and broad ; base truncate. Petioles semiuncial. Branches with rounded axils, pale, suffruticose. Pedicels 2–3 inches long, very slender. Heads smaller than in *S. macroglossus* with fewer and narrower bracteoles.

164. S. tamoides (DC.! l. c. 403); scandent, quite glabrous ; leaves petiolate, somewhat hastate, unequally and coarsely toothed or multilobulate, the lobules broadly deltoid, acute, with rounded interspaces ; corymb loose, fastigiate, many-headed, the pedicels long, slender, nearly nude ; inv. of 5–7 scales, half as long as the disc, cylindrical, truncate and scarcely calycled at base ; disc-fl. 10–12, rays 3–6, spreading ; achenes glabrous.

HAB. Omsamwubo, *Drege!* Natal, *Plant*, 105; *Gerr. & M'Ken*. 331. Katberg, *H. Hutton!* Fort Bowker, *H. Bowker*, 553. (Herb. D., Hk., Sd.)

A half herbaceous climber. Petioles slender, 1–2 inches long. Leaves 1½–2 inches long, 1–1½ inch wide, angle-lobulate, the angles salient, with rounded interspaces. Corymb many-headed, but loose, sessile or short-peduncled. Inv. 4 lines long ; disc-fl. fully 8 lines. Rays equalling the inv., spreading, 4-lined.

165. S. angulatus (Linn. f. suppl. *369*); suberect or scandent, quite glabrous, branching ; leaves petiolate, cuneate at base, ovate, acute, angle-lobed, the lobes few, short and broad or tooth-like ; corymb repeatedly forked, many-headed, pedicels nearly nude, spreading ; inv. sparingly calycled, of 10–12 scales, shorter than disc ; disc-fl. about 20, rays 4–6, oblong, 4-striate; achenes hispidulous. *Th.! Cap.* 382. *DC. l. c.* 404. *Zey.!* 196. *Cineraria lævis, Spr. in Zey.! coll.* 251. *S. macropodus, DC. l. c.* 403. *S. Natalensis, Sch. B. in Walp. Rep.* 6, p. 264. Also *S. tredecimsquamosus, Sch. B.*

HAB. Woods in Uitenhage and Albany, *Thunberg! E. & Z.! Dr. Atherstone!* Near Natal, *Drege! Krauss,* 279! (Herb. Th., D., Hk., Sd.)

A scrambling, half-climbing suffrutex, with stems many feet long, supported among shrubs in woods. Leaves on 1–1½ in. petioles, 1½–2½ in. long, 1–1½ in. broad, remarkably cuneate at base, somewhat fleshy, drying thin, more or less angle-lobed or repand, the lobes sometimes obsolete. Corymb 3–5 in. wide. Inv. 2–2½ lines long. A small specimen from *Drege* (Hb. Sd.) of *S. macropodus*, seems to me not different from this ; and those from *Krauss* (Hb. D., Hk.) of *S. Natalensis* are very similar.

166. S. brachypodus (DC. l. c. 403); herbaceous? quite glabrous; stem (or branches) elongate, terete, leafy; leaves tapering at base into a short petiole, broadly ovate, acute, subentire or repando-denticulate, penninerved, membranous and netted-veined when dry; corymb compound, much-branched, shortly pedunculate or subsessile, pedicels scaly, scarcely longer than the head; inv. of 8 scales, shorter than the disc, truncate and sparingly calycled at base; disc-fl. 6–8, rays 3–4; achenes glabrous. *S. macròpodus, Eckl.! in Hb. (non DC.)*

HAB. Philipstown, Katriver, *Ecklon!* (Herb. Sd., Cap.)
2–3 feet or more high; stems (or branches?) simple, weak, flexuous and perhaps scandent, laxly leafy throughout. Leaves 3–4 inches long, 2–2½ inches wide, a little fleshy perhaps when fresh, but drying thin, distantly and minutely denticulate. Corymb fastigiate. Inv. 2–3 lines long. Rays yellow.

§ 11. KLEINOIDEI. (Sp. 167–176.)

167. S. subsinuatus (DC.! l. c. 401); suffruticose, erect, somewhat fleshy, quite glabrous; branches virgate, leafy below, nude and loosely corymbose at the apex; leaves thickish, flat, oblongo-lanceolate or linear, tapering to the base, obtuse, entirely or obsoletely or very bluntly tooth-lobed; pedicels sparsely scaly; inv. scarcely calycled, of 10–12 acute scales; disc-fl. 15–20, rays 5; achenes minutely hispidulous.

HAB. Stellenbosch, *Ecklon!* (Herb. Sd., Cap.)
1–1½ ft. high. Leaves 2 inches long, 3–5 lines wide, with 1–2 shallow lobes or entire. Inv 2 lines long. Heads small and few fl.

168. S. crassiusculus (DC.! l. c. 401); suffruticose, erect, glabrous; stems terete, striate, loosely corymbose or panicled above; leaves thickish, flat, half-clasping, *variable* in shape, either linear, oblong, lanceolate or spathulate, sparingly pinnate-lobed and toothed, or unequally toothed, the lobes and teeth acute; corymb few-headed, pedicels nearly nude; inv. sparingly calycled, of 10–12 acuminate scales; disc-fl. 30–35, rays 8–10; achenes minutely hispidulous. *S. pinnulatus, Sieb. Cap.* 264, *non Th.*

HAB. Adow; Zuureberg, and betw. Coega and Zondag R., *Drege!* Adow and Zwartekops R., *Zey.!* 2979. Also said to be found on the Oudeberg, and betw. Grasberg R. and Water Val, Tulbagh, *Drege*, and near Stellenbosch and Capetown (?) by *Ecklon.* (Herb. D., Sd., Hk.)
1–2-ft. high, the stem and leaves somewhat fleshy. Leaves polymorphous, with a cartilaginous edge and sharp rigid teeth or lobes. Inv. 4–5 lines long. Heads much larger than in *S. subsinuatus.*

169. S. pyramidatus (DC.! l. c. 402); stem shrubby, fleshy, erect, simple, closely leafy below, lengthened upwards into a long, leafless, tomentose, pedunculoid flowering stem; leaves sessile, sub-terete, fleshy, acute, glabrous or cobwebbed; heads many, in a long, thyrsoid raceme, the pedicels tomentose, scaly; inv. sub-calyculate, campanulate, woolly, of 10–12 very broad scales; disc-fl. 50–60, rays 10–12, broad, spreading; achenes glabrous. *Bot. Mag. t.* 5396.

HAB. Uitenhage, *E. & Z.! Zey.!* 937. Sneeuweberg, *Drege!* (Herb. D., Sd., Hk.)
Stem short, thick and strong. Leaves 3 inches long, 2–3 lines diam. Flowering stem 2 ft. high thyrsus 6–12 inches long, white-woolly. Heads ¾ inch diameter, exclusive of the rays. Pappus copious, discoloured. A very fine species.

170. S. scaposus (DC.! l. c. 403); stem shrubby, fleshy, either very short or elongate and branched; leaves crowded at the apex of the stem or branches, fleshy, broadly linear, very obtuse, terete, often flattened near the tip and subspathulate, the younger cobwebbed, the older glabrous; pedunc. scapelike, nude or sparsely scaly, many times longer than the leaves, cobwebbed, one or several headed, the pedicels elongate; inv. cobwebbed, scarcely calycled, of 10–12 scales; disc-fl. very many, rays about 12; achenes nearly glabrous.

VAR. *a, acaulis*; stem very short or scarcely any. *S. scaposus, DC.*

VAR. *β, caulescens*; stem a foot high, branched. *S. calamifolius, Hook! Bot. Mag. t.* 4011.

HAB. Eastern Districts, *Bowie, Burchell.* Zwartkops R., *Drege! E. & Z.! Zey.!* 2983. (Herb. D., Hk., Sd.)

Wild specimens generally stemless, or with very short stems. Leaves 2–3 inches long, 2–3 lines diam. Pedunc. 1–1½ f. long, occasionally 1 headed, more frequently with 3–5, long pedicelled heads. Young parts more or less cobwebbed or thinly woolly.

171. S. aloides (DC.! l.c. 402); "a fleshy shrub, glabrous; branches terete; leaves thick, broadly linear, subobtuse, quite entire; fl.-branches naked, twice as long as the leaves, 1–2 headed; inv. calycled, of 10–11 scales, cylindrical, scarcely shorter than disc; rays 6–7; disc-fl. 20; achenes hispidulous on the striæ; pappus equalling the disc-fl." *DC.l.c.*

HAB. Hol. R., Kl. Namaqualand, *Drege.*

"A dwarf, ascending shrub. Lvs. 2 in. long, 2 l. wide. Heads 8–10 l. long." *DC.* [One of the sheets of "*Cacalia Arbuscula*" in Hb. Th., marked "No. 1," is a *Senecio,* seemingly very near *S. aloides,* if not the same. Its leaves are 2½ in. long, *acute,* about 1⅓ line wide. The pedunc. 3–4 in., bearing 3–4 subsessile heads.]

172. S. bulbinefolius (DC.? l. c. 402); a fleshy, glabrous shrub; branches terete; leaves thick and fleshy, (subterete?), broadly linear, obtuse, mucronate, narrowed at base, quite entire; pedunc. terminal, nude, 2–4 times as long as the leaves, 1-headed; inv. scarcely calycled, oblong, of 8–10 scales, shorter than the disc; rays few, narrow, revolute; disc-fl. 30–40; achenes densely pubescent.

HAB. Betw. Zilverfont. and Kaus, *Drege.* Modderfontein, *Rev. H. Whitehead.* (Herb. D.)

Not having seen *Drege's,* I describe from *Mr. Whitehead's* specimen, which agrees pretty nearly with *DC.'s* description. Leaves 1½–2 in. long, 2 l. diam. Pedunc. 5–6 in. long, sometimes (fide *DC.*) 2-headed.

173. S. corymbiferus (DC.! l. c. 402); shrubby, succulent, erect, quite glabrous; branches at the summit nude and corymbose; leaves fleshy, subterete, linear-elongate, acute, tapering to the base, quite entire; pedicels short, 2–3-scaled; corymb fastigiate, many-headed; inv. sparingly calycled, cylindrical, of about 8 scales; disc-fl. 15–20, rays 6–8; achenes striate, minutely hispidulous.

HAB. Silverfontein, Namaqualand, *Drege!* Bosjesmansland, *Zey.!* 953. (Herb. Hk., Sd.)

Stems ¼–¾ inch diameter. Leaves 2–3–4 inches long, 2–3 lines wide, probably subterete, flattened and shrinking in drying. Inv. 6 lines long, in age nodose at base. The habit is that of an *Othonna.* Pappus copious, dirty white.

174. S. succulentus (DC. l. c. 402); "shrubby, erect, glabrous; stem terete, somewhat fleshy; leaves crowded, linear-oblong, thick, *obtuse*, quite entire; pedunc. twice as long as the leaves, corymbose, few-headed; pedicels scaly, longer than the head; inv. cylindrical, scarcely calycled, of 7–8 scales; rays 4–5, small; disc-fl. 15–16; achenes hispidulous on the striæ." *DC. l. c.*

HAB. Kendo, on the Karroo, *Drege*.
Of this I have seen but a leaf, and fl.-head, in Hb. Sd. Leaves 1½ in. long, 2–1. wide. Heads 6 lines long, pale yellow.

175. S. cotyledonis (DC.! l. c. 402); "a fleshy shrub, quite glabrous; branches thick, terete, marked with leaf-scars; leaves crowded round the apices, fleshy, linear, tapering at base, callous-mucronate, somewhat keeled; pedicels shorter than the leaves, scaly, one-headed; inv. cylindrical, scarcely calycled, of 10–12 narrow, acuminate scales; disc-fl. abt. 25, rays 5–7; achenes puberulous.

HAB. Kendo and Zwaanepoelspoortberg, Karroo, *Drege!* (Herb. Hk., Sd.)
A thick-stemmed, fleshy bush. Leaves uncial, many in the tuft. Pedicel and head together not equalling the leaf. Fl. bright yellow.

176. S. acutifolius (DC. l. c. 402); "shrubby, erect, branched, quite glabrous; branches terete; leaves sessile, fleshy, linear-terete, mucronate; heads at the ends of the branches *solitary, sessile;* inv. calycled, cylindrical, of 7–9 scales; rays 5–6, *scarcely longer than their style;* disc-fl. 12–16; achenes terete, downy." *DC. l. c.*

HAB. Graaf Reynet, *Ecklon.*
This seems to be marked by its sessile fl.-heads and very short rays.

§ 12. APHYLLI. (Sp. 177.)

177. S. junceus (Harv.); quite glabrous; stems herbaceous, succulent, striate, erect, leafless, with a few distant, scattered, minute, deciduous, subulate scales; branches long and virgate; corymb terminal, slightly compound, few or several-headed; pedicels slender, sparingly scaly; inv. cylindrical, scarcely calyculate, 9–10 leaved, shorter than the disc; heads about 12–15 fl., rays about 5, oblong, flat; achenes glabrous, long and slender, rib-striate, slightly beaked. *Brachyrhynchos junceus, Less.! Syn. p. 393. DC.! l. c. 437. Cineraria coronata, Zey.! C. scirpina, E. Mey.!*

HAB. Langekloof, *Burchell,* 5008. Cape, *Mundt. & Maire!* Zwartkops R., *E. & Z. Zey.!* 2984, *Cooper,* 1575, Lislap, *Zey.!* 951. Nieuwveldt, *Drege!* Modderfontein, *Rev. H. Whitehead!* (Herb. D., Sd., Hk.)
2–3 ft. high, with many erect, rod-like, naked, green, soft branches, 2–3 lines diameter. Leaf-scales scarcely 1 line long. Pedicels ½–1½ inch long. Inv. 4–5 lines long. Ray and disc yellow. Achenes with an evident, cup-like epigynous disc.

Doubtful Species.

S. linoides (DC. l. c. 397); "shrubby, glabrous; branches erect, leafy, nude at the summit and panicled or rarely bifid; peduncles nearly nude, 1-headed; leaves linear, filiform, with revolute margins, quite entire, sessile, acute; inv. calycled, all the scales linear, subcarinate, wither-tipped; achenes terete, puberulous." *DC. l. c.*

HAB. About Table Mt., *Eck.* Betw. Gekau and Bashe, and on the Witberg, *Drege*. Unknown to me. Specimens from *Ecklon* (Hb. Sd.) under this name, in very bad order, to me seem referable to the narrow-leaved forms of *S. Albanensis;* and a fragment from *Drege* (also Herb. Sd.) to *S. umbellatus*. The widely sundered habitats assigned to the species suggest a doubt of the specific indentity of the specimens on which it was founded.

S. virgatus *(*Linn. Amoen. 6, Afr. 83*)*; "shrubby; leaves lyrate, tomentose beneath; pedunc. 1-headed; inv. sc. subulate. *DC. l. c.* 434.

S. barbareæfolius (Turcz. Bull. Mosc. xxiv. 2, p. 90); "herbaceous, glabrous; stem angular, furrowed, leafy at base, simple, from the middle branched, nude; branches long, corymbose, 1–3-headed; lower leaves petioled, lyrate-pinnatifid, the end lobe very large, cordate or deltoid, the lateral adnate-decurrent, trapeziform, denticulate; upper leaves half-clasping, coarsely toothed; uppermost minute; pedicels 2 or more times longer than the head, scaly; heads many-fl. [purple]; inv. calycled, of 8 glabrous scales." *Turcz. Walp. Ann. vol. 5, p.* 338.

HAB. Cape, *Zeyher*, 2966.

Brachyrhynchos trachycarpus (DC. l. c. 438, excl. syn. Th.); "stem herbaceous, erect, sparingly branched, glabrous or cobwebby; lower leaves petioled, lyrato-pinnatifid, on both sides glabrous, the lobes toothed, the lower ovate, small, the terminal ample, incised, toothed; uppermost leaves few, scale-like, acuminate, depressed, like the invol. scales; heads solitary, discoid; achenes muricate, scabrid, tapering." *DC. Cineraria incisa, Houtt. fide DC.*

HAB. Cape (in Herb. Deless., fide *DC.*).
I have not seen this, but by the above description it can hardly be different from *S. tuberosus*. Thunberg's *Doria incisa*, here quoted by *DC.*, is *Senecio erosus*.

Brachyrhynchos eupatorioides (DC.); "herbaceous, erect, quite glabrous; lowest leaves on long petioles, oblong, tapering to each end; upper semi-amplexicaul, lanceolate, dilated at base, acuminate, all coarsely and irregularly toothed; corymb compound, many-headed; heads discoid; inv. subcalyculate, with few acuminate scales, the inner scales not sphacelate at tip, nearly equalling the flowers." *DC. l. c.* 439. (excl. syn. Thunb.! & Less.)

HAB. Cape, *Zey.* 272, fide *DC.*
" *Br. eupatorioides*," Less.! is a synonym of *Senecio othonnæflorus, DC.*

CXVIII. **EURYOPS**, Cass.

Heads many-fl., radiate; the *ray-fl.* ligulate, female; *disc-fl.* tubular, 5-toothed, perfect. *Recept.* convex or conical, mostly honeycombed. *Inv.* scales uniseriate, their margins more or less concrete, valvate in æstivation. *Style*-branches truncate, pencilled at the summit only. *Achenes* roundish or subcompressed, wingless, beakless. *Pappus* pluriseriate, caducous, of rough, brittle, flexuous bristles, the outer hairs often deflexed or decurrent. *DC. Prodr.* 6, p. 443.

Small shrubs or shrubby plants, almost all S. African; one from Arabia. Leaves alternate, crowded, coriaceous or fleshy, entire or 3–5-fid, or pinnati-partite. Pe-

duncles nude, one-headed, terminal or axillary. Flowers yellow. Name from
ευρυωψ, *having large eyes;* alluding to the conspicuous fl. heads of many. Formerly
the species were included in *Othonna,* from which they differ in pappus and especially
in the fertile disc-fl. From *Gamolepis* they are only known by having pappus.

A. Leaves pinnate-parted, pectinate or inciso-serrate :
    All parts tomentose ; leaves pectinate ... ... (1) **pectinatus.**
    Glabrous ; leaves pinnati-partite :
      Lvs. 1–2 in. long ; inv. scales connate at base only (2) **abrotanifolius.**
      Lvs. 4–6 inches long ; inv. scales concrete in a
        broad-based cup ; rays very long and numerous (3) **Athanasiæ.**
    Glabrous ; leaves long, sharply inciso-serrate ... ... (4) **Serra.**

B. Leaves 3–5-fid, or 3–5-toothed or lobed at the extremity :
    Leaves *flat, cuneate,* 3–5-toothed or lobed :
      Lvs. powdery-tomentose ; pedunc. 8–10 inches
        long, terminal ... ... (5) **Dregeanus.**
      Lvs. glabrous : pedunc. 1–3 inches long, axillary :
        Lvs. sharply cut, the teeth subulate ; achen.
         glabrous ... ... (6) **virgineus.**
        Lvs. bluntly cut, teeth oblong ; achen. villous (7) **Algoensis.**
    Leaves *linear,* trifid or 5-fid ; the lobes linear :
      Heads conspicuously peduncled ; achenes villous :
        Pedunc. terminal, 3–4 times longer than lvs. (8) **trilobus.**
        Pedunc. lateral, filiform :
          Pedunc. about equalling the 2–3 in. long lvs. (9) **trifurcatus.**
          Pedunc. 2–3 times longer than the short leaves :
            Lvs. fleshy, 3-fid below the middle,
              obtuse ... ... (10) **trifidus.**
            Lvs. filiform, 3-5-7-fid *above* the middle (11) **multifidus.**
      Heads very shortly peduncled, terminal ; achenes
        glabrous ... ... (12) **calvescens.**

C. Leaves linear-terete, some simple, some 3-fid, some alternately 4–5-fid *on the
same plant.* (Intermediate between sec. B. and D.)
    Peduncles elongate :
      Inv. scales 12–13, linear-lanceolate, 3-nerved ; achenes
        glabrous, ribbed ; lvs. impunctate, 1–1½ inch ... (13) **diversifolius.**
      Inv. sc. 7–10, ovato-lanceolate, smooth ; achenes gla-
        brescent ; lvs. impunctate ... ... (14) **longipes.**
      Inv. sc. 10–12, acuminate ; achenes pubescent ; lvs.
        impress-dotted, ½–¾ inch, *slender* ... ... (15) **punctatus.**
    Peduncles 1–2ce as long as leaves ; inv. sc. 5–6 ... ... (16) **subcarnosus.**

D. Leaves quite entire, either filiform or flat :
    Heads radiate ; rays few or many :
      Leaves narrow-linear, spreading :
        Ovaries and achenes glabrous :
          Leaves *flat ;* inv. scales concrete beyond
            middle ... ... (17) **linearis.**
          Lvs. *filiform ;* inv. sc. connate at base only (18) **linifolius.**
        Ovaries pubescent ; achenes villous :
          Leaves 1½–3 inches long :
            Inv. scales 12–20, lanceolate ... ... (19) **tenuissimus.**
            Inv. scales 4–5, very broad, concrete (20) **spathaceus.**
          Leaves ½–¾ inch long, 3-angled, keeled (21) **oligoglossus.**
      Leaves *minute* (2 lines long), oblong or oval : achenes glabrous :
        Lvs. nerve-keeled, appressed ; rays 6–8 ... (22) **Candollei.**
        Lvs. nerveless, spreading ; rays 1–2 ... ... (23) **empetrifolius.**
      Leaves ¼–1 inch long, oblong or obovate : achenes villous.
        Leaves quite entire, smooth-edged ... ... (25) **lateriflorus.**
        Leaves cartilagineo-ciliate, recurved ... ... (26) **sulcatus.**
    Heads discoid, minute ; leaves 2–3 lines long, linear ;
      pedunc. very short ... ... (24) **asparagoides.**

1. E. pectinatus (Cass.) ; in all parts tomentose-canescent ; leaves petiolate, pectinato-pinnatifid, lobes in many pairs, linear, obtuse, quite entire (or few-toothed near the apex on the outside); pedunc. terminal, twice or thrice as long as the leaves; inv. scales 12–14, lanceolate, concrete ; rays equalling the involucre. *DC. l. c.* 443. *Othonna pectinata, Linn. sp.* 1309. *Th.! Cap.* 723. *Bot. Mag. t.* 306. *Mill. ic. t.* 149, *f.* 2.

VAR. β. **discoideus** (*DC.*) ; rays none.

HAB. Rocks about Table Mountain, frequent. Simon's Bay, *C. Wright*, 342. β. at Drakensteinberg, *Drege.* (Herb. Th., D., Hk.)
A strong, ashen-grey bush, 2–3 ft. high. Leaves 2–3 inches long, including the petiole ; lobes in 8–10 pairs, ¼–¾ inch long, 1–1½ line wide. Peduncle 3–6 inches long. Readily known by its copious, soft, whitish pubescence.

2. E. abrotanifolius (DC.! l. c. 443); glabrous or with tufts of wool in the axils of the floral leaves ; leaves densely crowded, pinnati-partite, the lobes mostly alternate, distant, entire, linear-filiform, subacute ; pedunc. terminal, as long as or 3–4 times longer than the leaves ; inv. scales 12–20, lanceolate, connate at base only. *E. comosus, Cass. Less. Syn.* 394.

VAR. α. **abrotanifolius**; pedunc. 3–4 times as long as the leaves. *DC. l. c. Othonna abrotanifolia, Linn. Th.! Cap.* 723. *Lodd. Cab. t.* 1698. *Bot. Reg. t.* 108.

VAR. β. **brachypus** ; pedunc. about twice as long as the shorter leaves; inv. scales 20. *Euryops Athanasiæ, DC.! l. c.* 444, *not of Less.*

VAR. γ. **brachypus**; pedunc. very short. *E. intermedius, DC. l. c.*

HAB. About Capetown and the W. districts. β. and γ. Stellenbosch, *E. & Z.!* Olifant's Hoek, *Dr. Pappe! Zey.!* 2994. (Herb. Th., D., Hk., Sd.)
A strong bush, 2–3 ft. high. Leaves and peduncles variable in length. Leaves 1–2 inches long, the lobes 2–9 lines long, spreading or recurved. Pedunc. 1–6 inches long. Inv. scales and size of heads varying in all the varieties.

3. E. Athanasiæ (Less.! Syn. 394) ; glabrous ; leaves petiolate, pinnati-partite, lobes opposite or alternate, 4–7 pair, *very long*, entire, linear-filiform, acute; pedunc. 3–4 times as long as the leaves, terminal; inv. scales 20–25, incurved, concrete for more than ¾ their length; ligules 20–30, twice or thrice as long as the heads. *Othonna Athanasiæ, Linn. f. Thunb.! Cap.* 722. *Jacq. Sch. t.* 242. *E. speciosissimus, DC.! l. c.*

HAB. Piquetberg, *Thunberg!* Cape, Herb. *Hook.* Olifant's R., near Tulbagh, *E. & Z.!* Piquetberg and Zeederberg, *Drege!* (Herb. Hk., D., Sd.)
A strong growing, resiniferous bush ("Resin Bush" of Colonists), 2–4 ft. high. Leaves 4–6 inches long, the lobes 2–3 inches. Pedunc. 10–12 inches long. Inv. very broad-based, broader than its height, with incurved scales, before and after flowering. Rays inch and half long. This is the true *Othonna Athanasiæ* of Herb. Thunb.!

4. E. Serra (DC. l. c. 444) ; glabrous; leaves sessile, crowded, rigid, erect, elongate, sharply inciso-serrate, the lobules (or teeth) on each side 8–12, shortly subulate, acute; pedunc. 2–3 times longer than the leaves, terminal; inv. sc. 20–25, concrete for their greater length; rays 20–25.

HAB. Cape, *Niven !* near Tulbagh, *Drege!* (Herb. D., Hk., Sd.)
This has the flat-bottomed, broad involucres of *E. Athanasiæ*, but very different foliage. Leaves 2–3 inches long ; their lobes or teeth 2–3 lines long.

5. E. Dregeanus (Sch. Bip. in Flora, xxviii. 51) ; leaves crowded at the ends of the branches, powdery-tomentulose, subcanescent, cuneate, much attenuated at base, deeply 3-5-dentato-crenate at the truncate apex, the uppermost sometimes entire ; pedunc. terminal and lateral, many times longer than the leaves ; invol. scales 12-16, ovate, acuminate, glabrous, connate at base only ; ovaries pilose. *Walp. Repert. 6, p. 272.*

HAB. Mierenkasteel, *Drege!* 2727. (Herb. Hk., D., Sd.)
A dwarf, woody scrub, under a foot high (?), ramulous, the bases of the branchlets denuded of leaves. Leaves 1-1¼ inch long, 3-4 lines wide at the extremity, grey with very minute pubescence. Pedunc. 8-10 inches long. Heads many-fl. Rays several.—A very distinct species.

6. E. virgineus (Less.! Syn. 394); glabrous ; leaves short, sessile, crowded, wedge-shaped, sharply and deeply 3-5-toothed ; peduncles axillary, filiform, 2-3 times as long as the leaves ; inv. scales 5-7, oblong, subacute, connate below ; recept. conical ; rays 6-7; achenes quite glabrous, smooth. *DC. l. c.* 445. *Othonna virginea, Linn. f. Th.! Cap.* 720. *Oth. flabellifolia, Lodd. Bot. Cab. t.* 728. *Euryops flabelliformis, Cass.*

HAB. Frequent in Uitenhage and Albany, *Thunberg, E. & Z.! Drege,* &c. Betw. the Knysna and Plettenbergbay, *Pappe!* Karregariver, *Zey.!* 2993. (Herb. Th., D., Hk., Sd.)
A virgate shrub, 1-2 feet high ; branches long, slender, imbricate with leaves to their apices. Leaves ½ inch long, flat, cuneate, 3-4 lines wide at top, deeply cleft. Heads as large as peas ; pedunc. 1-1½ inch long.

7. E. Algoensis (DC.! l. c. 445) ; glabrous ; leaves short, sessile, crowded, wedge-shaped, coriaceous, deeply 3-5-toothed, the teeth broad, oblong, blunt or mucronulate ; pedunc. axillary, filiform, 2-4 times as long as the leaves ; invol. scales 10-12, connate beyond the middle, acute ; recept. flattish ; rays 10-12 ; achenes densely villous. *Zey.!* 2991, 2992.

HAB. Districts of Uitenhage and Albany, frequent. (Herb. D., Hk., Sd. Cap.)
1-2 ft. high, robust, closely imbricated with leaves throughout. Leaves ⅓-⅔ inch long, 2-4 lines wide, the lowest sometimes longer and narrower, deeply but bluntly incised. Ped. 1-2½ inches long. Heads 4 lines diam.—Allied to *E. virgineus,* but with thicker, more bluntly cut leaves, larger fl. heads, more connate inv. scales, and hairy achenes.

8. E. trilobus (Harv.); glabrous ; leaves linear to beyond their middle, at the apex 3 lobed, the lobes linear, simple, obtuse, fleshy ; pedunc. terminal 3-4 times longer than the leaves ; inv. scales 12-15, concrete to the middle, acute ; achenes villous.

HAB. Spitzkop, Somerset, *Dr. Atherstone!* (Herb. Hook.)
Stems dwarf, tufted, 3-5 inches high, flexuous, ramulous. Leaves crowded, ¾ inch long, nearly 1 line wide ; the lobes 2 lines long. Pedunc. solitary, terminal ; 2-2½ inches long. Heads more than ½ inch across.—Only a single specimen seen by me.

9. E. trifurcatus (Cass.); glabrous ; leaves linear-filiform to beyond their middle, at the apex trifid, lobes linear, simple, acute ; pedunc. axillary, or tufted, filiform, not much longer than the leaves ; inv. scales 12-15, ovate, acute, connate at base only ; achenes villous. *Less! Syn. p.* 394. *DC. l. c.* 444. *Othonna trifurcata, Linn. f. Th. Cap.* 722.

HAB. Piquetberg, *Thunberg!* Winterhoek, Tulbagh, *Pappe!* Swellendam, *Drege.*
(Herb. Th., D.)
A strong growing bush, 2–3 feet high, with long branches, leafy to the summit.
Leaves 2–2½ inches long, the undivided portion 1½ inch, the lobes ½–¾ inch long,
½ line wide. Pedunc. 2–2½ inches long, rigid, very slender, pale. Achenes densely
villous ! (not '*glabrous,*' as stated by *DC.* l. c.)

10. E. trifidus (Less.! Syn. 394); glabrous, much branched, flexuous;
leaves short, fleshy, thick, linear, trifid to below the middle or entire,
imbricated, obtuse or mucronulate; pedunc. axillary, 2–3 times as long
as the leaves; inv. scales 8–9 oblong, membr. edged; rays oblong;
achenes very villous. *DC. l. c.* 444. *Othonna trifida, Th.! Cap.* 721.
(Herb. Th., litt. a.)

HAB. Piquetberg, *Thunberg!* Karroo and near the Gariep, *Drege!* (Herb. Th., Sd.)
Allied to *E. multifidus,* but with much thicker, shorter and more deeply cut,
close-lying leaves and somewhat larger fl. heads. Leaves ½ inch long, the undivided
portion 2 lines long; lobes nearly 1 line diameter. Achenes exactly like those of
E. multifidus.

11. E. multifidus (DC.! l. c. 444); glabrous, much branched, flexuous;
leaves linear-filiform to beyond their middle, at the apex 3–5 or many-
fid, the lobes linear, simple or bifid; pedunc. axillary, numerous, about
twice as long as the leaves; inv. scales 6–7, broadly ovate, membrane-
edged; rays 5–7; achenes very villous. *Othonna multifida, Linn. f.
Th.! Cap.* 722.

HAB. Carroo, *Drege! Mundt.!* Herrelogement, *Zey.!* 959. Namaqualand, *A. Wyley!*
Thaba Bossieu, in Basuta Land, and in Albert, *T. Cooper!* 736, 567. (Herb. Th.,
D., Hk., Sd.)
A strong, shrubby, ramulous bush, a foot or more in height. Leaves ½–1½ inches
long, their lobes 2–6 lines long, sometimes pedatifid. Pedunc. slender, 1–1½ inches
long. Heads small. Inv. scales broad, and less concrete than in most. Achenes
clothed with copious, very long hairs.

12. E. calvescens (DC. l. c. 445); glabrous; the upper axils bearded;
leaves crowded, imbricate, trifid, lobes linear, acute, longer than the
undivided part; heads terminal, on very short pedicels; inv. scales about
9, concrete beyond the middle; rays 7–9, twice as long as the involucre; achenes quite glabrous; pappus (fide DC.) " of few bristles, the
outer reflexed."

HAB. Stormberg, *Drege!* (Herb. Sond.)
So closely resembling some varieties of *Gamolepis brachypoda,* that, without examining the pappus, which I have been unable to do, it may readily be mistaken for that
plant. The leaf-lobes, however, are longer in proportion ; the inv. sc. more concrete
and much less taper-pointed; and the achenes (judging by the ovaries) probably
different. I rely on *DC.* for the characters of *pappus,* as above given.

13. E. diversifolius (Harv.); glabrous; stem slender, forked; leaves
filiform, some simple, some 3-fid to the middle, some pinnately 4–5-fid
with alternate lobes (not impress-dotted); pedunc. terminal and lateral,
6–8 times longer than the leaves; inv. scales 12–13, linear lanceolate,
3-ribbed, connate at the base only; rays 12–15, twice as long as the
head; achenes glabrous, rib striate.

VAR. β. **integrifolius**; leaves mostly entire. *E. longipes* β. *lasiocladus, DC.! l. c.*

HAB. Sandyground, betw. Langevalley and Berg Valley, *Zey.! 954, Dr. Pappe!* (Herb. Cap., D., Sd.)

Stem 1–2 ft. high, slender, sparingly branched, leafy. Leaves 1–1½ inch long, filiform. Pedunc. 4–5 inches long. Like *E. longipes* but with larger fl.-heads, more numerous and longer rays, and dissimilar inv. scales. *DC's* var. *β.* of *longifolius*, belongs to this, by the chars. of inv., ray and achenes. It is more robust than *E. punctatus*, with longer and thicker, and less divided leaves and smaller fl.-heads.

OBS. In Herb. D. and Hk. are specimens from *Dr. Sutherland*, gathered near Natal, of a *Euryops* closely allied to this species, but with a dwarfer habit and broader leaves. It may be distinct, but the specimens are insufficient, not having mature flowers.

14. E. longipes (DC.! l. c. 445); glabrous, stems short, flexuous, branching, the axils of the upper leaves woolly-bearded; leaves filiform, about uncial, some quite entire, some trifid to the middle, the lobes entire, subacute; pedunc. axillary and terminal, very long, 5–12 times longer than the leaves; inv. scales 7–10, ovato-lanceolate, not ribbed or margined; recept. conical; ligules 7–10; achenes (fide DC.) glabrescent; ovaries puberulous.

VAR. *β.* **integrifolius**; all the leaves entire.

HAB. Gnadendahl, *Burchell.* Gauritz R., *E.♂.Z.!* Gamke R., *Zey.!830.* Caledon, *Dr. Pappe.* (Herb. D., Hk., Sd.)

A rather slender, branching shrub, 6–10 inches high. Leaves 1–1½ inch long, more frequently entire than trifid (at least var. *β.* seems the commonest form). Pedunc. 6–12 inches long or more.

15. E. punctatus (DC.! l. c. 445); glabrous; stems flexuous, branching, closely leafy; leaves slender, short, impress-dotted, some simple, some trifid to the middle, some pinnati-partite, 5–7-lobed, the lobes filiform, acute; pedunc. slender, terminal, 8–10 times as long as the leaves; inv. scales 12–13, lanceolate, acuminate, ribbed; rays as many, longer than the disc; achenes pubescent (DC.). *Also (fide sp. in Hb. Sond.) Osteospermum tenuilobum, DC.! l. c. 467.*

HAB. Gauritz R., *Ecklon!* (Herb. Sd.)

A very slender shrublet, not much branched, 6–12 inches high. Leaves ¼–¾ inch long, very slender, rarely undivided, almost all either trifid, or alternately 4–5–7-fid, all taper-pointed. The *impressed dots* are not conspicuous on the specimens I have seen. Pedunc. 4–5 inches long. Heads small, the inv. 4 lines across.—A much smaller and more slender plant than *E. diversifolius*, to which it comes nearest; from *E. longipus* it is known by its more slender, shorter, and more divided leaves, and narrower and more numerous inv. scales.

16. E. subcarnosus (DC.! l. c. 445); glabrous; leaves crowded at the ends of the branches, linear-semiterete, fleshy, acute, some trifid, others (or all) undivided; peduncles filiform, as long or twice as long as the leaves; inv. scales 5–6, broadly ovate, subacute, connate at base only; achenes?

HAB. On the Carroo; also Zwarteberg and Zwart Ruggens, *Drege!* (Herb. D., Hk., Sd.)

A robust, dwarf scrub, with flexuous branches and twigs, all naked and cicatrised at base, with tufted leaves at the extremity. Leaves ½–1 inch long. Pedunc. 1–1½ inch, thread-like. Heads like those of *E. multifidus.*

17. E. linearis (Harv. Thes. Cap. t. 153); glabrous; leaves linear, elongate, *flat*, acute, faintly 1-nerved, quite entire; pedunc. axillary, about twice as long as the leaves, filiform; inv. scales 10–12, *concrete nearly to the apex*, acute; achenes glabrous, compressed.

HAB. Cape, *Mundt!* Swellendam, *Dr. Pappe!* (Herb. Hk., D.)
A rather slender bush, 1 foot or more high. Leaves 1–1¼ inch long, 1 line wide, quite flat (not filiform). Invol. scales concrete into a cup for more than ¼ their length; by which character and the broader, flat leaves it is known from *E. linifolius.*

18. E. linifolius (DC.! l. c. 446); glabrous, or the young branches somewhat woolly; leaves linear-filiform, elongate, acute, quite entire; pedunc. axillary, filiform, longer than the leaves; inv. scales 7–9, connate at base, oblong, acute; rays 6–8; recept. honey-combed; achenes glabrous. *Cineraria linifolia, Linn. f. Jacq. Schoenb. t.* 308. *Othonna tenuissima, litt. α, Herb. Thunb. (not litt. β. & γ.)*

HAB. Cape. *Thunberg!* Stellenbosch, *E. & Z.!* Under the Tigerberg, *Mundt! W. H. H.! Zey.! No.* 958. (Herb. Th. D., Sd., Hk.)
Very like *E. tenuissimus*, but with shorter and more slender leaves, smaller fl. heads, and glabrous achenes.

19. E. tenuissimus (Less.! Syn. 394); glabrous, or the young branches somewhat woolly; leaves linear-filiform, elongate, acute, quite entire, glabrous or the younger cobwebbed; pedunc. axillary, rather longer than the leaves, filiform; inv. scales 12–20, lanceolate, concrete at base; recept. convex, honeycombed; achenes hairy! *DC.! l. c.* 446. *Othonna tenuissima, Linn. f. Th. Cap.* 718. *Jacq. Schoenb. t.* 61. *Euryops carnosus, Cass. E. strictus, DC. l. c.* 447.

HAB. Cape, *Thunberg!* Kochmanskloof, *Mundt!* Banks of the Zwartkops R. and in Adow, Uit., *E & Z.! Zey.!* 2997. Swellendam, *Drege! Burchell!* R. Zonderende, *Zey.!* 2990. (Herb. Th., D., Hk., Sd.)
A tall, coarse shrub, 2–3 feet high, closely resembling *E. spathaceus* in foliage, but with very different involucres. *DC.* says the achenes are *glabrous;* but I find them hairy both in *Thunberg's* and *E. & Z.'s* specimens: nor can I distinguish *E. strictus, DC.* (in Hb. Drege!) by any valid character.

20. E. spathaceus (DC.! l. c. 446); glabrous; leaves crowded, linear-spathulate or linear-filiform, elongate, acute, quite entire; pedicels axillary, filiform, longer than the leaves; inv. campanulate, the young spathe-like, formed of 4–5, very broad, short scales, concrete for ¾ their length; rays 6–8, longer than the ovary; achenes villous.

HAB. Eastern Districts, *Burchell.* Uitenhage, *E. & Z.! Zey.!* 2988, 2989. Grahamstown, *Sir C. Bunbury! Genl. Bolton!* (Herb. Hk., D., Sd.)
A tall, strong shrub, 2–3 feet high; the old branches nude, cicatricised; younger densely leafy. Leaves 1½–2 inches long, sometimes linear throughout, sometimes narrowed at base and flattened and broader upwards, ½–1½ line wide. Pedunc. numerous, 2–3 inches long. Invol. irregularly opening, the broad, many-nerved scales continuing long concrete.

21. E. oligoglossus (DC.! l. c. 446); glabrous; lvs. linear-trigonous, keeled, acute, closely set, spreading; pedicels lateral, numerous, scarcely longer than the leaf; inv. scales 5–6, broadly ovate, acute, nerve-keeled, connate at base; rays 2–5, short; achenes villous. *E. triqueter, Less.! Mss. in Herb. Th. Othonna trifida, litt. β., Herb. Thunb.! E. racemosus, DC.! l. c.*

HAB. Cape, *Thunberg! Burchell*, 2826. Nieuwe-Hantam, *Drege!* Africa's Hoogte, *Burke & Zey.!* Beaufort, *E. & Z.!* (Herb. Th., D., Hk., Sd.)
A much-branched, closely leafy, small shrub, about 1 foot high. Leaves ½–¾ inch long, 1 line wide, rigid, subpungent. Pedicels as long. Heads 2 lines diameter. I cannot separate *E. racemosus, DC.* by any valid character; the heads are perhaps a shade larger in *Ecklon's* specimen. *Lessing's* syn. seems never to have been published, or I should have adopted it.

22. E. Candollei (Harv.)

; glabrous, much-branched; leaves *minute,* sessile, closely imbricated, linear-oblong, nerve-keeled, with an inflexed mucro; pedicels lateral, short; inv. scales 6–8, broad, connate beyond the middle, subacute; rays 6–7, twice as long as the involucre; achenes (fide DC.) glabrous. *E. imbricatus, DC.! l. c. 446, not of Lessing!*

HAB. Witberg, 7–8000 f., *Drege!* (Herb. D., Hk.,)
Stems 6–8 inches high, much-branched. Leaves (like those of a *Brunia*) 2 lines long, scarcely 1 line wide, broad at base, narrower upwards, with a thick, rounded keel, closely pressed on the stem and on each other. Pedicels ½–¾ inch long. Heads small. A remarkable species, quite different from *E. imbricatus*, Less.! which I consider a mere var. of *E. lateriflorus.*

23. E. empetrifolius (DC. l. c. 446)

; "glabrous; leaves *(minute)* sessile, thick, subcoriaceous, oval-oblong, obtuse, subtrigonous, nerveless, the younger subimbricate, the adult spreading; heads very few, subterminal, subsessile; inv. broadly 5-lobed, the rays 1–2; achenes glabrous." *DC. l. c.*

HAB. Roggeveld, near Riet R., *Burchell.* (Leaves scarcely 2 lines long.)

24. E. asparagoides (Less.! Syn. 394)

; branches robust, terete, spreading, cinereous, alternate, closely beset with subsessile tufts of leaves and flowers (i. e. minute, undeveloped, leafy ramuli); axils woolly; leaves very short, linear, fleshy, obtuse, channelled beneath; pedunc. *shorter* than the leaves, fascicled; heads homogamous, 8–12-flowered; invol. turbinate, of 5 scales, slightly connate at base; recept. honeycombed; ovaries pilose. *Mikania asparagoides, Licht.! MSS.*

HAB. Cape, *Lichtenstein!* (Herb. Berol., Sd.)
Of this I have seen frustules of the authentic specimen (in Hb. Berol.) which is said to have branches as thick as a goosequill. Its fl.-heads are the smallest in the genus. Leaves 2–3 lines long. Pedunc. 2 lines long. Inv. 1½–2 lines.

25. E. lateriflorus (Less.! Syn. 394)

; glabrous; leaves sessile, imbricated, ovate, oblong, or obovate-oblong, acute or obtuse, rigid-coriaceous, quite entire, callous-edged, nearly nerveless; pedunc. axillary, filiform, 3–4 times longer than the leaves; inv. scales 7–8, broadly ovate, connate; recept. convex, honey-combed; ligules 6–8, oblong; achenes villous. *DC.! l. c. 445. Othonna lateriflora, Linn. f. Thunb.! Cap.* 718.

VAR. *β.* imbricatus; leaves obovate, obtuse, rugulose when dry (somewhat fleshy?). *Euryops imbricata, Less.! Syn.* 394, *not of DC. Othonna imbricata, Th.! Cap.* 719. (Hb. Thunb.)

HAB. Karroo, near Gouds R., *Thunberg!* Plettenbergs Bay, *Burchell.* Gauritz R., and Sneeuweberg, *Drege!* Jackalsfontein; Brack R., and Rhinosterkopf, *Burke & Zey.!* Kraamberg, Albert, *T. Cooper*, 595. (Herb. Th., D., Hk., Sd.)
1–3 ft. high, robust, not much branched, imbricated with leaves throughout, or

the old parts denuded and cicatricised.   Leaves 4 lines to 1 inch long, 2 lines to ½
inch wide, very rigid and thick, but not fleshy.   Pedunc. 1–2½ inches long, from
the upper axils, or along the stem.   Var. β. differs solely by its rather smaller, very
obtuse leaves ; but the point in the normal form varies greatly, and *Drege's* specimen
from Mierenjesfontein completely unites the two forms.

26. E. sulcatus (Harv.) ; glabrous; leaves sessile, closely imbricate,
erect or spreading, oblong, obtuse or subacute, rigid, thick, flattish,
faintly 3-nerved above, with a medial impressed line or furrow beneath,
cartilagineo-ciliolate; pedicels lateral, short; inv. sc. about 5, broadly
oblong, obtuse, 3–4-nerved, subconnate at base; rays 2–3, short; recept.
convex, honey-combed; ovaries hairy, achenes woolly. *Othonna sulcata,*
Thunb.! Cap. p. 719.

VAR. β. **densifolius** (Sd.); more slender; leaves *recurved at the point*, ovate-oblong,
acute, with a fainter medial line beneath.   *E. densifolius, Sond. MSS. E. ciliatus*
Harv. in Herb. Hook.

HAB. Cape, *Thunberg!* β. Rhinosterkop, *Burke & Zey.!* Sand R., *Zey.!* (Herb.
Th., Hk., Sd.)

A rigid, more or less robust shrub, with the aspect of *E. lateriflorus,* from which
it is at once known by its elegantly ciliolate leaves.   Leaves 6–7 l. long, 3 l. wide ;
in β. 4–6 l. long 2 l. wide.   Pedicels uncial.   Heads few-flowered.

Doubtful Species.

E. subsessilis (Sch. Bip. Bot. Zeit. xxviii. 51); "shrubby, branching,
glabrous; leaves undivided, obovate-oblong, obtuse, about 4 lines long,
2 l. wide, imbricate, flat, sessile, fleshy, nearly nerveless; heads at the
ends of the branches, subsessile; inv. sc. about 5, obtuse; rays 3;
achenes villous." *Walp. Rep.* 6. *p.* 272.

If not *E. lateriflorus* β. *imbricatus,* this must come very near it.

CXIX. RUCKERIA, DC.

Heads many-fl., hetero-monœcious; *ray fl.* ligulate, female; *disc-fl.*
5-toothed, mostly abortive (not strictly "*male*"). *Inv.* uniseriate, the
scales more or less concrete at base. *Recept.* subconvex, areolate. Style
of the disc-fl. branched (!), its arms truncate, hispid at the apex. *Ray-*
achenes oblong, terete; those of the disc slender, abortive, velvetty.
Pappus of all the flowers in many rows, similar, caducous, the bristles
exceedingly slender, slightly nodulose, jointed, smooth, at length aggregated in copious woolly tufts. *DC. Prodr.* 6. 483.

S. African suffrutices or herbs, with the aspect of *Euryops.* Stems leafy at base,
ending in long, nude, one-headed peduncles. Leaves pinnate-partite, with linear
lobes. Flowers yellow. Name in honour of J. F. Rucker, a collector of Cape plants.
I have only had the opportunity of examining the gen. char. in *R. tagetoides,* and in
it, owing to the state of the specimens, very imperfectly: in it I find the disc-styles
bifid, not "undivided"; and the pappus bristles *smooth,* not "*barbate,*" as described
by *DC.*

Achenes densely velvetty:
    Leaves compressed, lobes 2–3 on each side... ... ...   (1) **Euryopsides.**
    Leaves terete, lobes several on each side, lowest tooth-
    like   ... ...   (2) **tagetoides.**
Achenes glabrescent; leaf-lobes 2–3 on each side, terete ...   (3) **othonnoides.**

1. R. Euryopsidis (DC. l. c. 484); leaves compressed, pinnati-partite, the lobes on each side 2–3, often alternate, subulate, subfalcate; ray-achenes velvetty; rays twice as long as the involucre. *R. euryopoides, Drege! in Herb.*

HAB. Ezelbank and Dwars R., Zeederberg, *Drege!* (Herb. Hk.)
Leaves 2–2½ in. long, 1½ l. wide; the lobes ½ in. long, very acute. Rays ¾ in. long, 2 l. wide, 7-striate. *DC.* says, "leaves 9–11 l. long, flat, pinnatifid, the lobes linear-lanceolate," none of which characters accord with the specimen seen by me.

2. R. tagetoides (DC.! in Deless. Ic. Sel. vol. 4, t. 66); leaves terete, pinnati-partite, the lobes on each side 5–7, the lowest short and tooth-like, upper long, filiform, acute, quite entire or "toothed" *(DC.);* achenes velvetty; rays 4 times longer than the invol. or longer. *DC. Prodr. 6, p.* 484.

HAB. Sand hills, betw. Packuis & Bredow Zeederberg, *Drege!* (Hb. D., Hk., Sd.)
Stem 5–6 inches long; the peduncle a foot or more. Leaves 3 in. long, the upper lobes ½–¾ in. long. Rays, fide DC., 1½ in. long, 10-striate. Our specimens are very imperfect.

3. R. othonnoides (DC. Prodr. 6., 484); "leaves terete, pinnati-partite, the lobes on each side 2, terete, undivided; ray achenes glabrescent; rays twice as long as the involucre." *DC. l. c.*

HAB. S. Africa, *Drege.* (Unknown to me).
"Leaves uncial, lobes semi-uncial. Rays 3 l. wide, 6 l. long. Pedunc. 7 in. long," *DC.* A few leaves, from *Drege* (in Hb. Sd.) purporting to be of this plant, have lobes in 2, 3 or 4 pair: and are otherwise like those of *R. tagetoides.*

Sub-Tribe 6. CALENDULEÆ. (Gen. 120–124.)

CXX. **DIMORPHOTHECA**, Vaill.

Heads radiate, *ray-fl.* ligulate, female, *disc-fl.* 5-toothed, either all abortive, or all perfect, or (more usually) the outer ones perfect, the inner abortive. *Invol.* uniseriate, of linear, acuminate scales. *Recept.* flat, becoming convex, naked or with a few deciduous paleæ. *Style* of the perfect flowers shortly bifid, the branches diverging, round-topped, glandular at the margin, and piliferous externally; of the fem. fl. with long, glabrous branches. *Achenes* without pappus, straight, those of the ray wingless, obconic, 3-cornered, tuberculated, or sharply toothed, rarely smooth; of the disc flattened, 2-winged, the wings thickened, wider than the seed. *DC. Prodr.* 6, *p.* 70. *Acanthotheca, DC.! l. c.* 73.

Herbaceous or half-shrubby plants, natives of S. Africa, very frequently viscoso-glandular. Leaves alternate, toothed, lobed, or pinnati-partite, rarely entire, commonly scabrous and glandular. Heads terminal, solitary; disc-fl. yellow or brown, or rarely purple; rays either white with purple under surfaces, or purple or yellow. The armature of the ray achenes varies so much in different species that I cannot adopt *Acanthotheca, DC.,* without dividing the genus into almost as many genera as there are species. Name from δις, *two;* μορφη, *shape;* and θηκα, here meaning the *achenes;* the achenes are of two forms.

A. Rays white, or purple, or white above, purple beneath (Sp. 1–9.)
    Root annual ... ... (1) **annua.**
    Perennial. Stems short, simple, tufted, ending in peduncles:
        Disc-achenes obcordate; of ray mostly abortive (2) **nudicaulis.**
        Disc-achenes abortive; of ray minutely pubes-
          cent, smooth ... ... (5) **caulescens.**
    Branching shrubs or suffrutices:
        Disc-fl. with glabrous or sparsely pilose teeth:

Lvs. obovate-spathulate, denticulate, thin	(3) fruticosa.
Leaves lanceolate, acute, entire	(4) Ecklonis.
Leaves lanceolate or linear, *toothed :*	
Ray-achenes bluntly 3-angled, wrinkled and pitted	(7) Tragus.
Ray-ach. sub-trigonous, muricate-tubercular	(8) Dregei.
Ray-ach. slender, sparingly tubercled	(9) leptocarpa.
Disc-fl. biform; outer bearded, inner flat-topped, closed	(6) Barberiæ.

B. Rays yellow or orange. (Sp. 10–21.)

Perennial, with short, tufted stems, ending in peduncles.	
Leaves linear-lanceolate, entire or toothed ...	(10) Zeyheri.
Branching shrubs or suffrutices.	
Leaves oblong-cuneate, coarsely few-toothed	(12) cuneata.
Leaves linear-oblong, obtuse, entire	(11) aurantiaca.
Leaves oblong, sharply incised ; ray achenes smooth	(13) chrysanthemifolia.
Leaves bi-pinnati-partite, lobes linear	(14) multifida.
Annuals :	
Leaves pinnati- or bipinnati-parted :	
Lvs. bi-pinnate. Ray-ach. margined with cuneate lobes	(15) polyptera.
Lvs. pinnate. Ray-ach. margined with sharp teeth	(16) pinnata.
Leaves oblong, sinuate-dentate or entire :	
Branched from the base, diffuse :	
Glabrous or nearly so. Lvs. sinuate	(17) sinuata.
Glabrous. Leaves entire, obtuse ...	(20) integrifolia.
Glandular-scabrous. Inv. sc. pilose	(18) calendulacea.
Erect, single-stemmed, branched or simple	(19) dentata.

A. *Rays white, purple, or discoloured beneath.* (Sp. 1–9.)

1. D. annua (Less. ! Syn. 257) ; annual, erect or diffuse, simple or branched, scabrid with jointed and gland-tipped hairs ; leaves narrow-oblong or obovate-oblong, tapering much to the base, sinuate-toothed, pilose, the upper small and narrow ; pedunc. terminal, in fruit cernuous; outer disc-fl. fertile, inner sterile ; disc-achenes with smooth margin ; ray-ach. much tuberculated. *D. pluvialis, Moench. Meth. 585, DC. l. c. 70, and D. hybrida, DC. l. c. Calendula decurrens, Th.! Cap. 703. Cal. hybrida, L. Th.! Cap. 704. Sw. Br. Fl. Gard. t. 39. Cal. pluvialis, Linn. Th.! Cap. 704. Schk. Handb. t. 264. Meteorina gracilipes, and M. crassipes, Cass. Mill. Ic. t. 79. f. 2, and t. 75. f. 1.*

HAB. Western districts, in sandy soil, common. Simon's Bay, *C. Wright!* 389. (Herb. Th., D., Hk., Sd.)

Very variable in size and ramification. Leaves 1–3 inches long, mostly spathulato-obovate, variably hairy or rough, bluntly toothed or incised. Pedunc. 2–3 inches long, at first cylindrical, afterwards swelling upwards under the head and finally nodding. Disc-fl. with taper-pointed, glabrous lobes. Anthers dark-coloured, blue or purple. Rays uncial, white above, purple beneath, minutely 3-toothed.

2. D. nudicaulis (DC.) ; stems many from the crown, short, simple, densely leafy, prolonged upwards into a long, naked or sparsely leafy, scape-like peduncle ; leaves *polymorphous*, narrow-linear, linear-lanceolate, oblongo-lanceolate, spathulate-oblong or obovate, the narrow ones entire or subentire, the broader sinuate-dentate, hispido-scabrous or more or less glabrous, acute or obtuse ; pedunc. roughly pubescent or hirsute, erect; ray-fl. and outer disc-fl., also innermost disc-fl. mostly abortive; achenes of medial disc-fl. fertile, roundish-obcordate, with smooth edges. *D. perennis, Less. !*

VAR. α. **latifolia**; leaves spathulate-oblong or obovate, sinuate, tapering at base, subacute. *D. nudicaulis, DC.! l. c.* 71. *Calendula nudicaulis, Linn. Th.! Cap.* 704.

VAR. β. **intermedia**; leaves lanceolate-oblong or lanceolate, dentate, scabrous. *Calendula* 941, *Hb. Eckl.*

VAR. γ. **graminifolia**; leaves narrow-linear, entire or remotely toothed, glabrous, or scaberulous, or scabrous. *D. graminifolia, DC. l. c. Calend. graminifolia, Linn. Th.! Cap.* 703. *Mill. Dict. t.* 76. *f.* 1. *Bot. Reg. t.* 289.

HAB. Sidesand summit of Table Mt., and flats near Wynberg, frequent. Gnadendahl, *Dr. Roser.* HassagaisKlo of, and Berg River, *Zey.!* 834. β. Cape, *Ecklon!* Simon's Bay, *C. Wright,* 390. (Herb. Th., D., Hk., Sd.)

Rootstock thick and woody; stems forming tufts often a foot across; the leafy part 3–4 inches long, ending in a leafless pedunc. often a foot or more high. Shape of leaves and pubescence extremely variable; no limits between the above varieties. Heads large; the rays uncial, white above, purple beneath. Disc-fl. sparsely pilose externally, with shortly lanceolate, acute (not taper-pointed) lobes. In all my specimens the ray-achenes are abortive.

3. D. fruticosa(Less.! Syn. 257); suffruticose at base, diffusely branched or decumbent, leafy; leaves alternate, obovato-spathulate, subacute, tapering much at base, sparingly denticulate, thin, glanduloso-scaberulous; pedunc. short, terminal; disc-fl. all abortive; achenes of ray trigonous, slightly curved, hispidulous, the sides wrinkled, base acute, apex 3-tubercled. *DC.! l. c.* 71. *Calendula fruticosa, Linn. Mill. Dict. t.* 283. *Th.! Cap. p.* 704. *Blaxium decumbens, Cass.*

HAB. Cape, *Thunberg!* Sandy sea shores: Kamps Bay, Capetown, *W. H. H.* Uitenhage, *E. & Z.!* Near Port Natal, *Drege! T. Williamson!* (Herb. Th., D., Hk., Sd.)

Stems 2–4 ft. long, weak, trailing on the ground, almost herbaceous, save near the base, branching, laxly leafy. Leaves 2–3 inches long, tapering into almost a petiole at base, ¾–1 inch wide near the apex, of much thinner substance than in most species. Pedunc. 2–3 inches long. Rays white, reddish or purplish beneath. Disc-fl. glabrous.

4. D. Ecklonis (DC. l. c. 71); suffruticose, robust, erect, branching, scaberulous; leaves crowded, linear-lanceolate or lanceolate, entire or subrepand-denticulate, acute, narrowed at base, thickish, mid-ribbed, calloso-mucronulate; pedunc. terminal, longish, one-headed, pulverulent-scabrous; inv. scales sub-biseriate, much acuminate, those of the inner row scarious; achenes of the ray pyriform-oblong, sharply trigonous, the sides finely wrinkled, the outer side convex-backed, the others flat; outer disc-fl. with glabrous teeth.

HAB. Koega R., near Elandstroom, *Ecklon!* Zwartkops R., Uit., *Zey.!* 2813. Port Elizabeth, *Mrs. Holland,* 46. Aapjes R. and Magalisberg, *Burke & Zey.! Zey.!* 831. (Herb. D., Hk., Sd.)

A strong-growing, half-herbaceous plant, 2 f. or more high, simple below, many-branched at the top, the stem and branches closely leafy, all parts minutely powdery-scabrous, drying pale. Leaves 2–3 inches long, ¼ to nearly ½ inch wide, rarely sub-calloso-denticulate. Pedunc. 5–8 inches long, erect. Heads resembling those of *D. nudicaulis;* the rays 1½ inch long, white above, purplish beneath, glandularly scabrous on the tube. Achenes not seen on the Magalisberg specimens, which are otherwise similar.

5. D. caulescens (Harv.); stems short, tufted, simple, sparsely leafy, prolonged into naked peduncles; all parts glandularly-scaberulous; radical leaves spathulato-lanceolate, much attenuate at base, entire or repand, callous pointed, cauline linear-lanceolate, sessile, thickish, mid-ribbed; inv. scales about 12, lanceolate, pubescent; achenes of ray oblong, bluntly trigonous, in the upper part minutely echinato-puberulent, below quite smooth, glossy, not wrinkled; disc-fl. with glabrous, acuminate teeth. *E. & Z.! Comp. Z. n. N. n. E.* 112–11 *and* 117–11.

HAB. Cape, *Eck. & Zey.!* Elandsberg, Queenstown, *T. Cooper!* 216. Weenen Country, Natal, *Dr. Sutherland!* (Herb. Sd., Hk., D.)

Stems (including the peduncle) 12–15 inches high, leafy for about ⅔ their length. Radical leaves several, 2–4 inches long, cauline about 2 inches, all varying from 2 to 4–5 lines broad, the radical broader. Heads erect, on peduncles 3–5 inches long. Rays white above, purple beneath, with glandularly scabrous tubes. Allied to *D. Ecklonis* in foliage; differing in habit and in the achenes. *Dr. Sutherland's* specimens, so far as they go, agree with those of *E. & Z.* on which the species is founded.

6. D. Barberiæ (Harv.); suffruticose, straggling, minutely glandular-pubescent; leaves oblongo-lanceolate, the lower ones much attenuated at base, upper sessile, acute or subacute, distantly calloso-denticulate or quite entire; pedunc. elongate; inv. scales much acuminate, scaberulous, with membranous edges; disc-fl. sterile, of two forms, the outer open, with bearded, bluntish lobes; the inner closed, the lobes connivent, inflated, or cap-like, glabrous; ray achenes smooth. *Bot. Mag. t.* 5337·

HAB. Kreili's Country, *H. Bowker,* 274. Emyate, Natal, *Gerr. & M'K.* 1059. (Hb. D.)

Young plants herbaceous, older weakly suffruticose, with long, diffuse, laxly leafy stems. Lower leaves 3–4 inches long, upper 1½–2 inches, 4–6 lines wide. Pedunc. 6–8 inches long. Fl. handsome; the rays uncial, of a deep bright purple on the upper surface, dull purple beneath. Disc-fl. also purple, the central ones with curious puffy-concave lobes bent inwards and forming together a flattened, waxy, 5-angled table. The species is named after *Mrs. F. W. Barber,* to whom I am indebted for seeds and dried specimens. It has been introduced to English gardens, but the cultivated specimens do not show the brilliant purple of the wild ones.

7. D. Tragus (DC. l. c. 71); suffruticose, rigid, scabrous, diffusely branched, the branches leafy nearly to the summit; leaves lanceolate-linear or lanceolate-oblong, acute or acuminate, scabrous and glandular, distantly spinoso-denticulate, the teeth short, subulate, spreading or deflexed; heads shortly pedunculate, at length cernuous, few-flowered; achenes of the disc-fl. all abortive, of the ray bluntly 3-angled, the angles swollen at base into knobs, the sides wrinkled and pitted. *Calendula Tragus, Jacq. Hort. Schoenb. t.* 153· *Cal. Tragus β., Bot. Mag. t.* 1981.

VAR. β. **scabra**; leaves narrow-linear or subulate, often entire; pedunc. rather longer. *D. scabra, DC. l. c.* 72.

HAB. Cape, *Jacquin.* Drakensteenberg, *Drege!* Worcester, *E. & Z.!* Waterfall, Tulbagh, *Dr. Pappe!* Witsenberg, *Zey.!* 832. β. Worcester, *Drege!* R. Zonder Einde, *Zey.!* 2814. Zwarteberg, *Pappe!* Caledon, *E. & Z.!* (Herb. D., Sd., Hk.)

12–18 inches high, robust, copiously glandular and scabrous. Leaves sessile or shortly tapering at base, 1–2 inches long, from 1½ line to half an inch broad, thickish, nerved beneath. Pedunc. scarcely 2 inches long. Invol. campanulate, longer than the disc, the scales much acuminate, roughly pubescent. Rays uncial, white above, purplish beneath. β. is very diffuse or procumbent, often many-stemmed from a leafy crown, with much narrower leaves and longer peduncles. The pubescence varies in amount ; the marginal teeth when present are always sharp and patent. I find intermediate states connecting the typical *D. Tragus* with " *D. scabra,*" *DC. Calendula scabra, Hb. Th.* partly belongs to this species ; partly to *D. annua !*

8. D. Dregei (DC. l. c. 72) ; "stem shrubby, erect, branched, the branches and leaves rough with glandular [scabrous] pubescence; leaves crowded, sessile, linear-elongate, with 2–3 obtuse teeth on each side, sinuate-dentate, with subreflexed margin; pedunc. naked, one-headed; invol. scales 10–13, linear-lanceolate, acuminate, on the back glandular; achenes of ray oblong, sub-trigonous, muricato-tubercular, of the disc sterile, smooth." *DC. l. c.*

HAB. Zeederberg, *Drege.*
"A shrublet 4–5 inches high. Leaves 1½–2 inches long, 1–1½ line wide. Fl. unknown." *DC.* Only known to me by a frustule, without flowers, in Hb. Sond. It may be a mere var. of *D. Tragus.*

9. D. leptocarpa (DC. l. c. 72) ; "stem subherbaceous, branched, glabrous [scaberulous] ; branches glandularly-pubescent, naked at the summit, one-headed ; leaves oblong-cuneate, toothed, glandularly-puberulent, the upper linear, entire; achenes of the ray slender, subtrigonous, along the angles and at apex sparingly tuberculate." *DC. l. c.*

HAB. Between Kaus and the Gariep, *Drege.* (Herb. Sond.)
Leaves 1–2 inches long, tapering at base, the smaller often spathulate. Pedunc. 2–3 inches long. Rays uncial, white, with blueish tips ("sub-cœrulescent," *DC.*) Only known to me by a very imperfect specimen.

B. Rays yellow or orange. (Sp. 10–20.)

10. D. Zeyheri (Sond.) ; stems short, tufted, simple, scaberulous, ending in naked peduncles ; leaves linear-lanceolate or linear, either entire or coarsely few-toothed, or sinuate, subglabrous, calloso-ciliolate, thickish, one-nerved; inv. scales lanceolate-acuminate, dorsally scabrous, with white margins ; rays yellow ; achenes of ray triquetrous, the angles crenate-toothed, the faces rugulose and gland-dotted ; of disc flat, with wide, thickened rim, occasionally 3-winged.

HAB. Zuurepoost, Stormberg, Sep., *Zey.!* Near Burghersdorp, Albert, *T. Cooper!* 771. (Herb. Sond., D.)
Stems in dense tufts, 1–2 inches long, leafy, prolonged 4–5 inches into a sparsely leafy flowering stem, ending in a naked, scabrous, one-headed peduncle. Leaves 1½–2 inches long, 2–4 lines wide, the surface mostly glabrous. Heads of small size; rays scarcely uncial. *Mr. Cooper's* plant is smaller, less scabrous, with more entire leaves.

11. D. aurantiaca (DC.! l. c. 72) ; stem suffruticose or fruticose, erect, glabrous, pale ; branches rodlike, sparingly leafy, ending in short pedunc., pubescent beneath the fl. head; leaves linear-oblong or spathulate, obtuse, entire, narrowed at base, thick, faintly ribbed, the young ones minutely softly ciliolate; inv. scales linear-acuminate, longer than the disc, with a dorsal line of short hairs, and wide, pale margins; rays orange-yellow; achenes . . . . ? *Calendula flaccida, Vent. Malm. t.* 20. *fide DC. Castalis Ventenati, Cass.*

HAB. Little Namaqualand, near Haasenkraal's river ; betw. Koussie and Zilverfontein, and near Kooperberg, *Drege!* (Herb. D., Hk., Sd.)

Apparently a tallish, laxly branched, rigid shrub, with slender, rodlike flowering branches 12–15 inches long.   Leaves 2–3 inches long, 3–5 lines wide, tapering much to the base, the upper gradually smaller.   Heads large, erect.   Rays numerous, an inch or more long, bright orange-yellow.   I have not seen fruit.

12. **D. cuneata** (Less.! Syn. 257); shrubby, much-branched, glandular, viscidulous; leaves crowded, cuneate or oblongo-obovate, or lanceolate, coarsely or sharply few-toothed, thick, obscurely midribbed, glabrous or nearly so, gland-dotted, subdecurrent; pedunc. short, glanduloso-pubescent; invol. scales linear-acuminate, minutely puberulous, pale-edged, barbato-ciliate near the apex; rays yellow; disc-fl. with glabrous teeth, the outer fertile; achenes of the ray (?); of the disc flattened, obcordate, with a wide, thick-rimmed marginal wing.   *DC. l. c. p.* 72. *Calendula cuneata, Th.! Cap.* 705. *C. viscosa, Andr. Rep. t.* 412. *Arctotis glutinosa, Sims, Bot. Mag. t.* 1343.

HAB. Cape, *Thunberg!* Nieuwe Hantum and Graaf Reynet, *Drege!* Uitenhage and Albany, frequent, *E. & Z.! Mrs. F. W. Barber! Zey.!* 2812, 3066.   Natal, *Miss Owen!* (Herb. Th., D., Hk., Sd.)
Young plants suffruticose, with long, slender fl. branches; old ones rigidly shrubby, twiggy, much and closely branched, the old branches naked, and rough with the bases of old leaves, the younger short, leafy.   Leaves variable in shape and size, always much narrowed at base, and generally sharply toothed, from ¾ inch to 1½ inch long, the young ones minutely glanduloso-puberulous.   Stems and leaves exude viscid, strong-scented substance.   Pedunc. 2–3 inches long; in young plants 4–5 inches.   Rays yellow, uncial.   I have not seen any *ripe* achenes of the ray.

13. **D. chrysanthemifolia** (DC. l. c. 73); shrubby, minutely scaberulous, subglabrous; branches virgate; leaves obovate-oblong, sharply incised or unequally toothed, the lobes often again toothed; pedunc. naked, one-headed; inv. scales lanceolate, acuminate, scaberulous or pilose, margined with soft hairs; achenes of ray trigonous, smooth; of disc flattened, thick-margined.   *Calendula chrysanthemifolia, Vent. Malm. t.* 56. *Bot. Reg. t.* 40. *Bot. Mag. t.* 2218. *Arnoldia aurea, Cass. Osteospermum floridum, E. Mey.!*

HAB. Cape, *Ventenat; Krebs,* 193.   Dutoit's kloof and Simonsberg, *Drege!* (Hb. Hk., Sd., D.)
2 f. or more high, with long, simple, leafy branches.   Leaves 2–3 inches long, ½–1 inch wide, more or less petiolate, or tapering at base, sparingly and minutely scaberulous.   Heads large, rays 1¼–1½ inch long, yellow.

14. **D. multifida** (DC. l. c. 73); "suffruticose, erect, branching, glabrous; branches long, terete, ending in long, naked peduncles; leaves bipinnati-partite, the lobes linear, acute; inv. scales 1–2 seriate, linear, with membranous margins, the inner few and very narrow; achenes of the ray compressed-trigonous, transversely furrowed, wingless; of the disc compressed, sterile, the outer larger with callous margin, the inner narrow, pellucid.

HAB. Fisch River, *Drege.*   Karroo and in Albany, *E. & Z.* (fide *DC.*)

15. **D. polyptera** (DC. l. c. 73); annual, many stemmed, much branched, viscidulous; leaves bipinnati-partite, the lobes linear, obtuse, with revolute margins; pedunc. short, cernuous, glandular; inv. scales lanceolate, glandular-pubescent, subciliate, equalling the disc; rays yellow; achenes

of ray oblong, subtrigonous, the outer angles many-lobed, the lobes cuneate; of disc flat, the outer obovate, perfect, with thickened rim, the inner oblong, sterile, pellucid.

HAB. Zilverfontein, *Drege!* (Herb. D., Hk., Sd.)
A much branched, viscidly glandular annual, 4–12 inches long, diffuse (or prostrate?) Leaves 1–1½ inch long, their lobes or lobules not a line wide. Fl. heads small; the pedunc. 1–2 inches long. Rays seemingly a pale yellow.

16. D. pinnata (Harv.); annual, glandularly pubescent and viscous, many stemmed, branched; leaves pinnati-partite, the lobes linear, obtuse, entire or here and there lobulate; pedunc. short; inv. scales linear, acute, glandular; achenes of rays obconic, trigonous, the angles sharply 3–4-toothed, teeth rigid, deltoid; of the disc abortive. *Calendula pinnata, Th.! Cap. p.* 705. *Acanthotheca pinnatisecta, DC. l. c.* 73. *Chrysanthemum Lidbeckioides, Less.! Syn.* 253.

HAB. Cape, *Thunb.! Dr. Thom!* Olifants R., and Zilverfontein, *Drege! Zey.!* 836, Namaqualand, *A. Wyley!* (Herb. Th., D., Hk., Sd.)
A much branched annual, 6–12 inches high, glandularly pubescent and viscid in all parts. Leaves ½–1 inch long, the lobes 1 line wide or less, often quite entire, as often toothed or lobed. Pedunc. 1–2 inches long. Achenes minutely glandular on the sides. Rays uncial, orange yellow.

17. D. sinuata (DC.! l. c. 72); annual, branched from the base, nearly glabrous, the branches leafy, ending in short peduncles; leaves oblong, obtuse, sinuate, narrowed at base, the younger ones subciliate; inv. scales lanceolate-acuminate, quite glabrous, longer than disc; achenes of ray trigonous, everywhere tubercled; of disc flat, with thickened rim; rays orange.

HAB. Zilverfontein, *Drege!* (Herb. Hook.)
A diffuse annual. Lower leaves 2–3 inches long, 5–6 lines wide. Rays about an inch long.

18. D. calendulacea (Harv.); annual, branched from the base, glandularly scaberulous throughout, the young parts pubescent; branches leafy, ending in short peduncles; leaves oblong or linear-oblong, sinuate-toothed, narrowed at base, obtuse; inv. scales lanceolate-acuminate, black-dotted and albo-pilose, with membr. margins, longer than the disc; achenes of ray (young) trigonous, tuberculated; of disc flat, with thickened rim; rays orange.

HAB. Namaqualand, *A. Wyley!* (Herb. D.)
Nearly allied to *D. sinuata*, and perhaps a mere variety. It differs in its copious glandular pubescence and more evidently toothed leaves; black-dotted invol. scales and longer rays. Leaves 2–3 inches long, 4–5 lines wide, upper smaller. Rays 1¼ inch long, orange yellow. The fl. heads resemble those of a "marigold."

19. D. dentata (DC.); annual, erect, branched, glandularly pubescent; branches ending in naked peduncles; leaves oblong, narrowed at base, sinuate-toothed, puberulous, the upper narrower, sessile; inv. scales lanceolate, acuminate, narrowly membrane-edged; achenes unknown. *Acanthotheca? dentata, DC. l. c. p.* 74.

HAB. Brackfontein, Clanwilliam, *E. & Z.!* (Herb. Sond.)
So similar to *D. annua* that except for the yellow rays (if they be yellow?) I can-

not distinguish it.   Until the fruit shall have been seen, it must be considered a
doubtful species.

20. D. integrifolia (DC.); " annual, many-stemmed, branching, glabrous ; leaves oblong, linear, obtuse, entire ; rays 7–8, yellow, brown
at base ; achenes only near the summit aculeate; disc-fl. few." *Acanthotheca integrifolia, DC. l. c.* 74.

HAB. Oliphant's River, *Drege !* (Unknown to us.)

CXXI. TRIPTERIS, Less.

Heads many-fl., monœcious; *ray-fl.* ligulate, female ; *disc-fl.* tubular,
bisexual, but sterile.   *Invol.* 1–2-seriate, the scales free, often membr.-
edged.   *Recept.* naked, flat, at length subconvex.   *Anthers* minutely
setose.   *Styles* of ray bifid ; of disc undivided.   *Achenes* of ray 3-cornered,
the angles produced in mostly unequal wings, the sides smooth or echinate, straight, substipitate, beaked, the beak hollow, on one side closed
with a hyaline membrane ; sometimes (monstrously ?) terete, elongate,
wingless, with solid beak.   *DC. Prodr.* 6, *p.* 456.

Herbs, suffrutices, or small rigid shrubs, chiefly S. African, mostly glandularly
viscid and strongly scented.   Leaves incised or toothed or entire, alternate or opposite.   Heads panicled or ending the branches or stems.   Rays yellow, white or
purplish.   Name from τρις, *three,* and πτερος, a *wing;* from the 3-winged achenes.

§ 1. OPPOSITIFOLIÆ.   Small shrubs, with constantly *opposite* leaves. (Sp. 1–6.)

Quite glabrous in all parts ; leaves glaucous, quite entire.
  Lvs. spathulate or lanceolate; rays deep-orange, 12–14 (1) **glabrata.**
  Lvs. linear-sublanceolate ; rays yellow, about 10 ... (2) **pallescens.**
Thinly gland-pubescent, powdery or canous :
  Lvs. linear-spathulate. entire, thinly glandular ... (3) **Thomii.**
  Leaves oblong-lanceolate, *sinuate-toothed*, minutely
    powdery ... ... (4) **sinuata.**
  Lvs. narrow-linear, acute, entire or denticulate, gland-
    pubescent ... ... (5) **linearis.**
  Lvs. linear, mucronulate, entire, *canous* ... ... (6) **incana.**

§ 2. RIGIDÆ.   Rigid, scabrous herbs; stems leafy, panicled at the summit; leaves,
except the uppermost, *opposite,* sharply toothed.   (Sp. 7–9.)

Leaves lanceolate, tapering at base :
  Stem *quite glabrous and smooth;* lvs. papillate-scabrid (7) **rigida.**
  All parts scabrid and glandular... ... ... (8) **dentata.**
  Lvs. ovate or ovato-lanceolate, broad and clasping at base (9) **amplexicaulis.**

§ 3. PANICULATÆ.   Annual or perennial herbs; stems leafy, panicled or corymbose at summit ; all the leaves *alternate.* (Sp. 10–14.)

Leaves toothed or sinuate, glandular or viscidulous :
  Inv. scales lanceolate-acuminate, *narrowly* memb.-edged :
    Lvs. ear-clasping, sharply-toothed ; achenes 5–6
      lines long, amply winged ... ... (10) **amplectens.**
    Leaves subpetiolate; inciso-pinnatifid or entire,
      achenes not 2 lines long, narrow-winged ... (11) **microcarpa.**
  Inv. scales oblong-acute, with *very wide* memb. edges :
    Rays as long as inv., or scarcely longer ... ... (12) **clandestina.**
    Rays 2–3 times as long as involucre ... ... (13) **hyoseroides.**
  Leaves quite entire, lanceolate ; the adult quite glabrous (14) **polycephala.**

§ 4. CAULESCENTES.   Rootstock perennial, ligneous, leafy at the crown ; stems subherbaceous, leafy, simple or branched ; branches ending in 1-headed peduncles ; all the leaves *alternate*. (Sp. 15-22.)

Lvs. glabrous, *entire*, cuneate-spathulate, obtuse; achenes
   narrow-winged ... ... (15) **spathulata.**
Lvs. hispid or glandular, more or less *toothed* or repand-sinuate :
   Achenes amply-winged, smooth or nearly smooth :
     Lower lvs. cuneate, acute, coarsely 3-toothed at
      summit, or *entire* ... ... (16) **Aghillana.**
     Lower lvs. linear-lanceolate, few-toothed, hispid;
      inv. sc. linear-lanceolate, acute ... ... (17) **setifera.**
     Lower lvs. linear-oblong, sinuate-toothed or run-
      cinate ; invol. sc. oblong, obtuse, with a wide
      border ... ... (18) **flexuosa.**
   Achenes amply winged, rough-sided :
     Stems much branched, flexuous ; leaves oblong-
      lanceolate, subentire ... ... (19) **Natalensis.**
     Stems ascending, sub-simple; leaves oblong-lan-
      ceolate, subentire or sharply toothed ... ... (20) **glandulosa.**
     Dwarf ; rad. leaves rosulate, obtuse, undulato-
      sinuate, margined, rigid ... ... (21) **humilis.**
Leaves *woolly*, subradical, long-petioled, obovate or obo-
   vato-lanceolate, entire ; stems scape-like ... ... (22) **tomentosa.**

§ 5. FRUTICOSÆ. Small suffrutices or shrubs, often spiny, with *alternate* leaves. (Sp. 23-27.)

Unarmed ; stems suffruticose ; leaves scabrous :
   Lvs. petioled, cuneate at base, coarsely 2-4-toothed (23) **petiolata.**
   Lvs. sessile, oblong or lanceolate, entire or denticulate (24) **arborescens.**
Twigs spinous-tipped ; stems rigid, ligneous ; lvs. mostly smooth :
   Lvs. *glaucous*, flat, glabr., entire or 3-toothed; wings
     of fr. membranous ... ... (25) **spinescens.**
   Lvs. not glaucous, viscid, flat, entire or 3-toothed ;
     wings of fr. coriaceous ... ... (26) **pachypteris.**
   Leaves viscid, linear-semiterete, 3-5-7-lobed ; lobes
     blunt ; wings of fr. narrow ... ... (27) **leptoloba.**

§ 1. OPPOSITIFOLIÆ. (Sp. 1-6.)

1. T. glabrata (Harv.); shrubby, unarmed, erect, glabrous, glaucous ; branches terete, pale ; leaves opposite, sessile, linear-spathulate or lan-ceolate, obtuse or subacute, quite entire, thickish and somewhat fleshy, faintly 3-nerved ; pedunc. terminal and from the upper axils, long, nude, scaberulous below the head ; heads at length nodding ; inv. sc. biseriate, lanceolate, membr.-edged, glabrous, serrulate ; achenes smooth, amply 3-winged. *Calendula glabrata, Thunb.! Cap.* 702. *Osteosp. glabratum, Less.! Syn.* 89. *DC. l. c.* 466. *Tripteris macroptera, DC.! l. c.* 456.

HAB. Heerelogement, Clanw., *Thunberg!* Silverfontein, *Drege!* Little Namaqua-land, *A. Wyley!* (Herb. Th., D. Hk., Sd.)

1-2 ft. high, with mostly opposite, widely spreading branches. Leaves 1-2 inches long, 2-4 lines wide, tapering to the base, strictly opposite, the uppermost sometimes alternate. Pedunc. 3-5 inches long, erect, nodding in fruit. Rays 12-14, long and broad, rich orange-colour (like "marigolds"); disc-fl. purplish. I have seen a single *achene* in Hb. Thunb. whose specimens quite agree with *Drege's.*

2. T. pallescens (DC. l. c. 456); "shrubby, glabrous; stem terete, pale; branches ending in 3 nude, one-headed pedicels ; leaves opposite, linear-sublanceolate, quite entire, glaucous, callous-tipped ; inv. sc. uniseriate, linear, submarginate, shorter than disc; rays about 10, four times longer than the inv., yellow; mature achenes unknown." *DC. l. c.*

HAB. Olifants R., *Drege.* (Unknown to me.)
Said to differ from *T. glabrata* by its narrower, more-linear leaves, yellow, not orange rays, and narrower inv. scales.

3. T. Thomii (Harv.); shrubby, oppositely branched, thinly gland-pubescent; leaves opposite, linear-spathulate, tapering to the base, obtuse, mucronulate, thickish, quite entire; heads ending the branches; inv. sc. uniseriate, linear, white-edged, slightly scabrid; achenes?

HAB. Cape, *Dr. Thom!* (Herb. Hk.)
A ligneous, much-branched shrub. Leaves 1 in. long, 2–3 l. wide, perhaps a little fleshy when recent. Rays 8–10, yellow.

4. T. sinuata (DC. l. c. 456); shrubby, oppositely branched, minutely powdery-pubescent, glaucous; leaves opposite, oblong or lanceolate, tapering at base, subobtuse, sinuate-toothed; heads on terminal peduncles; inv. sc. sub-biseriate, powdery-scabrid at back, lanceolate, membrane-edged, rather longer than the disc; achenes smooth, amply 3-winged.

HAB. Namaqualand, and betw. Zwarteberg and Aasvogelberg; and Driekoppen, Karroo, *Drege!* Graaf Reynet, *Eckl.!* Bitterfontein, *Zey!* 1023, Kareeberge, *E. & Z.!* Dwyca R., *Zey!* 1022. (Herb. Hk., Sd.)
A small, rigid shrub. Leaves 1–1½ in. long, 3–4 l. wide, variable in shape and amount of toothing. Rays yellow. Heads at length nodding.

5. T. linearis (Harv.); shrubby, oppositely branched, minutely and thinly viscoso-puberulous; leaves opposite, narrow-linear, acute, entire or here and there denticulate, thick and somewhat fleshy, nerveless; pedunc. terminal, one-headed, powdery-puberulous; inv. scales lanceolate, powdery-scabrid at back, margined; achenes? *Osteosp. No.* 162, *Hb. Eckl.*

HAB. Cape, *Ecklon!* (Herb. Sond.)
This is nearest *T. sinuata*, but differs in foliage. Leaves ¾–1 in. long, ½-line diam. Pedunc. 2–3 in. long. Fl. yellow.

6. T. incana (Harv.); shrubby, the young parts canous with very minute, close-lying pubescence; branches slender, rodlike, striate, mostly opposite; leaves opposite, sessile, linear, thickish, nerveless, mucronulate, slightly concave above; pedicels terminal, nodding; inv. sc. linear, acute, canous, uniseriate; achenes? *Osteosp. incanum, Th.!* Cap. 714.

HAB. Bokkeveld Karroo, *Thunberg!* (Herb. Th.)
1–2 ft. high, ligneous, much branched; branches decussate, 10–12 in. long, with distant leaf-pairs, simple. Leaves ¾–1 in. long, 1-l. wide, erect. Mature fl.-heads unknown. This has so many characters in common with the opposite-leaved sp. of *Tripteris*, that, though the fruit be unknown, I venture to place it among them.

§ 2. RIGIDÆ. (Sp. 7–9.)

7. T. rigida (Harv.); herbaceous, rigid; stem *quite glabrous*, terete, smooth, erect, branching and panicled upwards; lower and medial leaves opposite, very rigid, lanceolate, acute, tapering somewhat at base, sharply toothed, 3-nerved above the base, papillato-scabrid on both sides with hard, raised points, but not pubescent; upper leaves alternate, small, linear or subulate; pedicels scaberulous, at length nodding; inv. sc. lance-oblong, acute, scabrous, narrow-edged; achenes?

HAB. Roadside, betw. Wynberg and Constantia, *Dr. Wallich!* (Herb. D.) 2–3 ft. high, much branched, pale, very rigid. Leaves 1½–2 in. long, 4–6 l. wide, scorbiculate. Old, empty inv. only seen: but the genus can hardly be doubtful.

8. T. dentata (Harv.); herbaceous, scabrid and glandular in all parts; stems ascending or erect, terete, corymbose at summit; lower and medial leaves opposite, oblongo-lanceolate, tapering at base, the lowest subpetiolate, on both sides scabrid, sinuate-toothed, the teeth sharply callous-mucronate ; uppermost leaves alternate, subentire ; corymb loosely forked, several-headed ; inv. sc. lanceolate, acuminate, scabrid, narrow-edged; achenes thinly hispid or smooth, amply 3-winged.

HAB. Kampground, near Capetown, *E. & Z.!* Muysenberg, *Dr. Wallich!* (Herb. Sd., D.)
Rootstock ligneous, probably perennial. Stems several from the crown, 1–1½ ft. high. Leaves 2–2½ in. long, 4–6 lines wide, tapering much at base; the mucronate teeth subrecurved, with rounded interspaces. Achenes 5–6 lines long; wings 1½ l. wide, often purple. Rays about 12, bright-yellow.

9. T. amplexicaulis (Less.! Syn. 90, non DC.); herbaceous, scabrid and glandular in all parts; stem erect, terete, simple, ending in a corymbose, leafless panicle; lower and medial leaves opposite, broad and ear-clasping at base, often connate, ovate or ovato-lanceolate, acute, coarsely toothed, on both sides rough with hard, raised points; uppermost leaves alternate, lanceolate or subulate ; corymb compound, many-headed ; inv. sc. oblong, acuminate, membr. edged ; achenes muricated, 3-winged. *Calendula amplexicaulis, Th.! Cap.* 705. *Osteosp. connatum, DC.! l. c.* 466.

HAB. Cape, *Thunberg!* Brakfontein, Clanw. and Olifant's R., *E. & Z.!* Zeederberg, *Drege!* (Herb. Th., D., Sd., Hk.)
Stem 2–3 feet high, but soft, not ligceous. Leaves 2–4 in. long, 1–2 in. wide. Pedicels at length nodding. Heads small. Fl. pale-yellow.

§ 3. PANICULATÆ. (Sp. 10–14.)

10. T. amplectens (Harv.); annual; stem herbaceous, erect, branching, terete, gland-scabrid; leaves alternate, lanceolate, coarsely and sharply toothed, narrowed towards the base, but at the base more or less dilated and stem-clasping, scabrid and glandular on both sides; upper lvs. small, subentire, clasping ; branches panicled at the summit; pedicels long, glandular, cernuous ; invol. scales lanceolate, acuminate, dorsally scabrous, membr. edged; rays 12–15, thrice as long as the inv.; achenes smooth or scaberulous, amply 3-winged. *T. amplexicaulis, DC.! l. c.* 457, *excl. syn.*

HAB. Namaqualand, *Drege!* Cape Flats, *E. & Z.! W.H.H.* Kruilsrivier, *Zey.!* 3342. Simon's Bay, *T. Wright,* 381. (Herb. D., Hk., Sd.)
Stems 1½–2 ft. high, loosely branched. Leaves 2–3 in. long, 4–6 l. wide, rather rigid and harsh to the touch, more or less amply eared at base. Achenes 6–7 l. long, the wings sometimes purple ; sometimes quite smooth, sometimes with a few bristles, and in other cases very scabrous !

11. T. microcarpa (Harv.); herbaceous, more or less glandular-pubescent; branches virgate, corymboso-paniculate at top; lower leaves petiolate, with or without small, basal auricles, lance-oblong, inciso-pinnatifid or very deeply cut, glandularly pubescent; upper sessile, half-clasping, toothed or entire; inv. sc. lanceolate, acuminate, ciliate, with a narrow membr. border, dorsally scabrid; achenes *(small)* about as long as the inv. scales, quite glabrous, with a narrow membranous wing.

VAR. β. **glabrescens**; much less glandular; lower leaves unknown; upper broad-based, stem-clasping, entire or toothed; fruit as in α.

HAB. Löwentans, *Zey!* 1026. Namaqualand, *A. Wyley!* β. Namaqualand, *Pappe!* Modderfontein, *Rev. H. Whitehead!* (Herb. Hk., Sd., D.)

The specimens seen are all imperfect, and perhaps the description of the foliage may need correction, or I may possibly confound two species. The strongest character lies in the achenes, which are scarcely 2 lines long, with wings ½ line.

12. T. clandestina (Less.! in Linn. 6. p. 97); annual, viscidulous; stem herbaceous, erect, sparsely pilose, leafy, paniculately branched; leaves oblong obtuse, sinuato-dentate or repand, subglabrous or sparsely pilose, the lowest much attenuated at base, subpetiolate, upper sessile and more or less stem-clasping at base; pedicels ending the branches, cernuous; inv. sc. oblongo-lanceolate, acuminate, with narrow, opaque, scabrous centre and very wide, white, membr. edges; rays scarcely longer than inv.; achenes amply 3-winged, either smooth, or more or less echinate. *Calendula hybrida, fol. 3, Thunb.! in Hb.*

VAR. α. **leiocarpa**; achenes quite smooth. *T. herbacea, and T. scariosa, DC. l. c.* 458. *Calendula hybrida, Th.! fol. 3.*

VAR. β. **echinocarpa**; achenes more or less bristled on the sides. *T. clandestina, DC. l. c., non Less. T. hyoseroides, β. echinocarpa, DC. l. c.*

HAB. A common weed round Capetown, and in the W. districts. Hassaquays-kloof, *Zey.!* 3067, *ex. pte.* (Herb. Th., D., Hk., Sd.)

Stem 8–18 inches high, more or less branched. Lower leaves 4–5 in. long, ¾–1 in. wide; upper 1–1½ in. long. *Thunberg's* specimen, on which *Lessing* founded his "*T. clandestina*," is precisely *T. herbacea, E. M.* I find the rough and smooth achenes a very variable character in specimens from the same localities.

13. T. hyoseroides (DC. l. c. 458); "stem herbaceous, erect, minutely gland-pubescent, panicled at the summit; lower leaves elliptic-oblong, obtuse, obtusely sinuate, tapering into a petiole; cauline oblong-linear, acute, sub-entire, bluntly eared at base, all very minutely downy; pedunc. nude, elongate; inv. sc. elliptic, acute, dorsally glandular, widely membr. edged; rays twice or thrice as long as the inv.; achenes 3-winged, smooth." *DC. l. c., excl. var. β.*

HAB. Betw. Kaus, Natvoet, and Doornport, *Drege.* (Unknown to me.)

This seems to differ from *T. clandestina* chiefly in its longer ray-flowers.

14. T. polycephala(DC.! l. c. 458); perennial?; stem herbaceous, erect, glabrous, corymbosely much branched, branches many-headed; twigs, pedicels and young leaves minutely glandular, viscidulous; leaves lanceolate, elongate, tapering to each end, entire, glabrous, glaucous, 3-nerved; inv. sc. linear-subulate, acuminate, narrow-edged; rays 12–15, twice as long as the inv.; achenes quite smooth, 3-winged, wings glossy, rather narrow.

HAB. Hills near Verleptpraam, by the Gariep, *Drege!* (Herb. Hk.)

A large plant, perhaps 2–3 ft. high, with robust stem and branches, drying pale. Leaves 4–5 in. long, 5–8 l. wide. Achenes scarcely 2 l. long. Inv. sc. not ½ l. wide, very sharp-pointed. The rays seem to have been red or dark-coloured.

§ 4. CAULESCENTES. (Sp. 15–22.)

15. T. spathulata (DC.? l. c. 458); rootstock ligneous; stems fruticose and leafy at base, herbaceous upwards, ascending-erect, prolonged into subsimple or branching peduncles; lower leaves cuneato- or obovato-spathulate, obtuse, tapering much at base, glabrous, entire ("or here and there subdentate" *DC.*), coriaceous; upper oblong, half-clasping at base; pedicels scabrous; inv. sc. linear, obtuse, dorsally scaberulous, margined; achenes smooth, with 3 narrow wings.

HAB. Fields near Zwartkops R., Uit., *E. & Z.! Zey.!* 3065. (Herb. D., Sd.)

Stems probably many from the crown; the ligneous portion sometimes very short, in which case the lower leaves are subrosulate and almost radical. Fl. stems 1–1½ ft. high, quite herbaceous. Leaves 2½–3 in. long, ½–¾ in wide, wide with immersed nerve, scarcely obvious when dry. The wings of the achenes are remarkably narrow. I have not seen an authentic sp. of *DC.'s* plant.

16. T. Aghillana (DC. l. c. 457); stem half-shrubby, dwarf, branching, closely leafy below; leaves cuneate, tapering much at base, acute, either entire or (commonly) coarsely and sharply three-toothed at summit, sometimes glabrous, sometimes thinly or copiously sprinkled with thick hairs, especially along the margins; fl. branches long, one-headed, pilose gland-scabrid upwards, bearing a few leaf-scales; inv. sc. biseriate, ovato-lanceolate, acute, dorsally scabrous, membr. edged; achenes "nearly smooth, 3 winged." *DC.*

VAR. β. integrifolia; leaves mostly quite entire, cuneato-lanceolate, acute. (Hb. Sd.)

HAB. Cape, L'Agulhas, *Drege!* Cape, *E. & Z.!* (Herb. Sond.)

Stems 2–4 inches high, mostly branched. Pubescence very variable in amount, sometimes nearly absent. Leaves 2–2¼ in. long, 3–4 l. wide: in β. quite entire. Rays 10–12, yellow, 2ce. as long as the inv.

17. T. setifera (DC. l. c. 457); "rootstock ligneous, branched; stems subherbaceous, simple, erect, tufted, leafy below, nude above, at length incurved, one-headed; lower leaves linear-lanceolate, here and there few-toothed; upper linear, quite entire, all acute, at margin and on both sides rough with thick bristles; inv. sc. linear-lanceolate, acute, rather longer than the disc, setulose, membr.-edged; rays 12–15, twice as long as the inv.; achenes smooth, amply and roundly 3-winged." *DC. l. c.*

HAB. Zwarteberg, *Drege.* (Unknown to me.)

18. T. flexuosa (Harv.); rootstock ligneous; stems scarcely suffruticose, erect, flexuous, striate, branched, sparsely hispid; lower and medial leaves tapering much at base, oblong or linear, coarsely sinuate-toothed or subruncinate, glaucous, one-nerved, sparsely hispid, upper sessile, small, linear or lance-linear, entire or toothed; pedicels ending the branches, glandular; heads cernuous; inv. sc. oblong or linear-oblong, obtuse, dorsally scabrous, with a wide, scarious border; achenes smooth-sided, amply 3-winged. *Osteospermum scariosum, DC.! l. c.* 464.

HAB. Zwart Ruggens, *Drege.* Graaf Reynet, *Ecklon! Sneeuweberg, Dr. Wallich!* Jackal's Fontein, *Burke and Zeyher! Zey.!* 1021. Vet. R., *Zey.!* 1008. Queenstown, *Mr. F. W. Barber,* 616, *T. Cooper,* 1333. (Herb. D., Sd., Hk.)

Rootstock simple or branched, the older ligneous, deeply descending. Stems 10–12 in. high, rigid, but scarcely ligneous, sparingly branched; the branches with wide axils, spreading. Leaves 2–3 in. long, 2–4 l. wide, mostly 2–4-toothed on each side, with rounded interspaces.

19. T. Natalensis (Harv.); rootstock ligneous; stems suffruticose, erect, flexuous, branching, angle-striate, sparsely hispid; lower and medial leaves tapering to the base, oblongo-lanceolate, margined, sub-acute or obtuse, entire or subrepand, glabrous or nearly so above, sprinkled with rough points and thick white bristles beneath; upper lvs. stem-clasping, semi-lanceolate, concave, acute, often unidentate at base; pedicels ending the branches, glandular, cernuous; inv. sc. ellip-tical, acute, dorsally scabrous, widely membr.-edged; achenes echinate, with undulate, broad wings.

HAB. Natal Country, *Miss Owen!* (Herb. D.)

With the habit of *T. flexuosa;* but not glaucous, with subentire leaves and rough-sided achenes. Lower leaves 2–2½ in. long, 3–4 l. wide; upper ½–1 in. long, broad-based, tapering to an acute point. Stem and foliage pale green, harsh to the touch. Rays about 12, long, 4-lined, yellow.

20. T. glandulosa (Turcz.! Bull. Mosc. xxiv. ii. p. 93); rootstock lig-neous, leafy at the crown; stems herbaceous, ascending, simple or branched, laxly leafy, ending in one-headed peduncles, hispid with swollen, white hairs and glandular; subradical leaves tapering much at base, subpetio-late, oblongo-lanceolate or spathulate, acute, copiously pilose, subentire or sharply sinuate-toothed; cauline sessile, linear or lance-linear, entire or toothed; inv. sc. elliptic-oblong, acute, dorsally scabrous, with *wide* membranous borders; achenes rough-sided, amply 3-winged.

VAR. β. **dentata**; radical and cauline leaves sharply toothed with rounded inter-spaces.

HAB. Zwarteberg and Riv. Zonderende, *Zey.!* 3067, pte. β. Langehoogte, Caledon, *Ecklon!* Babylon'stoornberg, *Zey.!* 3067. (Herb. Sd., D., Hk.)

This must come very near *T. setifera, DC.,* if it be distinct; it seems to differ by the broad inv. scales and the muricated fruit. Rad. lvs. 3–4 in. long, 4–6 l. wide. Stems 10–12 in. high. Part of *Zey.!* 3067 belongs to *T. clandestina,* Less.

21. T. humilis (Turcz! Bull. Moscw. (1851) p. 211; rootstock peren-nial, ligneous, leafy at the crown, simple or divided; subradical leaves many, rosulate, ligulate, narrowed to the base, sessile, obtuse, undulato-sinuate and repand, with a thickened, pale margin, rigidly coriaceous, one-nerved, glabrous or sparsely hispid, especially beneath; stems short, glandular and pilose, subsimple, laxly leafy, their leaves small, linear, subentire; heads cernuous; inv. sc. lanceolate, acute, membr. edged, dorsally scabrous; achenes on two sides wrinkled, on the third roughly cross-ridged, amply winged.

HAB. Cape, *Ecklon,* Comp. 52, 114, 10. (Herb. Sd., D.)

Rootstock 1–1½ in.; its branches semiuncial. Leaves 2 inches long, 2–3 l. wide. Stems 3–4 in. high, their leaves ½–¾ in.

22. T. tomentosa (Less.! Syn. 90); rootstock ligneous, simple or branched, leafy at the crown; leaves subradical (or crowded on a very short·axis), subrosulate, numerous, obovate or obovato-lanceolate, acute or obtuse, tapering below into a long, broad-based, half-clasping petiole, entire or repand, on each side densely woolly; fl. stems scapelike, one-headed, glandularly puberulous and viscidulous, nude or sparsely scaly; inv. sc. oblongo-lanceolate, scabrous, membranous and piloso-ciliate at margin; achenes rough-sided, 3-winged, glandular. *DC.l.c.457. Calendula tomentosa, Th. Cap.* 702.

VAR. β. **cuspidata** ; leaves narrower, more acuminate, less copiously woolly. *Osteospermum cuspidatum, DC.! l. c.* 463.

HAB. Cafferkuels River, *Th.!* Paarlberg and Breede R., *Drege* (fide *DC.*) Klynrivier'sberg and Zwarteberg, *Zey.!* 3068. β. Caledon, *Ecklon!* (Herb. Th., Sd., Hk.)

Branches of the rootstock, 1–2 in. long, closely leafy. Leaves on 1–1½ in. petioles, ½–¾ in. long, equally wide, their toment closely interwoven, white. Scapes 6–12 inches long. Half-ripe achenes are in Hb. Thunb.—A very distinct species, unlike any other.

§ 5. FRUTICOSÆ. (Sp. 23–27.)

23. T. petiolata (DC.! l. c. 457); shrubby, erect, scabrous with hard, raised points; branches terete, glandular and sub-paniculate at the summit; leaves oblong, cuneate at base and tapering into a petiole, sharply and coarsely 2–4 toothed on each side, faintly 3-nerved; the uppermost sessile, entire, lanceolate ; inv. sc. lanceolate, acute, membr.-edged, dorsally scabrid, eq. the disc; rays 12–15 thrice as long as the inv., their tubes very hairy; achenes unknown.

HAB. Little Namaqualand, *Drege!* (Herb. D., Hk., Sd.)

Branches 12 in. long or more. Petiole ½ in.; lamina ¾–1 in. long. Indument minute and rigid, not at all hairy. Rays orange.

24. T. arborescens (Less. in Linn. 6, p. 96); shrubby, erect, branching; stems and branches flexuous, pale, terete, minutely gland-scabrid, becoming glabrous; "lowest lvs. obovate, tapering much at base, opposite;" rameal leaves alternate, oblong or lanceolate, sparsely denticulate or entire, 3-nerved above the narrowed base, on both sides rough with hard points ; pedicels long, spreading ; heads in fr. cernuous; inv. sc. lanceolate, acute, membr. edged; achenes smooth, amply 3-winged. *DC. l. c.* 456. *Calendula muricata, Th.! Cap.* 703. *Calendula rigida, Ait. Kew.* 3. 271.

HAB. Cape, *Thunberg!* Olifants R., *E. & Z.!* (Herb. Th., Sd.)

A slender, straggling, half-ligneous shrublet, 2–3 ft. high. Lower leaves I have not seen. Rameal lvs. 1–1½ in. long, 3–4 l. wide, as often entire as toothed. Achenes 5–6 l. long, wing 2 l. wide.

25. T. spinescens (Harv.); shrubby, divaricately branched, the branches virgate ; the upper twigs and old peduncles spinous, glabrous or scabrous, *glaucous;* leaves linear-spathulate, narrowed to the base, thick, flat nerveless, mucronate, entire or 3-toothed at the point; fl. branches divaricate, one-headed; inv. sc. lanceolate or oblongo-lanceolate, membr.-edged; achenes rough-sided, 3-winged, the inner wing narrower. *Osteosp. spinescens, Th.! Cap.* 713, *non DC. O. tripteroideum, DC. l. c.* 463. *O. tridens, DC. l. c.* 497, *ex. pte.*

VAR. β. **muricata**; stem and leaves sprinkled with rough points. *Zey!* 1015.

HAB. Roggeveld, *Thunberg!* Silverfontein, *Drege!* Kamiesberg, *E. & Z.!* Spring-
bokkeel, *Zey.!* 1025. Cradock, *Zey.!* 1016. Fish R., *Zey.!* 1012, 1020. Albert,
T. Cooper, 663. (Herb. Th., D., Sd., Hk.)
  A very pale, rigid, scrubby bush, glabrous in all parts, perhaps "*viscid*" also?
Leaves ½–¾ in. long, 1–2 l. wide. *Drege's* "*O tridens*" partly (Hb. D., Hk.) belongs
to our *T. leptoloba*.

26. T. pachypteris (Harv.) ; shrubby, robust, divaricately much
branched, the old twigs and pedunc. spinous, glabrous, *viscid* to the
touch ; leaves linear-spathulate, with slightly recurved edges, thick,
narrowed to the base, mucronulate, some entire, some 3-toothed or
lobed; pedicels ending the twigs, short, glandular; inv. sc. uniseriate,
oblong, subacute, equalling the disc; rays 10–12, twice as long as the
inv. ; achenes rough-sided, amply 3-winged, the wings rigid, thickish,
undulate, at length transversely lacerate. *Osteosp. pachypteris, DC.? l.
c. 460, ex descr. T. gariepina, Sond.! in Herb.*

HAB. Sternbergspruit on the Gariep, *Drege*. Nieuwejaarspruit, *Eckl.!* Queenstown
District, on limestone flats, *Mrs. F. W. Barber.* (Herb. D., Sd.)
  A robust, much branched, scrubby bush, " the whole plant very glutinous to the
touch and strongly scented, from 1–2 ft. high." *(M.E.B.)* Leaves ½–¾ in. long,
1–2 l. wide. Heads ending short, lateral twigs, many-fl. Very near *T. spinescens*
in foliage, but differing in habit and not glaucous. One of the folia of "*O. spinescens*"
in Hb. Th., marked No. 3, may belong to this ; though slightly differing from our
specimens.

27. T. leptoloba (Harv.); shrubby, divaricately branched, the old
twigs spinescent, glabrous, *viscid ;* leaves semiterete, midribbed below,
thickish, glandular, either 3-lobed or pinnately 5–7-lobed, the lobes
very blunt, exactly linear, as is the rachis ; pedicels ending the twigs,
glandular; inv. sc. subuniseriate, oblong, subacute, margined; rays
6–8; achenes smooth-sided, with 3 rather narrow, membr. wings. *Ost.
tridens, β. demissum, Drege! ex pte. (in Hb. Hk., D.)*

HAB. Sneeuweberg, *Drege!* Rhinosterkop, Beaufort; and Uitvlught, Gr. Reinet,
Zey.! 1014, 1017, 1019, 1313. (Herb. D., Hk., Sd.)
  This is much more glutinous and glandular than *T. spinescens*, with exactly linear
leaves and leaf-lobes, all very blunt, and smooth-sided fruit, which I have seen
(perhaps not quite mature) in Hb. Hk. Some of *Drege's* distributed specimens of
" *O. tridens, β. demissa* " belong to this ; others to *T. spinescens*.

Doubtful species.

T. limonifolia (DC. l. c. 458); "stem herbaceous, glabrous, branching,
ascending-erect; radical and lower leaves oval, cuneate at base, tapering
into a petiole, quite entire, 3-nerved; cauline sessile, often obtusely
eared at base, linear-lanceolate, acute; heads corymbose, at length
cernuous, gland-scabrid; inv. sc. linear-acuminate, narrow-edged; rays
about 20, twice as long as the inv.; achenes 3-winged, quite smooth."
DC. l. c.

HAB. Near Verleptpraam, *Drege*. (Unknown to me).
  By description this seems to agree well with *T. polycephala*, with the addition of
radical and lower leaves, which in that species are *unknown*. Both were gathered
by *Drege* in the same neighbourhood.

CXXII. OLIGOCARPUS, Less.

Heads few-flowered, monœcious ; *ray-fl.* uniseriate, ligulate, female; *disc-fl.* tubular, 5-toothed, male. *Inv.* uniseriate. *Recept.* naked. *Achenes* of ray sessile, *polymorphous,* terete or trigonous, scabrous or nearly smooth, or transversely ridged and pitted, either wingless, minutely winged, or obviously 3-winged; either beaked or nearly, or quite beakless, the beak sometimes solid, as long as the achene, hornlike ; sometimes short, knoblike ; sometimes inflated, hollow and cuplike ! *DC. Prodr.* 6, *p.* 455.

A small, many-stemmed, hairy and glandular, strong-scented annual. Leaves alternate. Name from ολιγος, *few,* and καρπος, *fruit.* As a genus it is scarcely distinct from *Tripteris.*

O. calendulaceus (Less. Syn. p. 90); *DC. l. c.* 455. Also *O. tripter-oides, DC. l. c. Osteospermum calendulaceum, Linn. f. Suppl.* 386. *Calendula parviflora, Thunb. Cap.* 703.

HAB. Cape, *Thunberg!* Stellenbosch, Uitenhage, Graaf Reynet and Caffirland, *E. & Z.! Zey./* 3070. Zuureberg, *Drege!* Bosjesveld, *Mundt.* Hassaquaskloof, *Zey !* 3069. (Hb. Th., D., Sd., Hk.)

Root annual, simple, somewhat fibrous. Stems many from the crown, at first ascending, then diffuse, procumbent or trailing, much branched and widely spreading, pubescent and glandular, as well as the leaves ; young parts cobwebby. Leaves alternate, 1–2 inches long, 2–5 lines wide, tapering at base or subpetiolate, oblong or lanceolate, irregularly few-toothed, repand or subentire; the upper small, sessile, linear. Pedicels terminal and axillary, one-headed. Heads small, few-flowered. Inv. sc. lanceolate, acute, variably membr.-edged. Achenes of many forms, on the same plant or in the same head ; 1, very much pitted and ridged across, obconical, crowned with an inflated, hollow, cuplike beak ; 2, slightly wrinkled with a similar beak ; 3, very much cross-ridged and beakless or nearly so ; 4, scabrous, but scarcely wrinkled, with an obsolete beak ; 5, scabrous or smooth, terete or 3-cornered, with a long, hornlike, solid beak; 6, three-cornered, the angles minutely winged; 7, cross-ridged and furrowed, with 3 membranous wings, exactly as in *Tripteris!*

CXXIII. OSTEOSPERMUM, Linn.

Heads many-fl., radiate ; *ray-fl.* ligulate, female; *disc-fl.* tubular, 5-toothed, male, with abortive ovule and style. *Invol.* in few rows, the scales free (not concrete). *Recept.* naked, or rarely setigerous. *Achenes* of the ray drupaceous or nutlike, thick, very hard, glabrous, beakless, without pappus. *DC. Prodr.* 6, *p.* 459. Also *Xerothamnus, DC. l. c. vol. 5, p.* 311.

South African shrubs or half-shrubs, rarely sub-herbaceous. Leaves alternate (very rarely opposite), either entire, toothed or pinnatifid. Heads yellow. Achenes oblong or obovate, smooth or rugulose, bluntly or sharply 3-angled, or 3-winged ; sometimes longitudinally many-nerved. Name from οστεον, *bone* and σπερμα, *a seed.*

§ 1. OPPOSITIFOLIA. Stem herbaceous, erect; leaves *opposite* (the uppermost sometimes alternate). (Sp. 1.)

Leaves sessile, cuneate at base, ovate, coarsely toothed,
membranous ... ...   (1) **herbaceum.**

§ 2. MONILIFERA. Shrubs or suffrutices, glabrous, cobwebby or hispid, but not glandular or viscidulous; leaves alternate, mostly *coriaceous,* with a hard, thickened crenato-dentate or spinoso-dentate margin. Achenes smooth, more or less drupaceous. (Sp. 2–6.)

Leaves petiolate:
   Shrub, much branched and ramulous; achenes sub-
     subglobose, smooth ... ...   (2) **moniliferum.**
   Suffrutex, erect, *virgate,* panicled at the summit;
     achenes oblong, 3-ridged and furrowed at the apex   (3) **coriaceum.**
Leaves sessile:
   Leaves coriaceous, lanceolate, cuneate at base,
     ciliate ... ...   (4) **ciliatum.**
   Lvs. linear-lanceolate or subulate, pungent, spinous
     toothed ... ...   (5) **pungens.**
   Lvs. obovate-oblong, sub-membranaceous, ciliate;
     stem and large branches hispid ... ...   (6) **laxum.**

§ 3. HIRSUTA. Shrubs, suffrutices or herbs, densely and *softly hairy*; leaves
alternate, membranous, not thick-edged, entire or sparsely few-toothed. (Sp. 7–9.)

   Suffruticose, diffuse; lvs. cuneate, oblong, mucro-
     nate, hispid, cottony-white beneath, entire or
     toothed; rays 5–6 ... ...   (7) **hispidum.**
   Shrubby, erect; lvs. cuneate-oblong, acuminate,
     entire, on both sides scabrous and hairy, but
     not cottony; rays 12 ... ...   (8) **hirsutum.**
   Sub-herbaceous; lvs. erect, linquæform, silky, 3–5-
     nerved; pedunc. long, glandular; rays about 8   (9) **helichrysoides.**

§ 4. VISCOSA. Shrubs or suffrutices, rarely sub-herbaceous, more or less copiously
viscoso-pubescent and glandular; leaves alternate, mostly toothed or cut, sometimes
entire. (Sp. 10–24.)

Spinescentia: divaricately branched, twigs spinous-tipped:
   Achenes bluntly 3-cornered, minutely glandular...   (10) **spinosum.**
   Achenes sharply 3-angled, glabrous, smooth ...   (11) **trigonospermum.**
   Achenes broadly 3-winged, muricated (some wing-
     less!) ... ...   (12) **pterospermum.**
Inermia: unarmed, not spiny:
   Shrubs or suffrutices, with *ovate or oblong, broad,* toothed or cut leaves:
     A rigid shrub, scabrous and harsh to the touch;
       lvs. clasping, acute, sinuous-toothed, with
       recurved margins, rigid ... ...   (13) **ilicifolium.**
     Suffrutices, viscidulous, but scarcely scabrous; lvs. membranous:
       Leaves sessile, more or less clasping at base:
         Leaves sinuate-toothed, with wide,
          rounded, interspaces; rays 8–9...   (14) **grandidentatum.**
         Lvs. sharply cut or pinnatifid; rays more than 12:
          Inv. sc. biseriate, the outer sc.
           longest and broadest ... ...   (15) **sonchifolium.**
          Inv. sc. 3–4 seriate, outer sc.
           shortest and narrowest ... ...   (16) **calendulaceum.**
       Leaves tapering at base into a petiole, toothed or incised:
         Inv. sc. elliptical, white-edged,
          not longer than disc ... ...   (17) **picridioides.**
         Inv. sc. lance-acuminate, longer
          than disc ... ...   (18) **sonchifolium,** β.
   Suffrutices with *lanceolate or linear, narrow, small,* toothed leaves:
     Diffuse, scabrous; lvs. *squarrose-recurved,* broad
       based and clasping, sharply 3–6-toothed on
       each side; the teeth recurved ... ...   (18) **squarrosum.**
     Erect, glaucous, hispid, gland-viscid above;
       lvs. linear-lanceolate, erecto-patent, sharply
       toothed or inciso-pinnatifid ... ...   (19) **muricatum.**
     Erect, scabrous; lower lvs. linear-spathulate,
       obtuse, bluntly few-toothed; upper subulate,
       entire ... ...   (20) **asperum.**

Erect, scabrous; lvs. *clasping at base,* linear-
acuminate, few-toothed, erect; rays 2-ce as
long as inv. ... ...   (21) **subauritum.**
Erect, glabrescent; lvs. lin.-lanceolate, acum-
inate to each end, 2–3-toothed; rays 4 times
as long as inv.   ... ...   (22) **grandiflorum.**
Herbaceous, erect, branching; lvs. remote, lanceolate, sessile, sub-dentate, puberulous; achenes
minutely winged... ... ...   (23) **plebeium.**
Herbaceous, leafy at crown; stems erect, simple,
one-headed; lower lvs. oblongo-lanceolate, *petio-
late,* remotely denticulate, puberulous   ... ...   (24) **caulescens.**

§ 5. DIVERSIFOLIA. Stems herbaceous, rigid, branching, angled and furrowed;
radical and lowest lvs. linear-lanceolate, tapering much at base, few-toothed, cobwebby beneath; rameal leaves sparse, linear-subulate. (Sp. 25–26.)

Lower lvs. *smooth* above; achenes with smooth
angles ... ...   (25) **bidens.**
Lower lvs. *scabrous* above; achenes with tuber-
cled angles ... ...   (26) **lavandulaceum.**
§ 6. SUBULIFOLIA. Small, branching shrubs or suffrutices, all the leaves *linear-
subulate,* entire, acute, scabrous, glabrous or glandular. (Sp. 27–30.)
Rough with *glandular*-bristles   ... ...   (27) **glandulosum.**
Leaves close-set, keeled, rough with hard points
not glandular   ... ...   (28) **scabrum.**
Leaves close-set, sharply-keeled, smooth, margins
not revolute ... ...   (29) **triquetrum.**
Leaves close-set, with subrevolute margins, the
adult reflexed   ... ...   (30) **subulatum.**

§ 7. POLYGALINA. Small shrubs or suffrutices, with coriaceous, glabrous, quite
entire, margined leaves; the margin smooth or scabro-ciliolate; pedicels mostly gland-
hispid or viscid. (Sp. 31–38.)

Leaves *broad-based,* ovate, oblong, oblongo-lanceolate, strapshaped or linear:
Achenes smooth (neither rigid, furrowed nor sharp-angled):
Leaves acuminate, lower linguæform 4–5
in. long, upper lanceolate; pedicels
subracemose... ... ...   (31) **Burchellii.**
Leaves mucronate, linear-oblong or linear,
not an inch long; branches one-headed   (32) **polygaloides.**
Achenes with 9 blunt ridges, furrowed and
pitted between   ... ...   (33) **imbricatum.**
Achenes subangular, obtuse, ridge-netted across   (34) **retirugum.**
Achenes *sharply* 3-angled, cross-ridged and furrowed between the angles:
Lvs. ovate, oblong, or obovate, not much
longer than broad   ... ...   (35) **corymbosum.**
Lvs. linear-lingulate or linear-oblong, the
the lower ones many times longer than
broad, all *strongly* nerved and margined   (36) **nervatum.**
Leaves lanceolate, tapering at base, subacute; inv.
sc. acuminate   ... ...   (37) **lanceolatum.**
Leaves (*small*) lance-linear, acuminate, mucronate;
inv. sc. oblong, sub-obtuse   ... ...   (38) **microphyllum.**

§ 1. OPPOSITIFOLIA. (Sp. 1.)

1. **O. herbaceum** (Linn. f., Suppl. 385); stem herbaceous, erect, terete,
hispidulous; leaves mostly opposite (the uppermost rameal sometimes
alternate), sessile or subsessile, cuneate attenuate, and either acute or
sub-cordate at base, ovate or rhomboid, from the middle coarsely few-

toothed, acute, membranaceous, hispidulous, penninerved and veiny;
pedunc. nude, terminal, one-headed, gland-scabrid; inv. sc. lanceolate,
acuminate, glabrous; achenes obovate, acute at base, with a blunt dorsal
ridge, smooth. *Thunb.! Cap. p.* 716. *DC. l. c.* 468. *O. Zeyheri, Spr.!*
in Zey. Pl. No. 214. *DC. l. c.* 466.

HAB. Cape, *Thunberg!* Uitenhage, *E. & Z.! Zey.!* 3063., *Drege!* Buffalo R., *Dr.
Pappe!* Natal, *Gueinzius!* 341. (Herb. Th., D., Hk., Sd.)

Tall, slender, with the habit of a *Wedelia.* Branches often opposite, erecto-patent.
Leaves 2–4 inches long, 1–2 in. wide, thin, the younger slightly cobwebbed beneath.
Heads small.

§ 2. MONILIFERA. (Sp. 2–6.)

2. O. moniliferum (Linn. Sp. 1308); shrubby, mostly unarmed,
much-branched, erect or diffuse; leaves petioled, obovate or obovato-
lanceolate, acute or obtuse, coarsely toothed or subentire, coriaceous,
nerved, the young cobwebbed or cottony, the adult glabrous and glossy;
pedunc. corymbose or solitary, scaly, short; inv. glabrous or woolly,
its outer scales narrow, inner *varying* from ovate to lanceolate; achenes
obovate-globose, drupaceous, glabrous and smooth. *Th.! Cap.* 715. *DC.
l. c.* 460. *Dill. Elth. t.* 68, *f.* 79. *Lam. Ill. t.* 714.

VAR. α. verum; lvs. obovate, mostly acute; inner inv. sc. lanceolate, oblongo-
lanceolate or oblong, acute. *O, moniliferum, DC.!*

VAR. β. rotundatum; lvs. broadly obovate, very obtuse; inv, sc. lanceolate. *O.
rotundatum, DC. l. c.* 461.

VAR. γ. pisiferum ; lvs. mostly glabrous, obovate or lanceolate; inner inv. sc.
broadly ovate or oblong. *O. pisiferum, Linn., DC. l. c. O. piliferum, Th.! Cap.* 715.

VAR. δ. lanosum (*DC.*) ; apices spinous; young leaves woolly on both sides. *O.
spineseens, DC. l. c.* 439, *non. Th. Jacq. Schoenbr. t.* 377 (good!)

VAR. ε. angustifolium (*DC.*); diffuse or trailing, more or less cobwebbed; leaves
lanceolate, cuneate at base, toothed or entire.

HAB. Throughout the Colony, frequent near the sea. β. Natal, *Drege, Krauss*
277, *Gerr. & McK.* 383. *J. Sanderson,* 587. ε. Uitenhage, *E. & Z.!* &c. (Herb. Th.,
D., Hk., Sd.)

A large bush, 2–4 ft. high, much branched, very variable in the shape of its
leaves and in the cobwebby or woolly hairs that more or less persistently clothe
the young parts. Leaves 1½–2½ in. long, ½–1½ in. wide, rarely quite entire, the old
ones almost always glossy. Pedunc. 1–2 inches long, axillary and terminal, often
several at the ends of the branches. Rays bright yellow. "The ashes contain much
alkali, and are sometimes used in the manufacture of soap; the colonial name is
Bush-tick Berry, and the berries are eaten by birds and Hottentot children." *Mrs.
Barber,* 479.—I find the inv. scales so variable, in specimens from the same locality,
that I cannot employ them to separate *O. pisiferum* from *moniliferum* of authors.
A fragment from *Drege* (Hb. Sd.) marked "*O. spinescens*" is referable to our var.
δ. of which Jacq. Schoenbr. t. 377, quoted by *DC.*, is an excellent figure. *O. rotundatum, DC.* is only a var. with larger, broader and blunter leaves than usual.

3. O. coriaceum (DC.! l. c. 461); suffruticose, erect, virgate, cobwebbed
when young, at length glabrous and glossy; leaves petioled, coriaceous,
obovate, obovato-lanceolate or lanceolate, tapering much to the base,
more or less sharply toothed beyond the middle or subentire, the uppermost small, narrow, acute, quite entire; pedunc. corymboso-paniculate,
pedicels scaly and at first white-woolly; outer inv. sc. subulate, woolly,
becoming nude, inner broadly oblong or obovate, acute or cuspidate-

acuminate ; achenes oblong, somewhat indented at base, bluntly 3-ridged at the apex, more or less furrowed between the ridges. *O. junceum, Herb. Th.! fol.* 2, 3. *O. paniculatum, Less. MSS. in Hb. Th.*

VAR. β, latifolium; lvs. broadly obovate or obovate-lanceolate ; achenes 4-5 lines long. *O. macrocarpum, DC. l. c.* 461.

HAB. Cape, *Thunberg.* Swellendam, *Eckl.!* Sneeuweberg, *Drege!* Zwarteberg and near the Waterfall, Tulb., *Pappe!* R. Zondereinde, *Zey.!* 3053. Albany, *Mrs. F.W. Barber!* 392. β, Uitenhage, *Eckl.!* Van Staaden's Mts., *Zey.!* 3052. (Herb. Th., D., Hk., Sd.)

Stems 3-4 feet high, subsimple, ending in a panicled inflorescence. Leaves 4-5 inches long, 1-2, or in β. 2½ inches wide, soon glabrous, very thick and rigid, with immersed veins. Invol. at first snow-white, becoming gradually nude. There are good specimens of this in Herb. Thunb.; he confounded it with his *O. junceum. O. macrocarpum, DC.,* is a var. with broader lvs. and somewhat larger fruit than usual.

4. O. ciliatum (Berg. Cap. 332); shrubby, unarmed, diffusely much branched, often trailing, the young parts cobwebbed, older glabrous; branches angular, the angle scabrous or smooth ; leaves sessile, half-clasping, commonly cuneate at base, lanceolate or oblongo-lanceolate, coriaceous, with a cartilaginous, crenato-dentate margin, rigidly ciliate; heads small, ending the subpanicled twigs ; inv. sc. lanceolate, acuminate; achenes cylindrical, nearly smooth. *Th.! Cap.* 715. *DC. l. c.* 465. *O. triquetrum, DC.? l. c. (non Linn. f.)*

VAR. β, **oblongifolia** ; leaves oblong, *broad and subcordate at base. Sieb. Fl.* 251. (Herb. D.)

HAB. Common on road-sides and waste ground near Capetown. *C. Wright,* 339. Genadendahl, *Dr. Pappe!* (Herb. Th., D., Sd. Hk.)

Stems 3-4 feet long, paniculately much branched. Leaves 1-1½ inch long, 4 lines to ¾ inch wide, opaque. β. is scarcely distinguishable from var. β. of *O. laxum,* but has glossy, and thicker leaves.

5. O. pungens (Harv.) ; suffruticose, diffuse or decumbent, unarmed, the young parts cobwebbed, older glabrous; branches angular; leaves linear-lanceolate and linear-subulate, pungent-mucronate, coriaceous, spinous-toothed (and sometimes ciliolate), nerve-keeled beneath, the margins recurved or revolute ; pedicels short, scaly, 1-headed ; inv. sc. ovato-lanceolate, much acuminate; achenes ? *O. triquetrum, DC.? l. c.* 465 *(non Linn.)*

HAB. Caledon, *Dr. Pappe!* (Herb. D.)

Rootstock and lower part of stems distinctly ligneous. Leaves about 1 inch long, ½-1 line wide, more frequently subulate than lin.-lanceolate, close-set, at length glossy and dark-green. Notwithstanding its distinct-looking foliage, this may be merely an extreme form of *O. ciliatum.* Dr. *Pappe's* specimens, here described, are evidently young, but just coming into flower. I venture to refer *DC.'s O. triquetrum,* founded on *Ecklon's* specimens from the Devil's Mt., to this place ; judging by description.

6. O. laxum (DC.! l. c. 465); suffruticose, unarmed, diffusely much-branched, the young parts cobwebbed, older glabrous; branches angular, sparsely hispid, the panicled twigs mostly glabrous; leaves sessile, half-clasping, obovate-oblong, acute, sub-membranaceous with a cartilaginous, crenato-dentate margin, ciliate; heads small, ending the twigs ; inv. sc. lanceolate, acuminate, glabrous; achenes cylindrical, obtuse, smooth, furrowed at base.

VAR. β. **auriculatum** (DC.); leaves obovate, sub-cordate-clasping at base, rigid, hispid, cartilangineo-serrulate. (Herb. D., Hk.)

HAB. Drakenstein, *Drege!* both varieties. (Herb. D., Hk., S.)

With thinner, less cuneate, more hispid, less polished leaves than *O. ciliatum*, with which however var. β. nearly unites it. I retain it with much hesitation.

§ 3. HIRSUTA. (Sp. 7–9.)

7. O. hispidum (Harv.); suffruticose, unarmed, diffusely branched, all parts densely hispid with short spreading hairs, the under-sides of the leaves and the involucres *cobweb-cottony;* branches terete, twigs obsoletely angular; leaves sessile, cuneate at base, obovate-oblong, mucronate-acuminate, membranous, on both sides hispid, *thin-edged*, quite entire or irregularly 2–3-toothed; heads on short, hispid pedicels, few-fl.; rays 5–6; inv. sc. lanceolate, acuminate; achenes?

HAB. Caledon Baths, *Ecklon! Zey.!* 3011. *Pappe!* (Herb. D., Sd., Hk.)

Allied to *O. laxum, O. hirsutum,* and *O. ciliatum,* but can scarcely be united to any of them ; it is much the most hispid ; it wants the cartilaginous and dentate-ciliate leaf edge of *laxum* and *ciliatum,* and has much smaller heads than *O. hirsutum,* with a rougher pubescence. Leaves 1–1½ in. long, ½–¾ in. wide.

8. O. hirsutum (Thunb.! Cap. 715); stem shrubby, terete, densely and softly hairy on the young parts; leaves sessile, cuneate at base, oblong, acuminate-mucronate, sub-opaque, entire, on both sides scabrous and hairy, the older becoming sub-glabrate; heads on terminal and latural, short, hairy pedicels, many-fl.; inv. sc. lanceolate, subacute, densely hairy, numerous, pluriseriate; rays about 12; achenes?

HAB. Cape *Thunberg!* (Herb. Th.)

Old branches nude, cicatriced. Leaves 1½–2 in. long, ¾–1 in. wide. Pedicels 2 in. long. Inner inv. sc. 12 or more. The pubescence is silky-villous, not in the least woolly or cobwebby.

9. O. helichrysoides (DC. l. c. 466); "subherbaceous, erect, simple, corymbose at the summit, stem and leaves clothed with copious, long, silky hairs ; leaves sessile, erect, imbricate, linguæform, quite entire, delicately 3–5-nerved, mucronate, at length glabrate; pedunc. long, gland-hispid ; inv. sc. oval-lanceolate, acute ; rays about 8 ; achenes subterete, transversely ridge-furrowed, 2–3-angled, the angles small, with very narrow wings." *DC. l. c.*

HAB. Omsamculo and Natal, *Drege.* (Unknown to me.)

"1½–2 ft. high ; lower leaves 4 in. long, 8–10 l. wide." *DC.*

§ 4. VISCOSA. (Sp. 10–24.)

10. O. spinosum (Linn. Sp. 1308; non Jacq.); shrubby, all parts viscoso-pubescent; branches and twigs divaricate, ending in hard spines; leaves linear-lanceolate, either entire or coarsely toothed, or pinnatifid, flat, nerveless or faintly one-nerved; fl. branches short, divaricate, one-headed; inv. sc. acuminate, membr. edged ; rays about 8 ; disc-fl. hairy; achenes bluntly 3-cornered, minutely glandular. *Th.! Cap.* 713. *DC.! l. c.* 459. *O. spinescens, Willd.*

HAB. Round Capetown and in the Western Districts, common. **Hassaquay's** Kl., *Zey.!* 3064. (Herb. Th., D., Hk., Sd.)

A strongly and disagreeably scented, thorny bush, much and intricately branched. Leaves ½–1½ in. long, 1–2 l. wide, very variable ; the teeth or lobes, when present, spreading, mucronate. Pedunc. ½–1½ in. long. Fully ripe achenes with convex, unwrinkled sides.

11. O. trigonospermum (DC.! l. c. 460); suffruticose, erect, all parts viscoso-pubescent ; the upper branches and twigs occasionally spinous ; leaves from broadish, half-clasping, toothed bases, subulate, sharply inciso-pinnatifid, one-nerved, the lobes 2–3 on each side, acuminate; inv. sc. dorsally glandular, oval, acuminate, membr. edged ; rays 9–10 ; achenes glabrous, smooth, 3-angled.

HAB. Olifant's R. and Zeederberg, *Drege!* Clanw., *E. & Z.!* (Hb. D., Hk., Sd.) Less robust and rigid than *O. spinosum*, with broad-based, more sharply incised leaves, and a more rigid, less close pubescence; broader inv. scales and more angular fruit.

12. O. pterospermum (E. Mey!); shrubby, divaricately much-branched, spiniferous, more or less viscoso-pubescent and scabrous ; leaves narrow, either subentire, sharply toothed, or inciso-pinnatifid, the lobes on each side 2–6, long or short, or obsolete, acute ; inv. sc. dorsally gland-scabrid, lanceolate, acute ; achenes *dimorphous*, some smoothish and broadly 3-winged, some (in the same head !) wingless, muricated. *DC. l. c.* 460.

VAR. β. **spinosissimum**; much more spiny than usual; leaves mostly minute, subulate entire. *O. pterospermum, litt. c., Drege!* (Hb. Hk., Sd., D.)

HAB. Stellenbosch and the Paarl, *Drege!* Worcester and Tulbagh, *Ecklon!* β. Simonsbay, Stell., *Drege!* Hex River, *Burke!* (Herb. D., Hk., Sd.)

Very like *O. spinosum*, but with different achenes. I find two kinds of achenes (as above described) on *Ecklon's* specimens (Hb. Sd.) Our var. β. in its extreme state looks different, being nearly leafless and very thorny, with very short and small inv. scales ; but *Zeyher's* 1010, also from Hex Riv., is intermediate with the common form.

13. O. ilicifolium (Linn. Sp. *1308*); shrubby, robust, unarmed, much-branched, in all parts scabrous, glandular and viscidulous; leaves half-clasping, crowded, spreading, broadly ovate or oblong, acute, mid-ribbed, faintly 3-nerved at base, sinuous-dentate, with recurved margins ; pedicels lateral and terminal, short, very scabrid ; inv. sc. lanceolate, acuminate, longer than the disc ; achenes subreniform, concave, with a large areole on the inner face, dorsally rounded, longitudinally many-ridged, the lateral ridges cross-wrinkled. *Th.! Cap.* 716. *DC. l. c.* 465. *Burm. Afr. t. 62.*

VAR. β. **effusum**; stems trailing, slender, widely spreading. (Herb. Sd.)

HAB. Common about Table Mountain. β. Zwarteberg, *E. & Z.! Zey.!* 3062. (Herb. Th., D., Hk., Sd.)

2–3 ft. high, bushy, very rough to the touch, clammy and strong-scented. Leaves 1–1½ in. long, ½–¾ in. wide.

14. O. grandidentatum (DC. ! l. c. 465); herbaceous or suffruticose, diffusely branched, leafy, thinly gland-pubescent, subviscidulous ; leaves sessile, cuneate and half-clasping at base, oblong, obtuse, sinuate-toothed, the teeth callous, with rounded interspaces, margins minutely reflexed ; pedicels terminal and axillary, exceeding the leaves, slender, one-headed ;

inv. sc. biseriate, lanceolate, acuminate, membr.-edged, longer than disc, dorsally scabrous , rays 8–9 ; young achenes 3-angled.

VAR. β. **hispidum**; all parts rather coarsely hispid.   *O. Natalense, Sch. B. in Pl. Krauss!* 391.

HAB. Katberg, *Drege!* Ceded Territory, *Ecklon!* Howison's Poort, *H. Hutton!* Grahamstown, *Genl. Bolton! P. McOwan.* 277. β. Natal, *Krauss,* 391. γ. Nonoti R., *W. T. Gerrard!* (Herb. D., Hk., Sd.)

Stems 2–3 ft. long or more, supporting themselves among bushes. Leaves 2–2½ in. long, ½–¾ in. wide, the teeth very shallow, distant, with rounded interspaces. Pedunc. 2–3 in. long. Rays bright yellow above, coppery-red beneath. Ripe achenes not seen ; when half grown they are sharply 3-angled and smoothish.

15. **O.?** **sonchifolium** (DC.! l. c. 465); herbaceous or suffruticose, branching, leafy, all parts glandular-pubescent, viscidulous ; leaves sessile, ear clasping at base, membranous, netted-veined, inciso-pinnatifid; the lobes short, sharply few-toothed, all the teeth taper-pointed ; inv. sc. biseriate, lanceolate-acuminate, longer than the disc, the outer scales *longest and broadest ;* rays 15–20 ; achenes ?

VAR. β. **subpetiolatum** ; leaves cuneate or tapering at base into an imperfect, amplexicaul petiole, either coarsely and sharply toothed, or inciso-pinnatifid as above.

HAB. Betw. Omtata and Omsamwubo, *Drege!* β. Kreili's Country, *H. Bowker,* 337. Keiskamma Hoek, *T. Cooper!* 227. (Herb. D., Hk.)

At length much-branched and somewhat lignescent, the branches well covered with broad leaves. Leaves always stem-clasping, but variable in form and degree of incision ; in α. oblong, deeply cut, in β. more or less, sometimes very much, attenuated at base. Flowers according to *Mr. Cooper* " purple blueish."

16. **O. calendulaceum** (Harv. non L. f.); herbaceous or suffruticose, branching, all parts glandular-pubescent, viscidulous ; leaves sessile, cuneate at base, oblong, coarsely and sharply toothed or inciso-pinnatifid, the teeth taperpointed ; pedunc. terminal and axillary, one-headed; inv. scales imbricated in 3–4 rows, lance-oblong, acute, equalling the disc, the outer scales *shortest and narrowest ;* rays numerous ; achenes rather sharply 3-angled, minutely wrinkled across.

HAB. Natal, *Gueinzius!* (Herb. Sond.)

Very like *O. sonchifolium* in general aspect, but with a very different involucre.

17. **O. picridioides** (DC. l. c. 465); " suffruticose, erect, sparingly branched, glabrescent ; leaves *much attenuated into a cuneate petiole,* somewhat ovate in the limb, coarsely toothed, the teeth callous and acutely mucronate; pedunc. 6–8 *inches long,* nude, 1-headed; invol. sc. elliptical, scarcely shorter than the disc, white-membraned at the edge; rays twice as long as the inv.; achenes unknown." *DC. l. c.*

HAB. Karroo, *Drege.* (Said to resemble *O. grandidentatum.*)

18. **O. squarrosum** (Harv.); suffruticose, scabro-pubescent, glandular, viscidulous ; stem diffusely much branched and ramulous, subdichotomous ; leaves stem-clasping, broad-based, squarrose or recurved, linear-acuminate, on each side 3–6-toothed, the teeth deltoid, *recurved,* taper-pointed, very scabrid ; heads ending short twigs, few-fl. ; invol. scales ovate, acute, margined, dorsally scabrid ; rays about 6 ; achenes oblong, bluntly 3-sided, deeply pitted and ridged longitudinally on the facets.

HAB. Ravine, under the Sneeuweberg, *Dr. Wallich!* (Herb. D.)
Root ligneous. Stems 2 feet or more long, spreading widely, rigid, harsh to the touch, much and intricately branched. Leaves ½-¾ inch long, 1-2 lines wide, all remarkably recurvo-patent; their teeth runcinate. Achenes 2-3 in each head. This seems a very distinct species.

19. O. muricatum (E. M.!); stems suffruticose, erect, much branched, glaucous, striate, glabrous or hispid below, glandular and viscidulous above; leaves sessile, linear or linear-lanceolate, sharply and deeply toothed, glabrous or hispid-scabrous, acute; branches 1-headed; invol. scales acuminate; achenes obovoid, trigonous, acute at base, the inner angle keel-shaped, the two outer cross-furrowed and tubercled.

VAR. *a.* **asperum**; stem and leaves copiously hispid-scabrous. *Zey.!* 1005.

VAR. *β.* **glabratum**; stem and leaves glabrous or nearly so. *Zey.!* 1004.

HAB. Camdebo, *Drege!* Cape, *E. & Z.!* Vaal R. and Rhinoster Kop, *Burke & Zeyher!* Zululand, *Miss Owen!* Mova R., *Gerr. & M'K.* 1061. (Herb, D., Hk., Sd.)
Stems 6-12 inches high, very pale, as well as the foliage. Leaves ½-1½ inch long, 2-3 lines wide, narrowed to the base, 4-8-toothed on each side. Achenes 10-12 in each head, their tubercles and furrows more or less strongly marked. Pubescence very rigid, variable in amount, sometimes very copious.

20. O. asperum (Less.! Syn. 89); suffruticose at base, scabro-pubescent, subglandular and viscidulous; stem erect, leafy, corymbose at summit, terete; leaves sessile (but not ear-clasping at base), the lowermost linear-spathulate, obtuse, coarsely and bluntly few-toothed beyond the middle, the uppermost linear-subulate, acute, quite entire; branches nude at the apex, 1-headed; inv. sc. lanceolate, acuminate, margined, dorsally scabrid; achenes? *DC. l. c.* 468. *Calendula aspera, Th.! Cap.* 703.

HAB. Cape, *Thunberg!* (Herb. Th.)
Described from *Thunberg's* specimen, from which the achenes have disappeared; possibly they may afford an additional character between this sp. and *O. subauritum,* which it much resembles. Lower leaves ½-¾ inch long, 1-1½ line wide; upper ½ line wide.

21. O subauritum (DC.! l. c. 464); half-herbaceous, scabrous, glandular-pubescent and subviscid; stem erect, leafy, corymbose at the summit, striate, subterete; leaves sessile, linear-acuminate, sparingly toothed, scaberulous, the uppermost especially clasping, and somewhat eared at base; branches nude at the apex, 1-headed; inv. scales ovate, acute, margined, dorsally scabrid; rays twice as long as the involucre; achenes acute at base, bluntly 3-angled, thickly scaberulous above and on the dorsal facets.

HAB. Paarl and Drakenstein, *Drege!* Brakfontein, Clanw., *E. & Z.!* (Herb. Sd., D., Hk.)
1-1½ foot high, scarcely lignescent. Leaves 1-1½ inch long, 2-3 lines wide, 2-3-toothed at each side. Achenes seen on *Ecklon's* specimens, Herb. Sond. This has closely the habit of *O. asperum,* but differs in foliage.

22. O. grandiflorum (DC. l. c. 464); "shrubby, erect, glabrescent, sparingly corymbose at the apex; leaves linear-lanceolate, at each end acuminate, on each side with 2-3 recurved teeth, the uppermost linear-subulate, entire; twigs 1-headed; invol. scales lanceolate, acuminate,

membr.-edged, equalling the disc ; rays 10-12, *at least 4 times as long
as the involucre.*" *DC. l. c.* [Achenes unknown.]

HAB. Zeederberg, *Drege.* (Said to be allied to *O. subauritum.*)

23. O. plebeium (DC. l. c. 464) ; "stem herbaceous, erect, branching
slender ; leaves remote, sessile, lanceolate, here and there somewhat
toothed, sparingly and remotely, as well as the stem, puberulous ;
pedicels terminal, solitary, nodding, with the invol. gland-pubescent ;
achenes obovate, obtuse, rather smooth, gibbous dorsally at base, and
minutely winged on the angles." *DC. l. c.*

HAB. Cape, *Drege.* Namaqualand, *Ecklon.*

24. O. caulescens (Harv.) ; herbaceous, minutely glandular-pubescent,
leafy at the crown ; stems several, erect, subsimple, angular, laxly
leafy, one-headed ; radical and lower leaves tapering at base into a
petiole, oblongo-lanceolate, acute or obtuse, remotely denticulate, cauline
sessile, half-clasping, linear, entire or denticulate ; inv. sc. acuminate,
viscoso-puberulous ; achenes obovoid, obscurely 6-angled, thickly scaberulous and subglandular, especially on the dorsal facets.

HAB. Rhinoster River, Betchuana Land, *Zeyher!* 1009. (Herb. Hk. Sd.)
Rootstock perennial. Radical leaves 2½–3 in. long, including the petiole, midribbed, 3–4 l. wide ; cauline 1–1½ in. long, 1–2 l. wide. All parts minutely glandular. Stems 6–10 inches long, **simple,** or branched at base. **Habit nearly that of**
Dimorphotheca caulescens.

§ 5. DIVERSIFOLIA. (Sp. 25–26.)

25. O. bidens (Thunb.! Cap. 714) ; stem herbaceous, angled and
furrowed, much branched ; leaves dimorphous, the lowest lanceolate,
tapering at base, 2–3-toothed at each side, coriaceous, with revolute
margins, smooth, loosely cobweb-cottony, at length glabrate ; upper
and rameal leaves *very sparse,* linear and subulate, erect, entire, glabrous or cottony ; heads terminal, solitary, on *scabrous* twigs ; inv. sc.
oblong, subacute, dorsally hispid or scabrous ; achenes ovate-oblong,
tapering, smooth, 3-tipped. *O. ephedroides, DC.! l. c.* 463. *Senecio stria-
tus, Th.! Cap.* 678.

HAB. Bockland, *Thunberg!* Bushman's Land, *E. & Z.!* (Herb. Th., Sd.)
1–1½ ft. high. Lower leaves crowded near the base of the stem, 2–2½ in. long,
3–4 l. wide, tapering almost into a petiole ; upper l. ½–1 in. long, ¼ l. wide. Apices
of the branches and twigs scabrous.

26. O. lavandulaceum (DC.! l. c. 466) ; stem herbaceous, angled and
furrowed, much-branched, cobweb-woolly and leafy at base, glabrous
and nearly nude above ; lower leaves linear-lanceolate and linear, tapering at base, entire or distantly 2–3-toothed, with revolute margins, *very
scabrous* on the glabrate and upper side, cobweb-cottony beneath ; upper
and rameal leaves very sparse, small or scalelike, subulate, entire ; pedicels gland-scabrid ; inv. sc. lanceolate, acute, membr. edged, dorsally
scabrous ; achenes (fide DC.) "3-angled, tubercled along the angles,
3-tipped."

HAB. Giftberg, and betw. Ezelsbank and Dwars Riv., *Drege!* (Hb. D., Sd., Hk.)

Very similar in habit and foliage to *O. bidens*, but differing in the scabrid upper surfaces of the leaves, and the tubercled achenes.

§ 6. SUBULIFOLIA. (Sp. 27–30.)

27. O. glandulosum (DC. l. c. 463); "suffruticose, erect, branching, scabrous with spreading, short, glandular bristles ; branches filiform, leafy, one-headed ; lower leaves ? ; rameal lvs. sessile, linear-subulate, quite entire ; inv. sc. oblong-lanceolate, acute, membr. edged, dorsally subscabrid, equalling the disc ; rays 10." *DC. l. c.*

HAB. Wapperthal, Zederberg, *Drege*. (Unknown to me.)

28. O. scabrum (Thunb.! Cap. 714); shrubby, corymbosely branched, rigid; leaves close-set, erecto-patent, sessile, linear-subulate, mucronate, keeled, more or less thickly sprinkled with bristles and hard points, the younger ciliate; pedunc. terminal, one-headed, glandularly scaberulous ; inv. sc. lanceolate, acuminate, margined, dorsally scabrous ; rays 10–12 ; achenes bluntly 3-angled, oblique, slightly incurved, with a convex, longitudinally ribbed dorsal region, and a smooth, keeled inner face, the terminal areole bent inwards. *Xerothamnus Ecklonianus, DC. Prodr. 5, p. 311.*

HAB. Cape, *Thunberg!* Amsterdam flat and by the Zwartkops Riv., Uit., *E. & Z.!* (Herb. Th., D., Hk., Sd.)

A small, erect shrub, 1 foot or more high, the young twigs scabrous. Leaves ¾–1 in. long, ½–¾ l. wide, rigid, sometimes very thickly, sometimes thinly covered with hard, sharp points. Pedunc. 2–3 in. long. Rays yellow. Anthers sagittate at base, exactly as in other species. Disc cor. deeply 5-fid, scabrous on the tube, its style simple, truncate, pencilled at the apex only.

29. O. triquetrum (Linn. f. Suppl. 385); suffruticose, suberect or diffuse, branching, glabrous ; branches slender, leafy to the summit; leaves close-set, sessile, linear-subulate, acute, smooth, sharply keeled, the keel decurrent as a ridge on the twig, the margins flattish or incurved ; pedunc. about as long as leaves, one-headed, lateral and terminal ; inv. sc. lanceolate-acuminate, longer than the disc ; achenes cylindrical, quite smooth.

VAR. β. **aciphyllum**; margin and keel of leaves serrulato-ciliolate. *O. aciphyllum, DC. l. c.* 463. (Herb. Sd.)

HAB. Cape, *Thunberg!* Zwarteberg near Caledon's Baths, *Zey.!* 3061. β. at Piquetberg, *Drege*. (Herb. Th., Sd., Hk.)

The young plant suberect ; the older diffuse or trailing. Leaves ¾–1–2½ inches long, about a line wide, rigid, subdecurrent at the edges, and conspicuously so from the keel. Pedunc. 1–2 in. long. Heads few-fl. *Zeyher's* specimens have shorter leaves than *Thunberg's*, but in other respects quite agree. The description in DC. l. c. of "*O. aciphyllum*" well accords with this species ; but a frustule, in Hb. Sd., differs as above indicated. *O. triquetrum*, DC., p. 465, is probably our *O. pungens*.

30. O. subulatum (DC. l. c. 463); "shrubby, erect ; leaves very crowded, sessile, quite entire, linear, subulate-mucronate, with subrevolute margins, the younger erect, loosely cobwebbed, adult glabrous, reflexed; pedicels villous, one-headed, rather longer than the leaves ; inv. sc. linear, acuminate, twice as long as the disc, dorsally hispid ; rays 12–13, longer than inv. ; achenes unknown." *DC. l. c.*

HAB. Betw. Cape L'Agulhas and Potberg, *Drege*. (Unknown to me.)

§ 7. POLYGALINA. (Sp. 31–38.)

31. O. Burchellii (DC.! l. c. 461); stem robust, suffruticose, erect, virgate, leafy, branched at the summit, glabrous below or woolly in the axils ; leaves sessile, coriaceous, glabrous, acuminate, entire, smooth-edged or ciliolate, the lowest linguæform, cauline lanceolate, erect, imbricating, rameal smaller ; branches leafy, corymbose, each ending in several racemoso-corymbose, simple or branched, glandular-scabrid, elongate pedicels ; inv. sc. 7–8, dorsally glandular, ovato-lanceolate, acute ; achenes ellipsoidal, smooth, obsoletely furrowed at base.

HAB. Anteniqualand; *Burchell.* Near George, *Drege!* Swellendam, *E. & Z.!* Voormansbosch, *Zeyh.!* 3056. (Herb. Sd., Hk.)
Stem 2–3 ft. high, with the aspect of *O. coriaceum,* but in character nearer *O. corymbosum.* Lower leaves 4–5 inches long, 1 inch wide, upper gradually smaller, 1–1½ in. long, ⅓ in. wide. Corymb finally widely spreading, very glandular.

32. O. polygaloides (Linn. Mant. 480); stem shrubby, erect, forked or flexuous, branching, glabrous or woolly in the axils ; leaves sessile, linear-oblong or linear, glabrous, mucronate, more or less distinctly nerved, subimbricate, smooth or scabrous at margin ; fl. branches terminal, 1-headed, gland-scabrid ; inv. scales dorsally scabrous ; achenes (fide DC.) "obovate, very obtusely 6-angled, smooth." *Th. ! Cap.* 714. *DC. l. c.* 462.

HAB. Cape, *Thunberg!* Western Districts, *E. & Z. ! Drege !* Caledon, *Pappe !* near Capetown, *W.H.H.* (Herb. Th., D., Hk., Sd.)
I have not seen *fruit* on this ; the character by which chiefly it is known from *O. imbricatum,* var. γ.

33. O. imbricatum (Linn. Mant. 290); stem shrubby, dichotomous or flexuous, erect, imbricated with leaves, glabrous, or the axils woolly; leaves sessile, oblong or oblongo-lanceolate, obscurely nerved, coriaceous, scabro-ciliolate or smooth-edged, mostly with a recurved point ; fl. branches terminal, 1-headed, gland-scabrid ; invol. sc. dorsally glandular, oblongo-lanceolate ; achenes oblong, with 9 thick, blunt ridges, furrowed and pitted between. *Th. ! Cap.* 715. *DC. l. c.* 462.

VAR. β. **longifolium** ; leaves 1–2 inches long, 4–6 lines wide, varying from oblong to lance-lingulate.

VAR. γ. **angustifolium** ; leaves oblong-linear, mucronate, ¾–1 inch long, 2 lines wide. *O. foveolatum, DC.! l. c.*

HAB. Cape, *Thunberg !* About Table Mt., *E. & Z. !* Caledon, *Eckl.;* Hartebeest R., Cal., *Zey.!* 1006. β. Worcester and Caledon, *Ecklon !* near Wynberg, *W.H.H.* γ, Zwell. and Hott. Holl., *Drege !* (Herb., Th., D., Hk., Sd.)
Known from *O. corymbosum* and *O. polygaloides* by its achenes ; *O. foveolatum,* DC., differs merely by its narrower and longer leaves from the normal form, but the leaves are too variable in all of this section to be depended upon.

34. O. retirugum (DC. l. c. 462) ; "shrubby, erect, divaricately branched, glabrous, the axils here and there woolly ; leaves sessile, erect, oblong, mucronate, subrecurved, nerveless, the margin in the younger scabrous; pedicels solitary, 1-headed, long, glabrescent; invol. sc. lanceolate, acuminate ; achenes obovate, subangular, obtuse, transversely ridge-netted." *DC. l. c.*

HAB. Cape, *Drege.* (Unknown to me.)

35. O. corymbosum (Linn. Mant. 290); stem suffruticose, erect, simple or branched, leafy, glabrous or woolly-tomentose; leaves sessile, coriaceous, glabrous, nerved, ovate, oblong, or obovate, acute or mucronate, entire, either smooth or scabro-ciliolate at edge; fl.-branches more or less corymbose, long or short, one-headed, gland-scabrid; inv. sc. mostly dorsally glandular, rarely smooth; achenes oblong, *sharply* 3-angled or with *3 very narrow* wings, the interspaces transversely ridged and furrowed. *Thunb.! Cap.* 715. *DC.! l. c.* 461.

VAR. β. **rotundifolium** (DC.); leaves crowded, broadly ovate, 1½ in. long, 1 in. wide; stem smooth. *O.corymbosum, Th.! Herb. O.dichotomum, E.Mey! DC.l.c.*462.

VAR. γ. **parvifolium**; lvs. crowded, ovate or oblong, ½ in. long, 4–5 l. wide; stem smooth. *Zey.!* 3059.

VAR. δ. **lasiocaulon** (DC.); leaves sparse, obovate or oblong; stem clothed with deciduous, white wool. *Zey.!* 3058, *T. Cooper!* 1504.

HAB. Cape, *Thunberg! E. & Z.! Drege!* β. Hott. Holl., *Zey.!* 3057. β & γ., Uitenhage, *E. & Z.!* (Herb. Th., D., Hk. Sd.)
Very variable in habit and in the size and form of the leaves. Var. γ, which has achenes similar to the others, is in foliage very near *O. imbricatum.*

36. O. nervatum (DC.! l. c. 462); rootstock ligneous, woolly at the crown; stem suffruticose, erect or ascending, decrescently leafy upwards, branched at the summit; lower and subradical leaves lineari-lingulate, narrowed toward the base, upper sessile, oblong-linear, all conspicuously mid-nerved and margined, glabrous, coriaceous, mucronate; fl.-branches laxly panicled or sub-corymbose, gland-scabrid; inv. sc. oblong, nearly smooth; achenes sharply 3-angled, between the angles ridged and furrowed.

HAB. Albany, Omtata and Omsamwubo, *Drege!* Cape, *Ecklon!* Brit. Caffraria, *T. Cooper*, 175. Umzinto, Natal, *M. J. M'Ken.! Gerr. & M'K.!* 322. (Herb. D., Hk., Sd.)
Lowest leaves 5–6 in. long, 3–4 l. wide; upper 1½–2 in. long; uppermost ½–1 in.; all somewhat glaucous, with a very pale or whitish nerve and smooth margin.

37. O. lanceolatum (DC. l. c. 463); " suffruticose, erect, corymbose at the apex, the young parts clothed with lax, cobwebby hairs, the older nude; leaves lanceolate, tapering at base, sessile, subacute, quite entire, one-nerved; branches terete, one headed, nearly nude at the apex; inv. sc. linear lanceolate, acuminate, immarginate, equalling the disc. *DC. l. c.*

HAB. Natal, *Drege.* (Unknown to me.)

38. O.? microphyllum (DC.: l. c. 463); very dwarf, robust, shrubby, ramulous, glabrous, the old twigs subspinescent, axils somewhat woolly; leaves rigid, lance-linear, acuminate, mucronate, flat, 1-nerved, quite entire (or ciliolate), thick-edged; ped. scarcely longer than the leaves, glabrous; inv. sc. oblong, membr.-edged, subobtuse; achenes unknown.

HAB. On the Karroo, *Ecklon!* Zwartbulletje, *Drege.* (Herb. Sond., D.)
Stem thick and woody, with rough bark, 3–4 inches high, irregularly divided and twiggy. Leaves 4–6 lines long, ½–¾ l. wide, pale. Mature fl. heads not seen.

Doubtful and imperfectly known species.

O.? thymelæoides (DC.l.c.462); "shrubby, erect, branched; branches

subterete, the younger angular; leaves sessile, erect, oblong-linear, tapering at base, mucronate, quite entire, 1-nerved, on each side (as well
as the branches) silky with short, close-pressed hairs; branches nude
at the apex, 1-headed; invol. sc. linear-lanceolate, acute; *fl. and fruit
unknown.*" *DC. l. c.*

HAB. Nieuweveldsbergen, near Beaufort, *Drege.* (Herb. Sd.)

O. **subcanescens** (DC. l. c. 464); "suffruticose, erect, branched; twigs,
pedicels, and involucres, cobweb-woolly; leaves lanceolate, tapering at
base, subpetiolate, acute, here and there subdentate or entire, 1-nerved,
glabrous; heads at the ends of the branches solitary, subcernuous; inv.
scales ovate, membr.-edged; achenes obovate-oblong, smooth, scarcely
angular at base." *DC. l. c.*

HAB. Karroo, *Drege.*

O. **scabridum** (DC. l. c. 460); "suffruticose, unarmed, ascending;
branches angular below, glabrous, leafy, produced into long, nude, 1-
headed, subscabrid peduncles; leaves sessile, linear, here and there
sharply toothed, glabrous; invol. scales acuminate, dorsally scabrous,
membr.-edged, subdenticulate; achenes?" *DC. l. c. (excl. syn. Th.)*

HAB. Genadendahl, *Ecklon.* (Unknown to me.)

(Species removed to other Genera.)

O. connatum, DC.	= Tripteris amplexicaulis.
O. cuspidatum, DC.	= Tripteris tomentosa.
O. fallax, Spr.	= Gamolepis munita.
O. glabratum, Less.	= Tripteris glabrata.
O. heterophyllum, DC.	= Gamolepis brachypoda.
O. incanum, Th.	= Tripteris incana.
O. leiocarpum, DC.	= Gamolepedis sp.?
O. pachypteris	= Tripteris pachypteris.
O. pinnatipartitum	= Gamolepedia, sp.?
O. scariosum, DC.	= Tripteris flexuosa.
O. tenuilobum, ex pte.	= Gamolepis trifurcata.
O. tenuilobum, ex pte.	= Euryops punctatus.
O. tripteroideum, DC.	= Tripteris spinescens.
O. tridens, DC. ex pte.	= Tripteris spinescens.
O. tridens, DC. ex pte.	= Tripteris leptoloba.

CXXIV.? XENISMIA, DC.

"*Heads* monœcious; ray-fl. 5–7, female, ligulate; *disc-fl.* 10–12,
tubular, 5-toothed, male. *Invol.* scales oblong, in a single row. *Recept.*
without paleæ. *Rays* obtuse, ciliate at base. *Anth.* . . . *Style* . . .
Achenes of disc none; of ray thick, glabrous, without pappus, every
where bristled with thick, rigid thorns. *Seed* thickish, oblong, tapering at base." *DC. Prodr. 5, p. 509.*

A somewhat downy, annual, many-stemmed, herbaceous plant. Root slender.
Leaves alternate, oblong-cuneate, toothed at the point, tapering at base. Heads
small, solitary at the ends of the branches, pedicellate. Name, from ξενισμος, a

foreigner or *outlier*, because this plant is the only plant of the group *Millerieæ*, in which *DC.* arranges it, which is found in Africa, the rest being American.

1. X. acanthosperma (DC. l. c.)

HAB. Kaus mountains, near Goedemanskraal, Rustbank, and Kookfontein, 3000–4000 f., *Drege*.

We are unacquainted with this plant. By the description it seems allied either to *Dimorphotheca* or to *Oligocarpus.*

Tribe 5. CYNAREÆ. *Flower-heads* either homogamous, or heterogamous, the *ray-ft.* in 1 or more rows, female or neuter; sometimes diœcious. *Involucre* commonly of dry and rigid, or of prickly-toothed scales. *Recept.* often honeycombed, and fimbrilliferous. *Style* suddenly thickened toward the apex, and often hispid at the thickening; its branches convex, either partially cohering or separate, minutely downy on the outer surface, or glabrous. *Habit* various. (Gen. 125–146.)

Sub-tribe 1. ARCTOTIDEÆ. *Heads* commonly radiate, rarely homogamous and discoid; *rays* uniseriate, female or neuter. *Anthers* minutely tailed. *Achenes* beakless, turbinate, with a terminal, flattened disc, often villous, either with or without pappus. *Pappus*, when present, of broad or narrow, membranous scales (except in *Heterolepis*, where it consists of barbed bristles). (Gen. 125–140.)

1. *Arctoteæ : Inv. sc.* unarmed, separate, the outer herbaceous, inner scarious or membrane-edged, obtuse. *Heads* radiate.

Ray-flowers *female*, producing *achenes :*
   Achenes with two collateral *cavities* at back :
      Pappus of 8 or more membranous scales ... (125) **Arctotis.**
      Pappus *none*, or of few *minute* squamules ... (126) **Venidium.**
   Achenes *solid* (no dorsal cavities) :
      Pappus of several *very delicate*, narrow scales :
         Filaments of stamens *scabrous* ... ... (127) **Haplocarpha.**
         Filaments of stamens *smooth*... ... ... (128) **Landtia.**
      Pappus of 15–20, *rigid*, barbed bristles ... (132) **Heterolepis.**
Ray-flowers *neuter*, never having achenes :
   Pappus none ; achenes 4-sided ... ... (129) **Arctotheca.**
   Pappus of several scales, hidden among the long,
      silky hairs that closely cover the achene... ... (130) **Cryptostemma.**
   Pappus crown-like, minute, crenate : achenes
      tomentose ... ... (131) **Microstephium.**

2. *Gorterieæ : Inv. sc.*, at least the outer and medial, pungent, and mostly spinous at the sides, more or less concrete. *Heads* either radiate or discoid; rays always neuter. (Gen. 133–140.)

Inv. sc. concrete into an urceolate cup, toothed or lobed round the apex :
   Achenes subglabrous ; pappus short, crown-like (133) **Gorteria.**
   Achenes very villous ; pappus of many delicate,
      toothed scales ... ... (134) **Gazania.**
Inv. sc. in two or many rows, concrete *at base only :*
   Pappus none; *achenes* glabr., hidden in pits of the
      receptacle ... ... (135) **Cullumia.**
   Pappus crown-like, crenate ; *achenes* glabrous ... (137) **Stephanocoma.**
   Pappus crown-like, splitting into bristles; *achenes*
      woolly ... ... (136) **Hirpicium.**
   Pappus of many flat, separate scales :
      Inv. sc. in *two* rows; pappus uniseriate ; the
         sc. fimbriate-plumose ... ... (140) **Didelta.**
      Inv. sc. imbricated in many rows :
         Pappus scales *very much acuminate* ... (139) **Berkheya.**
         Pappus sc. obtuse or subacute, *not taper-*
           pointed ... ... (138) **Stobæa.**

Sub-tribe 2. MUTISIACEÆ. *Heads* commonly radiate, rarely discoid; rays in one or more rows. *Invol.* imbricate. *Recept.* nude or fimbrilliferous. *Corolla* of disc or ray, or both, *very generally* (but not constantly) bilabiate or irregularly cleft. *Anthers* rigid, in all the Cape genera *long-tailed* at base. *Pollen* mostly smooth. (Gen. 141–146.)

    * Shrubs, half-shrubs, or dwarf woody plants (none *herbaceous*) :
       Leaves *very thick and leathery*, entire, penninerved,
        woolly beneath: stem dwarf or tall ... ...   (141) **Oldenburgia.**
       Lvs. membranous, various ; small, branching shrubs or shrublets :—
        Pappus of many shortly plumose bristles :
          Rays female ; shrubs   ... ...   (142) **Printzia.**
          Rays neuter or none ; suffrutices ... ...   (143) **Dicoma.**
        Pappus none ; recept. fimbrillate ... ...   (146) **Arrowsmithia.**
    ** Stemless herbs ; *leaves* radical, petioled ; *scapes* 1-headed :
       Heads radiate ... ...   (144) **Gerbera.**
       Heads discoid ... ...   (145) **Perdicium.**

Sub-tribe 1. ARCTOTIDEÆ. (Gen. 125–140.)

CXXV. ARCTOTIS, L.

Heads radiate ; *ray-fl.* female, ligulate ; *disc-fl.* 5-toothed, perfect, the innermost often abortive. *Recept.* honey-combed, fimbrilliferous. *Inv.* campanulate, its scales in several rows, free, the outer small, somewhat leafy, the inner longer, obtuse, scarious-membranous. *Filaments* smooth. *Achenes* mostly pubescent, ovate, dorsally 3–5-winged or ridged, the *lateral* wings or ridges inflexed, either entire or toothed, the *medial* straight, narrower. Copious *silky hairs* arise (in most species) from near the base of the achene. *Pappus* in 2 rows, paleaceous, the scales of the inner row mostly 8, spirally twisted before the opening of the flowers, sometimes very small. *DC. Prodr. 6, p.* 484.

    Stemless or caulescent, unarmed herbs, natives of S. Africa. Leaves alternate, petioled, variously incised or subentire. Heads peduncled, solitary. Name from ἄρκτος, a *bear*, and ὄυς, ὠτὸς, an *ear;* pappus scales look like *ears*. The species are very difficult to characterise, and perhaps too many are here retained.

    I. EUARCTOTIS. *Achenes* (silky or rarely glabrous) furnished at base with a tuft of long, straight, silky hairs. (Sp. 1–27.)

A. *Subacaules: Stemless* or nearly so : *pedunc.* scape-like, 1-headed.
    Outer inv. sc. with long, *spreading, slender*, woolly points :
       Rhizome *perennial*, more or less woody :
          Lvs. pinnate-parted, woolly beneath, with many
            pair of *narrow*, linear, toothed lateral lobes ...   (1) **candida.**
          Lvs. oblong, incised, sinuate, or lyrato-pinnatifid :
            Lvs. *green* and scabrous above, white-woolly
              beneath   ... ...   (2) **acaulis.**
            Lvs. green and scabrous on both sides   ...   (3) **campanulata.**
          Lvs. on both sides *tomentose :*
            Achenes silky or pubescent on the surface :
              Rhizome top-shaped, emitting small
              branched fibres :
                Lvs. roughly tomentose ; ped.
                short ... ...   (3) **campanulata.**
                Lvs. *thickly felted* with white
                wool ... ...   (4) **canescens.**
             Rhizome abrupt, emitting many
             long, *thick, simple*, glabrous roots   (5) **diffusa.**

Achenes quite glabrous on the surface ... (6) **leiocarpa.**
Root slender, fibrous, annual or biennial : lvs. lyrato-
pinnatifid, green above, woolly beneath :
    Rays orange-yellow, coppery beneath ; ach.
     silky ... ... (7) **leptorhiza.**
    Rays purple or blue? achenes *thinly pubescent* (8) **Dregei.**
Inv. sc. all *appressed*, without linear points :
    Lvs. lyrate, *on both sides* white-woolly ... ... (9) **adpressa.**
    Lvs. ellipt.-oblong, sinuate, bristly above, woolly beneath (10) **oocephala.**
    Lvs. long-petioled, *broadly ovate*, subcordate at base,
     on both sides thickly woolly ... ... (11) **verbascifolia.**

B. *Caulescentes :* Stems elongate, leafy, simple or branched ; *pedunc.* terminal, long
   or short, 1-headed.

  * *Lvs.* either ovate, obovate, lanceolate, sinuate, lyrate or pinnatifid, *but not linear*.
    Lvs. scabrous or hispid on one or both surfaces.
     Diffuse ; leaves lanceolate or oblong, subpetioled,
      toothed, white beneath ... ... (12) **angustifolia.**
     Erect, herbaceous, very scabrous ; lvs. ear-clasp-
      ing, subpanduriform or sinuate-pinnatifid, green (13) **bellidifolia.**
     Erect, half-shrubby, very scabrous ; leaves *very*
      setose on both sides, oblong or lanceolate, sharply
      toothed, the upper clasping ... ... (14) **glandulosa.**
     Erect, half-shrubby, hispid and scabrous ; leaves
      pinnatifid or deeply incised, with broad or nar-
      row lobes, mostly tomentose beneath ... ... (15) **aspera.**
     Erect, *annual*, hollow-stemmed, pilose ; leaves
      petioled, elliptic-oblong or lanceolate, toothed,
      green beneath, the upper ear-clasping ... ... (16) **fastuosa.**
    Lvs. glabrous or softly tomentose, neither scabrous nor hispid :
     All parts silvery ; lvs. lanceolate, petiolate, sub-
      entire ... ... (22) **lanceolata.**
     All parts nearly glabrous ; lvs. pinnatifid ... (17) **lævis.**
     All parts tomentose ; lvs. pinnati-partite, with
      many narrow, toothed lobes ... ... (18) **revoluta.**
     Lvs. tomentose on one or both sides, oblong or
      obovate, or lyrate-pinnatifid, very variable ... (19) **stæchadifolia.**
     Herbaceous, diffuse ; lvs. *petioled*, green above,
      white beneath, oblong-obovate, sharply toothed;
      inv. nearly glabrous ... ... (20) **petiolata.**
     Suffr., ascending ; leaves petioled, roundish or
      broadly obovate, toothed, woolly on both sides (21) **ænea.**
  ** *Lvs.* narrow, linear or linear-oblong :—
    Outer inv. sc. with linear, *spreading* points :
     Lvs. petioled, lin.-oblong, crenate, 3–5-nerved (23) **elongata.**
     Lvs. lin.-lanceolate and linear, repand, 3-nerved (24) **virgata.**
     Lvs. linear, sessile, obtuse, closely crenate, with
      revolute margins ... ... (25) **pinnatifida.**
    All the inv. scales close-pressed :
     Shrubby ; lvs. linear or spathulate, subentire, flat (26) **argentea.**
     Suffr. ; lvs. linear-*involute*, quite entire ... ... (27) **linearis.**

II. PSEUDARCTOTIS. *Achenes* minutely pubescent or glabrous on the surface, not
having a basal tuft of silky hairs. *Pappus* short. (Sp. 28–30.)

Perennial, caulescent ; leaves oblong, entire or toothed ... (28) **venidioides.**
Annual, nearly stemless ; lvs. petioled, elliptic, sinuate ... (29) **pusilla.**
Annual, with long, weak, branching stems ; leaves ear-
  clasping at base, oblong, obtuse, sub-entire ... ... (30) **flaccida.**

I. Euarctotis. (Sp. 1–27).

1. A. candida (Th.! Cap. 710); herbaceous, short-stemmed, simple or branched; leaves crowded on the short stem or subradical, petioled, pinnate-parted, white-woolly beneath, cobwebby above, at length nude, the lobes in several pairs, linear-oblong, callous-mucronate, few-toothed, with reflexed margins; pedunc. elongate, tomentose; inv. nearly glabrous, the outer sc. produced into a long, recurved, subulate, tomentose point. *Less.! Syn.* 15. *DC. l. c.* 485. *A. glaucophylla, Jacq. Schoenbr. t.* 170?

Hab. Cape, *Thunberg!* Flats, *Drege*, fide DC. (Herb. Thunb.)
Root ligneous, deeply descending. Stem 1–2 inches long, closely leafy; branches from the lowest axils, equalling the stem. Leaves 4–5 in. long; their lobes ½–¾ in. long, 1–2 lines wide. Rays purplish beneath, either creamy or yellow above. I have only seen the single specimen in Hb. Th.

2. A. acaulis (Linn. Sp. ed. 2. p. 1306); stemless or nearly so, with a thick, ligneous rootstock; radical leaves long-petioled, *polymorphous* (either oblong-subsinuate, incised, lobed, or *lyrate*, with few or many lateral lobes), white-woolly and many-nerved beneath, scabrous or gland-hairy above, the margin reflexed; pedunc. scapelike, elongate, tomentose, setose and glandular; inv. nearly glabrous, the outer produced into a long, recurved, subulate, tomentose point. *Less. Syn.!* 16. *Th.! Cap. p.* 708. *A. scapigera, Th.! Cap.* 709. *Bot. Reg. t.* 122. Also, fide Less., *A. tricolor, Jacq. Sch. t.* 159. *A. undulata, Jacq. l. c. t.* 160, and *A. speciosa, Jacq. l. c. t.* 161. *Bot. Mag. t.* 2182.

Hab. Cape, *Thunb.! E. & Z.!* Simon's Bay, *C. Wright*, 287. Tulbagh, *Dr. Pappe!* Uitenhage, *Zey.!* Brackfontein, *Zey.!* Hassaquas Kl.; Zw. Bg. and R. Zondereinde; and Hott. Hollandsberg, *Zey.!* 2998, β; 2999; 3000. (Hb. Th., D., Sd., Hk.)
Rootstock ½–1 in. thick. Lvs. 6–8 in. long, including the petiole; 1–2 in. broad, rather coarse in texture and rough to the touch. Lobes and lobules blunt, callous-mucronate. Rays yellow above, purplish beneath.

3. A. campanulata (DC.! l. c. 415); nearly stemless, radical leaves petioled, on both sides either tomentose-canescent or setoso-pubescent, lyrate, the lower lobes smaller, the terminal large, ovate-oblong, toothed or incised; pedunc. about equalling the leaves, tomentose and hairy; outer inv. scales produced into a linear, tomentose, spreading or recurved point.

Var. *a.* **subtomentosa**, DC.; leaves canescent, lobes blunt.
Var. β. **puberula**, DC.; leaves setoso-pubescent, lobes acute.
Hab. Both vars. in little Namaqualand, *Drege!* (Herb. D., Sd., Hk.)
Very near *A. acaulis*, but the leaves are uniformly coloured; either *both* sides hoary or both green and pubescent; in *A. acaulis*, so far as I know, the underside is always and alone hoary. The dry specimens show no floral difference. I retain the species with much hesitation.

4. A. canescens (DC.! l. c. 485); nearly stemless, with a thickish rhizome; radical leaves crowded, petioled, on both sides white-woolly, lyrato-pinnatipartite, the lateral lobes few or many, oblong, obtuse, sinuate-toothed, the terminal larger, obtuse; pedunc. 1½–3ce as long as leaves, tomentose; outer inv scales produced into a linear, tomentose, spreading point.

HAB. Kamiesberg and Kamdebosberg, *Drege!* Modderfontein, *Rev. H. White-head!* (Herb. D., Hk., Sd.)

Thickly clothed with felted, whitish hairs. Lvs. 3–6 inches long, sometimes with a large lobulate terminal lobe and 2–3 pair of small laterals; sometimes with 5–6 pair of uncial lateral lobes, the terminal not conspicuously larger. The length of the pedunc. also varies considerably. Heads large and showy; rays yellow above, coppery beneath.

5. A. diffusa (Th.! Cap. 707); stemless, rhizome creeping? emitting at each node many thick, simple, glabrous rootlets; radical leaves from a broad, many nerved base, petiolate, on both sides white-woolly, oblong-obovate, either crenate, incised or imperfectly lyrate; scape longer than the leaves, woolly; outer inv. sc. with linear, woolly appendices. *Less. Syn.* 18. *DC. l. c.* 486.

HAB. Cape, *Thunberg!* (Herb. Th.)

Possibly stoloniferous, forming at the end of each stolon a depressed, imperfect tuber nearly an inch across, which emits from its lower surface 6–12 simple roots, 4–6 inches long, 1 line diameter. Leaves 2–3 inches long, ½–¾ in. wide, thickly felted with white woolliness. The flowers have perished in *Thunberg's* specimens; no other collector seems to have met with this species.

6. A. leiocarpa (Harv.); nearly stemless, subradical leaves petioled, on both sides thinly pubescent, subtomentose, lyrato-pinnatifid, tapering at base, with oblong blunt lobes; fl. branches bearing 1–2 leaves, not much longer than the radical leaves; outer inv. sc. hairy, with short linear points, inner oblong, membranous, glabrous; achenes *quite glabrous* on the surface, with a tuft of basal, silky hairs. *E. & Z. No.* 437, *DC.*, 88. *(in Hb. Sd.)*

HAB. Beaufort, *E. & Z.* Gamka R., *Burke & Zeyher!* Springbokkeel, *Zey!* Hopetown Distr., *A. Wyley!* (Herb. Sond., Hk., D.)

The ripe achenes are black, and glossy; the silky basal tuft rather scanty; and the pappus scales often shorter than the body of the achene. Foliage as in *A. leptorhiza.* Rays white above; coppery purple beneath.

7. A. leptorhiza (DC.! l. c. 486); root slender, fibrous, annual; radical leaves petioled, green and hispid above, white-woolly beneath, oblong-lanceolate, *variably incised* (either repand, coarsely and sharply toothed or more or less deeply lyrato-pinnatifid, or pinnatipartite); scapes long or short, tomentose and hispid; outer inv. scales linear, tomentose, spreading and reflexed; inner glabrous, scarious; achenes *silky*, with a basal tuft of silky hairs, the lateral wings subentire.

VAR. *a.* **breviscapa**, DC.; scape shorter than the pinnatifid leaves. *A. breviscapa, Th.! Cap. p.* 709. *Less. Syn.* 18.

VAR. *β.* **longiscapa**, DC.; scape longer than the toothed or pinnatifid leaves.

HAB. About Capetown and in the Western Districts, *Th.! E. & Z.! W. H. H.* Driefontein, *Zey!* Simon's Bay, *C. Wright,* 292. Paardeneiland, *E. & Z.!* (Herb. Th., D., Hk., Sd.)

Not unlike *A. acaulis* in miniature; but the root is annual and the heads much smaller and fewer-flowered. The foliage varies as in that species. Rays orange-yellow, coppery outside.

8. A. Dregei (Turcz.! Bull. Moscw. XXIV. (1851), p. 93); root slender, fibrous, annual? radical leaves petioled, green and hispid above, white-woolly beneath, oblongo-lanceolate, variably incised, some entire, some lyrate, some pinnatipartite, the lobes oblong, entire ; scapes 2–3 times longer than the leaves, tomentose and hispid ; outer inv. sc. hirsute, with dark, spreading points ; inner tomentose, with a wide, glabrous margin ; achenes *thinly pubescent*, with a basal tuft of *short*, silky hairs, the lateral wings strongly toothed. *Walp. Ann. vol. 5. p. 351. Dietr. Fl. Univ., Comp. t. 6. Arctotheca grandiflora, Drege, fide Turcz.*

HAB. Riv. Zondereinde, *Zeyher!* 3005. (Herb. Sd., Hk.)
Very similar in general aspect and foliage to *A. leptorhiza;* but with a different involucre and achene, and apparently *purple* ray-flowers.

9. A. adpressa (DC. l. c. 485) ; nearly stemless, stem very short, decumbent, dividing at the crown ; leaves petioled, lyrate, toothed, on both sides woolly, the lobes short and blunt ; pedunc. elongate, tomentose and hairy ; outer inv. scales short, ovate, pointless, *appressed*. *A. acaulis, Jacq. Sch. t. 158, fide DC.*

HAB. Zeederberg, *Drege!* Sneeuwekop, *Dr. Wallich!* (Herb. Sond., D.)
Leaves 2–3 in. long, wholly canescent and somewhat rusty, lyrate, with few lateral lobes. Inv. scales pale, wholly pointless! I omit the ' Uitenhage' hab. of *E. & Z.*, quoted by DC., because all the specimens from those collectors which I have seen (in Hb. Sd., D.) belong to *Haplocarpha lyrata.*

10. A. oocephala (DC. l. c. 486); "nearly stemless ; radical leaves petioled, setose or tubercularly very rough above, thickly white-woolly beneath, 3–nerved, elliptic-oblong, obtuse, sinuate-toothed ; pedunc. subtomentose, hairy, equalling or exceeding the leaves; inv. scales *appressed.*" *DC. l. c.*

HAB. Zeederberge, *Drege.* (Unknown to me.)
"Collum and petioles woolly. Leaves 3 in. long, 6–9 l. wide. Achenes striate at back, sparingly villous at base." *DC. l. c.*

11. A. verbascifolia (Harv.) ; nearly stemless, with a thick rhizome ; subradical leaves on long petioles, broadly ovate, obtuse, subcordate at base, repando-crenate, as well as the petiole thickly clothed on both sides with felted, woolly hairs ; pedunc. 1–2ce as long as leaves, woolly ; all the inv. scales ovate, obtuse, *appressed*, without appendix.

HAB. Skurfdeberg, *Zeyher!* (Herb. Hk., Sd.)
Rootstock ½–¾ in. thick. Petioles 2–3 in. long ; lamina 2–2½ in. long, nearly as wide at base, immersedly many nerved. Toment very thick and shaggy, whitish. Achenes densely silky ; rather shorter than the pappus. Well marked by its foliage and involucre.

12. A. angustifolia (Linn. Sp. 1306, non Jacq.); stem herbaceous, diffuse or erect, branching, roughly hispid and glandular ; leaves short-petioled or subsessile, lanceolate or elliptic-oblong, acute, remotely toothed or entire, scabrous or hispid above, white-woolly and 3-nerved beneath, not eared at base ; pedunc. terminal, pilose, short or longish ; inv. cobwebby, the outer sc. with long, subulate, squarrose points. *Less. Syn. 20. Th. Cap. 706. DC. l. c. 486. A. decumbens, Jacq. Schoenbr. t. 381. C. Wright, No. 288, 289. A. decurrens, Jacq. l. c. t. 165. A. Kraussii, Sch. B.? Bot. Zeit. 27. p. 771* (ex. descr. cl. auct.)

VAR. β. **latifolia** ; leaves *ovate*, coarsely toothed, short-petioled.

HAB. Cape flats and in the W. Districts, common. *Zey!* 3007. β. Gt. Howhoek and Kl. Riviersberge, *Zey!* 3007. (Herb. Th., D., Hk., Sd.)

A large, coarse-growing, much-branched, glandular and hairy plant. Branches curved. Leaves 2-2½ in. long, ½-¾ in. wide, with reflexed edges.

13. A. bellidifolia (Berg. Cap. 318, non Th.) ; stem herbaceous, ascending or erect, branching, striate, hispid and scabrous ; branches long, ending in nude peduncles ; leaves hispid and scabrous above, thinly tomentose beneath, especially the younger ones, oblong or subpanduriform, the upper ones eared and stem-clasping at base, toothed, incised or sinuate-pinnatifid ; pedunc. very scabrid ; outer invol. sc. scabrousbristled, with linear-subulate, recurved points. *Less. Syn.* 21. *DC. l. c.* 486. *A. muricata, Th.! Cap.* 707. *A. paniculata, Jacq. Schoenbr. t.* 380.

HAB. Sandy fields, Zwartland, *Th.!* Cape, *E. & Z.!* Tulbagh's Kloof, *Zey.!* 969, Paarlberg, *Drege!* (Herb. Th., Hk., Sd.)

1-2 ft. high, sometimes much branched ; all parts more or less rough with short, glandular bristles. Leaves 2-3 inches long, the lowermost petioled, the rest strongly clasping with toothed ear-lobes at base. Heads rather small. The name "*bellidifolia*" is very inappropriate.

14. A. glandulosa (Thunb. Cap. 706) ; stem suffruticose, branched, furrowed, extremely rough with bristles, hairs, and glands ; leaves on both sides very setose, scabrous and glandular, oblong or lanceolate, the lowermost tapering much at base, subpetiolate, upper sessile, half-clasping, all sharply and unequally toothed, the young parts sometimes cobwebbed ; pedunc. short, very bristly ; outer inv. scales with subulate, scabrous points. *Less. Syn.* 23. *DC. l. c.* 487.

HAB. Piquetberg, *Thunberg!* Witsenberg and Windhoek, *Zey.!* 970. (Herb. Th., Hk., Sd.)

As strong-growing as *A. angustifolia*, exceedingly rough and bristly in all parts. Lower leaves 3-4 in. long, 1 in. wide ; upper 2-3 in. long, ½-¾ in. wide. Young plants thinly cobwebbed ; old quite nude. Heads large and showy, rays purplish ? perhaps white above.—Very near the simple-leaved forms of *A. aspera*.

15. A. aspera (Linn. Sp. 1307) ; stem half-shrubby, branching, hispid and scabrous, sometimes tomentose, leaves commonly tomentose beneath (sometimes nude), hispid and setose above, pinnatifid or incisopinnatifid, the lobes toothed, broad or narrow, the cauline dilated and clasping at base ; outer inv. scales linear-subulate, hispid. *Less. Syn. p.* 24. *DC. l. c.* 487.

VAR. *a.* ? **angustifolia**, *Less. ;* leaves subentire, much attenuate at base and obsoletely eared. *A. angustifolia, Jacq. Schoenbr. t.* 168, *fide Less.*

VAR. β, **incisa** ; leaves inciso-pinnatifid. *A. incisa, Th.! Cap.* 707. *A. melanocycla, Willd. A. auriculata, Jacq. t. c. t.* 169.

VAR. γ, **cichoracea**, Berg. ; leaves pinnatifid, segments short, blunt, spreading. *A. formosa, Th.! Cap.* 708. *A. lyrata, Willd. A. bicolor, Willd. A. caulescens, Th.! Cap.* 708.

VAR. δ, **scabra**, Berg. ; lvs. pinnatifid, the lobes long, spreading, flat and broadish. *A. maculata, Jacq. Schoenbr. t.* 379. *A. arborescens, Willd.*

VAR. ε, **undulata**, Berg. ; leaves pinnatifid, the lobes long, spreading, narrow, toothed, and undulate. *A. undulata, Th.! Cap.* 710, *nec. Jacq. A. cuprea, Jacq. t.* 176. *A. aureola, Edw. Bot. Reg. t.* 32.

HAB. Western districts in many places, frequent. (Herb. Th., D., Hk., Sd.)

Very variable in foliage, but generally known by its rough and glandular surface, the more or less pinnatifid char. of leaves, and their usually whitish, woolly under-side; this last char., however, is very variable, and in var. δ. especially there is sometimes no toment. Sometimes the leaf-lobes are ½-¾ in. wide, sometimes 1 line. In var. ε. they are most divided, and much crisped and curled.

16. A. fastuosa (Jacq.? Schoenbr. t. 166); stem herbaceous, fistular, erect, branched, pilose with jointed hairs; leaves scattered, petioled, elliptic-oblong or oblongo-lanceolate, coarsely toothed or sinuous, pilose on both sides, the medial half-clasping, the uppermost sessile, clasping; pedunc. ending the branches, very pilose; outer inv. sc. with long, subulate, pilose points, inner glabrous and glossy; achenes silky, with a very copious basal tuft. *Less. Syn.* 22. *DC. l. c.* 487.

VAR. β. **spinulosa**, Less.; lvs. undulate, the upper more auricled and clasping; pedunc. short. *A. spinulosa, Jacq. l. c. t.* 167.

HAB. Cape, *Jacquin*. Modderfontein, *Rev. H. Whitehead!* (Herb. D.)
Root said to be annual. Stem 2–3 ft. high, hollow, as thick as a swan's quill. Lower lvs. I have not seen; medial, including the petiole, 3–4 in. long, 1–1½ in. wide. Heads very large and showy; rays numerous, bright orange. I describe from *Mr. Whitehead's* specimen, which, so far as it goes, agrees well with *Jacquin's* figure, above quoted. I give var. β. wholly on *Lessing's* authority.

17. A. lævis (Thunb.! Cap. 708); stem suffruticose, branching, gla-brous or cobwebbed; leaves either glabrous and smooth on both sides, or thinly tomentose beneath, becoming glabrous in age, more or less incised or pinnatifid, the lateral lobes narrow, horizontal, mostly toothed, upper lvs. half-clasping at base; pedunc. terminal, nude or sparsely hispid; outer inv. sc. with glabrous or scabrous, linear, squarrose points. *Less.! Syn.* 22. *DC. l. c.* 487. *A. denudata, Th.! Cap.* 710. *A. gla-brata, Jacq. Schoenbr. t.* 175. *A. grandiflora, Jacq. l. c.* 378. *A. squarrosa, Jacq. l. c.* 177. *fide Less. l. c.*

HAB. Olifant's R., *Thunb.!* Leliefontein, and the Giftberg, *Drege!* (Herb. Th., D., Hk., Sd.)
Pale green, slightly glandular, mostly smooth in all parts; the young parts some-times with deciduous toment. Leaves mostly pinnatifid, the lateral lobes 1–3 in. long, 1–5 l. wide. Pedunc. 4–6 in. long, heads at length nodding.

18. A. revoluta (Jacq. ? Schoenbr. t. 173, fide DC.); stem suffruti-cose, rooting, branches furrow-striate, tomentose, leaves on both sides white-tomentose, pinnatipartite, the lobes numerous, narrow, linear or lanceolate, toothed, with revolute margins; pedicels tomentose (or "nigro-pilose"); outer inv. sc. with long, subulate, tomentose, subsquarrose points. *A. revoluta β. fruticosa, DC.! l. c.* 488.

HAB. Breedriver, *Drege!* (Herb. Sond.)
Of this I have seen but a fragment. The leaves are more like those of *A. candida* than of any of the fruticose species. Lf. lobes ½-¾ in. long, 1–2 l. wide. Rays yellow, fulvous externally. It is very doubtful whether this be Jacquin's plant.

19. A. stæchadifolia (Berg. Cap. 324); stem half-shrubby at base, diffuse or ascending, branches elongate, tomentose; leaves not scabrous, either tomentose on both surfaces or nude on the upper, whitish, taper-ing much at base, either oblong or obovate and toothed, or lyrate, or pinnatifid with broad, blunt lobes, sessile or half-clasping, and somewhat eared, outer inv. sc. linear, squarrose, tomentose. *Less. Syn.* 26. *DC. l. c.* 488.

VAR. α, **grandis**, Less.; lvs. 3–4 in. long, obovate-oblong, toothed, concolourous; fl. branches very long. *A. grandis, Th.! Cap.* 706.

VAR. β, **decumbens**, Less.; lvs. 2–3 in., repand or lyrate, with few lobes; stem decumbent; fl. branches shorter. *A. decumbens, Thunb. ! Cap.* 707. *A. rosea, Jacq. Schoenbr. t.* 162.

VAR. γ, **Bergii**, DC.; leaves lyrate, toothed, concolourous.

VAR. δ, **discolor**; lvs. sinuate or lyrate, green and glabrate above, white beneath. *Zey. ! 2996. A. cuneata, DC. l. c.* 489.

VAR. ε, **rosea**, Less.; leaves sinuate-pinnatifid, lobes short, obtuse, equal. *A. auriculata, DC. l. c.* 487, *non. Jacq. ?*

HAB. Cape, *Thunberg ! α*, Zwartland, *Th.! β*, Cape Flats, *W.ʰ.H.* Caledon R., *Zey.! 963.* Albert, *Cooper,* 666. Saldanha Bay, *E. & Z.!* Kreilsriver, *Zey.! 967.* δ, Uitenhage, *E. & Z.!* Zwartkops R., *Zey.! 2996. ε*, Hassaqua's Kl., *Zey.! 2997.* Kaus and Natvoet, *Drege !* Zwart Kei, *Mrs. F. W. Barber,* 417. (Herb. Th. D., Hk., Sd.)

Leaves very variable, but generally more or less lyrate. The toment. is usually copious, soft and white, rarely deficient; the surface is always smooth. Fl. heads large and showy. *A. cuneata, DC.,* in its narrow-leaved forms looks different, but gradually passes into one or other var.; its leaves are sometimes thinly cobwebbed above. *Zeyher's* 968 (Hb. Sd. Hk.) seems to be an extremely narrow form of var. δ, with almost linear leaves; but the leaves vary to obovate-spathulate on the same specimen (Hb. Sd.)

20. A. petiolata (Thunb. ! Cap. 708); stem herbaceous, diffuse or decumbent, striate, tomentose, leafy; leaves petioled, white-woolly and many-nerved beneath, green, smooth, and at length glabrous above, oblong-obovate, sharply toothed or pinnatifid, the lower ones lyrate, on longer petioles, simple at base, the cauline half-amplexicaul; invol. nearly glabrous, the outer sc. with narrow, linear, spreading points. *Less. Syn.* 19. *DC. l. c.* 486.

HAB. Sandy places, Zwartland, *Thunberg !* (Herb. Th.)
Stem rooting at intervals, afterwards ascending-erect. Petioles of the lower lvs. 1½–2 inches long; lamina 1–1½ inch. The incision of the leaves varies much. Very near *A. stœchadifolia,* var. *discolor,* if distinct.

21. A. ænea (Jacq. f. eclog. 1. t. 52, fide DC.); stem suffruticose, ascending, branched, branches striate, tomentose, ending in nude peduncles; leaves petiolate, half-clasping at base, roundish, elliptical or broadly obovate, sharply and unequally toothed, whitish-woolly on both sides, with a few teeth or small lobules on the petiole; outer inv. sc. with linear, tomentose points. *DC. l. c.* 489.

HAB. Brakfontein, Clanw., *E. & Z.!* (Herb. Sd., D.)
Allied to *A. stœchadifolia,* but with different foliage. Petioles 1½–2 in. long; lamina 1½ in. long and nearly as broad; the teeth almost spinulose. I describe from *E. & Z.'s* specimens, seen by De Candolle; of Jacquin's plant I know nothing.

22. A. lanceolata (Harv.); suffruticose at base only, stems ascending-erect, subsimple, laxly leafy, angular, thinly tomentose, ending in long nude peduncles; subradical leaves petioled, lanceolate or rarely lyrate, tapering much to each end, remotely denticulate, cauline not clasping, lanceolate or linear-lanceolate, subpetiolate, attenuate at base and apex, all thinly silvery; outer inv. sc. with linear, subtomentose points; achenes silky, with ample basal tuft. *E. & Z.* 108–5, 1836.

Hab. Cape, *E. & Z.!* (Herb. Sond.)
Stems probably several, 12–18 in. high, rather slender. Leaves, with the petiole, 4–5 in. long, 4–5 lines wide, some of the radical occasionally lyrate. All parts thinly silvery. Allied to *A. elongata* but much more slender, with smaller heads and petioled leaves tapering much to each end.

23. A. elongata (Th.! Cap. 706); stem herbaceous, branched, silvery, the flowering branches ending in long, nude, one-headed peduncles; leaves simple and tapering at base, petioled, linear-oblong, repando-crenate, with sometimes a pair of small lateral lobes, 3–5-nerved, cob-webby or glabrous above, silvery tomentose beneath; inv. cobwebby, the outer scales with linear, tomentose points. *Less! Syn.* 19. *DC. l. c.* 486. *A. tricolor, Willd.*

Hab. Cape, *Thunberg!* (Herb. Th., Sd.)
Leaves 3 in. long; ½ in. wide, the petiole broad and channelled. Peduunc. 8–12 in. long, whole plant thinly canescent.

24. A. virgata (Jacq.? Schoenbr. t. 307, fide DC.); stem herbaceous, branched, branches thinly silvery, striate, ending in long, nude pedun-cles; leaves linear-lanceolate or linear, distantly repand-toothed or lobu-late, glabrous above, thinly silvery beneath, somewhat 3-nerved; outer inv. scales cobwebby, with spreading, linear points. *DC. l. c.* 489.

Hab. Cape, *Burchell, E. & Z.!* (Herb. Sd.)
I describe from *Ecklon's* specimen, No. 1097, which was seen by *DC.*, but I think the reference to *Jacquin* rather doubtful. Leaves 1–1½ in. long, 1–2 l. wide.

25. A. pinnatifida (Th.! Cap. 705); suffruticose, branching, branches albo-tomentose, virgate, ending in peduncles; leaves tomentose, linear, thickish, obtuse, crenate, with reflexed margins, half-clasping at base; outer inv. sc. with linear, tomentose points. *Less.! Syn.* 27. *DC. l.c.* 489.

Hab. Cape, *Thunberg!* (Herb. Th.)
An erect, rigid suffrutex, hoary in all parts. Leaves ¾–1 in. long, 1–2 l. wide, bluntly toothed, *not pinnatifid*. The sp. name, as *DC.* observes, is very inappropriate, and tends to mislead.

26. A. argentea (Th! Cap. 705); stem shrubby, branching, erect, tomentose; branches long, virgate, ending in long, nude, tomentose peduncles; leaves tomentose, the rameal linear or spathulate, narrowed to the base, subentire; inv. sc. ovate or ovate-oblong, appressed, without excurrent points. *Less.! Syn.* 27. *DC. l. c.* 489. *Centaurea incana, Burm., fide DC.*

Hab. Cape, *Thunberg!* (Herb. Th.)
A slender, tall, rigid plant, albo-tomentose in all parts. Lower leaves not seen; upper 1–1½ in. long, 2 lines wide, not clasping at base. Peduunc. 8–12 in. long, rigid. Inv. sc. cobwebbed, then glabrous.

27. A. linearis (Thunb.! Cap. 705); stem suffruticose, branched, the young parts thinly tomentose, branches ending in nude peduncles; leaves thinly silvery, *linear-involute*, acute, very narrow, quite entire, half-clasping at base; outer inv. sc. taperpointed, appressed, glabrous. *Less. Syn.* 27. *DC. l. c. p.* 489.

HAB. Cape, *Thunberg!* (Herb. Th.)

With the habit of *A. argentea* and *A. pinnatifida*, this is known by its very narrow, *involute* leaves. Lvs. 1-1½ in. long, ½-1 line wide. DC's var. *β.*, denticulata, which I have seen in Hb. Sd., has decidedly spathulate, expanded leaves, and seems to me to be a narrow leaved form of *A. cuneata, DC. (A. stœchadifolia, δ.)*

II. PSEUDARCTOTIS. (Sp. 28-30.)

28. A. venidioides (DC.! l. c. 489); herbaceous, cæspitose-erect, thinly pubescent and glandular, branched from the base; leaves oblong, membranous, entire or toothed, acute, the lowest tapering to the base, the upper half-clasping; outer inv. sc. lanceolate, hairy, taper-pointed, erect, inner glabrous, ovate, obtuse; achenes shortly *pubescent;* pappus very short.

HAB. Olifants R., *Drege!* (Herb. Hk., Sd.)

The specimens seen by me are very imperfect, and I have chiefly drawn my description from *DC. l. c.*

29. A. pusilla (DC. l. c. 489); "root slender, annual; stem short, simple, one-headed, sparingly leafy, striate, puberous; radical leaves petioled, elliptic, sinuate-toothed, somewhat 3-nerved at base, pubescent on both sides; cauline sessile, lanceolate, erect; outer inv. sc. linear, hairy; inner glabrous, obtuse, scarious; achenes *glabrous,* crowned with 6-8 pappus scales." *DC. l. c.*

HAB. Olifant's R., *Drege!* (Unknown to me.)

30. A. flaccida (Jacq. Schoenbr. t. 163); stem herbaceous, ascending, weak, fistular, angle-furrowed, branched, leafy, hispidulous; cauline leaves ear-clasping at base, oblong or ovate-oblong, obtuse, subentire or repand, the lowermost tapering at base, all thinly puberulous or glabrescent, the youngest minutely cobwebbed beneath; uppermost lvs. linear; pedunc. terminal, very short, setose; outer inv. sc. in about 3 rows, setose, with linear, obtuse, erect, thinly canous points; achenes minutely puberulous, with strongly toothed wings, pappus of very short, semicircular scales. *DC. l. c.* 490.

HAB. Cape, *Jacquin.* Namaqualand, *Von Schlicht.* (Herb. Sond.)

Root annual. Radical leaves, fide *Jacquin,* petioled; all the cauline, save the lowest, strongly clasping. Lvs. 2-3 in. long, ½-¾ in. wide, pale green, membranous. Rays according to *Jacquin* white, yellow at base; in our sp. they seem to have been yellow, each with a dark spot. Our plant in other respects so nearly agrees with *Jacquin's* figure, that I can hardly think it different. It seems naturally allied to *Venid. semipapposum,* but has a more evident and complete, though minute, pappus.

(Garden species, unknown to us.)

A. decurrens (Jacq. Sch. t. 165); " stem suffruticose branched ; branches hairy, here and there subtomentose, leaves on both sides hairy, undivided, obovate-oblong, somewhat toothed, the limb decurrent along the half-clasping petiole ; outer inv. sc. with linear, spreading points." *DC. l. c.* 487.

HAB. Cape, *Jacquin.*

Drege's plant (Hb. Sd.) referred to this species by DC. seems to me to belong to *A. angustifolia.*

A. reptans (Jacq. l. c. t. 382); "stem herbaceous, ascending, branched; branches hispid leafy; leaves ear-clasping at base, hispid, canescent beneath, oblong-obovate, sinuate-incised; outer inv. sc. acuminate, spreading." *DC. l. c.* 487.

HAB. Cape, *Jacquin.*

A. cineraria (Jacq. Sch. t. 174); "stem half shrubby below, branching; branches albo-tomentose; leaves long, petioled, not eared at base, pinnatifid, cano-tomentose, the lobes elongate, patent, narrow, obtuse, toothed; outer inv. sc. Squarrose." *DC. l. c. p.* 488.

HAB. Cape, *Jacquin.*
I have (Herb. D.) a garden specimen which agrees well with *Jacquin's* figure, except that the petioles are auricled at base.

A. elatior (Jacq. l. c. t. 172); "stem shrubby, branched; branches tomentose-hairy, leafy, scarcely nude at the summit; leaves pinnatifid, hispid above, tomentose beneath, the lobes linear-lanceolate, angle-toothed; petioles scarcely dilated at base; outer inv. sc. squarrose." *DC. l. c.* 488.

HAB. Cape, *Jacquin.*

A. leucanthemoides (Jacq. l. c. t. 164); "root fibrous; stem herbaceous, hispidulous, branches sparingly leafy; leaves pubescent-hairy, the lower petioled, obovate, repand-toothed, upper lanceolate, sessile, quite entire; outer inv. sc. leafy, obtuse, hairy, squarrose." *DC. l. c.* 490.

HAB. Cape, *Jacquin.*

A. amplexicaulis (Less. Syn. 25); "stem shrubby, branched; flowering-branches elongate, flocculent-tomentose; leaves deeply pinnatifid, ear-clasping at base, glabrate above, smooth, cano-tomentose beneath; outer inv. sc. longish, squarrose." *DC. l. c.* 488. *A. elatior, Herb. Willd. No.* 16711, *fide Less.*

HAB. Cape.
A specimen in Hb. Sond. under this name seems to me not to differ from a var. of *A. aspera*, common on Table Mt. The leaves are neither *smooth* above nor *tomentose* beneath.

CXXVI. **VENIDIUM**, Less.

Heads radiate; *ray-fl.* female, ligulate; *disc-fl.* 5-toothed, perfect. *Recept.* honeycombed, mostly nude. *Inv. sc.* imbricate in several rows, the outer narrower, herbaceous, inner scarious. *Fil.* smooth. *Achenes* glabrous (rarely subpubescent), dorsally 3–5 winged or ridged, the *lateral* ridges inflexed, often toothed, the *medial* straight, narrower. No hairs from the base of the achene. *Pappus* either none, or of 4 *very minute*, unilateral scales. *DC. Prodr.* 6, *p.* 491.

Herbs with the aspect of *Arctotis*, from which genus this differs in the glabrous achenes, destitute of pappus. The section *Pseud-arctotis* is however, nearly intermediate with *Venidium.* Name?

A. Perennial; stems elongate, branching, leafy; leaves *subentire* (not lyrate), *sessile,* more or less *stem-clasping* :

<table>
<tr><td>Outer inv. scales *lanceolate-acuminate,* very long, with spreading points ... ...</td><td>(1) **semipapposum.**</td></tr>
<tr><td>Outer inv. sc. shorter than the inner, erect or suberect :
Lvs. linear or lance-linear, smooth above; achenes 1 line long ... ...</td><td>(2) **angustifolium.**</td></tr>
<tr><td>Lvs. oblong, bristly above; ach. 2 long, very smooth</td><td>(3) **macrospermum.**</td></tr>
</table>

B. Perennial; stems elongate, leafy; leaves *lyrate, lobed* or *pinnatifid, petioled :*

<table>
<tr><td colspan="2">Petioles (or the *upper* ones) conspicuously *eared* and clasping at base :</td></tr>
<tr><td>Stem *pilose* on the striæ; lvs. green, and *scabrous* above ... ...</td><td>(4) **discolor.**</td></tr>
<tr><td colspan="2">Stem and lvs. more or less albo-tomentose, not pilose :</td></tr>
<tr><td>Lvs. long-petioled, the petiole with a *decurrent* ear, lateral lobes small, or none, terminal roundish, sinuate ... ...</td><td>(5) **decurrens.**</td></tr>
<tr><td>Leaves short-petioled, *not* decurrent-eared; lat. lobes oblong, terminal ovate, toothed ...</td><td>(9) **arctotoides.**</td></tr>
<tr><td colspan="2">Petioles not conspicuously ear-clasping at base :</td></tr>
<tr><td>Outer inv. sc. leafy, large, and spathulate ... ...</td><td>(7) **spathuligerum.**</td></tr>
<tr><td colspan="2">Outer inv. sc. small, acute or taper-pointed :</td></tr>
<tr><td>Lvs. long-petioled, ovate or lyrate, tomentose; inv. sc. woolly.
Slender; leaves ovate or cordate, repand, green and hispidulous above ... ...</td><td>(6) **perfoliatum.**</td></tr>
<tr><td>Stem angle-ridged; leaves mostly lyrate, with few lat. lobes, canescent on one or both sides, *smooth* above ... ...</td><td>(5) **decurrens,** β.</td></tr>
<tr><td>Lvs. lyrato-pinnatifid or pinnati-partite, hispidulous on both sides; inv. sc. scabrous ...</td><td>(8) **hispidulum.**</td></tr>
</table>

C. Perennial : stems short and tufted; leaves chiefly radical and subradical, lyrate or pinnati-partite :

<table>
<tr><td>Leaves lyrato-pinnatifid, green above, white beneath ...</td><td>(11) **microcephalum.**</td></tr>
<tr><td>Leaves pinnati-partite, lobes small, very blunt, in many pairs, the upper lobulate, all thinly cobwebby on both sides ... ...</td><td>(10) **erosum.**</td></tr>
</table>

D. Annual, with fibrous roots; stem either long and leafy, or scarcely any; leaves various :

<table>
<tr><td>Nearly stemless; lvs. lyrate, subcanescent ... ...</td><td>(16) **subacaule.**</td></tr>
<tr><td colspan="2">More or less caulescent; stem simple or branched :</td></tr>
<tr><td>Lvs. ovato-lanceolate, subentire, pubescent ... ...</td><td>(17) **fugax.**</td></tr>
<tr><td>Lvs. ovate-sublyrate, tomentose beneath ... ...</td><td>(18) **Kraussii.**</td></tr>
<tr><td colspan="2">Lvs. lyrato-pinnatifid or pinnatifid :</td></tr>
<tr><td>Cobwebby-tomentose; leaves lyrato-pinnatifid :
fl. heads very large and showy ... ...</td><td>(15) **Wylei.**</td></tr>
<tr><td colspan="2">More or less hirsute, with long, jointed, soft hairs :</td></tr>
<tr><td colspan="2">Achenes cross-ridged and tubercled :</td></tr>
<tr><td>Stem branched, *thinly pilose;* heads very large ... ...</td><td>(13) **macrocephalum.**</td></tr>
<tr><td>Stem sub-simple, one-headed; leaves chiefly radical ... ...</td><td>(14) **aureum.**</td></tr>
<tr><td>Achenes smooth, coronate; all parts very hirsute ... ...</td><td>(12) **hirsutum.**</td></tr>
</table>

1. **V. semipapposum** (DC.! l. c. 491); stem herbaceous, erect, branched, furrow-striate, tomentose and more or less piloso-scabrid; cauline leaves oblong or subpanduriform, ear-clasping at base, subentire or sinuous-

toothed, rarely runcinate, scabrous above, albo-tomentose beneath, the uppermost narrower, often linear; outer inv. sc. lanceolate-acuminate, elongate, spreading, green and scabrous above, beneath either pilose or tomentose; *pappus of about 4 minute scales, at one side of the glabrous achene;* sometimes obsolete.

VAR. *a.* **scabrum**; outer inv. scales pilose beneath, scarcely tomentose. *V. scabrum, Less.! Syn. p.* 29. *V. semipapposum, DC. l. c. Arctotis scabra, Th. Cap.* 707.

VAR. *β.* **plantagineum**; outer inv. sc. white-woolly beneath. *V. plantagineum, Less.! Syn. p.* 30. *DC. l. c.* 492. *V. subcalvum, DC.! l. c.* 492. *Arctotis plantaginea, L. Th.! Cap.* 706. *Arct. tomentosa, Th.! in Herb.*

HAB. Var. *a.* Zwartland, *Thunberg!* Caledon and Gnadenthal, *Eckl.!* Drakenstein, *Drege!* *β.* Paarl. *Thunb.!* Draakenstein, *Drege!* Riv. Zon. Einde, *Zey.!* Zwartriver, *Zey.!* 3007. Tulbagh, *Pappe!* Hott. Holl., *Zey.!* (Herb. Th., D., Hk., Sd.)

Stems 1–2 ft. high, leafy throughout. Leaves variable in relative length and breadth; in *Thunbery's* specimens of var. *a.* much narrower than usual. Lvs. commonly 2–3 in. long, ¾–1 in. wide, green above. The toment. varies much, both on the leaves, stem, and inv. scales, and I have no hesitation in uniting the two species of *Lessing.* DC.'s *V. subcalvum* is absolutely identical with *Arctotis plataginea*, Th.! I adopt DC.'s name '*semipapposum*' for the united species, as it expresses the character by which this differs from other *Venidia.*

2. V. angustifolium (DC.! l. c. 492, excl. syn. Jacq.); stem herbaceous, erect, branched, striate, thinly tomentose; leaves linear or lanceolate-linear, remotely toothed or entire, with reflexed edges, more or less ear-clasping at base, the younger cobwebby above, at length glabrate, smooth, or rarely scabrous, all white-woolly beneath; outer inv. sc. taper-pointed, suberect, woolly externally; achenes smooth.

HAB. Dutoitskloof, *Drege!* Brakfontein, *Eckl.!* (Herb. D., Sd., Hk.)

1–2 ft. high, straggling. Lvs. 2–2½ in. long, 2–4 l. wide, mostly clasping at base, green above, white beneath. Like the narrow-leaved forms of *V. semipapposum*, but less scabrous. DC. quotes *Arct. angustifolia, Jacq. Sch. t.* 168, as a synonym, but that figure ill accords with our specimens. I have not seen the achenes. Some of *Drege's* distributed specimens (Hb. Hk.) have *scabrous* leaves!

3. V. macrospermum (DC. l. c. 492); " stem herbaceous, erect, branched, furrowed, tomentose in the furrows, scabrous-bristled along the ridges in the lower part, smooth above; leaves somewhat ear-clasping, oblong, toothed, setulous-bristly above, cano-tomentose beneath, 3-nerved, the lower wider, tapering at base, upper nearly linear, with revolute margins; pedicels tomentose; outer inv. sc. linear, erect, closely pubescent; achenes very smooth, 2 lines long." *DC. l. c.*

HAB. Draakensteinberg, *Drege.* (Unknown to me).

4. V. discolor. (Less. Syn. 31); stem herbaceous, erect, with wide-spreading, diffuse branches, furrowed, tomentose in the furrow, pilose on the ridges; leaves long-petioled, ovate or subrotund, sinuate-toothed or bluntly angle-lobed, scabro-pubescent above, albo-tomentose except on the nerves beneath; petiole amply eared at base, and sometimes 1–2 lobed in the middle; inv. campanulate, the outer scales setose, with reflexed, subulate points; "achenes somewhat cross-ridged." *(DC.)* *DC. l. c.* 492. *Osteosp. perfoliatum, ex. pte., Th. ! Herb. n.* 2.

HAB. Cape, *Thunberg.* Swellendam, *Mundt.! Ecklon!* Voormansbosch, *Zey!* 3008. (Herb. Th., Hk., Sd.)

Root fibrous. Stem 1–1½ ft. high, chiefly branched from near the base, the branches spreading subhorizontally, 1–2 ft. long. Petioles 1½–2½ inches long; lamina 1½ in. long, about as broad, subcordate at base, angle-lobed or repand, green above, white beneath. Heads small, or slender; pedunc. rather longer than the leaves; fl. yellow.

5. V. decurrens (Less.! Syn. 32); stem herbaceous, diffusely branched, tomentose, angular, minutely wing-ridged; leaves long petioled, mostly lyrate, the terminal lobe ovate or subrotund, sinuate-lobed or repand, at first cobwebbed, afterwards nude and punctate above, albo-tomentose beneath; lateral lobes small, sometimes obsolete, or altogether wanting; petiole (except in var. β.) amply eared at base, *the ear decurrent along the stem;* outer inv. sc. very woolly, erect, taper-pointed; achenes cross-ridged. *DC. l. c.* 492; also *V. canescens, DC.! l. c.* 493. *Arctotis micrantha, Th.! Herb.*

VAR. β. **calendulaceum**; petioles not eared at base, or with a very small ear! *V. calendulaceum, Less.! Syn.!* 32. *DC. l. c.* 493.

HAB. Cape, *Thunberg!* Albany, *Eckl.! Drege!* Vanstaadensberg, *Zey.!* 3009, 3010. β. Olifants Hoek, *Zey.!* Kaffraria, *T. Cooper!* 128, Plettenbergsbay, *Pappe!* (Herb. Th., D., Sd., Hk.)

A diffuse, canescent, soft plant, 1–2 ft. long. Petioles 2–2½ inches long. Lateral leaf-lobes often small or wanting; terminal 1½ in. long, and broad. *V. calendulaceum*, Less., differs merely by the want of ear-lobes to the petiole; but in *Dr. Pappe's* specimen I find some leaves with ears, some wanting them, on the same branch! *Lessing* described from garden specimens.

6. V. perfoliatum (Less. Syn. 30); stem herbaceous, slender, diffusely branched from the base, striate, thinly tomentose and setose on the striæ; leaves long-petioled, not eared at base, ovate or subcordate, repand, green and hispidulous above, cano-tomentose beneath; pedicels slender, axillary, setulose, rather longer than the leaves; heads small, few-flowered; inv. sc. sub-biseriate, outer woolly, linear-oblong, erect; achenes cross-wrinkle-striate. *V. cinerarium, DC.! l. c.* 493. *Osteosp. perfoliatum. No.* 1. *Herb. Th.! Cap.* 716.

HAB. Cape, *Thunberg!* Ataquaskloof, *Drege!* (Herb. Th. D., Hk., Sd.)

Stem and branches 6–12 inches long, laxly leafy. Petioles 1–1½ in. long; lamina 1 in. long, ¾ in. wide, thin and membranous. Heads 3 lines across, with very few flowers.

7. V. spathuligerum (DC. l. c. 493); "stem suberect; petioles, pedicels, involucres and the lower surfaces of leaves white-woolly; leaves glabrescent above, their petioles not eared at base, lyrato-pinnatifid, the lobes obtuse, subsinuate; pedicels longer than the leaves; *outer inv. sc. leafy, large, spathulate;* achenes cross-wrinkled." *DC. l. c.*

HAB. Near the Basche R., Caffraria, *Drege!* (Unknown to me.)

8. V. hispidulum (Less. Syn. 34, non DC.); stem diffusely branched, herbaceous, furrow-striate, glabrescent or thinly cobwebbed; leaves petioled, lyrato-pinnatifid or pinnati-partite, the lobes oblong, obtuse, toothed, the upper confluent, on both surfaces thinly and shortly scabro-pubescent, the younger thinly canescent beneath; pedunc. slender, scabrous; outer inv. sc. ovate, acuminate, erect, scabrous; achenes

minutely cross-wrinkled, small. *V. puberulum, DC.! l. c.* 493. *Osteosp.
arctotoides, Th. Herb. fol.* 1, *nec* 2.

HAB. Cape, *Thunberg.* Betw. Los-Tafelberg and Zwartkey, *Drege.* Zwartkops
R., *E. & Z.!* (Herb. Th., D., Sd.)
1–2-ft. high, much branched, weak-stemmed. Leaves 3–5 inches long. Petioles
mostly simple at base, rarely small-eared and clasping. Heads rather small, dark-
lined beneath.

9. V. arctotoides (Less.! Syn. 33); stem diffusely branched, herbaceous, striate, thinly white woolly; leaves petioled, lyrato-pinnatifid,
the lateral lobes few, oblong, obtuse, terminal large, ovate, coarsely
toothed or repand, the upper leaves ear-clasping at base, all thinly cobwebbed and punctate, becoming nude above, more or less albo-tomentose
beneath; pedunc. tomentose; outer inv. sc. ovate-lanceolate, obtusely
acuminate, woolly; recept. honey-combed; achenes finely cross-wrinkled.
DC. l. c. 493. *Osteospermum arctotoides, Linn. f. Th.! Cap.* 717, *Herb.
β.,* 2.

HAB. Cape, *Thunberg!* Uitenhage, *E. & Z.!* Albany, *Mrs. F. W. Barber!* near
Beaufort, *T. Cooper!* 556. (Herb. Th., D., Sd.)
Of this I have seen 6 specimens, two from each of the collectors quoted; of these
Thunberg's are the most woolly, *Cooper's* the least so, and *E. & Z.'s* nearly interme-
diate. The ears at the bases of the petioles are not constant; they are largest and
most frequent on the older plants, when the branches have more fully developed.
Stems 6–12 inches long, branching. Leaves 3–5 in. long, the terminal lobe 1 in.
wide. Heads small. Rays yellow, or pale.

10. V. erosum (Harv.); rootstock woody; stems short, tufted, leafy,
the fl.-branches shortly nude at the summit; subradical and lower leaves
crowded, petioled, pinnatiparted, with a narrow rachis, lobes in several
pairs, very short and blunt, the lowermost subrotund, upper inciso-
lobulate, terminal small, all cobwebby-canescent on both sides, with
recurved margins; cauline simple, sessile, lance-linear or subulate, entire
or the lowermost crenate; pedunc. shorter than the leaves; outer inv.
sc. linear, blunt, cobwebby; achenes not seen.

HAB. Bitterfonteyn, *Zeyher!* (Herb. Sond.)
Only one specimen seen. Leaves 3–4 in. long, the rachis 1 l. wide; lobes in 6–8
pair, the lowest 1–2 l. long, upper 4–6 l. long, mostly 3–4-lobuled. Whole plant
with a whitish aspect, but the hairs are scanty and probably deciduous. Rays
yellow on both sides.

11. V. microcephalum (DC.! l. c. 493); rootstock ligneous; stem
branched from the base, the branches short, tufted, closely leafy; leaves
green or greyish above, pubescent or punctate, beneath albo-tomentose,
crowded, petioled, lyrato-sinuate or pinnatifid, many lobed, the lobes
mostly blunt, subentire, the terminal not much larger, confluent; petiole
not eared at base; pedicels longer or shorter than the leaves, tomentose;
outer inv. sc. woolly; achenes cross-ridged.

HAB. Nieuweveld and Hex R., Kloof, *Drege!* Queenstown, *Cooper,* 425. Buffalo
R., Natal, *Gerr. & M·K.* 1014. (Herb. Sd., D.)
Stem very short, branches 1–2 in. long, or less. Leaves 3–5 in. long, green and
either nearly smooth or roughish above, white beneath, variably incised or sinuate.
Flowers yellow, of small size. *Cooper's* specimen seems quite the same as *Drege's;*—
Gerr. & M·K.'s is slightly different, and may be distinct.

12. V. hirsutum (Harv.); herbaceous, annual, the stem, branches, petioles, peduncles and leaves (both surfaces) hairy with long, jointed, spreading, soft hairs; stems diffuse, weak, hollow, simple or branched, laxly leafy, the branches pedunculoid, one-headed; radical leaves petioled, obovate-oblong, more or less *lyrate* or *sinuate*, the terminal lobe large, toothed or irregularly sinuate, the lateral small, narrow-oblong or toothlike; cauline lvs. ear-clasping, the lowermost petioled, the upper sessile, small, oblong or linear; outer inv. sc. in several rows, oblong, obtuse, pilose, with recurved, short points; achenes smooth, crowned with a minute appressed, 8-crenate pappus. *V. arctotoides, Hort. Hamb.* 1853, *nec Less. Arc. calendulacea, Hb. Th.! No. 6?*

HAB. Near Capetown, *Mundt.!* At Greenpoint, *W. H. H., Dr. Pappe!* (Herb. D., Hk., Sd.)

Nearly stemless when it begins to blossom; the stem afterwards lengthening by successive lateral branches, and becoming 10–12 inches long or more. Radical lvs. numerous, rosulate, 4–5 inches long, 1–1½ in. wide, variably incised. Pubescence very copious, especially on the young parts. Rays pale-yellow, concolourous. A specimen in Hb. Sond., from Hamburg Bot. Gard., called *V. arctotoides*, evidently belongs to this plant; its leaves, &c. are enlarged by cultivation, but the principal characters preserved.

13. V. macrocephalum (DC. l. c. 495); "annual; thinly sprinkled with jointed hairs; stem erect, branched, terete, striate; cauline leaves bluntly eared at base, half-clasping, pinnatifid, the lobes oblong, often sinuate; inv. scales nearly concrete at base into a disc, in many rows, hairy, the outer linear; achenes cross-ridged and tubercled." *DC. l. c.*

HAB. Betw. Kaus and the Gariep, *Drege.* (Unknown to me.)

14. V. aureum (DC.! l. c. 494); stem short, erect, subsimple, one-headed, pilose with long, spreading, jointed hairs, especially in the upper part; radical leaves numerous, lyrato-pinnatifid or pinnati-partite, the lobes ovate, toothed, the uppermost confluent, all thinly cobwebbed, subcanescent beneath, their petioles pilose with long, jointed hairs; cauline leaves sessile, pinnatifid, the uppermost stem-clasping, subentire, linear-acuminate; outer inv. scales linear-subulate, elongate, pilose; achenes cross-ridged and tubercled.

HAB. Betw. Zwartdoorn R. and Groen R., Little Namaqualand, *Drege!* (Hb. Hk.)
Root slender, annual. Stem 4–6 inches high, its leaves depauperated upwards. Rad. leaves 2½–3 inches long, ¾–1 inch wide, pilose on the nerves beneath. Hairs long, soft, spreading, glandular. "Rays 9–10 l. long, yellow, without spot." *DC.*

15. V. Wyleyi (Harv.); root fibrous, annual; stem subsimple or branched from the base, erect, thinly cobwebbed, laxly leafy; radical leaves numerous, petioled, lyrato-pinnatifid, on both sides cobwebby-tomentose, the lobes short, blunt, the terminal confluent; lower cauline leaves ear-clasping, petioled, oblong, subentire or sinuate, upper sessile, linear-oblong, obtuse; outer inv. sc. hirsute, with linear, blunt points; achenes cross-ridged and tubercled.

HAB. Namaqualand, *A. Wyley!* (Herb. D.)
Whole plant hoary, with long, interwoven, white hairs, not gland-hispid. Rad. leaves 3–4 in. long, ½ in. wide, sinuate-pinnatifid, the lobes and sinuses very blunt. Stem in our sp. 6–7 in. high, simple. Heads of large size, very showy; disc 1 inch

across, brownish-purple; rays 30 or more, 1–1¼ inch long, golden-yellow, with rich brown bases. A very handsome species.

16. V. subacaule (DC. l. c. 493); " annual; stem scarcely any; branches scarcely leafy at base, nude above, scapelike, pubescent, one-headed; leaves subradical, petioled, lyrate, subcanescent, the terminal lobe larger, rameal lvs. irregularly pinnate-lobed; outer inv. sc. linear, spreading, inner scarious, very obtuse." *DC. l. c.*

HAB. Betw. Hexriver and Bokkeveld, and near Holriver, *Drege.* (Unknown to me.) A specimen from *Drege,* in Hb. Hook., under this name, in no respect agrees with the above diagnosis; I have named it *V. fugax.*

17. V. fugax (Harv.); root fibrous, annual; stem erect, simple, or branched at base, laxly leafy, hispid, one-headed; subradical and lower leaves petioled, ovato-lanceolate, subacute, 3-nerved, repando-denticulate or subentire, hispidulous on both sides, the youngest slightly cobwebbed beneath; upper cauline leaves oblong, stem-clasping, sessile, entire or nearly so; outer inv. sc. taper-pointed, pilose; achenes smooth (not wrinkled), thinly glandular. *V. subacaule, Drege!* in *Hb. Hk., non DC.*

HAB. Cape, *Drege!* (Herb. Hk., D.) Stem 4–6 inches high, bearing 4–8 alternate leaves. Petioles of lower leaves 1–1½ in. long; lamina 1½ in. long, ¾ in. wide. Upper leaves scarcely uncial. Rays yellow, with dark spots at base.

18. V. (Antrospermum) Kraussii (Sch. Bip.); " root annual; stem slender, furrow-striate, branched, cobwebby-pubescent, leafy to the summit; lower leaves long petioled, ovate, 3-nerved, sub-lyrate, thinly tomentose beneath, rough above; medial amplexicaul, sessile, upper-most cordate-amplexicaul, linear-lanceolate, entire; inv. campanulate, sc. pluriseriate, the outer linear, acute, hispid and cobwebby; rays con-colourous, yellow; achenes pubescent, crowned on the inner side with 5, minute, rounded, close-pressed pappus scales." *Sch. Bip., in Walp. Rep. 6, p. 277, abbrev. V. hispidulum, DC., non Less., fide Sch. B. l. c.*

HAB. Betw. Groenekloof and Saldanha Bay, *Krauss.* (Unknown to me).

(Garden species, unknown to us.)

V. Schraderi (DC. l. c. 494); " stemless, glandularly pilose; leaves petioled, not eared at base, not decurrent, lyrato-pinnatipartite, lobes on each side 4–6, ovate, repando-dentate, obtuse, green, on the nerves beneath and beyond the nerves above hispid-scabrous; pedunc. as long as leaves; achenes very smooth." *DC. l. c.*

HAB. Raised from Cape Seeds, in Bot. Gard. Göttingen. (Unknown to us.)

CXXVII. HAPLOCARPHA, Less.

Heads radiate; *ray-fl.* ligulate, female; *disc-fl.* 5-toothed, perfect. *Recept.* nude, flat. *Inv. sc.* imbricate, pluriseriate, the outer acuminate, the inner scarious. *Filaments* granulated! *Achenes* wingless, turbinate, silky or glabrous with a tuft of silky, basal hairs. *Pappus* uniseriate, of many narrow, tapering, very delicate, diaphanous, nerved scales. *DC. Prodr.* 6, 494.

Stemless perennials, with a woody rhizome. Radical leaves numerous, short-petioled, entire or lyrato-pinnatifid, albo-tomentose beneath. Scapes one-headed, longer than the leaves. Cor. yellow. Name from ἁπαλος, *soft ;* χαρφος, a *scale ;* alluding to the pappus.

Leaves smooth above, lyrato-pinnatifid ; pappus tawny ... (1) **lyrata.**
Leaves scabrous above, oblong or obovate, subentire :
　Papp. sc. tapering to a very slender, bristle-like point :
　　Pedunc. 2-3 times longer than the lvs., hairy ; pappus tawny ... ... (2) **Thunbergii.**
　　Pedunc. many times longer than lvs., woolly ; pappus white ... ... (3) **scaposa.**
　Papp. sc. obtuse or subacute, not taper-pointed ... ... (4) **lanata.**

1. H. lyrata (Harv.) ; leaves smooth, at first cobwebbed, then glabrous above, albo-tomentose beneath, petioled, lyrato-pinnatifid (rarely some entire), the terminal lobe large, ovate, the lateral in 2-3 pair, shortly oblong or deltoid, blunt ; pedunc. 2-3 times as long as the leaves, tomentose ; inv. sc. in several rows, more or less woolly, the outer *taper-pointed*, subsquarrose, inner obtuse ; achenes *silky* and with a silky basal tuft ; pappus-sc. tawny, tapering into a very slender, bristle-shaped, scabrous point. *Arctotis adpressa, Eckl.! Herb.* An *Alloizonium arctotideum, Kze.? Linn.* 17, 572.

HAB. At Adow and near the Zwartkops R., *E. & Z.! Zey.!* 2995. Hassaqua's Kl., *Zey.!* 3003. Albany, *T. Williamson!* (Herb. D., Sd., Hk.)
Very near *H. Thunbergii*, but the leaves in the many specimens seen are lyrate, with scarcely an odd exception, and *smooth* on the upper surface. Lvs. 3-6 in. long, the terminal lobe ½-1½ in. wide. Pedunc. 4-12 in. long. Ray-fl. from the dry plant seem to have been red or purple, perhaps white or yellowish above ?

2. H. Thunbergii (Less.! Syn. 36) ; leaves scabrous above, white woolly beneath, oblong or ovate-oblong, obtuse, tapering to the base, subpetiolate, subentire or crenate ; pedunc. 2-3 times longer than the leaves, *hirsute* ; inv. sc. in few rows, glabrous or thinly pubescent, the outer ovato-lanceolate, acute ; achenes (young only seen) thinly pilose, with a copious tuft of basal, silky hairs ; pappus scales tawny, tapering into a very slender, bristle-shaped point. *Arctotis lanata, Thunb.! Cap.* 708.

HAB. Cape, *Thunberg!* (Herb. Th.)
Closely resembling *H. lanata*, from which its taper-pointed pappus at once distinguishes it. Leaves, with their petioles 2-3 in. long, ¾-1½ in. wide. Pedunc. 3-6 in. long, hairy with jointed, purple hairs, and subtomentose. I have only seen the 3 specimens preserved in Hb. Thunb. The rays are seemingly whitish above, and coppery-purplish beneath.

3. H. scaposa (Harv.) ; leaves scabrous above, white-woolly beneath, oblong or obovate-oblong, obtuse, tapering into a long, cuneate, thick-ribbed base, subentire or repand ; pedunc. *many times* longer than the lvs., striate, thickly-*woolly;* inv. sc. in several rows, more or less woolly, all linear or linear-oblong, *obtuse, especially the outer ones;* achenes quite glabrous, with a copious tuft of basal, silky hairs ; pappus-sc. white, tapering into a very slender, hair-like point. *H. Thunbergii, DC.! l. c.* 494, *non Less.*

HAB. Throughout the Eastern districts, and in Kaffraria and Natal, common. Gauritz R., *Burchell.* Ceded Terr. *E. & Z.!* Katriver, *Drege! H. Hutton!* Wolvekop

and Magalisberg, *Burke & Zey.! Zey.!* 962. Winterberg, *Mrs. F. W. Barber!* 531.
Old Ant Hills, Featherston's Kl. *P. McOwan,* 117. Bigarsberg, *Gerr. & McK.* 1041.
(Herb. D., Hk., Sd.)

Much larger than *H. Thunbergii,* with greatly longer peduncles and a very different
involucre. Leaves 3–10 in. long, 1½–4 in. wide. Pedunc. 12–18 in. long, even where
the leaves are but 3–4 in. Heads 2 in. across, with many long, yellow, concolourous
rays.

4. H. lanata (Less.! Linn. 6, 90); leaves scabrous above, white-woolly
beneath, oblong or obovate, acute or obtuse, tapering to the base, subpetiolate, subentire or crenate ; pedunc. 2–3 times longer than leaves,
hirsute ; outer inv. sc. lanceolate, hairy ; achenes *silky over the whole
surface;* pappus-scales lance-linear, obtuse or *subacute,* but not taper-
pointed. *Less. Syn. p. 36. H. Lessingii, DC. l. c.* 494.

HAB. Platteklip, on Table Mt. Cape, *Bergius, E. & Z.! Zey.!* 966. (Hb. Sd., Hk.)
Rootstock thick and woody. Leaves radical, 2–3 inches long, ¾–1½ in. wide, very
white beneath, thickish. Pedunc. 2–6 inches high. Rays white above, dark purple
beneath. *E. & Z.'s* specimens were confounded by *DC.* with *Arctotis acaulis, var. β*;
besides the generic characters, they are easily known from that species by the invol.

CXXVIII. LANDTIA, Less.

Heads radiate ; *ray-fl.* ligulate, female ; *disc-fl.* 5-toothed, perfect.
Recept. nude, flat. *Inv. sc.* imbricate in few rows, many-nerved, the
outer acuminate, inner scarious, obtuse. *Filaments* smooth ! *Achenes*
wingless, 4-sided, sulcate, hairy at base. *Pappus* uniseriate, of many
delicate, diaphanous, nerveless, obtuse scales. *DC. Prodr.* 6, *p.* 494.

Stemless perennials, with the habit of *Haplocarpha* (from which genus this chiefly
differs by its *smooth* filaments). Scapes *shorter* than the leaves. Name in honour of
Rev. M. Landt, who wrote on the Nat. Hist. of the Faroe Islands.

Upper surface of leaves and nerves beneath *hirsute* ...   ...   (1) **hirsuta.**
Upper surface of leaves and nerves beneath *quite glabrous* ...   (2) **nervosa.**

1. L. hirsuta (Less. Syn. 37); leaves hirsute, with jointed hairs above
white-woolly beneath, and hirsute on the nerves; outer inv. sc. pilose
DC. l. c. 495. Also *Arctotis echinata, DC.! l. c.* 486. *Landtia media,.
Drege!* in *Hb. Hk.*

HAB. Cape, *Thunberg !* Caffirland, *E. & Z.!* (Herb. D., Hk., Sd.)
Leaves radical, tapering at base into a petiole, elliptic-oblong, irregularly toothed,
repand or subruncinate, the teeth mucronate, green and very hairy above ; penninerved, with woolly interspaces beneath, 4–6 inches long, 1½–2½ in. wide. Pedunc.
hirsute, much shorter than leaves. Heads small, few-fl. ; outer inv. sc. lanceolate-
acuminate, hirsute ; inner glabrous, obtuse. Rays yellow.

2. L. nervosa (Less. Syn. 38); leaves quite glabrous and smooth
above, thinly tomentose-canous beneath and smooth on the nerves ;
outer inv. sc. glabrous. *DC. l. c.* 495. Also *L. media, DC.! l. c. Per-
dicium nervosum, Th. Cap.* 689. *Leria nervosa, Spr.*

HAB. Cape, *Thunberg !* Cape, *Mundt !* Katberg, *E. & Z., Drege !* (Hb. Hk., Sd.)
Similar to the last, except in pubescence ; very distinct-looking, but perhaps a
mere local var.

CXXIX. ARCTOTHECA, Wendl.

Heads radiate ; *ray-fl.* ligulate, neuter ; *disc-fl.* 5-toothed, perfect.

Recept. honey-combed, fimbrilliferous.   *Inv.* scales imbricate in many rows, the outer linear, foliaceous, inner larger, scarious, very obtuse. *Filaments* papillose !   *Achenes* ovate, somewhat 4-sided, without wings or *pappus. DC. Prod. 6, p. 495.*

A caulescent, perennial, creeping or decumbent herb. Leaves petioled, lyrato-pinnatifid, green and mostly smooth above, white-woolly beneath. Pedicels axillary, 1-headed. Fl. yellow. Name from αρκτος, a *bear*, and θηκα, a *receptacle?*

1. A. repens (Wendl. Hort. Herrenh. p. 8, t. 6). *R. Br. Hort. Kew. ed.* 2, *vol.* 5, *p.* 141. *Less. Syn. p.* 35. *DC. l. c.* 495. *Arctotis repens, Jacq. Schoenbr. t.* 306. *A. interrupta, Th.! Cap.* 708.

VAR. β, **grandiflora**; achenes puberulous ; leaves and fl. larger. *A. grandiflora, Schrad.! DC. l. c.* 495.

HAB. About Capetown and in the W. districts, common. β, raised from Cape seeds in Bot. Gard., Goett., *Schrader!* (Herb. Th., D., Hk., Sd.)

Stems prostrate, rooting at intervals. Lvs. 4–8 inches long, pinnatisect, the lat. lobes in 6–8 pair, acute, toothed, terminal larger. Pedunc. 3–12 in. long, according to soil ; varies much in size. β, of which I have seen an authentic specimen in Herb. Sond., is merely a garden variety. Without examining the *fruit*, this may readily be confounded with *Cryptostemma.*

CXXX. **CRYPTOSTEMMA**, R. Br.

Heads radiate ; *ray-fl.* ligulate, neuter, often cleft or irregularly cut, or biligulate ; *disc-fl.* 5-toothed, perfect. *Recept.* honeycombed. *Invol.* scales in many rows, imbricated, the outer narrow, herbaceous ; inner membranous, obtuse. *Filaments* scabrous. *Ach.* wingless, very thickly clothed with long, silkly, soft hairs. *Pappus* uniseriate, paleaceous, scarious, hidden among the hairs of the achene. *DC. Prodr. 6, p.* 495.

Stemless or caulescent herbs, more or less tomentose. Leaves very variable in size and incision, lyrato-pinnatifid, runcinate, or rarely undivided. Rays yellow ; disc dark-coloured. Name from κρυπτω, to *hide*, and στεμμα, a *crown;* alluding to the hidden pappus.

Leaves lyrate or runcinate, scabrous above ; lobes toothed   (1) **calendulaceum.**
Lvs. pinnati-partite, glabr. above, lobes linear-lanceolate,
    entire, with revolute margins ... ...   (2) **Forbesianum.**

1. C. calendulaceum (R. Br. Hort. Kew. 2, Vol. 5, p. 141); stemless or caulescent ; leaves scabrous above, hoary beneath, lyrate or runcinate, the lateral lobes variously cut or toothed, margins flat. *Less. Syn.* 39. *DC. l. c.* 495. *Arctotis calendulacea, Willd. Th.! Cap.* 709. *Jacq. Schoenbr. t.* 157.

VAR. a, **verum**; rays entire, or 2–3-toothed. *A. calend. R. Br.! l. c. Bot. Mag. t.* 2252.

VAR. β. **hypochondriacum**; rays deeply 3–5 cleft; lvs. lyrate-tomentose or cob-webbed. *Crypt. hypochondriacum, R. Br. l. c. DC. l. c.* 496. *Arct. hypochondriaca, A. tristis, A. corruscans,* and *A. superba, Linn.*

VAR. γ. **runcinatum**; rays 3–5 parted; lvs. runcinate, tomentose beneath. *Crypt. runcinatum, R. Br. l. c. DC. l. c.* 496.

HAB. Roadsides and waste places throughout the Colony, very common. (Hb. Th., D., Hk., Sd.)

Extremely variable in habit; sometimes stemless, with rosulate radical leaves and scape-like pedunc.; sometimes prostrate or creeping, with exactly the aspect of *Arcto-theca repens;* or erect, branched, 1–2 ft. high, like some *Arctotis.* Leaves as variable in cutting and pubescence ; all degrees of lyrate or pinnatifid, sometimes sinuate or entire ; either green above and white woolled beneath, or equally tomentose on both sides. The most constant character by which to recognize this *Proteus* is the achenes, which when ripe are hidden in their very thick coat of silky wool ; but alas ! I have found (Herb. Hk.!), *in the same head,* perfectly *glabrous* fruits mixed with woolly ones!! But for their pappus, such resemble the achenes of *Arctotheca.*

2. C. Forbesianum (Harv.) ; stemless ; leaves quite glabrous above, snow-white beneath, pinnatipartite (some simple), the lobes linear-lanceolate, quite entire, with revolute margins, some with an accessory lobule at base. *Gazania Forbesiana, DC. l. c.* 508.

HAB. Cape, *Forbes.* Betw. Zwarteberg and R. Zondereinde, *Zey.!* 3016. (Herb. Hk., Sd.)

Crown of root ¾ in. diam., emitting many strong, simple roots. Leaves all radical, on long petioles, 3–5 inches long. Leaflobes ½–¾ in. long, 1 l. wide, sometimes all quite simple, sometimes almost all with an accessory, basal lobule. Scapes about as long as the leaves. Inv. sc. 3–4 seriate, glabrous, the outer ovate-oblong, with a short linear appendage, the inner scarious, very obtuse. Filaments of stamens *scabrous!* A very distinct species, with the foliage of a *Gazania,* but all the generic characters of a *Cryptostemma.*

Doubtful Species.

Gazania subbipinnata (DC. l. c. 508) ; " crown of root woody ; rad. leaves long-petioled, glabrous above, cano-tomentose beneath, pinnatisect, the segments remote, narrow, some linear, quite entire, some sinuate or pinnatipartite, with quite entire margins ; scape not longer than the leaves, tomentose above ; inv. campanulate, with scarcely any tube, the scales free nearly to the base, the outer acute, inner obtuse." *DC. l. c.*

HAB. Knakerberg, at foot of hills, under 1000 f., *Drege.* (Unknown to us.)
By description this seems scarcely different from *C. Forbesianum.*

CXXXI. **MICROSTEPHIUM,** Less.

Heads radiate ; ray-fl. ligulate, neuter ; *disc-fl.* 5-toothed, perfect. *Recept.* slightly honey-combed. *Inv. sc.* pluriseriate, imbricate, unarmed, the inner membrane-edged. *Filaments* scabrous. *Achenes* wingless, tomentose. *Pappus* uniseriate, crown-like, crenate, callous at base, membranous at the apex, after flowering turned inwards. *DC. Prodr.* 6, *p.* 496.

A decumbent or creeping, branched herb, white-woolly in most parts. Leaves long petioled, roundish ovate, repand. Pedunc. 1-headed. Flowers yellow, not showy. Name from μικρος, *small,* and στεφος, a *crown;* the crown-like pappus.

1. M. niveum (Less. Syn. 55); *DC. l. c.* 496. *Osteospermum niveum, Linn. f. Th.! Cap.* 716. *Arctotis populifolia, Berg. Cap.* 323 ?

HAB. Sandy sea shores and places near the sea, from Capetown to Natal. *Thunberg! Mundt.! E. & Z.! W. H. H.,* Natal, *T. W., Gueinzius! Sanderson,* 507. *Gerr. & M'K.,* 352. (Herb. Th., D., Hk., Sd.)

Stem herbaceous, robust, decumbent or prostrate, branching, 1–2 ft. long. All parts clothed with thick, close-lying, very white woolliness, the leaves in age becoming nude above. Leaves on long petioles, ovate, cordate or subrotund, many-nerved, obtuse, entire or repand. Pedunc. often bifid, each branch bearing a head, axillary. Inv. very woolly. Flowers yellow, not very conspicuous.

CXXXII. HETEROLEPIS, Cass.

Heads radiate; *ray-fl.* female, ligulate; *disc-fl.* 5-toothed, perfect. *Recept.* honeycombed, villous in the middle, *Inv. sc.* free, in 2–3 rows, the outer lanceolate-acuminate, dorsally woolly; inner longer, oval, with a membranous, fringed apex. *Ray-flower* with a very minute, cirrhiform inner lobe; the outer ample, 4-toothed. *Filaments* smooth. *Anthers* shortly tailed. *Style* smooth, bifid; more deeply in the ray-fl. *Achenes* very villous, oblong. *Pappus* of 15–20 thick, closely barbed, unequal bristles, in two rows; the 10 inner ones longer. *DC. Prodr.* 6, 496.

Suffrutices; the younger branches woolly. Leaves crowded, sessile, linear, rigid, glabrous above, tomentose beneath, the margins revolute. Heads terminal, solitary. Cor. yellow. Name from ετεϱος, *different*, and λεπις, a *scale;* the inv. is of two sorts of scales.

Leaves with strongly revolute margins:
  Lvs. crowded nearly to the tops of the branches;
    pedunc. short ... ... (1) **decipiens.**
    Lvs. laxly set; pedunc. *very long* (6–10 inches) ... (2) **peduncularis.**
  Leaves quite flat, midnerved, glutinous ... ... (3) **mitis.**

1. H. decipiens (Cass.); branches crowded with leaves nearly to the summit, ending in a short peduncle; leaves linear with revolute margins, subentire or here and there denticulate. *Less. Syn.* 58. *DC. l. c.* 497. *Oedera aliena, Linn. f. Jacq. Schoenbr. t.* 154. *Leyssera arctotoides, Th. !* *Cap.* 691.

HAB. In the Karroo, *Thunb.!* Gordon's Slakte, *Mundt.* Tulbagh's Kl., French Hoek, and in Hott. Holl., *E. Z.!* Winterhoeksberg, *Pappe!* Stellenbosch Mt., *W.H.H.* Gr. Howhoek, *Zey.!* 3017, (977). (Herb. Th., D., Hk., Sd.)
A much branched, densely leafy bush, 1–2-ft. high. Leaves 1–1½ in. long, 1 line wide, acute, the younger canous, older glabrate. Pedunc. scabrous, ½–3 in. long. Heads large and showy; rays many, long, bright golden yellow.

2. H. peduncularis (DC.! l. c. 497); branches leafy below, ending in *very long* peduncles; leaves laxly set, linear, with revolute margins, quite entire.
HAB. Cape, *Burchell,* 6828. Zwellendam, *E. & Z.! Mundt.!* Buffeljagds R., *Zey.!* 3300. Genadendahl, *Dr. Prior!* (Herb. D., Hk., Sd.)
A half shrub, variably tomentose, 1–1½ feet high. Leaves much less closely set than in *H. decipiens,* but otherwise similar. Pedunc. 6–10 in. long, scabrous.

3. H. mitis (DC.! l. c. 497); branches leafy nearly to the summit; leaves linear, *flat,* nerved, viscidulous. *Gorteria mitis, Burm. Prodr. p.* 28, *fide DC. Burm. Afr. t.* 54, *f.* 2.
HAB. Zondags R. & Zuureberg, *Drege.* (Herb. Sond.)
Leaves crowded, quite flat! Heads smaller than in *H. decipiens,* with shorter rays, and less membr. inner inv. sc. Outer pappus very short.

CXXXIII. GORTERIA, Gaertn.

Heads many-fl., heterogamous; *ray-fl.* ligulate, neuter; *disc-fl.* very sharply 5-toothed, some of the marginal ones fertile, having a bifid style and abortive stamens; the central ones sterile, with simple style and perfect stamens. *Inv. sc.* pluriseriate, concrete into an urceolate, at length closed, tube, the apices linear-subulate, free, at length squarrose. *Recept.* slightly honeycombed or nude.   *Fil.* smooth.   *Achenes* obovate-3-angled, narrowed at base, barbed at the apex, otherwise subglabrous. surmounted by a short, crownlike pappus; outer skin of the achene membranous, easily peeling off.   *DC. Prodr. 6, p. 500.*

Hispid annuals; the seed germinating in the closed involucre, which remains like a bulb, through which the fibrous root pierces. Leaves alternate, entire or toothed, scabrous above and bristly, white beneath, one-nerved. Heads solitary or subcorymbose, ending the branches. Fl. yellow.—Name in honour of David Gorter, a Dutch botanist and author of a *Flora Belgica.*

Stems diffuse, hispid; heads ending the branches, solitary:
Rays *broadly obovate*, obtuse, rather shorter than the inv.   (1) **calendulacea.**
Rays linear, just equalling the involucre   ...   .   (2) **personata.**
Rays linear or linear-oblong, *longer* than the inv.   ...   (3) **diffusa.**
Stem erect, sparsely setose, corymbose; heads crowded, corymbose, bracteated; rays twice as long as inv. ...   ...   (4) **corymbosa.**

1. G. calendulacea (DC.! l. c. 501); stems diffuse, hispid; leaves entire, spathulate, closely setose above; rays broadly obovate, obtuse, minutely denticulate, scarcely equalling the involucre.

HAB. Betw. the Paarl and Paardeberg; also on the Lion Mt., Capetown, *Drege!* (Herb. D., Hk., Sd.).
Only known from *G. diffusa* by its very broad, short, blunt rays. It has similar inv. scales.

2. G. personata (Linn. Sp. 1283); stems diffuse, hispid; lower leaves tapering into a petiole, entire, toothed or pinnatifid, cauline sessile, oblong, entire or toothed; inv. sc. needle-shaped, round-backed, ciliate at base only; rays linear, not longer than the inv.   *Th.! Cap.* 698.   *Less. Syn.* 51.   *DC. l. c.* 501.   *Jacq. Coll.* 4, *t.* 21, *f.* 1.   *Pluk. phyt. t.* 273, *f.* 6.

HAB. Sandy ground round Capetown and in the Western districts. (Herb. Th. D., Hk., Sd.)
Stems 3–6 in. long. Rad. leaves 2–4 in. long, 4–5 l. wide above, tapering greatly to the base, hispid above, white beneath. Heads 3–4 l. diam., inv. sc. after flowering very rigid and pungent, spreading widely.

3. G. diffusa (Thunb.! Cap. 697); stems diffuse, hispid; lower leaves tapering into a petiole, entire, toothed or pinnatifid, cauline sessile, entire or toothed; inv. sc. narrow-subulate, mostly mid-ribbed or needle-shaped, copiously ciliate; rays longer than the involucre. *Less. Syn.* 52. *DC. l. c.* 501.

VAR. β. **intermedia**; leaves entire; rays linear-oblong, narrowed at base, rather longer than the involucre. *G. affinis, DC. l. c.* (Herb. D., Sd., Hk.)
HAB. Western districts, Cape, *Thunb.!* Caledon, *E. & Z.!* Riv. Zondereinde, *Zey..* 3018. β. Namaqualand, *Drege! A. Wyley!*
Very similar to *G. personata,* but with longer rays, and much more copiously ciliate, usually broader inv. scales. β. has shorter rays, but longer than in *G. personata:* in *Mr. Wyley's* specimens nearly or quite as long as in normal *G. diffusa.*

4. G. corymbosa (DC.! l. c. 501); stem erect, very sparsely setose, pale, corymbosely branched; cauline leaves roughly echinato-setose above and on the nerve beneath, lin.-spathulate, entire; heads crowded, subcorymbose, bracteate; involucre copiously clothed with long, white, straight hairs, the inner sc. ciliate with the same; rays twice as long as the involucre.

HAB. Near the Gariep, *Drege!* Namaqualand, *A.Wyley!* (Herb. D., Hk., Sd.) Stem 1–2 ft. high, fistular. Radical leaves not seen; cauline 1½–2½ inches long, 3–4 lines wide, very bristly. Rays beneath green in the middle, with a yellow border; above yellow. White hairs of invol. very conspicuous, 2–3 lines long.

CXXXIV. GAZANIA, Gaertn.

Heads radiate; *ray-fl.* ligulate, neuter; *disc-fl.* 5-toothed, perfect. *Recept.* honeycombed, the cells shallow. *Inv. sc.* in 2 or several rows, concrete below into an urceolate cup, toothed round the apex. *Fil.* smooth. *Achenes* wingless, very villous. *Pappus* biseriate, of very delicate, scarious, toothed scales, often hidden in the wool of the achene. *DC. Prodr. 6, p.* 508.

Herbaceous, mostly perennial, rarely annual plants, stemless or caulescent. Lvs. either crowded at the crown of the root, or scattered along the stem, variable in shape on the same plant. Pedunc. nude, 1-headed. Heads frequently of large size, and very showy; the rays yellow or orange, dark brown or reflecting peacock-colours at base; disc-fl. dark. Name said to be from γαζα, *riches,* from the brightness of the flowers. The species are extremely variable and difficult to define; I have rejected several that seem to me to be insufficiently defined; perhaps further evidence may alter the limits of several here retained.

A. **Perennial, caulescent and branching; branches alternately leafy :—**
　　Inv. woolly; fl.-heads small; rays *yellow,* not spotted　　(1) **uniflora.**
　　Inv. glabrous; fl.-heads large; rays *orange,* eye-spotted
　　　　at base ... ... (2) **rigens.**

B. **Perennial, stemless or nearly so; *rootstock* woody, simple or multifid; leaves radical or tufted at the ends of the short branches :**
　　Leaves quite glabrous on both surfaces, glaucous　　... (3) **Othonnites**
　　Leaves *glabrous,* smooth or rough with hard points on
　　　　the upper, hoary on the under surface :
　　　　Inv. sc. as long as the tube or longer, *very much* acuminate :
　　　　　　Inv. truncate and *ringed* above the conical base :
　　　　　　　　Lvs. linear-subulate, ciliate at base; ped.
　　　　　　　　　　pilose ... ... (4) **subulata.**
　　　　　　　　Lvs. lin.-lanceolate or pinnatisect; ped.
　　　　　　　　　　long, glabrous ... ... (5) **longiscapa.**
　　　　　　Inv. subtruncate, not *ringed;* pedunc. hairy;
　　　　　　　　lvs. pinnate ... ... (12) **pinnata,** η.
　　　　Inv. sc. shorter than the tube, the inner acute or acuminate :
　　　　　　Leaf-lobes spinous-mucronate or bristle-pointed :
　　　　　　　　Dwarf; leaf-lobes linear, spinous-mucronate (6) **jurineæfolia.**
　　　　　　　　Leaf-lobes lanceolate, tipped with a long
　　　　　　　　　　bristle ... ... (7) **mucronata.**
　　　　　　Leaf-lobes not conspicuously mucronate :
　　　　　　　　Lf.-lobes scaberulous above, obovate or
　　　　　　　　　　oblong, serrulate; pedunc. as long as
　　　　　　　　　　leaves, or longer ... ... (8) **varians.**
　　　　　　　　Leaf-lobes linear, acute, spinoso-ciliate;
　　　　　　　　　　outer inv. scales linear-filiform, medial
　　　　　　　　　　deltoid, inner semi-lanceolate ... ... (9) **oxyloba.**
　　　　　　　　Lf.-lobes (few) linear, acute; lvs. mostly
　　　　　　　　　　linear-lanceolate, ciliato-serrulate; inv.
　　　　　　　　　　sc. semi-lanceol., acuminate, the outer
　　　　　　　　　　ciliate ... ... (10) **serrulata.**

Lf.-lobes few, oblong, decurrent, not ser-
rulate; pedunc. shorter than leaves   (11) **Krebsiana.**
Inv. scales much shorter than the tube, the inner
obtuse or subobtuse :
   Lvs. mostly pinnate, the lobes short, obovate,
   cartilagineo-serrulate ; inv. few-flowered   (14) **arctotoides.**
   Lvs. either spathulate or pinnatifid, with linear
   segments ; margin callous-ciliate   ...   ...   (15) **longifolia.**
   Leaves either narrow-linear or pinnatifid, the
   lobes narrow, short   ...   .   (16) **leptophylla.**
Leaves tomentose or cobwebbed on the upper, hoary on
the lower surface :
   Lvs. pinnatisect, lobes short, linear, blunt, chan-
   nelled   ...   .   (20) **nivea.**
   Lvs. pinnatisect, lobes obovate or oblong, scaberu-
   lous   ...   .   (8) **varians.**
   Lvs. all narrow-linear, obtuse, channelled   ...   ...   (21) **canescens.**
Leaves *hispid* or roughly hairy above, hoary beneath :
   Inv. sc. much *acuminate*, especially the inner :
      Lvs. commonly pinnate (some simple) lobes oblong
      or linear in several pairs ; pedunc. longer than
      lvs. ...   .   (12) **pinnata.**
      Lvs. lanceolate or pinnatifid ; pedunc. short   (11) **Krebsiana.**
      Lvs. obovate, obtuse, sublyrate or pinnatisect,
      the margin toward the base ciliate with long
      bristles   ...   .   (17) **heterochæta.**
      Lvs. mostly *linear*, with strongly revolute mar-
      gins, tipped with a rigid bristle   ...   ...   (19) **armerioides.**
   Inv. sc. *short*, the inner broad, acute or subacute (not acuminate):
      Lvs. mostly pinnate or sub-bipinnate, fl. heads
      large, rays orange, eye-spotted at base   ...   (13) **Pavonia.**
      Leaves spathulate, dwarf ; heads small ; rays
      white, purplish beneath   ...   .   (18) **pygmæa.**
C. Annuals ; stems diffuse, leafy ;
   Invol. glabrous, truncate at base :
      Lvs. obovate-spathulate, subentire, cobwebby, ser-
      rato-ciliate, mucronate   ...   .   (22) **Lichtensteinii.**
      Lvs. mostly pinnatisect, lobes narrow-linear, entire,
      scaberulous above   ...   .   (23) **tenuifolia.**
   Inv. hispid, obovate : leaves very scabrous, pinnatisect,
   with linear setoso-mucronate lobes, and ciliate petioles   (24) **Burchellii.**

1. G. uniflora (Sims. Bot. Mag. t. 2270); stem suffruticose at base,
diffuse or ascending, leaves tapering at base into a petiole, oblongo-
obovate or spathulate, obtuse, smooth and glabrous (or cobwebbed)
above, white-woolly save the midrib beneath, some quite entire and
simple, some deeply 3-lobed, some *pinnately 5-7 lobed!*, the margins
revolute; pedunc. not much longer than the leaves, glabrous or woolly;
inv. obovate campanulate, obtuse or truncate at base, variably tomentose, lobes biseriate, short, the outer linear, obtuse, inner acuminate.
Less.! Syn. 41. *Lodg. Bot. Cab.* 795. *DC. l. c.* 512. *Gorteria uniflora,*
Linn. f. Th.! Cap. 698.

VAR. β. **leucolæna** ; more woolly than usual ; some leaves 3-lobed. *G. leucolæna,*
DC. l. c. 509.

VAR. γ. **pinnata** ; many of the lower leaves pinnately 5-7 lobed !

HAB. Sandy ground near the Coast. Zeeko R., *Thunberg!* Plettenberg Bay.
Mundt and Maire, sandhills on sea shore, Uit., *E. & Z.!* Natal, *Drege! T. W.! San-*
derson, 541. *Gerr. and M'K.,* 351. β. Mossel Bay, *Burchell,* 6233. Algoa Bay,

Drege! Mouth of Zwartkops R., with α, *Zey!* 3013. γ. Mth. of Zwartkops, *Zey.!* (Herb. Th., D., Hk., Sd.)

Stems spreading to 6-12 inches or more from a centre ; leaves very inconstant in form ; frequently *all* the leaves are quite simple ; in other specimens some leaves are deeply 3-lobed, the rest simple ; and in our var γ., which grows intermixed with the other varieties, the upper leaves are quite simple, the lower either 3-lobed or pinnately 5-7 lobed, all on the same branch ! In Herb. Th. both α and β occur. The woolliness varies greatly ; sometimes the whole plant is snowy white ; sometimes the whiteness is confined to the under sides of the leaves. Heads rather small ; fl. yellow, not spotted.

2. G. rigens (R. Br. l. c.) ; stems short, densely leafy, or diffuse, laxly leafy with ascending branches ; leaves petioled, either entire and spathulate, or sparingly pinnatifid, green and glabrous above, white beneath, except the midrib ; pedunc. glabrous, mostly longer than the leaves ; inv. glabrous, campanulate, sc. 3–4-seriate, the outer linear, acute, medial and inner semilanceolate, innermost with narrow membr. margins, subacute. *Less. Syn.* 47. *DC. l. c.* 510. *Gorteria rigens, Linn. Sp.* 1214, *non. Th. Bot. Mag. t.* 90. *Mill. Ic.* 1. *t.* 49.

HAB. Cult. in Europe ; wild plant unknown to us. (v. v. cult.)
Stems many from the crown, 6-12 in. long, spreading on the ground. Leaves 4-5 in. long, mostly entire, oblongo-spathulate, obtuse, tapering much at base, 5-6 l. wide in the broadest part, with reflexed edges. Pedunc. 4-8 in. long, incurved. Heads large and showy ; rays 1½ in. long, orange, with a black, eye-spotted spot at base. A common border and "bedding" plant in English gardens, requiring protection in winter.

3. G. Othonnites (Less. ! Syn. 45) ; rhizome multifid ; subradical leaves crowded, clasping at base, petioled, quite glabrous on both sides, glaucous, some simple, linear or lanceolate, some deeply pinnatifid, with linear-oblong, erecto-patent, acute lobes, all along the petiole and margin of leaf and lobes *spinoso-ciliate*, the *apices* of leaf and of lobes with an *entire* (not ciliate) cartilaginous margin, mucronate ; pedunc. glabrous, longer than the leaves ; invol. glabrous, turbinate, truncate and intruse at base, lobes 2–3 seriate, the outer deltoid, acute, inner ovato-semi-lanceolate-acuminate ; pappus-sc. long, tapering. *DC. l. c.* 510. Also *G. intrusa, E. Mey.! DC. l. c.* 511. *Gorteria Othonnites, Th.! Cap.* 699.

HAB. Verlooren Valley and Roggeveld, *Thunberg!* Little Namaqualand, *Drege! V. Schlicht! A. Wyley!* (Herb. Th., D., Hk., Sd.)
Rootstocks thick and woody, sometimes projecting 2-3 inches above the soil. Leaves 2-3 inches long, 1-3 l. wide. Pedunc. 2-8 in. long. All parts glabrous and glaucous. *Drege's* specimens quite agree with *Thunberg's.*

4. G. subulata (R. Br.) ; collum woody, multifid ; leaves subradical, numerous, glabrous and furrowed above, albo-tomentose beneath (the toment hidden under the margin), rigidly ciliate in the lower part, either all linear-subulate with strongly revolute margins, or some pinnatisect with narrow linear lobes ; pedunc. robust, sparsely pilose ; inv. glabrous, truncate and ridged above the conical base ; lobes all linear-subulate, much acuminate, the outer ciliate. *Less.! Syn.* 42. *DC. l. c.* 12. *Gorteria linearis, Th.! Cap.* 697.

HAB. Langekloof, *Thunb.!* Elandsriviersberge, *E. & Z.!* Betw. Vanstaadensberg and Bethelsdorp, and near Onzer, on a stony mountain cliff, *Drege!* (Herb. Th., Hk., Sd.)

Stems 1-1½ in. long, densely leafy. Leaves 4-6 in. long, about 1 line wide, pale, mucronate, quite glabrous. Pedunc. 10-12 in. long, 1½ l. diam., hollow. Lobes of invol. uncial, much longer than the tube.

5. G. longiscapa (DC. l. c. 513); collum divided; subradical leaves petioled, either quite glabrous or sparsely hispidulous above, white-woolly except the nerve beneath, some lanceolate-acuminate [or "elliptic oblong," *DC.*] undivided, others pinnatisect, with linear-lanceolate acute lobes, all with the margins reflexed, *cartilagineo-ciliate;* pedunc. glabrous, longer than the leaves; inv. glabrous, truncate at base, obovoid, the lobes linear-subulate, bi-triseriate, longer than the tube, much acuminate, the outer ciliate.

VAR. β. ovalis; lvs. long-petioled, elliptical or oblong, *obtuse at both ends,* 2-2½ in. long, 1-1½ in. wide. (Herb. Hk.)

HAB. Betw. Vanstaadensberg and Bethelsdorp, *Drege.* Port Elizabeth, *Dr. Pappe!* Albany, *T. W.* near Grahamstown, *P. MacOwan,* 150. Natal, *M. J. McKen in litt.* Var. β. Somerset, *Mrs. F. W. Barber!* (Herb. D., Sd., Hk.)

Except in foliage, which is very variable, and in the glabrous peduncle, this scarcely differs from *G. subulata.* I do not however find the petiole rigidly ciliate as in that species. *Mr. McKen's* specimen has very narrow leaves, almost like those of *G. subulata;* he tells me that the Kaffir name is '*Bensli,*' and that the small fringe which constitutes the dress of the young girls is made from the under surface of the leaves, scraped off.

6. G. jurineæfolia (DC.! l. c. 511); very dwarf; rootstock ligneous; subradical leaves petioled, pinnatipartite, the lobes linear, spinoso-mucronate, entire or 1-2-lobuled on the lower margin, glabrous and either smooth or scabrid above, white except the nerve beneath, with revolute margins; petiole rigidly ciliate; pedunc. glabrous, short; inv. glabrous, oval, rounded or truncate at base, its lobes 3-seriate, short, ovato-semi-lanceolate, acuminate, mucronate, the outer ciliolate or fringed at margin.

HAB. Zeekoe R., *Drege!* Queenstown flats, in dry, stony places, *Mrs. F. W. Barber!* 638. (Herb. D., Hk., Sd.)

Rootstock simple or divided, slender. Leaves "lying flat upon the ground" *(M. E. B.),* 1-2 in. long, fully half the length occupied by petiole; lobes in 2-5 pairs, close or subdistant, in the stronger specimens again lobuled. Pedunc. mostly shorter than the leaves. "Rays white; disc yellow" *(M. E. B.)*

7. G. mucronata (DC. l. c. 513), "crown woody, multifid; radical leaves petioled, glabrous above, cano-tomentose beneath, some entire, the majority pinnatisect, the lobes few, lanceolate, *tipped with a long bristle;* petiole here and there ciliate; pedunc. glabrous, twice as long as the leaves; inv. ovate, subtruncate at base, glabrous, the lobes 3-ranked, acuminate, shorter than the tube." *DC. l. c.*

HAB. Camiesberg, *Drege.* (Unknown to us).

Said to be allied to *G. jurineæfolia. DC. l. c.* also refers to "*G. rigens*" Eckl.; but a specimen (Hb. Sond.) so named by *Ecklon,* belongs to *G. pinnata,* Less. "Rays 10 l. long, brown spotted at base."

8. G. varians (DC.! l. c. 513); crown woody, multifid; subradical leaves petioled, flexuous, at first thinly cobwebbed, afterwards glabrous, but *scaberulous* above, beneath save on the nerve white-woolly, some undivided, oblongo- or lineari-lanceolate, the majority pinnatisect, the lobes in 1–3–5-pair, obovate or oblong, mucronate or pointless, the margin serrulate, strongly revolute; pedunc. as long as leaves or longer; inv. campanulate or turbinate, subtruncate at base, lobes 2–3-seriate, *variable* in length, the outer linear, acute, herbaceous, long or short, inner varying from ovate-acute to lanceolate-acuminate. *G. humilis, E. M.! DC. l. c. 513.*

HAB. Little Namaqualand and Olifant's R., and Nieuweveld, Beaufort, *Drege!* Namaqualand, *A. Wyley!* (Herb. Hk., D., Sd.)

The young plant looks quite hoary; even the old is pale and glaucescent. Leaves 3–4 in. long, in all that I have seen *scaberulous* with minute, sharp points. In *Drege's* specimen (Hb. Hk.) the outer and inner inv. sc. are *equal*, as described by DC. and about ⅔ as long as the inv., the inner being semilanceolate; in Mr. Wyley's (Hb. D.) which otherwise agrees, the outer are much shorter, and the inner ovate-acute, not ½ as long as inv. In *G. humilis,* E. M., undoubtedly the same sp., the *outer* sc. are very short; the *inner* exactly as in *Drege's* ' *G. varians.*'

9. G. oxyloba (DC.! l. c. 512); crown woody, multifid; subradical leaves above glabrous or minutely scaberulous, beneath, save the nerve, albo-tomentose, leathery, some linear-lanceolate, acute, tapering at base, quite entire, some pinnatisect above the middle, the lobes linear, acute, 2–4 pair, the margins of all reflexed and spinoso-ciliate; pedunc. glabrous, about equalling the leaves; inv. cylindr., subtruncate at base, glabrous, the scales pluriseriate, the outer herbaceous, linear-filiform, the medial deltoid with subulate points, the inner semilanceolate, acute, membr. edged.

HAB. Zwellendam, *Ecklon!* Gamka R., *Burke & Zey.!* (Herb. Sd., Hk.)

Glaucous (when dry); leaves 4 in. long, 2–3 l. wide, the lobes 1 l. wide. Outer inv. sc. in *Ecklon's* specimen dispersed over the tube, in the dry state easily escape notice, and are not described by DC. In *Burke & Zeyher's* plant they are much fewer, and sometimes wanting.

10. G. serrulata (DC.? l. c. 512); crown woody; radical leaves glabrous above, beneath, save on the nerve, tomentose, some linear-sublanceolate, acute, tapering at base into a petiole, others *(few)* pinnatisect, with linear, acute lobes, all with the reflexed margins minutely ciliato-serrulate; pedunc. shorter than the lvs., glabrous; inv. cylindr., truncate at base, glabrous, lobes sub-triseriate, semilanceolate-acuminate, rather shorter than the tube, the outer herbaceous, ciliate (few or many), the inner with widely membranous edges, taper-pointed.

HAB. Witberg, *Drege.* Natal country, *Dr. Sutherland!* Satohi, in bogs, *Dr. Kirk!* (Herb. D., Hk.)

I have not seen *Drege's* plant, and describe from *Dr. Sutherland's,* which may be different. Leaves 4–5 in. long, 1½–3 l. wide, in one specimen pinnatisect. Scapes 2½–3 in. long. Rays about 15, yellow, with a brown band beneath.

11. G. Krebsiana (Less.! Syn. 44); nearly stemless; subradical lvs. white woolly beneath, glabrous above, with reflexed edges, petioled, linear-lanceolate, entire or some sparingly pinnatifid, the lobes decurrent; pedunc. shorter than leaves, subglabrous; inv. obconic, intruse and truncate at base, glabrous, segments 2-seriate, the outer leafy, narrow-linear, acuminate, ciliate, inner semilanceolate, much acuminate, nerved, glabrous, membr. edged. *DC. l. c. 509.*

VAR. β. hispidula; leaves *hispidulous* above, and *glabrous* on the same root!
(Herb. D.)

HAB. Caffirland, *Krebs! E. & Z.!* β. Fort Beaufort, *Cooper,* 417. (Herb. Sd.)
Leaves 2½-3½ in. long, 2-4 l. wide, occasionally pinnatifid, the lobes 2-3 pair.
Pedunc. 2 in. long. Invol. 6-7 l. long. β. varies with hispidulous and glabrous
leaves; in other respects it agrees pretty well with α. *Zeyher's* 3014, from the
Zwartkop R., is also probably the same; it varies with rough or smooth leaves.

12. G. pinnata (Less.! Syn. 43); rootstock perennial, emitting many
fibres; leaves radical, petioled, white beneath, above and on the reflexed
margins and often on the midrib beneath hispid (rarely *glabrous*),
pinnatipartite (some entire), lobes oval-oblong or linear, 2-6-8 pair,
subdecurrent; pedunc. not much longer than the leaves, *hirsute or
smooth*; inv. campanulate, truncate at base, *hispid or glabrous,* the
segments in 2-3 rows, all lanceolato-subulate, much acuminate, nerved,
the outer leafy, rigidly ciliate, inner glabrous or softly ciliate. *Gorteria
pinnata,* Th., *Act. Soc. Hafn.* 4. *p. 6, t. 5. Fl.! Cap.* 699. *G. incisa,*
Th. l. c. *Gaz. ciliaris, DC.! l. c.* 513.

VAR. β. **speciosa**; lvs. glabrous or hispid above; pedunc. long; inv. subcylindrical,
glabrous or obsoletely tomentose, the lobes semilanceolate-linear, acuminate, his-
pido-ciliate, in 2-3 rows. *Less. G. speciosa, Less. Syn.* 43. *DC. l. c.* 510. *G. rigens*
β, *Th.* and *G. pectinata, Th. Cap.* 699, *fide Less.*

VAR. γ. **grandis**; lvs. sparsely setose above, pinnati-partite, lobes in 8-9 pair,
long and linear, obtuse, remote; pedunc. long, glabrous; inv. obovate-campanulate,
glabrous, lobes linear-subulate, longer than the tube, outer ciliate. *G. grandis, DC.
l. c.* 509.

VAR. δ. **multijuga**; lvs. setose above, lobes 7-9 pair, oblong, acute; pedunc. glabrous, long; inv. glabrous, its lobes linear-subulate, twice as long as tube. *G.
multijuga; DC. l. c.* 509.

VAR. ε? **scabra**; lvs. rigidly setoso-scabrous above, lobes in about 4 pair, oblong
lanceolate, scape short, glabrous; inv. subscabrid, the lobes acuminate, the outer
dorsally scabrid, inner longer. *G. scabra, DC. l. c.* 509.

VAR. ζ. **serrata**; hispido-scabrous, the terminal leaf-lobe *often* (not always) serrate; pedunc. and invol. hairy; lobes as in α. *G. serrata, DC. l. c.* 510.

VAR. η, **leiophylla**; leaves on rigidly ciliate petioles, entire or pinnati-partite,
quite *glabrous* above; pedunc. very hairy; inv. glabrous, the lobes pluriseriate, lanceolate-acuminate, the outer rigidly ciliate. (Hb. Hook.)

HAB. About Capetown and Stellenbosch, *E. & Z.!* Caledon's Baths, *Zeyher!*
Zwarteberg, *Beil.* Simonstown, *C. Wright,* 371. Paarl, *Drege!* β. Cape Flats,
Mundt. Stenberg, *Bergius.* γ, Saldanha Bay, *Drege.* δ, Zwartekops R., *Drege.*
ε, Betw. Kaus and Natvoet, Gariep, *Drege.* ς, Drakensteinberg, *Drege!* (Herb.
Th., Sd., D., Hk.)
Leaves 3-5-8 in. long, more or less pinnate, very generally hispid above, rarely
quite glabrous; lobes 3-10 l. long, 1-3 l. wide. In an original specimen marked
by *Lessing* (Hb. Sd.) the inv. is quite glabrous and the pedunc. roughly setose; *Mr.
Wright's* specimens vary with glabrous or hispid involucres. I regret that I have
not seen specimens of our first four varieties, which I venture to refer here, from
the descriptions given of their involucres by their authors: foliage is a most uncertain
character in this genus.

13. G. Pavonia (R. Br. Hort. Kew. 2, vol. 5, p. 140); rhizome dividing, ligneous; leaves subradical, petioled, hispid and green above, albo-
tomentose beneath, entire or pinnatisect, the segments oblong, acute,
subdecurrent, with revolute margins; pedunc. longer or shorter than
the leaves, pilose or glabrous; invol. campanulate, hispid or glabrous,

segments 2–3-seriate, *short*, the outer linear, leafy, strigose, inner from
an ovate base semi-lanceolate, acute or obtuse, memb.-edged, glabrous.
*Less.! Syn. 46. DC. l. 510. Bot. Reg. t. 35. Gorteria rigens, Th.! Cap.
698 (Herb.! ex pte). Gort. Pavonia, Andr. Rep. t. 523. Gazania, 6132,
Drege ? G. pinnata, DC. l. c. 509 (sp. ex Drege).*

VAR. β, **hirtella**; smaller, with less compound leaves; petiole rigidly ciliate;
inv. hairy. *G. hirtella, DC.! l. c. 511.* (Herb. D., Hk., Sd.)

VAR. γ.? **Zeyheri**; stem branching at base, branches short, diffuse; cauline and
subradical leaves long-petioled, scabrous above, *sub-bipinnatisect*, the lobes in several
pairs, very obtuse, the upper ones often *with* 1–2 *obtuse, spreading* lobules, the margins reflexed, ciliolate; petiole ciliate; inv. glabrous. (Herb. Sond.)

HAB. Roadsides, &c. near Capetown, also in Zwartland and Groenekloof, *Thunb.!
E. & Z.! Dr. Thom ! &c.* β, Paarl, *Drege!* γ, R. Zondereinde, near Appel's kraal,
Zey.! 3015. (Herb. Th., D., Sd., Hk.)

Stems short, ascending or procumbent, leafy at base. Leaves 3–9 inches long,
variably pinnate, sometimes undivided and lanceolate. Flowers large and showy,
the rays bright orange, with a black or eyed spot at base. Best known from *G.
pinnata* by its inv. scales. Var. γ, founded on a single specimen in Hb. Sond., has
remarkably compound foliage, and may perhaps be a distinct species.

14. G. arctotoides (Less.! Syn. 48); rhizome slender, multifid; subradical leaves petioled, glabrous (or cobwebbed) and smooth above,
white beneath, mostly pinnatipartite, the lobes short, *obovate*, obtuse or
acute, cartilagineo-serrulate along the margin ; pedunc. cobwebbed ;
inv. cylindrical, few-flowered, subtomentose; scales 2-seriate, the outer
deltoid, acute, inner ovate, subobtuse, all very short. *DC. l. c. 510.*

HAB. Gamko, *Mundt and Maire!* (Herb. Sond.)

Leaves 2–2½ inches long, lobes 4–5-pair, 3 lines long, 1 line wide. Inv. ⅓ inch
long, 3–4 l. wide at summit, 1½ l. at base. I have only seen a very poor specimen.

15. G. longifolia (Less. Syn. 48); rhizome multifid; subradical lvs.
thickish, unequal, flexuous, acute or obtuse, callous-mucronate, elongate, shortly petioled, tapering much at base, smooth and glabrous above,
except on the nerve white beneath, the margin reflexed and *callous-
ciliate*, some leaves quite entire, spathulate, some sinuate and some
deeply pinnatifid, with linear divergent segments; inv. cylindr., smooth,
at length glabrate, its lobes 2-seriate, the inner ovate, obtuse, membr.-
edged, the outer much smaller. *Less. l. c. abbrev. DC. l. c. 510.*

HAB. Schietfontein, *Lichtenstein.* Olifant's R., and near Gamko, *Mundt and
Maire.* Cape, *E. & Z. ?* (Herb. Sd. ?)

Of this I have seen no authentic specimens; but 2 specimens in Hb. Sd., collected
by *E. & Z.*, agree pretty well with the above character. The name "*longifolia*" is
unhappy, as many have much longer leaves. Lvs. 2–4 inches long. Heads rather
small; rays orange, with a black spot at base.

16. G. leptophylla (DC.! l. c. 512); crown woody, multifid; subradical leaves narrow-linear, flexuous, tapering into a ciliolate petiole,
either quite entire or pinnatifid toward the apex, the lobes narrow,
short, quite glabrous above, tomentose with revolute margins beneath ;
pedunc. glabrous, longer than the leaves; inv. cylindrical, truncate at
base, glabrous; lobes 3-seriate, the outer few, narrow-linear, reflexed,
inner broadly ovate, obtuse, membr.-edged, biseriate, all very short.

HAB. Zeekoe R., *Drege!* (Herb. Hk.)

A small species. Leaves 1½–2 inches long, 1 line wide. Heads small and few-fl.; rays narrow. Allied to *G. longifolia*, Less., with which its inv. nearly agrees.

17. G. heterochæta (DC. l. c. 511); "stem half-woody, short; sub-radical leaves tapering into a petiole, obovate, obtuse, sublyrate or variously pinnatisect, setose above, white-woolly beneath, the margin, chiefly near the base, ciliate with long bristles; pedunc. scarcely longer than the leaves, glabrous or sparsely bristled; inv. campanulate, gla-brescent, its lobes 3-seriate, acuminate, shorter than the tube." *DC. l. c.*

HAB. Betw. Zilverfontein, Kooperberg, and Kaus, *Drege!* (Unknown to me.)

18. G. pygmæa (Sond.! Linn. 23, p. 69); crown woody, multifid; sub-radical leaves hispid above, except the nerve white beneath, linear-spathulate, obtuse, tapering at base into a ciliate petiole, the margins revolute; pedunc. glabrous, shorter than the leaves; invol. glabrous, obconic, subtruncate at base, the lobes 3-seriate, outer herbaceous, linear-acute, ciliate, inner ovate, subacute, short, with broadly membra-nous margins.

HAB. Magallisberg, *Burke & Zeyher!* (Herb. Hk., Sd.)
Leaves (on the few specimens yet seen) *all* undivided, 2–3 inches long, 1½–2 lines wide, rigid, erect, pale. Scape and inv. in Herb. Hook. *quite glabrous;* in Hb. Sd. minutely puberulous. Heads small: rays about 12, white, with a purplish band beneath. A specimen (Hb. Hk.) found by *Burke* on the Aapjes Riv. differs in the shorter and less ciliate outer inv. sc.

19. G. armerioides (DC.! l. c. 511); rootstock woody, many-cleft; leaves subradical, linear (or a few pinnately 1–2-lobed), subsessile, with strongly revolute margins, setoso-mucronate, bristly above and on the margins, white tomentose beneath; pedunc. longer than the leaves, hairy; invol. broadly campanulate, very hairy, the sc. 3–4-seriate, the outer oblong or lanceolate, mucronate, inner much acuminate.

VAR. β, **Sutherlandi**; outer inv. sc. lin.-lanceolate, acuminate; inner hair-pointed.
HAB. Witberg, 7–8000 feet, *Drege!* β, at Klip-river, Natal, *Dr. Sutherland!* (Herb. D., Hk., Sd.)
Forming dense tufts. Leaves crowded, shortly petioled, 1–2 inches long, 1–1½ l. wide, in the stronger specimens sometimes pinnate-lobed, roughly hispid. Pedunc. 2–4 inches long. Invol. ¾-inch across. Rays numerous, pale, with a violet band beneath. In β. (Herb. Hook.) all the inv. sc., but particularly the outer ones, are more taper-pointed.

20. G. nivea (Less.! Syn. 49); very dwarf; rhizome woody, thick, divided at the summit; leaves crowded (subradical), petioled, hoary-tomentose on both sides, pinnatisect, the segments short, obtuse, linear, channelled; pedunc. not exceeding the leaves; inv. tomentose, truncate at base, lobes 2-seriate, the teeth very short, deltoid, the inner round-ish, glabrate, membr.-edged. *DC. l. c.* 509.

HAB. Gamko-karroo, *Mundt & Maire!* (Herb. Sond.)
Stems ½ inch long, densely leafy. Leaves, with their petioles, 1–1½ inch long; lobes 1–2 lines long. A very dwarf plant, with the aspect of *G. jurineæfolia,* but canous in all parts.

21. G. canescens (Harv.); rootstock thick and woody, multifid; sub-radical leaves narrow-linear, obtuse, above *thinly canescent*, smooth, channelled, beneath cano-tomentose, with entire, scarcely revolute margins ; pedunc. about as long as leaves, at first canous, then glabrate ; invol. truncate at base, cylindro-turbinate, *thinly canescent,* the lobes 2–3-seriate, the outer linear-filiform, mucronulate, longer than the glabrous, ovate-acute, membranous inner ones.

HAB. Basutu Land, *T. Cooper !* 719. (Herb. D.)

A dwarf plant, probably from high land, and dry, rocky ground. Leaves 1½–2½ in. long, not a line wide, none pinnatisect on our specimen. All the young parts pale, with very minute, white, downy pubescence, which rubs off in age. Heads of small size ; invol. ½ inch long, 3–4 lines wide at top ; rays about 12, yellow, with a central band beneath.

22. G. Lichtensteinii (Less.! Syn. 50) ; root slender, perpendicular, annual ! stems many from the crown, short, diffuse, leafy, cobwebbed ; leaves obovato-spathulate, tapering into a petiole, entire or bluntly 3–5-lobed, cobwebby-tomentose, becoming nude above, as well as the petiole serrato-ciliate, mucronate ; pedunc. slender, glabrous, longer than the leaves ; inv. glabrous, cylindro-turbinate, at base produced into a hollow, truncate, reversed cup, the lobes biseriate, outer very short, deltoid, inner longer, ovate-acuminate or ovato-semilanceolate, acute. *DC. l. c. p.* 514. Also, *G. araneosa, DC.! l. c.* 512.

HAB. Cape, *Lichtenstein.* Zilverfontein, and by the Gariep, *Drege !* Springbokkeel, *Zey.!* 972. Namaqualand, *A. Wyley !* (Herb. D., Hk., Sd.)

Root tapering, 2–4 inches long, with lateral fibres. Stems 2–4 in. long, spreading. Leaves 1½–2½ in. long, 2–4 l. wide, often crenate-lobed. Heads small ; inv. ¾ inch long, 3–5 l. wide at top, rays 8–10, broad, golden-yellow, with a dark spot at base, and on the under side a medial, greenish band.

23. G. tenuifolia (Less. Syn. 50) ; root slender, annual ; stems many from the crown, short, decumbent, leafy ; cauline leaves petioled, subscaberulous above, albo-tomentose beneath, some narrow-linear, entire, the majority pinnatisect above the middle, the lobes 3–4-pair, narrow-linear, acute, entire, the rachis and petiole narrow-margined ; pedunc. glabrous, little longer than the leaves ; inv. glabrous, at base truncate-ridged and obconic, short-tubed, lobes in several rows, much acuminate, the outer short, squarrose, inner lanceolate, rather longer than the short rays ; pappus scales elongate, much longer than the ovarian wool. *DC. l. c.* 513.

HAB. Cape, *Lichtenstein.* Zilverfont. and Olifant's R., *Drege !* (Hb. D., Hk., Sd.)

Stems 1–3 inches long. Leaves 2–2½ inches long, the rachis and lobes not a line wide, but evidently white beneath. Heads small, and rays very short. The habit is that of this genus, but the ach. and pappus-sc. are very differently proportioned.

24. G. Burchellii (DC.! l. c. 514) ; root perpendicular, slender, annual ; stem leafy at the crown, branched from the base, the branches spreading, hispid, with alternate leaves which are crowded toward the apices ; leaves petioled, very scabrous above, nearly nude beneath, with strongly revolute margins, either linear or pinnatisect, the apices of the lobes tipped, and the petiole ciliate with long, white, rigid bristles ; pedunc. terminal, longer than the leaves, pubescent ; inv. obovate, hispid, the scales pluriseriate, outer subulate, inner semi-lanceolate, all much acuminate, and tipped with rigid, subpungent bristles. *Hirpi-*

cium Echinus, Less.! Syn. 54. Gorteria hispida, Licht. MSS., fide Less.

HAB. Sak-rivier, *Lichtenstein.* Near the Gariep, *Burchell,* 2500. Zilverfontein, *Drege!* Springbokkeel, *Zey.!* 976. (Herb. D., Hk., Sd.)

Root seemingly annual. Stems 3-6 inches long, spreading from a leafy crown. Leaves 1-1½ inch long, not 1 line wide, almost always pinnatisect above the middle, with linear lobes. 1 am indebted to *Dr. Sonder,* who examined *Lichtenstein's* original specimens (Herb. Berol.), for the synonym of *Lessing.* This plant appears to me to be a true *Gazania,* having the habit, involucre, and *double pappus* of this genus; the outer scales about 10, long and denticulate at the extremity, the inner much shorter, entire. I retain *De Candolle's* specific name, as he was the first to refer it to its proper genus.

(Uncertain Species.)

G. coronopifolia (DC. l. c. 511); "subherbaceous, branching at the crown, leaves crowded at the ends of the branches, petioled, pinnati-partite, the lobes linear, obtuse, quite entire, with revolute margins, above glabrous, white-tomentose beneath; pedunc. glabrous, twice as long as the leaves; invol. obovate at base, the lobes rather longer than the tube, much acuminate, subrigid." *DC. l. c. Arnica coronopifolia, Linn. Sp.* 1247? *Burm. Cap. p.* 27.

HAB. Cape, *Burman,* in Herb. Delessert, fide DC. (Unknown to us.)
By description, this must come near *G. longiscapa.*

G. Kraussii (Sch. Bip. Flora. xxvii. 777); "radical leaves linear-lanceolate, quite entire, petioled, or pinnatipartite, with 1-2 pair of lobes, glabrous above, with revolute margins, white beneath; scapes a span long, cobwebbed; invol. cylindrical, truncate at base, the scales narrow-linear, the inner subulate, 10 lines long." *Sch. B., abbrev. Walp. Repert.* 6, 279.

HAB. Zitzikamma, Uit., *Krauss.* (Unknown to us.)
By description this is probably *G. longiscapa,* DC.

G. lineariloba (DC. l. c. 512); "crown subherbaceous; rad. leaves long-petioled, glabrous above, beneath, save the nerve, cano-tomentose, some undivided, oblong-linear, the majority pinnatipartite, the lobes in 1-2 pair, linear, acute, the margins of all subrevolute, entire; invol. truncate at base, shortly cylindrical, the lobes biseriate, acute, equal, shorter than the tube." *DC. l. c.*

HAB. Drakensteinberg, *Drege.* (Unknown to us.)

CXXXV. **CULLUMIA**, R. Br.

Heads radiate; ray-fl. ligulate, neuter; *disc-fl.* 5-toothed, perfect. *Recept.* very deeply honey-combed, the walls of the cells raised and enclosing the achenes, apices setigerous. *Inv. sc.* pluriseriate, concrete at base, the outer pectinate-spinous, similar to the leaves, the inner entire or fimbriate, pungent. *Achenes* angular, glabrous, without pappus. *DC. Prod.* 6, *p.* 497.

Small shrubs or suffrutices, all S. African. Branches leafy to the very summit. Leaves alternate, sessile or decurrent, margined with slender, spinous cilia : otherwise quite entire. Heads terminal, solitary. Fl. yellow. Named in honour of *Sir T. G. Cullum, Bart.,* a Norfolk botanist.

Leaves linear-subulate, with strongly reflexed or revolute margins, bordered with uniseriate spinous cilia :

Inv. sc. *all* erect and appressed (not *leaf-like*) :
  Leaves very erect, closely imbricate; outer inv. scales
    minutely serrulate   ... ...   (1) **sulcata.**
  Lvs. spreading ; outer inv. sc. *pectinate*   ... ...   (2) **bisulca.**
  Outer inv. scales *squarrose*, leaf-like and spine-bordered :
    Branches glabrous; lvs. falcate-recurved, cilia 2-3, long   (3) **decurrens.**
    Branches cobwebbed or tomentose ; cilia 4-5, or more.
      Lvs. spreading, *sessile;* cilia longer than breadth of
        leaf   ... ...   (4) **patula.**
      Lvs. strongly reflexed, subdecurrent; cilia equal-
        ling the breadth of leaf   ... ...   (5) **squarrosa.**

Leaves (small) closely *imbricated*, oblong or semi-lanceolate, glabrous ; the margin thickened and *bifariously* pectinate, with rigid cilia :
  Lvs. mucronate; inv. sc. erect   ... ...   (6) **pectinata.**
  Lvs. and inv. sc. tipped with a long, rigid, deflexed spine   (7) **ciliaris.**

Leaves lanceolate or oblong, erecto-patent, *spreading* or reflexed, flat or flattish, the margin bristling with solitary or geminate rigid cilia :
  Leaves widely spreading, recurved or deflexed :
    Lvs. more or less *decurrent:*
      Lvs. ovate-acuminate, *subrecurved*, cilia 2-3   ...   (8) **rigida.**
      Lvs. oblong, *strongly* recurved ; cilia 3-4-6   ...   (9) **setosa.**
    Lvs. *sessile*, broad-based, lanceolate, deflexed or squar-
      rose   ... ...   (10) **cirsioides.**
  Leaves erecto-patent (not reflexed) :
    Leaves more or less decurrent: inv. sc. spinous-ciliate:
      Lvs. glabrous, lin.-lanceolate ; cilia geminate, with
        a minute woolliness in their axils   ... .   (11) **floccosa.**
      Lvs. viscidulous, oblong or ellipt.; cilia very short,
        without a flocculent tuft   ... ...   (12) **micracantha.**
    Leaves sessile, not decurrent ; inv. sc. entire-edged :
      Lvs. and inv. quite glabrous ...   .   (13) **hispida.**
      Lvs. white-woolly beneath ; inv. cobwebby   ...   (14) **sublanata.**

1. C. sulcata (Less.! Syn. 84, vix DC) ; branches glabrous, virgate; leaves closely imbricated, straight and erect, shortly subulate or semi-lanceolate, with strongly revolute margins, 1-furrowed beneath, cilia much shorter than the breadth of the leaf ; invol. scales all dry, erect-appressed, all minutely ciliato-serrulate. *Cullumia adpressifolia, Sond.! Linn.* 23, *p.* 69. *Rohria sulcata, Th.! Cap.* 614.

HAB. Cape, *Thunberg!* R. Zondereinde, *Zey.!* 3024, var. *a.*   (Hb. Th., Sd., Hk.)
Allied to *C. bisulca*, but not so branching, with much shorter, very erect, and close-lying leaves ; smaller involucres, destitute of pectinate outer scales. *Zeyher's* plant above quoted is perfectly identical with *Thunberg's*. *Drege's*, relied on apparently by *De Candolle*, is *C. bisulca*, Less. (Herb. Hk., Sd.)

2. C. bisulca (Less.! Syn. 82); branches glabrous ; leaves crowded, subulate, subdecurrent, recurved at the points or squarrose, glabrous and nerved above, with revolute margins, generally nerved and bisul-cate beneath, cilia numerous, as long as the breadth of the leaf, or longer; inv. sc. all dry, erect-appressed, the outer pectinato-ciliate, the inner ciliato-serrulate or entire. *DC. l. c.* 498. *Rohria bisulca, Th.! Cap.* 615. *C. sulcata, Drege! in Herb.*

HAB. Cape, *Thunberg!* Swellendam, and Brackfontein, Clanw. *E. & Z.!* Hassa-quaskloof, *Zey!* 3024, var. *β.*   (Herb. Th., Sd.)

A much-branched bush 1-2 ft. high. Branches spreading. Leaves 4-7 lines long, 1 l. wide, sometimes simply falcate-recurved, sometimes curled backwards. The nerve on the lower side is sometimes quite obvious, sometimes concealed under the revolute margins, and such leaves are unisulcate, as in *C. sulcata.*

3. C. decurrens (Less.! Syn. 83); branches diffuse, glabrous; leaves patent-recurved, decurrent, subulate, flat above; cilia on each side 2-3, longer than the breadth of the leaf; outer inv. sc. squarrose, spinous-ciliate like the leaves. *DC. l. c.* 498; also *C. intermedia, DC.! l. c. Rohria setosa, Th.!* in *Hb., expte. Gorteria squarrosa, Linn. sp.* 1284, *excl. syn.*

HAB. W. and E. districts. Cape, *Thunb.!, Drege!* Langekloof, *Mundt & Maire.* Muysenberg; and Olifants Hoek, *Uit., E. & Z.!* Knysna, *Dr. Pappe!* (Herb. Th., D., Hk., Sd.)
Smaller and more diffuse and slender than any of this section, with very squarrose leaves. Lvs. 3-4 l. long, ½ l. wide. *DC.* attributes an "*inv. with erect scales*" to his *C. intermedia,* but in *Drege's* distributed specimens (Hb. D., Hk.. Sd.) they are quite squarrose; nor can I find any character to distinguish "*C. intermedia.*"

4. C. patula (Less.! Syn. 82); branches whitish tomentose; leaves spreading or recurved, sessile, subulate, with revolute margins and a prominent nerve beneath, glabrous above; cilia on each side 4-5, mostly longer than the breadth of the leaf; outer inv. sc. squarrose, spinoso-ciliate like the leaves, inner lanceolate, serrulato-ciliate. *DC. l. c.* 498. *Rohria patula, Th.! Cap.* 615, *and Gorteria ciliata, Th.! Cap.* 700.

HAB. Cape, *Thunberg! Burchell,* 5041, *Drege! Dr. Thom!* (Herb. Th., Hk., Sd.)
A much branched, robust bush. Leaves 1 in. or more in length, not 1 l. wide, very rigid, and commonly widely spreading, the upper sometimes erecto-patent. Outer inv. sc. like the leaves, more or less tomentose on the outside.

5. C. squarrosa (R. Br.); twigs robust, cobwebbed; leaves very densely crowded, squarrose-reflexed, subdecurrent, subulate, striate *(when dry)* above; cilia on each side 4-5, about equalling the breadth of the leaf; outer inv. sc. similar to the leaves, spreading or squarrose. *Less. Syn.* 84. *DC.! l. c.* 498. *Rohria squarrosa, Th.! Cap.* 615. *Gorteria squarrosa, Berg. Cap.* 301. *(nec L.)*

HAB. Sea shores, False Bay, *Thunb.! E. & Z.!* &c. Muysenberg, foot of hill, *W. H. H.* Simon's Bay, *C. Wright,* 338. (Herb. Th., Sd., D., Hk.)
Very robust, with curved branches, the lesser ones umbellate or corymbose, all most closely covered with reflexed leaves. Lvs. ¾-1 in. long, 1 l. wide, somewhat fleshy, when recent quite smooth above, but, in drying, the upper surface shrinks unequally and thus becomes striate.

6. C. pectinata (Less.! Syn. 85); branches glabrous; leaves erect-incurved, imbricated, semi-lanceolate, mucronate, (small), with a thickened reflexed margin which is pectinated on its outer edge with long, slender cilia, and on its inner (or under) with short ones; inv. sc. all erect, the outer pectinated like the leaves. *DC. l. c.* 499. *Rohria pectinata, Th.! Cap.* 615.

HAB. Cape, *Thunberg!* Onderbokkeveld, Clanw., *E. & Z.!* Betw. Grasberg River and the Waterfall, Tulbagh, *Drege!* 2896. *(Herb. Th., Sd.)*
A much branched, closely leafy bush, related to *C. ciliaris.* Leaves 2-3 lines long, 1½ l. wide, with a short, straight, erect mucro. Heads rather small.

7. C. ciliaris (R. Br.); branches glabrous ; leaves erect-appressed, imbricated, subdecurrent, elliptic-oblong (small), tipped with a long, rigid, deflexed spine, the thickened margin bifariously pectinate with many long, slender, spinous cilia; inv. sc. shorter than the rays, the outer leaf-like, tipped with long, deflexed bristles. *Less.! Syn.* 86. *DC.! l. c.* 499. *Rohria ciliaris, Th.! Cap.* 617.

HAB. Mountains round Capetown, *Thunberg!* and succeeding collectors, frequent. (Herb. Th., D., Hk., Sd.)

A straggling, small shrub, much or little branched. Leaves 3-4 lines long, their spinous mucro nearly as long and always strongly deflexed; marginal cilia in 8-10 pairs, those of both series nearly equally long. Rays twice as long as the involucre.

8. C. rigida (DC.! l. c. 500); branches and leaves glabrous ; leaves small, flattish, slightly concave above, subdecurrent, ovate-acuminate, tapering into a long, rigid bristle, horizontally spreading or subrecurved, the thickened margin bearing 2-3 spinous cilia; outer inv. sc. recurved, leaf-like, inner erect, oblong, spine-tipped, fringed with soft, short hairs,

HAB. Camiesberg, *Drege!* (Herb. D., Hk., Sd.)

A small, closely-branched species. Leaves 2-3 l. long, 1½ l. wide. By *DC.* it is considered most akin to *C. pectinata;* but I rather place it next *C. setosa,* which in many respects it resembles in miniature.

9. C. setosa (R. Br.); branches glabrous, curved, spreading; leaves oblong or lance-oblong, decurrent, strongly curved backwards, ending in a long, spinous point, the thickened margin bearing on each side 3-4 long, rigid, spinous cilia, which are occasionally geminate or bifarious; inv. sc. squarrose, resembling the leaves, but narrower. *Less.! Syn.* 86. *DC.! l. c.* 499. *Rohria setosa, Th.! Cap.* 616, *and Herb. ex pte.*

VAR. β. **adnata**; more robust, with larger leaves, more strongly decurrent. *C. adnata, DC. l. c.* (Herb. D., Hk., Sd.)

HAB. Mts. about Capetown and Simon's Bay, frequent. Langekloof, *Mundt.* Kleinriviersberg, *E. & Z.!* Zwarteberg, *Zey.!* 3021. Var. β. Berg R., *Drege!* (Herb. Th., D., Hk., Sd.)

A very straggling, weak, much or little branched bush. Leaves 3-5 lines long, 1-2 lines wide, the cilia long and stiff. Heads small. Some of *Drege's* distributed specimens under this name belong to *C. decurrens. C. adnata,* DC., is of stronger growth than usual, with larger and consequently more strongly decurrent leaves (often with 5-6 solitary or geminate spines); but the ordinary *C. setosa* varies considerably in size and luxuriance. All the specimens I have seen of "*C. adnata,*" from *E. & Z.!* belong to common *C. setosa.*

10. C. cirsioides (DC.! l. c. 499); branches robust, thinly cobwebby or glabrous; leaves crowded, lanceolate from a *broad* base, sessile, strongly deflexed, inversely imbricated, spinous mucronate, somewhat 3-nerved, the margin bearing long and slender, but rigid geminate or solitary spinous cilia; inv. sc. like the leaves, with spreading or deflexed points.

VAR. β. **carlinoides**; dwarfer, with rather smaller leaves, the marginal spines solitary. *C. carlinoides, DC.! l. c.* (Herb. Sd., Hk.)

VAR. γ. **squarrosa**; leaves more laxly set, smaller, hooked backwards; twigs crowded. (Herb. D., Hk.)

HAB. Near the Gauritz R., *Burchell,* 4643. Vanstaadensberg, Uit., *E. & Z.!*

Zey.! 3022. Var. β. Hoogekraal, betw. Kafferkuils R., and Zoetmelks R., *Drege!*
Var. γ, intermixed with α, at Vanstaadensberg, *Zeyher!* (Herb. D., Hk., Sd.)

A strong growing plant; β. smaller in all parts. Leaves ¾–1½ in. long, 2–4 l.
wide, the uppermost spreading, all the lower strongly deflexed. Var. γ. has quite
the look of strong growing specimens of *C. setosa*, var. β., but the leaves are not in
the least decurrent. *Zeyher*, who saw it in a fresh state, referred it to the present
species; the ordinary form of which he collected at the same time and place.

11. C. floccosa (E. Mey.!); branches glabrous; leaves erecto-patent,
decurrent, linear-lanceolate, with a thickened margin bearing on each
side 2–3 pair of geminate cilia, glabrous, except *a minute flocculent tuft
in the axils of the marginal cilia;* inv. sc. erect, the outer spinoso-ciliate
like the leaves, cobwebby, inner shorter than the ray. *DC. l. c.* 499.

HAB. Betw. Bergvalei and Langevalei, *Drege!* Kruisrivier, *Zey.!* 979. Elandsberg, *Dr. Wallich!* (Herb. D., Hk., Sd.)

An erect, rigid, much-branched and twiggy bush. Leaves 4–5 l. long, 1½–2 l.
wide. Heads of small size.

12. C. micracantha (DC.! l. c. 499); twigs and leaves minutely
viscidulous and glandular; leaves oblong or elliptic-oblong, erecto-
patent, mucronate, somewhat 3-nerved, decurrent, the margin bearing
very short and slender, geminate, or sometimes *solitary,* spinous cilia;
inv. sc. erect, like the leaves but narrower, the inner nearly equalling
the rays.

HAB. Piquetberg, *Drege!* Bergvalley, *E. & Z.!* *Zey.!* 980. (Herb. D., Hk., Sd.)
This resembles *C. hispida*, but is smaller and more slender, and specially differs
in the decurrent, glandular leaves, and small, weak, ciliary spines.

13. C. hispida (Less.! Syn. 87); the branches and leaves glabrous;
leaves oblongo- or lineari-spathulate, erecto-patent, spinous mucronate,
narrowed at base, sessile, (not decurrent), the margin bearing long and
rigid, solitary or geminate, spinous cilia; inv. sc. erect, oblong-elliptical,
entire, tapering into a long spinous point. *DC. l. c.* 499. *Rohria hispida,*
Th.! Cap. 616.

VAR. β. **angustifolia**; leaves narrower. *C. pauciflora, DC. l. c.* 500. (Herb. D.,
Sd., Hk.)

HAB. In the Karroo, *Thunberg!* (Hb. Thunb.) Var. β. **Eastern Distr.** *Burchell.*
Swellendam, *Mundt., E. & Z.!*
Branches robust, straight, 12–15 in. long. Leaves 1½ in. long; 4–6 l. wide; the
marginal cilia frequently geminate, 3–4 l. long. Heads rather large in the genus.
β. merely differs in its somewhat narrower leaves, and rather more bushy habit.
The name "*hispida*" is very unfortunate for a *glabrous* plant.

14. C? sublanata (DC. l. c. 500); "branches cobweb-woolly; leaves
sessile, half-spreading, linear lanceolate, glabrous above, white-woolly
beneath, the margin scarcely revolute, tipped with a spine, and furnished
with 6–8 spinous cilia; inv. cobwebby at base, the sc. erect, the outer
subciliate." *DC. l. c.*

HAB. Eastern Distr. *Burchell*, 6876. Ataquaskloof, *Drege.* (Unknown to us.)
This is said to have the habit and inflorescence of *C. hispida.* A leaf from *Drege*
(Hb. Sd.) said to belong to this is *oblong-obovate* (!) very obtuse: there must be
some strange blunder.

CXXXVI. HIRPICIUM, Cass.

Heads radiate; *ray-fl.* neuter; *disc-fl.* 5-toothed, perfect. *Recept.* shortly honeycombed. *Inv. sc.* concrete at base, in many rows. *Filam.* smooth. *Achenes* wingless, completely covered with long soft hairs. *Pappus* uniseriate, coroniform, splitting into bristles. *DC. Prodr. 6, p. 502.*

Much branched, erect, dwarf-shrubs. Leaves sessile, linear, leathery with revolute margins, white beneath. Heads terminal, solitary. Name ?

Leaves scabrous above, bristle-pointed ; inv. sc. subulate (1) **echinulatum.**
Leaves hispid above ; inv. sc. bristle-shaped, ciliate ... (2) **integrifolium.**

1. H. echinulatum (Cass.); leaves alternate, often fascicled, linear or oblong, tipped with a bristle, and ciliate near the base, with strongly revolute margins, scabrous above, albo-tomentose beneath; heads subsessile; inv. sc. linear-subulate, rigid, erect, pungent, minutely ciliolate. *Less.! Syn. 54. DC. l. c.* 502. *Œdera alienata, Th.! Cap. 725, excl. syn.*

HAB. Cape, *Thunberg!* Namaqualand and the Karroo, *E. & Z.!* Modderfontein, *Rev. H. Whitehead!* Kamiesberg, *Drege!* (Herb. Th., D., Hk., Sd.)
A rigid, twiggy, small shrub. Leaves pale-green.

2. H. integrifolium (Less.! Syn. 53); leaves approximate, alternate, or opposite, lance-linear, acute, with strongly revolute margins, hispid above, albo-tomentose beneath ; heads subsessile ; inv. sc. very narrow linear, with long, bristle-shaped, ciliate points. *DC. l. c.* 502. *Gorteria integrifolia, Th.! Cap.* 699.

HAB. Cape, *Thunb.!* Caledons Hoogte, *Mundt & Maire.* Hexrivierskloof, *Drege!* 9079. (Herb. Th., Sd.)
A much-branched, twiggy, small shrub ; young twigs hispid. Lvs. sub-petiolate, the rudimentary petiole half-clasping at base and membrane-edged ; lamina $\frac{1}{2}$–$\frac{3}{4}$ in. long, scarcely 1 l. wide, narrowed to both ends. Bristle-points of the inv. sc. nearly $\frac{1}{4}$ in. long. *Lessing* describes the leaves as "decussate and *connate;*" in *Thunberg's* specimen, though occasionally so, they are much more frequently alternate.

CXXXVII. STEPHANOCOMA, Less.

Heads discoid, all the fl. 5-toothed, perfect. *Recept.* honey-combed, the cells bearing long, rigid bristles. *Inv. sc.* pluriseriate, concrete at base, margined with spinous cilia. *Filaments* smooth. *Achenes* glabrous, obconic. *Pappus* sc. uniseriate, very short, at first concrete into a crenate, cup-like crown, afterwards partially separating. *DC. Prodr.* 6, 497.

A thistle-like herb, with the habit of a *Stobæa* (sec. *Apuleia*), from which genus this is only distinguishable by its crown-like pappus. This character appears to me insufficient, and I retain *Stephanocoma* merely in deference to *Lessing's* and *DC.'s* authority. I am obliged, however, to reject *St. Berkheyoides,* Less., because, on examination of the original specimens in Hb. Th., I find that it has the ordinary scaly pappus of a *Stobæa,* and I believe it to be identical with *St. scolymoides,* DC. The generic name is compounded of στεφανος, a *crown,* and κομη, *hair;* alluding to the crown-like pappus.

1. S. carduoides (Less.! Syn. 56); *Stobæa decurrens, Th.! Cap.* 621. *Stobæa crispa, DC.! l. c.* 517. *Berkheya carduiformis, DC.! l. c.* 507.

VAR. β. **latifolia**; rad. lvs. 2$\frac{1}{4}$–3 in. wide above the middle. (Herb. D., Sd.)

HAB. Hautniquas and Krom. R., *Th. /* Auteniqualand, *Mundt.* Uitenhage and Albany, *E. & Z. Zey.!* 3026. Knysna and Plettenbergs Bay, *Pappe!* Grahamstown, *P. MacOwan!* Howisons Poort, *H. Hutton*; and generally on the eastern frontier. β. Fort Beaufort and Kat R., *E. & Z.!* Winterberg, *Mrs. F. W. Barber!* (Herb. Th., D., Sd., Hk.)

Stem herbaceous, erect, striate, subglabrous. Leaves sparsely setulose or glabrous on both sides, the midrib beneath pilose, rigid, pinnatipartite or deeply pinnatifid, the lobes in many pairs, ovate-subrotund, spinous-toothed, with smaller spines between, curly undulate; the *radical* very long, tapering much at base, subsessile; cauline shorter and the upper very short, decurrent in long, spiny, curly-sinuate stem-wings. Heads small, discoid, subcorymbose or panicled; pedicels and base of inv. more or less woolly. Invol. sc. shorter than the disc, in many rows, subulate, margined with solitary slender spines. Achenes glabrous. Pappus sc. uniseriate, very short, at first concrete into a crenate, cup-like crown, afterwards separating. A specimen from *Drege* (Hb. Hk.) of *Berkheya carduiformis*, DC. is in all respects similar to *Steph. carduoides*, Less. *DC.* seems to have carelessly taken the long bristles of the receptacle for pappus !

CXXXVIII. STOBÆA, Thunb.

Heads many fl., either discoid, all the fl. tubular and perfect, or radiate, the *ray-fl.* neuter. *Inv. sc.* in many rows, concrete at base and tapering into a spinous point, bordered with 1–2 spines. *Recept.* honey-combed, the cells deep, nearly enclosing the achenes, cleft at the apex into bristle-shaped fimbrils. *Stam.* smooth. *Achenes* turbinate or inversely pyramidal, furrowed, either glabrous, downy, or villous. *Pappus* scales in 1–2 rows, equal or the alternate narrower, oval, obtuse, toothed at the apex. *DC. prodr. 6, p. 514.*

S. African thistle-like herbs or suffrutices, with much of the habit and characters of *Berkheya,* but differing in the *bluntish* not *much acuminate* pappus. Name in honour of Dr. Stobæus of Lund, one of Linnæus's earliest patrons.

Sect. 1. APULEIA. *Achenes* in all stages of growth *perfectly glabrous.* (Sp. 1–21.)

(*a.*) Cauline-leaves *conspicuously* decurrent, in spinous stem-wings:

Heads radiate: lower or radical tapering at base, but *not petioled:*

Leaves nearly glabrous on *both* sides, pinnatifid, with round lobes ... ... (1) **glabriuscula.**

Lvs. tomentose or canescent beneath:

Lower leaves deeply *pinnatifid,* with spinous lobes:

Pedicels and inv. sc. cobwebby or tomentose:

Leaves thinly *cobwebbed,* or subcanous beneath... ... ... (2) **polyacantha.**

Leaves *densely* white-woolly beneath, rigid:

Papp. sc. oblong, obtuse:

Inv. sc. with slender, solitary marg. spines ... (3) **scolymoides.**

Inv. sc. with binate marg. spines ... ... (4) **helianthiflora.**

Papp. sc. *very small* ; inv. sc. *pubescent* above... ... ... (5) **cirsiifolia.**

Pedicels and inv. sc. minutely *gland-pubescent*:

Stem wings long and narrow ; lvs. glabrous above:

Inv. scales gland-pubescent *above; lf. and wing spines 3–4 l. long* ... ... (6) **microcephala.**

Inv. sc. glabrous *above* ; lf. and wing-spines 1–2 l. long ... ... (7) **discolor.**

Stem wings short and broad ;
lvs. glandular above... ... (9) **onopordifolia.**
Lower (medial) lvs. obovate-oblong, re-
pand, *very rough* above ... ... (10) **Radula.**
Heads radiate ; radical and lower leaves on spine-bordered *petioles :*
Radical lvs. *rigid,* pinnatifid or sinuous lobed, tapering at base:
Rad. lvs. sparingly-cobwebbed or
nude beneath ... ... (11) **petiolata.**
Rad. lvs. white-woolly beneath ... (12) **echinopoda.**
Radical lvs. *thin,* oblong or elliptical,
repand ; spines minute ... ... (13) **membranifolia.**
Heads discoid :
Lvs. sinuato-pinnatifid, *spinellose* above, cob-
webbed beneath ... ... (8) **sphærocephala.**
Lvs. lanceolate, sharply 1-toothed at each
side, tomentose canescent ... ... (21) **M'Kenii.**
(*b.*) Cauline-leaves sessile, very *shortly subdecurrent* or adnate ; stem not winged :
Radical lvs. conspicuously petioled :
Rad. lvs. cordate, oblong, or oval, entire or sinuous-lobed ; heads
radiate :
Stem gland-setose ; lvs. thin, scabrous above,
with very small marginal spines ... ... (14) **speciosa.**
Stem quite glabrous, lvs. rigid, glabr. above,
with strong marg. spines and interposed
spinules ... ... (15) **acanthopoda.**
Rad. lvs. pinnati-partite or deeply pinnatifid : heads discoid :
Lvs. *spinellose* above, deeply pinnatisect ... (16) **epitrachys.**
Lvs. *smooth* above, pinnati-partite ; base of
petiole and crown of root very woolly ... (17) **eriobasis.**
Radical lvs. not petioled :
Stem shrubby, much branched ; lvs. pinnatipartite, the lobes 4–5 pair, rigid, nearly entire,
spine-tipped ... ... (18) **cardopatifolia.**
Herbaceous, branching *gland-puberulent* on all
young parts ; lvs. concolourous, the radical
sinuato-pinnatifid, with rounded lobes ... ... (19) **onobromoides.**
Herb., simple-stemmed ; rad. lvs. glabr. above,
white-woolled beneath, pinnatisect, lobes 18-20
pair, simple or 3–5-fid, spine-pointed ... ... (20) **multijuga.**

Sect. 2. EU-STOBÆA. Ovaries and young achenes pubescent or silky ; ripe achenes
sometimes becoming glabrous. (Sp. 22-43.)

(*a.*) Cauline leaves decurrent in long, spine-bordered stem wings :
Radical lvs. large, obovate or oblong, entire or pinnatifid ; cauline small :
Rad. lvs. cobwebbed-cottony or woolly beneath :
Rad. lvs. elongate, sinuous-pinnatifid, cob-
webbed ; fl. *purple* (or *white*) ... ... (22) **purpurea.**
Rad. lvs. obovate-oblong, subentire, white
beneath ; fl. yellow ... ... (23) **rhapontica.**
Rad. lvs. concolorous on both sides, nearly glabrous :
Stem wings narrow, their marg. spines
4-6 lines long ... ... (24) **aristosa.**
Stem wings broad, their marg. spines 1-2
lines long ... ... (25) **platyptera.**
No large rad. lvs. ; stem leafy to the summit ; lvs. spinoso-setose above :
Lvs. close-set, lanceolate, 2-3 in. long, 4-6 lines
wide ... ... (26) **echinacea.**
Lvs. subdistant, linear, 1-1½ in. long, 2 l. wide (27) **Gerrardi.**

(*b.*) Cauline leaves sessile, or *very shortly* decurrent ; stem not winged :

Heads radiate:

  Leaves alternate, linear or subulate, quite entire : bristles very long:

    Lvs. linear, *white* beneath; stems glandular
      pubescent ... ... **(28) insignis.**

    Lvs. subulate, with *strongly revolute* edges ;
      stem glabrous ... ... **(29) Zeyheri.**

  Lvs. alternate, pinnatifid or runcinate:

    Lvs. membranous, the lower on spine-bor-
    dered petioles, runcinato-pinnatifid, lobes
    3 pair, broad, with very small, weak marg.
    spines   ... ... **(30) sonchifolia.**

    Lvs. rigid, *gland-pubescent*, concolourous :

      Rad. lvs. 12–18 in. long, petioled,
      glandular *on both sides* ... ... **(31) grandifolia.**
      Rad. lvs. sessile? glabrous *beneath* ... **(32) viscosa.**

    Lvs. rigid, white-woolly beneath, softly
      pubescent above ... ... **(33) buphthalmoides**

  Lvs. *opposite!* connate, glabrous, roundish-oblong,
    spinous   ... ... **(34) oppositifolia.**

Heads discoid :

  Leaves concolourous, glabrous or nearly so :

    Glaucous, quite glabrous ; lvs. cordate at
    base, acuminate, many nerved, tipped and
    margined with long spines ... ... **(35) cruciata.**

    Not glaucous: subglabrous or puberulous:

      Upper surface of lvs. glabrous; lvs.
      cordate, clasping at base, netted-
      veined, acuminate; stem compressed **(36) glabrata.**

      Upper surface of lvs. *spinellose*; lvs.
      pinnatifid ... ... **(37) pinnatifida.**

  Leaves woolly or cottony beneath (or on both sides):

    Lvs. cordate-acuminate, spine-lobed at base **(36) glabrata, β.**

    Lvs. more or less deeply pinnatifid :

      Cauline lvs. subpetiolate, broadly-oblong,
      inciso-pinnatifid, white beneath, lobes
      oblong-lanceolate ; inv. sc. shorter
      than disc ... ... **(38) seminivea.**

      Lvs. sessile, pinnatipartite, with narrow lobes and revolute
      margins :

        Lvs. simply pinnatipartite, lobes
        lance-subulate ... ... **(39) rigida.**
        Lvs. pinnatipartite, lobes deeply
        cut or pinnatifid ... ... **(40) bipinnatifida.**

      Rad. lvs. short petioled; cauline clasp-
      ing, sinuate-pinnatifid and undulate:

        Stem and leaves copiously woolly-
        canescent ... ... **(41) heterophylla.**
        Stem-subglabrous; lvs. glabrous or
        thinly hoary beneath ... ... **(42) atractyloides.**

      Rad. lvs. *long petioled;* cauline ½-clasp-
      ing, pinnatifid, the segments 2-lobed,
      white beneath ... ... **(43) biloba.**

Sect. 1. APULEIA (Sp. 1–21).

1. S. glabriuscula (DC. l. c. 519) ; stem herbaceous, erect, striate, thinly pilose below, cobwebbed toward the summit; leaves on both sides glabrous or nearly so, deeply pinnatifid, the lobes roundish, tooth-lobed, set with long spines, the interspaces ciliato-spinellose, radical elongate, tapering to the base, cauline decurrent in long very spinous stem-wings, continued up to the head; inv. sc. longer than the disc, linear-lanceolate, bearing solitary spines and ciliato-spinellose; achenes glabrous; papp. sc. biseriate, oblong, toothed.

HAB. Betw. Hoffman'skloof and Driefontein, Albany, *Drege.* Uitenhage, *Ecklon!* (Herb. Sd.)

This seems to be little more than a glabrate form of *S. polyacantha*; I have not seen *Drege's* plant.

2. S. polyacantha (DC.! l. c. 519); stem herbaceous, erect, furrow-striate, glabrate below, cobwebby above; leaves glabrescent above, the lower ones beneath cano-tomentose, the upper cobwebby-canescent, cauline pinnatifid, with subovate, 3–5 spined lobes, multispinellose between the spines, decurrent in spinous stem wings; heads solitary, radiate; inv. sc. longer than the disc, cobwebbed beneath, margined with solitary spines; achenes glabrous; papp. sc. biseriate, oblong, toothed.

HAB. Cape, *Drege.* Uitenhage, *Ecklon!* (Herb. Sond.)

I have only seen *Ecklon's* plant, which is scarcely at all cobwebbed, certainly neither 'cano-tomentose' nor 'canescent;' seems nearly the same as *S. glabriuscula*.

3. S. scolymoides (DC.! l. c. 520); stem herbaceous, erect, glabrous, cobwebbed above; cauline leaves glabrous above, white-woolly beneath, oblong, narrowed to the base, decurrent in sinuous, spinous and ciliate stem wings, sinuous-lobed, the lobes short, 3–5 spined, with rounded ciliato-spinellose interspaces; heads solitary, sub-pedunculate, ebracteate; inv. sc. tomentose externally, longer than the disc, ciliato-spinose, whilst in flower patent-reflexed, in fruit suberect; achenes glabrous, turbinate, papp. sc. oblong, obtuse, denticled. *Rohria decurrens,* Th.! *Cap.* 620. *Stephanocoma Berkheyoides, Less.! Syn.* 57.

HAB. Cape, *Thunberg!* Klipplaat River near Schiloh, and on to Windvogelberg. *Drege!* Slaay Kraal, *Burke!* Boschman's and Karraga R., *Zey!* 3027. (Herb Th., Hk., Sd.)

A tall strong-growing species. Lower leaves 8–10 in. long, 1–2 in. wide, the younger sometimes cobwebbed above, afterwards nude. Spines strong and yellow-ish. Heads of large size; rays many, yellow. In Hb. Hk. is a specimen from *Bowie* that chiefly differs in having a much more woolly stem, and cobwebbed upper-surfaces to most of the leaves. *Thunberg's* specimens of "*Rohria decurrens,*" though imperfect, agree very well with this; their *pappus* is identical, of *separate*, toothed scales! by no means '*coroniform,*' as described by *Lessing.*

4. S. helianthiflora (DC. l. c. 520); "stem herbaceous, erect, fur-rowed, cobwebbed above; leaves glabrescent above, white-tomentose beneath, the lower tapering at base, sessile, pinnatifid, the lobes rounded, 7-spined, sparingly spinellose between the spines; cauline less sinuate, decurrent in spinoso-ciliate wings; heads solitary, ebracteate; inv. spreading, the sc. longer than the disc, cobwebby canescent beneath, with geminate marginal spines; achenes glabrous, turbinate; papp. sc. oval, obtuse, denticled." *DC. l. c.*

HAB. Zuureberg, 2–3000 f., *Drege.* (Unknown to me).

This seems to border very closely on *S. scolymoides.*

5. S. cirsiifolia (DC.! l. c. 519); stem herbaceous, erect, cobwebby at the summit; lower leaves elongate, pinnatifid, tapering at base, subsessile, the lobes roundish, bearing 1-3-5 long spines and ciliato-spinel-lose, glabrous and glossy above, white-woolly beneath, rigid; upper leaves oblong, sinuate, spinous-toothed, decurrent in spinous stem-wings, uppermost cobwebbed above; inv. sc. tomentose beneath, pubescent above, longer than the disc, spine-pointed and set with long, strong marginal, often geminate, spines; achenes glabrous, turbinate; papp. sc. very small, separate, denticled.

HAB. Witbergen, 6-7000 f., *Drege!* (Herb. D., Hk., Sd.)
Stem 2-3 ft. high, pale, leafy to the summit, ending in a corymb of fl. heads. Cauline leaves 4-6 inches long, smaller upwards; the uppermost 1-2 in. long. Young leaves cobwebbed above. Very like *S. scolymoides* in general aspect, but with *pubescent* inv. sc. bordered with much stronger spines than in that species.

6. S. microcephala (DC.! l. c. 520); stem erect, rib-furrowed, mi-nutely gland-pubescent, especially near the summit; leaves glabrous above, white woolly beneath, the lowest elongate, tapering at base, toward the apex shortly and roundly sinuate-lobed, the lobes about 3-spined, with ciliato-spinellose interspaces; cauline decurrent in sinu-ous, ciliato-spinous stem wings, heads solitary or sub-corymbose, ending nude, elongate, densely gland-pubescent branches, not bracteate; inv. sc. somewhat longer than the disc, lanceolate, gland-pubescent outside, margined with slender solitary spines and ciliato-spinulate between; achenes glabrous, turbinate; papp. sc. broadly ovate, 2-seriate, denticled.

HAB. Witbergen, 4-5000 f., *Drege!* Cape, *E. & Z.!* Winterberg, common, *Mrs. F. W. Barber!* 632. (Herb. D., Sd.)
Like other "common" plants, this has been rarely and very imperfectly collected.

7. S. discolor (DC.! l. c. 520); stem herbaceous, erect, furrowed, chiefly toward the apex glandular pubescent; leaves glabrous or minutely puberulous and green above, albotomentose beneath, the lower oblong-subobovate, tapering at base, sessile, sinuate-lobed, the lobes roundish, 3-5 spined, and spinellose between, cauline broad-based, decurrent in spinous-toothed stem-wings; branches corymbose, leafy, many-headed; heads (nude or) bracteate; inv. sc. glandular outside, ciliate with rigid, simple or geminate spines; achenes glabrous; papp. sc. obtuse, denticled.

HAB. Betw. the Key and Gekau, *Drege!* Natal, *Gerr. and M'K.!* 305. *Gueinzius!* 346. *Dr. Sutherland. Dr. W. B. Grant!* (Herb., D., Sd., Hk.)

8. S. sphærocephala (DC.! l. c. 518); "stem herbaceous, erect, cobwebbed above; leaves on the upper surface *spinelloso*-scabrid, cob-webby beneath, the younger sub-canous, at length glabrate, sinuato-pinnatifid, the margin spinuloso-ciliate, the lobes roundish, the radical lvs. tapering at base, cauline decurrent along the stem, in sinuate, spi-nuloso-ciliate wings; heads discoid, corymbose, pedicels tomentose; inv. much shorter than the disc, tomentose at base, sc. multiseriate, with several slender marginal spines." *DC. l. c.*

HAB. Gekau, and Omsamwubo, *Drege!* (Herb. Sd.)
I have seen but a leaf and a flowering twig. Achenes quite glabrous. Pappus of many, subbiseriate, separate, sharp-pointed, toothed scales.

9. S. onopordifolia (DC.! l. c. 521); stem herbaceous, erect, branched, glandular-pubescent and softly pilose; leaves on the upper surface
and on the nerves beneath glandular-pubescent, albo-tomentose beneath,
the cauline decurrent in *short*, broad, ciliato-spinous stem-wings, sinuato-pinnatifid, undulate, the lobes ample, roundish-reniform, strongly
many-spined, and ciliato-spinellose; heads solitary, bracteate; inv. sc.
glandular-puberulent, broad, 3-nerved, longer than the rays, bearing
slender, simple or geminate marginal spines; recept. covered with very
long, slender bristles; achenes glabrous, obconic; papp. sc. acute,
toothed.

HAB. Betw. Buffelvalei and Kraai Riv., Albert, 4500–5000 f. *Drege!* Beaufort
Distr., *Cooper!* 551. Near the Orange R., *Burke!* Zuureberg, Cradock, *Zey!*
(Herb. D., Hk., Sd.)
Readily known among its allies by its short, broad stem wings, the *very long*,
slender bristles of the receptacle, and the glandular foliage.

10. S. Radula (Harv.); stem herbaceous, erect, angle-furrowed, rough
with hard points, racemoso-corymbose at the summit; radical leaves
obovate-oblong, obtuse, tapering much at base, some subpetiolate, repand
or sinuate, rigid, unequally spinoso-ciliate, at first roughly setose, then scabrous with hard points above, white woolly beneath; cauline small, narrow-oblong or linear, very spiny, decurrent in long, narrow, *very spinous*
stem-wings, the stem winged to the summit; heads shortly radiate, racemose; inv. sc. linear-lanceolate, rather shorter than the disc, glabrous or
cobwebbed, closely spinous and ciliato-spinellose; achenes glabrous, obpyramidal; papp. sc. truncate, denticled.

HAB. Magalisberg, *Burke and Zeyher!* Zey.! 981. (Herb. D., Hk., Sd.)
This has much of the habit and foliage of *S. Rhapontica;* but besides the want of
woolliness on the stem, and very scabrous foliage, the achenes are glabrous, &c.

11. S. petiolata (DC. l. c. 520); stem herbaceous, erect, terete, glabrous below, cobwebbed above; leaves sparingly setulose above, cobwebby or nude beneath, the radical and lower *on long spine-bordered
petioles,* obovate-oblong, tapering to the base, pinnatifid, the lobes
rounded, with a sub-revolute, unequally many-spined margin, cauline
decurrent in long, narrow, ciliato-spinous stem wings; heads peduncled,
ebracteate, inv. woolly at base, the sc. longer than the disc. bearing
frequently geminate, marginal spines; achenes glabrous, turbinate;
papp. sc. oblong truncate, denticled.

HAB. Zwartkops R. and Vanstaadens R., *Drege, Zeyher!* Adow, Zey.! 3028.
(Herb. D., Hook., Sd.)
Rad. leaves, including the petiole, sometimes 15–16 in. long. The stem-wings
are sometimes very narrow, sometimes broadish, always very spiny. Pubescence
scanty and deciduous.

12. S. echinopoda (DC. l. c. 521); "stem herbaceous, erect, cobwebbed; leaves setuloso-scabrid above, snowy-white beneath, the radical
on petioles bordered with slender spines, oblong, tapering to both ends,
sinuate lobed, the lobes rounded, 7–9 spined and ciliato-spinellose, the
cauline decurrent in spinellose stem-wings; heads short-pedicelled,
pedicels tomentose, leafy; inv. sc. cobwebbed beneath, eq. the disc,

margined with solitary spines; achenes glabrous; papp. sc. oblong obtuse." *DC. l. c.*

HAB. Near the Zwartkops and Vanstaaden Rs., Uit. *Drege.* (Herb. Sd.)
Of this I have only seen 2 leaves and a fl. head in Hb. Sd. ; it seems to differ from *S. petiolata* chiefly in the copious snow-white wool of the under surfaces of the leaves.

13. S. membranifolia (DC. ! l. c. 521); stem herbaceous, erect, furrowed, cobwebby above; leaves *thinly membranous*, glabrous above, albo-tomentose beneath, the radical on long petioles bordered with slender spines, oblong or elliptical, repand or sinuate-lobed, ciliate with slender, short, but unequal spinules; cauline more copiously woolly beneath, furnished with longer and stronger spines and ciliato-spinellose between, more or less sinuous or pinnatifid, decurrent in ciliato-spinose, sinuous stem-wings; heads solitary, ebracteate; inv. sc. longer than the disc, lanceolate, cobwebbed beneath, margined with many, slender, often geminate spines; ach. glabrous, obpyramidal; papp. uniseriate, toothed.

HAB. Uitenhage Distr., *E. & Z. !* (Herb. Sond.)
Known from all allied species by the thinly membranous leaves, and slender weak spines.

14. S. speciosa (DC. ! l. c. 518); stem herbaceous, erect, striate, glanduloso-setose, simple or corymbose at the summit, chiefly leafy below; radical and lower leaves long petioled, the petiole aristoso-ciliate and sometimes winged or lobulate, the lamina either cordate, ovate or oblong, acute, scabrous above, white-tomentose beneath, ciliate with small, unequal bristles; lower cauline shortly decurrent, upper sessile, (small), few, passing from lanceolate to linear, aristoso-ciliate; heads 1–5, long-pedicelled, ebracteate, radiate; inv. sc. linear-lanceolate, acuminate, glandular, ciliate with long spines; achenes glabrous, furrowed, obconic; papp. sc. biseriate, truncate, toothed.

HAB. Sandplaat and Komga; Morley; Omsamwubo and Omtata, *Drege!* Katberg, *Drege! Ecklon! Hutton!* Caffraria, *Cooper!* 290. Fort Bowker, *H. Bowker!* 562. Natal, *T. Williamson!* Natal, *Gerr. and M'K.* 302, 303, 304. (Herb. D., Hk., Sd.)

15. S. acanthopoda (DC. l. c. 521); "stem herbaceous, erect, quite glabrous, flowering branches subcorymbose, the younger tomentose, at length glabrous; leaves rigid, glabrous above (the younger cobwebbed), white-woolly beneath, the radical and lower cauline petioled, oval, sinuous-lobed, the lobes strongly 1–3 spined and ciliato-spinellose; petiole armed with rigid, spreading spines; cauline lvs. sessile, wider at base, adnate or very shortly decurrent; heads pedicelled, ebracteate; inv. sc. tomentose beneath, linear, bearing geminate, marginal spines; achenes turbinate, glabrous; papp. sc. oblong.

HAB. Betw. the Key, Gekau and Basche; and betw. Omtata and Omsamwubo, *Drege.*
Unknown to me ; seemingly a very distinct species.

16. S. epitrachys (DC. ! l. c. 515); stems erect, subsimple, herbaceous, cobweb-woolly, terete; leaves *spinoso-setose* above and along the

margins, cobweb-woolly beneath, the radical long petioled, deeply pinnatisect, the lobes toothed or lobed, tipped with slender spines and
spine-ciliate, cauline sessile, not clasping, similarly pinnatisect; heads
discoid, subracemose, short-stalked; inv. sc. lanceolate, longer than the
disc, *often* spinoso-setose above, spinous-ciliate and with a pair of basal,
geminate spines; achenes "glabrous, shining." *DC.;* pappus of many
short, oblong, obtuse, narrow scales.

HAB. Kamiesberg, *Drege!* (Herb. D., Hk., Sd.)
This agrees with *S. pinnatifida* in the spinose upper surface of leaves, but differs
in the copious woolly covering, and the glabrous achenes. The specimens seen by
me are in very poor condition, insufficient clearly to verify all the characters.

17. **S. eriobasis** (DC. l. c. 516); "crown of root woody and woolly;
stems ascending woolly below, cobwebby above, simple, sparingly leafy,
radical leaves petioled, pinnatipartite, the petiole cobwebbed, very
woolly at base, the segments in many pairs, narrow, slightly cobwebby,
deeply lobed, lobules linear-lanceolate, tipped with a rigid spine, their
revolute margins spinoso-ciliate; cauline leaves few, sessile; heads
discoid, very short-pedicelled, subracemose; inv. sc. cobweb woolly,
equalling the disc, bearing subgeminate spines. *DC. l.c.*

HAB. Zeederberg, *Drege!* (Herb. Sd.)
For this I chiefly adopt *DC.'s* description, having seen but a radical leaf and a
detached involucre, in Hb. Sd.—"Achenes glabrous. Papp. sc. oval, in 1-2 rows."
DC.

18. **S. cardopatifolia** (DC. l. c. 514); "stem shrubby, erect, branched;
branches terete, the younger ones, with the undersides of the leaves,
woolly, the older glabrate; leaves above glabrous, sessile, pinnatipartite,
the lobes rigid, linear-lanceolate, produced into a spine, nearly entire,
in about 4–5 pair, sometimes solitary or geminate; heads discoid,
ending the branches, subracemose; outer inv. sc. spreading, leafy,
with solitary spines, inner erect; achenes glabrous; pappus sc. oval,
obtuse, subdenticulate." *DC. l. c.*

HAB. Near Graaf Reynet, *Drege.*

19. **S. onobromoides** (DC. l. c. 516); minutely *glandular*-puberulous
on all young parts; stem herbaceous, erect, terete, branched, thinly
pilose or glabrate; leaves on both sides more or less gland-pubescent,
becoming glabrate, concolourous; the radical sessile, very long, sinuato-
pinnatifid, with short, rounded, spinous-toothed lobes, and wide, spinous-
toothed interspaces; lower cauline similar, but shorter; upper cordate-
clasping at base, ovate or oblong, acuminate, bordered with rigid spines,
entire or sinuato-pinnatifid; heads discoid; inv. sc. ovato-lanceolate,
veiny, gland-dotted, spinous-edged, longer than the disc.

HAB. Boschesman's Doorn R., *Drege!* Cape, *Wallich!* Winterhoek, Tulbagh,
Dr. Pappe! (Herb. D., Sd., Hk.)
The young plant is probably viscidulous when recent, being covered with glandular
hairs, the bases of which are obvious on older leaves and inv. scales as small, raised
points. The heads in *Dr. Wallich's* specimens are smaller and more numerous; in
them the achenes are glabrous, and the pappus of many, small, narrow scales. All
the sp. of *Drege's* seen by me have the heads worm-eaten. *Dr. Pappe's* plant has
sub-pinnatifid cauline leaves; heads as in *Drege's.*

20. S. multijuga (DC.! l. c. 521); rhizome ligneous, horizontal; stem erect, simple, terete, cobweb-cottony; leaves glabrous above (the younger cobwebbed), softly and copiously white-woolly beneath, the lower elongate, sessile, cauline adnate or shortly decurrent, all deeply pinnatisect, the lobes in very many pair, close together, the lowermost short and simple or toothlike, the upper 3–5-fid, spinous pointed, base of the leaf (or winged petiole) broadly linear, with an entire margin; heads solitary; inv. sc. about as long as the rays, cobweb-cottony beneath, linear-lanceolate, bordered with simple spines; ach. glabrous; papp. sc. obtuse, uniseriate.

HAB. Witberg, 6–7000-f., *Drege!* (Herb. D., Hk., Sd.)
A very remarkable species, with foliage unlike that of any other. Lower leaves 8–12 in. long, 1–2 in. wide, with a very wide midrib, quite entire and smooth-edged for 1–2 in. above the base, thence upwards more and more deeply incised and pinnatisect.

21. S. M'Kenii (Harv.); suffruticose, much branched, all young parts cobwebby-canescent; branches closely leafy to the summit; leaves decurrent in slender, spinoso-ciliate stem-wings, lanceolate, acute, all but the uppermost, sharply one-toothed or tooth-lobed at each side below the middle, apex of leaf and lobes spine-bearing, the margin ciliato-spinellose, upper surface cobwebbed, becoming glabrous and netted-veined, lower tomentose; inv. sc. lanceolate, tomentose, longer than the disc, with a few, small, simple, marginal spines; heads discoid; achenes glabrous, obconic, striate; papp. sc. very short.

HAB. Highlands of Natal, *Gerr. & M'K.*, 1047. (Herb. D.)
Stems "3–4 ft. high." I have only seen flowering twigs, and cannot say what the lower leaves may be like. It seems a very distinct species. Leaves on the branchlets 1–1½ in. long; 3–5 l. wide.

Sect. 2. EU-STOBÆA, (Sp. 22–43.)

22. S. purpurea (DC.! l. c. 518); stem herbaceous, erect, simple, cobwebby, leafy to the summit; subradical and lower leaves sessile, elongate, tapering at base, more or less deeply sinuato-pinnatifid, above sprinkled with jointed, soft hairs, beneath cobweb-cottony; cauline gradually smaller upwards, and more and more decurrent, the stem-wings of the lower lvs. narrow, of the upper broad, and very spiny; heads radiate, racemose; inv. sc. ovato-lanceolate, spine-pointed and margined with slender, simple spines; achenes silky; papp. sc. oblong-toothed, biseriate.

VAR. β. albiflora; flowers *white* (Hb. Sd.)
HAB. Witberg, *Drege!* Winterberg, *Mrs. F. W. Barber*, 530. (Herb. D., Hk., Sd.)
This plant, *Mrs. Barber* tells me, "grows in the vallies of the Winterberg in patches, many thousands growing close together as if they had been planted in beds; flowers purple, showy." Lower leaves 6–8 in. long; uppermost very small, but decurrent as broad stem wings.

23. S. rhapontica (DC.! l. c. 519); stem herbaceous, erect, terete, cobwebby, racemose at the summit; radical leaves obovate or obovate-oblong, obtuse, tapering much at base, shortly petioled, subentire or repand, unequally spinelloso-ciliate, glabrous or setulose above, white-woolly beneath; cauline (small), narrow-oblong, subsinuate, decurrent in long, lobed, ciliato-spinellose stem wings; stem and branchlets winged to the summit; heads radiate, racemose, on short or long peduncles; inv. sc. linear-lanceolate, shorter than the disc, bordered with simple, short spines, tomentose at back, innermost subentire; achenes silky; pappus biseriate.

HAB. Port Natal, *Drege! T. W.! Gueinzius! Dr. Grant!* (Herb. D., Hk., Sd.) Stem 2–3 ft. high, winged to the summit. Rad. lvs. 5–12 in. long, 2–3 in. wide; the petiole 1–4 in. long, very white beneath, green above. Raceme, with few or many rather small heads; rays longer than the inv.

24. S. aristosa (DC. l. c. 518); stem herbaceous, erect, simple, cobwebby; radical leaves broadly obovate, cuneate at base, sessile, penninerved, on both sides minutely scabro-setulose, the subentire or repand margin unequally ciliate with long and short bristle-spines; cauline lvs. decurrent, erect, smaller upwards, passing from oblong to linear, they and the narrow stem-wings also ciliato-aristate; heads racemose, subsessile; outer inv. sc. margined with *long* and short bristles, like the leaves, spreading, inner entire, lanceolate, nearly as long as the rays; achenes silky; papp. sc. biseriate, toothed.

HAB. Kaffirland, *Ecklon!* (Herb. Sond.) Rad. leaves 5–6 in. long, 3–4 in. wide, very scabrid. Longer cilia of leaves, stem-wings and inv. sc. 3–5 lines long, slender and awn-like. Stem 1–1½ ft. high.

25. S. platyptera (Harv.); stem herbaceous, erect, simple, glabrous; radical leaves obovate-oblong, cuneate at base, sessile? sparsely setulose above, glabrous except on the midrib beneath, the subsinuate margin unequally ciliate with slender bristles; upper cauline leaves (lower unknown) decurrent in very broad stem-wings, erect, oblong or linear, they and the wings ciliate with *short*, unequal spines; heads racemose, subsessile; outer inv. sc. squarrose-reflexed, narrow, ciliate with *short*, simple spines; inner erect, entire, subulate, shorter than the rays; achenes silky, obconic; papp. subacute, toothed.

HAB. Umvoti, Natal, *Gerr. & M'K.* 1049. (Herb. D.) "Stem 3–4 feet high." Rad. lvs. 5–6 in. long, 2½–3 in. wide. Nearly allied to *S. aristosa*, but with much shorter bristles, wider stem wings, and squarrose outer inv. sc., &c.

26. S. echinacea (Harv.); glabrous; stems closely leafy throughout, simple, herbaceous; leaves much longer than the internodes, lanceolate, acute, densely spinoso-setose above, and closely ciliate with unequal, slender spines, pilose on the midrib beneath, decurrent in long, sinuous, very spinous stem-wings; heads shortly radiate, subcorymbose; inv. sc. shorter than the rays, subulate, bordered with mostly binate spines; ach. silky; papp. sc. obovate, toothed, biseriate.

HAB. Dry plains of Zululand, *Gerr. & M'K.* 1045. (Herb. D.) Stem 1–2 ft. high. Leaves 2½–3 in. long, 4–6 l. wide. Heads in our specimens 3. Nearly allied to *S. Gerrardi*; but with much longer and broader leaves, and larger flower-heads.

27. S. Gerrardi (Harv.); rootstock ligneous; all parts quite *glabrous*;

stem herbaceous, erect, simple, leafy throughout, one-headed; leaves
narrow-linear, erect, laxly set, densely spinoso-setose above, and ciliate
with similar, slender spines, nude and strong midribbed beneath, decurrent in long, narrow, very spinous, sinuous, stem-wings, which
nearly conceal the whole stem to the summit; heads shortly radiate;
invol. sc. longer than the rays, subulate, bordered with slender, simple
or binate spines; ach. silky; papp. scales obovate, toothed, 2-seriate.

HAB. Emyati, Natal, *Gerr. & M'K.* 1046. (Herb. D.)
This has much of the habit of a *Cullumia.* The free portion of the leaf is scarcely
as long as the decurrent wing, 1–1½ inches long, 2 l. wide; spines all slender, very
unequal, small alternating with longer. In essential character it comes near *S.
echinacea.*

28. S. insignis (Harv.); stems ascending-erect, closely leafy throughout, simple, 1-headed, glandular-pubescent; leaves sessile, cobwebbed,
becoming sparsely hispidulous or glabrous above, albo-tomentose beneath, linear, acute, spine-pointed, with sub-revolute margins, quite
entire, ciliate with long, slender bristles, and stipelled at base with
similar; invol. sc. linear-subulate, shorter than the ample rays, setoso-
ciliate and white beneath, like the leaves; achenes turbinate, silky;
pappus twice as long as the achene, of many biseriate, cuneato-spathulate, obtuse, sharply toothed scales.

HAB. Attercliffe, Natal, *J. Sanderson!* 234, 264; *W. T. Gerrard, in litt.* (Herb. D.)
A very remarkable plant, nearly allied to *S. Zeyheri,* Sond., but with broader leaves,
white beneath; glandular, closely leafy stems, and much larger fl. heads.

29. S. Zeyheri (Sond. and Harv., non Less.); stems ascending-erect,
closely leafy below, laxly-leafy and subpedunculoid upwards, 1-headed,
glabrous, cobwebbed above; leaves sessile, cobwebbed, becoming glabrous, linear-subulate, spine-pointed, with strongly revolute margins,
quite entire, laxly ciliate and stipelled at base with long, slender bristles; inv. sc. reflexed during anthesis, subulate, rather longer than the
rays, setoso-ciliate, cobwebby beneath; achenes turbinate, silky; papp.
twice as long as the achene, of many biseriate, cuneate, sharply toothed
scales. *Hirpicium ? Zeyheri, Sond. in Hb.*

HAB. Wonderfontyn, *Zey.!* 975. Schoonström, *Burke!* (Herb. D., Hk., Sd.)
St. Zeyheri, Less., is a trifling variety of *S. atractyloides.*

30. S. sonchifolia (Harv.); stem herbaceous, terete, glabrous or minutely pulverulent; leaves thinly membranous, sparsely and minutely
hispidulous above, very thinly tomentose-canescent beneath, the upper
clasping (but not decurrent), oblong or lanceolate, more or less sinuate,
lower (and radical ?) on spine-bordered petioles, runcinato-pinnatifid, the
lobes very ample, in about 3 pair, oblong, repando-sinuate, the angles
or teeth tipped with small, slender spines, the interspaces minutely
spinellose, ciliolate, ribs and veins glabrous; heads radiate, pedunculate; invol. scales lanceolate, glabrous, bordered with slender, simple
spines, and ciliate; ovaries pubescent; papp. sc. short, obtuse.

HAB. Albany, *Mrs. F. W. Barber!* (Herb. Hk.)
This has the delicately membranous texture of leaf of *S. membranifolia,* but the
lower leaves are very differently shaped, and the upper not decurrent, &c. I have

seen but one specimen, consisting of 2 detached lower leaves and a flowering branch. It was collected some 20 years ago.

31. S. grandifolia (DC. l. c. 517); stem herbaceous, tall, branching, terete, glandularly pubescent, as are also the leaves on both sides; radical leaves very long (12–18 inches), petioled, deeply pinnatifid or pinnatipartite, the lobes ovate, acute, unequally and coarsely spinous-toothed, the lower smaller, nerves 4–5 in each lobe, subparallel, running from the midrib through the lobe; cauline lvs. clasping at base, slightly adnate; heads on long pedicels, radiate; inv. sc. linear lanceolate, attenuate, about equal to the rays, margined with 6–7 solitary (or sometimes geminate) spines, glandular; achenes pubescent; papp. sc. 2-seriate, oblong, acute, dentate.

HAB. Witberg, *Drege!* (Herb. Hk.)

32. S. viscosa (DC. l. c. 517); "stem herbaceous, erect, branched, terete, glandularly puberulous and viscidulous; leaves viscid, gland-pubescent above, glabrous beneath, half-clasping, the lower sessile, pinnatifid, the lobes palmately spine-toothed, the cauline shortly decurrent, dilated at base, less incised; heads subpaniculate; inv. sc. spreading, rigid, lanceolate, acuminate, longer than the rays, bearing toward the base solitary or geminate spines; achenes pubescent; pappus sc. very short, separate, 1-seriate." *DC. l. c.*

HAB. Blaauwberg, *Drege!* (Herb. Sond.)
I have only seen a frustule; and therefore chiefly copy DC.'s description.

33. S. ? buphthalmoides (DC. l. c. 517); "stem herbaceous, erect, somewhat angled, sparsely setose, branches glabrous or cobwebbed; leaves above closely covered with jointed, soft, short hairs, white-woolly beneath, oblong, pinnatifid, tapering at base, the lobes rounded, ciliato-spinose, cauline sessile; inv. sc. spreading, bearing geminate marginal spines, longer than the disc; *rays* about equalling the disc, the inner-most narrower and shorter; achenes inversely prismatic, villous at the apex; papp. sc. 2-seriate, downy." *DC. l. c.*

VAR. β, **sublanata**; stems and upper sides of leaves cobweb-woolly; heads rather larger; inv. sc. rather longer than the leaves. *DC. l. c.*

HAB. Witberg, *Drege.* β, Blesboksvlagte; Los Tafelberg; and Stormberg, *Drege.*
Of this I have only seen 3 leaves and a fl.-head, in Hb. Sd.; I therefore nearly transcribe *DC.'s* sp. char.

34. S. oppositifolia (DC.! l. c. 517); stem herbaceous, erect, glabrous (or thinly cobwebbed); rad. leaves . . . . ; cauline leaves *opposite*, connate or perfoliate, glabrous, roundish-ovate or oblong, sinuate-toothed, the teeth bearing very long spines, the margins not ciliate; heads pedicellate, radiate; invol. sc. in few rows, broad, oblong, nearly equalling the rays, spinous-pointed and bordered with long, solitary spines; ach. silky; papp. sc. oblong, toothed, 2-seriate. *Rohria spinosa, Hb. Thb.!*

HAB. Cape, *Thunberg!* Silverfontein, *Drege!* Modderfontein, *Rev. H. Whitehead!* (Herb. Th., D., Hk., Sd.)
Readily known by its opposite and mostly connate leaves. Lvs. 1½–2½ in. long, 1–2 in. wide. Spines very long, yellow, slender. Lower and radical lvs. unknown. There are excellent specimens of this in Hb. Th.; why omitted by *Lessing* I know not.

35. S. cruciata (Harv.); herbaceous, erect, glaucous, quite glabrous; stem and branches terete; cauline leaves cordate-clasping at base, many-nerved and netted-veined, tapering into a subulate, spinous point, with similar, subulate, long marginal spine-lobes, and rounded, perfectly smooth interspaces; heads terminal, discoid; invol. sc. longer than the disc, spinoso-subulate, bearing 1–3 pair of marginal, simple or geminate, long, rigid spines; achenes silky-silvery; pappus sc. biseriate, oblong, toothed. *Stobœa glauca, DC. l. c.* 516. *Rohria cruciata, Th. Cap.* 619. *Berkheya cruciata, Willd. Less. Syn.* 76. *DC. l. c.* 506.

HAB. Oliphant's-rivier, *Thunberg !* Zwarteberg, *Drege !* (Herb. Th., Sond.)

Apparently a tall, divaricately branched plant, quite herbaceous. Cauline leaves 1½–2 in. long, ½–1 inch wide at base, very rigid. It is very different from *S. glabrata.* Why *Lessing* should have left it in *Berkheya* is unaccountable, as his own account of the pappus, "paleis obovatis, *obtusis,* denticulatis," clearly indicates a *Stobœa.*

36. S. glabrata (Th.! Cap. 620); stem herbaceous, robust, angularly compressed and ridged, thinly flocculent-tomentose or nearly glabrous; cauline leaves cordate-clasping at base, netted-veined, glabrous or cob-webbed, acuminate and spine-tipped, with 3–4 lanceolate-subulate, spine-tipped lobes at base, the rest of the margin unequally spinous, with erose or denticled interspaces; heads terminal, discoid, glabrous or cobweb-woolly; inv. sc. longer than the disc, lanceolate, acuminate, spine-margined at base; achenes silky-silvery; pappus scales 2-seriate, oblong, toothed. *Less.! Syn.* 67. *DC. l. c.* 516.

VAR. β, **lanata**; all parts cobweb-cottony. *St. lanata, Th. Cap.* 621.

HAB. Cape, *Thunberg !* (Herb. Th.)

A large and coarse-growing species, probably much branched. Branches filled with pith, strongly compressed and angular. Leaves 2–3–4 in. long (lower leaves probably much larger), 1½–2 in. wide, tapering from a broad base. Disc 1 inch or more in diameter. The sp. in Herb. Th. are flowering tops; the lower and medial cauline leaves are unknown. No recent collector has sent this plant.

37. S. pinnatifida (Th. Cap. 622); stem erect, terete, glandular-puberulous; cauline leaves narrowed toward the earclasping base, deeply pinnatifid, concolourous, glabrous or beneath puberulous, the lobes semi-lanceolate, spine-pointed, the whole margin and the *upper surface* (in a *greater* or *less* degree) set with slender, pungent spinules; heads discoid, clustered or corymbose; inv. sc. longer than the disc, semi-lanceolate, netted-veined, punctate, with binate basal spines and few or many marginal spinules; achenes thinly silky; pappus biseriate, short, denticled. *Less.! Syn.* 66. *DC. l. c.* 515.

VAR. β, **angustata**; inv. sc. lance-*subulate*, nar. ; lvs. v. spiny above. (Hb. Sd. Hk.)

HAB. Cape, *Thunberg !* Lion's Mt., *Eckl.!* Capetown, Dr. Alexander Prior! VAR. β. Voormansbosch, Swell.! *Zey.!* 3029. (Herb. Th., D., Sd.)

Very variable in the size of leaves and fl. heads, and also in the *spinules* on the upper surface of the leaves ; these last are most abundant on the dwarfest specimens ; in luxuriant plants they are few and far between. The expanded invol., in *Thunberg's* sp. is about ¾ in. from tip to tip diameter; in *Ecklon's* from 1–1½ in.; and in *Dr. Prior's,* fully 2 inches and upwards! Yet all these specimens probably grew within the circuit of a couple of miles, and there is quite a gradation in other respects between them. Var. β. has narrower inv. sc.; in foliage it precisely tallies with *Thunberg's* original specimen.

38. S. seminivea (DC ! l. c. 518); stem herbaceous, erect, terete, branching, cobwebbed above; leaves not decurrent, tapering at base into a winged, more or less spine-bordered petiole, oblong, pinnatifid or deeply incised, cobwebbed at length nude, or minutely pubescent (green) above, snow-white and woolly beneath, the lobes oblong or lanceolate, few toothed, the teeth and lobes spinous-mucronate, margins minutely revolute, spinello-denticulate; heads discoid, panicled, (small) inv. sc. rather shorter than the disc, cobwebby, margined with 2–3 spines; achenes thinly pilose; papp. sc. biseriate, obtuse, subdenticled.

HAB. Betw. Omtata and Omsamwubo, *Drege!* Natal, *T. Williamson! Gueinzius*, 604! Sutherland! *Grant! Gerr. and M'K.* 306. (Herb. D., Sd., Hk.)

A tall plant. Cauline leaves often 6–10 in. long, including the petiole, 4–5 in. wide; the lower leaves more deeply incised. Heads ½ in. diam.

39. S. rigida (Th.! Cap. 521); stem herbaceous, branching, terete, cobwebby-canescent; leaves rigid, sessile, half-clasping, the younger cobwebbed, the adult glabrate above, all white-woolly beneath, deeply pinnatifid or pinnatipartite, the lobes narrow, lanceolate or subulate, with revolute margins, spinous-tipped, the margin sometimes spinellose; heads with few *very short* rays, or discoid, aggregated, subracemose, short stalked; inv. sc. lanceolate, pungent, concave and woolly beneath, with 1–2 pair of sub-basal, geminate spines: achenes (Hb. Th.!) *glabrous* or (Hb. Eckl.?) very thinly silky.

HAB. Near the sea shore, about Capetown and Simon's bay, *Thunberg! Ecklon! McGillavray! Milne!* Cape Flats, *Wallich!* (Herb. Th., D., Hk., Sd.)

A hoary, much-branched plant, with narrow-lobed leaves, which are at first white on both sides, and afterwards green above. Leaves 2–4 in. long, spreading; lobes 1–2 lines wide. Heads small. I find perfectly glabrous ovaries on one of *Thunberg's* specimens; and on one from *Milne* (Hb. Hk.); but in *Ecklon's*, marked by *Lessing*, they are sparingly silky. In other respects the plants seem to agree. I have only seen *rays* on one sp. (Hb. Th.); are they generally present? They scarcely exceed the disc flowers in length.

40. S. bipinnatifida (Harv.); stem erect, suffruticose, terete, cobwebby-canescent, branching at the summit; cauline leaves sub-petiolate, clasping and stipell-spined at base, pinnatipartite, the lobes 3–5 pair, narrow, decurrent, deeply incised or pinnatifid, with revolute, entire margins, the lobes and lobules spinous-tipped, upper surface cobwebbed, becoming glabrous, lower white-woolly; heads discoid, small, subcorymbose at the ends of the branches; inv. sc. equalling the disc, narrow-lanceolote, bordered with simple spines; achenes silky; pappus short, denticled.

HAB. Howison's Poort, Albany, *H. Hutton!* (Herb. D.)

A tall, virgate plant, 2–3 ft. or more high, all parts more or less canescent. Leaves more divided than in any other; the lower truly bipinnatifid; lobes 1–2 lines wide. Heads very small, 4–5 lines diameter.

41. S. heterophylla (Th.! Cap. 622); stem herbaceous, erect or ascending, cobweb-woolly; leaves coriaceous, the adult glabrous above, albo-tomentose beneath, the radical and lower ones tapering at base into a petiole, coarsely sinuate-toothed, the teeth tipped with spines, cauline clasping, sinuato-pinnatifid, undulate, the lobes short, simple or 3-lobuled, the lobules tipped with stronger spines ; heads discoid, ending the branches, subcorymbose ; inv. sc. spreading or deflexed, concave and woolly beneath, cobwebby becoming glabrate above, generally bearing a pair of geminate spines near the base (sometimes also having 1-2 pair of solitary spines as well); young achenes covered with short, white, swollen hairs becoming in age glabrate; pappus sc. roundish, obtuse, sub-denticled. *DC.! l. c. p.* 516, also *S. adenocarpa, DC.! l. c.* 515. *Apuleia heterophylla, Less.! Syn. p.* 65.

HAB. Cape, *Thunberg!* Kochman'skloof, *Mundt.* Langekloof, Uitenhage and Albany, *Drege!* Waste places and cult. ground, Uitenhage, *E. & Z.!* Grasrugg, *Zey.!* 986. Zontag's R., *Zey!* 987. (Herb. Th., D., Hk., Sd.)

Varying from erect, 2-3 ft. high, to decumbent 10-15 in. long ; all the young parts hoary. I cannot distinguish *S. adenocarpa*, DC. ; its achenes are not covered with real *glands*, but with precisely similar swollen hairs to those on *S. heterophylla*, Th. ; these hairs are sometimes copious, sometimes scanty, but that is all : there is no difference in foliage or fl. heads.

42. S. atractyloides (Thunb.! Fl. Cap. 621); stem herbaceous, erect, glabrous, or the young parts very thinly cobwebbed ; cauline leaves soon becoming glabrous and glossy above, more or less tomentose or sometimes *glabrous* beneath, ear-clasping and adnate at base, rigid, inciso-pinnatifid, the lobes semilanceolate, with wide interspaces, tapering into a strong spine, the margin of lobes and interspaces ciliate with slender spinules; heads corymbose or panicled, discoid ; inv. scales widely spreading, ovato-lanceolate, or lanceolate, concave beneath, bearing strong geminate spines at base, and mostly ciliate with slender spinules; achenes copiously silky with white hairs ; pappus sc. short, subbiseriate, irregularly subconcrete. *DC. l. c.* 515? also *S. rubricaulis, DC.! l. c.* *Apuleia atractyloides, Less.! Syn.* 64. *Apuleia Zeyheri, Less. Syn. l. c.* *S. Zeyheri, DC. l. c.*

VAR. β. **carlinoides**, Less.! ; smaller and more glabrous in all parts, with more slender stems. *S. carlinoides, Th.! Cap.* 620. (Herb. Th.)

HAB. Roggeveld, *Th.!* About Capetown and in Zwellandam, *Eckl.!* Cape Flats, *Wallich!* Caledon, *Pappe!* (Herb. Th., D., Hk., Sd.)

Stem often red or reddish, becoming glabrous. Leaves either *green* or *white* beneath, varying much in this respect ; in Th. original specimen, though 'concolorous' in a dry state, a close inspection shows toment. beneath, which was probably once white. DC. says the achenes are *glabrous;* they are certainly not so in *Thunberg's* plant. I have not seen *Drege's* Namaqualand plant, quoted by DC. *S. Zeyheri*, Less.! of which I have seen the original, but now half-rotten specimen (in Hb. Sd.) marked by *Lessing* himself, does not seem to me to have any stable characters ; those he enumerates as separating it from *S atractyloides* are certainly variable. Of *Drege's S. Zeyheri*, I can say little, having seen but a rad. lf. and a fl. twig ; it is much more tomentose than *S. atractyloides* usually is.

43. S. biloba (DC. l. c. 516); "stem herbaceous, erect, terete, simple, cobwebbed ; leaves sparingly setulose above, cano-tomentose beneath, the radical obovate-oblong, tapering into a long petiole ; cauline oblong, half-clasping, all pinnatifid, *the lobes mostly* 2-*lobed*, lobules tapering into a spine, the margins here and there spinellose ; heads 1-3, pedicelled, discoid ; inv. sc. linear-lanceolate, having solitary, rarely geminate, marginal spines, the outer reflexed, inner erect, rather longer than the disc ; ach. subvillose ; papp. very short, uniseriate." *DC. l. c.*

HAB. Near Liefde, on hills, 1000 ft., *Drege.* (Unknown to me.)

(Imperfectly known Species.)

S. acarnoides (DC. l. c. 519); " stem herbaceous, erect, glabrous, furrowed ; leaves subglabrous above, white-woolly beneath, sinuato-pinnatifid, with rounded lobes, each carrying 3–5 spines and ciliato-spinellose in the interspaces, the radical tapering at base, subpetiolate, the cauline decurrent in lobed, ciliate and spiny stem-wings ; pedicels subnude, 1-headed; invol. sc. equalling the disc, glabrescent, margined with solitary spines; achenes glabrous, turbinate ; pappus sc. oval, *very minute*, denticled." *DC. l. c.*

HAB. Witberg, 6000–7000 ft., *Drege.*
This grows with *S. cirsiifolia,* to which it seems closely allied. In Hb. Sd. are two leaves from *Drege:* they tell but little of the plant.

S. erysithales (DC. l. c. 519); stem herbaceous, erect, branched, terete, glabrate, cobwebby above; leaves sparsely setoso-scabrid above, cano-tomentose beneath, scabrous on the midrib, the lowest pinnatifid, cobwebbed beneath, tapering to the base, the lobes ample, ovate, bearing 5–7 slender spines, and spinelloso-ciliolate in the interspaces, the cauline decurrent in spinous stem-wings, the uppermost half-clasping ; heads radiate, on nude, pubescent pedicels ; inv. sc. equalling the disc, scarcely cobwebbed beneath, the margin ciliato-spinellose ; achenes glabrous; pappus uniseriate.

HAB. Omsamcaba, near the river's mouth, *Drege.*
Of *Drege's* plant I have seen merely a leaf-lobe (Hb. Sd.). Judging by description, it must come very near *S. discolor.*

S. carlinæfolia (DC. l. c. 516); " stem herbaceous, erect, straight, substriate, puberulous ; radical leaves . . . . ; cauline shortly decurrent, erect, oblong, glabrescent on both sides, sinuato-pinnatifid, long-spinous-toothed ; heads subracemose, small, discoid ; inv. sc. lanceolate, spinous-toothed and ciliate, twice as long as the disc ; achenes silky ; papp. sc. 1-seriate, linear-oblong, subacute." *DC. l. c.*

HAB. Nieuweveld, betw. Zakrivierspoort and Leewenfont, *Drege.*

CXXXIX. BERKHEYA, Ehrh.

Heads radiate, rarely discoid; *ray-fl.* ligulate, neuter; *disc-fl.* 5-toothed, perfect. *Recept.* honeycombed or deeply pitted. *Inv. sc.* free or more or less concrete, in few or many rows, spinous-pointed. *Fil.* smooth. *Achenes* mostly silky or pubescent, rarely glabrous. *Pappus sc.* biseriate, lanceolate or subulate, acute or much acuminate, serrato-fimbriate or ciliato-serrulate. *DC. Prodr.* 6, *p.* 504.

Herbs, often thistle-like, or small shrubs. Leaves rigid, more or less ciliate or spinous-toothed. Heads solitary, or corymbose, terminal. Fl. yellow. Named in honour of *Lefranq von Berkhey.* It is only known from *Stobæa* by its acute or taper-pointed pappus scales.

Sect. 1. EVOPIS (Cass.). Herbs, with *large*, oblong or obovate, subpetiolate radical and lower leaves, and a subsimple, 1- or few-headed stem, clothed with *small* stem-clasping, spinoso-ciliate, cauline leaves. *Achenes* silky. *Pappus-sc.* lanceolate, serrato-fimbriate. (Sp. 1-6.)

Inv. sc. quite glabrous :
 All the inv. sc. concave, erect, entire-edged ... ... (1) **cynaroides.**
 Outer inv. sc. rigidly ciliate or pectinate :
  Inner inv. sc. lanceolate, *concave,* entire ... ... (2) **scolymoides.**
  Inner inv. sc. lance-linear, *flat,* much acuminate,
   ciliate or subentire ... ... (3) **carthamoides.**
Inv. sc. pubescent :
 Subradical leaves *glabrous* above, tomentose beneath :
  Outer invol. scales *squarrose,* sinuous, cordate or
   oblong ... ... (4) **carlinoides.**
  Outer inv. sc. *straight,* ovato-lanceolate, acumi-
   nate ... ... (5) **Ecklonis.**
 Subradical and cauline leaves *on both sides green,*
  softly hairy ... ... (6) **pungens.**

Sect. 2. BASTERIA (Less.). Branching shrubs or rigid herbs, uniformly leafy.
Leaves all pinnatipartite, alternate. *Achenes* sparingly pubescent or glabrous. *Papp.
sc.* lanceolate, serrato-fimbriate. (Sp. 7–10.)

Shrubs : leaf-lobes smooth above :
 Leaf-lobes approximate, the upper long, lower *suddenly* shorter :
  Lf.-lobes subglabrous beneath, with *reflexed* edges;
   inner invol. sc. short, broadly oblong, concave,
   mucronate ... ... (7) **eryngiifolia.**
  Leaf-lobes woolly beneath, with *revolute* edges;
   inner inv. sc. lanceolate-acuminate ... ... (8) **palmata.**
 Leaf-lobes distant, the lower *gradually* shorter ... (9) **pinnata.**
Herb : leaf-lobes rib-striate, spinous on the striæ above ... (10) **stobæoides.**

Sect. 3. EU-BERKHEYA (Less.). Shrubs, suffrutices, or herbs, uniformly leafy.
Leaves undivided, spinoso- or setoso-ciliate. *Achenes* silky. *Pappus-sc.* short, lanceolate, acute or subacuminate, serrate or lacerate. (Sp. 11–19.)

Leaves opposite :
 Leaves linear or lance-linear, with revolute margins (11) **lanceolata.**
 Lvs. oblong-elliptical, narrowed at base, flat ... ... (12) **grandiflora.**
Leaves alternate :
 Shrubs, with woolly or canous branches :
  Lvs. leathery with recurved edges, glabr. above,
   woolly beneath, subpetiolate, obovate, distantly
   spine-toothed ... ... (13) **coriacea.**
  Lvs. thin, flat, cobwebbed above, canous beneath,
   on spine-bordered petioles, oval or oblong, dis-
   tantly spine-toothed ... ... (14) **incana.**
  Lvs. on both sides woolly, cuneate-obovate, ta-
   pering at base, 3–5-spine-toothed at apex ... (15) **cuneata.**
  Leaves green and veiny above, woolly beneath,
   rigid, sessile, obovate, undulato-sinuate, spinous-toothed ... ... (16) **Dregei.**
 Rigid, subsimple herbs, with glabrous, striate stems :
  Lvs. ovate-oblong, green above, snow-white beneath ... ... (17) **seminivea.**
  Lvs. linear-subulate, with revolute margins, glabrous .. .. (18) **subulata.**
 A flaccid herb, with setose stems; lvs. obl.-lanceolate,
  white beneath, setose above, rigidly ciliate, but not
  spinous ... ... (19) **gazanioides.**

Sect. 4. TRICHODES (DC.). Shrubs, with alternate, undivided, or sinuoso-pinnatifid leaves, the teeth and lobes spinigerous. *Achenes* silky. *Papp. sc.* bristle-
shaped, tapering to a hairlike point. (Sp. 20–22).

Glabrous : lvs. obovate, sessile ... ... (20) **obovata.**

Branches (at least) tomentose or canous; lvs. generally so:
　Leaves petioled, elliptic or ovate, spine-stipelled ... ... (21) **canescens.**
　Lvs. sinuoso-pinnatifid, the lobes bearing long spines ... ... (22) **corymbosa.**

SECT. 5. TRICHOCOMA (DC.). Herbs with large, petiolate, spinulous on the upper
surface, radical and subradical leaves. Stem simple, nude or winged with the
decurrent bases of depauperated stem-leaves. *Achenes* pubescent. *Papp. sc.* bristle-
shaped, tapering to a hairlike point. (Sp. 23-24).

Leaves obovate-obl., tapering into a long petiole; stem nude,
　few-headed ... ... (23) **setifera.**
Lvs. lyrato-pinnatifid, on a lobulate petiole; stem spinous-winged;
　heads umbellate, on woolly pedicels ... ... (24) **umbellata.**

Sect. 6. PSEUDO-STOBÆA. Rigid, branching herbs, uniformly leafy. Leaves
stem-clasping, sinuoso-pinnatifid, spinous toothed. *Heads* discoid. *Achenes* minutely
pubescent. *Papp. sc.* very short and narrow, lanceolate, acute, serrate. (Sp. 25).
Stem cobwebbed; lvs. glabr. above, canous beneath; inv. sc. broad (25) **annectens.**

Sect. 1. EVOPIS. (Sp. 1-6).

1. B. cynaroides (Willd. Sp. 3. p. 2275); stem herbaceous, simple
or racemose at the summit, angular, subglabrous; lower leaves oblong
or lanceolate, subacute, tapering much at base, subpetioled, entire or
remotely denticled, glabrous above, woolly beneath; upper (small) amplexicaul, oblong or lanceolate, spinoso-ciliate; inv. sc. glabrous, erect,
imbricate, ovato-lanceolate, concave, shining, pungent, entire-edged.
Less. Syn. 69. *DC. l. c.* 504. *Rohria cynaroides, Vahl., Th. Cap.* 619.
B. scolymoides, Drege! in *Hb. Hk., nec DC.*

HAB. Cape, *Thunberg! Mundt!* Babylonscher Toornberg, *E. & Z.! Zey.!* 3030.
(Herb. Th., Hk., Sd.)
　Stem 1-1½ ft. high. Lower lvs. 5-6 in. long, ½-¾ in. wide, frequently quite entire.
Heads racemose, on 1-1½ in. long pedicels, erect; outer inv. sc. short, ovate, inner
more and more lanceolate, all rigid, dry, fulvous. Recept. deeply pitted, the cells
with jagged, membranous edges. Achenes clothed with long, silky hairs. Papp.
sc. lanceolate, acuminate, serrato-ciliate. A specimen from *Drege*, in Hb. Hk.,
marked "*B. scolymoides*," precisely agrees with *B. cynaroides*, Less.!

2. B. scolymoides (DC. l. c. 504); stem herbaceous, simple or racemoso-paniculate at the summit, striate, minutely puberulous; lower leaves
oblongo-lanceolate, pungent, tapering at base into a petiole, glabrous
above, woolly beneath, sinuous, the short lobes armed with 1-2 strong
spines; medial lvs. similar, but sessile; upper clasping, glabrous on
both sides, many-nerved at base and veiny, closely pectinate with
spinous cilia; inv. sc. glabrous, erect, imbricate, concave, the outer
ovate, rigidly ciliate, the inner lanceolate, equalling the rays, quite
entire.

HAB. Tulbagh, *Dr. Thom!* Paarlberg, *Drege*, fide DC., Hott. Holland, *E. & Z.!*
(Herb. Sd., Hk.)
　I describe from *E. & Z.*'s specimen, which quite agrees with *DC.*'s description;
whereas a sp. from *Drege* (in Hb. Hk.) marked "*scolymoides*, DC." is unquestionably *B. cynaroides*. The present is nearly intermediate between the last named
species and *B. carthamoides*. In *Dr. Thom's* sp. (Hb. Hk.) the heads are shortly
racemose.

3. B. carthamoides (Willd. Sp. 3. p. 2274); stem herbaceous, simple
or corymbose at the summit, angular, glabrous or cobwebbed; lower

leaves oblong or oblongo-lanceolate, 3-nerved, tapering much at base, subglabrous above, woolly beneath, repand or sinuous, undulate, more or less bordered with strong spines; upper stem-clasping, oblong or cordate, many-nerved and veiny, glabrous or canous beneath, closely pectinate with spinous cilia; inv. sc. glabrous (or the outer cobwebbed), erect, imbricate, the outer shorter and broader, pectinato-ciliate, inner lance-linear, flat, much acuminate, ciliate or subentire. *Less. Syn.* 70, *DC. l. c.* 505. *Rohria carthamoides, Th.!* Cap. 618. *Rohria armata, Vahl.*

VAR. β. **uniflora**; stem one-headed, outer inv. sc. ovato-lanceolate or lanceolate. *B. uniflora, Willd. l. c. Less.! Syn.* 71. *DC. l. c. Rohria monanthos, Th.!* Cap. 618. (Hb. Th.)

HAB. Round Capetown and in the W. districts, common. (Herb. Th., D., Hk., Sd.)

Known from the preceding by the very long, narrow, flat and often ciliate inner inv. scales. Var. β. appears to me to be a very trifling variety.

4. B. carlinoides (Willd. Sp. 3. p. 2275); stem herbaceous, simple or racemose at summit, angular, pubescent; lower leaves obovate-oblong or oblong, tapering much at base, glabrous above, woolly beneath, repando-sinuous, spinous toothed; upper cordate-clasping, acuminate, recurved, hispidulous on both sides, strongly spinous-toothed, undulate; inv. sc. pubescent, the outer squarrose, undulate (leaflike), sinuous, spinous-toothed, cordate or oblong, inner lance-subulate, much acuminate, bordered with *long* spines. *Less. Syn.* 72. *DC. l. c.* 505. *Rohria carlinoides, Th.!* Cap. 619.

HAB. Devil's Mt., Capetown, *Thunb.!* Lion's Mt., *Mundt,* fide Less. (Herb. Th.)

I have only seen the single specimen in Hb. Thunb.; one from *Ecklon* (Hb. Sd.) marked '*carlinoides,*' is certainly only a var. of *B. carthamoides.* *Lessing* does not notice the *broad,* leafy outer inv. sc., which are very conspicuous in *Thunberg's* specimen, and quite unlike the outer sc. of any other of this section.

5. B. Ecklonis (Harv.); stem herbaceous, racemose at summit, pubescent; lower leaves obovate-oblong, tapering much at base, glabrous above, tomentose (becoming glabrate) beneath, sinuous and spinous-toothed; upper cordate-clasping, hispidulous on both sides, strongly spinous-toothed; inv. sc. pubescent, spreading, the outer straight, ovato-lanceolate, acuminate, inner lance-subulate, much attenuate, all bordered with long, slender spines.

HAB. Swellendam? *Ecklon!* (Herb. Sond.)

This has the lower leaves of *B. carlinoides;* and the involucre of *B. pungens;* it is almost exactly intermediate therefore between these 'species.' It may be the plant quoted by *Lessing,* as from *Ecklon,* under *B. carlinoides.*

6. B. pungens (Willd. l. c.); stem, leaves on both sides, and inv. sc. softly pubescent; lower leaves oblong, narrowed to the base, pungent, somewhat sinuous and spinous-toothed, upper cordate-ovate, clasping, many-nerved, veiny, unequally spinous-toothed; inv. sc. spreading, the outer lanceolate, the inner lance-linear, acuminate, all ciliate with long spines. *Less.! Syn.* 73. *DC. l. c. Rohria pungens, Th.!* Cap. 618.

HAB. Cape, *Thunberg!* (Herb Th.)

Known from all the section by its copious pubescence, which is probably viscidulous when fresh. The lower leaves, in the solitary specimen seen by me, are not much longer than the upper; there is no *woolliness* on any part of the plant.

Sect. 2. BASTERIA. (Sp. 7–10.)

7. B. eryngiifolia (Less.! Syn. 73); shrubby, branches closely leafy to the summit, pubescent (not woolly); leaves very rigid, squarrose, puberulous on both sides, pinnatisect, upper segments 2–3 pair, lance-subulate, pungent, with reflexed margins, lower short toothlike; inv. sc. puberulous (not woolly), the outer spine-toothed at base, lanceolate, acuminate, the inner much shorter, broadly oblong, concave, mucronate. *DC. l. c.* 505. *Rohria palmata, litt. β, Th.! in Herb.*

HAB. Cape, *Thunberg!* (Herb. Th.)
Very similar to *B. palmata* in foliage and general aspect, but without woolliness, with rather broader leaves, and especially differing in the shape of the inner inv. scales. *Ecklon's* plant (quoted with doubt by DC.) belongs to *B. palmata.*

8. B. palmata (Willd. l. c. 2271); shrubby, branches closely leafy to the summit, woolly or flocculent; leaves very rigid, squarrose, glabrous and glossy above, cobweb-cottony beneath, pinnatisect, the segments close-set, upper much longer, subulate, pungent, keeled above, with revolute margins, lower short, spinelike; outer inv. sc. flocculent, spine-toothed at base, inner entire, lanceolate-acuminate, longer than the rays; achenes sparsely hispid; papp. sc. lanceolate, fimbriate. *Less. Syn.* 74. *DC. l. c.* 505. *Rohria palmata, Th.! Cap.* 620. *Heterorachis spinosissima, Sch. B. in Bot. Zeit. xxvii.* 775. *Walp. Rep.* 6, 278.

HAB. Cape, *Thunberg!* Tigerberg, *Mundt.!* Bergriver, *Ecklon!* (Herb. Th., D., Hk., Sd.)
An exceedingly rigid, coarse-growing shrub, 1–3 ft. high. Leaves 1–2 in. long, their upper lobes ¾–1 in. long, 1–1½ l. wide, strongly revolute.

9. B. pinnata (Less.! Syn. 75); shrubby, branches closely leafy to the summit, woolly; leaves very rigid, erecto-patent, woolly beneath, becoming glabrous and glossy above, pinnatisect, the segments *remote*, in several pair, the upper subulate, with revolute margins, pungent, lower gradually shorter and narrower, lowest spinelike; inv. sc. in. few rows, spreading, lanceolate-acuminate, the outer spine-toothed at base; ovaries villous; papp. sc. fimbriate. *DC. l. c.* 505. *Stobæa pinnata, Th.! Cap.* 622.

VAR. β. **minor**; smaller in all parts, with shorter and narrower leaves. (Hb. Hk.)
HAB. Bockland, *Thunb.!* Onderbokkeveld and Zeederbergen, *Drege!* VAR. β, Mountain-peaks in Graaf-Reynet, *Mrs. F. W. Barber.* (Herb. Th., Sd.)
Habit of *B. palmata*, but with more erect and longer leaves, and more numerous and distant leaflobes, &c. Leaves 2½–3½ in. long, the upper lobes 1–1½ in long. Var. β, has much smaller leaves, with narrower lobes, and smaller fl.-heads, but does not differ by a tangible character: its lvs. are 1–1½ in. long, lobes ½–¾ in. long, ½ l. wide.

10. B. stobæoides (Harv.); herbaceous, rigid, paniculately branched; stem striate, subglabrous, viscidulous; leaves alternate, spreading, half-clasping, and spine-stipelled at base, pinnatisect, the segments in about 3 pair, broadly subulate, with revolute margins, pungent and spinoso-

ciliate, closely rib-striate, with spinules on the striæ above, glabrous
beneath, often 1-lobed at base ; heads small, corymbose, discoid ; inv.
sc. subulate, longer than the disc, with revolute, spinoso-ciliate margins;
ovaries glabrous; papp. sc. lanceolate, acuminate, fimbriate.

HAB. Rhinoster River, *Burke and Zeyher! Zey.! 982.* (Herb. Hk., D., Sd.)
A very distinct species, with the general habit of a *Stobœa*, but the pappus of a
Berkheya. Leaves 2-2½ in. long, their lobes ¾-1 in. long, 1-2 lines wide. Heads
few-flowered; bristles of the receptacle ½ as long as the flowers.

Sect. 3. EU-BERKHEYA. (Sp. 11–18.)

11. B. lanceolata (Willd. l. c. 2270); branches tomentose; leaves
opposite, or the upper alternate, sessile, convex and becoming quite
glabrous above, tomentose beneath, linear or lance-linear, pungent, with
revolute margins, distantly spinous-toothed; heads terminal, subpedunculoid; inv. sc. linear, leaflike, spine-bordered, shorter than the rays ;
achenes silky; papp. sc. short, lanceolate-acuminate. *Less.! Syn.* 76.
DC. l. c. 506. *Rohria lanceolata, Th.! Cap.* 616. *R. revoluta, Vahl.*
B. grandiflora, var. angustifolia, DC.! in Hb. Drege.

HAB. Cape, *Thunberg!* Albany, *E. & Z.!* Algoa Bay, *Forbes!* Near Port Elizabeth, *Zey.!* 3023. Zuureberg, *Drege.* (Herb. Th., D., Sd., Hk.)
1–2 ft. high. Leaves 1-1½ in. long, 1-3 l. wide, subconnate, not narrowed at base.

12. B. grandiflora (Willd. l. c. 2271); branches tomentose; leaves
opposite or the upper sometimes alternate, flat, glabrous above, woolly
beneath, oblong or elliptical, narrowed to the base, spinous-toothed;
outer inv. sc. tomentose beneath, oblong or ovate, inner narrow, subglabrous, all bordered with *very long* spines ; achenes silky ; papp. sc.
short, lanceolate, acuminate. *Less.! Syn.* 77. *DC. l. c.* 506. *Rohria*
grandiflora, Th.! Cap. 619. *R. ilicifolia, Vahl.*

HAB. Rietvalley and Buffeljagts R., *Thunb.!* Hott. Holl. and Houw Hoek
Pass; and Pikenierskloof, *E. & Z.!* Goerre's Hoogte, *Mundt.!* Camp's Bay Hills,
W. H. H. Simonsbay, *C. Wright!* 337. (Herb. Th., D., Hk., Sd.)
Allied to *B. lanceolata,* but with much broader leaves and larger fl. heads. Leaves
1-1½ in. long, 4-8 lines wide, conspicuously narrowed to the base, the margin very
minutely recurved.

13. B. coriacea (Harv.); shrubby, the branches tomentose ; leaves
alternate, crowded, leathery, soon glabrous above, woolly beneath, obovate, tapering into a nude or few-spined petiole, with recurved margins,
distantly spine-toothed; inv. sc. oblongo-lanceolate, the outer dorsally
woolly, bordered with strong spines, inner narrower and shorter, subglabrous, ciliolate ; achenes silky; papp. sc. short, lanceolate, subacute,
toothed.

HAB. Port Beaufort, *Mundt!* (Herb. D., Hk.)
Allied in character to *B. incana,* but of much stronger growth, with tough and
thickish leaves and flower-heads as large as those of *B. grandiflora.* The leaves are
closely set, but all alternate.

14. B. incana (Willd. l. c. 2269); branches canescent; leaves alternate, oval or oblong, flat, on very short, half-clasping, spine-bordered
petioles, cobwebby at length glabrous above, thinly cano-tomentose
beneath; distantly spinous-toothed ; inv. sc. oblong or broadly linear,

bordered with slender spines, tomentose beneath; achenes silky; papp.
sc. lanceolate, acute, toothed. *B. fruticosa, Ehr. Less.! Syn.* 78. *DC.
l. c.* 506. *Rohria incana, Th.! Cap.* 617. *Gorteria asteroides, Lin. f.
Jacq. Ic. rar. t.* 591.

HAB. Cape, *Thunberg! Mundt!* Heerelogement, *E. & Z.!* Olifants R., and
Litt. Namaqualand, *Drege!* (Herb. Th., D., Hk., Sd.)
　　Rather a straggling suffrutex, with an ashen grey colour, the branches divaricate,
especially the flowering ones. Heads small ; the disc 4 l. across ; the expanded inv.
about 1 in. diam., in fruit.

15. B. cuneata (Willd. l. c. 2270); branches tomentose ; leaves
alternate, thickish, on both sides woolly, cuneate obovate, tapering at
base, 3–5 spined, toothed at the apex ; outer inv. sc. hoary, ovate-
oblong, acute, spine-bordered, inner longer and narrower; ovaries silky ;
papp. sc. *(fide Less.)* "much acuminate, unequal, plumose." *Less.!
Syn.* 79. *DC. l. c.* 506. *Rohria cuneata, Th.! Cap.* 617.

HAB. Karroo, *Thunb!* Gamko-karroo, *Mundt!* Zwellendam, *Dr. Thom! E. & Z!*
(Herb. Th., Sd., Hk.)
　　Leaves about 1 in. long, 3–4 l. wide at the apex, tapering much to the base.
Heads subsessile, small; the rays shorter than the inv.

16. B. Dregei (Harv.); shrubby, the branches tomentose ; leaves
alternate, rigid, thinly cobwebbed and veiny, becoming glabrate above,
woolly beneath, sessile, obovate, repando-sinuate, subundulate, and
unequally spinous toothed; inv. sc. lanceolate, dorsally woolly bordered,
with strong spines. *Berkh.* 2901, *α, Hb. Drege!*

HAB. Cape, *Drege!* (Herb. Sd.)
　　Of this seemingly distinct species I have seen but a small fragment, the fl. head
of which is completely worm-eaten. Leaves 2 in. long, 1–1½ in. wide.

17. B. seminivea (Harv. and Sond.); stem herbaceous, subsimple,
striate, minutely puberulous or glabrous, leafy; leaves alternate, sessile,
ovate-oblong or elliptical, green and minutely scaberulous above, snow-
white beneath, spine-stipelled at base and ciliate with rigid spines ;
heads 1–3, ending the nude, pedunculoid apex of the stem or branches ;
inv. sc. linear-lanceolate, glabrous, acuminate, closely spinoso-ciliate ;
ach. silky ; papp. sc. lance-oblong, subacute, toothed.

HAB. Magalisberg, *Burke and Zeyher, Zey.!* 794. (Herb. D., Hk., Sd.)
　　A very handsome species. Stems 1–2 ft. high, leafy for upwards of ¼ their length,
thence pedunculoid, 1–3 headed, pale green, the young ones very minutely puberulous. Leaves 1–1½ in..long, 5–8 lines wide, delicately penninerved and veiny above.
Heads of large size, radiate. The pappus scales are almost intermediate in character between those of *Stobæa* and *Berkheya*, but the habit is more of the latter genus.

18. B. subulata (Harv.); stem herbaceous, subsimple, terete, striate,
glabrous or near the summit gland-puberulous, leafy ; leaves approximate, alternate, glabrous, linear-subulate, with revolute margins, remotely spinoso-ciliate and spine-stipelled ; heads ending the laxly leafy
or nude apex of the stem ; inv. sc. subulate, acuminate, minutely glandular, spinoso-ciliate ; papp. sc. lanceolate, acuminate, incised, twice as
long as the silky achene.

HAB. Dry plains, Zululand, *Gerr. & McK.!* 1048. (Herb. D.)

A remarkable species, unlike any other *Berkheya*, but with somewhat the aspect of *Stobœa Zeyheri*, S. and H. Stem 1-2 ft. high ; the barren stems very densely, the flowering more laxly leafy. Leaves 1½-2 in. long, not 1 line wide.

19. B. ? gazanioides (Harv.) ; stem herbaceous, rigidly setose ; leaves alternate, subdistant, oblongo-lanceolate, subacute, tapering at base, entire, rigidly setose above and on the midrib beneath, tomentose beneath, rigidly-ciliate especially toward the base, but not spinous ; outer inv. sc. lineari-subulate, attenuate, sub-biseriate, pectinato-ciliate, inner much shorter, lanceolate, membr.-edged and purple-ribbed, ciliolate ; ovaries very silky ; papp. sc. nearly as long as disc-corolla, lanceolate, incised.

Hab. Damaraland, *Mrs. Kolbe!* (Herb. D.)
Of this most remarkable plant I have only seen a single flowering branch, about 6 inches long, bearing 3 distant leaves and a fl.-head closely subtended by 3 others. Leaves 2-2¼ in. long, 3-4 l. wide. Outer inv. sc. 1-1¼ in. long, ¼ l. wide. The *setæ* on all parts of the plant are *rigid*, but none are spinous. Rays many, longer than the inv. sc., yellow. In gen. char. it oscillates between *Gazania* and *Berkheya ;* but the inv. sc. are concrete at the very base only.

Sect. 4. TRICHODES. (Sp. 20-22).

20. B. obovata (Willd. l. c. 2269) ; *glabrous ;* leaves alternate, sessile, obovate or oblong-obovate, distantly spinous-toothed, pungent ; inv. sc. oblong or lanceolate, few-spined, the inner shorter and narrower; achenes silky ; papp. sc. very long, subulate, very much acuminate. *Less.! Syn.* 80. *DC. l. c.* 507. *Rohria obovata, Th.! Cap.* 617.

Hab. Karroo, *Thunberg! Drege!* Kamnario, *Mundt.* (Hb. Th., Hk., Sd.)
Readily known among its allies by the perfectly glabrous foliage.

21. B. canescens (DC.! l. c. 507) ; branches cano-tomentose ; leaves alternate, petiolate, elliptical or ovate, distantly spinous-toothed, on both sides thinly canous ; petiole at the base 3-5-spine-stipelled ; heads subcorymbose, pedicelled ; inv. sc. glabrescent, narrow, shorter than the rays, spine-bordered ; achenes silky ; papp. sc. bristle-shaped, tapering to a hair-like point.

Hab. Gariep, near Verleptpram, *Drege!* (Herb. D., Hk., Sd.)
Very like *B. incana* in general aspect ; but with larger leaves, smaller fl. heads, more glabrous inv. sc., and a very different pappus. A specimen from *Drege* (Hb. Sd.) marked " B. fruticosa, c." belongs to this species.

22. B. corymbosa (DC.! l. c. 507) ; branches cano-tomentose ; leaves alternate, sessile or subpetiolate, on both sides cano-tomentose (or *glabrate!*) sinuoso-pinnatifid or tooth-lobed, the lobes produced into long spines ; corymb compound, level-topped, many-headed ; inv. sc. about equalling the rays, lanceolate, glabrate, few-spined ; ach. silky ; papp. sc. bristle-shaped, tapering to a hair-like point.

Var. β. **glabrata** ; leaves becoming glabrous on both sides. (Herb. Hk., Sd.)
Hab. Zilverfontein, *Drege!* Modderfontein, *Rev. H. Whitehead!* β. Kamos, *Zey.!* 983. (Herb. D., Hk., Sd.)
Leaves 1½-2 in. long, ½ in. wide ; in more luxuriant specimens the lower leaves are subpetioled, the petiole with a pair of spine-toothed ears at base. *Zeyher's* 983 has nearly glabrous foliage !

Sect. 5. TRICHOCOMA. (Sp. 23–24).

23. B. setifera (DC. l. c. 507); leaves confined to the base of the stem, obovate or obovate-oblong, tapering much at base into a long, half-clasping petiole, subentire or distantly spinous-toothed, the margin and upper surface thickly covered with pungent bristles, otherwise glabrous, the under thinly pilose ; peduncular-stem nearly nude, elongate, terete, cobwebbed or glabrous, bearing about 3 pedicelled fl. heads ; inv. sc. linear-lanceolate, closely ciliate with slender spinules ; ach. pubescent ; papp. sc. very long, straw-colour, subulate-acuminate, tapering to hair-like points.

HAB. Omtata, *Drege.* Doornkop, *Burke!* (Herb. Hk.)
Of *Drege's* plant I have only seen a leaf, in Hb. Sd. ; I describe from *Burke's* (Hb. Hk.) which agrees well with *DC.'s* character. Leaves (including their petiole) 12–15 in. long, 3–4 in. wide. Peduncular-stem 2–2½ ft. high.

24. B. umbellata (DC. l. c. 507); lower and subradical leaves lyrato-pinnatifid, tapering at base into a lobulate and spine-toothed petiole, the leaf-lobes broad, longer or shorter, the margin and upper surface covered with short, pungent bristles, the under scaberulous ; cauline leaves small, linear, spine bordered, decurrent in long, sinuous, spinous stem-wings which clothe the stem to its summit; heads subumbellate, 10–12, on woolly pedicels ; heads discoid ; inv. sc. linear-lanceolate, ciliolate and spinous at base ; ach. pubescent ; papp. sc subulate-acuminate, tapering to hair-like points.

HAB. Omtata, and betw. Omtendo and Omsamculo, *Drege.* Natal, *Gueinzius!* 318, 596. Bullies' Grave. Brit. Kaffraria, *W. S. M. D'Urban,* 58. (Herb. Sond., D.)
With much of the habit of *B. setifera,* but readily known by its winged-stem and more or less deeply pinnatifid lower leaves, &c.

Sect. 6. PSEUDO-STOBÆA. (Sp. 25)

25. B.? annectens (Harv.) ; stem herbaceous, branching, terete, cobwebbed; leaves half-clasping at base, glabrous above, canescent beneath, sinuoso-pinnatifid, spinoso-acuminate, the lobes semi-lanceolate, spine-tipped, the interspaces sparingly spinellate ; heads small, few-fl., discoid ; inv. sc. in few series, concrete, ovate-acuminate, dorsally canous, few-spined at base, widely spreading, the innermost longest; recept. deeply pitted, the margins of the cells fimbriate; achenes ribbed, thinly pubescent; papp. sc. *very short, lanceolate, acute,* serrate.

HAB. Bitterfontein, *Zey.!* 984. (Herb. Hk., Sd.)
In habit, and every other character save *pappus,* this is like a *Stobæa* : but as the papp. scales, though small, are assuredly *acute,* I place it in *Berkheya.* It must be confessed however that it almost unites the two genera, being even more a *Stobæa* than *B. stobæoides.*

(Imperfectly known species.)

B.? hirpicioides (Sond.!); root-stock creeping?; stems erect, sub-simple, leafy, setose ; leaves alternate, linear-lanceolate, tapering at base into a setoso-ciliate petiole, closely echinate with sharp points above, tomentose beneath, with reflexed, distantly spine-toothed margins, setose on the midrib beneath; heads terminal, subsessile ; inv.

sc. lanceolate, gland puberulous, spinoso-ciliate at base only ; achenes and pappus unknown.

HAB. Aapjes R., *Zeyher!* (Herb. Sd.)
A small plant, very different in foliage from any other *Berkheya.* The only fl. head I have seen is completely worm-eaten; the genus is therefore at present doubt-ful. It may be a *Stobæa.*

CXL. **DIDELTA**, L'Her.

Heads mostly radiate ; the ray-fl., when present, ligulate, neuter; *disc-fl.* 5-toothed, perfect. *Recept.* honey-combed, the margins of the cells rigidly fimbrilliferous. *Inv. sc.* concrete at base, in two rows; the sc. of the rows very unequal, sometimes the outer, sometimes the inner scales largest. *Filam.* smooth. *Achenes* wingless. *Pappus* uniseriate, palea-ceous, the paleæ fimbriato-plumose. *DC. Prodr. 6, p. 503.*

Half-shrubs or herbs. Leaves opposite or alternate, entire or sinuate-toothed, unarmed or spinous. Heads terminal, peduncled, solitary. Fl. yellow. Name from δις, *double,* and δελτα, the Greek letter Δ (D) ; the inv. is *double,* and the inv. sc. in the original species *triangular.*

Sect. 1. CHORISTEA (Th.). Outer invol. sc. few, broadly ovate or cordate, inner small.

Shrubby ; leaves opposite, cordate or elliptical, veiny ... (1) **spinosum.**
Herbaceous ; lvs. alternate, oblongo-spathulate, veinless :

    Stem and leaves glabrous or nearly so ... ... (2) **carnosum.**
    Stem and leaves densely white-woolly ... ... (3) **tomentosum.**

Sect. 2. CUSPIDIA (Less.). Outer inv. scales numerous, very small, spine-tipped, inner larger, ovate-oblong or lanceolate, ciliate. Annuals.

Stems elongate ; lvs. clasping, obovate-oblong, not greatly
    tapering at base; ovaries nearly glabrous ... . (4) **cernuum.**
Stems short ; lvs. *greatly* tapering at base ; ovaries villous (5) **annuum.**

Sect. 1. CHORISTEA. (Sp. 1–3.)

1. D. spinosum (Ait. Kew. 3, p. 256); shrubby, glabrous; leaves op-posite, decussate, sessile, broadly cordate or elliptical, mucronate, quite entire or distantly spiniferous, midribbed and netted-veined beneath ; outer inv. sc. 4–5, very large, cordate-ovate. *Less.! Syn.* 60. *DC. l. c. 503. Choristea spinosa, Thunb. Cap.* 702.

HAB. Olifant's R., *Thunberg!* Tulbagh, *Dr. Thom!* Bergvalley, *Drege!* Heere-logement, Clanw., *E. & Z.! Zey.!* 978. Namaqualand, *A. Wyley!* Modderfontein, *Rev. H. Whitehead!* (Herb. Th., D., Hk., Sd.)
Young parts sometimes tomentose, the toment. soon wearing off. Leaves some-times clasping, sometimes simply sessile, 2–2½ in. long, 1–2 in. wide, of thin sub-stance, but rigid. Outer and medial invol. sc. leaf-like ; innermost very small and narrow, acute.

2. D. carnosum (Ait. l. c. 256); herbaceous, the young parts floccu-lent, then glabrous; leaves mostly alternate, fleshy, with immersed veins, oblongo-spathulate, tapering much at base; outer inv. sc. 3–5, deltoid-ovate, subserrulate, veiny, inner lanceolate, acuminate. *Less.! Syn.* 61. *DC. l. c.* 503 ? *D. tetragoniæfolium, L'Her. St. Nov. t.* 28. *Choristea carnosa, Th.! Cap.* 703.

HAB. Cape, *Thunberg!* Heerelogement, Clanw., *E. & Z.!* Mierenkasteel and Holriver, *Drege!* Sea shore, St. Helena Bay, *Dr. Wallich!* Namaqualand, *A. Wyley!* (Herb. Th., D., Hk., Sd.).

Root seemingly annual. Stem much branched from the base, erect or diffuse. Lower leaves sometimes opposite. Leaves 2–3 in. long, 3–5 l. wide, obtuse. Rays golden-yellow.

3. **D. tomentosum** (Less.! Syn. 61); herbaceous, stem and leaves on both surfaces *persistently* white-woolly; leaves alternate (or sometimes opposite), fleshy, veinless, subspathulate-oblong, not much narrowed at base; outer inv. sc. *3–5*, deltoid-ovate, subserrulate, veiny, inner lanceolate, acuminate. *DC. l. c. 503. Arctotis lancea, Hb. Th.! D. carnosum, litt. c. Hb. Drege!*

HAB. Cape, *Thunberg!* Rietvalley, Cape, *E. & Z.!* Near the mouth of the Gariep, *Drege!* Namaqualand, *A. Wyley!* (Herb. Th., D., Hk., Sd.)

Root seemingly annual. Stem branching, erect or diffuse. Leaves 1½–2½ in long, 3–4 l. wide. Rays deep orange. Very similar to *D. carnosum*, except that all parts are thickly covered with white woolliness.

Sect. 2. CUSPIDIA. (Sp. 4–5.)

4. **D. cernuum** (Less.! Syn. 62.); herbaceous, glabrous or nearly so, branching, diffuse; leaves alternate, half-clasping, oblong or obovate, acute, coarsely spinous-toothed, the margin spinelloso-ciliate; outer inv. sc. numerous, short, spine-tipped, inner leafy, oblong-lanceolate, ciliate-pungent, achenes glabrous, at the apex pubescent. *DC. l. c. 503. Gorteria cernua, Linn. f. Th.! Cap. 697. Berkheya cernua, R. Br., Hort. Kew. 2, v. 5, p. 139.*

HAB. Cape, *Thunbery!* *Krebs, Burchell,* &c., Uitenhage, *E. & Z.!* *Zey.!* 3019. Zuureberg, *Drege!* Somerset, *Mrs. F. W. Barber.* (Herb. Th., D., Hk., Sd.)

Root seemingly annual. Stems 6 inches to 1–2 ft. long, spreading or ascending, pale; leaves membranous, 1–1½ in. long, 4–6 l. wide, variably tooth-lobed. Inv. sc. concrete, as in *Gorteria*, imbricated in several rows, but of two kinds. Rays pale yellow. Pappus sc. narrow, lance-subulate, plumose.

5. **D. annuum** (Less. Syn. 61); herbaceous, cobwebby, simple or branched at base; leaves alternate, half-clasping, obovate-oblong, acute, *very much attenuated to the base,* and petiolate, coarsely spinous-toothed, the margin spinelloso-ciliate; outer inv. sc. few, short, spine-tipped; medial ovate-oblong, inner lanceolate, both leafy, spine-toothed, ciliate and pungent; ovaries villous.

HAB. Cape, *Mundt and Maire.* Betw. Beaufort and Rhinosterkops, and betw. Zwarteberg and Kendo, *Drege!* (Herb. Sond.)

Annual. Allied to *D. cernuum,* but smaller and less branched; with the leaves greatly tapering to the base, and broader inner inv. scales. Leaves 2–2½ in. long, 4–5 l. wide.

(Doubtful genera of Arctotideæ.)

DAMATRIS, Cass.

Heads radiate; *rays* female, uniseriate; *disc-fl.* male, 5-toothed. *Inv.* longer than the disc, hemispherical, scales imbricate, outer ovate, inner linear, subulate at apex. *Recept.* convex, with a row of scarious, 3-lobed, externally concave paleæ between the ray and disc. *Ray-achenes* subcylindrical, hairy, with long, soft, red hairs. *Pappus* 2-seriate, scaly. *Disc-achenes* abortive. *DC. Prodr. 7, p. 255.*

D. pudica (Cass.) A short-stemmed annual, with pedunculoid branches. Leaves alternate, half-clasping, linear-lanceolate, sinuate, albo-tomentose beneath. Heads solitary, terminal. Fl. yellow.

GIBBARIA, Cass.

Heads radiate ; *rays* female, uniseriate ; *disc-fl.* male, 5-toothed. *Inv.* equalling disc, hemispherical, scales in few rows, imbricated, spine-tipped. *Recept.* flat, nude. *Achenes* of ray short, thick, dorsally gibbous ; of disc sterile, compressed, striate, with a coroniform, dimidiate, very short, irregularly torn pappus. *DC. Prodr.* 7, 257.

G. bicolor (Cass.) A branching plant. Leaves alternate, long, semiterete, acute, 1-nerved, half-clasping, glabrous. Heads terminal, solitary ; disc yellow ; rays white above, yellow beneath.

Sub-tribe 2, MUTISIACEÆ. (Gen. 141–145).

CXLI. OLDENBURGIA, Less.

Heads many-fl., radiate, homogamous (all the fl. perfect) ; corolla of *disc-fl.* nearly regular, very deeply 5-fid ; of the *ray* bilabiate, the outer lip long, strapshaped ; the inner minute, bifid. *Inv. sc.* unarmed, linear, acuminate, several nerved, the inner herbaceous. *Recept.* nude. *Cor.* glabrous. *Filaments* smooth. *Anthers* tailed. *Style* glabrous, its branches very short, obtuse. *Achenes* turbinate, beakless. *Pappus* of many shortly plumose, equal bristles. *DC. Prodr.* 7, p. 12.

South African shrubs or suffruticose, dwarf plants. Rootstock very woolly. Leaves rosulate, sessile, coriaceous, one-nerved, obtuse, glabrous above, very hairy beneath. Heads solitary, of large size. Cor. purplish. Name in honour of *Oldenburg,* a companion of Thunberg in his S. African herborizations : he died afterwards in Madagascar of fever.

Inv. sc. standing loosely in many rows, very much acuminate :
    Nearly stemless, densely tufted ; heads immersed among the
      narrow fl. leaves ... ...   (1) **paradoxa.**
    Shrubby, branching ; heads on terminal peduncles, sub-
      corymbose ... ...   (2) **Arbuscula.**
Inv. sc. closely imbricated, connivent, scarcely acuminate ; pedunc.
    long ... ...   (3) **Papionum.**

1. O. paradoxa (Less. Linn. v. p. 252, f. 69–75) ; nearly stemless, from a thick woody rootstock ; leaves, on *barren* shoots, tufted at the end of the rudimentary branch, obovate-oblong, cuneate at base, with subrevolute margins, glabrous and impress-nerved above, tomentose and strongly ribbed and penninerved beneath ; *fertile* stems simple or multifid, very short, closely leafy, their leaves tongue-shaped, narrow, with revolute margins, tapering into a very densely shaggy petiole ; heads sessile among the leaves ; inv. sc. very shaggy at base, very much acuminate and subglabrous at apex ; achenes silky-villous, short. *DC. l. c.* 12.

HAB. Cape, *Mundt and Maire.* Eastern parts, *Burchell.* Swellendam, *E. & Z.!* R. Zondereinde, *Zey.!* 3073. Top of Georgetown Mt., forming dense tufts, *Dr. Alexander Prior!* (Herb. D., Sd., Hk.)

Stems 1–1½ in. long. The leaves on *barren* stems are very similar in size and character to those of *O. Arbuscula :* those on *flowering* stems (which alone have been described by *Lessing* and *De Candolle*) are 2–2¼ in. long, 3–5 lines wide, nearly veinless, glabrous above, woolly beneath.

2. O. Arbuscula (DC.! l. c.) ; stem elongate, shrubby or arborescent, branched, very shaggy ; leaves crowded at the ends of the branches, obovate-oblong, cuneate at base, convex above, with subrevolute margins, glabrous and impress-nerved above, tomentose and strongly ribbed and penninerved beneath; flowering branches elongate, subpedunculoid, laxly leafy, their leaves alternate, oblongo-lanceolate or tongue-shaped, subacute, with scarcely recurved margins, and a short, woolly petiole ; heads pedicellate, subcorymbose; inv. sc. in very many rows, tomentose at base, very much acuminate, and becoming subglabrous ; achenes glabrous, elongate. *Arnica grandis, Thunb.! Cap. p.* 668.

HAB. Cape, *Thunberg!* Zuureberg, *Drege!* Uitenhage and Albany, *E. & Z.! Cooper,* 1543, &c. (Herb. Th., D., Hk., Sd.)

A coarse shrub or small tree, 12–15 ft. high, the principal stem measuring 1½ ft. circumference. Leaves 6–9 in. long, 3–5 in. wide, very rigid and leathery. Fl. branch or peduncle 1–2 ft. long, ¾–1 in. thick. Heads very large, in fruit 4–5 in. diameter.

3. O. Papionum (DC.! l. c.); stem very short, thick and woody, tomentose, closely marked with leaf-scars ; leaves crowded at the apex of the stem, obovate-oblong, cuneate-attenuate at base, with subrevolute margins, glabrous and nerveless above, tomentose, strongly ribbed and penninerved, with netted interspaces beneath ; fl. stem pedunculoid, elongate, glabrous, very laxly leafy, its lower leaves similar to the cauline, but smaller, upper much depauperated ; heads 1–2, long-pedicelled ; inv. very tomentose, connivent, urceolate, its scales densely tomentose, mucronate or subacuminate, closely imbricate; achenes ? *Scytala Papionum, E. Mey.! in Hb. Drege.*

HAB. Mts. near Tulbagh and Nieuwekloof, *Drege!* (Herb. D., Hk, Sd.)

Leaves 5–7 in. long, 2–2½ in. wide, much narrower than in *O. Arbuscula.* Flowering stem 2–3 ft. high, striate, minutely tomentulose, becoming glabrous. I have only seen immature fl. heads and old, denuded receptacles. Flowers unknown. The habit is completely similar to that of the other species ; the inv. very different.

CXLII. **PRINTZIA**, Cass.

Heads heterogamous, radiate ; *ray-fl.* ligulate (rarely bilabiate), female; *disc-fl.* regular, deeply 5-lobed, the lobes revolute. *Inv.* scales imbricate, lanceolate. *Recept.* nude, honeycombed. *Fil.* glabrous. *Anth.* exserted, tailed at base. *Style-branches* obtuse or acute ; in the disc-fl. short, equal, downy ; in the ray longer. *Achenes* beakless, oblong, villous. *Pappus* copious (except in *P. Huttoni*), in many rows, of shortly plumose or barbed bristles. *DC. Prodr.* 7, *p.* 13.

Small, branching, S. Afr. shrubs, more or less tomentose. Leaves alternate, sessile, crowded, cobwebbed or glabrate above, white-woolly beneath. Heads terminal, solitary. Rays white, blue or purple ; disc yellow, like those of an *Aster.*

Leaves *decurrent* in long, narrow stem wings ; pappus rufous ... (1) **Bergii.**
Leaves *sessile,* but not decurrent :
   Leaves obovate, amply auricled at base ... ... (2) **auriculata.**
   Leaves narrow-linear, with revolute margins ... ... (3) **aromatica.**
Leaves *petiolate :*
   Lvs. ovate or subrotund, toothed, woolly beneath ; rays
     elongate ... ... (4) **pyrifolia.**
   Lvs. obl.-lanceolate, denticled, green beneath ; rays very short (5) **Huttoni.**

1. P. Bergii (Cass.); branches tomentose, winged with the narrow, decurrent bases of the oblong or oblong-obovate, undulate, mucronate leaves, which are cobwebby, becoming glabrate above, tomentose beneath; inv. sc. 2–3 seriate, narrow, subequal; rays elongate; pappus rufous. *Less. Syn.* 108. *DC. l. c. 13. Inula cernua, Berg. Cap.* 288. *Leyssera polifolia, Th.! Cap.* 692.

HAB. Hills round Capetown, *Bergius, E. & Z.!. W. H. H.* Algoa Bay, *Forbes.* Auteniqualand, *Burchell.* Caledon, *E. & Z.* Rietkuil, *Zey.!* 3074. Zwarteberg, *Drege.* Paarlberg, *W.H.H.* Genadenthahl, *Dr. Roser!* Steendaal, Tulb., *Dr. Pappe!* (Herb. Thunb., D., Hk., Sd.)

A much-branched bush, 2–3 ft. high and in diameter. Leaves ¾–1½ in. long, 3–6 l. wide, either quite entire or remotely callous-toothed. Rays blue. Flowers like those of an *Aster*, in which genus Linnæus placed it.

2. P. auriculata (Harv.); branches tomentose; leaves sessile, *amply auricled*, obovate, cuneate at base, entire or 5–7-toothed beyond the middle, cobwebbed, becoming glabrous above, tomentose beneath; inv. sc. imbricated in 4–5 rows, the outer shorter, all lanceolate, subsilky; rays elongate; pappus white.

HAB. Bashee R., Fort Bowker, *H. Bowker!* 370. (Herb. D.)

In the shape and pubescence of the leaves this is like *P. Bergii;* but the leaves are *not* decurrent, but auricled and stem-clasping; the inv. sc. are pluriseriate, and the pappus is white. The rays may have been either blue or white.

3. P. aromatica (Less. Syn. 108); branches tomentose; leaves sessile, narrow-linear, obtuse, with revolute margins, entire, on both sides densely tomentose; inv. sc. imbricated in 3–4 rows, the outer shorter, all oblong, obtuse, tomentose; rays elongate; pappus white. *DC. l. c. 13. Inula aromatica, Linn. Amoen. 6, p.* 103. *Th.! Cap.* 667.

HAB. Cape, *Thunberg, E. & Z.!* Paarlberg, *Drege!* French Hoek, *W. H. H.* Winterhoek, Tulbagh, and Genadenthahl, *Dr. Pappe!* (Herb. Th., D., Sd., Hk.)

A very much-branched and leafy undershrub, 6 inches to 2 ft. or more high, subsimple below, the branches corymbose. All parts tomentose. Leaves ½–¾ in. long, 1 l. wide. Heads subcorymbose; rays purple (sometimes white?).

4. P. pyrifolia (Less. Syn. 108); branches tomentulose, in age glabrate; leaves shortly petioled, ovate or subrotund, broad and subcordate at base, sharply many-toothed, glabrous and netted above, rigid. tomentose beneath; inv. sc. imbricate, in 5–6 rows, the outer shorter, all fringed with woolly hairs; rays elongate; pappus whitish. *DC. l. c. 13.*

HAB. Kaffirland, *Krebs, E. & Z.!* Witberg, *Drege!* Albany, *T. Williamson!* Basutuland, *T. Cooper,* 691. Tongaat R., Natal, *Mrs. Saunders!* (Hb. D., Hk., Sd.)

A shrub, 3–4 ft. high. Leaves varying much in size; in *E. & Z.* sp. 1¼ in. long, 1 inch wide; in *Drege's* about ¼ inch long and wide, or less, in all very rigid. Fl. purple?

5. P. Huttoni (Harv. Thes. Cap. t. 158); stem diffusely branched, subscandent; branches flexuous, terete, scabro-puberulous; lvs. shortly-petioled, oblong-lanceolate, membranaceous, penninerved, on both sides scaberulous, distantly denticulate; petiole ear-clasping at base; heads ending the branches and axillary twigs, solitary, short-pedicelled; inv. scales very narrow, acuminate, pubescent, 2–3-seriate; rays *very short*, spoon-shaped; pappus of few, unequal bristles; achenes hispidulous.

HAB. Katberg, *H. Hutton!* (Herb. D., Hk., Sd.)

A tall, scrambling or half-climbing, slender shrub. Leaves 2–3 in. long, ¾–1 in. wide; petiole ¼–½ in. long.

CXLIII. **DICOMA**, Less.

Heads many-fl., homo- or hetero-gamous; *ray-fl.*, when present, neuter, ligulate, bilabiate or terete-tubular; *disc-fl.* regular, 5-parted, the lobes longer than the tube: *corollas* very generally pubescent externally. *Inv.* campanulate, equalling the disc ; the scales in many rows, coriaceous, obsoletely many-nerved, entire, acuminate, sometimes pungent, broad or narrow. *Recept.* honey-combed. *Anthers* long-tailed, the tails barbed; *fil.* glabrous. *Style-branches* short, erect, obtuse, externally hispid at the apex. *Achene* turbinate, villous, beakless, often 10-ridged. *Pappus* in 2 or many rows, of equal or unequal, shortly plumose bristles. *DC. Prodr.* 7, *p. 36.*

Small suffrutices or half-herbs. Leaves alternate, entire or toothed. Heads solitary. Name from δις, *twice*, and κομη, *hair;* the double pappus of the first described species.

Sect. 1. STEIROCOMA. *Heads* discoid, heterogamous ; *marg. fl.* uniseriate, sterile, very short, tubular or bilabiate. *Pappus* of the marg. fl. of scabrous bristles ; of the disc. fl. biseriate, the outer bristle-shaped, inner broad-based, more or less plumose. Procumbent, tomentose herbs. (Sp. 1-2.)

Lvs. on both sides woolly ; inner inv. sc. acute ; disc-papp. plumose beyond the middle ... ... (1) **Capensis.**
Lvs. becoming glabrous above; inner inv. sc. taper-pointed ; disc-papp. ciliato-barbellate from base to apex ... ... (2) **macrocephala.**

Sect. 2. RHIGIOTHAMNUS. *Heads* radiate ; *rays* neuter. *Disc*-pappus multiseriate, bristle shaped, serrated. A much branched, erect, rigid half-shrub. (Sp. 3.)

Lvs. obovate or spathulate, thinly canous ; branches rib-striate (3) **radiata.**

Sect. 3. MACLEDIUM. *Heads* discoid, homogamous, *Pappus* of broadish, *flat*, subulate, acuminate, serrulate scales. Small, much-branched shrubs. (Sp. 4-6.)

Leaves nerveless, broad-based, sessile, lanceolate, subpungent, concave ... ... (4) **relhanioides.**
Leaves 1-nerved, tapering at base, subpetiolate :
   Lvs. obovate, woolly beneath ; no subulate bracts on the pedicels ... ... (5) **diacanthoides.**
   Lvs. narrow, with *revolute* margins, glabr. or tomentose beneath ; many subulate bracts on the pedicels ... ... (6) **Burmanni.**

Sect. 4. PSILOCOMA. Heads discoid, homogamous. *Pappus* multiseriate, of very slender, ciliato-barbellate bristles. *Inv. sc.* multiseriate, linear-subulate, midribbed, erect or squarrose. A diffuse half-shrub.

Leaves linear or lanceolate, tomentose beneath, serrulate, one-nerved ... ... (7) **anomala.**

Sect. 5. PTEROCOMA. Heads discoid, homogamous. *Pappus* bristles conspicuously plumose. Erect herbs ; stem simple or corymbose, one or several-headed. (Sp. 8-11.)

Fl.-heads pedicellate, corymbose or subpaniculate (or terminal, solitary).
   Leaves *tapering* at base :
      Lvs. lanceolate, acuminate, soon quite glabrous on both sides ... ... (8) **speciosa.**
      Lvs. oblongo-lanceolate, white-woolly beneath ... (9) **Zeyheri.**
   Leaves *broad-based*, lanceolate, thinly canous beneath, with revolute margins ... ... (10) **Kirkii.**
Fl.-heads *sessile* in the axils of the uppermost leaves ; lvs. thinly canous beneath, tapering at base ... ... (11) **sessiliflora.**

Sect. 1. STEIROCOMA. (Sp. 1–2.)

1. D. Capensis (Less. in Linn. v. p. 277); root perpendicular, simple; stems many from the crown, diffuse or decumbent, tomentose; leaves tapering at base into a petiole, oblongo-lanceolate, on both sides tomentose, undulate, minutely calloso-serrulate; heads subsessile, discoid, terminal or lateral; outer invol. sc. cobwebbed or glabrous, very much acuminate, squarrose, nerved, inner lanceolate, acute, nerveless, erect; disc-achenes furrowed, marginal sterile; pappus of the marg. fl. bristle-shaped, serrate, of the disc double, the outer bristle-shaped, inner plumose beyond the middle, expanded at base. *DC. l. c. 36.*

HAB. Cape, *Lichtenstein.* Karroo, *Drege!* Nieweveld, Beaufort, *E. & Z.!* Springbokkeel, *Zey.!* 1030. Namaqualand, *A. Wyley!* Aapje's and Vaal R., *Burke!* (Herb. D., Hk., Sd.)

Probably an annual or biennial. Stems 6–12 inches long or more, laxly leafy. Leaves 1–1½ inch long, 2–4 lines wide, like those of some *Aizoon.* Invol. in the majority of specimens seen by me quite glabrous, as in DC.'s var. "*leiolepis.*" Inv. scales white or straw-colour.

2. D. macrocephala (DC.! l. c. 36); stems many from the crown, diffuse, tomentose; leaves tapering at base into a more or less distinct petiole, oblong or oblongo-lanceolate, at first cobwebbed, then glabrate above, thickly woolly beneath; heads equalling or sub exceeding the leaves, sessile, discoid; inv. sc. glabrous or nearly so, outer and inner alike tapering to a fine, bristle point, subsquarrose; papp. of marg. fl. very long, bristle-shaped; of disc double, the outer of few, deciduous scales, inner subdilated at base, much acuminate, ciliato-barbellate from base to apex.

HAB. Klipplaat R., *Drege!* Magalisberg, *Zey!* 1029, Vaal R., *Burke.* (Herb. Hk., D., Sd.)

Allied to *D. Capensis,* but with broader and less woolly leaves, much larger fl. heads, more taper-pointed inner inv. sc. and dissimilar pappus. Heads 1½ in. long. Lvs. 1½–2 in. long, 4–5 l. wide.

Sect. 2. RHIGIOTHAMNUS. (Sp. 3.)

3. D. radiata (Less. in Linn. v., 278); stem suffruticose, erect, shrubby, much-branched, branches rib-striate, tomentulose in the furrows; lvs. obovate or spathulate, tapering into a petiole, mucronate, onenerved, entire, on both sides thinly canous; heads terminal, radiate; inv. sc. glabrous, rigid, linear-lanceolate, acuminate, erect, nerved, the inner gradually longer; ray achenes sterile, glabrescent, with fewer, setiform papp. bristles; disc-ach. very silky, with multiseriate, serrate papp. bristles. *DC. l. c. 37. Leyssera picta, Th.! Cap.* 692.

HAB. Cape, *Thunberg! Mundt and Maire,* Uitenhage, *E. & Z.! Dr. Alex. Prior!* (Herb. Th., D., Sd.)

A broom-like bush, often nearly leafless, with very rigid twigs. Leaves 1–1½ in. long, 2–4 l. wide, scattered or tufted. Inv. sc. marked with two purple-bands, one at each side of the slender midrib. It varies (fide *DC.*) with unilabiate or bilabiate ray-fl., and equally cut or bilabiate disc-fl.

Sect. 3. MACLEDIUM. (Sp. 4–6.)

4. D. relhanioides (Less.! l. c. 279); shrubby, twigs minutely subcanescent; leaves nerveless, broad-based, sessile, lanceolate-acuminate, sub-pungent, with inflexed edges, quite glabrous; inv. sc. ovate, much acuminate, imbricate. *DC. l. c. 37.*

HAB. Cape, *Mundt.!* (Herb. Sond.)

I have seen an authentic fragment in Hb. Sd.; in it the leaves are quite glabrous on *both* sides, not "tomentose above." Achenes shortly turbinate, silky, fulvous. Papp. bristles subulate-attenuate, pluriseriate, serrulate, white.

5. D. diacanthoides (Less. l. c. 279); much branched, shrubby, twigs canescent; leaves one-nerved, obovate or obovato-spathulate, tapering at base, subpetioled, the younger cobweb-tomentose above, then nearly glabrous, all tomentose beneath; outer inv. sc. widely spreading, elongate, subulate-acuminate, pungent, inner ovate, acuminate, glossy, subconcave, connivent. *DC.l.c. 37. Helichrysum spinosum, Th.! Cap. 664. Stobœa Bergeri, Spr.! Syst. 3, 394.*

HAB. Cape, *Thunberg!* Karroo, *E. & Z.!* Langekloof, *Drege!* Uitenhage, *Dr. Alex. Prior!* Somerset, *Mrs. F. W. Barber!* (Herb. Th., D., Hk., Sd.)

A coarse-growing, robust, dwarf, much-branched shrub, under 1 foot high. Lvs. and young parts hoary. Outer invol. sc. straw-colour; inner purplish, tipped with white. Leaves varying much in breadth.

6. D. Burmanni (Less. l. c.); much branched, shrubby, twigs thinly canescent or glabrous; leaves one-nerved, lanceolate or obovate-spathulate, tapering at base, often 2–3-toothed, soon glabrous above, tomentose or glabrous beneath; outer inv. sc. widely spreading, long, subulate-acuminate, pungent, gradually passing below into pungent bracts, continued down the petiole; inner inv. scales ovate, much acuminate, glossy, connivent. *DC. l. c. 37. Burm. Afr. t. 67, f. 3.*

VAR. β, **glabrata**; leaves (adult) glabrous on both sides. *Cryptostephane Kraussii, Sch.! B. Bot. Zeit.* xxvii., 782. *Walp. Rep.* 6, *p.* 318.

HAB. Cape, *Burman.* Swellendam, *E. & Z.!* Langekloof, *Echl.* Hassaquaskloof, *Zey.!* 3075. Var. β, Swellendam, *Dr. Thom! Krauss!* Groote Howhoek, *Zey.!* 3076. (Herb. D., Hk., Sd.)

Very similar to *D. diacanthoides*, and perhaps a mere variety, with narrower lvs. and more numerous empty bracts on the pedicels. Var. β. merely differs by its more glabrous foliage: I possess an authentic specimen from *Dr. Krauss* himself.

Sect. 4. PSILOCOMA. (Sp. 7.)

7. D. anomala (Sond.! in Linn. 23, p. 71); rootstock woody; stems suffruticose, diffuse or ascending-erect, simple or branched, flexuous, angular, glabrate or canescent; leaves narrowed at base, linear or lanceolate, acute or subobtuse, serrulate, with subrevolute margins, one-nerved, soon glabrous above, albo-tomentose beneath; invol. sc. multiseriate, linear-subulate, acuminate, subpungent, midribbed, the inner gradually longer; papp. bristles multiseriate, very slender, ciliato-barbellate.

VAR. α, **Sonderi**; leaves narrow-linear, 1–2 lines wide; invol. sc. erect, straight. *Sond.! l. c.*

VAR. β, **cirsioides**; leaves lanceolate, 4–5 l. wide; inv. sc. squarrose. *D. cirsioides, Harv. MS.*

VAR. γ, **microcephala**; leaves as in α.; heads much smaller, ½ in. long, 4–5 lines wide. *D. Gerrardi, Harv. MS.*

HAB. Dornkop, Laayspruit and Magalisberg, *Burke & Zey.! Zey.!* 1028. North

Aliwal, *Cooper!* 1369. Var. β. Ingoma, *Gerr. & M'K.* 1003. Dargle Farm, *G. F. Fannin*, 88. Var. γ. Tugela River, *W. T. Gerrard!* 1058. (Herb. D., Hk., Sd.)

Very variable in the breadth of the leaves and in the size of the fl. heads. At first I had regarded the 3 vars. given above as distinct species, but a comparison of numerous specimens shows so many intermediate conditions that I cannot keep them separate. Leaves in α. and β. 2–3 in. long, 1–2 l. wide; in β. 2–2½ in. long, 4–5 l. wide, with lateral veins. Heads in α. and γ. 1 in. long, 1¼ in. diam.; in γ. ½ in. long, 4–5 l. diam. Flowers purple. Pappus white.

Sect. 5. PTEROCOMA. (Sp. 8–11.)

8. D. speciosa (DC.! l. c. 37); rootstock woody; stem herbaceous, erect, tall, angle-striate, cobwebbed, panicled or corymbose at the summit; leaves linear-lanceolate and lanceolate, elongate, acute or acuminate, entire, flat, at first thinly cobwebbed, then quite glabrous, netted-veined; heads pedicelled; inv scales multiseriate, shining, lanceolate-acuminate, concave, pungent, the outer spreading, gradually passing below into pungent bracts, continued down the petiole.

HAB. Omsamculo and Omcomas, *Drege!* Natal, *Gueinzius!* 352. Umgena and Umcomas, *Gerr. & M'K.*, 1027, 1581. (Herb. Th., D., Sd.)

Stem 1–2½ ft. high, mostly several-headed. Leaves 4–6 in. long, ¼–1 inch wide, tapering much to both ends, with very deciduous cobweb.

9. D. Zeyheri (Sond.! in Linn. xxiii. p. 71); stem herbaceous, erect, cobwebby-canescent, simple or subcorymbose at the summit; leaves oblongo-lanceolate or oblong, acute, narrowed at base, soon quite glabrous above and netted-veined, white-woolly beneath, with slightly reflexed edges, entire or distantly subdenticled; heads pedicelled, ovate; invol. sc. multiseriate, shining, broadly lanceolate-acuminate, concave, pungent, the outer spreading or reflexed. *Harv. Thes. Cap. t.* 68.

HAB. Magalisberg, *Burke & Zeyher!* *Zey.!* 1031. Zululand Plains, *Gerr. & M'K.* 1028. (Herb. Hook., D., Sd.)

Very like *D. speciosa*, but much more woolly, with broader leaves and inv. scales, and no empty bracts on the pedicels.

10. D. Kirkii (Harv.); rootstock woody; stem herbaceous, erect, terete, cobwebbed, corymbose at the summit; leaves sessile, *broad-based*, lanceolate-acuminate, soon quite glabrous and netted-veined above, thinly canescent beneath, with revolute margins, entire; heads pedicelled, ovate; inv. sc. multiseriate, shining, lanceolate-acuminate, flattish, pungent, the outer spreading or reflexed.

HAB. Highlands of Bakota Country, *Dr. Kirk!* (Herb. Hk.)

Intermediate in appearance between *D. speciosa* and *D. Zeyheri*, but differing from both in the broad-based, half-clasping leaves, with evidently *revolute* margins. Inv. sc. nearly as in *D. speciosa*, but no bracts on the pedicel.

11. D. sessiliflora (Harv.); stem herbaceous, erect, terete, cobweb-canescent, several-headed, simple; leaves oblong or lance-oblong, narrowed at base, acute, soon quite glabrous and netted above, thinly canescent beneath, subserrulate, flat or with subrevolute margins; heads *sessile* in the axils of the upper leaves, ovate; inv. sc. multiseriate, lanceolate-acuminate, pungent, spreading.

HAB. Manganja Hills, *C. J. Meller!* (Herb. Hk.)

Known by its sessile, spicate fl.-heads. Leaves 4–6 in. long, 1–1½ in. wide. Native name *Sachamburi.* A very similar plant, gathered by *Barter* on the Niger expedition, is also in Hb. Hk.

CXLIV. **GERBERA**, Gron.

Heads many-fl., heterogamous, radiate ; *ray-fl.* in one or two rows, those of the inner row, when present, very short, subtubular; of the outer bilabiate, the outer lip ligulate, elongate, tridentate, the inner minute, bifid ; *disc-fl.* also sub-bilabiate, the outer lip 3, the inner 2-fid. *Inv.* scales oblong or lanceolate, closely imbricate, subherbaceous. *Recept.* nude. *Anthers* tailed ; *fl.* glabrous. *Style-branches* short, obtuse. *Achenes* beaked. *Pappus* in two or more rows, of rough bristles. *DC. Prodr.* 7, *p.* 15. Also LASIOPUS, *Cass. DC. l. c. p.* 18.

Stemless, perennial herbs. Leaves radical, petioled, entire or pinnatilobed, coriaceous or membranous. Scapes one-headed. Fl. yellow or orange, the rays often red or coppery outside. Named after *Gerber*, a German naturalist who travelled in Russia.

Sect. 1. EUGERBERA. *Inv. sc.* much shorter than the ray-fl. *Rays* uniseriate, uniform. *Achenes* papulose, not (or scarcely) tapering upwards. *Leaves* coriaceous, rigid. *Scape* mostly scale-bearing. (Sp. 1–7.)

Pappus rufous :
　Leaves linear, incised, pinnatifid or pinnatipartite ... ...　(1) **asplenifolia.**
　Leaves oblong, ovate or cordate, entire or sinuate :
　　Leaves rusty-tomentose beneath :
　　　Leaves oblong, sinuous ; scapes scale-bearing ;
　　　　inv. subglabrous　...　...　...　...　...　...　(2) **ferruginea.**
　　　Lvs. cordate, subentire ; scapes without scales ;
　　　　inv. rufo-tomentose ...　...　...　...　...　...　(4) **tomentosa.**
　　Leaves on both sides *glabrous,* quite entire　...　...　(3) **integralis.**
　　Lvs. cordate, thinly albo-tomentose beneath ; inv.
　　　subglabrous　...　...　...　...　...　...　...　(5) **Wrightii.**
Pappus milk-white :
　　Lvs. cordate at base, inciso-sinuous, rufo-tomentose
　　　beneath ...　...　...　...　...　...　...　...　(6) **leucothrix.**
　　Lvs. oblong or obovate, entire or sinuous, thinly
　　　albo-tomentose or glabrous beneath　...　...　...　(7) **Burmanni.**

SECT 2. LEPTICA. *Inv. sc.* not much shorter than ray-fl. *Rays* uniseriate, uniform. *Achenes* minutely papulose, when *mature* tapering into a long, slender beak. *Leaves* membranaceous. *Scapes* not scaly. (Sp. 8–9.)

　Lvs. long-petioled, cordate at base, albo-tomentose beneath　(8) **cordata.**
　Lvs. short-petioled, elliptic-oblong, cuneate at base, softly
　　hairy beneath ...　...　...　...　...　...　...　...　(9) **piloselloides.**

Sect. 3. LASIOPUS. *Rays* biseriate, female ; *outer* strap-shaped, elongate ; *inner* short, subtubular, slender, with a very short, outer lip and a much exserted style. Other characters as in *Leptica.* (Sp. 10–15.)

Pappus rufous or foxy :
　Leaves albo-tomentose beneath :
　　Lvs. elliptic or oblong, obtuse, scarcely narrowed at
　　　base ...　...　...　...　...　...　...　...　...　(10) **ambigua.**
　　Lvs. oblongo-lanceolate, tapering very much at base　(11) **discolor.**
　Leaves green on both sides :
　　Lvs. obovate-oblong, tapering very much at base ...　(12) **plantaginea.**
　　Lvs. elliptical or oblong, scarcely tapering at base ...　(13) **viridifolia.**
Pappus violet-purple :
　　Leaves oblongo-lanceolate, green on both sides　...　...　(14) **aurantiaca.**
　　Leaves ovate-oblong, tomentose and strongly penninerved
　　　beneath　...　...　...　...　...　...　...　...　(15) **Kraussii.**

Sect. 1. EU-GERBERA. (Sp. 1–7.)

1. G. asplenifolia (Spr. Syst. 3, 576); leaves pinnatifid or pinnati-sect, rufo-tomentose beneath, lobes roundish, concave, with revolute margins, the terminal not larger than the rest; scape cobwebbed or woolly, bracteolate; pappus rufous. *DC. l. c.* 15. *Arnica Gerbera, Linn. Sp.* 1246. *Th.! Cap.* 669. *Burm. Afr. t.* 56. *f.* 1. *Lam. Ill. t.* 679, *f.* 5.

VAR. β. **linearis**; leaves. linear, with revolute margins, *quite entire* or runcinate-toothed (on the same root).

HAB. Round Capetown and Simonstown, frequent. Drakenstein, *Drege!* Caledon Baths, *E. & Z.! Pappe!* β. Voorman's Bosch, Swell., *Zey.!* 3078, pte. (Herb. Th., D., Hk., Sd.)
Leaves narrow, 4–6 in. long, petioled ; lobes glossy above, 2–4 l. long and wide. The leaves vary much in degree of incision, and the inv. sc. in comparative breadth, and in indument. Rays purple. A very handsome plant, with fern-like leaves.

2. G. ferruginea (DC.! l. c. 15); leaves long-petioled, oblong, re-motely denticulate, subsinuate, with narrow, revolute margins, glabrous above, rusty-tomentose beneath; scape minutely velvetty-pubescent above the middle, squamellose; inv. sc. about 3-seriate, lanceolate, subglabrous; pappus rufous. *Gerb. sinuata, Less.! in Linn.* 1830, 292. *Arnica serrata, Th.! Cap.* 669.

HAB. Cape, *Thunberg!* Swellendam and Outeniqualand, *Drege! E. & Z.!* Voormansbosch, *Zey.!* 3079. (Herb. Th., D., Hk., Sd.)
The leaves in *Thunberg's* specimen are narrower than in *Drege's ;* in other respects the plants seem identical. Petioles 1–3 in. long, glabrous : lamina 2½–3 in. long, ½–1½ in. wide, obtuse at base. Possibly merely a broad and short, subundivided leaved form of *G. asplenifolia ;* its rufous pappus distinguishes it from *G. sinuata.*

3. G. integralis (Sond.!); leaves long-petioled, ovate-oblong, obtuse, cuneate or tapering at base, entire, on both sides quite glabrous, with minutely revolute edges; scapes scaly, glabrous; inv. sc. 3-seriate, glabrous ; pappus rufous.

HAB. Cape Flats, *E. & Z.!* Near Capetown, *W. H. H.* (Herb. Hk., Sd.)
Though the leaves, at a glance, seem to be *perfectly* entire, a lens shows indications of distant, very minute, *reversed* denticles, which would become runcinate teeth if developed ; these, taken with the *rufous* pappus, indicate an affinity with *G. ferruginea,* rather than *G. Burmanni.* But if a variety of that species, it is at least so strongly characterised as to deserve a name.

4. G. tomentosa (DC.! l. c. 16); leaves petioled, cordate or oblong, subcordate at base, entire or repando-denticled, glabrous above, per-sistently rufo-tomentose beneath; petioles, scapes and inv. sc. rufo-tomentose, the scapes *mostly* without bracts, inv. sc. in 3 or several rows, acuminate, more or less distinctly blackened at the tip; pappus rufous. *Gerbera hirsuta, Spr.,* non *Th.*

VAR. β. **lanata**; toment very thick and copious; the scapes and inv. densely woolly.

HAB. Hott. Holl. and Caledon, *E. & Z.!* Langekloof and Drakensteinberg, *Drege!* Genadendahl, *Dr. Roser!* Zwarteberg, *Pappe!* β. Hott. Holl. *Bowie! Zey.* 3077. (Herb. D., Hk., Sd.)
Much more tomentose than *G. ferruginea,* with woolly involucres. It varies considerably in the size of the fl. heads, and in the number of rows of inv. scales. Leaves 2½–4 in. long, 1½–2 in. wide ; petioles 2–3 in. long.

5. G. Wrightii (Harv.); leaves petioled, cordate, repando-denticulate with subrevolute margins, soon glabrous above, thinly *albo*-tomentose beneath; petioles, scapes and inv. sc. thinly cobwebbed, becoming nearly glabrous, the scapes scale-bearing; inv. sc. in 4–5 rows, lanceolate, acute; pappus rufous.

HAB. Near Simonstown, *C. Wright*, 343. (Herb. D.)
In the shape of the leaves this agrees with *G. tomentosa*, but the scanty tomentum is *white*, not rufous, and the involucre is nearly glabrous. Heads of large size, with 35–40 rays.

6. G. leucothrix (Harv.); leaves oblong, cordate at base, deeply inciso sinuous, and subdentate, with revolute margins, glabrous above, persistently rufo-tomentose beneath; scapes squamellose, cobwebby or nude below, tomentellose at the summit; inv. sc. 4–5 seriate, minutely tomentose or glabrate; pappus *white*.

HAB. Cape, *E. & Z!* (Herb. Sond.)
This may possibly be a natural hybrid between *G. ferruginea* of which it has the foliage and pubescence, and *G. Burmanni* which it resembles in involucre and pappus. I have seen but a solitary specimen.

7. G. Burmanni (Cass. Dict. 18, p. 461); leaves petioled, elliptical, oblong, or obovate, more or less cuneate or tapering at base, either minutely denticulate, toothed, or sinuous-toothed (very variable in this respect), glabrous above, either glabrous or thinly canous beneath; scapes scale-bearing, glabrous or cobwebbed, or minutely velvetty; inv. sc. about 3-seriate, lanceolate, subglabrous; pappus slender, snow-white, eq. the disc. *Less. in Linn.* 1830, *p.* 294.

VAR. *a*, Burmanni; lvs. minutely toothed, on both sides glabrous; inv. sc. cobwebby. *Ger. Burmanni, DC.! l. c.* 16. *Arnica crocea, Linn. Th.! Cap.* 668. *Burm. Afr. t.* 56, *f.* 2.

VAR. *β.* sinuata; lvs. more or less toothed or sinuate, generally thinly canous beneath, sometimes quite glabrous; inv. sc. glabrous. *G. sinuata, Spr. Syst.* 3, 756. *DC.! l. c.* 15. *Arnica sinuata, Th.! Cap.* 668.

HAB. Both varieties near Capetown and in the Western Districts generally. (Herb. Th., D., Hk., Sd.)
I cannot find either in pubescence or the toothing of the leaves any permanent character between the above varieties, which grow in the same districts and often intermixed. The pappus is constantly and conspicuously *white*, by which character those with sinuous leaves are known from subsimilar forms of *G. ferruginea*.

Sect. 2. LEPTICA. (Sp. 8–9.)

8. G. cordata (Less. in Linn. v. p. 297); leaves long-petioled, membranous, elliptical or oblong-ovate, cordate at base, entire or denticulate, sparsely pilose becoming nude above, softly and persistently albotomentose beneath, and when young also clothed with long, brown, silky, deciduous hairs; inv. sc. at length reflexed; ripe-achenes long-beaked. *DC.! l. c.* 16. *Arnica cordata, Th.! Cap.* 628.

HAB. Cape, *Thunberg!* Zuureberg, *Drege!* Uitenhage, *E. & Z.!* Vanstaadensberg, *Zeyl* 3080. Tzitsikamma, *Pappe!* Albany, *H. Hutton! T. Cooper!* 241, 1554. Grahamstown, *R. W. Reade*, 10. (Herb. Th., D., Hk., Sd.)
Petioles 3–5 in. long; lamina 2½–5 in. long, 2–3 in. wide, obtuse or acute. Scape 12–15 in. long, tomentose. Inv. very woolly. Rays yellow, not much longer than the inv. scales.

9. G. piloselloides (Cass. Dict. 18, p. 461); leaves elliptic-oblong, obtuse or subacute, tapering into a short petiole, entire or repando-denticulate, sparsely setose or nude above, beneath and on the margins softly hairy ; scape shaggy and somewhat thickened at the apex ; inv. sc. erect in fruit ; ripe achenes long-beaked, papulose, striate. *DC. l. c.* 16. *Arnica piloselloides, Linn. Amoen.* 6, 105. *Gerbera Schimperi, Sch. Bip., in Hb. Schimp! No.* 185. *G. ovalifolia, DC.! l. c.* 17.

VAR. β. **discolor**; (DC.); leaves albo-tomentose beneath. *Arnica nudiflora, Th.! in Herb.*

HAB. Round Capetown and in the W. districts, *Burchell, W.H.H.* and Draken-stein, *W.H.H.* Caledon, *E. & Z.* Uitenhage, *Eckl.* Witberg, Katberg, and Omtata and Basche R., *Drege!* Br. Kaffraria, *T. Cooper,* 172. Kreilis Country, *H. Bowker!* 243. Bigarsberg, Natal, *Gerr. & McK.* 1042. Var. β. Zuureberg, *Drege, Thunberg!* (Herb. D., Hk., Sd.)

Leaves 3–5 in. long, 1½–2¼ wide. Pedunc. 4–12 in. long. Rays turning dark-brown in drying. I cannot distinguish *G. ovalifolia,* DC.!, which is widely spread through temperate India ; our Cape plant has certainly papulose achenes.

Sect. 3. LASIOPUS. (Sp. 10–15.)

10. G. ambigua (Sch. Bip.) ; leaves short-petioled, subcoriaceous, elliptical or oblong, obtuse, scarcely tapering at base, entire or inversely-subsinuate, pilose, becoming glabrous above, albo-tomentose beneath, midribbed and penninerved ; achenes hispidulous; pappus fulvous ; inv. sc. longer than the disc, villous. *Lasiopus ambiguus, Cass. Dict.* 25, p. 299. *DC.! l. c.* 18. *Lasiopus coriaceus, DC.! l. c.* 19.

HAB. Zuureberg, *Drege!* Uitenhage, *E. & Z.!* Bushman's R., Albany, *Zeyher!* (Herb. D., Hk., Sd.)

Petiole 1–2 in. long ; lamina 2½–4 in. long, 1–2 in. wide, all but the very old leaves tomentose beneath. There are three specimens in Hb. Sd. from *Ecklon* marked "Las. coriaceus," *DC.:* one of them is precisely identical with an authentically marked sp. from *Drege* (Hb. D.) of "L. ambiguus;" the other two are undistinguishable from "Las. viridifolius." I am therefore compelled to reject "L. coriaceous" altogether.

11. G. discolor (Sond.! in Hb.); leaves oblongo-lanceolate or lanceolate, subacute, tapering much at base into a longish petiole, entire or subdenticulate, glabrous and green above, tomentose beneath ; inv. sc. villous ; ovaries hispidulous ; pappus fulvous.

HAB. Magalisberg, *Burke & Zeyher !* *Zey.!* 1033. Aapjes R., *Zey.!* 1032. (Hb. Hk., Sd., D.)

Petioles 2–5 in. long ; lamina 4–6 in. long, ¾–1 inch wide. Scapes tomentose, 8–16 in. long. Heads like those of *G. ambigua.*

12. G. plantaginea (Harv.); leaves obovate-oblong, subacute, tapering much at base into a variable petiole, entire, glabrous above, at first cobwebbed, then thinly pubescent beneath, on both sides green ; inv. scales villous; marg. ovaries hispid, of disc thinly pubescent ; pappus fulvous.

HAB. Magalisberg, *Burke & Zeyher !* (Herb. Hk., Sd.)

Crown of root not copiously silky. Petioles 1–3 in. long ; lamina 2–5 in. long, 1–1½ in. wide. Near *G. viridifolia,* but differing in aspect, with much longer lvs. It differs from *G. discolor* in having both sides of leaf green.

13. G. viridifolia (Sch. Bip.); leaves short-petioled, subcoriaceous, elliptical or oblong, obtuse, slightly tapering at base, entire or denticulate, soon glabrous above, either quite glabrous or thinly pilose beneath, on both sides green; inv. sc. subglabrous; marginal achenes hispidulous; pappus fulvous. *Lasiopus viridifolius, DC.! l. c. 19. Gerbera Natalensis, Sch.! Bip. Bot. Zeit. xxvii. p. 778.*

HAB. Keiskamma; Buffel R.; Klipplaat R.; and Zwartekey, *Drege!* Tambukiland, *E. & Z.! Cooper!* 173. Albany, *T. Williamson!* Grahamstown, *P. Mac Owan!* 56; *Genl. Bolton!* Cradock, *Burke!* Near Shiloh, Queenstown, *Mrs. F. W. Barber!* 318. Natal, *Krauss!* 452; *Gueinzius!* 338. Near Maritzberg, *Dr. Sutherland!* (Herb. D., Hk., Sd.)

Crown of root very copiously silky. Petioles 1–2 in. long; lamina 1½–2½ inches wide. Rays "white above, red beneath" *(M.E.B.).* I have examined *Krauss's* original specimen in Hb. D. of " *G. Natalensis,*" Sch. B., and find it a "*Lasiopus;*" in other respects it agrees with the pilose forms of *G. viridifolia.* I have not seen ripe achenes; ovaries of the disc nearly glabrous; of the marg. fem. fl. hispidulous.

14. G. aurantiaca (Sch. B. l. c.); leaves membranous, lanceolate or oblongo-lanceolate, acute, tapering much at base into a short petiole, entire or denticulate, setose, becoming nude above, cobwebby-canescent, becoming nude and netted-veined beneath; invol. sc. cobwebbed, subbiseriate, acuminate; achenes hispid; pappus violet-purple. *Walp. Rep. 6, 317.*

HAB. Near Maritzburg, Natal, *Krauss! Dr. Sutherland! J. Sanderson,* 154. (Hb. D., Hk.)

Leaves 5–6 in. long, 1–2 in. wide, the pubescence of both sides deciduous. Scape woolly, 8–15 inches high, without bracts. Rays (when dry) rich red-brown above, pale beneath; said to be orange when fresh. A very handsome species.

15. G. Kraussii (Sch. Bip.! Bot. Zeit. XXVII. 781); leaves petioled, membranous, ovate-oblong, denticulate, acute, either rounded or cuneate-attenuate at base, at first setose, afterwards glabrous above, persistently tomentose, strongly midribbed and penni-nerved beneath; inv. sc. silky, sub-biseriate; achenes hispid; pappus violet-purple. *Walp. Rep. 6, 317.*

VAR. *a.* **Kraussii**; leaves mostly cuneate-attenuate at base; toment. fulvous. *Sch.! B. l. c.*

VAR. *β.* **Gueinzii**; leaves mostly rounded at base; toment. whitish, *G. nervosa, Sond. in Linn.* 23, *p.* 70.

HAB. Both forms grow at Natal. Var. *a.* sent by *Krauss!* 402. *Sanderson!* 81, 127. Var. *β. Gueinzius!* 356. *Sanderson!* 58. *Plant!* 37. *Gerr. & M'K.!* (Herb. D., Hk., Sd.)

Solitary specimens of our two varieties look very distinct, but I possess an intermediate form from *Mr. Sanderson,* which quite connects them. The strongly marked rib and nerves on the lower side characterize both forms equally.

CXLV. PERDICIUM, Lag.

Heads many-fl., heterogamous, discoid; *marginal*-fl. female, uniseriate, tubular, bilabiate, the outer lip shortly ligulate, 3-toothed, inner shorter, bipartite, with linear lobes; *disc*-fl. also bilabiate, the outer lip unequally 3-toothed, inner bipartite. *Inv.* sc. leafy, imbricate, lanceolate, appressed. *Anthers* tailed; *fil.* smooth. *Style* shortly 2-lobed, the branches scarcely divergent, semiterete, obtuse, pubescent. *Achenes* ovate-oblong, rostrate. Terminal callus dilated. *Pappus* pluriseriate, of scabrous bristles, falling off with the epigynous disc or annulus. *DC. Prodr. 7. p. 38.*

Small S. Afr. herbs, with the aspect of *Taraxacum* or some other Cichoracea. Leaves radical, runcinate, glabrous or cobwebby-canescent. Scapes 1-headed, about equalling the leaves.

Leaves glabrous on both sides ... ... (1) **Taraxaci.**
Leaves cobweb-cottony beneath ... ... (2) **leiocarpum.**

1. P. Taraxaci (Vahl.); leaves on both sides green, glabrous. sinuato-runcinate ; achenes densely papulose, shortly beaked; pappus not much longer than the inv. scales. *DC.! l. c. 38. Perd. semiflosculare, Linn. Thunb.! Cap. 690.*

HAB. Cape, *Burman, Thunberg! Chamisso!* Breede R., *Drege!* Hott. Holl., *E. & Z.!* (Herb. Th., D., Hk., Sd.)
Leaves 2-4 in. long, short-petioled, rigid. Scape about as long as leaves. Pappus straw coloured.

2. P. leiocarpum (DC.! l. c. 39); leaves cobwebby-canescent beneath, sinuato-runcinate, the lobes toothed; achenes quite smooth, long beaked; pappus much longer than the involucre.

HAB. Camiesberg, *Drege!* (Herb. D., Hk., Sd.)
Rather larger than the preceding, with discoloured leaves, larger fl.-heads and very conspicuous, straw-coloured pappus.

(Genus of uncertain affinity.)

CXLVI. ARROWSMITHIA, DC.

Heads many-fl., radiate; *ray-fl.* ligulate, female, in one row; *disc-fl.* campanulate, 5-toothed, sterile; all the corollas woolly on the tube. *Inv.* scales scarious, imbricate in several rows, the inner membrane-tipped. *Recept.* (fide DC.) "flat, clothed with linear-setaceous fimbrils or paleæ." *Anthers* cuspidate at base, produced into a very slender, partly adnate, bristle-shaped tail. *Style* thickened upwards, its branches short, convex-backed, obtuse, equally pubescent outside. *Achenes* without pappus, those of the ray compressed; of the disc terete, glabrous, sterile. *DC. Prodr. 7. p. 254.*

A rigid, small shrub or half-shrub, with the aspect of a *Relhania*. Branches terete, thinly cobwebbed, closely leafy to the summit. Leaves rigid, alternate, spreading or reflexed, sessile, half-clasping, lanceolate, pungent-mucronate, with recurved edges, quite entire, glabrous and glossy above, tomentose beneath, veinless. Heads terminal, solitary or in pairs, sessile; fl. yellow. Named in honour of the well known geographer of Atlas celebrity. Its proper place in the system uncertain.

1. A. styphelioides (DC.! l. c. 254). *Deless. Ic. Sel.* 4, *t.* 100.

HAB. Katberg, *Drege!* (Herb. Hook., Sond.)
Leaves 4-5 lines long, 1 line wide. Heads 6-7 lines long, 5 lines wide, campanulate.

TRIBE VI.—CICHORACEÆ.

Flower-heads semi-flosculose, that is, having *all* the flowers ligulate and bisexual. *Style-branches* long, subobtuse, filiform, equally pubescent on the outer surface.—Herbaceous plants, with alternate leaves, and milky, very bitter juice. (Gen. CXLVII.–CLIV.)

Pappus plumose :
  Invol. imbricate ; recept. paleaceous ; central achenes
    beaked ... ... (147) **Hypochoeris.**
  Inv. uniseriate ; recept. nude ; all the achenes beaked (148) **Urospermum.**
Pappus bristle- or hair-like :
  Achenes long-beaked :
    Ach. flattened, smooth or striate ; stem branched (149) **Lactuca.**
    Achenes oblong, with hard, sharp points at apex ;
      lvs. all radical ; stem scape-like, nude, 1-headed (150) **Taraxacum.**
    Achenes oblong, compressed, the outer with a short,
      inner with a long beak ; leaves radical ; stem
      branched, nearly nude ... ... (154) **Anisoramphus.**
  Achenes beakless, or very shortly beaked :
    Achenes 4–5-angled, slightly beaked ; stems trail-
      ing, bearing tufted leaves and axillary fl.-heads (151) **Microrhynchus.**
    Ach. compressed, beakless ; pappus very soft and
      white ; branching herbs ... ... (152) **Sonchus.**
    Ach. 5-angled, beakless ; pappus rigid, discoloured ;
      branching herbs ; leaves chiefly radical, rosulate (153) **Hieracium.**

CXLVII. HYPOCHŒRIS, Vaill.

Heads many-flowered. *Invol.* ovate-oblong or subcampanulate, the scales imbricated. *Recept.* bearing paleæ between the flowers. *Achenes* glabrous, rough with sharp points, the marginal ones *(in the Cape species)* beakless, those of the centre with a long, slender beak. *Pappus* biseriate, the outer of short bristles, the inner long, plumose. *DC. Prodr.* 7, *p.* 90.

Annual or biennial herbs, glabrous or pilose. Leaves mostly radical, rosulate. Flowers yellow. Name, ὑπο, *for*, and χοιρος, *a pig;* pigs eat the roots greedily.

1. H. glabra (Linn. sp. 1141); leaves radical, numerous, oblong or spathulate, sinuate-toothed or runcinate, glabrous or hispidulous; scapes glabrous, leafless, mostly branched, in dwarf specimens simple; invol. glabrous; outer achenes beakless, inner long-beaked; pappus double. *DC. l. c.* 90. *Fl. Lond.! t.* 149. *Fl. Dan. t.* 424. *E. Bot. t.* 575. *H. Capensis, Less. Syn.* 130. *Hyoseris tenella, Th.! Cap.* 613.

HAB. Table Mt. summit, common. Simonstown, *C. Wright*, 349; *E. & Z.! Dr. Wallich*, &c. Riv. Zondereinde, *Zey.!* 3081. Zwartkops R., *Zey.!* Fish R., *Zey.!* (Herb. Th. D., Hk., Sd.)

CXLVIII. UROSPERMUM, Scop.

Heads many-fl. *Invol.* campanulate, composed of about 8 uniseriate scales, concrete at base into a tube. *Recept.* without paleæ, piloso-fim-brilliferous. *Corolla* hairy at the apex of tube and base of limb. *Achenes* sessile, rough with sharp points, beaked, beak very long, hollow, in-flated at base. *Pappus* uniseriate, plumose. *DC. Prodr.* 7, *p.* 116.

Annual or biennial herbs. Leaves more or less pinnatifid or lyrate, the cauline amplexicaul. Heads on long, nude peduncles. Flowers yellow. Name, ουρα, a *tail*, and σπερμη, *seed.*

1. U. picroides (Desf.); leaves runcinate, toothed, the cauline with sagittate, toothed ears; involucres rough with rigid bristles. *DC. l. c.* 116. *Tragopogon picroides, Linn. Sp.* 1111. *Lam. Ill. t.* 646, *f.* 3. *T. Capensis, Jacq. Ic. Rar. t.* 577. *Urospermum Capense, Spr.*

HAB. Waste ground near cultivation, throughout the Colony and in Kaffraria : introduced from the south of Europe. (Herb. D., Hk., Sd.)

1–2 ft. high, the stems smooth or bristly, hollow, pale. Leaves very variable in their cutting.

CXLIX. LACTUCA, Tourn.

Heads several or few-flowered. *Inv.* cylindrical, imbricate, calycled, 2–4-seriate, the outer scales short. *Recept.* nude. *Achenes* flattened, wingless, abruptly produced into a slender beak. *Pappus* hair-like, very soft, soon falling off. *DC. Prodr.* 7, *p.* 133.

Annual, biennial, or perennial herbs, mostly glabrous. Heads generally panicled. *Flowers* yellow (or blue or purple). *Pappus* white or yellow. Name from *lac*, milk ; from the milky juices of these plants.

Mid-rib of leaves smooth ; achenes faintly one-nerved ... ... (1) **Capensis.**
Mid-rib of leaves aculeate ; achenes strongly many-striate ... ... (2) **Dregeana.**

1. L. Capensis (Thunb.! Cap. 614); quite glabrous ; stem erect, terete, smooth, simple at base, loosely panicled above ; leaves smooth on the midrib, the lower runcinate-pinnatifid or quite entire, narrow, acuminate, upper sagittate at base, linear-lanceolate, acuminate, entire ; achenes with a single, slender or faint central rib, rather longer than the beak ; pappus pale-yellowish, rarely white. *Less.! Syn.* 135. *DC. l. c.* 136. Also *Scorzonera Capensis, Th.! Cap.* 141, *et Th.! Herb. fol.* 1.

HAB. Cape, *Thunberg!* Uitenhage, *E. & Z.! Zey.!* 3083. Zuureberg, Kipplaat R., and Zw. Key, *Drege!* Howison's Poort, *H. Hutton!* Albert and Br. Kaffr., *Cooper!* 653, 654. Thaba Unka, *Burke!* Near D'Urban, Natal, *Gerr. & McK.!* 317. Magalisberg, *Zey.!* 1035. Wienen Country, *Dr. Sutherland.* (Herb. Th., D., Hk., Sd.)

Stem rigid, 1–2 feet high, leafy chiefly near the base, the upper leaves small and sparse, much-branched above. Leaves 3–5 in. long, seldom more than 4–5 lines wide, much acuminate ; frequently quite entire and then linear-lanceolate. In Herb. Th., sheets 1 of " *Scorzonera Capensis*" and sh. 2, 4, of " *Lact. Capensis*" belong to this ; but sheet 1, 3 of *Lact. Capensis* belong to *Sonchus Ecklonianus.* Sheet 2 of " *Scorz. Capensis*" is also a *Sonchus,* but without its leaves.

2. L. Dregeana (DC. l. c. 137); stem erect, terete, simple at base, and more or less rough with bristles, panicled and glabrous above ; leaves with a prickly midrib, sagittate-half-clasping, the lower runcinate or runcinate or sinuous-toothed and rigidly ciliate, the upper quite entire, linear-lanceolate, much acuminate ; achenes multistriate, about equalling their beak. *Drege,* 3784. *L. virosa, Thunb.! Fl. Cap. p.* 614.

HAB. Cannaland, *Thunberg!* Zuureberg, *Drege!* Caledon R., *Burke & Zeyher! Zey!* 1037. (Herb. Th., D., Hk., Sd.)

Stem 2–3 ft. high, pale ; the taller specimens very much branched above. Cauline leaves 4–9 in. long, 4–8 l. wide. Very near *L. saligna* and *L. virosa,* if sufficiently distinct from the latter. Cultivated specimens from Hort. Kew. and Hort. Hamburg have broadly oblong or obovate, obtuse, toothed cauline leaves ! altogether unlike those of the wild plant, as above described.

CL. TARAXACUM, Hall.

Heads many-fl. *Inv.* double, the *outer* sc. (or *calycle*) small, either appressed, spreading or reflexed, *inner* uniseriate, erect, all frequently callous-tipped. *Recept.* nude, *Achenes* oblong, striate, along the striæ muricate, or spinellose near the apex, produced into a long, slender beak. *Pappus* pilose, pluriseriate. *DC. Prodr.* 7, *p.* 145.

Stemless herbs. Leaves all radical, entire or runcinate, mostly glabrous. Scapes 1-headed, fistular, longer than the leaves. Heads yellow. The well known "*Dandelion*" is the type of this genus.

1. **T. fulvipilis** (Harv.); glabrous; leaves very narrow, runcinate, sparingly toothed; outer inv. sc. erect; achenes compressed, multistriate, the striæ spinulose near the apex, smooth below; *papus tawny !* *Hypochœris ?* 6176, *Drege! in Hb.*

HAB. On the flats between Los Tafelberg and Wildschutsberg, *Drege!* (Herb. Sond.)

Leaves 2–3 in. long, the rachis 1–2 l. wide. Scapes 4–5 in. high. Fl. heads small. This seems to be quite a *Taraxacum;* but its *pale-fulvous* pappus is unlike that of any other known species.

CLI. **MICRORHYNCUS**, Less.

Heads several-fl. *Inv.* cylindrical, the scales membrane-edged, the outer much shorter, imbricate, forming a calycle. *Recept.* nude. *Achenes* 4, rarely 5-angled, sub-rostrate, (the beak very short, not obvious in immature fruit), ribbed and furrowed, the ribs either smooth or cross-ridged, furrows narrow. *Pappus* multiseriate, pilose. *DC. Prodr.* 7. *p.* 180.

Glabrous, perennial herbs. Leaves coarsely toothed or pinnatifid. Flowers yellow. Name, μιχρος, *small,* ρυγχος, *a beak.*

1. **M. Dregeanus** (DC. l. c. 181); stems trailing, flagelliform; leaves fascicled at the nodes, obovate, or obovato-spathulate, tapering at base, sub-petiolate, closely calloso-denticulate; pedicels among the tufts of leaves, short, bracteate.

HAB. Omsamculo and Omcomas, *Drege!* Natal, *Rev. Mr. Hewetson! T. Williamson! J. Sanderson! Gerr. & McK.* 1016. (Herb. D., Hk., Sd.)

Trailing stems "sometimes 8–10 ft. long" (*Gerr. & McK*). Lower leaves, 2–5 in. long, 1–1½ in. broad; upper seldom 1 in. long, ½ in. wide. Pedicels scarcely uncial.

CLII. **SONCHUS**, Linn.

Heads many-fl. *Inv.* imbricate. *Achenes* similar, wingless, compressed, beakless, longitudinally ribbed, the ribs often cross-ridged or muricate. *Pappus* soft, very white, multiseriate, the hairs very slender. *DC. Prodr.* 7, *p.* 184.

Herbaceous (rarely half-shrubby) plants, of very various aspect. Leaves entire or pinnatisect, runcinate or lyrate. Flowers yellow.—Name, σονχος, the sow-thistle, said to be from σομφος, spongy or hollow.

Root perennial :

1. S. Dregeanus (DC.! l. c. 184); perennial, glabrous,. glaucous; stem erect, terete; leaves linear-elongate, acuminate, the radical and lower either quite entire or variably runcinate, the lobes strongly recurved, taper-pointed, entire or toothed; upper cauline lvs. sagittate at base, stem-clasping, toothed or entire; heads subcorymbose, on a long, nude peduncle; inv. tomentose at base, the scales membrane-edged; achenes slightly tapering upwards, longitudinally many rib-striate, striæ smooth.

HAB. On the flats near Los Tafelberg; between Buffelvalei and Kraai rivier; near Gaatje; on the Sneeuweberg; and by the banks of the Zwartkops R., *Drege!* Zwartkops R., and near Smalldeel, Ceded Terr., *Zey.!* Near Grahamstown, *P. MacOwan*, 531. (Hb. D., Hk., Sd.)

Mr. *Mac Owan's* specimen is much more luxuriant than any I have seen from *Drege* or *Zeyher*, and the leaves, besides being *runcinate*, are toothed on and between the lobes; in the common forms the margin is quite entire betw. the lobes. Leaves 3–10 in. long, 3–8 l. wide.

2. S. Ecklonianus (DC. l. c. 184); perennial, glabrous, not glaucous; stem nearly nude, terete, subsimple or sparingly branched; leaves chiefly radical, linear-elongate, either entire, runcinate, or runcinato-pinnatifid; cauline few, *sessile* (not sagittate at base); heads few, long-pedicelled; inv. subtomentose at base, the sc. membr.-edged; achenes slightly tapering upwards, longitudinally rib-striate, striæ smooth. *Lactuca Capensis, Sheets* 1, 3, *in Hb. Thunb.!*

HAB. Cape, *Thunberg!* *Ecklon!* Sides of Table Mt., Capetown, rare, *W. H. H.* Simon's Bay, *C. Wright*, 350. (Herb. Th., D., Sd.)

Very close indeed to *S. Dregeanus*, but not glaucous, of smaller size, and with fewer and less clasping cauline leaves. Radical leaves rosulate, 4–6 in. long. Stem 6–12 in. high, few-headed.

3. S. integrifolius (Harv.); perennial, glabrous, not glaucous; stem erect, terete, leafy, subsimple; leaves oblong or oblong-linear, acute, entire or the lower subruncinate, all strongly stem-clasping, and slightly adnate at base, the margin minutely callous-denticled; heads on very long pedicels; inv. quite glabrous, sc. 3-seriate, the inner membr. edged; achenes nearly linear, longitudinally ribstriate, the striæ smooth.

HAB. Natal, *T. Williamson!* *J. Sanderson!* Near D'Urban, *Gerr. & M'K.!* 315. (Herb. D.)

This has much broader, less tapering and more entire leaves than any of this section; the cauline leaves especially more thoroughly amplexicaul, and partly adnate. Leaves 4–5 in. long, 1–1½ in. wide.

4. S. nanus (Sond.! in Herb.); perennial, glabrous, nearly stemless; radical leaves runcinate or pinnate-lobed, tapering at base, rosulate; heads subsessile among the radical leaves; inv. sc. broad, membr.-edged; achenes tapering upwards, much compressed, longitudinally rib-striate, the striæ smooth. *Compos. Z. n. N. n. E.*, 114. 10, *E. & Z.!*

HAB. Cape, *Ecklon and Zeyher!* (Herb. Sond.)

A very dwarf plant, probably from some alpine height. Leaves 4–5 in. long, mostly runcinate, like those of a *Dandelion*. Heads several, subsessile on a woody crown of root.

5. S. oleraceus (Linn. Sp. 1116); annual; stem erect, terete, glabrous, or sprinkled with gland-tipped bristles near the summit and on the involucres; cauline leaves ear-clasping at base, sharply ciliate-toothed, *polymorphous*, either undivided, runcinate, lyrate or deeply pinnatifid; inv. and pedicels glabrous or rarely glandular; inv. after flowering conical, finally reflexed; achenes compressed or flattened, finely striate, longitudinally the striæ rough or smooth. *E. Bot. t. 843.*

VAR. *a*, ciliatus; achenes compressed, narrowed to the base, finely multistriate, the striæ cross-ridged. *S. ciliatus. Lam. DC.! l. c.* 185. *S. oleraceus, a & β., Linn. l. c. S. asper, Blackw. Herb. t.* 30. *S. umbellifer, Th.! Cap.* 614. *S. zacinthoides, DC.! l. c.* 184. *S. reversus, E. Mey.! DC. l. c.* 186.

VAR. *β*, fallax; achenes flattened, narrowed at base, with 3–5 subdistant, smooth striæ on each face. *S. fallax, Wallr. DC! l. c.* 185. *S. oleraceus, γ. & δ., Linn. S. spinosus, Lam. S. glaber, Thunb.! Cap.* 614.

HAB. A weed in cultivated ground throughout the Colony, introduced from Europe. (Herb. Th., D., Sd.)

The common "*Sowthistle.*" There are evidently two "races" of this plant, known by the characters of fruit above described; but they grow often intermixed, nor is there any permanent character of foliage to distinguish them. I therefore perfer to leave the species as Linnæus had it.

CLIII. HIERACIUM, L.

Heads many-fl. *Inv.* ovate, often cylindrical, composed of linear obtuse, often acuminated, rarely 2-ranked, commonly multiseriate, imbricated scales. *Recept.* nude, areolate, the areoles pentagonal, often with a minutely fimbriate margin. *Achenes* 5-angled, substriate, mostly clavate, beakless or nearly so; or oblong or fusiform. *Pappus* persistent, uniseriate, simple, sessile, often dirty-whiteish, the bristles rigid, scabrid. *DC. Prodr.* 7, *p.* 199.

A vast, cosmopolitan genus; very few from the Southern hemisphere. Perennial, caulescent herbs, with alternate leaves, variously pubescent; the hairs often glandular or stellate. Name from ἱεραξ, a *hawk;* "Hawk weed." *Engl.*

Root leaves obovate-oblong obtuse; stem and pedicels glabrous;
inv. powdery ... ... (1) **Capense.**
Root leavesl anceolate, acute; stem hispid; pedicels and inv. *gland-hispid* ... ... (2) **polyodon.**

1. H. Capense (Linn. Amoen. 6, p. 96); stem nearly nude, scape-like, glabrous, striate, panicled or racemoso-corymbose at the summit; radical leaves rosulate, obovate-oblong, obtuse, toothed, denticulate, or subentire, on both sides sparsely pubescent; cauline leaves reduced to scales; pedicels elongate, straight, alternate; inv. sc. subtomentose at base, sc. thinly powdery, in few rows, linear, very acute; ligules pubescent externally; achenes cylindrical; pappus dirty-white. *DC. l. c.* 218. *Fries, Epicr. Hier.* 141. *Crepis striata, Th.! Cap.* 613. *Picris striata, Spr. Syst.* 3, 264. *Schmidtia Capensis, Reichb.*

VAR. *β*. microcephala; rad. lvs. quite glabrous, lanceolate, entire; stem very much branched; heads ¼ smaller; achenes delicately striate.

HAB. Cape, *Thunberg!* About Table Mt., especially the East side, and at Newlands, in woods. *E. & Z.! W.H.H.* &c. Drakensteinberg, *Drege!* Zwartkops R., *Zey!* 3082. Slaayekraal, *Burke!* Grahamstown, *H. Hutton!* Br. Kaffraria, *Cooper!* 149. Natal, *Krauss!* 345. *Gerr. & McK.!* 310. VAR. *β.* Magalisberg, *Zey!* 1038. (Herb. Th., D., Sd., Hk.)

Root leaves 3–4 in. long, ½–1 in. wide, much attenuate at base. Stem 1–2 ft. high or more, simple for ¾ its length. "Used by the natives about Port Natal in treating snake-bites. They scarify the bitten limb and rub in the powder into the cuts, and give the patient a strong hot decoction of the roots to drink. It is also given to cattle, with success in cases of internal inflammation." *Dr. Callaway, in litt. to D. Hanbury.* β. may be a distinct species, but the specimens seen are in imperfect condition.

2. H. polyodon (Fries, Epicr. Hier. p. 67); stem nearly nude, scape-like, striate, glabrous below, above flocculent and rough with rigid, subulate, short bristles, few-headed; radical leaves rosulate, linear-lanceolate, acute, glabrous, toothed, the teeth reversed, subruncinate; pedicels and involucres glandularly setose, the scales in few rows, linear, obtuse, dark-coloured, with pale, membr. margins; achenes columnar; papp. dirty whitish.

HAB. Cape, *Eklon!* (Herb. Sond.)

Leaves 4–5 in. long, ¼–½ in. wide. Stem 12–15 in. high, in the specimens seen 2 headed. Very distinct in aspect and pubescence from *H. Capense.*

CLIV. ANISORAMPHUS, DC.

Heads many-fl. *Inv.* calyculate-imbricate, the scales broadly linear, the outer about 2-ranked, short, inner sub-biseriate, elongate. *Recept.* nude. *Achenes* oblong, *(immature only seen)* compressed, striate, ros-trate; the beak of the outer ones short, of the inner longer. *Pappus* pilose, multiseriate, yellowish. *DC. Prodr.* 7, *p.* 251.

"A small plant with the aspect of a *Hieracium* or *Hypochæris.* Root thick, cylindrical. Radical leaves oblong, subattenuate at base, subretrorsely toothed, glabrous; the cauline very few, linear, quite entire. Stem nearly nude, sulcate and glabrous at base, sparingly branched above, about 3-headed, hairy with black, rigid bristles. Inv. sc. more copiously hairy with similar bristles. Fl. yellow." *DC. l. c.* Name, ανισος, *unequal,* and ραμφος, a *beak.*

1. A. hypochoerideus (DC. l. c.)

HAB. Windvogelberg, in rocky and stony places, 4–5000 f. *Drege.* (Unknown to us.)

ORDER LXXVII. CAMPANULACEÆ, Juss.

(By W. SONDER.)

Calyx mostly 5-lobed (varying from 3 to 10-lobed), rarely truncate, adhering to the ovary, or partially or wholly free; lobes equal or unequal. *Corolla* monopetalous, rarely cleft nearly to the base or polypetalous, regular or irregular, the lobes valvate or induplicate in æstivation. *Stam.* as many as the lobes of the corolla, alternate with them, epigy-nous, free from the corolla, very rarely inserted on its tube; *filaments* expanded at base; *anth.* 2-celled, splitting longitudinally. *Ovary* inferior (or half-inferior, rarely superior), 2–3–5–6–8–10-celled; *ovules* many or few, rarely definite or solitary; *style* simple. *Fruit* a capsule or berry; rarely a drupe or nut. *Seeds* albuminous; *embryo* straight.

Herbaceous, rarely shrubby or arborescent plants, often with milky acrid juice. Leaves alternate or rarely opposite, mostly toothed, sometimes lobed or much cut, without stipules. *Flowers* racemose, or panicled, or solitary, terminal or axillary. A large order, chiefly from the temperate zones of both hemispheres.

Tribe 1. **LOBELIEÆ.** *Corolla* monopetalous, irregular, 1-2-lipped, with a split or rarely an entire tube, persistent. *Anthers* syngenesious, cohering in a tube round the stigma ; pollen ovoid, smooth. *Style* glabrous ; *stigma* girt with a circle of hairs, mostly 2-lobed. *Juice milky.* (Gen. 1-9.)

Sub-tribe 1. *Clintonieæ* : Capsule 1-celled, prismatic, elongate, 3-valved.

I. **Grammatotheca.**

Sub-tribe 2. *Lobelieæ* : Caps. 2-celled, 2-valved, the valves in the middle septi-ferous, rarely opening by pores.

* *Tube of the corolla cleft down one side to the base.*

II. **Metzleria.**—*Cor.* subregular, 5-parted, segments lanceolate. 2 lower anthers tipped with bristles.
III. **Monopsis.**—*Cor.* rotate, 5-lobed, lobes roundish-ovate. Anth. all bearded.
IV. **Isolobus.**—*Cor.* 5-fid, unilabiate ; lobes equal, spreading. 2 lower anth. bearded.
V. **Parastranthus.**—*Cor.* 5-fid, 2-lipped, upp. lip 3, lower 2-fid. Anth. all bearded.
VI. **Lobelia.**—*Cor.* 5-fid, 2-lipped ; upp. 2-, lower 3-fid. Anth. all, or the 2 lower bearded.
VII. **Dobrowskya.**—*Cor.* 3-parted, 2 upper petals separate, clawed ; 3 lower connate into a 3-fid lip. Anth. all bearded.

** *Tube of the corolla funnelshaped, not cleft at one side.*

VIII. **Enchysia.**—*Cor.* sub-regular, the lobes sub-equal, erect.
IX. **Laurentia.**—*Cor.* bilabiate, 2 upp. lobes small, 3 lower larger, reflexed.

Tribe 2. **CAMPANULEÆ.** *Corolla* monopetalous (rarely 4-5-petaled), regular. *Anthers* separate ; *pollen* spherical, echinulate. *Style* pubescent ; *stigma* nude (not in a cup), mostly 2-3-5-lobed. *Juice commonly milky.* (Gen. 10-17.)

Sub-tribe 1. *Wahlenbergieæ.* Capsule opening at the apex. Ovules many.

* *Capsule opening with valves regularly.*

X. **Lightfootia.**—*Cor.* 4-5-parted to the base or nearly so.
XI. **Microcodon.**—*Cor.* cylindrical, small, 5-lobed at apex. *Caps.* 5-celled, cells *alternating* with the cal.-lobes.
XII. **Wahlenbergia.**—*Cor.* funnel or bell-shaped, 5-lobed at the apex, or 5-fid to the middle. Caps. half superior, 2-5-celled ; cells, when 5, *opposite* the cal.-lobes.
XIII. **Leptocodon.**—*Cor.* cylindrical, small, 5-lobed at the apex. *Caps.* 2 celled, *wholly* inferior.

** *Capsule elongated, bursting at the top or opening by a terminal pore. Ovary quite inferior.*

XIV. **Prismatocarpus.**—Caps. 2-celled, long and slender, *naked*, at length splitting from the top nearly to the bottom into 5 segments.
XV. **Roella.**—Caps. 2-celled, crowned by the persistent calyx-lobes, opening by a terminal pore.

Sub-tribe 2. *Merciereæ.* *Stamens* free. *Ovary* one-celled, with an incomplete septum. *Ovules* 4 in the bottom of the ovary.

XVI. **Merciera.** *Cor.* with a very long, narrow tube.

Sub-tribe 3. *Siphocodeæ.* *Stam.* inserted in the tube of the corolla. *Ovary* 3-celled ; cells 2-ovuled.

XVII. **Siphocodon.** *Corolla* tubular, 5-lobed at apex.

Doubtful Genus.

Rhigiophyllum. Cor. with a long tube, 5-lobed. St. inserted near the apex of the tube of corolla. Ovary 3-celled ; cells many ovuled.

Tribe 3. **CYPHIEÆ.** Petals 5, separate or partially cohering by their claws above

the base, spreading more or less irregularly in the form of a labiate corolla. *Anthers* separate, rigid, mostly hispid at back ; *pollen* globose. *Style* glabrous ; *stigma* obtuse, simple, nude, or with an imperfect indusium. *Caps.* 2-celled ; many seeded.
XVIII. **Cyphia.** Character of the tribe.

Tribe 4. **GOODENOVIEÆ.** *Cor.* monopetalous, irregular, the tube split in front, limb 5-parted, 1–2-lipped, with induplicate æstivation. *Anthers* separate or cohering ; pollen simple or compound. *Style* simple (very rarely 2) ; *stigma* fleshy, girt with a cup-like indusium.

XIX. **Scævola.**—*Cor.* unilabiate. *Anth.* free. Drupe fleshy or dry ; seeds solitary. *(A sea-coast shrub.)*

TRIBE I.—**LOBELIEÆ.** (Gen. I–IX.)

I. **GRAMMATOTHECA**, Presl.

Calyx-tube elongate, linear-triquetrous ; lobes 5-parted, spreading or reflexed. *Corolla* tubulose ; tube cleft the whole length ; limb 5-parted, 2-labiate, the 2 upper lobes linear, erect, lower lip larger, reflexed, 3-lobed. *Anthers* all bearded. *Stigma* bilobed. *Capsule* linear, triquetrous, 1-celled, 3 valved, many-seeded ; valves coherent at the base and apex, 2 of them placentiferous in the middle. *Seeds* ovoid. *Prodr. Lob. p.* 43. *DC. Prodr.* 7, 348, *Lobeliæ sect. Clintonia, Chamiss. in Linnæa*, 8, *p.* 217. *Endl. Gen. n.* 3051.

Herbaceous, glabrous plants ; leaves sessile, linear or lanceolate ; flowers axillary, sessile, blue. Name from γραμμα, a *line*, and θηχα, *capsule*, in reference to the linear capsule.

1. G. erinoides (Sond.). *Lobelia erinoides, Thunb. Prodr. p.* 40. *Fl. Cap. p.* 180.

VAR. *a.* **Thunbergiana** ; leaves linear-lanceolate, remotely denticulate ; flowers about twice shorter than the leaf. *L. erinoides, Thunb.! l. c. Lob. (Clintonia) Bergiana Cham. l. c. Gram. Bergiana et Mundtiana, Presl, l. c. p.* 44. *DC. l. c. E. & Z.̅ n.* 2410.

VAR. *β,* **Dregeana** ; leaves linear-lanceolate or linear, remotely denticulate ; flowers about as long or a little longer than the leaf. *G. Dregeana, Presl, l. c. Deless! Icon. v.* 5, *t.* 6. *G. Meyeriana, Presl! l. c. (a small specimen). G. Eckloniana, Presl! l. c. E. & Z. n.* 2489. *Rapuntium, Zeyh. n.* 1051.

HAB. Wet or boggy places, and by rivulets near Capetown, Hanglip, Paarl, Bergriver, Dutoitskloof, Zwartkopsriver, and Port Natal (*Lob. Anceps.* Plant. n. 31). Dec.–Jan. (Herb. Thunb. Reg., Berol., Hook., D., Sd.)
Root perennial, creeping. Stem ascending, erect or procumbent, sometimes rooting and stoloniferous, ½–2 feet long, flexuous, simple or branched, compressed or triangular above. Leaves alternate, lower ones often broader, obovate or oblongcuneate, but generally linear-lanceolate or linear, with a callous point, remotely serrulate or denticulate, 1–2 in. long, 1–2 lines wide, spreading. Flowers in the upper axils of the leaves sometimes racemose. Ovary sessile, with 2 linear bracteæ at the base, 4 lines to 1 inch long, erect, crowned by the linear, acute, serrulate or ciliolate calyx-lobes, which are unequal, about 1 line long. Corolla 4–4½ lines long, bright blue, the tube striated, the lobes of the lower lip ovate or ovate-oblong, acute. Capsule incurved, dehiscing at the sides. *G. Eckloniana* is more diffuse than *G. Dregeana*, but in no other respect different ; the flowers are not longer. *G. Meyeriana* is founded on a small and narrow-leaved specimen ; the flowers are not pedicellate.

II. **METZLERIA**, Presl.

Calyx-tube hemispherical ; *limb* 5-parted. *Corolla* split the whole length on back, 5-parted, subregular, lobes lanceolate, the lower connate at the base, spreading, the upper lobes free. *Anthers* subincurved, the two lower terminated by a single bristle, rarely bearded. *Capsule* globose, bivalved. *Prod. Mon. Lob. p. 7. Endl. Gen. n. 3055.* Also *DC.! in Prodr.* 7, 350. *Lobeliæ species, Thunb.*

Small, annual, decumbent herbs, with alternate leaves, and minute, axillary, solitary, pedicellate flowers. Named after Giseke-Metzler, professor of natural history in England, 1813.

Leaves serrulate ... ... (4) **humifusa.**
Leaves entire :
　　Leaves petiolate, obovate ... . (1) **depressa.**
　　Leaves sessile, ovate, narrowed to the base ... ... (2) **Dregeana.**
　　Leaves sessile, ovate, acute, upper lanceolate ... ... (3) **filicaulis.**

1. M. depressa (Sd); glabrous; stems decumbent, filiform, branched; leaves petiolate, obovate, obtuse, entire ; peduncles axillary, solitary, twice longer than the leaf. *Lobelia depressa, Thunb. prod. p. 39. Fl. Cap. p.* 178. *Linn. f. Suppl. p.* 395. *E. Z. n.* 2443.

HAB. Sandy places near Capetown, Greenpoint, Tablemountain, *Thunb., Pappe, W.H.H.* Swellendam, *E. & Z.* Platteklipp, *Zey. n.* 1049, b. Nov.–Feb. (Herb. Thunb., Hook., D., Sd.)

Habit of *Elatine Hydropiper.* Stems or branches diffuse, 1–3 inches long, white. Leaves obovate, or ovate-oblong, 1–2 lines long ; petioles as long or shorter than the leaf. Pedicels 3–4 lines long. Flowers minute, white ? or pale rose-coloured. Corolla 1 line long, lobes ovato-lanceolate. Capsule ½ line long.

2. M. Dregeana (Sond.); glabrous ; stems prostrate, branched, rooting ; branches angular, very leafy ; leaves ovate, tapering at the base, acutish or obtuse, entire ; peduncles axillary, as long or shorter than the leaf, biglandular at the base. *M. depressa, Presl ! l. c. DC.! Prod. p.* 350, *not Lob. depressa, Thunb.*

HAB. Witbergen, 6–7000 ft. *Drege.* Aapjesriver, *Burke & Zeyh. n.* 1047, October. (Herb. Hook., D., Sd.)

Stems 2–5 inches long. Leaves on a very short petiole, or subattenuated at the base, subfleshy, 2–3 lines long, 1½ line wide. Calyx-lobes twice shorter than the tube. Corolla rose-coloured ? 1½ line long, lobes lanceolate or acuminate. Capsule 1 line long. Distinguished from *M. depressa* by very numerous broader, nearly sessile leaves, shorter peduncles, and larger flowers.

3. M. filicaulis (Presl! l. c.); glabrous ; stems humifuse, angular, filiform, branched ; leaves sessile, ovate, acute, entire, upper ones lanceolate ; peduncles axillary, solitary, shorter or longer than the leaf. *DC. l. c. p.* 351. *E. Z.! n.* 2444.

HAB. In waterpools on the fields near the Zwartkopsriver, Uitenhage, *E. & Z.* Port Natal, *Gerr. & M'K.,* n. 1483. Near Capetown, *W. H. H.* October. (Herb. Hook., D., Sd.)

Larger specimens resemble *Anagallis phaenicea.* Leaves not fleshy, commonly distant, 3 lines long, 2 lines wide, the upper 3–4 lines long, 1 line wide. Pedicels 2–6 lines long. Flowers very minute as in *L. depressa,* of which it is perhaps a variety. *Presl.* has only seen an imperfect specimen, communicated by *Ecklon.*

4. M. humifusa (A. DC.! l. c. p. 351); quite glabrous ; stem and branches capillary, angular, diffuse ; leaves ovate, very shortly petiolate, serrulate, serratures 3–4 on each side, callously mucronulate as well as the terminal, obtuse lobe; pedicels 2–3 times longer than the leaf, naked; calyx-tube subturbinate, lobes ovate, acute; corolla 3 times longer than the calyx-lobes. *Rapuntium pygmaeum, Presl.! excl. syn. E. Z.! n.* 2486.

HAB. Top of Table Mt., *E. & Z., Dreye, Pappe, W.H.H.* Dec.–Jan. (Herb. D.,Sd.)
Stems 1–4 inches long. Leaves 2–3 lines long, 1½–2 lines wide, or smaller, on very short, broad petioles. Peduncles erect. Calyx-lobes longer than the very short tube. Corolla 1½ line long, white; in one specimen they appear blue, lobes ovate-lanceolate, the 2 upper a little narrower, nearly free. Anthers glabrous or hairy on back, the lower setaceous or bearded at top. It varies with more fleshy leaves, of which the upper are entire.

III. MONOPSIS, Salisb.

Calyx-tube obconical or hemispherical; lobes equal, linear, acute. *Corolla* funnel-, or salver-shaped, with a tube cleft the whole length; *limb* rotate, 5-lobed, lobes equal, roundish. *Anthers* all bearded. *Stigmas* 2. *Trans. Hort. Soc. Lond.* 2, *p.* 37. *A. DC. Prodr.* 7, *p.* 35 ¹. *Lobeliæ spec. Linn. et Thunb.*

Small annual herbs with diffuse branches, linear-lanceolate leaves, very long, axillary, 1-flowered, naked peduncles, and showy deep blue flowers with yellow anthers. Name from μονος, *one*, and οψις, *a face*, in reference to the regular, not bilabiate flowers.

Cal. tube obconical; capsule acute or tapering at base ... ... (1) **debilis.**
Cal. tube hemispherical; caps. roundish... ... ... (2) **campanulata.**

1. M. debilis (Presl! Prod. Lob. p. 11); branches diffuse; leaves linear, acute, remotely serrate, subpilose or glabrous; pedicels erect or ascending, 2–4 times longer than the leaf; calyx-tube obconical, lobes linear-acuminate; tube of corolla a little longer than the calyx, with roundish, mucronulate lobes; capsule a little curved, acute or attenuate at the base. *Comm. pl. Dreg. p.* 284. *E. & Z. n.* 2452. *Deless. Icon.* 5, *v. t.* 7. *Lobelia debilis, Linn. fil.! Suppl. p.* 395. *L. Erinus, Thunb.! fol.* β. *et* γ. *fl. Cap. p.* 179, *ex. pte. L. debilis, Thunb. in Herb. Holm., non in Herb. Thunb. Zey. n.* 1052-3114.

VAR. β. **conspicua**; smaller; leaves linear or linear-lanceolate; tube of calyx shortly obconical; corolla often larger. *M. conspicua, Salisb. l. c. p.* 40. *t.* 2. *Presl. l. c. p.* 10, *E. & Z. n.* 2451. *M. Speculum A. DC.! Lob. Speculum Andr. Bot. Rep. t.* 644. *Sims Bot. Mag. t.* 1499. *L. simplex Thunb.! Prod. p.* 39. *A. DC. Prodr. p.* 371. *Rap. simplex Presl. prodr. Lob. p.* 17. *Lob. campanulata, Lam. dict.* 3. *p.* 588. *Zey. n.* 1054, 1056.

HAB. Moist places in the distr. of Cape, Stellenbosch and Worcester, var. β. near Capetown, Klypfontein, Brackfontein, etc. also near Magalisberg, Sept. Nov. (Herb. Thunb., Holm., Hook., D., Sd.)
Very variable in size, from 2 inches to 1 foot, leafy, commonly branched from the base, the branches filiform or capillary. Leaves entire at the base, 6–12 lines long, ½–1 line wide. Peduncles 1–2½ inches long. Corolla 4–6 lines long, lobes rotate. Capsule subincurved or horizontal, in var. α., 2–4 lines long, evidently 10-nerved, in var. β. usually shorter, 1–3 lines long, hispid or adpressed hairy, sometimes quite glabrous. The calyx-lobes are erect or spreading in the same specimen.

2. M. campanulata (Sond.); stem branched, glabrous; leaves linear-lanceolate or lanceolate, remotely serrate, with cartilaginous margins; pedicels erect, subpilose, 2–4-times longer than the leaf; calyx-tube hemispherical, hispid, shorter than the linear, acuminate lobes ; tube of corolla equalling the calyx, with roundish mucronulate lobes ; capsule roundish. *M. conspicua, var. gracilis Presl.! Comm. pl. Drege, p.* 284. *M. gracilis, A. DC.! l. c. p. 352. Lobelia campanulata, Lam.! dict. 3. p.* 588.

HAB. Draakensteenbergen, 2–300 ft., *Drege;* Modderfontyn, Namaqualand, *V. Schlicht. Rev. H. Whitehead.* Cape, *Vaillant,* Herb. *Lamark.* Ort. (Herb. Roeper., D., Sd.)

In habit and size exactly *M. conspicua,* Sal. and only differing by the capsule, roundish at the base. Lower leaves ovate-lanceolate, entire or a few toothed cauline leaves 6–8 lines long, 1–2 lines wide. Calyx-lobes soon reflexed, 2 lines long. Tube of corolla 4 lines long, limb rotate. Capsule the size of a small pea.

IV. **ISOLOBUS**, A. DC.

Calyx-tube turbinate or obconical, elongate. *Corolla* split the whole length down the back, unilabiate, 5-fid; lobes equal, spreading. *Anthers* of the two lower stamens terminated by a single seta. *Stigma* bilobed. *DC. Prod.* 7, 2, *p.* 352. *Lobeliæ, Spec. Auct. Pratiæ, Spec. G. Don. Gen. Syst. Gard. 3, p.* 699. *Monopsis, Sp. Presl. Prod. Lob. p.* 11. *Rapunt. Spec. Presl. Comm. pl. Drege, p.* 287.

Perennial herbs or sub-shrubs. Leaves alternate, serrate. Flowers solitary, axillary or aggregated at the apex, white or purplish. Name from ισος, *equal,* and λοβος, *lobe,* in reference to the equal lobes of stigma.

Ascending or erect ; fl. in dense, terminal corymbs ... ... (1) **corymbosus.**
Diffuse, procumbent ; fl. axillary, on thread-like pedicels ... (2) **Ecklonianus.**

1. I. corymbosus (A. DC.! l. c.) ; glabrous ; stem ascending or erect, angular, branched ; leaves inciso-serrate, lower ones obovate or suborbicular, petiolate, upper sessile, lanceolate-spathulate or sublinear; *flowers disposed in dense, terminal corymbs;* bracteæ narrow-linear, toothed, longer than the pedicels ; calyx-tube turbinate with acumi-nate lobes. *Monopsis corymbosa, Presl! l. c.*

VAR. *a.* **foliosus**; branches leafy at the apex ; upper leaves lanceolate or linear-spathulate. *I. corymbosus, A. DC. Lob. corymbosa, Grah. in Jameson's Edinb. New Philos. Jour. Oct.* 1826, *p.* 385. *Hook. Bot. Mag. t.* 2693. *Pratia corymbosa, G. Don. l. c. E. Z. n.* 2453. *Herb. Un. itin. n.* 489.

VAR. *β.* **jasionoides**; branches sparingly leafy or nearly naked at the apex ; upper leaves linear-lanceolate or linear. *I. jasionoides, A. DC.! l. c. p.* 353. *Deless. Icon. vol.* 5, *t.* 8. *Rap. corymbosum, Presl.!* E. *Meyer, comm. pl. Drege, p.* 287. *Zeyh. n.* 3117.

VAR. *γ.* **sparsiflorus**; branches sparingly leafy or nearly naked at the apex ; upper leaves sublinear ; flowers axillary, solitary, the terminal subcorymbose. *Zey. n.* 3118.

HAB. Sandy and rocky places, Tablemountain ; Hott. Holl. ; Kleinriviersberg, Caledon, *E.& Z.;* var. *β.* Baviansberg, Genadenthal, *Dr. Pappe;* Riverzonder Einde, *E. & Z.;* between Ebenezar and Giftberg, and in Caffraria, *Drege;* var. *γ.* near Voormansbosh, Swellendam, *Zey., W. H. H.* Oct.–Jan. (Herb. Hook., D., Sd.)

Perennial herb, ½–1½ foot high, much-branched. Leaves decurrent, sharply serrate, 6–8 lines long ; the lower suborbicular, ovate or obovate, 3 lines wide ; the middle oblong-cuneate or spathulate, the upper sublinear, about 1 line wide ; peti-oles of the lower leaves as long or shorter than the limb. Corymb or condensed raceme ½–1 inch long and broad, sometimes smaller. Pedicels and calyx hairy or glabrous. Corolla 2 lines long, 5-fid, purplish or white with purplish spots. Anthers shorter or as long as the corolla, often subexserted. Capsule bivalved.

In var. β. the upper part of the branches is leafless or nearly so, the flowers are aggregated to a dense corymb; var. γ. has the same habit, but the pedicels are solitary in the axil of the leaves, and only a few (4–6) form a terminal leafy corymb.

2. I. Ecklonianus (Sond.); quite glabrous; stems diffuse, procumbent, axillary, branched; leaves tapering into a short petiole, dentate or serrate; the lower obovate or spathulate; the middle oblong-cuneate or lanceolate; the upper sessile, sublinear, bidentate; *pedicels axillary, capillary,* half as long as the leaves, bibracteolate at the base; calyx-tube turbinate with subulate lobes. *Lobelia Eckloniana, A. DC. l. c. p.* 368. *Rap. Ecklonianum, Presl! E. Z. n.* 2464. *E. Mey. Comm. pl. Drege, p.* 286. *Isol. sparsiflorus, A. DC.! l. c. p.* 353.

HAB. On Kasteelberg, Tablemountain, *E. & Z., W. H. H.;* Drakensteenbergen, *Drege.* Jan. (Herb. D., Sd.)

Nearly allied to var. γ. of the preceding, but the stems diffuse, more capillary, weak, and all the flowers axillary, distant. Root perennial. Stems 4–8 in. long, prostrate, trailing. Leaves 4–8 lines long, the lowest 2–3 lines wide, coarsely serrate, the serratures 3–4, mucronate; uppermost leaves about 1 line wide, narrowed, at the base. Pedicels 4 lines long. Corolla 2 lines long, white, with purplish lobes, lanceolate. Ovary with 2 ovules in each cell. Capsule with an inflated, subbilobed apex, opening by 2 broad, apiculate valves bearing the septa in their middle. Seeds "elliptic." *Harv.* in Herb. D.

V. **PARASTRANTHUS,** G. Don.

Calyx 5-cleft, tube cylindrical or obconical. *Corolla* cleft the whole length, bilabiate, upper lip of 3 lobes, low r one of 2 lobes, smaller. *Anthers* all bearded. *Stigma* bifid. *Seeds* roundish, very minute. *Gen. Hist. Gard.* 3, *p.* 716. *DC. Prodr.* 7, *p.* 354. *Endl. Gen. n.* 3055. *Lobeliæ Spec. Linn. Thunb. Rapuntii, sect. Xanthomeria. Presl, Prodr. Lob. p.* 11.

Perennial or suffruticulose herbs, with sessile, toothed leaves, and terminal racemes of nearly sessile or pedicellate flowers. Name from παραστρεφω, to *invert,* and ανθος, a *flower;* the flowers are upside down, compared with those of *Lobelia.*

Flowers yellow:
Leaves lanceolate, serrate, acute ... ...   (1) **luteus.**
Leaves linear, the lower entire, upper ones serrate or toothed   (2) **variifolius.**
Flowers blue or white; leaves ovate, attenuate to the base ... ...   (3) **thermalis.**

1. P. luteus (A. DC. Prodr. p. 354); glabrous or slightly hairy; stems simple or branched, ascending; leaves lanceolate, serrate, acute; flowers almost sessile at the tops of the stems, forming a loose spike; bracteæ toothed, shorter than their calycine segments; corolla hairy. *Burm. Afr. p.* 101, t. 39. *Lobelia lutea, Linn. Spec. p.* 1322. *Thunb.! Fl. Cap. p.* 180. *Sims. Bot. Mag. t.* 1319. *Rapunt. luteum, Presl, l. c. E. & Z. n.* 2454. *P. simplex, Don. l. c. Sieb. ext. Cap.* 249. *Herb. Un. Itin. n.* 18, 484. *Zeyh. n.* 1056.

VAR. β, **capitatus**; flowers subcapitate. *Rap. capitatum, Presl! l. c. E. & Z. n.* 2456. *Parast. capitatus, A. DC. l. c.*

VAR. γ, **euphrasioides**; leaves linear-lanceolate or linear, serrate. *Rap. euphrasioides, Presl! l. c.; E. & Z. n.* 2457. *Parast. euphrasioides, A. DC. l. c. p.* 355.

VAR. δ, **ericoides**; stems diffuse, filiform; leaves very small, alternate or subopposite; flowers spiked. *Rap. ericoides, Presl, l. c. p.* 12.; *E. & Z.* 2458. *Parast. ericoides, A. DC. l. c. Lob. lutea, Thunb.! Herb. var. a, ex pte.*

HAB. Sandy wet places, Cape flats, and in mountains near Capetown, Zoutrivier, and Rietvalley, Winterhoek, Tulbaghskloof, and Witsenberg. Oct.–April. (Herb. Thunb., Holm., Hook., D., Sd.)

Stems many from the root, filiform, $\frac{1}{2}$–1 foot high. Leaves numerous, often secundate, glabrous, $\frac{1}{4}$–1 inch long, 1–1$\frac{1}{2}$, rarely 2–3 lines wide; in var. δ. 2–3 lines long, $\frac{1}{2}$ line wide. Spike $\frac{1}{2}$ to several inches. Calyx obconical, glabrous or hairy; lobes lanceolate, acuminate. Corolla deep yellow. Capsule cylindrical, 4 lines long. *Rapunt. flavum. Presl.! in E. Z. n.* 2455, is *P. luteus*, with a few-flowered raceme, in which 1 or 2 flowers are more or less pedicellate and bracteated by a narrower leaf.

2. P. variifolius (Don. l. c.); stem erect, simple, rarely branched above; leaves numerous, erect, linear, glabrous, upper ones toothed or serrate, the lower entire; flowers solitary, axillary, nearly terminal, almost sessile; bracteæ as long as the flowers; corolla hairy. *Lobelia variifolia, Sims. Bot. Mag. t.* 1692. *Rap. variifolium, Presl, Prodr. p.* 12. *comm. pl. Dreg. p.* 285, *E. Z. n.* 2459.

HAB. Wet places on hills, Waterfall, Tulbagh, *E. & Z., Pappe.* Nieuwekloof Vierentwintigrivier and Pickenierskloof, *Drege.* Betw. Roux and Bergrivier, *Zeyh.* n. 1055. Nov.–Dec. (Herb. Hook., D., Sd.)

2–4 inches high, very leafy from the base to the apex, glabrous or slightly hairy. Upper leaves 4–6 lines long, $\frac{1}{2}$ line wide, concave or sulcate above, acute, with 2–3 sharp teeth on each side, the lower leaves 3–2 lines long. 1 or 3 deep yellow flowers at top of the branches. Calyx-tube hairy, obconical; lobes lanceolate, acuminate. Stigma always bifid, not trifid.

3. P. thermalis (Sond.); stems decumbent or creeping, simple or branched, filiform, hispid or glabrous; leaves sessile, ovate, acute, narrowed at the base, serrate, glabrous or hairy beneath; flowers axillary, shortly pedicellate; calyx-tube clavate, hispid, 2–3 times longer than the lanceolate lobes; corolla minutely downy, hispid at the lobes; capsule cylindraceous. *A. DC. Prod. p.* 354. *Lob. leptocarpa, Grissel. in Linnaea,* 5, *p.* 419. *DC. l. c. p.* 358. *Lob. thermalis, Thunb.! prod. p.* 40. *Fl. Cap.* 181. *Rap. thermale, Presl, prod. p.* 12. *L. pubescens, Dietr. ex Steud. L. Mundtiana Chamiss! Linnaea,* 8, *p.* 215. *L. glabrifolia, E. Meyer in Herb. Drege. Rap. glabrifolium, Presl! Prod. p.* 12. *E. & Z. n.* 2460.

HAB. Watercourses and damp situations near the baths of Oliphantsriver, *Thunb.;* near Winterhoeksberg, Uitenhage, *E. & Z.,* Nieuweveld; on the Brakrivier, and between Keiskamma and Basche, *Drege;* Gamkariver, *Mundt;* Magalisberg, *Burke & Zey.* n. 1071; Distr. of Albert, *T. Cooper, n.* 1391; Queenstown D., *Mrs. F. W. Barber;* Grahamstown, *Dr Atherstone;* Transvaal and Port Natal, *I. Sanderson, Miss Owen;* Roggeveld, *A. Wyley.* Dec.–Feb. (Herb. Th., Holm., Hk., D., Sd.)

Stems $\frac{1}{2}$–1 foot long, subangular. Leaves alternate, distant by intervals of 3–4 lines, ovate or elliptical or sublanceolate, sharply serrated, 4–6 lines long, 2–3 lines wide. Flowers rare in the axils; peduncles 1–2, rarely 3–4 lines long, hispid. Calyx-lobes 1$\frac{1}{2}$ line long. Corolla about 8 lines long; tube white, dilated above; lower lip bifid; lobes erect, acute, blue; upper lip 3-fid, blue. Capsule hairy or subglabrous, 6–8 lines long. It is a true *Parastranthus;* the flower is incorrectly described by Chamisso and others. In the dry flowers the blue colour of the lobes has generally disappeared.

VI. **LOBELIA**, Linn.

Calyx 5-lobed, sub-obconical, ovoid, or hemispherical. *Corolla* irregular, bilabiate, tube cylindraceous or infundibuliform, cleft on the upper side; upper lip often smaller and erect; lower lip often spreading or

pendulous, 3-fid. *Anthers* cohering, the 2 lower ones or sometimes all
bearded. *Ovary* inferior or half-superior. *Capsule* 2-celled, 2 valved,
many seeded, dehiscing at the apex. *Lobeliæ spec. Linn. sp. p.* 1317,
non Presl. Rapuntium Tourn. Presl, prod. Lob. p. 11, *excl. Sect.* 1. *A.
DC. in DC. Prod.* 7, 2, *p.* 35, *f. Endl. gen. n.* 3058.

Erect or procumbent herbs or sub-shrubs. Leaves alternate. Flowers pedicellate,
usually disposed in terminal leafy spikes or racemes, rarely solitary and axillary,
blue, white, violaceous or red. Named in honour of *Matthew Lobel*, author of various
botanical works.

Sect. 1. All the anthers bearded. (Sp. 1–15.)

Cal.-tube acute at base, turbinate or linear-obconic:
　Leaves cordate or ovate:
　　　Lvs. cordate-orbicular, obtuse and lobed   ...   ... (1) **Preslii.**
　　　Lvs. ovate, acute, subserrate; raceme terminal;
　　　　pubescent   ... ... (2) **Zeyheri.**
　　　Lvs. ovate or sub-cordate, mucronate, toothed;
　　　　pedicels axillary; *glabrous* ... ... (3) **alsinoides.**
　Leaves linear or subulate:
　　　Lvs. linear-subulate, keeled, ciliate at base; ped.
　　　　and cal. woolly ... ... (4) **spartioides.**
　　　Lvs. linear, glabr.; pedicels glabr.; cal. appressed
　　　　hairy ... . (5) **linearis.**
Cal.-tube obtuse at base, ovoid or hemispherical:
　Herbaceous:
　　Leaves linear or setaceous:
　　　Pubescent scabrous. Pedicels sub-solitary,
　　　　terminal ... ... (6) **decipiens.**
　　　Glabrous. St. terete; leaves setaceous; raceme;
　　　　ped. as long as bracts ... ... (7) **setacea.**
　　　Glabr. St. angular; lvs. setaceous; raceme
　　　　branched at base; ped. shorter than the
　　　　bracts ... ... (8) **capillifolia.**
　　　Stem and branches glabr. Lvs. ciliate; raceme
　　　　loose, secund; ped. and cal. *hispid* ... ... (9) **lasiantha.**
　　Leaves petioled, ovate or obovate:
　　　St. and branches glabrous. Leaves ovate,
　　　　acute ... ... (14) **patula.**
　　　St. and branches hairy. Leaves obovate, obtuse (15) **chamædryfolia.**
　Suffruticose, somewhat woody:
　　　Leaves ovate or oblong, obtuse, entire or bluntly
　　　　toothed; plant *hairy* ... ... (10) **hirsuta.**
　　　Lvs. linear, keeled, entire, rigid; pedunc. few-fl.,
　　　　terminal ... ... (11) **pinifolia.**
　　　Lvs. linear, with 2–4 blunt teeth, tomentose ... (12) **tomentosa.**
　　　Lvs. lanceolate, toothed or pinnatifid, hairy or
　　　　glabrous ... ... (13) **coronopifolia.**

Sect. 2. The lower anthers alone bearded or tipped with bristles. (Sp. 16–26.)

Leaves all radical, rosulate, (obovate); spike racemose   ... (26) **cyphioides.**
Leaves scattered along the stem and branches:
　Stem terete or subangular (not compressed or winged):
　　Leaves entire or toothed (not lobed):
　　　Prostrate, creeping; with axillary, solitary
　　　　flowers   ... ... (20) **Boivini.**

Erect or ascending; fl. in terminal racemes :
<table>
<tr><td>Glabrous. Lower lvs. toothed, petioled,
obovate; upp. lanceolate; cal.-tube as
as long as the ovate, acute lobes ...</td><td>(18) Dregeana.</td></tr>
<tr><td>Hairy at base. Lower lvs. obovate,
obtuse, toothed; upp. lanceol. or linear;
cal.-tube shorter than the linear acumi-
nate lobes ... ...</td><td>(16) Erinus.</td></tr>
<tr><td>Glabrous. St. 3-angled; lower lvs. oblong;
upp. linear, remotely serrulate; raceme
loose; cal.-tube shorter than lobes ...</td><td>(17) Natalensis.</td></tr>
<tr><td>Glabr. or subpilose. St. 3-angled ; lvs.
lin.-lanceol., coarsely and unequally
toothed ; raceme dense ; cal.-tube as
long as lobes ... ...</td><td>(19) triquetra.</td></tr>
<tr><td>St. angular, pubescent. Lvs. ovate, acute,
serrate; upp. lanceol. ; cal.-lobes linear,
subulate ... ...</td><td>(21) pubescens.</td></tr>
<tr><td>St. subangular, glabrous. Lvs. obovate-
oblong ; medial lanceol. ; upp. linear ;
cal. tube ovoid, lobes linear, denticled</td><td>(22) linarioides.</td></tr>
<tr><td>Leaves roundish, deeply 5–7 cleft or lobed ... ...</td><td>(25) muscoides.</td></tr>
</table>

Stem winged, or compressed, sharply 2-edged :
<table>
<tr><td>Cauline lvs. linear-oblong, decurrent. Cal. lobes
triangular, much shorter than the cylindrical tube</td><td>(23) anceps.</td></tr>
<tr><td>Cauline lvs. lanceolate, tapering at base. Cal. lobes
linear, as long as the obconic cal. tube ... ...</td><td>(24) fervens.</td></tr>
</table>

§ 1. Anthers all bearded at the apex or tipped with setæ. (Sp. 1–15.)

1. L. Preslii (A. DC. Prod. p. 358) pubescent; stem ascendent, angular, branched at the base; leaves longish, petiolate, cordate, subor-bicular, obtuse and lobed; petiole flat, broader at the apex; peduncle terminal, elongated, glabrous, few-flowered; flowers racemose, secun-date; pedicels as long as the setaceous bractea; lobes of calyx linear, acute, erect, rather longer than the turbinate 10-nerved tube; limb of corolla hairy outwards. *Rapuntium cordatum, Presl.! in E .Mey. Comment. pl. Dreg. p.* 285, *non. L. cordata, Willd.*

HAB. Rocky Mts., Witbergen near Bambooshoek, 6–7000 ft. *Drege.* (Herb. Sd.) Perennial. Petioles 1–2½ inches long, with spreading hairs. Leaves 6–8 lines in diameter with blunt teeth or lobes. Flower blue, not 1 inch long.

2. L. Zeyheri (Sond.) shortly pubescent; stems erect, terete, simple ; leaves sessile, ovate, acute, with a few acute serratures; upper ones much smaller; raceme terminal, few-flowered; pedicels elongated, often unilateral, bibracteolate at the base; calyx pubescent; lobes linear about equal in length to the tube, nearly twice shorter than the corolla ; capsule turbinate.

HAB. Wet rocky places, Aapjesriver, *Burke & Zey. n.* 1046. Oct. (Herb. Hook., D., Sd.)
A small perennial, stems slender, numerous, 3–5 inches high, leafy in the lower part, and nearly naked at the apex. Leaves 2–3 lines long, 1–2 lines wide, with thickened margins, and 2–4 short, acute teeth. Raceme 4–7-flowered. Pedicels erect, spreading, 6–8 lines long. Flowers very small. Calyx 1–1½ line long, hairy at the corolla. Anthers bearded. Capsule erect, crowned by the short erect calycine lobes.

3. L. alsinoides (Lam.! dict. 3. p. 588); glabrous ; branches diffuse, ascendent or erect, trigonal as well as the stem; leaves subsessile, ovate or subcordate, the lower rotundate, dentate, teeth mucronulate; pedicels slender, 2–3 times longer than the leaves, bibracteolate at the base; calyx-tube obconical, lobes linear, acuminate, equalling the tube; corolla glabrous, somewhat longer than the calyx-lobes; anthers all bearded at the apex; capsule obovoid. *L. trigona Roxb. fl. ind.* 2, *p.* 111. *DC. Prod. l. c. p.* 359. *L. stipularis, Rott. and L. Heyniana, R. & Sch. Syst. Veg. Rap. Zeilanicum, Presl ! in Herb. Sd. non L. Zeylanica, Linn. L. trialata Hamilt. in Don. prod. fl. nepal. p.* 157. *Rap. trialatum, Presl !*

HAB. Cape of Good Hope, *Sonnerat* in Herb. Lam. (Herb. Lam.)
Stem creeping at the base, 4–10 inches, weak. Leaves 3–4 lines long, membranous, obtuse, or acute, serrulate; the upper or bracteæ scarcely smaller. Bracteolæ setaceous, minute. Flowers small, blue. Calyx-lobes 1½ line long. The specimen from the herbarium Lamarck, communicated by Prof. Roeper, agrees exactly with those from East India communicated by Wallich and Drs. Hooker and Thomson. *L. Zeilanica Linn.,* of which I have seen a specimen in Herb. Thunberg, is the same as *L. affinis* Wallich! It is similar to *L. alsinoides,* but stronger and different by a dense pubescence.

4. L. spartioides (A. DC. l. c. p. 358); stem much-branched, branches divaricate, terete, glabrous ; leaves sessile, linear-subulate, carinate, ciliate at the base; flowers terminal, racemose; pedicels and calyx-tube woolly; bracteolæ subulate ; lobes of calyx lanceolate, acuminate, longer than the tube, but shorter than the villous corolla; capsule subturbinate, 10-costate. *Rapuntium spartioides, Presl ! in E. & Z.! enum. n.* 2462. *E. Mey. Comment. p.* 286.

HAB. Hills in Karroo like ground near the Gauritzriver and in Langekloof, *E. & Z.; Zwartbergen, Drege;* Dec. (Herb. D., Sd.)
Stem shrubby, 2 ft. and more in height, nearly leafless ; branches rugulose-striated. Leaves 4–6 lines long, or smaller. Racemes few-flowered. Bractea often equal in length to the pedicels. Calyx-tube in flower 1 line, lobes 2–3 lines long. Corolla more than ¼ inch long, blue; the lobes ovate-oblong. It varies with glabrate pedicel and calyx.

5. L. linearis (Thunb.! Prod. p. 39); stems erect, panicled-branched, as well as the branches terete, striate, glabrous; leaves narrow-linear, entire or obsoletely denticulated; pedicels axillary, longer than the leaves, glabrous; lobes of calyx lanceolate, acuminate, a little longer than the turbinate, appressed hairy tube; corolla pubescent; capsule subturbinate, glabrous, 10-costate. *Fl. Cap. p.* 178. *Herb. Thunb. fol. α. non β. Rapunt. tenuifolium, Presl ! E. & Z.! n.* 2461. *Lob. ericetorum. A. DC. Prod. p.* 784. *Rap. genistoides, Presl ! expte.*

HAB. Paardeberg, *Thunb.; Capeflats near Doornhoogte, E. & Z. Zey. n.* 3104; Genadenthal, *Drege.* Dec. (Herb. Thunb., D., Sd.)
1–1½ foot high, more slender than *L. spartioides* and leafy. Leaves about 1 inch long, not ½ line wide, acute, quite entire or minutely denticulated towards the apex. Racemes forming a terminal panicle. Pedicels 4–5 lines long. Calyx acute at the base 10-nerved, 1 line long ; lobes 4-times shorter than the blue corolla, which has oval lobes. Capsule broadly ovate-turbinate, longer than the erect calyx-lobes.

6. L. decipiens (Sond.) pubescent-scabrous ; stem simple or somewhat branched, subangulate; leaves sessile, linear, acute, entire or 1–

2-dentate, with revolute margins; pedicels few or solitary, terminal, calyx scabrous, tube hemispherical, shorter than the linear lobes; corolla pubescent, 4-times longer than the calyx; lobes elliptical or oval, apiculate; capsule hemispherical. *Rapuntium Breynii, Presl! Prod. Lob. p. 19, excl. synon.*

HAB. Katriviersberg, Ceded Territory, *E. & Z. Drege;* Port Natal, *Gueinzius n.* 135. *J. Sanderson;* Magalisberg, *Burke & Zey. n.* 1072.

Habit of *Dobrowsk. scabra;* 3 inches to 1 foot high, very slender, few-branched. Leaves alternate or sub-opposite, 3-4 lines long, ½ line wide, rarely broader and smaller, and only in the lower part of stem. Calyx-tube 1 line, lobes 1½-2 lines long. Corolla blue, yellow in the throat; the lobes of the upper lip oval, twice smaller than those of the lower lip. Capsule as large as a small pea, scabrous.

7. L. setacea (Thunb.! fl. Cap. p. 178) glabrous; stems diffuse or erect, terete, striate, branched; leaves sessile, linear-setaceous, quite entire or obsoletely ciliato-scabrous; pedicels axillary, racemose, about as long as the leafy bractea, bibracteolate at the base; calyx scabrous, tube sub-hemispherical, shorter than the linear-acuminate lobes; corolla downy, 3-4 times longer than the calyx; capsule hemispherical, 10-costate. *Rapuntium setaceum, Presl! Prod. Lob. p. 19. E. & Z. n.* 2477. *Lob. nematophylla Spreng. in Zey. pl. Cap. II. n.* 252. *Sieber! pl. exs. cap. n. 139.*

VAR. β. **parviflora**; flowers smaller; leaves often secundate.

HAB. Capeflats, *Thunb. E. & Z. Sieber, W.H.H.; Simonsbay, I. M'Gillivray and Milne;* Zwartland and Hottentott's, Holland, *Ecklon;* var. β. Rivierzonder Einde, *Zey. n.* 3105. Dec. Mart. (Herb. Thunb., Hook, D., Sd.)

Many perennial stems, ½-1 foot high, more or less branched, sometimes simple. Leaves ½-1 inch long, in var. β. filiform, sub-unilateral, ⅔ line long. Pedicels 6-8 lines long. Flowers blue, ½ inch, in var. β. 3 lines long; lobes oval or elliptical. Anthers hirsute at the apex, hairy on the back. Very near *L. linearis* and only distinguished by the hemispherical, obtuse calyx-tube.

8. L. capillifolia (A. DC. Prod. p. 362); stem diffuse or erect, angulate, glabrous; leaves sessile, setaceous, entire; raceme loose, branched at the base; pedicels shorter than the linear, capillaceous bractea, shortly hairy as the calyx and corolla; calyx-tube hemispherical, twice shorter than the linear-setaceous lobes; corolla 3-4 times longer than the calyx; capsule hemispherical, 10-striate. *Rapunt. capillifolium, Presl! l. c. p. 19. E. Mey. Comm. pl. Dreg. p. 288. Rap. genistoides, Presl! in E. Mey. Comm. p. 285. Lob. genistoides, A. DC. Prod. p. 358. ex pte. E. Z. n. 2478.*

VAR. β, **velutina** (A. DC. l. c.); branches, leaves and calyx minutely downy.

HAB. Mountains near Waterfall and Tulbagh, Mount Zwarteberg, Caledon, *E. Z., Dr. Pappe, Zeyh. n.* 3101, β, and 3102; Hassaquaskloof, *Zeyh. n.* 3101, a; Stellenbosch, Roodesand, Klein-Drakensteen, and Onderbokkeveld, *Drege;* var. β, near Waterfall and Tulbagh. Nov.-Feb. (Herb. Hook., D., Sd.)

Perennial, rigid, stronger than *L. setacea,* 1-2 feet and more high, branched in the upper part. Leaves filiform, callous at the apex, 4-6 lines long. Pedicels 2-3 lines long, often shorter, rarely longer, erect. Calyx-lobes 1½-2 lines long. Corolla blue; lobes oblong. Anthers hairy, hirsute at the apex. Capsule as large as a pea.

9. L. lasiantha (A. DC. l. c. p. 362); stem and branches erect, terete, glabrous; leaves sessile, linear, narrow, ciliate, callous at the apex;

racemes loose, few-flowered, secundate ; pedicels hairy, as long as the
linear, leafy bractea, in the middle or near the base with a linear bracteola ; calyx hemispherical, hispid ; lobes linear-subulate, longer than
the 10-striated tube ; corolla pubescent, hirsute at the apex, 3 times
longer than the calyx. *Rapunt. lasianthum, Presl, E. Mey. Comm. pl.
Dreg. p.* 288.

HAB. In mountains, Giftberg, 1500–2000 ft., *Drege.* (Herb. Sd.)
Distinguished from *L. setacea* by the hispid calyx-tube and the bracteola ; from
L. spartioides by the hemispherical, not woolly calyx. Leaves ½–1 inch long, ½ line
wide. Flowers somewhat larger than in *L. setacea.* Anthers bearded at the apex,
not hairy.

10. **L. hirsuta** (Linn. Spec. p. 1322) ; stem suffruticulose, woody,
much branched, as well as the branches, hairy ; leaves sessile, ovate or
oblong, obtuse at both ends, entire, or with a few obtuse teeth, coriaceous, ciliate and subhirsute beneath ; peduncles terminal, naked, elongated, glabrous, 1–2-flowered, bibracteated at the apex ; flowers drooping ; tube of calyx appressed-hairy or hispid, hemispherical, shorter
than the linear-lanceolate lobes ; corolla 4 times longer than the calyx.
Thunb. Fl. Cap. p. 179. *L. longipes, A. DC. l. c. p.* 361. *Rapunt. pedunculare, Presl ! Prod. Lob. p.* 19. *E. & Z.* 2478, *Burm. Afr. p.* 105,
t. 40, *f.* 2 *(a bad figure). Zeyh.* 3108.

HAB. Mountains in the distr. of Swellendam, George, Uitenhage, and Albany.
Oct.–Feb. (Herb. Thunb., Hook., D., Sd.)
Stems or branches 6–12 in. or more high, purplish, with spreading or adpressed
hairs, rarely glabrous. Leaves with revolute margins, glabrous above, sulcate by
the impressed, middle nerve, 4–6 lines long, in small-leaved specimens 1–1½ line, in
broad-leaved 2–3 lines wide. Peduncles filiform, glabrous or subpilose, 3–4 inches
long, sometimes in lateral branches scarcely longer than the leaves. Pedicels 1–3
lines long, with 1 or 2 bracteolæ. Calyx-tube ¾ line, lobes 1½ line long. Corolla
blue, hairy. Capsule the size of a small pea.

11. **L. pinifolia** (Linn. Spec. Sp. 1318) ; suffruticose ; branches erect ;
leaves approximate, erect, linear, carinate, subtrigonal, acute, quite entire, with subrevolute margins, glabrous or pubescent ; peduncles terminal, few, equalling the leaves or somewhat longer, bracteated ; tube
of calyx hemispherical, silky as the peduncle by appressed hairs ; lobes
of calyx lanceolate, acuminate, 4 times shorter than the pubescent
corolla. *Thunb. ! Fl. Cap. p.* 178. *Breyn. Cent. p.* 173, *t.* 187. *Burm.
Afr. p.* 111, *t.* 41, *f.* 2. *Andr. Bot. Rep. t.* 273. *Sieb. Herb. Cap. n.* 83.
Herb. Un. Itin. 481. *Rapuntium pinifolium, Presl, Prod. p.* 19. *E. & Z.*
2475. *Zeyh.* 3099.

HAB. Mountains near Capetown, in Caledon, Stellenbosch, etc. Sept.–Jan.
(Herb. Thunb., Hook., D., Sd.)
A rigid shrub, 1–1½ ft. Stem nodulose, leafless at the base ; branches closely leafy.
Leaves imbricate, ½–1 inch long, nearly 1 line wide, upper ones smaller. Peduncles
1–2 inches long, rarely longer, 1–3-flowered. Calyx-tube short ; lobes 1½ lines long.
Corolla blue ; lobes of the lower lip oval or obovate.

12. **L. tomentosa** (Linn. suppl. p. 394) ; stems erect or ascending,
closely leafy at the base as well as the leaves, tomentose ; leaves linear
or linear-lanceolate, with revolute margins, and 2–4 obtuse teeth on

each side; peduncles elongated, glabrous, naked, 2–4-flowered; bracteæ linear, ciliated; pedicels bibracteolate; calyx-tube appressed-hairy, 3 times shorter than the lanceolate lobes; corolla subpilose, 6 times longer than the calycine lobes; lips very unequal, the lower deeply trifid; lobes oblong; ovary partly free. *Thunb.! Fl. Cap. p.* 182. *DC. Prod. p.* 363. *Rapunt. tomentosum, Presl! Prodr. Lob. p.* 19. *E. Mey. comm. pl. Dreg. p.* 288. *E. Z. n.* 2480. *L. Chamæpitys, Lam. Dict.* 3, *p.* 590. *A. DC.! Prodr. p.* 362. *Rap. Chamæpitys, Presl! non Spreng. E. & Z.* 2482.

VAR. β, **ceratophylla**; leaves pinnatifid-lobed. *L. ceratophylla, A. DC. l. c. p.* 362. *Rapunt. ceratophyllum, Presl! in E. Z.* 2484. *L. Simsii, Sweet. H. Brit. p.* 247. *L. pedunculata, Sims, Bot. Mag. t.* 2251. *Rapunt. Simsii, Presl, l. c. p.* 20, *ex icone.*

VAR. γ, **paucidentata**; stem, branches, calyx, and corolla pubescent; leaves with few teeth or lobes. *Lob. paucidentata, A. DC. l. c. Rapunt. paucidentatum, Presl! Prod. p.* 20.

VAR. δ, **multiflora** (A. DC. l. c.); peduncles often 3–4-flowered. *Rap. coeruleum, Presl! in E. Z.* 2481, *ex pte.*

HAB. Stony places, in mountains, in the districts of Uitenhage, George, Caledon, Albany, and in Caffraria. Var. β, Hottentottsholld., *Zey.* 3106, var. 6. Hemel en Aarde, *Zeyh.* 3107, var. 6. Var. γ, Uitenhage. Var. δ, Grahamstown and Port Elizabeth. Oct.–Feb. (Herb. Thunb., Hook., D., Sd.)

Stem shrubby, terete; branches fastigiate. Leaves 6 lines long, toothed near the apex, rarely towards the base, densely tomentose, greyish. Peduncles filiform, glabrous or with appressed hairs, 3 inches to 1 foot long. Corolla blue or violaceous, 6–8 lines long.

13. L. coronopifolia (Linn. spec. p. 1322); stem decumbent at the base, very leafy, hairy; leaves lanceolate, dentately-pinnatifid, hairy; teeth opposite, acute; peduncles elongated, naked, glabrous, 1–5-flow-ered; bractea linear-acuminate; pedicels bibracteolate; calyx tube ovoid-turbinate, hispid or appressed-hairy, nearly twice shorter than the lan-ceolate lobes; corolla 5–6 times longer than the calycine lobes; lips very unequal, lobes of the upper lanceolate, of the lower deeply trifid, segments obovate, mucronulate; ovary partly free. *Thunb.! Fl. Cap. p.* 134. *Sieb. Fl. Cap. exs. n,* 137. *Breyn. Cent. t.* 88. *Burm. Afr. p.* 98, *t.* 38, *f.* 1. *L. Thunbergii Sweet. Hort. Brit. p.* 247. *Bot. Mag. t.* 644. *Reich. Ic. exot.* 1, *p.* 61, *t.* 89. *Rap. coronopifol. Presl! Prod. Lob. p.* 20. *E. & Z.! n.* 2483. *Herb. un. itin. n.* 38, *et.* 485. *Rap. Simsii, Presl! E. & Z. n.* 2485, *non Lob. pedunculata Sims. Zey.! n.* 1057, 3107.

VAR. β. **coerulea**; peduncle 3–4-flowered, flowers large. *Lob. coerulea, Hook. Bot. Mag. t.* 2701. *Rap. coeruleum. Presl, Prod. p.* 20. *E. & Z. n.* 2481, *ex pte.*

VAR. γ. **glabrescens** (Presl.); leaves glabrous or nearly so.

VAR. δ. **macularis**; peduncles 2–5-flowered; leaves linear-lanceolate, with few very short teeth. *L. macularis, A. DC. l. c. p.* 364. *Rap. maculare, Presl! Prod. Lob. p.* 20. *E. Mey. Comm. pl. Afr. Austr. p.* 288.

HAB. In flats and mountains throughout the colony; var. δ, rear Port Natal, *Drege, R. W. Plant, n.* 34. *Gueinzius, n.* 466. *J. Sanderson, Dr. Sutherland, Gerr. & M'K. n.* 686. Sept.–Feb. (Herb. Thunb., Hook., D., Sd.)

Sub-shrub, very like the preceding, but not tomentose, and the leaves usually broader and serrato-pinnatifid, spreading, hairy or glabrous, flat or concave above. Peduncles 1–5-flowered. Flowers large, sometimes 1 inch long, blue. Valves of capsule as long or longer than the base.

14. L. patula (Linn. fil. Suppl. p. 395); stem and branches ascending or prostrate, elongate, glabrous; leaves petiolate, coarsely toothed or lobate-dentate, hairy, lobes obtuse, mucronulate; lower leaves on longish petioles, ovate, subcordate, acute; upper ones lanceolate; flowers terminal, loosely racemose; pedicels capillary, much longer than the linear bractea; calyx appressed-hairy, tube ovoid or subhemispherical, nearly as long as the subulate lobes; corolla 4–5 times longer than the calycine lobes. *Thunb.! Prod. p. 40. Fl. Cap. p.* 181.

VAR. β. **pteropoda**; petioles winged; stem subangular. *L. pteropoda, A. DC.! l. c. p.* 364. *Rap. pteropodum, Presl! Prod. p.* 21.

HAB. Mountains, Langekloof, *Thunb.* Draakensberg, Orange Free State, *T. Cooper,* 1067. Var. β. near Port Natal, *Drege, Gueinzius,* 420, *Dr. Sutherland.* November. (Herb. Thunb., D.)

Perennial 1 foot or more high. Branches filiform. Leaves ½–1 inch long and wide, in var β. usually larger and subglabrous. Petiole as long or longer than the leaf, ciliolate. Raceme 2-6-flowered. Calyx-tube 1 line long. Corolla blue or whitish. Capsule hemispherical, half superior.

15. L. chamædrifolia (A. DC.! l. c. p. 364); stem ascending, subangular, branched, hairy; leaves obovate, obtuse, tapering into a petiole, coarsely toothed; teeth obtuse, mucronulate, with revolute margins; upper leaves oblong-lanceolate; flowers terminal, loosely racemose; pedicels much longer than the linear-bracteæ, pubescent as the subhemispherical tube of calyx; corolla 3-times longer than the subulate calycine lobes; capsule half superior. *Rap. chamaedryfolium, Presl! Prod. p.* 21. *E. Mey. comm. pl. Dreg. p.* 290.

HAB. Rocky places near Omsamcaba, 1500-2000 ft. *Drege.* (Herb. D., Sd.)

Habit of *L. patula,* but different by the obovate, cuneate leaves, and smaller flowers. Larger leaves 1 inch long, 4-5 lines wide, with 2-4 equal teeth on each side, glabrous or hairy on the nerve beneath. Pedicels 3-4 lines long. Corolla blue, lobes of the lower lip oblong, glabrous. Capsule hemispherical.

§ 2. The 2 lower anthers only bearded at the apex or terminated by setæ. (Sp. 16–26.)

16. L. Erinus (Linn. spec. p. 1321); stem herbaceous, erect or ascending, as well as the filiform branches, angular or subterete, hairy at the base; leaves alternate, lower ones petiolate, obovate, obtuse, dentate or serrato-dentate; the middle lanceolate, serrate or denticulate, upper ones linear, often entire; flowers racemose; pedicels axillary and terminal, longer than the leaf or bractea; calyx-tube obconical, shorter than the linear, acuminate, toothed or entire lobes; corolla 3 times longer than the calyx; anthers hairy on the back; the two lower bearded.

VAR. *a,* **Erinus**; glabrous or hairy at the base; stem ascending; leaves glabrous or subpilose; lower ones petiolate, obovate, subdentate; upper ones lanceolate, sessile; corolla 2–3 times longer than the linear calyx-lobes. *L. Erinus, L. Thunb.! Prod. p.* 40. *herbar.* δ. *Sims' Bot. Mag. t.* 901. *Rap. Erinus, Moench. Presl! Prod. p.* 17. *Lob. bicolor, Sims' Bot. Mag. t.* 514. *Rap. bicolor, Presl! p.* 17. *Sieb. ext.* 138. *Rap. bellidifolium, var. a, Presl! R. flaccidum, Presl! Prod. p.* 13; *E. & Z.!* 2466, 2473. *Lob. flaccida, Erinus et bicolor, A. DC. Prod. L. Breynii, Spreng.! in Zeyh. Fl. Cap. II.* 21, *non Lam. L. linearis, var. β, Herb. Thunb. Zeyh.* 3110, 1052. *Herb. Un. Itin.* 30, 486, 487.

VAR. β, **bellidifolia**; stem and leaves subhirsute by spreading hairs; branches glabrous; leaves dentate, lower ones obovate or elliptical, middle and upper ones lanceolate, serrate or entire; corolla 3 times longer than the linear calyx lobes. *L. bellidifolia, Thunb.! Fl. Cap. p.* 180. *Rap. bellidifol. var.* β, *hirsutum, Presl, l. c. p.* 15. *R. scabripes, Presl! l. c. p.* 17. *Lob. algoensis, A. DC. l. c. p.* 369. *L. Krebsiana, A. DC. l. c. p.* 385. *E. & Z.* 2465, 2472. *Krebs.* 137.

VAR. γ, **microdon**; stems erect, pubescent at the base; radical leaves aggregated, obovate-oblong, shortly petiolate, denticulate or callously serrated. *L. microdon, A. DC. l. c. p.* 369. *Rap. microdon, Presl.! in E. Mey. Comm. pl. Dreg. p.* 287. *R. bellidifol. var.* β, *brevidens, Presl! in E. & Z.* 2465.

HAB. Stony places in flats and on mountains. Var. *a.* and β. in the distr. of Cape, Swellendam, George, Uitenhage, Albany, and Caffraria. Var. γ, Paarlberg, *Drege.* Puspasvalley, Swellendam, *E. & Z.* Zwarteberg, Caledon, *Zey.* 3113. Aapjes River, *Burke & Zeyher.* Port Natal, *Miss Owen.* July–Dec. (Herb. Thunb., Hook., Reg. Berol., D., Sd.)

Annual herb. Small specimens 3–4 inches high, simple; larger ones 1–2 feet, with filiform branches. Lower leaves ½–1½ inch long, 3–6 lines wide, often smaller and roundish, toothed or subserrated. Petioles short. Upper leaves sessile. Flowers racemose or subpanicled. Bracteæ linear. Pedicels 4–8 lines long, pubescent or glabrous, compressed. Calyx-lobes 1½–2¼ lines long, entire or with 1 or 2 acute teeth. Corolla blue or violet, glabrous. . Lobes of the upper lip linear; the 3 segments of the lower lip oval or obovate, with 2 white lines or spots near the base. Capsule 2–4 lines long. Var. γ. is about a span high, with radical leaves 6–8 lines long, 2–3 lines wide, minutely toothed; the stem-leaves are lanceolate, 1 l. wide, serrulate; the flowers are not different. Our var. *a.* is often cultivated in gardens as *L. heterophylla.*

17. L. Natalensis (A. DC.! Prod. p. 369); annual, quite glabrous; stem erect, branched, subtriquetrous, flexuous; leaves sessile, linear, obtuse, remotely and callously serrulate; lower ones oblong or oblong-lanceolate, narrowed at the base.; flowers loosely racemose; pedicels filiform, 2–3 times longer than the linear, denticulate or entire bractea; calyx-tube turbinate, shorter than the linear lobes; corolla 2–3 times longer than the calyx; anthers hairy on the back, the 2 lower bearded. *Rapuntium flexuosum, Presl! Prod. p.* 16, *non p.* 23. *E. Mey. comm. pl. Dreg. p.* 287.

VAR. β. **subulifolia**; leaves linear-subulate.

HAB. Port Natal, *Drege, J. Sanderson, Dr. Grant, Dr. Sutherland.* Delagoa-bay, *Capt. Speke.* Var. β. Port Natal, *T. Cooper,* 1126. Sept. (Herb. Hook, D., Sd.)

Possibly this is also a variety of *L. Erinus,* but I have not seen any intermediate form. It is 1 foot and more high. Leaves 1–2 inches long, 1–2 lines wide, the lower and upper ones smaller. Raceme 4–10-flowered. Lobes of calyx 1–2 lines long, narrow. Corolla blue, upper lip linear, narrow, often hairy, about 1 line long, lobes of lower lip oval or obovate. Capsule 3–4 lines long. Var. β. is distinct by very narrow leaves (1 inch long, ⅓ line wide), and a 3–4-flowered raceme.

18. L. Dregeana (A. DC. Prod. p. 371); quite glabrous; stems prostrate or ascending, weak, angular; leaves dentate, lower ones suborbicular or obovate, obtuse, petiolate; upper ones subsessile, narrowed at the base, oblong, acute or sublanceolate; flowers loosely racemose; pedicels filiform, 2–3 times longer than the lanceolate-cuneate 2–4-toothed bractea; calyx-tube shortly obconical, about as long as the ovate, acute lobes; corolla 2–3 times longer than the calycine lobes; anthers hairy on the back, the 2 lower bearded.

HAB. Little Namaqualand, Nieuweveld, between Rhinosterkop and Ganzefontein, Sternbergspruit, 3–5000 ft., *Drege.* Dec. (Herb. D., Sd.)

A small annual, 4 inches high. Radical leaves suborbicular, 2–3 lines long and wide, equal in length to the petiole; stem leaves 4–10 lines long, 2–3 lines wide, with 6–8 short teeth above the middle, obtuse, sessile or narrowed into a petiole; upper leaves and bractea 4 lines long, 1 line wide. Calyx at flowering time 1 line long; lobes acute. Corolla blue; lobes of the lower lip oblong. Capsule obconical.

19. L. triquetra (Linn. Mant. p. 120); glabrous or subpilose at the base; stem erect, as well as the branches smooth, obsoletely trigonous, very leafy, but naked above; leaves sessile, linear-lanceolate, coarsely and unequally dentate; raceme coarctate, terminal; pedicels twice longer than the lanceolate bractea; tube of calyx obconical, as long as the linear lobes; corolla 3 times longer than the calycine lobes, glabrous; capsule ovoid, elongate. *Thunb.! Fl. Cap. p.* 184. *L. capitata, Burm. Prod.* 25. *L. heteromalla, Schrad.! Hort. Goett. Herb. Un. itin.* 482, 483. *Rapunt. heteromallum et triquetrum, Presl! l. c. p.* 15, 16. *E. Z.* 2467, 2468.

VAR. β. **comosa** (A. DC.! Prod. p. 370); leaves linear; spike oblong, leafy; bracteæ about as long as the flowers. *L. comosa, L. spec. p.* 1323.

VAR. γ. **secundata**; leaves remotely dentate, glabrous or ciliolate; raceme elongate, subunilateral. *L. acutangulum, Presl, prod. p.* 16. *Lob. lavandulacea, Klotzsch! in Peters Mozamb.*

HAB. Cape flats and mountains near Capetown, Muysenberg, Piquetberg, Hexriver, Klynriviersberg and Babylonschetoornberg, *Zeyh.* 3100. Var. γ. in the Cape flats, *Zeyh.* 1053. Nov.–Dec. (Herb. Thunb., Hook., D., Sd.)

Annual, 1–2 feet high, erect or decumbent at the base, simple or few-branched above. Leaves alternate, sessile, toothed, sometimes subpinnatif.-dentate; teeth acute, 1 inch or more long, 1–2 lines wide. Flowers forming a coarctate, fastigiate raceme. Var. γ, elongate, 4–9 inches long. Corolla small, blue. Anthers hirsute on the back, the 2 lower bearded. Capsule 4 lines long, 10–ribbed.

20. L. Boivini (Sond.); stems prostrate, elongate, sometimes rooting, hairy; branches filiform; leaves shortly petiolate, oval, elliptical or suborbicular, bluntish, denticulate, hairy and ciliate or subglabrous; peduncles axillary, 1-flowered, 3–4 times longer than the leaf; calyxtube obconical; lobes lanceolate or ovate, acute, longer than the tube, one-half shorter or nearly as long as the corolla.

HAB. Cape of Good Hope, *Boivin*, 624. (Herb. D., Sd.)

A very distinct species, by the creeping stems, the habit of *Veronica agrestis*, and the variable size of calyx-lobes. Root annual. Stems 1 ft. and more long; branches shorter. Leaves 3–5 lines long, 3–4 lines wide, obtuse, crenate; upper sometimes oblong. Peduncles 1 inch and more long, hairy or glabrous. Calyx-lobes lanceolate-subulate, 1½ line long, and ovate, 2–3 lines long, 1 line wide in the same branch. Corolla blue; tube pale or whitish. Anthers hairy, the 2 lower bearded.

21. L. pubescens (Ait. H. Kew, ed. 1, v. 3, p. 498); stem and branches angular, pubescent or subhirsute, or the branches glabrous above; leaves petiolate, ovate, acute or subcordate, dentato-serrate; upper ones sessile, ovato-lanceolate and lanceolate, serrate or entire; flowers racemose; pedicels 2–3 times longer than leaf; tube of calyx ovoid, obconical, as long or shorter than the linear-subulate lobes; corolla 3 times longer than the calycine lobes.

VAR. a, **Thunbergiana**; hirsute or villous; lower leaves larger, subcordate, roundish, on longish petioles; flowers pale blue. *L. pubescens, Thunb.! Fl. Cap. p.* 181. *L. cuneifolia, Link. & Otto! Abbild. h. Berol. t.* 39. *DC. Prod. p.* 370. *Rapuntium ovatum, Presl! Prod. p.* 16. *E. & Z.* 2469.

VAR. β, **Jacquiniana**; minutely pubescent; lower leaves ovate or obovate, crenate or dentate, upper ones serrate; flowers white, with reddish or blueish tube. *Jacq. Hort. Schoenb.* 2, *t.* 178. *DC. l. c. Rap. pubescens, Presl! l. c. p.* 17. *E. & Z.* 2471.

VAR. γ, **incisa**; pubescent; leaves petiolate, ovate, duplicate-dentate; teeth acute or acuminate. *Rap. incisum, Presl! in E. & Z.* 2470.

HAB. Stony places and among shrubs. Var. *a*, near Capetown, *Thunb.* Lange-kloof, George, *E. & Z.* Simonsbay, *J. McGillivray & Milne.* Buffeljagtriver, *Bowie.* Var. β, near Genadenthal and Winterhoek, *Drege!* Rivier Zondereinde, *E. & Z.* Zwartland, *Brehm.* Zwarteberg, Caledon, *Dr. Pappe.* Breederivier and near Riet-kuil, *Zeyh.* 3111. Var. γ, Puspasvalley, Swellendam, *E. & Z.* Oct.–Feb. (Herb. Thunb., Hook., D., Sd.)

Annual (not perennial, as indicated by authors), very similar to the largest forms of *L. bellidifolia,* and only distinguished by the pubescence, commonly larger, roundish leaves, and white flowers. Stem 1 foot or more high, flexuous; branches ascending, filiform. Leaves ½–1 inch long, the lower 6–10 lines wide, on broad petioles 4–8 lines long, petioled; middle leaves usually somewhat narrower and shorter, petiolate; upper ones lanceolate or linear, toothed or entire. Pedicels drooping or spreading, 1–1½ inch long, minutely pubescent, or villous as the calyx. Corolla with a narrow tube; lobes of the lower lip oblong. Capsule 4 lines long.

22. L. linarioides (A. DC. Prodr. p. 371); annual, glabrous; stem erect, much-branched, angular as the branches; leaves remotely serru-lated, the lowest obovate-oblong, petiolate, middle ones lanceolate, subsessile; bracteæ linear, subentire; raceme loose; pedicels filiform, 2–3 times longer than the bracteæ; calyx-tube ovoid-obconical, twice shorter than the linear-acute, denticulate lobes; corolla 2–3 times longer than the calyx; capsule obovoid. *Rap. linarioides, Presl! Prod. p.* 22. *Comm. pl. Drege, p.* 290.

HAB. Grassy hills between Key and Gekau, and near Omtata, *Drege.* George, *Dr. Pappe.* Jan.–Oct. (Herb. D., Sd.)

4–5 inches high, lateral branches often ascending. Leaves 4–8 lines long, the lower 2 lines, the middle 1 line wide, acute. Pedicels ½ inch long, erect. Flowers small, blue. Capsule 1–2 lines long.

23. L. anceps (Thunb. Prod. p. 40); glabrous; stem erect, or as-cending, branched, winged-trigonal; radical and lower leaves obovate, crenated, petiolate; cauline leaves linear-oblong or lanceolate, denticu-late, decurrent; pedicels axillary, shorter than the leaf; calyx-tube cylindrical, much longer than the triangular lobes; corolla 2 or 3 times longer than the calycine lobes; capsule cylindraceous. *Thunb. Fl. Cap. p.* 180. *DC. l. c. p.* 375. *Rap. anceps, Presl! Prod. p.* 14. *E. Z.* 2463. *Dobrowskya anceps, Presl! l. c. p.* 10. *Lob. decumbens, Sims, Bot. Mag. t.* 2277. *L. rhizophyta, Spreng.! nov. Prov. Hort. Hal.* 1818. *Sims, Bot. Mag. t.* 2519. *L. alata, Lab.! nov. Holl.* 1, *p.* 51, *t.* 72. *Rap. alatum, Presl, l. c. p.* 14. *Sieb. pl. ext. nov. Holl. n.* 559. *Zeyh.* 1050, *a.*

VAR. β, **minor**; stems numerous, prostrate, often rooting; branches short; leaves subpetiolate, obovate, obtuse, pubescent or glabrous, upper ones cuneate or lanceo-late; raceme few-flowered. *L. repens, Thunb.! Prod. p.* 40. *Fl. Cap. p.* 181.

HAB. Wet places and in rivulets throughout the colony; near Port Natal, *Drege, J. Sanderson.* (New Holland, New Zealand, Chili.) Var. β. in the Cape downs, *E. & Z., Drege,* 957. *Zeyh.* 1050, *b.* Dec.–Jan. (Herb. Thunb., Hook., D., Sd.)

A very variable perennial, erect or prostrate, easily recognised by the winged stem and branches, and the cylindraceous capsule. Leaves subfleshy, 1–2 inches long, the upper much smaller. Flowers blue, small. Calyx glabrous or pubescent. Capsule 4–6 lines long.

24. L. fervens (Thunb.! Fl. Cap. p. 183); glabrous; stem erect or
ascending, angular, rather branched; branches compressed, narrowly
winged; leaves serrulated; lower ones obovate or ovate-oblong, shortly
petiolate; upper ones ovate-lanceolate or lanceolate, attenuate at the
base, sessile; pedicels axillary, racemose, shorter or about as long as the
linear bracteæ; calyx-tube ovoid-obconical, equalling the linear lobes;
corolla 3 times longer than the calycine lobes; capsule obovóid, acute
at the base. *Rap. fervens, Presl, Prod. p. 30. Lobelia Petersiana, humilis, pterocaulon, et subalata, Klotzsch! in Peters. Mozamb.*

VAR. β, **asperulata** ; branches and leaves with scattered hairs; pedicels and calyx
hispid. *Lobelia asperulata, Klotzch! l. c.*

HAB. Cape of Good Hope, *Thunb.* (Mozambique, and var. β. Zanzibar, *Dr. Peters.*)
(Herb. Thunb., reg. Berol.)
Root annual. Stem ½–1 foot and more high. Branches filiform. Leaves alternate, 1 inch long, 2–3 lines wide, the lower ones smaller, but evidently petiolate.
Bracteæ linear-lanceolate, the uppermost linear, minutely toothed or entire. Pedicels
3–6 lines long. Flowers blue. Calyx-lobes 1 line, corolla 4 lines long. Capsule
3–4 times shortѧr than the calyx-lobes. It has some resemblance to *L. anceps.*

25. L.? muscoides (Chamiss! in Linnæa, v. 8, p. 215); herb small,
slender, trailing, quite glabrous; stems filiform, rooting at the nodes;
leaves roundish, deeply 5–7-cleft, truncate or cuneated at the base;
lobes elliptic or obovate, acute, mucronulate; petioles as long or shorter
than the leaves; pedicels axillary, ebracteate, erect, much longer than
the leaves; calyx-tube shortly obconical; lobes ovate, acuminate. *DC.
Prod. p. 378. Rap. muscoides, Presl! Prod. p. 22, excl. syn. L. minuta.*

HAB. Highest mts. of Zwellendam, *Mundt, Zey.* April. (Herb. reg. Berol., Sd.)
Herb densely tufted, pale green; branches 2–4 inches long. Leaves 2–4 lines
long or smaller. Calyx scarcely 1 line long. Corolla purplish, 2 lines long. I
have only seen a few and not very perfect flowers, in one of which I observed a
3-parted corolla, with 2 distinct petals, representing the upper lip; the lower lip is
3-fid. It may be the type of a new genus or a *Dobrowskya.*

26. L. cyphioides (Harv.! in Thes. Cap. t. 162); stem scape-form,
striated, glabrous; leaves radical, obovate, narrowed at the base or on
very short petioles, obtuse, appressed-hairy, paler beneath; flowers
forming an elongate, secundate raceme; pedicels very short, equalling
the linear bracteæ; calyx-tube obconical, shorter than the linear lobes;
corolla one-half longer than the calycine lobes.

HAB. Cape of Good Hope, *Lahaie.* (Herb. D.)
Annual, exactly resembling *Cyphia Phyteuma,* with the inflorescence of *L. spicata.*
Leaves 1½–2 inches long, 1–1½ inch wide, entire or obscurely crenulated. Scape
with one small leaf near the base, 1 foot high, the spiked raceme 4 inches long.
Flower 3 lines long. Lobes of the upper lip erect, linear-acuminated; lobes of the
lower lip ovate, acute.

(Doubtful species.)

L. Breynii (Lam. Dict. 3. p. 588); glabrous; stems procumbent,
simple; leaves sessile, lanceolate, serrate, acute, thickish, usually unilateral; flowers axillary, racemose; pedicels short, topped by the
ovate-lanceolate bracteæ; tube of calyx hemispherical, shorter than
the linear lobes; corolla (blue) 4 times longer than the calycine lobes,

longitudinally split; segments linear-lanceolate; anthers . . . *A. DC. l. c. p. 361. Rapuntium œthiopicum cœruleo galeato flore foliolis dentatis, Breyn. Cent. p. 175, t. 89. Rapuntium Breynii, Presl, Prod. Lob. p. 19, but not comm. pl. Afr Austr. Dreg. p. 288.*

HAB. Cape of Good Hope.

VII. **DOBROWSKYA**, Presl.

Calyx-tube turbinate or hemispherical. *Corolla* split the whole length on back, 3-partite, the 2 upper petals distinct, unguiculate, erect; the 3 lower connate to a 3-fid lip. *Filaments* and *anthers* connate; all the anthers bearded at apex. *Stigmas* 2, linear, revolute. *Prod. Lob. p. 8, Comm. pl. Dreg. p. 283, Endl. gen. n. 3056. Lobeliæ spec. Thunb. Parastranthi spec. G. Don.*

Annual or perennial herbs. Lvs. alternate, opposite, or 3–5-verticillate. Flowers axillary on elongate pedicels, blue. Named in honour of James Dobrowsky, professor of philosophy at Prague.

Leaves all linear, with revolute, entire margins; perennial ... (1) **scabra.**
Leaves all lanceolate, remotely serrulated; annual ... ... (4) **stellarioides.**
Lower lvs. ovate or roundish, 1–2-dentate on each side; upper
  ones lanceolate or linear; calyx-lobes longer than the tube,
  2–3 times shorter than the corolla ... ... (2) **tenella.**
Lower lvs. obovate-oblong or ovate-lanceolate, 1–2-dentate, upper
  lanceolate; calyx-lobes equalling the tube, and nearly as long
  as the corolla ... ... (3) **aspera.**
Lower lvs. oblong-linear, obtuse, 1–2-dentate on each side, upper
  serrate; calyx-lobes shorter than the longish, obconical tube (5) **serratifolia.**

1. D. scabra (A. DC. Prod. 7, 2, p. 355); stem erect, scabrous, branched above, very leafy; leaves sessile, subopposite or ternate, linear, subpilose, scabrous with revolute margins, spreading; pedicels axillary, 3–4 times longer than the leaf, appressed, hispid as well as the calyx-tube; lobes of calyx linear, acute, longer than the tube, 2–3 times shorter than the corolla. *Deless. Icon. v. 5, t. 9. Lobelia scabra, Thunb.! Fl. Cap. p. 178. D. Thunbergiana, Presl, l. c. p. 9. Comm. pl. Dreg. p. 283. D.! polyphylla, Presl! l. c.*

VAR. β, **Dregeana**; lower leaves lanceolate or linear-lanceolate, sometimes with 1 tooth on each side. *D. Dregeana, Presl! l. c. E. & Z. 2446.*

VAR. γ, **glabrata**; stem and branches glabrous, or with a few appressed hairs; leaves ciliate-scabrous; the lower ones rarely 1–2-dentate. *D. Zeyheriana, Presl ! l. c. p. 10. Cyphia simplex, Spreng. in Zey. pl. Cap. ext. p. 18. E. & Z. 2447. D. lævicaulis, Presl! l. c. Comm. pl. Dreg. p. 284.*

HAB. Grassy places in mountains. Var. a. and β. near Grahamstown, *E. & Z., Genl. Bolton.* Howison's Poort, *H. Hutton.* Georgetown, *Pappe.* Zuurebergen and near Gekau, *Drege.* Var. γ, Zwartkopsriver, *E. & Z., Drege, Zey.* 3116. Port Natal, *Drege.* Oct.–Jan. (Herb. Thunb., Holm., Hook., D., Sd.)

Perennial, ½–1 ft., scabrous-hairy, simple or a few-branched. Leaves longer than the intervals, mucronulate or acute, 4–6 lines long, rarely 1 line wide, the lower smaller, often broader, the uppermost sometimes 1 in. long, not 1 l. wide. Peduncles 1–3 inches long. Calyx-tube obconical or obovoid; lobes reflexed, at flowering time 2 lines long. Segments of corolla ovate, acute. Capsule obovoid or hemispherical, rarely obconical, attenuate at the base, 10-nerved.

2. D. tenella (Sond.); stems diffuse, angular, scabrous, branched;

leaves subsessile, alternate, opposite or ternate, with cartilaginous margins, scabrous; lower ones ovate or roundish, 1–2-toothed on each side; the upper lanceolate or linear-lanceolate, bidentate or entire; pedicels subterminal, much longer than the leaf; calyx-tube scabrous or hairy; lobes linear, reflexed, longer than the tube, 2–3 times shorter than the corolla. *Lobelia tenella, Thunb.! Prod. p.* 40. *Linn. Mant. p.* 120. *L. unidentata, H. Kew. ed.* 2, *v.* 1, *p.* 356. *Bot. Mag. t.* 1484. *L. scapigera, Willd. Parastr. unidentata, G. Don. Dob. Massoniana, Presl! Prod. Lob. p.* 9. *Comm. pl. Dreg. p.* 283. *E. & Z.* 2445. *D. unidentata, A. DC. l. c. p.* 355. *D. stricta, Presl! E. & Z.* 2448.

HAB. Hills near Musselbay, *Th.* Grootvadersbosch, *E. & Z., Zey.,* 3109. Knysna, George, *Pappe;* and in Langekloof, *Ecklon.* Glenfilling, *Drege.* Oct.–Nov. (Herb. Thunb., Holm., Hook., D., Sd.)

Perennial, very variable in habit and size, 3–4 inches or 2 feet long. Lower leaves 2–4 lines long and 2–3 lines wide; the middle 4–6 lines long, 1½–2 l. wide; the uppermost sublinear, acute or mucronulate, the margins revolute. Peduncles at the base with 2 linear bracteæ, minutely downy as the calyx and corolla. Capsule as in *D. scabra,* from which it is distinguished by the more distant, broader, and toothed leaves, and prostrate stem with elongate branches. It varies also with smooth stems.

3. D. aspera (A. DC.! l. c. p. 356); annual, scabrous by retrorse, rigid hairs; stem erect, angular, branched; leaves alternate and opposite, scabrous on margins, and prominent middle nerve; the lower obovate-oblong or ovate-lanceolate, mucronulate or acute, very shortly petiolate; with 1 or 2 teeth on each side, the middle and upper lanceolate or linear, often entire; pedicels spreading, 3–4 times longer than the leaf, forming a loose raceme; flowers minutely pubescent; calyx-lobes linear, equalling the tube, and nearly as long as the corolla. *Lob. aspera, Spr. neue Entd.* 3, *p.* 222. *L. Chamæpitys, Spreng. in Zey. pl. Cap.* 280. *Lob. Herb. Un. Itin.* 16. *Wahlenbergia chondrophylla, Buek! E. & Z.* 2388. *Dob. Eckloniana, Presl! Prod. Lob. p.* 9. *Comm. pl. Dreg. p* 283. *E. Z.* 2449.

HAB. Wet sandy places and on rivulets, Cape flats and on Devilsmountain, *E. & Z.* *W. H. H.* Zuurebergen and Quaggasvlakte, Bontjesriver, *Drege.* Howison's Port, *H. Hutton.* Oct.–Dec. (Herb. D., Sd.)

Commonly much branched from the base; branches erect or ascending, the lower often opposite. Leaves with revolute margins, 4–6, rarely 8 lines long, 1–2 lines wide, with a few serratures. Primary branches terminated by a 6–8-flowered, very loose raceme; peduncles 1–2 inches long. Corolla blue, small. Capsule obovoid or obconical, 2–3 lines long, twice longer than the rigid, erect, calycine lobes. By the annual root, weak, more-branched stem, racemose, spreading peduncles, and smaller flowers, it differs sufficiently from *D. tenella.*

4. D. stellarioides (Presl.! Prod. Lob. p. 10); annual; stem erect, angular, as well as the margins and middle nerve of the leaves, scabrous by retrorse bristles, simple or few-branched; branches weak, opposite; leaves very shortly petiolate, opposite, lanceolate, remotely serrulated; pedicels axillary, as long or longer than the leaf, erect, scabrous; calyx-lobes linear, spreading or reflexed, longer than the tube, and one-half or nearly twice shorter than the minutely pubescent corolla. *E. & Z.* 2450, *DC.! l. c. p.* 356.

HAB Mountains in Caffraria, *I. Wahlberg* Winterberg, near Katriver, *E. & Z.*

Zululand, Port Natal, *T. Cooper*, 1137, *J. Sanderson, Gerr. & M'K.*, 1438. March. (Herb. Hook., D., Sd.)

2-3 feet long, habit of *Galium*. Leaves 1-1½ inch long, 2 lines wide, with 3-6 minute serratures on each side, mucronulate. Peduncles 1-2 inches long, with leaf-like, 6-8 lines long bracteæ. Flowers blue, nodding, nearly ½ inch long. Calyx scabrous-hairy. Capsule 2½ lines long, obovoid or obconical.

5. D. serratifolia (A. DC. l. c. p. *356*); annual; stem erect, as well as the branches angular, smooth, glabrous; leaves glabrous, the lower oblong-linear, obtuse, tapering in a short petiole, with 1-2 teeth on each side; the middle and upper linear, sessile, acute, serrate, serratures 3-5, subulate, on each side; pedicels axillary, glabrous, somewhat longer than the leaf; calyx-tube elongate-obconical, glabrous, longer than the erect, linear, scabrously-margined lobes. *Burch. Catal. Pl. Afr. Austr. extratrop.* 796.

HAB. Cape of Good Hope, *Burchell*.

4 inches high. Root fibrous. Leaves erect, the middle longer, 4-8 lines long, scarcely ½ line wide. Pedicels erect, naked, in fruit ½ longer than the leaf. Corolla unknown. Capsule erect, longish, obconical, glabrous, 2 lines long.

VIII. **ENCHYSIA**, Presl.

Calyx 5-fid, tube ovoid. *Corolla* infundibuliform, subregular, tube entire, lobes sub-equal, suberect. The 2 lower *anthers* aristate by 4 setæ. *Prod. Mon. Lob. p.* 40. *A. DC. l. c. p.* 408. *Lobelia spec. Linn. et Thunb. Monopsis sp. Salisb. Laurentia sect. Enchysia, Endl gen. n.* 3060, *b*.

Small annual herbs with racemose, white or pale purplish flowers. Name from ενχυσις, *funnel;* the corolla is funnel-shaped.

1. E. secunda (Sond.) stems decumbent or ascending, minutely pubescent; radical leaves petiolate, obovate or oblong-spathulate, obtuse, denticulate, lower cauline leaves oblong-lanceolate, upper linear, sessile, entire; pedicels axillary, 2-3 times shorter than the leaf, minutely pubescent as the calyx; lobes of calyx as long as the ovoid tube; corolla tubulose-infundibuliform, 5-fid, lobes acute, subequal. *Lobelia secunda Linn. fil./ Suppl. p.* 395. *Thunb./ prod. p.* 40. *fl. Cap. p.* 183. *L. depressa, var. γ. Herb. Thunb./ Lob. pubescens, Willd. enum. non Ait. E. repens, Presl, l. c. excl. syn. A. DC. l. c. E. erinoides, Presl, excl. syn. Hook. Bot. Mag. t.* 3609. *E.&Z.* 2487, 2488. *Herb. Un. Itin. p.* 77.

VAR. β. **glabrata**; glabrous or minutely pubescent above; stems erect or decumbent. *E. erecta, A. DC./ l. c. p.* 409. *Deless. Icon. v.* 5. *t.* 13.

VAR. γ. **dentata**; glabrous, pedicels and calyx pubescent; leaves dentate or serrate, the upper entire or denticulate. *E. dentata, A. DC. l. c.*

VAR. δ. **reflexa**; quite glabrous, or pedicels and calyx pubescent; pedicels reflexed. *E. repens var. reflexa, Kunze/ ind. sem. Hort. Lips.* 1846.

HAB. Sandy wet places and by rivulets near Capetown, in the flats, on Devil's and Tablemountains, Hottholld., Kuilsriver, Bergriver and Dutoitskloof. Jan.—April. (Herb. Thunb., Holm., Hook., D., Sd.)

Root fasciculate, fibrous. Stems numerous, 2 inches to a span long, filiform. Radical leaves aggregated, about 1 inch long, 2-3 lines wide, subentire or denticulate; petiole short or as long as the leaf, cauline-leaf 5-6 lines long, gradually smaller, entire or toothed, the upper 2-3 lines long, linear, rarely spathulate. Raceme long, secundate. Pedicels commonly 2-3 times, rarely 4-5 times shorter than the leaf; in large, loose-flowered specimens (*E. erinoides, var. β. laxa, Presl*) the pedicel equals the leaf. Calyx tube 1 line long. Corolla white, 2 lines long to ⅜, 5-fid, pubescent or glabrous. Capsule 1¼ line long.

IX. LAURENTIA. Mich.

Calyx 5-fid or 5-dentate. *Cor.-tube* entire, cylindraceous, straight; *limb* bilabiate, the 2 upper lobes smaller, usually erect, the lower lip larger, 3-lobed, reflexed, *Filaments* free at the base, the middle connate, anthers included, glabrous, the 2 lower somewhat hairy, or terminated by setae. *Capsule* obovoid, inferior. Seeds minute. *Nov. Gen. p.* 18. *t,* 14. *A. DC. Prod. p.* 410. *Lobeliæ spec. Linn. Solenopsis, Presl, prod. mon. Lob. p.* 32. *Enchysiæ spec. Presl, Isotomæ, sp. G. Don. and Benth. Laurentiæ, sect. Solenopsis Endl. gen.* 1, *p.* 512.

Small glabrous herbs, with small blue or whitish flowers on terminal and axillary pedicels.—Named after M. A. Laurenti, Botanist at Bologna.

Glabrous. Stem 1 foot, erect. Radical leaves aggregated, oblong, entire or sinuate-denticulate; raceme elongate, many-flowered   (1) **arabidea.**

Minutely pubescent at base. Stem 1-3 inches, erect. Radical leaves subaggregated, obovate, unequally and coarsely toothed; pedicels 2-4, subracemose ... ...   (2) **bifida.**

Pubescent or glabrous, branches decumbent. Leaves roundish, acutely 5-9-toothed ... ...   (3) **pygmæa.**

Densely pubescent. Branches ascending. Leaves reniform, 3-5-lobed, lobes toothed ... ...   (5) **hederacea.**

1. L. arabidea (A. DC.! l. c.); annual, glabrous; stem mostly solitary, erect, terete, slender; radical leaves aggregated, oblong, acute or obtuse, sinuato-denticulate or entire, shortly petiolate or sessile, tapering at the base; cauline and floral leaves rare, linear, denticulate; raceme elongate; pedicels one-half or twice longer than the bractea; calyx-tube ovoid, narrowed at the base, terete, equalling the linear-acuminate lobes; corolla narrow, cylindraceous, 3 times longer than the calycine lobes; segments linear, the 2 upper narrower, the 2 lower of the subglabrous anthers terminated by a seta. *Deless. Icon. v.* 5, *t.* 14 *Rap. arabideum, Presl! Prod. p.* 18. *Comm. pl. Dreg. p.* 288. *E. & Z.* 2474. *Herb. Un. Itin.* 488, *ex pte.*

Hab. Wet places near Waterfall, Tulbagh, and near Capetown, *E. & Z.* Wynberg, *W. H. H.* Bergriver, Paarlberg, and Dutoitskloof, *Drege.* Nov.-Dec. (Herb. Hook., D., Sd.)

Radical leaves about 1 inch long, 2 lines wide. Stem 1 foot, simple, straight, or with a few erect branches. Cauline leaves 12-6 lines long, 1 line wide. Raceme 3-6 inches; pedicels often secundate, 4-6 lines long, erect, spreading. Corolla light blue, 3 lines long. Capsule ovoid, when ripe cylindrical, acute at the base.

2. L. bifida (Sond.); annual, minutely pubescent at the base; stem solitary, erect, simple, terete, very slender; radical leaves 2-4, obovate, shortly petiolate, unequally and coarsely toothed, rarely subentire; cauline leaves few, obverse-lanceolate or oblong-lanceolate, sessile; pedicels 2-4, racemose, 2-3 times longer than the linear bractea; calyx-tube terete, narrow, a little shorter than the linear-acuminate lobes; corolla narrow, cylindraceous, 4 times longer than the calycine lobes; segments oval, subequal, the 2 upper narrower; the 2 lower of the dor-

sally glabrous anthers bearded. *Lobelia bifida, Thunb.! Prod. p.* 40.
Fl. Cap. p. 183. *Solenopsis Dregeana, Presl! Comm. pl. Dreg. p.* 290.
Laurentia Dregeana, A. DC.! l. c. p. 411.

HAB. Mount Bockland, *Thunb.* Giftberg, *Drege.* Nov. (Herb. Thunb., D., Sd.)
Stem filiform, 1–3 in. high. Radical leaves 4–6 lines long, 2 lines wide. Cauline
leaves 2–3, distant, denticulate. Pedicels 4–6 lines long. Corolla blue, 3 lines long.
Capsule linear-turbinate.

3. L. pygmæa (Sond.) ; annual; branches decumbent, filiform, pubescent; leaves petiolate, rotundate, roundish or obtuse at the base,
acutely 5–9-dentate, subpilose; petiole hairy, as long or longer than
the leaf; pedicels erect, capillary, glabrous, naked, 4–5 times longer
than the leaf; calyx glabrous; tube obovoid, acute at the base, terete,
equalling the lanceolate, acuminate lobes; corolla 5–6 times longer
than the calycine lobes; tube narrow; the 2 upper lobes a little smaller;
anthers glabrous, the 2 inferior terminated by a seta. *Lobelia pygmæa,*
Thunb.! Prod. p. 40. *Fl. Cap. p.* 181. *Rap. pygmæum, Presl Prod. p.*
22. *Wahlenbergia micrantha, E. Mey. in Herb. Drege. Laurentia mi-*
crantha, A. DC. l. p. 411.

VAR. β, **glabra**; quite glabrous, or the pedicels with a few hairs ; corolla 3–4
times longer than the calycine lobes. *Wahl. micrantha, c, Herb. Drege.*

VAR. γ, **obtusiloba**; glabrous or the petioles with a few hairs ; leaves on longish
petioles, rotundate or elliptical, sinuato-3-7-dentate, teeth obtuse ; corolla 2–3 times
longer than the calycine lobes. *Lobelia minuta, Thunb.! Prod. p.* 40. *Fl. Cap. p.*
182, *excl. syn. L. minuta, Linn. L. pusilla, Don. Gen. Syst.* 3, *p.* 710.

HAB. Mount Ribeck Castel, *Thunb.* Paarlberg and Dutoitskloof, *Drege.* Lam-
bertskloofberg, *Wallich.* Var. β, Blauwberg, *Drege.* Var. γ, summit of Tablemountain, *Thunb., Ecklon, Milne.* Dec.–April. (Herb. Thunb., Hook., D., Sd.)
A small herb, with the habit of *Wahlenbergia hederacea.* Branches 2–4 inches
long. Leaves 2–4 lines wide, membranaceous, angulate-dentate; teeth acute, unequal. Peduncles 1–1½ inch long. Calyx minute. Corolla white, 1½–2 lines long.
Capsule ovoid. Var. γ is at first sight distinct by obtusely-lobed leaves, but the
same leaves are also found in young specimens of var. β. The tube of corolla is
very long in var. α.; in var. γ about half the length ; var. β. is an intermediate

4. L. hederacea (Sond.) ; annual, minutely pubescent ; branches
ascending, filiform ; leaves petiolate, reniform-cordate, 5–9-dentate or
lobed ; petiole as long as the leaf; pedicels axillary, solitary, longer
than the leaf, naked ; calyx ovoid, acute at the base, pubescent; lobes
lanceolate, twice shorter than the glabrous corolla; anthers glabrous,
the 2 inferior terminated by 1 or 2 setæ.

HAB. Eastern frontier, *Henry Hutton.* (Herb. D., Sd.)
Near *L. pygmæa* in habit and size, differing by a dense but minute pubescence in
all parts, and cordate leaves, broader than long. Branches densely leafy. Leaves
4–5 lines long, 3–4 lines wide, obscurely 3–5-lobed; lobes unequally toothed; teeth
subacute. Pedicels 1 inch long. Calyx-lobes about as long as the tube. Corolla
white, 2 lines long. Capsule oval.

———

Lobelia minuta (Linn. Mant. p. 292). "Plant one inch high. Leaves radical,
obovate, obtuse, scarcely crenate, longish-petiolate. Scape setaceous, 2 inches long,
without bracteæ, 1-flowered. Calyx oblong, lanceolate. Corolla infundibuliform.
Capsule oblong." Is possibly not different from *L. minuta,* Thunb.

Laurentia minuta (A. DC. l. c. p. 410); with blue flowers and hemispherical
calyx-tube. Founded on : *Sims, Bot. Mag. t.* 2590, and *Lob. minima, Sims' Bot.*
Mag. t. 2077 ; is not a Cape species, but *L. tenella,* Biv., perfectly agreeing with
Sicilian specimens collected by Prof. *Parlatore.*

Tribe 2. CAMPANULEÆ. (Gen. X.–XVII.)

X. LIGHTFOOTIA, L'Herit.

Calyx 5-cleft. *Corolla* usually 5-parted, rarely deeply 5-fid. *Stamens* 5; filaments broad, ciliated; anthers free, caducous. *Ovary* 3-, rarely 5- or 2-celled. *Style* thicker at the apex. *Stigmas* 2, 3, or 5, short, fili- form. *Capsule* usually half-superior, 2–5-celled, dehiscing by valves at the apex; cells, when 5, opposite the calycine segments. *Seeds* ovoid-trigonal, minute. *Sert. Angl. p. 3. Endl. Gen.* 3072.

Small shrubs, rarely perennial or annual herbs. Leaves alternate, sometimes opposite, usually with fascicles of smaller ones in the axils, sessile, small, scattered equally over the whole plant. Flowers usually racemose; pedicels erect. Named after the *Rev. J. Lightfoot,* author of the first Flora Scotica.

A. Capsule 5-celled. Lvs. subulate; fl. subsessile ... ...   (1) **sessiliflora.**
B. Capsule 3–celled :
  Annuals :
    Lvs. spiny-toothed, linear; fl. sessile ... ...   (2) **albicaulis.**
    Lvs. entire, linear ; fl. peduncled ... ...   (3) **Namaquana.**
  Small shrubs or perennial herbs :
    1. Fl. terminal, subsessile, densely glomerate :
      Leaves, *reflexed,* ovate, thick :
        Fl. in a globose tuft; corolla hardly longer
          than calyx ... ...   (5) **fasciculata.**
        Fl. in an oblong tuft; cor. 2–3-ce as long
          as calyx... ... ...   (21) **tenella,** var. δ.
      Leaves erect :
        Lvs. lanceolate, elongated, flat, uncial; fl.
          in a dense terminal spike ... ...   (4) **Huttoni.**
        Lvs. lin.-lanceol. (2–4 l. long), setaceously-
          toothed ; tufts of fl. small, racemed or
          panicled ; cal. glabr.; cor. hairy ...   (6) **rubens.**
        Lvs. appressed, lin.-subulate (3–4 l. long),
          greyish-tomentose; fl. sessile, tufted or
          solitary; cal. tomentose; cor. hairy ...   (8) **cinerea.**
      Leaves spreading, subulate (4–6 l. long); spike
        ovate, 6–8-fl.; cal. tomentose; cor. glabrous   (7) **Albanensis.**
    2. Fl. axillary, sessile or subsessile; spike elongated :
      Lvs. reflexed, ovato-lanceolate, denticulate ...   (9) **spicata.**
      Lvs. erect-appressed, lin.-lanceolate, entire ...   (10) **axillaris.**
    3. Fl. racemose or panicled; or subsolitary, *pedicelled,* subterminal :
      *Leaves alternate (some opp. in No. 11) :
        Cal.-tube turbinate or obconical, acute at base :
          Racemes elongate, many-fl., unilateral   (13) **Caledonica.**
          Racemes few-fl., or fl. axillary ; pedi-
            cels spreading ... ...   (17) **Buekii.**
        Cal.-tube hemispherical :
          Leaves erect or spreading *(not reflexed) :*
            Raceme long ; lvs. erect, subulate,
              subdenticled ... ...   (11) **longifolia.**
            Rac. long or short; lvs. lanceol.
              or lin.-lanc., flat, setaceously 1-
              2-toothed at base, erect; pedi-
              cels longer than cal. ; cor. 5-
              parted ... ...   (12) **unidentata.**
            Rac. corymbose, or fl. subsolitary;
              lvs. lin.-lanceol., mucronate,
              convex, ciliate-toothed at base,
              spreading; ped. as long as cal.;
              cor. dceply 5-fid ... ...   (18) **ciliata.**

Rac. panicled or with spreading
pedicels ; lvs. suberect, linear-
acuminate, denticled; cal.-lobes
denticled ; stem herbaceous.   (14) **denticulata**.

Rac. lax ; lvs. spreading, linear-
subulate, entire, convex; cal.-
lobes entire ; stem woody ;
branches downy ... ...   (16) **albens**.

Panicle large ; branches divaricate; lvs. erect, subulate, channelled, entire; cal.-lobes entire;
stem woody. Quite glabrous   (15) **paniculata**.

Leaves reflexed :
  Fl. solitary, or a few fl. raceme :
    Lvs. lin.-lanceolate, flat; ped.
    rigid; cor. 2-3 times longer
    than the calyx   ... ...   (19) **Thunbergiana**.
    Lvs. ovato-lanceolate, flat ;
    ped. capillary ; cor. but
    little longer than the calyx   (20) **oxycoccoides**.
  Fl. racemose or panicled ; leaves
    ovate, concave   ... ...   (21) **tenella**.
**Leaves opposite : (linear-lanceolate)
  Lvs. acuminate, entire; branches erect   (22) **oppositifolia**.
  Leaves acute, remotely denticled ;
    branches diffuse ... ...   (23) **rubioides**.
C. Capsule 2-celled, elongate, conical:
  Leaves ovate-lanceolate or lanceolate; fl. spiked ...   (24) **juncea**.
  Leaves linear-elongate; fl. racemose   ... ...   (25) **laxiflora**.

§ 1. Capsule 5-celled.

1. L. sessiliflora (Sond.) ; stem ascending or erect, rather woody,
simple or branched ; leaves fascicled, alternate, erectish, subulate, narrow, subdenticulated ; flowers nearly sessile, racemosely-spiked ; corolla
5-parted ; segments longer than the calyx ; valves of the capsule nearly
as long as the calyx lobes. *Campanula sessiliflora Linn.! Suppl. p.* 139.
Thunb.! Prodr. p. 38. *Fl. cap. p.* 171. *Mem. acad. Petersb. T.* iv. *p.*
367. *t.* 5. *f.* 1. *C. subulata, Spreng. Syst.* 1. *p.* 728, *non. Thunb. Herb.*
Un. itin. 1 *et* 158. *Lightf. subulata, l'Herit. Sert. angl.* 1, *t.* 5, *A. DC.*
Monog. p. 107. *E. Z.* 2340. *Dreg. pl. exs.* 6312, *etc. Zey.* 3124. *Burch.*
cat. 284.

VAR. β. **foliosissima** ; stem shorter, much branched, branches and ramuli densely
beset with appressed leaves.

HAB. Sandy and stony places on the mountains about Capetown and in Hottentottsholland. Var. β. in Zwartland, *Zey.* 1083. Nov. to Jan. (Herb. Th., Holm.,
Sd., Hk., D.)

Root perpendicular, stem ½-1½ foot, usually much branched, branches reddish,
mostly simple, filiform, in var. β. with numerous short pubescent ramuli ; leaves
alternate, rarely opposite, numerous, very narrow, 3-6 lines long, erectish or spreading. Flowers forming an elongated spike or raceme, blue, peduncles shorter than
the calyx. In var. β. the flowers are sessile and solitary on top of the short ramuli ;
the calyx quite covered by leaves. Tube of calyx hemispherical, 1 line long, shorter
than the subulate lobes, segments of corolla very narrow, scarcely one half longer
than the calyx, usually hairy outwardly. Style as long as the corolla, thickened
upwards, or densely pubescent. It varies with glabrous or somewhat hairy or densely
pubescent branches or calyx.

§ 2 Capsule 3-celled. (Sp. 2-23).

2. L. albicaulis (Sond.) annual ; stems erect-spreading, white, glabrous ; leaves alternate, sessile, linear, narrow, spiny-toothed, recurved
at the apex, greyish, hairy or nearly glabrous ; flowers sessile, solitary,
or aggregated ; calyx tube hairy, shorter than the denticulated subulate
lobes ; corolla 5-parted, longer than the calyx. *Lobelia cinerea, Thunb.!*
Prodr. p. 40. *Fl. Cap., p.* 182. *DC. Prodr.* 7, 2, *p.* 385. *Rapuntium*
cinereum, Presl, Prodr. Lob. p. 30.

HAB. Roggeveld, *Thunb.;* Namaqualand, *A. Wyley,* Nov.-Dec. (Herb. Th., D.)
Root fusiform, simple, white, 2-4 inches long. Stems many from the same root,
erect or ascending, 1-3 inches long, terete, smooth, simple or a few branched.
Leaves 4-6 lines long, acute, with thickened margins, carinate above ; the teeth
short but rigid. Calyx about 3 lines long, tube one-half shorter than the lobes.
Corolla white, a little longer than the calyx lobes, 5-fid, at length 5-parted, segments
lanceolate. Ovary hairy, elliptical, stigma trifid.

3. L. Namaquana (Sond); annual; stems erect, branched, white,
terete, glabrous or shortly hairy ; leaves distant, alternate, spreading,
linear, narrow, entire or obsoletely denticulate, canaliculate above ; flowers
solitary, pedicellate, disposed in a terminal, racemose panicle ; pedicels
and calyx minutely downy ; tube of calyx hemispherical, equalling the
linear, carinate, ciliolate lobes ; corolla sub-5-parted, 3-4 times longer
than the calyx ; capsule 3-valved, valves as long as the reflexed calyx-
lobes.

HAB. Modderfontein, Rev. H. Whitehead (Herb. D.)
Habit of *L. denticulata.* Root long, subsimple, white as the whole plant. Stems
a span high. Branches few, or very numerous ones in the axil of the leaves, bearing
1 or 2 pedicellate flowers. Leaves 4-6 lines long, ⅓ line broad. Pedicels of the
panicle 2-3 lines long. Calyx when in flower very small. Segments of corolla
narrow, 3 lines long. Capsule hemispherical, with 5 prominent and 5 nearly obsolete ribs ; the tube one half longer than the valves.

4. L. Huttoni (Sond.); stem erect, glabrous or a little downy ;
leaves numerous, erect, alternate, lanceolate, elongate, calloso-mucronate, flat, with thickened, entire or subdenticled margins, glabrous ;
flowers in dense terminal spikes ; tube of calyx glabrous, a little shorter
than the subulate lobes ; corolla 5-parted, much longer than the calyx.

HAB. Kreili's Country, Caffraria, *H. Bowker* ; Katberg, Eastern Frontier, *H.*
Hutton. (Herb. D., Sd.)
Stem 1 foot, stiff, simple, with some short branches in the upper part, closely leafy
from the base. Leaves about one inch long, at the base 1½-2 lines broad, the upper
ones gradually smaller, glabrous or ciliated near the base, 1-nerved. Flowers 4-12,
sessile at top, in the axils of the upper leaves. Calyx 3-4 lines long with 2 bracts
at the base. Corolla with linear segments, longer than the trifid stigma.

5. L. fasciculata (A. DC. Monog. p. 112) stem erect, woody, branched,
hairy ; leaves numerous, alternate, reflexed, ovate, acute, thickish, remotely denticulated, glabrous or hairy beneath ; flowers in round fascicles at the top of the branches ; tube of calyx nearly equalling the subulate lobes ; corolla 5-parted, hardly longer than the calyx, hairy. *Cam-*
panula fasciculata, Linn. Suppl. p. 139. *Thunb.! Prodr. p.* 39. *Mem.*
Acad. Petersb. 4, 372, *t.* 6, *f.* 1. *Roella reflexa, Banks herb. L. oxycoccoides,*
Litt. e. Dreg. ext. (fide DC. non in Herb. Sd.)

HAB. Between Knysna and Plettenbergsbay, Jan. *Dr. Pappe!* (Herb. Th., D., Sd.) Plant 1-2 feet, simple or branched at the top. Leaves sessile, coriaceous, fascicled, about 2 lines long, 1 line broad, the lower ones often acuminate and 3 lines long. Inflorescence composed of a glomerate panicle, 6-8 lines broad, the flowers sessile or on very short hairy peduncle, bracteated by acuminate leaves, nearly as long as the calyx. Calyx 3-3½ lines long, tube ovate, 10-nerved, with short hairs. Corolla with linear segments. Valves of the capsule 3-4 times shorter than the calyx.

6. L. rubens (Buek! in E. Z. Enum. p. 373); stem erect, woody; branches spreading, hairy; leaves alternate, glabrous, erect, linear-lanceolate, acute; margins revolute, near the base with 4-6 setaceous teeth; flowers fascicled at top of the branches; fascicles 3-12-flowered, densely racemose or laxly panicled; pedicels shorter than the hemispherical calyx tube; calyx lobes subulate, twice shorter than the 5-parted, hairy corolla. *P. angustifolia. A. DC.! Prodr. p. 418.*

HAB. Near Kromrivier, *Drege,* 6320; near Capetown, *W. H. H.*; between Krakakamma and Vanstaadensberg, *E. & Z.; Zey.!* 3133; *T. Cooper,* 1465. Winterhoek-*Zey.* 3121; Port Elizabeth, *Zeyh.* 3132. Bethelsdorp, *Zey.* 312 f. Dec.–May. (Herb. Sd., D.)

An erect, very leafy shrub, 1-2 feet high, with the habit of *L. fasciculata,* but differing by much narrower, erect leaves, with setaceous teeth and revolute margins and branches, reddish, with greyish pubescence. Leaves often reddish when old, mostly in 4 rows, erect or erect-spreading, 2-4 lines long, ½-1 line broad, rigid. Fascicles of flowers in some specimens sessile, aggregated in the upper part of the branches, in others divided by spreading peduncles and then forming a panicle of few-flowered fascicles. Calyx 2 lines long, glabrous; the lobes equalling the tube, but longer than the valves of capsule. Style as long as the corolla.

7. L. Albanensis (Sond.); perennial, stem and branches erect, terete, rather tomentose; leaves erect-spreading, subulate, nearly terete by the revolute margins, quite entire or denticulate near the base, glabrous; flowers sessile, forming an ovate spike at top of the branches; tube of calyx tomentose, lobes subulate, 3 times shorter than the 5-parted glabrous corolla.

HAB. Grassy places in Howisonspoort, Albany. *Zey.* May. (Herb Sd.) Stem nearly 1 foot high, with few erectish branches. Leaves with fascicles of smaller ones in the axils, 4-6 lines long, sessile, with broader woolly base. Spike composed of 6-8 flowers, leafy-bracteated. Calyx 3 lines long, tube shorter than the lobes. Segments of the 5-parted corolla linear, a little longer than the minutely pubescent style. Stigma trifid. Perhaps a variety of the following.

8. L. cinerea (Sond.); stem and branches erect, woody, as well as the base of the leaves, greyish, tomentose; leaves numerous, sessile, erect, appressed, linear-subulate, nearly terete by the revolute margins, quite entire or denticulate near the base, greyish; flowers sessile at top of the branches, solitary or 4-6 aggregated; calyx woolly, tube obconical, nearly equalling the subulate lobes; corolla 5-parted, twice longer than the calyx, hairy. *Campanula cinerea, Linn.! Suppl. p. 139. Thunb.! Prodr. p. 38. Mem. Acad. Petrop. 4, p. 368, t. 6, f. 4. Fl. Cap. p. 172. Roella cinerea, A. DC. Monog. p. 175. Lightf. grisea, Buek! in E. & Z. Enum. p. 374.*

HAB. Grassy fields near Swellendam, *Thunb.!* on the Vanstadensriviersberg and near Eladsrivier, *E. & Z.* Nov.–May. (Herb. Th., Holm., D., Sd., Hk.)

Many stems from the woody rootstock, 1-2 feet high, more or less branched ; branches virgate or short. The whole plant of a greyish colour. Leaves with fascicles in the axils, 3-4 lines long, often imbricated, glabrous, but tomentose at the base. Calyx 3 lines long. Corolla white, segments linear, longer than the 3-fid style. Capsule 3-valved, valves shorter than the calyx lobes.

9. L. spicata (Buek.! l. c. p. 374); stem erect, woody ; branches virgate, angular, shortly pubescent ; leaves alternate, reflexed, ovato-lanceolate, coriaceous, glabrous, concave, with thickened, denticulate margins ; spike elongate; flowers sessile, the lower ones shortly pedicellate ; calyx glabrous or subpilose ; tube hemispherical; lobes lanceolate, 2-3 times shorter than the hairy corolla. *L. spicata, DC.! Prod.* p. 418. *L. macrostachys. DC. l. c. p.* 787. *L. subulata c. et L. oxycoccoides, e. Herb. Drege.*

HAB. Stony sandy places near the cataract of Tulbagh, *E. & Z.* In mountains, Giftsberg and Cederbergen, *Drege.* Nov. (Herb. D., Sd.)
Stem 1-2 feet high, reddish as the branches. Leaves numerous, with small ones in the axils, 3-4 lines long, 1 line broad. Calyx lobes longer than the tube, entire, with thickened margins. Corolla 5-parted, segments linear, shorter than the incrassate style, with trifid stigma. It resembles *L. tenella* in the shape of its leaves, but is distinguished by the sessile, axillary flowers.

10. L. axillaris (Sond.) ; branches woody, erect, terete, shortly pubescent ; leaves alternate, erect, appressed, linear-lanceolate, quite entire, glabrous, with revolute margins ; spike elongate ; flowers shortly pedicellate ; calyx tube hemispherical, pubescent ; lobes lanceolate, a little shorter than the glabrous corolla.

HAB. Mts. between Hottholld. and Caledon Stellenbosch, *Ecklon.* in (Herb. Sd.)
Branches as in the preceding. Leaves 4 lines long, 1 line broad, with slightly revolute not denticulate margins. Flowers solitary ; the pedicels equalling the calyx tube. Calyx 2 lines long. Corolla as in *L. spicata,* but not much longer than the calyx.

11. L. longifolia (DC.! Monog. p. 107, excl. syn. Thunb.) ; stem woody, erect, simple or branched ; branches virgate, undivided, minutely downy ; leaves alternate, erect, linear-subulate, narrow, rigid, subdenticulated, glabrous; raceme terminal, elongate ; pedicels solitary or ternate, as long or a little longer than the calyx ; tube of calyx hemispherical ; lobes subulate, twice shorter than the hairy corolla. *Prodr. VII.* 2, *p.* 417. *E. Z.* 2341. *L. linearis, Eckl. Herb. Un. itin.* 157.

VAR. β. **oppositifolia** ; leaves opposite, or the upper ones alternate.

HAB. Sandy places amongst shrubs in the Capeflats near Doornhoogde & Zeekoovalley, and near Grooteport, *E. & Z., Wallich ;* Vaalfontyn near Eersterivier. *Zeyher,* 3123. Var. β. in the Cape flats, *Ecklon, W. H. H.* Nov.–Janr. (Herb. D., Sd., Hk.)
A virgate sub-shrub, 1-1½ foot high. Leaves 6-10 lines long, ½ line broad, sub-carinate, minutely denticulated, stiff. Raceme 2-4 inches long ; pedicels in the axil of the leaves, glabrous or minutely downy. Calyx 2-3 lines long ; lobes longer than the glabrous or downy tube. Corolla 5-parted, segments very narrow, equalling the 3-lobed, rarely 4-lobed stigma. Capsule 3-valved, the valves equal in length to the base of the capsule. It is nearly allied in habit and characters to *L. sessiliflora,* from which it differs by longer leaves, racemose inflorescence, and 3-valved capsule.

12. L. unidentata (A. DC. Monog. p. 109); stem erect, woody, simple, or branched at the base ; leaves alternate, erect, subappressed, linear-lanceolate, acuminate, with 1–2 setaceous teeth on each side near the base; flowers racemose; pedicels solitary, longer than the calyx; tube of calyx hemispherical, shorter than the acuminate lobes ; corolla 5-parted, 2–3 times longer than the calyx, hairy.

VAR. *a.* **glabra** ; raceme narrow elongated. *Camp. unidentata, Thunb.! Prodr. p.* 39. *Mem. Acad. Petersb.* 4, *p.* 371, *t.* 5, *f.* 2.

VAR. *β.* **pubescent** ; raceme narrow, mostly elongated ; branches, pedicels, and calyx shortly pubescent ; leaves glabrous, or hairy beneath. *C. unidentata, Herb. Holm, specimen a Thunb. lectum.*

VAR. *γ.* **lycopodioides** ; leaves smaller, 1–2 dentate or entire ; flowers on shorter pedicels, or disposed in a loose raceme. *L. lycopodioides, A. DC. Monog. p.* 114.

HAB. Var. *a.* Cape of Good Hope, *Thunb.* Var. *β.* Wittedrift, Plettenberg's Bay, *Dr. Pappe.* Var. *γ.* Cape, *Lalande,* in Herb. *Kunth.* In Zwarteberg, Caledon, *Dr. Pappe, Zeyher,* 3122. Dec.–Janr. (Herb. Thb., Hlm., Kth., Sd., D., Hk.)

Stem 1 foot or more in height virgate as the branches, often reddish, leafy, Leaves 4–6 lines long, 1 line broad ; in var. *γ.* 3–4 lines long, ½ line wide, rigid, acute, with slightly revolute margins. Pedicels capillary, 2–6 lines long, with bracteas and sometimes a second flower in the middle, in var. *γ.* not disposed in a regular raceme, but more distant, but sometimes more approximate at the top of the branchlets. Calyx 2 lines long. Corolla blue (*Thunb.*) Valves of the capsule shorter than the calyx lobes. Var. *γ.* differs in habit by more branched stem, filiform branches, smaller leaves and flowers ; the leaves are appressed ; the flowering branches exactly as in the original branches in Herb. Kunth., but in young sterile twigs they are generally spreading, or erect spreading.

13. L. Caledonica (Sond.) ; branches herbaceous, erect, glabrous, or shortly pubescent ; leaves alternate, erect, linear, narrow, glabrous, rigid, remotely denticulate on the thickened margins ; flowers disposed in an elongate raceme ; peduncles leafy, 1–3 flowered ; pedicels short; calyx-tube turbinate; lobes denticulate, as long as the tube ; corolla 5-parted, glabrous.

HAB. Caledonriver, *Zeyher.* Janr. (Herb. Sd.)

Branches 1 foot, terete. Leaves about 4 lines long, ½ line broad, with a fascicle of smaller ones in the axil. Raceme secundate. Peduncles 4–6 lines long, with small leaflike bracts. Pedicels equalling the calyx tube. Calyx glabrous. Corolla with narrow segments. Capsule with 3 valves, twice shorter than the turbinate base of the capsule. It has the leaves of the following species, but a very different capsule.

14. L. denticulata (Sond.) ; stem herbaceous, much branched as well as the branches glabrous or subscabrous; leaves alternate, linear, narrow, acuminate, glabrous or subpilose, with thickened remotely denticulate margins; flowers racemose-panicled ; tube of calyx hemispherical ; lobes linear, denticulate ; corolla deeply 5-fid, longer than the calyx. *Camp. denticulata, Burch. Trav., vol.* 1, *p.* 538. *Wahlenb. denticulata, A. DC. ! Monog. p.* 152, *t.* 16. *E. & Z.* 2378. *Lightf. capillaris, Buek. in E. & Z. Enum.* 2342. *W. corymbosa, E. Mey. in Herb. Drege,* and 4005, 6300, *a. Zeyher,* 1058, 1059.

HAB. Eastern part of the Colony, Zwartkops and Krumrivier in Caffraria, near the Aapges, Caledon and Mooyeriver, and Port Natal, *Burchell, Drege, Krauss, Eckl. and Zeyh.* Nov.–Dec. (Herb. Sd., D., Hk.)

A perennial herb, ½–1 foot, very leafy. Leaves alternate, ternate, rarely oppo-site, 4–6 lines long, ½ line broad, the upper smaller. Flowers numerous towards the tops of the branches. Pedicels unequal, spreading, capillary, 2–6, in *L. capillaris*, 6–10 lines long. Panicle few or many flowered. Calyx 2 lines long. Corolla 5-fid, at length nearly 5-parted ; segments very narrow. Capsule hemispherical, 10-nerved ; the 3 valves a little shorter than the base of the capsule. The plant I possess from *Drege*, under the name "*Wahlenb. Caffra, DC.*" agrees perfectly with *L. denticulata.*

15. L. paniculata (Sond.) ; quite glabrous ; stems erect, woody, much branched, striate-angulate; branches filiform, divaricate, divided, forming a large, nearly leafless panicle; leaves rare, alternate, linear-subulate, canaliculate, entire ; upper ones or the bracteæ gradually smaller; flowers pedicellate; pedicels 2–6 times longer than the calyx ; calyx obconical, lobes linear, twice shorter than the 5-parted corolla with narrow segments; valves one half, or nearly doubly shorter than the base of the capsule.

HAB. Magalisberg. Nov. *Burke & Zey.*, 1067. (Herb. Hook., D., Sd.)
Root-stalk with several slender stems, 1 foot or more in height. Branches alter-nate, 4–6 lines long ; the ultimate ramuli or the pedicels 3–10 lines long. Leaves erect, 4–6 lines, upper ones more spreading, 2–1 line long. Calyx tube terete, 1 line long, equal in length to the lobes.

16. L. albens (Spreng. ! in Zeyh. fl. cap. ex. 266); stem erect, woody, branched ; branches rigid, minutely downy ; leaves alternate, in fas-cicles, erect-spreading, linear, very narrow, entire, with revolute mar-gins, thickish, pale ; flowers disposed in a loose raceme; pedicels divari-cate, glabrous, rigid ; calyx tube hemispherical, lobes erect, subulate, entire ; corolla 5-parted, 3 times longer than the calyx ; valves acute, equal in length to the base of the capsule. *A. DC. Monog. p.* 110. *DC. Prodr. VII. 2, p.* 417. *Burch. Catal.* 1720. *E. & Z.* 2350. *L. laricina, Buek! in E. & Z. Enum.* 2344, *ex parte.*

HAB. Sandy places on the Zwartkopsriver, *E. & Z. Zeyh.* 3129. Klein Fisch-rivier, *Drege,* 3619. Rivier Zonder Einde, and in Hassaquaskloof, *Zey.* 3120. Dis-trict of Albert, *T. Cooper,* 1861. Sanddrift, Orange River, *Burke.* Oct.–Dec. (Herb. Hk., D., Sd.)
Shrub 1 foot, much branched. Leaves 2–4 lines long, ½ line broad, rigid, cori-aceous, convex above by the revolute margins, glabrous. Pedicels 4–6 lines long. Calyx glabrous, 1½ line long. Segments of corolla very narrow. Capsule hemis-pherical, small.

17. L. Buekii (Sond.) ; stem erect, woody ; branches virgate, filiform, minutely downy ; leaves alternate, rarely opposite, often in fascicles, spreading, linear, very narrow, quite entire or with a short tooth at each side at the base ; flowers few, at the top of the branches terminal and axillary, racemose ; pedicels and calyx downy ; tube of calyx obconi-cal, lobes linear, spreading, twice shorter than the 5-parted corolla ; valves acute, 3 times shorter than the base of the capsule. *L. laricina Buek! in E. Z. Enum.* 2344, *ex pte.*

HAB. Hills between the Zondagsriver and Adow, and on the Zwartkopsriver, *E. & Z.; Zey.* 5129, ex pte. Jul. (Herb. Sd.)
Distinguished from *L. albens* by the elongated, nearly filiform, 1–1½ foot long branches, subfiliform leaves, capillary, downy pedicels, and obconical capsule with longer lobes and shorter valves. Whole plant of a pale green colour. Leaves 3–4

lines long, much spreading.   Pedicels 2-4 lines long.   Lobes of calyx in the flowering plant erect, in the fruitbearing divergent.   Segments of corolla very narrow. Capsule 2 lines long.

18. L. ciliata (Sond.); stem woody, erect; branches simple or divided, glabrous or pubescent; leaves alternate, rarely opposite, spreading, linear-lanceolate, mucronate; margins revolute, ciliated near the base by 2-4 setaceous teeth; flowers solitary or in few-flowered loose racemes at the top of the branches; pedicels short, capillary; tube of calyx hemispherical, lobes acuminate, setaceous-ciliate at the base, 2-3 times shorter than the deeply 5-fid corolla. *Campan. ciliata, Thunb.! Fl. Cap. p.* 172, *C. Thunbergii, R. et Sch., Syst. Veg. v. p.* 135. *Roella ? Thunbergii, A. DC. Monog. p.* 174. *Prod. p.* 446. *Lightf. thymifolia Buek! l. c.* 2343. *divaricata, Buek!* 2352. *mucronulata, Buek!* 2353. *L. Loddigesii, A. DC. Monog. p.* 114. *L. tenella, Lodd. Bot. Cab. t.* 1038, *non A. DC. Zey.* 3130.

VAR. β. **major**; floriferous branches longer, forming a terminal racemose panicle. *L. Uitenhagensis, Buek!* 2345. *L. intermedia, Buek!* 2351.

VAR. γ. **debilis**; branches procumbent, diffuse, filiform; leaves spreading or a little reflexed, alternate or often opposite, with hooked apex, mucronate; flowers terminal, subsolitary; calyx glabrous or pilose. *L. rubioides Buek! b. c.* 2363. *non A. DC.? L. tenella, A. DC.! var. in Herb. Drege. Zey.* 3131.

VAR. δ. **pubescent**; branches, leaves and calyx pubescent; leaves opposite or alternate with hooked apex. *L. pubescens. DC.! Prod. p.* 419.

HAB. Mountains between Hassagaybosch and Grahamstown, Albany; near Uitenhage, Adow, and Port Elizabeth. *E. & Z., Drege*; var. β. on the Zwartkopsrivier and Van Staadensrivier mountains; *E. & Z., Drege*, 6317, 9138; var. γ, Zuureberge, Winterhoeksberge, *E. Z. Drege!* Knysna, *Dr. Pappe*; var. δ. between Omtata and Omsamwubo, *Drege*, 7496. Oct.-Mai. (Herb. Th., Hlm., D., Sd. Hk.)

Stem ½-1 foot or more high; branches reddish, woody, erect, very leafy, terete, in var. γ, very long, filiform or capillary, trailing. Branchlets mostly short. Leaves in fascicles, sessile, much spreading or subreflexed, glabrous or hairy, pale green, 2-3, rarely four lines long, ⅓ line broad, pointed by a recurved mucro; the teeth-like cilia, short, spreading. Pedicels about as long as the calyx, in some flowers shorter, in others a little longer. Lobes of calyx acute or mucronulate, longer than the tube. Corolla variable in size, 5-fid, the segments lanceolate, blue? Style as long as the corolla, stigma 3-fid. Capsule hemispherical or subdepressed, 10-nerved, spreading or recurved; the calyx lobes twice longer than the valves of capsule. Seeds ovoid, trigonal. VAR. β. is a more rigid shrub with panicled inflorescence, the calyx glabrous or hairy. Var. γ. is characterized by very long, nearly capillary slightly divided branches; the leaves 1-2 lines long, incurved or hooked, evidently mucronulate; the flowers solitary, in size nearly as in var. α. Var. δ. is only distinguished by the pubescence.

19. L. Thunbergiana (Buek! in E. Z. Enum. 2359.); glabrous branches woody, erect-spreading; leaves, alternate, reflexed, linear-lanceolate, acute, with thickened margins, quite entire or remotely denticulated, flat; peduncles divergent, rigid, disposed in a few-flowered, loose raceme; the terminal pedicels 3-4 lines longer than the lanceolate-acuminate bractea; tube of calyx hemispherical; lobes linear-acuminate, entire; corolla deeply 5-fid, 2-3 times longer than the calyx. *L. anomala, A. DC.! Prodr. p.* 418.

HAB. Karroolike hills, near the Gauritz river, Swellendam, *E. & Z.*; near Leliefontein, 3-4000 ft., and between Houtbay and Wynberg, *Drege.* Nov.-Dec. (Herb. Hook, Sd.)

More woody and erect than the following, to which it has a great resemblance, and nearly the same but smaller leaves. Stem 1 foot and more, as well as the primary branches leafless. Ultimate branches often finely-pubescent. Leaves, 3–6 lines long, ¾–1 line broad. Pedicels 2–4 lines long. Calyx with spreading lobes, 2 lines long. Segments of corolla very narrow, nearly 5-parted. Valves of capsule, shorter than the calyx lobes.

20. L. oxycoccoides (L'Her. Sert. Angl. t. 4); stem glabrous or hairy, erectish or ascending, woody at the base, much branched, branches diffuse, filiform or capillary; leaves alternate, reflexed, ovato-lanceolate, acute, flat, thin, quite entire or denticulated at the base; peduncles capillary, solitary, terminal, sometimes loosely racemose, equalling or a little longer than the leaves; tube of calyx hemispherical, lobes subulate, entire; corolla 5-parted, one-half longer than the calyx. *A. DC. Monog. p.* 113, *E. Z.* 2360. *Campanula tenella, Thunb.! Herb. ex pte. act. acad. Petersb.* 4, 1, 7, *f.* 3. *Herb. Un. itin,* 151. *C. Ottoniana, R. and Sch. Syst.* 5, *p.* 113, *fide DC. Drege,* 6306, 6315.

HAB. Rocks on Table and Devil's Mountain; Drakensteensberg, and near Simon's bay. Decb.–Feb. (Herb. Thunb., Holm., D., Hk., Sd.)

Stem short, perennial, with many prostrate, longish branches, resembling *Oxycoccos*. Leaves, glabrous, 1–2 lines long. Calyx 1 line long. Corolla white, with rose-coloured nerves.

21. L. tenella (A. DC. Monog. p. 111, t. 3. f. B.); stem erect, woody, branched; branches minutely downy; leaves alternate, usually in fascicles, reflexed, ovate, obtuse or acute, concave or canaliculate, thick, entire or denticulate, with incrassate margins; flowers racemose; corolla sub-5-parted, 2–3 times longer than calyx; valves of capsules equal or a little shorter than the linear-lanceolate calyx-lobes. *DC. Prodr. p.* 418. *E. Z.* 2355. *Campanula tenella Linn.! suppl. p.* 141. *Thunb.! herb. ex parte. Herb. Un. itin.* 156. *Roella filiformis et glabra. Poir. dict. Lightf. oxycoccoides, Dreg. ex s. ex pte. et* 6316, 6318.

VAR. β. **microphylla**; leaves minute, glomerate; flowers racemose-corymbose or subpanicled; valves of the capsule longer than the calyx lobes. *L. nodosa, Buek! l. c.* 2358. *L. tenella, var. longivalvis, A. DC. Drege,* 6308. *L. oxycoccoides, Drege b.*

VAR. γ. **tenerrima**; branches elongated, filiform; leaves small; raceme remote-flowered; valves of capsule equal or a little longer than the calyx-lobes. *L. tenerrima, Buek ! l. c.* 2356.

VAR. δ. **fasciculata**; leaves very numerous, ovate-acuminate; flowers shortly pedicellate, densely aggregated in an ovate or oblong raceme. *L. fasciculata, E. Z.!* 2354. *Camp. fasciculata, Thunb.! Herb. var. β.*

VAR. ε. **rigida**; more rigid, branches woody, diffused, sparingly leafed; leaves small, pedicels disposed in a loose raceme; valves of capsule longer or shorter than the calyx lobes. *L. diffusa Buek ! l. c.* 2357.

HAB. Cape Flats, and on the seashore from Cape town to Port Natal (*Miss Owen*), *Zey.* 3126, Nov.–Dec.; var. β. Port Elizabeth and on the Zwartskops river, *Zey.* 3119 ex parte, 3128, Sept.; var. γ. distr. of Swellendam and Hassaquaskloof, *E. & Z. Dr. Pappe, Zey.* 3125. Nov.–Dec.; var. δ. near Vankampsbay, and in Uitenhage, *Thunb. E. & Z.*, Oct.–Nov.; var. ε. on Zondags and Coegariver, Uitenhage, *E. & Z.*; Rhinosterkop, near Beaufort, *Burke and Zey.* 1062. Africas-hoogde, *Zey.* 1063, Mar. (Herb. Thunb., Holm, D., Hook, Sd.)

Very variable, from 4 inches to several feet high, but always shrubby and woody. Leaves when small about as long as broad, and usually 1 line; in var. δ. 2–3 lines long, coriacous, often in four rows, very rarely in the upper branches opposite.

Raceme ½-2 inches long, in large specimens panicled and then resembling in habit the inflorescence of *Wahlenbergia adpressa;* in very small specimens, as distributed by Drege *(L. oxycoccoides, b.),* the branches are terminated by one or two flowers. Pedicels 1-2 lines ; var. ε. often 3-4 lines long. Calyx with hemispherical tube and acuminate lobes 1½-2 lines long. Segments of corolla linear or lanceolate, blue, glabrous, or shortly hairy.

22. L. oppositifolia (A. DC. Monog. p. 115); branches erect or ascending, slender, rigid, simple ; leaves opposite, spreading or reflexed, linear-lanceolate, acuminate, subdenticulated or entire, with thickish margins ; flowers few, usually terminal ; calyx glabrous, tube hemispherical ; lobes erect, lanceolate-acuminate, denticulate; corolla sub-5-partite, twice longer than the calyx. *Prod. p.* 419. *E. & Z.* 2362. *P. lanceolata et L. muscosa, Link. Enum. Hort. Berol.* 1, *p.* 217. *E. & Z.* 2361. *L. oxycoccoides, var. Spreng. Syst.* 1, *p.* 809.

HAB. Sandy places in the Cape flats, near Doornhoogde, and on the Table mountain, *E. & Z.* Dutoitskloof, *Drege* (A. DC.). Nov. (Herb. Reg. Berol., Sd.)

Plant ½-1 foot, much branched ; branches reddish, filiform, glabrous or a little hairy. Leaves 3-4 lines long, ½ line broad, glabrous. Pedicels capillary, 1-3 at top of the branches, 2-6 lines long. Calyx 1 line long ; lobes longer than the tube. Drege's specimen, 6314 in Herb. Sond., is not *L. oppositifolia,* but *L. capillaris,* Buek. = *denticulata,* Sd.

23. L. rubioides, (A. DC. Monog. p. 116); branches procumbent diffuse ; leaves opposite, spreading or subreflexed, lanceolate, acute, remotely denticulated; flowers few, terminal, and axillary; tube of calyx hairy, spherical, lobes spreading, linear-lanceolate, remotely denticulate; corolla 5-parted, longer than the calycine lobes. *Prodr. p.* 419.

HAB. Cape of Good Hope, Herb. *Banks.* (Unknown to us.)

Plant trailing, much branched ; branches divaricate, 3-4 inches long, a little hairy, reddish, more or less leafy. Leaves sessile, 2-4 lines long, scarcely 1 line broad, glabrous or furnished with a few white hairs. Flowers pedicellate; pedicels filiform, subpilose, 3-5 lines long. Tube of calyx about 1 line long ; lobes acute, 1½-2 lines long. Style a little longer than the corolla, thicker and 3-dentate at the top. Capsule spherical.

§ 3. Capsule 2-celled, elongate-conical. (Sp. 24-25.)

24. L. juncea (Sond.) ; stem and branches virgate, striate, herbaceous, glabrous, hairy at the base ; lower leaves ovato-lanceolate, acute, erect, alternate, with revolute, incrassate, entire or denticulate margins, glabrous, sometimes hairy beneath ; upper ones distant, gradually smaller, lanceolate or linear-lanceolate ; flowers disposed in an elongate spike, solitary or ternate, sessile; tube of calyx obconical, glabrous, longer than the triangular lobes; corolla deeply 5-fid; segments 3-4 times longer than the calyx; capsule obconical; valves equalling the calyx-lobes. *Prismatocarpus junceus, Buek! in E. & Z. Enum.* 2400. *Wahlenbergia spicata, E. Mey.! DC. Prod. p.* 441.

HAB. Stony places and the Klipplaatriver, near Silo, Tambukiland, *E.&Z.* Port Beaufort, Caffraria, *Drege.* Dec.–Jan. (Herb. Thunb., D., Sd.)

Perennial, 1-1½ foot high. Root woody, with many terete, pale green stems ; branches similar. Lower leaves approximate, 4-6 lines long, 2 lines broad ; the intermediate 1 line broad ; the uppermost 2-1 line long. Flowers in the axil of

3 small bracts. Spike 1–2 inches long, remoti-flowered. Lobes of calyx acute, nearly 1 line long. Segments of corolla narrow, glabrous; stamens 5; anthers linear. Style with 2, rarely 3, linear, revolute stigmas. Capsule terete, 10-nerved, 4 lines long. Seeds oblong, acute at both ends, sub-3-angular.

25. L. laxiflora (Sond.); branches erect, virgate, striate, subsimple, glabrous; leaves alternate, distant, elongate-linear, narrow, quite entire, keeled by the middle nerve, glabrous; upper ones smaller; flowers disposed in a lax, often unilateral raceme; peduncles capillary, with 2 subulate bracts in the middle; tube of calyx obconical, glabrous; lobes subulate, as long as the flowering tube; corolla deeply 5-fid, 3 times longer than the calyx; capsule linear-obconical, terete; valves 3 times shorter than the calyx lobes.

HAB. In Caffraria, *Drege.* Katberg, Eastern Frontier, *H. Hutton.* (Hb. D., Sd.) Well distinguished from *L. juncea* by longer branches, narrower leaves, and pedunculate flowers. Branches in the specimen 1½ foot. Leaves ¾–1 inch long, 1 line broad. Raceme 4–8-flowered, 3–4 inches long. Lower peduncle 1 inch, the upper shorter. Lobes of calyx erect, ciliated at the base, 2–2½ lines long. Segments of corolla lanceolate-linear. Base of filaments ciliated. Ovary 2-celled. Style with 2 stigmas. Ripe capsule about half an inch long, striate.

XI. MICROCODON, A. DC.

Calyx 5-cleft, tube ovoid or spherical. *Corolla* 5-lobed at the apex, small, cylindrical. *Stamens* 5, free; filaments very slender, not expanded at the base. *Style* filiform. *Stigmas* 5. *Capsule* 5-celled, half-superior, dehiscing by 5 valves at the apex; cells alternating with the calycine lobes. *Seeds* ovoid, small. *Monog. camp. p. 127, t. 19. DC. Prod. p. 421. Endl. Gen. 3078. Wahlenbergiæ, Spec. Thunb.*

Small, annual herbs. Leaves alternate or subopposite, sessile, small, narrow; upper ones rather the longest. Flowers terminal, sessile or pedicellate. Tube of calyx hispid, rarely glabrous. Name from μικρος, *small,* and κωδων, a *bell;* in reference to the shape and smallness of the flowers.

Sect. 1. EUMICROCODON. Ovary conical and free above. Capsule dehiscing by the elevated valves.

Flowers solitary, on longish pedicels ... ...   (1) **lineare.**
Flowers subsessile, or shortly peduncled :
    Flowers glomerate, terminal ... .   (2) **glomeratum.**
    Flowers solitary in the axils of the upper leaves ... ...   (3) **sparsiflorum.**

Sect. 2. CŒLOTHECA. Ovary flat above, quite inferior. Capsule
dehiscing by depressed valves ... ...   (4) **hispidulum.**

Sect. 1. EUMICROCODON, A. DC., Prod. (Sp. 1–3.)

1. M. lineare (Buek! in E. & Z. enum. 2368); stem flexuous, erect, simple or branched above, terete, glabrous at the base; leaves linear, acuminate, narrow, subdenticulate with cartilaginous margins; flowers pedicellate, often subpanicled; tube of calyx hairy or glabrous, spherical; lobes linear, nearly as long as the corolla; capsule spherical, obtusely 5-angled, 3 times shorter than the calycine lobes. *Campanula linearis, Linn. f. Suppl. p. 140. Thunb. Mem. Acad. Petersb. 4, t. 5, f. 3. Fl. Cap. p. 171. Wahlenb. linearis, A. DC.! Monog. p. 137.*

VAR. β. **diffusa**; stem much branched; branches diffuse. *Wahlenb. diffusa, A. DC. l.c.*

HAB. Sandy places near Heerenlogement, *Thunb.* Near Waterfall, Tulbagh, *E. & Z.* Bergrivier, *Drege.* Oct.-Nov. (Herb. Thunb., D., Sd.)

Stem ½-1 foot high, filiform, purplish, more or less dichotomously branched at the apex. Leaves glabrous or ciliate at the base, about 6-8 lines long, ½ line wide. Flowers white. Peduncles spreading, ½-1 inch long. Bracteæ leafy, linear. Corolla narrow. Capsule hispid or quite glabrous, the size of a small pea ; valves 1 line long, erect.

2. M. glomeratum (A. DC.! Monog. p. 127, t. 19); stem erect, simple or branched; branches spreading, simple, hairy; leaves linear, acuminated, hairy and ciliated ; flowers subsessile, glomerate, terminal; bracteæ leafy, longer than the flowers; tube of calyx ovoid, hispid ; lobes acuminate, long; capsule ovoid, terete, valves erect, twice shorter than their base. *DC. Prod. p. 421. M. Candolleanum, Buek. l. c.* 2364, *ex pte. Herb. Un. Itin.* 149.

VAR. β, **brevibracteatum** ; smaller; leaves rare; flowers densely glomerate; bracteæ as long as the flowers; calyx-lobes shorter. *M. brevibracteatum, Buek! l. c.* 2366.

VAR. γ, **pygmæum** ; dwarf, 1 inch high, mostly simple, with 2-4 terminal flowers. *M. pygmæum, Buek! l. c.* 2365.

VAR. δ, **singuliflorum** ; flowers solitary or subsolitary.

HAB. Stony sandy places in the flats and mountains near Capetown. Driefontein, *Zey.* 1065. Var. γ, near Waterfall, Tulbagh, *E. & Z.* Oct.-Nov. (Hb. Hk., D., Sd.)

2 inches to 1 foot high. Leaves alternate, or lower ones opposite, in small specimens 3-4 lines, in the largest 8-10 lines long, 1 line wide, minutely denticulated ; floral ones or bracteæ 1 inch long, but in the varieties not exceeding the flowers. Flowers glomerate to a head, rarely solitary. Calyx-tube 1-1½ line long, ovoid or subsphæroid ; lobes 2-3 times longer, or in var. β about as long as the tube. Corolla 2-3 lines long. Capsule very hispid. Var. δ resembles *M. lineare*, but differs by more acuminate leaves and very hispid, terete, not obtusely-5-angular capsule.

3. M. sparsiflorum (A. DC. Monog. p. 128) ; stem branched from the base ; branches subdivided, hairy ; leaves linear-lanceolate, ciliate, denticulated; flowers shortly peduncled, solitary in the axils of the upper leaves; tube of calyx spheroid, densely hairy ; lobes linear, long ; capsule spherical, valves erect. *DC. Prod. p. 421. M. Candolleanum, Buek, l. c. ex parte. Wahlenb. hispidula, Schrad. Campanula hispidula, Link. enum. p.* 215.

HAB. Cape of Good Hope. (Unknown to me.)

Root annual, filiform. Stem erect, 2-3 inches high ; branches divergent, terete, hairy, often subdichotomously divided, alternate, lower ones opposite. Leaves opposite or alternate, sessile, linear-lanceolate, 2-8 lines long, nearly 1 line wide, rather hairy, incrassate on the margins ; upper ones rather longer. Flowers on rigid, 3-6-lines-long peduncles. Calyx 5-fid ; tube very hispid, 1 line long ; lobes spreading, acuminate, leafy, subdenticulate, 2 lines long. Corolla tubular, 2½ lines long, 5-lobed at the apex, lobes erect, at length twisted. Style equalling the half corolla. Capsule 5-celled, ½ superior ; inferior part spherical, 1½ line long ; valves the same length, opposite the calyx-lobes. I have not seen an authenticated specimen ; all I have found in several herbariums are not different from small specimens of *M. hispidulum.* Can it be the same ?

Sect. 2. CŒLOTHECA, A. DC., Prod. (Sp. 4.)

4. M. hispidulum (Sond.); stems much branched; branches diffuse, hispid, very leafy ; leaves linear-lanceolate, acuminate, marginate, subdenticulated, subciliated ; flowers subsessile or shortly pedicellate, terminal and axillary; calyx-tube obovoid, hispid; lobes linear-acuminate, denticulate, leafy, longer than the tube; corolla as long or shorter than the calycine lobes, 5-lobed at the apex; capsule obconical, hispid. *Campanula hispidula, Thunb.! Prod. p. 38. Fl. Cap. p. 172. Willd.! herbar. Wahlenb. hispidula, A. DC. Monog. p. 155. Prod. p. 430. Microcodon depressum, A. DC.! Prod. p. 422. Deless. Icon. Sel. vol. 5, t. 16. Herb. Un. Itin. 150.*

HAB. Sandy hills in Swartland, *Thunb., Sparrman, Bergius.* Near Capetown, *W. H.H.* Tablemountain, *E. & Z.* Betw. Elephantrivier and Bergrivier, *Drege.* Breederivierpont, near Zwellendam, *Zey.* 3134. Sept.–Oct. (Hb.Th., Holm., Hk., D., Sd.)
     The annual root very long, white. Stem 2–4 inches high; branches 2–5 inches long, lower ones opposite, dichotomously divided. Leaves 3–6 lines long, 1 line wide; floral ones or bracteæ rather longer. Calyx-tube in flower 1 line, lobes 2–4, sometimes 6–7 lines long. Corolla blue and white (Thb.), tubular-infundibuliform, 4–6 lines long. Stigma 5-lobed. Capsule coriaceous, 2–2½ lines long, crowned by the rigid, elongated lobes, wholly inferior; the valves are depressed, and not elevated and superior, as in *M. glomeratum.* In some specimens, however, and chiefly in the small ones, which I think to be *M. sparsiflorum,* A. DC., there are rudimentary or perfectly developed superior valves, about ½ line long.

XII. **WAHLENBERGIA**, Schrad.

Calyx 5–3-fid. *Corolla* 5–3-lobed at the apex, rarely divided to the middle, infundibuliform, subcampanulate, or tubulose. *Stamens* 5–3, free; filaments broad at the base. *Style* inclosed, pilose, particularly in the upper part. *Stigmas* 5–2, often linear, short. *Capsule* 5–3–2-celled, each opening by so many valves at the apex; valves with a dissepiment in the middle; cells (when 5) opposite to the calycine lobes. *Seeds* numerous, minute. *Schrad. in Catal. Hort. Goetting.* 1814. *A. DC. Monog. Camp. p. 129. Endl. gen. 3079. Campanula spec. Linn.*

Herbs for the most part annual, sometimes perennial, rarely subshrubs. Leaves alternate, rarely opposite, generally most numerous towards the lower part of the plant. Peduncles terminal and axilary, often elongated, dichotomous. Pedicels often elongated, filiform, terminal or opposite to a leaf. Flowers at first drooping; the capsule erect. Name in honour of George Wahlenberg, professor of botany in Upsala.

Sect. 1. CAPSULE 5-celled. (Sp. 1–8.)

Cor. 5-fid, spreading, 2-coloured. Pedunc. long, 1 fl. (stem
    1–2 ft. high) ... ...   (1) **Capensis.**
Cor. 5-lobed at the apex, one coloured :
    Glabrous (or pilose, at base only or rarely downy above) :
        Cor. nearly as long as calyx. Dichotomously branched   (7) **debilis.**
        Cor. 2–3ce as long as calyx :
            Lvs. narrow-linear, denticled; branches erect
                (2–4 in. high) ... ...   (2) **costata.**
            Lvs. linear, acute, denticled; br. divaricate (2–3
                in. high) ... ...   (3) **divergens.**
            Lvs. linear, acuminate, entire or subdenticled;
                stem ½–1 ft. high, glabr. above; cal. lobes
                much longer than the tube ... ...   (6) **Ecklonii.**
            Lvs. lanceolate, subpungent, denticled; stem
                4–6 in. high, downy above; cal. lobes *a little*
                longer than the tube ... ...   (5) **oxyphylla.**

Cor. 6-8 times longer than the calyx; lvs. linear;
   stem 3-4 in. high ... ...   (4) **decipiens.**
Hispid, or covered in all parts with spreading hairs   ...   (8) **pilosa.**

Sect. 2. CAPSULE 3, rarely 2-celled. (Sp. 9-46.)

A. Herbaceous. Leaves *opposite* (rarely some alternate). Caps 3-celled :
  Tufted, depressed ; lvs. obovate or oblong lanceolate   ...   (14) **montana.**
  Stem erect, or ascending, or prostrate ; branched :
    Lvs. ovate, obovate, or ovato-lanceolate :
      Cal. tube hairy. Stem erect. Pedunc. termi-
        nal, 1-2-fl. leafy ... ...   (9) **flaccida.**
      Cal. hispid. Stem ascending. Pedicels axillary,
        about as long as leaf ... ...   (12) **oppositifolia.**
      Cal. glabrous or appressed-pubescent. Lvs. ob-
        tuse. Stem creeping. Ped. twice as long as
        leaf ....   (13) **procumbens.**
    Lvs. linear or lanceolate ; cal. glabrous :
      Lvs. linear, rigid, pungent, *denticled.* Branches
        terete ... ...   (11) **spinulosa.**
      Lvs lanceolate, *quite entire*, Branches 4-angled   (10) **stellarioides.**
B. Herbaceous. Leaves alternate (the lowest sometimes opposite) :
  *Fl. sessile or subsessile. Lvs. scattered, not *rosulate* or tufted at the base of
    stem :
    Leaves linear, denticled ... ...   (15) **inconspicua.**
    Leaves petioled, obovate, obtuse ... ...   (16) **acaulis.**
  **Fl. peduncled ; pedunc. and pedicels short. Lvs. scattered,
    not rosulate or tufted :
  Annuals :
    Leaves flattish (not channelled), with revolute or thickened margins, with
      acute but not spinous teeth ; caps. hemispherical or turbinate :
      Patently hairy ; leaves ovate or ovate-lanceol. ;
        cor. 4 times as long as calyx. Caps. hemi-
        spherical ... ...   (21) **polyclada.**
      Shortly hairy or downy ; lvs. lanceol. or lin.-
        lanceol. ; cor. 2-ce as long as calyx; caps. tur-
        binate ... ...   (20) **paniculata.**
      Quite glabrous, or subpilose near the base :
        Glabr. ½-1 ft. ; lvs. linear, denticled, 4-6 l.
          long ; fl. 1-3, terminal, shortly pedicelled ;
          cor. 2-3 times longer than calyx ... ...   (27) **Bowkeri.**
        Glabr. or sub-pilose, ½-1 ft. ; lvs. narrow-
          linear, 3-4 l. long; fl. panicled ; cor. 2-ce
          as long as the calyx ... ...   (19) **Banksiana.**
        Subpilose, ½ ft. ; lvs. linear, acuminate,
          serrulate ; panicle few-fl. ; cor. ½ longer
          than the cal. ... ...   (26) **Caffra.**
        Minutely downy, 3-4 in. ; lvs. linear, 2-4 l.
          long ; panicle dichotomous, with long
          pedicels ; cor. 3-4 times as long as calyx   (18) **ramulosa.**
    Leaves linear, channelled, toothed ; teeth 2-spinous :
      Cal. tube turbinate. Cor. campanulate, 6 times
        longer than cal. ... ...   (22) **prostrata.**
      Cal. tube hemispheric. Cor. tubulose, 4-5 times
        longer than cal. ... ...   (23) **ingrata.**
    Leaves linear, *flat*, entire or subentire, not thickened at margin :—
      Cal. tube obconical, hispid. Pedicels solitary.
        Cor. 3-4 times longer than cal. ... ...   (17) **tenuis.**
      Capsule ovoid. Panicle dichot. Cor. 2-3 times
        as long as calyx ... ...   (24) **oocarpa.**

Caps. obovoid.   Pedicels opp. leaves, the upper
   panicled ; cor. ½ longer than the cal.   ...   ...   (25) **exilis.**
Perennial.   Leaves linear-filiform, fascicled   ...   .   (28) **capillacea.**
***Pedunc. and ped. elongate.   Leaves crowded or often
   rosulate at the base of the stem.   Caps. 3 or 2 celled :—
   Style naked, or with some glands below the stigma :
      Pedicels erect :
         Quite glabrous.   Leaves linear lanceolate   (29) **denudata.**
         Hairy at the base :
            Annuals :—
               Caps. 2-celled, ovoid. Cor. 5-lobed
                  at apex   ...   .   (30) **Meyeri.**
               Caps. 2–3 celled, obconic.   Cor. 5-
                  fid. Lvs. ovate, oblong, or sub-
                  lanceol. undulated, crenate   ...   (32) **undulata.**
               Caps. 3-celled, obovoid or spheroidal :
                  Lower lvs. obovate, obtuse ;
                     upp. lanceol. irreg. toothed ;
                     pedunc. simple or forked ...   (31) **cernua.**
                  Lvs. subrosulate, obovate or
                     oblong-lanceol. undulated,
                     toothed ; pedunc. forked ;
                     cal. lobes ovate, acute   ...   (36) **arenaria.**
                  Lvs. crowded at base, and un-
                     dulately crenate, obovate
                     or lanceol. ; cal. lobes sub-
                     ulate, subinequal   ...   ...   (35) **nudicaulis.**
                  Lvs. rosulate, ovato-lanceo-
                     late, denticled ...   .   (38) **androsacea.**
               Perennials : caps. 3-celled
                  Caps. obovoid ; lvs. lanceol. mar-
                     gined, subentire   ...   .   (33) **Caledonica.**
                  Caps. globose ; lvs. lanceol. ser-
                     rate, or toothed   ...   .   (34) **Zeyheri.**
               Pedunc. and pedicels divaricate ; caps. 3-celled :
                  Caps. obovoid ; lvs. ovato-lanceol. leathery,
                     minutely denticled ; cal. tube as long as
                     lobes ...   .   (42) **Wyleyana.**
                  Caps. ovoid ; lvs. ovate, acute, membra-
                     nous ; cal. tube ½ as long as the lobes ...   (43) **patula.**
      Style with a tumid ring below the stigma :
         Caps. obconical.   Leaves oblongo-lanceolate :
            Lvs. denticled, hairy ; cor. 4 times longer
               than cal. ...   .   (37) **dichotoma.**
            Lvs. petioled, undulate-crenate, villous; cor.
               6 times as long as cal.   ...   .   (41) **Namaquana.**
         Caps. globose.   Lvs. lin.-lanceol. denticled, pu-
            bescent ; cor. 6 times as long as cal.   ...   (39) **annularis.**
         Caps. elongate, obovoid.   Lvs. lin.-lanceol. cre-
            nulate ; cor. 1½ as long as the cal.   ...   ...   (40) **pauciflora.**
C. Half-shrubs or shrubs.   Leaves alternate, or subopposite.   Caps. 3-celled.
   Half-shrub. Lvs. reflexed, lanceolate, acuminate, den-
      ticled. Panicle large   ...   .   (44) **adpressa.**
   Shrub. Lvs. erect or spreading, oblong, tapering to base,
      entire. Fl. axillary, solitary...   ...   .   (45) **robusta.**
   Half-shrub. Lvs. broadly ovate, acute, imbricate.   Spike
      leafy ...   .   (46) **epacridea.**

§ 1. Capsule 5-celled.   Seeds shining.

1. W. Capensis (A. DC. l. c. p. 136, t. 18) ; stem erect, herbaceous, simple or branched, hairy at the base ; leaves ovate-lanceolate, or lanceolate, pilose, irregularly toothed ; peduncles elongated, 1-flowered, glabrous ; tube of calx ovoid, very hairy ; lobes linear-lanceolate, ciliate, one half shorter than the 5-fid spreading corolla ; capsule obovoid, hairy. *Campanula Capensis, Linn. spec. p.* 240. *Thunb. ! Prod. p.* 39. *Herbar. ex parte. E. Z.!* 2369. *Mem. acad. Petersb.* 4, *p.* 374, *t.* 6, *f.* 3. *Bot. Mag. t.* 782. *Roella decurrens, Andr. bot. rep. t.* 238, *non l'Her. Camp. elongata, Willd. enum. h. Berol. suppl. p.* 10, *Wahlenb. elongata, Schrad ! cat. h. Goett.* 1814. *Herb. Unitin.* 146. *et* 147. *Zey.* 1073, 3135.

HAB. Grassy places in the Cape Flats, and in mountains near Capetown, Simon's-bay, Kampsbay, Rivierzonder Einde, in Zwartland, and near Tulbagh, Worcester. Nov.-Dec. (Herb. Thunb., Ð., Hook, Holm., Sd.)

Stem 1-1½ foot. Leaves often opposite in the lower part of the stem, about 1-2 inch. long. Flowers at first drooping, but, at length, nearly erect. Calyx covered with recurved white hair. Calyx lobes 3 lines long. Corolla bluish-green on the outside, dark blue inside at the bottom, but greenish at the origin of the lobes, spotted with black within the lobes at the recesses, the lobes violaceous. Capsule 4 lines long.

2. W. costata (A. DC. ! Prodr. p. 427); stem branched from the base ; branches erect, subpilose by short hairs ; leaves linear, narrow, acute, with thick, denticulate margins, the lower opposite, upper ones alternate ; raceme elongate ; peduncles shorter than the leaves ; calyx scabrous, tube hemispherical, shorter than the linear-acuminate, ciliolate lobes ; corolla cylindraceous, twice longer than the calyx ; capsule broadly turbinate, 5-costate or 5-angular.

HAB. Bergvalei, Novb., *Drege,* 6296. (Herb. D., Sd.)

An annual, 2-4 inches high. Stem and branches reddish-brown. Leaves glabrous, or a little hairy, 4-6 lines, the upper ones often 8 lines long, ½-¾ line wide. Peduncles in the axils of the leaves, from the base of the branches, 1-3-flowered, pedicels as long or shorter than the calyx. Corolla rarely 3 times longer than the calyx. Lower part of capsule 1 line long, valves erect, a little shorter.

3. W. divergens (A. DC. ! Prodr. p. 427); stem branched from the base ; branches divergent, scabrous-hairy ; hairs minute vesiculiform ; leaves linear, acute, glabrous, with thick, denticulated margins ; lower ones opposite, the upper alternate ; peduncles divaricate, dichotomous ; pedicels longer than the leaf ; calyx scabrous ; tube hemispherical, shorter than the linear-acuminate, denticulate lobes ; corolla cylindraceous, twice longer than the calyx ; capsule broadly turbinate, 5-costate.

HAB. Sandy hills near Ebenezar, Nov., *Drege,* 6299. (Also *W. ramulosa,* b. *in Herb., Sd.*) (Herb. D., Sd.)

Distinct by the much spreading branches and pedicels. Annual, 2-3 inches high ; branches opposite, upper ones alternate, ending in a divaricate racemose panicle. Leaves 3-6 lines long. Pedicels 2-4 lines long. Calyx lobes minutely toothed. Capsule 1 line long, the base twice longer than the valves, but nearly twice shorter than the calyx lobes.

4. W. decipiens (A. DC. ! l. c.); stem much-branched, erect, finely scabrous ; leaves opposite and alternate, linear, acuminate, subspinous, rigid, subdenticulate ; peduncles dichotomous, divaricate, disposed in a terminal racemose-panicle ; bracts a little shorter than the pedicels ;

calyx glabrous, with very short 5-gonous tube ; lobes reflexed, spreading, rigid, cuspidate, as long as the tube ; corolla tubulose, 5-lobed at
the apex, 6–8 times longer than the calyx ; capsule minute, 5-angular.

HAB. Boschkloof, Clanwilliam, Dec., *Drege*, 6310.   (Herb. Sd.)
Perennial, rigid, 3–4 inches high.   Leaves approximate on the lower part of the
stem and branches.   Leaves 2–3 lines long, $\frac{1}{4}$ line wide. Pedicels rigid, 3–4 times
longer than the calyx, spinelike after the capsules have fallen. Corolla 3 lines long,
with linear lobes, white. Capsule hemispherical, not 1 line long, 5-celled, rarely
3-celled ; valves as long as the base of the capsule.

5. W. oxyphylla (A. DC.! l. c.); stem much branched from the
base ; branches subsimple, densely hairy, much leaved, ending in a
divaricate, dichotomous panicle ; leaves opposite, verticillate and alternate, longer than the internodes, spreading, ovate-acuminate or lanceolate, rigid, subspinous, marginated, subdenticulate, rather canaliculate
above, hairy on both sides ; panicle finely scabrous, pedicels thick,
rigid, longer than the leafy bracts ; calyx hemispherical, tube shorter
than the subulate, entire lobes ; corolla tubulose, 5-lobed at the apex,
3-times longer than the calyx ; capsule 5-angular.

HAB. Mountains at Liliefontyn, Kamiesbergen, and near Kaus, *Drege*, 6313 ;
Namaqualand, *Rev. H. Whitehead.* Oct.–Nov.   (Herb. D., Sd.)
A perennial, very rigid herb, approaching *W. decipiens*, but distinguished by the
dense, broader leaves, large panicle, with a very short, greyish indument, and larger
calyx and fruit. Root thick, with very long branches. Stem or its primary divisions 3
inches, terminated by an equal or longer diffuse panicle. Leaves 4–6 lines long,
1 line wide, with a white, spinous point. Primary divisions of the panicle 1–$\frac{3}{4}$
inch long, the upper shorter. Calyx in flower 1–$\frac{1}{2}$ line long. Corolla white, with
short lobes. Capsule hemispherical, tube 1 line long ; lobes widely spreading, from
the same length, but twice longer than the valves of the capsule.

6. W. Ecklonii (Buek ! in E. Z. enum. 2379); glabrous or hairy at
the base ; branches erect or ascending, terminated by a dichotomous
panicle ; leaves aggregated on the lower part, alternate or opposite,
linear, acuminate, entire or subdenticulated on the incrassate margins ;
panicle few, or many flowered ; peduncles short or elongate, capillary,
glabrous or minutely downy ; tube of calyx hemispherical ; lobes
linear, setaceous, much longer than tube ; corolla tubulose, 5-lobed at
the apex, 2–3 times longer than the calyx ; capsule hemispherical or
subturbinate, 5-costate or 5-angular, *W. Swellendamensis, Buek! l. c.*
2383. *W. turbinata et saxifraga, DC. ! Prodr. p.* 427. *Drege*, 6313. *a.*
Camp. paniculata, Thunb. in Mem. acad. Petersb., IV. t. 7, *f.* 4, *excl.*
descript. p. 370, *et Herb. Thunb. fol.* β. 2, *Camp. paniculata, Herb. Ber-*
gius et Swartz. ex parte.

VAR. β **gracilis** ; panicle loose, pedicels elongate, *W. paniculata, E. Z.!* 2882.
W. oxyphylla ; var. β. *DC.! l. c., W. Dregeana, DC.! l. c. p.* 428.

VAR. γ. **pilosa** ; tube of calyx, pilose, *W. pentamera, DC. l. c.*

HAB. Rocky places, Winterhoeksberg, near Tulbagh, between Puspasvalley and
Kochmannskloof, *E. & Z.*, Swellendam, *W. H. Harvey* ; Gnadenthal, *Dr. Pappe ;*
Witsenberg, *Zey.* 1066. Rivier Zonder Einde, *Zey.* 3142 ; Piquetberg and Dutoitskloof, *Drege* ; var. β. Tulbagh, *E. & Z.*, Namaqualand, *Drege, Rev. W. Whitehead ;*
var. γ. Wupperthal, *v. Wurmb.* Oct.–Dec.   (Herb. Holm., Thunb., Hook, Sd.)
Habit of *Lightf. denticulata*, $\frac{1}{2}$–1 foot high, quite glabrous or rather hairy on the
lower and very leafy part of the branches. Leaves 4–8 lines long, $\frac{1}{2}$–1 line wide,

glabrous, or a little hairy. Peduncles commonly longer (2-4 times), rarely as long, or a little shorter than the calyx. Corolla 4–5 lines long. Capsule hemispherical, or roundish or broadly-turbinate, glabrous or minutely downy, 2–3 times shorter than the spreading lobes ; valves about equal to the base of capsule. It seems to be an annual plant.

7. W. debilis (Buek ! l. c. 2370); stem herbaceous, much branched from the base ; branches filiform, dichotomous, diffused, minutely hairy ; leaves alternate or opposite at the base, lanceolate, with incrassate, denticulated margins, glabrous ; peduncles elongate, axillary, and terminal, capillary, 1-flowered ; tube of calyx hemispherical, twice shorter than the setaceous lobes ; corolla tubulose, 5-lobed at the apex, nearly as long as the calyx ; capsule subglobose.

HAB. Sandy places in Zwartland, *E. & Z.*, Sept. (Herb. Sd.)

A small annual, 3–4 inches high, with opposite, lower, and alternate upper branches. Leaves 3–4 lines-long, 1 line wide, upper ones narrow, 2–1 line wide, spreading or recurved. Peduncles ½ inch long. Flower 1–1½ line. Style glabrous, as long as the corolla. Capsule, which I have not seen in a ripe state, is described by Dr. Buek as 5-celled, scarcely 1 line in diameter.

8. W. pilosa (Buek ! l. c. 2385); stem herbaceous, much branched from the base ; branches ascendent, hispid by white hairs ; leaves alternate, lanceolate, or linear-lanceolate, denticulated, hairy or subglabrous ; flowers panicled ; peduncles as long or longer than the calyx ; tube of caylx hemispherical, hispid, a little shorter than the linear lobes ; corolla infundibuliform, 5-lobed at the apex, 3-times longer than the calyx ; capsule spherical, 5-costate, pilose.

HAB. Sandy places near Olifants river and Brackfontein, Clanwilliam, *E. & Z.*, Sept. (Herb. D., Sd.)

Annual, ¼ foot to a span high, quite similar to *W. polyclada*, DC., from which it differs by smaller calyx-lobes and 5-costate, 5-valved, not terete, 3-valved capsule. Leaves 4–6 lines long, nearly 1 line wide. Flowering calyx 2 lines, corolla 5–6 lines long. Capsule with spreading recurved calyx lobes, equal in length, but a little longer than the valves.

§ 2. Capsule 3-celled, rarely 2-celled. (Sp. 9–46.)

9. W. flaccida (A. DC. Monog. p. 138); rather hairy ; stem erect, simple, rather leafy ; leaves opposite and alternate, ovate, acute entire; peduncle terminal, 1–2-flowered, leafy ; tube of calyx hairy, ovoid ; lobes lanceolate-acuminate, erect, equal ; corolla 5-lobed, one half longer than the calyx lobes ; capsule ovoid-spherical, st. angular. *DC. Prod. p. 428. Camp. flaccida, Masson. in Herb. Banks.*

HAB. Cape of Good Hope, *Masson.* Unknown to me.

Stem 3–4 inches high ; hairs short, rigid. Leaves sessile, 3–4 lines long, about 2 lines broad. Calyx about 3 lines long, 5-fid ; lobes flat. Capsule 2½ lines long, 10-nerved.

10. W. stellarioides (Cham.! in Linnæa, vol. 8, p. 196); glabrous or nearly so ; stems ascendent, subcæspitose, as well as the branches filiform, tetragonal ; leaves opposite, approximate at the base, ovato-lanceolate, or linear lanceolate, acute, with incrassate, revolute, quite entire margins, glabrous or hairy on the middle nerve beneath ; peduncles terminal or axillary, solitary, naked ; tube of calyx glabrous, obconical; lobes lanceolate-acuminate, erect; corolla infundibuliform, 5-fid, longer than the calyx-lobes; capsule obconical.

VAR. α. **angusta**; leaves smaller; pedicels longer than the flowers. *W. stellarioides, Cham. l. c. E. & Z.* 2387. *DC.! Prod. p.* 428.

VAR. β. **major**; leaves larger, pedicels shorter or as long as the flower. *W. integrifolia, DC.! l. c.*

HAB. Hills between Bosjesman and Karregarivier near Grahamstown, and on the Fishrivier, Albany, *E. & Z., Drege.* Caffraria, *Krebs.* 136. Capeflats, *Ecklon.* near Salem, *Zey.* 3115. Var. β. on the Keyriver, *Drege.* Albany, *Mrs. F. W. Barber.* Oct.–Jan. (Herb. Sd., D., Hook.)

Perennial, ½ foot and longer; branches glabrous, or a little hairy near the leaves. Internodes short at the base, but ½–1 inch long in the middle of the branches. Leaves sessile, lower ones more approximate and shorter, the upper about 4–5 lines long, nearly 1 line wide. Pedicels erect, ½–1 inch long. Calyx about 3 lines long. Corolla ½ or twice longer than the calyx, broad, 5-fid to the middle; lobes acute. Var. γ. is more robust, the leaves are somewhat longer (6–10 lines), more rigid; the flowers larger (6 lines long), on a shorter pedicel; the capsule the same. Specimens connecting both forms are collected by *Mrs. F. W. Barber.*

11. **W. spinulosa** (A. DC. Monog. p. 155); plant humble, much branched; branches stiff, glabrous, or rather scabrous; leaves narrow-linear, glabrous or pilose, denticulated, rigid, ending in a hard point; tube of calyx glabrous, spherical; lobes with revolute ciliated margins; corolla narrow, 5-lobed at the apex; capsule spherical. *DC. Prod. p. 429. Camp. spinulosa, Bank's Herb.*

HAB. Cape of Good Hope. (Unknown to me.)

Stem and branches diffuse, 1–2 inches high. Leaves opposite, in the upper part alternate, greyish, about 2–3 lines long. Flowers terminal and axillary, solitary; the pedicels 3–4 lines long, equal, spreading, rigid, and appearing like spines after the capsules have fallen. Calyx 5-fid. Corolla 3–4 lines long. Style at top 3-lobed. Capsule 1½ line long, opening by 3 valves. It comes near *W. (Lightfoot.) denticulata,* but is smaller, very much branched, and distinguished by shorter, ciliated leaves, and rigid, shorter and ciliated calyx lobes (DC.)

12. **W. oppositifolia** (A. DC. Prod. p. 429); glabrous or subpilose; stem ascending, flexuous; branches tetragonal; leaves opposite, ovate or ovato-lanceolate, subacute at both ends, mucronulate, denticulated; pedicels axillary, as long or shorter than the leaf; tube of calyx hispid, hemispherical, nearly as long as the lanceolate, erect lobes; corolla a little longer than the calyx; capsule subspherical.

VAR. β. **crispa**; (DC. l. c.) leaves crispate and whitish on the margins; pedicels as long or a little longer than the leaf.

HAB. Woods near Port Natal, *Drege, Dr. Krauss, Gueinzius,* 421. *J. Sanderson, Gerr. & M'K.* Albany, near the coast, *Mrs. F. Barber.* Var. β. on the same locality. Jan.–Feb. (Herb. Sd., D., Hook.)

An annual herb, 1–2 feet long, ascending, or the filiform branches procumbent. Leaves variable in size, in some specimens 4 lines long, 2 lines broad, in others nearly 1 inch long, 4 lines wide; the margins with obtuse but mucronulate teeth. Pedicels capillary. Calyx in flower, 1½–2 lines long. Corolla white. Capsule hairy, rarely glabrous, as long as the spreading calyx lobes. Valves very short.

13. **W. procumbens** (A. DC. Monog. p. 140, t. 15); glabrous or subpilose; stem creeping, branched; leaves opposite or verticillate, ovate, elliptic or obovate, obtuse, nearly entire; pedicels terminal and axil-

lary, twice longer than the leaves; tube of calyx ovoid, as long as the lobes, glabrous, or with short appressed hairs; corolla 3 times longer than the calyx; capsule subspherical. *Campanula procumbens, Thunb.! fl. cap. p.* 174, *Mem. Acad. Petersb.* 4, *p.* 375. *Herb. Un. Itin.* 154. *E. & Z.* 2371.

VAR. β. **diversifolia**; leaves of the branches ovate, acute, or ovato-lanceolate. *W. diversifolia, A. DC. l. c. p.* 139.

VAR γ. **foliosa** (A. DC. Prod. p. 429); stem and branches short, leaves smaller; pedicels often shorter.

HAB. Watery places through the whole colony; very common in the Cape flats, Jan.–April. (Herb. Thunb., Holm., Hook, D., Sd.)

Very similar to the preceding, but the whole plant is prostrate, the branches usually shorter, the leaves roundish, obtuse, 2–3 lines long, 2 lines wide, only in var. β. narrower and acute, obsoletely crenate, the pedicels 1–4 times longer and the corolla larger. The capsule the size of a small pea, glabrous, terminated by the lanceolate calyx-lobes.

14. W. montana (DC. Prodr. p. 430); caespitose, dwarf, much branched, branches short, villous, diffuse; leaves crowded, opposite, and alternate, obovate or oblong-lanceolate, acute, remotely denticulate, marginate, ciliate at the base; flowers solitary, terminal, shortly pedunculate; tube of calyx obconical, hispid; lobes glabrous, lanceolate-acuminate, marginate, entire or subdenticulate, as long as the tube; corolla infundibuliform, 2–3–times longer than the calyx.

VAR. β. **glabrata**; branches glabrous, leaves oblongo-lanceolate, sub-acuminate, scarcely ciliate at the base.

HAB. Mount Katberg, 5–6000 ft., *Drege;* var β. Draakensteenberg, *T. Cooper.* Nov. (Herb. Sd., Hook, D.)

A perennial, resembling the Alpine *Edraianthus*, prostrate, the twigs very numerous, short, leafy. Leaves angustate at the base, 4–6 lines long, 1½–2 lines wide, glabrous, except the base. Peduncle 1–3 lines long, hairy. Corolla ½ inch long, 5-lobed at the apex. Capsule crowned by the erect calyx-lobes.

15. W. inconspicua (A. DC. Prodr., p. 430); dwarf, glabrous; stem erect, terete, branched at the base; leaves erect, alternate, linear, denticulate, lower ones opposite; flowers terminal and lateral, sessile; tube of calyx obconical, lobes linear, denticulate, acute, longer than the tube; corolla tubulose, 2–3 times longer than the calyx-lobes; capsule turbinate.

HAB. Cape of Good Hope, *Burmann.* (Unknown to me.)

Root annual, simple, 2 inches long. Stem rigid, thickish, 2 inches high. Leaves very narrow, 4–5 lines long. Flowers few, 3 lines long. Valves of the capsule 3, a little shorter than the calyx-lobes.

16. W. acaulis (E. Mey.! in Herb. Drege); root elongate, simple; stem or branches very short, condensate; leaves petiolate, hairy, lower ones obovate, obtuse, subdentate, angustate in a longer ciliate petiole; the others lanceolate or linear-lanceolate, dentate, attenuate in a shorter petiole; flowers sessile, terminal and lateral; tube of calyx obovoid, terete, hairy, with nearly glabrous lanceolate, acute lobes; corolla infundibuliform, 5-lobed at the apex, 3 times longer than the calyx lobes. *DC. Prodr. p.* 430.

HAB. Rocky places near Zilverfontein, 2000 ft. ! *Drege*. (Herb. D., Sd.)

The annual root, several inches, the stems or branches ½–1 inch long. Larger leaves 3–6 lines long, 1–1½ line wide, the petiole as long or somewhat longer ; upper or interior leaves smaller and narrower, sometimes nearly linear. Flowers 3 lines long, lobes of corolla short, acute. All Drege's specimens are in an insufficient state.

17. W. tenuis (A. DC. ! Prodr. p. 440) ; stems erect, as well as the filiform branches shortly hairy, leafy ; leaves sessile, alternate or the lower ones opposite, linear-lanceolate or linear, entire, flat ; pedicels solitary, capillary, hairy, as long or somewhat longer than the leaf ; calyx tube obconical, hispid, shorter than the linear-acuminate, hairy lobes ; corolla infundibuliform, 5-lobed at the apex, 3–4 times longer than the calycine lobes ; ovary 3-celled.

HAB. Little Namaqualand, *Drege*, 6293. (Herb. Sd.)

A small annual, with elongate, simple roots. Stems 4–6 inches long, the middle erect, lateral ones ascending ; branches short. Leaves scattered from the base to the apex, nearly glabrous, 4–6 lines long, ¾–1 line wide. Pedicels 3–4 lines long, sub-terminal on the upper branches. Calyx 1 line long. Corolla blue. Style with 3-linear, revolute stigmas. Ripe capsules are wanting. I examined 7 flowers, but never found a 2-lobed stigma, as indicated by Decandolle.

18. W. ramulosa (E. Mey.! in Herb. Drege) ; stem erect, much branched, branches terete, filiform, minutely downy at the base, glabrous and dichotomous at the apex ; leaves sessile, linear, acute, with revolute margins, entire or subdenticulated, the lowest opposite ; pedicels of the dichotomous panicle capillary, much longer than the opposite bractea ; calyx glabrous, 5–costate, hemispherical or acute at the base ; lobes subulate, as long as the tube ; corolla 3–4 times longer than the calyx ; capsule roundish, 5-costate. *DC. Prodr. p.* 432.

HAB. Between Bergrivier and Nieuwekloof, and on Piquetberg. *Drege ;* Hassaquaskloof and on Breederiviersport, *Zey.* 3141 ; Steendaal, Tulbagh, *Dr. Pappe.* Oct.–Nov. (Herb. Sd., Hook, D.)

Annual, 3–4 inches high, very slender. Leaves in some specimens 2–3. in others 3–4 lines long, ½-line wide, glabrous or minutely rough. Pedicels 4–12 lines long, glabrous. Calyx in flower 1 line long. Corolla 5-lobed at the apex, 2–3 lines long. Capsule 1 line long, with filiform, erect calyx lobes, a little longer than the valves.

19. W. Banksiana (A. DC. Monog., p. 154) ; stem erect, paniculate, glabrous ; leaves erect, linear, very narrow, remotely denticulated ; panicle spreading ; pedicels erect-spreading, 2–3 times longer than the bractea ; calyx glabrous ; tube hemispherical, nearly as long as the linear lobes ; corolla infundibuliform, 5-lobed at the apex, twice longer than the calyx ; capsule hemispherical. *DC. Prod. p.* 430. *Camp. capillacea. Herb. Holm. ex pte.*

VAR. β. ? *Zey.* ; stem leafless above the middle ; panicle few-flowered ; calyx-lobes ciliate.

HAB. Cape. *Thunberg, Wahlberg.* VAR. β. Crocodile River, *Burke & Zey.,* 1070. Nov. (Herb. Holm., Sd., Hook.)

Perennial ? ½–1 foot high. Stems several from the simple root ; branched from the middle. Leaves alternate, erect or subappressed, 3–4 lines long, scarcely ½ line wide, acute or acuminate. Panicle ending each of the branches, racemose, 6–12-flowered. Pedicels 2–6 lines long. Flowering calyx 1½–2 lines long. Corolla 3 lines long, with short, acute lobes. Capsule 5-costate, 1 line long ; the valves nearly as long as the spreading calyx-lobes. Var. β may be a different species, it is only beset with alternate leaves on the lower part of the stem, which is a little hairy near the base, and 2–3 times forked in the upper parts ; the leaves are quite the same, 3–5 lines long ; calyx, corolla and capsule the same size.

20. W. paniculata (A. DC. Monog. p. 153) ; stem much branched, hairy ; leaves linear-lanceolate or lanceolate-acuminate, entire or sub-denticulated, with incrassate margins ; peduncles dichotomous, longer than the bractea and calyx, paniculate ; tube of calyx hairy, obovate or turbinate, shorter than the linear, acuminate entire lobes ; corolla infundibuliform, 5-lobed at the apex, twice longer than the calyx ; capsule turbinate or obovate.

VAR. *a.* **Thunbergii** ; stem with short indument ; branches subflexuous, spreading, leaves lanceolate-acuminate ; pedicels longish, forming a remotiflowered panicle. *Campanula paniculata, Linn. fil! suppl., p.* 139 *in Herbar. Alstroemer. & Bergius. Thunb.! Prod. p.* 38. *Flor. Cap. p.* 172. *Herbar. fol.* γ. *Mem. acad. Petersb. IV., p.* 370 ; *description, but not the figure, t.* 7, *f.* 4. *Wahl. rudis, a, E. Mey. in Herb. Drege, partim.*

VAR. β. **rudis** ; branches very numerous, often filiform, pilose or pubescent ; leaves lanceolate, acuminate ; pedicels numerous, shorter than in var. *a*, forming an irregular, many-flowered panicle. *Camp. paniculata Thunb., herb. fol.* β. 1. *Wahlenb. hispidula, E. Z.!* 2384, *excl. syn. W. rudis. E. Mey. b. W. divaricata, E. Mey. a. W. Dunantii A. DC. Monog. p.* 152, *E. Z.!* 2384 (panicle more regular).

VAR. γ. **Massonii** ; plant small ; leaves smaller, linear ; calyx pubescent or hispid, hemispherical ; capsule broadly cuneate or hemispherical. *Camp. paniculata Thunb. herb., fol. a. C. linearis, herb. Swartz., non Thunb., Wahlenb. Massonii, A.DC. l. c. p.* 153, *E. Z.!* 2381. *W. rudis, E. Mey. c.*

VAR. δ. **glabrata** ; stem and leaves nearly glabrous, calyx minutely rough or with a few appressed, white hairs.

HAB. var. *a.* Cape Flats, *Ecklon,* near Devilsmountain, *Drege ;* var. β. near Table Mountain, Tulbagh and in Zwartland, *E. & Z.,* Langevallei and Olifantsriver, *Drege ;* Prekstael, *Zey.* 1076 ; var. γ. Bergvallei, Clanwilliam, *E. & Z. ;* Ebenezer, *Drege ;* Vygekraal, *Zey.* 3410 ; var. δ. near Capetown, *W. H. Harvey, Zey., Wallich.* Sept.–Nov. Herb. Thunb., Holm., Sd., Hook, D.)

Annual, very variable in size, from 3-inches to 1 foot high, generally branched from the base only, simple to the middle. Branches erect or spreading. Leaves scattered, 4–5 lines long, about 1-line wide, often smaller, 2–3 lines long and nearly linear, hairy or subglabrous, often quite entire. Peduncles 2–6 ; but in var. *a.* 6–10 lines long, and more rigid. Calyx 2–2½ lines long, tube usually cuneate, but sometimes hemispherical in the same specimens, pubescent or hispid, the lobes very narrow, erect. Corolla 3–4 lines long ; in var. γ. with a violet or blue tube. Capsule 1½–2 lines long ; the valves short.

21. W. polyclada (A. DC. ! Prod. p. 789) ; whole plant spreading, hairy ; stem branched from the base, branches divergent, subangular above ; leaves scattered, ovate-acute or ovate-lanceolate, with thick, subrevolute, denticulate margins ; flowers panicled ; pedicels shorter than the bractea or the calyx ; tube of calyx hemispherical, shorter than the acuminate lobes ; corolla infundibuliform, 5-lobed at the apex, 4-times longer than the calyx ; capsule hemispherical, *W. pilosa, DC. Prodr. p.* 431.

HAB. Sandy places near Ebenezar, *Drege,* Novb. (Herb., Hook, Sd.)
Except the larger leaves, shorter pedicels and roundish capsule, it is not to be distinguished from *W. paniculata, var.* β. *rudis.* Root annual. Branches 2–4 inches long, 3–4 lines long, leaves alternate. 1 line wide. Panicle leafy. Pedicels

1–2 lines long. Calyx-lobes erect, a little broader than in *W. rudis.* Capsule hairy, the valves twice shorter than the calyx lobes.

22. **W. prostrata** (A. DC. ! Prodr. p. 431); branches divergent from the base, simple, prostrate, terete, nearly glabrous; leaves scattered, alternate and opposite, thickish, linear, incurved, channelled above, acute, denticulated, teeth terminated by a minute bifid mucro, panicle 3–5 flowered; pedicels as long or longer than the bractea or the calyx; tube of calyx turbinate, hairy or subglabrous; a little shorter than the ovate, acute, ciliate lobes; corolla 6-times longer than the calyx, infundibuliform-campanulate, 5-lobed at the apex; capsule hemispherical, glabrous, *W.? prostrata, E. Mey.*

HAB. Between Zilverfontein, Kooperbergen and Kaus, Namaqualand, 2–3000 ft. *Drege,* Sept.–Oct. (Herb. Hook, Sd.)
Annual. Root 3–4 inches long, white, simple. Stems or branches scarcely longer than the root. Leaves sessile, 3–4 lines long, 1 line wide, spreading-recurved, rather fleshy, on each side with 6–8 minute bispinous teeth, glabrous or subpilose near the base. Pedicels 1–2 lines long, minutely hairy. Calyx 1 line long. Corolla nearly half an inch long, the lobes obtuse or apiculate. Stamens and trifid style shorter than the corolla. Capsule crowned by the spreading recurved calyx-lobes.

23. **W. ingrata** (A. DC. ! Prodr. p. 432); branches ascendent from the base, glabrous, simple, ending in a 7–12-flowered panicle; leaves scattered, alternate and opposite, thickish, linear, channelled above, acute, denticulated; teeth terminated by a minute, bifid mucro; pedicels as long or shorter than the bractea or the calyx; tube of calyx hemispherical, hairy, nearly as long as the lanceolate denticulate, ciliate lobes; corolla 4–5 times longer than the calyx, tubulose, 5-lobed at the apex; capsule hemispherical, hairy. *W. divaricata, b., E. Mey. in herb. Drege (ex DC.) non litt. a.*

HAB. Sandy plains near the mouth of the Gariep, *Drege.* Oct. (Herb. Sd.)
Scarcely to be separated from the preceding, with which it has the long, white roots, simple stems or branches and leaves. The difference lies only in the more roundish calyx, with spreading hairs, and the narrow, not campanulate corolla. Leaves 4–6 lines long, subpilose near the base. Pedicels glabrous. Capsule 1 line long, equalling the recurved-spreading calyx-lobes and the valves of capsule.

24. **W. oocarpa** (Sond.); stems erect, subpilose; branches filiform, paniculated; leaves alternate, erect, linear, narrow, entire or indistinctly denticulate, glabrous or subpilose; panicle dichotomous; pedicels erect-spreading, capillary, 3–4 times longer than the bractea; calyx tube oval, glabrous; lobes subulate, twice shorter than the tube; corolla tubulose, 5-lobed at the apex, 2–3 times longer than the calyx; capsule ovoid, terete.

HAB. Sandhills on the Zwartkopsriver, *Zey.* 3140. Octob. (Herb. Sd.)
A slender annual, 4–6 inches high. Root longish, simple. Stem solitary, or 2–4, often flexuous, dichotomously branched. Leaves 3–4 lines long, ½ line wide, acute, sessile; the lowermost opposite. Pedicels 4–6 lines long, glabrous. Flowers small, calyx in flower 1 line, corolla about 2 lines long. Capsule 2 lines long, 3-, rarely 4-valved; valves equaling the short, erect calyx lybes.

25. **W. exilis** (A. DC.! Monog. p. 151, t. xvi.); stem much-branched, subpilose at the base; branches glabrous, filiform; leaves scattered,

linear-acuminate, entire, with revolute margins, subpilose; pedicels capillary, opposite the leaf, often paniculate at top of the branches; tube of calyx glabrous or minutely rough, obconical, lobes linear, acute, erect, shorter than the tube; corolla tubular, 5-cleft at the apex, one half longer than the calyx; capsule obovoid, narrow. *DC. Prodr. p.* 432, *E. Z.! 2377. W. claviculata, E. Mey. in herb. Drege.*

VAR. β. **major**; plant stronger, stems numerous from the roots, leaves broader, pedicels more distant, not panicled.

HAB. Sandy places near Capetown, *E. & Z., W. H. Harvey;* Paarl and Paarlberg, Bergriver, Dutoitskoof and Gnadenthal. *Drege;* Stellenbosch, *Ecklon;* Kampsbay and near Caledon, *Zey.* 3143; Tulbagh, *Dr. Pappe;* var. β. Cape flats, *Ecklon;* hills near Rivier Zonder Einde, *Zey.* 3142. Nov. (Herb. Hook, D., Sd.)

Annual, very slender, and commonly much branched, 2–4; var. β. 6–12 ins. high. Leaves 4–8 lines long, ½–1 line wide, alternate, or opposite at the base of the stem. Pedicels numerous, 4–6 lines long, reddish, glabrous, erect or ascending. Calyx in flower 1 line, in fruit oblong-cuneate 2 lines long, terete, striate, crowned by the very short calyx-lobes, which are longer than the valves of capsule. The corolla is scarcely longer than a line. Var β. has the habit and size of *W. paniculata*, A. DC. from which it differs by the leaves, flowers and narrow, longish capsule.

26. W. Caffra (A. DC. Prod. p. 432); stems erect, subpilose, simple at the base, branched above, glabrous; leaves scattered, alternate, opposite or ternate, numerous, linear-acuminate, serrulate, glabrous; panicle few-flowered, coarctate; pedicels as long as the bractea; tube of calyx glabrous, obovoid, terete; lobes linear-acuminate, spreading at the apex, reflexed on the margins, longer than the tube; corolla 5-fid, tubulose, one half longer than the calycine-lobes; capsule hemispherical, 10-nerved. *Lightfootia,* 6307, *Hb. Drege, fide, A. DC.*

HAB. Caffraria, *Drege.*
Root elongate, simple. Stems 3–4, leafy, ½ foot high. Leaves 3–4 lines long, 1 line wide. Flowers small. Capsule 3-locular, valves acute, a little shorter than the calycine-lobes. I have not seen any authentic specimen. Drege's, 6307, as quoted by Decandolle, is *L. denticulata* in herb. Sond.

27. W. Bowkeriæ (Sond.); quite glabrous, stem erect, few-branched, terete; leaves scattered, alternate, linear, acute or apiculate, denticulate on the incrassate margins; flowers terminal, 1–3, shortly pedicellate; tube of calyx hemispherical, shorter than the lanceolate denticulate lobes; corolla infundibuliform, 2–3 times longer than the calyx, deeply 5-lobed; capsule hemispherical, 10-nerved.

HAB. Somerset, *Miss Bowker.* (Herb. Hook.)
The herbaceous (annual?) stems 1 foot or more high, leafy from the base to the flowers. Leaves 4–6 lines long, 1 line wide, erect, acute or obtuse, with a minute mucro; the margins whitish, with 3–6 very small teeth. Peduncles 2–4 lines long, with 1 or 2 minute bracts. Calyx 1½ line long. Corolla with ovate, acute lobes. capsule with erect calycine lobes, longer than the valves.

28. W. capillacea (A. DC.! Monog. p. 156); stems erect, simple, rarely branched, glabrous, but minutely hairy at the base; leaves numerous, alternate, fasciculate, linear-filiform, entire, with revolute margins; panicle terminal, leafless; pedicels filiform, 2–6 times longer than the bractea; tube of calyx hemispherical, lobes linear-acuminate longer than the tube; corolla infundibuliform, 3–4 times longer than the calyx, 5-lobed at the apex; capsule hemispherical, *E. & Z.* 2386. *Campanula capillacea, Thunb.! Prod. p.* 38, *Mem. Acad. Petersb. T. IV. p. 366, t. 5, f.* 4, *Fl. Cap. p.* 170, *excl. syn., Willd.*

HAB. Stony places amongst grasses on the Van Stadesmountains, Uitenhage, Zuureberge, and near Grahamstown, Albany, *E. & Z., Dreje, Sir C. Bunbury, Miss Bowker, Zey.* 3144. May–Jul. (Herb. Thunb. ! Holm., Hook., D., Sd.)

A perennial herb (Zey.), 1-1½ foot. Leaves approximate on the lower and the middle part of the stems, not at the base, 3–6 lines long, nearly terete, incurved or a little hooked at the apex, with many smaller ones in the axils ; glabrous, rarely pubescent. Racemose panicle 2–4 inches long. Flowers ½ inch long, blue. Capsule 1 line long, crowned by erect calycine lobes.

29. W. denudata (A. DC. Monog. p. 147); quite glabrous; branches ascendent, slender, few-flowered; leaves linear-lanceolate, with undulated, whitish, nearly entire margins; peduncles elongated, nearly naked; tube of calyx obconical, elongate ; lobes linear-acuminate, entire, glabrous, as long as the tube; capsule elongated, pyriform, 3-celled. *DC. Prod. p.* 435.

HAB. Cape of Good Hope, *Burchell,* 2769. (Unknown to me.)

Habit of *W. Capensis,* perhaps annual. Stem 1 foot high, leafy at the base. Leaves nearly as in *W. lobelioides,* 6 lines long, 1 line broad, greyish green, with white edges. Flower solitary, terminal, with a small axillary alabastrum in the specimen collected by Burchell. Calyx-tube in flower about 2 lines long, the lobes 2–3 lines long, with revolute margins. Corolla unknown. Capsule 4 lines long, scarcely 1½ line broad, probably 3-valved. Seems only a variety of *W. cernua.*

30. W. Meyeri (DC. Prod. p. 439); glabrous, hairy below ; stem branched from the base ; branches as well as the stem erect, naked, dichotomous ; leaves aggregate or subrosulate at the base, obovate or obverse-lanceolate, acute, with undulate whitish margins, membranaceous; bracteæ linear-acuminate, minute, entire ; peduncles dichotomous, capillary; tube of calyx obovoid, as long as the ovate acute lobes; corolla infundibuliform, 5-lobed at the apex, 3 times longer than the calyx-lobes; capsule obovoid, 2-celled. *W. cernua, E. Meyer in Herb. Drege, non. A. DC.*

HAB. Rocks on the Kamiesbergen near Leliefontein, Ezelskop and Roodeberg, 3–5000 ft. *Drege.* Nov. (Herb. D., Sd.)

A very slender annual, with the habit of *W. androsacea.* Stem filiform, ½-1 foot high ; branches generally smaller, the outer ascendent. Leaves 1–1½ inch long, 3–5 lines wide, ciliate, hairy at the angustate base, on the surfaces with few white hairs ; bracteæ about 1 line long. Pedicels at flowering time 1–3 lines, fruit bearing 4–6 lines long. Calyx not 1 line, corolla 2–2½ lines long. Capsule striate, 1 line long, terminated by the very minute calyx-teeth.

31. W. cernua (A. DC. Monog. p. 148); branched from the hairy base, glabrous above; branches erect, slender; leaves irregularly toothed; lower ones opposite, obovate or ovate-obtuse, middle ones alternate, ovate-lanceolate, acute, angustate at the base ; bracteæ minute, linear ; peduncles simple or dichotomous ; tube of calyx glabrous, obovoid, as long as the lanceolate, ciliolate lobes ; corolla infundibuliform, sub-5-fid, 3 times longer than the calyx ; capsule obovoid, 3-celled. *Campanula cernua, Thunb.! Prod. p.* 39. *Mem. Acad. Petersb. t.* 4, *p.* 374, *t.* 7, *f.* 1. *Herb. Un. Itin.* 148, *Sieb. cap.* 243. *E. & Z.* 2372. *W. gracilis, E. Mey. in Herb. Drege, non A. DC. Zeyh.* 1074, 3136, 3139.

VAR. β. **minor**; branches filiform, the lateral ascending; lower leaves petiolato-attenuated; flowers and capsule smaller.   *W. ciliolata, A. DC. Prod. p.* 436, *ex pte. W. variabilis, var. pumila, E. Mey. Drege,* 6286.

VAR. γ. erect, leaves ovate-lanceolate or lanceolate, often margined, capsule elongate, pyriform, *W. ciliolata, A. DC. l. c. ex pte, Drege,* 6282.

HAB. Sandy stony places on mountains near Capetown, in Caledon, and in Tulbaghskloof. Var. β. Dutoitskloof and Paarlberg, *Drege.* Var. γ. Paarlberg, *Drege.* Tulbagh, *Dr. Pappe.* Hassaquaskloof, *Zeyher.* Port Natal, *Drege, Ger. & M'K.,* 1501. Oct.–Nov.   (Herb. Thunb., Holm., D., Hk., Sd.)

Very variable in habit and size; small specimens with ascendent, filiform branches 2–4 inches long, the largest more erect, 1½–2 feet high, more or less branched. Leaves numerous at the base, but not rosulate, 4 lines to 1 inch long, 2–5 lines broad, the middle and upper ones denticulate or nearly entire. Peduncles much elongated, 1-flowered, filiform. Calyx 1–2 lines long. Corolla drooping in the bud state, white at the base; the lobes blue. Capsule in var. α. 2–3 lines long; in var. γ. 4 lines long, 1½–2 lines broad.

32. W. undulata (A. DC. Monog. p. 148); stem ascendent or erect, branched, hairy at the base, naked and glabrous at the apex; leaves sessile or somewhat clasping, alternate, rarely opposite, ovato-oblong or ovato-lanceolate, coarsely crenated or toothed, sinuated and undulated, marginate; peduncles elongated, few flowered, rarely repetito-dichotomous; calyx glabrous; tube obconical; lobes ovate, acute, or acuminate, ciliolate; corolla large, 5-fid, infundibuliform, 3–4 times longer than the calyx; capsule obconical, 2 or 3-celled, striate. *Campanula undulata, Linn. fil.! Suppl. p.* 142. *Thunb. Prod. p.* 39, *fl. Cap. p.* 173. *E. & Z.* 2373. *Wahlenbergia, Cham.! in Linnæa* 8, *p.* 194. *W. Chamissoniana, G. Don. Gen. Syst.* 3, *p.* 740. *DC. Prod. p.* 439. *W. bilocularis, A. DC.! Prod. p.* 439. *Zeyh.* 1068, 3137. *Drege,* 4003, 6280, 6281.

VAR. β. **glabrata**; glabrous, or the leaves hairy on the middle, now beneath. *W. striata, A. DC. l. c. Krebs! Herb. Reg. Berol.* 133.

VAR. γ. **macrantha**; (Cham. l. c.) flower nearly an inch broad.

HAB. Zwartland, near Bergriver, *Thunb.* Tulbagh, *Zeyh.* District of George, *Dr. Pappe, Mundt.* Albany, *E. & Z., Drege, H. Hutton.* Uitenhage, *E. & Z.* Orange Free State, district of Albert. *T. Cooper,* Caffraria, *Krebs. Drege,* to Port Natal, *Dr. Grant, Williamson, Plant, Gueinzius. Gerrard,* 871. *Sanderson,* 102, 380, 458. Sept.–Jan.   (Herb. Thunb., Holm., Hk., Dr. Berol. Sd.)

A rigid annual, well known from all the other species by the undulate, crenated, coriaceous, margined leaves. Stem ½ 1 feet high, subangulate by the decurrent leaves; much leafy from the base, and generally clothed with white spreading hairs; branches as long, or often much longer than the stem, naked or leafy at the base, terminated by the quite naked, simple, or near the apex, 2–3-times forked, peduncles, each of which bears 2–5 pedicellate flowers. Leaves spreading, about 1 inch long, 3–4 lines wide, larger or smaller in different specimens, hairy, especially in the young, not flowering branches, but sometimes quite glabrous; upper ones gradually smaller and narrower, those at the base of the peduncles 3–1 line long, entire. Pedicels in flower short, in fruit ½–1 inch long, glabrous as the peduncles and flowers. Calyx variable in size, 1–3 lines long; lobes mostly broad, acute, shorter than tube, but sometimes as long or a little longer, acuminate. Corolla blue, ½–1 inch in diameter, with acute lobes. Style with two glands below the stigma. Capsule 4–6 lines long, 10-nerved, crowned by the erect short calycine lobes. Valves 2 or 3 in the same peduncle. Seeds oblong or ovoid. It varies also, but rarely with 1 flowered, much elongate peduncles, and flattish, but always distinctly undulate leaves.

33. W. Caledonica (Sond.); stem erect, branched, leafy, hairy, naked and glabrous at the apex; leaves sessile, alternate, lanceolate or linear-

lanceolate, marginate, quite entire or obsoletely denticulate, coriaceous; peduncles elongated, dichotomous, pedicels longish, 1-flowered;
tube of calyx hemispherical, glabrous ; lobes ovate, lanceolate, ciliolate;
corolla infundibuliform, 5-fid, twice longer than the calyx-lobes ; capsule obovoid.

HAB. Caledon river, *Burke & Zey.* 1069. Dec. (Herb Hook, D., Sd.)
Perennial (Zey.), 2-3 feet high, near the preceding, but stem and branches more
slender, leaves smaller, not undulated nor crenated, and calyx-tube larger. Stems
with spreading hairs. Leaves scattered, not much longer than the internodes, ¾ inch
long, 1½-2 lines wide, with 4-8 very minute teeth on the thickened margins ; upper
ones very small and linear. Peduncles 3-4 times forked at the apex. Pedicels
1 inch long or longer. Calyx 1-1½ line long, lobes broad at the base. Corolla
smaller than in smallest-flowered forms of *W. undulata.* Capsule 2 lines long, obovoid or nearly hemispherical ; the valves short.

34. W. Zeyheri (E. Z.! Enum, 2374); stem erect, simple or branched,
hairy or glabrous above ; leaves alternate, sessile, ovate-lanceolate or
lanceolate, lower ones angustate at the base, petiolate, marginate, dentate or serrate, acute; peduncles elongate, leafless, simple or dichotomous, few-flowered; pedicels filiform ; tube of calyx hemispherical,
lobes lanceolate-acuminate, twice larger than the tube ; corolla infundibuliform, twice larger than the calyx ; capsule globose. *W. variabilis, E. Meyer, in herb. Drege, excl. var. pumila. DC. Prodr. p.* 436.

VAR. β. **Krebsiana** (DC.! l. c.); hairy all over or glabrous above; stem subsimple ;
leaves oblong-cuneate or oblong-lanceolate, acute, coarsely dentate, glabrous or hairy
beneath. *W. Krebsii, Cham.! in Linnæa* 8, p. 195.

VAR. γ. **linearis** ; glabrous or hairy near the base ; stem erect, slender: leaves
scattered, linear, ½-¾ inch long, 1 line wide, ciliate-serrate ; peduncles elongate.

VAR. δ. **Natalensis** ; stem very leafy, hairy ; leaves elongated, linear-lanceolate,
sharply serrate, hairy beneath. 2 inches long, 2 lines wide ; peduncles elongate, glabrous, dichotomously panicled.

HAB. Mount Winterberg, near Philipstown, *E. & Z. ;* Mount Katberg, *Drege ;*
district of Albert, *T. Cooper;* var. β. in Caffraria, *Krebs.* 134, *T. Cooper;* Kreili's
Country, *H. Bowker;* Transkei, *Mrs. F. W. Barber;* Albany, *Col. Bolton, H.
Hutton, T. Williamson;* var. γ. Katberg, *Zey.;* var δ. Port Natal.
Sanderson, W. F. Gerrard, 785. Oct.–Feb. (Herb. D., Hook, Sd.)
Perennial herb. Small specimens, 3-4 inches high, mostly simple, larger ones
1-1½ foot, more or less, but not much branched. Leaves in var. *a. et β.* more
numerous near the base ; the lowest much attenuated to the base, about 1 inch long,
3-4 lines broad ; in other specimens smaller. Middle leaves 1-½ inch long, 2-1 line
wide. In var. γ. the lower leaves are not broader than the upper ones. Peduncles
terete, varying with 3-flowered to a dichotomously many-flowered loose panicle.
Pedicels short, or 1 inch long. Calyx-tube 1 line, lobes 2 lines long. Corolla blue,
½ inch long and broad, with acute lobes. Style without glands near the stigma.
Capsule the size of a small pea.

35. W. nudicaulis (A. DC. Monog. p. 149); stems erect, branched,
hairy at the base, glabrous and nearly naked above; leaves approximate at the base, obovate or lanceolate, undulately crenated, subpilose,
whitish on the margins, peduncles dichotomous; calyx glabrous, with
a nearly spherical tube, lobes subulate, subunequal, entire, as long as
the tube; corolla broadly funnelshaped, deeply 5-fid, 3 or 4 times
longer than the calycine lobes; capsule spherical. *Camp. diffusa,
herb.; Banks.*

Var. β. flowers larger. *Camp. elongata, herb. Banks.*

Hab. Cape of Good Hope. (Unknown to me.)

Root cylindrical, whitish. Stems very numerous, 1, in var. β. 1½ foot high, often dichotomous. Leaves nearly all at the base of the plant, obtuse or acute, 1-2 inches long, 4-6 lines broad. Flowers very numerous, terminating the slender, naked pedicels. Tube of calyx scarcely 1 line long. Corolla at top about 4 lines broad. Capsule 2 lines long, glabrous, 3-valved. It is larger than *W. androsacea,* and differs by the diffuse branches, DC. According to the description it must be very near *W. arenaria.*

36. W. arenaria (A. DC.! Prodr. p. 436); stems simple or branched from the hairy base; branches naked, dichotomous and glabrous above; leaves rosulate or approximate at the base, obovate or oblong-lanceolate, acute or acuminate, irregularly toothed, glabrous or hairy; bracteæ minute, acuminate; calyx glabrous, tube obovoid, about as long as the lobes; corolla infundibuliform, 5-fid, 2-3 times longer than the calyx; capsule obovoid or subspherical, 3-celled. *W. androsacea, E. Z.! 2376. W. nudicaulis, E. Z.! 2375, Zey. 3138.*

Var. β. **Verreauxii**; corolla one half or twice longer than the calyx, *W. Verreauxii, A. DC.! l. c. W. semiglabra.* DC.! l. c. p. 437. *W. nudicaulis, E. Z. ex pte.*

Var. γ. **sphaerocarpa**; plant dwarf; corolla one half longer than the calyx; capsule spherical. *W. sphaerocarpa, A. DC.! l. c.*

Var. δ. **lasiocarpa**; calyx tube and capsule ellipsoid, densely beset with short hairs; corolla one half longer than the calyx-lobes.

Hab. Sandy places, var. α. β. et γ., near Capetown, in Zwartland, Caffraria, Namaqualand, and in the districts of Albany, Uitenhage, to Port Natal; var. δ. in Namaqualand, *Drege.* Sept.-Nov. (Herb. Hook, D., Sd.)

Annual, 3 inches to 1 foot and more high, variable in habit. Stem or branches erect, hairy at the lower part, rarely to the middle. Leaves in some specimens all rosulate, in others radical, and on the lower part of the stem 1-2, sometimes 3 inches long, 3-4 lines broad, more or less petiolated, crenulate. dentate, rarely pinnatifid-toothed at the base, hairy on both sides, or glabrous above or quite glabrous. Peduncle in the small specimen simple, few-flowered, in the largest 4-5 times forked; pedicels filiform, ½-2 inches long. Calyx 1-1¼ line long, lobes ovate acute or lanceolate-acuminate. Corolla blue, with ovate, acute lobes. Style with 4 glands at the base of the stigma. Capsule 2-3 lines long, 1½-2 lines broad, sometimes only 1 line long and broad, crowned by the erect calycine-lobes, which are generally very short in var. β., a little longer in var. δ, as long as the capsule. The small specimens with rosulate leaves are scarcely to be distinguished from *W. androsacea,* DC., with which *W. nudicaulis* and *W. arenaria* perhaps must be united.

37. W. dichotoma (A. DC.! Prodr. p. 437); stems many from the base, erect, dichotomous, glabrous, or a little hairy near the base, nearly naked; leaves approximate or subrosulate at the base, oblongo-lanceolate or lanceolate, or sublinear, shortly petiolate, denticulated, hairy; bracteæ minute, glabrous; peduncles dichotomous; calyx glabrous, tube obconical, as long as the erect linear lobes; corolla 5-fid, 4 times longer than the calyx; capsule obconical, 3, rarely 2-celled.

Var. β. **simplex**; stem simple, slender; leaves densely hairy. *W. parviflora, A. DC. l. c.*

Hab. Hills near Ebenezar, *Drege,* 6285; var. β. Bergriver, *Drege,* 6290. (Hb. Sd.)

Root simple, several inches long. Stems 3-5 inches in var. β., to 1 foot high. Leaves ½-1 inch long, 1-3 lines wide, spreading-hairy. Peduncles 3-6 times dichotomous. Flowers very small. Calyx 1 line; corolla 3 lines long, deeply 5-fid. Capsule 2 lines long; valves obtuse, a little shorter than the calyx-lobes. Style,

with a ring below the 3-fid stigma. The corolla is not tubulose in *W. dichotoma*, as
described by Decandolle, but quite the same as in *W. parviflora.*

38. W. androsacea (A. DC. Monog. p. 150, t. 19, f. 1); stem simple,
naked, few-flowered, glabrous; radical leaves rosulate, ovato-lanceolate,
shortly petiolate, acute, denticulated, hairy; peduncles dichotomous;
bracteæ glabrous; calyx glabrous, tube hemispherical, lobes triangular,
entire, a little shorter than the tube; corolla broadly infundibuliform,
deeply 5-fid, 2–3 times longer than the calyx; capsule spheroid, 3-celled.
DC. Prodr. p. 437. Camp. Burch. catal. pl. Afr. 2273.

HAB. Cape of Good Hope, *Burchell.* (Unknown to me.)
    Root simple. Stem erect, 5 inches high, twice dichotomous at top, and nearly
naked, except the bracteæ at the base of the pedicels. Flowers blue ? Stigmas 3,
at the base with some glands.

39. W. annularis (A. DC. ! Prodr. p. 437); stems erect, branched,
pubescent at the base, glabrous above; leaves aggregated at the base,
linear-lanceolate, denticulated, pubescent; bracteæ linear, glabrous;
peduncles dichotomous; calyx glabrous, tube turbinate, lobes ovate,
acute, about as long as the tube; corolla broadly infundibuliform,
deeply 5-fid, 6 times longer than the calyx; capsule globose, 3-celled.

HAB. Bergvalei. Nov. *Drege*, 6288. (Herb. D., Sd.)
    Annual. 1½–2 feet, branches terete, stem-like, forked at top. Leaves numerous
at the base, scattered on the lower part of stem and branches, 1–2 inches long, 1–2
lines wide, narrower at the base, minutely toothed; upper ones on the base of the
forks, smaller, bract-like, glabrous. Pedicels 1–1½ inch long. Corolla 8–10 lines
broad, blue, lobes broadly ovate, spreading. Style with a tumid ring at the base of
the rather large 3-fid stigma. Capsule the size of a pea, 10–ribbed; tube scarcely
longer than the ovate, acuminate, spreading recurved, calycine-lobes, which are a
little longer than the obtuse valves of capsule.

40. W. pauciflora (A. DC. ! Prodr. p. 437); stem erect, simple,
terete, glabrous, ending in a few dichotomous peduncles; leaves aggregated at the base, linear-lanceolate, obtuse or apiculate, crenulate, hairy;
bracteæ linear, glabrous; peduncle elongated; few-flowered; calyx glabrous, tube narrow-obovoid, lobes linear, acuminate, corolla broadly
infundibuliform, 5-fid, one half longer than the calyx-lobes; capsule
elongate obovoid.

HAB. Rocky places, Pedroskloof to Liliefontein, 3–4000 ft. Nov. *Drege.* (Hb. Sd.)
    Distinguished from *W. annularis* by the ovary and elongated calyx-lobes. Root
annual, perpendicular, stem 1 foot. Leaves 1½ inch long, 2 lines wide, radical and
on the lower part of the stem, attenuated at the base, subpetiolated, glabrous
above, with rigid hairs beneath. Peduncle 1–2 inches long. Flowers as in *W.
annularis,* but the calyx-lobes nearly 3 lines long. Ovary 3-celled. Stigmas 3, at
the base with a tumid ring. Ripe capsule unknown.

41. W. Namaquana (Sond.); stems erect, terete, simple, villous at
the base, glabrous and naked above; leaves radical, oblong-lanceolate,
attenuated or petiolate at the base, undulately crenate, acute, villous
or hairy at both sides; bracteæ minute, glabrous; peduncles dichotomous, elongated, few-flowered; calyx glabrous, tube broadly obconical, lobes lanceolate, acuminate, longer than the tube; corolla broadly
infundibuliform, deeply 5-fid, 6-times longer than the calyx-lobes.

HAB. Modderfontyn, Namaqualand, *Rev. H. Whitehead.* (Herb. D.)

Annual, nearly 1 foot high. Leaves 1½–2 inches long, 3–4 lines wide, or smaller, all radical, rosulate, or a few lanceolate at the base of the shortly pubescent stem ; hairs on the leaves, white. Peduncle 2–3 times forked ; pedicels 1–3 inches long. Tube of calyx 1 line, lobes 1½ line long. Corolla 8–12 lines broad, lobes ovate, acute. Style shorter than the corolla, with a tumid ring below the 3-fid stigma.

42. W. Wyleyana (Sond.) ; stems many, dichotomously branched, glabrous, branches spreading, divaricate ; leaves aggregated at the base, ovato-lanceolate, acute, attenuated at the base, subpilose, coriaceous, minutely denticulated on the thickened margins; bracteæ lanceolate or linear, upper ones minute ; peduncles dichotomous; calyx glabrous, tube broadly obconical, about as long as the acuminate lobes ; corolla broadly infundibuliform, 5-lobed at the apex, 3-times longer than the calyx-lobes ; capsule obovoid, 3-celled.

HAB. Namaqualand, *A. Wyley.* (Herb. D., Sd.)

Near the following and preceding ; from the last it differs by much branched stems with divaricate peduncles, smaller, coriaceous leaves and much smaller flowers. From *W. patula* it is distinguished by glabrous branches, coriaceous, smaller leaves and flowers and shorter calyx-lobes. Stems 1–1½ foot high. Radical leaves 1 inch, 2–3 lines wide, ciliated-hairy at the base and on the middlenerve beneath ; stem leaves none, or a few smaller. Peduncles 3–5-times dichotomous, filiform ; pedicels 1–1½ inch. Calyx 1 line, corolla 3 lines long. Style without gland or ring. Capsule 2 lines long, crowned by the twice shorter spreading calyx-lobes ; valves obtuse.

43. W. patula (A. DC.! Prod. p. 436); stem much branched from the base, diffuse, hairy at the base, glabrous above; branches divaricate, dichotomous ; leaves petiolate, ovate, acute, membranous, glabrous, coarsely dentate, upper ones lanceolate ; bracteæ minute, linear or wanting ; calyx glabrous, tube hemispherical, twice shorter than the linear acuminate lobes ; corolla infundibuliform, sub 5-fid, twice longer than the calycine-lobes ; capsule ovoid, 3-celled.

HAB. Rocky places on the Gariep, *Drege,* 3070; Namaqualand, *A. Wyley.* (Hb. Hook, D., Sd.)

Root annual, simple. Branches leafy, leaves alternate, shortly petiolate, 1–2 inches long, ½–1 inch wide, with unequal, acute or subulate teeth. Peduncle and pedicels much divaricate. Corolla blue, 3 lines long. Style without gland or ring. Capsule 2 lines long, terminated by the erect calyx-teeth of the same length.

44. W. adpressa (Sond.) ; stem erect, woody, simple or branched ; branches ending in a large, compound, spreading panicle ; leaves reflexed, lanceolate-acuminate or linear-lanceolate, denticulate on the incrassate margins, glabrous, ciliate at the base ; bracteæ very small ; pedicels filiform ; calyx tube hemispherical; lobes subulate, longer ; corolla tubulose, glabrous, semi 5-cleft, 3-times longer than the calyx-lobes ; capsule hemispherical. *Camp. adpressa Thunb.! Prodr. p.* 38, *Mem. acad. Petersb. T. IV., p.* 368, *t.* 7, *f.* 2. *Lightfootia ? adpressa, A. DC. Monog. p.* 110. *Prod. p.* 417.

HAB. Cape of Good Hope, *Thunb.,* Ebenezer, *Drege, Lightfootia,* 6309 ; Brandenburg, Capedowns, *Zey.* 1064. Nov.–Jan. (Herb. Thunb., Hook, D., Sd.)

Subshrub 2–4 feet high, more or less branched, branches angular by the decurrent leaves, greyish. Leaves aggregate on the lower part, opposite or ternate, about half an inch long, 1 line wide, rigid or subspinous toothed on the margins ;

upper leaves scattered, gradually smaller. Panicle sometimes 1 foot or more long, branches spreading or divaricate, glabrous; pedicels 2–4 lines long. Calyx 1 line long; corolla 3–4 lines long, with lanceolate lobes. Capsule 1 line long; valves equalling the erect, rigid calycine-lobes. Seeds ovoid. This species connects the genera *Wahlenbergia*, and *Lightfootia;* the habit is that of the last, but the corolla and ovoid seeds belong to *Wahlenbergia.*

45. W. robusta (Sond.); shrubby, glabrous, branches woody, diffuse, rigid, terete, sub-spinescent; leaves fasciculate or solitary, very shortly petiolate, oblong, attenuate at the base, obtuse or mucronulate, quite entire, marginated; flowers in the axil of the leaves, solitary, rarely geminate or ternate; pedicels shorter than the leaf, tube of calyx hemispherical, lobes subulate; corolla tubular, 4–5 times longer than the calyx; capsule hemispherical, 5-costate. *Lightfootia? robusta, DC.! Prodr. p.* 427. *Wahlenb. rigida, Bernh.! in pl. Krauss Flora,* 1845.

HAB. Hills near Zwaanepoelpoortberg, 2–3000 ft., *Drege;* Grootrivier, Uitenhage, *E. & Z.;* Mount Winterhoek, *Dr. Krauss.* April–Aug. (Herb. Sd.)

Branches alternate, reddish-brown; twigs short, divergent. Leaves 4–6 lines long, 1–1½ line wide, coriaceous, mostly fasciculate. Pedicels 2–3 lines, calyx 1 line long. Corolla semi-5-cleft, lobes lanceolate. Style as long as the corolla; stigma 3-fid, revolute. Capsule 1 line long; the valves a little longer than the calyx-lobes. Seeds obovoid.

46. W. epacridea (Son.); stem woody; branches erect, simple, minutely pubescent, leaves appressed, imbricated, broadly ovate, acutish, entire or with a few short, bluntish teeth, pubescent; flowers aggregated in a terminal, oblong, leafy spike; calyx-lobes pubescent, ovate, acuminate; corolla 5-lobed at the apex, twice or three times longer than the calyx.

VAR. β. **glabrata**; branches and leaves glabrous or nearly so, leaves ovate, acuminate.

HAB. Port Natal, Zulu plains, *Gerrard and M'Ken,* 1502; var. β. Transvaal, *Sanderson.* April. (Herb. Hook, D.)

Root woody. Stem in our specimens short, branches ½–1 foot long, densely leafy from the base. Leaves 1½–2 lines long, 1¼ line wide, in var. β. about 3 lines long, 1 line wide, denticulate, 1-nerved. Spikes 1–2 inches long, 10-20-flowered. Tube of calyx 1 line, lobes 1½ line long. Corolla dark blue, infundibuliform, ovary 3-locular. Style shorter than the corolla, conically incrassate at the base; stigma tri-fid.

Wahlenbergia chondrophylla, Buek! in E. Z., Enum, 2388, *is Dobrowskya aspera.*

XIII. **LEPTOCODON**, Sond.

Calyx 5-lobed, with ovate tube. *Corolla* tubular, 5-lobed at the apex. *Stamens* 5, free; *anthers* linear. *Ovary* 2-celled. *Style* filiform, the base dilated, permanent. *Stigmas* 2, recurved. *Capsule* 2-celled, many-seeded, wholly inferior, cylindrical, covered by the superior, solid, conical base of the style. *Dissepiment* complete, the whole length in the middle, placentiferous. *Seeds* ovoid, subangular, reticulate-punctate.

An annual, prostrate herb, with the habit of *Microcodon.* Branches hairy, subdivided, sparingly leafed. Leaves alternate, linear, narrow, denticulated, ciliated at the base. Flowers sessile, terminal, densely glomerated, involucrated and bracteated by elongate, linear leaves. Calyx hairy. Name from λεπτος, *slender,* and κωδων, *a bell.*

1. L. longebracteatum (Sond.); *Microcodon longebracteatum, Buek!*
in E. Z. Enum. 2367.

HAB. Stony places near Tokay, Tablemountain, and near Babylonschetoorn,
E. & Z. Dec. (Herb. Sd.)

Root 1–2 inches long, white. Stem divided at the base, branches opposite and
alternate, diffuse, 2–4 inches long, with filiform, often very short ramuli, terminated by the glomerate flowers. Leaves 4–6 lines long, 1 line wide, floral ones on
bracteæ 6–8 lines long, keeled by the middle nerve, on the margins with minute but
sharp teeth. Calyx 1½ line long, lobes nearly 3-times shorter than the 10-nerved
tube, ovate, cuspidate, aristate, ciliated, bimucronulate at top. Corolla tubular,
narrow, 2 lines long; lobes short, acute, with some hairs outside. Capsule 1 line
long, operculated by the conical base of style, which is nearly equal in length to
the capsule. Dehiscence of capsule unknown.

XIV. **PRISMATOCARPUS**, A. DC.

Calyx 5-lobed, with a cylindrical, elongated, 5-nerved tube. *Corolla*
infundibuliform, or subcampanulate, 5-lobed or 5-fid. *Stamens* 5,
anthers free. *Ovary* inferior, 2-celled. *Style* persistent at the base.
Stigmas 2. *Capsule* prismatic or cylindrical, elongated, 2-celled, splitting from the top towards the base into 5 segments. *Seeds* ovoid, subcompressed, obtuse, dotted, *A. DC. Monog. camp. p.* 164. *Prismatocarpus spec. L'Her. sert angl. p.* 1, *excl. the Europ. species. Campanula
spec. auct. Polemonium spec., Thunb. End. Gen.* 3080.

Subshrubs, rarely perennial or annual herbs. Leaves alternate, often narrow,
linear. Flowers terminal, solitary, or 2–3 approximate, sessile at the top of the
peduncles or shortly petiolate, rarely in the axils of the leaves. Name from πρισμα,
a prism, and καρπος, a fruit, in reference to the long, prismatical form of the fruit.

§. 1. Style exserted (longer than corolla) :
    Lvs. linear-acuminate, remotely denticled ; fl. crowded
      at the ends of the branches ... ...   (1) **fastigiatus.**
    Lvs. narrow-linear, entire ; fl. panicled ... ...   (2) **diffusus.**
§. 2. Style included (shorter than corolla) :
    * Leaves numerous at the base, few or none in the upper part :
      Suffrutices, with woody stems :
        Fl. panicled or racemose :
          Lvs. *spreading*, lin.-subulate, ciliate :
            Fl. loosely panicled or aggregated;
            cal.-lobes glabrous, as long as the
            tube of corolla   ... ...   (3) **roelloides.**
            Fl. few, near together; cal.-lobes vel-
            vetty   ... ...   (4) **altiflorus.**
            Panicle long ; cal.-lobes glabrous, lan-
            ceol., ¼ as long as the tube of
            corolla   ... ...   (5) **subulatus.**
          Lvs. *erect*, linear, ciliate at base :
            Panicle long ; cal.-lobes glabr. lanceol.-
            acuminate, half as long as the tube
            of corolla   ... ...   (6) **brevilobus.**
            Fl. few, racemose ; cal.-lobes glabrous,
            ovate, acute, as long as cor.-tube   (7) **fruticosus.**
        Fl. sessile, spiked :
          Lvs. linear, entire ; bracts entire... ...   (8) **Ecklonii.**
          Lvs. lanceolate, 2–4 toothed ; bracts pin-
           natifid   ... ...   (9) **Candolleanus.**
      Perennial or annual herbs :
        Perennial. Lvs. linear, flat ; fl. sessile, spiked   (11) **campanuloides**

Annual. Lvs. long, linear, undulate-toothed ;
     pedunc. |divaricate... ... ...   (10) **crispus.**
** Leaves uniformly scattered over the whole plant :
Lvs. ovate-oblong or lanceolate, flat, serrulated   ...   (12) **nitidus.**
Lvs. narrow-linear, entire :
     Lvs. channelled, ciliate at base ; fl. axillary,
     crowded, shorter than the leaves   ...   ...   (13) **sessilis.**
     Lvs. flat, glabr. ; bracts ciliated ; fl. sessile,
     spiked, longer than the leaves   ...   .   (14) **tenerrimus.**

§. 1. Style exserted. (Sp. 1–2.)

1. **P. fastigiatus** (Presl ! in Herb. Drege) ; . glabrous ; stem woody at the base, branches erect, angulate-striate, simple ; leaves scattered, erect, linear-acuminate, with revolute, remotely denticulated margins ; flowers approximate at top of the branches, sessile ; bracteæ ovate, acute, subciliated ; lobes of calyx ovate, acuminate, connate at the base ; corolla 2–3 times longer than the calycine lobes. *DC. Prod. p.* 442.

HAB. Rocky hills, Uienvalei, 2000–2500 ft., *Drege.* Dec. (Herb. D. Sd.)
Habit of *P. strictus.* Branches leafy from the base nearly to the apex. Lower leaves more numerous, 6–8 lines long, ¾-line wide, upper ones 4–3 lines long, terminated by a rigid point ; the teeth on the margins very minute, 2–3 on each side. 2–8 flowers forming a spike. Lobes of calyx 1 line long, 2–3 times shorter than the tube. Corolla infundibuliform, tube twice longer than the calyx ; lobes short. Style exserted, glabrous : stigmas incrassate.

2. **P. diffusus** (A. DC. Monog. p. 164) ; stem woody at the base, erect, terete, glabrous ; branches leafy at the base ; leaves linear, very narrow, entire, glabrous or subpilose ; flowers in loose, nearly naked panicles ; peduncles diverging, stiff ; bracteæ subulate, lobes of calyx ovate, acute, or acuminate, glabrous ; corolla 3–4 times longer than the calycine lobes. *Trachelium diffusum, Linn. fil. suppl. p.* 143. *Thunb.! Prod. p.* 38, *Fl. Cap. p.* 176. *P. diffusus, E. Z.,* 2389. *P. laricinus Presl! Monog. lob. p.* 52. *Herb. Zey.* 357.

HAB. Mountains in Hottentottsholld. *Thunb. E. & Z.* ; Winterhoeksberg, *Drege ;* Tulbaghskloof, *E. Z., Dr. Pappe, Zey.,* 1081. Kardow, *Zey.,* 1079. Alexanderskloof and Sneuwkop, Cederbergen, *Wallich.* Nov.–Feb. (Herb. Thunb., Holm., Hook, D., Sd.)
Stem ½–1½ foot high, much branched, branches filiform, subflexuous. Leaves 6–10 lines long, ¼ line wide, sulcate, rigid, pointed. Panicle in small specimens 2–3 inches, in others 1 foot long ; peduncles much-spreading or divaricate, 3-flowered. Calyx 1 line long. Corolla with longish tube and lanceolate lobes. Anthers subexserted. Style very long. Capsule 6–8 lines long.

§. Style included. (3–14.)

2. **P. roelloides** (Sond.) ; branches woody at the base, terete, glabrous or scabrous-hairy, very leafy, dichotomous and naked above ; leaves fascicled, lineari-subulate, entire, suberect or spreading, setaceously ciliated, incrassate on the margins ; panicle loose, often fewflowered ; bracteæ subulate ; flowers sessile, lobes of calyx acuminate, incrassate on the margins, glabrous ; corolla 5-fid, broadly infundibuliform, subcampanulate, tube as long as the calyx-lobes. *Polemonium roelloides Linn. suppl. p.* 139. *Thunb.! Fl. Cap. p.* 177. *Campanula ericoides Lam. ill.,* 2574. *Roella ericoides Spreng. syst.* 1. p. 723. *P.*

paniculatus L'Her, sert. angl. p. 1, *A..DC. Monog. p.* 165, *t.* 20. *Prodr. p.* 443, *E. Z.,* 2390.

VAR. β. **grandiflorus** (Sond.); flowers larger. *Roella pedunculata, Berg.! pl. cap. p.* 42. *P. pedunculatus, DC.! Prod. p.* 443. *P. grandiflorus, E. Mey.! in Herb. Drege, DC. l. c. P. interruptus, E. Z.* 2391.

VAR. **ß. interruptus** (Sond.); stem leafy nearly to the apex; flowers approximate at the tops of the peduncles. *P. interruptus L'Her. sert. angl. p.* 1, *A. DC. Monog. p.* 166. *Prod. p.* 443. *Camp. interrupta Pers. ench.* 1, *p.* 192.

HAB. Mountains, Nieuwekloof, *Drege;* Waterfall, Tulbagh, Rivierzonder Ende and Kochmanns Kloof, *E. & Z., Pappe;* Kardow, Hassaquaskloof and Buffeljagd-rivier, *Zey.* 3145; var. β. Breederivier, *E. & Z.,* Piquetberg, *Drege;* Zwarteberg and Rivierzonder Ende, *Zey.* 3146; Mitchell's Pass, *A. Wyley,* Genadenthal, *Dr. Roser;* var. γ. Klipfontyn, *Zey.* 1077, Oct.–Jan. (Herb. Thunb., Holm., Hook, D., Sd.)

Stem erect, 1–2 feet or more high; branches simple, greyish. Leaves longer than the internodes. sessile, linear-lanceolate, entire, 4–8 lines long, ½ line wide, pedunculoid, apex of branches dichotomously divided; rarely in var. β 1 or 2-flowered; as long or shorter, rarely longer than the leafy branch; peduncles 1–2 inches long, spreading, 1-flowered, or with 2–3 sessile approximate flowers at top. Calyx-lobes 2–3 lines long, in var. β. broader and 4–5 lines long. Corolla about 6–8 lines; in var. β. 1 inch long and broad; lobes ovate, as long or longer than the tube. Base of the style incrassate, at length larger, ovate, obtuse, persistent (ovary, *Thunb.*); stigmas 2 linear, recurved. Capsule 1–1½ inch long, 4 angular, striate. Var. γ. seems to be a distinct species at first sight, but I have intermediate specimens collected by *E. Z.* The stem and branches are 3 feet high, leafy also in the pedunculoid part at the base of the shorter and more condensate panicle. Calyx and corolla as in var. α.

4. P. altiflorus (L'Her. sert. angl. p. 1); suffrutescent; leaves linear-subulate, entire, ciliated; flowers few, approximate at the tops of the peduncles; bracteæ subulate; lobes of calyx ovate, acute, velvetty. *A. DC. Monog. p.* 167. *Prod. p.* 444. *Roella erecta, Banks, herb. Camp. altiflora Poir. suppl.* 2, *p.* 66.

HAB. Cape of Good Hope. *Masson* in herb. Banks; unknown to us.

Leaves and habit of *P. roelloides,* and probably a variety. Stems nearly simple, 1 foot, glabrous, naked at the base and top, but very leafy in the middle, divided at the top into 2 erect 2–4-flowered peduncles. Leaves often with tufts of smaller ones in their axils. Flowers sessile. Calyx greyish; the tube filiform, 3–4 lines long, lobes 3 lines long, 1 line broad.

5. P. subulatus (A. DC. Monog. p. 166); stem erect, woody, branched, minutely pubescent, very leafy from the base to the middle, glabrous and naked above; leaves fascicled, linear-subulate, narrow, entire, ciliate, erectish or spreading-recurved; panicle elongate, loose, naked, flowers pedicellate and sessile, bracteæ subulate; lobes of calyx acuminate, glabrous; corolla infundibuliform, subcampanulate, twice longer than the calyx-lobes. *Prod.! p.* 444, *E. & Z.!* 2393. *Campanula subulata, Thunb.! Prod. p.* 38, *Mem. Acad. Petersb.* 4, *p.* 369, *t.* 6, *Fl. Cap. p.* 171. *Lightf. ciliata Spreng. syst.* 1, *p.* 809. *P. altiflorus, E. & Z.!* 2394.

VAR. β. **pauciflorus**; branches very slender, velvety; panicle few-flowered. *P. lycopodioides, A. DC.! Prod. p.* 453.

VAR. γ.**Bergianus;** more robust, branches minutely pubescent; leaves subracemose, canaliculate, recurved, ciliate at the base; panicle much elongated. *P. Bergianus! Cham.! Linn. vol.* 8, *p.* 199, *E. & Z.!* 2392.

VAR. δ. **penicillata** (A. DC.! Prod. p.444); lobes of corolla pubescent at the apex.

HAB. Mountains near Hexriver, *Thunb.;* distr. of Cape, Worcester and Caledon, *E. & Z. ; Zey.* 1078. Dutoitskloof, *Drege ;* Alexanderskloof, *Dr. Wallich ;* var. β. Dutoitskloof, *Drege ;* var. γ. in the Capeflats, *Bergius, Ecklon, W. H. Harvey, Zey.* 3147. Dec.–Jan. (Herb. Thunb., Holm., Hook, D., Sd.)

Habit of *P. paniculatus.* It differs by the elongate, more divided panicle, with filiform or capillary, erectish peduncles, smaller calyx and corolla, with longer tube. Leaves about 6 lines long, ⅓ line wide; in var. γ. more rigid and subglabrous. Flowers mostly pedunculate. Calyx-lobes 1–1½ line long. Corolla with ovate, acute lobes, 3–4 times longer than the calycine lobes. Capsule ¾–1 inch long.

6. P. brevilobus (A. DC.! Prod. p. 443); branches woody at the base, minutely downy, paniculated and glabrous above; leaves erect, linear, narrow, subconcave, entire, mucronulate, subciliated; peduncles naked, erect, slender, dichotomously panicled; bracteæ minute, linear, subciliate; calyx glabrous; lobes lanceolate, acute, much shorter than the tube; corolla 3–4 times longer than the calycine lobes.

HAB. Hills near Bergrivier, *Drege.* Paarl. *Rev. W. Elliott.* Aug. (Hb. Hk., D. Sd.)

Habit and panicle as in *P. subulatus,* but the leaves are erect, subappressed, not fascicled, about 1 inch long, not ⅓ line wide, subtrigonous, beset from the base with very short cilia. The flowers or the whole purplish panicle is the same as in *P. subulatus,* in which the tube of calyx is also constricted under the lobes. Tube of corolla twice longer than the calyx. Capsule ½–1 inch. I have not seen a specimen with quite glabrous branches.

7. P. fruticosus (L'Her Sert. Angl. p. 2); stem minutely downy, subwoody at the base, glabrous above; leafy in the middle; leaves narrow, linear, acute, erect, flat above, ciliate at the base; margins revolute; flowers few, racemose, or dichotomously subpanicled; bracteæ minute; calyx glabrous; lobes ovate, acute, or subacuminate; corolla infundibuliform, 5-cleft to the middle. *A. DC. Monog. p. 169. Prod. p.* 444. *E. & Z.* 2395. *Campanula fruticosa, Herb. Banks.*

HAB. Table Mountain and in Hottentottsholland, *E. & Z.* Dec. (Herb. Sd.)

Stem 1 foot, branches simple, ending in a naked few-flowered raceme. Leaves 8–10 lines long, nearly 1 line wide, nearly glabrous, with a few short cilia at the base. Pedicels 2–4 lines long (DC.), in our specimens ½–1 inch. Tube of calyx in the specimen of Herb. Banks only 2 lines, in ours 3–4 lines long. Lobes erect, entire, 1 line long. Lobes of corolla acute. Capsule cylindrical, angulate-striate. It may be regarded as a few-flowered, broad leafed variety of *P. brevilobus.*

8. P. Ecklonii (A. D.C. ! Monog. p. 168); stem woody, branched; branches minutely downy, very leafy at the base, ending in a nearly naked, loose, glabrous spike; leaves erect, linear-acuminate, entire, glabrous, with revolute margins ; flowers sessile, remotely spiked ; bracteæ subulate, at the base larger than the leaves, quite entire; calyx glabrous, lobes linear-lanceolate, shorter than the 5-fid, broadly infundibuliform corolla. *Prod. p.* 444, *E. & Z.* 2396. *P. fruticosus, Eckl. pl. exs. olim, not of L'Her.*

HAB. Stony places in the Zwarteberg near Caledon, and in Mount Baviansberg near Genadenthal, *E. & Z.* Feb. (Herb. Sd.)

A subshrub, 1–2 feet high ; branches erect, simple, straight, terete, greyish, velvety. Leaves numerous at the base, rare from the middle, but not smaller, rigid, 6–8 lines long, 1 line wide at the base ; raceme 6–8 flowered, flowers solitary in the axil of a leafy, glabrous bractea, which incloses 3 smaller bracteolæ ; upper flowers subapproximate, the lower ones distant about 1 inch. Calyx-tube at flowering time equalling the bractea ; the lobes of the same length. Corolla 5-fid to the middle ; the tube a little shorter than the calyx lobes. Capsule 6 lines long, 4-angular.

9. P. Candolleanus (Chamiss.! Linnaea, vol. 8, p. 197) ; stem woody, branched ; branches erect, minutely downy, very leafy, ending in a nearly naked or somewhat leafy, remoti-flowered, glabrous spike ; leaves erect, lanceolate, or linear-lanceolate, acuminate, acute, 2–4 dentate at the base, with revolute margins, glabrous, or the lower ones subpilose beneath ; flowers sessile, remotely spiked ; bracteæ subulate-acuminate, at the base larger than the leaves, pinnatifid ; calyx glabrous, lobes linear-lanceolate, one half shorter than the 5-fid corolla. *E. & Z.!* 2401.

HAB. Near Ruiterbosch, *Mundt.* Jan. (Herb. Hook., Sd.)
Nearly allied to *P. Ecklonii* in size and habit ; stem and branches quite similar, the leaves broader at the base, with 2 or rarely 4 acute or subspinous teeth, rigid, 6–8 lines long, 1 line or somewhat broader at the base. Spike often leafy at the lower part; inferior flowers 1 inch distant, the upper approximate. Bracteæ leaf-like, as long as the calyx tube, pinnatifid-dentate, middle lobes longer ; upper ones smaller. Calyx lobes 4 lines long, acuminate, glabrous. Corolla 5 fid to the middle, lobes acute. Capsule ½ inch long, prismatical, glabrous. The flowers are generally solitary, rarely geminate or ternate, and then on a very short pedicel or branch.

10. P. crispus (L'Her. Sert. Angl. p. 2) ; stem herbaceous, terete, simple or branched, and hairy at the base, quite glabrous, dichotomously branched and naked above ; leaves linear, acute, elongate, undulately toothed, whitish margined ; peduncles divaricate, 1-flowered ; bracteæ subulate ; calyx glabrous or subpilose, tube elongate ; lobes linear, acuminated, entire ; corolla infundibuliform, deeply 5-fid, twice longer than the calyx. *A. DC. Monog. p.* 168. *Prod.!* p. 444. *E. & Z.* 2397. *Camp. crispa, Herb. Banks. C. plicata, Pers. enchir.* 1, p. 193.

HAB. Sandy stony places near Waterfall, Tulbagh, *E. Z.* Rivier Zonder Einde, *Zey.* Glenfilling and Grahamstown, Albany, *Drege.* Oct.–Dec. (Herb. Hook., D. Sd.)
Annual herb, 1–1¼ feet high, subflexuous ; branches alternate, the lower some times opposite, ascending. Leaves 1–1½ inch long, 1–2 lines wide, curled on the margins. Calyx lobes very narrow, 4–5 lines long, erect. Corolla to ¾-5-fid, lobes oblong, acute. Capsule 1–3 inches long.

11. P. campanuloides (Sond.) ; stem herbaceous, simple or branched, glabrous or pilose at the base ; leaves erect or spreading, linear, acute, glabrous or subciliate, with revolute remotely denticulate or entire margins ; flowers spiked, solitary or geminate ; bracteæ ovate, acumi-nate ; calyx glabrous, tube elongate, lobes linear, acute, entire ; corolla infundibuliform, 5-fid, one half longer than the calycine lobes. *Pole-monium campanuloides. Linn. suppl. p.* 139. *Thunb.! fl. cap. p.* 177. *Lightfootia sessiliflora, Spreng. in Zeyh. fl. cap. exs.* 203. *Prismatoc. lina-riaefolius et P. strictus, A. DC. Monog. p.* 169. *Prod. p.* 444. *E. & Z.* 2398, 2399.

HAB. Sandy places on the Zwartkopsrivier, and on Winterhoeksberg near Elands-rivier, *E. & Z. Zeyh.* 3148. Tulbagh near Waterfall, *Dr. Pappe.* Ataquaskloof and between Kovi and Katrivier, *Drege.* Cafferland, *Dr. Gill.* Dec.–Apr. (Herb. Thunb., Hook., D., Sond.)
Perennial, 1–2 feet or more in height, angular-striated by the decurrent leaves ; branches alternate, virgate, glabrous. Leaves numerous, sometimes few and distant, alternate, ½–1 inch long, 1 line wide, flat, with more or less revolute margins, cili-

ated at the base or to the apex, or quite entire, or with a few minute teeth. Branches ending in a few-flowered, nearly naked spike, rarely terminated by a single flower. Bracts 4-6 lines long, enclosing 1-3 flowers. Calyx lobes 5-6 lines long, at flowering time as long or shorter than the tube. Corolla subcampanulate, with ovate, acute, lobes, blueish. Stigmas 2. Capsule 1 inch long, 5-valved. It has a great affinity to *P. Ecklonii*, from which it is distinguished by herbaceous stem, angulate, striate, glabrous branches, and less rigid leaves.

12. P. nitidus (L'Her. sert. angl. p. 2, t. 3); plant rather woody, branched; leaves ovate-oblong or lanceolate, flat, serrulated, spreading, shining; flowers 2-4, sessile, and approximate at top of the branches; bracteæ leafy; calyx glabrous, tube longer than the bracteæ, lobes lanceolate, shorter than the tube; corolla infundibuliform, subcampanulate, twice longer than the calyx. *A. DC.! Monog. p. 170. Prod. p. 445. E. & Z. 2403. Camp. Prismatocarpus, Ait. Hook. Bot. Mag. t. 2733. Herb. Un. Itin. 155.*

HAB. Table Mountain. Jan. *E. & Z. W. H. Harvey.* (Herb. Hook., D., Sd.) A small glabrous perennial, with short or elongated diffuse reddish branches. Leaves 4-6 times longer than broad, commonly 4-6 lines long, sharply serrate, acute. Flowers 2-4 in a cluster, rarely solitary in the axils of the bracteæ. Calyx 1½ line long. Corolla white, 5 lobed. Capsule 1 inch long.

13. P. sessilis (Eckl.! A. DC. Monog. p. 171); stem much branched; branches simple or dichotomously divided, filiform, diffuse, glabrous; leaves fascicled, spreading, narrow-linear, canaliculate above, entire, subciliate at the base; flowers sessile, axillary, aggregated, rarely solitary, pedicellate, opposite the leaf; calyx glabrous, lobes minute, ovate-lanceolate; corolla infundibuliform, 2-3 times longer than the calyx; capsule shorter than the leaf. *Prod. p. 445. E. & Z. 2404. Herb. Un. Itin. 152, et 194.*

HAB. Grassy places on the Table Mountain and in Capeflats, *E. & Z., W. H. Harvey, Zey.* 1080, near Simon's Bay, *Milne,* 169. Nov.-Jan. (Herb. Hk., D., Sd.) Branches some inches to 1 foot and more long, purplish, shining. Leaves or fascicles often secundate, the outer or larger ½-1 inch long, not ½ line wide, channelled above, subtrigonous, longer than flower and fruit. Generally 2-8 flowers are aggregated in a fascicle, sometimes they are pedicellate and solitary. Calyx lobes ⅓ line long. Corolla white, 5-lobed, 2-3 lines long. Capsule prismatical, 3 lines long It is indicated as a perennial by Ecklon and Zeyher; small specimens seem to be annual.

14. P. tenerrimus (Buek.! in E. Z. enum. 2402); stem much branched; branches filiform, subsimple, velvety or sub-glabrous, flowering secundate, remotely leafy; leaves alternate, erect, linear, acuminate, glabrous; bracteæ ovate, acuminate, ciliate; flowers axillary, sessile, solitary or geminate, spiked; calyx glabrous, lobes acuminate; corolla infundibuliform, twice longer than the calyx; capsule longer than the leaf.

HAB. Mts. near Rivier Zonder Einde, Swellendam. Feb.-Mar. *E. & Z.* (Hb. Sd.) A subshrub, woody at the base, in characters very near the foregoing, but in habit very distinct. Branches reddish. Leaves ½-1 inch long, ½ line wide, not fascicled, flattish, the margins scarcely revolute, not ciliate at the base. The lateral branches form a distant-flowered spike; the inferior or smaller branches bear sometimes 1 or 2 single flowers at the top. Calyx lobes 2 lines long. Corolla white, 5-lobed. Capsule 4-6 lines long.

XV. ROELLA, Linn.

Calyx 5-fid, tube cylindraceous.   *Corolla* infundibuliform, tubular or campanulate, large, 5-lobed.   *Stamens* 5 ; anthers free.   *Ovary* 2-celled. *Stigmas* 2, thick.   *Capsule* cylindraceous, 2-celled, wholly inferior, perforated by the permanent base of the style, and therefore appearing as covered by a concave operculum, and at length dehiscing by a large hole at the apex, without valves.   *Seeds* numerous, angular, scabrous, thick. *Hort. Cliff. p.* 492, *t.* 16, *f.* 5. *Lam. ill.* 346, *t.* 123. *ADC. Monog. camp. p.* 172. *Endl. gen.* 3081.

Subshrubs, rarely herbs.   Leaves numerous, alternate, scattered, usually narrow and stiff.   Flowers sessile, terminal, solitary or glomerate.   Named in honour of *W. Roell,* professor of anatomy in the academy of Amsterdam.

Flowers solitary, terminal ; branches woody :
Cal. lobes and floral leaves *recurved,* squarrose ... ...   (6) **recurvata.**
Cal. lobes erect :
   Cal. lobes ovate, acute, broad   ... ...   (4) **latiloba.**
   Cal. lobes lanceol., tapering to a long, hair-like point ...   (5) **leptosepala.**
   Cal. lobes lanceolate, acuminate :
      Lvs. erect or spreading, lin.-acuminate ; caps. gla-
      brous ... ...   (1) **ciliata.**
      Lvs. erect, appressed, lin.-subulate ; caps. pubescent   (2) **reticulata.**
      Lvs. reflexed, linear-acuminate ... ...   (3) **bryoides.**
Flowers solitary, terminal ; stem and branches herbaceous ; lvs.
ovato-lanceolate, reflexed, serrulate ... ...   (11) **muscosa.**
Flowers aggregate or glomerate ; branches woody :
   Lvs. narrow-linear ; branches minutely pubescent :
      Bracts laciniate, with a long point ; cal.-lobes recurved
      at apex ... ...   (7) **Ecklonii.**
      Bracts toothed, middle tooth larger, recurved ; cal.-lobes
      not recurved ... ...   (8) **spicata.**
   Lvs. linear-lanceolate ; branches glabrous ... ...   (9) **glomerata.**
   Lvs. ovate, acute, reflexed ... ...   (10) **squarrosa.**

1. R. ciliata (Linn.! Spec. 1, p. 241) ; stem branched, hairy ; branches spreading ; leaves fascicled, erect, or spreading, linear, acuminated, ciliated, upper ones longer or larger ; flowers solitary, terminal ; lobes of calyx lanceolate or acuminated, longer than the leaves, ciliated or laciniately toothed, glabrous or hairy outwards ; corolla infundibuliform, variegated ; capsule glabrous.

VAR. *a.* **Linnaeana** ; stem erect ; leaves linear, rigid ; floral ones longer and larger, acuminated, laciniately toothed and ciliated ; corolla large, twice longer than the calycine lobes. *R. ciliata L. l. c. Hort. Cliffort. t.* 35. *Berg. cap. p.* 41, *ex pte. Thunb.! fl. cap. p.* 175. *A. DC.! Monog.* p. 173. *Chamiss.! Linnæa, vol.* 8, *p.* 200. *E. Z.!* 2405. *Herb. Un. itin.* 694, 695. *Lam. ill. t.* 123, *f.* 1. *Curt. bot. mag. t.* 378. *Lodd. bot. cab. t.* 1156. *R. reticulata Linn. ex parte fide specim. Burmanniani in herb. Thunb.!*

VAR. *β.* **incurva** ; stem erect ; leaves shorter, incurved, floral ones densely imbricated, scarcely longer, rigid-ciliated or laciniately toothed ; corolla as long or longer than the glabrous calycine lobes. *R. incurva, Banks herb. ex A. DC. mon. p.* 172. *R. ciliata, Bergius! cap. p.* 41, *ex pte.* E. & Z.! 2405, *ex pte.*

VAR. *γ.* **Dregeana** ; stem erect ; leaves shorter, incurved, floral ones densely imbricated, longer ; calyx lobes laciniately toothed and ciliated, acuminated or elongated, hairy or pubescent outwards, shorter than the corolla. *R. Dregeana, A. DC. Prod.! p.* 446.

VAR. *δ.* **minor** ; stem short, slender, with erect or diffuse branches ; leaves incurved or straight, ciliated, floral ones longer, glabrous, acuminated, ciliated and toothed ; corolla as long or shorter than the calyx lobes. *R. minor Eckl.! enum.* 2407 (branches ascendent, leaves somewhat spreading) ; *R. gracilis Buek.! l. c.* 2409, *et R. ciliata, β. elongata Bk. l. c.* (branches and smaller leaves erect).

HAB. Sandy and rocky places in the districts of Cape, Stellenbosch, Worcester, and Caledon ; var. *β.* in Klynhowhoek, *Zeyh.,* 3149 ; var. *γ.* near Capetown, Paarlberg, Simonsbay, Driekop, Hemel en Aarde ; var. *δ.* Cape flats, Palmiets river and Zwarteberg, near Caledon. Oct.-April. (Herb. Thunb. Holm. Hook., D., Sd.)

Root perennial. Stem ½-1 foot, much leafy. Leaves scattered, with numerous smaller ones in the axils, glabrous, ciliated by white, rigid, subcartilaginous hairs, 2-5 lines long, ½-1 line wide, calyx lobes 6-8 lines long, 1-2 lines wide. Corolla in var. *a.* one inch and more long, in var. *β.* and *γ.* often smaller, in var. *γ.* 3-4 lines long, blue or rose coloured, with a deep purple or blue circle at the base of the lobes. Capsule cylindrical.

2. R. reticulata (A. DC. Monog. p. 174) ; stem branched, hairy ; branches diffuse or ascendent, upper ones very short, secundate ; leaves fascicled, erect, appressed, linear-subulate, ciliated, upper ones not longer; flowers terminating the lateral branches, solitary, rarely subaggregated ; lobes of calyx subulate, entire or remotely toothed, ciliated, longer than the leaves, hairy or pubescent inside, glabrous or hairy at the apex outside ; corolla infundibuliform, concolorous ; capsule pubescent.

VAR. *a.* **ternifolia**; leaves very narrow, appressed ; lobes of calyx subulate, ciliated ; corolla as long or one half longer than the clyx. *R. reticulata, A. DC. l. c. ex. cl. syn. Linn. Chamisso! Linnæa,* v. 8, p. 200. *R. ternifolia, Willd. herb.!* 3875, *but not of Thunb. R. gracilis, E. & Z.!* 2409, *ex pte. R. ciliata, Dreg. herb. ex pte.*

VAR. *β.* **Dunantii**; leaves linear-acuminated, erect, not appressed ; lobes of calyx acuminated, denticulated ; corolla nearly one half longer than the calyx. *R. Dunantii, A. DC.! Monog.* p. 175. *Prod.* p. 446.

VAR. *γ.* **prostrata** ; leaves spreading erect, subulate ; lobes of calyx spreading of subrecurved ; corolla as long or a little shorter than the calyx. *R. prostrata, E. Meyer.! DC. Prod. p.* 447.

HAB. Sandy places in the Cape flats, *W. H. Harvey, Ecklon., Zeyh.* 3417. Brederiver, *Mundt. ;* var. *β.* Cape flats, *E. & Z., Dr. Wallich* ; var. *γ.* near Cape town, *Drege, W. H. Harvey. Carmichael,* Predikstael, *Zeyher.* Febr.-May. (Herb. reg. Berol., Holm., Hook., D., Sd.)

A much branched and ramulous, densely leafy, perennial or subshrub, 1-1½ foot long. The lateral branches generally very short, and forming at flowering time a secundate spike, but in other specimens elongate, virgate. Leaves imbricated, mucronate, in var. *γ.* spreading; they are as long as in *R. ciliata,* but twice narrower and hairy-ciliated. Calyx lobes always pubescent inside, about ⅓ inch long, ½ line wide, ciliated or remotely ciliate-toothed. Corolla white or perhaps yellowish, smaller than in common forms of *R. ciliata,* deeply 5-fid, the lobes narrower. The var. *γ.* resembles in aspect the var. *δ. minor* of *R. ciliata. R. reticulata* Linn. spec. 1, p. 241 ; foliis ciliatis, mucrone-reflexo, with the quotations ; *Petiver. Vaill. and Pluken. amalth.* is *Gorteria ciliaris.*

3. R. bryoides (Buek ! in E. Z. enum. 386) ; stem branched woody ; branches erect, virgate, ramulous, hairy ; leaves alternate, reflexed, linear-acuminated, ciliated ; floral ones erect, nearly as long as the calyx lobes, toothed ; flowers solitary, terminal ; lobes of calyx subulate, hairy inside, with 2 or 4 longish lateral teeth, as long or one half shorter than the corolla.

HAB. Sandy places near Zeekoevalley, and on Mount Kamiesberg, Clanwilliam, *E. & Z. ; Zeyh.* 1082. Dec.-Feb. (Herb., Hook., Sd.)

Not unlike *R. reticulata,* but more woody and erect, and at first sight distinguished by the reflexed, not densely fascicled, 1½-2 lines long, rigid leaves. Flowers at top of the primary and secondary very short branches. Calyx-lobes 3-4 lines long, the lateral teeth ¼ or nearly 1 line long. Corolla in dried state white or pale yellowish. The name is a very bad one, but the 2 specimens in herb. *E. & Z.* are very incomplete. Those collected by *Zeyher* are much better, but also without fruit.

4. R. latiloba (A. DC. Prod. p. 447); stem woody ; branches erect, pubescent; leaves spreading, linear, acute, setaceously-ciliated, floral ones erect, scarcely longer; flowers solitary, terminal ; lobes of calyx ovate, acute, broad, nearly entire, scabrous outside, pubescent inside ; corolla twice longer than the calycine lobes.

HAB. Bergvalei and Langevalei. Nov. *Drege.* (Herb., Hook., Sd.)
Habit of *R. bryoides*, with the same horizontally spreading, but not reflexed leaves. Calyx-lobes 3 lines long, 1½ line wide, muricately asperous, sometimes with some short teeth on the margin. Corolla deeply 5-fid, lobes oblong.

5. R. leptosepala (Sond.) ; stem woody ; branches erect, minutely pubescent, the ultimate lateral ones very short and terminated by the flower ; leaves linear-acuminated, glabrous, remotely ciliate-toothed; floral ones longer, with 2 or 3 subulate teeth on each side, and minutely ciliated at the base ; calyx-lobes lanceolate, attenuated in a long, hair-like acumen, entire, or with 4-6 tooth-like cilia, and minutely pubescent near the base ; corolla deeply 5-cleft, as long as the calycine lobes.

HAB. Dutoitskloof. Drege. (Herb., Sd.)
Stem 1 foot and more high, branches slender, in the upper part beset with many ¼-1 inch long ramuli, each of which bears a flower. Leaves 3-4 lines long, at the base ¼ line wide, much acuminated, not ciliated in the axils with many smaller ones, which are bluntish and mucronulate. Floral leaves not densely imbricated, 5-6 lines long, filiform at the apex, bractlike, not green ; the lateral filiform teeth long, spreading. Calyx-lobes 6-8 lines long, greenish, flat, attenuated in a long point, equalling the oblong lobes of corolla. Ovary minutely downy.

6. R. recurvata (A.DC.! Prod. p.447); stem woody, much branched ; branches short, diffuse, very leafy, spreading, hairy ; leaves spreading, recurved, linear, mucronate, remotely ciliated-toothed; floral ones a little longer, as well as the lobes of calyx recurvate, setaceous-ciliated at the base, laciniately-toothed at the apex ; flowers solitary, terminal.

HAB. Cape. *Drege,* 6323. (Herb., Sd.)
Near *R. bryoides* and *R. reticulata, var. prostrata ;* distinguished from both by the squarrose-recurved floral leaves and calyx-lobes. Leaves 2 lines long. Flowering branches 1-2 inches long. Flowers small, bracteæ numerous, imbricate, 3-4 lines long. Corolla wanting.

7. R. Eckloniana (Buek ! in E. & Z. enum. 2408) ; stem branched, woody, glabrous ; branches ascending, often secundate, simple or divided, minutely pubescent, leafy ; leaves fascicled, spreading, linear, acuminated, entire, ciliated at the base ; flowers terminal, aggregated, each of which is subtended by 3 bracteæ, covering the calyx-tube ; bracteæ broad, cuneate, ciliated at top, laciniately-toothed with a long linear acumen ; calyx-lobes lineari-acuminated, entire, glabrous, subrecurved at the apex, longer than the bracteæ; corolla 5-lobed, as long or shorter than the calyx-lobes ; capsule minutely downy.

VAR. β. **pubescens**; calyx-lobes and bracteæ downy.

Hab. Sandy stony places between Krum and Camtoursrivier, Uitenhage, *E. & Z.*; var. β. near Genadenthal. *Ecklon.* Dec. (Herb., D., Sd.)

Stem diffuse, 1–1½ foot long. Branches ascending or erect, 2–4 inches long, much leafy ; leaves 3–4 lines, at the base ½ line wide, 3–4 flowers at the top of the branches. Branches 2 lines long and broad, carinate, with a subulate, 3 lines long, spreading or recurved acumen. Calyx-lobes 4–5 lines long, subscarious. Lobes of corolla acute. Stigma 2-fid

8. R. spicata (Linn.! suppl. p. 143); stem branched, woody ; branches ascending or erect, ramulous, sometimes secundate, very leafy, naked at the base, minutely pubescent ; leaves fasicled, linear-acuminate, spreading, ciliate at the base; flowers terminal, aggregated, rarely solitary ; fascicles globose or spiked ; bracteæ broad, cuneate, ciliolate, at top dentate, middle tooth larger, recurvate ; calyx-lobes lanceolate, acuminated, entire, glabrous, longer than the bracteæ, not recurved; corolla 5-lobed, one half or twice longer than the calyx-lobes; capsule minutely downy. *Thunb.! Prod. p. 38. Fl. cap. p. 175. R. secunda Buek! in E. & Z. enum.* 2414. *R. ericoides, Buek.! l. c.* 2410. *R. glauca et prostrata Eckl.! l. c.* 2415, 2416. *R. campestris, A. DC.! Prod.* 447. *Camp. bracteata Thunb. fl. cap. p.* 171. *R. bracteata A. DC. l. c. p.* 446.

Hab. Hills on the Zwartkoprivier and in Krakakamma. Gauritzrivier, *E. & Z.* Zeyh. 3151, at Bethelsdorp, *Drege, Zey.* 3150. Dec.–Feb. (Herb. Thunb., Holm., Hook., D., Sd.)

Variable in size and habit. Stem decumbent or erect, 1–2 feet. Branches short or long. Leaves 1–3 lines long. Flowers sessile, aggregated to a head, nearly as large as a hazelnut, or forming an oblong spike. Bracteæ more numerous than in *R. Eckloniana,* 1–1¼ line long ; the larger tooth ½ line long. Calyx lanceolate, shortly acuminated, 2½ lines long. Corolla white ; tube short, lobes acute. It differs from *R. Eckloniana* by the smaller leaves, smaller bracts without elongate acumen, shorter calyx lobes without recurved apex and smaller flowers. *R. ericoides,* Bk. is founded on a single bad specimen in Herb. Ecklon. *R. glauca* and *prostrata* are not distinguished by Dr. Buek, and quite the same as *R. secunda.*

9. R. glomerata (A. DC.! Prod. p. 447); erect, or branches ascendent, glabrous, angular, naked at the base, very leafy; leaves linear-lanceolate or linear, acute, entire or toothed on the incrassate margins, glabrous or subciliated, spreading-erect; floral ones often longer, coarsely toothed; flowers capitate; bracteæ broadly ovate, 5–7-lobate, lobes acute; calyx lobes lanceolate, setaceously acuminate, entire, glabrous ; corolla 5-lobed, longer than the calycine lobes.

Hab. On the Omsamcaba, *Drege.* Wentworth Bluff, near Natal, *J. Sanderson,* near D'Urban, *Ger. & M'K.* 872. Feb. (Herb. D., Sd.)

A shrub, 1–2 feet, sometimes simple or with short branches at the summit. Leaves ½ inch long, 1 line wide, fascicled, generally with 2–3 short but rigid teeth on the margins. Branches terminated by a round many flowered head, 6–9 lines in diameter. Bracteæ often broader than long, with setaceously acuminated, short lobes, not much surpassed by the calyx. Corolla white, tubular, deeply 5-lobed, lobes acutate. Stigma broad, 2-lobed. Capsule (unripe) glabrous.

10. R. squarrosa (Berg.! cap. p. 42); stem erect, woody; branched above or from the base ; branches diffuse, herbaceous, filiform, simple cr ramulous, glabrous or subpilose ; leaves decurrent, ovate, acute or cuspidate, reflexed, remotely ciliated and toothed; flowers terminal, glomerate, rarely solitary; calyx lobes lanceolate, acute, ciliate; corolla one half or twice longer than the calyx. *Thunb. / fl. cap. p.* 175. *A. DC. Prod. p.* 447. *E. & Z.* 2412. *Herb. Un. Itin.* 696. *R. filiformis, Lam. Ill.* 2580, *t.* 123, *f.* 2.

HAB. In the Table Mountain and Kasteelberg, Dec.–Jan. (Herb. Thunb. Holm., Hook., D., Sond.)

Shrubby, ½–1 foot and more high. Branches simple or subpanicled, reddish. Leaves glabrous, approximate, subquadrifarious, not fascicled, 1 line long, in larger specimens sometimes 2 lines long and acuminated, setaceously toothed. Bracteæ a little longer than the leaves, ciliated, with recurved apex. Calyx lobes 1 line long. Corolla white, 5-fid; lobes acute. Capsule cylindrical, glabrous. Seeds ovate.

11. R. muscosa (Thunb.! in Linn. f. Suppl. p. 143); herbaceous, humble, trailing; branches diffuse, filiform, glabrous, leafy; leaves ovate acute, or ovato-lanceolate, spreading reflexed, remotely serrated, glabrous; flowers terminal, solitary; calyx lobes lanceolate, entire, glabrous or subciliated; corolla infundibuliform, 3–4 times longer than the calyx lobes. *Flor. Cap. p.* 175. *A. DC. Monog. p.* 177. *Prod. p.* 447. *E. & Z. enum.* 2413.

HAB. In the Poort, in front of the Table Mountain. Jan. (Herb. Thunb., Holm., Hook., D., Sd.)

Perennial, much branched from the base, branches aggregated, 1-several inches long. Leaves 1–4 lines long, 1 line wide, acute, subattenuated at the base. Calyx tube beset with some leaves. Corolla blue, 4 lines long.

Doubtful Species.

R. decurrens (L'Her. Sert. Angl. p. 2, t. 6), stem erect, branched, herbaceous, leafy, pilose; leaves spreading, decurrent, ovate-oblong, ciliated at the base, nearly entire; flowers solitary, sessile, terminal; tube of calyx ovoid, glabrous, lobes acuminated, ciliated, as long as the tube; corolla campanulate, 5-fid, 3 times longer than the calyx lobes *R. decurrens, A. DC. Monog. p.* 178.

HAB. Cape of Good Hope. Fl. July–Sept. (Unknown to us.)

Leaves 4–8 lines long, 2 lines wide, sessile, flat, paler beneath. Flowers terminal, sessile. Calyx 5-cleft. Corolla 3–4 lines long, half 5-fid, lobes ovate, acute, spreading. Style filiform, equal in length to the tube of corolla; stigma 2-lobed.

XVI. MERCIERA, A. DC.

Calyx 4–5 cleft, tube ovoid. *Corolla* tubular, 4–5 lobed; tube very narrow. *Stamens* 4–5, free; filaments very slender at the base, much longer than the anthers. *Ovary* inferior, with an incomplete dissepiment, 1-celled. *Ovules* 4 (or 2 ?) in the bottom of the ovary. *Style* filiform, usually exserted, glabrous; stigmas 2, very short. *Capsule* unknown, indehiscent? *Monog. Camp. p.* 369. *DC. Prod.* 496. *Endl. Gen.* 3091. *Trachelii et Roellae spec. Thunberg.*

Small subshrubs, with the habit of *Roella ciliata.* Stem erect, branched; branches very leafy, downy, rather woody. Leaves sessile, alternate, erect, crowded, linear-subulate, rigid, ciliated, with a fascicle of smaller ones in the axils. Flowers sessile, solitary, spiked, axillary. Calyx with a very hispid tube, and lanceolate, acute, smoothish segments. Corolla like that of *Trachelium*, long and narrow. Named in honour of *Philip Mercier*, who has written a monograph on the order *Polemoniaceæ*.

Corolla blue ... ... (1) tenuifolia.
Corolla white ... ... (2) brevifolia.

1. M. tenuifolia (A. DC.! l. c. p. 370, t. 5); leaves ciliated; corolla *blue*, as long or longer than the leaves, 5-lobed at the very apex; lobes spreading; style a little exserted.

VAR. *a.* **Candolleana**; leaves elongated, much ciliated; corolla longer than the leaves, pilose on the back; lobes lanceolate. *M. tenuifolia, A. DC. l. c. and DC. Prod. p.* 496. *E. & Z.* 2417.

VAR. *β.* **Thunbergiana**; leaves subciliated; corolla as long as the leaves, glabrous; lobes sublinear. *Trachelium tenuifolium, Thunb. herbar. var. a. Prod. fl. cap. p.* 38, *et Roella tenuifolia, fl. cap. p.* 174, *ex pte.*

VAR. *γ.* **Eckloniana**; leaves short, more or less ciliated; corolla one half or twice longer than the leaves; lobes lanceolate. *M. Eckloniana, Buek! in E. & Z. enum.* 2420.

HAB. Mountains Hauhoeksberg, and in Hottentotsholland, *Thunb. E. & Z. Zey*, 3152. Var. *γ.* near the cataract of Tulbagh, *E. & Z.* Dec.–Jan. (Herb. Thunb., Holm., Hook., D., Sd.)

Stems solitary or several from the root, simple or branched, 4–6 inches high. Leaves in var. *a.* nearly 1 inch long, at the base ½ line wide, recurved at the ciliated apex. In var. *β.* 6–8 lines, in var. *γ.* 3–4 lines long. Calyx lobes ¼ line long. Lobes of corolla 1–1¼ line long. Style not much exceeding the corolla, rarely twice longer than the lobes. Ovary minute, at the base with 2 linear bracts. Var. *γ.* has exactly the habit of *M. brevifolia*, from which it differs by the elongated blue corolla.

2. M. brevifolia (A. DC.! l. c. p. 371); leaves ciliated; corolla white; as long or a little shorter than the leaves, 4–5-lobed at the apex; lobes spreading; style as long or doubly longer than the corolla. *DC. Prod. p.* 496. *Deless. Icon. Select. vol.* 5, *p.* 7, *t.* 17. *M. brevifolia et leptoloba; E. & Z.* 2418, 2419. *Trachelium tenuifolium, var. β. Thunb.! herb. Roella tenuifolia, fl. cap. ex part.*

VAR. *γ.* **leptoloba**; leaves longer; lobes of corolla linear, nearly as long as the tube. *M. leptoloba, A. DC. l. c. p.* 371.

HAB. Babylon's Toornberg, *Thunb. Zey.* 3154. Dutoitskloof, *Drege.* Zwarteberg, Caledon, and on the Steenberge near Muysenberg, *E. & Z.* Genadenthal, *Dr. Roser.* Var. *β.* Cape, *Dr. Thom,* In district Caledon, *Zey.* Dec.–Feb. (Herb. Thunb., Holm., Hook., D., Sd.)

Habit of the preceding, but branches more slender. Leaves 4–6 lines long, in some specimens rarely 3 lines long, and very rigid-ciliated. Corolla 4–5 lines long. Style not exceeding the lobes of corolla, or twice longer. *M. heteromorpha, Buek! in E. & Z. enum.* 2421, *is a Rubiacea.*

XVII. SIPHOCODON, Turczan. (Emend.)

Calyx 5-parted, tube ovoid, *Corolla* tubular, 5-lobed at the apex; tube narrow. *Stamens* 5, filaments short, inserted in the middle of tube of the corolla; anthers free, 2-celled, longer than the filaments. *Ovary* circumscissile below the calyx lobes, 3-celled; cells 2-ovulate; placenta affixed to the vertex. *Style* filiform; stigmas 3, revolute. *Capsule* opening transversely, the upper part of the calyx falling off; the cup-like remains 3-celled, cells 1-seeded. *Seeds* ovoid, with very loose, thin, rugose testa. *Turczaninoff decas. f. gen. nov. in. Bullet Mosc.* 1852, 3, *p.* 38.

A slender, much branched, quite glabrous subshrub, with the habit of *Thesium*. Branches subangulate, filiform, virgate. Leaves alternate, linear, trigonal, minute, squamiform, appressed. Flowers blue, shortly peduncled, racemose; racemes often panicled.

1. S. spartioides (Turcz. l. c.).

HAB. Sandy grassy places on Klynhowhoek, *Zey.* 3103. Rocky places, Zwarteberg, Caledon, *Zey.* 3103, *b, Dr. Pappe.* Nov.-Dec. (Herb. D., Sd.)

Stem 1-2 feet high. Leaves 1 line long, distant. Flowers at top of the branches. Peduncles 1-2 lines long. Calyx lobes linear-lanceolate, acute, twice longer than the tube, but 3 times shorter than the corolla. Corolla about ½ an inch long. Capsule, when ripe and opened, 1 line long.

(*Doubtful genus.*)

RHIGIOPHYLLUM, Hochst.

Calyx-tube ovate, short ; limb 5-parted, lobes linear-lanceolate, much longer than the tube. *Corolla* elongated, tubular, tube very narrow, much longer than the calyx, limb 5-parted with spreading, oblong, obtuse lobes. *Stamens* 5, subinclosed, filaments inserted below the apex of the tube (but decurrent nearly to the base). *Ovary* inferior, 3-locular, cells with many ovules. *Style* very long, filiform, exserted ; stigmas 3-lobed, lobes spreading, recurved. *Capsule* 3-celled, as it seems operculate by the persistent base of style ; ripe capsule unknown. *Regensb. Bot. Zeitg.* 1842, 13.

A very rigid shrub, with the habit of *Roella.* Leaves imbricated. squarrose, quite entire. Flowers capitate, bracteolate. Name from ριγιος, stiff, and φυλλον, leaf.

1. R. squarrosum (Hochst. in Krauss, pl. ext. 1088).

HAB. Stony sandy hills near Elin, Zwellendam, *Dr. Krauss.* Dec.

1-1½ foot high, branches erect. Leaves rhombic-obovate, acute, coriaceous, quite smooth, subquadrifarious-imbricated, appressed to the stem by a cuneate base, the apex spreading, squarrose, quite entire, shining, the middle nerve prominent below. Flowers purple, aggregated to a head, separated by lanceolate, rigid bracteæ, and the pedicels with 2 similar bracteolæ.

Campanula stellata, Thunb. ! Mem. Acad. Petrop. 4, p. 373 is Viola scrotiformis, DC. in fruit. Campanula integra, var. δ. herb. Thunb. is Tenaris rubella, E. Mey. an Asclepiadea.

TRIBE 3.—CYPHIEÆ (Gen. XVIII.)

XVIII. CYPHIA, Berg.

Calyx-tube turbinate, adhering to the ovary, limb 5-parted, the segments subunequal. *Petals* 5, their claws conniving in a split tube, or slightly cohering above or below ; limbs subequal, spreading as a 2-lipped corolla. *St.* 5, *fil.* pilose ; *anthers* free, often hispid on the connective at back. *Ovary* inferior or half inferior, 2-celled, many ovuled ; *style* simple, *stigma* with an obsolete, ciliate indusium. *Caps.* 2-celled, many-seeded. *Endl. Gen.* 3041. *DC. Prodr.* 7, *p.* 498.

Herbaceous, perennial plants, with alternate, undivided, pinnatifid or pinnate leaves, and blue, white, or red flowers. Name from κυφος, curved ; in reference to the stigma being gibbous.

   * *Eucyphia.* Corolla hypocrateriform or tubulose, bilabiato-pentapetalous.

    1. Stem erect, scape-like ; leaves all radical.

Leaves oblong or obovate-oblong, denticulate	(1) **Phyteuma.**
Leaves obovate, incised, or subpinnatifid	(2) **incisa.**
Leaves deeply pinnatifid or pinnate	(3) **Cardamines.**

    2. Stem erect, leafy, radical leaves none.

Leaves palmatipartite, lobes linear-oblong, entire or divided ...	(8) **bulbosa.**
Leaves linear-lanceolate, remotely serrulated ; pedicels shorter than the bracteæ	(4) **persicifolia.**
Leaves narrow-linear (4-6 lines long) ; pedicels longer than the ovato-acuminate bracteæ ; calyx-lobes triangular, acute, longer than the tube	(5) **campestris.**

Leaves narrow-linear (6–12 lines long) ; pedicels longer than
the capillary bracteæ ; calyx-lobes triangular, acute, as
long as the tube ... ...   (6) **linarioides.**
Leaves linear, elongate (2–3 inch. long) ; pedicels shorter than
the capillary bracteæ ; calyx-lobes lanceolate, subulate ...   (7) **assimilis.**
Leaves oblongo-lanceolate, acute, denticulate ; petals longer
than the calyx-lobes... ... ...   (9) **elata.**
Leaves oblong, obtuse, apiculate, crenulate ; petals bearded,
nearly as long as the calyx-lobes ... ...   (10) **oblongifolia.**
Leaves cordate, acute, serrato-dentate . ...   (11) **Gerrardi.**

3. Stem twining.

Leaves similar, broadly-ovate, cuneate at the base, serrato-
dentate ... ...   (12) **corylifolia.**
Leaves ovato- or oblongo-lanceolate, lobate at the base ; the
upper lanceolate; racemes axillary, few-flowered ; petals equalling the style ... ...   (13) **crenata.**
Leaves ovate or oblong, acute, subtrilobed at the base ; the
upper 3-lobed, lobes lanceolate ; flowers axillary, solitary ;
petals twice longer than the style ... ...   (14) **heterophylla.**
Leaves petiolate, lanceolate, elongate, entire or denticulate ...   (15) **sylvatica.**
Leaves sessile, lanceolate, elongate, undulately and coarsely
dentate or incised ... ...   (16) **undulata.**
Leaves sessile, linear, callously dentate with reflexed margins ;
bracteæ 3-partite or entire ... ...   (19) **volubilis.**
Leaves ternate or deeply 3-partite, segments lanceolate (1–1½
inch long, 2–3 lines wide), entire, or the lateral 2-parted ;
bracteæ 3-partite or entire ... ...   (17) **dentariæfolia.**
Leaves digitately 3–5-partite ; segments linear (1½–1 inch long,
¼ line wide); bracteæ 3-partite or entire ... ...   (18) **digitata.**

** *Cyphiella.* Corolla hypocrateriform, tube at length irregularly split at the base,
not bilabiato-5-petalous.

Stems twining ... ...   (20) **Zeyheriana.**

1. **C. Phyteuma** (Willd. Spec. 1, p. 953); perennial; stem simple,
erect, striate, glabrous or hairy, naked or with some narrow scales ;
leaves oblong or obovate-oblong, remotely denticulated, tapering in a
short petiole, hairy or glabrous ; flowers subspicate ; pedicels shorter
than the ovate-acuminate, ciliate bracteæ, glabrous ; calyx glabrous,
lobes lanceolate-acuminate, serrate, at length reflexed, longer than the
tube ; corolla 3-times longer than the calyx-lobes. *Ker. Bot. Reg. t.*
625. Presl in comm. pl. Drege, p. 292, E. & Z. 2423. Lobelia Phy-
teuma, Linn. spec. p. 1319. *Thunb. Fl. Cap. p.* 184. *L. nudicaulis*
Lam. dict. 3, p. 590. *Cyphia serrata Spr. syst.* 1, *p.* 809. *Herb. Un. itin.*
230. *Zey.* 3098.

HAB. Sandy flats and hills near Capetown, Hotthld., Rivierzonder Ende, Buffeljagdrivier, Caledon, Paarl, Klipfontein, Roodesand, Giftberg, etc. Sept.–Oct.
(Herb. Thunb., Holm., Hook, D., Sd.)
Root tuberous. Stem about 1 foot, spike 3–6 inches long. Leaves spreading,
1–2 inches long, obtuse or acute. Calyx of the upper flowers nearly sessile, lobes
rather longer than the tube. Bracteæ 3. Corolla ⅓ inch long, 2-labiate, pink or
flesh-coloured ; petals at length separated, lanceolate. Anthers hirsute. Capsule
half superior ; the base hemispherical, 3 lines long, 10-nerved and reticulated ;
2-celled ; the upper part free, a little longer than the base, ovate, acute, bisulcate ;
at length dehiscent by 2 bipartite valves. Seeds winged ; perfect not seen.

2. **C. incisa** (Willd. Spec. 1, p. 953); perennial ; stem simple, erect,

sulcate, hirsute naked; leaves petiolate, hirsute, as long or rather shorter than the petiole, obovate, obtuse, incised, subpinnatifid; flowers race-mose; pedicels longer than the bracteæ, hirsute as well as the calyx with erect, linear-lanceolate lobes; corolla 3–4 times longer than the calyx-lobes. *A. DC. l. c. p.* 449, *E. & Z.,* 2424. *Lobelia incisa, Thunb. Fl. Cap. p.* 185.

HAB. Sandy places near Capetown and in Swellandam, *Thunb., E. & Z., W. H. Harvey.* Oct.–Nov. (Herb. Th., Hook, D., Sd.)

Very similar in habit to the preceding, but a distinct species. The whole plant is hirsute; the leaves sinuato-pinnatifid, particularly at the base, 1 inch long; the pedicels are not very short, as indicated by *Thunberg,* but 4–6 lines long, at the base with 3 lanceolate or linear bracteæ. Raceme 4–8 flowers; the calyx-lobes ciliated but not serrate; corolla 2-labiate, petals lanceolate; anthers hirsute. Capsule as in *C. Phyteuma;* the upper part longer than the tube, splitting in 4 (or in 2, at length bipartite) valves. Seeds compressed, flat, broadly winged, orbicular, 2 lines in diameter; wings thin, scarious, nearly as wide as the elliptical body of the seed. Embryo straight, lying in the middle of fleshy albumen; radicle next the hilum.

3. C. Cardamines (Willd. Spec. 1, p. 953); perennial; stem simple, erect, sulcate, hirsute, naked; leaves petiolated, pinnatifid or pinnate; leaflets ovate, toothed or 2–3 lobed, pilose; flowers racemose, pedicels as long or longer than the bracteæ, subhirsute as well as the calyx with erect, linear-lanceolate lobes; corolla 3–4 times longer than the calyx-lobes. *DC. l. c., E. & Z.* 2425. *Lobelia Cardamines, Thunb.! Fl. Cap. p.* 185.

HAB. Sandy fields near Capetown and at Greenpoint, *Thunb., E. & Z., W. H. Harvey.* Aug.–Sept. (Herb. Holm., D., Sd.)

Probably a variety of *C. incisa,* with more rigid leaves. Petioles 1½ inch, narrow, as long or longer than the limb, with 3-lobed terminal lobes and toothed or lobed lateral segments. Raceme 4–6-flowered; pedicels 4–2 lines long. Corolla and anthers as in the preceding; capsule a little broader, the seeds quite the same.

3. C. persicifolia (Presl! comm. pl. Drege, p. 296); perennial, glabrous; stem erect, simple, angulate-striate, leafy; leaves sessile, linear-lanceolate, subacute, remotely serrulated, with revolute margins; upper ones much smaller, subappressed; flowers subspicate; pedicels shorter than the lanceolate bracteæ, calyx-lobes linear-lanceolate, serrate, erect, longer than the tube, 4 times shorter than the corolla; anthers glabrous. *DC. l. c., Harv. Thes. cap. t.* 159.

HAB. Grassy hills near Morley, *Drege;* Transkey, *Mrs. F. W. Barber.* (Herb. D., Sd.)

Stem 1½–2 feet high. Leaves 2–3 inches long, about two lines wide, uninerved, the upper 8–4 lines long, ½ line wide. Spike 4–6 inches, sometimes branched at the base, many-flowered, pedicels ½–1 line long. Flowers ½-inch, incurved. Corolla bilabiate, petals free. Filaments hairy at apex. Ovary half superior. Stigma obtuse.

5. C. campestris (E & Z.! 2438); glabrous; stem erect, terete, striate, branched at the base; branches divergent; leaves sessile, linear, narrow, remotely denticulate and entire; racemes flexuous; pedicels one half longer than the ovato-acuminate, entire or 3-lobed bractea, bibracteolate at the apex; calyx-lobes triangular, acute, erect, longer than the very short turbinate tube. *C. linarioides Presl! Comm. pl. Drege,* 1, p. 294, *ex parte.*

HAB. Hills near Zwarteberg, Caledon, *E. & Z.*, *Drege.* Aug. (Herb. Sond.)
Root tuberous, edible. Stem 6-12 inches, often flexuous. Leaves 4-6 lines long,
½ line wide. Raceme simple, 2-4 inches long, sometimes branched. Pedicels 2-3
lines long. Calyx 1 line long ; corolla 4-5 lines. Filaments hairy ; anthers hirsute.

6. C. linarioides (Presl! E. & Z. ! 2437) ; glabrous or minutely
pubescent above ; stem erect, branched at the base ; branches divergent ; leaves sessile, linear, narrow, remotely denticulate ; raceme flexuous ; pedicels longer than the capillaceous entire bracteæ, bibractcolate
at the apex ; calyx-lobes triangular, acute, erect, as long as the short
turbinate tube ; filaments three times longer than the anthers.   *C.
tenuifolia, A. DC. l. c. p.* 498.   *Lightfootia sessilifolia Zeyh. pl. exs.* 227,
non 203.   *C. linarioides β. micrantha Presl! in Comm. pl. Dreg. p.* 294.

HAB. On the fields near the Zwartkopsriver, *E. & Z. ;* Rivierzonder Einde, near
Appelskraal, *Zeyher,* 3088. Grahamstown, *Col. Bolton ;* distr. of Albert, *T. Cooper ;*
on the Garip, *Drege.* Feb.-March. (Herb., Hook., D., Sd.)
Quite similar to *C. campestris,* and only distinct by shorter pedicels, capillaceous
bracts and calyx-lobes. Leaves 6-12 lines long, about ½ line wide. Raceme unilateral, 1-4 inches long. Pedicels 1-1½ line long, fruit-bearing longer. Flowers
usually a little smaller than in the preceding, 4 lines long, not much longer than the
stamens. Capsule half superior, the size of a pea ; the base hemisphaerical, 10-
nerved ; the upper free part acute, as long or somewhat longer than the base, dehiscent by 4 valves. Seeds small, triquetrous. The tuberous, eatable root is called by
the Hottentotts "Barroe."

7. C. assimilis (Sond.) ; glabrous ; stem erect, simple ; leaves sessile, linear, elongate, narrow, remotely denticulate ; raceme evidently
pedunculate ; pedicels shorter than the capillaceous, entire bractea ;
calyx-lobes lanceolate, subulate, as long or longer than the turbinate
tube ; filaments as long as the anthers.   *C. linarioides γ. major, Presl.!
in E. & Z., enumer. p.* 391.

HAB. Hills near Klipplaat river, Tambukiland, *E. & Z. ;* Alice, Victoria East,
Dr. Pappe. April. (Herb., Hook., D., Sd.)
Perennial like *C. linarioides,* 1-1½ foot high, leafy in the lower and middle part,
naked above. Leaves 2-3 inches long, nearly 1 line wide. Raceme 4-8 flowered.
Pedicels 1 line long. Petals lanceolate, 6-8 lines long. Stamens short, anthers
hirsute.

8. C. bulbosa (Berg. cap p. 172) ; root tuberous ; stem erect, simple,
minutely pubescent or subglabrous ; lower leaves palmatipartite, tapering in a short petiole ; lobes linear-oblong, entire or irregularly 2-fid
and toothed ; upper leaves sessile, deeply 3-7-fid ; flowers spicate-
racemose ; pedicels shorter than the leaf, bibracteolate at apex ; calyx-
lobes lanceolate, 3 times longer than the turbinate tube.   *Willd. Spec.
1, p.* 953, *E. & Z.* 2426. *Herb. Un. itin.* 231. *Zeyh.* 3096. *Sieb. pl.
cap.* 136. *Lobelia bulbosa Linn. Spec. p.* 1319. *Thunb. fl. cap. p.* 184.
Burm. Afr. l. 38, *f.* 2. *Cyphia bulbosa, var. α. et β. Presl in Comm. pl.
Dreg. C. Botrys Willd.! in Roem. and Sch. syst.* 5, *p.* 477.

HAB. Stony hills of the mountains in the districts of Cape and Stellenbosch.
Aug.-Sept. (Herb., Thunb., Holm., Willd., Hook., D., Sd.)
Stem simple, rarely with a few branches, ½-1 foot, striate or angulate-striate.
Leaves commonly 1 inch long and wide, on a shorter cuneate petiole ; lobes 1 line,
sometimes ½ line wide (var. angustiloba DC.), the middle pinnatifid. Uppermost
leaves often entire. Raceme many flowered. Flowers about ½ inch long. Capsule

4 lines long, half superior, the upper free part ovate acute, longer than the hemispherical, 10-nerved lower part or tube, dehiscing by 4 acute valves. Seeds compressed, angular, subtriquetrous, glabrous, minutely punctate, nearly 1 line long.

9. C. elata (Harv. Thes. Cap. t. 160); minutely pubescent or glabrous; stem erect, simple, leafy; leaves sessile, oblongo-lanceolate, acute, denticulate; flowers subsessile, disposed into a dense-flowered terminal leafy spike; calyx-lobes lanceolate, deeply incised or pinnatifid, separated by the lanceolate minutely pubescent or subglabrous petals.

VAR. γ. **glabra**; (Harv.), stem, leaves, bracteæ, and calyx glabrous; petals subpubescent inside.

HAB. Draakensberg, Orange Free State, *T. Cooper*, 1087. Var. β. Port Natal, *Gerrard and M'Ken.*, 1462. (Herb., D.)

Stem 3-4 feet high. Leaves coriaceous, veined beneath, with subrevolute acutely denticulate margins, 2-2½ inches long, 6-8 lines wide, the upper smaller. Bracteæ ovate-acuminate, including 1 or several flowers or a few-flowered spike. Calyx tube turbinate with 2-4 linear-lanceolate ciliated bracteæ, lobes of calyx 3 times longer than the tube. Petals narrow, 4 lines long. Filaments hairy, anthers hispid.

10. C. oblongifolia (Sond. and Harv.); minutely hairy, subglabrous at the base; stem erect, simple, leafy; leaves sessile, oblong, obtuse, apiculate, crenulate-toothed; flowers disposed in a terminal spike; bracteæ serrato-incised; calyx-lobes lanceolate, with 2 or 4 tooth-like cilia, scarcely longer than the lanceolate, bearded petals.

HAB. Ingoma, Port Natal, *Ger. and M'K.* 1463. (Herb., D.)

Very nearly allied in habit and characters to *C. elata.* Lower leaves 1½ inch long, 6 lines wide, the upper gradually smaller and more acutely denticulate, leaves alternate, including one flower. Bracteæ equalling the sessile flower. Petals very narrow, 3-4 lines long, bearded by white hairs inside.

11. C. Gerrardi (Harv. Mss.); pubescent all over; stem erect, simple or branched above, leafy; leaves sessile, cordate or ovate cordate, acute, serrato-dentate; flowers aggregated in the axils of the bractea, forming an elongated, leafy, many flowered spike, sometimes branched at the base; bracteolae linear, longer than the pedicels; calyx-lobes linear acuminate, with some cilia-like teeth, much longer than the tube; petals narrow, scarcely longer than the calyx-lobes.

HAB. Ingoma, Zulu, Port Natal, *Ger. and M'K.* 1461, 1464. (Herb., D.)

4-6 feet high. Cauline leaf—only one seen—3 inches long, 2 inches wide, coriaceous, veined beneath, multidentate. Upper leaves much smaller, cuspidate. Bracteæ ovato-acuminate, about as long as the fascicles of flowers. Calyx tube 1 line, the lobes 2-2½ lines long. Petals lanceolate, acuminate, and recurved at apex, hairy.

12. C. corylifolia (Harv., Thes. Cap. t. 161); branches climbing, angulate-striate, minutely pubescent; leaves petiolate, broadly ovate, acute, cuneate at the base, serrato-dentate, glabrous; flowers disposed into an elongated remote-flowered spike, the lower fascicled in the axil of the leaves, the upper geminate or solitary with lanceolate or linear bracteæ; calyx-tube turbinate, hairy, 3-times shorter than the linear-lanceolate 1-3-dentate lobes; petals lanceolate, subglabrous, one half longer than the calyx.

HAB. Ingoma, Port Natal, *Ger. and M'K.* 1465. Feb. (Herb., D.)

Perennial.   Branches 4–6 feet.   Leaves about 1–1¼ inch long and broad, much
veined and paler beneath; petiole 4–6 lines long.   Flowers shortly pedicellate, the
lower fascicles 6–8 flowered, 3–4 times shorter than the leaf, distant from the following fascicle by an interval of 2 inches.   Uppermost flowers more approximte, as
long or longer than the bracteæ.   Corolla 4–5 lines long.   Ovary half superior.

13. C. crenata (Sond.); quite glabrous; stem and branches filiform
striate, twining; leaves shortly petiolate, ovato- or oblongo-lanceolate,
dentato-lobate at the base, remotely denticulated at the apex; upper
ones lanceolate crenato-serrate; racemes axillary, few-flowered, subsessile, about as long as the leafy bracteæ; pedicels much longer than the
linear bracteolæ; calyx-lobes linear-lanceolate, erect, spreading, acute,
4–5 times longer than the very short tube, ½ shorter than the narrow
spreading petals. *Lobelia crenata. Thunb.! Prod. p. 39. Fl. Cap. p. 180.
Cyphia Dregeana Presl.! Comm. pl. Dreg. p. 294. C. atriplicifolia, Presl!
E. & Z. 2428.*

Hab. Sandy places, Rietvalley and Simonsbay, *E. & Z.*, near Saldanhabay,
Thunb. Drege. (Herb., Thunb., Sd.)
Habit of *Convolvulus* like the preceding.   Leaves 2–3 inches long, 6–10 lines wide,
membranaceous, remotely and bluntish-toothed, the upper ones about 1 inch long,
3–4 lines wide, the bracteæ entire or nearly so.   Petiole 2–3 lines long.   Raceme
shortly peduncled or subsessile, 3–5 flowered, pedicels unequal, 4–8 lines long, with
2 bracteolæ at the base.   Calyx-lobes entire, 2 lines long.   Petals acuminate, rather
longer than the calyx, equalling the style, with incrassate apex.   Capsule subglobose, the size of a pea.   It varies but rarely with solitary pedicels in the uppermost axils.   The raceme, if indistinctly-peduncled, has the aspect of an umbel.
Drege's specimens are more perfect than those of *Thunberg.*   *C. crenata Presl* is a
doubtful plant; it seems not different from   *C. sylvatica.*

14. C. heterophylla (Presl ! E. & Z. 2429); glabrous; stem and
branches twining, terete; leaves petiolate, the lower ovate or oblong,
acute, unequally denticulate or serrate, angulate or subtrilobed at the
base, upper ones trilobed, lobes lanceolate, acute, serrate and entire,
middle lobe 2–3-times longer; flowers axillary, solitary, disposed in an
elongate raceme; pedicels bibracteolate in the middle, equalling the
ovato-lanceolate bracteæ; calyx-lobes lanceolate, acute, 3-times longer
than the tube, 4–5-times shorter than the petals. *E. Mey. comm. pl.
Drege, p. 297. DC. l. c. p. 500.*

Hab. Woods at Ganzevalei, near Doukamma, *Drege;* George, *E. & Z.;* Albany,
Mrs. F. W. Barber. Sept. (Herb. Hook, Sd.)
Branches very long.   Leaves very variable in size, the lower and middle 1–2 inches
long, the upper 6–8 lines long, with very short lateral lobes.   Petiole 4–6 lines long.
Bracteæ and pedicels 2–3 lines long.   Calyx-lobes acuminate.   Corolla 6–8 lines
long, twice as long as the stamens and style.   It varies with sessile and petiolate
bracteæ, and shorter pedicels.

15. C. sylvatica (Eckl.! in South Afr. Quart Journ. 1830, p. 373);
glabrous; stem subangulate, with the branches twining; leaves shortly
petiolate, alternate, distant, narrow, lanceolate, elongate, entire or remotely denticulate, acute, callous, attenuate at the base, floral ones
sublinear; flowers axillary, solitary, peduncled; bracteæ minute, capillary, much shorter than the peduncle; calyx-lobes lanceolate, acute,
a little longer than the turbinate tube. *E. & Z. 2427. C. salicifolia,
E. & Z.! 2432.*

HAB. In thickets near the Zwartkopsriver, *E. & Z., Zey.* 3090 ; *Dr. Pappe, T. Cooper,* 1466 ; Albany, *T. Williamson;* Boshmansrivier, *E. & Z.;* Brackfontein, *E. & Z.* April–Aug. (Herb. D., Sd.)

In habit like *C. crenata,* but the leaves are narrower, entire, and the flowers solitary. Leaves 1½–3 inches long, 2–3 lines wide ; petiole 1 line long, sometimes nearly wanting. Pedicels 3–4 lines long, twice shorter than the corolla. Capsule the size of a large pea, half superior, dehiscing in 4 valves ; the tube hemispherical, 10-nerved. Seeds angular, nearly 1 line long. The large root is eatable, and called *Bosh-barroe* by the Hottentotts.

16. C. undulata (Eckl. ! in South Afric., Quart. Journ. 1830, p. 373) ; glabrous ; stem subangulate, simple, erect, at length subbranched and twining ; leaves sessile, the radical aggregate, oblong, or ovato-lanceolate, undulated ; cauline alternate, lanceolate or elongate-lanceolate, undulately and coarsely dentate or subincised, floral ones linear-lanceolate or linear, entire or denticulate ; pedicels axillary, solitary, racemose ; bracteæ capillary, minute, shorter than the pedicels ; calyx-lobes lanceolate, acute, longer than the urceolate tube. *C. anomala, E. & Z.!* 2434. *C. angustifolia, E. & Z.* 2436.

HAB. Amongst grasses in the hills between Coega and Zondagsriver, near Trompeterspoort and Grootriver, Winterhoeksberg and Zuureberg, *E. & Z.;* above Elandsriver, *Zey.* 3087. March.–June. (Herb. Hook, D., Sd.)

Very near *C. Sylvatica* ; the stem is often simple, scarcely twining, more leafy and the sessile leaves wider and eroso-dentate or subincised. Root fleshy, eatable, called *Veld-barroe.* Leaves about 1 inch long, 3–4 lines wide, in the upper part longer and narrower, but near the base smaller and wider. Flowers forming a short or elongate, secundate raceme, usually leafless. Corolla half an inch long. Capsule half superior, the size of a pea, 4-valved. Seeds ½ line long, 3-quetrous. *C. undulata* is the simple-stemmed plant, *C. anomala* the same with elongated raceme, *C. angustifolia* much-branched.

17. C. dentariæfolia (Presl ! in E. Mey. Comm. pl. Drege, p. 295) ; glabrous, stem terete, striate, branched ; leaves ternate or deeply tripartite, lower ones petiolate ; leaflets or segments tapering at the base or on a short petiolule, lateral ones bipartite, the segments as well as the middle lobe or leaflets lanceolate, acuminate, entire or remotely serrulate ; upper leaves sessile or shortly petiolate, smaller 3-fid, with linear, entire lobes ; bracteæ linear, entire ; pedicels axillary, solitary, shorter than the leaf, bibracteolate ; calyx-lobes lanceolate, acute, twice longer than the tube. *C. digitata,* var. β. *major, E. & Z.!* 2439.

HAB. Among shrubs near Driefontein, *Drege,* 6276 ; on the side of Table Mountain, near Constantia, *E. & Z.* ; Bergrivierspoort, *Zey.* 1048. July–Aug. (Herb. Sd.)

Stronger than *C. digitata.* Petioles of the lower leaves 4–6 lines long. Middle lobe or leaflet 1–1½ inch long, 2–3 lines wide, the lateral lobes or leaflets one half shorter. Floral leaves equalling or longer, rarely shorter than the pedicels. Calyxlobes acuminate, nearly 1½ line, corolla 6–8 lines long. Filaments hairy ; anthers hirsute.

18. C. digitata (Willd. spec. 1, p. 953) ; glabrous ; stem terete, striate, branched ; leaves shortly petiolate or subsessile, digitately 3–5-parted, segments linear-oblong or linear, entire or bifid, equal or the middle longer ; upper leaves or bracteæ 3-parted, rarely entire ; pedicels axillary, solitary, as long or longer than the bractea ; calyx-lobes ovato-lanceolate, acute, longer than the tube, *E. and Z.* 2439, *excl. var.* β *Lo-belia digitata Thunb. Fl. Cap. p.* 185. *C. digitata et polydactyla Presl! in E. Mey. Comm. pl. Drege, p.* 296, *DC. l. c. p.* 500. *C. angustiloba Presl! in E. & Z.* 2430.

VAR. β. **tomentosa** ; pubescent or leaves, pedicels and calyx subglabrous. *C. to-mentosa, Presl ! l. c. p.* 295, *C. digitata, E. & Z.* 2439, *ex pte.*

HAB. Hills and sandy places near Capetown and Drakenstein, Thunb. ; Ronde-bosh, Brackfontein, and in Zwarteberg, Caledon, *E. & Z.; * Tulbagh, *Dr. Pappe; * Hassaquaskloof and Rivierzonder Einde, *Zey.* 3091, 3095 ; Dutoitskloof, Wupper-thal and Modderfontainsberg, *Drege;* var. β. Paarl and on Hexrivier, *Drege.* Jun.-Aug. (Herb. Holm., Hook, Sd.)
Root fleshy, bulbous, eatible. Stem twining, filiform. Subradical leaves ovate or ovato-lanceolate, serrate or incised ; the cauline digitate, lobes ½-1 inch long, ½ line wide, the middle often somewhat longer and a little wider. Bractea 3-lobed, middle lobe longer. Pedicels with two very small bracteolæ. Calyx-lobes 1 line long. Corolla 4-5 lines long.

19. C. volubilis (Willd. spec. 1, p. 952) ; glabrous; stem terete, branch-ed ; leaves sessile, distant, linear, callously dentate with incrassate re-flexed margins ; upper leaves and bracteæ entire or tripartite ; pedicels axillary, solitary, as long or longer than the bracteæ ; calyx-lobes ovato-acuminate, longer than the turbinate tube. *Lobelia volubilis, Thunb.! Prod. p.* 39, *fl. cap. p.* 179. *Herb. Un. Itin.* 232, *C. volubilis, longipetala and latipetala, Presl ! E. & Z.* 2433, 2431, 2440.

HAB. Hills and mountains, Lions and Table Mountain, *Thunb., W. H. Harv.* Palmietriver and Zwarteberg, Caledon, *E. & Z.; * Paarlberg, Sparrbosch and Du-toitskloof, *Drege;* Simonsbay, *C. Wright; * Rivierzonder Einde and on the foot of Babylonisches Toornsberg, *Zeyh.* 3093, 3094. Jul.-Sept. (Herb. Thunb., Holm., Hook., D., Sd.)
Perennial. Leaves 1 inch or more long, ½-1 line wide. Corolla 6-8 lines long, bilabiato-pentapetalous. It varies with subentire leaves and hairy or subglabrous anthers, and is distinguished from *C. digitata* by linear not digitate lower leaves and larger flowers, with usually longer lobes of corolla.

20. C. Zeyheriana (Presl ! E. & Z. 2442) ; glabrous, stem twining, terete, striate, somewhat branched ; leaves sessile, distant, digitately 3-5-parted ; segments unequal, linear, denticulate ; floral ones smaller, tripartite ; flowers axillary, solitary, shortly pedicellate ; calyx-lobes linear-lanceolate, acute, longer than the tube ; corolla hypocrateriform, tube 3-times longer than the calyx.

VAR. γ. **Eckloniana** ; flowers smaller. *C. Eckloniana, Presl ! E. & Z.* 2441.

HAB. Stony hills near Brackfontein, Clanwilliam ; var. β. on the River Zonde Einde, Swellendam, *E. & Z.* July-Sept. (Herb., Sd.)
Very similar to *C. digitata,* and only differing by mostly tripartite leaves and gamopetalous, not bilabiato-pentapetalous corolla. Cauline leaves 1-1¼ inch long, about 1 line wide. Corolla 6 lines long, lobes ovate or oblong-lanceolate. It varies but rarely with undivided leaves at the base of the branches.

TRIBE 4.—GOODENOVIEÆ (Gen. XIX.)

XIX. **SCAEVOLA**, Linn.

Calyx 5-cleft or 5-toothed, rarely subentire. *Corolla* cleft longitu-dinally on the upper side, limb 5-parted, all to one side ; lobes winged, about equal in size and shape. *Anthers* free. *Indusium* of stigma ciliated. *Drupe* fleshy or dry, 1–4-locular; cells 1-seeded. *R. Brown Prod. p. 582. Endl. Gen. 3038.*

Shrubs and herbs very variable in habit. Leaves alternate, rarely opposite, quite entire or toothed or subincised. Inflorescence axillary. Flowers bibracteate. Corolla white or blue, rarely yellow. Name from *scaeva*, the left hand, in reference to the form of the corolla.

1. S. Thunbergii (E. & Z. ! 2422); shrubby, glabrous; leaves obovate, obtuse, quite entire, subpetiolate, fleshy, with bearded axils; peduncles axillary, angulate, dichotomous, cymose-spiked; bracteæ linear, opposite; calyx campanulate, bluntish, denticulate; tube of corolla very villous inside, twice longer than the calyx, lobes marginate; style villous at the base; drupe globose, smooth, bilocular. *Presl in E. Mey. Comm. pl. Dreg. p. 292. S. Lobelia Thunb.! fl. cap. p. 186.*

HAB. Sea shore near Mosselbay, *Thunb.*; Albany, *H. Hutton;* Algoabay and mouth of the Zwartkops river, *E. & Z. Zeyh,* 3086; Port Natal, *Drege, T. Williamson.* Nov.–Dec. (Herb., Thunb., Holm., Hook., D., Sd.)

Leaves 2–3 inches long, 1½ in. wide. Flowers nearly 1 in. long. Drupe the size of a small cherry. [Scarcely, if at all, different from *S. Plumieri*, Vahl. W.H.H.]

ADDENDA AND CORRIGENDA.

Page 1, line 35, from the char. of **Stylocoryne** remove the words "or funnel-shaped," and transfer them to that of **Gardenia**, after the words " salver-shaped."

Page 17, line 24, for **parviflora** read **pauciflora.**

Page 47, for the synonyms of **Ethulia conyzoides**, instead of *D. Kraussii, D. gracilis, D. angustifolia*, read *E. Kraussii, E. gracilis, E. angustifolia.*

Page 60, line 46, for (Gen. 1-24.) read (Gen. 11-24).

Page 86, in the gen. char. of **Erigeron**, insert "or" between the words "honey-combed" "fimbrilliferous."

Page 92, in the gen. char. of **Garuleum**, for "abortive ovules" read "abortive ovaries."

Page 94, for **2. C. tenuifolia** read **3. C. tenuifolia.**

Page 130, in the table, char. of (100) **Leyssera**, for "pappus in the ray of short *bristles*" read "pap. in the ray of short *scales.*"

Page 153, immediately before **Eumorphia**, introduce :

Sub-tribe 3. **ANTHEMIDEÆ.** (Gen. 50-75.)

Page 158, under **G. trifurcata**, add the syn. *Osteospermum tenuifolium, DC. l. c. 467, ex pte.*

Page 159, after **G. ericoides**, insert :

12. G. polytrichoides (Harv.); stems shrubby, dichotomous, closely set with slender leaves, glabrous ; leaves setaceo-subulate, mucronate, recurvo-patent or squarrose; pedunc. filiform, elongate; inv. sc. subconnate at base, lanceolate.

HAB. Shady slopes, E. of Woert's Hill, Grahamstown, *P. Mac Owan*, 517. (Hb. D.) Like *G. ericoides*, but with much longer and more slender and tapering, spreading or recurved leaves. Lvs. 4–6 l. long, not ½ l. diameter. Pedunc. 2–3 in. long.

(Imperfectly known species.)

G.? pinnatipartita (DC.); " glabrous; stem shrubby, erect, simple, or scarcely branched, densely leafy ; leaves crowded, pinnatipartite, the lobes on each side 2–3 pair, together with the terminal lobe and rachis linear, entire, scarcely acute; pedicels nude, 1-headed, 3–4 times longer than the leaves; inv. sc. concrete at base only; rays 12–13; ray-ach. muricate-echinulate (young)." *Osteospermum pinnatipartitum, DC. l. c. p.* 467.

HAB. Langekloof, *Drege.*

G.? leiocarpa (DC.); "glabrous; stem shrubby, erect, slender, sub-simple, densely leafy; leaves pinnatipartite, lobes and rachis linear, entire, mucronate, in 3-pair, the lower lobes shorter, 3-terminal longer, equal; pedicels 4 times as long as the leaves, nude, 1-headed; inv. 8-fid; rays 8, oval; young achenes smooth." *Osteospermum leiocarpum, DC. l. c.* 467.

Page 162, after **C. segetum,** add :

1.* C. coronarium (Linn. Sp. 1264); herbaceous, glabrous; stem erect, diffuse, branched; leaves ear-clasping, bipinnatipartite, the lobes lanceolate, dilated upwards, serrate; branches nude, 1-headed; all the inv. sc. membrane-edged. *DC. l. c.* 64.

HAB. Cape flats, *Ecklon!* (Herb. Sd.)
A weed, introduced from the S. of Europe. Rays yellow.

1.** C. frutescens (Linn. Sp. 1251, non Th.); ligneous, mostly gla-brous; leaves somewhat fleshy, pinnatipartite, the lobes few, linear, toothed, acute; upper leaves linear, entire or trifid; flowering branches shortish; ray-achenes narrow-winged; rays white. *DC. l. c. p.* 65. *Argyranthemum frutescens, Schultz. B.*

HAB. Cape flats, *Ecklon!* (Herb. Sd.)
A native of the Canary Islands; probably escaped from some garden.

Page 171, line 5, for "*less* subsessile" read "leaves subsessile."

Page 171, in gen. char. of **Cotula,** insert, after the words "Recept. flat," the words "or conical."

Pages 212, 213, in the table :

> For (107) **cymosum** read (106) **cymosum;**
> „ (106) **tenuiculum** „ (107) **tenuiculum;**
> „ (115) **maritimum** „ (116) **maritimum;**
> „ (116) **pannosum** „ (115) **pannosum.**

Page 230, line 38, for **59 H. coriaceum** (Harv.) read **59 H. ful-vellum** (Harv.).

The name *coriaceum* is preoccupied ; see No. 85, p. 239.

Page 275, line 25, for *Amphidoxa triflora* read *Amphiglossa triflora.*

Page 277, line 28, for *Helichr. arctioides* read *Hel. aretioides.*

Page 293, in gen. char. of **Antithrixia,** for the words "pubescent *style*" read "pubescent stipe."

Page 352, for

> (137) **paniculatus** read (138) **paniculatus;**
> (138) **multicaulis** „ (137) **multicaulis.**

Page 435, line 33, for "rigid" read "ridged."

The following COMPOSITÆ are found apparently "wild" in many parts of the Colony, but being undoubtedly of exotic origin, have been omitted in this work :—

Zinnia multiflora (L. DC. Prod. 5, p. 535) ; common on the Eastern frontier and at Natal ; escaped from cultivation. A native of Mexico.

Centaurea Cyanus (L., DC. l. c. 6, p. 578) ; found in cornfields and the borders of gardens. Native of S. of Europe.

Centaurea Melitensis (L., DC. l. c. p. 593) ; found on Cape flats and about Greenpoint by *E. & Z.* Native of Malta and Sicily, but widely dispersed over the globe.

Cnicus benedictus (L., DC. l. c. p. 606) ; found about Capetown by *W. H. H.* (1838) and sent from the slopes of the Wittberg by *Mrs. F. W. Barber !* (1863). Native of Persia and Tauria, from whence it has found its way to the S. of Europe, and thence to the Andes of Chili ! *(DC.)*

Kentrophyllum lanatum (DC. l. c. p. 610) ; found about Capetown by *E. & Z.* Native of S. of Europe.

Cichorium Intybus (L., DC. l. c. 7, p. 84) ; sent from Natal by *Mr. Sanderson.* Common "*chicory,*" a well-known European weed, whose cultivated roots are used to adulterate coffee ; and its blanched leaves eaten as winter salad.

INDEX.

630

MADE AND PRINTED IN GREAT BRITAIN BY
WILLIAM CLOWES AND SONS, LIMITED, LONDON AND BECCLES

Printed in the United States
By Bookmasters